STUDENT'S SOLUTIONS MANUAL

JUDITH A. PENNA

Indiana University Purdue University Indianapolis

PREALGEBRA AND INTRODUCTORY ALGEBRA

THIRD EDITION

Marvin L. Bittinger

Indiana University Purdue University Indianapolis

David J. Ellenbogen

Community College of Vermont

Judith A. Beecher

Indiana University Purdue University Indianapolis

Barbara L. Johnson

Indiana University Purdue University Indianapolis

Addison-Wesley
is an imprint of

The author and publisher of this book have used their best efforts in preparing this book. These efforts include the development, research, and testing of the theories and programs to determine their effectiveness. The author and publisher make no warranty of any kind, expressed or implied, with regard to these programs or the documentation contained in this book. The author and publisher shall not be liable in any event for incidental or consequential damages in connection with, or arising out of, the furnishing, performance, or use of these programs.

Reproduced by Pearson Addison-Wesley from electronic files supplied by the author.

Publishing as Addison-Wesley, 75 Arlington Street, Boston, MA 02116.

ISBN-13: 978-0-321-73169-2
ISBN-10: 0-321-73169-7

9 10 11 12 EBM 17 16 15 14

Addison-Wesley
is an imprint of

www.pearsonhighered.com

Contents

Chapter 1 1

Chapter 2 39

Chapter 3 61

Chapter 4 95

Chapter 5 147

Chapter 6 187

Chapter 7 233

Chapter 8 247

Chapter 9 275

Chapter 10 295

Chapter 11 341

Chapter 12 383

Chapter 13 413

Chapter 14 463

Chapter 15 517

Chapter 16 561

Chapter 17 583

Chapter 1

Whole Numbers

Exercise Set 1.1

1. 2 3 $\boxed{5}$, 8 8 8

The digit 5 means 5 thousands.

3. 1, 4 8 8, $\boxed{5}$ 2 6

The digit 5 means 5 hundreds.

5. 1 2 1, 6 2 $\boxed{9}$, 2 7 0

The digit 9 names the number of thousands.

7. 1 2 1, 6 2 9, 2 $\boxed{7}$ 0

The digit 7 names the number of tens.

9. 5702 = 5 thousands + 7 hundreds + 0 tens + 2 ones, or 5 thousands + 7 hundreds + 2 ones

11. 93,986 = 9 ten thousands + 3 thousands + 9 hundreds + 8 tens + 6 ones

13. 2058 = 2 thousands + 0 hundreds + 5 tens + 8 ones, or 2 thousands + 5 tens + 8 ones

15. 1576 = 1 thousand + 5 hundreds + 7 tens + 6 ones

17. 1,424,161,948 = 1 billion + 4 hundred millions + 2 ten millions + 4 millions + 1 hundred thousand + 6 ten thousands + 1 thousand + 9 hundreds + 4 tens + 8 ones

19. 99,886,568 = 9 ten millions + 9 millions + 8 hundred thousands + 8 ten thousands + 6 thousands + 5 hundreds + 6 tens + 8 ones

21. 617,249 = 6 hundred thousands + 1 ten thousand + 7 thousands + 2 hundreds + 4 tens + 9 ones

23. A word name for 85 is eighty-five.

25. 88,000

Eighty-eight thousand → 88

27. 123,765

One hundred twenty-three thousand, → 123

seven hundred sixty-five → 765

29. 7, 754, 211, 577

Seven billion, → 7

seven hundred fifty-four million, → 754

two hundred eleven thousand, → 211

five hundred seventy-seven → 577

31. 700, 634

Seven hundred thousand, → 700

six hundred thirty-four → 634

33. 3 , 048, 005

Three million, → 3

forty-eight thousand, → 048

five → 005

35. Two million,

two hundred thirty-three thousand,

eight hundred twelve

Standard notation is 2, 233, 812.

37. Eight billion

Standard notation is 8,000,000,000.

39. Fifty thousand,

three hundred twenty-four

Standard notation is 50,324.

41. Six hundred thirty-two thousand,

eight hundred ninety-six

Standard notation is 632, 896.

43. One billion,

six hundred million,

Standard notation is 1,600,000,000.

45. Sixty-four million,

one hundred eighty-six thousand,

Standard notation is 64, 186, 000.

47. First consider the whole numbers from 100 through 199. The 10 numbers 102, 112, 122, ... , 192 contain the digit 2. In addition, the 10 numbers 120, 121, 122, ... , 129 contain the digit 2. However, we do not count the number 122 in this group because it was counted in the first group of ten numbers. Thus, 19 numbers from 100 through 199 contain the digit 2. Using the same type of reasoning for the whole numbers from 300 to 400, we see that there are also 19 numbers in this group that contain the digit 2.

Finally, consider the 100 whole numbers 200 through 299. Each contains the digit 2.

Thus, there are $19 + 19 + 100$, or 138 whole numbers between 100 and 400 that contain the digit 2 in their standard notation.

Exercise Set 1.2

1. $$\begin{array}{r} 364 \\ +\ \ 23 \\ \hline 387 \end{array}$$ Add ones, add tens, then add hundreds.

3. $$\begin{array}{r} {\scriptstyle 1}\ \ \\ 86 \\ +78 \\ \hline 164 \end{array}$$ Add ones: We get 14 ones, or 1 ten + 4 ones. Write 4 in the ones column and 1 above the tens. Add tens: We get 16 tens.

5. $$\begin{array}{r} {\scriptstyle 1}\ \ \ \ \ \ \\ 1716 \\ +3482 \\ \hline 5198 \end{array}$$ Add ones: We get 8. Add tens: We get 9 tens. Add hundreds: We get 11 hundreds, or 1 thousand + 1 hundred. Write 1 in the hundreds column and 1 above the thousands. Add thousands: We get 5 thousands.

7. $$\begin{array}{r} {\scriptstyle 1}\ \ \\ 99 \\ +\ \ 1 \\ \hline 100 \end{array}$$ Add ones: We get 10 ones, or 1 ten + 0 ones. Write 0 in the ones column and 1 above the tens. Add tens: We get 10 tens.

9. $$\begin{array}{r} {\scriptstyle 1}\ \ \ \ \ \\ 8113 \\ +\ \ 390 \\ \hline 8503 \end{array}$$ Add ones: We get 3. Add tens: We get 10 tens, or 1 hundred + 0 tens. Write 0 in the tens column and 1 above the hundreds. Add hundreds: We get 5. Add thousands: We get 8.

11. $$\begin{array}{r} {\scriptstyle 1}\ \ \ \ \ \ \\ 356 \\ +4910 \\ \hline 5266 \end{array}$$ Add ones: We get 6. Add tens: We get 6. Add hundreds: We get 12 hundreds, or 1 thousand + 2 hundreds. Write 2 in the hundreds column and 1 above the thousands. Add thousands: We get 5.

13. $$\begin{array}{r} {\scriptstyle 1\,2\,1}\ \ \\ 3870 \\ 92 \\ 7 \\ +\ \ 497 \\ \hline 4466 \end{array}$$ Add ones: We get 16 ones, or 1 ten + 6 ones. Write 6 in the ones column and 1 above the tens. Add tens: We get 26 tens, or 2 hundreds + 6 tens. Write 6 in the tens column and 2 above the hundreds. Add hundreds: We get 14 hundreds, or 1 thousand + 4 hundreds. Write 4 in the hundreds column and 1 above the thousands. Add thousands: We get 4.

15. $$\begin{array}{r} {\scriptstyle 1\,1}\ \ \ \ \\ 4825 \\ +1783 \\ \hline 6608 \end{array}$$ Add ones: We get 8. Add tens: We get 10 tens. Write 0 in the tens column and 1 above the hundreds. Add hundreds: We get 16 hundreds. Write 6 in the hundreds column and 1 above the thousands. Add thousands: We get 6 thousands.

17. $$\begin{array}{r} {\scriptstyle 1\,1\,1}\ \ \\ 23,443 \\ +10,989 \\ \hline 34,432 \end{array}$$ Add ones: We get 12 ones, or 1 ten + 2 ones. Write 2 in the ones column and 1 above the tens. Add tens: We get 13 tens. Write 3 in the tens column and 1 above the hundreds. Add hundreds: We get 14 hundreds. Write 4 in the hundreds column and 1 above the thousands. Add thousands: We get 4 thousands. Add ten thousands: We get 3 ten thousands.

19. $$\begin{array}{r} {\scriptstyle 1\,1\,1\,1}\ \ \\ 77,543 \\ +23,767 \\ \hline 101,310 \end{array}$$ Add ones: We get 10 ones, or 1 ten + 0 ones. Write 0 in the ones column and 1 above the tens. Add tens: We get 11 tens. Write 1 in the tens column and 1 above the hundreds. Add hundreds: We get 13 hundreds. Write 3 in the hundreds column and 1 above the thousands. Add thousands: We get 11 thousands. Write 1 in the thousands column and 1 above the ten thousands. Add ten thousands: We get 10 ten thousands.

21. We look for pairs of numbers whose sums are 10, 20, 30, and so on.

$$\begin{array}{r} {\scriptstyle 2}\ \ \\ 45 \\ 25 \\ 36 \\ 44 \\ +80 \\ \hline 230 \end{array}$$

23. $$\begin{array}{r} {\scriptstyle 1\,1}\ \ \ \ \\ 12,070 \\ 2954 \\ +\ \ 3400 \\ \hline 18,424 \end{array}$$ Add ones: We get 4. Add tens: We get 12 tens, or 1 hundred + 2 tens. Write 2 in the tens column and 1 above the hundreds. Add hundreds: We get 14 hundreds, or 1 thousand + 4 hundreds. Write 4 in the hundreds column and 1 above the thousands. Add thousands: We get 8 thousands. Add ten thousands: We get 1 ten thousand.

25. $$\begin{array}{r} {\scriptstyle 3\,1\,2}\ \ \\ 4835 \\ 729 \\ 9204 \\ 8986 \\ +7931 \\ \hline 31,685 \end{array}$$ Add ones: We get 25. Write 5 in the ones column and 2 above the tens. Add tens: We get 18 tens. Write 8 in the tens column and 1 above the hundreds. Add hundreds: We get 36 hundreds. Write 6 in the hundreds column and 3 above the thousands. Add thousands: We get 31 thousands.

27. Perimeter = 50 yd + 23 yd + 40 yd + 19 yd

We carry out the addition.

$$\begin{array}{r} {\scriptstyle 1}\ \ \\ 50 \\ 23 \\ 40 \\ +19 \\ \hline 132 \end{array}$$

The perimeter of the figure is 132 yd.

29. Perimeter = 402 ft + 298 ft + 196 ft + 100 ft + 453 ft + 212 ft

We carry out the addition.

$$\begin{array}{r} {}^{2\;2}\\ 4\,0\,2\\ 2\,9\,8\\ 1\,9\,6\\ 1\,0\,0\\ 4\,5\,3\\ +\,2\,1\,2\\ \hline 1\,6\,6\,1 \end{array}$$

The perimeter of the figure is 1661 ft.

31. Perimeter = 200 ft + 85 ft + 200 ft + 85 ft

We carry out the addition.

$$\begin{array}{r} {}^{1\;1}\\ 2\,0\,0\\ 8\,5\\ 2\,0\,0\\ +\quad 8\,5\\ \hline 5\,7\,0 \end{array}$$

The perimeter of the hockey rink is 570 ft.

33. 4 [8] 6, 2 0 5

The digit 8 tells the number of ten thousands.

35. One method is described in the answer section in the text. Another method is: 1 + 100 = 101, 2 + 99 = 101, . . . , 50 + 51 = 101. Then the sum of 50 101's is 5050.

Exercise Set 1.3

1. $$\begin{array}{r} 6\,5\\ -\,2\,1\\ \hline 4\,4 \end{array}$$

Subtract ones, then subtract tens.

3. $$\begin{array}{r} 8\,6\,6\\ -\,3\,3\,3\\ \hline 5\,3\,3 \end{array}$$

Subtract ones, subtract tens, then subtract hundreds.

5. $$\begin{array}{r} {}^{7\;16}\\ \not{8}\;\not{6}\\ -\,4\;7\\ \hline 3\;9 \end{array}$$

We cannot subtract 7 ones from 6 ones. Borrow 1 ten to get 16 ones. Subtract ones, then subtract tens.

7. $$\begin{array}{r} {}^{4\;11}\\ \not{5}\;\not{1}\\ -\,3\;7\\ \hline 1\;4 \end{array}$$

We cannot subtract 7 ones from 1 one. Borrow 1 ten to get 11 ones. Subtract ones, then subtract tens.

9. $$\begin{array}{r} {}^{15}\\ {}^{4\;\not{5}\;13}\\ \not{5}\;\not{6}\;\not{3}\\ -\,1\;9\;4\\ \hline 3\;6\;9 \end{array}$$

We cannot subtract 4 ones from 3 ones. Borrow 1 ten to get 13 ones. Subtract ones. We cannot subtract 9 tens from 5 tens. Borrow 1 hundred to get 15 tens. Subtract tens, then subtract hundreds.

11. $$\begin{array}{r} {}^{8\;11}\\ 3\;\not{9}\;\not{1}\\ -\,3\;6\;5\\ \hline 2\;6 \end{array}$$

We cannot subtract 5 ones from 1 one. Borrow 1 ten to get 11 ones. Subtract ones, then tens, then hundreds. Note that 3 hundreds − 3 hundreds = 0 hundreds, but we do not write the 0 when it is the first digit of a difference.

13. $$\begin{array}{r} {}^{7\;11}\\ 9\;\not{8}\;\not{1}\\ -\,7\;4\;7\\ \hline 2\;3\;4 \end{array}$$

We cannot subtract 7 ones from 1 one. Borrow 1 ten to get 11 ones. Subtract ones, subtract tens, then subtract hundreds.

15. $$\begin{array}{r} {}^{7\;13}\\ 6\;\not{8}\;\not{3}\\ -\,2\;6\;6\\ \hline 4\;1\;7 \end{array}$$

We cannot subtract 6 ones from 3 ones. Borrow 1 ten to get 13 ones. Subtract ones, then tens, then hundreds.

17. $$\begin{array}{r} {}^{6\;16}\\ 7\;\not{7}\;\not{6}\;9\\ -\,2\;3\;8\;7\\ \hline 5\;3\;8\;2 \end{array}$$

Subtract ones. We cannot subtract 8 tens from 6 tens. Borrow 1 hundred to get 16 tens. Subtract tens, subtract hundreds, then subtract thousands.

19. $$\begin{array}{r} {}^{14\;10}\\ {}^{3\;\not{4}\;\not{0}\;12}\\ \not{4}\;\not{5}\;\not{1}\;\not{2}\\ -\,1\;7\;3\;4\\ \hline 2\;7\;7\;8 \end{array}$$

We cannot subtract 4 ones from 2 ones. Borrow 1 ten to get 12 ones. Subtract ones. We cannot subtract 3 tens from 0 tens. Borrow 1 hundred to get 10 tens. Subtract tens. We cannot subtract 7 hundreds from 4 hundreds. Borrow 1 thousand to get 14 hundreds. Subtract hundreds, then thousands.

21. $$\begin{array}{r} {}^{10}\\ {}^{2\;\not{0}\;18}\\ 5\;\not{3}\;\not{1}\;\not{8}\\ -\,2\;2\;4\;9\\ \hline 3\;0\;6\;9 \end{array}$$

We cannot subtract 9 ones from 8 ones. Borrow 1 ten to get 18 ones. Subtract ones. We cannot subtract 4 tens from 0 tens. Borrow 1 hundred to get 10 tens. Subtract tens, then hundreds, then thousands.

23. $$\begin{array}{r} {}^{13}\\ {}^{8\;\not{3}\;17}\\ 3\;\not{9}\;\not{4}\;\not{7}\\ -\,2\;8\;5\;8\\ \hline 1\;0\;8\;9 \end{array}$$

We cannot subtract 8 ones from 7 ones. Borrow 1 ten to get 17 ones. Subtract ones. We cannot subtract 5 tens from 3 tens. Borrow 1 hundred to get 13 tens. Subtract tens, then hundreds, then thousands.

25. $$\begin{array}{r} {}^{11\;15\;13}\\ {}^{\not{1}\;\not{5}\;\not{3}\;17}\\ \not{1}\;\not{2},\;\not{6}\;\not{4}\;7\\ -\quad 4\;8\;9\;9\\ \hline 7\;7\;4\;8 \end{array}$$

27.
13
4 11 2 3 12
5 1, 3 4 2
− 4 7, 1 9 8
4 1 4 4

29.
7 10
8 0
− 2 4
5 6

31.
8 10
6 9 0
− 2 3 6
4 5 4

33.
6 16 3 10
7 6 4 0
− 3 8 0 9
3 8 3 1

35.
7 9 18
6 8 0 8
− 3 0 5 9
3 7 4 9

We have 8 hundreds or 80 tens. We borrow 1 ten to get 18 ones. We then have 79 tens. Subtract ones, then tens, then hundreds, then thousands.

37.
2 9 10
2 3 0 0
− 1 0 9
2 1 9 1

We have 3 hundreds or 30 tens. We borrow 1 ten to get 10 ones. We then have 29 tens. Subtract ones, then tens, then hundreds, then thousands.

39.
5 9 9 17
6 0 0 7
− 1 5 8 9
4 4 1 8

We have 6 thousands, or 600 tens. We borrow 1 ten to get 17 ones. We then have 599 tens. Subtract ones, then tens, then hundreds, then thousands.

41.
8 10 2 17
9 0, 2 3 7
− 4 7, 2 0 9
4 3, 0 2 8

43.
10 16
9 0 6 13
1 0 1, 7 3 4
− 5 7 6 0
9 5, 9 7 4

45.
6 9 9 10
7 0 0 0
− 2 7 9 4
4 2 0 6

We have 7 thousands or 700 tens. We borrow 1 ten to get 10 ones. We then have 699 tens. Subtract ones, then tens, then hundreds, then thousands.

47.
8 9 9 10
3 9, 0 0 0
−3 7, 6 9 5
1 3 0 5

We have 9 thousands or 900 tens. We borrow 1 ten to get 10 ones. We then have 899 tens. Subtract ones, then tens, then hundreds, then thousands. Note that 3 ten thousands − 3 ten thousands = 0 ten thousands, but we do not write the 0 when it is the first digit of a difference.

49.
9 9 9 18
1 0, 0 0 8
− 1 9
9 9 8 9

We have 1 ten thousand, or 1000 tens. We borrow 1 ten to get 18 ones. We then have 999 tens. Subtract ones, then tens, then hundreds, then thousands.

51.
4 9 9 9 11
5 0, 0 0 1
− 1 9 8 4
4 8, 0 1 7

We have 5 ten thousands, or 5000 tens. We borrow 1 ten to get 11 ones. We then have 4999 tens. Subtract ones, then tens, then hundreds, then thousands.

53.
1 1
9 4 6
\+ 7 8
1 0 2 4

Add ones: We get 14. Write 4 in the ones column and 1 above the tens. Add tens: We get 12. Write 2 in the tens column and 1 above the hundreds. Add hundreds: We get 10 hundreds.

55.
1 1 1
5 7, 8 7 7
\+ 3 2, 4 0 6
9 0, 2 8 3

Add ones: We get 13. Write 3 in the ones column and 1 above the tens. Add tens: We get 8. Add hundreds: We get 12. Write 2 in the hundreds column and 1 above the thousands. Add thousands: We get 10. Write 0 in the thousands column and 1 above the ten thousands. Add ten thousands: We get 9 ten thousands.

57.
1 1
5 6 7
\+ 7 7 8
1 3 4 5

Add ones: We get 15. Write 5 in the ones column and 1 above the tens. Add tens: We get 14. Write 4 in the tens column and 1 above the hundreds. Add hundreds: We get 13 hundreds.

59.
1 1 1
1 2, 8 8 5
\+ 9 8 0 7
2 2, 6 9 2

Add ones: We get 12. Write 2 in the ones column and 1 above the tens. Add tens: We get 9. Add hundreds: We get 16. Write 6 in the hundreds column and 1 above the thousands. Add thousands: We get 12. Write 2 in the thousands column and 1 above the ten thousands. Add ten thousands. We get 2 ten thousands.

61.
6, 375, 602

Six million,
three hundred seventy-five thousand,
six hundred two

63.
9, _ 4 8, 6 2 1
− 2, 0 9 7, _ 8 1
7, 2 5 1, 1 4 0

To subtract tens, we borrow 1 hundred to get 12 tens.

5 12
9, _ 4 8, 6 2 1
− 2, 0 9 7, _ 8 1
7, 2 5 1, 1 4 0

In order to have 1 hundred in the difference, the missing digit in the subtrahend must be 4 $(5 - 4 = 1)$.

$$\begin{array}{r} \scriptstyle 5\ 12 \\ 9,\_\,48,\not{6}\,\not{2}\,1 \\ -\,2,097,481 \\ \hline 7,251,140 \end{array}$$

In order to subtract ten thousands, we must borrow 1 hundred thousand to get 14 ten thousands. The number of hundred thousands left must be 2 since the hundred thousands place in the difference is 2 $(2 - 0 = 2)$. Thus, the missing digit in the minuend must be $2 + 1$, or 3.

$$\begin{array}{r} \scriptstyle 2\ 14\quad 5\ 12 \\ 9,\not{3}\,\not{4}\,8,\not{6}\,\not{2}\,1 \\ -\,2,097,481 \\ \hline 7,251,140 \end{array}$$

Exercise Set 1.4

1.
$$\begin{array}{rl} \scriptstyle 4 & \\ 65 & \\ \times\ 8 & \text{Multiplying by 8} \\ \hline 520 & \end{array}$$

3.
$$\begin{array}{rl} \scriptstyle 2 & \\ 94 & \\ \times\ 6 & \text{Multiplying by 6} \\ \hline 564 & \end{array}$$

5.
$$\begin{array}{rl} \scriptstyle 2 & \\ 509 & \\ \times\ 3 & \text{Multiplying by 3} \\ \hline 1527 & \end{array}$$

7.
$$\begin{array}{rl} \scriptstyle 1\,2\,6 & \\ 9229 & \\ \times\ 7 & \text{Multiplying by 7} \\ \hline 64,603 & \end{array}$$

9.
$$\begin{array}{rl} \scriptstyle 2 & \\ 53 & \\ \times\ 90 & \text{Multiplying by 9 tens (We write 0 and then multiply 53 by 9.)} \\ \hline 4770 & \end{array}$$

11.
$$\begin{array}{rl} \scriptstyle 2 & \\ \scriptstyle 3 & \\ 85 & \\ \times 47 & \\ \hline 595 & \text{Multiplying by 7} \\ 3400 & \text{Multiplying by 40} \\ \hline 3995 & \text{Adding} \end{array}$$

13.
$$\begin{array}{rl} 87 & \\ \times 10 & \text{Multiplying by 1 ten (We write 0 and then multiply 87 by 1.)} \\ \hline 870 & \end{array}$$

15.
$$\begin{array}{rl} 96 & \\ \times\ 20 & \text{Multiplying by 2 tens (We write 0 and then multiply 96 by 2.)} \\ \hline 1920 & \end{array}$$

17.
$$\begin{array}{rl} \scriptstyle 3\,2 & \\ 643 & \\ \times\ 72 & \\ \hline 1286 & \text{Multiplying by 2} \\ 45010 & \text{Multiplying by 70} \\ \hline 46,296 & \text{Adding} \end{array}$$

19.
$$\begin{array}{rl} \scriptstyle 1\,1 & \\ \scriptstyle 1\,1 & \\ 444 & \\ \times\ 33 & \\ \hline 1332 & \text{Multiplying by 3} \\ 13320 & \text{Multiplying by 30} \\ \hline 14,652 & \text{Adding} \end{array}$$

21.
$$\begin{array}{rl} \scriptstyle 2\,1 & \\ \scriptstyle 3\,2 & \\ \scriptstyle 5\,3 & \\ 564 & \\ \times 458 & \\ \hline 4512 & \text{Multiplying by 8} \\ 28200 & \text{Multiplying by 50} \\ 225600 & \text{Multiplying by 400} \\ \hline 258,312 & \text{Adding} \end{array}$$

23.
$$\begin{array}{rl} \scriptstyle 4\,2 & \\ \scriptstyle 1\ \ & \\ \scriptstyle 3\,1 & \\ 853 & \\ \times 936 & \\ \hline 5118 & \text{Multiplying by 6} \\ 25590 & \text{Multiplying by 30} \\ 767700 & \text{Multiplying by 900} \\ \hline 798,408 & \text{Adding} \end{array}$$

25.
$$\begin{array}{rl} \scriptstyle 1\ \ 2 & \\ \scriptstyle 1 & \\ \scriptstyle 1 & \\ \scriptstyle 1\,1\,3\ \ & \\ 6428 & \\ \times 3224 & \\ \hline 25712 & \text{Multiplying by 4} \\ 128560 & \text{Multiplying by 20} \\ 1285600 & \text{Multiplying by 200} \\ 19284000 & \text{Multiplying by 3000} \\ \hline 20,723,872 & \text{Adding} \end{array}$$

27.
$$\begin{array}{rl} \scriptstyle 1\,3\ \ \ & \\ 3482 & \\ \times\ 104 & \\ \hline 13928 & \text{Multiplying by 4} \\ 348200 & \text{Multiplying by 1 hundred (We write 00 and then multiply 3482 by 1.)} \\ \hline 362,128 & \end{array}$$

29.
$$\begin{array}{rl} \scriptstyle 2\,1 & \\ \scriptstyle 3\,2 & \\ \scriptstyle 3\,3 & \\ 876 & \\ \times 345 & \\ \hline 4380 & \text{Multiplying by 5} \\ 35040 & \text{Multiplying by 40} \\ 262800 & \text{Multiplying by 300} \\ \hline 302,220 & \text{Adding} \end{array}$$

31.
$$\begin{array}{rl}
{\scriptstyle 5\,5\,5} & \\
{\scriptstyle 1\,1\,1} & \\
{\scriptstyle 1\,1\,1} & \\
{\scriptstyle 3\,3\,3} & \\
7\,8\,8\,9 & \\
\times\ 6\,2\,2\,4 & \\
\hline
3\,1\,5\,5\,6 & \text{Multiplying by 4} \\
1\,5\,7\,7\,8\,0 & \text{Multiplying by 20} \\
1\,5\,7\,7\,8\,0\,0 & \text{Multiplying by 200} \\
4\,7\,3\,3\,4\,0\,0\,0 & \text{Multiplying by 6000} \\
\hline
4\,9,1\,0\,1,1\,3\,6 & \text{Adding}
\end{array}$$

33.
$$\begin{array}{rl}
{\scriptstyle 2\ \ 3} & \\
{\scriptstyle 3\ \ 4} & \\
5\,6\,0\,8 & \\
\times\ 4\,5\,0\,0 & \\
\hline
2\,8\,0\,4\,0\,0\,0 & \text{Multiplying by 5 hundreds (We write 00 and then multiply 5608 by 5.)} \\
2\,2\,4\,3\,2\,0\,0\,0 & \text{Multiplying by 4000} \\
\hline
2\,5,2\,3\,6,0\,0\,0 & \text{Adding}
\end{array}$$

35.
$$\begin{array}{rl}
{\scriptstyle 2} & \\
{\scriptstyle 4} & \\
5\,0\,0\,6 & \\
\times\ 4\,0\,0\,8 & \\
\hline
4\,0\,0\,4\,8 & \text{Multiplying by 8} \\
2\,0\,0\,2\,4\,0\,0\,0 & \text{Multiplying by 4 thousands (We write 000 and then multiply 5006 by 4.)} \\
\hline
2\,0,0\,6\,4,0\,4\,8 &
\end{array}$$

37. $A = 728\text{ mi} \times 728\text{ mi} = 529,984$ square miles

39. $A = l \times w = 90\text{ ft} \times 90\text{ ft} = 8100$ square feet

41.
$$\begin{array}{r}
{\scriptstyle 1\ \ 1\ \ \ \ \ } \\
4\,9\,0\,8 \\
5\,6\,6\,7 \\
+\ 2\,1\,1\,0 \\
\hline
1\,2,6\,8\,5
\end{array}$$

Add ones: We get 15. Write 5 in the ones column and 1 above the tens. Add tens: We get 8. Add hundreds: We get 16. Write 6 in the hundreds column and 1 above the thousands. Add thousands: We get 12 thousands.

43.
$$\begin{array}{r}
{\scriptstyle 1\ \ \ \ 1\ 1\ 1\ \ } \\
3\,4\,0,7\,9\,8 \\
+\ \ 8\,6,6\,7\,9 \\
\hline
4\,2\,7,4\,7\,7
\end{array}$$

Add ones: We get 17. Write 7 in the ones column and 1 above the tens. Add tens: We get 17. Write 7 in the tens column and 1 above the hundreds. Add hundreds: We get 14. Write 4 in the hundreds column and 1 above the thousands. Add thousands: We get 7. Add ten thousands: We get 12. Write 2 in the ten thousands column and 1 above the hundred thousands. Add hundred thousands: We get 4 hundred thousands.

45.
$$\begin{array}{r}
{\scriptstyle 8\ 10\ \ } \\
4\,\not 9\,\not 0\,8 \\
-\ 3\,6\,6\,7 \\
\hline
1\,2\,4\,1
\end{array}$$

Subtract ones. We cannot subtract 6 tens from 0 tens. We have 9 hundreds or 90 tens. We borrow 1 hundred to get 10 tens. We have 8 hundreds. Subtract tens, hundreds, and thousands.

47.
$$\begin{array}{r}
{\scriptstyle 13} \\
{\scriptstyle 2\ \not 3\ 10\ \ 8\ 18} \\
\not 3\,\not 4\,\not 0,7\,\not 9\,\not 8 \\
-\ \ 8\,6,6\,7\,9 \\
\hline
2\,5\,4,1\,1\,9
\end{array}$$

We cannot subtract 9 ones from 8 ones. Borrow 1 ten to get 18 ones. Subtract ones. Then subtract tens and hundreds. We cannot subtract 6 thousands from 0 thousands. We have 4 ten thousands or 40 thousands. We borrow 1 ten thousand to get 10 thousands. Subtract thousands. We cannot subtract 8 ten thousands from 3 ten thousands. We borrow 1 hundred thousand to get 13 ten thousands. Subtract ten thousands and then hundred thousands.

49. Use a calculator to perform the computations in this exercise.

First find the total area of each floor:

$A = l \times w = 172 \times 84 = 14,448$ square feet

Find the area lost to the elevator and the stairwell:

$A = l \times w = 35 \times 20 = 700$ square feet

Subtract to find the area available as office space on each floor:

$14,448 - 700 = 13,748$ square feet

Finally, multiply by the number of floors, 18, to find the total area available as office space:

$18 \times 13,748 = 247,464$ square feet

Exercise Set 1.5

1.
$$\begin{array}{r}
1\,2 \\
6\,\overline{)\,7\,2} \\
6\ \ \\
\hline
1\,2 \\
1\,2 \\
\hline
0
\end{array}$$

Think: 7 tens ÷ 6. Estimate 1 ten.
Think: 12 ones ÷ 6. Estimate 2 ones.

The answer is 12.

3. $\dfrac{23}{23} = 1$ Any nonzero number divided by itself is 1.

5. $22 \div 1 = 22$ Any number divided by 1 is that same number.

7. $\dfrac{0}{7} = 0$ Zero divided by any nonzero number is 0.

9. $\dfrac{16}{0}$ is not defined, because division by 0 is not defined.

11. $\dfrac{48}{8} = 6$ because $48 = 8 \cdot 6$.

13.
$$\begin{array}{r}
5\,5 \\
5\,\overline{)\,2\,7\,7} \\
2\,5\ \ \\
\hline
2\,7 \\
2\,5 \\
\hline
2
\end{array}$$

Think: $27 \div 5$. Try 5.
Think: $27 \div 5$ again. Try 5.

The answer is 55 R 2.

15.

```
   1 0 8
8 ⟌8 6 4     8 ÷ 8 = 1
  8
    6 4     There are no groups of 8 in 6.
    6 4     Write 0 above 6. Bring down 4.
      0     64 ÷ 8 = 8
```

The answer is 108.

17.

```
   3 0 7
4 ⟌1 2 2 8    12 ÷ 4 = 3
   1 2
       2 8    There are no groups of 4 in 2.
       2 8    Write 0 above the second 2.
         0    Bring down 8. 28 ÷ 4 = 7
```

The answer is 307.

19.

```
   7 5 3
6 ⟌4 5 2 1    Think: 45 ÷ 6. Try 7.
   4 2
     3 2      Think: 32 ÷ 6. Try 5.
     3 0
       2 1    Think: 21 ÷ 6. Try 3.
       1 8
         3
```

The answer is 753 R 3.

21.

```
   7 4
4 ⟌2 9 7    Think: 29 ÷ 4. Try 7.
   2 8
     1 7    Think 17 ÷ 4. Try 4.
     1 6
       1
```

The answer is 74 R 1.

23.

```
   9 2
8 ⟌7 3 8    Think: 73 ÷ 8. Try 9.
   7 2
     1 8    Think: 18 ÷ 8. Try 2.
     1 6
       2
```

The answer is 92 R 2.

25.

```
   1 7 0 3
5 ⟌8 5 1 5    Think: 8 ÷ 5. Try 1.
   5
   3 5        35 ÷ 5 = 7
   3 5
       1 5    Bring down 1. There are no
       1 5    groups of 5 in 1. Write 0 above
         0    1. Bring down 5. 15 ÷ 5 = 3
```

The answer is 1703.

27.

```
   9 8 7
9 ⟌8 8 8 8    Think: 88 ÷ 9. Try 9.
   8 1
     7 8      Think: 78 ÷ 9. Try 8.
     7 2
       6 8    Think: 68 ÷ 9. Try 7.
       6 3
         5
```

The answer is 987 R 5.

29.

```
     1 2,7 0 0
1 0 ⟌1 2 7,0 0 0    Think: 12 ÷ 10. Try 1.
     1 0
       2 7          Think: 27 ÷ 10. Try 2.
       2 0
         70         70 ÷ 10 = 7. Since the last two
         70         numbers in the dividend are 0
          0         and 0 ÷ 7 = 0, the last two
                    digits of the quotient are 0.
```

The answer is 12,700.

31.

```
               1 2 7
1 0 0 0 ⟌1 2 7, 0 0 0    Think: 1270 ÷ 1000. Try 1.
          1 0 0 0
            2 7 0 0      Think: 2700 ÷ 1000. Try 2.
            2 0 0 0
              7 0 0 0    7000 ÷ 1000 = 7
              7 0 0 0
                    0
```

The answer is 127.

33.

```
         5 2
7 0 ⟌3 6 9 2    Think: 369 ÷ 70. Try 5.
     3 5 0
       1 9 2    Think: 192 ÷ 70. Try 2.
       1 4 0
         5 2
```

The answer is 52 R 52.

35.

```
       2 9
3 0 ⟌8 7 5    Think: 87 ÷ 30. Try 2.
     6 0
     2 7 5    Think: 275 ÷ 30. Try 9.
     2 7 0
         5
```

The answer is 29 R 5.

37.

```
       4 0
2 1 ⟌8 5 2    21 is close to 20. Think: 85 ÷ 20. Try 4.
     8 4
       1 2    There are no groups of 21 in 12.
              Write a 0 above 2.
```

The answer is 40 R 12.

39.

```
         8
8 5 ⟌7 6 7 2    85 is close to 90. Think: 767 ÷ 90.
     6 8 0      Try 8.
     [8 7]
```

Since 87 is larger than the divisor 85, the estimate is too low. We try 9.

```
       9 0
8 5 ⟌7 6 7 2
     7 6 5
         2 2    There are no groups of 85 in 22.
                Write 0 above 2.
```

The answer is 90 R 22.

41.
```
        3
111 ⟌3219    111 is close to 100. Think: 321 ÷ 100.
      333    Try 3.
```
Since we cannot subtract 333 from 321, 3 is too large. Try 2.
```
       29
111 ⟌3219
     222
      999    Think: 999 ÷ 100. Try 9.
      999
        0
```
The answer is 29.

43.
```
     105
 8 ⟌843    8 ÷ 8 = 1
    8
     43    There are no groups of 8 in 4.
     40    Write 0 above 4. Bring down 3.
      3    Think: 43 ÷ 8. Try 5.
```
The answer is 105 R 3.

45.
```
    1609
 5 ⟌8047   Think: 8 ÷ 5. Try 1.
    5
    30     30 ÷ 5 = 6
    30
      47   There are no groups of 5 in 4. Write
      45   0 above 4. Bring down 7.
       2   Think: 47 ÷ 5. Try 9.
```
The answer is 1609 R 2.

47.
```
    1007
 5 ⟌5036   5 ÷ 5 = 1. The next number in the
    5      dividend is 0, and 0 ÷ 5 = 0.
      36   There are no groups of 5 in 3. Write
      35   0 above 3. Bring down 6.
       1   Think: 36 ÷ 5. Try 7.
```
The answer is 1007 R 1.

49.
```
       22
 46 ⟌1058   46 is close to 50. Think: 105 ÷ 50.
      92    Try 2.
      138   Think: 138 ÷ 50. Try 2.
       92
       46
```
Since the difference, 46, is not smaller than the divisor, 46, 2 is too small. Try 3.
```
       23
 46 ⟌1058
      92
      138
      138
        0
```
The answer is 23.

51.
```
      107
 32 ⟌3425   32 is close to 30. Think: 34 ÷ 30.
     32     Try 1.
      225   There are no groups of 32 in 22.
      224   Write 0 above 2. Bring down 5.
        1   Think: 225 ÷ 30. Try 7.
```
The answer is 107 R 1.

53.
```
       4
 24 ⟌8880   24 is close to 20. Think: 88 ÷ 20.
     96     Try 4.
```
Since we cannot subtract 96 from 88, 4 is too large. Try 3.
```
      38
 24 ⟌8880
     72
     168    Think: 168 ÷ 20. Try 8.
     192
```
Since we cannot subtract 192 from 168, 8 is too large. Try 7.
```
      370
 24 ⟌8880
     72
     168
     168
       0    The last digit in the dividend is 0,
            and 0 ÷ 24 = 0.
```
The answer is 370.

55.
```
        5
 28 ⟌17,067   28 is close to 30. Think: 170 ÷ 30.
     140      Try 5.
      30
```
Since 30 is larger than the divisor, 28, 5 is too small. Try 6.
```
        608
 28 ⟌17,067
     168
        267   There are no groups of 28 in 26.
        224   Write 0 above 6. Bring down 7.
         43   Think 267 ÷ 30. Try 8.
```
Since 43 is larger than the divisor, 28, 8 is too small. Try 9.
```
        609
 28 ⟌17,067
     168
        267
        252
         15
```
The answer is 609 R 15.

57.
```
        304
 80 ⟌24,320
     240
        320
        320
          0
```
The answer is 304.

59.

$$\begin{array}{r} 3508 \\ 285\overline{)999,999} \\ \underline{855} \\ 1449 \\ \underline{1425} \\ 2499 \\ \underline{2280} \\ 219 \end{array}$$

The answer is 3508 R 219.

61.

$$\begin{array}{r} 8070 \\ 456\overline{)3,679,920} \\ \underline{3648} \\ 3192 \\ \underline{3192} \\ 0 \end{array}$$

The answer is 8070.

63. The distance around an object is its perimeter.

65. For large numbers, digits are separated by commas into groups of three, called periods.

67. In the sentence $10 \times 1000 = 10,000$, 10 and 1000 are called factors and 10,000 is called the product.

69. The sentence $3 \times (6 \times 2) = (3 \times 6) \times 2$ illustrates the associative law of multiplication.

71.

a	b	$a \cdot b$	$a+b$
	68	3672	
84			117
		32	12

To find a in the first row we divide $a \cdot b$ by b:

$3672 \div 68 = 54$

Then we add to find $a+b$:

$54 + 68 = 122$

To find b in the second row we subtract a from $a+b$:

$117 - 84 = 33$

Then we multiply to find $a \cdot b$:

$84 \cdot 33 = 2772$

To find a and b in the last row we find a pair of numbers whose product is 32 and whose sum is 12. Pairs of numbers whose product is 32 are 1 and 32, 2 and 16, 4 and 8. Since $4 + 8 = 12$, the numbers we want are 4 and 8. We will let $a = 4$ and $b = 8$. (We could also let $a = 8$ and $b = 4$).

The completed table is shown below.

a	b	$a \cdot b$	$a+b$
54	68	3672	122
84	33	2772	117
4	8	32	12

73. We divide 1231 by 42:

$$\begin{array}{r} 29 \\ 42\overline{)1231} \\ \underline{84} \\ 391 \\ \underline{378} \\ 13 \end{array}$$

The answer is 29 R 13. Since 13 students will be left after 29 buses are filled, then 30 buses are needed.

Chapter 1 Mid-Chapter Review

1. The statement is false. For example, $8 - 5 = 3$, but 5 is not equal to $8 + 3$.

2. The statement is true. See page 19 in the text.

3. The statement is true. See page 19 in the text.

4. The statement is false. For example, $3 \cdot 0 = 0$ and 0 is not greater than 3. Also, $1 \cdot 1 = 1$ and 1 is not greater than 1.

5. It is true that zero divided by any nonzero number is 0.

6. The statement is false. Any number divided by 1 is the number itself. For example, $\frac{27}{1} = 27$.

7. 95, 406, 237

Ninety-five million,
four hundred six thousand,
two hundred thirty-seven

8.

$$\begin{array}{r} \overset{5}{\cancel{6}}\,\overset{9}{\cancel{0}}\,\overset{14}{\cancel{4}} \\ -\ 4\ 9\ 7 \\ \hline 1\ 0\ 7 \end{array}$$

9. 2 [6] 9 8

The digit 6 names the number of hundreds.

10. [6] 1, 2 0 4

The digit 6 names the number of ten thousands.

11. 1 4 [6], 2 3 7

The digit 6 names the number of thousands.

12. 5 8 [6]

The digit 6 names the number of ones.

13. 3 0 6, 4 5 8, 1 [2] 9

The digit 2 names the number of tens.

14. 3 0 [6], 4 5 8, 1 2 9

The digit 6 names the number of millions.

15. 3 0 6, 4 [5] 8, 1 2 9

The digit 5 names the number of ten thousands.

16. 3 0 6, 4 5 8, $\boxed{1}$ 2 9

The digit 1 names the number of hundreds.

17. 5602 = 5 thousands + 6 hundreds + 0 tens + 2 ones, or 5 thousands + 6 hundreds + 2 ones

18. 69,345 = 6 ten thousands + 9 thousands + 3 hundreds + 4 tens + 5 ones

19. A word name for 136 is one hundred thirty-six.

20. A word name for 64,325 is sixty-four thousand, three hundred twenty-five.

21. Standard notation for three hundred eight thousand, seven hundred sixteen is 308,716.

22. Standard notation for four million, five hundred sixty-seven thousand, two hundred sixteen is 4,567,216.

23.
$$\begin{array}{r} 316 \\ +\,482 \\ \hline 798 \end{array}$$

24.
$$\begin{array}{r} {\scriptstyle 1\,1} \\ 593 \\ +\,437 \\ \hline 1030 \end{array}$$

25.
$$\begin{array}{r} {\scriptstyle 1\,1} \\ 2638 \\ +\,5284 \\ \hline 7922 \end{array}$$

26.
$$\begin{array}{r} {\scriptstyle 1\,1\,1} \\ 4617 \\ 2436 \\ +\quad 481 \\ \hline 7534 \end{array}$$

27.
$$\begin{array}{r} 786 \\ -\,321 \\ \hline 465 \end{array}$$

28.
$$\begin{array}{r} {\scriptstyle 11} \\ {\scriptstyle 5\;\;1\;\;14} \\ \not{6}\;\not{2}\;\not{4} \\ -\,2\;8\;5 \\ \hline 3\;3\;9 \end{array}$$

29.
$$\begin{array}{r} {\scriptstyle 15} \\ {\scriptstyle 2\;\;5\;\;9\;12} \\ \not{3}\;\not{6}\;\not{0}\;\not{2} \\ -\,1\;7\;4\;8 \\ \hline 1\;8\;5\;4 \end{array}$$

30.
$$\begin{array}{r} {\scriptstyle 4\;\;9\;\;9\;14} \\ \not{5}\;\not{0}\;\not{0}\;\not{4} \\ -\quad 6\;7\;6 \\ \hline 4\;3\;2\;8 \end{array}$$

31.
$$\begin{array}{r} {\scriptstyle 3} \\ 36 \\ \times\;\;6 \\ \hline 216 \end{array}$$

32.
$$\begin{array}{r} {\scriptstyle 1\,1} \\ {\scriptstyle 5\,5} \\ 567 \\ \times\;\;28 \\ \hline 4536 \\ 11340 \\ \hline 15,876 \end{array}$$

33.
$$\begin{array}{r} {\scriptstyle 2} \\ {\scriptstyle 1} \\ {\scriptstyle 3} \\ 407 \\ \times\,325 \\ \hline 2035 \\ 8140 \\ 122100 \\ \hline 132,275 \end{array}$$

34.
$$\begin{array}{r} {\scriptstyle 2\;2\;3} \\ {\scriptstyle 1} \\ 9435 \\ \times\quad 602 \\ \hline 18870 \\ 5661000 \\ \hline 5,679,870 \end{array}$$

35.
$$\begin{array}{r} 253 \\ 4\,\overline{)1012} \\ \underline{8} \\ 21 \\ \underline{20} \\ 12 \\ \underline{12} \\ 0 \end{array}$$

The answer is 253.

36.
$$\begin{array}{r} 112 \\ 38\,\overline{)4261} \\ \underline{38} \\ 46 \\ \underline{38} \\ 81 \\ \underline{76} \\ 5 \end{array}$$

The answer is 112 R 5.

37.
$$\begin{array}{r} 23 \\ 60\,\overline{)1399} \\ \underline{120} \\ 199 \\ \underline{180} \\ 19 \end{array}$$

The answer is 23 R 19.

38.
$$\begin{array}{r} 144 \\ 56\,\overline{)8095} \\ \underline{56} \\ 249 \\ \underline{224} \\ 255 \\ \underline{224} \\ 31 \end{array}$$

The answer is 144 R 31.

39. Perimeter = 10 m + 4 m + 8 m + 3 m = 25 m

40. $A = 4$ in. $\times$ 2 in. $= 8$ sq in.

41. When numbers are being added, it does not matter how they are grouped.

42. Subtraction is not commutative. For example, $5 - 2 = 3$, but $2 - 5 \neq 3$.

43. Answers will vary. Suppose one coat costs \$150. Then the multiplication $4 \cdot \$150$ gives the cost of four coats.

Suppose one ream of copy paper costs \$4. Then the multiplication $\$4 \cdot 150$ gives the cost of 150 reams.

44. Using the definition of division, $0 \div 0 = a$ such that $a \cdot 0 = 0$. We see that a could be *any* number since $a \cdot 0 = 0$ for any number a. Thus, we cannot say that $0 \div 0 = 0$. This is why we agree not to allow division by 0.

Exercise Set 1.6

1. Round 48 to the nearest ten.

4 [8]
↑

The digit 4 is in the tens place. Consider the next digit to the right. Since the digit, 8, is 5 or higher, round 4 tens up to 5 tens. Then change the digit to the right of the tens digit to zero.

The answer is 50.

3. Round 463 to the nearest ten.

4 6 [3]
↑

The digit 6 is in the tens place. Consider the next digit to the right. Since the digit, 3, is 4 or lower, round down, meaning that 6 tens stays as 6 tens. Then change the digit to the right of the tens digit to zero.

The answer is 460.

5. Round 731 to the nearest ten.

7 3 [1]
↑

The digit 3 is in the tens place. Consider the next digit to the right. Since the digit, 1, is 4 or lower, round down, meaning that 3 tens stays as 3 tens. Then change the digit to the right of the tens digit to zero.

The answer is 730.

7. Round 895 to the nearest ten.

8 9 [5]
↑

The digit 9 is in the tens place. Consider the next digit to the right. Since the digit, 5, is 5 or higher, we round up. The 89 tens become 90 tens. Then change the digit to the right of the tens digit to zero.

The answer is 900.

9. Round 146 to the nearest hundred.

1 [4] 6
↑

The digit 1 is in the hundreds place. Consider the next digit to the right. Since the digit, 4, is 4 or lower, round down, meaning that 1 hundred stays as 1 hundred. Then change all digits to the right of the hundreds digit to zeros.

The answer is 100.

11. Round 957 to the nearest hundred.

9 [5] 7
↑

The digit 9 is in the hundreds place. Consider the next digit to the right. Since the digit, 5, is 5 or higher, round up. The 9 hundreds become 10 hundreds. Then change all digits to the right of the hundreds digit to zeros.

The answer is 1000.

13. Round 9079 to the nearest hundred.

9 0 [7] 9
↑

The digit 0 is in the hundreds place. Consider the next digit to the right. Since the digit, 7, is 5 or higher, round 0 hundreds up to 1 hundred. Then change all digits to the right of the hundreds digit to zeros.

The answer is 9100.

15. Round 32,839 to the nearest hundred.

3 2, 8 [3] 9
↑

The digit 8 is in the hundreds place. Consider the next digit to the right. Since the digit, 3, is 4 or lower, round down, meaning 8 hundreds stays as 8 hundreds. Then change all digits to the right of the hundreds digit to zero.

The answer is 32,800.

17. Round 5876 to the nearest thousand.

5 [8] 7 6
↑

The digit 5 is in the thousands place. Consider the next digit to the right. Since the digit, 8, is 5 or higher, round 5 thousands up to 6 thousands. Then change all digits to the right of the thousands digit to zeros.

The answer is 6000.

19. Round 7500 to the nearest thousand.

7 [5] 0 0
↑

The digit 7 is in the thousands place. Consider the next digit to the right. Since the digit, 5, is 5 or higher, round 7 thousands up to 8 thousands. Then change all the digits to the right of the thousands digit to zeros.

The answer is 8000.

21. Round 45,340 to the nearest thousand.

$$4\,5,\boxed{3}\,4\,0$$

The digit 5 is in the thousands place. Consider the next digit to the right. Since the digit, 3, is 4 or lower, round down, meaning that 5 thousands stays as 5 thousands. Then change all the digits to the right of the thousands digit to zeros.

The answer is 45,000.

23. Round 373,405 to the nearest thousand.

$$3\,7\,3,\boxed{4}\,0\,5$$

The digit 3 is in the thousands place. Consider the next digit to the right. Since the digit, 4, is 4 or lower, round down, meaning that 3 thousands stays as 3 thousands. Then change all the digits to the right of the thousands digit to zeros.

The answer is 373,000.

25. Rounded to the nearest ten

$$\begin{array}{rr} 78 & 80 \\ +92 & +\ 90 \\ \hline & 170 \leftarrow \text{Estimated answer} \end{array}$$

27. Rounded to the nearest ten

$$\begin{array}{rr} 8074 & 8070 \\ -2347 & -2350 \\ \hline & 5720 \leftarrow \text{Estimated answer} \end{array}$$

29. Rounded to the nearest ten

$$\begin{array}{rr} 45 & 50 \\ 77 & 80 \\ 25 & 30 \\ +56 & +60 \\ \hline 343 & 220 \leftarrow \text{Estimated answer} \end{array}$$

The sum 343 seems to be incorrect since 220 is not close to 343.

31. Rounded to the nearest ten

$$\begin{array}{rr} 622 & 620 \\ 78 & 80 \\ 81 & 80 \\ +111 & +110 \\ \hline 932 & 890 \leftarrow \text{Estimated answer} \end{array}$$

The sum 932 seems to be incorrect since 890 is not close to 932.

33. Rounded to the nearest hundred

$$\begin{array}{rr} 7348 & 7300 \\ +9247 & +\ 9200 \\ \hline & 16,500 \leftarrow \text{Estimated answer} \end{array}$$

35. Rounded to the nearest hundred

$$\begin{array}{rr} 6852 & 6900 \\ -1748 & -1700 \\ \hline & 5200 \leftarrow \text{Estimated answer} \end{array}$$

37. Rounded to the nearest hundred

$$\begin{array}{rr} 216 & 200 \\ 84 & 100 \\ 745 & 700 \\ +595 & +600 \\ \hline 1640 & 1600 \leftarrow \text{Estimated answer} \end{array}$$

The sum 1640 seems to be correct since 1600 is close to 1640.

39. Rounded to the nearest hundred

$$\begin{array}{rr} 750 & 800 \\ 428 & 400 \\ 63 & 100 \\ +205 & +200 \\ \hline 1446 & 1500 \leftarrow \text{Estimated answer} \end{array}$$

The sum 1446 seems to be correct since 1500 is close to 1446.

41. Rounded to the nearest thousand

$$\begin{array}{rr} 9643 & 10,000 \\ 4821 & 5000 \\ 8943 & 9000 \\ +7004 & +\ 7000 \\ \hline & 31,000 \leftarrow \text{Estimated answer} \end{array}$$

43. Rounded to the nearest thousand

$$\begin{array}{rr} 92,149 & 92,000 \\ -\ 22,555 & -\ 23,000 \\ \hline & 69,000 \leftarrow \text{Estimated answer} \end{array}$$

45. Rounded to the nearest ten

$$\begin{array}{rr} 45 & 50 \\ \times\ 67 & \times\ 70 \\ \hline & 3500 \leftarrow \text{Estimated answer} \end{array}$$

47. Rounded to the nearest ten

$$\begin{array}{rr} 34 & 30 \\ \times\ 29 & \times\ 30 \\ \hline & 900 \leftarrow \text{Estimated answer} \end{array}$$

49. Rounded to the nearest hundred

$$\begin{array}{rr} 876 & 900 \\ \times\ 345 & \times\ 300 \\ \hline & 270,000 \leftarrow \text{Estimated answer} \end{array}$$

51. Rounded to the nearest hundred

$$\begin{array}{r} 432 \\ \times\ 199 \\ \hline \end{array} \qquad \begin{array}{r} 400 \\ \times\ 200 \\ \hline 80,000 \end{array} \leftarrow \text{Estimated answer}$$

53. Rounding to the nearest ten, we have $347 \div 73 \approx 350 \div 70$.

$$\begin{array}{r} 5 \\ 70\overline{)350} \\ 350 \\ \hline 0 \end{array}$$

55. $8452 \div 46 \approx 8450 \div 50$

$$\begin{array}{r} 169 \\ 50\overline{)8450} \\ 50 \\ \hline 345 \\ 300 \\ \hline 450 \\ 450 \\ \hline 0 \end{array}$$

57. $1165 \div 236 \approx 1200 \div 200$

$$\begin{array}{r} 6 \\ 200\overline{)1200} \\ 1200 \\ \hline 0 \end{array}$$

59. $8358 \div 295 \approx 8400 \div 300$

$$\begin{array}{r} 28 \\ 300\overline{)8400} \\ 600 \\ \hline 2400 \\ 2400 \\ \hline 0 \end{array}$$

61. We round the cost of each option to the nearest hundred and add.

$$\begin{array}{r} 7450 \\ 1595 \\ 1540 \\ +\ \ 625 \\ \hline \end{array} \qquad \begin{array}{r} 7500 \\ 1600 \\ 1500 \\ +\ \ 600 \\ \hline 11,200 \end{array}$$

The estimated cost is \$11,200.

63. We round the cost of each option to the nearest hundred and add.

$$\begin{array}{r} 8820 \\ 2870 \\ 6245 \\ +\ \ 985 \\ \hline \end{array} \qquad \begin{array}{r} 8800 \\ 2900 \\ 6200 \\ +\ 1000 \\ \hline 18,900 \end{array}$$

The estimated cost is \$18,900. Since this is more than Sara and Ben's budget of \$17,700, they cannot afford their choices.

65. Answers will vary depending on the options chosen.

67. a) First we round the cost of the car and the destination charges to the nearest hundred and add.

$$\begin{array}{r} 21,160 \\ +\ \ \ 670 \\ \hline \end{array} \qquad \begin{array}{r} 21,200 \\ +\ \ \ 700 \\ \hline 21,900 \end{array}$$

The number of sales representatives, 112, rounded to the nearest hundred is 100. Now we multiply the rounded total cost of a car and the rounded number of representatives.

$$\begin{array}{r} 21,900 \\ \times\ \ \ 100 \\ \hline 2,190,000 \end{array}$$

The cost of the purchase is approximately \$2,190,000.

b) First we round the cost of the car to the nearest thousand and the destination charges to the nearest hundred and add.

$$\begin{array}{r} 21,160 \\ +\ \ \ 670 \\ \hline \end{array} \qquad \begin{array}{r} 21,000 \\ +\ \ \ 700 \\ \hline 21,700 \end{array}$$

From part (a) we know that the number of sales representatives, rounded to the nearest hundred, is 100. We multiply the rounded total cost of a car and the rounded number of representatives.

$$\begin{array}{r} 21,700 \\ \times\ \ \ 100 \\ \hline 2,170,000 \end{array}$$

The cost of the purchase is approximately \$2,170,000.

69. $2716 \div 28 \approx 2700 \div 30$

$$\begin{array}{r} 90 \\ 30\overline{)2700} \\ 270 \\ \hline 0 \\ 0 \\ \hline 0 \end{array}$$

We estimate that 90 people attended the banquet.

71. 0 5 10 17

Since 0 is to the left of 17, $0 < 17$.

73. 5 12 15 20 25 30 34

Since 34 is to the right of 12, $34 > 12$.

75. 1000 1001

Since 1000 is to the left of 1001, $1000 < 1001$.

77. 132 133

Since 133 is to the right of 132, $133 > 132$.

79. 0 17 100 200 300 400 460 500

Since 460 is to the right of 17, $460 > 17$.

81. 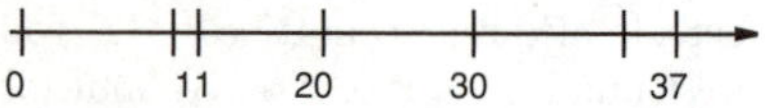

Since 37 is to the right of 11, $37 > 11$.

83. Since 284,370 lies to the right of 172,000 on the number line, we can write $284,370 > 172,000$.

Conversely, since 172,000 lies to the left of 284,370 on the number line, we could also write $172,000 < 284,370$.

85. Since 1663 lies to the left of 9453 on the number line, we can write $1663 < 9453$.

Conversely, since 9453 lies to the right of 1663 on the number line, we could also write $9453 > 1663$.

87.
$$\begin{array}{r} \scriptstyle 1\ 1\ 1\ 1 \\ 6\,7,\,7\,8\,9 \\ +\ 1\,8,\,9\,6\,5 \\ \hline 8\,6,\,7\,5\,4 \end{array}$$

Add ones. We get 14. Write 4 in the ones column and 1 above the tens. Add tens: We get 15 tens. Write 5 in the tens column and 1 above the hundreds. Add hundreds: We get 17 hundreds. Write 7 in the hundreds column and 1 above the thousands. Add thousands: We get 16 thousands. Write 6 in the thousands column and 1 above the ten thousands. Add ten thousands: We get 8 ten thousands.

89.
$$\begin{array}{r} \scriptstyle 16 \\ \scriptstyle 5\ \not 6\ 17 \\ \not 6\,\not 7,\,\not 7\,8\,9 \\ -\ 1\,8,\,9\,6\,5 \\ \hline 4\,8,\,8\,2\,4 \end{array}$$

Subtract ones: We get 4. Subtract tens: We get 2. We cannot subtract 9 hundreds from 7 hundreds. We borrow 1 thousand to get 17 hundreds. Subtract hundreds. We cannot subtract 8 thousands from 6 thousands. We borrow 1 ten thousand to get 16 thousands. Subtract thousands, then ten thousands.

91.
$$\begin{array}{r} \scriptstyle 1 \\ \scriptstyle 4 \\ 4\,6 \\ \times\ 3\,7 \\ \hline 3\,2\,2 \\ 1\,3\,8\,0 \\ \hline 1\,7\,0\,2 \end{array}$$

93.
$$\begin{array}{r} 5\,4 \\ 6\,\overline{)\,3\,2\,8} \\ 3\,0 \\ \hline 2\,8 \\ 2\,4 \\ \hline 4 \end{array}$$

The answer is 54 R 4.

95. Using a calculator, we find that the sum is 30,411. This is close to the estimated sum found in Exercise 41.

97. Using a calculator, we find that the difference is 69,594. This is close to the estimated difference found in Exercise 43.

Exercise Set 1.7

1. $x + 0 = 14$

We replace x by different numbers until we get a true equation. If we replace x by 14, we get a true equation: $14 + 0 = 14$. No other replacement makes the equation true, so the solution is 14.

3. $y \cdot 17 = 0$

We replace y by different numbers until we get a true equation. If we replace y by 0, we get a true equation: $0 \cdot 17 = 0$. No other replacement makes the equation true, so the solution is 0.

5. $x = 12,345 + 78,555$

To solve the equation we carry out the calculation.

$$\begin{array}{r} 1\,2,\,3\,4\,5 \\ +\ 7\,8,\,5\,5\,5 \\ \hline 9\,0,\,9\,0\,0 \end{array}$$

We can check by repeating the calculation. The solution is 90,900.

7. $908 - 458 = p$

To solve the equation we carry out the calculation.

$$\begin{array}{r} 9\,0\,8 \\ -\ 4\,5\,8 \\ \hline 4\,5\,0 \end{array}$$

We can check by repeating the calculation. The solution is 450.

9. $16 \cdot 22 = y$

To solve the equation we carry out the calculation.

$$\begin{array}{r} 2\,2 \\ \times\ 1\,6 \\ \hline 1\,3\,2 \\ 2\,2\,0 \\ \hline 3\,5\,2 \end{array}$$

We can check by repeating the calculation. The solution is 352.

11. $t = 125 \div 5$

To solve the equation we carry out the calculation.

$$\begin{array}{r} 2\,5 \\ 5\,\overline{)\,1\,2\,5} \\ 1\,0 \\ \hline 2\,5 \\ 2\,5 \\ \hline 0 \end{array}$$

We can check by repeating the calculation. The solution is 25.

13.
$$\begin{aligned} 13 + x &= 42 \\ 13 + x - 13 &= 42 - 13 && \text{Subtracting 13 on both sides} \\ 0 + x &= 29 && \text{13 plus } x \text{ minus 13 is } 0 + x. \\ x &= 29 \end{aligned}$$

Check: $13 + x = 42$

$13 + 29 \;?\; 42$

42 TRUE

The solution is 29.

15. $12 = 12 + m$

$12 - 12 = 12 + m - 12$ Subtracting 12 on both sides

$0 = 0 + m$ 12 plus m minus 12 is 0 + m.

$0 = m$

Check: $12 = 12 + m$

$12 \;?\; 12 + 0$

12 TRUE

The solution is 0.

17. $10 + x = 89$

$10 + x - 10 = 89 - 10$

$x = 79$

Check: $10 + x = 89$

$10 + 79 \;?\; 89$

89 TRUE

The solution is 79.

19. $61 = 16 + y$

$61 - 16 = 16 + y - 16$

$45 = y$

Check: $61 = 16 + y$

$61 \;?\; 16 + 45$

61 TRUE

The solution is 45.

21. $3 \cdot x = 24$

$\frac{3 \cdot x}{3} = \frac{24}{3}$ Dividing by 3 on both sides

$x = 8$ 3 times x divided by 3 is x.

Check: $3 \cdot x = 24$

$3 \cdot 8 \;?\; 24$

24 TRUE

The solution is 8.

23. $112 = n \cdot 8$

$\frac{112}{8} = \frac{n \cdot 8}{8}$ Dividing by 8 on both sides

$14 = n$

Check: $112 = n \cdot 8$

$112 \;?\; 14 \cdot 8$

112 TRUE

The solution is 14.

25. $3 \cdot m = 96$

$\frac{3 \cdot m}{3} = \frac{96}{3}$ Dividing by 3 on both sides

$m = 32$

Check: $3 \cdot m = 96$

$3 \cdot 32 \;?\; 96$

96 TRUE

The solution is 32.

27. $715 = 5 \cdot z$

$\frac{715}{5} = \frac{5 \cdot z}{5}$ Dividing by 5 on both sides

$143 = z$

Check: $715 = 5 \cdot x$

$715 \;?\; 5 \cdot 143$

715 TRUE

The solution is 143.

29. $8322 + 9281 = x$

$17,603 = x$ Doing the addition

The number 17,603 checks. It is the solution.

31. $47 + n = 84$

$47 + n - 47 = 84 - 47$

$n = 37$

The number 37 checks. It is the solution.

33. $45 \times 23 = x$

To solve the equation we carry out the calculation.

$$\begin{array}{r} 4\,5 \\ \times\, 2\,3 \\ \hline 1\,3\,5 \\ 9\,0\,0 \\ \hline 1\,0\,3\,5 \end{array}$$

The number 1035 checks. It is the solution.

35. $x + 78 = 144$

$x + 78 - 78 = 144 - 78$

$x = 66$

The number 66 checks. It is the solution.

37. $6 \cdot p = 1944$

$\frac{6 \cdot p}{6} = \frac{1944}{6}$

$p = 324$

The number 324 checks. It is the solution.

39. $5 \cdot x = 3715$

$\frac{5 \cdot x}{5} = \frac{3715}{5}$

$x = 743$

The number 743 checks. It is the solution.

41. $x + 214 = 389$

$x + 214 - 214 = 389 - 214$

$x = 175$

The number 175 checks. It is the solution.

43. $567 + x = 902$

$567 + x - 567 = 902 - 567$

$x = 335$

The number 335 checks. It is the solution.

45. $234 \cdot 78 = y$

$18,252 = y$ Doing the multiplication

The number 18,252 checks. It is the solution.

47.
$$\begin{aligned} 18 \cdot x &= 1872 \\ \frac{18 \cdot x}{18} &= \frac{1872}{18} \\ x &= 104 \end{aligned}$$

The number 104 checks. It is the solution.

49.
$$\begin{aligned} 40 \cdot x &= 1800 \\ \frac{40 \cdot x}{40} &= \frac{1800}{40} \\ x &= 45 \end{aligned}$$

The number 45 checks. It is the solution.

51.
$$\begin{aligned} 2344 + y &= 6400 \\ 2344 + y - 2344 &= 6400 - 2344 \\ y &= 4056 \end{aligned}$$

The number 4056 checks. It is the solution.

53. $m = 7006 - 4159$

To solve the equation we carry out the calculation.

$$\begin{array}{r} {\scriptstyle 6\ 9\ 9\ 16} \\ \not7\ \not0\ \not0\ \not6 \\ -\ 4\ 1\ 5\ 9 \\ \hline 2\ 8\ 4\ 7 \end{array}$$

The number 2847 checks. It is the solution.

55.
$$\begin{aligned} 165 &= 11 \cdot n \\ \frac{165}{11} &= \frac{11 \cdot n}{11} \\ 15 &= n \end{aligned}$$

The number 15 checks. It is the solution.

57.
$$\begin{aligned} 58 \cdot m &= 11,890 \\ \frac{58 \cdot m}{58} &= \frac{11,890}{58} \\ m &= 205 \end{aligned}$$

The number 205 checks. It is the solution.

59. $491 - 34 = y$

To solve the equation we carry out the calculation.

$$\begin{array}{r} {\scriptstyle 8\ 11} \\ 4\ \not9\ \not1 \\ -\ \ 3\ 4 \\ \hline 4\ 5\ 7 \end{array}$$

The number 457 checks. It is the solution.

61.
$$\begin{array}{r|l} 1\ 4\ 2 & \\ 9\ \overline{)\,1\ 2\ 8\ 3} & \text{Think } 12 \div 9. \text{ Try 1.} \\ 9 & \\ \hline 3\ 8 & \text{Think } 38 \div 9. \text{ Try 4.} \\ 3\ 6 & \\ \hline 2\ 3 & \text{Think } 23 \div 9. \text{ Try 2.} \\ 1\ 8 & \\ \hline 5 & \end{array}$$

The answer is 142 R 5.

63.
$$\begin{array}{r|l} 3\ 3\ 4 & \\ 1\ 7\ \overline{)\,5\ 6\ 7\ 8} & 17 \text{ is close to 20. Think: } 56 \div 20. \text{ If we} \\ 5\ 1 & \text{try 2, we find that it is too small. Try 3.} \\ \hline 5\ 7 & \text{Think } 57 \div 20. \text{ Again, 2 is too small.} \\ 5\ 1 & \text{Try 3.} \\ \hline 6\ 8 & \text{Think } 68 \div 20. \text{ If we try 3, we find that} \\ 6\ 8 & \text{it is too small. Try 4.} \\ \hline 0 & \end{array}$$

The answer is 334.

65. Since 123 is to the left of 789 on the number line, $123 < 789$.

67. Since 688 is to the right of 0 on the number line, $688 > 0$.

69. Round 6, 3 7 5, $\boxed{6}$ 0 2 to the nearest thousand.
↑

The digit 5 is in the thousands place. Consider the next digit to the right. Since the digit 6 is 5 or higher, round 5 thousands to 6 thousands. Then change all digits to the right of the thousands digit to zero.

The answer is 6,376,000.

71.
$$\begin{aligned} 23,465 \cdot x &= 8,142,355 \\ \frac{23,465 \cdot x}{23,465} &= \frac{8,142,355}{23,465} \\ x &= 347 \quad \text{Using a calculator to divide} \end{aligned}$$

The number 347 checks. It is the solution.

Exercise Set 1.8

1. ***Familiarize.*** We visualize the situation. Let f = the number of feet by which the length of Kingda Ka exceeds the length of Top Thrill Dragster.

Length of Top Thrill Dragster 2800 ft	Excess length of Kingda Ka f
Length of Kingda Ka 3118 ft	

Translate. We translate to an equation.

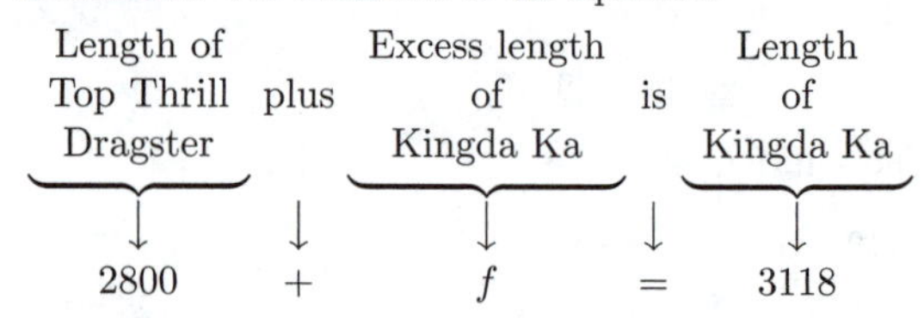

Solve. We subtract 2800 on both sides of the equation.

$$\begin{aligned} 2800 + f &= 3118 \\ 2800 + f - 2800 &= 3118 - 2800 \\ f &= 318 \end{aligned}$$

Check. We can add the difference, 318, to the length of Top Thrill Dragster, 2800: $2800 + 318 = 3118$. We can also estimate:

$$3118 - 2800 \approx 3100 - 2800 = 300 \approx 318.$$

The answer checks.

State. Kingda Ka is 318 ft longer than Top Thrill Dragster.

3. ***Familiarize.*** We visualize the situation. Let l = the length of The Ultimate, in feet.

Length of The Ultimate l	Excess length of Steel Dragon 683 ft
Length of Steel Dragon 8133 ft	

Translate. We translate to an equation.

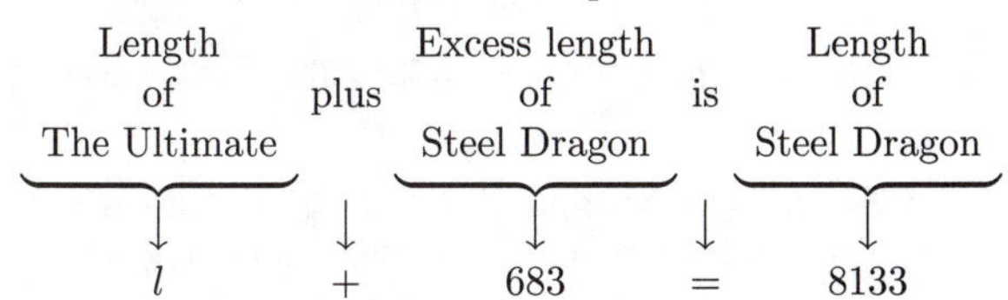

Solve. We subtract 683 on both sides of the equation.

$$l + 683 = 8133$$
$$l + 683 - 683 = 8133 - 683$$
$$l = 7450$$

Check. We can add:

$$7450 + 683 = 8133.$$

We can also estimate:

$$7450 + 683 \approx 7500 + 700 = 8200 \approx 8133.$$

The answer seems reasonable.

State. The length of The Ultimate is 7450 ft.

5. ***Familiarize.*** We visualize the situation. Let c = the amount of caffeine in an 8-oz serving of coffee, in milligrams.

Caffeine in Red Bull 76 milligrams	Excess caffeine in coffee 19 milligrams
Caffeine in coffee c	

Translate. We translate to an equation.

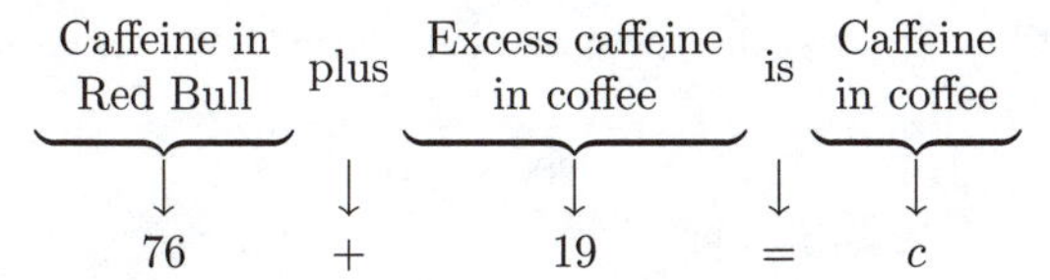

Solve. We carry out the calculation.

$$\begin{array}{r} \scriptstyle 1 \\ 76 \\ +\,19 \\ \hline 95 \end{array}$$

Thus, $95 = c$.

Check. We can repeat the calculation. We can also estimate: $76 + 19 \approx 80 + 20 = 100 \approx 95$. The answer seems reasonable.

State. An 8-oz serving of coffee contains 95 milligrams of caffeine.

7. ***Familiarize.*** We first make a drawing. Let r = the number of rows.

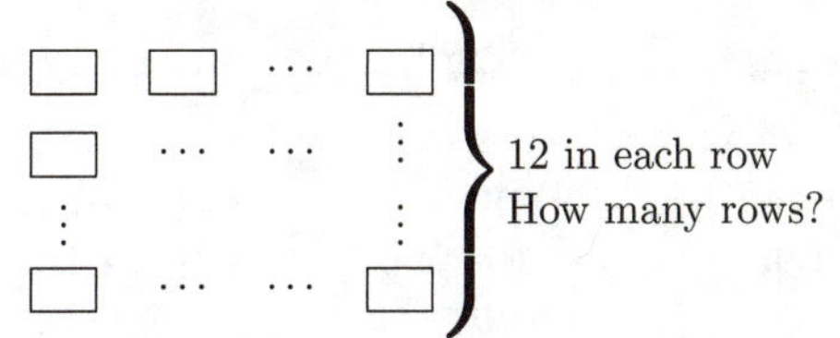

Translate. We translate to an equation.

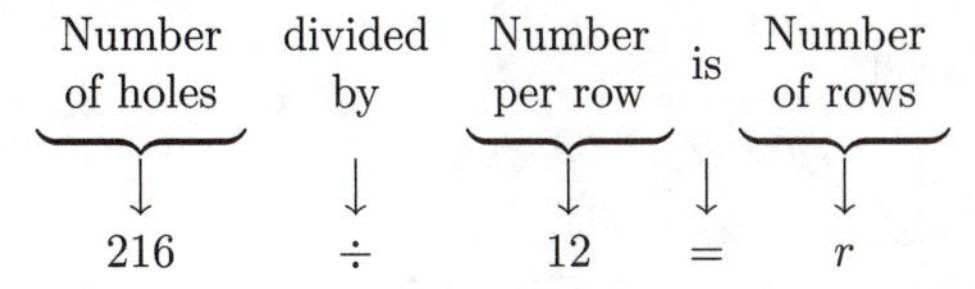

Solve. We carry out the division.

$$\begin{array}{r} 18 \\ 12\overline{)216} \\ 12 \\ \hline 96 \\ 96 \\ \hline 0 \end{array}$$

Thus, $18 = r$, or $r = 18$.

Check. We can check by multiplying: $12 \cdot 18 = 216$. Our answer checks.

State. There are 18 rows.

9. ***Familiarize.*** We visualize the situation. Let h = the number of hours that people age 12 and older watched TV, on average, in 2000.

TV-watching hours in 2000 h	Excess hours in 2008 202
TV-watching hours in 2008 1704	

Translate. We translate to an equation.

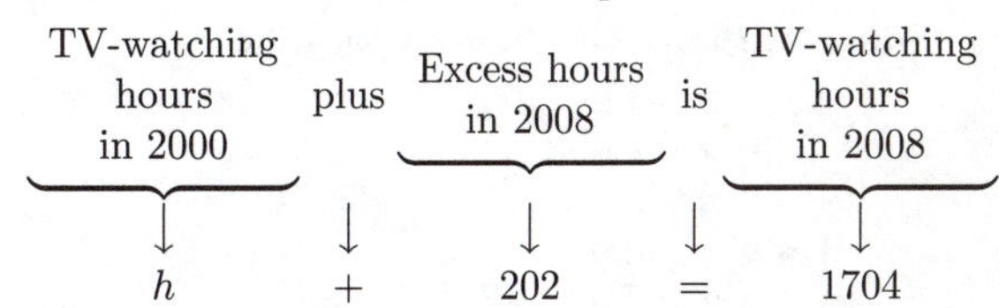

Solve. We subtract 202 on both sides of the equation.

$$h + 202 = 1704$$
$$h + 202 - 202 = 1704 - 202$$
$$h = 1502$$

Check. We can add the difference, 1502, to 202: $202 + 1502 = 1704$. We can also estimate:

$$1704 - 202 \approx 1700 - 200 = 1500 \approx 1502.$$

The answer checks.

State. On average, people age 12 and older spent 1502 hours watching TV in 2000.

11. ***Familiarize.*** We visualize the situation. Let m = the number of miles by which the Canadian border exceeds the Mexican border.

Mexican border 1933 mi	Excess miles in Canadian border m
Canadian border 3987 mi	

Translate. We translate to an equation.

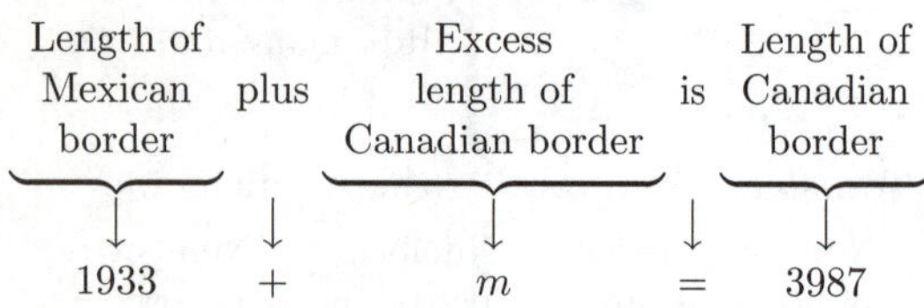

Solve. We subtract 1933 on both sides of the equation.

$$1933 + m = 3987$$
$$1933 + m - 1933 = 3987 - 1933$$
$$m = 2054$$

Check. We can add the difference, 2054, to the subtrahend, 1933: $1933 + 2054 = 3987$. We can also estimate:

$$3987 - 1933 \approx 4000 - 2000$$
$$\approx 2000 \approx 2054$$

The answer checks.

State. The Canadian border is 2054 mi longer than the Mexican border.

13. ***Familiarize.*** We visualize the situation. Let p = the number of pixels on the screen. Repeated addition works well here.

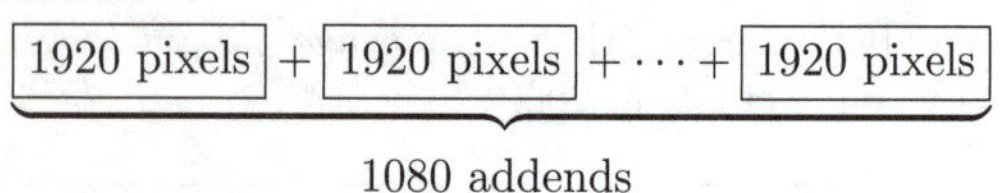

Translate. We translate to an equation.

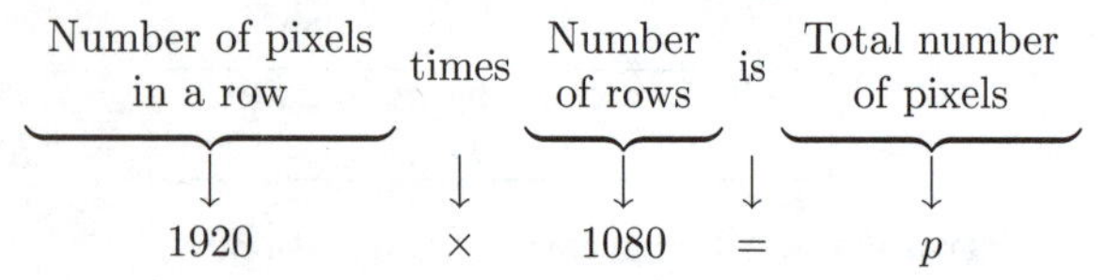

Solve. We carry out the multiplication.

$$\begin{array}{r} 1\,9\,2\,0 \\ \times\ 1\,0\,8\,0 \\ \hline 1\,5\,3\,6\,0\,0 \\ 1\,9\,2\,0\,0\,0\,0 \\ \hline 2,0\,7\,3,6\,0\,0 \end{array}$$

Thus, $2,073,600 = p$.

Check. We can repeat the calculation. The answer checks.

State. There are 2,073,600 pixels on the screen.

15. ***Familiarize.*** We visualize the situation. Let m = the number of associate's degrees earned by men in 2007.

Number of degrees earned by women 453,000	Number of degrees earned by men m
Total number of degrees earned 728,000	

Translate. We translate to an equation.

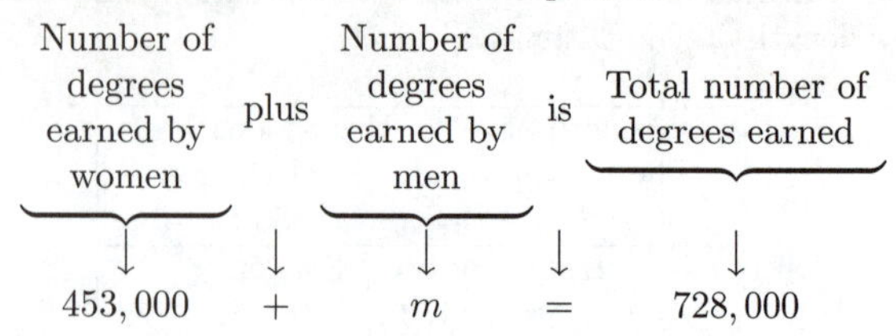

Solve. We subtract 453,000 on both sides of the equation.

$$453,000 + m = 728,000$$
$$453,000 + m - 453,000 = 728,000 - 453,000$$
$$m = 275,000$$

Check. We can add the number of degrees earned by women and the number of degrees earned by men.

$$453,000 + 275,000 = 728,000$$

We get the total number of degrees earned, so the answer checks.

State. 275,000 associate's degrees were earned by men in 2007.

17. ***Familiarize.*** We first draw a picture. Let h = the number of hours in a week. Repeated addition works well here.

24 hours + 24 hours + ··· + 24 hours

7 addends

Translate. We translate to an equation.

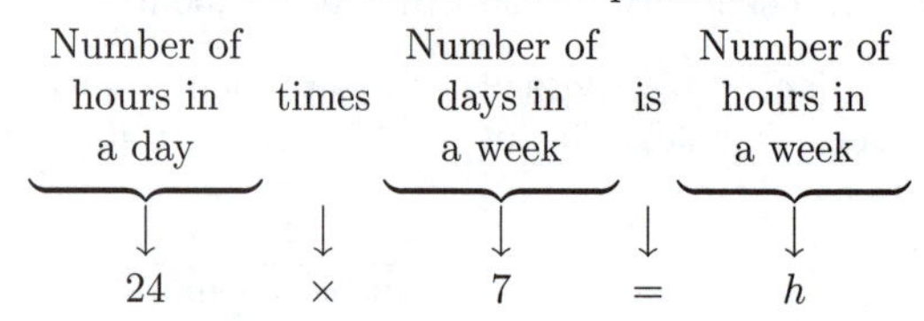

Solve. We carry out the multiplication.

$$\begin{array}{r} 2\,4 \\ \times\ \ 7 \\ \hline 1\,6\,8 \end{array}$$

Thus, $168 = h$, or $h = 168$.

Check. We can repeat the calculation. We an also estimate:

$$24 \times 7 \approx 20 \times 10 = 200 \approx 168$$

Our answer checks.

State. There are 168 hours in a week.

19. ***Familiarize.*** We visualize the situation. Let r = the number of dollars by which the average monthly rent in Houston exceeds the average monthly rent in Dallas/Fort Worth.

Rent in Dallas/Fort Worth $755	Excess rent in Houston r
Rent in Houston $778	

Translate. We translate to an equation.

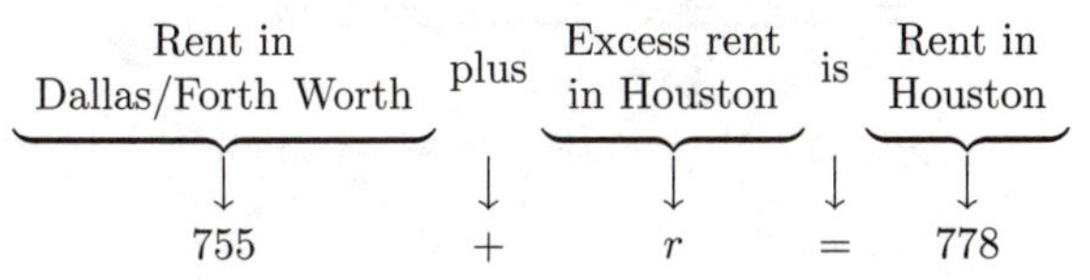

Solve. We subtract 755 on both sides of the equation.

$$755 + r = 778$$
$$755 + r - 755 = 778 - 755$$
$$r = 23$$

Check. We can add: $755 + 23 = 778$. We get the rent in Houston, so the answer checks.

State. The average monthly rent in Houston is \$23 higher than in Dallas/Fort Worth.

21. ***Familiarize.*** We first draw a picture. We let r = the average monthly rent each person will pay.

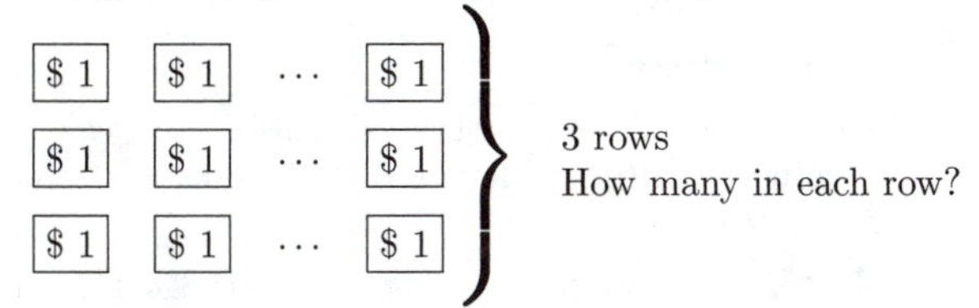

Translate. We translate to an equation.

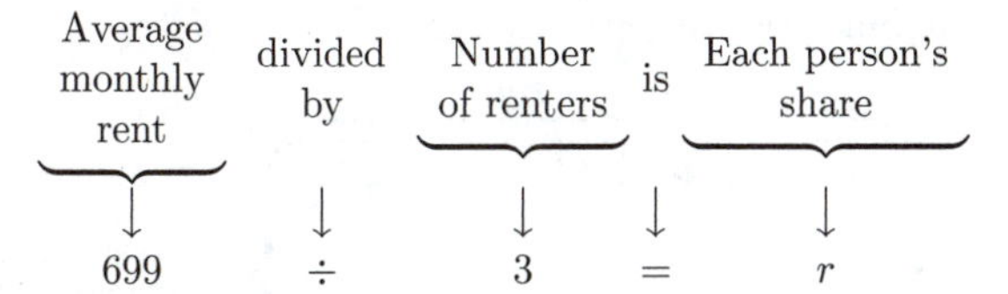

Solve. We carry out the division.

$$\begin{array}{r} 233 \\ 3\overline{)699} \\ \underline{6} \\ 9 \\ \underline{9} \\ 9 \\ \underline{9} \\ 0 \end{array}$$

Thus, $233 = r$.

Check. We can check by multiplying: $233 \cdot 3 = 699$. The answer checks.

State. Each person can expect to pay an average monthly rent of \$233.

23. ***Familiarize.*** We draw a picture. Let t = the total amount of rent a person would pay in Atlanta during a 12-month period. Repeated addition works well here.

\$773 + \$773 + · · · + \$773

12 addends

Translate. We translate to an equation.

Average monthly rent	times	Number of months	is	Total rent paid
↓	↓	↓	↓	↓
773	×	12	=	t

Solve. We carry out the multiplication.

$$\begin{array}{r} 773 \\ \times\ \ 12 \\ \hline 1546 \\ 7730 \\ \hline 9276 \end{array}$$

Thus, $9276 = t$.

Check. We repeat the calculation. The answer checks.

State. On average, a tenant would pay \$9276 in rent during a 12-month period in Atlanta.

25. ***Familiarize.*** We visualize the situation. Let p = the Colonial population in 1680.

Population in 1680 p	Increase in population 2,628,900
Population in 1780 2,780,400	

Translate. We translate to an equation

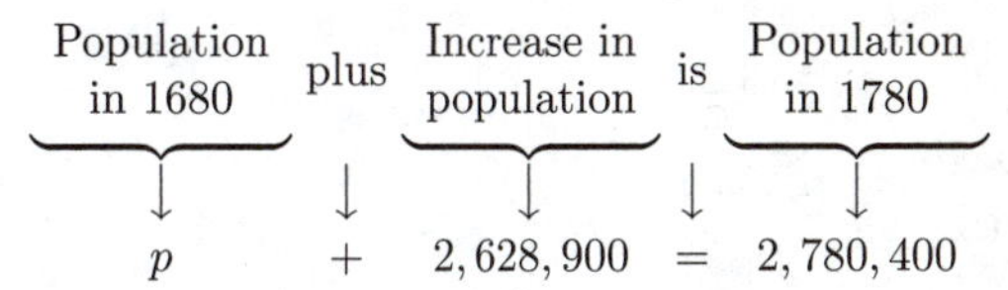

Solve. We subtract 2,628,900 on both sides of the equation.

$$p + 2,628,900 = 2,780,400$$
$$p + 2,628,900 - 2,628,900 = 2,780,400 - 2,628,900$$
$$p = 151,500$$

Check. Since $2,628,900 + 151,500 = 2,780,400$, the answer checks.

State. In 1680 the Colonial population was 151,500.

27. ***Familiarize.*** We visualize the situation. Let m = the number of motorcycles that were sold in 2006.

Motorcycles sold in 2008 521,000	Excess sales in 2006 669,000
Motorcycles sold in 2006 m	

Translate. We translate to an equation.

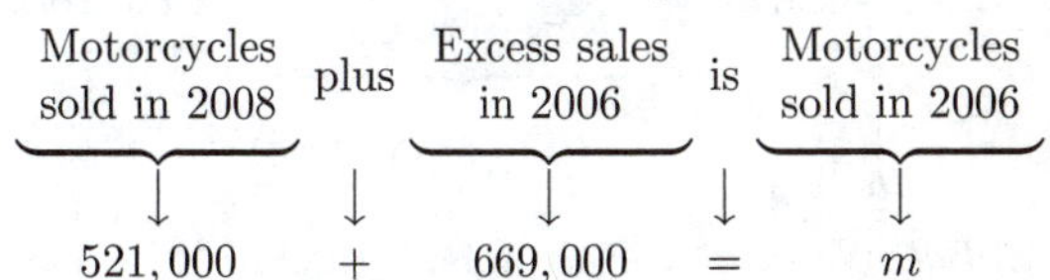

Solve. We carry out the addition.

$$\begin{array}{r} \scriptstyle 1 \\ 521,000 \\ +\,669,000 \\ \hline 1,190,000 \end{array}$$

Thus, $1,190,000 = m$.

Check. We can repeat the addition. We can also estimate: $521,000 + 669,000 \approx 500,000 + 700,000 = 1,200,000 \approx 1,190,000$

The answer seems reasonable.

State. 1,190,000 motorcycles were sold in 2006.

29. ***Familiarize.*** We draw a picture. Let g = the amount each grandchild received.

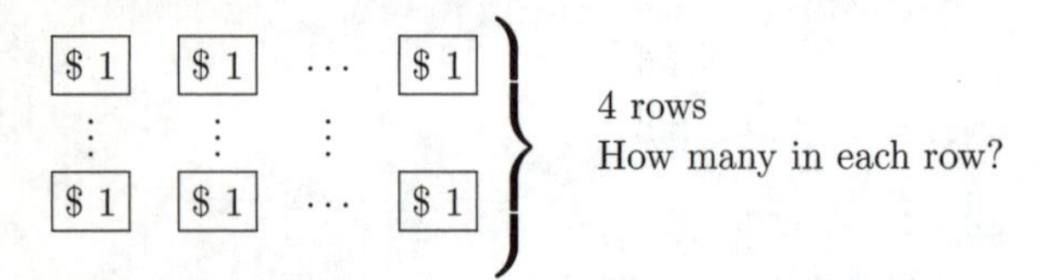

Translate. We translate to an equation.

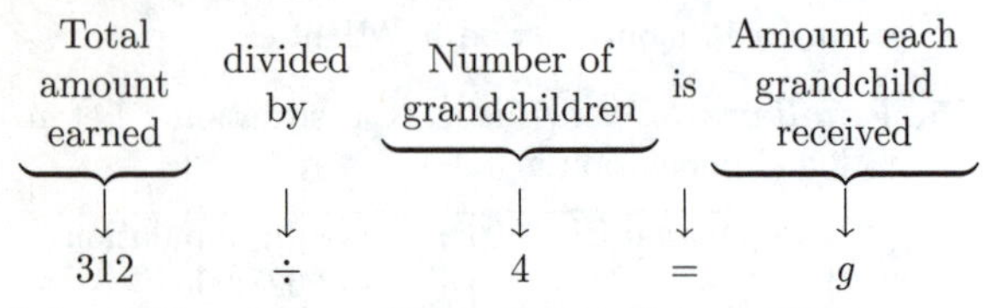

Solve. We carry out the division.

$$\begin{array}{r} 78 \\ 4\overline{)312} \\ \underline{28} \\ 32 \\ \underline{32} \\ 0 \end{array}$$

Thus, $78 = g$.

Check. We can check by multiplying the amount each grandchild received by the number of grandchildren:

$\$78 \cdot 4 = \$312$. The answer checks.

State. Each grandchild received $78.

31. ***Familiarize.*** We visualize the situation. Let b = the average monthly parking rate in Bakersfield.

Rate in Bakersfield b	Additional cost in New York City \$545
Rate in New York City \$585	

Translate. We translate to an equation.

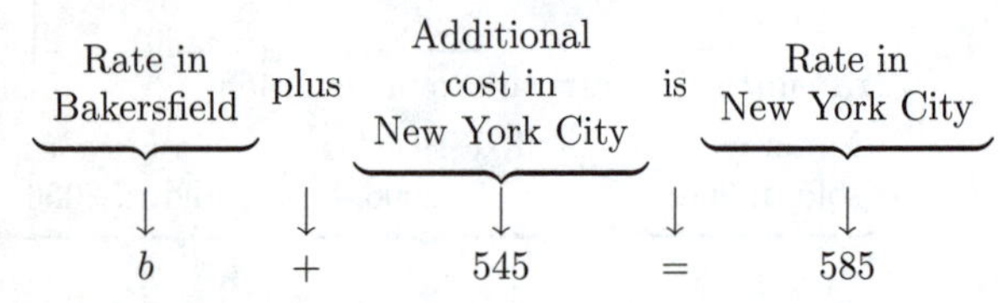

Solve. We subtract 545 on both sides of the equation.

$$b + 545 = 585$$
$$b + 545 - 545 = 585 - 545$$
$$b = 40$$

Check. We can add the rate in Bakersfield to the additional cost in New York City: $\$40 + \$545 = \$585$. The answer checks.

State. The average monthly parking rate in Bakersfield is $40.

33. ***Familiarize.*** We draw a picture of the situation. Let c = the total cost of the purchase. Repeated addition works well here.

$1019 + $1019 + ⋯ + $1019

24 addends

Translate. We translate to an equation.

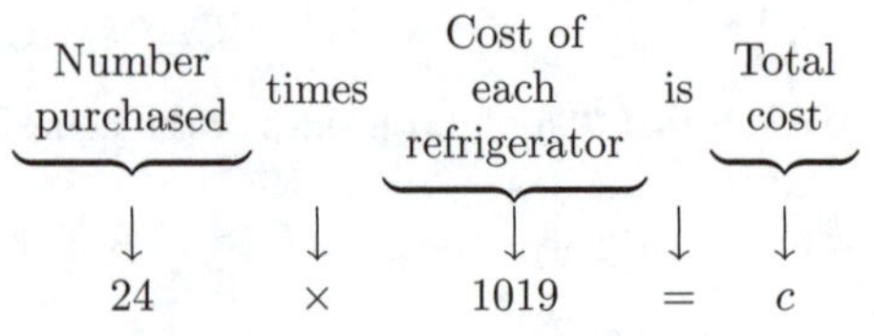

Solve. We carry out the multiplication.

$$\begin{array}{r} {\scriptstyle 1} \\ {\scriptstyle 3} \\ 1\,0\,1\,9 \\ \times \quad 2\,4 \\ \hline 4\,0\,7\,6 \\ 2\,0\,3\,8\,0 \\ \hline 2\,4,4\,5\,6 \end{array}$$

Thus, $24,456 = c$.

Check. We can repeat the calculation. We can also estimate: $24 \times 1019 \approx 24 \times 1000 \approx 24,000 \approx 24,456$. The answer checks.

State. The total cost of the purchase is $24,456.

35. ***Familiarize.*** We first draw a picture. Let w = the number of full weeks the episodes can run.

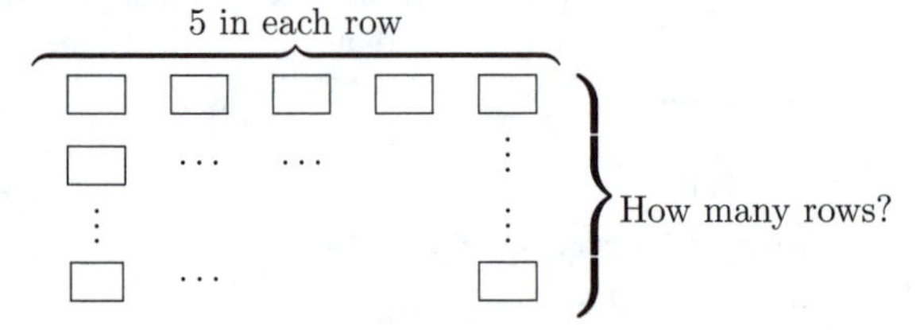

Translate. We translate to an equation.

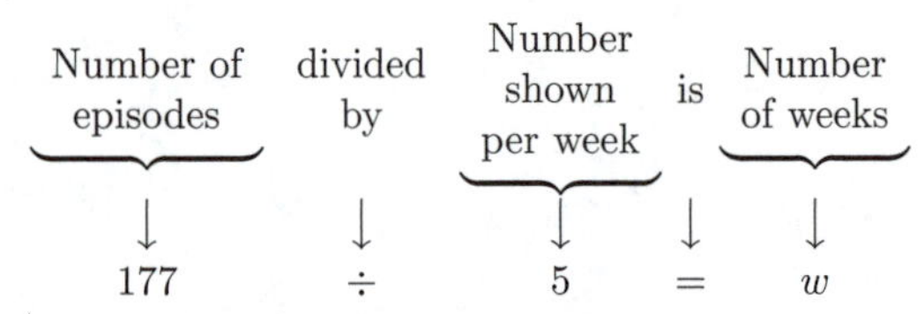

Solve. We carry out the division.

$$\begin{array}{r} 35 \\ 5\overline{)177} \\ \underline{15} \\ 27 \\ \underline{25} \\ 2 \end{array}$$

Check. We can check by multiplying the number of weeks by 5 and adding the remainder, 2:

$5 \cdot 35 = 175, \qquad 175 + 2 = 177$

State. 35 full weeks will pass before the station must start over. There will be 2 episodes left over.

37. ***Familiarize.*** We first draw a picture of the situation. Let g = the number of gallons that will be used in 7750 mi of city driving.

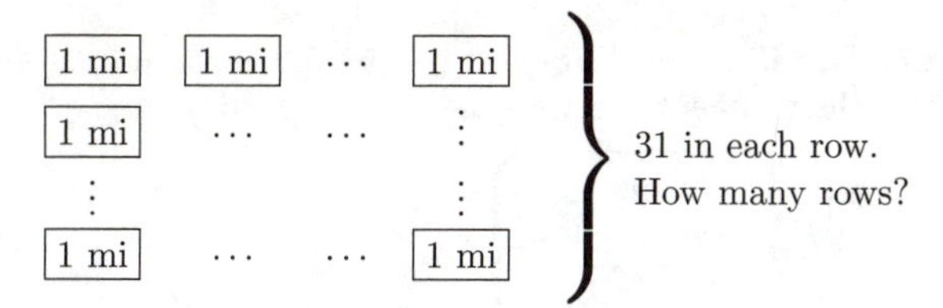

Translate. We translate to an equation.

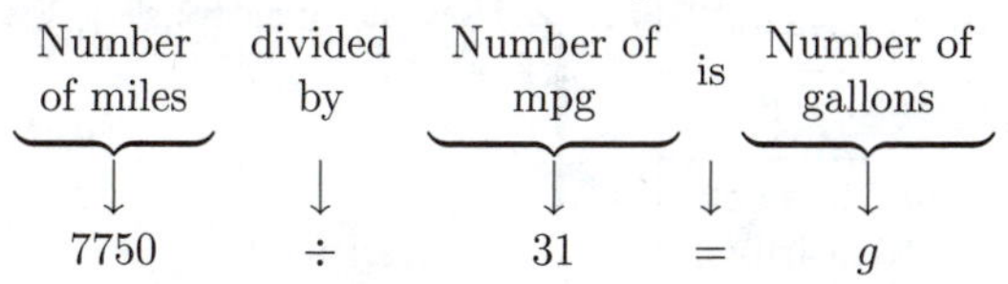

Solve. We carry out the division.

```
       2 5 0
  3 1 ⟌7 7 5 0
       6 2
       1 5 5
       1 5 5
           0
           0
           0
```

Thus, $250 = g$.

Check. We can check by multiplying the number of gallons by the number of miles per gallon: $31 \cdot 250 = 7750$. The answer checks.

State. The 2010 Hyundai Tucson will use 250 gal of gasoline in 7750 mi of city driving.

39. ***Familiarize.*** First we draw a picture. Let c = the number of columns. The number of columns is the same as the number of squares in each row.

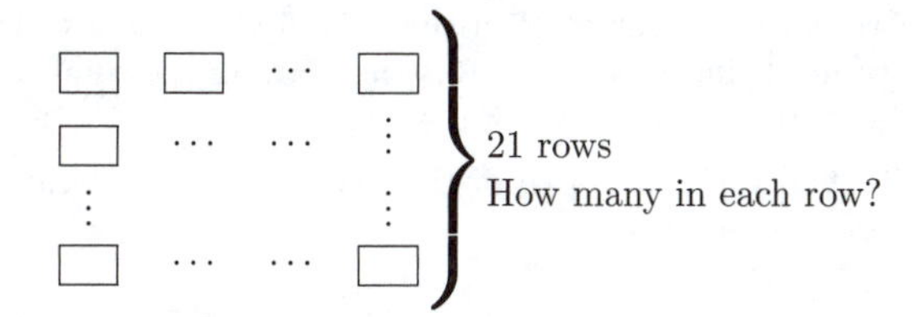

Translate. We translate to an equation.

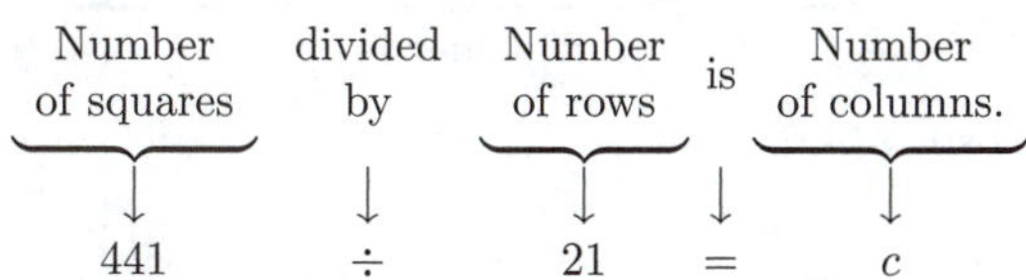

Solve. We carry out the division.

```
       2 1
  2 1 ⟌4 4 1
       4 2
         2 1
         2 1
           0
```

Thus, $21 = c$.

Check. We can check by multiplying the number of rows by the number of columns: $21 \cdot 21 = 441$. The answer checks.

State. The puzzle has 21 columns.

41. ***Familiarize.*** We first draw a picture. Let A = the area and P = the perimeter of the court, in feet.

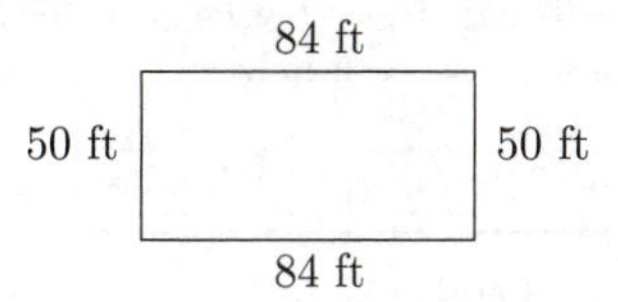

Translate. We write one equation to find the area and another to find the perimeter.

a) Using the formula for the area of a rectangle, we have

$$A = l \cdot w = 84 \cdot 50$$

b) Recall that the perimeter is the distance around the court.

$$P = 84 + 50 + 84 + 50$$

Solve. We carry out the calculations.

a)

```
      5 0
    × 8 4
    2 0 0
  4 0 0 0
  4 2 0 0
```

Thus, $A = 4200$.

b) $P = 84 + 50 + 84 + 50 = 268$

Check. We can repeat the calculation. The answers check.

State. a) The area of the court is 4200 square feet.

b) The perimeter of the court is 268 ft.

43. ***Familiarize.*** We first draw a picture. We let x = the amount of each payment.

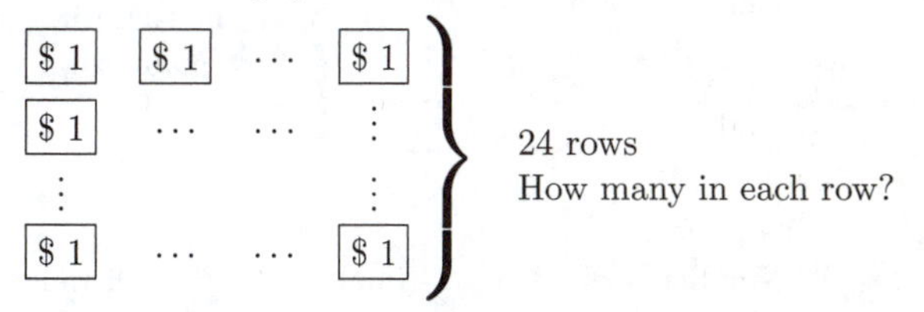

Translate. We translate to an equation.

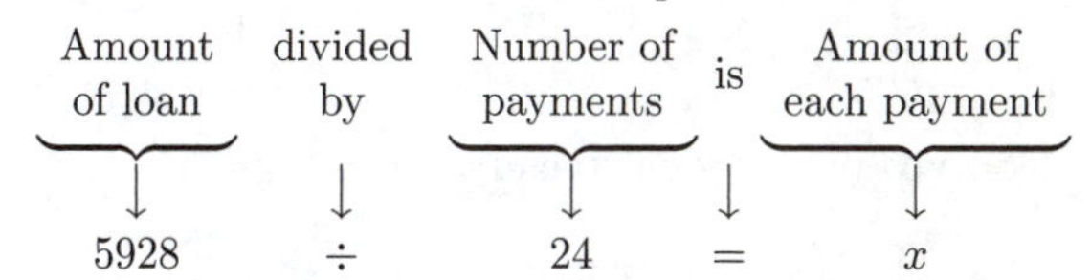

Solve. We carry out the division.

```
       2 4 7
  2 4 ⟌5 9 2 8
       4 8
       1 1 2
         9 6
         1 6 8
         1 6 8
             0
```

Thus, $247 = x$, or $x = 247$.

Check. We can check by multiplying 247 by 24: $24 \cdot 247 = 5928$. The answer checks.

State. Each payment is $247.

45. ***Familiarize.*** First we find the distance in reality between two cities that are 3 in. apart on the map. We make a drawing. Let $d =$ the distance between the cities, in miles. Repeated addition works well here.

$$\underbrace{\boxed{215 \text{ miles}} + \boxed{215 \text{ miles}} + \boxed{215 \text{ miles}}}_{3 \text{ addends}}$$

Translate.

$$\underbrace{\text{Number of miles per inch}}_{215} \quad \underbrace{\text{times}}_{\times} \quad \underbrace{\text{Number of inches}}_{3} \quad \underbrace{\text{is}}_{=} \quad \underbrace{\text{Distance, in miles}}_{d}$$

Solve. We carry out the multiplication.

$$\begin{array}{r} 2\,1\,5 \\ \times \quad 3 \\ \hline 6\,4\,5 \end{array}$$

Thus, $645 = d$.

Check. We can repeat the calculation or estimate the product. Our answer checks.

State. Two cities that are 3 in. apart on the map are 645 miles apart in reality.

Next we find the distance on the map between two cities that, in reality, are 1075 mi apart.

Familiarize. We visualize the situation. Let $m =$ the distance between the cities on the map.

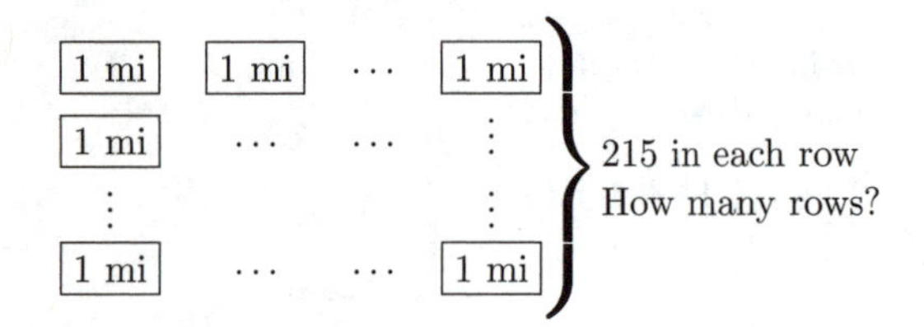

Translate.

$$\underbrace{\text{Number of miles}}_{1075} \quad \underbrace{\text{divided by}}_{\div} \quad \underbrace{\text{Number of miles per inch}}_{215} \quad \underbrace{\text{is}}_{=} \quad \underbrace{\text{Distance, in inches.}}_{m}$$

Solve. We carry out the division.

$$\begin{array}{r} 5 \\ 2\,1\,5\overline{)\,1\,0\,7\,5} \\ 1\,0\,7\,5 \\ \hline 0 \end{array}$$

Thus, $5 = m$.

Check. We can check by multiplying: $215 \cdot 5 = 1075$.

Our answer checks.

State. The cities are 5 in. apart on the map.

47. ***Familiarize.*** We draw a picture of the situation. Let $c =$ the number of cartons that can be filled.

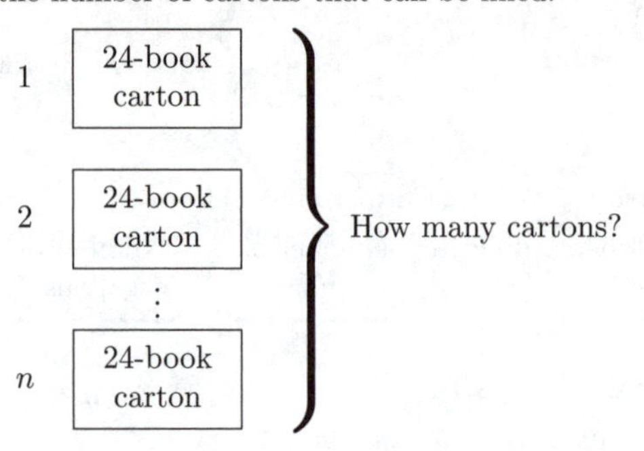

Translate.

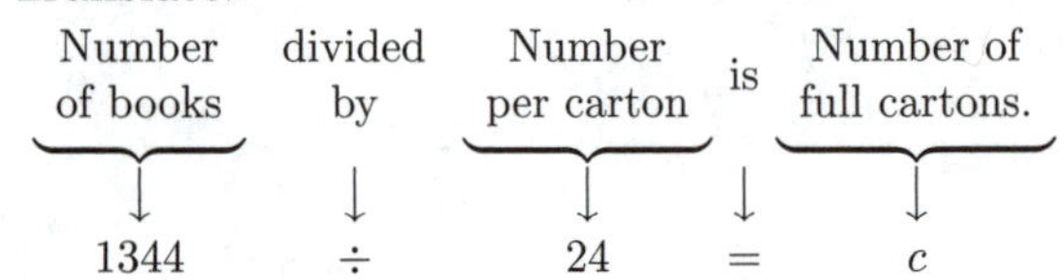

Solve. We carry out the division.

$$\begin{array}{r} 5\,6 \\ 2\,4\overline{)\,1\,3\,4\,4} \\ 1\,2\,0 \\ \hline 1\,4\,4 \\ 1\,4\,4 \\ \hline 0 \end{array}$$

Check. We can check by multiplying the number of cartons by 24: $24 \cdot 56 = 1344$. The answer checks.

State. 56 cartons can be filled.

49. ***Familiarize.*** This is a multistep problem.

We must find the total price of the 5 video games. Then we must find how many 10's there are in the total price. Let $p =$ the total price of the games.

To find the total price of the 5 video games we can use repeated addition.

$$\underbrace{\boxed{\$64} + \boxed{\$64} + \boxed{\$64} + \boxed{\$64} + \boxed{\$64}}_{5 \text{ addends}}$$

Translate.

$$\underbrace{\text{Price per game}}_{64} \quad \underbrace{\text{times}}_{\cdot} \quad \underbrace{\text{Number of games}}_{5} \quad \underbrace{\text{is}}_{=} \quad \underbrace{\text{Total price of games}}_{p}$$

Solve. First we carry out the multiplication.

$$64 \cdot 5 = p$$
$$320 = p$$

The total price of the 5 video games is \$320. Repeated addition can be used again to find how many 10's there are in \$320. We let $x =$ the number of \$10 bills required.

\$320			
\$10	\$10	···	\$10

Translate to an equation and solve.

$$10 \cdot x = 320$$
$$\frac{10 \cdot x}{10} = \frac{320}{10}$$
$$x = 32$$

Check. We repeat the calculations. The answer checks.

State. It takes 32 ten dollar bills.

51. ***Familiarize***. This is a multistep problem. We must find the total amount of the debits. Then we subtract this amount from the original balance and add the amount of the deposit. Let a = the total amount of the debits. To find this we can add.

Translate.

$$\underbrace{\text{First debit}} \text{ plus } \underbrace{\text{Second debit}} \text{ plus } \underbrace{\text{Third debit}} \text{ is } \underbrace{\text{Total amount}}$$
$$46 + 87 + 129 = a$$

Solve. First we carry out the addition.

$$\begin{array}{r} \scriptstyle 1\,2 \\ 46 \\ 87 \\ +\,129 \\ \hline 262 \end{array}$$

Thus, $262 = a$.

Now let b = the amount left in the account after the debits.

$$\underbrace{\text{Amount left}} \text{ is } \underbrace{\text{Original amount}} \text{ minus } \underbrace{\text{Amount of debits}}$$
$$b = 568 - 262$$

We solve this equation by carrying out the subtraction.

$$\begin{array}{r} 568 \\ -\,262 \\ \hline 306 \end{array}$$

Thus, $b = 306$.

Finally, let f = the final amount in the account after the deposit is made.

$$\underbrace{\text{Final amount}} \text{ is } \underbrace{\text{Amount after debits}} \text{ plus } \underbrace{\text{Amount of deposit}}$$
$$f = 306 + 94$$

We solve this equation by carrying out the addition.

$$\begin{array}{r} \scriptstyle 1\,1 \\ 306 \\ +\,94 \\ \hline 400 \end{array}$$

Thus, $f = 400$.

Check. We repeat the calculations. The answer checks.

State. There is $400 left in the account.

53. ***Familiarize***. This is a multistep problem. We begin by visualizing the situation.

One pound 3500 calories			
100 cal 15 min	100 cal 15 min	...	100 cal 15 min

Let x = the number of hundreds in 3500. Repeated addition applies here.

Translate. We translate to an equation.

$$\underbrace{\text{100 calories}} \text{ times } \underbrace{\text{How many 100's}} \text{ is 3500?}$$
$$100 \cdot x = 3500$$

Solve. We divide by 100 on both sides of the equation.

$$100 \cdot x = 3500$$
$$\frac{100 \cdot x}{100} = \frac{3500}{100}$$
$$x = 35$$

From the chart we know that doing aerobic exercise for 15 min burns 100 calories. Thus we must do 15 min of exercise 35 times in order to lose one pound. Let t = the number of minutes of aerobic exercise required to lose one pound. We translate to an equation.

$$\underbrace{\text{Number of times}} \text{ times } \underbrace{\text{Number of minutes}} \text{ is } \underbrace{\text{Total time}}$$
$$35 \times 15 = t$$

$$\begin{array}{r} 15 \\ \times\,35 \\ \hline 75 \\ 450 \\ \hline 525 \end{array}$$

Thus, $525 = t$.

Check. $525 \div 15 = 35$, so there are 35 15's in 525 min, and $35 \cdot 100 = 3500$, the number of calories that must be burned in order to lose one pound. The answer checks.

State. You must do aerobic exercise for 525 min, or 8 hr, 45 min, in order to lose one pound.

55. ***Familiarize***. This is a multistep problem. We begin by visualizing the situation.

One pound 3500 calories			
100 cal 20 min	100 cal 20 min	...	100 cal 20 min

From Exercise 53 we know that there are 35 100's in 3500. From the chart we know that golfing for 20 min will burn 100 calories. This must be done 35 times in order to lose one pound. Let t = the time it takes to lose one pound. We have:

$$t = 35 \times 20$$
$$t = 700$$

Check. $700 \div 20 = 35$, so there are 35 20's in 700 min, and $35 \cdot 100 = 3500$, the number of calories that must be burned in order to lose one pound. The answer checks.

State. You must golf for 700 min, or 11 hr, 40 min, walking, in order to lose one pound.

57. ***Familiarize***. This is a multistep problem. We will find the number of seats in each class and then add to find the total seating capacity. Let $F =$ the number of first-class seats, $E =$ the number of economy-class seats, and $T =$ the total number of seats.

Translate. We translate to three equations.

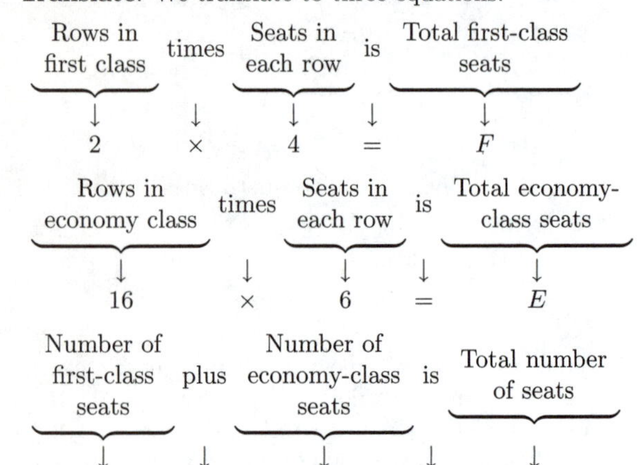

Solve. We solve the first two equations.

$2 \times 4 = F \qquad 16 \times 6 = E$

$8 = F \qquad 96 = E$

Now we substitute 8 for F and 96 for E in the third equation and add to find T.

$F + E = T$

$8 + 96 = T$

$104 = T$

Check. We repeat the calculations. The answer checks.

State. The total seating capacity of the plane is 104.

59. ***Familiarize***. This is a multistep problem. We will find the number of bones in both hands and the number in both feet and then the total of these two numbers. Let $h =$ the number of bones in two human hands, $f =$ the number of bones in two human feet, and $t =$ the total number of bones in two hands and two feet.

Translate. We translate to three equations.

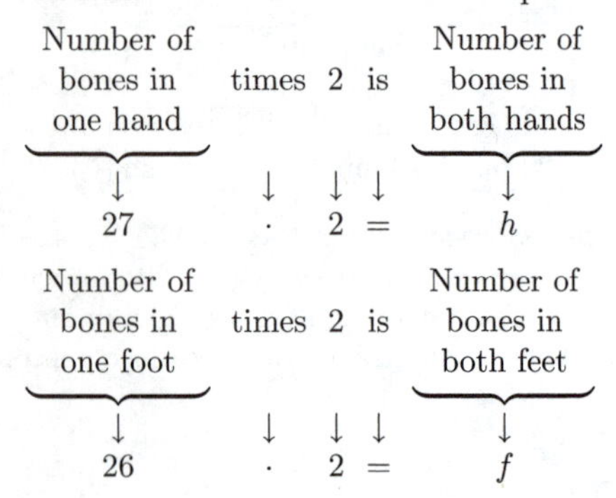

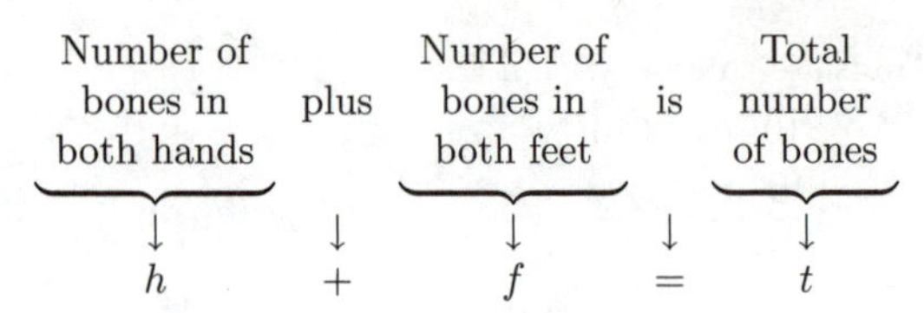

Solve. We solve each equation.

$27 \cdot 2 = h \qquad 26 \cdot 2 = f$

$54 = h \qquad 52 = f$

$h + f = t$

$54 + 52 = t$

$106 = t$

Check. We repeat the calculations. The answer checks.

State. In all, a human has 106 bones in both hands and both feet.

61. Round 234,562 to the nearest hundred.

2 3 4, 5 [6] 2
↑

The digit 5 is in the hundreds place. Consider the next digit to the right. Since the digit, 6, is 5 or higher, round 5 hundreds up to 6 hundreds. Then change all digits to the right of the hundreds place to zeros.

The answer is 234,600.

63. Round 234,562 to the nearest thousand.

2 3 4, [5] 6 2
↑

The digit 4 is in the thousands place. Consider the next digit to the right. Since the digit, 5, is 5 or higher, round 4 thousands up to 5 thousands. Then change all digits to the right of the thousands place to zeros.

The answer is 235,000.

65.

	Rounded to the nearest thousand
28,430	28,000
− 11,977	− 12,000
	16,000 ← Estimated answer

67.

	Rounded to the nearest thousand
2100	2000
+ 5800	+ 6000
	8000 ← Estimated answer

69.

	Rounded to the nearest hundred
799	800
× 887	× 900
	720,000 ← Estimated answer

71. ***Familiarize***. This is a multistep problem. First we will find the differences in the distances traveled in 1 second. Then we will find the differences for 18 seconds. Let $d =$ the difference in the number of miles light would travel per second in a vacuum and in ice. Let $g =$ the difference in the number of miles light would travel per second in a vacuum and in glass.

Translate. We translate to two equations.

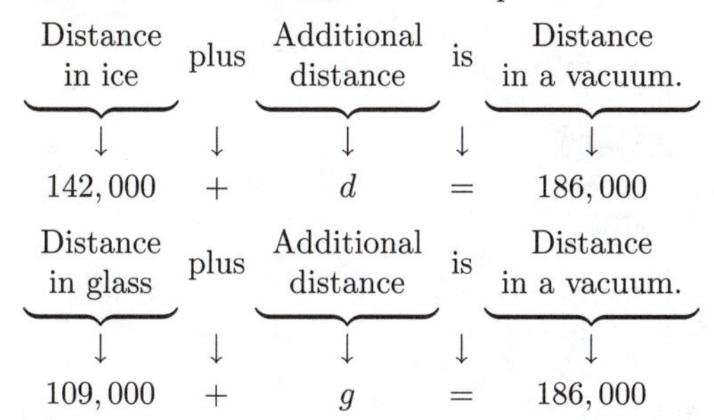

Solve. We begin by solving each equation.

$$142,000 + d = 186,000$$
$$142,000 + d - 142,000 = 186,000 - 142,000$$
$$d = 44,000$$

$$109,000 + g = 186,000$$
$$109,000 + g - 109,000 = 186,000 - 109,000$$
$$g = 77,000$$

Now to find the differences in the distances in 18 seconds, we multiply each solution by 18.

For ice: $18 \cdot 44,000 = 792,000$

For glass: $18 \cdot 77,000 = 1,386,000$

Check. We repeat the calculations. Our answers check.

State. In 18 seconds light travels 792,000 miles more in a vacuum than in ice and 1,386,000 miles more in a vacuum than in glass.

Exercise Set 1.9

1. Exponential notation for $3 \cdot 3 \cdot 3 \cdot 3$ is 3^4.

3. Exponential notation for $5 \cdot 5$ is 5^2.

5. Exponential notation for $7 \cdot 7 \cdot 7 \cdot 7 \cdot 7$ is 7^5.

7. Exponential notation for $10 \cdot 10 \cdot 10$ is 10^3.

9. $7^2 = 7 \cdot 7 = 49$

11. $9^3 = 9 \cdot 9 \cdot 9 = 729$

13. $12^4 = 12 \cdot 12 \cdot 12 \cdot 12 = 20,736$

15. $3^5 = 3 \cdot 3 \cdot 3 \cdot 3 \cdot 3 = 243$

17. $12 + (6 + 4) = 12 + 10$ Doing the calculation inside the parentheses
$= 22$ Adding

19. $52 - (40 - 8) = 52 - 32$ Doing the calculation inside the parentheses
$= 20$ Subtracting

21. $1000 \div (100 \div 10)$
$= 1000 \div 10$ Doing the calculation inside the parentheses
$= 100$ Dividing

23. $(256 \div 64) \div 4 = 4 \div 4$ Doing the calculation inside the parentheses
$= 1$ Dividing

25. $(2 + 5)^2 = 7^2$ Doing the calculation inside the parentheses
$= 49$ Evaluating the exponential expression

27. $16 \cdot 24 + 50 = 384 + 50$ Doing all multiplications and divisions in order from left to right
$= 434$ Doing all additions and subtractions in order from left to right

29. $83 - 7 \cdot 6 = 83 - 42$ Doing all multiplications and divisions in order from left to right
$= 41$ Doing all additions and subtractions in order from left to right

31. $10 \cdot 10 - 3 \times 4$
$= 100 - 12$ Doing all multiplications and divisions in order from left to right
$= 88$ Doing all additions and subtractions in order from left to right

33. $4^3 \div 8 - 4$
$= 64 \div 8 - 4$ Evaluating the exponential expression
$= 8 - 4$ Doing all multiplications and divisions in order from left to right
$= 4$ Doing all additions and subtractions in order from left to right

35. $17 \cdot 20 - (17 + 20)$
$= 17 \cdot 20 - 37$ Carrying out the operation inside parentheses
$= 340 - 37$ Doing all multiplications and divisions in order from left to right
$= 303$ Doing all additions and subtractions in order from left to right

37. $5(6 - 3) - (18 - 4)$
$= 5(3) - 14$ Doing the calculations inside the parentheses
$= 15 - 14$ Multiplying
$= 1$ Subtracting

39. $15(7 - 3) \div 12$
$= 15(4) \div 12$ Doing the calculation inside the parentheses
$= 60 \div 12$ Doing all multiplications and
$= 5$ divisions in order from left to right

41. $(11-8)^2-(18-16)^2$

$=3^2-2^2$ Doing the calculations inside the parentheses

$=9-4$ Evaluating the exponential expressions

$=5$ Subtracting

43. $4^2+8^2\div 2^2=16+64\div 4$

$=16+16$

$=32$

45. $10^3-10\cdot 6-(4+5\cdot 6)=10^3-10\cdot 6-(4+30)$

$=10^3-10\cdot 6-34$

$=1000-10\cdot 6-34$

$=1000-60-34$

$=940-34$

$=906$

47. $6\times 11-(7+3)\div 5-(6-4)=6\times 11-10\div 5-2$

$=66-2-2$

$=64-2$

$=62$

49. $120-3^3\cdot 4\div(5\cdot 6-6\cdot 4)$

$=120-3^3\cdot 4\div(30-24)$

$=120-3^3\cdot 4\div 6$

$=120-27\cdot 4\div 6$

$=120-108\div 6$

$=120-18$

$=102$

51. $3\cdot(2+8)^2-5(4-3)^2$

$=3\cdot 10^2-5\cdot 1^2$ Doing the calculations inside the parentheses

$=3\cdot 100-5\cdot 1$ Evaluating the exponential expressions

$=300-5$ Doing all multiplications and divisions in order from left to right

$=295$ Subtracting

53. $2^3\cdot 2^8\div 2^6=8\cdot 256\div 64$ Evaluating the exponential expressions

$=2048\div 64$ Doing all multiplications and divisions in order from left to right

$=32$

55. $3(5^2)-6(9-2)+4(7-1)$

$=3\cdot 25-6\cdot 7+4\cdot 6$ Doing the calculations inside the parentheses

$=75-42+24$ Doing all multiplications and divisions in order from left to right

$=33+24$ Doing all additions and subtractions in order from left to right

$=57$

57. $24\div(7-5)^2\cdot 3^2$

$=24\div 2^2\cdot 3^2$ Doing the calculation inside the parentheses

$=24\div 4\cdot 9$ Evaluating the exponential expressions

$=6\cdot 9$ Doing all multiplications and divisions in order from left to right

$=54$

59. We add the numbers and then divide by the number of addends.

$$\frac{\$64+\$97+\$121}{3}=\frac{\$282}{3}=\$94$$

61. We add the numbers and then divide by the number of addends.

$$\frac{320+128+276+880}{4}=\frac{1604}{4}=401$$

63. $2[3+5(9+1)]$

$=2[3+5\cdot 10]$

$=2[3+50]$

$=2\cdot 53$

$=106$

65. $6+3[15-2(7-3)+1]$

$=6+3[15-2\cdot 4+1]$

$=6+3[15-8+1]$

$=6+3[7+1]$

$=6+3\cdot 8$

$=6+24$

$=30$

67. $25\div 5+10[3(9-2)-2(10-7)]$

$=25\div 5+10[3\cdot 7-2\cdot 3]$

$=25\div 5+10[21-6]$

$=25\div 5+10\cdot 15$

$=5+150$

$=155$

69. $30-\{50-[4+5(3+4)]\}$

$=30-\{50-[4+5\cdot 7]\}$

$=30-\{50-[4+35]\}$

$=30-\{50-39\}$

$=30-11$

$=19$

71. $15-2\cdot(9-7)-3[5-2(3-1)]$

$=15-2\cdot 2-3[5-2\cdot 2]$

$=15-2\cdot 2-3[5-4]$

$=15-2\cdot 2-3\cdot 1$

$=15-4-3$

$=11-3$

$=8$

73. $6(7-2)\div 5[8-(7-1)]$

$=6\cdot 5\div 5[8-6]$

$=6\cdot 5\div 5\cdot 2$

$=30\div 5\cdot 2$

$=6\cdot 2$

$=12$

75. $8\times 13+\{42\div[18-(6+5)]\}$

$=8\times 13+\{42\div[18-11]\}$

$=8\times 13+\{42\div 7\}$

$=8\times 13+6$

$=104+6$

$=110$

77. $[14-(3+5)\div 2]-[18\div(8-2)]$

$=[14-8\div 2]-[18\div 6]$

$=[14-4]-3$

$=10-3$

$=7$

79. $(82-14)\times[(10+45\div 5)-(6\cdot 6-5\cdot 5)]$
$= (82-14)\times[(10+9)-(36-25)]$
$= (82-14)\times[19-11]$
$= 68\times 8$
$= 544$

81. $2+(4+3^2)-2\{18-3[7-(8-6)]+1\}$
$= 2+(4+9)-2\{18-3[7-2]+1\}$
$= 2+13-2\{18-3\cdot 5+1\}$
$= 2+13-2\{18-15+1\}$
$= 2+13-2\{3+1\}$
$= 2+13-2\cdot 4$
$= 2+13-8$
$= 15-8$
$= 7$

83. $1+\{7[4(5-2)^3-8(2^4-3\cdot 5)]-3-2\}$
$= 1+\{7[4\cdot 3^3-8(16-3\cdot 5)]-3-2\}$
$= 1+\{7[4\cdot 27-8(16-15)]-3-2\}$
$= 1+\{7[4\cdot 27-8\cdot 1]-3-2\}$
$= 1+\{7[108-8]-3-2\}$
$= 1+\{7\cdot 100-3-2\}$
$= 1+\{700-3-2\}$
$= 1+\{697-2\}$
$= 1+695$
$= 696$

85.
$$x+341=793$$
$$x+341-341=793-341$$
$$x=452$$

The solution is 452.

87.
$$7\cdot x=91$$
$$\frac{7\cdot x}{7}=\frac{91}{7}$$
$$x=13$$

The solution is 13.

89.
$$3240=y+898$$
$$3240-898=y+898-898$$
$$2342=y$$

The solution is 2342.

91.
$$25\cdot t=625$$
$$\frac{25\cdot t}{25}=\frac{625}{25}$$
$$t=25$$

The solution is 25.

93. $1+5\cdot 4+3=1+20+3$
$= 24$ Correct answer

To make the incorrect answer correct we add parentheses:

$1+5\cdot(4+3)=36$

95. $12\div 4+2\cdot 3-2=3+6-2$
$= 7$ Correct answer

To make the incorrect answer correct we add parentheses:

$12\div(4+2)\cdot 3-2=4$

Chapter 1 Concept Reinforcement

1. The statement is true. See page 42 in the text.

2. $a\div a=\dfrac{a}{a}=1$, $a\neq 0$; the statement is true.

3. $a\div 0$ is not defined, so the statement is false.

4. The statement is false. For example, $1+1=3$ is not a true equation.

5. The statement is true. See page 72 in the text.

6. The statement is false. For example, the average of 3, 6, and 12 is $\dfrac{3+6+12}{3}=\dfrac{21}{3}=7$ rather than the middle number, 6.

Chapter 1 Important Concepts

1. 4 3 [2], 0 7 9

The digit 2 names the number of thousands.

2.
$$\begin{array}{r} \scriptstyle 1\ \ \ 1\ 1\ \\ 36{,}047 \\ +\ 29{,}255 \\ \hline 65{,}302 \end{array}$$

3.
$$\begin{array}{r} \scriptstyle 7\ 9\ 15 \\ 4\ \not{8}\ \not{0}\ \not{5} \\ -1\ 5\ 6\ 8 \\ \hline 3\ 2\ 3\ 7 \end{array}$$

4.
$$\begin{array}{rl} \scriptstyle 2\ 1\ \ & \\ \scriptstyle 1\ \ \ & \\ \scriptstyle 7\ 3\ \ & \\ 6\ 8\ 4 & \\ \times\ 3\ 2\ 9 & \\ \hline 6\ 1\ 5\ 6 & \text{Multiplying by 9} \\ 1\ 3\ 6\ 8\ 0 & \text{Multiplying by 20} \\ 2\ 0\ 5\ 2\ 0\ 0 & \text{Multiplying by 300} \\ \hline 2\ 2\ 5{,}0\ 3\ 6 & \end{array}$$

5.
$$\begin{array}{r} 3\ 1\ 5 \\ 27\overline{)\,8\ 5\ 1\ 9} \\ 8\ 1\ \ \ \ \\ \hline 4\ 1\ \ \\ 2\ 7\ \ \\ \hline 1\ 4\ 9 \\ 1\ 3\ 5 \\ \hline 1\ 4 \end{array}$$

The answer is 315 R 14.

6. Round 36,468 to the nearest hundred.

3 6, 4 [6] 8
↑

The digit 4 is in the hundreds place. Consider the next digit to the right. Since the digit, 6, is 5 or higher, round 4 hundreds up to 5 hundreds. Then change the digits to the right of the hundreds digit to zeros.

The answer is 36,500.

7. Round 36,468 to the nearest thousand.

3 6, [4] 6 8
↑

The digit 6 is in the thousands place. Consider the next digit to the right. Since the digit, 4, is 4 or lower, round down, meaning that 6 thousands stays as 6 thousands. Then change the digits to the right of the thousands digit to zeros.

The answer is 36,000.

8. Since 78 is to the left of 81 on the number line, $78 < 81$.

9. $24 \cdot x = 864$

$$\frac{24 \cdot x}{24} = \frac{864}{24} \quad \text{Dividing by 24}$$

$$x = 36$$

Check: $24 \cdot x = 864$

$24 \cdot 36 \; ? \; 864$

$864 \;|\;$ TRUE

The solution is 36.

10. $6^3 = 6 \cdot 6 \cdot 6 = 216$

Chapter 1 Review Exercises

1. 4, 6 7 [8] , 9 5 2

The digit 8 means 8 thousands.

2. 1 [3] , 7 6 8, 9 4 0

The digit 3 names the number of millions.

3. 2793 = 2 thousands + 7 hundreds + 9 tens + 3 ones

4. 56,078 = 5 ten thousands + 6 thousands + 0 hundreds + 7 tens + 8 ones, or 5 ten thousands + 6 thousands + 7 tens + 8 ones

5. 4,007,101 = 4 millions + 0 hundred thousands + 0 ten thousands + 7 thousands + 1 hundred + 0 tens + 1 one, or 4 millions + 7 thousands + 1 hundred + 1 one

6.

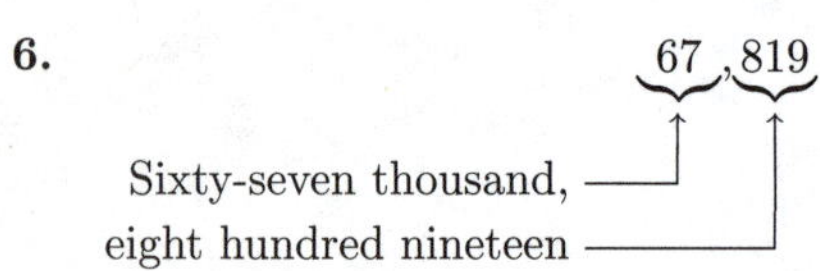

7.

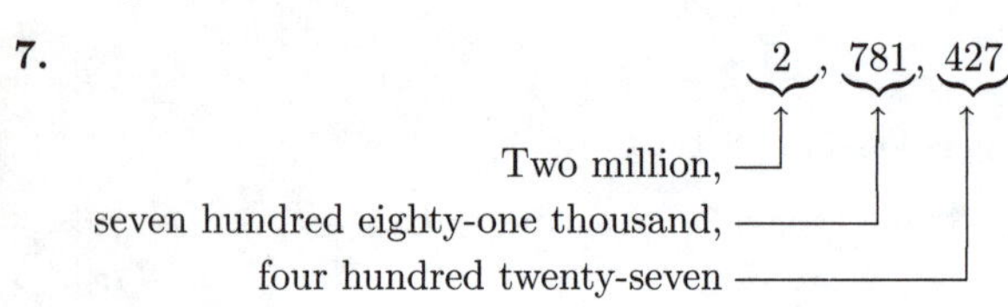

8.

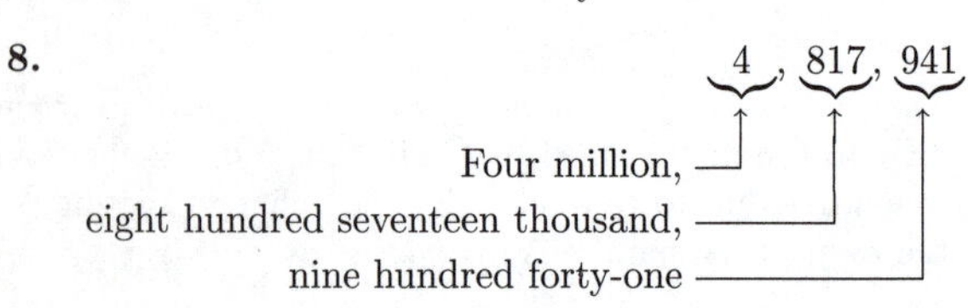

9. Four hundred seventy-six thousand, five hundred eighty-eight

Standard notation is 476, 588.

10. One billion, six hundred twenty million,

Standard notation is 1 ,620,000,000.

11.

$$\begin{array}{r} {}^{1\;\;1} \\ 7\,3\,0\,4 \\ +\;6\,9\,6\,8 \\ \hline 1\,4,2\,7\,2 \end{array}$$

12.

$$\begin{array}{r} 2\,7,6\,0\,9 \\ +\;3\,8,4\,1\,5 \\ \hline 6\,6,0\,2\,4 \end{array}$$

13.

$$\begin{array}{r} 2\,7\,0\,3 \\ 4\,1\,2\,5 \\ 6\,0\,0\,4 \\ +\;8\,9\,5\,6 \\ \hline 2\,1,7\,8\,8 \end{array}$$

14.

$$\begin{array}{r} 9\,1,4\,2\,6 \\ +\;\;7,4\,9\,5 \\ \hline 9\,8,9\,2\,1 \end{array}$$

15. (borrow marks: 7 9 13 15)

$$\begin{array}{r} 8\,0\,4\,5 \\ -\;2\,8\,9\,7 \\ \hline 5\,1\,4\,8 \end{array}$$

16. (borrow marks: 8 9 9 11)

$$\begin{array}{r} 9\,0\,0\,1 \\ -\;7\,3\,1\,2 \\ \hline 1\,6\,8\,9 \end{array}$$

17. (borrow marks: 5 9 9 13)

$$\begin{array}{r} 6\,0\,0\,3 \\ -\;3\,7\,2\,9 \\ \hline 2\,2\,7\,4 \end{array}$$

18. (borrow marks: 2 16 13 9 15)

$$\begin{array}{r} 3\,7,4\,0\,5 \\ -\;1\,9,6\,4\,8 \\ \hline 1\,7,7\,5\,7 \end{array}$$

19.

$$\begin{array}{r} 1\,7,0\,0\,0 \\ \times\;\;\;3\,0\,0 \\ \hline 5,1\,0\,0,0\,0\,0 \end{array}$$

Multiplying by 300 (Write 00 and then multiply 17,000 by 3.)

20.

$$\begin{array}{r} {}^{6\;3\;4} \\ 7\,8\,4\,6 \\ \times\;\;8\,0\,0 \\ \hline 6,2\,7\,6,8\,0\,0 \end{array}$$

Multiplying by 800 (Write 00 and then multiply 7846 by 8.)

21.
$$\begin{array}{r l}
{}^{1\,3} & \\
{}^{2\,5} & \\
{}^{2\,4} & \\
7\,2\,6 & \\
\times\ 6\,9\,8 & \\
\hline
5\,8\,0\,8 & \text{Multiplying by 8}\\
6\,5\,3\,4\,0 & \text{Multiplying by 9}\\
4\,3\,5\,6\,0\,0 & \text{Multiplying by 6}\\
\hline
5\,0\,6,7\,4\,8 &
\end{array}$$

22.
$$\begin{array}{r l}
{}^{3\,2} & \\
{}^{6\,4} & \\
5\,8\,7 & \\
\times\ \ 4\,7 & \\
\hline
4\,1\,0\,9 & \text{Multiplying by 7}\\
2\,3\,4\,8\,0 & \text{Multiplying by 4}\\
\hline
2\,7,5\,8\,9 &
\end{array}$$

23.
$$\begin{array}{r}
8\,3\,0\,5\\
\times\ \ \ 6\,4\,2\\
\hline
1\,6\,6\,1\,0\\
3\,3\,2\,2\,0\,0\\
4\,9\,8\,3\,0\,0\,0\\
\hline
5,3\,3\,1,8\,1\,0
\end{array}$$

24.
$$\begin{array}{r}
1\,2\\
5\,\overline{)6\,3}\\
\underline{5}\\
1\,3\\
\underline{1\,0}\\
3
\end{array}$$

The answer is 12 R 3.

25.
$$\begin{array}{r}
5\\
1\,6\,\overline{)8\,0}\\
\underline{8\,0}\\
0
\end{array}$$

The answer is 5.

26.
$$\begin{array}{r}
9\,1\,3\\
7\,\overline{)6\,3\,9\,4}\\
\underline{6\,3}\\
9\\
\underline{7}\\
2\,4\\
\underline{2\,1}\\
3
\end{array}$$

The answer is 913 R 3.

27.
$$\begin{array}{r}
3\,8\,4\\
8\,\overline{)3\,0\,7\,3}\\
\underline{2\,4}\\
6\,7\\
\underline{6\,4}\\
3\,3\\
\underline{3\,2}\\
1
\end{array}$$

The answer is 384 R 1.

28.
$$\begin{array}{r}
4\\
6\,0\,\overline{)2\,8\,6}\\
\underline{2\,4\,0}\\
4\,6
\end{array}$$

The answer is 4 R 46.

29.
$$\begin{array}{r}
5\,4\\
7\,9\,\overline{)4\,2\,6\,6}\\
\underline{3\,9\,5}\\
3\,1\,6\\
\underline{3\,1\,6}\\
0
\end{array}$$

The answer is 54.

30.
$$\begin{array}{r}
4\,5\,2\\
3\,8\,\overline{)1\,7,1\,7\,6}\\
\underline{1\,5\,2}\\
1\,9\,7\\
\underline{1\,9\,0}\\
7\,6\\
\underline{7\,6}\\
0
\end{array}$$

The answer is 452.

31.
$$\begin{array}{r}
5\,0\,0\,8\\
1\,4\,\overline{)7\,0,1\,1\,2}\\
\underline{7\,0}\\
1\,1\,2\\
\underline{1\,1\,2}\\
0
\end{array}$$

The answer is 5008.

32.
$$\begin{array}{r}
4\,3\,8\,9\\
1\,2\,\overline{)5\,2,6\,6\,8}\\
\underline{4\,8}\\
4\,6\\
\underline{3\,6}\\
1\,0\,6\\
\underline{9\,6}\\
1\,0\,8\\
\underline{1\,0\,8}\\
0
\end{array}$$

The answer is 4389.

33.
$$\begin{array}{r}
3\,2\,0\\
1\,0\,0\,\overline{)3\,2,0\,0\,0}\\
\underline{3\,0\,0}\\
2\,0\,0\\
\underline{2\,0\,0}\\
0\\
\underline{0}\\
0
\end{array}$$

The answer is 320.

34. Round 345,759 to the nearest hundred.

$$3\,4\,5,7\,\boxed{5}\,9$$
$$\uparrow$$

The digit 7 is in the hundreds place. Consider the next digit to the right. Since the digit, 5, is 5 or higher, round 7 hundreds up to 8 hundreds. Then change the digits to the right of the hundreds digit to zero.

The answer is 345,800.

35. Round 345,759 to the nearest ten.

3 4 5, 7 5 $\boxed{9}$
↑

The digit 5 is in the tens place. Consider the next digit to the right. Since the digit, 9, is 5 or higher, round 5 tens up to 6 tens. Then change the digit to the right of the tens digit to zero.

The answer is 345,760.

36. Round 345,759 to the nearest thousand.

3 4 5, $\boxed{7}$ 5 9
↑

The digit 5 is in the thousands place. Consider the next digit to the right. Since the digit, 7, is 5 or higher, round 5 thousands up to 6 thousands. Then change the digits to the right of the thousands digit to zero.

The answer is 346,000.

37. Since 67 is to the right of 56 on the number line, $67 > 56$.

38. Since 1 is to the left of 23 on the number line, $1 < 23$.

39.

	Rounded to the nearest hundred
4 1, 3 4 8	4 1, 3 0 0
+ 1 9, 7 4 9	+ 1 9, 7 0 0
	6 1, 0 0 0 ← Estimated answer

40.

	Rounded to the nearest hundred
3 8, 6 5 2	3 8, 7 0 0
− 2 4, 5 4 9	− 2 4, 5 0 0
	1 4, 2 0 0 ← Estimated answer

41.

	Rounded to the nearest hundred
3 9 6	4 0 0
× 7 4 8	× 7 0 0
	2 8 0, 0 0 0 ← Estimated answer

42.
$$46 \cdot n = 368$$
$$\frac{46 \cdot n}{46} = \frac{368}{46}$$
$$n = 8$$

Check: $46 \cdot n = 368$
$46 \cdot 8 \ ? \ 368$
$368 \ |$ TRUE

The solution is 8.

43.
$$47 + x = 92$$
$$47 + x - 47 = 92 - 47$$
$$x = 45$$

Check: $47 + x = 92$
$47 + 45 \ ? \ 92$
$92 \ |$ TRUE

The solution is 45.

44.
$$1 \cdot y = 58$$
$$y = 58 \qquad (1 \cdot y = y)$$

The number 58 checks. It is the solution.

45.
$$24 = x + 24$$
$$24 - 24 = x + 24 - 24$$
$$0 = x$$

The number 0 checks. It is the solution.

46. Exponential notation for $4 \cdot 4 \cdot 4$ is 4^3.

47. $10^4 = 10 \cdot 10 \cdot 10 \cdot 10 = 10,000$

48. $6^2 = 6 \cdot 6 = 36$

49. $8 \cdot 6 + 17 = 48 + 17$ Multiplying
$= 65$ Adding

50. $10 \cdot 24 - (18 + 2) \div 4 - (9 - 7)$
$= 10 \cdot 24 - 20 \div 4 - 2$ Doing the calculations inside the parentheses
$= 240 - 5 - 2$ Multiplying and dividing
$= 235 - 2$ Subtracting from
$= 233$ left to right

51. $(80 \div 16) \times [(20 - 56 \div 8) + (8 \cdot 8 - 5 \cdot 5)]$
$= 5 \times [(20 - 7) + (64 - 25)]$
$= 5 \times [13 + 39]$
$= 5 \times 52$
$= 260$

52. We add the numbers and divide by the number of addends.
$$\frac{157 + 170 + 168}{3} = \frac{495}{3} = 165$$

53. ***Familiarize.*** Let x = the additional amount of money, in dollars, Natasha needs to buy the desk.

Translate.

Money available plus Additional amount is Price of desk
↓ ↓ ↓ ↓ ↓
$196 + x = 698$

Solve. We subtract 196 on both sides of the equation.
$$196 + x = 698$$
$$196 + x - 196 = 698 - 196$$
$$x = 502$$

Check. We can estimate.
$$196 + 502 \approx 200 + 500 \approx 700 \approx 698$$

The answer checks.

State. Natasha needs $502 dollars.

54. ***Familiarize.*** Let b = the balance in Toni's account after the deposit.

Translate.

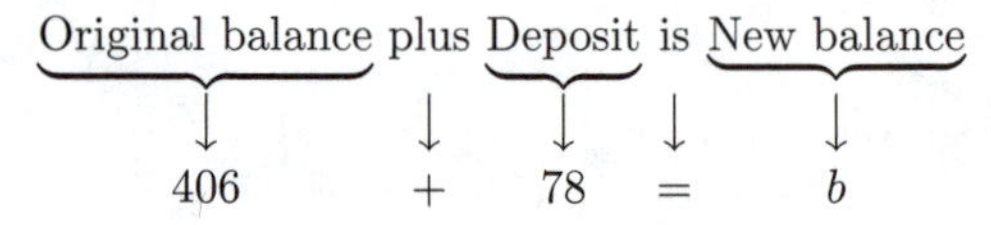

Solve. We add on the left side.

$$406 + 78 = b$$
$$484 = b$$

Check. We can repeat the calculation. The answer checks.

State. The new balance is $484.

55. ***Familiarize***. Let $y =$ the year in which the copper content of pennies was reduced.

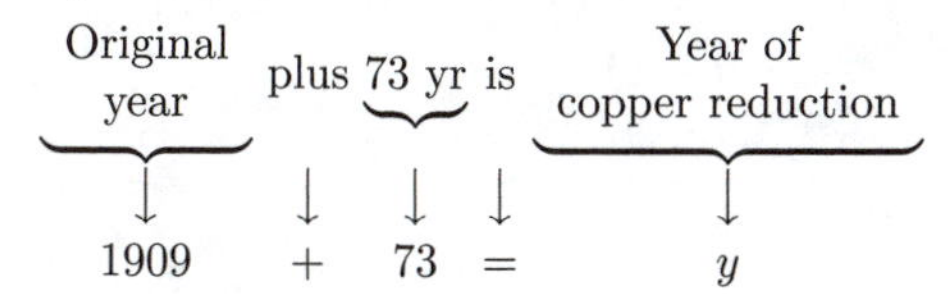

Solve. We add on the left side.

$$1909 + 73 = y$$
$$1982 = y$$

Check. We can estimate.

$$1909 + 73 \approx 1910 + 70 \approx 1980 \approx 1982$$

The answer checks.

State. The copper content of pennies was reduced in 1982.

56. ***Familiarize***. We first make a drawing. Let $c =$ the number of cartons filled.

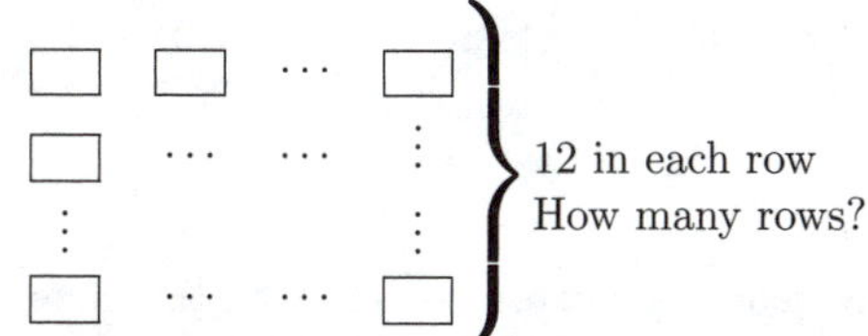

Translate.

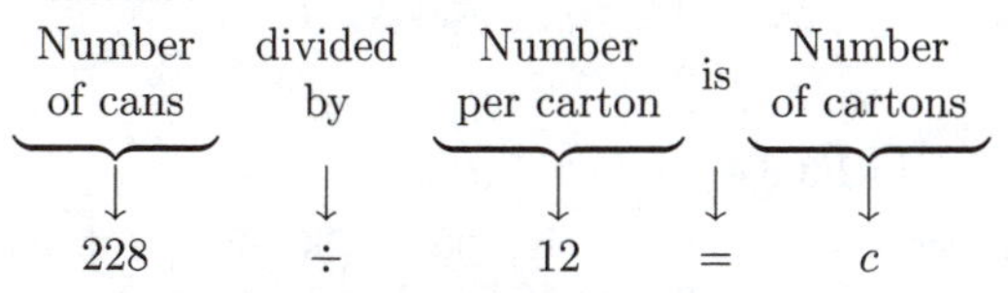

Solve. We carry out the division.

$$\begin{array}{r} 19 \\ 12\overline{)228} \\ 12 \\ \hline 108 \\ 108 \\ \hline 0 \end{array}$$

Thus, $19 = c$, or $c = 19$.

Check. We can check by multiplying: $12 \cdot 19 = 228$. Our answer checks.

State. 19 cartons were filled.

57. ***Familiarize***. This is a multistep problem. Let $s =$ the cost of 13 stoves, $r =$ the cost of 13 refrigerators, and $t =$ the total cost of the stoves and refrigerators.

Translate.

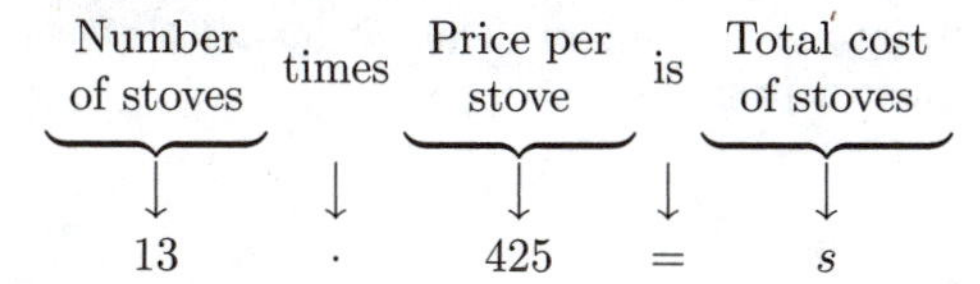

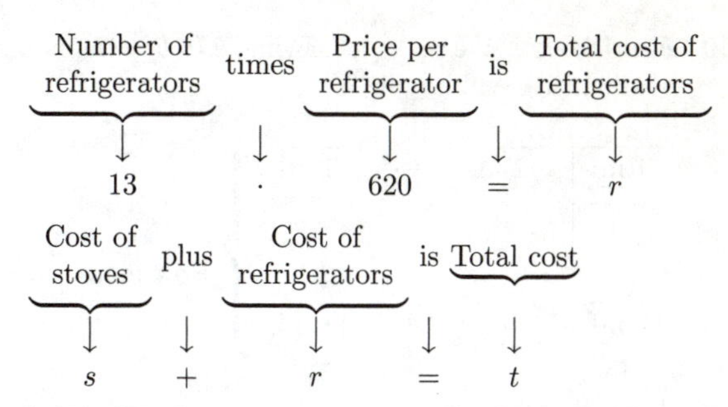

Solve. We first carry out the multiplications in the first two equations.

$$13 \cdot 425 = s \qquad 13 \cdot 620 = r$$
$$5525 = s \qquad 8060 = r$$

Now we substitute 5525 for s and 8060 for r in the third equation and then add on the left side.

$$s + r = t$$
$$5525 + 8060 = t$$
$$13,585 = t$$

Check. We repeat the calculations. The answer checks.

State. The total cost was $13,585.

58. ***Familiarize***. Let $b =$ the number of beehives the farmer needs.

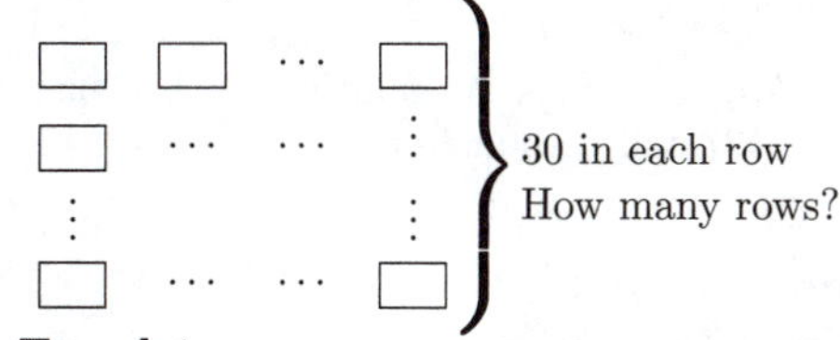

Translate.

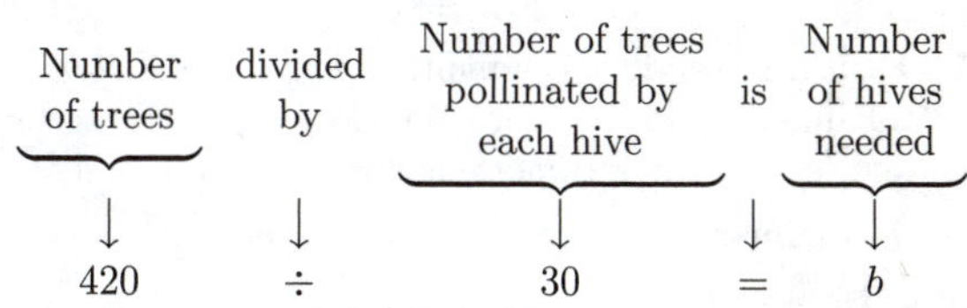

Solve. We carry out the division.

$$\begin{array}{r} 14 \\ 30\overline{)420} \\ 30 \\ \hline 120 \\ 120 \\ \hline 0 \end{array}$$

Thus, $14 = b$, or $b = 14$.

Check. We can check by multiplying: $30 \cdot 14 = 420$. The answer checks.

State. The farmer needs 14 beehives.

59. $A = l \cdot w = 14 \text{ ft} \cdot 7 \text{ ft} = 98$ square ft

Perimeter $= 14 \text{ ft} + 7 \text{ ft} + 14 \text{ ft} + 7 \text{ ft} = 42 \text{ ft}$

60. ***Familiarize.*** We make a drawing. Let b = the number of beakers that will be filled.

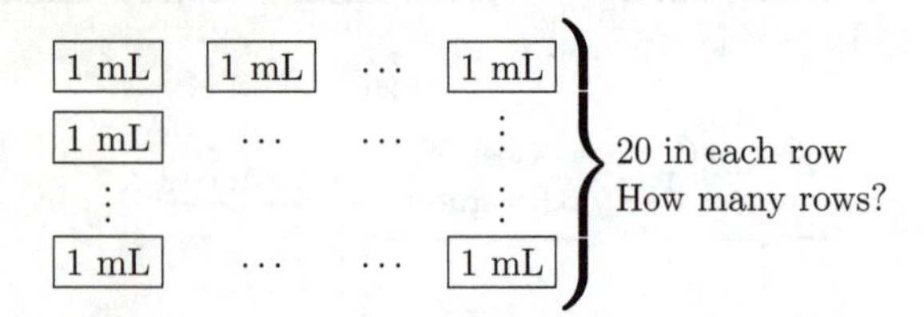

Translate.

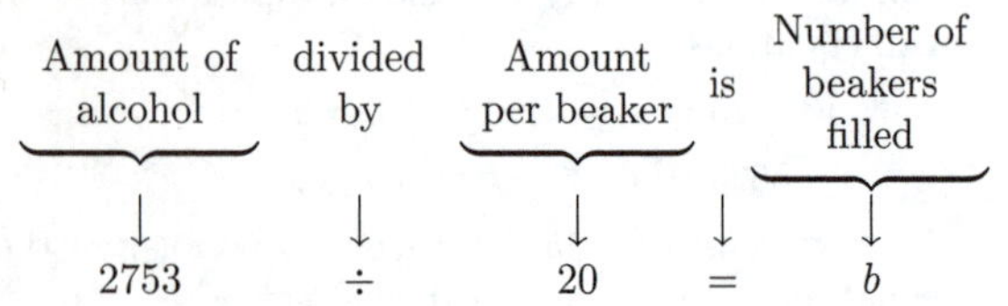

Solve. We carry out the division.

```
      1 3 7
 2 0 |2 7 5 3
      2 0
        7 5
        6 0
        1 5 3
        1 4 0
          1 3
```

Thus, 137 R 13 = b.

Check. We can check by multiplying the number of beakers by 137 and then adding the remainder, 13.

$137 \cdot 20 = 2740$ and $2740 + 13 = 2753$

The answer checks.

State. 137 beakers can be filled; 13 mL will be left over.

61. ***Familiarize.*** This is a multistep problem. Let b = the total amount budgeted for food, clothing, and entertainment and let r = the income remaining after these allotments.

Translate.

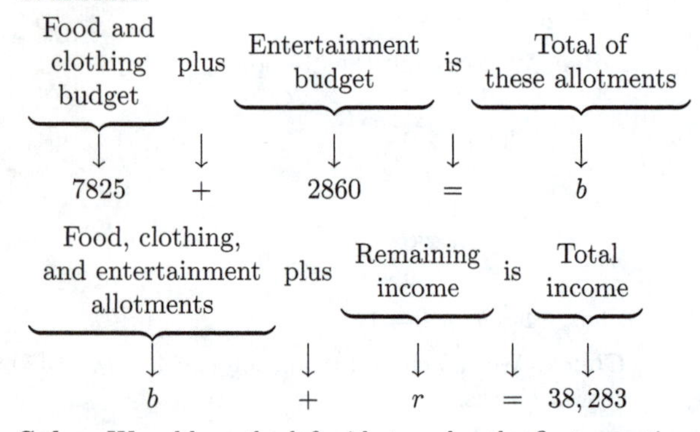

Solve. We add on the left side to solve the first equation.

$$7825 + 2860 = b$$
$$10,685 = b$$

Now we substitute 10,685 for b in the second equation and solve for r.

$$b + r = 38,283$$
$$10,685 + r = 38,283$$
$$10,685 + r - 10,685 = 38,283 - 10,685$$
$$r = 27,598$$

Check. We repeat the calculations. The answer checks.

State. After the allotments for food, clothing, and entertainment, $27,598 remains.

62. $7 + (4+3)^2 = 7 + 7^2$
$= 7 + 49$
$= 56$

Answer B is correct.

63. $7 + 4^2 + 3^2 = 7 + 16 + 9$
$= 23 + 9$
$= 32$

Answer A is correct.

64.
$[46 - (4-2) \cdot 5] \div 2 + 4$
$= [46 - 2 \cdot 5] \div 2 + 4$
$= [46 - 10] \div 2 + 4$
$= 36 \div 2 + 4$
$= 18 + 4$
$= 22$

Answer D is correct.

65.
```
     9 d
  ×  d 2
 8 0 3 6
```

By using rough estimates, we see that the factor $d2 \approx 8100 \div 90 = 90$ or $d2 \approx 8000 \div 100 = 80$. Since $99 \times 92 = 9108$ and $98 \times 82 = 8036$, we have $d = 8$.

66.
```
          9 a 1
 2 b 1 |2 3 6,4 2 1
```

Since $250 \times 1000 = 250,000 \approx 236,421$ we deduce that $2b1 \approx 250$ and $9a1 \approx 1000$. By trial we find that $a = 8$ and $b = 4$.

67. At the beginning of each day the tunnel reaches 500 ft − 200 ft, or 300 ft, farther into the mountain than it did the day before. We calculate how far the tunnel reaches into the mountain at the beginning of each day, starting with Day 2.

Day 2: 300 ft

Day 3: 300 ft + 300 ft = 600 ft

Day 4: 600 ft + 300 ft = 900 ft

Day 5: 900 ft + 300 ft = 1200 ft

Day 6: 1200 ft + 300 ft = 1500 ft

We see that the tunnel reaches 1500 ft into the mountain at the beginning of Day 6. On Day 6 the crew tunnels an additional 500 ft, so the tunnel reaches 1500 ft + 500 ft, or 2000 ft, into the mountain. Thus, it takes 6 days to reach the copper deposit.

Chapter 1 Discussion and Writing Exercises

1. No; if subtraction were associative, then $a - (b - c) = (a - b) - c$ for any a, b, and c. But, for example,

   $$12 - (8 - 4) = 12 - 4 = 8,$$

   whereas

   $$(12 - 8) - 4 = 4 - 4 = 0.$$

   Since $8 \neq 0$, this example shows that subtraction is not associative.

2. By rounding prices and estimating their sum a shopper can estimate the total grocery bill while shopping. This is particularly useful if the shopper wants to spend no more than a certain amount.

3. Answers will vary. Anthony is driving from Kansas City to Minneapolis, a distance of 512 miles. He stops for gas after driving 183 miles. How much farther must he drive?

4. The parentheses are not necessary in the expression $9 - (4 \cdot 2)$. Using the rules for order of operations, the multiplication would be performed before the subtraction even if the parentheses were not present.

   The parentheses are necessary in the expression $(3 \cdot 4)^2$; $(3 \cdot 4)^2 = 12^2 = 144$, but $3 \cdot 4^2 = 3 \cdot 16 = 48$.

Chapter 1 Test

1. $\boxed{5}\,4\,6,\,7\,8\,9$

   The digit 5 tells the number of hundred thousands.

2. $8843 = 8$ thousands $+ 8$ hundreds $+ 4$ tens $+ 3$ ones

3. 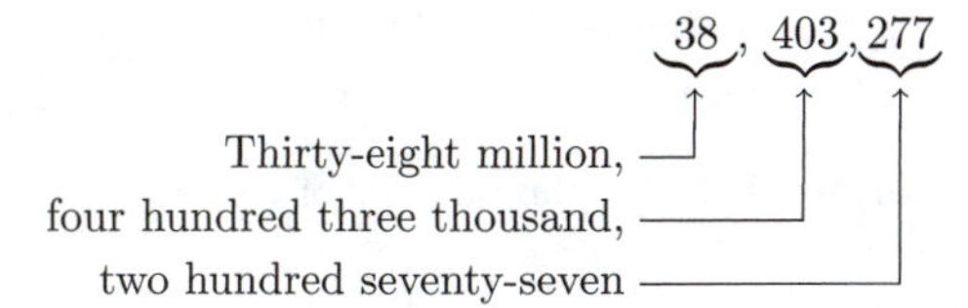

4. $$\begin{array}{r} 6811 \\ +\,3178 \\ \hline 9989 \end{array}$$ Add ones, add tens, add hundreds, and then add thousands.

5. $$\begin{array}{r} {\scriptstyle 1\ 1\quad 1} \\ 45,889 \\ +\,17,902 \\ \hline 63,791 \end{array}$$

6. $$\begin{array}{r} {\scriptstyle 2\,1\,1\ } \\ 1239 \\ 843 \\ 301 \\ +\ \ 782 \\ \hline 3165 \end{array}$$

7. $$\begin{array}{r} 6203 \\ +\ \ 4312 \\ \hline 10,515 \end{array}$$

8. $$\begin{array}{r} 7983 \\ -\,4353 \\ \hline 3630 \end{array}$$ Subtract ones, subtract tens, subtract hundreds, and then subtract thousands.

9. $$\begin{array}{r} {\scriptstyle 6\ 14} \\ 29\not{7}\,\not{4} \\ -\,1935 \\ \hline 1039 \end{array}$$

10. $$\begin{array}{r} {\scriptstyle 8\ \ 9\ \ 17} \\ 8\,\not{9}\,\not{0}\,\not{7} \\ -\,2059 \\ \hline 6848 \end{array}$$

11. $$\begin{array}{r} {\scriptstyle 12} \\ {\scriptstyle 1\ \ \not{2}\ \ 9\ 16} \\ \not{2}\,\not{3},\not{0}\,\not{6}\,7 \\ -\,17,892 \\ \hline 5175 \end{array}$$

12. $$\begin{array}{r} {\scriptstyle 5\ 6\ 7\ } \\ 4568 \\ \times\qquad 9 \\ \hline 41,112 \end{array}$$

13. $$\begin{array}{r} {\scriptstyle 5\ 4\ 3\ } \\ 8876 \\ \times\,600 \\ \hline 5,325,600 \end{array}$$ Multiply by 6 hundreds (We write 00 and then multiply 8876 by 6.)

14. $$\begin{array}{r} 65 \\ \times\,37 \\ \hline 455 \\ 1950 \\ \hline 2405 \end{array}$$ Multiplying by 7; Multiplying by 30; Adding

15. $$\begin{array}{r} 678 \\ \times\,788 \\ \hline 5424 \\ 54240 \\ 474600 \\ \hline 534,264 \end{array}$$

16. $$\begin{array}{r} 3 \\ 4\overline{)15} \\ 12 \\ \hline 3 \end{array}$$

    The answer is 3 R 3.

17. $$\begin{array}{r} 70 \\ 6\overline{)420} \\ 42 \\ \hline 0 \\ 0 \\ \hline 0 \end{array}$$

    The answer is 70.

18.
$$\begin{array}{r} 97 \\ 89\overline{)8633} \\ \underline{801} \\ 623 \\ \underline{623} \\ 0 \end{array}$$

The answer is 97.

19.
$$\begin{array}{r} 805 \\ 44\overline{)35,428} \\ \underline{352} \\ 228 \\ \underline{220} \\ 8 \end{array}$$

The answer is 805 R 8.

20. Round 34,528 to the nearest thousand.

3 4, $\boxed{5}$ 2 8
↑

The digit 4 is in the thousands place. Consider the next digit to the right, 5. Since 5 is 5 or higher, round 4 thousands up to 5 thousands. Then change all digits to the right of thousands to zeros.

The answer is 35,000.

21. Round 34,528 to the nearest ten.

3 4, 5 2 $\boxed{8}$
↑

The digit 2 is in the tens place. Consider the next digit to the right, 8. Since 8 is 5 or higher, round 2 tens up to 3 tens. Then change the digit to the right of tens to zero.

The answer is 34,530.

22. Round 34,528 to the nearest hundred.

3 4, 5 $\boxed{2}$ 8
↑

The digit 5 is in the hundreds place. Consider the next digit to the right, 2. Since 2 is 4 or lower, round down, meaning that 5 hundreds stays as 5 hundreds. Then change all digits to the right of hundreds to zero.

The answer is 34,500.

23.

	Rounded to the nearest hundred	
2 3, 6 4 9	2 3, 6 0 0	
+ 5 4, 7 4 6	+ 5 4, 7 0 0	
	7 8, 3 0 0	← Estimated answer

24.

	Rounded to the nearest hundred	
5 4, 7 5 1	5 4, 8 0 0	
− 2 3, 6 4 9	− 2 3, 6 0 0	
	3 1, 2 0 0	← Estimated answer

25.

	Rounded to the nearest hundred	
8 2 4	8 0 0	
× 4 8 9	× 5 0 0	
	4 0 0, 0 0 0	← Estimated answer

26. Since 34 is to the right of 17 on the number line, $34 > 17$.

27. Since 117 is to the left of 157 on the number line, $117 < 157$.

28.
$$28 + x = 74$$
$$28 + x - 28 = 74 - 28 \quad \text{Subtracting 28 on both sides}$$
$$x = 46$$

Check:
$$\begin{array}{c|c} 28 + x = 74 & \\ \hline 28 + 46 \ ? \ 74 & \\ 74 & \text{TRUE} \end{array}$$

The solution is 46.

29. $169 \div 13 = n$

We carry out the division.

$$\begin{array}{r} 13 \\ 13\overline{)169} \\ \underline{13} \\ 39 \\ \underline{39} \\ 0 \end{array}$$

The solution is 13.

30.
$$38 \cdot y = 532$$
$$\frac{38 \cdot y}{38} = \frac{532}{38} \quad \text{Dividing by 38 on both sides}$$
$$y = 14$$

Check:
$$\begin{array}{c|c} 38 \cdot y = 532 & \\ \hline 38 \cdot 14 \ ? \ 532 & \\ 532 & \text{TRUE} \end{array}$$

The solution is 14.

31.
$$381 = 0 + a$$
$$381 = a \quad \text{Adding on the right side}$$

The solution is 381.

32. ***Familiarize***. Let s = the number of calories in an 8-oz serving of skim milk.

Translate.

Number of calories in skim milk	plus	How many more calories	is	Number of calories in whole milk
↓	↓	↓	↓	↓
s	$+$	63	$=$	146

Solve. We subtract 63 on both sides of the equation.

$$s + 63 = 146$$
$$s + 63 - 63 = 146 - 63$$
$$s = 83$$

Check. Since 63 calories more than 83 calories is $83 + 63$, or 146 calories, the answer checks.

State. An 8-oz serving of skim milk contains 83 calories.

33. ***Familiarize.*** Let s = the number of staplers that can be filled. We can think of this as repeated subtraction, taking successive sets of 250 staples and putting them into s staplers.

Translate.

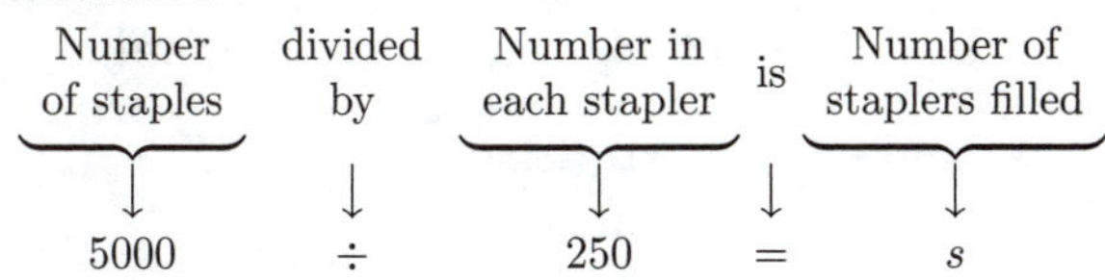

Solve. We carry out the division.

```
        2 0
2 5 0 ⟌5 0 0 0
       5 0 0
           0
           0
           0
```

Then $20 = s$.

Check. We can multiply the number of staplers filled by the number of staples in each one.

$20 \cdot 250 = 5000$

The answer checks.

State. 20 staplers can be filled from a box of 5000 staples.

34. ***Familiarize.*** Let a = the total land area of the five largest states, in square meters. Since we are combining the areas of the states, we can add.

Translate.

$571,951 + 261,797 + 155,959 + 145,552 + 121,356 = a$

Solve. We carry out the addition.

```
    2 1  3 3 2
    5 7 1, 9 5 1
    2 6 1, 7 9 7
    1 5 5, 9 5 9
    1 4 5, 5 5 2
  + 1 2 1, 3 5 6
 1, 2 5 6, 6 1 5
```

Then $1,256,615 = a$.

Check. We can repeat the calculation. We can also estimate the result by rounding. We will round to the nearest ten thousand.

$$571,951 + 261,797 + 155,959 + 145,552 + 121,356$$
$$\approx 570,000 + 260,000 + 160,000 + 150,000 + 120,000$$
$$= 1,260,000$$

Since $1,260,000 \approx 1,256,615$, we have a partial check.

State. The total land area of Alaska, Texas, California, Montana, and New Mexico is 1,256,615 sq mi.

35. a) We will use the formula Perimeter $= 2 \cdot$ length $+$ $2 \cdot$ width to find the perimeter of each pool table in inches. We will use the formula Area = length · width to find the area of each pool table, in sq in.

For the 50 in. by 100 in. table:

$$\text{Perimeter} = 2 \cdot 100 \text{ in.} + 2 \cdot 50 \text{ in.}$$
$$= 200 \text{ in.} + 100 \text{ in.}$$
$$= 300 \text{ in.}$$
$$\text{Area} = 100 \text{ in.} \cdot 50 \text{ in.} = 5000 \text{ sq in.}$$

For the 44 in. by 88 in. table:

$$\text{Perimeter} = 2 \cdot 88 \text{ in.} + 2 \cdot 44 \text{ in.}$$
$$= 176 \text{ in.} + 88 \text{ in.}$$
$$= 264 \text{ in.}$$
$$\text{Area} = 88 \text{ in.} \cdot 44 \text{ in.} = 3872 \text{ sq in.}$$

For the 38 in. by 76 in. table:

$$\text{Perimeter} = 2 \cdot 76 \text{ in.} + 2 \cdot 38 \text{ in.}$$
$$= 152 \text{ in.} + 76 \text{ in.}$$
$$= 228 \text{ in.}$$
$$\text{Area} = 76 \text{ in.} \cdot 38 \text{ in.} = 2888 \text{ sq in.}$$

b) Let a = the number of square inches by which the area of the largest table exceeds the area of the smallest table. We subtract to find a.

$$a = 5000 \text{ sq in.} - 2888 \text{ sq in.} = 2112 \text{ sq in.}$$

36. ***Familiarize.*** Let n = the number of 12-packs that can be filled. We can think of this as repeated subtraction, taking successive sets of 12 snack cakes and putting them into n packages.

Translate.

Number of cakes	divided by	Number in each package	is	Number of 12-packs
↓	↓	↓	↓	↓
22,231	÷	12	=	n

Solve. We carry out the division.

```
        1 8 5 2
  1 2 ⟌2 2, 2 3 1
       1 2
       1 0 2
         9 6
           6 3
           6 0
             3 1
             2 4
               7
```

Then $1852 \text{ R } 7 = n$.

Check. We multiply the number of packages by 12 and then add the remainder, 7.

$12 \cdot 1852 = 22,224$

$22,224 + 7 = 22,231$

The answer checks.

State. 1852 twelve-packs can be filled. There will be 7 cakes left over.

37. ***Familiarize***. This a multistep problem. Let b = the total cost of the black cartridges, p = the total cost of the photo cartridges, and t = the total cost of the entire purchase.

Translate.

For the black ink cartridges:

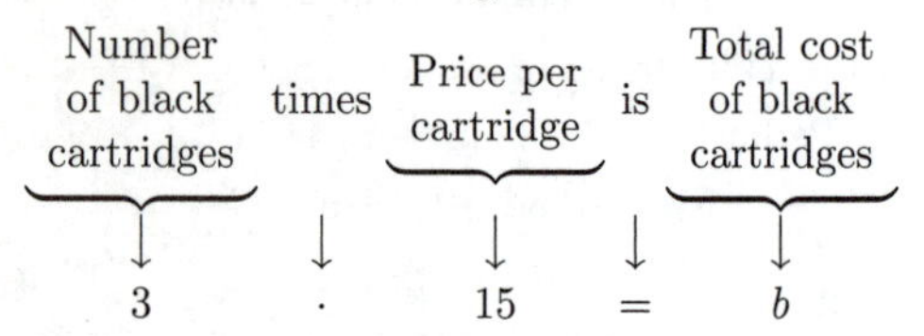

For the photo cartridges:

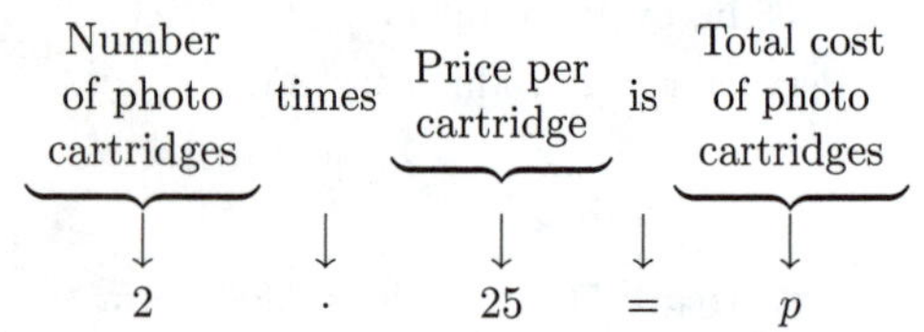

For the total cost of the order:

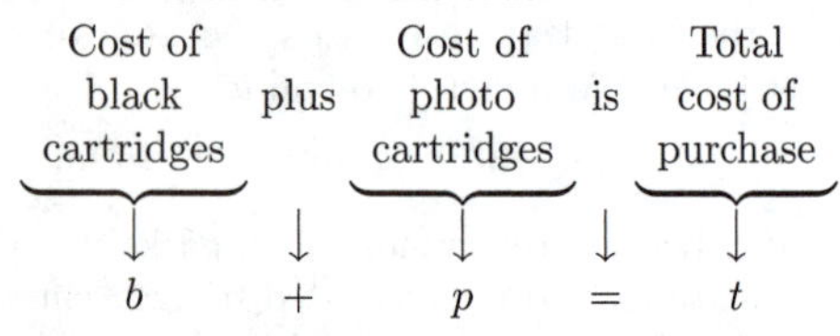

Solve. We solve the first two equations and then add the solutions.

$3 \cdot 15 = b$

$45 = b$

$2 \cdot 25 = p$

$50 = p$

$b + p = t$

$45 + 50 = t$

$95 = t$

Check. We repeat the calculations. The answer checks.

State. The total cost of the purchase was \$95.

38. Exponential notation for $12 \cdot 12 \cdot 12 \cdot 12$ is 12^4.

39. $7^3 = 7 \cdot 7 \cdot 7 = 343$

40. $10^5 = 10 \cdot 10 \cdot 10 \cdot 10 \cdot 10 = 100,000$

41. $35 - 1 \cdot 28 \div 4 + 3$

$= 35 - 28 \div 4 + 3$ Doing all multiplications and

$= 35 - 7 + 3$ divisions in order from left to right

$= 28 + 3$ Doing all additions and subtractions

$= 31$ in order from left to right

42. $10^2 - 2^2 \div 2$

$= 100 - 4 \div 2$ Evaluating the exponential expressions

$= 100 - 2$ Dividing

$= 98$ Subtracting

43. $(25 - 15) \div 5$

$= 10 \div 5$ Doing the calculation inside the parentheses

$= 2$ Dividing

44. $2^4 + 24 \div 12$

$= 16 + 24 \div 12$ Evaluating the exponential expression

$= 16 + 2$ Dividing

$= 18$ Adding

45. $8 \times \{(20 - 11) \cdot [(12 + 48) \div 6 - (9 - 2)]\}$

$= 8 \times \{9 \cdot [60 \div 6 - 7]\}$

$= 8 \times \{9 \cdot [10 - 7]\}$

$= 8 \times \{9 \cdot 3\}$

$= 8 \times 27$

$= 216$

46. We add the numbers and then divide by the number of addends.

$$\frac{97 + 99 + 87 + 89}{4} = \frac{372}{4} = 93$$

Answer A is correct.

47. ***Familiarize***. We make a drawing.

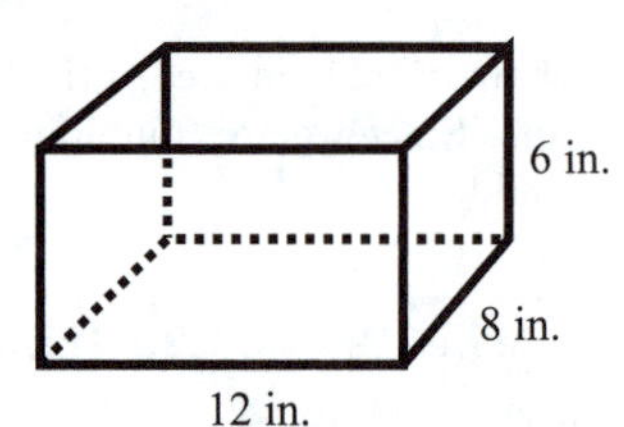

Observe that the dimensions of two sides of the container are 8 in. by 6 in. The area of each is 8 in. · 6 in. and their total area is 2 · 8 in. · 6 in. The dimensions of the other two sides are 12 in. by 6 in. The area of each is 12 in. · 6 in. and their total area is 2 · 12 in. · 6 in. The dimensions of the bottom of the box are 12 in. by 8 in. and its area is 12 in. · 8 in. Let c = the number of square inches of cardboard that are used for the container.

Translate. We add the areas of the sides and the bottom of the container.

$2 \cdot 8 \text{ in.} \cdot 6 \text{ in.} + 2 \cdot 12 \text{ in.} \cdot 6 \text{ in.} + 12 \text{ in.} \cdot 8 \text{ in.} = c$

Solve. We carry out the calculation.

$2 \cdot 8 \text{ in.} \cdot 6 \text{ in.} + 2 \cdot 12 \text{ in.} \cdot 6 \text{ in.} + 12 \text{ in.} \cdot 8 \text{ in.} = c$

$96 \text{ sq in.} + 144 \text{ sq in.} + 96 \text{ sq in.} = c$

$336 \text{ sq in.} = c$

Check. We can repeat the calculations. The answer checks.

State. 336 sq in. of cardboard are used for the container.

48. We can reduce the number of trials required by simplifying the expression on the left side of the equation and then using the addition principle.

$$359 - 46 + a \div 3 \times 25 - 7^2 = 339$$
$$359 - 46 + a \div 3 \times 25 - 49 = 339$$
$$359 - 46 + \frac{a}{3} \times 25 - 49 = 339$$
$$359 - 46 + \frac{25 \cdot a}{3} - 49 = 339$$
$$313 + \frac{25 \cdot a}{3} - 49 = 339$$
$$264 + \frac{25 \cdot a}{3} = 339$$
$$264 + \frac{25 \cdot a}{3} - 264 = 339 - 264$$
$$\frac{25 \cdot a}{3} = 75$$

We see that when we multiply a by 25 and divide by 3, the result is 75. By trial, we find that $\frac{25 \cdot 9}{3} = \frac{225}{3} = 75$, so $a = 9$. We could also reason that since $75 = 25 \cdot 3$ and $9/3 = 3$, we have $a = 9$.

49. ***Familiarize***. First observe that a 10-yr loan with monthly payments has a total of $10 \cdot 12$, or 120, payments. Let $m =$ the number of monthly payments represented by \$9160 and let $p =$ the number of payments remaining after \$9160 has been repaid.

Translate. First we will translate to an equation that can be used to find m. Then we will write an equation that can be used to find p.

Payments to date	divided by	Amount of each payment	is	Number of payments made
↓	↓	↓	↓	↓
9160	$\div$	229	$=$	m

Payments already made	plus	Remaining payments	is	Total number of payments
↓	↓	↓	↓	↓
m	$+$	p	$=$	120

Solve. To solve the first equation we carry out the division.

```
        4 0
2 2 9 ) 9 1 6 0
        9 1 6
        -----
            0
            0
            -
            0
```

Thus, $m = 40$.

Now we solve the second equation.

$$m + p = 120$$
$$40 + p = 120 \quad \text{Substituting 40 for } m$$
$$40 + p - 40 = 120 - 40$$
$$p = 80$$

Check. We can approach the problem in a different way to check the answer. In 10 years, Cara's loan payments will total $120 \cdot \$229$, or \$27,480. If \$9160 has already been paid, then $\$27,480 - \$9160$, or \$18,320, remains to be paid. Since $80 \cdot \$229 = \$18,320$, the answer checks.

State. 80 payments remain on the loan.

Chapter 2

Introduction to Integers and Algebraic Expressions

Exercise Set 2.1

1. The integer -282 corresponds to 282 ft below sea level.

3. The integer 820 corresponds to receiving an \$820 refund; the integer -541 corresponds to owing \$541.

5. The integer 950,000,000 corresponds to a temperature of 950,000,000°F above zero; the integer -460 corresponds to a temperature of 460°F below zero.

7. The integer 40 corresponds to receiving \$40 for a ton of paper; the integer -15 corresponds to paying \$15 to get rid of a ton of paper.

9. Since -8 is to the left of 0, we have $-8 < 0$.

11. Since 9 is to the right of 0, we have $9 > 0$.

13. Since 8 is to the right of -8, we have $8 > -8$.

15. Since -6 is to the left of -4, we have $-6 < -4$.

17. Since -8 is to the left of -5, we have $-8 < -5$.

19. Since -13 is to the left of -9, we have $-13 < -9$.

21. Since -3 is to the right of -4, we have $-3 > -4$.

23. The distance from 57 to 0 is 57, so $|57| = 57$.

25. The distance from 0 to 0 is 0, so $|0| = 0$.

27. The distance from -24 to 0 is 24, so $|-24| = 24$.

29. The distance from 53 to 0 is 53, so $|53| = 53$.

31. This distance from -8 to 0 is 8, so $|-8| = 8$.

33. To find the opposite of x when x is -7, we reflect -7 to the other side of 0. We have $-(-7) = 7$. The opposite of -7 is 7.

35. To find the opposite of x when x is 7, we reflect 7 to the other side of 0. We have $-(7) = -7$. The opposite of 7 is -7.

37. When we try to reflect 0 to the other side of 0, we go nowhere. The opposite of 0 is 0.

39. $-(-21) = 21$

41. $-(53) = -53$

43. $-(-1) = 1$

45. We replace x by 7. We wish to find $-(-7)$. Reflecting 7 to the other side of 0 gives us -7 and then reflecting back gives us 7. Thus, $-(-x) = 7$ when x is 7.

47. We replace x by -9. We wish to find $-(-(-9))$. Reflecting -9 to the other side of 0 gives us 9 and then reflecting back gives us -9. Thus, $-(-x) = -9$ when x is -9.

49. We replace x by -17. We wish to find $-(-(-17))$. Reflecting -17 to the other side of 0 gives us 17 and then reflecting back gives us -17. Thus $-(-x) = -17$ when x is -17.

51. We replace x by 23. We wish to find $-(-23)$. Reflecting 23 to the other side of 0 gives us -23 and then reflecting back gives us 23. Thus, $-(-x) = 23$ when x is 23.

53. We replace x by -1. We wish to find $-(-(-1))$. Reflecting -1 to the other side of 0 gives us 1 and then reflecting back gives us -1. Thus, $-(-x) = -1$ when x is -1.

55. We replace x by 85. We wish to find $-(-85)$. Reflecting 85 to the other side of 0 gives us -85 and then reflecting back gives us 85. Thus, $-(-x) = 85$ when x is 85.

57. We have $-|-x| = -|-345|$. Since $|-345| = 345$, it follows that $-|-345| = -345$.

59. We have $-|-x| = -|-0|$. Since $|-0| = |0| = 0$, it follows that $-|-0| = -|0| = -0 = 0$.

61. We have $-|-x| = -|-(-8)| = -|8|$. Since $|8| = 8$, it follows that $-|-(-8)| = -8$.

63.
$$\begin{array}{r} {\scriptstyle 1\ 1\ \ } \\ 3\,2\,7 \\ +\ 4\,9\,8 \\ \hline 8\,2\,5 \end{array}$$

65.
$$\begin{array}{rl} {\scriptstyle 2} & \\[-4pt] {\scriptstyle 3} & \\ 2\,0\,9 & \\ \times\ \ 3\,4 & \\ \hline 8\,3\,6 & \text{Multiplying 209 by 4} \\ 6\,2\,7\,0 & \text{Multiplying 209 by 30} \\ \hline 7\,1\,0\,6 & \text{Adding} \end{array}$$

67. $7(9-3) = 7 \cdot 6 = 42$

69. $|-5| = 5$ and $|-2| = 2$. Since 5 is to the right of 2, we have $|-5| > |-2|$.

71. $|-8| = 8$ and $|8| = 8$, so $|-8| = |8|$.

73. The integers whose distance from 0 is less than 2 are -1, 0, and 1. These are the solutions.

75. First note that $2^{10} = 1024$, $|-6| = 6$, $|3| = 3$, $2^7 = 128$, $7^2 = 49$, and $10^2 = 100$. Listing the entire set of integers in order from least to greatest, we have -100, -5, 0, $|3|$, 4, $|-6|$, 7^2, 10^2, 2^7, 2^{10}.

Exercise Set 2.2

1. Add: $-7+2$

Start at -7. Move 2 units to the right.

$-7+2=-5$

3. Add: $-9+5$

Start at -9. Move 5 units to the right.

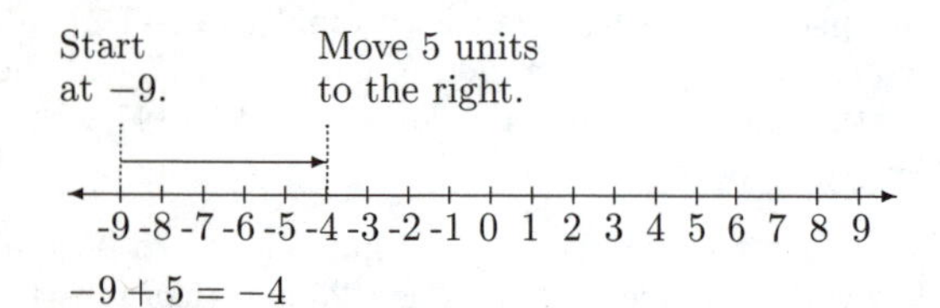

$-9+5=-4$

5. Add: $-3+9$

Start at -3. Move 9 units to the right.

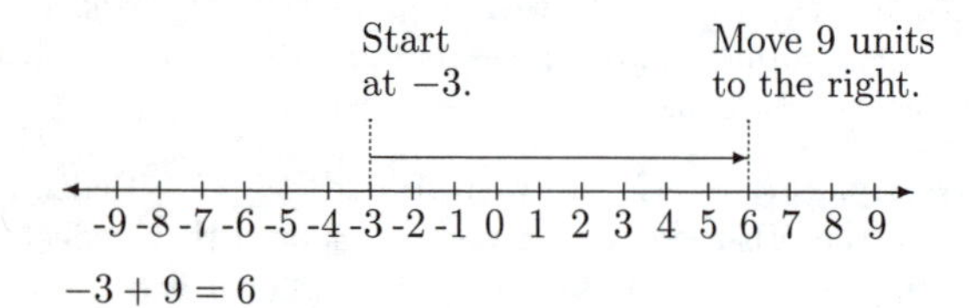

$-3+9=6$

7. $-7+7$

Start at -7. Move 7 units to the right.

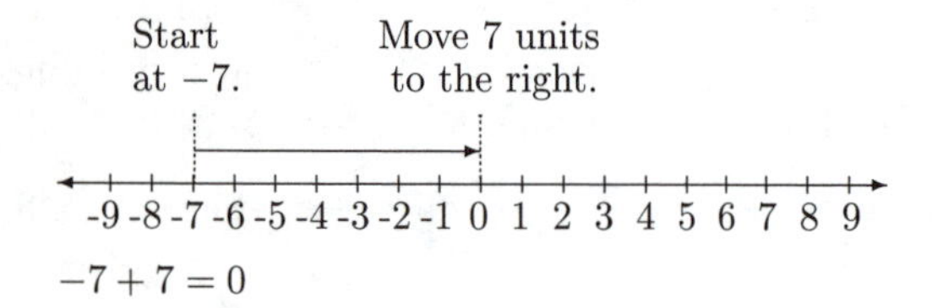

$-7+7=0$

9. $-3+(-1)$

Move 1 unit to the left. Start at -3.

-5 -4 -3 -2 -1 0 1 2 3 4 5

$-3+(-1)=-4$

11. $-3+(-9)$ Two negative integers

Add the absolute values: $3+9=12$
Make the answer negative: $-3+(-9)=-12$

13. $-6+(-5)$ Two negative integers

Add the absolute values: $6+5=11$
Make the answer negative: $-6+(-5)=-11$

15. $-15+0$ One number is 0.

The sum is the other number.
$-15+0=-15$

17. $0+42$ One number is 0.

The sum is the other number.
$0+42=42$

19. $9+(-4)$ The absolute values are 9 and 4. The difference is 5. The positive number has the larger absolute value, so the answer is positive. $9+(-4)=5$

21. $-10+6$ The absolute values are 10 and 6. The difference is $10-6$, or 4. The negative number has the larger absolute value, so the answer is negative. $-10+6=-4$

23. $5+(-5)=0$

For any integer a, $a+(-a)=0$.

25. $-2+2=0$

For any integer a, $-a+a=0$.

27. $-4+(-5)$ Two negatives. Add the absolute values, 4 and 5, getting 9. Make the answer negative. $-4+(-5)=-9$

29. $13+(-6)$ The absolute values are 13 and 6. The difference is $13-6$, or 7. The positive number has the larger absolute value, so the answer is positive. $13+(-6)=7$

31. $-25+25$ A positive and a negative number. The numbers have the same absolute value. The sum is 0. $-25+25=0$

33. $63+(-18)$ The absolute values are 63 and 18. The difference is $63-18$, or 45. The positive number has the larger absolute value, so the answer is positive.
$63+(-18)=45$

35. $-11+8$ The absolute values are 11 and 8. The difference is 3. Since the negative number has the larger absolute value, the answer is negative. $-11+8=-3$

37. $-19+19=0$

For any integer a, $-a+a=0$.

39. $-16+6$ The absolute values are 16 and 6. The difference is 10. Since the negative number has the larger absolute value, the answer is negative. $-16+6=-10$

41. $-17+(-7)$ Two negative integers

Add the absolute values: $17+7=24$
Make the answer negative: $-17+(-7)=-24$

43. $11+(-16)$ The absolute values are 11 and 16. The difference is 5. Since the negative number has the larger absolute value, the answer is negative. $11+(-16)=-5$

45. $-15+(-6)$ Two negative integers

Add the absolute values: $15+6=21$
Make the answer negative: $-15+(-6)=-21$

47. $-15+(-15)$ Two negative integers

Add the absolute values: $15+15=30$
Make the answer negative: $-15+(-15)=-30$

49. $-11+17$ The absolute values are 11 and 17. The difference is 6. The positive number has the larger absolute value, so the answer is positive. $-11+17=6$

51. We will add from left to right.

$$\begin{aligned}-15+(-7)+1 &= -22+1\\ &= -21\end{aligned}$$

53. We will add from left to right.

$$30 + (-10) + 5 = 20 + 5 \\ = 25$$

55. We will add from left to right.

$$-23 + (-9) + 15 = -32 + 15 \\ = -17$$

57. We will add from left to right.

$$40 + (-40) + 6 = 0 + 6 \\ = 6$$

59. $12 + (-65) + (-12)$

Note that $12 + (-12) = 0$. Then we have $0 + (-65)$, so the sum is -65.

61. We will add from left to right.

$$\begin{aligned} & -24 + (-37) + (-19) + (-45) + (-35) \\ = & -61 + (-19) + (-45) + (-35) \\ = & -80 + (-45) + (-35) \\ = & -125 + (-35) \\ = & -160 \end{aligned}$$

63. $28 + (-44) + 17 + 31 + (-94)$

a) $28 + 17 + 31 = 76$ Adding the positive numbers

b) $-44 + (-94) = -138$ Adding the negative numbers

c) $76 + (-138) = -62$ Adding the results

65. $-19 + 73 + (-23) + 19 + (-73)$

a) $-19 + 19 = 0$ Adding one pair of opposites

b) $73 + (-73) = 0$ Adding the other pair of opposites

c) We have $0 + (-23) = -23$

67.
$$\begin{array}{r} \scriptstyle 1\,1 \\ 587 \\ +\,6094 \\ \hline 6681 \end{array}$$

69. 3 ten thousands + 9 thousands + 4 hundreds + 1 ten + 7 ones

71.
$$\begin{array}{r} 32 \\ 9\overline{)288} \\ \underline{27} \\ 18 \\ \underline{18} \\ 0 \end{array}$$

The answer is 32.

73. $-|27| + (-|-13|) = -27 + (-13) = -40$

75. We use a calculator.

$-3496 + (-2987) = -6483$

77. If $-x$ is positive, it is the reflection of a negative number x across 0 on the number line. Thus, $-x$ is positive for all negative numbers x.

79. If n is positive, $-n$ is negative. Then $-n + m$, the sum of two negative numbers, is negative.

81. If n is negative and m is less than n, then m is also negative. Then $n + m$, the sum of two negative numbers, is negative.

Exercise Set 2.3

1. $3 - 7 = 3 + (-7) = -4$

3. $0 - 7 = 0 + (-7) = -7$

5. $-8 - (-2) = -8 + 2 = -6$

7. $-10 - (-10) = -10 + 10 = 0$

9. $12 - 16 = 12 + (-16) = -4$

11. $20 - 27 = 20 + (-27) = -7$

13. $-9 - (-3) = -9 + 3 = -6$

15. $-11 - (-11) = -11 + 11 = 0$

17. $7 - 7 = 7 + (-7) = 0$

19. $7 - (-7) = 7 + 7 = 14$

21. $8 - (-3) = 8 + 3 = 11$

23. $-6 - 8 = -6 + (-8) = -14$

25. $-4 - (-9) = -4 + 9 = 5$

27. $2 - 9 = 2 + (-9) = -7$

29. $-6 - (-5) = -6 + 5 = -1$

31. $8 - (-10) = 8 + 10 = 18$

33. $0 - 10 = 0 + (-10) = -10$

35. $-5 - (-2) = -5 + 2 = -3$

37. $-7 - 14 = -7 + (-14) = -21$

39. $0 - (-5) = 0 + 5 = 5$

41. $-8 - 0 = -8 + 0 = -8$

43. $7 - (-5) = 7 + 5 = 12$

45. $6 - 25 = 6 + (-25) = -19$

47. $-42 - 26 = -42 + (-26) = -68$

49. $-72 - 9 = -72 + (-9) = -81$

51. $24 - (-92) = 24 + 92 = 116$

53. $-50 - (-50) = -50 + 50 = 0$

55. $-30 - (-85) = -30 + 85 = 55$

57. $7 - (-5) + 4 - (-3) = 7 + 5 + 4 + 3 = 19$

59.
$$\begin{aligned} & -31 + (-28) - (-14) - 17 \\ = & -31 + (-28) + 14 + (-17) \\ = & -31 + (-28) + (-17) + 14 \quad \text{Using a commutative law} \\ = & -76 + 14 \quad \text{Adding the negative numbers} \\ = & -62 \end{aligned}$$

61. $-34 - 28 + (-33) - 44 = (-34) + (-28) + (-33) + (-44) = -139$

63. $-93-(-84)-41-(-56)$
$= -93+84+(-41)+56$
$= -93+(-41)+84+56$ Using a commutative law
$= -134+140$ Adding negatives and adding positives
$= 6$

65. $-5-(-30)+30+40-(-12)$
$= -5+30+30+40+12$
$= -5+112$ Adding the positive numbers
$= 107$

67. $132-(-21)+45-(-21) = 132+21+45+21 = 219$

69. We subtract the beginning page number from the final page number.

$62-37=25$

Alicia read 25 pages.

71. The integer 8 corresponds to 8 lb above the ideal weight, and -9 corresponds to 9 lb below it. We subtract the lower weight from the higher weight:

$8-(-9)=8+9=17$

Rod lost 17 lb.

73. We start with the original temperature, add the rise in temperature, and subtract the drop in temperature.

$32+15-50 = 32+15+(-50) = 47+(-50) = -3$

The final temperature was $-3°$.

75. The integer -5000 represents a loss of \$5000, and the integer 8000 represents a profit of \$8000. We subtract the amount of the loss from the profit.

$8000-(-5000) = 8000+5000 = 13,000$

The store made \$13,000 more in 2010 than in 2009.

77. Let T = the amount by which the temperature dropped, in degrees Fahrenheit.

Temperature drop	is	Higher temperature	minus	Lower temperature
↓	↓	↓	↓	↓
T	$=$	44	$-$	(-56)

We carry out the subtraction.

$T = 44-(-56) = 44+56 = 100$

The temperature dropped 100°F.

79. Let E = the difference between the elevations, in feet.

Difference in elevations	is	Higher elevation	minus	Lower elevation
↓	↓	↓	↓	↓
E	$=$	5672	$-$	(-4)

We carry out the subtraction.

$E = 5672-(-4) = 5672+4 = 5676$

The difference between the elevations is 5676 ft.

81. Let B = the final balance.

Final balance	=	Original balance	−	Amount of first check	+	Deposit	−	Amount of second check
↓	↓	↓	↓	↓	↓	↓	↓	↓
B	$=$	460	$-$	530	$+$	75	$-$	90

We carry out the computation.

$B = 460-530+75-90$
$= -70+75-90$
$= 5-90$
$= -85$

The balance in the account is $-\$85$. (That is, the account is \$85 overdrawn.)

83. To find the elevation that is 2293 ft deeper than -7718 ft, we subtract the additional depth from the original depth.

$-7718-2293 = -7718+(-2293) = -10,011$

In 2005 the elevation of the deepwater drilling record was $-10,011$ ft.

85. To find by how much the Ramones overspent their balance, we subtract the amount of the tolls from the original balance.

$\$12-\$15 = \$12+(-\$15) = -\$3$

To find the total debt, we represent the \$80 fine as $-\$80$ and add this to the amount overspent.

$-\$3+(-\$80) = -\$83$

The Ramones were \$83 in debt.

87. $4^3 = 4\cdot4\cdot4 = 64$

89. $1^7 = 1\cdot1\cdot1\cdot1\cdot1\cdot1\cdot1 = 1$

91. ***Familiarize***. Let n = the number of 12-oz cans that can be filled. We think of an array consisting of 96 oz with 12 oz in each row.

The number n corresponds to the number of rows in the array.

Translate and Solve. We translate to an equation and solve it.

$96 \div 12 = n$

$$\begin{array}{r} 8 \\ 12\overline{)96} \\ \underline{96} \\ 0 \end{array}$$

Check. We multiply the number of cans by 12: $8\cdot12=96$. The result checks.

State. Eight 12-oz cans can be filled.

93. $5+4^2+2\cdot7$
$= 5+16+2\cdot7$
$= 5+16+14$
$= 21+14$
$= 35$

95. $(9+7)(9-7) = (16)(2) = 32$

97. Use a calculator to do this exercise.

$123,907 - 433,789 = -309,882$

99. False; $3 - 0 \neq 0 - 3$.

101. True

103. True

105. a is the number we add to -57 to get -34. If we think of starting at -57 on the number line and moving to -34, we move 17 units to the right, so $a = 17$.

107. The changes during weeks 1 to 5 are represented by the integers -13, -16, 36, -11, and 19, respectively. We add to find the total rise or fall:

$$-13 + (-16) + 36 + (-11) + 19 = 15$$

The market rose 15 points during the 5 week period.

Exercise Set 2.4

1. $-2 \cdot 8 = -16$

3. $10 \cdot (-6) - 60$

5. $8 \cdot (-6) = -48$

7. $-10 \cdot 3 = -30$

9. $-3(-5) = 15$

11. $-9 \cdot (-2) = 18$

13. $(-6)(-7) = 42$

15. $-10(-2) = 20$

17. $12(-10) = -120$

19. $-23 \cdot 0 = 0$

21. $(-72)(-1) = 72$

23. $(-20)17 = -340$

25. $-8(-50) = 400$

27. $0(-14) = 0$

29. $3 \cdot (-8) \cdot (-1)$
$= -24 \cdot (-1)$ Multiplying the first two numbers
$= 24$

31. $7(-4)(-3)5$
$= 7 \cdot 12 \cdot 5$ Multiplying the negative numbers
$= 84 \cdot 5$
$= 420$

33. $-2(-5)(-7)$
$= 10 \cdot (-7)$ Multiplying the first two numbers
$= -70$

35. $(-5)(-2)(-3)(-1)$
$= 10 \cdot 3$ Multiplying the first two numbers and the last two numbers
$= 30$

37. $(-15)(-29) \cdot 0 \cdot 8$
$= 435 \cdot 0$ Multiplying the first two numbers and the last two numbers
$= 0$

(We might have noted at the outset that the product would be 0 since one of the numbers in the product is 0.)

39. $(-7)(-1)(7)(-6)$
$= 7(-42)$ Multiplying the first two numbers and the last two numbers
$= -294$

41. $(-6)^2 = (-6)(-6) = 36$

43. $(-5)^3 = (-5)(-5)(-5)$
$= 25(-5)$
$= -125$

45. $(-10)^4 = (-10)(-10)(-10)(-10)$
$= 100 \cdot 100$
$= 10,000$

47. $-2^4 = -1 \cdot 2^4$
$= -1 \cdot 2 \cdot 2 \cdot 2 \cdot 2$
$= -1 \cdot 4 \cdot 4$
$= -1 \cdot 16$
$= -16$

49. $(-3)^5 = (-3)(-3)(-3)(-3)(-3)$
$= 9 \cdot 9 \cdot (-3)$
$= 81(-3)$
$= -243$

51. $(-1)^{12}$
$= (-1) \cdot (-1) \cdot (-1) \cdot (-1) \cdot (-1) \cdot (-1) \cdot (-1) \cdot (-1) \cdot (-1) \cdot (-1) \cdot (-1) \cdot (-1)$
$= 1 \cdot 1 \cdot 1 \cdot 1 \cdot 1 \cdot 1$
$= 1 \cdot 1 \cdot 1$
$= 1 \cdot 1$
$= 1$

53. $-11^2 = -1 \cdot 11^2$
$= -1 \cdot 11 \cdot 11$
$= -1 \cdot 121$
$= -121$

55. $-4^3 = -1 \cdot 4^3$
$= -1 \cdot 4 \cdot 4 \cdot 4$
$= -4 \cdot 16$
$= -64$

57. -8^4 is read "the opposite of eight to the fourth power."

59. $(-9)^{10}$ is read "negative nine to the tenth power."

61. a) Locate the digit in the hundreds place.

5 3 2, 4 [5] 1
↑

b) Then consider the next digit to the right.

c) Since that digit is 5 or higher, round 4 hundreds up to 5 hundreds.

d) Change all digits to the right of hundreds to zeros.

The answer is 532,500.

63.
$$\begin{array}{r} 80 \\ 36\overline{)2880} \\ \underline{288} \\ 0 \\ \underline{0} \\ 0 \end{array}$$

The answer is 80.

65. $10 - 2^3 + 6 \div 2$

$= 10 - 8 + 6 \div 2$ Evaluating the exponential expression

$= 10 - 8 + 3$ Dividing

$= 2 + 3$ Adding and subtracting in

$= 5$ order from left to right

67. ***Familiarize***. We first make a drawing.

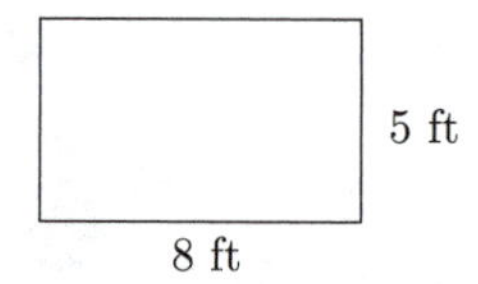

Let $A =$ the area.

Translate. Using the formula for area, we have $A = l \cdot w = 8 \cdot 5$.

Solve. We carry out the multiplication.

$A = 8 \cdot 5 = 40$

Check. We repeat our calculation.

State. The area of the rug is 40 ft^2.

69. ***Familiarize***. Let $n =$ the number of trips the ferry will make carrying 12 cars. We can think of this problem as repeated subtraction.

Translate.

Number of cars	divided by	Number per trip	is	Number of 12-car trips
↓	↓	↓	↓	↓
53	÷	12	=	n

Solve. We carry out the division.

$$\begin{array}{r} 4 \\ 12\overline{)53} \\ \underline{48} \\ 5 \end{array}$$

$53 \div 12 = n$

$4 \text{ R } 5 = n$

After the ferry makes 4 trips, carrying 12 cars on each trip, 5 cars remained to be ferried. Thus, a fifth trip will be required to ferry the remaining cars.

Check. We can check by multiplying the number of full trips by 12 and then adding the remainder, 5.

$4 \cdot 12 = 48$

$48 + 5 = 53$

Since 53 cars were to be ferried, the answer checks.

State. A total of 5 trips will be required.

71. $(-3)^5(-1)^{379} = -243(-1) = 243$

73. $-9^4 + (-9)^4 = -6561 + 6561 = 0$

75.
$$\begin{aligned} &|(-2)^5 + 3^2| - (3-7)^2 \\ &= |-32+9| - (-4)^2 \\ &= |-23| - 16 \\ &= 23 - 16 \\ &= 7 \end{aligned}$$

77. Use a calculator. On many scientific calculators the keystrokes are [4] [7] [x^2] [+/−]. We get −2209.

79. Use a calculator. On many scientific calculators the keystrokes are [1] [9] [+/−] [x^y] [4] [=]. We get 130,321. (Some calculators have a [y^x] key rather than an [x^y] key.)

81. Use a calculator. On many scientific calculators the keystrokes are [(] [7] [3] [−] [8] [6] [)] [x^y] [3] [=]. We get −2197. (Some calculators have a [y^x] key rather than an [x^y] key.)

83. Use a calculator. On many scientific calculators the keystrokes are [9] [3] [5] [+/−] [×] [5] [+/−] [x^y] [3] [=]. We get 116,875. (Some calculators have a [y^x] key rather than an [x^y] key.)

85. The new balance will be $\$68 - 7(\$13) = \$68 - \$91 = -\$23$.

87. a) If $[(-5)^m]^n$ is to be negative, first m must be an odd number so that $(-5)^m$ is negative. Similarly, n must also be odd in order for $[(-5)^m]^n$ to be negative. Thus, both m and n must be odd numbers.

b) If $[(-5)^m]^n$ is to be positive, at least one of m and n must be an even number. For example, if m is even then $(-5)^m$ is positive and so is $[(-5)^m]^n$ regardless of whether n is even or odd. If m is odd, then $(-5)^m$ is negative and n must be even in order for $[(-5)^m]^n$ to be positive.

Exercise Set 2.5

1. $36 \div (-6) = -6$ Check: $-6 \cdot (-6) = 36$

3. $\dfrac{26}{-2} = -13$ Check: $-13 \cdot (-2) = 26$

5. $\dfrac{-16}{8} = -2$ Check: $-2 \cdot 8 = -16$

7. $\dfrac{-48}{-12} = 4$ Check: $4(-12) = -48$

9. $\dfrac{-72}{8} = -9$ Check: $-9 \cdot 8 = -72$

11. $-100 \div (-50) = 2$ Check: $2(-50) = -100$

13. $-108 \div 9 = -12$ Check: $9(-12) = -108$

15. $\dfrac{200}{-25} = -8$ Check: $-8(-25) = 200$

17. $\dfrac{-56}{0}$ is undefined.

19. $\dfrac{88}{-11} = -8$ Check: $-8(-11) = 88$

21. $-\dfrac{276}{12} = \dfrac{-276}{12} = -23$ Check: $-23 \cdot 12 = -276$

23. $\dfrac{0}{-2} = 0$ Check: $0 \cdot (-2) = 0$

25. $\dfrac{19}{-1} = -19$ Check: $-19(-1) = 19$

27. $-41 \div 1 = -41$ Check: $-41 \cdot 1 = -41$

29. $8 - 2 \cdot 3 - 9 = 8 - 6 - 9$ Multiplying
$= 2 - 9$ Doing all additions and subtractions in order from left to right
$= -7$

31. $(8 - 2 \cdot 3) - 9 = (8 - 6) - 9$ Multiplying inside parentheses
$= 2 - 9$ Subtracting inside parentheses
$= -7$ Subtracting

33. $16 \cdot (-24) + 50 = -384 + 50$ Multiplying
$= -334$ Adding

35. $40 - 3^2 - 2^3$
$= 40 - 9 - 8$ Evaluating the exponential expressions
$= 31 - 8$ Doing all additions and subtractions
$= 23$ in order from left to right

37. $4 \cdot (6 + 8)/(4 + 3)$
$= 4 \cdot 14/7$ Adding inside parentheses
$= 56/7$ Doing all multiplications and divisions
$= 8$ in order from left to right

39. $4 \cdot 5 - 2 \cdot 6 + 4 = 20 - 12 + 4$ Multiplying
$= 8 + 4$
$= 12$

41. $1 - (-2)^2 \cdot 3 \div 6$
$= 1 - 4 \cdot 3 \div 6$ Evaluating the exponential expression
$= 1 - 12 \div 6$ Doing all multiplications and
$= 1 - 2$ divisions in order from left to right
$= -1$

43. $18 - (-3)^3 - 3^2 \cdot 5$
$= 18 - (-27) - 9 \cdot 5$ Evaluating the exponential expressions
$= 18 - (-27) - 45$ Multiplying
$= 18 + 27 - 45$
$= 45 - 45$
$= 0$

45. $\dfrac{9^2 - 1}{1 - 3^2}$
$= \dfrac{81 - 1}{1 - 9}$ Evaluating the exponential expressions
$= \dfrac{80}{-8}$ Subtracting in the numerator and in the denominator
$= -10$

47. $8(-7) + 6(-5) = -56 - 30$ Multiplying
$= -86$

49. $20 \div 5(-3) + 3 = 4(-3) + 3$ Dividing
$= -12 + 3$ Multiplying
$= -9$ Adding

51. $18 - 0(3^2 - 5^2 \cdot 7 - 4)$
Observe that $a \cdot 0 = 0$ for an integer a. Then $0(3^2 - 5^2 \cdot 7 - 4) = 0$, so the result is $18 - 0$, or 18.

53. $-4^2 + 6 = -16 + 6$
$= -10$

55. $-8^2 - 3 = -64 - 3$
$= -67$

57. $4 \cdot 5^2 \div 10$
$= 4 \cdot 25 \div 10$ Evaluating the exponential expression
$= 100 \div 10$ Multiplying
$= 10$ Dividing

59. $(3 - 8)^2 \div (-1)$
$= (-5)^2 \div (-1)$ Subtracting inside parentheses
$= 25 \div (-1)$ Evaluating the exponential expression
$= -25$

61. $12 - 20^3 = 12 - 8000$
$= -7988$

63. $2 \times 10^3 - 5000 = 2 \times 1000 - 5000$
$= 2000 - 5000$
$= -3000$

65. $6[9 - (3 - 4)] = 6[9 - (-1)]$ Subtracting inside the innermost parentheses
$= 6[9 + 1]$
$= 6[10]$
$= 60$

67. $-1000 \div (-100) \div 10 = 10 \div 10$ Doing the divisions in order from left to right

$= 1$

69. $-7 - 3[-80 \div (2-10)]$

$= -7 - 3[-80 \div (-8)]$

$= -7 - 3 \cdot 10$

$= -7 - 30$

$= -37$

71. $-2[3-(7-9)^3] = -2[3-(-2)^3]$

$= -2[3-(-8)]$

$= -2[3+8]$

$= -2 \cdot 11$

$= -22$

73. $8 - |7-9| \cdot 3 = 8 - |-2| \cdot 3$

$= 8 - 2 \cdot 3$

$= 8 - 6$

$= 2$

75. $9 - |7-3^2| = 9 - |7-9|$

$= 9 - |-2|$

$= 9 - 2$

$= 7$

77. $\dfrac{6^3 - 7 \cdot 3^4 - 2^5 \cdot 9}{(1-2^3)^3 + 7^3}$

$= \dfrac{216 - 7 \cdot 81 - 32 \cdot 9}{(1-8)^3 + 343}$

$= \dfrac{216 - 567 - 288}{(-7)^3 + 343}$

$= \dfrac{-351 - 288}{-343 + 343}$

$= -\dfrac{639}{0}$

Since division by 0 is not defined, this expression is not defined.

79. $\dfrac{2 \cdot 3^2 \div (3^2 - (2+1))}{5^2 - 6^2 - 2^2(-3)}$

$= \dfrac{2 \cdot 3^2 \div (3^2 - 3)}{25 - 36 - 4(-3)}$

$= \dfrac{2 \cdot 3^2 \div (9-3)}{25 - 36 + 12}$

$= \dfrac{2 \cdot 3^2 \div 6}{-11 + 12}$

$= \dfrac{2 \cdot 9 \div 6}{1}$

$= \dfrac{18 \div 6}{1}$

$= \dfrac{3}{1}$

$= 3$

81. $\dfrac{(-5)^3 + 17}{10(2-6) - 2(5+2)}$

$= \dfrac{-125 + 17}{10(2-6) - 2(5+2)}$ Evaluating the exponential expression

$= \dfrac{-125 + 17}{10(-4) - 2 \cdot 7}$ Doing the calculations within parentheses

$= \dfrac{-125 + 17}{-40 - 14}$ Multiplying

$= \dfrac{-108}{-54}$ Adding and subtracting

$= 2$

83. $\dfrac{2 \cdot 4^3 - 4 \cdot 32}{19^3 - 17^4}$

$= \dfrac{2 \cdot 64 - 4 \cdot 32}{6859 - 83,521}$ Evaluating the exponential expressions

$= \dfrac{128 - 128}{6859 - 83,521}$ Multiplying

$= \dfrac{0}{-76,662}$ Subtracting

$= 0$ Dividing

85. ***Familiarize.*** We first make a drawing.

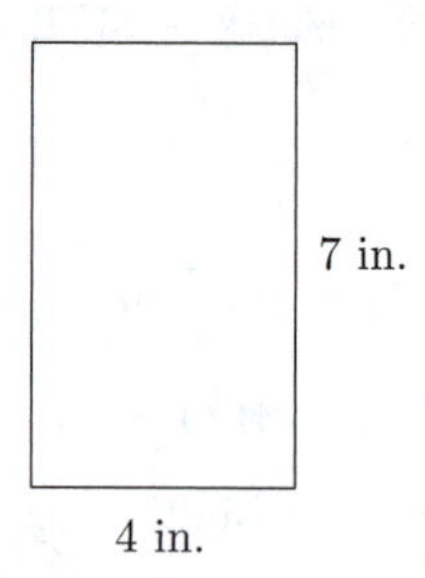

Let $A =$ the area.

Translate. Using the formula for area, we have $A = l \cdot w = 7 \cdot 4$.

Solve. We carry out the multiplication.

$A = 7 \cdot 4 = 28$

Check. We repeat the calculation.

State. The area of the ad was 28 in^2.

87. ***Familiarize.*** We let $g =$ the number of gallons needed to travel 384 miles. Think of a rectangular array consisting of 384 miles with 32 miles in each row. The number g is the number of rows.

Translate.

The number of miles	divided by	the number of miles per gallon	is	the number of gallons needed.
↓	↓	↓	↓	↓
384	$\div$	32	$=$	g

Solve. We carry out the division.

$$\begin{array}{r} 12 \\ 32\overline{)384} \\ \underline{320} \\ 64 \\ \underline{64} \\ 0 \end{array}$$

Check. We multiply the number of gallons by the number of miles per gallon:

$12 \cdot 32 = 384$

We get the number of miles to be traveled, so the answer checks.

State. It will take 12 gallons of gasoline to travel 384 miles.

89. *Familiarize*. We let $c =$ the number of calories in a 1-oz serving. Think of a rectangular array consisting of 1050 calories arranged in 7 rows. The number c is the number of calories in each row.

Translate.

Total calories	divided by	number of ounces	is	number of calories in 1 oz.
↓	↓	↓	↓	↓
1050	÷	7	=	c

Solve. We carry out the division.

$$\begin{array}{r} 150 \\ 7\overline{)1050} \\ \underline{7} \\ 35 \\ \underline{35} \\ 0 \\ \underline{0} \\ 0 \end{array}$$

Check. We multiply the number of calories in 1 oz by 7:

$150 \cdot 7 = 1050$

The result checks

State. There are 150 calories in a 1-oz serving.

91. *Familiarize*. Let $p =$ the number of whole pieces of gum each person will receive. We can think of this problem as repeated subtraction.

Translate.

Total number of sticks	divided by	number of people	is	number of whole pieces per person.
↓	↓	↓	↓	↓
18	÷	4	=	p

Solve. We carry out the division.

$$\begin{array}{r} 4 \\ 4\overline{)18} \\ \underline{16} \\ 2 \end{array}$$

$18 \div 4 = p$

$4 \text{ R } 2 = p$

Check. We multiply the number of whole pieces of gum per person by the number of people and then add the number of remaining pieces.

$4 \cdot 4 = 16$

$16 + 2 = 18$

We got the number of sticks of gum in a package, 18, so the answer checks.

State. Each person will receive 4 pieces of gum, and there will be 2 extra pieces remaining.

93.
$$\begin{aligned} &\frac{9-3^2}{2\cdot 4^2 - 5^2\cdot 9 + 8^2\cdot 7} \\ &= \frac{9-9}{2\cdot 16 - 25\cdot 9 + 64\cdot 7} \\ &= \frac{0}{32-225+448} \\ &= \frac{0}{-193+448} \\ &= \frac{0}{255} \\ &= 0 \end{aligned}$$

95.
$$\begin{aligned} \frac{(25-4^2)^3}{17^2-16^2}\cdot((-6)^2-6^2) &= \frac{(25-16)^3}{289-256}\cdot(36-36) \\ &= \frac{9^3}{289-256}\cdot 0 \\ &= 0 \end{aligned}$$

97. Use a calculator.

$$\frac{19-17^2}{13^2-34} = -2$$

99. Use a calculator.

$28^2 - 36^2/4^2 + 17^2 = 992$

101. [( ] [1] [5] [x^2] −5 [x^y] [3] [ )] [÷] [( ] [3] [x^2] [+] [4] [x^2] [ )] [=] (Some calculators have a [y^x] key rather than an [x^y] key.)

103. Entering the given keystrokes and then pressing [=], we get 5.

105. $-n$ and m are both negative, so $\dfrac{-n}{m}$ is the quotient of two negative numbers and, thus, is positive.

107. $\dfrac{-n}{m}$ is positive (see Exercise 105), so $-\left(\dfrac{-n}{m}\right)$ is the opposite of a positive number and, thus, is negative.

109. $-n$ is negative and $-m$ is positive, so $\dfrac{-n}{-m}$ is the quotient of a negative and a positive number and, thus, is negative. Then $-\left(\dfrac{-n}{-m}\right)$ is the opposite of a negative number and, thus, is positive.

Chapter 2 Mid-Chapter Review

1. The statement is false. The integer 0 is neither positive nor negative.

2. If $a > b$, then a lies to the right of b on the number line. Thus, the given statement is false.

3. The absolute value of a number is its distance from zero on the number line. Since distance is always nonnegative, the absolute value of a number is always nonnegative. The given statement is true.

4. $-x = -(-4) = 4$
   $-(-x) = -(-(-4)) = -(4) = -4$

5. $5 - 13 = 5 + (-13) = -8$

6. $-6 - (-7) = -6 + 7 = 1$

7. The integer 450 corresponds to a \$450 deposit; the integer -79 corresponds to writing a check for \$79.

8. $-(9) = -9$

9. Since -6 is to the left of 6, we have $-6 < 6$.

10. Since -5 is to the left of -3, we have $-5 < -3$.

11. Since -10 is to the left of 0, we have $-10 < 0$.

12. Since -20 is to the right of -30, we have $-20 > -30$.

13. The distance of 38 from 0 is 38, so $|38| = 38$.

14. The distance of -18 from 0 is 18, so $|-18| = 18$.

15. The distance of 0 from 0 is 0, so $|0| = 0$.

16. The distance of -12 from 0 is 12, so $|-12| = 12$.

17. The additive inverse of -56 is 56 because $-56 + 56 = 0$.

18. The additive inverse of 3 is -3 because $3 + (-3) = 0$.

19. The additive inverse of 0 is 0 because $0 + 0 = 0$.

20. The additive inverse of -49 is 49 because $-49 + 49 = 0$.

21. If $x = -19$, then $-x = -(-19) = 19$.

22. If $x = 23$, then $-(-x) = -(-23) = 23$.

23. $7+(-9)$ The absolute values are 7 and 9. The difference is $9-7$, or 2. The negative number has the larger absolute value, so the answer is negative. $7 + (-9) = -2$

24. $-6 + (-10)$ Two negative numbers

    Add the absolute values, 6 and 10, getting 16. Make the answer negative.

    $-6 + (-10) = -16$

25. $36 + (-36)$ A positive and a negative number

    The numbers have the same absolute value. The sum is 0.

    $36 + (-36) = 0$

26. $-8 + (-9)$ Two negative numbers

    Add the absolute values, 8 and 9, getting 17. Make the answer negative. $-8 + (-9) = -17$

27. $-9+10$ The absolute values are 9 and 10. The difference is $10-9$, or 1. The positive number has the larger absolute value, so the answer is positive.
    $-9 + 10 = 1$

28. $19 + (-17)$ The absolute values are 19 and 17. The difference is $19-17$, or 2. The positive number has the larger absolute value, so the answer is positive.
    $19 + (-17) = 2$

29. $2 - 28 = 2 + (-28) = -26$

30. $-8 - (-4) = -8 + 4 = -4$

31. $-3 - 10 = -3 + (-10) = -13$

32. $5 - (-11) = 5 + 11 = 16$

33. $0 - (-6) = 0 + 6 = 6$

34. $12 - 24 = 12 + (-24) = -12$

35. $-12 \cdot 3 = -36$

36. $6(-9) = -54$

37. $(-13)(-2) = 26$

38. $(-2)(-41) = 82$

39. $(-9)^2 = (-9)(-9) = 81$

40. $-9^2 = -1 \cdot 9^2 = -1 \cdot 9 \cdot 9 = -81$

41. $-75 \div (-3) = 25$

42. $-20 \div 4 = -5$

43. $17 - (-25) + 15 - (-18) = 17 + 25 + 15 + 18 = 75$

44. $-9 + (-3) + 16 - (-10) = -9 + (-3) + 16 + 10 = 14$

45. $(-7)(-2)(-1)(-3) = 14 \cdot 3 = 42$

46. $$\begin{aligned} 3 - 6 \cdot 5 - 11 &= 3 - 30 - 11 \\ &= -27 - 11 \\ &= -38 \end{aligned}$$

47. $$\begin{aligned} -5^2 + 6[1 - (3 - 4)] &= -5^2 + 6[1 - (-1)] \\ &= -5^2 + 6[1 + 1] \\ &= -5^2 + 6 \cdot 2 \\ &= -25 + 6 \cdot 2 \\ &= -25 + 12 \\ &= -13 \end{aligned}$$

48. $$\begin{aligned} \frac{6^2 - 3(5 - 9)}{7^2 - (-5)^2} &= \frac{36 - 3(5 - 9)}{49 - 25} \\ &= \frac{36 - 3(-4)}{24} \\ &= \frac{36 + 12}{24} \\ &= \frac{48}{24} \\ &= 2 \end{aligned}$$

49. Let T = the difference in the temperatures, in degrees Celsius.

$$\underbrace{\text{Difference in temperatures}}_{\downarrow\ T} \ \overset{\text{is}}{\downarrow =} \ \underbrace{\text{Higher temperature}}_{\downarrow\ 25} \ \overset{\text{minus}}{\downarrow -} \ \underbrace{\text{Lower temperature}}_{\downarrow\ (-8)}$$

We carry out the subtraction.

$T = 25 - (-8) = 25 + 8 = 33$

The difference in the two temperature is 33°C.

50. Let S = the final value of the stock.

$$\underbrace{\text{Final value}}_{\downarrow\ S} \ \downarrow = \ \underbrace{\text{Beginning price}}_{\downarrow\ 56} \ \downarrow + \ \underbrace{\text{First change}}_{\downarrow\ (-6)} \ \downarrow + \ \underbrace{\text{Second change}}_{\downarrow\ 2} \ \downarrow + \ \underbrace{\text{Third change}}_{\downarrow\ (-8)}$$

We carry out the addition.

$S = 56 + (-6) + 2 + (-8) = 44$

The final value of the stock was \$44.

51. The student is confusing the absolute values of the numbers with the numbers themselves.

52. No; for example, $(10 - 5) - 2 = 5 - 2 = 3$, but $10 - (5 - 2) = 10 - 3 = 7$.

53. Answers may vary. If we think of the addition on the number line, we start at a negative number and move to the left. This always brings us to a point on the negative portion of the number line.

54. Yes; consider $m - (-n)$ where both m and n are positive. Then $m - (-n) = m + n$. Now $m + n$, the sum of two positive numbers, is positive.

Exercise Set 2.6

1. $10n = 10 \cdot 2 = 20¢$

3. $\dfrac{x}{y} = \dfrac{6}{-3} = -2$

5. $\dfrac{2d}{c} = \dfrac{2 \cdot 3}{6} = \dfrac{6}{6} = 1$

7. $\dfrac{72}{r} = \dfrac{72}{4} = 18$ yr

9. $3 - 5 \cdot x = 3 - 5 \cdot 2 = 3 - 10 = -7$

11. $2l + 2w = 2 \cdot 3 + 2 \cdot 4 = 6 + 8 = 14$ ft

13. $2(l + w) = 2(3 + 4) = 2 \cdot 7 = 14$ ft

15. $7a - 7b = 7(-1) - 7 \cdot 2 = -7 - 14 = -21$

17. $7(a - b) = 7(-1 - 2) = 7(-3) = -21$

19. $16t^2 = 16 \cdot 5^2 = 16 \cdot 25 = 400$ ft

21.
$$\begin{aligned} a + (b - a)^2 &= 6 + (10 - 6)^2 \\ &= 6 + (4)^2 \\ &= 6 + 16 \\ &= 22 \end{aligned}$$

23.
$$\begin{aligned} 9a + 9b &= 9 \cdot 13 + 9(-13) \\ &= 117 - 117 \\ &= 0 \end{aligned}$$

25.
$$\begin{aligned} \frac{n^2 - n}{2} &= \frac{9^2 - 9}{2} \\ &= \frac{81 - 9}{2} \\ &= \frac{72}{2} \\ &= 36 \end{aligned}$$

27.
$$\begin{aligned} 1 - x^2 &= 1 - (-2)^2 \\ &= 1 - 4 \\ &= -3 \end{aligned}$$

29.
$$\begin{aligned} m^2 - n^2 &= 6^2 - 5^2 \\ &= 36 - 25 \\ &= 11 \end{aligned}$$

31.
$$\begin{aligned} a^3 - a^2 &= (-10)^3 - (-10)^2 \\ &= -1000 - 100 \\ &= -1100 \end{aligned}$$

33. $-\dfrac{a}{b}$, $\dfrac{-a}{b}$, and $\dfrac{a}{-b}$ all represent the same number. Thus we can also write $-\dfrac{5}{t}$ as $\dfrac{-5}{t}$ and $\dfrac{5}{-t}$.

35. $-\dfrac{a}{b}$, $\dfrac{-a}{b}$, and $\dfrac{a}{-b}$ all represent the same number. Thus we can also write $\dfrac{-n}{b}$ as $-\dfrac{n}{b}$ and $\dfrac{n}{-b}$.

37. $-\dfrac{a}{b}$, $\dfrac{-a}{b}$, and $\dfrac{a}{-b}$ all represent the same number. Thus we can also write $\dfrac{9}{-p}$ as $-\dfrac{9}{p}$ and $\dfrac{-9}{p}$.

39. $-\dfrac{a}{b}$, $\dfrac{-a}{b}$, and $\dfrac{a}{-b}$ all represent the same number. Thus we can also write $\dfrac{-14}{w}$ as $-\dfrac{14}{w}$ and $\dfrac{14}{-w}$.

41. $\dfrac{-a}{b} = \dfrac{-45}{9} = -5$;
$\dfrac{a}{-b} = \dfrac{45}{-9} = -5$;
$-\dfrac{a}{b} = -\dfrac{45}{9} = -5$

43. $\dfrac{-a}{b} = \dfrac{-81}{3} = -27$;
$\dfrac{a}{-b} = \dfrac{81}{-3} = -27$;
$-\dfrac{a}{b} = -\dfrac{81}{3} = -27$

45. $(-3x)^2 = (-3 \cdot 2)^2 = (-6)^2 = 36$;
$-3x^2 = -3(2)^2 = -3 \cdot 4 = -12$

47. $5x^2 = 5(3)^2 = 5 \cdot 9 = 45$;
$5x^2 = 5(-3)^2 = 5 \cdot 9 = 45$

49. $x^3 = 6^3 = 6 \cdot 6 \cdot 6 = 216$;
$x^3 = (-6)^3 = (-6) \cdot (-6) \cdot (-6) = -216$

51. $x^8 = 1^8 = 1 \cdot 1 \cdot 1 \cdot 1 \cdot 1 \cdot 1 \cdot 1 \cdot 1 = 1$;
$x^8 = (-1)^8 =$
$(-1) \cdot (-1) \cdot (-1) \cdot (-1) \cdot (-1) \cdot (-1) \cdot (-1) \cdot (-1) = 1$

53. $a^5 = 2^5 = 2 \cdot 2 \cdot 2 \cdot 2 \cdot 2 = 32$;
$a^5 = (-2)^5 = (-2)(-2)(-2)(-2)(-2) = -32$

55. $5(a+b) = 5 \cdot a + 5 \cdot b = 5a + 5b$

57. $4(x+1) = 4 \cdot x + 4 \cdot 1 = 4x + 4$

59. $2(b+5) = 2 \cdot b + 2 \cdot 5 = 2b + 10$

61. $7(1-t) = 7 \cdot 1 - 7 \cdot t = 7 - 7t$

63. $6(5x-2) = 6 \cdot 5x - 6 \cdot 2 = 30x - 12$

65. $8(x+7+6y) = 8 \cdot x + 8 \cdot 7 + 8 \cdot 6y = 8x + 56 + 48y$

67. $-7(y-2) = -7 \cdot y - (-7) \cdot 2 = -7y - (-14) = -7y + 14$

69. $(x+2)3 = x \cdot 3 + 2 \cdot 3 = 3x + 6$

71. $-4(x - 3y - 2z) = -4 \cdot x - (-4)3y - (-4)2z =$
$-4x - (-12y) - (-8z) = -4x + 12y + 8z$

73. $8(a - 3b + c) = 8 \cdot a - 8 \cdot 3b + 8 \cdot c =$
$8a - 24b + 8c$

75. $4(x - 3y - 7z) = 4 \cdot x - 4 \cdot 3y - 4 \cdot 7z =$
$4x - 12y - 28z$

77. $(4a - 5b + c - 2d)5 = 4a \cdot 5 - 5b \cdot 5 + c \cdot 5 - 2d \cdot 5 =$
$20a - 25b + 5c - 10d$

79. $-1(3m + 2n) = -1 \cdot 3m - 1 \cdot 2n = -3m - 2n$

81. $-1(2a - 3b + 4) = -1 \cdot 2a - 1(-3b) - 1 \cdot 4 = -2a + 3b - 4$

83. $-(x - y - z) = -1 \cdot x - 1 \cdot (-y) - 1 \cdot (-z) = -x + y + z$

85. Twenty-three million, forty-three thousand, nine hundred twenty-one

87.
$$\begin{array}{r} 5\,2\,8\,3 \\ -\,2\,4\,7\,5 \\ \hline \end{array} \qquad \begin{array}{r} 5\,2\,8\,0 \\ -\,2\,4\,8\,0 \\ \hline 2\,8\,0\,0 \end{array}$$

89. ***Familiarize***. Since we are finding the difference in snowfall amounts, subtraction can be used. We let $s =$ the number of inches of snow remaining.

Translate. We translate to an equation.

$12 - 7 = s$

Solve. We carry out the subtraction.

$12 - 7 = 5$

Thus, $5 = s$, or $s = 5$.

Check. We repeat the calculation. The answer checks.

State. 5 in. of snow remained.

91. We substitute 370 for C in the formula in Example 5.

$$\frac{9C}{5} + 32 = \frac{9 \cdot 370}{5} + 32 = \frac{3330}{5} + 32 = 666 + 32 = 698$$

The temperature 370°Celsius corresponds to a Fahrenheit temperature of 698°.

93. Use a calculator.

$a - b^3 + 17a = 19 - (-16)^3 + 17 \cdot 19 = 4438$

95. Use a calculator.

$r^3 + r^2t - rt^2 = (-9)^3 + (-9)^2 \cdot 7 - (-9) \cdot 7^2 = 279$

97.
$$\begin{aligned} a^{1996} - a^{1997} &= (-1)^{1996} - (-1)^{1997} \\ &= 1 - (-1) \\ &= 1 + 1 \\ &= 2 \end{aligned}$$

99.
$$\begin{aligned} (m^3 - mn)^m &= (4^3 - 4 \cdot 6)^4 \\ &= (64 - 4 \cdot 6)^4 \\ &= (64 - 24)^4 \\ &= 40^4 \\ &= 2,560,000 \end{aligned}$$

101. $-32 \boxed{\times} (88 \boxed{-} 29) = -1888$

103. True

105. True

Exercise Set 2.7

1. $2a + 5b - 7c = 2a + 5b + (-7c)$

The terms are $2a$, $5b$, and $-7c$.

3. $mn - 6n + 8 = mn + (-6n) + 8$

The terms are mn, $-6n$, and 8.

5. $3x^2y - 4y^2 - 2z^3 = 3x^2y + (-4y^2) + (-2z^3)$

The terms are $3x^2y$, $-4y^2$, and $-2z^3$.

7. $5x + 9x = (5+9)x = 14x$

9. $10a - 15a = (10 - 15)a = -5a$

11.
$$\begin{aligned} 2x + 6y + x &= 2x + x + 6y \\ &= 2x + 1 \cdot x + 6y \\ &= (2+1)x + 6y \\ &= 3x + 6y \end{aligned}$$

13.
$$\begin{aligned} 27a + 70 - 40a - 8 &= 27a - 40a + 70 - 8 \\ &= (27 - 40)a + 70 - 8 \\ &= -13a + 62 \end{aligned}$$

15.
$$\begin{aligned} &9 + 5t + 7y - t - y - 13 \\ &= 9 - 13 + 5t - 1 \cdot t + 7y - 1 \cdot y \\ &= (9 - 13) + (5 - 1)t + (7 - 1)y \\ &= -4 + 4t + 6y \end{aligned}$$

17.
$$\begin{aligned} &a + 3b + 5a - 2 + b \\ &= a + 5a + 3b + b - 2 \\ &= (1+5)a + (3+1)b - 2 \\ &= 6a + 4b - 2 \end{aligned}$$

19.
$$\begin{aligned} &-8 + 11a - 5b - 10a - 7b + 7 \\ &= 11a - 10a - 5b - 7b - 8 + 7 \\ &= (11 - 10)a + (-5 - 7)b + (-8 + 7) \\ &= a - 12b - 1 \end{aligned}$$

21.
$$\begin{aligned} 8x^2 + 3y - x^2 &= 8x^2 - x^2 + 3y \\ &= (8-1)x^2 + 3y \\ &= 7x^2 + 3y \end{aligned}$$

23. $11x^4 + 2y^3 - 4x^4 - y^3 = 11x^4 - 4x^4 + 2y^3 - y^3$
$= (11-4)x^4 + (2-1)y^3$
$= 7x^4 + y^3$

25. $9a^2 - 4a + a - 3a^2 = 9a^2 - 3a^2 - 4a + a$
$= (9-3)a^2 + (-4+1)a$
$= 6a^2 - 3a$

27. $x^3 - 5x^2 + 2x^3 - 3x^2 + 4$
$= x^3 + 2x^3 - 5x^2 - 3x^2 + 4$
$= (1+2)x^3 + (-5-3)x^2 + 4$
$= 3x^3 - 8x^2 + 4$

29. $9x^3y + 4xy^3 - 5xy^3 + 3xy$
$= 9x^3y + (4-5)xy^3 + 3xy$
$= 9x^3y - xy^3 + 3xy$

31. $3a^6 - b^4 + 2a^6b^4 - 7a^6 - 2b^4$
$= 3a^6 - 7a^6 - b^4 - 2b^4 + 2a^6b^4$
$= (3-7)a^6 + (-1-2)b^4 + 2a^6b^4$
$= -4a^6 - 3b^4 + 2a^6b^4$

33. $P = 2 \cdot (l + w)$ Perimeter of a rectangle
$P = 2 \cdot (3 \text{ ft} + 2 \text{ ft})$
$P = 2 \cdot (3+2) \text{ ft}$
$P = 2 \cdot 5 \text{ ft}$
$P = 10 \text{ ft}$

35. Perimeter
$= 7 \text{ km} + 7 \text{ km} + 7 \text{ km} + 7 \text{ km} + 7 \text{ km} + 7 \text{ km}$
$= (7+7+7+7+7+7) \text{ km}$
$= 42 \text{ km}$

37. Perimeter $= 3 \text{ m} + 1 \text{ m} + 3 \text{ m} + 1 \text{ m}$
$= (3+1+3+1) \text{ m}$
$= 8 \text{ m}$

39. A singles court is 78 ft by 27 ft.
$P = 2l + 2w = 2 \cdot 78 \text{ ft} + 2 \cdot 27 \text{ ft}$
$= 156 \text{ ft} + 54 \text{ ft}$
$= 210 \text{ ft}$

41. The rectangle formed by the services lines and the singles sideline is 42 ft by 27 ft.
$P = 2l + 2w = 2 \cdot 42 \text{ ft} + 2 \cdot 27 \text{ ft}$
$= 84 \text{ ft} + 54 \text{ ft}$
$= 138 \text{ ft}$

43. $P = 2(l + w) = 2(10 \text{ ft} + 8 \text{ ft})$
$= 2 \cdot 18 \text{ ft} = 36 \text{ ft}$

45. $P = 4s$
$= 4 \cdot 14 \text{ in.} = 56 \text{ in.}$

47. $P = 4s$
$= 4 \cdot 65 \text{ cm} = 260 \text{ cm}$

49. $P = 2(l + w) = 2(20 \text{ ft} + 12 \text{ ft})$
$= 2 \cdot 32 \text{ ft} = 64 \text{ ft}$

51. ***Familiarize.*** Let s = the number of servings of Shaw's Corn Flakes in one box. Visualize a rectangular array consisting of 510 grams with 30 grams in each row. Then s is the number of rows.

Translate.

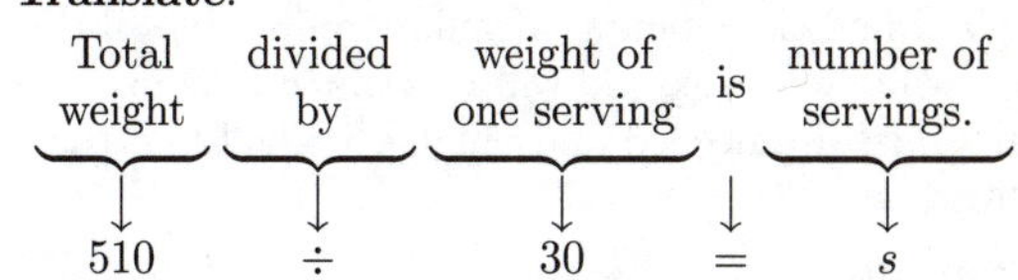

Solve. We carry out the division.

$$\begin{array}{r} 17 \\ 30\overline{)510} \\ \underline{30} \\ 210 \\ \underline{210} \\ 0 \end{array}$$

We have $17 = s$, or $s = 17$.

Check. We multiply the number of servings by the weight of a serving.

$17 \cdot 30 = 510$

We get 510 oz, the total weight of the corn flakes, so the answer checks.

State. There are 17 servings in a box of Shaw's Corn Flakes.

53. $5 + 3 \cdot 2^3$
$= 5 + 3 \cdot 8$ Evaluating the exponential expression
$= 5 + 24$ Multiplying
$= 29$ Adding

55. $12 \div 3 \cdot 2$
$= 4 \cdot 2$ Dividing and multiplying in order
$= 8$ from left to right

57. $15 - 3 \cdot 2 + 7$
$= 15 - 6 + 7$ Multiplying
$= 9 + 7$ Subtracting and adding in order
$= 16$ from left to right

59. $25 = t + 9$
$25 - 9 = t + 9 - 9$
$16 = t$

The solution is 16.

61. $45 = 3x$
$\dfrac{45}{3} = \dfrac{3x}{3}$
$15 = x$

The solution is 15.

63. $5(x+3) + 2(x-7) = 5x + 15 + 2x - 14 = 7x + 1$

65. $2(3-4a) + 5(a-7) = 6 - 8a + 5a - 35 = -3a - 29$

67. $-5(2 + 3x + 4y) + 7(2x - y) =$
$-10 - 15x - 20y + 14x - 7y = -10 - x - 27y$

69. ***Familiarize.*** First we will find the amount of sealant needed to caulk each door and each window, keeping in mind that the bottom of each door requires no caulk. Thus, for each door we add the lengths of the other three sides and for each window we find the perimeter. Then we will find the amount of caulk required for all the doors and windows. Next we will determine how many sealant cartridges are needed and, finally, we will find the cost of the sealant.

Translate.

The amount of caulk required for each door is given by

$7 \text{ ft} + 3 \text{ ft} + 7 \text{ ft}.$

The perimeter of each window is given by

$P = 2(l + w) = 2(3 \text{ ft} + 4 \text{ ft}).$

Solve. First we do the calculations in the Translate step.

For each door: $7 \text{ ft} + 3 \text{ ft} + 7 \text{ ft} = 17 \text{ ft}$

For each window: $P = 2(3 \text{ ft} + 4 \text{ ft}) = 2 \cdot 7 \text{ ft} = 14 \text{ ft}$

We multiply to find the amount of caulk required for 3 doors and of 13 windows.

Doors: $3 \cdot 17 \text{ ft} = 51 \text{ ft}$

Windows: $13 \cdot 14 \text{ ft} = 182 \text{ ft}$

We add to find the total of the perimeters:

$51 \text{ ft} + 182 \text{ ft} = 233 \text{ ft}$

Next we divide to determine how many sealant cartridges are needed:

$$\begin{array}{r} 4 \\ 56\overline{)233} \\ \underline{224} \\ 9 \end{array}$$

The answer is 4 R 9. Since 9 ft will be left unsealed after 4 cartridges are used, Andrea should buy 5 sealant cartridges.

Finally, we multiply to find the cost of 5 sealant cartridges.

$5 \cdot \$6 = \$30$

Check. We repeat the calculations. The result checks.

State. It will cost Andrea \$30 to seal the windows and doors.

71. ***Familiarize.*** The inside of the rack is a square whose side has a length that is the total diameter of 4 balls, or $4 \cdot 57$ mm, or 228 mm. We find the perimeter of a square with side 228 mm.

Translate.

$P = 4s = 4 \cdot 228 \text{ mm}$

Solve. We calculate the perimeter.

$P = 4 \cdot 228 \text{ mm} = 912 \text{ mm}$

Check. We repeat the calculation. The answer checks.

State. The inside perimeter of the storage rack is 912 mm.

Exercise Set 2.8

1. First note that $x = -1$ and $x + 5 = 4$ are equations. The solution of $x = -1$ is -1. We substitute to determine if -1 is also the solution of $x + 5 = 4$.

$$\begin{array}{rl} x + 5 = 4 & \\ -1 + 5 = 4 & \text{TRUE} \end{array}$$

We see that the equations are equivalent.

3. First note that $7a - 3$ and $13 + 7a - 16$ are expressions, not equations. To determine if they are equivalent, we combine like terms in the second expression.

$$13 + 7a - 16 = 7a + (13 - 16) = 7a - 3$$

We see that the expressions are equivalent.

5. First note that $4r + 3$ and $9 - r + 5r - 6$ are expressions, not equations. To determine if they are equivalent, we combine like terms in the second expression.

$$9 - r + 5r - 6 = (-1 + 5)r + (9 - 6) = 4r + 3$$

We see that the expressions are equivalent.

7. First note that $x - 9 = 8$ and $x = 20 - 3$ are equations. To find the solution of $x = 20 - 3$, we carry out the subtracting, getting $x = 17$. We substitute to determine if 17 is also a solution of $x - 9 = 8$.

$$\begin{array}{rl} x - 9 = 8 & \\ 17 - 9 = 8 & \text{TRUE} \end{array}$$

We see that the equations are equivalent.

9. First note that $3(t + 2)$ and $5 + 3t + 1$ are expressions, not equations. We simply each expression.

$$3(t + 2) = 3t + 6$$
$$5 + 3t + 1 = 3t + (5 + 1) = 3t + 6$$

We see that the expressions are equivalent.

11. First note that $x + 4 = -8$ and $2x = -24$ are equations. We solve each equation.

$$\begin{array}{rcl} x + 4 = -8 & \qquad & 2x = -24 \\ x + 4 - 4 = -8 - 4 & & \frac{2x}{2} = \frac{-24}{2} \\ x = -12 & & x = -12 \end{array}$$

The equations have the same solution, so they are equivalent.

13.

$$\begin{array}{rl} x - 6 = -9 & \\ x - 6 + 6 = -9 + 6 & \text{Adding 6 to both sides} \\ x + 0 = -3 & \\ x = -3 & \end{array}$$

Check:
$$\begin{array}{c|c} x - 6 & -9 \\ \hline -3 - 6 \ ? & -9 \\ -9 & -9 \quad \text{TRUE} \end{array}$$

The solution is -3.

15. $x - 4 = -12$

$x - 4 + 4 = -12 + 4$ Adding 4 to both sides

$x + 0 = -8$

$x = -8$

Check:

$$\begin{array}{c|c} x - 4 & -12 \\ \hline -8 - 4 \ ? & -12 \\ -12 & -12 \ \text{TRUE} \end{array}$$

The solution is -8.

17. $a + 7 = 25$

$a + 7 - 7 = 25 - 7$ Subtracting 7 from both sides

$a + 0 = 18$

$a = 18$

The solution is 18.

19. $-8 = n + 7$

$-8 - 7 = n + 7 - 7$ Subtracting 7 from both sides

$-15 = n$

The solution is -15.

21. $24 = t - 8$

$24 + 8 = t - 8 + 8$ Adding 8 to both sides

$32 = t + 0$

$32 = t$

The solution is 32.

23. $-12 = x + 5$

$-12 - 5 = x + 5 - 5$ Subtracting 5 from both sides

$-17 + 0 = x$

$-17 = x$

The solution is -17.

25. $-5 + a = 12$

$5 - 5 + a = 5 + 12$ Adding 5 to both sides

$0 + a = 17$

$a = 17$

The solution is 17.

27. $-8 = -8 + t$

$8 - 8 = 8 - 8 + t$ Adding 8 to both sides

$0 = 0 + t$

$0 = t$

The solution is 0.

29. $6x = 60$

$\frac{6x}{6} = \frac{60}{6}$ Dividing both sides by 6

$x = 10$

The solution is 10.

31. $-3t = 42$

$\frac{-3t}{-3} = \frac{42}{-3}$ Dividing both sides by -3

$t = -14$

The solution is -14.

33. $-7n = -35$

$\frac{-7n}{-7} = \frac{-35}{-7}$ Dividing both sides by -7

$n = 5$

The solution is 5.

35. $0 = 6x$

$\frac{0}{6} = \frac{6x}{6}$ Dividing both sides by 6

$0 = x$

The solution is 0.

37. $55 = -5t$

$\frac{55}{-5} = \frac{-5t}{-5}$ Dividing both sides by -5

$-11 = t$

The solution is -11.

39. $-x = 56$

$\frac{-x}{-1} = \frac{56}{-1}$ Dividing both sides by -1

$x = -56$

The solution is -56.

41. $n(-4) = -48$

$\frac{n(-4)}{-4} = \frac{-48}{-4}$ Dividing both sides by -4

$n = 12$

The solution is 12.

43. $-x = -390$

$\frac{-x}{-1} = \frac{-390}{-1}$ Dividing both sides by -1

$x = 390$

The solution is 390.

45. $t - 6 = -2$

To undo the addition of -6, or the subtraction of 6, we subtract -6, or simply add 6, to both sides.

$$t - 6 = -2$$
$$t - 6 + 6 = -2 + 6$$
$$t + 0 = 4$$
$$t = 4$$

The solution is 4.

47. $6x = -54$

To undo multiplication by 6, we divide both sides by 6.

$$6x = -54$$
$$\frac{6x}{6} = \frac{-54}{6}$$
$$x = -9$$

The solution is -9.

49. $15 = -x$

$-1 \cdot 15 = -1 \cdot (-x)$ Multiplying both sides by -1

$-15 = x$

The solution is -15.

51. $-21 = x + 5$

$-21 - 5 = x + 5 - 5$ Subtracting 5 from both sides

$-26 = x$

The solution is -26.

53. $35 = -7t$

To undo multiplication by -7, we divide both sides by -7.

$$35 = -7t$$
$$\frac{35}{-7} = \frac{-7t}{-7}$$
$$-5 = t$$

The solution is -5.

55. $-17x = 68$

To undo multiplication by -17, we divide both sides by -17.

$$-17x = 68$$
$$\frac{-17x}{-17} = \frac{68}{-17}$$
$$x = -4$$

The solution is -4.

57. $12 + t = -160$

To undo the addition of 12, we subtract 12 from both sides.

$$12 + t = -160$$
$$12 + t - 12 = -160 - 12$$
$$t + 0 = -172$$
$$t = -172$$

The solution is -172.

59. $-27 = x - 23$

To undo the subtraction of 23, we add 23 to both sides.

$$-27 = x - 23$$
$$-27 + 23 = x - 23 + 23$$
$$-4 = x + 0$$
$$-4 = x$$

The solution is -4.

61. $5x - 1 = 34$

$5x - 1 + 1 = 34 + 1$ Adding 1 to both sides

$5x + 0 = 35$

$5x = 35$

$\frac{5x}{5} = \frac{35}{5}$ Dividing both sides by 5

$x = 7$

Check: $5x - 1 = 34$

$5 \cdot 7 - 1$?	34
$35 - 1$	
34	34 TRUE

The solution is 7.

63. $4t + 2 = 14$

$4t + 2 - 2 = 14 - 2$ Subtracting 2 from both sides

$4t + 0 = 12$

$4t = 12$

$\frac{4t}{4} = \frac{12}{4}$ Dividing both sides by 4

$t = 3$

Check: $4t + 2 = 14$

$4 \cdot 3 + 2$?	14
$12 + 2$	
14	14 TRUE

The solution is 3.

65. $6a + 1 = -17$

$6a + 1 - 1 = -17 - 1$ Subtracting 1 from both sides

$6a + 0 = -18$

$6a = -18$

$\frac{6a}{6} = \frac{-18}{6}$ Dividing both sides by 6

$a = -3$

The solution is -3.

67. $2x - 9 = -23$

$2x - 9 + 9 = -23 + 9$ Adding 9 to both sides

$2x + 0 = -14$

$2x = -14$

$\frac{2x}{2} = \frac{-14}{2}$ Dividing both sides by 2

$x = -7$

The solution is -7.

69. $-2x + 1 = 17$

$-2x + 1 - 1 = 17 - 1$ Subtracting 1 from both sides

$-2x + 0 = 16$

$-2x = 16$

$\frac{-2x}{-2} = \frac{16}{-2}$ Dividing both sides by -2

$x = -8$

The solution is -8.

71. $-8t - 3 = -67$

$-8t - 3 + 3 = -67 + 3$ Adding 3 to both sides

$-8t + 0 = -64$

$-8t = -64$

$\frac{-8t}{-8} = \frac{-64}{-8}$ Dividing both sides by -8

$t = 8$

The solution is 8.

73.
$$
\begin{aligned}
-x+9 &= -15 \\
-x+9-9 &= -15-9 \quad \text{Subtracting 9 from both sides} \\
-x+0 &= -24 \\
-x &= -24 \\
-1(-x) &= -1(-24) \quad \text{Multiplying both sides by } -1 \\
x &= 24
\end{aligned}
$$
The solution is 24.

75.
$$
\begin{aligned}
7 &= 2x-5 \\
7+5 &= 2x-5+5 \quad \text{Adding 5 to both sides} \\
12 &= 2x+0 \\
12 &= 2x \\
\frac{12}{2} &= \frac{2x}{2} \quad \text{Dividing both sides by 2} \\
6 &= x
\end{aligned}
$$
The solution is 6.

77.
$$
\begin{aligned}
13 &= 3+2x \\
13-3 &= 3+2x-3 \quad \text{Subtracting 3 from both sides} \\
10 &= 2x+0 \\
10 &= 2x \\
\frac{10}{2} &= \frac{2x}{2} \quad \text{Dividing both sides by 2} \\
5 &= x
\end{aligned}
$$
The solution is 5.

79.
$$
\begin{aligned}
13 &= 5-x \\
13-5 &= 5-x-5 \quad \text{Subtracting 5 from both sides} \\
8 &= 0-x \\
8 &= -x \\
-1\cdot 8 &= -1(-x) \quad \text{Multiplying both sides by } -1 \\
-8 &= x
\end{aligned}
$$
The solution is -8.

81. A polygon is a closed geometric figure.

83. Numbers we multiply together are called factors.

85. The result of an addition is a sum.

87. The absolute value of a number is its distance from zero on a number line.

89.
$$
\begin{aligned}
2x-7x &= -40 \\
-5x &= -40 \quad \text{Collecting like terms} \\
\frac{-5x}{-5} &= \frac{-40}{-5} \\
x &= 8
\end{aligned}
$$
The solution is 8.

91.
$$
\begin{aligned}
2x-7+x &= 5-12 \\
3x-7 &= -7 \quad \text{Collecting like terms} \\
3x-7+7 &= -7+7 \\
3x &= 0 \\
\frac{3x}{3} &= \frac{0}{3} \\
x &= 0
\end{aligned}
$$
The solution is 0.

93.
$$
\begin{aligned}
n+n &= -2-3\cdot 6\div 2+1 \\
2n &= -2-18\div 2+1 \\
2n &= -2-9+1 \\
2n &= -10 \\
\frac{2n}{n} &= \frac{-10}{2} \\
n &= -5
\end{aligned}
$$
The solution is -5.

95.
$$
\begin{aligned}
17-3^2 &= 4+t-5^2 \\
17-9 &= 4+t-25 \\
8 &= t-21 \quad \text{Collecting like terms} \\
8+21 &= t-21+21 \quad \text{Adding 21 to both sides} \\
29 &= t
\end{aligned}
$$
The solution is 29.

97.
$$
\begin{aligned}
(-7)^2-5 &= t+4^3 \\
49-5 &= t+64 \\
44 &= t+64 \\
44-64 &= t+64-64 \quad \text{Subtracting 64 from both sides} \\
-20 &= t
\end{aligned}
$$
The solution is -20.

99.
$$
\begin{aligned}
x-(19)^3 &= -18^3 \\
x-6859 &= -5832 \\
x-6859+6859 &= -5832+6859 \\
x &= 1027
\end{aligned}
$$
The solution is 1027.

101.
$$
\begin{aligned}
35^3 &= -125t \\
42,875 &= -125t \\
\frac{42,875}{-125} &= \frac{-125t}{-125} \\
-343 &= t
\end{aligned}
$$
The solution is -343.

103.
$$
\begin{aligned}
529-143x &= -1902 \\
529-143x-529 &= -1902-529 \\
-143x &= -2431 \\
\frac{-143x}{-143} &= \frac{-2431}{-143} \\
x &= 17
\end{aligned}
$$
The solution is 17.

Chapter 2 Concept Reinforcement

1. The statement is true. See page 91 in the text.

2. The statement is true. See page 108 in the text.

3. The statement is false. For example, let $x = 1$. Then $2(x+3) = 2(1+3) = 2 \cdot 4 = 8$ and $2 \cdot x + 3 = 2 \cdot 1 + 3 = 5$. We see that $2(x+3) \neq 2 \cdot x + 3$ for an allowable replacement for x, so the expressions are not equivalent.

Chapter 2 Important Concepts

1. a) The number is negative, so we make it positive.
   $|-17| = 17$

   b) The number is positive, so the absolute value is the same as the number.

   $|300| = 300$

2. $37 + (-16)$ A positive number and a negative number

   The difference of the absolute values is $37 - 16$, or 21. The positive number has the larger absolute value, so the answer is positive.

   $37 + (-16) = 21$

3. $6 - (-8) = 6 + 8 = 14$

4. $6(-15) = -90$

5. $99 \div (-9) = -11$

6. $$\begin{aligned} 4 - 8^2 \div (10-6) &= 4 - 8^2 \div 4 \\ &= 4 - 64 \div 4 \\ &= 4 - 16 \\ &= -12 \end{aligned}$$

7. $5(6x - 8y - z) = 5 \cdot 6x - 5 \cdot 8y - 5 \cdot z = 30x - 40y - 5z$

8. $$\begin{aligned} 8a - b + 9a - 6b &= 8a + (-1 \cdot b) + 9a + (-6b) \\ &= 8a + 9a + (-1 \cdot b) + (-6b) \\ &= (8+9)a + (-1-6)b \\ &= 17a - 7b \end{aligned}$$

9. $$\begin{aligned} -19 &= 5x + 11 \\ -19 - 11 &= 5x + 11 - 11 \\ -30 &= 5x \\ \frac{-30}{5} &= \frac{5x}{5} \\ -6 &= x \end{aligned}$$

   The solution is -6.

Chapter 2 Review Exercises

1. The integer -45 corresponds to a debt of \$45; the integer 72 corresponds to having \$72 in a savings account.

2. Since 0 is to the right of -5, we have $0 > -5$.

3. Since -7 is to the left of 6, we have $-7 < 6$.

4. Since -4 is to the right of -19, we have $-4 > -19$.

5. The distance from -39 to 0 is 39, so $|-39| = 39$.

6. The distance from 23 to 0 is 23, so $|23| = 23$.

7. The distance from 0 to 0 is 0, so $|0| = 0$.

8. When $x = -72$, $-x = -(-72) = 72$.

9. When $x = 59$, $-(-x) = -(-59) = 59$.

10. $-14+5$ The absolute values are 14 and 5. The difference is 9. The negative number has the larger absolute value, so the answer is negative. $-14 + 5 = -9$

11. $-5 + (-6)$

    Add the absolute values: $5 + 6 = 11$

    Make the answer negative: $-5 + (-6) = -11$

12. $14 + (-8)$ The absolute values are 14 and 8. The difference is 6. The positive number has the larger absolute value, so the answer is positive. $14 + (-8) = 6$

13. $0 + (-24) = -24$

    When 0 is added to any number, that number remains unchanged.

14. $17 - 29 = 17 + (-29) = -12$

15. $9 - (-14) = 9 + 14 = 23$

16. $-8 - (-7) = -8 + 7 = -1$

17. $-3 - (-3) = -3 + 3 = 0$

18. $$\begin{aligned} & -3 + 7 + (-8) \\ &= -3 + (-8) + 7 \quad \text{Using a commutative law} \\ &= -11 + 7 \\ &= -4 \end{aligned}$$

19. $$\begin{aligned} & 8 - (-9) - 7 + 2 \\ &= 8 + 9 + (-7) + 2 \\ &= 19 + (-7) \quad \text{Adding the positive numbers} \\ &= 12 \end{aligned}$$

20. $-23 \cdot (-4) = 92$

21. $7(-12) = -84$

22. $$\begin{aligned} & 2(-4)(-5)(-1) \\ &= -8 \cdot 5 \quad \text{Multiplying the first two numbers and the last two numbers} \\ &= -40 \end{aligned}$$

23. $15 \div (-5) = -3$ Check: $-3(-5) = 15$

24. $\frac{-55}{11} = -5$ Check: $-5 \cdot 11 = -55$

25. $\frac{0}{7} = 0$ Check: $0 \cdot 7 = 0$

26. $625 \div (-25) \div 5 = -25 \div 5 = -5$

27. $$\begin{aligned} -16 \div 4 - 30 \div (-5) &= -4 - (-6) \\ &= -4 + 6 \\ &= 2 \end{aligned}$$

28. $9[(7-14)-13] = 9[-7-13] = 9[-20] = -180$

29. $$\begin{aligned} &(-3)|4-3^2|-5 \\ &= (-3)|4-9|-5 \\ &= (-3)|-5|-5 \\ &= -3 \cdot 5 - 5 \\ &= -15-5 \\ &= -20 \end{aligned}$$

30. $$\begin{aligned} &[-12(-3)-2^3]-(-9)(-10) \\ &= [-12(-3)-8]-(-9)(-10) \\ &= [36-8]-(-9)(-10) \\ &= 28-(-9)(-10) \\ &= 28-90 \\ &= -62 \end{aligned}$$

31. $3a+b = 3 \cdot 4 + (-5) = 12 + (-5) = 7$

32. $$\frac{-x}{y} = \frac{-30}{5} = -6$$
$$\frac{x}{-y} = \frac{30}{-5} = -6$$
$$-\frac{x}{y} = -\frac{30}{5} = -6$$

33. $4(5x+9) = 4 \cdot 5x + 4 \cdot 9 = 20x + 36$

34. $3(2a-4b+5) = 3 \cdot 2a - 3 \cdot 4b + 3 \cdot 5 = 6a - 12b + 15$

35. $-10(2x+y) = -10 \cdot 2x + (-10) \cdot y = -20x - 10y$

36. $5a + 12a = (5+12)a = 17a$

37. $-7x + 13x = (-7+13)x = 6x$

38. $$\begin{aligned} &9m+14-12m-8 \\ &= 9m-12m+14-8 \\ &= (9-12)m+(14-8) \\ &= -3m+6 \end{aligned}$$

39. $$\begin{aligned} P = 2l + 2w &= 2 \cdot 10 \text{ in.} + 2 \cdot 8 \text{ in.} \\ &= 20 \text{ in.} + 16 \text{ in.} = 36 \text{ in.} \end{aligned}$$

40. $P = 4s = 4 \cdot 25 \text{ cm} = 100 \text{ cm}$

41. $$\begin{aligned} x - 9 &= -17 \\ x - 9 + 9 &= -17 + 9 \\ x &= -8 \end{aligned}$$
The solution is -8.

42. $$\begin{aligned} -4t &= 36 \\ \frac{-4t}{-4} &= \frac{36}{-4} \\ t &= -9 \end{aligned}$$
The solution is -9.

43. $$\begin{aligned} 13 &= -x \\ -1 \cdot 13 &= -1 \cdot (-x) \\ -13 &= x \end{aligned}$$
The solution is -13.

44. $$\begin{aligned} 56 &= 6x - 10 \\ 56 + 10 &= 6x - 10 + 10 \\ 66 &= 6x \\ \frac{66}{6} &= \frac{6x}{6} \\ 11 &= x \end{aligned}$$
The solution is 11.

45. $$\begin{aligned} -x + 3 &= -12 \\ -x + 3 - 3 &= -12 - 3 \\ -x &= -15 \\ \frac{-x}{-1} &= \frac{-15}{-1} \\ x &= 15 \end{aligned}$$
The solution is 15.

46. $$\begin{aligned} 18 &= 4 - 2x \\ 18 - 4 &= 4 - 2x - 4 \\ 14 &= -2x \\ \frac{14}{-2} &= \frac{-2x}{-2} \\ -7 &= x \end{aligned}$$
The solution is -7.

47. Let t = the total gain or loss. We represent the gains as positive numbers and the loss as a negative number. We add the gains and the loss to find t.

$t = 5 + (-12) + 15 = -7 + 15 = 8$

There is a total gain of 8 yd.

48. Let a = Kaleb's total assets after he borrows \$300.

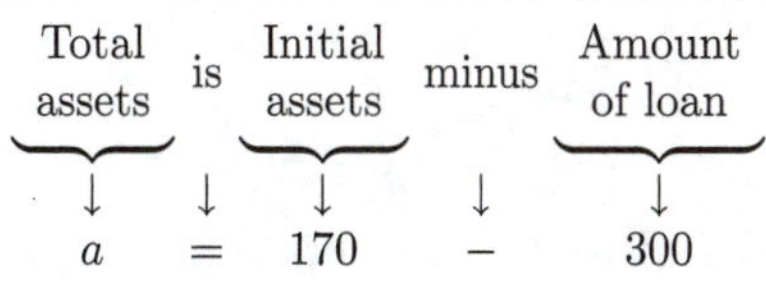

We carry out the subtraction.

$a = 170 - 300 = -130$

Kaleb's total assets were $-\$130$.

49. $-|-(-10)| = -|10| = -10$

Answer A is correct.

50. $-3 \cdot 4 - 12 \div 4 = -12 - 3 = -15$

Answer B is correct.

51. ***Familiarize.*** Let x = the larger number. Then $800 - x$ = the smaller number.

Translate.

$$\underbrace{\text{Larger number}}_{x} \text{ minus } \underbrace{\text{Smaller number}}_{(800-x)} \text{ is } 6.$$
$$x - (800 - x) = 6$$

Solve.

$$x-(800-x)=6$$
$$x-800+x=6$$
$$2x-800=6$$
$$2x-800+800=6+800$$
$$2x=806$$
$$\frac{2x}{2}=\frac{806}{2}$$
$$x=403$$

If $x=403$, then $800-x=800-403=397$.

Check. $403+397=800$ and $403-397=6$, so the answer checks.

State. The numbers are 403 and 397.

52. a) $-7+(-6)+(-5)+(-4)+(-3)+(-2)+(-1)+0+1+2+3+4+5+6+7+8$

b) Since one of the factors is 0, the product is 0.

53.
$$\begin{aligned}&87\div 3\cdot 29^3-(-6)^6+1957\\&=87\div 3\cdot 24,389-46,656+1957\\&=29\cdot 24,389-46,656+1957\\&=707,281-46,656+1957\\&=660,625+1957\\&=662,582\end{aligned}$$

54.
$$\begin{aligned}&1969+(-8)^5-17\cdot 15^3\\&=1969+(-32,768)-17\cdot 3375\\&=1969+(-32,768)-57,375\\&=-30,799-57,375\\&=-88,174\end{aligned}$$

55.
$$\begin{aligned}\frac{113-17^3}{15+8^3-507}&=\frac{113-4913}{15+512-507}\\&=\frac{-4800}{527-507}\\&=\frac{-4800}{20}\\&=-240\end{aligned}$$

56. $8+x^3$ will be negative for all values of x for which x^3 is less than -8. Thus, $8+x^3$ will be negative for $x<-2$.

57. $|x|>x$ for all negative values of x, or for $x<0$.

Chapter 2 Discussion and Writing Exercises

1. We know that the product of an even number of negative numbers is positive, and the product of an odd number of negative numbers is negative. Since $(-7)^8$ is equivalent to the product of eight negative numbers, it will be a positive number. Similarly, since $(-7)^{11}$ is equivalent to the product of eleven negative numbers, it will be a negative number.

2. No; when x is a negative number or 0, $-x$ is nonnegative. For example, when x is -3, $-x=-(-3)=3$ and when x is 0, $-x=-(0)=0$.

3. Jake is expecting the multiplication to be performed before the division.

4. The expression $-x^2$ represents a negative number, except for $x=0$. For all other values of x, x^2 is positive, and thus the opposite of x^2 is negative.

Chapter 2 Test

1. The integer -542 corresponds to selling 542 fewer shirts than expected; the integer 307 corresponds to selling 307 more shirts than expected.

2. Since -14 is to the right of -21, we have $-14>-21$.

3. The distance from -739 to 0 is 739, so $|-739|=739$.

4. When $x=-19$, $-(-x)=-(-(-19))=-(19)=-19$.

5. $6+(-17)$ The absolute values are 6 and 17. The difference is 11. The negative number has the larger absolute value, so the answer is negative. $6+(-17)=-11$

6. $-9+(-12)$

Add the absolute values: $9+12=21$

Make the answer negative: $-9+(-12)=-21$

7. $-8+17$ The absolute values are 8 and 17. The difference is 9. The positive number has the larger absolute value, so the answer is positive. $-8+17=9$

8. $0-12=0+(-12)=-12$

When 0 is added to any number, that number remains unchanged.

9. $7-22=7+(-22)=-15$

10. $-5-19=-5+(-19)=-24$

11. $-8-(-27)=-8+27=19$

12.
$$\begin{aligned}&31-(-3)-5+9\\&=31+3+(-5)+9\\&=43+(-5)\quad\text{Adding the positive numbers}\\&=38\end{aligned}$$

13. $(-4)^3=-4(-4)(-4)=16(-4)=-64$

14. $27(-10)=-270$

15. $-9\cdot 0=0$

16. $-72\div(-9)=8$ Check: $8(-9)=-72$

17. $\frac{-56}{7}=-8$ Check: $-8\cdot 7=-56$

18.
$$\begin{aligned}8\div 2\cdot 2-3^2&=8\div 2\cdot 2-9\\&=4\cdot 2-9\\&=8-9\\&=-1\end{aligned}$$

19. $29 - (3-5)^2 = 29 - (-2)^2$
$= 29 - 4$
$= 25$

20. We subtract the lower temperature from the higher temperature.

$$-67 - (-81) = -67 + 81 = 14$$

The average high temperature is 14°F higher than the average low temperature.

21. $\frac{a-b}{6} = \frac{-8-10}{6} = \frac{-18}{6} = -3$

22. $7(2x + 3y - 1) = 7 \cdot 2x + 7 \cdot 3y - 7 \cdot 1 = 14x + 21y - 7$

23. $9x - 14 - 5x - 3 = 9x - 5x - 14 - 3$
$= (9-5)x + (-14-3)$
$= 4x - 17$

24. $P = 4s = 4 \cdot 5 \text{ ft} = 20 \text{ ft}$

25. $-7x = -35$
$\frac{-7x}{-7} = \frac{-35}{-7}$
$x = 5$

The solution is 5.

26. $a + 9 = -3$
$a + 9 - 9 = -3 - 9$
$a = -12$

The solution is -12.

27. $95 = -x$
$95 = -1 \cdot x$
$\frac{95}{-1} = \frac{-1 \cdot x}{-1}$
$-95 = x$

The solution is -95.

28. $3t - 7 = 5$
$3t - 7 + 7 = 5 + 7$
$3t = 12$
$\frac{3t}{3} = \frac{12}{3}$
$t = 4$

The solution is 4.

29. $-2(n - 6m) = -2 \cdot n - (-2) \cdot 6m = -2n - (-12m) = -2n + 12m$

Answer C is correct.

30. The amount of trim needed is given by the perimeter of the room, less the 3 ft width of the door, plus the lengths of the three sides of the door that will get trim.

Perimeter of room: $P = 2(l + w)$
$= 2(14 \text{ ft} + 12 \text{ ft})$
$= 2(26 \text{ ft})$
$= 52 \text{ ft}$

Subtract the width of the door: $52 \text{ ft} - 3 \text{ ft} = 49 \text{ ft}$

Trim on door: $7 \text{ ft} + 3 \text{ ft} + 7 \text{ ft} = 17 \text{ ft}$

Total length of trim: $49 \text{ ft} + 17 \text{ ft} = 66 \text{ ft}$

31. $9 - 5[x + 2(3 - 4x)] + 14$
$= 9 - 5[x + 6 - 8x] + 14$
$= 9 - 5(-7x + 6) + 14$
$= 9 + 35x - 30 + 14$
$= 35x - 7$

32. $15x + 3(2x - 7) - 9(4 + 5x)$
$= 15x + 6x - 21 - 36 - 45x$
$= -24x - 57$

33. $49 \cdot 14^3 \div 7^4 + 1926^2 \div 6^2$
$= 49 \cdot 2744 \div 2401 + 3,709,476 \div 36$
$= 134,456 \div 2401 + 3,709,476 \div 36$
$= 56 + 3,709,476 \div 36$
$= 56 + 103,041$
$= 103,097$

34. $3487 - 16 \div 4 \cdot 4 \div 2^8 \cdot 14^4$
$= 3487 - 16 \div 4 \cdot 4 \div 256 \cdot 38,416$
$= 3487 - 4 \cdot 4 \div 256 \cdot 38,416$
$= 3487 - 16 \div 256 \cdot 38,416$
$= 3487 - 2401$ Dividing and then multiplying
$= 1086$

Chapter 3

Fraction Notation: Multiplication and Division

Exercise Set 3.1

1.

$1 \cdot 7 = 7$	$6 \cdot 7 = 42$
$2 \cdot 7 = 14$	$7 \cdot 7 = 49$
$3 \cdot 7 = 21$	$8 \cdot 7 = 56$
$4 \cdot 7 = 28$	$9 \cdot 7 = 63$
$5 \cdot 7 = 35$	$10 \cdot 7 = 70$

3.

$1 \cdot 20 = 20$	$6 \cdot 20 = 120$
$2 \cdot 20 = 40$	$7 \cdot 20 = 140$
$3 \cdot 20 = 60$	$8 \cdot 20 = 160$
$4 \cdot 20 = 80$	$9 \cdot 20 = 180$
$5 \cdot 20 = 100$	$10 \cdot 20 = 200$

5.

$1 \cdot 3 = 3$	$6 \cdot 3 = 18$
$2 \cdot 3 = 6$	$7 \cdot 3 = 21$
$3 \cdot 3 = 9$	$8 \cdot 3 = 24$
$4 \cdot 3 = 12$	$9 \cdot 3 = 27$
$5 \cdot 3 = 15$	$10 \cdot 3 = 30$

7.

$1 \cdot 12 = 12$	$6 \cdot 12 = 72$
$2 \cdot 12 = 24$	$7 \cdot 12 = 84$
$3 \cdot 12 = 36$	$8 \cdot 12 = 96$
$4 \cdot 12 = 48$	$9 \cdot 12 = 108$
$5 \cdot 12 = 60$	$10 \cdot 12 = 120$

9.

$1 \cdot 10 = 10$	$6 \cdot 10 = 60$
$2 \cdot 10 = 20$	$7 \cdot 10 = 70$
$3 \cdot 10 = 30$	$8 \cdot 10 = 80$
$4 \cdot 10 = 40$	$9 \cdot 10 = 90$
$5 \cdot 10 = 50$	$10 \cdot 10 = 100$

11.

$1 \cdot 25 = 25$	$6 \cdot 25 = 150$
$2 \cdot 25 = 50$	$7 \cdot 25 = 175$
$3 \cdot 25 = 75$	$8 \cdot 25 = 200$
$4 \cdot 25 = 100$	$9 \cdot 25 = 225$
$5 \cdot 25 = 125$	$10 \cdot 25 = 250$

13. We divide 83 by 3.

$$\begin{array}{r} 27 \\ 3\overline{)83} \\ \underline{6} \\ 23 \\ \underline{21} \\ 2 \end{array}$$

Since the remainder is not 0 we know that 83 is not divisible by 3.

15. We divide 525 by 7.

$$\begin{array}{r} 75 \\ 7\overline{)525} \\ \underline{49} \\ 35 \\ \underline{35} \\ 0 \end{array}$$

The remainder of 0 indicates that 525 is divisible by 7.

17. We divide 8127 by 9.

$$\begin{array}{r} 903 \\ 9\overline{)8127} \\ \underline{81} \\ 27 \\ \underline{27} \\ 0 \end{array}$$

The remainder of 0 indicates that 8127 is divisible by 9.

19. Because $8 + 4 = 12$ and 12 is divisible by 3, 84 is divisible by 3.

21. 5553 is not divisible by 5 because the ones digit is neither 0 nor 5.

23. 671,500 is divisible by 10 because the ones digit is 0.

25. Because $1 + 7 + 7 + 3 = 18$ and 18 is divisible by 9, 1773 is divisible by 9.

27. 21,687 is not divisible by 2 because the ones digit is not even.

29. 32,109 is not divisible by 6 because it is not even.

31. 6825 is not divisible by 2 because the ones digit is not even.

Because $6 + 8 + 2 + 5 = 21$ and 21 is divisible by 3, 6825 is divisible by 3.

6825 is divisible by 5 because the ones digit is 5.

6825 is not divisible by 6 because it is not even.

Because $6 + 8 + 2 + 5 = 21$ and 21 is not divisible by 9, 6825 is not divisible by 9.

6825 is not divisible by 10 because the ones digit is not 0.

33. 119,117 is not divisible by 2 because the ones digit is not even.

Because $1 + 1 + 9 + 1 + 1 + 7 = 20$ and 20 is not divisible by 3, then 119,117 is not divisible by 3.

119,117 is not divisible by 5 because the ones digit is neither 0 nor 5.

119,117 is not divisible by 6 because it is not even.

Because $1+1+9+1+1+7=20$ and 20 is not divisible by 9, then 119,117 is not divisible by 9.

119,117 is not divisible by 10 because the ones digit is not 0.

35. 127,575 is not divisible by 2 because the ones digit is not even.

Because $1+2+7+5+7+5=27$ and 27 is divisible by 3, then 127,575 is divisible by 3.

127,575 is divisible by 5 because the ones digit is 5.

127,575 is not divisible by 6 because it is not even.

Because $1+2+7+5+7+5=27$ and 27 is divisible by 9, then 127,575 is divisible by 9.

127,575 is not divisible by 10 because the ones digit is not 0.

37. 9360 is divisible by 2 because the ones digit is even.

Because $9+3+6+0=18$ and 18 is divisible by 3, 9360 is divisible by 3.

9360 is divisible by 5 because the ones digit is 0.

We saw above that 9360 is even and that it is divisible by 3, so it is divisible by 6.

Because $9+3+6+0=18$ and 18 is divisible by 9, 9360 is divisible by 9.

9360 is divisible by 10 because the ones digit is 0.

39. A number is divisible by 3 if the sum of the digits is divisible by 3.

46 is not divisible by 3 because $4+6=10$ and 10 is not divisible by 3.

224 is not divisible by 3 because $2+2+4=8$ and 8 is not divisible by 3.

19 is not divisible by 3 because $1+9=10$ and 10 is not divisible by 3.

555 is divisible by 3 because $5+5+5=15$ and 15 is divisible by 3.

300 is divisible by 3 because $3+0+0=3$ and 3 is divisible by 3.

36 is divisible by 3 because $3+6=9$ and 9 is divisible by 3.

45,270 is divisible by 3 because $4+5+2+7+0=18$ and 18 is divisible by 3.

4444 is not divisible by 3 because $4+4+4+4=16$ and 16 is not divisible by 3.

85 is not divisible by 3 because $8+5=13$ and 13 is not divisible by 3.

711 is divisible by 3 because $7+1+1=9$ and 9 is divisible by 3.

13,251 is divisible by 3 because $1+3+2+5+1=12$ and 12 is divisible by 3.

254,765 is not divisible by 3 because $2+5+4+7+6+5=29$ and 29 is not divisible by 3.

256 is not divisible by 3 because $2+5+6=13$ and 13 is not divisible by 3.

8064 is divisible by 3 because $8+0+6+4=18$ and 18 is divisible by 3.

1867 is not divisible by 3 because $1+8+6+7=22$ and 22 is not divisible by 3.

21,568 is not divisible by 3 because $2+1+5+6+8=22$ and 22 is not divisible by 3.

41. A number is divisible by 10 if its ones digit is 0.

Of the numbers under consideration, only 300 and 45,270 have one digits of 0. Therefore, only 300 and 45,270 are divisible by 10.

43. For a number to be divisible by 6, the sum of the digits must be divisible by 3 and the ones digit must be 0, 2, 4, 6 or 8 (even). It is most efficient to determine if the ones digit is even first and then, if so, to determine if the sum of the digits is divisible by 3.

46 is not divisible by 6 because 46 is not divisible by 3.

$4+6=10$
↑
Not divisible by 3

224 is not divisible by 6 because 224 is not divisible by 3.

$2+2+4=8$
↑
Not divisible by 3

19 is not divisible by 6 because 19 is not even.

19
↑
Not even

555 is not divisible by 6 because 555 is not even.

555
↑
Not even

300 is divisible by 6.

300 ↑ Even $3+0+0=3$ ↑ Divisible by 3

36 is divisible by 6.

36 ↑ Even $3+6=9$ ↑ Divisible by 3

45,270 is divisible by 6.

45,270 ↑ Even $4+5+2+7+0=18$ ↑ Divisible by 3

4444 is not divisible by 6 because 4444 is not divisible by 3.

$4+4+4+4=16$
↑
Not divisible by 3

85 is not divisible by 6 because 85 is not even.

85
↑
Not even

711 is not divisible by 6 because 711 is not even.

711
↑
Not even

13,251 is not divisible by 6 because 13,251 is not even.

13,251
↑
Not even

254,765 is not divisible by 6 because 254,765 is not even.

254,765
↑
Not even

256 is not divisible by 6 because 256 is not divisible by 3.

$2+5+6=13$
↑
Not divisible by 3

8064 is divisible by 6.

8064 — $8+0+6+4=18$
↑ — ↑
Even — Divisible by 3

1867 is not divisible by 6 because 1867 is not even.

1867
↑
Not even

21,568 is not divisible by 6 because 21,568 is not divisible by 3.

$2+1+5+6+8=22$
↑
Not divisible by 3

45. A number is divisible by 2 if its ones digit is even.

56 is divisible by 2 because 6 is even.
324 is divisible by 2 because 4 is even.
784 is divisible by 2 because 4 is even.
55,555 is not divisible by 2 because 5 is not even.
200 is divisible by 2 because 0 is even.
42 is divisible by 2 because 2 is even.
501 is not divisible by 2 because 1 is not even.
3009 is not divisible by 2 because 9 is not even.

75 is not divisible by 2 because 5 is not even.
812 is divisible by 2 because 2 is even.
2345 is not divisible by 2 because 5 is not even.
2001 is not divisible by 2 because 1 is not even.
35 is not divisible by 2 because 5 is not even.
402 is divisible by 2 because 2 is even.
111,111 is not divisible by 2 because 1 is not even.
1005 is not divisible by 2 because 5 is not even.

47. A number is divisible by 5 if the ones digit is 0 or 5.

56 is not divisible by 5 because the ones digit (6) is not 0 or 5.

324 is not divisible by 5 because the ones digit (4) is not 0 or 5.

784 is not divisible by 5 because the ones digit (4) is not 0 or 5.

55,555 is divisible by 5 because the ones digit is 5.

200 is divisible by 5 because the ones digit is 0.

42 is not divisible by 5 because the ones digit (2) is not 0 or 5.

501 is not divisible by 5 because the ones digit (1) is not 0 or 5.

3009 is not divisible by 5 because the ones digit (9) is not 0 or 5.

75 is divisible by 5 because the ones digit is 5.

812 is not divisible by 5 because the ones digit (2) is not 0 or 5.

2345 is divisible by 5 because the ones digit is 5.

2001 is not divisible by 5 because the ones digit (1) is not 0 or 5.

35 is divisible by 5 because the ones digit is 5.

402 is not divisible by 5 because the ones digit (2) is not 0 or 5.

111,111 is not divisible by 5 because the ones digit (1) is not 0 or 5.

1005 is divisible by 5 because the ones digit is 5.

49. A number is divisible by 9 if the sum of the digits is divisible by 9.

56 is not divisible by 9 because $5+6=11$ and 11 is not divisible by 9.

324 is divisible by 9 because $3+2+4=9$ and 9 is divisible by 9.

784 is not divisible by 9 because $7+8+4=19$ and 19 is not divisible by 9.

55,555 is not divisible by 9 because $5+5+5+5+5=25$ and 25 is not divisible by 9.

200 is not divisible by 9 because $2+0+0=2$ and 2 is not divisible by 9.

42 is not divisible by 9 because $4+2=6$ and 6 is not divisible by 9.

501 is not divisible by 9 because $5+0+1=6$ and 6 is not divisible by 9.

3009 is not divisible by 9 because $3+0+0+9=12$ and 12 is not divisible by 9.

75 is not divisible by 9 because $7+5=12$ and 12 is not divisible by 9.

812 is not divisible by 9 because $8+1+2=11$ and 11 is not divisible by 9.

2345 is not divisible by 9 because $2+3+4+5=14$ and 14 is not divisible by 9.

2001 is not divisible by 9 because $2+0+0+1=3$ and 3 is not divisible by 9.

35 is not divisible by 9 because $3+5=8$ and 8 is not divisible by 9.

402 is not divisible by 9 because $4+0+2=6$ and is not divisible by 9.

111,111 is not divisible by 9 because $1+1+1+1+1+1=6$ and 6 is not divisible by 9.

1005 is not divisible by 9 because $1+0+0+5=6$ and 6 is not divisible by 9.

51. $16 \cdot t = 848$

$$\frac{16 \cdot t}{16} = \frac{848}{16} \quad \text{Dividing by 16 on both sides}$$

$$t = 53$$

The solution is 53.

53.

$$23 + x = 15$$
$$23 + x - 23 = 15 - 23 \quad \text{Subtracting 23 on both sides}$$
$$x = -8$$

The solution is -8.

55. ***Familiarize.*** We visualize the situation. Let g = the number of gallons of gasoline the automobile will use to travel 1485 mi.

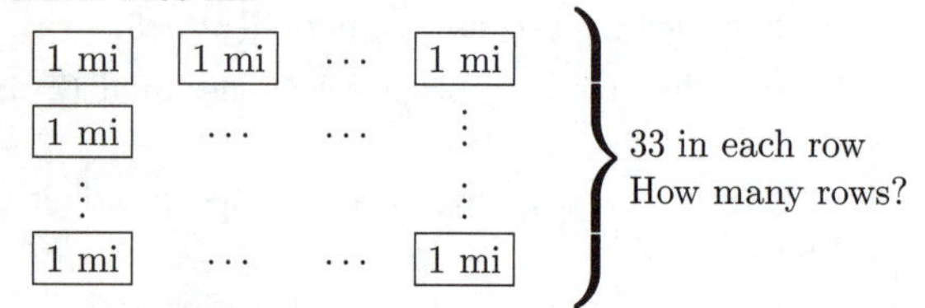

Translate. We translate to an equation.

Number of miles	divided by	Miles per gallon	is	Number of gallons
↓	↓	↓	↓	↓
1485	÷	33	=	g

Solve. We carry out the division.

```
      4 5
3 3 ) 1 4 8 5
      1 3 2
        1 6 5
        1 6 5
            0
```

Thus $45 = g$, or $g = 45$.

Check. We can repeat the calculation. The answer checks.

State. The automobile will use 45 gallons of gasoline to travel 1485 mi.

57. $5^3 = 5 \cdot 5 \cdot 5 = 125$

59. $\underbrace{9 \cdot 9 \cdot 9}_{\text{3 factors}} = 9^3$

61. When we use a calculator to divide the largest five-digit number, 99,999, by 47 we get 2127.638298. This tells us that 99,999 is not divisible by 47 but that 2127×47, or 99,969, is divisible by 47 and that it is the largest such five-digit number.

63. We list multiples of 2, 3, and 5 and find the smallest number that is on all 3 lists.

Multiples of 2: $2, 4, 6, 8, 10, 12, 14, 16, 18, 20, 22, 24, 26, 28, \underline{30}, 32, \cdots$

Multiples of 3: $3, 6, 9, 12, 15, 18, 21, 24, 27, \underline{30}, 33, \cdots$

Multiples of 5: $5, 10, 15, 20, 25, \underline{30}, 35, \cdots$

The smallest number that is simultaneously a multiple of 2, 3, and 5 is 30.

65. We list the multiples of 6, 10, and 14 and then find the smallest number that is on all three lists.

For 6: $6, 12, 18, 24, 30, 36, 42, 48, 54, 60, 66, 72, 78, 84, 90, 96, 102, 108, 114, 120, 126, 132, 138, 144, 150, 156, 162, 168, 174, 180, 186, 192, 198, 204, \underline{210}, 216, \cdots$

For 10: $10, 20, 30, 40, 50, 60, 70, 80, 90, 100, 110, 120, 130, 140, 150, 160, 170, 180, 190, 200, \underline{210}, 220, \cdots$

For 14: $14, 28, 42, 56, 70, 84, 98, 112, 126, 140, 154, 168, 182, 196, \underline{210}, 224, \cdots$

The smallest number that is simultaneously a multiple of 6, 10, and 14 is 210.

67. First note that 120 is a multiple of 30 ($120 = 4 \cdot 30$). Thus, any multiple of 120 is also a multiple of 30. Now we list multiples of 70 and 120 and find the smallest number that is on both lists.

Multiples of 70: 70, 140, 210, 280, 350, 420, 490, 560, 630, 700, 770, $\underline{840}$, 910, ...

Multiples of 120: 120, 240, 360, 480, 600, 720, $\underline{840}$, 960, ...

The smallest number that is simultaneously a multiple of 30, 70, and 120 is 840.

69. a) $999a + 99b + 9c = 9(111a + 11b + c)$, so it is divisible by 9 and $9(111a + 11b + c) = 3 \cdot 3(111a + 11b + c)$ so it is also divisible by 3.

b) $abcd$ is divisible by 9 if $a + b + c + d$ is divisible by 9, because $abcd = (999a + 99b + 9c) + (a + b + c + d)$ and $999a + 99b + 9c$ is divisible by 9. Similarly, $abcd$ is divisible by 3 if $a + b + c + d$ is divisible by 3, because $abcd = (999a + 99b + 9c) + (a + b + c + d)$ and $999a + 99b + 9c$ is divisible by 3.

71. 332,986,412 is divisible by 4 because the number named by its last two digits, 12, is divisible by 4.

332,986,412 is not divisible by 8 because the number named by its last three digits, 412, is not divisible by 8.

To test for divisibility by 7, we start with the three digits on the right, 412, and we subtract the next group of three digits to the left, 986:

$$412 - 986 = -574.$$

Now we add the next group to the left, 332:

$$-574 + 332 = -242.$$

Since -242 is not divisible by 7, the number 332,986,412 is not divisible by 7.

To test for divisibility by 11, we first find the sum of the odd-numbered digits:

$$2 + 4 + 8 + 2 + 3 = 19.$$

Next we find the sum of the even-numbered digits:

$$1 + 6 + 9 + 3 = 19.$$

Now we subtract the sums found above:

$$19 - 19 = 0.$$

Since 0 is divisible by 11, the number 332,986,412 is divisible by 11.

Exercise Set 3.2

1. One is a factor of every number. Since 18 is even, we know that 2 is a factor. Since the sum of the digits is 9 and 9 is divisible by both 3 and 9, we know that 3 and 9 are both factors. Also, since 2 and 3 are factors, 6 is a factor as well. We write a list of factorizations.

   $18 = 1 \cdot 18$ $\quad$ $18 = 3 \cdot 6$
   $18 = 2 \cdot 9$

   Factors: 1, 2, 3, 6, 9, 18

3. One is a factor of every number. Since 54 is even, we know that 2 is a factor. Since the sum of the digits is 9 and 9 is divisible by both 3 and 9, we know that 3 and 9 are both factors. Also, since 2 and 3 are factors, 6 is a factor as well. We write a list of factorizations.

   $54 = 1 \cdot 54$ $\quad$ $54 = 3 \cdot 18$
   $54 = 2 \cdot 27$ $\quad$ $54 = 6 \cdot 9$

   Factors: 1, 2, 3, 6, 9, 18, 27, 54

5. One is a factor of every number. Since 9 is divisible by 3, we know that 3 is a factor. We write a list of factorizations.

   $9 = 1 \cdot 9$
   $9 = 3 \cdot 3$

   Factors: 1, 3, 9

7. The number 13 is prime. It has only 1 and 13 as factors.

9. One is a factor of every number. Since 98 is even, 2 is a factor: $98 = 2 \cdot 49$. We continue to check positive integers to determine if they are factors of 98. We find that 7 is a factor: $98 = 7 \cdot 14$. The next factor we find is 49, but it is already listed in the factorization $2 \cdot 49$. Thus the factors of 98 are 1, 2, 7, 14, 49, and 98.

11. By checking positive integers to determine if they are factors of 255, we find that the factors are 1, 3, 5, 15, 17, 51, 85, and 255.

13. The number 17 is prime. It has only the factors 1 and 17.

15. The number 22 has factors 1, 2, 11, and 22. Since it has at least one factor other than itself and 1, it is composite.

17. The number 48 has factors 1, 2, 3, 4, 6, 8, 12, 16, 24, and 48. Since it has at least one factor other than itself and 1, it is composite.

19. The number 53 is prime. It has only the factors 1 and 53.

21. 1 is neither prime nor composite.

23. The number 81 has factors 1, 3, 9, 27, and 81.

    Since it has at least one factor other than itself and 1, it is composite.

25. The number 47 is prime. It has only the factors 1 and 47.

27. The number 29 is prime. It has only the factors 1 and 29.

29. $\begin{array}{r|l} & 3 \\ 3 & 9 \\ 3 & 27 \end{array}$ $\leftarrow$ 3 is prime.

    $27 = 3 \cdot 3 \cdot 3$

31. $\begin{array}{r|l} & 7 \\ 2 & 14 \end{array}$ $\leftarrow$ 7 is prime.

    $14 = 2 \cdot 7$

33. $\begin{array}{r|l} & 5 \\ 2 & 10 \\ 2 & 20 \\ 2 & 40 \\ 2 & 80 \end{array}$ $\leftarrow$ 5 is prime.

    $80 = 2 \cdot 2 \cdot 2 \cdot 2 \cdot 5$

    We can also use a factor tree.

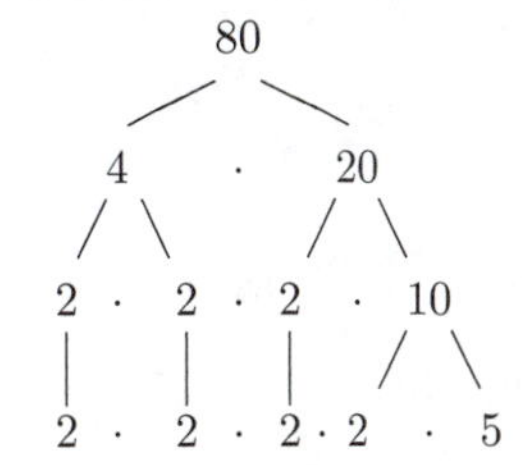

35. $\begin{array}{r|l} & 5 \\ 5 & 25 \end{array}$ $\leftarrow$ 5 is prime. (25 is not divisible by 2 or 3. We move to 5.)

    $25 = 5 \cdot 5$

37. $\begin{array}{r|l} & 31 \\ 2 & 62 \end{array}$ $\leftarrow$ 31 is prime.

    $62 = 2 \cdot 31$

39. $\begin{array}{r|l} & 5 \\ 5 & 25 \\ 2 & 50 \\ 2 & 100 \end{array}$ $\leftarrow$ 5 is prime.

    $100 = 2 \cdot 2 \cdot 5 \cdot 5$

    We can also use a factor tree.

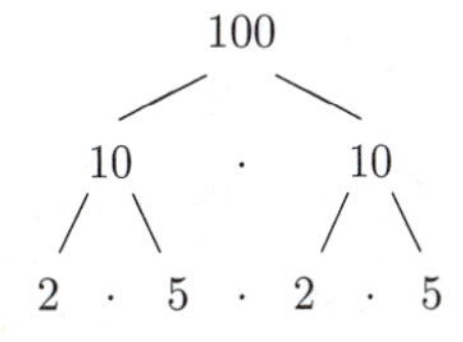

41. $\begin{array}{r|l} & 13 \\ 11 & 143 \end{array}$ $\leftarrow$ 13 is prime. (143 is not divisible by 2, 3, 5, or 7. We move to 11.)

    $143 = 11 \cdot 13$

43. $\begin{array}{r|l} & 11 \\ 11 & 121 \end{array}$ $\leftarrow$ 11 is prime. (121 is not divisible by 2, 3, 5, or 7. We move to 11.)

    $121 = 11 \cdot 11$

45. $\begin{array}{r|l} & 13 \\ 7 & 91 \\ 3 & 273 \end{array}$ $\leftarrow$ 13 is prime. (273 is not divisible by 2. We move to 3.)

    $273 = 3 \cdot 7 \cdot 13$

47.
$$\begin{array}{r|l} & 7 \\ \cline{2-2} 5 & 35 \\ \cline{2-2} 5 & 175 \end{array}$$
$\leftarrow$ 7 is prime. (175 is not divisible by 2 or 3. We move to 5.)

$175 = 5 \cdot 5 \cdot 7$

49.
$$\begin{array}{r|l} & 19 \\ \cline{2-2} 11 & 209 \end{array}$$
$\leftarrow$ 19 is prime. (209 is not divisible by 2, 3, 5, or 7. We move to 11.)

$209 = 11 \cdot 19$

51.
$$\begin{array}{r|l} & 5 \\ \cline{2-2} 5 & 25 \\ \cline{2-2} 3 & 75 \\ \cline{2-2} 2 & 150 \\ \cline{2-2} 2 & 300 \\ \cline{2-2} 2 & 600 \\ \cline{2-2} 2 & 1200 \end{array}$$
$\leftarrow$ 5 is prime

$1200 = 2 \cdot 2 \cdot 2 \cdot 2 \cdot 3 \cdot 5 \cdot 5$

53.
$$\begin{array}{r|l} & 11 \\ \cline{2-2} 7 & 77 \\ \cline{2-2} 3 & 231 \\ \cline{2-2} 3 & 693 \end{array}$$
$\leftarrow$ 11 is prime

$693 = 3 \cdot 3 \cdot 7 \cdot 11$

55.
$$\begin{array}{r|l} & 103 \\ \cline{2-2} 7 & 721 \\ \cline{2-2} 2 & 1442 \\ \cline{2-2} 2 & 2884 \end{array}$$
$\leftarrow$ 103 is prime.

$2884 = 2 \cdot 2 \cdot 7 \cdot 103$

We can also use a factor tree.

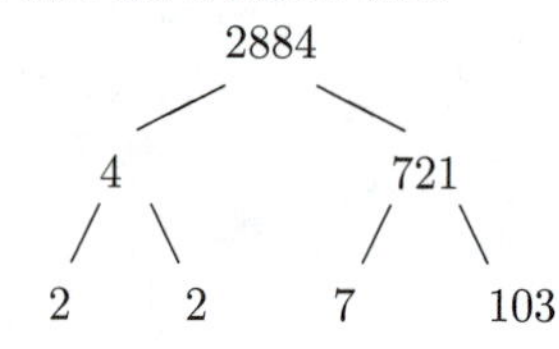

57.
$$\begin{array}{r|l} & 17 \\ \cline{2-2} 11 & 187 \\ \cline{2-2} 3 & 561 \\ \cline{2-2} 2 & 1122 \end{array}$$
$\leftarrow$ 17 is prime.

$1122 = 2 \cdot 3 \cdot 11 \cdot 17$

59. $-2 \cdot 13 = -26$ (The signs are different, so the answer is negative.)

61. $-17 + 25$ The absolute values are 17 and 25. The difference is 8. The positive number has the larger absolute value, so the answer is positive. $-17 + 25 = 8$

63. $53 \div 53 = 1$

65. $0 \div 22 = 0$ (0 divided by a nonzero number is 0.)

67. ***Familiarize***. Let $t =$ the total amount Kate took in.

Translate. We write an equation.

$$\underbrace{\text{Number of pounds}}_{43} \;\; \underset{\cdot}{\text{times}} \;\; \underbrace{\text{Price per pound}}_{22} \;\; \underset{=}{\text{is}} \;\; \underbrace{\text{Total amount taken in}}_{t}$$

Solve. We carry out the multiplication.

$$\begin{array}{r} 43 \\ \times\ 22 \\ \hline 86 \\ 860 \\ \hline 946 \end{array}$$

Thus, $946 = t$.

Check. We can repeat the calculation. The answer checks.

State. Kate took in a total of \$946.

69. We use a calculator to divide 136,097 by prime numbers.

$136,097 = 13 \cdot 19 \cdot 19 \cdot 29$

71. We use a calculator to divide 473,073,361 by prime numbers.

$473,073,361 = 23 \cdot 31 \cdot 61 \cdot 73 \cdot 149$

73. Answers may vary. One arrangement is a 3-dimensional rectangular array consisting of 2 tiers of 12 objects each where each tier consists of a rectangular array of 4 rows with 3 objects each.

75. The factors of 63 whose sum is 16 are 7 and 9.

The factors of 36 whose sum is 20 are 2 and 18.

The factors of 72 whose sum is 38 are 2 and 36.

The factors of 140 whose sum is 24 are 10 and 14.

The factors of 96 whose sum is 20 are 8 and 12.

The factors of 48 whose sum is 14 are 6 and 8.

The factors of 168 whose sum is 29 are 8 and 21.

The factors of 110 whose sum is 21 are 10 and 11.

The factors of 90 whose sum is 19 are 9 and 10.

The factors of 432 whose sum is 42 are 18 and 24.

The factors of 63 whose sum is 24 are 3 and 21.

Exercise Set 3.3

1. The top number is the numerator, and the bottom number is the denominator.

$$\frac{3}{4} \begin{array}{l} \leftarrow \text{Numerator} \\ \leftarrow \text{Denominator} \end{array}$$

3. $\dfrac{7}{-9}$ $\begin{array}{l} \leftarrow \text{Numerator} \\ \leftarrow \text{Denominator} \end{array}$

5. $\dfrac{2x}{3y}$ $\begin{array}{l} \leftarrow \text{Numerator} \\ \leftarrow \text{Denominator} \end{array}$

7. The dollar is divided into 4 parts of the same size, and 2 of them are shaded. This is $2 \cdot \frac{1}{4}$ or $\frac{2}{4}$. Thus, $\frac{2}{4}$ (two-fourths) of the dollar is shaded.

9. The yard is divided into 8 parts of the same size, and 1 of them is shaded. Thus, $\frac{1}{8}$ (one-eighth) of the yard is shaded.

11. The window is divided into 9 parts of the same size, and 4 of them are shaded. Thus, $\frac{4}{9}$ (four-ninths) of the window is shaded.

13. The acre is divided into 4 parts of the same size, and 3 of them are shaded. This is $3 \cdot \frac{1}{4}$ or $\frac{3}{4}$ of the acre.

15. The pie is divided into 8 equal parts. The unit is $\frac{1}{8}$. The denominator is 8. We have 4 parts shaded. This tells us that the numerator is 4. Thus, $\frac{4}{8}$ is shaded.

17. The rectangle is divided into 12 equal parts. The unit is $\frac{1}{12}$. The denominator is 12. All 12 parts are shaded. This tells us that the numerator is 12. Thus, $\frac{12}{12}$ is shaded.

19. We have 2 quarts, each divided into 3 equal parts. We take 4 of those parts. This is $4 \cdot \frac{1}{3}$, or $\frac{4}{3}$. Thus, $\frac{4}{3}$ of a quart is shaded.

21. We have 2 spools, each divided into 5 parts. We take 7 of those parts. This is $7 \cdot \frac{1}{5}$, or $\frac{7}{5}$. Thus, $\frac{7}{5}$ of a spool is shaded.

23. There are 8 circles, and 5 are shaded. Thus, $\frac{5}{8}$ of the circles are shaded.

25. There are 7 objects in the set, and 4 of the objects are shaded. Thus, $\frac{4}{7}$ of the set is shaded.

27. Each inch on the ruler is divided into 16 equal parts. The shading extends to the 12th mark, so $\frac{12}{16}$ is shaded.

29. Each inch on the ruler is divided into 16 equal parts. The shading extends to the 38th mark, so $\frac{38}{16}$ is shaded.

31. The gas gauge is divided into 8 equal parts.

a) The needle is 2 marks from the E (empty) mark, so the amount of gas in the tank is $\frac{2}{8}$ of a full tank.

b) The needle is 6 marks from the F (full) mark, so $\frac{6}{8}$ of a full tank of gas has been burned.

33. The gas gauge is divided into 8 equal parts.

a) The needle is 3 marks from the E (empty) mark, so the amount of gas in the tank is $\frac{3}{8}$ of a full tank.

b) The needle is 5 marks from the F (full) mark, so $\frac{5}{8}$ of a full tank of gas has been burned.

35. a) There are 8 people in the set and 5 are women, so the desired ratio is $\frac{5}{8}$.

b) There are 5 women and 3 men, so the ratio of women to men is $\frac{5}{3}$.

c) There are 8 people in the set and 3 are men, so the desired ratio is $\frac{3}{8}$.

d) There are 3 men and 5 women, so the ratio of men to women is $\frac{3}{5}$.

37. a) In Minnesota there are 1068 registered nurses per 100,000 residents, so the ratio is $\frac{1068}{100,000}$.

b) In Hawaii there are 680 registered nurses per 100,000 residents, so the ratio is $\frac{680}{100,000}$.

c) In Florida there are 797 registered nurses per 100,000 residents, so the ratio is $\frac{797}{100,000}$.

d) In Kentucky there are 962 registered nurses per 100,000 residents, so the ratio is $\frac{962}{100,000}$.

e) In New York there are 866 registered nurses per 100,000 residents, so the ratio is $\frac{866}{100,000}$.

f) In the District of Columbia there are 1561 registered nurses per 100,000 residents, so the ratio is $\frac{1561}{100,000}$.

39. a) The ratio is $\frac{4}{15}$.

b) The number of orders not delivered is $15 - 4$, or 11. The ratio is $\frac{4}{11}$.

c) From part (b) we know that the number of orders not delivered is 11. The ratio is $\frac{11}{15}$.

41. The ratio of voters in the 18-24 age group in the 2004 presidential election to all voters in the election is $\frac{16}{100}$.

The ratio of voters in the 18-24 age group in the 2008 presidential election to all voters in the election is $\frac{18.5}{100}$.

43. Remember: $\frac{0}{n} = 0$, for any whole number n that is not 0.

$\frac{0}{8} = 0$

Think of dividing an object into 8 parts and taking none of them. We get 0.

45. $\frac{8-1}{9-8} = \frac{7}{1}$ Remember: $\frac{n}{1} = n$.

$\frac{7}{1} = 7$

Think of taking 7 objects and dividing each into 1 part. (We do not divide them.) We have 7 objects.

47. Remember: $\frac{n}{n} = 1$, for any integer n that is not 0.

$\frac{20}{20} = 1$

If we divide an object into 20 parts and take 20 of them, we get all of the object (1 whole object).

49. Remember: $\frac{n}{n} = 1$, for any integer n that is not 0.

$\frac{-45}{-45} = 1$

51. Remember: $\frac{0}{n} = 0$, for any integer n that is not 0.

$\frac{0}{-238} = 0$

53. Remember: $\frac{n}{1} = n$.

$\frac{19x}{1} = 19x$

55. Remember: $\frac{n}{n} = 1$, for any integer n that is not 0.

$\frac{13t}{13t} = 1$

57. Remember: $\frac{n}{1} = n$.

$\frac{-87}{1} = -87$

59. Remember: $\frac{0}{n} = 0$, for any integer n that is not 0.

$\frac{0}{2a} = 0$

61. Remember: $\frac{n}{0}$ is not defined.

$\frac{52}{0}$ is undefined.

63. Remember: $\frac{n}{1} = n$,

$\frac{7n}{1} = 7n$

65. $\frac{5}{6-6} = \frac{5}{0}$

Remember: $\frac{n}{0}$ is not defined. Thus, $\frac{5}{6-6}$ is undefined.

67. $-7(30) = -210$
(The signs are different, so the answer is negative.)

69. $(-71)(-12)0 = -71 \cdot 0 = 0$
(We might have observed at the outset that the answer is 0 since one of the factors is 0.)

71. ***Familiarize.*** Let m = the number by which the membership in 2008 exceeded the membership in 1983. We visualize the situation.

Members in 1983 22,190	Additional members in 2008 m
Members in 2008 46,655	

Translate.

$$\underbrace{\text{Members in 1983}}_{22,190} + \underbrace{\text{Additional members}}_{m} \text{ is } \underbrace{\text{Members in 2008}}_{46,655}$$

$$22,190 + m = 46,655$$

Solve. We subtract 22,190 on both sides of the equation.

$$22,190 + m = 46,655$$
$$22,190 + m - 22,190 = 46,655 - 22,190$$
$$m = 24,465$$

Check. We can add the difference, 24,465, to the number of members in 1983, 22,190: $22,190 + 24,465 = 46,655$. We get the number of members in 2008, so the answer checks.

State. In 2008 the New York Road Runners had 24,465 more members than in 1983.

73. We can think of the object as being divided into 6 sections, each the size of the area shaded. Thus, $\frac{1}{6}$ of the object is shaded.

75. We can think of the object as being divided into 16 sections, each the size of one of the shaded sections. Since 2 sections are shaded, $\frac{2}{16}$ of the object is shaded. We could also express this as $\frac{1}{8}$.

77. The set contains 5 objects, so we shade 3 of them.

79. The figure has 5 rows, so we shade 3 of them.

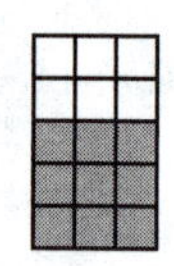

81. $365 = 52 \cdot 7 + 1$, so in one year there are 52 full weeks plus one additional day. Since 2006 began on a Sunday, the additional day is not a Monday. (It is a Sunday.) Thus, of the 365 days in 2006, 52 were Mondays, so $\frac{52}{365}$ were Mondays.

83. The surface of Earth is divided into 3+1, or 4 parts. Three of them are taken up by water, so $\frac{3}{4}$ is water. One of them is land, so $\frac{1}{4}$ is land.

Exercise Set 3.4

1. $3 \cdot \frac{1}{8} = \frac{3 \cdot 1}{8} = \frac{3}{8}$

3. $(-5) \times \frac{1}{6} = \frac{-5 \times 1}{6} = \frac{-5}{6}$, or $-\frac{5}{6}$

5. $\frac{2}{3} \cdot 7 = \frac{2 \cdot 7}{3} = \frac{14}{3}$

7. $(-1)\frac{7}{9} = \frac{(-1)7}{9} = \frac{-7}{9}$, or $-\frac{7}{9}$

9. $\frac{5}{6} \cdot x = \frac{5 \cdot x}{6} = \frac{5x}{6}$

11. $\frac{2}{5}(-3) = \frac{2(-3)}{5} = \frac{-6}{5}$, or $-\frac{6}{5}$

13. $a \cdot \frac{2}{7} = \frac{a \cdot 2}{7} = \frac{2a}{7}$

15. $-17 \times \frac{m}{6} = \frac{-17 \times m}{6} = \frac{-17m}{6}$, or $-\frac{17m}{6}$

17. $-3 \cdot \frac{-2}{5} = \frac{-3}{1} \cdot \frac{-2}{5} = \frac{-3(-2)}{1 \cdot 5} = \frac{6}{5}$

19. $-\frac{2}{7}(-x) = \frac{-2}{7} \cdot \frac{-x}{1} = \frac{-2(-x)}{7 \cdot 1} = \frac{2x}{7}$

21. $\frac{2}{5} \cdot \frac{2}{3} = \frac{2 \cdot 2}{5 \cdot 3} = \frac{4}{15}$

23. $\left(-\frac{1}{4}\right) \times \frac{1}{10} = -\frac{1 \times 1}{4 \times 10} = -\frac{1}{40}$, or $\frac{-1}{40}$

25. $\frac{2}{3} \times \frac{1}{5} = \frac{2 \times 1}{3 \times 5} = \frac{2}{15}$

27. $\frac{2}{y} \cdot \frac{x}{9} = \frac{2 \cdot x}{y \cdot 9} = \frac{2x}{9y}$

29. $\left(-\frac{3}{4}\right)\left(-\frac{3}{4}\right) = \frac{(-3)(-3)}{4 \cdot 4} = \frac{9}{16}$

31. $\frac{2}{3} \cdot \frac{7}{13} = \frac{2 \cdot 7}{3 \cdot 13} = \frac{14}{39}$

33. $\frac{1}{10}\left(\frac{-3}{5}\right) = \frac{1(-3)}{10 \cdot 5} = \frac{-3}{50}$, or $-\frac{3}{50}$

35. $\frac{7}{8} \cdot \frac{a}{8} = \frac{7 \cdot a}{8 \cdot 8} = \frac{7a}{64}$

37. $\frac{1}{y} \cdot 100 = \frac{1 \cdot 100}{y} = \frac{100}{y}$

39. $\frac{-21}{4} \cdot \frac{7}{5} = \frac{-21 \cdot 7}{4 \cdot 5} = \frac{-147}{20}$, or $-\frac{147}{20}$

41. ***Familiarize***. Let c = the portion of a cheesecake that corresponds to $\frac{1}{2}$ piece.

Translate. We are finding $\frac{1}{2}$ of $\frac{1}{12}$, so the multiplication sentence $\frac{1}{2} \cdot \frac{1}{12} = c$ corresponds to this situation.

Solve. We multiply.

$$\frac{1}{2} \cdot \frac{1}{12} = \frac{1 \cdot 1}{2 \cdot 12} = \frac{1}{24}$$

Thus, $\frac{1}{24} = c$.

Check. We repeat the calculation. The answer checks.

State. $\frac{1}{2}$ piece is $\frac{1}{24}$ of a cheesecake.

43. ***Familiarize***. Let b = the fractional part of high school basketball players who play professional basketball.

Translate. We are finding $\frac{1}{75}$ of $\frac{1}{35}$, so the multiplication sentence $\frac{1}{75} \cdot \frac{1}{35} = b$ corresponds to this situation.

Solve. We multiply.

$$\frac{1}{75} \cdot \frac{1}{35} = \frac{1 \cdot 1}{75 \cdot 35} = \frac{1}{2625}$$

Thus, $\frac{1}{2625} = b$.

Check. We repeat the calculation. The answer checks.

State. $\frac{1}{2625}$ of high school basketball players play professional basketball.

45. ***Familiarize***. Let r = the number of yards of ribbon needed to make 8 bows.

Translate.

Ribbon for 1 bow	times	Number of bows	is	Total amount needed
$\frac{5}{3}$	$\cdot$	8	=	r

Solve. We multiply.

$$\frac{5}{3} \cdot 8 = \frac{5 \cdot 8}{3} = \frac{40}{3}$$

Thus, $\frac{40}{3} = r$.

Check. We repeat the calculation. The answer checks.

State. $\frac{40}{3}$ yd of ribbon is needed to make 8 bows.

47. ***Familiarize***. A picture of the situation appears in the text. Let f = the fraction of the floor that has been tiled.

Translate. The multiplication sentence $\frac{3}{5} \cdot \frac{3}{4} = f$ corresponds to the situation.

Solve. We multiply.

$$\frac{3}{5} \cdot \frac{3}{4} = \frac{3 \cdot 3}{5 \cdot 4} = \frac{9}{20}$$

Check. We repeat the calculation. The answer checks.

State. $\frac{9}{20}$ of the floor has been tiled.

49.
```
      2 0 4
 3 5 ) 7 1 4 0
       7 0
       ---
         1 4 0
         1 4 0
         -----
             0
```

The answer is 204.

51. $\frac{-65}{-5} = 13$

(The signs are the same, so the answer is positive.)

53. $8 \cdot 12 - (63 \div 9 + 13 \cdot 3)$

$= 8 \cdot 12 - (7 + 13 \cdot 3)$ Dividing inside the parentheses

$= 8 \cdot 12 - (7 + 39)$ Multiplying inside the parentheses

$= 8 \cdot 12 - 46$ Adding inside the parentheses

$= 96 - 46$ Multiplying

$= 50$ Subtracting

55. Use a calculator.

$$\frac{341}{517} \cdot \frac{209}{349} = \frac{341 \cdot 209}{517 \cdot 349} = \frac{71,269}{180,433}$$

57. $\left(\frac{2}{5}\right)^3\left(-\frac{7}{9}\right) = \frac{8}{125}\left(-\frac{7}{9}\right)$ Evaluating the exponential expression

$$= -\frac{8(7)}{125 \cdot 9}$$

$$= -\frac{56}{1125}, \text{ or } \frac{-56}{1125}$$

59. ***Familiarize***. Let t = the number of gallons of two-cycle oil in a freshly filled chainsaw.

Translate. The multiplication sentence $\frac{1}{16} \cdot \frac{1}{5} = t$ corresponds to this situation.

Solve. We carry out the multiplication.

$$\frac{1}{16} \cdot \frac{1}{5} = \frac{1 \cdot 1}{16 \cdot 5} = \frac{1}{80}$$

Check. We repeat the calculation. The answer checks.

State. There is $\frac{1}{80}$ gal of two-cycle oil in a freshly filled chainsaw.

Exercise Set 3.5

1. Since $10 \div 2 = 5$, we multiply by $\frac{5}{5}$.

$$\frac{1}{2} = \frac{1}{2} \cdot \frac{5}{5} = \frac{1 \cdot 5}{2 \cdot 5} = \frac{5}{10}$$

3. Since $-48 \div 4 = -12$, we multiply by $\frac{-12}{-12}$.

$$\frac{3}{4} = \frac{3}{4}\left(\frac{-12}{-12}\right) = \frac{3(-12)}{4(-12)} == \frac{-36}{-48}$$

5. Since $50 \div 10 = 5$, we multiply by $\frac{5}{5}$.

$$\frac{7}{10} = \frac{7}{10} \cdot \frac{5}{5} = \frac{7 \cdot 5}{10 \cdot 5} = \frac{35}{50}$$

7. Since $5t \div 5 = t$, we multiply by $\frac{t}{t}$.

$$\frac{11}{5} = \frac{11}{5} \cdot \frac{t}{t} = \frac{11 \cdot t}{5 \cdot t} = \frac{11t}{5t}$$

9. Since $4 \div 1 = 4$, we multiply by $\frac{4}{4}$.

$$\frac{5}{1} = \frac{5}{1} \cdot \frac{4}{4} = \frac{5 \cdot 4}{1 \cdot 4} = \frac{20}{4}$$

11. Since $54 \div 18 = 3$, we multiply by $\frac{3}{3}$.

$$-\frac{17}{18} = -\frac{17}{18} \cdot \frac{3}{3} = -\frac{17 \cdot 3}{18 \cdot 3} = -\frac{51}{54}$$

13. Since $-40 \div -8 = 5$, we multiply by $\frac{5}{5}$.

$$\frac{3}{-8} = \frac{3}{-8} \cdot \frac{5}{5} = \frac{3 \cdot 5}{-8 \cdot 5} = \frac{15}{-40}$$

15. Since $132 \div 22 = 6$, we multiply by $\frac{6}{6}$.

$$\frac{-7}{22} = \frac{-7}{22} \cdot \frac{6}{6} = \frac{-7 \cdot 6}{22 \cdot 6} = \frac{-42}{132}$$

17. Since $8x \div 8 = x$, we multiply by $\frac{x}{x}$.

$$\frac{1}{8} = \frac{1}{8} \cdot \frac{x}{x} = \frac{x}{8x}$$

19. Since $7a \div 7 = a$, we multiply by $\frac{a}{a}$.

$$\frac{-10}{7} = \frac{-10}{7} \cdot \frac{a}{a} = \frac{-10a}{7a}$$

21. Since $9ab \div 9 = ab$, we multiply by $\frac{ab}{ab}$.

$$\frac{4}{9} \cdot \frac{ab}{ab} = \frac{4ab}{9ab}$$

23. Since $27b \div 9 = 3b$, we multiply by $\frac{3b}{3b}$.

$$\frac{4}{9} = \frac{4}{9} \cdot \frac{3b}{3b} = \frac{12b}{27b}$$

25. $\frac{2}{4} = \frac{1 \cdot 2}{2 \cdot 2}$ ⟵ Factor the numerator ⟵ Factor the denominator

$= \frac{1}{2} \cdot \frac{2}{2}$ ⟵ Factor the fraction

$= \frac{1}{2} \cdot 1$ ⟵ $\frac{2}{2} = 1$

$= \frac{1}{2}$ ⟵ Removing a factor of 1

27. $-\frac{6}{9} = -\frac{2 \cdot 3}{3 \cdot 3}$ ⟵ Factor the numerator ⟵ Factor the denominator

$= -\frac{2}{3} \cdot \frac{3}{3}$ ⟵ Factor the fraction

$= -\frac{2}{3} \cdot 1$ ⟵ $\frac{3}{3} = 1$

$= -\frac{2}{3}$ ⟵ Removing a factor of 1

29. $\frac{10}{25} = \frac{2 \cdot 5}{5 \cdot 5}$ ⟵ Factor the numerator ⟵ Factor the denominator

$= \frac{2}{5} \cdot \frac{5}{5}$ ⟵ Factor the fraction

$= \frac{2}{5} \cdot 1$ ⟵ $\frac{5}{5} = 1$

$= \frac{2}{5}$ ⟵ Removing a factor of 1

31. $\frac{24}{8} = \frac{3 \cdot 8}{1 \cdot 8} = \frac{3}{1} \cdot \frac{8}{8} = \frac{3}{1} \cdot 1 = 3$

33. $\frac{27}{36} = \frac{9 \cdot 3}{9 \cdot 4} = \frac{9}{9} \cdot \frac{3}{4} = 1 \cdot \frac{3}{4} = \frac{3}{4}$

35. $-\frac{24}{14} = -\frac{12 \cdot 2}{7 \cdot 2} = -\frac{12}{7} \cdot \frac{2}{2} = -\frac{12}{7}$

37. $\frac{3n}{4n} = \frac{3 \cdot n}{4 \cdot n} = \frac{3}{4} \cdot \frac{n}{n} = \frac{3}{4}$

39. $\dfrac{-17}{51} = \dfrac{-1 \cdot 17}{3 \cdot 17} = \dfrac{-1}{3} \cdot \dfrac{17}{17} = \dfrac{-1}{3}$

41. $\dfrac{-100}{20} = \dfrac{-5 \cdot 20}{1 \cdot 20} = \dfrac{-5}{1} \cdot \dfrac{20}{20} = -5$

43.
$$\begin{aligned}\frac{420}{480} &= \frac{2 \cdot 2 \cdot 3 \cdot 5 \cdot 7}{2 \cdot 2 \cdot 2 \cdot 2 \cdot 2 \cdot 3 \cdot 5} \\ &= \frac{2}{2} \cdot \frac{2}{2} \cdot \frac{3}{3} \cdot \frac{5}{5} \cdot \frac{7}{2 \cdot 2 \cdot 2} \\ &= \frac{7}{2 \cdot 2 \cdot 2} \\ &= \frac{7}{8}\end{aligned}$$

45.
$$\begin{aligned}\frac{-540}{810} &= \frac{-2 \cdot 2 \cdot 3 \cdot 3 \cdot 3 \cdot 5}{2 \cdot 3 \cdot 3 \cdot 3 \cdot 3 \cdot 5} \\ &= \frac{2}{2} \cdot \frac{3}{3} \cdot \frac{3}{3} \cdot \frac{3}{3} \cdot \frac{5}{5} \cdot \frac{-2}{3} \\ &= \frac{-2}{3}\end{aligned}$$

47. $\dfrac{12x}{30x} = \dfrac{2 \cdot 2 \cdot 3 \cdot x}{2 \cdot 3 \cdot 5 \cdot x} = \dfrac{2}{2} \cdot \dfrac{3}{3} \cdot \dfrac{x}{x} \cdot \dfrac{2}{5} = \dfrac{2}{5}$

49.
$$\begin{aligned}\frac{153}{136} &= \frac{3 \cdot 3 \cdot 17}{2 \cdot 2 \cdot 2 \cdot 17} \\ &= \frac{3 \cdot 3}{2 \cdot 2 \cdot 2} \cdot \frac{17}{17} \\ &= \frac{3 \cdot 3}{2 \cdot 2 \cdot 2} \\ &= \frac{9}{8}\end{aligned}$$

51. $\dfrac{132}{143} = \dfrac{11 \cdot 12}{11 \cdot 13} = \dfrac{11}{11} \cdot \dfrac{12}{13} = \dfrac{12}{13}$

53. $\dfrac{221}{247} = \dfrac{13 \cdot 17}{13 \cdot 19} = \dfrac{13}{13} \cdot \dfrac{17}{19} = \dfrac{17}{19}$

55. $\dfrac{3ab}{8ab} = \dfrac{3 \cdot a \cdot b}{8 \cdot a \cdot b} = \dfrac{3}{8} \cdot \dfrac{a}{a} \cdot \dfrac{b}{b} = \dfrac{3}{8}$

57. $\dfrac{9xy}{6x} = \dfrac{3 \cdot 3 \cdot x \cdot y}{2 \cdot 3 \cdot x} = \dfrac{3 \cdot y}{2} \cdot \dfrac{3}{3} \cdot \dfrac{x}{x} = \dfrac{3y}{2}$

59. $\dfrac{-18a}{20ab} = \dfrac{-9 \cdot 2 \cdot a}{10 \cdot 2 \cdot a \cdot b} = \dfrac{-9}{10 \cdot b} \cdot \dfrac{2}{2} \cdot \dfrac{a}{a} = \dfrac{-9}{10b}$

61. We multiply these two numbers: We multiply these two numbers:

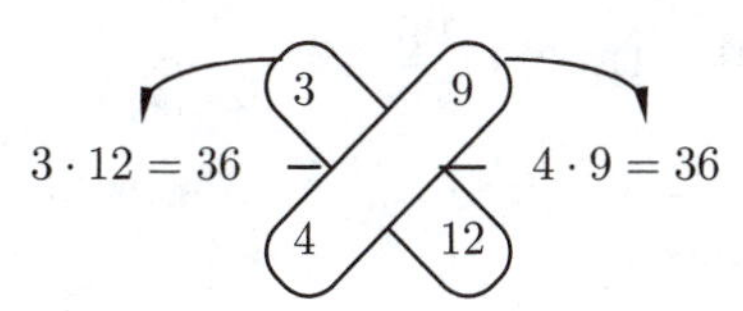

Since $36 = 36$, $\dfrac{3}{4} = \dfrac{9}{12}$.

63. We multiply these two numbers: We multiply these two numbers:

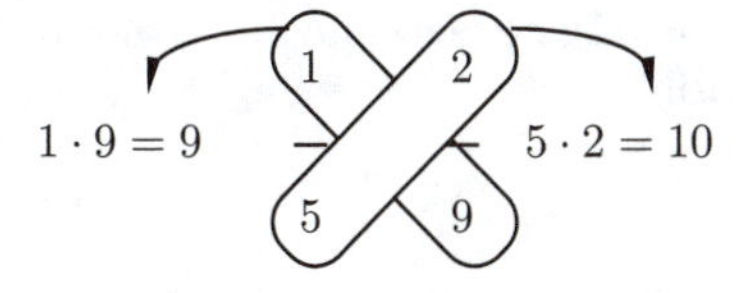

Since $9 \neq 10$, $\dfrac{1}{5}$ and $\dfrac{2}{9}$ do not name the same number. Thus, $\dfrac{1}{5} \neq \dfrac{2}{9}$.

65. We multiply these two numbers: We multiply these two numbers:

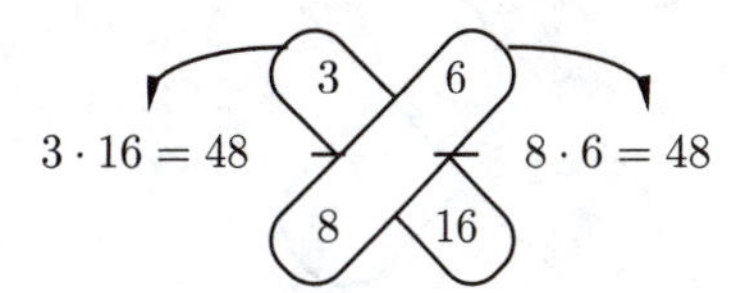

Since $48 = 48$, $\dfrac{3}{8} = \dfrac{6}{16}$.

67. We multiply these two numbers: We multiply these two numbers:

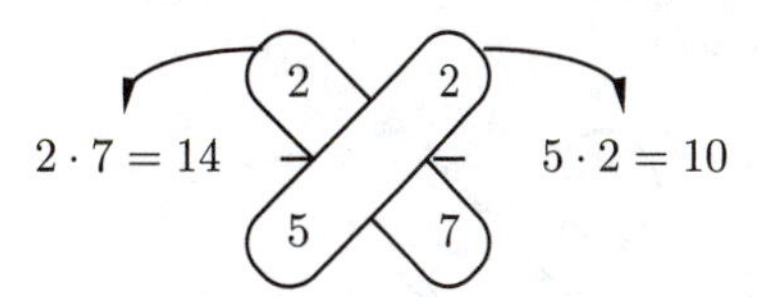

Since $14 \neq 10$, $\dfrac{2}{5}$ and $\dfrac{2}{7}$ do not name the same number. Thus, $\dfrac{2}{5} \neq \dfrac{2}{7}$.

69. We multiply these two numbers: We multiply these two numbers:

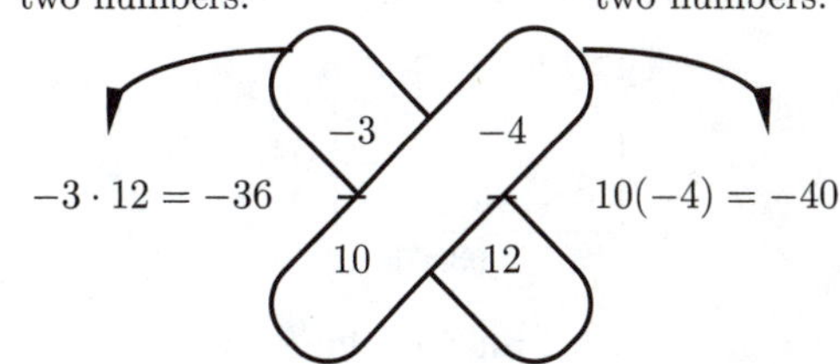

Since $-36 \neq -40$, $\dfrac{-3}{10}$ and $\dfrac{-4}{12}$ do not name the same number. Thus, $\dfrac{-3}{10} \neq \dfrac{-4}{12}$.

71. We rewrite $-\dfrac{12}{9}$ as $\dfrac{-12}{9}$ and check cross products.

We multiply these two numbers: We multiply these two numbers:

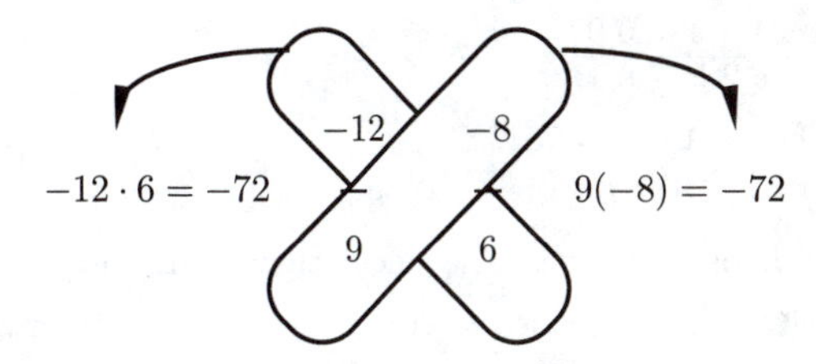

Since $-72 = -72$, $-\dfrac{12}{9} = \dfrac{-8}{6}$.

73. We rewrite $-\frac{17}{7}$ as $\frac{17}{-7}$ and check cross products.

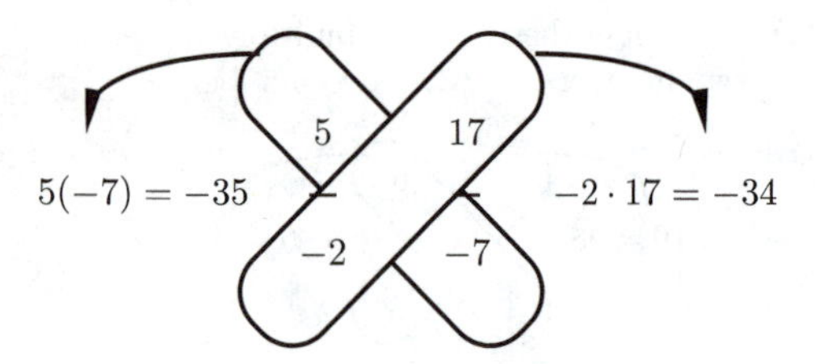

Since $-35 \neq -34$, $\frac{5}{-2}$ and $\frac{17}{-7}$ do not name the same number. Thus, $\frac{5}{-2} \neq \frac{17}{-7}$, or $\frac{5}{-2} \neq -\frac{17}{7}$.

75.

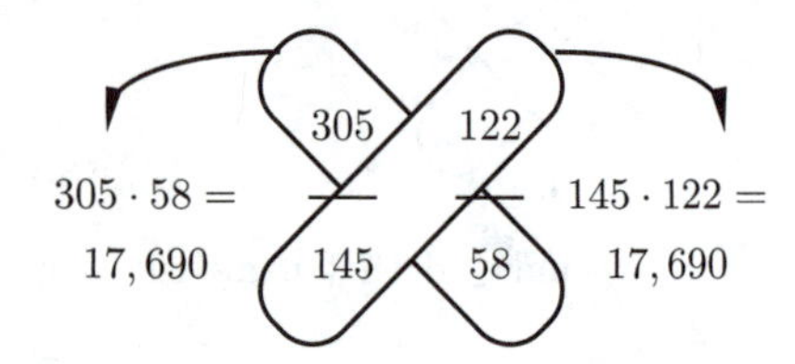

Since $17,690 = 17,690$, $\frac{305}{145} = \frac{122}{58}$.

77. ***Familiarize***. We make a drawing. We let $A =$ the area.

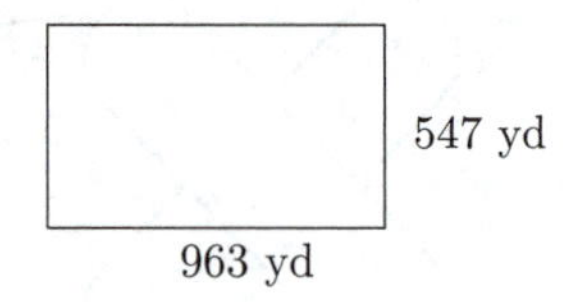

Translate. Using the formula for area, we have

$A = l \cdot w = 963 \cdot 547.$

Using the formula for perimeter, we have

$P = 2l + 2w = 2 \cdot 963 + 2 \cdot 547$

Solve. We carry out the computations.

$$\begin{array}{r} 963 \\ \times\ 547 \\ \hline 6741 \\ 38520 \\ 481500 \\ \hline 526,761 \end{array}$$

Thus, $A = 526,761$.

$P = 2 \cdot 963 + 2 \cdot 547 = 1926 + 1094 = 3020$

Check. We repeat the calculations. The answers check.

State. The area is 526,761 yd^2. The perimeter is 3020 yd.

79. $-12(-5) = 60$

(The signs are the same, so the product is positive.)

81. $-9 \cdot 7 = -63$

(The signs are different, so the product is negative.)

83. $30 \cdot x = 150$

$\frac{30 \cdot x}{30} = \frac{150}{30}$ Dividing both sides by 30

$x = 5$

The solution is 5.

85. $5280 = 1760 + t$

$5280 - 1760 = 1760 + t - 1760$ Subtracting 1760 from both sides

$3520 = t$

The solution is 3520.

87. $\frac{391}{667} = \frac{17 \cdot 23}{23 \cdot 29} = \frac{17}{29} \cdot \frac{23}{23} = \frac{17}{29}$

89. $-\frac{1073x}{555y} = -\frac{29 \cdot 37 \cdot x}{15 \cdot 37 \cdot y} = -\frac{29 \cdot x}{15 \cdot y} \cdot \frac{37}{37} = -\frac{29x}{15y}$

91. $\frac{4247}{4619} = \frac{31 \cdot 137}{31 \cdot 149} = \frac{31}{31} \cdot \frac{137}{149} = \frac{137}{149}$

93. a) The part of the population that is shy is $\frac{4}{10}$. We simplify:

$\frac{4}{10} = \frac{2 \cdot 2}{5 \cdot 2} = \frac{2}{5} \cdot \frac{2}{2} = \frac{2}{5}$

b) Since 4 out of 10 people are shy, then $10 - 4$, or 6, are not shy. The part of the population that is not shy is $\frac{6}{10}$. We simplify:

$\frac{6}{10} = \frac{3 \cdot 2}{5 \cdot 2} = \frac{3}{5} \cdot \frac{2}{2} = \frac{3}{5}$

95. Hanley Ramirez's batting average was $\frac{197}{576}$; Joe Mauer's batting average was $\frac{191}{523}$. We test these fractions for equality:

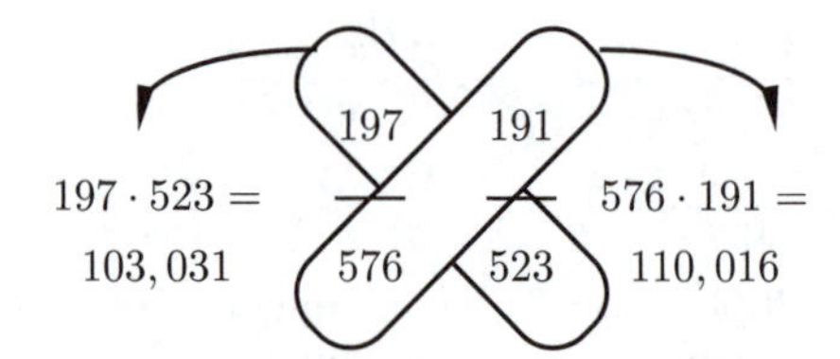

Since $103,031 \neq 110,016$, $\frac{197}{576}$ and $\frac{191}{523}$ do not name the same number. Thus, $\frac{197}{576} \neq \frac{191}{523}$ and the batting averages are not the same.

Chapter 3 Mid-Chapter Review

1. The statement is true. See page 159 in the text.

2. The statement is false. For example 15 is not divisible by 6 but it is divisible by 3.

3. We multiply these two numbers: We multiply these two numbers:

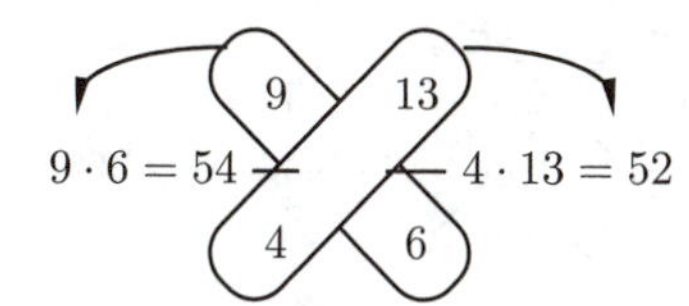

Since $54 \neq 52$, $\frac{9}{4} \neq \frac{13}{6}$. The given statement is false.

4. The statement is true. See page 160 in the text.

5. $\frac{n}{n} = 1$ for any whole number n that is not 0, so $\frac{25}{25} = 1$.

6. $\frac{0}{n} = 0$ for any whole number n that is not 0, so $\frac{0}{9} = 0$.

7. $\frac{n}{1} = n$ for any whole number n, so $\frac{8}{1} = 8$.

8. $\frac{6}{13} = \frac{6}{13} \cdot \frac{3}{3} = \frac{18}{39}$

9.
$$\begin{aligned}\frac{70}{225} &= \frac{2 \cdot 5 \cdot 7}{3 \cdot 3 \cdot 5 \cdot 5}\\ &= \frac{5}{5} \cdot \frac{2 \cdot 7}{3 \cdot 3 \cdot 5}\\ &= 1 \cdot \frac{14}{45}\\ &= \frac{14}{45}\end{aligned}$$

10. A number is divisible by 2 if it has an even ones digit. A number is divisible by 10 if its ones digit is 0. Thus, we find the numbers with a ones digit of 2, 4, 6, or 8. Those numbers are 84, 17,576, 224, 132, 594, 504, and 1632. These are the numbers divisible by 2 but not by 10.

11. A number is divisible by 4 if the number named by its last two digits is divisible by 4. A number is divisible by 8 if the number named by its last three digits is divisible by 8. The numbers that are divisible by 8 are 17,576, 224, 120, 504, and 1632. Of the remaining numbers, those divisible by 4 are 84, 300, 132, 500, and 180. These are the numbers that are divisible by 4 but not by 8.

12. A number is divisible by 4 if the number named by its last two digits is divisible by 4. A number is divisible by 6 if its ones digit is even and the sum of its digits is divisible by 3. The numbers divisible by 4 are 84, 300, 17,576, 224, 132, 500, 180, 120, 504, and 1632. Of these, the numbers for which the sum of the digits is not divisible by 3 are 17,576, 224, and 500. These are the numbers that are divisible by 4 but not by 6.

13. A number is divisible by 3 if the sum of its digits is divisible by 3. A number is divisible by 9 if the sum of its digits is divisible by 9. The numbers divisible by 3 are 84, 300, 132, 180, 351, 594, 120, 1125, 495, 14,850, 504, and 1632. Of these, the numbers that are not divisible by 9 are 84, 300, 132, 120, and 1632. These are the numbers that are divisible by 3 but not by 9.

14. A number is divisible by 4 if the number named by its last two digits is divisible by 4. A number is divisible by 5 if its ones digit is 0 or 5. A number is divisible by 6 if its ones digit is even and the sum of its digits is divisible by 3. From Exercise 12, we know that the numbers that are divisible by 4 are 84, 300, 17,576, 224, 132, 500, 180, 120, 504, and 1632. Of these, 300, 500, 180, and 120 have a ones digit of 0 or 5. Of these, 300, 180, and 120 are divisible by 6. These are the numbers that are divisible by 4, 5, and 6.

15. The number 61 is prime. It has only the factors 1 and 61.

16. The number 2 is prime. It has only the factors 1 and 2.

17. The number 91 has factors 1, 7, 13, and 91. Since 91 is not 1 and not prime, it is composite.

18. The number 1 is neither prime nor composite.

19. To find all the factors of 160, we find all the two-factor factorizations.

$160 = 1 \cdot 160 \quad 160 = 5 \cdot 32$

$160 = 2 \cdot 80 \quad 160 = 8 \cdot 20$

$160 = 4 \cdot 40 \quad 160 = 10 \cdot 16$

Factors: 1, 2, 4, 5, 8, 10, 16, 20, 32, 40, 80, 160

Now we find the prime factorization of 160.

```
      5  ← 5 is prime
   2 ⌈ 1 0
   2 ⌈ 2 0
   2 ⌈ 4 0
   2 ⌈ 8 0
 2 ⌈ 1 6 0
```

$160 = 2 \cdot 2 \cdot 2 \cdot 2 \cdot 2 \cdot 5$

20. To find all the factors of 222, we find all the two-factor factorizations.

$222 = 1 \cdot 222 \quad 222 = 3 \cdot 74$

$222 = 2 \cdot 111 \quad 222 = 6 \cdot 37$

Factors: 1, 2, 3, 6, 37, 74, 111, 222

Now we find the prime factorization.

```
     3 7  ← 37 is prime
 3 ⌈ 1 1 1
 2 ⌈ 2 2 2
```

$222 = 2 \cdot 3 \cdot 37$

21. To find all the factors of 98, we find all the two-factor factorizations.

$98 = 1 \cdot 98 \quad 98 = 7 \cdot 14$

$98 = 2 \cdot 49$

Factors: 1, 2, 7, 14, 49, 98

Now we find the prime factorization.

```
     7  ← 7 is prime
 7 ⌈ 4 9
 2 ⌈ 9 8
```

$98 = 2 \cdot 7 \cdot 7$

22. To find all the factorizations of 315, we find all the two-factor factorizations.

$315 = 1 \cdot 315 \qquad 315 = 7 \cdot 45$

$315 = 3 \cdot 105 \qquad 315 = 9 \cdot 35$

$315 = 5 \cdot 63 \qquad 315 = 15 \cdot 21$

Factors: 1, 3, 5, 7, 9, 15, 21, 35, 45, 63, 105, 315

Now we find the prime factorization.

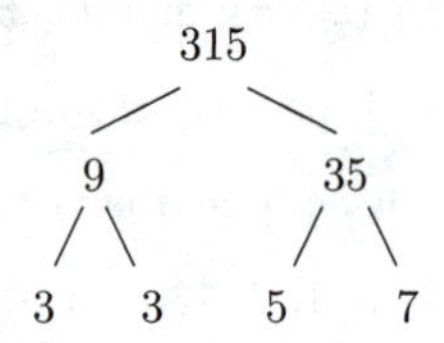

$315 = 3 \cdot 3 \cdot 5 \cdot 7$

23. The rectangle is divided into 24 equal parts. The unit is $\frac{1}{24}$. The denominator is 24. We have 8 parts shaded. This tells us that the numerator is 8. Thus, $\frac{8}{24}$ is shaded. We can simplify $\frac{8}{24}$:

$$\frac{8}{24} = \frac{8 \cdot 1}{8 \cdot 3} = \frac{8}{8} \cdot \frac{1}{3} = \frac{1}{3}.$$

24. Each circle is divided into 4 equal parts. The unit is $\frac{1}{4}$. The denominator is 4. We see that 5 of the parts are shaded, so the numerator is 5. Thus, $\frac{5}{4}$ is shaded.

25. $7 \cdot \frac{1}{9} = \frac{7 \cdot 1}{9} = \frac{7}{9}$

26. $\frac{4}{15} \cdot \frac{2}{3} = \frac{4 \cdot 2}{15 \cdot 3} = \frac{8}{45}$

27. $\frac{5}{11} \cdot (-8) = \frac{5 \cdot (-8)}{11} = \frac{-40}{11}$, or $-\frac{40}{11}$

28. $\frac{24}{60} = \frac{2 \cdot 2 \cdot 2 \cdot 3}{2 \cdot 2 \cdot 3 \cdot 5} = \frac{2 \cdot 2 \cdot 3}{2 \cdot 2 \cdot 3} \cdot \frac{2}{5} = \frac{2}{5}$

29. $\frac{220n}{60n} = \frac{2 \cdot 10 \cdot 11 \cdot n}{2 \cdot 3 \cdot 10 \cdot n} = \frac{2 \cdot 10 \cdot n}{2 \cdot 10 \cdot n} \cdot \frac{11}{3} = \frac{11}{3}$

30. $\frac{17x}{17x} = 1$

31. $\frac{0}{-23} = 0$

32. $\frac{54}{186} = \frac{2 \cdot 3 \cdot 3 \cdot 3}{2 \cdot 3 \cdot 31} = \frac{2 \cdot 3}{2 \cdot 3} \cdot \frac{3 \cdot 3}{31} = \frac{9}{31}$

33. $\frac{-36}{20} = \frac{4 \cdot (-9)}{4 \cdot 5} = \frac{-9}{5}$

34. $\frac{75}{630} = \frac{3 \cdot 5 \cdot 5}{2 \cdot 3 \cdot 3 \cdot 5 \cdot 7} = \frac{3 \cdot 5}{3 \cdot 5} \cdot \frac{5}{2 \cdot 3 \cdot 7} = \frac{5}{42}$

35. $\frac{315}{435} = \frac{3 \cdot 3 \cdot 5 \cdot 7}{3 \cdot 5 \cdot 29} = \frac{3 \cdot 5}{3 \cdot 5} \cdot \frac{3 \cdot 7}{29} = \frac{21}{29}$

36. $\frac{14}{0}$ is undefined.

37. We multiply these two numbers: We multiply these two numbers:

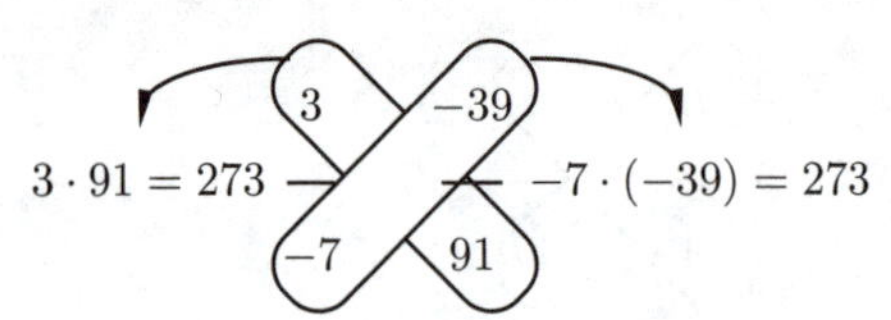

Since $273 = 273$, $\frac{3}{-7} = \frac{-39}{91}$.

38. We multiply these two numbers: We multiply these two numbers:

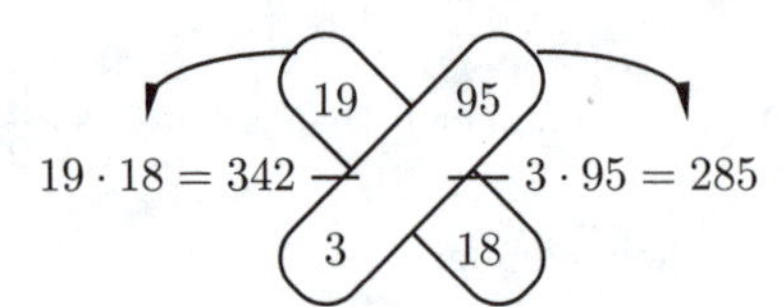

Since $342 \neq 285$, $\frac{19}{3} \neq \frac{95}{18}$.

39. The ratio is $\frac{60}{500}$. This can be simplified:

$$\frac{60}{500} = \frac{2 \cdot 3 \cdot 10}{2 \cdot 5 \cdot 5 \cdot 10} = \frac{2 \cdot 10}{2 \cdot 10} \cdot \frac{3}{5 \cdot 5} = \frac{3}{25}.$$

40. ***Familiarize.*** We will use the formula $A = l \times w$.

Translate. We substitute in the formula.

$$A = l \times w = \frac{7}{100} \times \frac{3}{100}$$

Solve. We carry out the multiplication.

$$A = \frac{7}{100} \times \frac{3}{100} = \frac{7 \times 3}{100 \times 100} = \frac{21}{10,000}.$$

Check. We repeat the calculation. The answer checks.

State. The area of the rink is $\frac{21}{10,000}$ m^2.

41. Find the product of two prime numbers.

42. Using the divisibility tests, it is quickly clear that none of the even-numbered years is a prime number. In addition, the divisibility tests for 5 and 3 show that 2000, 2001, 2005, 2007, 2013, 2015, and 2019 are not prime numbers. Then the years 2003, 2009, 2011, and 2017 can be divided by prime numbers to determine if they are prime. When we do this, we find that 2011 and 2017 are prime numbers.

If the divisibility tests are not used, each of the numbers from 2000 to 2020 can be divided by prime numbers to determine if they are prime.

43. It is possible to cancel only when identical *factors* appear in the numerator and denominator of a fraction. Situations in which it is not possible to cancel include the occurrence of identical *terms* or *digits* in the numerator and denominator.

44. No; since the only factors of a prime number are the number itself and 1, two different prime numbers cannot contain a common factor (other than 1).

Exercise Set 3.6

1. $\frac{2}{3}\cdot\frac{1}{2}=\frac{2\cdot 1}{3\cdot 2}=\frac{2}{2}\cdot\frac{1}{3}=1\cdot\frac{1}{3}=\frac{1}{3}$

3. $\frac{7}{8}\cdot\frac{-1}{7}=\frac{7(-1)}{8\cdot 7}=\frac{7}{7}\cdot\frac{-1}{8}=\frac{-1}{8}$, or $-\frac{1}{8}$

5. $\frac{2}{3}\cdot\frac{6}{7}=\frac{2\cdot 6}{3\cdot 7}=\frac{2\cdot 2\cdot 3}{3\cdot 7}=\frac{3}{3}\cdot\frac{2\cdot 2}{7}=\frac{4}{7}$

7. $\frac{2}{9}\cdot\frac{3}{10}=\frac{2\cdot 3}{9\cdot 10}=\frac{2\cdot 3\cdot 1}{3\cdot 3\cdot 2\cdot 5}=\frac{2}{2}\cdot\frac{3}{3}\cdot\frac{1}{3\cdot 5}=\frac{1}{15}$

9. $\frac{9}{-5}\cdot\frac{12}{8}=\frac{9\cdot 12}{-5\cdot 8}=\frac{9\cdot 4\cdot 3}{-5\cdot 2\cdot 4}=\frac{4}{4}\cdot\frac{9\cdot 3}{-5\cdot 2}=\frac{9\cdot 3}{-5\cdot 2}=$
$\frac{27}{-10}$, or $-\frac{27}{10}$

11. $\frac{5x}{9}\cdot\frac{4}{5}=\frac{5x\cdot 4}{9\cdot 5}=\frac{5\cdot x\cdot 4}{9\cdot 5}=\frac{5}{5}\cdot\frac{x\cdot 4}{9}=\frac{4x}{9}$

13. $9\cdot\frac{1}{9}=\frac{9\cdot 1}{9}=\frac{9}{9}=1$

15. $\frac{7}{10}\cdot\frac{10}{7}=\frac{7\cdot 10}{10\cdot 7}=\frac{7\cdot 10}{7\cdot 10}=1$

17. $\frac{1}{4}\cdot 12=\frac{1\cdot 12}{4}=\frac{12}{4}=\frac{4\cdot 3}{4\cdot 1}=\frac{4}{4}\cdot\frac{3}{1}=\frac{3}{1}=3$

19. $21\cdot\frac{1}{3}=\frac{21\cdot 1}{3}=\frac{21}{3}=\frac{3\cdot 7}{3\cdot 1}=\frac{3}{3}\cdot\frac{7}{1}=\frac{7}{1}=7$

21. $-16\left(-\frac{3}{4}\right)=\frac{16\cdot 3}{4}=\frac{4\cdot 4\cdot 3}{4\cdot 1}=\frac{4}{4}\cdot\frac{4\cdot 3}{1}=$
$\frac{4\cdot 3}{1}=\frac{12}{1}=12$

23. $\frac{3}{8}\cdot 8a=\frac{3\cdot 8a}{8}=\frac{3\cdot 8\cdot a}{8\cdot 1}=\frac{8}{8}\cdot\frac{3\cdot a}{1}=\frac{3a}{1}=3a$

25. $\left(-\frac{3}{8}\right)\left(-\frac{8}{3}\right)=\frac{3\cdot 8}{8\cdot 3}=\frac{3\cdot 8}{3\cdot 8}=1$

27. $\frac{a}{b}\cdot\frac{b}{a}=\frac{a\cdot b}{b\cdot a}=\frac{a\cdot b}{a\cdot b}=1$

29. $\frac{4}{10}\cdot\frac{5}{10}=\frac{4\cdot 5}{10\cdot 10}=\frac{2\cdot 2\cdot 5\cdot 1}{2\cdot 5\cdot 2\cdot 5}=\frac{2\cdot 2\cdot 5}{2\cdot 2\cdot 5}\cdot\frac{1}{5}=\frac{1}{5}$

31. $\frac{8}{10}\cdot\frac{45}{100}=\frac{8\cdot 45}{10\cdot 100}=\frac{2\cdot 2\cdot 2\cdot 3\cdot 3\cdot 5}{2\cdot 5\cdot 2\cdot 2\cdot 5\cdot 5}=\frac{2\cdot 2\cdot 2\cdot 5}{2\cdot 2\cdot 2\cdot 5}=$
$\frac{3\cdot 3}{5\cdot 5}=\frac{9}{25}$

33. $\frac{1}{6}\cdot 360n=\frac{1\cdot 360n}{6}=\frac{1\cdot 6\cdot 60n}{1\cdot 6}=\frac{6}{6}\cdot\frac{1\cdot 60n}{1}=60n$

35. $20\left(\frac{1}{-6}\right)=\frac{20\cdot 1}{-6}=\frac{2\cdot 10\cdot 1}{-3\cdot 2}=\frac{2}{2}\cdot\frac{10\cdot 1}{-3}=$
$\frac{10}{-3}$, or $-\frac{10}{3}$

37. $-8x\cdot\frac{1}{-8x}=\frac{8x\cdot 1}{8x\cdot 1}=1$

39. $\frac{2x}{9}\cdot\frac{27}{2x}=\frac{2x\cdot 27}{9\cdot 2x}=\frac{2x\cdot 9\cdot 3}{9\cdot 2x\cdot 1}=$
$\frac{2x\cdot 9}{2x\cdot 9}\cdot\frac{3}{1}=3$

41. $\frac{7}{10}\cdot\frac{34}{150}=\frac{7\cdot 34}{10\cdot 150}=\frac{7\cdot 2\cdot 17}{2\cdot 5\cdot 150}=\frac{2}{2}\cdot\frac{7\cdot 17}{5\cdot 150}=\frac{119}{750}$

43. $\frac{36}{85}\cdot\frac{25}{-99}=\frac{9\cdot 4\cdot 5\cdot 5}{5\cdot 17\cdot 9(-11)}=\frac{9\cdot 5}{9\cdot 5}\cdot\frac{4\cdot 5}{17(-11)}=$
$\frac{20}{-187}$, or $-\frac{20}{187}$

45. $\frac{-98}{99}\cdot\frac{27a}{175a}=\frac{7(-14)\cdot 9\cdot 3\cdot a}{9\cdot 11\cdot 7\cdot 25\cdot a}=$
$\frac{7\cdot 9\cdot a}{7\cdot 9\cdot a}\cdot\frac{-14\cdot 3}{11\cdot 25}=\frac{-42}{275}$, or $-\frac{42}{275}$

47. $\frac{110}{33}\cdot\frac{-24}{25x}=-\frac{2\cdot 5\cdot 11\cdot 3\cdot 8}{3\cdot 11\cdot 5\cdot 5\cdot x}=-\frac{2\cdot 8}{5\cdot x}\cdot\frac{3\cdot 5\cdot 11}{3\cdot 5\cdot 11}=$
$\frac{-16}{5x}$, or $-\frac{16}{5x}$

49. $\left(-\frac{11}{24}\right)\frac{3}{5}=-\frac{11\cdot 3}{24\cdot 5}=-\frac{11\cdot 3}{3\cdot 8\cdot 5}=\frac{3}{3}\left(-\frac{11}{8\cdot 5}\right)=$
$-\frac{11}{40}$

51. $\frac{10a}{21}\cdot\frac{3}{8b}=\frac{10a\cdot 3}{21\cdot 8b}=\frac{2\cdot 5\cdot a\cdot 3}{3\cdot 7\cdot 2\cdot 4\cdot b}=\frac{2\cdot 3}{2\cdot 3}\cdot\frac{5\cdot a}{7\cdot 4\cdot b}=\frac{5a}{28b}$

53. ***Familiarize***. Let n = the number of inches the screw will go into the piece of oak when it is turned 10 complete rotations.

Translate. We write an equation.

Total distance	is	Distance for one revolution	times	Number of revolutions
↓	↓	↓	↓	↓
n	$=$	$\frac{1}{16}$	$\cdot$	10

Solve. We carry out the multiplication.

$$n=\frac{1}{16}\cdot 10=\frac{1\cdot 10}{16}$$
$$=\frac{1\cdot 2\cdot 5}{2\cdot 8}=\frac{2}{2}\cdot\frac{1\cdot 5}{8}$$
$$=\frac{5}{8}$$

Check. We can repeat the calculation. We can also determine that the answer seems reasonable since we multiplied 10 by a number less than 10 and the result is less than 10. The answer checks.

State. The screw will go $\frac{5}{8}$ in. into the piece of oak when it is turned 10 complete rotations.

55. ***Familiarize***. Let a = the median income of people with associate's degrees.

Translate.

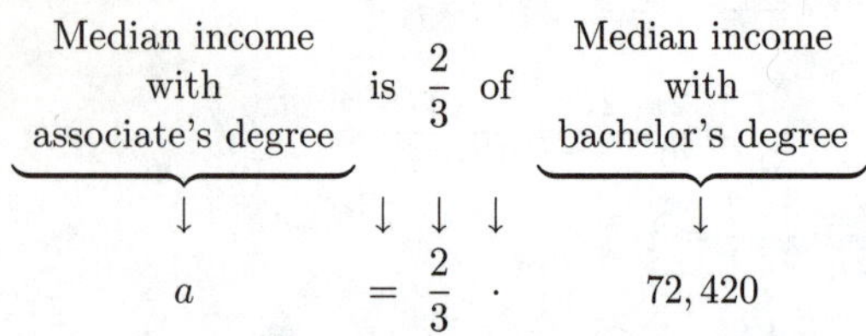

$$a \quad = \quad \frac{2}{3} \quad \cdot \quad 72,420$$

Solve. We carry out the multiplication.

$$a = \frac{2}{3} \cdot 72,420 = \frac{2 \cdot 72,420}{3} = \frac{2 \cdot 3 \cdot 24,140}{3 \cdot 1}$$
$$= \frac{3}{3} \cdot \frac{2 \cdot 24,140}{1} = 48,280$$

Check. We can repeat the calculation. We can also observe that the answer seems reasonable since we multiplied 72,420 by a number less than 1 and the result is less than 72,420. The answer checks.

State. The median income of people with associate's degrees is $48,280.

57. ***Familiarize.*** We visualize the situation. We let n = the number of addresses that will be incorrect after one year.

Mailing list 2500 addresses			
1/4 of the addresses n			

Translate.

$$\underbrace{\text{Number incorrect}}_{\downarrow} \text{ is } \frac{1}{4} \text{ of } \underbrace{\text{Number of addresses}}_{\downarrow}$$
$$n \quad = \quad \frac{1}{4} \quad \cdot \quad 2500$$

Solve. We carry out the multiplication.

$$n = \frac{1}{4} \cdot 2500 = \frac{1 \cdot 2500}{4} = \frac{2500}{4}$$
$$= \frac{4 \cdot 625}{4 \cdot 1} = \frac{4}{4} \cdot \frac{625}{1}$$
$$= 625$$

Check. We can repeat the calculation. We can also determine that the answer seems reasonable since we multiplied 2500 by a number less than 1 and the result is less than 2500. The answer checks.

State. After one year 625 addresses will be incorrect.

59. ***Familiarize.*** We draw a picture.

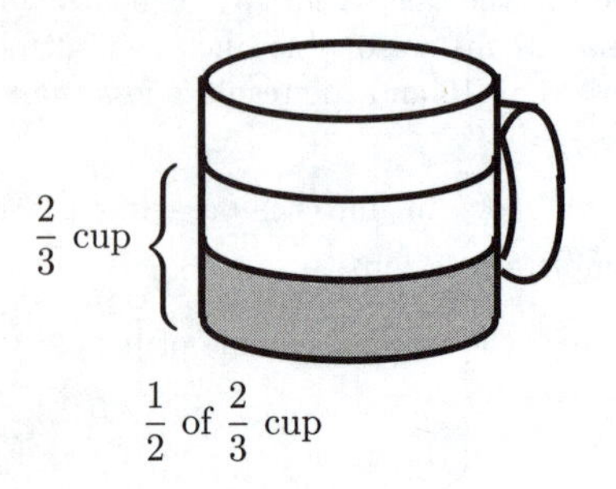

We let n = the amount of flour the chef should use.

Translate. The multiplication sentence

$$\frac{1}{2} \cdot \frac{2}{3} = n$$

corresponds to the situation.

Solve. We multiply and simplify:

$$n = \frac{1}{2} \cdot \frac{2}{3} = \frac{1 \cdot 2}{2 \cdot 3} = \frac{2}{2} \cdot \frac{1}{3} = \frac{1}{3}$$

Check. We can repeat the calculation. We can also determine that the answer seems reasonable since we multiplied $\frac{2}{3}$ by a number less than 1 and the result is less than $\frac{2}{3}$. The answer checks.

State. The chef should use $\frac{1}{3}$ cup of flour.

61. ***Familiarize.*** We visualize the situation. Let a = the assessed value of the house.

Value of house $154,000	
3/4 of the value $a	

Translate. We write an equation.

$$\underbrace{\text{Assessed value}}_{\downarrow} \text{ is } \frac{3}{4} \text{ of } \underbrace{\text{the value of the house}}_{\downarrow}$$
$$a \quad = \quad \frac{3}{4} \quad \cdot \quad 154,000$$

Solve. We carry out the multiplication.

$$a = \frac{3}{4} \cdot 154,000 = \frac{3 \cdot 154,000}{4}$$
$$= \frac{3 \cdot 4 \cdot 38,500}{4 \cdot 1} = \frac{4}{4} \cdot \frac{3 \cdot 38,500}{1}$$
$$= 115,500$$

Check. We can repeat the calculation. We can also determine that the answer seems reasonable since we multiplied 154,000 by a number less than 1 and the result is less than 154,000. The answer checks.

State. The assessed value of the house is $115,500.

63. ***Familiarize.*** We draw a picture.

$\frac{2}{3}$ in.

1 in.
240 miles

We let n = the number of miles represented by $\frac{2}{3}$ in.

Translate. The multiplication sentence

$$n = \frac{2}{3} \cdot 240$$

corresponds to the situation.

Solve. We multiply and simplify:

$$n = \frac{2}{3} \cdot 240 = \frac{2 \cdot 240}{3} = \frac{2 \cdot 3 \cdot 80}{1 \cdot 3}$$
$$= \frac{3}{3} \cdot \frac{2 \cdot 80}{1} = \frac{2 \cdot 80}{1}$$
$$= 160$$

Check. We can repeat the calculation. We can also determine that the answer seems reasonable since we multiplied 240 by a number less than 1 and the result is less than 240.

State. $\frac{2}{3}$ in. on the map represents 160 miles.

65. ***Familiarize***. This is a multistep problem. First we find the amount of each of the given expenses. Then we find the total of these expenses and take it away from the annual income to find how much is spent for other expenses.

We let f, h, c, s, and t represent the amounts spent on food, housing, clothing, savings, and taxes, respectively.

Translate. The following multiplication sentences correspond to the situation.

$\frac{1}{5} \cdot 42{,}000 = f$ $\quad$ $\frac{1}{14} \cdot 42{,}000 = s$

$\frac{1}{4} \cdot 42{,}000 = h$ $\quad$ $\frac{1}{5} \cdot 42{,}000 = t$

$\frac{1}{10} \cdot 42{,}000 = c$

Solve. We multiply and simplify.

$$f = \frac{1}{5} \cdot 42{,}000 = \frac{42{,}000}{5} = \frac{5 \cdot 8400}{5 \cdot 1} = \frac{5}{5} \cdot \frac{8400}{1} = 8400$$

$$h = \frac{1}{4} \cdot 42{,}000 = \frac{42{,}000}{4} = \frac{4 \cdot 10{,}500}{4 \cdot 1} = \frac{4}{4} \cdot \frac{10{,}500}{1} = 10{,}500$$

$$c = \frac{1}{10} \cdot 42{,}000 = \frac{42{,}000}{10} = \frac{10 \cdot 4200}{10 \cdot 1} = \frac{10}{10} \cdot \frac{4200}{1} = 4200$$

$$s = \frac{1}{14} \cdot 42{,}000 = \frac{42{,}000}{14} = \frac{14 \cdot 3000}{14 \cdot 1} = \frac{14}{14} \cdot \frac{3000}{1} = 3000$$

$t = \frac{1}{5} \cdot 42{,}000$; this is the same computation we did to find f above. Thus, $t = 8400$.

We add to find the total of these expenses.

```
   $ 8 4 0 0
   1 0, 5 0 0
       4 2 0 0
       3 0 0 0
       8 4 0 0
  -----------
 $ 3 4, 5 0 0
```

We let m = the amount spent on other expenses and subtract to find this amount.

Annual income	minus	Total of itemized expenses	is	Total spent on other expenses	
↓	↓	↓	↓	↓	
$42,000	−	$34,500	=	m	
		$7500	=	m	Subtracting

Check. We repeat the calculations. The results check.

State. $8400 is spent for food, $10,500 for housing, $4200 for clothing, $3000 for savings, $8400 for taxes, and $7500 for other expenses.

67. $A = \frac{1}{2} \cdot b \cdot h$ $\quad$ Area of a triangle

$A = \frac{1}{2} \cdot 15 \text{ in.} \cdot 8 \text{ in.}$ $\quad$ Substituting 15 in. for b and 8 in. for h

$A = \frac{15 \cdot 8}{2} \text{ in}^2$

$A = 60 \text{ in}^2$

69. $A = \frac{1}{2} \cdot b \cdot h$ $\quad$ Area of a triangle

$A = \frac{1}{2} \cdot 5 \text{ mm} \cdot \frac{7}{2} \text{ mm}$ $\quad$ Substituting 5 mm for b and $\frac{7}{2}$ mm for h

$A = \frac{5 \cdot 7}{2 \cdot 2} \text{ mm}^2$

$A = \frac{35}{4} \text{ mm}^2$

71. $A = \frac{1}{2} \cdot b \cdot h$ $\quad$ Area of a triangle

$A = \frac{1}{2} \cdot \frac{9}{2} \text{ m} \cdot \frac{7}{2} \text{ m}$ $\quad$ Substituting $\frac{9}{2}$ m for b and $\frac{7}{2}$ m for h

$A = \frac{9 \cdot 7}{2 \cdot 2 \cdot 2} \text{ m}^2$

$A = \frac{63}{8} \text{ m}^2$

73. ***Familiarize***. We look for figures whose areas we can calculate using area formulas we already know.

Translate. The figure consists of a rectangle with a length of 10 mi and a width of 8 mi and of a triangle with a base of 13 − 10, or 3 mi, and a height of 8 mi. We use the formula $A = l \cdot w$ for the area of a rectangle and the formula $A = \frac{1}{2} \cdot b \cdot h$ for the area of a triangle and add the two areas.

Solve. For the rectangle: $A = l \cdot w = 10 \text{ mi} \cdot 8 \text{ mi} = 80 \text{ mi}^2$

For the triangle: $A = \frac{1}{2} \cdot b \cdot h = \frac{1}{2} \cdot 3 \text{ mi} \cdot 8 \text{ mi} = 12 \text{ mi}^2$

Then we add: $80 \text{ mi}^2 + 12 \text{ mi}^2 = 92 \text{ mi}^2$

Check. We repeat the calculations.

State. The area of the figure is 92 mi^2.

75. ***Familiarize***. We look for figures whose areas we can calculate using formulas we already know.

Translate. The figure consists of two triangles. Each has a base of $2 \cdot \frac{7}{2}$, or 7 m. The heights are 7 mm and 9 mm.

We use the formula $A = \frac{1}{2} \cdot b \cdot h$ twice.

Solve. Top triangle:

$$A = \frac{1}{2} \cdot b \cdot h = \frac{1}{2} \cdot 7 \text{ mm} \cdot 7 \text{ mm} = \frac{49}{2} \text{ mm}^2$$

Bottom triangle:

$$A = \frac{1}{2} \cdot b \cdot h = \frac{1}{2} \cdot 7 \text{ mm} \cdot 9 \text{ mm} = \frac{63}{2} \text{ mm}^2$$

Total area: $\frac{49}{2}\text{ mm}^2 + \frac{63}{2}\text{ mm}^2 = \frac{112}{2}\text{ mm}^2 = 56\text{ mm}^2$

Check. We repeat the calculations.

State. The area is 56 mm^2.

77. $48 \cdot t = 1680$

$\frac{48 \cdot t}{48} = \frac{1680}{48}$

$t = 35$

The solution is 35.

79. $3125 = 25 \cdot t$

$\frac{3125}{25} = \frac{25 \cdot t}{25}$ Dividing by 25 on both sides

$125 = t$

The solution is 125.

81. $t + 28 = 5017$

$t = 5017 - 28$

$t = 4989$

The solution is 4989.

83. $8797 = y + 2299$

$8797 - 2299 = y + 2299 - 2299$ Subtracting 2299 on both sides

$6498 = y$

The solution is 6498.

85. $$\frac{201}{535} \cdot \frac{4601}{6499} = \frac{201 \cdot 4601}{535 \cdot 6499} = \frac{3 \cdot 67 \cdot 43 \cdot 107}{5 \cdot 107 \cdot 67 \cdot 97} = \frac{67 \cdot 107}{67 \cdot 107} \cdot \frac{3 \cdot 43}{5 \cdot 97} = \frac{129}{485}$$

87. ***Familiarize***. We are told that $\frac{2}{3}$ of $\frac{7}{8}$ of the students are high school graduates who are older than 20, and $\frac{1}{7}$ of this fraction are left-handed. Thus, we want to find $\frac{1}{7}$ of $\frac{2}{3}$ of $\frac{7}{8}$. We let f represent this fraction.

Translate. The multiplication sentence

$$f = \frac{1}{7} \cdot \frac{2}{3} \cdot \frac{7}{8}$$

corresponds to this situation.

Solve. We multiply and simplify.

$$f = \frac{1}{7} \cdot \frac{2}{3} \cdot \frac{7}{8} = \frac{1 \cdot 2}{7 \cdot 3} \cdot \frac{7}{8} = \frac{1 \cdot 2 \cdot 7}{7 \cdot 3 \cdot 8} = \frac{1 \cdot 2 \cdot 7}{7 \cdot 3 \cdot 2 \cdot 4} =$$
$$\frac{2 \cdot 7}{2 \cdot 7} \cdot \frac{1}{3 \cdot 4} = \frac{1}{3 \cdot 4} = \frac{1}{12}$$

Check. We repeat the calculation. The result checks.

State. $\frac{1}{12}$ of the students are left-handed high school graduates over the age of 20.

89. Area of each triangular end:

$A = \frac{1}{2} \cdot b \cdot h = \frac{1}{2} \cdot 30\text{ mm} \cdot 26\text{ mm} = 390\text{ mm}^2$

Area of each rectangular side:

$A = l \cdot w = 140\text{ mm} \cdot 30\text{ mm} = 4200\text{ mm}^2$

Total area: $2 \cdot 390\text{ mm}^2 + 3 \cdot 4200\text{ mm}^2 = 780\text{ mm}^2 + 12,600\text{ mm}^2 = 13,380\text{ mm}^2$

Exercise Set 3.7

1. $\frac{7}{3}$ Interchange the numerator and denominator.

The reciprocal of $\frac{7}{3}$ is $\frac{3}{7}$. $\left(\frac{7}{3} \cdot \frac{3}{7} = \frac{21}{21} = 1\right)$

3. Think of 9 as $\frac{9}{1}$.

$\frac{9}{1}$ Interchange the numerator and denominator.

The reciprocal of 9 is $\frac{1}{9}$. $\left(\frac{9}{1} \cdot \frac{1}{9} = \frac{9}{9} = 1\right)$

5. $\frac{1}{7}$ Interchange the numerator and denominator.

The reciprocal of $\frac{1}{7}$ is 7. $\left(\frac{7}{1} = 7; \frac{1}{7} \cdot \frac{7}{1} = \frac{7}{7} = 1\right)$

7. $-\frac{8}{9}$ Interchange the numerator and denominator.

The reciprocal of $-\frac{8}{9}$ is $-\frac{9}{8}$. $\left[-\frac{8}{9}\left(-\frac{9}{8}\right) = \frac{72}{72} = 1\right]$

9. $\frac{a}{c}$ Interchange the numerator and denominator.

The reciprocal of $\frac{a}{c}$ is $\frac{c}{a}$. $\left(\frac{a}{c} \cdot \frac{c}{a} = \frac{ac}{ac} = 1\right)$

11. $\frac{-3n}{m}$ Interchange the numerator and denominator.

The reciprocal of $\frac{-3n}{m}$ is $\frac{m}{-3n}$.

$\left(\frac{-3n}{m} \cdot \frac{m}{-3n} = \frac{-3mn}{-3mn} = 1\right)$

13. $\frac{8}{-15}$ Interchange the numerator and denominator.

The reciprocal of $\frac{8}{-15}$ is $\frac{-15}{8}$.

$\left[\frac{8}{-15}\left(\frac{-15}{8}\right) = \frac{-120}{-120} = 1\right]$

15. Think of $7m$ as $\frac{7m}{1}$.

$\frac{7m}{1}$ Interchange the numerator and denominator.

The reciprocal of $7m$ is $\frac{1}{7m}$. $\left(\frac{7m}{1} \cdot \frac{1}{7m} = \frac{7m}{7m} = 1\right)$

17. $\dfrac{1}{4a}$ Interchange the numerator and denominator.

The reciprocal of $\dfrac{1}{4a}$ is $\dfrac{4a}{1}$, or $4a$.

$\left(\dfrac{1}{4a}\cdot\dfrac{4a}{1}=\dfrac{4a}{4a}=1\right)$

19. The reciprocal of $-\dfrac{1}{3z}$ is $-\dfrac{3z}{1}$, or $-3z$.

$\left(-\dfrac{1}{3z}\cdot(-3z)=\dfrac{3z}{3z}=1\right)$

21. $\dfrac{3}{7}\div\dfrac{3}{4}=\dfrac{3}{7}\cdot\dfrac{4}{3}$ Multiplying by the reciprocal of the divisor

$=\dfrac{3\cdot 4}{7\cdot 3}$ Multiplying numerators and denominators

$=\dfrac{3}{3}\cdot\dfrac{4}{7}=\dfrac{4}{7}$ Removing a factor equal to 1

23. $\dfrac{3}{5}\div\dfrac{9}{4}=\dfrac{3}{5}\cdot\dfrac{4}{9}$ Multiplying by the reciprocal of the divisor

$=\dfrac{3\cdot 4}{5\cdot 9}$

$=\dfrac{3\cdot 4}{5\cdot 3\cdot 3}$

$=\dfrac{3}{3}\cdot\dfrac{4}{5\cdot 3}$

$=\dfrac{4}{15}$

25. $\dfrac{4}{3}\div\dfrac{1}{3}=\dfrac{4}{3}\cdot 3=\dfrac{4\cdot 3}{3}=\dfrac{3}{3}\cdot 4=4$

27. $\left(-\dfrac{1}{3}\right)\div\dfrac{1}{6}=-\dfrac{1}{3}\cdot 6=-\dfrac{1\cdot 2\cdot 3}{1\cdot 3}=$

$-\dfrac{1\cdot 3}{1\cdot 3}\cdot 2=-2$

29. $\left(-\dfrac{10}{21}\right)\div\left(-\dfrac{2}{15}\right)=\left(-\dfrac{10}{21}\right)\cdot\left(-\dfrac{15}{2}\right)=\dfrac{10\cdot 15}{21\cdot 2}=$

$\dfrac{2\cdot 5\cdot 3\cdot 5}{3\cdot 7\cdot 2}=\dfrac{2\cdot 3}{2\cdot 3}\cdot\dfrac{5\cdot 5}{7}=\dfrac{25}{7}$

31. $\dfrac{3}{8}\div 3=\dfrac{3}{8}\cdot\dfrac{1}{3}=\dfrac{3\cdot 1}{8\cdot 3}=\dfrac{3}{3}\cdot\dfrac{1}{8}=\dfrac{1}{8}$

33. $\dfrac{12}{7}\div 16=\dfrac{12}{7}\cdot\dfrac{1}{16}=\dfrac{4\cdot 3\cdot 1}{7\cdot 4\cdot 4}=\dfrac{4}{4}\cdot\dfrac{3\cdot 1}{7\cdot 4}=\dfrac{3}{28}$

35. $(-12)\div\dfrac{3}{2}=-12\cdot\dfrac{2}{3}=-\dfrac{3\cdot 4\cdot 2}{3\cdot 1}$

$=-\dfrac{3}{3}\cdot\dfrac{4\cdot 2}{1}=-\dfrac{8}{1}=-8$

37. $\dfrac{x}{8}\div\dfrac{1}{4}=\dfrac{x}{8}\cdot\dfrac{4}{1}=\dfrac{x\cdot 2\cdot 2}{2\cdot 2\cdot 2\cdot 1}=\dfrac{2\cdot 2}{2\cdot 2}\cdot\dfrac{x}{2\cdot 1}=\dfrac{x}{2}$

39. $\dfrac{2}{3}\div(6x)=\dfrac{2}{3}\cdot\dfrac{1}{6x}=\dfrac{2\cdot 1}{3\cdot 2\cdot 3\cdot x}=\dfrac{2}{2}\cdot\dfrac{1}{3\cdot 3\cdot x}=\dfrac{1}{9x}$

41. $28\div\dfrac{4}{5a}=28\cdot\dfrac{5a}{4}=\dfrac{28\cdot 5a}{4}=\dfrac{4\cdot 7\cdot 5\cdot a}{4\cdot 1}=\dfrac{4}{4}\cdot\dfrac{7\cdot 5\cdot a}{1}$

$=35a$

43. $\left(-\dfrac{5}{8}\right)\div\left(-\dfrac{5}{8}\right)=-\dfrac{5}{8}\left(-\dfrac{8}{5}\right)=\dfrac{5\cdot 8}{8\cdot 5}=\dfrac{5\cdot 8}{5\cdot 8}=1$

45. $\dfrac{-8}{15}\div\dfrac{4}{5}=\dfrac{-8}{15}\cdot\dfrac{5}{4}=\dfrac{-8\cdot 5}{15\cdot 4}=\dfrac{-2\cdot 4\cdot 5}{3\cdot 5\cdot 4}=$

$\dfrac{4\cdot 5}{4\cdot 5}\cdot\dfrac{-2}{3}=\dfrac{-2}{3}$, or $-\dfrac{2}{3}$

47. $\dfrac{77}{64}\div\dfrac{49}{18}=\dfrac{77}{64}\cdot\dfrac{18}{49}=\dfrac{7\cdot 11\cdot 2\cdot 9}{2\cdot 32\cdot 7\cdot 7}=$

$\dfrac{2\cdot 7}{2\cdot 7}\cdot\dfrac{11\cdot 9}{32\cdot 7}=\dfrac{99}{224}$

49. $120a\div\dfrac{45}{14}=120a\cdot\dfrac{14}{45}=\dfrac{8\cdot 15\cdot a\cdot 14}{3\cdot 15}=$

$\dfrac{15}{15}\cdot\dfrac{8\cdot a\cdot 14}{3}=\dfrac{112a}{3}$

51. $\dfrac{\frac{2}{5}}{\frac{3}{7}}=\dfrac{2}{5}\div\dfrac{3}{7}=\dfrac{2}{5}\cdot\dfrac{7}{3}=\dfrac{2\cdot 7}{5\cdot 3}=\dfrac{14}{15}$

53. $\dfrac{-\frac{7}{20}}{-\frac{8}{5}}=-\dfrac{7}{20}\div\left(-\dfrac{8}{5}\right)=-\dfrac{7}{20}\cdot\left(-\dfrac{5}{8}\right)=\dfrac{7\cdot 5}{20\cdot 8}=$

$\dfrac{7\cdot 5}{4\cdot 5\cdot 8}=\dfrac{5}{5}\cdot\dfrac{7}{4\cdot 8}=\dfrac{7}{32}$

55. $\dfrac{-\frac{15}{8}}{\frac{9}{10}}=-\dfrac{15}{8}\div\dfrac{9}{10}=-\dfrac{15}{8}\cdot\dfrac{10}{9}=-\dfrac{15\cdot 10}{8\cdot 9}=$

$-\dfrac{3\cdot 5\cdot 2\cdot 5}{2\cdot 4\cdot 3\cdot 3}=-\dfrac{5\cdot 5}{4\cdot 3}\cdot\dfrac{3\cdot 2}{3\cdot 2}=-\dfrac{25}{12}$

57. The equation $14+(2+30)=(14+2)+30$ illustrates the associative law of addition.

59. A natural number that has exactly two different factors, itself and 1, is called a prime number.

61. Since $a+0=a$ for any number a, the number 0 is the additive identity.

63. The sum of 6 and -6 is 0; we say that 6 and -6 are opposites of each other.

65. $\left(\dfrac{4}{15}\div\dfrac{2}{25}\right)^2=\left(\dfrac{4}{15}\cdot\dfrac{25}{2}\right)^2$

$=\left(\dfrac{4\cdot 25}{15\cdot 2}\right)^2$

$=\left(\dfrac{2\cdot 2\cdot 5\cdot 5}{3\cdot 5\cdot 2}\right)^2$

$=\left(\dfrac{2\cdot 5}{2\cdot 5}\cdot\dfrac{2\cdot 5}{3}\right)$

$=\left(\dfrac{10}{3}\right)^2$

$=\dfrac{100}{9}$

67. $\left(\frac{9}{10}\div\frac{2}{5}\div\frac{3}{8}\right)^2=\left(\frac{9}{10}\cdot\frac{5}{2}\div\frac{3}{8}\right)^2$

$=\left(\frac{9\cdot 5}{10\cdot 2}\div\frac{3}{8}\right)^2$

$=\left(\frac{9\cdot 5}{2\cdot 5\cdot 2}\div\frac{3}{8}\right)^2$

$=\left(\frac{9}{2\cdot 2}\div\frac{3}{8}\right)^2$

$=\left(\frac{9}{2\cdot 2}\cdot\frac{8}{3}\right)^2$

$=\left(\frac{9\cdot 8}{2\cdot 2\cdot 3}\right)^2$

$=\left(\frac{3\cdot 3\cdot 2\cdot 2\cdot 2}{2\cdot 2\cdot 3\cdot 1}\right)^2$

$=\left(\frac{3\cdot 2}{1}\right)^2$

$=6^2$

$=36$

69. $\left(\frac{14}{15}\div\frac{49}{65}\cdot\frac{77}{260}\right)^2=\left(\frac{14}{15}\cdot\frac{65}{49}\cdot\frac{77}{260}\right)^2$

$=\left(\frac{2\cdot 7\cdot 5\cdot 13\cdot 7\cdot 11}{3\cdot 5\cdot 7\cdot 7\cdot 2\cdot 2\cdot 5\cdot 13}\right)^2$

$=\left(\frac{2\cdot 5\cdot 7\cdot 7\cdot 13}{2\cdot 5\cdot 7\cdot 7\cdot 13}\cdot\frac{11}{2\cdot 3\cdot 5}\right)^2$

$=\left(\frac{11}{30}\right)^2$

$=\frac{121}{900}$

71. Use a calculator.

$\frac{711}{1957}\div\frac{10,033}{13,081}=\frac{711}{1957}\cdot\frac{13,081}{10,033}$

$=\frac{711\cdot 13,081}{1957\cdot 10,033}$

$=\frac{3\cdot 3\cdot 79\cdot 103\cdot 127}{19\cdot 103\cdot 79\cdot 127}$

$=\frac{79\cdot 103\cdot 127}{79\cdot 103\cdot 127}\cdot\frac{3\cdot 3}{19}$

$=\frac{9}{19}$

73. Use a calculator.

$\frac{451}{289}\div\frac{123}{340}=\frac{451}{289}\cdot\frac{340}{123}$

$=\frac{451\cdot 340}{289\cdot 123}$

$=\frac{11\cdot 41\cdot 17\cdot 20}{17\cdot 17\cdot 3\cdot 41}$

$=\frac{41\cdot 17}{41\cdot 17}\cdot\frac{11\cdot 20}{17\cdot 3}$

$=\frac{220}{51}$

Exercise Set 3.8

1. $\frac{4}{5}x=12$

$\frac{5}{4}\cdot\frac{4}{5}x=\frac{5}{4}\cdot 12$ The reciprocal of $\frac{4}{5}$ is $\frac{5}{4}$.

$1x=\frac{5\cdot 4\cdot 3}{4}$

$x=15$ Removing the factor $\frac{4}{4}$

Check: $\frac{4}{5}x=12$

$\frac{4}{5}\cdot 15\ ?\ 12$

$\frac{4\cdot 3\cdot \not{5}}{\not{5}\cdot 1}$

12 | 12 TRUE

The solution is 15.

3. $\frac{7}{3}a=21$

$\frac{3}{7}\cdot\frac{7}{3}a=\frac{3}{7}\cdot 21$ The reciprocal of $\frac{7}{3}$ is $\frac{3}{7}$.

$1a=\frac{3\cdot 3\cdot 7}{7}$

$a=9$ Removing the factor $\frac{7}{7}$

Check: $\frac{7}{3}a=21$

$\frac{7}{3}\cdot 9\ ?\ 21$

$\frac{7\cdot 3\cdot \not{3}}{\not{3}\cdot 1}$

21 | 21 TRUE

The solution is 9.

5. $\frac{2}{9}y=-10$

$\frac{9}{2}\cdot\frac{2}{9}y=\frac{9}{2}(-10)$ The reciprocal of $\frac{2}{9}$ is $\frac{9}{2}$.

$1y=-\frac{9\cdot 2\cdot 5}{2\cdot 1}$

$y=-45$ Removing the factor $\frac{2}{2}$

Check: $\frac{2}{9}y=-10$

$\frac{2}{9}(-45)\ ?\ -10$

$-\frac{2\cdot 5\cdot \not{9}}{\not{9}\cdot 1}$

-10 | -10 TRUE

The solution is -45.

7.
$$6t = \frac{12}{17}$$
$$\frac{1}{6} \cdot 6t = \frac{1}{6} \cdot \frac{12}{17} \quad \text{The reciprocal of 6 is } \frac{1}{6}.$$
$$1t = \frac{2 \cdot 6}{6 \cdot 17}$$
$$t = \frac{2}{17} \quad \text{Removing the factor } \frac{6}{6}$$

Check: $6t = \frac{12}{17}$

$$6 \cdot \frac{2}{17} \ ? \ \frac{12}{17}$$
$$\frac{12}{17} \ \Big| \ \frac{12}{17} \quad \text{TRUE}$$

The solution is $\frac{2}{17}$.

9.
$$\frac{1}{4}x = \frac{3}{5}$$
$$\frac{4}{1} \cdot \frac{1}{4}x = \frac{4}{1} \cdot \frac{3}{5}$$
$$x = \frac{12}{5}$$

$\frac{12}{5}$ checks and is the solution.

11.
$$\frac{3}{2}t = -\frac{8}{7}$$
$$\frac{2}{3} \cdot \frac{3}{2}t = \frac{2}{3}\left(-\frac{8}{7}\right)$$
$$t = -\frac{16}{21}$$

$-\frac{16}{21}$ checks and is the solution.

13.
$$\frac{4}{5} = -10a$$
$$-\frac{1}{10} \cdot \frac{4}{5} = -\frac{1}{10}(-10a)$$
$$-\frac{2 \cdot 2}{2 \cdot 5 \cdot 5} = a$$
$$-\frac{2}{25} = a$$

$-\frac{2}{25}$ checks and is the solution.

15.
$$x \cdot \frac{9}{5} = \frac{3}{10}$$
$$x \cdot \frac{9}{5} \cdot \frac{5}{9} = \frac{3}{10} \cdot \frac{5}{9}$$
$$x = \frac{5 \cdot 3 \cdot 1}{3 \cdot 3 \cdot 2 \cdot 5}$$
$$x = \frac{1}{6}$$

$\frac{1}{6}$ checks and is the solution.

17.
$$-\frac{1}{10}x = 8$$
$$-\frac{10}{1}\left(-\frac{1}{10}x\right) = -\frac{10}{1} \cdot 8$$
$$x = -\frac{10 \cdot 8}{1}$$
$$x = -80$$

-80 checks and is the solution.

19.
$$a \cdot \frac{9}{7} = -\frac{3}{14}$$
$$a \cdot \frac{9}{7} \cdot \frac{7}{9} = -\frac{3}{14} \cdot \frac{7}{9}$$
$$a \cdot 1 = -\frac{3 \cdot 7 \cdot 1}{2 \cdot 7 \cdot 3 \cdot 3}$$
$$a = -\frac{1}{6}$$

$-\frac{1}{6}$ checks and is the solution.

21.
$$-x = \frac{7}{13}$$
$$-1(-x) = -1 \cdot \frac{7}{13}$$
$$x = -\frac{7}{13}$$

$-\frac{7}{13}$ checks and is the solution.

23.
$$-x = -\frac{27}{31}$$
$$-1(-x) = -1\left(-\frac{27}{31}\right)$$
$$x = \frac{27}{31}$$

$\frac{27}{31}$ checks and is the solution.

25.
$$7t = 6$$
$$\frac{1}{7} \cdot 7t = \frac{1}{7} \cdot 6$$
$$t = \frac{6}{7}$$

$\frac{6}{7}$ checks and is the solution.

27.
$$-24 = -10a$$
$$-\frac{1}{10}(-24) = -\frac{1}{10}(-10a)$$
$$\frac{2 \cdot 12}{2 \cdot 5} = a$$
$$\frac{12}{5} = a$$

$\frac{12}{5}$ checks and is the solution.

29.
$$-\frac{14}{9} = \frac{10}{3}t$$
$$\frac{3}{10}\left(-\frac{14}{9}\right) = \frac{3}{10}\cdot\frac{10}{3}t$$
$$-\frac{3\cdot 2\cdot 7}{2\cdot 5\cdot 3\cdot 3} = t$$
$$-\frac{7}{15} = t$$

$-\frac{7}{15}$ checks and is the solution.

31.
$$n\cdot\frac{4}{15} = \frac{12}{25}$$
$$n\cdot\frac{4}{15}\cdot\frac{15}{4} = \frac{12}{25}\cdot\frac{15}{4}$$
$$n = \frac{4\cdot 3\cdot 3\cdot 5}{5\cdot 5\cdot 4}$$
$$n = \frac{9}{5}$$

$\frac{9}{5}$ checks and is the solution.

33.
$$-\frac{7}{20}x = -\frac{21}{10}$$
$$-\frac{20}{7}\left(-\frac{7}{20}x\right) = -\frac{20}{7}\left(-\frac{21}{10}\right)$$
$$x = \frac{2\cdot 10\cdot 3\cdot 7}{7\cdot 10}$$
$$x = 6$$

6 checks and is the solution.

35.
$$-\frac{25}{17} = -\frac{35}{34}a$$
$$-\frac{34}{35}\left(-\frac{25}{17}\right) = -\frac{34}{35}\left(-\frac{35}{34}a\right)$$
$$\frac{2\cdot 17\cdot 5\cdot 5}{5\cdot 7\cdot 17} = a$$
$$\frac{10}{7} = a$$

$\frac{10}{7}$ checks and is the solution.

37. ***Familiarize***. We draw a picture. Let c = the number of extension cords that can be made from 2240 ft of cable.

$\frac{7}{3}$ ft	$\frac{7}{3}$ ft	⋯	$\frac{7}{3}$ ft

c cords

Translate. The multiplication that corresponds to the situation is

$$\frac{7}{3}\cdot c = 2240.$$

Solve. We solve the equation by dividing on both sides by $\frac{7}{3}$ and carrying out the division:

$$c = 2240 \div \frac{7}{3} = 2240\cdot\frac{3}{7} = \frac{2240\cdot 3}{7} = \frac{7\cdot 320\cdot 3}{7\cdot 1}$$
$$= \frac{7}{7}\cdot\frac{320\cdot 3}{1} = \frac{320\cdot 3}{1} = 960$$

Check. We repeat the calculation. The answer checks.

State. 960 $\frac{7}{3}$-ft extension cords can be made from 2240 ft of cable.

39. ***Familiarize***. Let g = the number of gallons of gasoline the tanker holds when it is fully loaded.

Translate. We translate to an equation.

1400 gal	is	$\frac{7}{9}$	of	a full load
↓	↓	↓	↓	↓
1400	$=$	$\frac{7}{9}$	$\cdot$	g

Solve. We solve the equation.

$$1400 = \frac{7}{9}\cdot g$$
$$1400 \div \frac{7}{9} = g$$
$$1400\cdot\frac{9}{7} = g$$
$$\frac{1400\cdot 9}{7} = g$$
$$\frac{7\cdot 200\cdot 9}{7\cdot 1} = g$$
$$\frac{7}{7}\cdot\frac{200\cdot 9}{1} = g$$
$$1800 = g$$

Check. $\frac{7}{9}$ of 1800 gal is $\frac{7}{9}\cdot 1800 = \frac{7\cdot 1800}{9} = \frac{7\cdot 9\cdot 200}{9\cdot 1} = \frac{9}{9}\cdot\frac{7\cdot 200}{1} = 1400$ gal. The answer checks.

State. The tanker holds 1800 gal of gasoline when it is full.

41. ***Familiarize***. Let w = the number of worker bees it takes to produce $\frac{3}{4}$ tsp of honey.

Translate.

Amount produced by one bee	times	Number of bees	is	$\frac{3}{4}$ tsp
↓	↓	↓	↓	↓
$\frac{1}{12}$	$\cdot$	w	$=$	$\frac{3}{4}$

Solve. We solve the equation.

$$\frac{1}{12}\cdot w = \frac{3}{4}$$
$$w = \frac{3}{4}\div\frac{1}{12} = \frac{3}{4}\cdot\frac{12}{1} = \frac{3\cdot 12}{4\cdot 1}$$
$$= \frac{3\cdot 3\cdot 4}{4\cdot 1} = \frac{3\cdot 3}{1}\cdot\frac{4}{4} = \frac{3\cdot 3}{1}$$
$$= 9$$

Check. Since $\frac{1}{12}\cdot 9 = \frac{9}{12} = \frac{3}{4}$, the answer checks.

State. It takes 9 worker bees to produce $\frac{3}{4}$ tsp of honey.

43. ***Familiarize***. We make a drawing. Let $p =$ the number of packages that can be made from 15 lb of cheese.

$\boxed{\frac{3}{4}\text{ lb}}\ \boxed{\frac{3}{4}\text{ lb}}\ \cdots\ \boxed{\frac{3}{4}\text{ lb}}$

p packages

Translate. The problem translates to the following equation:

$$p = 15 \div \frac{3}{4}.$$

Solve. We carry out the division.

$$\begin{aligned} p &= 15 \div \frac{3}{4} \\ &= 15 \cdot \frac{4}{3} \\ &= \frac{3 \cdot 5 \cdot 4}{3 \cdot 1} = \frac{3}{3} \cdot \frac{5 \cdot 4}{1} \\ &= 20 \end{aligned}$$

Check. If 20 packages, each containing $\frac{3}{4}$ lb of cheese, are made, a total of

$$20 \cdot \frac{3}{4} = \frac{5 \cdot 4 \cdot 3}{4 \cdot 1} = \frac{4}{4} \cdot \frac{5 \cdot 3}{1} = 15,$$

or 15 lb of cheese is used. The answer checks.

State. 20 packages can be made.

45. ***Familiarize***. Let $c =$ the amount of clay each art department will receive, in tons.

Translate. The problem translates to the following equation:

$$c = \frac{3}{4} \div 6.$$

Solve. We carry out the division.

$$\begin{aligned} c &= \frac{3}{4} \div 6 \\ &= \frac{3}{4} \cdot \frac{1}{6} \\ &= \frac{3 \cdot 1}{4 \cdot 2 \cdot 3} = \frac{3}{3} \cdot \frac{1}{4 \cdot 2} \\ &= \frac{1}{8} \end{aligned}$$

Check. If each of 6 art departments get $\frac{1}{8}$ T of clay, the total amount of clay is

$$6 \cdot \frac{1}{8} = \frac{6 \cdot 1}{8} = \frac{2 \cdot 3 \cdot 1}{2 \cdot 4} = \frac{3}{4}\text{ T}.$$

The answer checks.

State. Each art department will receive $\frac{1}{8}$ T of clay.

47. ***Familiarize***. This is a multistep problem. First we find the length of the total trip. Then we find how many kilometers were left to drive. We draw a picture. We let $n =$ the length of the total trip.

$\frac{5}{8}$ of the trip

180 km

n km

Translate. We translate to an equation.

Fraction of trip completed	times	Total length of trip	is	Amount already traveled
↓	↓	↓	↓	↓
$\frac{5}{8}$	$\cdot$	n	$=$	180

Solve. We solve the equation as follows:

$$\frac{5}{8} \cdot n = 180$$

$$\begin{aligned} n &= 180 \div \frac{5}{8} = 180 \cdot \frac{8}{5} = \frac{5 \cdot 36 \cdot 8}{5 \cdot 1} \\ &= \frac{5}{5} \cdot \frac{36 \cdot 8}{1} = \frac{36 \cdot 8}{1} = 288 \end{aligned}$$

The total trip was 288 km.

Now we find how many kilometers were left to travel. Let $t =$ this number.

Length of total trip	minus	Distance traveled	is	Distance left to travel
↓	↓	↓	↓	↓
288	$-$	180	$=$	t

We carry out the subtraction:

$$\begin{aligned} 288 - 180 &= t \\ 108 &= t \end{aligned}$$

Check. We repeat the calculation. The results check.

State. The total trip was 288 km. There were 108 km left to travel.

49. ***Familiarize***. We make a drawing. Let $c =$ the number of customers that can be accommodated with a 30 yd batch of mulch.

$\boxed{\frac{2}{3}\text{ yd}}\ \boxed{\frac{2}{3}\text{ yd}}\ \cdots\ \boxed{\frac{2}{3}\text{ yd}}$

c customers

Translate. The problem translates to the following situation.

$$c = 30 \div \frac{2}{3}.$$

Solve. We carry out the division.

$$\begin{aligned} c &= 30 \div \frac{2}{3} \\ &= 30 \cdot \frac{3}{2} \\ &= \frac{2 \cdot 15 \cdot 3}{2 \cdot 1} = \frac{2}{2} \cdot \frac{15 \cdot 3}{1} \\ &= 45 \end{aligned}$$

Check. If each of 45 customers gets $\frac{2}{3}$ yd of mulch, a total of

$$45 \cdot \frac{2}{3} = \frac{3 \cdot 15 \cdot 2}{3 \cdot 1} = \frac{3}{3} \cdot \frac{15 \cdot 2}{1} = 30,$$

or 30 yd of mulch is used. The answer checks.

State. 45 customers can be accommodated with 30 yd of mulch.

51. ***Familiarize***. We draw a picture.

□ □ □ ... □ } 24 yd makes how many pairs?

$\frac{3}{4}$ yd per pair

We let $s =$ the number of pairs of basketball shorts that can be made.

Translate. The problem translates to the following equation:

$$s = 24 \div \frac{3}{4}$$

Solve. We carry out the division.

$$\begin{aligned} s &= 24 \div \frac{3}{4} \\ &= 24 \cdot \frac{4}{3} \\ &= \frac{3 \cdot 8 \cdot 4}{1 \cdot 3} = \frac{3}{3} \cdot \frac{8 \cdot 4}{1} \\ &= 32 \end{aligned}$$

Check. If each of 32 pairs of shorts requires $\frac{3}{4}$ yd of nylon, a total of

$$32 \cdot \frac{3}{4} = \frac{32 \cdot 3}{4} = \frac{4 \cdot 8 \cdot 3}{4} = 8 \cdot 3,$$

or 24 yd of nylon is needed. Our answer checks.

State. 32 pairs of basketball shorts can be made from 24 yd of nylon.

53. ***Familiarize***. Let $p =$ the pitch of the screw, in inches. The distance the screw has traveled into the wallboard is found by multiplying the pitch by the number of complete rotations.

Translate. We translate to an equation.

Pitch of screw	times	Number of rotations	is	Distance traveled
↓	↓	↓	↓	↓
p	$\cdot$	8	$=$	$\frac{1}{2}$

Solve. We divide on both sides of the equation by 8 and carry out the division.

$$p = \frac{1}{2} \div 8 = \frac{1}{2} \cdot \frac{1}{8} = \frac{1 \cdot 1}{2 \cdot 8} = \frac{1}{16}$$

Check. We repeat the calculation. The answer checks.

State. The pitch of the screw is $\frac{1}{16}$ in.

55. $-23 + 49 = 26$

(Find the difference of the absolute values. The positive integer has the larger absolute value, so the answer is positive.)

57. $-38 - 29 = -67$

(Add the absolute values. The answer is negative.)

59.
$$\begin{aligned} 36 \div (-3)^2 \times (7-2) &= 36 \div (-3)^2 \times 5 \\ &= 36 \div 9 \times 5 \\ &= 4 \times 5 \\ &= 20 \end{aligned}$$

61. $13x + 4x = (13+4)x = 17x$

63.
$$\begin{aligned} 2a + 3 + 5a &= 2a + 5a + 3 \\ &= (2+5)a + 3 \\ &= 7a + 3 \end{aligned}$$

65.
$$\begin{aligned} 2x - 7x &= -\frac{10}{9} \\ -5x &= -\frac{10}{9} \\ -\frac{1}{5}(-5x) &= -\frac{1}{5}\left(-\frac{10}{9}\right) \\ x &= \frac{2 \cdot 5}{5 \cdot 9} \\ x &= \frac{2}{9} \end{aligned}$$

67. ***Familiarize***. Let $w =$ the weight of the package when it is completely filled.

Translate.

$\frac{3}{4}$	of	total weight	is	$\frac{21}{32}$ lb
↓	↓	↓	↓	↓
$\frac{3}{4}$	$\cdot$	w	$=$	$\frac{21}{32}$

Solve. We solve the equation.

$$\begin{aligned} \frac{3}{4} \cdot w &= \frac{21}{32} \\ \frac{4}{3} \cdot \frac{3}{4} \cdot w &= \frac{4}{3} \cdot \frac{21}{32} \\ w &= \frac{4 \cdot 3 \cdot 7}{3 \cdot 4 \cdot 8} = \frac{4 \cdot 3}{4 \cdot 3} \cdot \frac{7}{8} \\ w &= \frac{7}{8} \end{aligned}$$

Check. We find $\frac{3}{4}$ of $\frac{7}{8}$ lb.

$$\frac{3}{4} \cdot \frac{7}{8} = \frac{3 \cdot 7}{4 \cdot 8} = \frac{21}{32} \text{ lb}$$

The answer checks.

State. The package could hold $\frac{7}{8}$ lb of coffee beans when it is completely filled.

69. ***Familiarize***. Let $x =$ the number of slices yielded by the $\frac{3}{32}$-in. cuts and $y =$ the number of slices yielded by the $\frac{5}{32}$-in. cuts. Half the block is $\frac{1}{2} \cdot 12$ in., or 6 in.

Translate. The problem translates to the following situations.

$x = 6 \div \frac{3}{32}$ and $y = 6 \div \frac{5}{32}$

Solve. We carry out the division.

$$\begin{aligned} x &= 6 \div \frac{3}{32} \\ &= 6 \cdot \frac{32}{3} \\ &= \frac{2 \cdot 3 \cdot 32}{3 \cdot 1} = \frac{3}{3} \cdot \frac{2 \cdot 32}{1} \\ &= 64 \end{aligned}$$

$$\begin{aligned} y &= 6 \div \frac{5}{32} \\ &= 6 \cdot \frac{32}{5} \\ &= \frac{192}{5} \\ &= 38 \text{ R } 2 \end{aligned}$$

The $\frac{3}{32}$-in. cuts yield 64 slices. The $\frac{5}{32}$-in. cuts yield 38 slices that are $\frac{5}{32}$ in. thick and an additional slice (indicated by the remainder) that is less than $\frac{5}{32}$ in., so this cutting yields 39 slices. Then the total number of slices is $64 + 39$, or 103.

Check. We repeat the calculations. The answer checks.

State. The cutting will yield 103 slices of cheese.

71. ***Familiarize.*** First we will find the number of walkways that can be covered by 6 yd of gravel. Then we will find the amount Eric received. We make a drawing. Let $w =$ the number of walkways that can be covered with 6 yd of gravel.

$\frac{3}{5}$ yd | $\frac{3}{5}$ yd | ... | $\frac{3}{5}$ yd

w walkways

Also, let $t =$ the total amount Eric was paid for a full load of gravel.

Translate. The problem translates to the following situations:

$w = 6 \div \frac{3}{5}$

$t = 85 \cdot w$

Solve. We carry out the division first.

$$\begin{aligned} w &= 6 \div \frac{3}{5} \\ &= 6 \cdot \frac{5}{3} \\ &= \frac{2 \cdot 3 \cdot 5}{3 \cdot 1} = \frac{3}{3} \cdot \frac{2 \cdot 5}{1} \\ &= 10 \end{aligned}$$

Now we substitute 10 for w in the second equation and multiply to find t.

$t = 85 \cdot w = 85 \cdot 10 = 850$

Check. If each of 10 walkways is covered with $\frac{3}{5}$ yd of gravel, a total of

$$10 \cdot \frac{3}{5} = \frac{2 \cdot 5 \cdot 3}{5 \cdot 1} = \frac{5}{5} \cdot \frac{2 \cdot 3}{1} = 6,$$

or 6 yd of gravel is used. Also, if Eric charges \$85 for the gravel for each cottage and he earns \$850, then $850 \div 85$, or 10 cottages have received gravel. The answer checks.

State. Eric will receive \$850 for a full load of gravel.

Chapter 3 Concept Reinforcement

1. For any nonzero integer n, $\frac{n}{n} = 1$ and $\frac{0}{n} = 0$. Since $1 > 0$, the statement is true.

2. A number is divisible by 10 only if its ones digit is 0. The statement is false.

3. Since 3 is a factor of 9, any number that is divisible by 9 is also divisible by 3. The statement is true.

4. Since $13 > 11$, the $\frac{13}{6} > \frac{11}{6}$. The statement is true.

Chapter 3 Important Concepts

1. We find as many two-factor factorizations as we can.

$104 = 1 \cdot 104$ $\quad$ $104 = 4 \cdot 26$

$104 = 2 \cdot 52$ $\quad$ $104 = 8 \cdot 13$

Factors: 1, 2, 4, 8, 13, 26, 52, 104

2.

$$\begin{array}{r} 13 \leftarrow 13 \text{ is prime} \\ 2 \,\overline{)\, 26} \\ 2 \,\overline{)\, 52} \\ 2 \,\overline{)\, 104} \end{array}$$

$104 = 2 \cdot 2 \cdot 2 \cdot 13$

3. $\frac{0}{18} = 0$; $\frac{18}{18} = 1$; $\frac{18}{1} = 18$

4. Since $12 \cdot 8 = 96$, we multiply by $\frac{8}{8}$.

$$\frac{7}{12} = \frac{7}{12} \cdot \frac{8}{8} = \frac{7 \cdot 8}{12 \cdot 8} = \frac{56}{96}$$

5. $\frac{100}{280} = \frac{2 \cdot 5 \cdot 10}{2 \cdot 2 \cdot 7 \cdot 10} = \frac{2 \cdot 10}{2 \cdot 10} \cdot \frac{5}{2 \cdot 7} = 1 \cdot \frac{5}{2 \cdot 7} = \frac{5}{14}$

6. We multiply these two numbers: $\quad$ We multiply these two numbers:

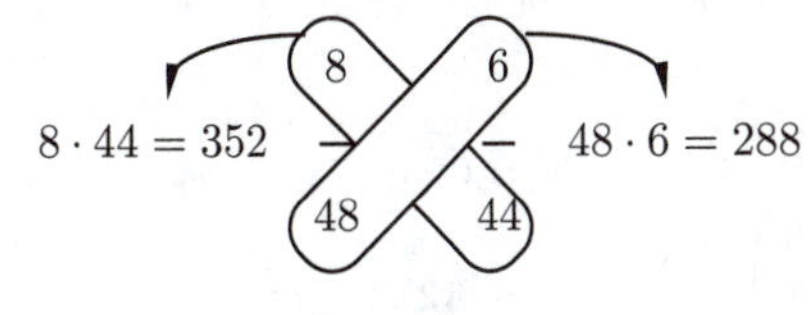

Since $352 \neq 288$, $\frac{8}{48} \neq \frac{6}{44}$.

7. $\frac{80}{3}\cdot\frac{21}{72}=\frac{80\cdot21}{3\cdot72}=\frac{2\cdot5\cdot8\cdot3\cdot7}{3\cdot9\cdot8}=\frac{3\cdot8}{3\cdot8}\cdot\frac{2\cdot5\cdot7}{9}=$
$1\cdot\frac{2\cdot5\cdot7}{9}=\frac{70}{9}$

8. $\frac{9}{4}\div\frac{45}{14}=\frac{9}{4}\cdot\frac{14}{45}=\frac{9\cdot14}{4\cdot45}=\frac{9\cdot2\cdot7}{2\cdot2\cdot9\cdot5}=\frac{9\cdot2}{9\cdot2}\cdot\frac{7}{2\cdot5}=$
$1\cdot\frac{7}{2\cdot5}=\frac{7}{10}$

9. ***Familiarize.*** Let w = the number of cups of water the vase can hold when full.

Translate.

$$\underbrace{\frac{7}{4}\text{ cups}}\ \text{is}\ \frac{3}{4}\ \text{of}\ \underbrace{\text{full amount}}$$
$$\downarrow\quad\downarrow\quad\downarrow\quad\downarrow\qquad\downarrow$$
$$\frac{7}{4}\quad=\quad\frac{3}{4}\quad\cdot\qquad w$$

Solve. We multiply both sides by $\frac{4}{3}$.

$$\frac{4}{3}\cdot\frac{7}{4}=\frac{4}{3}\cdot\frac{3}{4}w$$
$$\frac{\not{4}\cdot7}{3\cdot\not{4}}=w$$
$$\frac{7}{3}=w$$

Check. We find $\frac{3}{4}$ of $\frac{7}{3}$: $\frac{3}{4}\cdot\frac{7}{3}=\frac{3\cdot7}{4\cdot3}=\frac{3}{3}\cdot\frac{7}{4}=\frac{7}{4}$. The answer checks.

State. The vase holds $\frac{7}{3}$ cups of water when full.

Chapter 3 Review Exercises

1.
$1\cdot8=8$ $\qquad$ $6\cdot8=48$
$2\cdot8=16$ $\qquad$ $7\cdot8=56$
$3\cdot8=24$ $\qquad$ $8\cdot8=64$
$4\cdot8=32$ $\qquad$ $9\cdot8=72$
$5\cdot8=40$ $\qquad$ $10\cdot8=80$

2. 3920 is even because the ones digit is even; $3+9+2+0=14$ and 14 is not divisible by 3, so 3920 is not divisible by 3. Since 3920 is not divisible by 3, it is not divisible by 6.

3. Because $6+8+5+3+7=29$ and 29 is not divisible by 3, then 68,537 is not divisible by 3.

4. 673 is not divisible by 5 because the ones digit is neither 0 nor 5.

5. 4936 is divisible by 2 because the ones digit is even.

6. Because $5+2+3+8=18$ and 18 is divisible by 9, then 5238 is divisible by 9.

7. Since the ones digit of 60 is 0 we know that 2, 5, and 10 are factors. Since the sum of the digits is 6 and 6 is divisible by 3, then 3 is a factor. Since 2 and 3 are factors, 6 is also a factor. We write a list of factorizations.
$60=1\cdot60$ $\qquad$ $60=4\cdot15$
$60=2\cdot30$ $\qquad$ $60=5\cdot12$
$60=3\cdot20$ $\qquad$ $60=6\cdot10$
Factors: 1, 2, 3, 4, 5, 6, 10, 12, 15, 20, 30, 60

8. 176 is even so 2 is a factor. None of the other tests for divisibility yields additional factors, so we find as many two-factor factorizations as we can:
$176=1\cdot176$ $\qquad$ $176=8\cdot22$
$176=2\cdot88$ $\qquad$ $176=11\cdot16$
$176=4\cdot44$
Factors: 1, 2, 4, 8, 11, 16, 22, 44, 88, 176

9. The only factors of 37 are 1 and 37, so 37 is prime.

10. 1 is neither prime nor composite.

11. The number 91 has factors 1, 7, 13, and 91, so it is composite.

12.
$$\begin{array}{rl} & 7 \quad\leftarrow\ \text{7 is prime.}\\ 5 & \overline{\smash{)}35}\\ 2 & \overline{\smash{)}70}\end{array}$$
$70=2\cdot5\cdot7$

13.
$$\begin{array}{rl} & 3 \quad\leftarrow\ \text{3 is prime.}\\ 3 & \overline{\smash{)}9}\\ 2 & \overline{\smash{)}18}\\ 2 & \overline{\smash{)}36}\\ 2 & \overline{\smash{)}72}\end{array}$$
$72=2\cdot2\cdot2\cdot3\cdot3$

14.
$$\begin{array}{rl} & 5 \quad\leftarrow\ \text{5 is prime.}\\ 3 & \overline{\smash{)}15}\\ 3 & \overline{\smash{)}45}\end{array}$$
$45=3\cdot3\cdot5$

15.
$$\begin{array}{rl} & 5 \quad\leftarrow\ \text{5 is prime.}\\ 5 & \overline{\smash{)}25}\\ 3 & \overline{\smash{)}75}\\ 2 & \overline{\smash{)}150}\end{array}$$
$150=2\cdot3\cdot5\cdot5$

16.
$$\begin{array}{rl} & 3 \quad\leftarrow\ \text{3 is prime.}\\ 3 & \overline{\smash{)}9}\\ 3 & \overline{\smash{)}27}\\ 3 & \overline{\smash{)}81}\\ 2 & \overline{\smash{)}162}\\ 2 & \overline{\smash{)}324}\\ 2 & \overline{\smash{)}648}\end{array}$$
$648=2\cdot2\cdot2\cdot3\cdot3\cdot3\cdot3$

17.
$$\begin{array}{rl} & 5 \quad\leftarrow\ \text{5 is prime.}\\ 5 & \overline{\smash{)}25}\\ 3 & \overline{\smash{)}75}\\ 2 & \overline{\smash{)}150}\\ 2 & \overline{\smash{)}300}\\ 2 & \overline{\smash{)}600}\\ 2 & \overline{\smash{)}1200}\end{array}$$
$1200=2\cdot2\cdot2\cdot2\cdot3\cdot5\cdot5$

18. The top number is the numerator, and the bottom number is the denominator.
$$\frac{9}{7}\ \begin{array}{l}\leftarrow\text{Numerator}\\ \leftarrow\text{Denominator}\end{array}$$

19. The object is divided into 5 equal parts. The unit is $\frac{1}{5}$. The denominator is 5. We have 3 parts shaded. This tells us that the numerator is 3. Thus, $\frac{3}{5}$ is shaded.

20. Each object is divided into 6 equal parts. The unit is $\frac{1}{6}$. The denominator is 6. We have 7 parts shaded. This tells us that the numerator is 7. Thus, $\frac{7}{6}$ is shaded.

21. a) The ratio is $\frac{3}{5}$.

b) The ratio is $\frac{5}{3}$.

c) There are $3+5$, or 8, members of the committee. The desired ratio is $\frac{3}{8}$.

22. $\frac{0}{n} = 0$, for any number n that is not 0.

$\frac{0}{6} = 0$

23. $\frac{n}{n} = 1$, for any number n that is not 0.

$\frac{74}{74} = 1$

24. $\frac{n}{1} = n$, for any number n.

$\frac{48}{1} = 48$

25. Remember: $\frac{n}{n} = 1$ for any number n that is not 0.

$\frac{7x}{7x} = 1$

26. $-\frac{10}{15} = -\frac{2\cdot 5}{3\cdot 5} = -\frac{2}{3}\cdot\frac{5}{5} = -\frac{2}{3}$

27. $\frac{7}{28} = \frac{7\cdot 1}{7\cdot 4} = \frac{7}{7}\cdot\frac{1}{4} = \frac{1}{4}$

28. $\frac{-42}{42} = \frac{-1\cdot 42}{1\cdot 42} = \frac{-1}{1}\cdot\frac{42}{42} = \frac{-1}{1} = -1$

29. $\frac{9m}{12m} = \frac{3\cdot 3\cdot m}{3\cdot 4\cdot m} = \frac{3\cdot m}{3\cdot m}\cdot\frac{3}{4} = \frac{3}{4}$

30. $\frac{-12}{-30} = \frac{2\cdot 6}{5\cdot 6} = \frac{2}{5}\cdot\frac{6}{6} = \frac{2}{5}$

31. Remember: $\frac{n}{0}$ is not defined.

$\frac{-27}{0}$ is undefined.

32. $\frac{140}{490} = \frac{2\cdot 7\cdot 10}{7\cdot 7\cdot 10} = \frac{2}{7}\cdot\frac{7\cdot 10}{7\cdot 10} = \frac{2}{7}$

33. $\frac{288}{2025} = \frac{2\cdot 2\cdot 2\cdot 2\cdot 2\cdot 3\cdot 3}{3\cdot 3\cdot 3\cdot 3\cdot 5\cdot 5} = \frac{3\cdot 3}{3\cdot 3}\cdot\frac{2\cdot 2\cdot 2\cdot 2\cdot 2}{3\cdot 3\cdot 5\cdot 5} = \frac{32}{225}$

34. Since $21 \div 7 = 3$, we multiply by $\frac{3}{3}$.

$\frac{5}{7} = \frac{5}{7}\cdot\frac{3}{3} = \frac{5\cdot 3}{7\cdot 3} = \frac{15}{21}$

35. Since $55 \div 11 = 5$, we multiply by $\frac{5}{5}$.

$\frac{-6}{11} = \frac{-6}{11}\cdot\frac{5}{5} = \frac{-6\cdot 5}{11\cdot 5} = \frac{-30}{55}$

36. $\frac{15}{100} = \frac{3\cdot 5}{20\cdot 5} = \frac{3}{20}\cdot\frac{5}{5} = \frac{3}{20}$

$\frac{38}{100} = \frac{2\cdot 19}{2\cdot 50} = \frac{2}{2}\cdot\frac{19}{50} = \frac{19}{50}$

23 and 100 have no prime factors in common, so $\frac{23}{100}$ cannot be simplified.

$\frac{24}{100} = \frac{4\cdot 6}{4\cdot 25} = \frac{4}{4}\cdot\frac{6}{25} = \frac{6}{25}$

37. We multiply these two numbers: We multiply these two numbers:

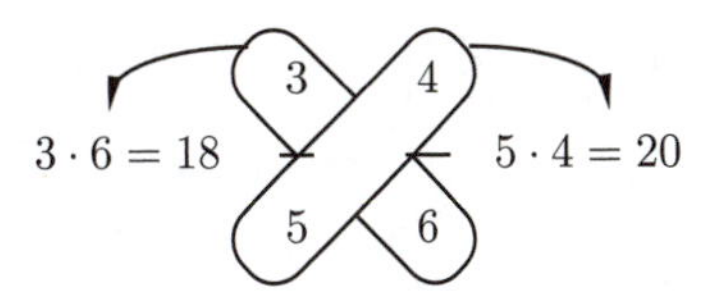

Since $18 \neq 20$, $\frac{3}{5}$ and $\frac{4}{6}$ do not name the same number. Thus, $\frac{3}{5} \neq \frac{4}{6}$.

38. We multiply these two numbers: We multiply these two numbers:

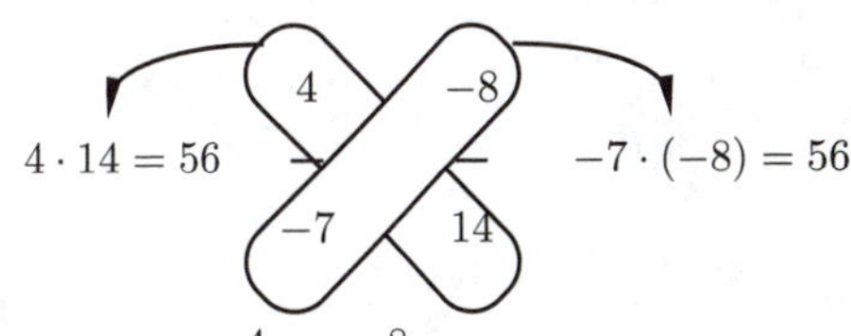

Since $56 = 56$, $\frac{4}{-7} = \frac{-8}{14}$.

39. We multiply these two numbers: We multiply these two numbers:

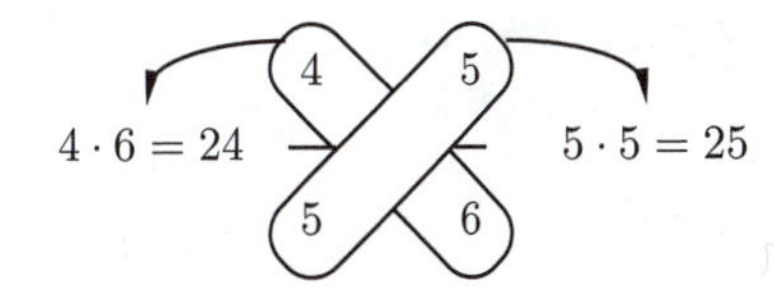

Since $24 \neq 25$, $\frac{4}{5}$ and $\frac{5}{6}$ do not name the same number. Thus, $\frac{4}{5} \neq \frac{5}{6}$.

40. We multiply these two numbers: We multiply these two numbers:

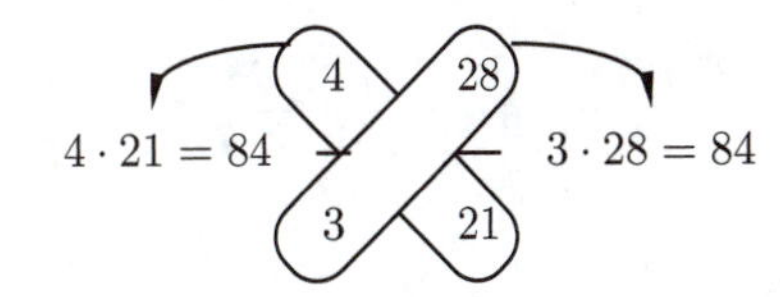

Since $84 = 84$, $\frac{4}{3} = \frac{28}{21}$.

41. Interchange the numerator and denominator.

The reciprocal of $\frac{2}{13}$ is $\frac{13}{2}$. $\left(\frac{2}{13}\cdot\frac{13}{2}=\frac{26}{26}=1\right)$

42. Think of -7 as $\frac{-7}{1}$. Interchange the numerator and denominator.

The reciprocal of -7 is $\frac{1}{-7}$, or $\frac{-1}{7}$, or $-\frac{1}{7}$.

$\left(-7\cdot\left(-\frac{1}{7}\right)=\frac{7}{7}=1\right)$

43. Interchange the numerator and denominator.

The reciprocal of $\frac{1}{8}$ is $\frac{8}{1}$, or 8. $\left(\frac{1}{8}\cdot 8=\frac{8}{8}=1\right)$

44. Interchange the numerator and denominator.

The reciprocal of $\frac{3x}{5y}$ is $\frac{5y}{3x}$. $\left(\frac{3x}{5y}\cdot\frac{5y}{3x}=\frac{15xy}{15xy}=1\right)$

45. $\frac{2}{9}\cdot\frac{7}{5}=\frac{2\cdot 7}{9\cdot 5}=\frac{14}{45}$

46. $\frac{3}{x}\cdot\frac{y}{7}=\frac{3\cdot y}{x\cdot 7}=\frac{3y}{7x}$

47. $\frac{3}{4}\cdot\frac{8}{9}=\frac{3\cdot 8}{4\cdot 9}=\frac{3\cdot 2\cdot 4}{4\cdot 3\cdot 3}=\frac{3\cdot 4}{3\cdot 4}\cdot\frac{2}{3}=\frac{2}{3}$

48. $-10\cdot\frac{7}{5}=\frac{-10\cdot 7}{5}=-\frac{2\cdot 5\cdot 7}{5\cdot 1}=-\frac{2\cdot 7}{1}\cdot\frac{5}{5}=-14$

49. $\frac{11}{3}\cdot\frac{30}{77}=\frac{11\cdot 3\cdot 10}{3\cdot 7\cdot 11}=\frac{3\cdot 11}{3\cdot 11}\cdot\frac{10}{7}=\frac{10}{7}$

50. $\frac{4a}{7}\cdot\frac{7}{4a}=\frac{4a\cdot 7}{4a\cdot 7}=1$

51. $\frac{6}{5}\cdot 20x=\frac{6\cdot 20x}{5}=\frac{6\cdot 4\cdot 5\cdot x}{5\cdot 1}=\frac{5}{5}\cdot\frac{6\cdot 4\cdot x}{1}=24x$

52. $\frac{3}{14}\div\frac{6}{7}=\frac{3}{14}\cdot\frac{7}{6}=\frac{3\cdot 7}{14\cdot 6}=\frac{3\cdot 7\cdot 1}{2\cdot 7\cdot 2\cdot 3}=\frac{3\cdot 7}{3\cdot 7}\cdot\frac{1}{2\cdot 2}=$
$\frac{1}{2\cdot 2}=\frac{1}{4}$

53. $20\div\frac{3}{4}=20\cdot\frac{4}{3}=\frac{20\cdot 4}{3}=\frac{80}{3}$

54. $-\frac{5}{36}\div\left(-\frac{25}{12}\right)=-\frac{5}{36}\cdot\left(-\frac{12}{25}\right)=\frac{5\cdot 12}{36\cdot 25}=$
$\frac{5\cdot 12\cdot 1}{3\cdot 12\cdot 5\cdot 5}=\frac{5\cdot 12}{5\cdot 12}\cdot\frac{1}{3\cdot 5}=\frac{1}{15}$

55. $21\div\frac{7}{2a}=\frac{21}{1}\cdot\frac{2a}{7}=\frac{3\cdot 7\cdot 2a}{1\cdot 7}=\frac{7}{7}\cdot\frac{3\cdot 2a}{1}=6a$

56. $-\frac{23}{25}\div\frac{23}{25}=-\frac{23}{25}\cdot\frac{25}{23}=-\frac{23\cdot 25}{25\cdot 23}=-1$

57. $\frac{\frac{21}{30}}{\frac{14}{15}}=\frac{21}{30}\cdot\frac{15}{14}=\frac{21\cdot 15}{30\cdot 14}=\frac{3\cdot 7\cdot 3\cdot 5}{2\cdot 3\cdot 5\cdot 2\cdot 7}=$
$\frac{3\cdot 5\cdot 7}{3\cdot 5\cdot 7}\cdot\frac{3}{2\cdot 2}=\frac{3}{4}$

58. $\frac{-\frac{2}{3}}{-\frac{3}{2}}=-\frac{2}{3}\cdot\left(-\frac{2}{3}\right)=\frac{2\cdot 2}{3\cdot 3}=\frac{4}{9}$

59.
$$\begin{aligned}\frac{2}{3}x&=160\\ \frac{3}{2}\cdot\frac{2}{3}x&=\frac{3}{2}\cdot 160\\ 1x&=\frac{3\cdot 2\cdot 80}{2}\\ x&=240\end{aligned}$$

The solution is 240.

60.
$$\begin{aligned}\frac{3}{8}&=-\frac{5}{4}t\\ -\frac{4}{5}\cdot\frac{3}{8}&=-\frac{4}{5}\left(-\frac{5}{4}t\right)\\ -\frac{4\cdot 3}{5\cdot 2\cdot 4}&=1t\\ -\frac{3}{10}&=t\end{aligned}$$

The solution is $-\frac{3}{10}$.

61.
$$\begin{aligned}-\frac{1}{7}n&=-4\\ -7\left(-\frac{1}{7}n\right)&=-7(-4)\\ n&=28\end{aligned}$$

The solution is 28.

62.
$$\begin{aligned}y\cdot\frac{1}{2}&=\frac{1}{3}\\ y\cdot\frac{1}{2}\cdot\frac{2}{1}&=\frac{1}{3}\cdot\frac{2}{1}\\ y&=\frac{2}{3}\end{aligned}$$

The solution is $\frac{2}{3}$.

63.
$$\begin{aligned}A&=\frac{1}{2}\cdot b\cdot h\\ A&=\frac{1}{2}\cdot 14\text{ m}\cdot 6\text{ m}\\ A&=\frac{14\cdot 6}{2}\text{ m}^2\\ A&=42\text{ m}^2\end{aligned}$$

64.
$$\begin{aligned}A&=\frac{1}{2}\cdot b\cdot h\\ A&=\frac{1}{2}\cdot\frac{7}{2}\text{ ft}\cdot 10\text{ ft}\\ A&=\frac{7\cdot 10}{2\cdot 2}\text{ ft}^2\\ A&=\frac{35}{2}\text{ ft}^2\end{aligned}$$

65. ***Familiarize.*** Let d = the number of days it will take to repave the road.

Translate.

$$\underbrace{\text{Number of miles repaved each day}}_{\downarrow\ \frac{1}{12}}\ \underset{\downarrow\ \cdot}{\text{times}}\ \underbrace{\text{Number of days}}_{\downarrow\ d}\ \underset{\downarrow\ =}{\text{is}}\ \underbrace{\text{Total number of miles repaved}}_{\downarrow\ \frac{3}{4}}$$

Solve. We divide by $\frac{1}{12}$ on both sides of the equation.

$$d = \frac{3}{4} \div \frac{1}{12}$$

$$d = \frac{3}{4} \cdot \frac{12}{1} = \frac{3 \cdot 12}{4 \cdot 1} = \frac{3 \cdot 3 \cdot 4}{4 \cdot 1}$$

$$= \frac{4}{4} \cdot \frac{3 \cdot 3}{1} = \frac{3 \cdot 3}{1} = 9$$

Check. We repeat the calculation. The answer checks.

State. It will take 9 days to repave the road.

66. ***Familiarize.*** Let c = the number of bales of cotton produced in the United States in 2008.

Translate.

$$\underbrace{\text{U.S. cotton production}}_{\downarrow\ c}\ \underset{\downarrow\ =}{\text{was}}\ \underset{\downarrow\ \frac{3}{25}}{\frac{3}{25}}\ \underset{\downarrow\ \cdot}{\text{of}}\ \underbrace{\text{World cotton production}}_{\downarrow\ 113{,}000{,}000}$$

Solve. We carry out the multiplication.

$$c = \frac{3}{25} \cdot 113{,}000{,}000 = \frac{3 \cdot 113{,}000{,}000}{25} =$$

$$\frac{3 \cdot 25 \cdot 4{,}520{,}000}{25 \cdot 1} = \frac{25}{25} \cdot \frac{3 \cdot 4{,}520{,}000}{1} = 13{,}560{,}000$$

Check. We can repeat the calculation. The answer checks.

State. The United States produced 13,560,000 bales of cotton in 2008.

67. ***Familiarize.*** Let t = the total length of the trip, in km.

Translate.

$$\underbrace{\text{Distance driven}}_{\downarrow\ 600}\ \underset{\downarrow\ =}{\text{is}}\ \underset{\downarrow\ \frac{3}{5}}{\frac{3}{5}}\ \underset{\downarrow\ \cdot}{\text{of}}\ \underbrace{\text{Total distance}}_{\downarrow\ t}$$

Solve. We divide by $\frac{3}{5}$ on both sides of the equation.

$$t = 600 \div \frac{3}{5}$$

$$t = 600 \cdot \frac{5}{3} = \frac{600 \cdot 5}{3} = \frac{3 \cdot 200 \cdot 5}{3 \cdot 1}$$

$$= \frac{3}{3} \cdot \frac{200 \cdot 5}{1} = \frac{200 \cdot 5}{1} = 1000$$

Check. We repeat the calculation. The answer checks.

State. The trip is 1000 km long.

68. ***Familiarize.*** Let x = the number of cups of peppers needed for $\frac{1}{2}$ recipe.

Translate. We want to find $\frac{1}{2}$ of $\frac{2}{3}$ cup, so we have the multiplication sentence $x = \frac{1}{2} \cdot \frac{2}{3}$.

Solve. We carry out the multiplication.

$$x = \frac{1}{2} \cdot \frac{2}{3} = \frac{1 \cdot 2}{2 \cdot 3} = \frac{2}{2} \cdot \frac{1}{3} = \frac{1}{3}$$

Check. We repeat the calculation. The answer checks.

State. For $\frac{1}{2}$ recipe, $\frac{1}{3}$ cup of peppers is needed.

69. ***Familiarize.*** Let d = the distance each person will swim, in miles.

Translate.

$$\underbrace{\text{Number of swimmers}}_{\downarrow\ 4}\ \underset{\downarrow\ \cdot}{\text{times}}\ \underbrace{\text{Distance each swims}}_{\downarrow\ d}\ \underset{\downarrow\ =}{\text{is}}\ \underbrace{\text{Total distance}}_{\downarrow\ \frac{2}{3}}$$

Solve. We solve the equation.

$$4 \cdot d = \frac{2}{3}$$

$$\frac{1}{4} \cdot 4 \cdot d = \frac{1}{4} \cdot \frac{2}{3}$$

$$d = \frac{1 \cdot 2}{4 \cdot 3} = \frac{1 \cdot 2}{2 \cdot 2 \cdot 3}$$

$$= \frac{2}{2} \cdot \frac{1}{2 \cdot 3} = \frac{1}{6}$$

Check. Since $4 \cdot \frac{1}{6} = \frac{4}{6} = \frac{2}{3}$, the answer checks.

State. Each person will swim $\frac{1}{6}$ mi.

70. ***Familiarize.*** Let b = the number of bags that can be made from 48 yd of fabric.

Translate.

$$\underbrace{\text{Fabric for one bag}}_{\downarrow\ \frac{4}{5}}\ \underset{\downarrow\ \cdot}{\text{times}}\ \underbrace{\text{Number of bags}}_{\downarrow\ b}\ \underset{\downarrow\ =}{\text{is}}\ \underbrace{\text{Total amount of fabric}}_{\downarrow\ 48}$$

Solve. We divide by $\frac{4}{5}$ on both sides of the equation.

$$\frac{4}{5} \cdot b = 48$$

$$b = 48 \div \frac{4}{5}$$

$$b = 48 \cdot \frac{5}{4} = \frac{48 \cdot 5}{4} = \frac{4 \cdot 12 \cdot 5}{4 \cdot 1}$$

$$= \frac{4}{4} \cdot \frac{12 \cdot 5}{1} = \frac{12 \cdot 5}{1} = 60$$

Check. Since $\frac{4}{5} \cdot 60 = \frac{4 \cdot 60}{5} = \frac{4 \cdot 5 \cdot 12}{5 \cdot 1} = \frac{5}{5} \cdot \frac{4 \cdot 12}{1} = 48$, the answer checks.

State. 60 book bags can be made from 48 yd of fabric.

71. $\frac{2}{13} \cdot x = \frac{1}{2}$

We divide by $\frac{2}{13}$ on both sides and carry out the division.

$$x = \frac{1}{2} \div \frac{2}{13} = \frac{1}{2} \cdot \frac{13}{2} = \frac{1 \cdot 13}{2 \cdot 2} = \frac{13}{4}$$

Answer D is correct.

72. $$\frac{15}{26} \cdot \frac{13}{90} = \frac{15 \cdot 13}{26 \cdot 90} = \frac{3 \cdot 5 \cdot 13 \cdot 1}{2 \cdot 13 \cdot 2 \cdot 3 \cdot 3 \cdot 5} = \frac{3 \cdot 5 \cdot 13}{3 \cdot 5 \cdot 13} = \frac{1}{2 \cdot 2 \cdot 3} = \frac{1}{12}$$

Answer B is correct.

73.
$$\begin{aligned}
&\frac{15x}{14z} \cdot \frac{17yz}{35xy} \div \left(-\frac{3}{7}\right)^2 \\
&= \frac{15x}{14z} \cdot \frac{17yz}{35xy} \div \frac{9}{49} \\
&= \frac{15x \cdot 17yz}{14z \cdot 35xy} \div \frac{9}{49} \\
&= \frac{15x \cdot 17yz}{14z \cdot 35xy} \cdot \frac{49}{9} \\
&= \frac{15x \cdot 17yz \cdot 49}{14z \cdot 35xy \cdot 9} \\
&= \frac{3 \cdot 5 \cdot x \cdot 17 \cdot y \cdot z \cdot 7 \cdot 7}{2 \cdot 7 \cdot z \cdot 5 \cdot 7 \cdot x \cdot y \cdot 3 \cdot 3} \\
&= \frac{3 \cdot 5 \cdot 7 \cdot 7 \cdot x \cdot y \cdot z}{3 \cdot 5 \cdot 7 \cdot 7 \cdot x \cdot y \cdot z} \cdot \frac{17}{2 \cdot 3} \\
&= \frac{17}{6}
\end{aligned}$$

74. The digit must be even and the sum of the digits must be divisible by 3. Let d = the digit to be inserted. Then $5+7+4+d$, or $16+d$, must be divisible by 3. The only even digits for which $16+d$ is divisible by 3 are 2 and 8.

75. $\frac{19}{24} \div \frac{a}{b} = \frac{19}{24} \cdot \frac{b}{a} = \frac{19 \cdot b}{24 \cdot a} = \frac{187,853}{268,224}$

Then, assuming the quotient has not been simplified, we have

$$\begin{aligned}
19 \cdot b &= 187,853 \quad \text{and} \quad 24 \cdot a = 268,224 \\
b &= \frac{187,853}{19} \quad \text{and} \quad a = \frac{268,224}{24} \\
b &= 9887 \quad \text{and} \quad a = 11,176.
\end{aligned}$$

76. 13 and 31 are both prime numbers, so 13 is a palindrome prime.

19 is prime but 91 is not $(91 = 7 \cdot 13)$, so 19 is not a palindrome prime.

16 is not prime $(16 = 2 \cdot 8 = 4 \cdot 4)$, so it is not a palindrome prime.

11 is prime and when its digits are reversed we have 11 again, so 11 is a palindrome prime.

15 is not prime $(15 = 3 \cdot 5)$, so it is not a palindrome prime.

24 is not prime $(24 = 2 \cdot 12 = 3 \cdot 8 = 4 \cdot 6)$, so it is not a palindrome prime.

29 is prime but 92 is not $(92 = 2 \cdot 46 = 4 \cdot 23)$, so 29 is not a palindrome prime.

101 is prime and when its digits are reversed we get 101 again, so 101 is a palindrome prime.

201 is not prime $(201 = 3 \cdot 67)$, so it is not a palindrome prime.

37 and 73 are both prime numbers, so 37 is a palindrome prime.

Chapter 3 Discussion and Writing Exercises

1. The student is probably multiplying the divisor by the reciprocal of the dividend rather than multiplying the dividend by the reciprocal of the divisor.

2. $9432 = 9 \cdot 1000 + 4 \cdot 100 + 3 \cdot 10 + 2 \cdot 1 = 9(999+1) + 4(99+1) + 3(9+1) + 2 \cdot 1 = 9 \cdot 999 + 9 \cdot 1 + 4 \cdot 99 + 4 \cdot 1 + 3 \cdot 9 + 3 \cdot 1 + 2 \cdot 1$. Since 999, 99, and 9 are each a multiple of 9, $9 \cdot 999$, $4 \cdot 99$, and $3 \cdot 9$ are multiples of 9. This leaves $9 \cdot 1 + 4 \cdot 1 + 3 \cdot 1 + 2 \cdot 1$, or $9+4+3+2$. If $9+4+3+2$, the sum of the digits, is divisible by 9, then 9432 is divisible by 9.

3. Taking $\frac{1}{2}$ of a number is equivalent to multiplying the number by $\frac{1}{2}$. Dividing by $\frac{1}{2}$ is equivalent to multiplying by the reciprocal of $\frac{1}{2}$, or 2. Thus taking $\frac{1}{2}$ of a number is not the same as dividing by $\frac{1}{2}$.

4. We first consider some object and take $\frac{4}{7}$ of it. We divide it into 7 parts and take 4 of them as shown by the shading below.

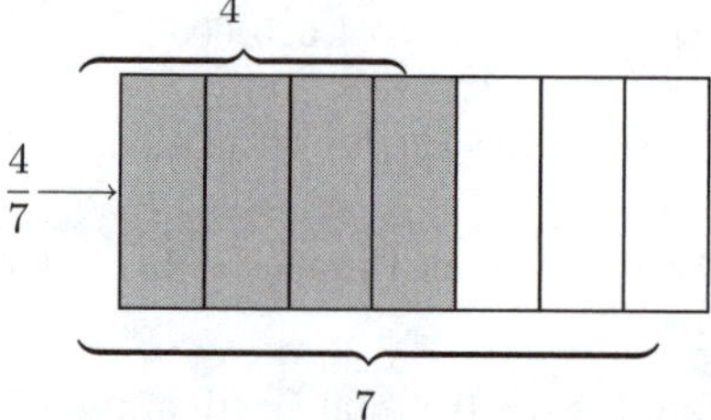

Next we take $\frac{2}{3}$ of the shaded area above. We divide it into 3 parts and take two of them as shown below

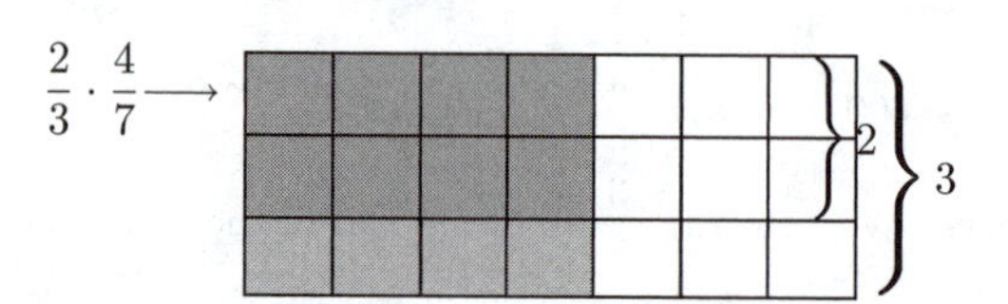

The entire object has been divided into 21 parts, 8 of which have been shaded. Thus, $\frac{2}{3} \cdot \frac{4}{7} = \frac{8}{21}$.

5. Since $\frac{1}{7}$ is a smaller number than $\frac{2}{3}$, there are more $\frac{1}{7}$'s in 5 than $\frac{2}{3}$'s. Thus, $5 \div \frac{1}{7}$ is a bigger number than $5 \div \frac{2}{3}$.

6. No; in order to simplify a fraction, we must be able to remove a factor of the type $\frac{n}{n}, n \neq 0$, where n is a factor that the numerator and denominator have in common.

Chapter 3 Test

1. Because $5+6+8+2=21$ and 21 is divisible by 3, then 5682 is divisible by 3.

2. 7018 is not divisible by 5 because the ones digit is neither 0 nor 5.

3. Since the ones digit of 90 is 0 we know that 2, 5, and 10 are factors. Since the sum of the digits is 9 and 9 is divisible by both 3 and 9, we know that 3 and 9 are factors. Since 2 and 3 are factors, 6 is also a factor. We write a list of factorizations.

$90 = 1 \cdot 90 \quad 90 = 5 \cdot 18$
$90 = 2 \cdot 45 \quad 90 = 6 \cdot 15$
$90 = 3 \cdot 30 \quad 90 = 9 \cdot 10$

Factors: 1, 2, 3, 5, 6, 9, 10, 15, 18, 30, 45, 90

4. The number 93 has factors 1, 3, 31, and 93. Since it has at least one factor other than itself and 1, it is composite.

5.

```
        3   ←  3 is prime.
    3 ⌈ 9
  2 ⌈ 1 8
  2 ⌈ 3 6
```

$36 = 2 \cdot 2 \cdot 3 \cdot 3$

6. We use a factor tree.

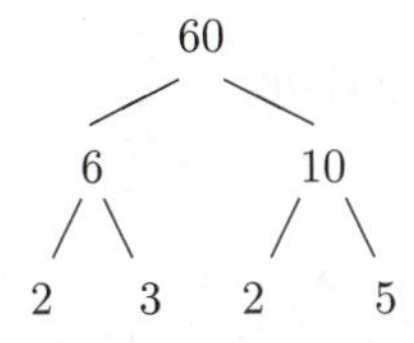

$60 = 2 \cdot 3 \cdot 2 \cdot 5$, or $2 \cdot 2 \cdot 3 \cdot 5$

7. $\frac{4}{9}$ ← Numerator, ← Denominator

8. The figure is divided into 4 equal parts, so the unit is $\frac{1}{4}$ and the denominator is 4. Three of the units are shaded, so the numerator is 3. Thus, $\frac{3}{4}$ is shaded.

9. There are 7 objects in the set, so the denominator is 7. Three of the objects are shaded, so the numerator is 3. Thus, $\frac{3}{7}$ of the set is shaded.

10. a) The ratio is $\frac{180}{47}$.

b) The ratio is $\frac{47}{93}$.

11. Remember: $\frac{n}{1} = n$.

$\frac{32}{1} = 32$

12. Remember: $\frac{n}{n} = 1$ for any integer n that is not 0.

$\frac{-12}{-12} = 1$

13. Remember: $\frac{0}{n} = 0$ for any integer n that is not 0.

$\frac{0}{16} = 0$

14. $\frac{-8}{24} = \frac{-1 \cdot 8}{3 \cdot 8} = \frac{-1}{3} \cdot \frac{8}{8} = \frac{-1}{3}$

15. $\frac{42}{7} = \frac{6 \cdot 7}{7 \cdot 1} = \frac{7}{7} \cdot \frac{6}{1} = 6$

16. $\frac{9x}{45x} = \frac{9 \cdot x \cdot 1}{5 \cdot 9 \cdot x} = \frac{9 \cdot x}{9 \cdot x} \cdot \frac{1}{5} = \frac{1}{5}$

17. Remember: $\frac{n}{0}$ is not defined.

$\frac{-62}{0}$ is undefined.

18. $\frac{72}{108} = \frac{2 \cdot 2 \cdot 2 \cdot 3 \cdot 3}{2 \cdot 2 \cdot 3 \cdot 3 \cdot 3} = \frac{2 \cdot 2 \cdot 3 \cdot 3}{2 \cdot 2 \cdot 3 \cdot 3} \cdot \frac{2}{3} = \frac{2}{3}$

19. We multiply these two numbers: We multiply these two numbers:

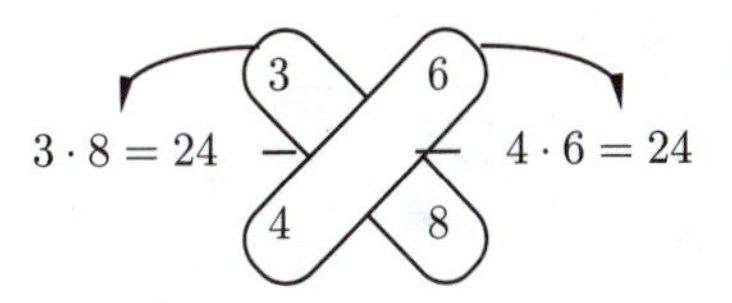

Since $24 = 24$, $\frac{3}{4} = \frac{6}{8}$.

20. We multiply these two numbers: We multiply these two numbers:

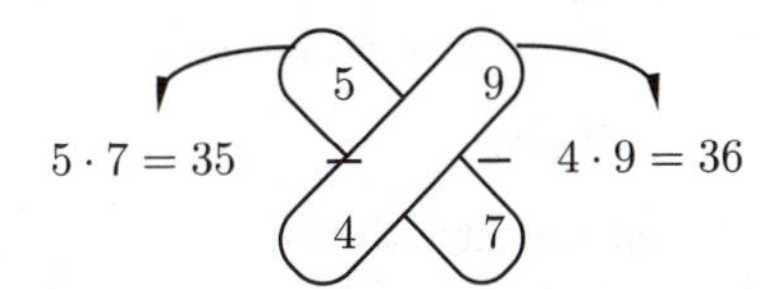

Since $35 \neq 36$, $\frac{5}{4}$ and $\frac{9}{7}$ do not name the same number. Thus, $\frac{5}{4} \neq \frac{9}{7}$.

21. Since $40 \div 8 = 5$, we multiply by $\frac{5}{5}$.

$\frac{3}{8} = \frac{3}{8} \cdot \frac{5}{5} = \frac{3 \cdot 5}{8 \cdot 5} = \frac{15}{40}$

22. Interchange the numerator and denominator.

The reciprocal of $\frac{a}{42}$ is $\frac{42}{a}$. $\left(\frac{a}{42} \cdot \frac{42}{a} = \frac{42a}{42a} = 1\right)$

23. Think of -9 as $\frac{-9}{1}$. Interchange the numerator and denominator.

The reciprocal of -9 is $\frac{1}{-9}$, or $-\frac{1}{9}$, or $\frac{-1}{9}$.

$\left[-9 \cdot \left(-\frac{1}{9}\right) = \frac{9 \cdot 1}{9} = 1\right]$

24. $\frac{2}{3}\cdot\frac{15}{4}=\frac{2\cdot 3\cdot 5}{3\cdot 2\cdot 2}=\frac{2\cdot 3}{2\cdot 3}\cdot\frac{5}{2}=\frac{5}{2}$

25. $\frac{2}{11}\div\frac{3}{4}=\frac{2}{11}\cdot\frac{4}{3}=\frac{2\cdot 4}{11\cdot 3}=\frac{8}{33}$

26. $3\cdot\frac{x}{8}=\frac{3\cdot x}{8}=\frac{3x}{8}$

27. $\frac{\frac{4}{7}}{-\frac{8}{3}}=\frac{4}{7}\cdot\left(-\frac{3}{8}\right)=-\frac{4\cdot 3}{7\cdot 8}=-\frac{4\cdot 3}{7\cdot 2\cdot 4}=$

$-\frac{3}{7\cdot 2}\cdot\frac{4}{4}=-\frac{3}{14}$

28. $12\div\frac{2}{3}=12\cdot\frac{3}{2}=\frac{12\cdot 3}{2}=\frac{2\cdot 6\cdot 3}{2\cdot 1}=\frac{2}{2}\cdot\frac{6\cdot 3}{1}=\frac{6\cdot 3}{1}=18$

29. $\frac{22c}{15}\cdot\frac{5}{33c}=\frac{2\cdot 11\cdot c\cdot 5}{3\cdot 5\cdot 3\cdot 11\cdot c}=\frac{5\cdot 11\cdot c}{5\cdot 11\cdot c}\cdot\frac{2}{3\cdot 3}=\frac{2}{9}$

30. ***Familiarize.*** Let $c=$ the number of pounds of cheese each person receives.

Translate. We write a division sentence.

$$c=\frac{3}{4}\div 5$$

Solve. We carry out the division.

$$c=\frac{3}{4}\div 5=\frac{3}{4}\cdot\frac{1}{5}=\frac{3\cdot 1}{4\cdot 5}=\frac{3}{20}$$

Check. Since $\frac{3}{20}\cdot 5=\frac{3\cdot 5}{20}=\frac{3\cdot 5}{4\cdot 5}=\frac{3}{4}$, the answer checks.

State. Each person receives $\frac{3}{20}$ lb of cheese.

31. ***Familiarize.*** Let $w=$ Monroe's weight, in pounds. We want to find $\frac{5}{7}$ of 175 lb.

Translate. We write a multiplication sentence.

$$w=\frac{5}{7}\cdot 175$$

Solve. We carry out the multiplication.

$$w=\frac{5}{7}\cdot 175=\frac{5\cdot 175}{7}=\frac{5\cdot 7\cdot 25}{7\cdot 1}$$
$$=\frac{7}{7}\cdot\frac{5\cdot 25}{1}$$
$$=125$$

Check. We can repeat the calculation. The answer checks.

State. Monroe weighs 125 lb.

32. $\frac{7}{8}\cdot x=56$

$x=56\div\frac{7}{8}$ Dividing by $\frac{7}{8}$ on both sides

$x=56\cdot\frac{8}{7}$

$=\frac{56\cdot 8}{7}=\frac{7\cdot 8\cdot 8}{7\cdot 1}=\frac{7}{7}\cdot\frac{8\cdot 8}{1}=\frac{8\cdot 8}{1}=64$

The solution is 64.

33. $\frac{7}{10}=\frac{-2}{5}\cdot t$

$\frac{5}{-2}\cdot\frac{7}{10}=\frac{5}{-2}\cdot\frac{-2}{5}\cdot t$

$\frac{5\cdot 7}{-2\cdot 10}=1t$

$\frac{5\cdot 7}{-2\cdot 2\cdot 5}=t$

$\frac{7}{-2\cdot 2}\cdot\frac{5}{5}=t$

$\frac{7}{-4}=t$

The solution is $\frac{7}{-4}$, or $-\frac{7}{4}$, or $\frac{-7}{4}$.

34. $A=\frac{1}{2}\cdot b\cdot h$

$A=\frac{1}{2}\cdot 13\text{ m}\cdot 7\text{ m}$

$A=\frac{13\cdot 7}{2}\text{ m}^2=\frac{91}{2}\text{ m}^2$

35. Only the figures in C and D are divided into 6 equal parts, so in each of these the unit is $\frac{1}{6}$. The denominator is 6. In C, 7 of the units are shaded, so the numerator is 7 and $\frac{7}{6}$ is shaded. In D, 5 of the units are shaded, so the numerator is 5 and $\frac{5}{6}$ is shaded. We see that the correct answer is C.

36. ***Familiarize.*** This is a multistep problem. First we will find the number of acres Karl received. Then we will find how much of that land Shannon received. Let $k=$ the number of acres of land Karl received.

Translate. We translate to an equation.

$$k=\frac{7}{8}\cdot\frac{2}{3}$$

Solve. We carry out the multiplication.

$$k=\frac{7}{8}\cdot\frac{2}{3}=\frac{7\cdot 2}{8\cdot 3}=\frac{7\cdot 2}{2\cdot 4\cdot 3}=\frac{2}{2}\cdot\frac{7}{4\cdot 3}=\frac{7}{12}$$

Karl received $\frac{7}{12}$ acre of land. Let $a=$ the number of acres Shannon received. An equation that corresponds to this situation is

$$a=\frac{1}{4}\cdot\frac{7}{12}.$$

We solve the equation by carrying out the multiplication.

$$a=\frac{1}{4}\cdot\frac{7}{12}=\frac{1\cdot 7}{4\cdot 12}=\frac{7}{48}$$

Check. We repeat the calculations. The answer checks.

State. Shannon received $\frac{7}{48}$ acre of land.

37. First we will evaluate the exponential expression; then we will multiply and divide in order from left to right.

$$\begin{aligned}
\left(-\frac{3}{8}\right)^2 \div \frac{6}{7} \cdot \frac{2}{9} \div (-5) &= \frac{9}{64} \div \frac{6}{7} \cdot \frac{2}{9} \div (-5) \\
&= \frac{9}{64} \cdot \frac{7}{6} \cdot \frac{2}{9} \div (-5) \\
&= \frac{9 \cdot 7}{64 \cdot 6} \cdot \frac{2}{9} \div (-5) \\
&= \frac{9 \cdot 7 \cdot 2}{64 \cdot 6 \cdot 9} \div (-5) \\
&= \frac{9 \cdot 7 \cdot 2}{64 \cdot 6 \cdot 9} \cdot \left(-\frac{1}{5}\right) \\
&= -\frac{9 \cdot 7 \cdot 2 \cdot 1}{64 \cdot 6 \cdot 9 \cdot 5} \\
&= -\frac{9 \cdot 7 \cdot 2 \cdot 1}{64 \cdot 2 \cdot 3 \cdot 9 \cdot 5} \\
&= -\frac{9 \cdot 2}{9 \cdot 2} \cdot \frac{7 \cdot 1}{64 \cdot 3 \cdot 5} \\
&= -\frac{7}{960}, \text{ or } \frac{-7}{960}
\end{aligned}$$

Chapter 4

Fraction Notation: Addition, Subtraction, and Mixed Numerals

Exercise Set 4.1

In this section we will find the LCM using the multiples method in Exercises 1 - 19 and the prime factorization method in Exercises 21 - 50.

1. 1. 4 is the larger number and is a multiple of 2, so it is the LCM.

The LCM = 4.

3. 1. 25 is the larger number, but it is not a multiple of 10.

2. Check multiples of 25:

$2 \cdot 25 = 50$ A multiple of 10

The LCM = 50.

5. 1. 40 is the larger number and is a multiple of 20, so it is the LCM.

The LCM = 40.

7. 1. 27 is the larger number, but it is not a multiple of 18.

2. Check multiples of 27:

$2 \cdot 27 = 54$ A multiple of 18

The LCM = 54.

9. 1. 50 is the larger number, but it is not a multiple of 30.

2. Check multiples of 50:

$2 \cdot 50 = 100$ Not a multiple of 30
$3 \cdot 50 = 150$ A multiple of 30

The LCM = 150.

11. 1. 40 is the larger number, but it is not a multiple of 30.

2. Check multiples of 40:

$2 \cdot 40 = 80$ Not a multiple of 30
$3 \cdot 40 = 120$ A multiple of 30

The LCM = 120.

13. 1. 24 is the larger number, but it is not a multiple of 18.

2. Check multiples of 24:

$2 \cdot 24 = 48$ Not a multiple of 18
$3 \cdot 24 = 72$ A multiple of 18

The LCM = 72.

15. 1. 70 is the larger number, but it is not a multiple of 60.

2. Check multiples of 70:

$2 \cdot 70 = 140$ Not a multiple of 60
$3 \cdot 70 = 210$ Not a multiple of 60
$4 \cdot 70 = 280$ Not a multiple of 60
$5 \cdot 70 = 350$ Not a multiple of 60
$6 \cdot 70 = 420$ A multiple of 60

The LCM = 420.

17. 1. 36 is the larger number, but it is not a multiple of 16.

2. Check multiples of 36:

$2 \cdot 36 = 72$ Not a multiple of 16
$3 \cdot 36 = 108$ Not a multiple of 16
$4 \cdot 36 = 144$ A multiple of 16

The LCM = 144.

19. 1. 20 is the larger number, but it is not a multiple of 18.

2. Check multiples of 20:

$2 \cdot 20 = 40$ Not a multiple of 18
$3 \cdot 20 = 60$ Not a multiple of 18
$4 \cdot 20 = 80$ Not a multiple of 18
$5 \cdot 20 = 100$ Not a multiple of 18
$6 \cdot 20 = 120$ Not a multiple of 18
$7 \cdot 20 = 140$ Not a multiple of 18
$8 \cdot 20 = 160$ Not a multiple of 18
$9 \cdot 20 = 180$ A multiple of 18

The LCM = 180.

21. 1. Write the prime factorization of each number. Because 2, 3, and 7 are all prime we write $2 = 2$, $3 = 3$, and $7 = 7$.

2. a) None of the factorizations contains the other two.

b) We begin with 2. Since 3 contains a factor of 3, we multiply by 3:

$2 \cdot 3$

Next we multiply $2 \cdot 3$ by 7, the factor of 7 that is missing:

$2 \cdot 3 \cdot 7$

The LCM is $2 \cdot 3 \cdot 7$, or 42.

3. To check, note that 2, 3, and 7 appear in the LCM the greatest number of times that each appears as a factor of 2, 3, or 7. The LCM is $2 \cdot 3 \cdot 7$, or 42.

23. 1. Write the prime factorization of each number.

$3 = 3$
$6 = 2 \cdot 3$
$15 = 3 \cdot 5$

2. a) None of the factorizations contains the other two.

b) We first consider 3 and 6. Since the factorization of 6 contains 3, we next multiply $2 \cdot 3$ by the factor of 15 that is missing, 5. The LCM is $2 \cdot 3 \cdot 5$, or 30.

3. To check, note that 2, 3, and 5 appear in the LCM the greatest number of times that each appears as a factor of 3, 6, or 15. The LCM is $2 \cdot 3 \cdot 5$, or 30.

25. 1. Write the prime factorization of each number.

$24 = 2 \cdot 2 \cdot 2 \cdot 3$
$36 = 2 \cdot 2 \cdot 3 \cdot 3$
$12 = 2 \cdot 2 \cdot 3$

2. a) None of the factorizations contains the other two.

b) We begin with the factorization of 24, $2 \cdot 2 \cdot 2 \cdot 3$. Since 36 contains a second factor of 3, we multiply by another factor of 3:

$$2 \cdot 2 \cdot 2 \cdot 3 \cdot 3$$

Next we look for factors of 12 that are still missing. There are none. The LCM is

$2 \cdot 2 \cdot 2 \cdot 3 \cdot 3$, or 72.

3. To check, note that 2 and 3 appear in the LCM the greatest number of times that each appears as a factor of 24, 36, or 12. The LCM is $2 \cdot 2 \cdot 2 \cdot 3 \cdot 3$, or 72.

27. 1. Write the prime factorization of each number.

$5 = 5$
$12 = 2 \cdot 2 \cdot 3$
$15 = 3 \cdot 5$

2. a) None of the factorizations contains the other two.

b) We begin with the factorization of 12, $2 \cdot 2 \cdot 3$. Since 5 contains a factor of 5, we multiply by 5:

$$2 \cdot 2 \cdot 3 \cdot 5$$

Next we look for factors of 15 that are still missing. There are none. The LCM is $2 \cdot 2 \cdot 3 \cdot 5$, or 60.

3. The result checks.

29. 1. Write the prime factorization of each number.

$9 = 3 \cdot 3$
$12 = 2 \cdot 2 \cdot 3$
$6 = 2 \cdot 3$

2. a) None of the factorizations contains the other two.

b) We begin with the factorization of 12, $2 \cdot 2 \cdot 3$. Since 9 contains a second factor of 3, we multiply by another factor of 3:

$$2 \cdot 2 \cdot 3 \cdot 3$$

Next we look for factors of 6 that are still missing. There are none. The LCM is $2 \cdot 2 \cdot 3 \cdot 3$, or 36.

3. The result checks.

31. 1. Write the prime factorization of each number.

$180 = 2 \cdot 2 \cdot 3 \cdot 3 \cdot 5$
$100 = 2 \cdot 2 \cdot 5 \cdot 5$
$450 = 2 \cdot 3 \cdot 3 \cdot 5 \cdot 5$

2. a) None of the factorizations contains the other two.

b) We begin with the factorization of 450, $2 \cdot 3 \cdot 3 \cdot 5 \cdot 5$. Since 180 contains another factor of 2, we multiply by 2:

$$2 \cdot 3 \cdot 3 \cdot 5 \cdot 5 \cdot 2$$

Next we look for factors of 100 that are still missing. There are none. The LCM is $2 \cdot 3 \cdot 3 \cdot 5 \cdot 5 \cdot 2$, or 900.

3. The result checks.

33. 1. Write the prime factorization of each number.

$8 = 2 \cdot 2 \cdot 2$
$48 = 2 \cdot 2 \cdot 2 \cdot 2 \cdot 3$

2. a) The factorization of 48 contains the factorization of 8. Thus, the LCM is $2 \cdot 2 \cdot 2 \cdot 2 \cdot 3$, or 48.

35. 1. Write the prime factorization of each number.

$10 = 2 \cdot 5$
$21 = 3 \cdot 7$

2. a) Neither factorization contains the other.

b) We begin with the factorization of 21, $3 \cdot 7$. Since 10 contains factors of 2 and 5, we multiply by 2 and 5:

$$3 \cdot 7 \cdot 2 \cdot 5$$

The LCM is $3 \cdot 7 \cdot 2 \cdot 5$, or 210.

3. The result checks.

37. 1. Write the prime factorization of each number.

$75 = 3 \cdot 5 \cdot 5$
$100 = 2 \cdot 2 \cdot 5 \cdot 5$

2. a) Neither factorization contains the other.

b) We begin with the factorization of 100, $2 \cdot 2 \cdot 5 \cdot 5$. Since 75 contains a factor of 3, we multiply by 3:

$$2 \cdot 2 \cdot 5 \cdot 5 \cdot 3$$

The LCM is $2 \cdot 2 \cdot 5 \cdot 5 \cdot 3$, or 300.

3. The result checks.

39. 1. Write the prime factorization of each number.

$12 = 2 \cdot 2 \cdot 3$
$15 = 3 \cdot 5$
$60 = 2 \cdot 2 \cdot 3 \cdot 5$

2. a) The factorization of 60 contains the factorization of 12 and the factorization of 15. Thus, the LCM is $2 \cdot 2 \cdot 3 \cdot 5$, or 60.

41. 1. We have the following factorizations:

$ab = a \cdot b$
$bc = b \cdot c$

2. a) Neither factorization contains the other.

b) Consider the factorization of ab, $a \cdot b$. Since bc contains a factor of c, we multiply by c.

$$a \cdot b \cdot c$$

The LCM is $a \cdot b \cdot c$, or abc.

3. The result checks.

43. 1. We have the following factorizations:

$3x = 3 \cdot x$
$9x^2 = 3 \cdot 3 \cdot x \cdot x$

2. a) One factorization, $3 \cdot 3 \cdot x \cdot x$, contains the other. Thus the LCM is $3 \cdot 3 \cdot x \cdot x$, or $9x^2$.

45. 1. We have the following factorizations:

$4x^3 = 2 \cdot 2 \cdot x \cdot x \cdot x$
$x^2y = x \cdot x \cdot y$

2. a) Neither factorization contains the other.

b) Consider the factorization of $4x^3$, $2 \cdot 2 \cdot x \cdot x \cdot x$. Since x^2y contains a factor of y, we multiply by y.

$2 \cdot 2 \cdot x \cdot x \cdot x \cdot y$

The LCM is $2 \cdot 2 \cdot x \cdot x \cdot x \cdot y$, or $4x^3y$.

3. The result checks.

47. 1. We have the following factorizations:

$6r^3st^4 = 2 \cdot 3 \cdot r \cdot r \cdot r \cdot s \cdot t \cdot t \cdot t \cdot t$
$8rs^2t = 2 \cdot 2 \cdot 2 \cdot r \cdot s \cdot s \cdot t$

2. a) Neither factorization contains the other.

b) Consider the factorization of $6r^3st^4$, $2 \cdot 3 \cdot r \cdot r \cdot r \cdot s \cdot t \cdot t \cdot t \cdot t$. Since $8rs^2t$ contains two more factors of 2 and one more factor of s, we multiply by $2 \cdot 2 \cdot s$.

$2 \cdot 3 \cdot r \cdot r \cdot r \cdot s \cdot t \cdot t \cdot t \cdot t \cdot 2 \cdot 2 \cdot s$

The LCM is $2 \cdot 3 \cdot 2 \cdot 2 \cdot r \cdot r \cdot r \cdot s \cdot s \cdot t \cdot t \cdot t \cdot t$, or $24r^3s^2t^4$.

3. The result checks.

49. 1. We have the following factorizations:

$a^3b = a \cdot a \cdot a \cdot b$
$b^2c = b \cdot b \cdot c$
$ac^2 = a \cdot c \cdot c$

2. a) No one factorization contains the others.

b) Consider the factorization of a^3b, $a \cdot a \cdot a \cdot b$. Since b^2c contains another factor of b and a factor of c, we multiply by $b \cdot c$.

$a \cdot a \cdot a \cdot b \cdot b \cdot c$

Now consider ac^2. Since ac^2 contains another factor of c, we multiply by c.

$a \cdot a \cdot a \cdot b \cdot b \cdot c \cdot c$

The LCM is $a \cdot a \cdot a \cdot b \cdot b \cdot c \cdot c$, or $a^3b^2c^2$.

3. The result checks.

51. We find the LCM of the number of years it takes Jupiter and Saturn to make a complete revolution around the sun.

Jupiter: $12 = 2 \cdot 2 \cdot 3$

Saturn: $30 = 2 \cdot 3 \cdot 5$

The LCM $= 2 \cdot 2 \cdot 3 \cdot 5$, or 60. Thus, Jupiter and Saturn will appear in the exact same direction in the night sky as seen from Earth once every 60 years.

53. We find the LCM of the number of years it takes Saturn and Uranus to make a complete revolution around the sun.

Saturn: $30 = 2 \cdot 3 \cdot 5$

Uranus: $84 = 2 \cdot 2 \cdot 3 \cdot 7$

The LCM is $2 \cdot 2 \cdot 3 \cdot 5 \cdot 7$, or 420. Thus, Saturn and Uranus will appear in the same direction in the night sky once every 420 years.

55. ***Familiarize***. Let t = the number of tornadoes occurring in 2008.

Translate.

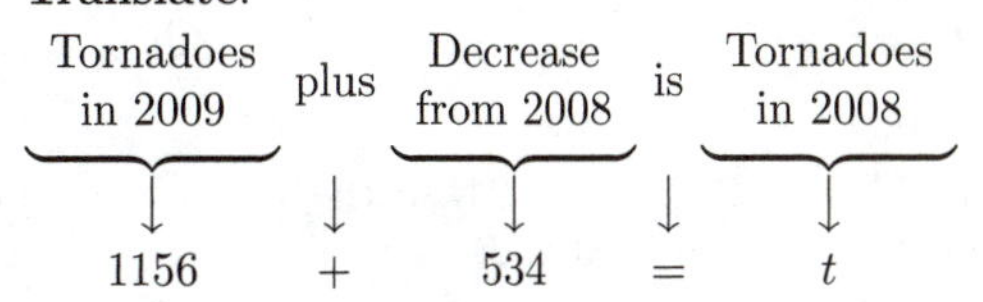

Solve. We carry out the addition.

$1156 + 534 = t$

$1690 = t$

Check. A decrease of 534 from 1690 is $1690 - 534$, or 1156. The answer checks.

State. There were 1690 tornadoes in 2008.

57. $-38 + 52$

The absolute values are 38 and 52. The difference is 14. The positive number has the larger absolute value, so the answer is positive.

$-38 + 52 = 14$

59.

$$\begin{array}{r} {\scriptstyle 1} \\ {\scriptstyle 1\,1} \\ 3\,4\,5 \\ \times\quad 2\,3 \\ \hline 1\,0\,3\,5 \\ 6\,9\,0\,0 \\ \hline 7\,9\,3\,5 \end{array}$$

61. $\frac{4}{5} \div \left(-\frac{7}{10}\right) = \frac{4}{5} \cdot \left(-\frac{10}{7}\right) = -\frac{4 \cdot 10}{5 \cdot 7} = -\frac{4 \cdot 2 \cdot 5}{5 \cdot 7} = -\frac{4 \cdot 2}{7} \cdot \frac{5}{5} = -\frac{8}{7}$

63. 1. 324 is not a multiple of 288.

2. Check multiples using a calculator.

$2 \cdot 324 = 648$	Not a multiple of 288
$3 \cdot 324 = 972$	Not a multiple of 288
$4 \cdot 324 = 1296$	Not a multiple of 288
$5 \cdot 324 = 1620$	Not a multiple of 288
$6 \cdot 324 = 1944$	Not a multiple of 288
$7 \cdot 324 = 2268$	Not a multiple of 288
$8 \cdot 324 = 2592$	A multiple of 288

The LCM = 2592.

65. 1. 18,011 is not a multiple of 7719.

2. Check multiples using a calculator.

$2 \cdot 18{,}011 = 36{,}022$	Not a multiple of 7719
$3 \cdot 18{,}011 = 54{,}033$	A multiple of 7719

The LCM = 54,033.

67. The desired length is the LCM of 6 ft and 8 ft.

$6 = 2 \cdot 3$

$8 = 2 \cdot 2 \cdot 2$

The LCM is $2 \cdot 2 \cdot 2 \cdot 3$, or 24.

The shortest aisle that can accommodate tables of either length is 24 ft.

69. The smallest number of strands that can be used is the LCM of 4, 6, and 8.

$4 = 2 \cdot 2$

$6 = 2 \cdot 3$

$8 = 2 \cdot 2 \cdot 2$

The LCM $= 2 \cdot 2 \cdot 2 \cdot 3$, or 24, so the smallest number of strands that can be used is 24.

71. a) This is not the LCM, because a^2b^5 is not a factor of a^3b^3.

b) This is not the LCM, because a^3b^2 is not a factor of a^2b^5.

c) This is the LCM, because both a^3b^2 and a^2b^5 are factors of a^3b^5 and this is the smallest such expression.

73. 1. From Example 6 we know that the LCM of 18 and 21 is $2 \cdot 3 \cdot 3 \cdot 7$. From Example 8 we know that the LCM of 24 and 36 is $2 \cdot 2 \cdot 2 \cdot 3 \cdot 3$. Also $63 = 3 \cdot 3 \cdot 7$, $56 = 2 \cdot 2 \cdot 2 \cdot 7$, and $20 = 2 \cdot 2 \cdot 5$.

2. a) None of the factorizations contains all of the others. Begin with the LCM of 18 and 21. We multiply by two factors of 2, the prime factors of the LCM of 24 and 36 that are missing. We have $2 \cdot 3 \cdot 3 \cdot 7 \cdot 2 \cdot 2$. There are no prime factors of either 63 or 56 that are missing in this factorization. We multiply by 5, the prime factor of 20 that is missing. The LCM is $2 \cdot 3 \cdot 3 \cdot 7 \cdot 2 \cdot 2 \cdot 5$, or 2520.

3. The result checks.

75. Answers may vary. Since $54 = 2 \cdot 3 \cdot 3 \cdot 3$, we use appropriate combinations of these factors to find the desired numbers. One pair is 2 and $3 \cdot 3 \cdot 3$, or 2 and 27. Another is $2 \cdot 3$ and $3 \cdot 3 \cdot 3$, or 6 and 27. A third pair is $2 \cdot 3 \cdot 3$ and $3 \cdot 3 \cdot 3$, or 18 and 27.

Exercise Set 4.2

1. $\frac{4}{9} + \frac{1}{9} = \frac{4+1}{9} = \frac{5}{9}$

3. $\frac{4}{7} + \frac{3}{7} = \frac{4+3}{7} = \frac{7}{7} = 1$

5. $\frac{7}{10} + \frac{3}{-10} = \frac{7}{10} + \frac{-3}{10} = \frac{7+(-3)}{10} = \frac{4}{10} = \frac{2 \cdot 2}{2 \cdot 5} = \frac{2}{2} \cdot \frac{2}{5} = 1 \cdot \frac{2}{5} = \frac{2}{5}$

7. $\frac{9}{a} + \frac{4}{a} = \frac{9+4}{a} = \frac{13}{a}$

9. $\frac{-1}{4} + \frac{-1}{4} = \frac{-1+(-1)}{4} = \frac{-2}{4} = \frac{-1 \cdot 2}{2 \cdot 2} = \frac{-1}{2} \cdot \frac{2}{2} = \frac{-1}{2}$, or $-\frac{1}{2}$

11. $\frac{2}{9}x + \frac{5}{9}x = \left(\frac{2}{9} + \frac{5}{9}\right)x = \frac{7}{9}x$

13.
$$\frac{3}{32}t + \frac{13}{32}t = \left(\frac{3}{32} + \frac{13}{32}\right)t = \frac{16}{32}t = \frac{16 \cdot 1}{16 \cdot 2}t = \frac{16}{16} \cdot \frac{1}{2}t = \frac{1}{2}t$$

15. $-\frac{2}{x} + \left(-\frac{7}{x}\right) = \frac{-2}{x} + \frac{-7}{x} = \frac{-2+(-7)}{x} = \frac{-9}{x}$, or $-\frac{9}{x}$

17. $\frac{1}{8} + \frac{1}{6}$ — $8 = 2 \cdot 2 \cdot 2$ and $6 = 2 \cdot 3$, so the LCD is $2 \cdot 2 \cdot 2 \cdot 3$, or 24

$= \frac{1}{8} \cdot \frac{3}{3} + \frac{1}{6} \cdot \frac{4}{4}$

Think: $6 \times \square = 24$. The answer is 4, so we multiply by 1, using $\frac{4}{4}$.

Think: $8 \times \square = 24$. The answer is 3, so we multiply by 1, using $\frac{3}{3}$.

$= \frac{3}{24} + \frac{4}{24}$

$= \frac{7}{24}$

19. $\frac{-4}{5} + \frac{7}{10}$ — 5 is a factor of 10, so the LCD is 10.

$= \frac{-4}{5} \cdot \frac{2}{2} + \frac{7}{10}$ ← This fraction already has the LCD as denominator.

Think: $5 \times \square = 10$. The answer is 2, so we multiply by 1, using $\frac{2}{2}$.

$= \frac{-8}{10} + \frac{7}{10}$

$= \frac{-1}{10}$, or $-\frac{1}{10}$

21. $\frac{7}{12} + \frac{3}{8}$ — $12 = 2 \cdot 2 \cdot 3$ and $8 = 2 \cdot 2 \cdot 2$, so the LCD is $2 \cdot 2 \cdot 2 \cdot 3$, or 24.

$= \frac{7}{12} \cdot \frac{2}{2} + \frac{3}{8} \cdot \frac{3}{3}$

Think: $8 \times \square = 24$. The answer is 3, so we multiply by 1, using $\frac{3}{3}$.

Think: $12 \times \square = 24$. The answer is 2, so we multiply by 1, using $\frac{2}{2}$.

$= \frac{14}{24} + \frac{9}{24} = \frac{23}{24}$

23. $\dfrac{3}{20}+4$

$=\dfrac{3}{20}+\dfrac{4}{1}$ Rewriting 4 in fractional notation

$=\dfrac{3}{20}+\dfrac{4}{1}\cdot\dfrac{20}{20}$ The LCD is 20.

$=\dfrac{3}{20}+\dfrac{80}{20}$

$=\dfrac{83}{20}$

25. $\dfrac{5}{-8}+\dfrac{5}{6}$

$=\dfrac{-5}{8}+\dfrac{5}{6}$ Recall that $\dfrac{m}{-n}=\dfrac{-m}{n}$. The LCD is 24. (See Exercise 17.)

$=\dfrac{-5}{8}\cdot\dfrac{3}{3}+\dfrac{5}{6}\cdot\dfrac{4}{4}$

$=\dfrac{-15}{24}+\dfrac{20}{24}$

$=\dfrac{5}{24}$

27. $\dfrac{3}{10}x+\dfrac{7}{100}x$

$=\dfrac{3}{10}\cdot\dfrac{10}{10}\cdot x+\dfrac{7}{100}x$ 10 is a factor of 100, so the LCD is 100.

$=\dfrac{30}{100}x+\dfrac{7}{100}x$

$=\dfrac{37}{100}x$

29. $\dfrac{5}{12}+\dfrac{8}{15}$ $12=2\cdot 2\cdot 3$ and $15=3\cdot 5$, so the LCM is $2\cdot 2\cdot 3\cdot 5$, or 60.

$=\dfrac{5}{12}\cdot\dfrac{5}{5}+\dfrac{8}{15}\cdot\dfrac{4}{4}$

$=\dfrac{25}{60}+\dfrac{32}{60}=\dfrac{57}{60}$

$=\dfrac{3\cdot 19}{3\cdot 20}=\dfrac{3}{3}\cdot\dfrac{19}{20}$

$=\dfrac{19}{20}$

31. $\dfrac{7}{8}+\dfrac{0}{1}$ 1 is a factor of 8, so the LCD is 8.

$=\dfrac{7}{8}+\dfrac{0}{1}\cdot\dfrac{8}{8}$

$=\dfrac{7}{8}+\dfrac{0}{8}=\dfrac{7}{8}$

Note that if we had observed at the outset that $\dfrac{0}{1}=0$, the computation becomes $\dfrac{7}{8}+0=\dfrac{7}{8}$.

33. $\dfrac{-7}{10}+\dfrac{-29}{100}$ 10 is a factor of 100, so the LCD is 100.

$=\dfrac{-7}{10}\cdot\dfrac{10}{10}+\dfrac{-29}{100}$

$=\dfrac{-70}{100}+\dfrac{-29}{100}=\dfrac{-99}{100}$, or $-\dfrac{99}{100}$

35. $-\dfrac{1}{10}x+\dfrac{1}{15}x$

$=-\dfrac{1}{2\cdot 5}x+\dfrac{1}{3\cdot 5}x$ The LCD is $2\cdot 5\cdot 3$.

$=-\dfrac{1}{2\cdot 5}\cdot\dfrac{3}{3}x+\dfrac{1}{3\cdot 5}\cdot\dfrac{2}{2}x$

$=-\dfrac{3}{30}x+\dfrac{2}{30}x$

$=-\dfrac{1}{30}x$

37. $-5t+\dfrac{2}{7}t$

$=\dfrac{-5}{1}t+\dfrac{2}{7}t$ The LCD is 7.

$=\dfrac{-5}{1}\cdot\dfrac{7}{7}\cdot t+\dfrac{2}{7}t$

$=\dfrac{-35}{7}t+\dfrac{2}{7}t$

$=\dfrac{-33}{7}t$, or $-\dfrac{33}{7}t$

39. $-\dfrac{5}{12}+\dfrac{7}{-24}$

$\dfrac{-5}{12}+\dfrac{-7}{24}$ 12 is a factor of 24, so the LCD is 24.

$=\dfrac{-5}{12}\cdot\dfrac{2}{2}+\dfrac{-7}{24}$

$=\dfrac{-10}{24}+\dfrac{-7}{24}$

$=\dfrac{-17}{24}$, or $-\dfrac{17}{24}$

41. $\dfrac{3}{16}+\dfrac{5}{16}+\dfrac{4}{16}=\dfrac{3+5+4}{16}=\dfrac{12}{16}=\dfrac{3}{4}$

43. $\dfrac{4}{10}+\dfrac{3}{100}+\dfrac{7}{1000}$ 10 and 100 are factors of 1000, so the LCD is 1000.

$=\dfrac{4}{10}\cdot\dfrac{100}{100}+\dfrac{3}{100}\cdot\dfrac{10}{10}+\dfrac{7}{1000}$

$=\dfrac{400}{1000}+\dfrac{30}{1000}+\dfrac{7}{1000}$

$=\dfrac{437}{1000}$

45. $\frac{3}{10}+\frac{5}{12}+\frac{8}{15}$

$=\frac{3}{2\cdot 5}+\frac{5}{2\cdot 2\cdot 3}+\frac{8}{3\cdot 5}$ Factoring the denominators

The LCD is $2\cdot 5\cdot 2\cdot 3$.

$=\frac{3}{2\cdot 5}\cdot\frac{2\cdot 3}{2\cdot 3}+\frac{5}{2\cdot 2\cdot 3}\cdot\frac{5}{5}+\frac{8}{3\cdot 5}\cdot\frac{2\cdot 2}{2\cdot 2}$

In each case we multiply by 1 to obtain the LCD.

$=\frac{3\cdot 2\cdot 3}{2\cdot 5\cdot 2\cdot 3}+\frac{5\cdot 5}{2\cdot 2\cdot 3\cdot 5}+\frac{8\cdot 2\cdot 2}{3\cdot 5\cdot 2\cdot 2}$

$=\frac{18}{2\cdot 5\cdot 2\cdot 3}+\frac{25}{2\cdot 5\cdot 2\cdot 3}+\frac{32}{2\cdot 5\cdot 2\cdot 3}$

$=\frac{75}{2\cdot 5\cdot 2\cdot 3}$

$=\frac{3\cdot 5\cdot 5}{2\cdot 5\cdot 2\cdot 3}=\frac{3\cdot 5}{3\cdot 5}\cdot\frac{5}{2\cdot 2}$

$=\frac{5}{4}$

47. $\frac{5}{6}+\frac{25}{52}+\frac{7}{4}$

$=\frac{5}{2\cdot 3}+\frac{25}{2\cdot 2\cdot 13}+\frac{7}{2\cdot 2}$ LCD is $2\cdot 3\cdot 2\cdot 13$.

$=\frac{5}{2\cdot 3}\cdot\frac{2\cdot 13}{2\cdot 13}+\frac{25}{2\cdot 2\cdot 13}\cdot\frac{3}{3}+\frac{7}{2\cdot 2}\cdot\frac{3\cdot 13}{3\cdot 13}$

$=\frac{5\cdot 2\cdot 13}{2\cdot 3\cdot 2\cdot 13}+\frac{25\cdot 3}{2\cdot 2\cdot 13\cdot 3}+\frac{7\cdot 3\cdot 13}{2\cdot 2\cdot 3\cdot 13}$

$=\frac{130}{2\cdot 3\cdot 2\cdot 13}+\frac{75}{2\cdot 3\cdot 2\cdot 13}+\frac{273}{2\cdot 3\cdot 2\cdot 13}$

$=\frac{478}{2\cdot 3\cdot 2\cdot 13}$

$=\frac{2\cdot 239}{2\cdot 3\cdot 2\cdot 13}=\frac{2}{2}\cdot\frac{239}{3\cdot 2\cdot 13}$

$=\frac{239}{78}$

49. $\frac{2}{9}+\frac{7}{10}+\frac{-4}{15}$

$=\frac{2}{3\cdot 3}+\frac{7}{2\cdot 5}+\frac{-4}{3\cdot 5}$ LCD is $3\cdot 3\cdot 2\cdot 5$.

$=\frac{2}{3\cdot 3}\cdot\frac{2\cdot 5}{2\cdot 5}+\frac{7}{2\cdot 5}\cdot\frac{3\cdot 3}{3\cdot 3}+\frac{-4}{3\cdot 5}\cdot\frac{3\cdot 2}{3\cdot 2}$

$=\frac{2\cdot 2\cdot 5}{3\cdot 3\cdot 2\cdot 5}+\frac{7\cdot 3\cdot 3}{2\cdot 5\cdot 3\cdot 3}+\frac{-4\cdot 3\cdot 2}{3\cdot 5\cdot 3\cdot 2}$

$=\frac{20}{3\cdot 3\cdot 2\cdot 5}+\frac{63}{3\cdot 3\cdot 2\cdot 5}+\frac{-24}{3\cdot 3\cdot 2\cdot 5}$

$=\frac{59}{3\cdot 3\cdot 2\cdot 5}$

$=\frac{59}{90}$

51. $-\frac{3}{4}+\frac{1}{5}+\frac{-7}{10}$

$=\frac{-3}{4}+\frac{1}{5}+\frac{-7}{10}$

$=\frac{-3}{2\cdot 2}+\frac{1}{5}+\frac{-7}{2\cdot 5}$ The LCD is $2\cdot 2\cdot 5$.

$=\frac{-3}{2\cdot 2}\cdot\frac{5}{5}+\frac{1}{5}\cdot\frac{2\cdot 2}{2\cdot 2}+\frac{-7}{2\cdot 5}\cdot\frac{2}{2}$

$=\frac{-15}{2\cdot 2\cdot 5}+\frac{4}{5\cdot 2\cdot 2}+\frac{-14}{2\cdot 5\cdot 2}$

$=\frac{-25}{2\cdot 2\cdot 5}=\frac{-5\cdot 5}{2\cdot 2\cdot 5}=\frac{-5}{2\cdot 2}\cdot\frac{5}{5}$

$=\frac{-5}{4}$, or $-\frac{5}{4}$

53. Since there is a common denominator, compare the numerators.

$3>2$, so $\frac{3}{8}>\frac{2}{8}$.

55. The LCD is 6. We multiply $\frac{2}{3}$ by 1 to make the denominators the same.

$\frac{2}{3}\cdot\frac{2}{2}=\frac{4}{6}$

The denominator of $\frac{5}{6}$ is the LCD.

Since $4<5$, it follows that $\frac{4}{6}<\frac{5}{6}$, so $\frac{2}{3}<\frac{5}{6}$.

57. Since there is a common denominator, compare the numerators.

$-2>-5$, so $\frac{-2}{7}>\frac{-5}{7}$.

59. The LCD is 30. We multiply by 1 to make the denominators the same.

$\frac{9}{15}\cdot\frac{2}{2}=\frac{18}{30}$

$\frac{7}{10}\cdot\frac{3}{3}=\frac{21}{30}$

Since $18<21$, it follows that $\frac{18}{30}<\frac{21}{30}$, so $\frac{9}{15}<\frac{7}{10}$.

61. Express $-\frac{1}{5}$ as $\frac{-1}{5}$. The LCD is 20. Multiply by 1 to make the denominators the same.

$\frac{3}{4}\cdot\frac{5}{5}=\frac{15}{20}$

$\frac{-1}{5}\cdot\frac{4}{4}=\frac{-4}{20}$

Since $15>-4$, it follows that $\frac{15}{20}>\frac{-4}{20}$, so $\frac{3}{4}>-\frac{1}{5}$. We might have observed at the outset that one number is positive and the other is negative and it follows that the positive number is greater than the negative number.

63. The LCD is 60. We multiply by 1 to make the denominators the same.

$$\frac{-7}{20}\cdot\frac{3}{3}=\frac{-21}{60}$$
$$\frac{-6}{15}\cdot\frac{4}{4}=\frac{-24}{60}$$

Since $-21>-24$, it follows that $\frac{-21}{60}>\frac{-24}{60}$, so $\frac{-7}{20}>\frac{-6}{15}$.

65. The LCD is 60. We multiply by 1 to make the denominators the same.

$$\frac{3}{10}\cdot\frac{6}{6}=\frac{18}{60}$$
$$\frac{5}{12}\cdot\frac{5}{5}=\frac{25}{60}$$
$$\frac{4}{15}\cdot\frac{4}{4}=\frac{16}{60}$$

Since $16<18$ and $18<25$, when we arrange $\frac{18}{60}$, $\frac{25}{60}$, and $\frac{16}{60}$ from smallest to largest we have $\frac{16}{60}$, $\frac{18}{60}$, $\frac{25}{60}$. Then it follows that when we arrange the original fractions from smallest to largest we have $\frac{4}{15}$, $\frac{3}{10}$, $\frac{5}{12}$.

67. ***Familiarize.*** Let d = the total distance Tate rode his Segway. This is the sum of the three distances he traveled.

Translate.

Distance to library	plus	Distance to class	plus	Distance to work	is	Total distance
↓	↓	↓	↓	↓	↓	↓
$\frac{5}{6}$	$+$	$\frac{3}{4}$	$+$	$\frac{3}{2}$	$=$	d

Solve. We carry out the addition. The LCM of the denominators is 12.

$$\frac{5}{6}\cdot\frac{2}{2}+\frac{3}{4}\cdot\frac{3}{3}+\frac{3}{2}\cdot\frac{6}{6}=d$$
$$\frac{10}{12}+\frac{9}{12}+\frac{18}{12}=d$$
$$\frac{37}{12}=d$$

Check. We can repeat the calculation. Also note that the sum is larger than any of the individual distances, so the answer seems reasonable.

State. Tate rode his Segway $\frac{37}{12}$ mi.

69. ***Familiarize.*** Let m = the number of miles of the highway that the students cleaned. This is the sum of the three individual distances cleaned.

Translate.

First day distance	plus	Second day distance	plus	Third day distance	is	Total distance
↓	↓	↓	↓	↓	↓	↓
$\frac{4}{5}$	$+$	$\frac{5}{8}$	$+$	$\frac{1}{2}$	$=$	m

Solve. We carry out the addition. The LCM of the denominators is 40.

$$\frac{4}{5}\cdot\frac{8}{8}+\frac{5}{8}\cdot\frac{5}{5}+\frac{1}{2}\cdot\frac{20}{20}=m$$
$$\frac{32}{40}+\frac{25}{40}+\frac{20}{40}=m$$
$$\frac{77}{40}=m$$

Check. We repeat the calculation. We also note that the sum is larger than any of the individual distances, so the answer is reasonable.

State. The students cleaned $\frac{77}{40}$ mi along the highway.

71. ***Familiarize.*** We draw a picture and let r = the total rainfall, in inches.

$\frac{1}{2}$ in.	$\frac{3}{8}$ in.
r	

Translate. The problem can be translated to an equation as follows:

Morning rain	plus	Afternoon rain	is	Total rainfall
↓	↓	↓	↓	↓
$\frac{1}{2}$	$+$	$\frac{3}{8}$	$=$	r

Solve. We carry out the addition. Since 8 is a multiple of 2, the LCM of the denominators is 8.

$$\frac{1}{2}+\frac{3}{8}=r$$
$$\frac{1}{2}\cdot\frac{4}{4}+\frac{3}{8}=r$$
$$\frac{4}{8}+\frac{3}{8}=r$$
$$\frac{7}{8}=r$$

Check. We repeat the calculations. We also note that the sum is larger than either of the individual amounts, so the answer seems reasonable.

State. Altogether it rained $\frac{7}{8}$ in.

73. ***Familiarize.*** We draw a picture and let d = the number of miles the naturalist hikes.

Lookout — Nest — Campsite

$\frac{3}{5}$ mi, $\frac{3}{10}$ mi, $\frac{3}{4}$ mi; total d

Translate. We translate to an equation as follows:

Miles to lookout	plus	Miles to nest	plus	Miles to campsite	is	Total miles
↓	↓	↓	↓	↓	↓	↓
$\frac{3}{5}$	$+$	$\frac{3}{10}$	$+$	$\frac{3}{4}$	$=$	d

Solve. We carry out the addition. Since $5 = 5$, $10 = 2 \cdot 5$, and $4 = 2 \cdot 2$, the LCM of the denominators is $2 \cdot 2 \cdot 5$, or 20.

$$\frac{3}{5} + \frac{3}{10} + \frac{3}{4} = d$$
$$\frac{3}{5} \cdot \frac{4}{4} + \frac{3}{10} \cdot \frac{2}{2} + \frac{3}{4} \cdot \frac{5}{5} = d$$
$$\frac{12}{20} + \frac{6}{20} + \frac{15}{20} = d$$
$$\frac{33}{20} = d$$

Check. We repeat the calculation. We also note that the sum is larger than any of the individual distances, as expected.

State. The naturalist hiked a total of $\frac{33}{20}$ mi.

75. ***Familiarize.*** We draw a picture and let f = the total amount of flour used.

$\frac{1}{2}$ lb	$\frac{1}{4}$ lb	$\frac{1}{3}$ lb
f		

Translate. The problem can be translated to an equation as follows:

Amount for rolls	plus	Amount for donuts	plus	Amount for cookies	is	Total amount
↓	↓	↓	↓	↓	↓	↓
$\frac{1}{2}$	$+$	$\frac{1}{4}$	$+$	$\frac{1}{3}$	$=$	f

Solve. We carry out the addition. Since $2 = 2$, $4 = 2 \cdot 2$, and $3 = 3$, the LCM of the denominators is $2 \cdot 2 \cdot 3$, or 12.

$$\frac{1}{2} + \frac{1}{4} + \frac{1}{3} = f$$
$$\frac{1}{2} \cdot \frac{6}{6} + \frac{1}{4} \cdot \frac{3}{3} + \frac{1}{3} \cdot \frac{4}{4} = f$$
$$\frac{6}{12} + \frac{3}{12} + \frac{4}{12} = f$$
$$\frac{13}{12} = f$$

Check. We repeat the calculation. We also note that the sum is larger than any of the individual amounts, as expected.

State. $\frac{13}{12}$ lb of flour was used.

77. ***Familiarize.*** Let t = the thickness of the iced brownies, in inches. We see from the drawing in the text that the total thickness will be the sum of the thicknesses of the brownie and the icing.

Translate.

Thickness of brownie	plus	Thickness of icing	is	Thickness of iced brownie
↓	↓	↓	↓	↓
$\frac{11}{16}$	$+$	$\frac{5}{32}$	$=$	t

Solve. We carry out the addition. Since 32 is a multiple of 16, the LCM of the denominators is 32.

$$\frac{11}{16} \cdot \frac{2}{2} + \frac{5}{32} = t$$
$$\frac{22}{32} + \frac{5}{32} = t$$
$$\frac{27}{32} = t$$

Check. We repeat the calculation. We also note that the sum is larger than either of the individual thicknesses, so the answer seems reasonable.

State. The thickness of the iced brownie is $\frac{27}{32}$ in.

79. ***Familiarize.*** First we will find the amount of liquid needed. Let l = this amount, in quarts.

Translate.

Amount of ginger ale	plus	Amount of strawberry soda	is	Total amount of liquid
↓	↓	↓	↓	↓
$\frac{1}{5}$	$+$	$\frac{3}{5}$	$=$	l

Solve. We carry out the addition.

$$\frac{1}{5} + \frac{3}{5} = l$$
$$\frac{4}{5} = l$$

Let d = the number of quarts of liquid needed if the recipe is doubled. We write a multiplication sentence and carry out the multiplication.

$$d = 2 \cdot \frac{4}{5}$$
$$d = \frac{2 \cdot 4}{5} = \frac{8}{5}$$

Let h = the number of quarts of liquid needed if the recipe is halved. We write another multiplication sentence and carry out the multiplication.

$$h = \frac{1}{2} \cdot \frac{4}{5}$$
$$h = \frac{1 \cdot 4}{2 \cdot 5} = \frac{1 \cdot 2 \cdot 2}{2 \cdot 5} = \frac{2}{2} \cdot \frac{1 \cdot 2}{5} = \frac{2}{5}$$

Check. We repeat the calculations. The answers check.

State. The recipe requires $\frac{4}{5}$ qt of liquid. If the recipe is doubled, $\frac{8}{5}$ qt is required; $\frac{2}{5}$ qt is required if the recipe is halved.

81. $-7 - 6 = -7 + (-6) = -13$

83. $9 - 17 = 9 + (-17) = -8$

85. $\dfrac{x-y}{3} = \dfrac{7-(-3)}{3} = \dfrac{7+3}{3} = \dfrac{10}{3}$

87. ***Familiarize.*** Let $p =$ the number of votes by which Gore's popular votes exceeded Bush's.

Translate.

Bush's votes	plus	Gore's excess votes	is	Gore's votes
↓	↓	↓	↓	↓
50, 459, 211	+	p	=	51, 003, 894

Solve. We subtract 50,459,211 on both sides of the equation.

$$50,459,211 + p = 51,003,894$$
$$50,459,211 + p - 50,459,211 = 51,003,894 - 50,459,211$$
$$p = 544,683$$

Check. Since $50,459,211 + 544,683 = 51,003,894$, the answer checks.

State. Albert A. Gore received 544,683 more popular votes than George W. Bush in 2000.

89. ***Familiarize.*** Let v = the number of votes by which Kennedy's electoral votes exceeded Nixon's.

Translate. This is a "how much more" situation.

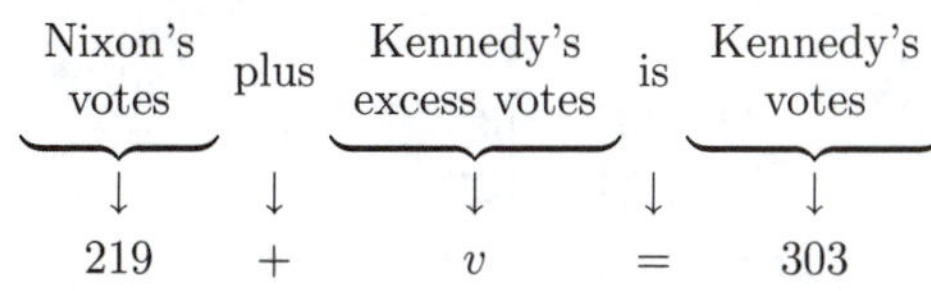

Solve. We subtract 219 on both sides of the equation.

$$219 + v = 303$$
$$219 + v - 219 = 303 - 219$$
$$v = 84$$

Check. Since $219 + 84 = 303$, the answer checks.

State. John F. Kennedy had 84 more electoral votes than Richard M. Nixon in 1960.

91. ***Familiarize.*** This is a multistep problem. Let $x =$ the total popular vote in 2000, $y =$ the total popular vote in 1976, and $n =$ the number of votes by which the 2000 total exceeded that in 1976.

Translate. First we add to find x and y.

$$x = 50,459,211 + 51,003,894$$
$$y = 40,830,763 + 39,147,793$$

To find n, we write a third equation.

1976 votes	plus	excess 2000 votes	is	2000 votes
↓	↓	↓	↓	↓
y	+	n	=	x

Solve. We first carry out the additions to find x and y.

$$x = 50,459,211 + 51,003,894$$
$$x = 101,463,105$$

$$y = 40,830,763 + 39,147,793$$
$$y = 79,978,556$$

Now we substitute 101,463,105 for x and $79,978,556$ for y in the third equation and solve for n.

$$y + n = x$$
$$79,978,556 + n = 101,463,105$$
$$79,978,556 + n - 79,978,556 = 101,463,105 - 79,978,556$$
$$n = 21,484,549$$

Check. Since $79,978,556 + 21,484,549 = 101,463,105$, the answer checks.

State. The total popular vote in 2000 exceeded that in 1976 by 21,484,549 votes.

93. $\dfrac{3}{10}t + \dfrac{2}{7} + \dfrac{2}{15}t + \dfrac{3}{5}$

$= \left(\dfrac{3}{10} + \dfrac{2}{15}\right)t + \left(\dfrac{2}{7} + \dfrac{3}{5}\right)$

$= \left(\dfrac{3}{10} \cdot \dfrac{3}{3} + \dfrac{2}{15} \cdot \dfrac{2}{2}\right)t + \left(\dfrac{2}{7} \cdot \dfrac{5}{5} + \dfrac{3}{5} \cdot \dfrac{7}{7}\right)$

$= \left(\dfrac{9}{30} + \dfrac{4}{30}\right)t + \left(\dfrac{10}{35} + \dfrac{21}{35}\right)$

$= \dfrac{13}{30}t + \dfrac{31}{35}$

95. $5t^2 + \dfrac{6}{a}t + 2t^2 + \dfrac{3}{a}t$

$= (5+2)t^2 + \left(\dfrac{6}{a} + \dfrac{3}{a}\right)t$

$= 7t^2 + \dfrac{9}{a}t$

97. Use a calculator to do this exercise. First, add on the left.

$$\frac{12}{169} + \frac{53}{103} = \frac{10,193}{17,407}$$

Now compare $\dfrac{10,193}{17,407}$ and $\dfrac{10,192}{17,407}$. The denominators are the same. Since $10,193 > 10,192$, it follows that $\dfrac{10,193}{17,407} > \dfrac{10,192}{17,407}$, so $\dfrac{12}{169} + \dfrac{53}{103} > \dfrac{10,192}{17,407}$.

99. ***Familiarize.*** First we find the fractional part of the band's pay that the guitarist received. We let f = this fraction.

Translate. We translate to an equation.

One-third	of	one-half	plus	one-fifth	of	one-half	is	fractional part
↓	↓	↓	↓	↓	↓	↓	↓	↓
$\frac{1}{3}$	$\cdot$	$\frac{1}{2}$	+	$\frac{1}{5}$	$\cdot$	$\frac{1}{2}$	=	f

Solve. We carry out the calculation.

$$\frac{1}{3} \cdot \frac{1}{2} + \frac{1}{5} \cdot \frac{1}{2} = f$$
$$\frac{1}{6} + \frac{1}{10} = f \quad \text{LCD is 30.}$$
$$\frac{1}{6} \cdot \frac{5}{5} + \frac{1}{10} \cdot \frac{3}{3} = f$$
$$\frac{5}{30} + \frac{3}{30} = f$$
$$\frac{8}{30} = f$$
$$\frac{4}{15} = f$$

Now we find how much of the \$1200 received by the band was paid to the guitarist. We let $p =$ the amount.

$$\underbrace{\text{Four-fifteenths}} \text{ of } \$1200 = \underbrace{\text{guitarist's pay}}$$
$$\frac{4}{15} \cdot 1200 = p$$

We solve the equation.

$$\frac{4}{15} \cdot 1200 = p$$
$$\frac{4 \cdot 1200}{15} = p$$
$$\frac{4 \cdot 3 \cdot 5 \cdot 80}{3 \cdot 5} = p$$
$$320 = p$$

Check. We repeat the calculations.

State. The guitarist received $\frac{4}{15}$ of the band's pay. This was \$320.

101. $\frac{a}{17} + \frac{1b}{23}$

$= \frac{a}{17} + \frac{10+b}{23}$

$= \frac{a}{17} + \frac{10}{23} + \frac{b}{23}$

$= \frac{a}{17} \cdot \frac{23}{23} + \frac{10}{23} \cdot \frac{17}{17} + \frac{b}{23} \cdot \frac{17}{17}$

$= \frac{23a}{391} + \frac{170}{391} + \frac{17b}{391}$

$= \frac{23a + 170 + 17b}{391}$

Writing $35a$ as $350 + a$, we have

$$\frac{23a + 170 + 17b}{391} = \frac{350 + a}{391}.$$

Thus, $23a + 170 + 17b = 350 + a$, or $22a + 17b = 180$. Using a calculator and trying values of a and b, where $a < 4$ and $b > 6$, we find that $a = 2$ and $b = 8$. (Using the addition and division principles, we could also find that $a = \frac{180 - 17b}{22}$ and then try values of b greater than 6, looking for one for which a is a natural number. Similarly, we could find that $b = \frac{180 - 22a}{17}$ and try values of a less than 4.)

Exercise Set 4.3

1. When denominators are the same, subtract the numerators and keep the denominator.

$$\frac{5}{6} - \frac{1}{6} = \frac{5-1}{6} = \frac{4}{6} = \frac{2 \cdot 2}{2 \cdot 3} = \frac{2}{2} \cdot \frac{2}{3} = \frac{2}{3}$$

3. When denominators are the same, subtract the numerators and keep the denominator.

$$\frac{9}{16} - \frac{13}{16} = \frac{9-13}{16} = \frac{-4}{16} = \frac{-1 \cdot 4}{4 \cdot 4} = \frac{-1}{4} \cdot \frac{4}{4} = \frac{-1}{4},$$

or $-\frac{1}{4}$

5. $\frac{8}{a} - \frac{6}{a} = \frac{8-6}{a} = \frac{2}{a}$

7. $-\frac{3}{8} - \frac{1}{8} = \frac{-3-1}{8} = \frac{-4}{8} = \frac{-1 \cdot 4}{2 \cdot 4} = \frac{-1}{2} \cdot \frac{4}{4} = \frac{-1}{2}$, or $-\frac{1}{2}$

9. $\frac{3}{5a} - \frac{7}{5a} = \frac{3-7}{5a} = \frac{-4}{5a}$, or $-\frac{4}{5a}$

11. $\frac{10}{3t} - \frac{4}{3t} = \frac{10-4}{3t} = \frac{6}{3t} = \frac{3}{3} \cdot \frac{2}{t} = \frac{2}{t}$

13. The LCM of 8 and 16 is 16.

$\frac{7}{8} - \frac{1}{16} = \frac{7}{8} \cdot \frac{2}{2} - \frac{1}{16}$ ⟵ This fraction already has the LCM as the denominator.

Think: $8 \times \square = 16$. The answer is 2, so we multiply by 1, using $\frac{2}{2}$.

$= \frac{14}{16} - \frac{1}{16} = \frac{13}{16}$

15. The LCM of 15 and 5 is 15.

$\frac{7}{15} - \frac{4}{5} = \frac{7}{15} - \frac{4}{5} \cdot \frac{3}{2}$

Think: $5 \times \square = 15$. The answer is 3, so we multiply by 1, using $\frac{3}{3}$.

This fraction already has the LCM as the denominator.

$= \frac{7}{15} - \frac{12}{15}$

$= \frac{-5}{15} = \frac{5}{5} \cdot \frac{-1}{3}$

$= \frac{-1}{3}$, or $-\frac{1}{3}$

17. The LCM of 4 and 20 is 20.

$\frac{3}{4} - \frac{1}{20} = \frac{3}{4} \cdot \frac{5}{5} - \frac{1}{20}$

$= \frac{15}{20} - \frac{1}{20} = \frac{14}{20}$

$= \frac{2 \cdot 7}{2 \cdot 10} = \frac{2}{2} \cdot \frac{7}{10}$

$= \frac{7}{10}$

19. The LCM of 15 and 12 is 60.

$$\frac{2}{15}-\frac{5}{12}=\frac{2}{15}\cdot\frac{4}{4}-\frac{5}{12}\cdot\frac{5}{5}$$
$$=\frac{8}{60}-\frac{25}{60}=\frac{8-25}{60}$$
$$=\frac{-17}{60},\text{ or }-\frac{17}{60}$$

21. The LCM of 10 and 100 is 100.

$$\frac{7}{10}-\frac{23}{100}=\frac{7}{10}\cdot\frac{10}{10}-\frac{23}{100}$$
$$=\frac{70}{100}-\frac{23}{100}=\frac{47}{100}$$

23. The LCM of 15 and 25 is 75.

$$\frac{7}{15}-\frac{3}{25}=\frac{7}{15}\cdot\frac{5}{5}-\frac{3}{25}\cdot\frac{3}{3}$$
$$=\frac{35}{75}-\frac{9}{75}=\frac{26}{75}$$

25. The LCM of 10 and 100 is 100.

$$\frac{-41}{100}-\frac{3}{10}=\frac{-41}{100}-\frac{3}{10}\cdot\frac{10}{10}$$
$$=\frac{-41}{100}-\frac{30}{100}=\frac{-41-30}{100}$$
$$=\frac{-71}{100},\text{ or }-\frac{71}{100}$$

27. The LCM of 3 and 8 is 24.

$$\frac{2}{3}-\frac{1}{8}=\frac{2}{3}\cdot\frac{8}{8}-\frac{1}{8}\cdot\frac{3}{3}$$
$$=\frac{16}{24}-\frac{3}{24}$$
$$=\frac{13}{24}$$

29. The LCM of 10 and 25 is 50.

$$-\frac{3}{10}-\frac{7}{25}=-\frac{3}{10}\cdot\frac{5}{5}-\frac{7}{25}\cdot\frac{2}{2}$$
$$=-\frac{15}{50}-\frac{14}{50}$$
$$=-\frac{29}{50},\text{ or }-\frac{29}{50}$$

31. The LCM of 8 and 12 is 24.

$$\frac{3}{8}-\frac{5}{12}=\frac{3}{8}\cdot\frac{3}{3}-\frac{5}{12}\cdot\frac{2}{2}$$
$$=\frac{9}{24}-\frac{10}{24}$$
$$=\frac{-1}{24},\text{ or }-\frac{1}{24}$$

33. The LCM of 18 and 24 is 72.

$$\frac{-5}{18}-\frac{7}{24}=\frac{-5}{18}\cdot\frac{4}{4}-\frac{7}{24}\cdot\frac{3}{3}$$
$$=\frac{-20}{72}-\frac{21}{72}$$
$$=\frac{-41}{72},\text{ or }-\frac{41}{72}$$

35. The LCM of 90 and 120 is 360.

$$\frac{13}{90}-\frac{17}{120}=\frac{13}{90}\cdot\frac{4}{4}-\frac{17}{120}\cdot\frac{3}{3}$$
$$=\frac{52}{360}-\frac{51}{360}$$
$$=\frac{1}{360}$$

37. The LCM of 3 and 9 is 9.

$$\frac{2}{3}x-\frac{4}{9}x=\frac{2}{3}\cdot\frac{3}{3}\cdot x-\frac{4}{9}x$$
$$=\frac{6}{9}x-\frac{4}{9}x$$
$$=\frac{2}{9}x$$

39. The LCM of 5 and 4 is 20.

$$\frac{2}{5}a-\frac{3}{4}a=\frac{2}{5}\cdot\frac{4}{4}\cdot a-\frac{3}{4}\cdot\frac{5}{5}a$$
$$=\frac{8}{20}a-\frac{15}{20}a$$
$$=\frac{-7}{20}a,\text{ or }-\frac{7}{20}a$$

41.

$$x-\frac{4}{9}=\frac{3}{9}$$
$$x-\frac{4}{9}+\frac{4}{9}=\frac{3}{9}+\frac{4}{9}\quad\text{Adding }\frac{4}{9}\text{ to both sides}$$
$$x+0=\frac{7}{9}$$
$$x=\frac{7}{9}$$

The solution is $\frac{7}{9}$.

43.

$$a+\frac{2}{11}=\frac{6}{11}$$
$$a+\frac{2}{11}-\frac{2}{11}=\frac{6}{11}-\frac{2}{11}\quad\text{Subtracting }\frac{2}{11}\text{ from both sides}$$
$$a+0=\frac{4}{11}$$
$$a=\frac{4}{11}$$

The solution is $\frac{4}{11}$.

45. $y+\frac{1}{30}=\frac{1}{10}$

$y+\frac{1}{30}-\frac{1}{30}=\frac{1}{10}-\frac{1}{30}$ Subtracting $\frac{1}{30}$ from both sides

$y+0=\frac{1}{10}\cdot\frac{3}{3}-\frac{1}{30}$ The LCD is 30. We multiply by 1 to get the LCD.

$y=\frac{3}{30}-\frac{1}{30}=\frac{2}{30}$

$y=\frac{2\cdot 1}{2\cdot 15}=\frac{2}{2}\cdot\frac{1}{15}$

$y=\frac{1}{15}$

The solution is $\frac{1}{15}$.

47. $a-\frac{3}{8}=\frac{3}{4}$

$a-\frac{3}{8}+\frac{3}{8}=\frac{3}{4}+\frac{3}{8}$ Adding $\frac{3}{8}$ on both sides

$a+0=\frac{3}{4}\cdot\frac{2}{2}+\frac{3}{8}$ The LCD is 8. We multiply by 1 to get the LCD.

$a=\frac{6}{8}+\frac{3}{8}=\frac{9}{8}$

The solution is $\frac{9}{8}$.

49. $\frac{2}{3}+x=\frac{4}{5}$

$\frac{2}{3}+x-\frac{2}{3}=\frac{4}{5}-\frac{2}{3}$ Subtracting $\frac{2}{3}$ on both sides

$x+0=\frac{4}{5}\cdot\frac{3}{3}-\frac{2}{3}\cdot\frac{5}{5}$ The LCD is 15. We multiply by 1 to get the LCD.

$x=\frac{12}{15}-\frac{10}{15}=\frac{2}{15}$

The solution is $\frac{2}{15}$.

51. $\frac{3}{8}+a=\frac{1}{12}$

$\frac{3}{8}+a-\frac{3}{8}=\frac{1}{12}-\frac{3}{8}$ Subtracting $\frac{3}{8}$ on both sides

$a+0=\frac{1}{12}\cdot\frac{2}{2}-\frac{3}{8}\cdot\frac{3}{3}$ The LCD is 24. We multiply by 1 to get the LCD.

$a=\frac{2}{24}-\frac{9}{24}=\frac{2-9}{24}$

$a=\frac{-7}{24}$, or $-\frac{7}{24}$

The solution is $-\frac{7}{24}$.

53. $n-\frac{3}{10}=-\frac{1}{6}$

$n-\frac{3}{10}+\frac{3}{10}=-\frac{1}{6}+\frac{3}{10}$ Adding $\frac{3}{10}$ to both sides

$n+0=-\frac{1}{6}\cdot\frac{5}{5}+\frac{3}{10}\cdot\frac{3}{3}$ The LCD is 30. We multiply by 1 to get the LCD.

$n=-\frac{5}{30}+\frac{9}{30}$

$n=\frac{4}{30}$

$n=\frac{2\cdot 2}{2\cdot 15}=\frac{2}{2}\cdot\frac{2}{15}$

$n=\frac{2}{15}$

The solution is $\frac{2}{15}$.

55. $x+\frac{3}{4}=-\frac{1}{2}$

$x+\frac{3}{4}-\frac{3}{4}=-\frac{1}{2}-\frac{3}{4}$ Subtracting $\frac{3}{4}$ on both sides

$x+0=-\frac{1}{2}\cdot\frac{2}{2}-\frac{3}{4}$ The LCD is 4. We multiply by 1 to get the LCD.

$x=-\frac{2}{4}-\frac{3}{4}=\frac{-2}{4}-\frac{3}{4}$

$x=\frac{-2-3}{4}$

$x=\frac{-5}{4}$, or $-\frac{5}{4}$

The solution is $-\frac{5}{4}$.

57. ***Familiarize***. We visualize the situation. Let t = the number of hours by which the time Kaitlyn spent on google.com exceeded the time she spent on chacha.com.

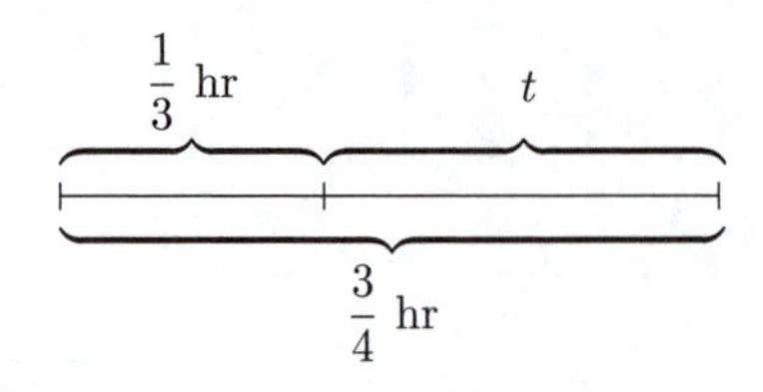

Translate.

Time spent on chacha.com	plus	Excess time spent on google.com	is	Time spent on google.com
↓	↓	↓	↓	↓
$\frac{1}{3}$	$+$	t	$=$	$\frac{3}{4}$

Solve. We subtract $\dfrac{1}{3}$ on both sides of the equation.

$$\frac{1}{3}+t-\frac{1}{3}=\frac{3}{4}-\frac{1}{3}$$

$$t+0=\frac{3}{4}\cdot\frac{3}{3}-\frac{1}{3}\cdot\frac{4}{4} \quad \text{The LCD is 12. We multiply by 1 to get the LCD.}$$

$$t=\frac{9}{12}-\frac{4}{12}=\frac{5}{12}$$

Check. We return to the original problem and add.

$$\frac{1}{3}+\frac{5}{12}=\frac{1}{3}\cdot\frac{4}{4}+\frac{5}{12}=\frac{4}{12}+\frac{5}{12}=\frac{9}{12}=\frac{3}{3}\cdot\frac{3}{4}=\frac{3}{4}$$

The answer checks.

State. Kaitlyn spent $\dfrac{5}{12}$ hr more on google.com than on chacha.com.

59. ***Familiarize***. Let $d=$ the number of inches by which the depth of the long-life tread exceeds the depth of the more typical tread.

Translate.

$$\underbrace{\text{Typical tread depth}}_{\frac{11}{32}} \ \underset{+}{\text{plus}} \ \underbrace{\text{Excess depth of long-life tire}}_{d} \ \underset{=}{\text{is}} \ \underbrace{\text{Depth of long-life tire}}_{\frac{3}{8}}$$

Solve. We subtract $\dfrac{11}{32}$ on both sides of the equation.

$$\frac{11}{32}+d=\frac{3}{8}$$

$$\frac{11}{32}+d-\frac{11}{32}=\frac{3}{8}-\frac{11}{32}$$

$$d+0=\frac{3}{8}\cdot\frac{4}{4}-\frac{11}{32} \quad \text{The LCD is 32.}$$

$$d=\frac{12}{32}-\frac{11}{32}$$

$$d=\frac{1}{32}$$

Check. We return to the original problem and add.

$$\frac{11}{32}+\frac{1}{32}=\frac{12}{32}=\frac{4\cdot 3}{4\cdot 8}=\frac{4}{4}\cdot\frac{3}{8}=\frac{3}{8}$$

The answer checks.

State. The long-life tire tread is $\dfrac{1}{32}$ in. deeper than the more typical $\dfrac{11}{32}$ in. tread.

61. ***Familiarize***. We visualize the situation. Let $c=$ the amount of cheese remaining, in pounds.

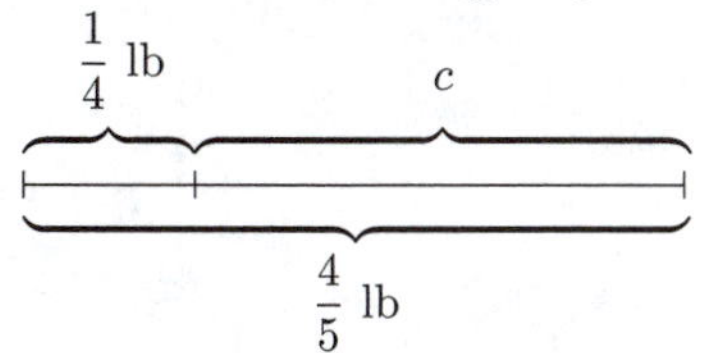

Translate. This is a "how much more" situation.

$$\underbrace{\text{Amount served}}_{\frac{1}{4}} \ \underset{+}{\text{plus}} \ \underbrace{\text{Amount remaining}}_{c} \ \underset{=}{\text{is}} \ \underbrace{\text{Original amount}}_{\frac{4}{5}}$$

Solve. We subtract $\dfrac{1}{4}$ on both sides of the equation.

$$\frac{1}{4}+c-\frac{1}{4}=\frac{4}{5}-\frac{1}{4}$$

$$c+0=\frac{4}{5}\cdot\frac{4}{4}-\frac{1}{4}\cdot\frac{5}{5} \quad \text{The LCD is 20.}$$

$$c=\frac{16}{20}-\frac{5}{20}$$

$$c=\frac{11}{20}$$

Check. Since $\dfrac{1}{4}+\dfrac{11}{20}=\dfrac{1}{4}\cdot\dfrac{5}{5}+\dfrac{11}{20}=\dfrac{5}{20}+\dfrac{11}{20}=\dfrac{16}{20}=\dfrac{4\cdot 4}{4\cdot 5}=\dfrac{4}{4}\cdot\dfrac{4}{5}=\dfrac{4}{5}$, the answer checks.

State. $\dfrac{11}{20}$ lb of cheese remains on the wheel.

63. ***Familiarize***. We visualize the situation. Let $t=$ the time spent on country driving, in hours.

$\frac{2}{5}$ hr t

$\frac{3}{4}$ hr

Translate. This is a "how much more" situation.

$$\underbrace{\text{City driving time}}_{\frac{2}{5}} \ \underset{+}{\text{plus}} \ \underbrace{\text{Country driving time}}_{t} \ \underset{=}{\text{is}} \ \underbrace{\text{Total time}}_{\frac{3}{4}}$$

Solve. We subtract $\dfrac{2}{5}$ from both sides of the equation.

$$\frac{2}{5}+t-\frac{2}{5}=\frac{3}{4}-\frac{2}{5}$$

$$t+0=\frac{3}{4}\cdot\frac{5}{5}-\frac{2}{5}\cdot\frac{4}{4} \quad \text{The LCD is 20.}$$

$$t=\frac{15}{20}-\frac{8}{20}=\frac{7}{20}$$

Check. We return to the original problem and add.

$$\frac{2}{5}+\frac{7}{20}=\frac{2}{5}\cdot\frac{4}{4}+\frac{7}{20}=\frac{8}{20}+\frac{7}{20}=\frac{15}{20}=\frac{3\cdot 5}{4\cdot 5}=\frac{3}{4}$$

The answer checks.

State. Jorge spent $\dfrac{7}{20}$ hr on country driving.

65. ***Familiarize***. Using the label on the drawing in the text, we let $r=$ the amount by which the board must be planed down, in inches.

Translate.

$$\underbrace{\text{Desired thickness}}_{\downarrow\ \frac{3}{4}} \ \text{plus}\ \underbrace{\text{Excess amount}}_{\downarrow\ r} \ \text{is}\ \underbrace{\text{Original thickness}}_{\downarrow\ \frac{15}{16}}$$

$$\frac{3}{4} + r = \frac{15}{16}$$

Solve. We subtract $\frac{3}{4}$ from both sides of the equation.

$$\frac{3}{4} + r - \frac{3}{4} = \frac{15}{16} - \frac{3}{4}$$

$$r + 0 = \frac{15}{16} - \frac{3}{4}\cdot\frac{4}{4} \quad \text{The LCD is 16. We multiply by 1 to get the LCD.}$$

$$r = \frac{15}{16} - \frac{12}{16}$$

$$r = \frac{3}{16}$$

Check. We return to the original problem and add.

$$\frac{3}{4} + \frac{3}{16} = \frac{3}{4}\cdot\frac{4}{4} + \frac{3}{16} = \frac{12}{16} + \frac{3}{16} = \frac{15}{16}$$

The answer checks.

State. The board must be planed down $\frac{3}{16}$ in.

67. ***Familiarize.*** We visualize the situation. Let s = the amount of syrup that should be added, in cups.

$\frac{1}{3}$ cup $\quad s$

$\frac{5}{8}$ cup

Translate.

$$\underbrace{\text{Original amount}}_{\downarrow\ \frac{1}{3}} \ \text{plus}\ \underbrace{\text{Additional amount}}_{\downarrow\ s} \ \text{is}\ \underbrace{\text{Total amount}}_{\downarrow\ \frac{5}{8}}$$

$$\frac{1}{3} + s = \frac{5}{8}$$

Solve. We subtract $\frac{1}{3}$ from both sides of the equation.

$$\frac{1}{3} + s - \frac{1}{3} = \frac{5}{8} - \frac{1}{3}$$

$$s + 0 = \frac{5}{8}\cdot\frac{3}{3} - \frac{1}{3}\cdot\frac{8}{8} \quad \text{The LCD is 24. We multiply by 1 to get the LCD.}$$

$$s = \frac{15}{24} - \frac{8}{24} = \frac{7}{24}$$

Check. We return to the original problem and add.

$$\frac{1}{3} + \frac{7}{24} = \frac{1}{3}\cdot\frac{8}{8} + \frac{7}{24} = \frac{8}{24} + \frac{7}{24} = \frac{15}{24} = \frac{3}{3}\cdot\frac{5}{8} = \frac{5}{8}$$

The answer checks.

State. Blake should add $\frac{7}{24}$ cup of syrup to the batter.

69.
$$\frac{3}{7} \div \frac{9}{4} = \frac{3}{7}\cdot\frac{4}{9} = \frac{3\cdot 4}{7\cdot 9} = \frac{3\cdot 4}{7\cdot 3\cdot 3} = \frac{3}{3}\cdot\frac{4}{7\cdot 3} = \frac{4}{21}$$

71.
$$7 \div \frac{1}{3} = 7\cdot\frac{3}{1} = \frac{7\cdot 3}{1} = \frac{21}{1} = 21$$

73. ***Familiarize.*** Let c = the number of crayons in the Crayola 64 box that were sold from 1958 to 2008.

Translate.

$$\underbrace{\text{Number of crayons in box}}_{\downarrow\ 64} \ \text{times}\ \underbrace{\text{Number of boxes sold}}_{\downarrow\ 200,000,000} \ \text{is}\ \underbrace{\text{Total number of crayons sold}}_{\downarrow\ c}$$

$$64 \cdot 200,000,000 = c$$

Solve. We carry out the multiplication.

```
      2 0 0, 0 0 0, 0 0 0
    ×                 6 4
    ---------------------
      8 0 0 0 0 0 0 0 0 0
  1 2 0 0 0 0 0 0 0 0 0 0
  -----------------------
  1 2, 8 0 0, 0 0 0, 0 0 0
```

Thus, $12,800,000,000 = c$.

Check. We repeat the calculation. The answer checks.

State. About 12,800,000,000, or 12.8 billion, crayons in the 64 box were sold from 1958 to 2008.

75.
$$3x - 8 = 25$$
$$3x - 8 + 8 = 25 + 8 \quad \text{Adding 8 to both sides}$$
$$3x + 0 = 33$$
$$3x = 33$$
$$\frac{3x}{3} = \frac{33}{3} \quad \text{Dividing both sides by 3}$$
$$x = 11$$

The solution is 11.

77. The LCM of 8, 4, and 16 is 16.

$$\frac{7}{8} - \frac{3}{4} - \frac{1}{16} = \frac{7}{8}\cdot\frac{2}{2} - \frac{3}{4}\cdot\frac{4}{4} - \frac{1}{16} = \frac{14}{16} - \frac{12}{16} - \frac{1}{16} = \frac{14-12-1}{16} = \frac{1}{16}$$

79. $\frac{2}{5}-\frac{1}{6}(-3)^2 = \frac{2}{5}-\frac{1}{6}\cdot 9$

$= \frac{2}{5}-\frac{9}{6}$

$= \frac{2}{5}\cdot\frac{6}{6}-\frac{9}{6}\cdot\frac{5}{5}$ The LCD is 30.

$= \frac{12}{30}-\frac{45}{30}$

$= \frac{-33}{30} = \frac{3}{3}\cdot\frac{-11}{10}$

$= \frac{-11}{10}$, or $-\frac{11}{10}$

81. $-4\cdot\frac{3}{7}-\frac{1}{7}\cdot\frac{4}{5} = \frac{-12}{7}-\frac{4}{35}$

$= \frac{-12}{7}\cdot\frac{5}{5}-\frac{4}{35}$ The LCD is 35.

$= \frac{-60}{35}-\frac{4}{35}$

$= \frac{-64}{35}$, or $-\frac{64}{35}$

83. $\left(-\frac{2}{5}\right)^3-\left(-\frac{3}{10}\right)^3$

$= -\frac{8}{125}-\left(-\frac{27}{1000}\right)$

$= -\frac{8}{125}+\frac{27}{1000}$

$= -\frac{8}{125}\cdot\frac{8}{8}+\frac{27}{1000}$ The LCD is 1000.

$= -\frac{64}{1000}+\frac{27}{1000}$

$= -\frac{37}{1000}$

85. ***Familiarize***. This is a multistep problem. First we find the portion of shoppers who stay for 1-2 hr. Let s = this portion of the shoppers.

Translate. This is a "missing addend" situation.

$$\underbrace{\text{Portions shown in graph}}_{\frac{25}{50}+\frac{5}{50}+\frac{2}{50}} \quad \underbrace{\text{plus}}_{+} \quad \underbrace{\text{Remaining portion}}_{s} \quad \underbrace{\text{is}}_{=} \quad \underbrace{\text{One entire group}}_{1}$$

Solve. First we collect like terms on the left.

$\frac{26}{50}+\frac{5}{50}+\frac{2}{50}+s = 1$

$\frac{33}{50}+s = 1$

$\frac{33}{50}+s-\frac{33}{50} = 1-\frac{33}{50}$ Subtracting $\frac{33}{50}$ from both sides

$s+0 = 1\cdot\frac{50}{50}-\frac{33}{50}$ The LCD is 50.

$s = \frac{50}{50}-\frac{33}{50}$

$s = \frac{17}{50}$

Now we add the portion of shoppers who stay less than one hour and the portion who stay 1-2 hr to find the portion of shoppers who stay 0-2 hr.

$\frac{26}{50}+\frac{17}{50} = \frac{43}{50}$

Check. We repeat the calculation.

State. $\frac{43}{50}$ of shoppers stay 0-2 hr when visiting a mall.

87. ***Familiarize***. Let d = the fractional portion of the dealership that Paul owns. Then Ella owns the same portion and together the Romanos and the Chrenkas own $\frac{7}{12}+\frac{1}{6}+d+d$.

Translate.

$$\underbrace{\text{Total portions}}_{\frac{7}{12}+\frac{1}{6}+d+d} \quad \underbrace{\text{are}}_{=} \quad \underbrace{\text{1 dealership}}_{1}$$

Solve. First we collect like terms on the left.

$\frac{7}{12}+\frac{1}{6}+d+d = 1$

$\frac{7}{12}+\frac{1}{6}\cdot\frac{2}{2}+2d = 1$ The LCD is 12.

$\frac{7}{12}+\frac{2}{12}+2d = 1$

$\frac{9}{12}+2d = 1$

$\frac{9}{12}+2d-\frac{9}{12} = 1-\frac{9}{12}$

$2d+0 = 1\cdot\frac{12}{12}-\frac{9}{12}$

$2d = \frac{3}{12}$

$\frac{1}{2}\cdot 2d = \frac{1}{2}\cdot\frac{3}{12}$

$d = \frac{3}{24} = \frac{3}{3}\cdot\frac{1}{8}$

$d = \frac{1}{8}$

Check. We return to the original problem and add.

$\frac{7}{12}+\frac{1}{6}+\frac{1}{8}+\frac{1}{8} = \frac{7}{12}\cdot\frac{2}{2}+\frac{1}{6}\cdot\frac{4}{4}+\frac{1}{8}\cdot\frac{3}{3}+\frac{1}{8}\cdot\frac{3}{3} =$

$\frac{14}{24}+\frac{4}{24}+\frac{3}{24}+\frac{3}{24} = \frac{24}{24} = 1$

The answer checks.

State. Paul owns $\frac{1}{8}$ of the dealership.

89. ***Familiarize***. First we find how far the athlete swam. We let s = this distance. We visualize the situation.

$\longrightarrow \longrightarrow \cdots \longrightarrow$ } 10 laps

$\frac{3}{80}$ km in each lap

Translate. We translate to the following equation:

$s = 10\cdot\frac{3}{80}$

Solve. We carry out the multiplication.

$$s = 10 \cdot \frac{3}{80} = \frac{10 \cdot 3}{80}$$
$$s = \frac{10 \cdot 3}{10 \cdot 8} = \frac{10}{10} \cdot \frac{3}{8}$$
$$s = \frac{3}{8}$$

Now we find the distance the athlete must walk. We let w = the distance.

$$\underbrace{\text{Distance swum}}_{\downarrow\ \frac{3}{8}} \quad \underset{\downarrow\ +}{\text{plus}} \quad \underbrace{\text{Distance walked}}_{\downarrow\ w} \quad \underset{\downarrow\ =}{\text{is}} \quad \underbrace{\frac{9}{10}\text{ km}}_{\downarrow\ \frac{9}{10}}$$

We solve the equation.

$$\frac{3}{8} + w = \frac{9}{10}$$
$$\frac{3}{8} + w - \frac{3}{8} = \frac{9}{10} - \frac{3}{8}$$
$$w + 0 = \frac{9}{10} \cdot \frac{4}{4} - \frac{3}{8} \cdot \frac{5}{5} \quad \text{The LCD is 40.}$$
$$w = \frac{36}{40} - \frac{15}{40}$$
$$w = \frac{21}{40}$$

Check. We add the distance swum and the distance walked:

$$\frac{3}{8} + \frac{21}{40} = \frac{3}{8} \cdot \frac{5}{5} + \frac{21}{40} = \frac{15}{40} + \frac{21}{40} = \frac{36}{40} = \frac{9 \cdot 4}{10 \cdot 4} = \frac{9}{10} \cdot \frac{4}{4} = \frac{9}{10}$$

State. The athlete must walk $\frac{21}{40}$ km after swimming 10 laps.

91. Use a calculator.

$$x + \frac{16}{323} = \frac{10}{187}$$
$$x + \frac{16}{323} - \frac{16}{323} = \frac{10}{187} - \frac{16}{323}$$
$$x + 0 = \frac{10}{11 \cdot 17} - \frac{16}{17 \cdot 19}$$
$$x = \frac{10}{11 \cdot 17} \cdot \frac{19}{19} - \frac{16}{17 \cdot 19} \cdot \frac{11}{11} \quad \text{The LCD is } 11 \cdot 17 \cdot 19.$$
$$x = \frac{190}{11 \cdot 17 \cdot 19} - \frac{176}{17 \cdot 19 \cdot 11}$$
$$x = \frac{14}{11 \cdot 17 \cdot 19}$$
$$x = \frac{14}{3553}$$

The solution is $\frac{14}{3553}$.

93. Use the two cuts to cut the bar into three pieces as follows: one piece is $\frac{1}{7}$ of the bar, one is $\frac{2}{7}$ of the bar, and then the remaining piece is $\frac{4}{7}$ of the bar. On Day 1, give the contractor $\frac{1}{7}$ of the bar. On Day 2, have him/her return the $\frac{1}{7}$ and give him/her $\frac{2}{7}$ of the bar. On Day 3, add $\frac{1}{7}$ to what the contractor already has, making $\frac{3}{7}$ of the bar. On Day 4, have the contractor return the $\frac{1}{7}$ and $\frac{2}{7}$ pieces and give him/her the $\frac{4}{7}$ piece. On Day 5, add the $\frac{1}{7}$ piece to what the contractor already has, making $\frac{5}{7}$ of the bar. On Day 6, have the contractor return the $\frac{1}{7}$ piece and give him/her the $\frac{2}{7}$ to go with the $\frac{4}{7}$ piece he/she also has, making $\frac{6}{7}$ of the bar. On Day 7, give him/her the $\frac{1}{7}$ piece again. Now the contractor has all three pieces, or the entire bar. This assumes that he/she does not spend any part of the gold during the week.

Exercise Set 4.4

1.

$$6x - 3 = 15$$
$$6x - 3 + 3 = 15 + 3 \quad \text{Using the addition principle}$$
$$6x + 0 = 18$$
$$6x = 18$$
$$\frac{1}{6} \cdot 6x = \frac{1}{6} \cdot 18 \quad \text{Using the multiplication principle}$$
$$1x = \frac{18}{6}$$
$$x = 3$$

Check:

$6x - 3 = 15$	
$6 \cdot 3 - 3$?	15
$18 - 3$	
15	15 TRUE

The solution is 3.

3.

$$5x + 7 = 10$$
$$5x + 7 - 7 = 10 - 7 \quad \text{Using the addition principle}$$
$$5x = 3$$
$$\frac{1}{5} \cdot 5x = \frac{1}{5} \cdot 3 \quad \text{Using the multiplication principle}$$
$$1x = \frac{3}{5}$$
$$x = \frac{3}{5}$$

Check:

$5x + 7 = 10$	
$\frac{3}{5} + 7$?	10
$3 + 7$	
10	10 TRUE

The solution is $\frac{3}{5}$.

5.

$$8 = 3x + 11$$
$$8 - 11 = 3x + 11 - 11 \quad \text{Using the addition principle}$$
$$-3 = 3x + 0$$
$$-3 = 3x$$
$$\frac{1}{3} \cdot (-3) = \frac{1}{3} \cdot 3x \quad \text{Using the multiplication principle}$$
$$-1 = 1x$$
$$-1 = x$$

Check:

$$\begin{array}{l|l} \multicolumn{2}{c}{8 = 3x + 11} \\ \hline 8 \ ? & 3 \cdot (-1) + 11 \\ & -3 + 11 \\ 8 & 8 \quad \text{TRUE} \end{array}$$

The solution is -1.

7.

$$\frac{2}{3}y - 8 = 1$$
$$\frac{2}{3}y - 8 + 8 = 1 + 8 \quad \text{Using the addition principle}$$
$$\frac{2}{3}y + 0 = 9$$
$$\frac{2}{3}y = 9$$
$$\frac{3}{2} \cdot \frac{2}{3}y = \frac{3}{2} \cdot 9 \quad \text{Using the multiplication principle}$$
$$1y = \frac{3 \cdot 9}{2}$$
$$y = \frac{27}{2}$$

Check:

$$\begin{array}{l|l} \multicolumn{2}{c}{\frac{2}{3}y - 8 = 1} \\ \hline \frac{2}{3} \cdot \frac{27}{2} - 8 \ ? & 1 \\ \frac{2 \cdot 3 \cdot 9}{3 \cdot 2 \cdot 1} - 8 & \\ \frac{2 \cdot 3}{2 \cdot 3} \cdot \frac{9}{1} - 8 & \\ 9 - 8 & \\ 1 & 1 \quad \text{TRUE} \end{array}$$

The solution is $\frac{27}{2}$.

9.

$$\frac{3}{2}t - \frac{1}{4} = \frac{1}{2}$$
$$\frac{3}{2}t - \frac{1}{4} + \frac{1}{4} = \frac{1}{2} + \frac{1}{4} \quad \text{Adding } \frac{1}{4} \text{ to each side}$$
$$\frac{3}{2}t + 0 = \frac{2}{4} + \frac{1}{4}$$
$$\frac{3}{2}t = \frac{3}{4}$$
$$\frac{2}{3} \cdot \frac{3}{2}t = \frac{2}{3} \cdot \frac{3}{4} \quad \text{Multiplying both sides by } \frac{2}{3}$$
$$1t = \frac{\not{2} \cdot \not{3}}{\not{3} \cdot \not{2} \cdot 2}$$
$$t = \frac{1}{2}$$

Check:

$$\begin{array}{l|l} \multicolumn{2}{c}{\frac{3}{2}t - \frac{1}{4} = \frac{1}{2}} \\ \hline \frac{3}{2} \cdot \frac{1}{2} - \frac{1}{4} \ ? & \frac{1}{2} \\ \frac{3}{4} - \frac{1}{4} & \\ \frac{2}{4} & \\ \frac{1}{2} & \frac{1}{2} \quad \text{TRUE} \end{array}$$

The solution is $\frac{1}{2}$.

11.

$$\frac{1}{5}x + \frac{3}{10} = \frac{3}{5}$$
$$\frac{1}{5}x + \frac{3}{10} - \frac{3}{10} = \frac{3}{5} - \frac{3}{10} \quad \text{Subtracting } \frac{3}{10} \text{ from both sides}$$
$$\frac{1}{5}x + 0 = \frac{6}{10} - \frac{3}{10}$$
$$\frac{1}{5}x = \frac{3}{10}$$
$$\frac{5}{1} \cdot \frac{1}{5}x = \frac{5}{1} \cdot \frac{3}{10} \quad \text{Multiplying both sides by } \frac{5}{1}$$
$$1x = \frac{\not{5} \cdot 3}{1 \cdot \not{5} \cdot 2}$$
$$x = \frac{3}{2}$$

Check:

$$\begin{array}{l|l} \multicolumn{2}{c}{\frac{1}{5}x + \frac{3}{10} = \frac{3}{5}} \\ \hline \frac{1}{5} \cdot \frac{3}{2} + \frac{3}{10} \ ? & \frac{3}{5} \\ \frac{1 \cdot 3}{5 \cdot 2} + \frac{3}{10} & \\ \frac{3}{10} + \frac{3}{10} & \\ \frac{6}{10} & \\ \frac{3}{5} & \frac{3}{5} \quad \text{TRUE} \end{array}$$

The solution is $\frac{3}{2}$.

13.
$$
\begin{aligned}
5-\frac{3}{4}x&=6\\
5-\frac{3}{4}x-5&=6-5 \quad \text{Subtracting 5 from both sides}\\
-\frac{3}{4}x&=1\\
-\frac{4}{3}\left(-\frac{3}{4}x\right)&=-\frac{4}{3}\cdot 1 \quad \text{Multiplying both sides by } -\frac{4}{3}\\
1x&=-\frac{4}{3}\\
x&=-\frac{4}{3}
\end{aligned}
$$

The number $-\frac{4}{3}$ checks and is the solution.

15.
$$
\begin{aligned}
-1+\frac{2}{5}t&=-\frac{4}{5}\\
-1+\frac{2}{5}t+1&=-\frac{4}{5}+1 \quad \text{Adding 1 to both sides}\\
\frac{2}{5}t&=-\frac{4}{5}+\frac{5}{5}\\
\frac{2}{5}t&=\frac{1}{5}\\
\frac{5}{2}\cdot\frac{2}{5}t&=\frac{5}{2}\cdot\frac{1}{5}\\
1t&=\frac{\cancel{5}\cdot 1}{2\cdot\cancel{5}}\\
t&=\frac{1}{2}
\end{aligned}
$$

The number $\frac{1}{2}$ checks and is the solution.

17.
$$
\begin{aligned}
12&=8+\frac{7}{2}t\\
12-8&=8+\frac{7}{12}t-8 \quad \text{Subtracting 8 from both sides}\\
4&=\frac{7}{2}t\\
\frac{2}{7}\cdot 4&=\frac{2}{7}\cdot\frac{7}{2}t \quad \text{Multiply both sides by } \frac{2}{7}\\
\frac{8}{7}&=t
\end{aligned}
$$

The number $\frac{8}{7}$ checks and is the solution.

19.
$$
\begin{aligned}
-11&=\frac{2}{3}x-7\\
-11+7&=\frac{2}{3}x-7+7 \quad \text{Adding 7 to both sides}\\
-4&=\frac{2}{3}x\\
\frac{3}{2}\cdot(-4)&=\frac{3}{2}\cdot\frac{2}{3}x \quad \text{Multiplying both sides by } \frac{3}{2}\\
-\frac{3\cdot\cancel{2}\cdot 2}{\cancel{2}\cdot 1}&=x\\
-6&=x
\end{aligned}
$$

The number -6 checks and is the solution.

21.
$$
\begin{aligned}
7&=a+\frac{14}{5}\\
7-\frac{14}{5}&=a+\frac{14}{5}-\frac{14}{5} \quad \text{Using the addition principle}\\
7\cdot\frac{5}{5}-\frac{14}{5}&=a+0\\
\frac{35}{5}-\frac{14}{5}&=a\\
\frac{21}{5}&=a
\end{aligned}
$$

The number $\frac{21}{5}$ checks and is the solution.

23.
$$
\begin{aligned}
\frac{2}{5}t-1&=\frac{7}{5}\\
\frac{2}{5}t-1+1&=\frac{7}{5}+1 \quad \text{Using the addition principle}\\
\frac{2}{5}t+0&=\frac{7}{5}+1\cdot\frac{5}{5}\\
\frac{2}{5}t&=\frac{7}{5}+\frac{5}{5}\\
\frac{2}{5}t&=\frac{12}{5}\\
\frac{5}{2}\cdot\frac{2}{5}t&=\frac{5}{2}\cdot\frac{12}{5} \quad \text{Using the multiplication principle}\\
1t&=\frac{5\cdot 12}{2\cdot 5}\\
t&=\frac{5\cdot 2\cdot 6}{2\cdot 5\cdot 1}=\frac{5\cdot 2}{5\cdot 2}\cdot\frac{6}{1}\\
t&=6
\end{aligned}
$$

The number 6 checks and is the solution.

25.
$$
\begin{aligned}
\frac{39}{8}&=\frac{11}{4}-\frac{1}{2}x\\
\frac{39}{8}-\frac{11}{4}&=\frac{11}{4}-\frac{1}{2}x-\frac{11}{4} \quad \text{Using the addition principle}\\
\frac{39}{8}-\frac{11}{4}\cdot\frac{2}{2}&=-\frac{1}{2}x+0\\
\frac{39}{8}-\frac{22}{8}&=-\frac{1}{2}x\\
\frac{17}{8}&=-\frac{1}{2}x\\
-2\cdot\frac{17}{8}&=-2\cdot\left(-\frac{1}{2}x\right) \quad \text{Using the multiplication principle}\\
\frac{2\cdot 17}{8}&=1x\\
-\frac{\cancel{2}\cdot 17}{\cancel{2}\cdot 4}&=x\\
-\frac{17}{4}&=x
\end{aligned}
$$

The number $-\frac{17}{4}$ checks and is the solution.

27.

$$-\frac{13}{3}x+\frac{11}{2}=\frac{35}{4}$$

$$-\frac{13}{3}x+\frac{11}{2}-\frac{11}{2}=\frac{35}{4}-\frac{11}{2} \quad \text{Using the addition principle}$$

$$-\frac{13}{3}x+0=\frac{35}{4}-\frac{11}{2}\cdot\frac{2}{2}$$

$$-\frac{13}{3}x=\frac{35}{4}-\frac{22}{4}$$

$$-\frac{13}{3}x=\frac{13}{4}$$

$$-\frac{3}{13}\cdot\left(-\frac{13}{3}\right)x=-\frac{3}{13}\cdot\frac{13}{4} \quad \text{Using the multiplication principle}$$

$$1x=-\frac{3\cdot\cancel{13}}{\cancel{13}\cdot 4}$$

$$x=-\frac{3}{4}$$

The number $-\frac{3}{4}$ checks and is the solution.

29.

$$\frac{1}{2}x-\frac{1}{4}=\frac{1}{2}$$

$$4\left(\frac{1}{2}x-\frac{1}{4}\right)=4\cdot\frac{1}{2} \quad \text{Multiplying both sides by the LCD, 4}$$

$$\frac{4\cdot 1}{2}x-4\cdot\frac{1}{4}=\frac{4\cdot 1}{2}$$

$$\frac{\cancel{2}\cdot 2}{\cancel{2}}x-1=\frac{\cancel{2}\cdot 2}{\cancel{2}}$$

$$2x-1=2$$

$$2x-1+1=2+1 \quad \text{Adding 1 to both sides}$$

$$2x=3$$

$$\frac{2x}{2}=\frac{3}{2} \quad \text{Dividing both sides by 2}$$

$$x=\frac{3}{2}$$

The number $\frac{3}{2}$ checks and is the solution.

31.

$$7=\frac{4}{9}t+5$$

$$9\cdot 7=9\left(\frac{4}{9}t+5\right) \quad \text{Multiplying both sides by the LCD, 9}$$

$$63=\frac{\cancel{9}\cdot 4}{\cancel{9}}t+9\cdot 5$$

$$63=4t+45$$

$$63-45=4t+45-45 \quad \text{Subtracting 45 from both sides}$$

$$18=4t$$

$$\frac{18}{4}=\frac{4t}{4} \quad \text{Dividing both sides by 4}$$

$$\frac{9\cdot\cancel{2}}{2\cdot\cancel{2}}=1t$$

$$\frac{9}{2}=t$$

The number $\frac{9}{2}$ checks and is the solution.

33.

$$-3=\frac{3}{4}t-\frac{1}{2}$$

$$4(-3)=4\left(\frac{3}{4}t-\frac{1}{2}\right) \quad \text{Multiplying both sides by the LCD, 4}$$

$$-12=\frac{4\cdot 3}{4}t-\frac{4\cdot 1}{2}$$

$$-12=\frac{\cancel{4}\cdot 3}{\cancel{4}}t-\frac{\cancel{2}\cdot 2}{\cancel{2}}$$

$$-12=3t-2$$

$$-12+2=3t-2+2 \quad \text{Adding 2 to both sides}$$

$$-10=3t$$

$$\frac{-10}{3}=\frac{3t}{3} \quad \text{Dividing both sides by 3}$$

$$-\frac{10}{3}=t$$

The number $-\frac{10}{3}$ checks and is the solution.

35.

$$\frac{4}{3}-\frac{5}{6}x=\frac{3}{2}$$

$$6\left(\frac{4}{3}-\frac{5}{6}x\right)=6\cdot\frac{3}{2} \quad \text{Multiplying both sides by the LCD, 6}$$

$$\frac{6\cdot 4}{3}-\frac{6\cdot 5}{6}x=\frac{6\cdot 3}{2}$$

$$\frac{2\cdot\cancel{3}\cdot 4}{\cancel{3}}-\frac{\cancel{6}\cdot 5}{\cancel{6}}x=\frac{\cancel{2}\cdot 3\cdot 3}{\cancel{2}}$$

$$8-5x=9$$

$$8-5x-8=9-8 \quad \text{Subtracting 8 from both sides}$$

$$-5x=1$$

$$\frac{-5x}{-5}=\frac{1}{-5} \quad \text{Dividing both sides by } -5$$

$$x=-\frac{1}{5}$$

The number $-\frac{1}{5}$ checks and is the solution.

37.

$$-\frac{3}{4}=-\frac{5}{6}-\frac{1}{2}x$$

$$12\left(-\frac{3}{4}\right)=12\left(-\frac{5}{6}-\frac{1}{2}x\right) \quad \text{Multiplying both sides by the LCD, 12}$$

$$-\frac{12\cdot 3}{4}=-\frac{12\cdot 5}{6}-\frac{12\cdot 1}{2}x$$

$$-\frac{\cancel{4}\cdot 3\cdot 3}{\cancel{4}}=-\frac{2\cdot\cancel{6}\cdot 5}{\cancel{6}}-\frac{\cancel{2}\cdot 6}{\cancel{2}}x$$

$$-9=-10-6x$$

$$-9+10=-10-6x+10 \quad \text{Adding 10 to both sides}$$

$$1=-6x$$

$$\frac{1}{-6}=\frac{-6x}{-6} \quad \text{Dividing both sides by } -6$$

$$-\frac{1}{6}=x$$

The number $-\frac{1}{6}$ checks and is the solution.

39. $\frac{4}{3}-\frac{1}{5}t=\frac{3}{4}$

$60\left(\frac{4}{3}-\frac{1}{5}t\right)=60\cdot\frac{3}{4}$ Multiplying both sides by the LCD, 60

$\frac{60\cdot 4}{3}-\frac{60\cdot 1}{5}t=\frac{60\cdot 3}{4}$

$\frac{\not{3}\cdot 20\cdot 4}{\not{3}}-\frac{\not{5}\cdot 12}{\not{5}}t=\frac{\not{4}\cdot 15\cdot 3}{\not{4}}$

$80-12t=45$

$80-12t-80=45-80$ Subtracting 80 from both sides

$-12t=-35$

$\frac{-12t}{-12}=\frac{-35}{-12}$

$t=\frac{35}{12}$

The number $\frac{35}{12}$ checks and is the solution.

41. $39\div(-3)=-13$

Think: What number multiplied by -3 gives 39? The number is -13.

43. $(-72)\div(-4)=18$

Think: What number multiplied by -4 gives -72? The number is 18.

45. -200 represents the \$200 withdrawal, 90 represents the \$90 deposit, and -40 represents the \$40 withdrawal. We add these numbers to find the change in the balance.

$$-200+90+(-40)=-110+(-40)=-150$$

The account balance decreased by \$150.

47. $\frac{10}{7}\div(2m)=\frac{10}{7}\cdot\frac{1}{2m}=\frac{10\cdot 1}{7\cdot 2m}=\frac{2\cdot 5\cdot 1}{7\cdot 2\cdot m}=$
$\frac{2}{2}\cdot\frac{5\cdot 1}{7\cdot m}=\frac{5}{7m}$

49. Use a calculator.

$\frac{553}{2451}a-\frac{13}{57}=\frac{29}{43}$

$\frac{553}{2451}a-\frac{13}{57}+\frac{13}{57}=\frac{29}{43}+\frac{13}{57}$

$\frac{553}{2451}a=\frac{29}{43}+\frac{13}{57}$

$\frac{553}{2451}a=\frac{29}{43}\cdot\frac{57}{57}+\frac{13}{57}\cdot\frac{43}{43}$

$\frac{553}{2451}a=\frac{1653}{43\cdot 57}+\frac{559}{57\cdot 43}$

$\frac{553}{2451}=\frac{2212}{2451}$

$\frac{2451}{553}\cdot\frac{553}{2451}a=\frac{2451}{553}\cdot\frac{2212}{2451}$

$a=4$

The solution is 4.

51. Use a calculator.

$\frac{11}{17}=\frac{13}{41}-\frac{23}{29}t$

$\frac{11}{17}-\frac{13}{41}=\frac{13}{41}-\frac{23}{29}t-\frac{13}{41}$

$\frac{11}{17}\cdot\frac{41}{41}-\frac{13}{41}\cdot\frac{17}{17}=-\frac{23}{29}t$

$\frac{451}{697}-\frac{221}{697}=-\frac{23}{29}t$

$\frac{230}{697}=-\frac{23}{29}t$

$-\frac{29}{23}\left(\frac{230}{697}\right)=-\frac{29}{23}\left(-\frac{23}{29}t\right)$

$-\frac{29\cdot\not{23}\cdot 10}{\not{23}\cdot 697}=t$

$-\frac{290}{697}=t$

The solution is $-\frac{290}{697}$.

53. $\frac{47}{5}-\frac{a}{4}=\frac{44}{7}$

$\frac{47}{5}-\frac{a}{4}-\frac{47}{5}=\frac{44}{7}-\frac{47}{5}$

$-\frac{a}{4}=\frac{44}{7}\cdot\frac{5}{5}-\frac{47}{5}\cdot\frac{7}{7}$

$-\frac{a}{4}=\frac{220}{35}-\frac{329}{35}$

$-\frac{a}{4}=-\frac{109}{35}$

$-4\left(-\frac{a}{4}\right)=-4\left(-\frac{109}{35}\right)$

$a=\frac{436}{35}$

The solution is $\frac{436}{35}$.

55. $\frac{5}{4}x+x+\frac{5}{2}+6+2=15$

$\frac{5}{4}x+x=15-\frac{5}{2}-6-2$

$\frac{5}{4}x+x\cdot\frac{4}{4}=7-\frac{5}{2}$

$\frac{5}{4}x+\frac{4}{4}x=7\cdot\frac{2}{2}-\frac{5}{2}$

$\frac{9}{4}x=\frac{14}{2}-\frac{5}{2}$

$\frac{9}{4}x=\frac{9}{2}$

$\frac{4}{9}\cdot\frac{9}{4}x=\frac{4}{9}\cdot\frac{9}{2}$

$x=\frac{\not{2}\cdot 2\cdot\not{9}}{\not{9}\cdot\not{2}\cdot 1}$

$x=2$

The value of x is 2 cm.

Exercise Set 4.5

1. $7\frac{2}{3} = \frac{23}{3}$

a) Multiply: $3 \cdot 7 = 21$.

b) Add: $21 + 2 = 23$.

c) Keep the denominator.

3. $6\frac{1}{4} = \frac{25}{4}$

a) Multiply: $6 \cdot 4 = 24$.

b) Add: $24 + 1 = 25$.

c) Keep the denominator.

5. $-20\frac{1}{8} = -\frac{161}{8}$ $(20 \cdot 8 = 160; 160 + 1 = 161;$ include the negative sign)

7. $5\frac{1}{10} = \frac{51}{10}$ $(5 \cdot 10 = 50;\ 50 + 1 = 51)$

9. $20\frac{3}{5} = \frac{103}{5}$ $(20 \cdot 5 = 100;\ 100 + 3 = 103)$

11. $-33\frac{1}{3} = -\frac{100}{3}$ $(33 \cdot 3 = 99;\ 99 + 1 = 100;$ include the negative sign)

13. $1\frac{5}{8} = \frac{13}{8}$ $(1 \cdot 8 = 8;\ 8 + 5 = 13)$

15. $-12\frac{3}{4} = -\frac{51}{4}$ $(12 \cdot 4 = 48;\ 48 + 3 = 51;$ include the negative sign)

17. $5\frac{7}{10} = \frac{57}{10}$ $(5 \cdot 10 = 50;\ 50 + 7 = 57)$

19. $-5\frac{7}{100} = -\frac{507}{100}$ $(5 \cdot 100 = 500;\ 500 + 7 = 507;$ include the negative sign)

21. To convert $\frac{16}{3}$ to a mixed numeral, we divide.

$$\begin{array}{r} 5 \\ 3\overline{)16} \\ \underline{15} \\ 1 \end{array} \qquad \frac{16}{3} = 5\frac{1}{3}$$

23. To convert $\frac{45}{6}$ to a mixed numeral, we divide.

$$\begin{array}{r} 7 \\ 6\overline{)45} \\ \underline{42} \\ 3 \end{array} \qquad \frac{45}{6} = 7\frac{3}{6} = 7\frac{1}{2}$$

25.

$$\begin{array}{r} 5 \\ 10\overline{)57} \\ \underline{50} \\ 7 \end{array} \qquad \frac{57}{10} = 5\frac{7}{10}$$

27.

$$\begin{array}{r} 7 \\ 9\overline{)65} \\ \underline{63} \\ 2 \end{array} \qquad \frac{65}{9} = 7\frac{2}{9}$$

29.

$$\begin{array}{r} 5 \\ 6\overline{)33} \\ \underline{30} \\ 3 \end{array} \qquad \frac{33}{6} = 5\frac{3}{6} = 5\frac{1}{2}$$

Since $\frac{33}{6} = 5\frac{1}{2}$, we have $\frac{-33}{6} = -5\frac{1}{2}$.

31.

$$\begin{array}{r} 11 \\ 4\overline{)46} \\ \underline{4} \\ 6 \\ \underline{4} \\ 2 \end{array} \qquad \frac{46}{4} = 11\frac{2}{4} = 11\frac{1}{2}$$

33.

$$\begin{array}{r} 1 \\ 8\overline{)12} \\ \underline{8} \\ 4 \end{array} \qquad \frac{12}{8} = 1\frac{4}{8} = 1\frac{1}{2}$$

Since $\frac{12}{8} = 1\frac{1}{2}$, we have $\frac{-12}{8} = -1\frac{1}{2}$.

35.

$$\begin{array}{r} 61 \\ 5\overline{)307} \\ \underline{30} \\ 7 \\ \underline{5} \\ 2 \end{array} \qquad \frac{307}{5} = 61\frac{2}{5}$$

37.

$$\begin{array}{r} 8 \\ 50\overline{)413} \\ \underline{400} \\ 13 \end{array} \qquad \frac{413}{50} = 8\frac{13}{50}$$

Since $\frac{413}{50} = 8\frac{13}{50}$, we have $-\frac{413}{50} = -8\frac{13}{50}$.

39. We first divide as usual.

$$\begin{array}{r} 108 \\ 8\overline{)869} \\ \underline{8} \\ 69 \\ \underline{64} \\ 5 \end{array}$$

The answer is 108 R 5. We write a mixed numeral for the quotient as follows: $108\frac{5}{8}$.

41. We first divide as usual.

$$\begin{array}{r} 906 \\ 7\overline{)6345} \\ \underline{63} \\ 45 \\ \underline{42} \\ 3 \end{array}$$

The answer is 906 R 3. We write a mixed numeral for the quotient as follows: $906\frac{3}{7}$.

43.
$$\begin{array}{r} 4\,0 \\ 21\,\overline{)\,8\,5\,2} \\ \underline{8\,4} \\ 1\,2 \end{array}$$

We get $40\frac{12}{21}$. This simplifies as $40\frac{4}{7}$.

45. First we find $302 \div 15$.
$$\begin{array}{r} 2\,0 \\ 15\,\overline{)\,3\,0\,2} \\ \underline{3\,0} \\ 2 \\ \underline{0} \\ 2 \end{array} \qquad \frac{302}{15} = 20\frac{2}{15}$$

Since $302 \div 15 = 20\frac{2}{15}$, we have $-302 \div 15 = -20\frac{2}{15}$.

47. First we find $471 \div 21$.
$$\begin{array}{r} 2\,2 \\ 21\,\overline{)\,4\,7\,1} \\ \underline{4\,2} \\ 5\,1 \\ \underline{4\,2} \\ 9 \end{array} \qquad \frac{471}{21} = 22\frac{9}{21} = 22\frac{3}{7}$$

Since $471 \div 21 = 22\frac{3}{7}$, we have $471 \div (-21) = -22\frac{3}{7}$.

49. There are 5 humanitarian organizations in the list. We add the expenses for these organizations and then divide by 5.
$$\frac{\$1 + \$2 + \$4 + \$11 + \$14}{5} = \frac{\$32}{5} = \$6\frac{2}{5}$$

51. We add the expenses for the first five organizations in the list and then divide by 5.
$$\frac{\$1 + \$2 + \$3 + \$4 + \$6}{5} = \frac{\$16}{5} = \$3\frac{1}{5}$$

53.
$$\begin{aligned} \frac{7}{9}\cdot\frac{24}{21} &= \frac{7\cdot 24}{9\cdot 21} \\ &= \frac{7\cdot 3\cdot 8}{3\cdot 3\cdot 3\cdot 7} \\ &= \frac{3\cdot 7}{3\cdot 7}\cdot\frac{8}{3\cdot 3} \\ &= \frac{8}{9} \end{aligned}$$

55.
$$\frac{7}{10}\cdot\frac{5}{14} = \frac{7\cdot 5}{10\cdot 14} = \frac{7\cdot 5\cdot 1}{2\cdot 5\cdot 2\cdot 7} = \frac{7\cdot 5}{7\cdot 5}\cdot\frac{1}{2\cdot 2} = \frac{1}{2\cdot 2} = \frac{1}{4}$$

57.
$$\begin{aligned} -\frac{17}{25}\cdot\frac{15}{34} &= -\frac{17\cdot 15}{25\cdot 34} \\ &= -\frac{17\cdot 3\cdot 5}{5\cdot 5\cdot 2\cdot 17} \\ &= -\frac{17\cdot 5}{17\cdot 5}\cdot\frac{3}{5\cdot 2} \\ &= -\frac{3}{10} \end{aligned}$$

59. Use a calculator.
$$\frac{128{,}236}{541} = 237\frac{19}{541}$$

61.
$$\begin{aligned} \frac{56}{7} + \frac{2}{3} &= 8 + \frac{2}{3} \qquad (56 \div 7 = 8) \\ &= 8\frac{2}{3} \end{aligned}$$

63.
$$\frac{12}{5} + \frac{19}{15} = \frac{36}{15} + \frac{19}{15} = \frac{55}{15}$$

$$\begin{array}{r} 3 \\ 15\,\overline{)\,5\,5} \\ \underline{4\,5} \\ 1\,0 \end{array} \qquad \frac{55}{15} = 3\frac{10}{15} = 3\frac{2}{3}$$

Thus, $\frac{12}{5} + \frac{19}{15} = 3\frac{2}{3}$.

65.
$$\begin{array}{r} 5\,2 \\ 7\,\overline{)\,3\,6\,5} \\ \underline{3\,5} \\ 1\,5 \\ \underline{1\,4} \\ 1 \end{array} \qquad \frac{365}{7} = 52\frac{1}{7}$$

Chapter 4 Mid-Chapter Review

1. The statement is true. To determine which of two numbers is greater when there is a common denominator, we compare the numerators.

2. $1\frac{1}{2} = \frac{3}{2} = \frac{3}{2}\cdot\frac{2}{2} = \frac{6}{4}$; the statement is true.

3. The statement is false. The least common multiple of two natural numbers is the smallest number that is a multiple of both numbers.

4. The statement is false. To add fractions when denominators are the same, add the numerators and keep the denominator.

5.
$$\begin{aligned} \frac{11}{42} - \frac{3}{35} &= \frac{11}{2\cdot 3\cdot 7} - \frac{3}{5\cdot 7} \\ &= \frac{11}{2\cdot 3\cdot 7}\cdot\frac{5}{5} - \frac{3}{5\cdot 7}\cdot\frac{2\cdot 3}{2\cdot 3} \\ &= \frac{11\cdot 5}{2\cdot 3\cdot 7\cdot 5} - \frac{3\cdot 2\cdot 3}{5\cdot 7\cdot 2\cdot 3} \\ &= \frac{55}{2\cdot 3\cdot 5\cdot 7} - \frac{18}{2\cdot 3\cdot 5\cdot 7} \\ &= \frac{55 - 18}{2\cdot 3\cdot 5\cdot 7} = \frac{37}{210} \end{aligned}$$

6.
$$\begin{aligned} x + \frac{1}{8} &= \frac{2}{3} \\ x + \frac{1}{8} - \frac{1}{8} &= \frac{2}{3} - \frac{1}{8} \\ x + 0 &= \frac{2}{3}\cdot\frac{8}{8} - \frac{1}{8}\cdot\frac{3}{3} \\ x &= \frac{16}{24} - \frac{3}{24} \\ x &= \frac{13}{24} \end{aligned}$$

The solution is $\frac{13}{24}$.

7. For 45 and 50:

$45 = 3 \cdot 3 \cdot 5$
$50 = 2 \cdot 5 \cdot 5$

The LCM is $2 \cdot 3 \cdot 3 \cdot 5 \cdot 5$, or 450.

For 50 and 80:

$50 = 2 \cdot 5 \cdot 5$
$80 = 2 \cdot 2 \cdot 2 \cdot 2 \cdot 5$

The LCM is $2 \cdot 2 \cdot 2 \cdot 2 \cdot 5 \cdot 5$, or 400.

For 30 and 24:

$30 = 2 \cdot 3 \cdot 5$
$24 = 2 \cdot 2 \cdot 2 \cdot 3$

The LCM is $2 \cdot 2 \cdot 2 \cdot 3 \cdot 5$, or 120.

For 18, 24, and 80:

$18 = 2 \cdot 3 \cdot 3$
$24 = 2 \cdot 2 \cdot 2 \cdot 3$
$80 = 2 \cdot 2 \cdot 2 \cdot 2 \cdot 5$

The LCM is $2 \cdot 2 \cdot 2 \cdot 2 \cdot 3 \cdot 3 \cdot 5$, or 720.

For 30, 45, and 50:

$30 = 2 \cdot 3 \cdot 5$
$45 = 3 \cdot 3 \cdot 5$
$50 = 2 \cdot 5 \cdot 5$

The LCM is $2 \cdot 3 \cdot 3 \cdot 5 \cdot 5$, or 450.

8. $\frac{1}{5} + \frac{7}{45} = \frac{1}{5} \cdot \frac{9}{9} + \frac{7}{45} = \frac{9}{45} + \frac{7}{45} = \frac{16}{45}$

9. $\frac{5}{6} + \frac{2}{3} + \frac{7}{12} = \frac{5}{6} \cdot \frac{2}{2} + \frac{2}{3} \cdot \frac{4}{4} + \frac{7}{12} = \frac{10}{12} + \frac{8}{12} + \frac{7}{12} = \frac{25}{12}$

10. $\frac{2}{9} - \frac{1}{6} = \frac{2}{9} \cdot \frac{2}{2} - \frac{1}{6} \cdot \frac{3}{3} = \frac{4}{18} - \frac{3}{18} = \frac{1}{18}$

11.
$$\begin{aligned} \frac{1}{15} - \frac{5}{18} &= \frac{1}{3 \cdot 5} - \frac{5}{2 \cdot 3 \cdot 3} \\ &= \frac{1}{3 \cdot 5} \cdot \frac{2 \cdot 3}{2 \cdot 3} - \frac{5}{2 \cdot 3 \cdot 3} \cdot \frac{5}{5} \\ &= \frac{6}{3 \cdot 5 \cdot 2 \cdot 3} - \frac{25}{2 \cdot 3 \cdot 3 \cdot 5} \\ &= \frac{6 - 25}{2 \cdot 3 \cdot 3 \cdot 5} = -\frac{19}{90} \end{aligned}$$

12.
$$\begin{aligned} \frac{19}{48} - \frac{11}{30} &= \frac{19}{8 \cdot 6} - \frac{11}{5 \cdot 6} \\ &= \frac{19}{8 \cdot 6} \cdot \frac{5}{5} - \frac{11}{5 \cdot 6} \cdot \frac{8}{8} \\ &= \frac{95}{8 \cdot 6 \cdot 5} - \frac{88}{5 \cdot 6 \cdot 8} \\ &= \frac{95 - 88}{5 \cdot 6 \cdot 8} = \frac{7}{240} \end{aligned}$$

13.
$$\begin{aligned} -\frac{3}{8}x + \frac{1}{12}x &= -\frac{3}{8} \cdot \frac{3}{3} \cdot x + \frac{1}{12} \cdot \frac{2}{2} \cdot x \\ &= -\frac{9}{24}x + \frac{2}{24}x \\ &= \frac{-9}{24}x + \frac{2}{24}x \\ &= \frac{-9 + 2}{24}x \\ &= \frac{-7}{24}x, \text{ or } -\frac{7}{24}x \end{aligned}$$

14.
$$\begin{aligned} \frac{-3}{40} + \frac{-5}{24} &= \frac{-3}{40} \cdot \frac{3}{3} + \frac{-5}{24} \cdot \frac{5}{5} \\ &= \frac{-9}{120} + \frac{-25}{120} \\ &= \frac{-9 + (-25)}{120} \\ &= \frac{-34}{120} = \frac{-17 \cdot \cancel{2}}{60 \cdot \cancel{2}} \\ &= \frac{-17}{60}, \text{ or } -\frac{17}{60} \end{aligned}$$

15.
$$\begin{aligned} \frac{8}{65} - \frac{2}{35} &= \frac{8}{5 \cdot 13} - \frac{2}{5 \cdot 7} \\ &= \frac{8}{5 \cdot 13} \cdot \frac{7}{7} - \frac{2}{5 \cdot 7} \cdot \frac{13}{13} \\ &= \frac{56}{5 \cdot 13 \cdot 7} - \frac{26}{5 \cdot 7 \cdot 13} \\ &= \frac{30}{5 \cdot 7 \cdot 13} = \frac{5 \cdot 6}{5 \cdot 7 \cdot 13} \\ &= \frac{5}{5} \cdot \frac{6}{7 \cdot 13} = \frac{6}{91} \end{aligned}$$

16. ***Familiarize.*** Let d = the total distance Miguel jogs.

Translate.

$\underbrace{\text{First distance}}$ plus $\underbrace{\text{Second distance}}$ is $\underbrace{\text{Total distance}}$

$$\frac{4}{5} + \frac{2}{3} = d$$

Solve. We carry out the addition.

$$\frac{4}{5} + \frac{2}{3} = \frac{4}{5} \cdot \frac{3}{3} + \frac{2}{3} \cdot \frac{5}{5} = \frac{12}{15} + \frac{10}{15} = \frac{22}{15}$$

Thus, $\frac{22}{15} = d$.

Check. We can repeat the calculation. Also note that the result is greater than either of the individual distances, so the answer seems reasonable.

State. Miguel jogs $\frac{22}{15}$ mi in all.

17. ***Familiarize.*** Let t = the number of hours Kirby spent playing Brain Challenge.

Translate.

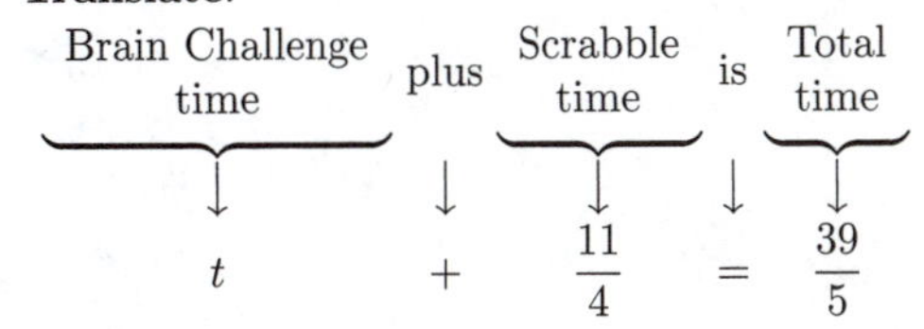

$$t + \frac{11}{4} = \frac{39}{5}$$

Solve. We subtract $\frac{11}{4}$ on both sides of the equation.

$$t + \frac{11}{4} = \frac{39}{5}$$
$$t + \frac{11}{4} - \frac{11}{4} = \frac{39}{5} - \frac{11}{4}$$
$$t + 0 = \frac{39}{5} \cdot \frac{4}{4} - \frac{11}{4} \cdot \frac{5}{5}$$
$$t = \frac{156}{20} - \frac{55}{20}$$
$$t = \frac{101}{20}$$

Check. We can add to perform a check.

$$\frac{101}{20} + \frac{11}{4} = \frac{101}{20} + \frac{11}{4} \cdot \frac{5}{5} = \frac{101}{20} + \frac{55}{20} = \frac{156}{20} =$$
$$\frac{4 \cdot 39}{4 \cdot 5} = \frac{4}{4} \cdot \frac{39}{5} = \frac{39}{5}$$

The answer checks.

State. Kirby played Brain Challenge for $\frac{101}{20}$ hr.

18. $\frac{4}{9} = \frac{4}{3 \cdot 3}$; $\frac{3}{10} = \frac{3}{2 \cdot 5}$; $\frac{2}{7}$; $\frac{1}{5}$

The LCD is $2 \cdot 3 \cdot 3 \cdot 5 \cdot 7$, or 630. We write each fraction with this denominator and then compare numerators.

$$\frac{4}{9} = \frac{4}{3 \cdot 3} \cdot \frac{2 \cdot 5 \cdot 7}{2 \cdot 5 \cdot 7} = \frac{280}{630}$$
$$\frac{3}{10} = \frac{3}{2 \cdot 5} \cdot \frac{3 \cdot 3 \cdot 7}{3 \cdot 3 \cdot 7} = \frac{189}{630}$$
$$\frac{2}{7} = \frac{2}{7} \cdot \frac{2 \cdot 3 \cdot 3 \cdot 5}{2 \cdot 3 \cdot 3 \cdot 5} = \frac{180}{630}$$
$$\frac{1}{5} = \frac{1}{5} \cdot \frac{2 \cdot 3 \cdot 3 \cdot 7}{2 \cdot 3 \cdot 3 \cdot 7} = \frac{126}{630}$$

Arranging the numbers in order from smallest to largest, we have

$$\frac{1}{5}, \frac{2}{7}, \frac{3}{10}, \frac{4}{9}.$$

19.
$$\frac{2}{5} + x = \frac{9}{16}$$
$$\frac{2}{5} + x - \frac{2}{5} = \frac{9}{16} - \frac{2}{5} \quad \text{Subtracting } \frac{2}{5}$$
$$x + 0 = \frac{9}{16} \cdot \frac{5}{5} - \frac{2}{5} \cdot \frac{16}{16} \quad \text{The LCD is } 16 \cdot 5 \text{, or } 80$$
$$x = \frac{45}{80} - \frac{32}{80}$$
$$x = \frac{13}{80}$$

The solution is $\frac{13}{80}$.

20.
$$\frac{3}{4}x + 1 = \frac{1}{3}$$
$$\frac{3}{4}x + 1 - 1 = \frac{1}{3} - 1$$
$$\frac{3}{4}x + 0 = \frac{1}{3} - 1 \cdot \frac{3}{3}$$
$$\frac{3}{4}x = \frac{1}{3} - \frac{3}{3}$$
$$\frac{3}{4}x = -\frac{2}{3}$$
$$\frac{4}{3} \cdot \frac{3}{4}x = \frac{4}{3} \cdot \left(-\frac{2}{3}\right)$$
$$1x = -\frac{8}{9}$$
$$x = -\frac{8}{9}$$

The solution is $-\frac{8}{9}$.

21.
$$\begin{array}{r} 17 \\ 15 \overline{)263} \\ \underline{15} \\ 113 \\ \underline{105} \\ 8 \end{array}$$

The answer is 17 R 8. A mixed numeral for the answer is written $17\frac{8}{15}$.

22. $9\frac{3}{8} = \frac{75}{8}$ $(8 \cdot 9 = 72,\ 72 + 3 = 75)$

Answer C is correct.

23.
$$\begin{array}{r} 9 \\ 4 \overline{)39} \\ \underline{36} \\ 3 \end{array}$$

$\frac{39}{4} = 9\frac{3}{4}$, so $-\frac{39}{4} = -9\frac{3}{4}$.

Answer C is correct.

24. No; if one number is a multiple of the other, for example, the LCM is the larger of the numbers.

25. We multiply by 1, using the notation n/n, to express each fraction in terms of the least common denominator.

26. Write $\frac{8}{5}$ as $\frac{16}{10}$ and $\frac{8}{2}$ as $\frac{40}{10}$ and use a drawing to show that $\frac{16}{10} - \frac{40}{10} \neq \frac{8}{3}$. You could also find the sum $\frac{8}{3} + \frac{8}{2}$ and show that it is not $\frac{8}{5}$.

27. No; $2\frac{1}{3} = \frac{7}{3}$ but $2 \cdot \frac{1}{3} = \frac{2}{3}$.

Exercise Set 4.6

1. $$\begin{array}{r} 6 \\ +5\,\frac{2}{5} \\ \hline 11\,\frac{2}{5} \end{array}$$

3. $$\begin{array}{r} 2\,\frac{7}{8} \\ +6\,\frac{5}{8} \\ \hline 8\,\frac{12}{8} \end{array} = 8 + \frac{12}{8} = 8 + 1\frac{1}{2} = 9\frac{1}{2}$$

To find a mixed numeral for $\frac{12}{8}$ we divide:

$$8\overline{)\,12}\quad \text{quotient } 1,\ \ 8,\ \text{remainder } 4 \qquad \frac{12}{8} = 1\frac{4}{8} = 1\frac{1}{2}$$

5. The LCD is 12.

$$\begin{array}{rcl} 4\,\boxed{\frac{1}{4}\cdot\frac{3}{3}} & = & 4\,\frac{3}{12} \\ +1\,\frac{1}{12} & = & +1\,\frac{1}{12} \\ \hline & & 5\,\frac{4}{12} = 5\frac{1}{3} \end{array}$$

7. The LCD is 12.

$$\begin{array}{rcl} 7\,\boxed{\frac{3}{4}\cdot\frac{3}{3}} & = & 7\,\frac{9}{12} \\ +5\,\boxed{\frac{5}{6}\cdot\frac{2}{2}} & = & +5\,\frac{10}{12} \\ \hline & & 12\,\frac{19}{12} = 12 + \frac{19}{12} \\ & & = 12 + 1\frac{7}{12} \\ & & = 13\frac{7}{12} \end{array}$$

9. The LCD is 10.

$$\begin{array}{rcl} 3\,\boxed{\frac{2}{5}\cdot\frac{2}{2}} & = & 3\,\frac{4}{10} \\ +8\,\frac{7}{10} & = & +8\,\frac{7}{10} \\ \hline & & 11\,\frac{11}{10} = 11 + \frac{11}{10} \\ & & = 11 + 1\frac{1}{10} \\ & & = 12\frac{1}{10} \end{array}$$

11. The LCD is 24.

$$\begin{array}{rcl} 6\,\boxed{\frac{3}{8}\cdot\frac{3}{3}} & = & 6\,\frac{9}{24} \\ +10\,\boxed{\frac{5}{6}\cdot\frac{4}{4}} & = & +10\,\frac{20}{24} \\ \hline & & 16\,\frac{29}{24} = 16 + \frac{29}{24} \\ & & = 16 + 1\frac{5}{24} \\ & & = 17\frac{5}{24} \end{array}$$

13. The LCD is 10.

$$\begin{array}{rcl} 18\,\boxed{\frac{4}{5}\cdot\frac{2}{2}} & = & 18\,\frac{8}{10} \\ +2\,\frac{7}{10} & = & +2\,\frac{7}{10} \\ \hline & & 20\,\frac{15}{10} = 20 + \frac{15}{10} \\ & & = 20 + 1\frac{5}{10} \\ & & = 21\frac{5}{10} \\ & & = 21\frac{1}{2} \end{array}$$

15. The LCD is 8.

$$\begin{array}{rcl} 14\,\frac{5}{8} & = & 14\,\frac{5}{8} \\ +13\,\boxed{\frac{1}{4}\cdot\frac{2}{2}} & = & +13\,\frac{2}{8} \\ \hline & & 27\,\frac{7}{8} \end{array}$$

17. $$\begin{array}{r} 8\frac{9}{10} \\ -1\frac{7}{10} \\ \hline 7\frac{2}{10} \end{array} = 7\frac{1}{5}$$

19. The LCD is 10.

$$\begin{array}{rcl} 9\,\boxed{\frac{3}{5}\cdot\frac{2}{2}} & = & 9\,\frac{6}{10} \\ -3\,\boxed{\frac{1}{2}\cdot\frac{5}{5}} & = & -3\,\frac{5}{10} \\ \hline & & 6\,\frac{1}{10} \end{array}$$

21. $\begin{array}{rcr} 4\frac{1}{5} & = & 3\frac{6}{5} \\ -2\frac{3}{5} & = & -2\frac{3}{5} \\ \hline & & 1\frac{3}{5} \end{array}$

Since $\frac{1}{5}$ is smaller than $\frac{3}{5}$, we cannot subtract until we borrow:

$4\frac{1}{5} = 3+\frac{5}{5}+\frac{1}{5} = 3+\frac{6}{5} = 3\frac{6}{5}$

23. $\begin{array}{rcr} 19 & = & 18\frac{4}{4} \\ -\ 5\frac{3}{4} & = & -\ 5\frac{3}{4} \\ \hline & & 13\frac{1}{4} \end{array}$ $\left(19 = 18+1 = 18+\frac{4}{4} = 18\frac{4}{4}\right)$

25. $\begin{array}{rcr} 34 & = & 33\frac{8}{8} \\ -\ 18\frac{5}{8} & = & -\ 18\frac{5}{8} \\ \hline & & 15\frac{3}{8} \end{array}$ $\left(34 = 33+1 = 33+\frac{8}{8} = 33\frac{8}{8}\right)$

27. The LCD is 12.

$\begin{array}{rcrcr} 21\left[\frac{1}{6}\cdot\frac{2}{2}\right] & = & 21\ \frac{2}{12} & = & 20\ \frac{14}{12} \\ -13\left[\frac{3}{4}\cdot\frac{3}{3}\right] & = & -13\ \frac{9}{12} & = & -13\ \frac{9}{12} \\ \hline & & \uparrow & & 7\ \frac{5}{12} \end{array}$

$\left(\text{Since } \frac{2}{12} \text{ is smaller than } \frac{9}{12}, \text{ we cannot subtract until we borrow: } 21\frac{2}{12} = 20+\frac{12}{12}+\frac{2}{12} = 20+\frac{14}{12} = 20\frac{14}{12}.\right)$

29. The LCD is 18.

$\begin{array}{rcrcr} 25\left[\frac{1}{9}\cdot\frac{2}{2}\right] & = & 25\ \frac{2}{18} & = & 24\ \frac{20}{18} \\ -13\left[\frac{5}{6}\cdot\frac{3}{3}\right] & = & -13\ \frac{15}{18} & = & -13\ \frac{15}{18} \\ \hline & & \uparrow & & 11\ \frac{5}{18} \end{array}$

$\left(\text{Since } \frac{2}{18} \text{ is smaller than } \frac{15}{18}, \text{ we cannot subtract until we borrow: } 25\frac{2}{18} = 24+\frac{18}{18}+\frac{2}{18} = 24+\frac{20}{18} = 24\frac{20}{18}.\right)$

31. $1\frac{1}{8}t + 7\frac{5}{8}t$

$= \left(1\frac{1}{8} + 7\frac{5}{8}\right)t$ Using the distributive law

$= 8\frac{6}{8}t$ Adding

$= 8\frac{3}{4}t$

33. $9\frac{1}{2}x - 7\frac{1}{2}x$

$= \left(9\frac{1}{2} - 7\frac{1}{2}\right)x$ Using the distributive law

$= 2x$ Subtracting

35. $5\frac{9}{10}y + 2\frac{2}{5}y$

$= \left(5\frac{9}{10} + 2\frac{2}{5}\right)y$ Using the distributive law

$= \left(5\frac{9}{10} + 2\frac{4}{10}\right)y$ The LCD is 10.

$= 7\frac{13}{10}y = 8\frac{3}{10}y$

37. $37\frac{5}{9}t - 25\frac{2}{3}t$

$= \left(37\frac{5}{9} - 25\frac{2}{3}\right)t$ Using the distributive law

$= \left(37\frac{5}{9} - 25\frac{6}{9}\right)t$ The LCD is 9.

$= \left(36\frac{14}{9} - 25\frac{6}{9}\right)t$

$= 11\frac{8}{9}t$

39. $2\frac{5}{6}x + 7\frac{3}{8}x$

$= \left(2\frac{5}{6} + 7\frac{3}{8}\right)x$ Using the distributive law

$= \left(2\frac{20}{24} + 7\frac{9}{24}\right)x$ The LCD is 24.

$= 9\frac{29}{24}x = 10\frac{5}{24}x$

41. $11a - 8\frac{2}{3}a$

$= \left(11 - 8\frac{2}{3}\right)a$ Using the distributive law

$= \left(10\frac{3}{3} - 8\frac{2}{3}\right)a$

$= 2\frac{1}{3}a$

43. ***Familiarize***. We draw a picture, letting x = the amount of pipe that was used, in inches.

$\vdash$ $10\frac{5}{16}$ in. $+$ $8\frac{3}{4}$ in. $\dashv$

$\vdash$ x $\dashv$

Translate. We write an addition sentence.

$\underbrace{\text{First length}}\ \text{plus}\ \underbrace{\text{Second length}}\ \text{is}\ \underbrace{\text{Total length}}$

$10\frac{5}{16} \quad + \quad 8\frac{3}{4} \quad = \quad x$

Solve. We carry out the addition. The LCD is 16.

$$
\begin{array}{rcl}
10\,\frac{5}{16} & = & 10\,\frac{5}{16} \\
+\ 8\boxed{\frac{3}{4}\cdot\frac{4}{4}} & = & +\ 8\,\frac{12}{16} \\
\hline
 & & 18\,\frac{17}{16} = 18+\frac{17}{16} \\
 & & = 18+1\frac{1}{16} \\
 & & = 19\frac{1}{16}
\end{array}
$$

Check. We repeat the calculation. We also note that the total length is larger than either of the individual lengths, so the answer seems reasonable.

State. The plumber used $19\frac{1}{16}$ in. of pipe.

45. ***Familiarize.*** We let $w =$ the total weight of the meat.

Translate. We write an equation.

Weight of one package	plus	Weight of second package	is	Total weight
↓	↓	↓	↓	↓
$1\frac{2}{3}$	$+$	$5\frac{3}{4}$	$=$	w

Solve. We carry out the addition. The LCD is 12.

$$
\begin{array}{rcl}
1\boxed{\frac{2}{3}\cdot\frac{4}{4}} & = & 1\,\frac{8}{12} \\
+5\boxed{\frac{3}{4}\cdot\frac{3}{3}} & = & +5\,\frac{9}{12} \\
\hline
 & & 6\,\frac{17}{12} = 6+\frac{17}{12} \\
 & & = 6+1\frac{5}{12} \\
 & & = 7\frac{5}{12}
\end{array}
$$

Check. We repeat the calculation. We also note that the answer is larger than either of the individual weights, so the answer seems reasonable.

State. The total weight of the meat was $7\frac{5}{12}$ lb.

47. ***Familiarize.*** Let $g =$ the number of feet by which the distance between the goalposts was reduced.

Translate.

Reduced distance	plus	Amount of reduction	is	Original distance
↓	↓	↓	↓	↓
$18\frac{1}{2}$	$+$	g	$=$	$23\frac{1}{3}$

Solve.

$$18\frac{1}{2}+g = 23\frac{1}{3}$$

$$18\frac{1}{2}+g-18\frac{1}{2} = 23\frac{1}{3}-18\frac{1}{2}$$

$$g = 23\frac{1}{3}-18\frac{1}{2}$$

We carry out the subtraction of the right side of the equation.

$$
\begin{array}{rcccl}
23\boxed{\frac{1}{3}\cdot\frac{2}{2}} & = & 23\,\frac{2}{6} & = & 22\frac{8}{6} \\
-18\boxed{\frac{1}{2}\cdot\frac{3}{3}} & = & -18\,\frac{3}{6} & = & -18\frac{3}{6} \\
\hline
 & & & & 4\frac{5}{6}
\end{array}
$$

Thus, $g = 4\frac{5}{6}$.

Check. We can add the reduced distance and the amount of the reduction to determine if we get the original distance.

$$18\frac{1}{2}+4\frac{5}{6} = 18\frac{3}{6}+4\frac{5}{6} = 22\frac{8}{6} = 23\frac{2}{6} = 23\frac{1}{3}$$

The answer checks.

State. The distance between the goalposts was reduced by $4\frac{5}{6}$ ft.

49. ***Familiarize.*** We let $h =$ Tara's excess height.

Translate. We have a "how much more" situation.

Tom's height	plus	How much more height	is	Tara's height
↓	↓	↓	↓	↓
$59\frac{7}{12}$	$+$	h	$=$	66

Solve. We solve the equation as follows:

$$h = 66-59\frac{7}{12}$$

$$
\begin{array}{rcl}
66 & = & 65\frac{12}{12} \\
-\ 59\frac{7}{12} & = & -\ 59\frac{7}{12} \\
\hline
 & & 6\frac{5}{12}
\end{array}
$$

Check. We add Tara's excess height to Tom's height:

$$6\frac{5}{12}+59\frac{7}{12} = 65\frac{12}{12} = 66$$

The answer checks.

State. Tara is $6\frac{5}{12}$ in. taller.

51. ***Familiarize.*** We make a drawing. We let $t =$ the number of hours Sue worked on the third day.

|— $2\frac{1}{2}$ hr —|— $4\frac{1}{5}$ hr —|— t —|

|———— $10\frac{1}{2}$ hr ————|

Translate. We write an addition sentence.

$$2\frac{1}{2}+4\frac{1}{5}+t = 10\frac{1}{2}$$

Solve. This is a two-step problem.

First we add $2\frac{1}{2}+4\frac{1}{5}$ to find the time worked on the first two days. The LCD is 10.

$$\begin{array}{rcl} 2\boxed{\frac{1}{2}\cdot\frac{5}{5}} & = & 2\,\frac{5}{10} \\ +\,4\boxed{\frac{1}{5}\cdot\frac{2}{2}} & = & +\,4\,\frac{2}{10} \\ \hline & & 6\,\frac{7}{10} \end{array}$$

Then we subtract $6\frac{7}{10}$ from $10\frac{1}{2}$ to find the time worked on the third day. The LCD is 10.

$$6\frac{7}{10}+t=10\frac{1}{2}$$

$$t=10\frac{1}{2}-6\frac{7}{10}$$

$$\begin{array}{rcccl} 10\boxed{\frac{1}{2}\cdot\frac{5}{5}} & = & 10\,\frac{5}{10} & = & 9\,\frac{15}{10} \\ -\,6\,\frac{7}{10} & = & -\,6\,\frac{7}{10} & = & -\,6\,\frac{7}{10} \\ \hline & & & & 3\,\frac{8}{10}=3\frac{4}{5} \end{array}$$

Check. We repeat the calculations.

State. Sue worked $3\frac{4}{5}$ hr the third day.

53. ***Familiarize.*** We let $P=$ the perimeter of the mirror.

Translate. We add the lengths of the four sides of the mirror.

$$P=30\frac{1}{2}+36\frac{5}{8}+30\frac{1}{2}+36\frac{5}{8}$$

Solve. We carry out the addition. The LCD is 8.

$$\begin{array}{rcl} 30\boxed{\frac{1}{2}\cdot\frac{4}{4}} & = & 30\,\frac{4}{8} \\ 36\,\frac{5}{8} & = & 36\,\frac{5}{8} \\ 30\boxed{\frac{1}{2}\cdot\frac{4}{4}} & = & 30\,\frac{4}{8} \\ +36\,\frac{5}{8} & = & +\,36\,\frac{5}{8} \\ \hline & & 132\,\frac{18}{8}=132+\frac{18}{8} \\ & & =132+2\frac{2}{8} \\ & & =134\frac{2}{8} \\ & & =134\frac{1}{4} \end{array}$$

Check. We repeat the calculation. The answer checks.

State. The perimeter of the mirror is $134\frac{1}{4}$ in.

55. ***Familiarize.*** Let $f=$ the total number of flats planted.

Translate. We write an equation.

$$\underbrace{\text{Flats of impatiens}}_{4\frac{1}{2}} + \underbrace{\text{Flats of snapdragons}}_{6\frac{2}{3}} + \underbrace{\text{Flats of phlox}}_{3\frac{3}{8}} \;\text{is}\; \underbrace{\text{Total flats}}_{f}$$

$$4\frac{1}{2}+6\frac{2}{3}+3\frac{3}{8}=f$$

Solve. We add. The LCD is 24.

$$\begin{array}{rcl} 4\boxed{\frac{1}{2}\cdot\frac{12}{12}} & = & 4\,\frac{12}{24} \\ 6\boxed{\frac{2}{3}\cdot\frac{8}{8}} & = & 6\,\frac{16}{24} \\ +3\boxed{\frac{3}{8}\cdot\frac{3}{3}} & = & +3\,\frac{9}{24} \\ \hline & & 13\,\frac{37}{24}=13+\frac{37}{24} \\ & & =13+1\frac{13}{24} \\ & & =14\frac{13}{24} \end{array}$$

Check. We can repeat the calculation. Also note that the answer is reasonable since it is larger than any of the individual number of flats.

State. The landscaper planted $14\frac{13}{24}$ flats of flowers.

57. The length of each of the five sides is $5\frac{3}{4}$ yd. We add to find the distance around the figure.

$$5\frac{3}{4}+5\frac{3}{4}+5\frac{3}{4}+5\frac{3}{4}+5\frac{3}{4}=25\frac{15}{4}=25+3\frac{3}{4}=28\frac{3}{4}$$

The distance is $28\frac{3}{4}$ yd.

59. We add to find the perimeter.

$$\begin{aligned} &1\frac{5}{12}+\frac{17}{24}+1+1+\frac{17}{24} \\ &=1\frac{10}{24}+\frac{17}{24}+1+1+\frac{17}{24} \\ &=3\frac{44}{24}=3+\frac{44}{24} \\ &=3+1\frac{20}{24}=3+1\frac{5}{6} \\ &=4\frac{5}{6} \end{aligned}$$

The perimeter is $4\frac{5}{6}$ ft.

61. We see that d and the two smallest distances combined are the same as the largest distance. We translate and solve.

$$2\frac{3}{4}+d+2\frac{3}{4}=12\frac{7}{8}$$
$$d=12\frac{7}{8}-2\frac{3}{4}-2\frac{3}{4}$$
$$=10\frac{1}{8}-2\frac{3}{4}\quad \text{Subtracting } 2\frac{3}{4} \text{ from } 12\frac{7}{8}$$
$$=7\frac{3}{8}\quad \text{Subtracting } 2\frac{3}{4} \text{ from } 10\frac{1}{8}$$

The length of d is $7\frac{3}{8}$ ft.

63. ***Familiarize***. Let f = the number of yards of fabric needed to make the outfit.

Translate. We write an equation.

$$\underbrace{\text{Fabric for dress}}_{1\frac{3}{8}} + \underbrace{\text{Fabric for band}}_{\frac{5}{8}} + \underbrace{\text{Fabric for jacket}}_{3\frac{3}{8}} \text{ is } \underbrace{\text{Total fabric}}_{f}$$

$$1\frac{3}{8}+\frac{5}{8}+3\frac{3}{8}=f$$

Solve. We add.

$$\begin{array}{r} 1\,\frac{3}{8} \\ \frac{5}{8} \\ +\,3\,\frac{3}{8} \\ \hline 4\,\frac{11}{8} \end{array}$$

$$4\frac{11}{8}=4+\frac{11}{8}=4+1\frac{3}{8}=5\frac{3}{8}$$

Check. We can repeat the calculation. Also note that the answer is reasonable since it is larger than any of the individual amounts of fabric.

State. The outfit requires $5\frac{3}{8}$ yd of fabric.

65. ***Familiarize***. The height of the window, $4\frac{5}{6}$ ft, will remain the same. The $8\frac{1}{4}$ ft will increase by $2\frac{1}{3}$ ft. Let w = the new width of the window.

Translate.

$$\underbrace{\text{Original width}}_{8\frac{1}{4}} \text{ plus } \underbrace{\text{Added width}}_{2\frac{1}{3}} \text{ is } \underbrace{\text{New width}}_{w}$$

$$8\frac{1}{4}+2\frac{1}{3}=w$$

Solve. We carry out the addition. The LCD is 12.

$$\begin{array}{rcr} 8\,\boxed{\frac{1}{4}\cdot\frac{3}{3}} & = & 8\,\frac{3}{12} \\ +\,2\,\boxed{\frac{1}{3}\cdot\frac{4}{4}} & = & +\;2\,\frac{4}{12} \\ \hline & & 10\,\frac{7}{12} \end{array}$$

Check. We repeat the calculation. The answer checks.

State. The new dimensions of the window are $4\frac{5}{6}$ ft $\times$ $10\frac{7}{12}$ ft.

67. $9-9\frac{2}{5}=9+\left(-9\frac{2}{5}\right)$

Since $9\frac{2}{5}$ is greater than 9, the answer will be negative.

The difference in absolute values is

$$\begin{array}{r} 9\,\frac{2}{5} \\ -9 \\ \hline \frac{2}{5} \end{array}$$

so $9-9\frac{2}{5}=-\frac{2}{5}$.

69. $3\frac{1}{2}-6\frac{3}{4}=3\frac{1}{2}+\left(-6\frac{3}{4}\right)$

Since $6\frac{3}{4}$ is greater than $3\frac{1}{2}$, the answer will be negative.

The difference in absolute values is

$$\begin{array}{rcrcr} 6\,\frac{3}{4} & = & 6\,\frac{3}{4} & = & 6\,\frac{3}{4} \\ -\,3\,\frac{1}{2} & = & -\,3\,\boxed{\frac{1}{2}\cdot\frac{2}{2}} & = & -\,3\,\frac{2}{4} \\ \hline & & & & 3\,\frac{1}{4} \end{array}$$

so $3\frac{1}{2}-6\frac{3}{4}=-3\frac{1}{4}$.

71. $3\frac{4}{5}+\left(-7\frac{2}{3}\right)$

Since $7\frac{2}{3}$ is greater than $3\frac{4}{5}$, the answer will be negative.

The difference in absolute values is

$$\begin{array}{rcrcrcr} 7\,\frac{2}{3} & = & 7\,\boxed{\frac{2}{3}\cdot\frac{5}{5}} & = & 7\,\frac{10}{15} & = & 6\,\frac{25}{15} \\ -\,3\,\frac{4}{5} & = & -\,3\,\boxed{\frac{4}{5}\cdot\frac{3}{3}} & = & -\,3\,\frac{12}{15} & = & -\,3\,\frac{12}{15} \\ \hline & & & & & & 3\,\frac{13}{15} \end{array}$$

so $3\frac{4}{5}-7\frac{2}{3}=-3\frac{13}{15}$.

73. $-3\frac{1}{5} - 4\frac{2}{5} = -3\frac{1}{5} + \left(-4\frac{2}{5}\right)$

We add the absolute values and make the answer negative.

$$\begin{array}{r} 3\,\frac{1}{5} \\ +\,4\,\frac{2}{5} \\ \hline 7\,\frac{3}{5} \end{array}$$

Thus, $-3\frac{1}{5} - 4\frac{2}{5} = -7\frac{3}{5}$.

75. $-4\frac{1}{12} + 6\frac{2}{3}$

Since $6\frac{2}{3}$ is greater than $4\frac{1}{12}$, the answer will be positive. The difference in absolute values is

$$\begin{array}{rcr} 6\,\boxed{\frac{2}{3}\cdot\frac{4}{4}} & = & 6\,\frac{8}{12} \\ -\,4\,\frac{1}{12} & = & -\,4\,\frac{1}{12} \\ \hline & & 2\,\frac{7}{12} \end{array}$$

so $-4\frac{1}{12} + 6\frac{2}{3} = 2\frac{7}{12}$.

77. $-6\frac{1}{9} - \left(-4\frac{2}{9}\right) = -6\frac{1}{9} + 4\frac{2}{9}$

Since $-6\frac{1}{9}$ has the greater absolute value, the answer will be negative. The difference in absolute values is

$$\begin{array}{rcr} 6\,\frac{1}{9} & = & 5\,\frac{10}{9} \\ -4\,\frac{2}{9} & = & -4\,\frac{2}{9} \\ \hline & & 1\,\frac{8}{9} \end{array}$$

so $-6\frac{1}{9} - \left(-4\frac{2}{9}\right) = -1\frac{8}{9}$.

79. ***Familiarize.*** We visualize the situation. Repeated subtraction, or division, works well here.

$$\underbrace{\boxed{\frac{3}{4}\text{ lb}}\ \boxed{\frac{3}{4}\text{ lb}}\cdots\boxed{\frac{3}{4}\text{ lb}}}$$

12 lb fills how many packages?

Let n = the number of packages that can be made.

Translate. We translate to an equation.

$$n = 12 \div \frac{3}{4}$$

Solve. We carry out the division.

$$\begin{aligned} n &= 12 \div \frac{3}{4} = 12 \cdot \frac{4}{3} = \frac{12\cdot 4}{3} \\ &= \frac{3\cdot 4\cdot 4}{3\cdot 1} = \frac{3}{3}\cdot\frac{4\cdot 4}{1} \\ &= 16 \end{aligned}$$

Check. If each of 16 packages contains $\frac{3}{4}$ lb of cheese, a total of

$$16\cdot\frac{3}{4} = \frac{16\cdot 3}{4} = \frac{4\cdot 4\cdot 3}{4} = 4\cdot 3,$$

or 12 lb of cheese is used. The answer checks.

State. 16 packages of cheese can be made from a 12-lb slab.

81. The sum of the digits is $9+9+9+3 = 30$. Since 30 is divisible by 3, then 9993 is divisible by 3.

83. The sum of the digits is $2+3+4+5 = 14$. Since 14 is not divisible by 9, then 2345 is not divisible by 9.

85. The ones digit of 2335 is not 0, so 2335 is not divisible by 10.

87. The ones digit of 18,888 is even. Thus, 18,888 is divisible by 2.

89. $$\begin{aligned} \frac{15}{9}\cdot\frac{18}{39} &= \frac{15\cdot 18}{9\cdot 39} = \frac{3\cdot 5\cdot 2\cdot 3\cdot 3}{3\cdot 3\cdot 3\cdot 13} \\ &= \frac{3\cdot 3\cdot 3}{3\cdot 3\cdot 3} = \frac{5\cdot 2}{13} \\ &= \frac{10}{13} \end{aligned}$$

91. Use a calculator.

$$\begin{array}{rcrcr} 3289\frac{1047}{1189} & = & 3289\ \ \frac{1047}{1189} & = & 3289\frac{1047}{1189} \\ +\,5278\frac{32}{41} & = & +\,5278\boxed{\frac{32}{41}\cdot\frac{29}{29}} & = & +\,5278\frac{928}{1189} \\ \hline & & & & 8567\frac{1975}{1189} = 8568\frac{786}{1189} \end{array}$$

93. $$\begin{aligned} 35\frac{2}{3} + n &= 46\frac{1}{4} \\ n &= 46\frac{1}{4} - 35\frac{2}{3} \\ n &= 46\frac{3}{12} - 35\frac{8}{12} \\ n &= 45\frac{15}{12} - 35\frac{8}{12} \\ n &= 10\frac{7}{12} \end{aligned}$$

95. $$\begin{aligned} -15\frac{7}{8} &= 12\frac{1}{2} + t \\ -15\frac{7}{8} - 12\frac{1}{2} &= t \\ -28\frac{3}{8} &= t \end{aligned}$$

97. The resulting rectangle will have length $\left(8\frac{1}{2}+1\frac{1}{8}+8\frac{1}{2}\right)$ in. and width $9\frac{3}{4}$ in. We add to find the perimeter.

$$8\frac{1}{2}+1\frac{1}{8}+8\frac{1}{2}+9\frac{3}{4}+8\frac{1}{2}+1\frac{1}{8}+8\frac{1}{2}+9\frac{3}{4}$$
$$=8\frac{4}{8}+1\frac{1}{8}+8\frac{4}{8}+9\frac{6}{8}+8\frac{4}{8}+1\frac{1}{8}+8\frac{4}{8}+9\frac{6}{8}$$
$$=52\frac{30}{8}=52+\frac{30}{8}$$
$$=52+3\frac{6}{8}=52+3\frac{3}{4}$$
$$=55\frac{3}{4}$$

The perimeter is $55\frac{3}{4}$ in.

Exercise Set 4.7

1. $8 \cdot 2\frac{5}{6}$

$=\frac{8}{1}\cdot\frac{17}{6}$ Writing fractional notation

$=\frac{8\cdot 17}{1\cdot 6}=\frac{2\cdot 4\cdot 17}{1\cdot 2\cdot 3}=\frac{2}{2}\cdot\frac{4\cdot 17}{1\cdot 3}=\frac{68}{3}=22\frac{2}{3}$

3. $6\frac{2}{3}\cdot\frac{1}{4}$

$=\frac{20}{3}\cdot\frac{1}{4}$ Writing fraction notation

$=\frac{20\cdot 1}{3\cdot 4}=\frac{4\cdot 5\cdot 1}{3\cdot 4}=\frac{4}{4}\cdot\frac{5\cdot 1}{3}=\frac{5}{3}=1\frac{2}{3}$

5. $20\left(-2\frac{5}{6}\right)=\frac{20}{1}\cdot\left(-\frac{17}{6}\right)=-\frac{20\cdot 17}{1\cdot 6}=-\frac{2\cdot 10\cdot 17}{2\cdot 3}=\frac{2}{2}\left(-\frac{10\cdot 17}{3}\right)=-\frac{170}{3}=-56\frac{2}{3}$

7. $3\frac{1}{2}\cdot 4\frac{2}{3}=\frac{7}{2}\cdot\frac{14}{3}=\frac{7\cdot 14}{2\cdot 3}=\frac{7\cdot 2\cdot 7}{2\cdot 3}=\frac{2}{2}\cdot\frac{7\cdot 7}{3}=\frac{49}{3}=16\frac{1}{3}$

9. $-2\frac{3}{10}\cdot 4\frac{2}{5}=-\frac{23}{10}\cdot\frac{22}{5}=-\frac{23\cdot 22}{10\cdot 5}=-\frac{23\cdot 2\cdot 11}{2\cdot 5\cdot 5}=\frac{2}{2}\left(-\frac{23\cdot 11}{5\cdot 5}\right)=-\frac{253}{25}=-10\frac{3}{25}$

11. $\left(-6\frac{3}{10}\right)\left(-5\frac{7}{10}\right)=\frac{63}{10}\cdot\frac{57}{100}=\frac{3591}{100}=35\frac{91}{100}$

13. $20\div 3\frac{1}{5}$

$=20\div\frac{16}{5}$ Writing fractional notation

$=20\cdot\frac{5}{16}$ Multiplying by the reciprocal

$=\frac{20\cdot 5}{16}=\frac{4\cdot 5\cdot 5}{4\cdot 4}=\frac{4}{4}\cdot\frac{5\cdot 5}{4}=\frac{25}{4}=6\frac{1}{4}$

15. $8\frac{2}{5}\div 7$

$=\frac{42}{5}\div 7$ Writing fractional notation

$=\frac{42}{5}\cdot\frac{1}{7}$ Multiplying by the reciprocal

$=\frac{42\cdot 1}{5\cdot 7}=\frac{6\cdot 7}{5\cdot 7}=\frac{7}{7}\cdot\frac{6}{5}=\frac{6}{5}=1\frac{1}{5}$

17. $6\frac{1}{4}\div 3\frac{3}{4}=\frac{25}{4}\div\frac{15}{4}=\frac{25}{4}\cdot\frac{4}{15}=\frac{5\cdot 5\cdot 4}{4\cdot 3\cdot 5}=\frac{5\cdot 4}{5\cdot 4}\cdot\frac{5}{3}=\frac{5}{3}=1\frac{2}{3}$

19. $-1\frac{7}{8}\div 1\frac{2}{3}=-\frac{15}{8}\div\frac{5}{3}=-\frac{15}{8}\cdot\frac{3}{5}=-\frac{3\cdot 5\cdot 3}{8\cdot 5}=-\frac{3\cdot 3}{8}\cdot\frac{5}{5}=-\frac{9}{8}=-1\frac{1}{8}$

21. $5\frac{1}{10}\div 4\frac{3}{10}=\frac{51}{10}\div\frac{43}{10}=\frac{51}{10}\cdot\frac{10}{43}=\frac{51\cdot 10}{10\cdot 43}=\frac{10}{10}\cdot\frac{51}{43}=\frac{51}{43}=1\frac{8}{43}$

23. $-20\frac{1}{4}\div(-90)=-\frac{81}{4}\div(-90)=-\frac{81}{4}\left(-\frac{1}{90}\right)=\frac{81\cdot 1}{4\cdot 90}=\frac{9\cdot 9\cdot 1}{4\cdot 9\cdot 10}=\frac{9}{9}\cdot\frac{9\cdot 1}{4\cdot 10}=\frac{9}{40}$

25. $lw=2\frac{3}{5}\cdot 9$

$=\frac{13}{5}\cdot 9$

$=\frac{117}{5}=23\frac{2}{5}$

27. $rs=5\cdot 3\frac{1}{7}$

$=5\cdot\frac{22}{7}$

$=\frac{110}{7}=15\frac{5}{7}$

29. $mt=6\frac{2}{9}\left(-4\frac{3}{8}\right)$

$=\frac{56}{9}\left(-\frac{35}{8}\right)$

$=-\frac{7\cdot 8\cdot 35}{9\cdot 8}=-\frac{7\cdot 35}{9}\cdot\frac{8}{8}$

$=-\frac{245}{9}=-27\frac{2}{9}$

31. $R \cdot S \div T = 4\frac{2}{3} \cdot 1\frac{3}{7} \div (-5)$

$= \frac{14}{3} \cdot \frac{10}{7} \div (-5)$

$= \frac{14 \cdot 10}{3 \cdot 7} \div (-5)$

$= \frac{2 \cdot 7 \cdot 10}{3 \cdot 7} \div (-5)$

$= \frac{7}{7} \cdot \frac{2 \cdot 10}{3} \div (-5)$

$= \frac{20}{3} \div (-5)$

$= \frac{20}{3} \cdot \left(-\frac{1}{5}\right)$

$= -\frac{20 \cdot 1}{3 \cdot 5}$

$= -\frac{4 \cdot 5}{3 \cdot 5}$

$= -\frac{4}{3}$

$= -1\frac{1}{3}$

33. $r + ps = 5\frac{1}{2} + 3 \cdot 2\frac{1}{4}$

$= 5\frac{1}{2} + \frac{3}{1} \cdot \frac{9}{4}$

$= 5\frac{1}{2} + \frac{27}{4}$

$= 5\frac{1}{2} + 6\frac{3}{4}$

$= 5\frac{2}{4} + 6\frac{3}{4}$

$= 11\frac{5}{4}$

$= 12\frac{1}{4}$

35. $m + n \div p = 7\frac{2}{5} + 4\frac{1}{2} \div 6$

$= 7\frac{2}{5} + \frac{9}{2} \div 6$

$= 7\frac{2}{5} + \frac{9}{2} \cdot \frac{1}{6}$

$= 7\frac{2}{5} + \frac{9 \cdot 1}{2 \cdot 6} = 7\frac{2}{5} + \frac{3 \cdot 3 \cdot 1}{2 \cdot 2 \cdot 3}$

$= 7\frac{2}{5} + \frac{3 \cdot 1}{2 \cdot 2} = 7\frac{2}{5} + \frac{3}{4}$

$= 7\frac{8}{20} + \frac{15}{20} = 7\frac{23}{20}$

$= 8\frac{3}{20}$

37. ***Familiarize.*** Let b = the number of beagles registered with The American Kennel Club.

Translate.

$3\frac{4}{9}$	times	Number of beagles registered	is	Number of Labrador retrievers registered
↓	↓	↓	↓	↓
$3\frac{4}{9}$	$\cdot$	b	$=$	$155,000$

Solve. We divide by $3\frac{4}{9}$ on both sides of the equation.

$b = 155,000 \div 3\frac{4}{9}$

$b = 155,000 \div \frac{31}{9}$

$b = 155,000 \cdot \frac{9}{31} = \frac{155,000 \cdot 9}{31}$

$b = \frac{31 \cdot 5000 \cdot 9}{31 \cdot 1} = \frac{31}{31} \cdot \frac{5000 \cdot 9}{1} = \frac{5000 \cdot 9}{1}$

$b = 45,000$

Check. Since $3\frac{4}{9} \cdot 45,000 = \frac{31}{9} \cdot 45,000 = 155,000$, the answer checks.

State. There are 45,000 beagles registered with The American Kennel Club.

39. ***Familiarize.*** Let s = the number of teaspoons of sodium the average American woman consumes in 30 days.

Translate. A multiplication corresponds to this situation.

$s = 30 \cdot 1\frac{1}{3}$

Solve. We carry out the multiplication.

$s = 30 \cdot 1\frac{1}{3} = 30 \cdot \frac{4}{3} = \frac{30 \cdot 4}{3} = \frac{3 \cdot 10 \cdot 4}{3 \cdot 1} = 10 \cdot 4 = 40$

Check. We repeat the calculation. The answer checks.

State. In 30 days the average American woman consumes 40 tsp of sodium.

41. ***Familiarize.*** Let t = the number of tiles to be used.

Translate.

Length of one tile	times	Number of tiles	is	Length of sidewalk
↓	↓	↓	↓	↓
$1\frac{1}{8}$	$\cdot$	t	$=$	$14\frac{2}{5}$

Solve. We divide by $1\frac{1}{8}$ on both sides of the equation.

$t = 14\frac{2}{5} \div 1\frac{1}{8}$

$t = \frac{72}{5} \div \frac{9}{8}$

$t = \frac{72}{5} \cdot \frac{8}{9} = \frac{72 \cdot 8}{5 \cdot 9}$

$t = \frac{8 \cdot 9 \cdot 8}{5 \cdot 9} = \frac{9}{9} \cdot \frac{8 \cdot 8}{5}$

$t = \frac{64}{5} = 12\frac{4}{5}$

Check. Since $1\frac{1}{8} \cdot 12\frac{4}{5} = \frac{9}{8} \cdot \frac{64}{5} = \frac{9 \cdot 8 \cdot 8}{8 \cdot 5} = \frac{72}{5} = 14\frac{2}{5}$, the answer checks.

State. $12\frac{4}{5}$ tiles will be used.

43. ***Familiarize***. Let $p =$ the population of Alaska.

Translate.

$$\underbrace{6\frac{4}{5}}_{} \text{ times } \underbrace{\text{Population of Alaska}} \text{ is } \underbrace{\text{Population of Alabama}}$$
$$6\frac{4}{5} \cdot p = 4,700,000$$

Solve. We divide by $6\frac{4}{5}$ on both sides.

$$6\frac{4}{5} \cdot p = 4,700,000$$
$$p = 4,700,000 \div 6\frac{4}{5}$$
$$p = 4,700,000 \div \frac{34}{5}$$
$$p = 4,700,000 \cdot \frac{5}{34}$$
$$p = \frac{4,700,000 \cdot 5}{34}$$
$$p \approx 691,176 \approx 690,000$$

Check. We repeat the calculation. The answer checks.

State. The population of Alaska is about 690,000.

45. ***Familiarize***. We let $t =$ the Fahrenheit temperature.

Translate.

$$\underbrace{\text{Celsius temperature}} \text{ times } 1\frac{4}{5} \text{ plus } 32^\circ \text{ is } \underbrace{\text{Fahrenheit temperature}}$$
$$20 \cdot 1\frac{4}{5} + 32 = t$$

Solve. We multiply and then add, according to the rules for order of operations.

$$t = 20 \cdot 1\frac{4}{5} + 32 = \frac{20}{1} \cdot \frac{9}{5} + 32 = \frac{20 \cdot 9}{1 \cdot 5} + 32 =$$
$$\frac{4 \cdot 5 \cdot 9}{1 \cdot 5} + 32 = \frac{5}{5} \cdot \frac{4 \cdot 9}{1} + 32 = 36 + 32 = 68$$

Check. We repeat the calculation.

State. 68° Fahrenheit corresponds to 20° Celsius.

47. ***Familiarize***. Let $c =$ the daily circulation of *USA Today*.

Translate.

$$3\frac{1}{5} \text{ times } \underbrace{\textit{Washington Post} \text{ circulation}} \text{ is } \underbrace{\textit{USA Today} \text{ circulation}}$$
$$3\frac{1}{5} \cdot 665,000 = c$$

Solve. We carry out the multiplication.

$$3\frac{1}{5} \cdot 665,000 = \frac{16}{5} \cdot 665,000 = \frac{16 \cdot 665,000}{5} =$$
$$\frac{16 \cdot 5 \cdot 133,000}{5 \cdot 1} = \frac{5}{5} \cdot \frac{16 \cdot 133,000}{1} = 2,128,000$$

Check. We repeat the calculation. The answer checks.

State. In 2008 the daily circulation of *USA Today* was 2,128,000.

49. ***Familiarize***. Let $A =$ the area of the mural, in square feet. Recall that the area of a rectangle is length times width.

Translate.

$$\text{Area} = \text{length} \times \text{width}$$
$$A = 9\frac{3}{8} \times 6\frac{2}{3}$$

Solve. We carry out the multiplication.

$$A = 9\frac{3}{8} \times 6\frac{2}{3} = \frac{75}{8} \cdot \frac{20}{3} = \frac{75 \cdot 20}{8 \cdot 3}$$
$$= \frac{3 \cdot 25 \cdot 4 \cdot 5}{4 \cdot 2 \cdot 3} = \frac{3 \cdot 4}{3 \cdot 4} \cdot \frac{25 \cdot 5}{2}$$
$$= \frac{125}{2} = 62\frac{1}{2}$$

Check. We repeat the calculation. The answer checks.

State. The area of the mural is $62\frac{1}{2}$ ft^2.

51. ***Familiarize***. Let $p =$ the population of the United States in 2008.

Translate.

$$3\frac{3}{4} \text{ times } \underbrace{\text{Population of U.S.}} \text{ is } \underbrace{\text{Population of India}}$$
$$3\frac{3}{4} \cdot p = 1,149,000,000$$

Solve. We divide by $3\frac{3}{4}$ on both sides.

$$3\frac{3}{4} \cdot p = 1,149,000,000$$
$$p = 1,149,000,000 \div 3\frac{3}{4}$$
$$p = 1,149,000,000 \div \frac{15}{4}$$
$$p = 1,149,000,000 \cdot \frac{4}{15}$$
$$p = \frac{1,149,000,000 \cdot 4}{15} = \frac{\cancel{15} \cdot 76,600,000 \cdot 4}{\cancel{15} \cdot 1}$$
$$p = 306,400,000$$

Check. We can find $3\frac{3}{4}$ times 306,400,000.

$$3\frac{3}{4} \cdot 306,400,000 = \frac{15}{4} \cdot 306,400,000 =$$
$$\frac{15 \cdot 306,400,000}{4} = \frac{15 \cdot \cancel{4} \cdot 76,600,000}{\cancel{4} \cdot 1} =$$
$$1,149,000,000$$

This is the population of India, so the answer checks.

State. The population of the United States was 306,400,000 in 2008.

53. ***Familiarize***. Let $f =$ the number of cups of flour and $s =$ the number of cups of sugar in the doubled recipe.

Translate. We write two equations. We multiply each of the original amounts by 2 to find the amounts in the doubled recipe.

$$f = 2 \cdot 2\frac{3}{4}$$
$$s = 2 \cdot 1\frac{1}{3}$$

Solve. We carry out the calculations.

$$f = 2 \cdot 2\frac{3}{4} = 2 \cdot \frac{11}{4} = \frac{2 \cdot 11}{4} = \frac{\cancel{2} \cdot 11}{\cancel{2} \cdot 2} = \frac{11}{2} = 5\frac{1}{2}$$
$$s = 2 \cdot 1\frac{1}{3} = 2 \cdot \frac{4}{3} = \frac{2 \cdot 4}{3} = \frac{8}{3} = 2\frac{2}{3}$$

Check. We repeat the calculations. The answer checks.

State. The chef will need $5\frac{1}{2}$ cups of flour and $2\frac{2}{3}$ cups of sugar for the doubled recipe.

55. ***Familiarize***. We let n = the number of cubic feet occupied by 25,000 lb of water.

Translate. We write an equation.

Total weight	÷	Weight per cubic foot	=	Number of cubic feet
↓	↓	↓	↓	↓
$25,000$	÷	$62\frac{1}{2}$	=	n

Solve. To solve the equation we carry out the division.

$$n = 25,000 \div 62\frac{1}{2} = 25,000 \div \frac{125}{2}$$
$$= 25,000 \cdot \frac{2}{125} = \frac{200 \cdot 125 \cdot 2}{125 \cdot 1}$$
$$= \frac{125}{125} \cdot \frac{200 \cdot 2}{1} = 400$$

Check. We repeat the calculation.

State. 400 cubic feet would be occupied.

57. ***Familiarize***. We draw a picture.

$\frac{1}{3}$ lb	$\frac{1}{3}$ lb	....	$\frac{1}{3}$ lb

⟵ $5\frac{1}{2}$ lb ⟶

We let s = the number of servings that can be prepared from $5\frac{1}{2}$ lb of flounder fillet.

Translate. The situation corresponds to a division sentence.

$$s = 5\frac{1}{2} \div \frac{1}{3}$$

Solve. We carry out the division.

$$s = 5\frac{1}{2} \div \frac{1}{3} = \frac{11}{2} \div \frac{1}{3}$$
$$= \frac{11}{2} \cdot \frac{3}{1} = \frac{33}{2}$$
$$= 16\frac{1}{2}$$

Check. We check by multiplying. If $16\frac{1}{2}$ servings are prepared, then

$$16\frac{1}{2} \cdot \frac{1}{3} = \frac{33}{2} \cdot \frac{1}{3} = \frac{3 \cdot 11 \cdot 1}{2 \cdot 3} = \frac{3}{3} \cdot \frac{11 \cdot 1}{2} = \frac{11}{2} = 5\frac{1}{2} \text{ lb}$$

of flounder is used. Our answer checks.

State. $16\frac{1}{2}$ servings can be prepared from $5\frac{1}{2}$ lb of flounder fillet.

59. ***Familiarize***. We let m = the number of miles per gallon the car got.

Translate. We write an equation.

Total number of miles traveled	÷	Number of gallons of gas used	=	Miles per gallon
↓	↓	↓	↓	↓
213	÷	$14\frac{2}{10}$	=	m

Solve. To solve the equation we carry out the division.

$$m = 213 \div 14\frac{2}{10} = 213 \div \frac{142}{10}$$
$$= 213 \cdot \frac{10}{142} = \frac{3 \cdot 71 \cdot 2 \cdot 5}{2 \cdot 71 \cdot 1}$$
$$= \frac{2 \cdot 71}{2 \cdot 71} \cdot \frac{3 \cdot 5}{1} = 15$$

Check. We repeat the calculation.

State. The car got 15 miles per gallon of gas.

61. ***Familiarize***. First we will find the total width of the columns and determine if this is less than or greater than $8\frac{1}{2}$ in. Then, if the total is less than $8\frac{1}{2}$ in., we will find the difference between $8\frac{1}{2}$ in. and the total width and then divide by 2 to find the width of each margin. Let w = the total width of the columns.

Translate. First we write an equation for finding the total width of the columns.

$$w = 2 \cdot 1\frac{1}{2} + 5 \cdot \frac{3}{4}$$

Solve. We solve the equation.

$$w = 2 \cdot 1\frac{1}{2} + 5 \cdot \frac{3}{4}$$
$$w = 2 \cdot \frac{3}{2} + 5 \cdot \frac{3}{4}$$
$$w = \frac{2 \cdot 3}{2} + \frac{5 \cdot 3}{4}$$
$$w = 3 + \frac{15}{4} = 3 + 3\frac{3}{4}$$
$$w = 6\frac{3}{4}$$

Since the total width of the columns is less than $8\frac{1}{2}$ in., the table will fit on a piece of standard paper. Let l = the number of inches by which the width of the paper exceeds the width of the table. Then we have:

$$l = 8\frac{1}{2} - 6\frac{3}{4} = 8\frac{2}{4} - 6\frac{3}{4}$$
$$l = 7\frac{6}{4} - 6\frac{3}{4}$$
$$l = 1\frac{3}{4}$$

This tells us that the total width of the margins will be $1\frac{3}{4}$ in. Since the margins are of equal width, we divide by 2 to find the width of each margin. Let m = this width, in inches. We have:

$$m = 1\frac{3}{4} \div 2$$
$$m = \frac{7}{4} \div 2$$
$$m = \frac{7}{4} \cdot \frac{1}{2} = \frac{7}{8}$$

Check. We repeat the calculations.

State. The table will fit on a standard piece of paper. Each margin will be $\frac{7}{8}$ in. wide.

63. *Familiarize*. We can refer to the drawing in the text. Let a = the total area of the sod.

Translate. The total area is the sum of the areas of the two rectangles.

$$a = 20 \cdot 15\frac{1}{2} + 12\frac{1}{2} \cdot 10\frac{1}{2}$$

Solve. We perform the multiplication and then add.

$$a = 20 \cdot 15\frac{1}{2} + 12\frac{1}{2} \cdot 10\frac{1}{2}$$
$$= 20 \cdot \frac{31}{2} + \frac{25}{2} \cdot \frac{21}{2}$$
$$= \frac{20 \cdot 31}{2} + \frac{25 \cdot 21}{4}$$
$$= \frac{\not{2} \cdot 10 \cdot 31}{\not{2} \cdot 1} + \frac{525}{4}$$
$$= 310 + 131\frac{1}{4}$$
$$= 441\frac{1}{4}$$

Check. We can perform a partial check by estimating the total area as $20 \cdot 16 + 13 \cdot 11 = 320 + 143 = 463 \approx 441\frac{1}{4}$. Our answer seems reasonable.

State. The total area of the sod is $441\frac{1}{4}$ ft^2.

65. *Familiarize*. The figure contains a square with sides of $10\frac{1}{2}$ ft and a rectangle with dimensions of $8\frac{1}{2}$ ft by 4 ft. The area of the shaded region consists of the area of the square less the area of the rectangle. Let A = the area of the shaded region, in square feet.

Translate. We write an equation.

$$A = 10\frac{1}{2} \cdot 10\frac{1}{2} - 8\frac{1}{2} \cdot 4$$

Solve. We multiply and then subtract.

$$A = 10\frac{1}{2} \cdot 10\frac{1}{2} - 8\frac{1}{2} \cdot 4$$
$$A = \frac{21}{2} \cdot \frac{21}{2} - \frac{17}{2} \cdot 4$$
$$A = \frac{441}{4} - \frac{68}{2}$$
$$A = \frac{441}{4} - \frac{68}{2} \cdot \frac{2}{2}$$
$$A = \frac{441}{4} - \frac{136}{4}$$
$$A = \frac{305}{4} = 76\frac{1}{4}$$

Check. We repeat the calculation.

State. The area of the shaded region is $76\frac{1}{4}$ ft^2.

67. *Familiarize*. The figure is a rectangle with dimensions $7\frac{1}{4}$ cm by $4\frac{1}{2}$ cm with a rectangular area cut out of it. One dimension of the cut-out area is $4\frac{1}{4}$ cm. We subtract to find the other dimension:

$4\frac{1}{2} - \left(1\frac{1}{2} + 1\frac{3}{4}\right) = 4\frac{1}{2} - \left(1\frac{2}{4} + 1\frac{3}{4}\right) = 4\frac{1}{2} - 2\frac{5}{4} = 4\frac{1}{2} - 3\frac{1}{4} = 4\frac{2}{4} - 3\frac{1}{4} = 1\frac{1}{4}$.

The area of the shaded region consists of the area of the larger rectangle less the area of the smaller rectangle. Let A = the area of the shaded region, in square centimeters.

Translate.

$$A = 7\frac{1}{4} \cdot 4\frac{1}{2} - 4\frac{1}{4} \cdot 1\frac{1}{4}$$

Solve.

$$A = 7\frac{1}{4} \cdot 4\frac{1}{2} - 4\frac{1}{4} \cdot 1\frac{1}{4}$$
$$A = \frac{29}{4} \cdot \frac{9}{2} - \frac{17}{4} \cdot \frac{5}{4}$$
$$A = \frac{29 \cdot 9}{4 \cdot 2} - \frac{17 \cdot 5}{4 \cdot 4}$$
$$A = \frac{261}{8} - \frac{85}{16}$$
$$A = \frac{261}{8} \cdot \frac{2}{2} - \frac{85}{16}$$
$$A = \frac{522}{16} - \frac{85}{16}$$
$$A = \frac{437}{16} = 27\frac{5}{16}$$

Check. We can perform a partial check by estimating the area as $7 \cdot 5 - 4 \cdot 1 = 35 - 4 = 31$. Our answer seems reasonable.

State. The area of the shaded region is $27\frac{5}{16}$ cm^2.

69. The set $\{\ldots, -3, -2, -1, 0, 1, 2, 3, \ldots\}$ is the set of integers.

71. The numbers 91, 95, and 111 are examples of composite numbers.

73. To add fractions with different denominators, we must first find the least common multiple of the denominators.

75. In the expression $\frac{c}{d}$, we call c the numerator.

77. $-8 \div \frac{1}{2}+\frac{3}{4}+\left(-5-\frac{5}{8}\right)^2 = -8 \div \frac{1}{2}+\frac{3}{4}+\left(-\frac{40}{8}-\frac{5}{8}\right)^2 =$
$-8 \div \frac{1}{2}+\frac{3}{4}+\left(-\frac{45}{8}\right)^2 = -8 \div \frac{1}{2}+\frac{3}{4}+\frac{2025}{64} =$
$-8 \cdot 2+\frac{3}{4}+\frac{2025}{64} = -16+\frac{3}{4}+\frac{2025}{64} =$
$-\frac{1024}{64}+\frac{48}{64}+\frac{2025}{64} = \frac{1049}{64} = 16\frac{25}{64}$

79.
$$\frac{1}{3} \div \left(\frac{1}{2}-\frac{1}{5}\right) \times \frac{1}{4}+\frac{1}{6}$$
$$=\frac{1}{3} \div \left(\frac{5}{10}-\frac{2}{10}\right) \times \frac{1}{4}+\frac{1}{6}$$
$$=\frac{1}{3} \div \frac{3}{10} \times \frac{1}{4}+\frac{1}{6}$$
$$=\frac{1}{3} \times \frac{10}{3} \times \frac{1}{4}+\frac{1}{6}$$
$$=\frac{10}{9} \times \frac{1}{4}+\frac{1}{6}$$
$$=\frac{2\times 5\times 1}{9\times 2\times 2}+\frac{1}{6} = \frac{2}{2}\times\frac{5\times 1}{9\times 2}+\frac{1}{6}$$
$$=\frac{5}{18}+\frac{1}{6} = \frac{5}{18}+\frac{3}{18} = \frac{8}{18} = \frac{4}{9}$$

81.
$$\frac{1}{r} = \frac{1}{40}+\frac{1}{60}+\frac{1}{80}$$
$$\frac{1}{r} = \frac{1}{40}\cdot\frac{6}{6}+\frac{1}{60}\cdot\frac{4}{4}+\frac{1}{80}\cdot\frac{3}{3}$$
$$\frac{1}{r} = \frac{6}{240}+\frac{4}{240}+\frac{3}{240}$$
$$\frac{1}{r} = \frac{13}{240}$$

Then r is the reciprocal of $\frac{13}{240}$, so $r = \frac{240}{13}$, or $18\frac{6}{13}$.

83. ***Familiarize***. Let w = the amount of hot water required for two showers and two loads of wash. Note that washing one load of clothes requires $2\frac{2}{3} \cdot 12$ gallons of hot water.

Translate. We write an equation.

$$w = 2 \cdot 20 + 2 \cdot 2\frac{2}{3} \cdot 20$$

Solve. We perform the multiplications and then we add.

$$w = 2 \cdot 12 + 2 \cdot 2\frac{2}{3} \cdot 12$$
$$w = 24 + 2 \cdot \frac{8}{3} \cdot 12$$
$$w = 24 + \frac{2 \cdot 8 \cdot 12}{3}$$
$$w = 24 + \frac{192}{3}$$
$$w = 24 + 64$$
$$w = 88$$

Check. We repeat the calculations. The answer checks.

State. Two showers and two loads of wash require 88 gallons of hot water.

Exercise Set 4.8

1.
$$\frac{1}{8}+\frac{1}{4}\cdot\frac{2}{3}$$
$$=\frac{1}{8}+\frac{1\cdot 2}{4\cdot 3} \quad \text{Multiplying}$$
$$=\frac{1}{8}+\frac{1\cdot \cancel{2}}{\cancel{2}\cdot 2\cdot 3}$$
$$=\frac{1}{8}+\frac{1}{6} \quad \text{Removing a factor of 1}$$
$$=\frac{1}{8}\cdot\frac{3}{3}+\frac{1}{6}\cdot\frac{4}{4} \quad \text{Multiplying by 1 to obtain the LCD}$$
$$=\frac{3}{24}+\frac{4}{24}$$
$$=\frac{7}{24}$$

3.
$$-\frac{1}{6}-3\left(-\frac{5}{9}\right)$$
$$=-\frac{1}{6}+\frac{3\cdot 5}{9} \quad \text{Multiplying}$$
$$=-\frac{1}{6}+\frac{\cancel{3}\cdot 5}{\cancel{3}\cdot 3}$$
$$=-\frac{1}{6}+\frac{5}{3} \quad \text{Removing a factor of 1}$$
$$=-\frac{1}{6}+\frac{5}{3}\cdot\frac{2}{2} \quad \text{Multiplying by 1 to obtain the LCD}$$
$$=-\frac{1}{6}+\frac{10}{6}$$
$$=\frac{9}{6}=\frac{\cancel{3}\cdot 3}{2\cdot\cancel{3}}$$
$$=\frac{3}{2}, \text{ or } 1\frac{1}{2}$$

5.
$$\frac{9}{10}-\left(\frac{2}{5}-\frac{3}{8}\right)$$
$$=\frac{9}{10}-\left(\frac{2}{5}\cdot\frac{8}{8}-\frac{3}{8}\cdot\frac{5}{5}\right)$$
$$=\frac{9}{10}-\left(\frac{16}{40}-\frac{15}{40}\right)$$
$$=\frac{9}{10}-\frac{1}{40}$$
$$=\frac{9}{10}\cdot\frac{4}{4}-\frac{1}{40}$$
$$=\frac{36}{40}-\frac{1}{40}$$
$$=\frac{35}{40}=\frac{\cancel{5}\cdot 7}{\cancel{5}\cdot 8}$$
$$=\frac{7}{8}$$

7. $\frac{5}{8} \div \frac{1}{4} - \frac{2}{3} \cdot \frac{4}{5}$

$= \frac{5}{8} \cdot \frac{4}{1} - \frac{2}{3} \cdot \frac{4}{5}$ Dividing

$= \frac{5 \cdot \cancel{4}}{2 \cdot \cancel{4} \cdot 1} - \frac{2}{3} \cdot \frac{4}{5}$

$= \frac{5}{2} - \frac{2}{3} \cdot \frac{4}{5}$ Removing a factor of 1

$= \frac{5}{2} - \frac{2 \cdot 4}{3 \cdot 5}$ Multiplying

$= \frac{5}{2} - \frac{8}{15}$

$= \frac{5}{2} \cdot \frac{15}{15} - \frac{8}{15} \cdot \frac{2}{2}$ Multiplying by 1 to obtain the LCD

$= \frac{75}{30} - \frac{16}{30}$

$= \frac{59}{30}$, or $1\frac{29}{30}$ Subtracting

9. $\frac{7}{8} \div \frac{1}{2} \cdot \frac{1}{4}$

$= \frac{7}{8} \cdot \frac{2}{1} \cdot \frac{1}{4}$ Dividing

$= \frac{7 \cdot \cancel{2}}{\cancel{2} \cdot 4 \cdot 1} \cdot \frac{1}{4}$

$= \frac{7}{4} \cdot \frac{1}{4}$ Removing a factor of 1

$= \frac{7}{16}$ Multiplying

11. $\frac{3}{4} - \frac{2}{3} \cdot \left(\frac{1}{2} + \frac{2}{5}\right)$

$= \frac{3}{4} - \frac{2}{3} \cdot \left(\frac{5}{10} + \frac{4}{10}\right)$ Adding inside

$= \frac{3}{4} - \frac{2}{3} \cdot \frac{9}{10}$ the parentheses

$= \frac{3}{4} - \frac{2 \cdot 9}{3 \cdot 10}$ Multiplying

$= \frac{3}{4} - \frac{2 \cdot 3 \cdot 3}{3 \cdot 2 \cdot 5}$

$= \frac{3}{4} - \frac{2 \cdot 3}{2 \cdot 3} \cdot \frac{3}{5}$

$= \frac{3}{4} - \frac{3}{5}$

$= \frac{15}{20} - \frac{12}{20}$

$= \frac{3}{20}$ Subtracting

13. $\frac{4}{5} \div \left(\frac{2}{9} \cdot \frac{1}{2}\right) \cdot \left(-\frac{5}{6}\right)$

$= \frac{4}{5} \div \left(\frac{2 \cdot 1}{9 \cdot 2}\right) \cdot \left(-\frac{5}{6}\right)$ Multiplying inside the parentheses

$= \frac{4}{5} \div \frac{1}{9} \cdot \left(-\frac{5}{6}\right)$

$= \frac{4}{5} \cdot \frac{9}{1} \cdot \left(-\frac{5}{6}\right)$ Dividing

$= \frac{4 \cdot 9}{5 \cdot 1}\left(-\frac{5}{6}\right)$

$= -\frac{4 \cdot 9 \cdot 5}{5 \cdot 1 \cdot 6}$ Multiplying

$= -\frac{\cancel{2} \cdot 2 \cdot \cancel{3} \cdot 3 \cdot \cancel{5}}{\cancel{5} \cdot 1 \cdot \cancel{2} \cdot \cancel{3}}$

$= -6$

15. $\left(\frac{2}{3}\right)^2 - \frac{1}{3} \cdot 1\frac{1}{4}$

$= \frac{4}{9} - \frac{1}{3} \cdot 1\frac{1}{4}$ Evaluating the exponental expression

$= \frac{4}{9} - \frac{1}{3} \cdot \frac{5}{4}$

$= \frac{4}{9} - \frac{5}{12}$ Multiplying

$= \frac{4}{9} \cdot \frac{4}{4} - \frac{5}{12} \cdot \frac{3}{3}$

$= \frac{16}{36} - \frac{15}{36}$

$= \frac{1}{36}$ Subtracting

17. $-\frac{12}{25}\left(\frac{3}{4} - \frac{1}{2}\right)^2$

$= -\frac{12}{25}\left(\frac{3}{4} - \frac{1}{2} \cdot \frac{2}{2}\right)^2$

$= -\frac{12}{25}\left(\frac{3}{4} - \frac{2}{4}\right)^2$

$= -\frac{12}{25} \cdot \left(\frac{1}{4}\right)^2$ Subtracting

$= -\frac{12}{25} \cdot \frac{1}{16}$ Evaluating the exponential expression

$= -\frac{12 \cdot 1}{25 \cdot 16}$ Multiplying

$= -\frac{3 \cdot \cancel{4} \cdot 1}{5 \cdot 5 \cdot \cancel{4} \cdot 4}$

$= -\frac{3}{100}$

19.
$$-\frac{3}{4} \div \left(\frac{2}{3}-\frac{1}{6}\right)+\frac{1}{2}$$
$$= -\frac{3}{4} \div \left(\frac{2}{3}\cdot\frac{2}{2}-\frac{1}{6}\right)+\frac{1}{2}$$
$$= -\frac{3}{4} \div \left(\frac{4}{6}-\frac{1}{6}\right)+\frac{1}{2}$$
$$= -\frac{3}{4} \div \frac{3}{6}+\frac{1}{2}$$
$$= -\frac{3}{4} \cdot \frac{6}{3}+\frac{1}{2}$$
$$= -\frac{3\cdot 6}{4\cdot 3}+\frac{1}{2}$$
$$= \frac{\cancel{3}\cdot\cancel{2}\cdot 3}{\cancel{2}\cdot 2\cdot\cancel{3}}+\frac{1}{2}$$
$$= -\frac{3}{2}+\frac{1}{2}$$
$$= -\frac{2}{2}$$
$$= -1$$

21.
$$\left(-\frac{3}{2}\right)^2-2\left(\frac{1}{4}-\frac{3}{2}\right)$$
$$= \left(-\frac{3}{2}\right)^2-2\left(\frac{1}{4}-\frac{3}{2}\cdot\frac{2}{2}\right)$$
$$= \left(-\frac{3}{2}\right)^2-2\left(\frac{1}{4}-\frac{6}{4}\right)$$
$$= \left(-\frac{3}{2}\right)^2-2\left(-\frac{5}{4}\right)$$
$$= \frac{9}{4}-2\left(-\frac{5}{4}\right)$$
$$= \frac{9}{4}+\frac{2\cdot 5}{4}$$
$$= \frac{9}{4}+\frac{10}{4}$$
$$= \frac{19}{4}, \text{ or } 4\frac{3}{4}$$

23. $\frac{1}{2}-\left(\frac{1}{2}\right)^2+\left(\frac{1}{2}\right)^3$

$= \frac{1}{2}-\frac{1}{4}+\frac{1}{8}$ Evaluating the exponental expressions

$= \frac{2}{4}-\frac{1}{4}+\frac{1}{8}$ Doing the additions and

$= \frac{1}{4}+\frac{1}{8}$ subtractions in order

$= \frac{2}{8}+\frac{1}{8}$ from left to right

$= \frac{3}{8}$

25.
$$\left(\frac{3}{5}-\frac{1}{2}\right)\div\left(\frac{3}{4}-\frac{3}{10}\right)$$
$$= \left(\frac{6}{10}-\frac{5}{10}\right)\div\left(\frac{15}{20}-\frac{6}{20}\right)$$
$$= \frac{1}{10}\div\frac{9}{20}$$
$$= \frac{1}{10}\cdot\frac{20}{9}$$
$$= \frac{1\cdot 20}{10\cdot 9}$$
$$= \frac{1\cdot 2\cdot\cancel{10}}{\cancel{10}\cdot 9}$$
$$= \frac{2}{9}$$

27. $\dfrac{\frac{3}{8}}{\frac{11}{8}} = \frac{3}{8}\div\frac{11}{8} = \frac{3}{8}\cdot\frac{8}{11} = \frac{3\cdot\cancel{8}}{\cancel{8}\cdot 11} = \frac{3}{11}$

29. $\dfrac{-4}{\frac{6}{7}} = -4\div\frac{6}{7} = -4\cdot\frac{7}{6} = -\frac{4\cdot 7}{6} = -\frac{\cancel{2}\cdot 2\cdot 7}{\cancel{2}\cdot 3} = -\frac{14}{3}$, or $-4\frac{2}{3}$

31. $\dfrac{\frac{1}{40}}{-\frac{1}{50}} = \frac{1}{40}\div\left(-\frac{1}{50}\right) = \frac{1}{40}\cdot\left(-\frac{50}{1}\right) = -\frac{1\cdot 50}{40\cdot 1} = -\frac{1\cdot 5\cdot\cancel{10}}{4\cdot\cancel{10}\cdot 1} = -\frac{5}{4}$, or $-1\frac{1}{4}$

33. $\dfrac{-\frac{1}{10}}{-10} = -\frac{1}{10}\div(-10) = -\frac{1}{10}\cdot\left(-\frac{1}{10}\right) = \frac{1}{100}$

35. $\dfrac{\frac{5}{18}}{-1\frac{2}{3}} = \dfrac{\frac{5}{18}}{-\frac{5}{3}} = \frac{5}{18}\div\left(-\frac{5}{3}\right) = \frac{5}{18}\cdot\left(-\frac{3}{5}\right) = -\frac{5\cdot 3}{18\cdot 5} = -\frac{\cancel{5}\cdot\cancel{3}\cdot 1}{\cancel{3}\cdot 6\cdot\cancel{5}} = -\frac{1}{6}$

37. $\dfrac{\frac{x}{28}}{\frac{5}{8}} = \frac{x}{28}\div\frac{5}{8} = \frac{x}{28}\cdot\frac{8}{5} = \frac{x\cdot 8}{28\cdot 5} = \frac{x\cdot 2\cdot\cancel{4}}{\cancel{4}\cdot 7\cdot 5} = \frac{2x}{35}$

39. $\dfrac{\frac{n}{14}}{-\frac{2}{3}} = \frac{n}{14}\div\left(-\frac{2}{3}\right) = \frac{n}{14}\cdot\left(-\frac{3}{2}\right) = -\frac{3n}{28}$

41. $\dfrac{-\frac{3}{35}}{-\frac{x}{10}} = -\frac{3}{35}\div\left(-\frac{x}{10}\right) = -\frac{3}{35}\cdot\left(-\frac{10}{x}\right) = \frac{3\cdot 10}{35\cdot x} = \frac{3\cdot 2\cdot\cancel{5}}{\cancel{5}\cdot 7\cdot x} = \frac{6}{7x}$

43. $\dfrac{-\dfrac{5}{8}}{\left(\dfrac{3}{2}\right)^2} = -\dfrac{5}{8} \div \left(\dfrac{3}{2}\right)^2 = -\dfrac{5}{8} \div \dfrac{9}{4} = -\dfrac{5}{8} \cdot \dfrac{4}{9} = -\dfrac{5 \cdot 4}{8 \cdot 9} =$

$-\dfrac{5 \cdot \not{4}}{2 \cdot \not{4} \cdot 9} = -\dfrac{5}{18}$

45. $\dfrac{\dfrac{1}{6} - \dfrac{5}{9}}{\dfrac{2}{3}} = \left(\dfrac{1}{6} - \dfrac{5}{9}\right) \div \dfrac{2}{3} = \left(\dfrac{3}{18} - \dfrac{10}{18}\right) \div \dfrac{2}{3} =$

$-\dfrac{7}{18} \div \dfrac{2}{3} = -\dfrac{7}{18} \cdot \dfrac{3}{2} = -\dfrac{7 \cdot 3}{18 \cdot 2} = -\dfrac{7 \cdot \not{3}}{\not{3} \cdot 6 \cdot 2} = -\dfrac{7}{12}$

47. $\dfrac{\dfrac{1}{4} - \dfrac{3}{8}}{\dfrac{1}{2} - \dfrac{7}{8}} = \dfrac{\dfrac{2}{8} - \dfrac{3}{8}}{\dfrac{4}{8} - \dfrac{7}{8}} = \dfrac{-\dfrac{1}{8}}{-\dfrac{3}{8}} = -\dfrac{1}{8} \div \left(-\dfrac{3}{8}\right) =$

$-\dfrac{1}{8} \cdot \left(-\dfrac{8}{3}\right) = \dfrac{1 \cdot \not{8}}{\not{8} \cdot 3} = \dfrac{1}{3}$

49. Add the numbers and divide by the number of addends.

$$\dfrac{\dfrac{2}{3} + \dfrac{7}{8}}{2}$$

$= \dfrac{\dfrac{16}{24} + \dfrac{21}{24}}{2}$ The LCD is 24.

$= \dfrac{\dfrac{37}{24}}{2}$ Adding

$= \dfrac{37}{24} \cdot \dfrac{1}{2}$ Dividing

$= \dfrac{37}{48}$

51. Add the numbers and divide by the number of addends.

$$\dfrac{\dfrac{1}{6} + \dfrac{1}{8} + \dfrac{3}{4}}{3}$$

$$= \dfrac{\dfrac{4}{24} + \dfrac{3}{24} + \dfrac{18}{24}}{3}$$

$$= \dfrac{\dfrac{25}{24}}{3}$$

$$= \dfrac{25}{24} \cdot \dfrac{1}{3}$$

$$= \dfrac{25}{72}$$

53. Add the numbers and divide by the number of addends.

$$\dfrac{3\frac{1}{2} + 9\frac{3}{8}}{2}$$

$$= \dfrac{3\frac{4}{8} + 9\frac{3}{8}}{2}$$

$$= \dfrac{12\frac{7}{8}}{2}$$

$$= \dfrac{\dfrac{103}{8}}{2}$$

$$= \dfrac{103}{8} \cdot \dfrac{1}{2}$$

$$= \dfrac{103}{16}, \text{ or } 6\frac{7}{16}$$

55. We add the numbers and divide by the number of addends.

$$\dfrac{15\frac{5}{32} + 20\frac{3}{16} + 12\frac{7}{8}}{3}$$

$$= \dfrac{15\frac{5}{32} + 20\frac{6}{32} + 12\frac{28}{32}}{3}$$

$$= \dfrac{47\frac{39}{32}}{3} = \dfrac{48\frac{7}{32}}{3}$$

$$= \dfrac{\dfrac{1543}{32}}{3} = \dfrac{1543}{32} \cdot \dfrac{1}{3}$$

$$= \dfrac{1543}{96} = 16\frac{7}{96}$$

The average of the distances was $16\frac{7}{96}$ mi.

57. We add the numbers and divide by the number of addends.

$$\dfrac{7\frac{1}{2} + 8 + 9\frac{1}{2} + 10\frac{5}{8} + 11\frac{3}{4}}{5}$$

$$= \dfrac{7\frac{4}{8} + 8 + 9\frac{4}{8} + 10\frac{5}{8} + 11\frac{6}{8}}{5}$$

$$= \dfrac{45\frac{19}{8}}{5} = \dfrac{47\frac{3}{8}}{5}$$

$$= \dfrac{\dfrac{379}{8}}{5} = \dfrac{379}{8} \cdot \dfrac{1}{5}$$

$$= \dfrac{379}{40} = 9\frac{19}{40}$$

The average weight was $9\frac{19}{40}$ lb.

59. $\dfrac{4}{5} \div \left(-\dfrac{3}{10}\right) = \dfrac{4}{5} \cdot \left(-\dfrac{10}{3}\right) = -\dfrac{4 \cdot 2 \cdot 5}{5 \cdot 3} =$

$\dfrac{5}{5} \cdot \left(-\dfrac{4 \cdot 2}{3}\right) = -\dfrac{8}{3}$

61. 1 is neither prime nor composite.

The only factors of 5 are 1 and 5, so 5 is prime.

The only factors of 7 are 1 and 7, so 7 is prime.

$9 = 3 \cdot 3$, so 9 is composite.

$14 = 2 \cdot 7$, so 14 is composite.

The only factors of 23 are 1 and 23, so 23 is prime.

The only factors of 43 are 1 and 43, so 43 is prime.

63. ***Familiarize***. We make a drawing.

◯ ◯ ◯ ··· ◯ } 6 lb feeds how many people?

$\frac{3}{8}$ lb per person

We let $p =$ the number of people who can attend the luncheon.

Translate. The problem translates to the following equation:

$$p = 6 \div \frac{3}{8}$$

Solve. We carry out the division.

$$\begin{aligned} p &= 6 \div \frac{3}{8} \\ &= 6 \cdot \frac{8}{3} = \frac{6 \cdot 8}{3} \\ &= \frac{2 \cdot 3 \cdot 8}{3 \cdot 1} = \frac{3}{3} \cdot \frac{2 \cdot 8}{1} \\ &= 16 \end{aligned}$$

Check. If each of 16 people is allotted $\frac{3}{8}$ lb of cold cuts, a total of

$$16 \cdot \frac{3}{8} = \frac{16 \cdot 3}{8} = \frac{2 \cdot 8 \cdot 3}{8} = 2 \cdot 3,$$

or 6 lb of cold cuts are used. Our answer checks.

State. 16 people can attend the luncheon.

65.

$$\begin{aligned} &\left(1\frac{1}{2} - 1\frac{1}{3}\right)^2 \cdot 144 - \frac{9}{10} \div 4\frac{1}{5} \\ &= \left(\frac{3}{2} - \frac{4}{3}\right)^2 \cdot 144 - \frac{9}{10} \div \frac{21}{5} \\ &= \left(\frac{9}{6} - \frac{8}{6}\right)^2 \cdot 144 - \frac{9}{10} \div \frac{21}{5} \\ &= \left(\frac{1}{6}\right)^2 \cdot 144 - \frac{9}{10} \div \frac{21}{5} \\ &= \frac{1}{36} \cdot 144 - \frac{9}{10} \div \frac{21}{5} \\ &= \frac{1}{36} \cdot 144 - \frac{9}{10} \cdot \frac{5}{21} \\ &= \frac{1 \cdot 144}{36} - \frac{9 \cdot 5}{10 \cdot 21} \\ &= \frac{1 \cdot \cancel{36} \cdot 4}{\cancel{36} \cdot 1} - \frac{\cancel{3} \cdot 3 \cdot \cancel{5}}{2 \cdot \cancel{5} \cdot \cancel{3} \cdot 7} \\ &= 4 - \frac{3}{14} = \frac{56}{14} - \frac{3}{14} \\ &= \frac{53}{14}, \text{ or } 3\frac{11}{14} \end{aligned}$$

67.

$$\begin{aligned} \frac{\frac{2}{3}x - \frac{1}{2}x}{\left(1\frac{1}{4} + \frac{1}{2}\right)^2} &= \frac{\frac{4}{6}x - \frac{3}{6}x}{\left(\frac{5}{4} + \frac{1}{2}\right)^2} \\ &= \frac{\frac{1}{6}x}{\left(\frac{5}{4} + \frac{2}{4}\right)^2} \\ &= \frac{\frac{1}{6}x}{\left(\frac{7}{4}\right)^2} \\ &= \frac{\frac{1}{6}x}{\frac{49}{16}} \\ &= \frac{1}{6}x \div \frac{49}{16} \\ &= \frac{1}{6}x \cdot \frac{16}{49} \\ &= \frac{x \cdot 16}{6 \cdot 49} \\ &= \frac{x \cdot \cancel{2} \cdot 8}{\cancel{2} \cdot 3 \cdot 49} \\ &= \frac{8x}{147}, \text{ or } \frac{8}{147}x \end{aligned}$$

69. Because 2 is very small compared to 99, $\frac{2}{99} \approx 0$.

71. Because $2 \cdot 13 = 26$ and 26 is close to 27, the denominator is about twice the numerator. Thus, $\frac{13}{27} \approx \frac{1}{2}$.

73. Because $2 \cdot 215 = 430$ and 430 is close to 429, the denominator is about twice the numerator. Thus, $\frac{215}{429} \approx \frac{1}{2}$.

Chapter 4 Concept Reinforcement

1. The statement is true; $5\frac{2}{3} = 5 + \frac{2}{3} = 5 \cdot 1 + \frac{2}{3} = 5 \cdot \frac{3}{3} + \frac{2}{3}$.
2. The statement is true. If one number is a multiple of the other, the LCM is the larger number. If one number is not a multiple of the other, then the LCM contains each prime factor that appears in either number the greatest number of times it occurs in any one factorization and, thus, is larger than both numbers.
3. The statement is true. See page 252 in the text.
4. The statement is true. Any mixed numeral is greater than 1, so the product of any two mixed numerals is greater than 1.

Chapter 4 Important Concepts

1. $52 = 2 \cdot 2 \cdot 13$
$78 = 2 \cdot 3 \cdot 13$

The LCM is $2 \cdot 2 \cdot 3 \cdot 13$, or 156.

2. $$\frac{19}{60} + \frac{11}{36} = \frac{19}{2 \cdot 2 \cdot 3 \cdot 5} + \frac{11}{2 \cdot 2 \cdot 3 \cdot 3}$$
$$= \frac{19}{2 \cdot 2 \cdot 3 \cdot 5} \cdot \frac{3}{3} + \frac{11}{2 \cdot 2 \cdot 3 \cdot 3} \cdot \frac{5}{5}$$
$$= \frac{57}{180} + \frac{55}{180} = \frac{112}{180}$$
$$= \frac{4 \cdot 28}{4 \cdot 45} = \frac{4}{4} \cdot \frac{28}{45} = \frac{28}{45}$$

3. The LCD is $13 \cdot 12$, or 156.

$$\frac{3}{13} = \frac{3}{13} \cdot \frac{12}{12} = \frac{36}{156}$$
$$\frac{5}{12} = \frac{5}{12} \cdot \frac{13}{13} = \frac{65}{156}$$

Since $36 < 65$, $\frac{36}{156} < \frac{65}{156}$, and thus $\frac{3}{13} < \frac{5}{12}$.

4. $$\frac{29}{35} - \frac{5}{7} = \frac{29}{35} - \frac{5}{7} \cdot \frac{5}{5} = \frac{29}{35} - \frac{25}{35} = \frac{4}{35}$$

5. $$\frac{2}{9} + \frac{2}{3}x = \frac{1}{6}$$
$$\frac{2}{9} + \frac{2}{3}x - \frac{2}{9} = \frac{1}{6} - \frac{2}{9}$$
$$\frac{2}{3}x = \frac{6}{36} - \frac{8}{36}$$
$$\frac{2}{3}x = -\frac{2}{36}$$
$$\frac{2}{3}x = -\frac{1}{18} \quad \left(-\frac{2}{36} = -\frac{\not{2} \cdot 1}{\not{2} \cdot 18} = -\frac{1}{18}\right)$$
$$\frac{3}{2}\left(\frac{2}{3}x\right) = \frac{3}{2}\left(-\frac{1}{18}\right)$$
$$x = -\frac{3 \cdot 1}{2 \cdot 18}$$
$$x = -\frac{\not{3} \cdot 1}{2 \cdot \not{3} \cdot 6}$$
$$x = -\frac{1}{12}$$

The solution is $-\frac{1}{12}$.

6. $8\frac{2}{3} = \frac{26}{3}$ $(3 \cdot 8 = 24, 24 + 2 = 26)$

7. $$\begin{array}{r} 5 \\ 9\overline{)\,47} \\ \underline{45} \\ 2 \end{array} \qquad \frac{47}{9} = 5\frac{2}{9}$$

8. $$\begin{array}{rcrcr} 10\,\frac{5}{7} & = & 10\,\frac{20}{28} & = & 9\,\frac{48}{28} \\ -2\,\frac{3}{4} & = & -2\,\frac{21}{28} & = & -2\,\frac{21}{28} \\ \hline & & & & 7\,\frac{27}{28} \end{array}$$

9. $$4\frac{1}{5} \cdot 3\frac{7}{15} = \frac{21}{5} \cdot \frac{52}{15} = \frac{21 \cdot 52}{5 \cdot 15} = \frac{3 \cdot 7 \cdot 52}{5 \cdot 3 \cdot 5} =$$
$$\frac{3}{3} \cdot \frac{7 \cdot 52}{5 \cdot 5} = \frac{364}{25} = 14\frac{14}{25}$$

10. ***Familiarize***. Let $p =$ the population of Louisiana.

Translate.

$$\underbrace{2\frac{1}{2}}\ \text{times}\ \underbrace{\text{Population of West Virginia}}\ \text{is}\ \underbrace{\text{Population of Louisiana}}$$
$$2\frac{1}{2} \quad \cdot \quad 1,800,000 \quad = \quad p$$

Solve. We carry out the multiplication.

$$2\frac{1}{2} \cdot 1,800,000 = \frac{5}{2} \cdot 1,800,000 = \frac{5 \cdot 1,800,000}{2} =$$
$$\frac{5 \cdot 2 \cdot 900,000}{2 \cdot 1} = \frac{2}{2} \cdot \frac{5 \cdot 900,000}{1} = \frac{5 \cdot 900,000}{1} =$$
$$4,500,000$$

Thus, $4,500,000 = p$.

Check. We repeat the calculation. The answer checks.

State. The population of Louisiana is 4,500,000.

11. $$\frac{3}{2} \cdot 1\frac{1}{3} \div \left(\frac{2}{3}\right)^2 = \frac{3}{2} \cdot 1\frac{1}{3} \div \frac{4}{9}$$
$$= \frac{3}{2} \cdot \frac{4}{3} \div \frac{4}{9}$$
$$= \frac{3 \cdot 4}{2 \cdot 3} \div \frac{4}{9}$$
$$= \frac{3 \cdot 4}{2 \cdot 3} \cdot \frac{9}{4}$$
$$= \frac{3 \cdot 4 \cdot 9}{2 \cdot 3 \cdot 4} = \frac{3 \cdot 4}{3 \cdot 4} \cdot \frac{9}{2}$$
$$= \frac{9}{2}, \text{ or } 4\frac{1}{2}$$

Chapter 4 Review Exercises

1. a) 18 is not a multiple of 12.

b) Check multiples:

$2 \cdot 18 = 36$ A multiple of 12

c) The LCM is 36.

2. 1.) 45 is not a multiple of 18.

2.) Check multiples:

$2 \cdot 45 = 90$ A multiple of 18

The LCM is 90.

3. Note that 3 and 6 are factors of 30. Since the largest number, 30, has the other two numbers as factors, it is the LCM.

4. a) Find the prime factorization of each number.

$$26 = 2 \cdot 13$$
$$36 = 2 \cdot 2 \cdot 3 \cdot 3$$
$$54 = 2 \cdot 3 \cdot 3 \cdot 3$$

b) Create a product by writing each factor the greatest number of times it occurs in any one factorization.

The greatest number of times 2 occurs in any one factorization is two times.

The greatest number of times 3 occurs in any one factorization is three times.

The greatest number of times 13 occurs in any one factorization is one time.

Since there are no other prime factors in any of the factorizations, the LCM is $2 \cdot 2 \cdot 3 \cdot 3 \cdot 3 \cdot 13$, or 1404.

5. $\frac{2}{9} + \frac{5}{9} = \frac{2+5}{9} = \frac{7}{9}$

6. $\frac{7}{x} + \frac{2}{x} = \frac{7+2}{x} = \frac{9}{x}$

7. The LCM of 5 and 15 is 15.

$$-\frac{6}{5} + \frac{11}{15} = -\frac{6}{5} \cdot \frac{3}{3} + \frac{11}{15} = -\frac{18}{15} + \frac{11}{15} = -\frac{7}{15}$$

8. The LCM of 16 and 24 is 48.

$$\begin{aligned}\frac{5}{16} + \frac{3}{24} &= \frac{5}{16} \cdot \frac{3}{3} + \frac{3}{24} \cdot \frac{2}{2} \\ &= \frac{15}{48} + \frac{6}{48} = \frac{21}{48} \\ &= \frac{3 \cdot 7}{3 \cdot 16} = \frac{3}{3} \cdot \frac{7}{16} = \frac{7}{16}\end{aligned}$$

9. $\frac{7}{9} - \frac{5}{9} = \frac{7-5}{9} = \frac{2}{9}$

10. The LCM of 4 and 8 is 8.

$$\frac{1}{4} - \frac{3}{8} = \frac{1}{4} \cdot \frac{2}{2} - \frac{3}{8} = \frac{2}{8} - \frac{3}{8} = -\frac{1}{8}$$

11. The LCM of 27 and 9 is 27.

$$\frac{10}{27} - \frac{2}{9} = \frac{10}{27} - \frac{2}{9} \cdot \frac{3}{3} = \frac{10}{27} - \frac{6}{27} = \frac{10-6}{27} = \frac{4}{27}$$

12. The LCM of 6 and 9 is 18.

$$\begin{aligned}\frac{5}{6} - \frac{7}{9} &= \frac{5}{6} \cdot \frac{3}{3} - \frac{7}{9} \cdot \frac{2}{2} \\ &= \frac{15}{18} - \frac{14}{18} = \frac{1}{18}\end{aligned}$$

13. The LCD is $7 \cdot 9$, or 63.

$$\frac{4}{7} \cdot \frac{9}{9} = \frac{36}{63}$$
$$\frac{5}{9} \cdot \frac{7}{7} = \frac{35}{63}$$

Since $36 > 35$, it follows that $\frac{36}{63} > \frac{35}{63}$, so $\frac{4}{7} > \frac{5}{9}$.

14. The LCD is $9 \cdot 13$, or 117.

$$\frac{-8}{9} \cdot \frac{13}{13} = \frac{-104}{117}$$
$$\frac{-11}{13} \cdot \frac{9}{9} = \frac{-99}{117}$$

Since $-104 < -99$, it follows that $\frac{-104}{117} < \frac{-99}{117}$, so $-\frac{8}{9} < -\frac{11}{13}$.

15.
$$\begin{aligned}x + \frac{2}{5} &= \frac{7}{8} \\ x + \frac{2}{5} - \frac{2}{5} &= \frac{7}{8} - \frac{2}{5} \\ x + 0 &= \frac{7}{8} \cdot \frac{5}{5} - \frac{2}{5} \cdot \frac{8}{8} \\ x &= \frac{35}{40} - \frac{16}{40} \\ x &= \frac{19}{40}\end{aligned}$$

The solution is $\frac{19}{40}$.

16.
$$\begin{aligned}7a - 3 &= 25 \\ 7a - 3 + 3 &= 25 + 3 \\ 7a &= 28 \\ \frac{1}{7} \cdot 7a &= \frac{1}{7} \cdot 28 \\ a &= 4\end{aligned}$$

The solution is 4.

17.
$$\begin{aligned}5 + \frac{16}{3}x &= \frac{5}{9} \\ 5 + \frac{16}{3}x - 5 &= \frac{5}{9} - 5 \\ \frac{16}{3}x &= \frac{5}{9} - \frac{45}{9} \\ \frac{16}{3}x &= -\frac{40}{9} \\ \frac{3}{16} \cdot \frac{16}{3}x &= \frac{3}{16}\left(-\frac{40}{9}\right) \\ x &= -\frac{3 \cdot 40}{16 \cdot 9} = -\frac{3 \cdot 5 \cdot 8}{2 \cdot 8 \cdot 3 \cdot 3} \\ x &= -\frac{5}{2 \cdot 3} \cdot \frac{3 \cdot 8}{3 \cdot 8} = -\frac{5}{6}\end{aligned}$$

The solution is $-\frac{5}{6}$.

18.
$$\begin{aligned}\frac{22}{5} &= \frac{16}{5} + \frac{5}{2}x \\ \frac{22}{5} - \frac{16}{5} &= \frac{16}{5} + \frac{5}{2}x - \frac{16}{5} \\ \frac{6}{5} &= \frac{5}{2}x \\ \frac{2}{5} \cdot \frac{6}{5} &= \frac{2}{5} \cdot \frac{5}{2}x \\ \frac{12}{25} &= x\end{aligned}$$

The solution is $\frac{12}{25}$.

19.
$$\frac{5}{3}x+\frac{5}{6}=\frac{3}{2}$$
$$6\left(\frac{5}{3}x+\frac{5}{6}\right)=6\cdot\frac{3}{2}\quad\text{The LCD is 6.}$$
$$\frac{6\cdot 5}{3}x+6\cdot\frac{5}{6}=\frac{18}{2}$$
$$\frac{2\cdot\not{3}\cdot 5}{\not{3}\cdot 1}x+5=9$$
$$10x+5=9$$
$$10x+5-5=9-5$$
$$10x=4$$
$$\frac{10x}{10}=\frac{4}{10}$$
$$x=\frac{2}{5}$$
The solution is $\frac{2}{5}$.

20. $7\frac{1}{2}=\frac{15}{2}\quad(7\cdot 2=14,\ 14+1=15)$

21. $8\frac{3}{8}=\frac{67}{8}\quad(8\cdot 8=64,\ 64+3=67)$

22. $4\frac{1}{3}=\frac{13}{3}\quad(4\cdot 3=12,\ 12+1=13)$

23. $-1\frac{5}{7}=-\left(1\frac{5}{7}\right)=-\frac{12}{7}$

$(1\cdot 7=7;\ 7+5=12;$ include the negative sign.)

24. To convert $\frac{7}{3}$ to a mixed numeral,we divide.

$$\begin{array}{r} 2 \\ 3\overline{)\,7} \\ \underline{6} \\ 1 \end{array}\qquad \frac{7}{3}=2\frac{1}{3}$$

25. First consider $\frac{27}{4}$.

$$\begin{array}{r} 6 \\ 4\overline{)\,27} \\ \underline{24} \\ 3 \end{array}\qquad \frac{27}{4}=6\frac{3}{4}$$

Since $\frac{27}{4}=6\frac{3}{4}$, we have $\frac{-27}{4}=-6\frac{3}{4}$.

26. To convert $\frac{63}{5}$ to a mixed numeral, we divide.

$$\begin{array}{r} 12 \\ 5\overline{)\,63} \\ \underline{5} \\ 13 \\ \underline{10} \\ 3 \end{array}\qquad \frac{63}{5}=12\frac{3}{5}$$

27.
$$\begin{array}{r} 3 \\ 2\overline{)\,7} \\ \underline{6} \\ 1 \end{array}\qquad \frac{7}{2}=3\frac{1}{2}$$

28. First we find $7896\div 9$.

$$\begin{array}{r} 877 \\ 9\overline{)\,7896} \\ \underline{72} \\ 69 \\ \underline{63} \\ 66 \\ \underline{63} \\ 3 \end{array}$$

Since $877\frac{3}{9}=877\frac{1}{3}$, we have $7896\div(-9)=-877\frac{1}{3}$.

29. $\frac{80+82+85}{3}=\frac{247}{3}=82\frac{1}{3}$

30.
$$\begin{array}{r} 7\,\frac{3}{5} \\ +2\,\frac{4}{5} \\ \hline 9\,\frac{7}{5} \end{array}=9+\frac{7}{5}$$
$$=9+1\frac{2}{5}$$
$$=10\frac{2}{5}$$

31.
$$\begin{array}{rcr} 6\,\boxed{\frac{1}{3}\cdot\frac{5}{5}} & = & 6\,\frac{5}{15} \\ +5\,\boxed{\frac{2}{5}\cdot\frac{3}{3}} & = & +5\,\frac{6}{15} \\ \hline & & 11\,\frac{11}{15} \end{array}$$

32. $-3\frac{5}{6}+\left(-5\frac{1}{6}\right)$

We add the absolute values and make the answer negative.

$$\begin{array}{r} 3\,\frac{5}{6} \\ +5\,\frac{1}{6} \\ \hline 8\,\frac{6}{6} \end{array}=8+1=9$$

Thus, $-3\frac{5}{6}+\left(-5\frac{1}{6}\right)=-9$.

33. $-2\frac{3}{4}+4\frac{1}{2}=4\frac{1}{2}-2\frac{3}{4}$

$$\begin{array}{rcrcr} 4\,\boxed{\frac{1}{2}\cdot\frac{2}{2}} & = & 4\,\frac{2}{4} & = & 3\,\frac{6}{4} \\ -2\,\frac{3}{4} & = & -2\,\frac{3}{4} & = & -2\,\frac{3}{4} \\ \hline & & & & 1\,\frac{3}{4} \end{array}$$

34.
$$\begin{array}{rcr} 14 & = & 13\frac{9}{9} \\ -\ 6\frac{2}{9} & = & -\ 6\frac{2}{9} \\ \hline & & 7\frac{7}{9} \end{array}$$

35.
$$\begin{array}{rcrcr} 9\boxed{\dfrac{3}{5}\cdot\dfrac{3}{3}} & = & 9\dfrac{9}{15} & = & 8\dfrac{24}{15} \\ -4\dfrac{13}{15} & = & -4\dfrac{13}{15} & = & -4\dfrac{13}{15} \\ \hline & & & & 4\dfrac{11}{15} \end{array}$$

36. $4\frac{5}{8} - 9\frac{3}{4} = 4\frac{5}{8} + \left(-9\frac{3}{4}\right)$

Since $9\frac{3}{4}$ is greater than $4\frac{5}{8}$, the answer will be negative. The difference of the absolute values is

$$\begin{array}{rcr} 9\boxed{\dfrac{3}{4}\cdot\dfrac{2}{2}} & = & 9\dfrac{6}{8} \\ -4\dfrac{5}{8} & = & -4\dfrac{5}{8} \\ \hline & & 5\dfrac{1}{8} \end{array}$$

so $4\frac{5}{8} - 9\frac{3}{4} = -5\frac{1}{8}$.

37. $-7\frac{1}{2} - 6\frac{3}{4} = -7\frac{1}{2} + \left(-6\frac{3}{4}\right)$

We add the absolute values and make the answer negative.

$$\begin{array}{rcl} 7\boxed{\dfrac{1}{2}\cdot\dfrac{2}{2}} & = & 7\dfrac{2}{4} \\ +6\dfrac{3}{4} & = & +6\dfrac{3}{4} \\ \hline & & 13\dfrac{5}{4} = 13 + \dfrac{5}{4} \\ & & \quad = 13 + 1\dfrac{1}{4} \\ & & \quad = 14\dfrac{1}{4} \end{array}$$

Thus, $-7\frac{1}{2} - 6\frac{3}{4} = -14\frac{1}{4}$.

38.
$$\begin{aligned} \frac{4}{9}x + \frac{1}{3}x &= \frac{4}{9}x + \frac{1}{3}\cdot\frac{3}{3}x \\ &= \frac{4}{9}x + \frac{3}{9}x \\ &= \frac{7}{9}x \end{aligned}$$

39.
$$\begin{aligned} 8\frac{3}{10}a - 5\frac{1}{8}a &= \left(8\frac{3}{10} - 5\frac{1}{8}\right)a \\ &= \left(8\frac{12}{40} - 5\frac{5}{40}\right)a \\ &= 3\frac{7}{40}a \end{aligned}$$

40. $6\cdot 2\frac{2}{3} = 6\cdot\frac{8}{3} = \frac{6\cdot 8}{3} = \frac{2\cdot 3\cdot 8}{3\cdot 1} = \frac{3}{3}\cdot\frac{2\cdot 8}{1} = 16$

41. $-5\frac{1}{4}\cdot\frac{2}{3} = -\frac{21}{4}\cdot\frac{2}{3} = -\frac{21\cdot 2}{4\cdot 3} = -\frac{3\cdot 7\cdot 2}{2\cdot 2\cdot 3} =$
$\frac{2\cdot 3}{2\cdot 3}\cdot\left(-\frac{7}{2}\right) = -\frac{7}{2} = -3\frac{1}{2}$

42. $2\frac{1}{5}\cdot 1\frac{1}{10} = \frac{11}{5}\cdot\frac{11}{10} = \frac{11\cdot 11}{5\cdot 10} = \frac{121}{50} = 2\frac{21}{50}$

43. $2\frac{2}{5}\cdot 2\frac{1}{2} = \frac{12}{5}\cdot\frac{5}{2} = \frac{12\cdot 5}{5\cdot 2} = \frac{2\cdot 6\cdot 5}{5\cdot 2\cdot 1} = \frac{2\cdot 5}{2\cdot 5}\cdot\frac{6}{1} = 6$

44. $-54 \div 2\frac{1}{4} = -54 \div \frac{9}{4} = -54\cdot\frac{4}{9} = \frac{-54\cdot 4}{9} =$
$\frac{-2\cdot 3\cdot 9\cdot 4}{9\cdot 1} = \frac{9}{9}\cdot\frac{-2\cdot 3\cdot 4}{1} = -24$

45. $2\frac{2}{5} \div \left(-1\frac{7}{10}\right) = \frac{12}{5} \div \left(-\frac{17}{10}\right) = \frac{12}{5}\cdot\left(-\frac{10}{17}\right) =$
$-\frac{12\cdot 10}{5\cdot 17} = -\frac{12\cdot 2\cdot 5}{5\cdot 17} = \frac{5}{5}\cdot\left(-\frac{12\cdot 2}{17}\right) = -\frac{24}{17} = -1\frac{7}{17}$

46. $3\frac{1}{4} \div 26 = \frac{13}{4} \div 26 = \frac{13}{4}\cdot\frac{1}{26} = \frac{13\cdot 1}{4\cdot 26} = \frac{13\cdot 1}{4\cdot 2\cdot 13} =$
$\frac{13}{13}\cdot\frac{1}{4\cdot 2} = \frac{1}{8}$

47. $4\frac{1}{5} \div 4\frac{2}{3} = \frac{21}{5} \div \frac{14}{3} = \frac{21}{5}\cdot\frac{3}{14} = \frac{21\cdot 3}{5\cdot 14} = \frac{3\cdot 7\cdot 3}{5\cdot 2\cdot 7} =$
$\frac{7}{7}\cdot\frac{3\cdot 3}{5\cdot 2} = \frac{9}{10}$

48.
$$\begin{aligned} 5x - y &= 5\cdot 3\frac{1}{5} - 2\frac{2}{7} \\ &= 5\cdot\frac{16}{5} - 2\frac{2}{7} \\ &= \frac{\not{5}\cdot 16}{1\cdot\not{5}} - 2\frac{2}{7} \\ &= 16 - 2\frac{2}{7} \\ &= 15\frac{7}{7} - 2\frac{2}{7} \\ &= 13\frac{5}{7} \end{aligned}$$

49.
$$\begin{aligned} 2a \div b &= 2\cdot 5\frac{2}{11} \div 3\frac{4}{5} \\ &= 2\cdot\frac{57}{11} \div \frac{19}{5} \\ &= \frac{2\cdot 57}{11} \div \frac{19}{5} \\ &= \frac{2\cdot 57}{11}\cdot\frac{5}{19} \\ &= \frac{2\cdot 57\cdot 5}{11\cdot 19} \\ &= \frac{2\cdot 3\cdot \not{19}\cdot 5}{11\cdot\not{19}} \\ &= \frac{30}{11} \\ &= 2\frac{8}{11} \end{aligned}$$

50. ***Familiarize***. Let f = the number of yards of fabric Gloria needs.

Translate.

Fabric for slacks	plus	Fabric for jacket	is	Total fabric needed
↓	↓	↓	↓	↓
$1\frac{5}{8}$	$+$	$2\frac{5}{8}$	$=$	f

Solve. We carry out the addition.

$$\begin{array}{r} 1\,\frac{5}{8} \\ +2\,\frac{5}{8} \\ \hline 3\,\frac{10}{8} \end{array}$$

$$3\frac{10}{8} = 3 + \frac{10}{8} = 3 + 1\frac{2}{8} = 4\frac{2}{8} = 4\frac{1}{4}$$

Check. We repeat the calculation. The answer checks.

State. Gloria needs $4\frac{1}{4}$ yd of fabric.

51. ***Familiarize***. Let $p =$ the number of pizzas that remained.

Translate.

$$p = \frac{3}{8} + 1\frac{1}{2} + 1\frac{1}{4}$$

Solve. We carry out the addition.

$$\begin{array}{rcl} \frac{3}{8} & = & \frac{3}{8} \\ 1\boxed{\frac{1}{2}\cdot\frac{4}{4}} & = & 1\frac{4}{8} \\ +1\boxed{\frac{1}{4}\cdot\frac{2}{2}} & = & +1\frac{2}{8} \\ \hline & & 2\frac{9}{8} \end{array}$$

$$2\frac{9}{8} = 2 + \frac{9}{8} = 2 + 1\frac{1}{8} = 3\frac{1}{8}$$

Check. We repeat the calculation. The answer checks.

State. Altogether, $3\frac{1}{8}$ pizzas remained.

52. ***Familiarize***. Let $t =$ the number of pounds of turkey needed for 32 servings.

Translate.

Servings per pound	times	Number of pounds	is	Number of servings
↓	↓	↓	↓	↓
$1\frac{1}{3}$	$\cdot$	t	$=$	32

Solve. We divide by $1\frac{1}{3}$ on both sides of the equation.

$$t = 32 \div 1\frac{1}{3}$$

$$t = 32 \div \frac{4}{3}$$

$$t = 32 \cdot \frac{3}{4} = \frac{32\cdot 3}{4}$$

$$t = \frac{4\cdot 8\cdot 3}{4\cdot 1} = \frac{4}{4}\cdot\frac{8\cdot 3}{1}$$

$$t = 24$$

Check. Since $1\frac{1}{3}\cdot 24 = \frac{4}{3}\cdot 24 = \frac{4\cdot 24}{3} = \frac{4\cdot 3\cdot 8}{3\cdot 1} = \frac{3}{3}\cdot\frac{4\cdot 8}{1} = 32$, the answer checks.

State. 24 pounds of turkey are needed for 32 servings.

53. We find the area of each rectangle and then add to find the total area. Recall that the area of a rectangle is length × width.

Area of rectangle A:

$12\times 9\frac{1}{2} = 12\times\frac{19}{2} = \frac{12\times 19}{2} = \frac{2\cdot 6\cdot 19}{2\cdot 1} = \frac{2}{2}\cdot\frac{6\cdot 19}{1} =$ 114 in^2

Area of rectangle B:

$8\frac{1}{2}\times 7\frac{1}{2} = \frac{17}{2}\times\frac{15}{2} = \frac{17\times 15}{2\times 2} = \frac{255}{4} = 63\frac{3}{4}$ in^2

Sum of the areas:

114 in^2 + $63\frac{3}{4}$ in^2 = $177\frac{3}{4}$ in^2

54. We subtract the area of rectangle B from the area of rectangle A.

$$\begin{array}{rcr} 114 & = & 113\,\frac{4}{4} \\ -63\,\frac{3}{4} & = & -63\,\frac{3}{4} \\ \hline & & 50\,\frac{1}{4} \end{array}$$

The area of rectangle A is $50\frac{1}{4}$ in^2 greater than the area of rectangle B.

55. ***Familiarize***. This is a multistep problem. First we find the perimeter of the room, in feet. Then we find what the artist charges to paint that many feet. Let $P =$ the perimeter, in feet, and C the amount the artist charges.

Translate. To find the perimeter, we add the lengths of the four sides of the room.

$$P = 11\frac{3}{4} + 9\frac{1}{2} + 11\frac{3}{4} + 9\frac{1}{2}$$

We multiply to find the cost.

Number of feet	times	Cost per foot	is	Total cost
↓	↓	↓	↓	↓
P	$\cdot$	20	$=$	C

Solve. First we find P.

$P = 11\frac{3}{4} + 9\frac{1}{2} + 11\frac{3}{4} + 9\frac{1}{2} = 11\frac{3}{4} + 9\frac{2}{4} + 11\frac{3}{4} + 9\frac{2}{4} =$

$40\frac{10}{4} = 40 + \frac{10}{4} = 40 + 2\frac{2}{4} = 42\frac{2}{4} = 42\frac{1}{2}$

Now we substitute $42\frac{1}{2}$ for P in the second equation and find C.

$$P \cdot 20 = C$$
$$42\frac{1}{2} \cdot 20 = C$$
$$\frac{85}{2} \cdot 20 = C$$
$$\frac{85 \cdot 20}{2} = C$$
$$\frac{85 \cdot \cancel{2} \cdot 10}{\cancel{2} \cdot 1} = C$$
$$850 = C$$

Check. We repeat the calculations. The answer checks.

State. The project will cost \$850.

56. ***Familiarize***. Let $s =$ the number of cups of shortening in the lower calorie cake.

Translate.

New amount of shortening	plus	Amount of prune puree	is	Original amount of shortening
↓	↓	↓	↓	↓
s	$+$	$3\frac{5}{8}$	$=$	12

Solve. We subtract $3\frac{5}{8}$ on both sides of the equation.

$$\begin{array}{rcr} 12 & = & 11\frac{8}{8} \\ -3\frac{5}{8} & = & -3\frac{5}{8} \\ \hline & & 8\frac{3}{8} \end{array}$$

Thus, $s = 8\frac{3}{8}$.

Check. $8\frac{3}{8} + 3\frac{5}{8} = 11\frac{8}{8} = 12$, so the answer checks.

State. The lower calorie recipe uses $8\frac{3}{8}$ cups of shortening.

57. ***Familiarize***. Let $s =$ the number of pies sold and let $l =$ the number of pies left over.

Translate.

Number of pies sold	times	Number of pieces per pie	is	Number of pieces sold
↓	↓	↓	↓	↓
s	$\cdot$	6	$=$	382

Number of pies sold	plus	Number left over	is	Number of pies donated
↓	↓	↓	↓	↓
s	$+$	l	$=$	83

Solve. To solve the first equation we divide by 6 on both sides.

$$s \cdot 6 = 382$$
$$s = \frac{382}{6} = 63\frac{2}{3}$$

Now we substitute $63\frac{2}{3}$ for s in the second equation and solve for l.

$$s + l = 83$$
$$63\frac{2}{3} + l = 83$$
$$l = 83 - 63\frac{2}{3}$$
$$l = 19\frac{1}{3}$$

Check. We repeat the calculations. The answer checks.

State. $63\frac{2}{3}$ pies were sold; $19\frac{1}{3}$ pies were left over.

58. For the $13\frac{1}{4}$ in. $\times$ $13\frac{1}{4}$ in. side:

Perimeter $= 13\frac{1}{4} + 13\frac{1}{4} + 13\frac{1}{4} + 13\frac{1}{4} = 52\frac{4}{4} = 52 + \frac{4}{4} = 52 + 1 = 53$ in.

Area $= 13\frac{1}{4} \times 13\frac{1}{4} = \frac{53}{4} \times \frac{53}{4} = \frac{53 \times 53}{4 \times 4} = \frac{2809}{16} = 175\frac{9}{16}$ in^2

For the $13\frac{1}{4}$ in. $\times$ $3\frac{1}{4}$ in. side:

Perimeter $= 13\frac{1}{4} + 3\frac{1}{4} + 13\frac{1}{4} + 3\frac{1}{4} = 32\frac{4}{4} = 32 + \frac{4}{4} = 32 + 1 = 33$ in.

Area $= 13\frac{1}{4} \times 3\frac{1}{4} = \frac{53}{4} \times \frac{13}{4} = \frac{53 \times 13}{4 \times 4} = \frac{689}{16} = 43\frac{1}{16}$ in^2

59.
$$\begin{aligned} \frac{1}{2} + \frac{1}{8} \div \frac{1}{4} &= \frac{1}{2} + \frac{1}{8} \cdot \frac{4}{1} \\ &= \frac{1}{2} + \frac{4}{8} \\ &= \frac{1}{2} + \frac{\cancel{4} \cdot 1}{\cancel{4} \cdot 2} \\ &= \frac{1}{2} + \frac{1}{2} \\ &= 1 \end{aligned}$$

60.
$$\begin{aligned} \frac{4}{5} - \frac{1}{2} \cdot \left(\frac{1}{4} - \frac{3}{5}\right) &= \frac{4}{5} - \frac{1}{2} \cdot \left(\frac{5}{20} - \frac{12}{20}\right) \\ &= \frac{4}{5} - \frac{1}{2} \cdot \left(-\frac{7}{20}\right) \\ &= \frac{4}{5} + \frac{7}{40} \\ &= \frac{32}{40} + \frac{7}{40} \\ &= \frac{39}{40} \end{aligned}$$

61. $20\frac{3}{4} - 1\frac{1}{2} \times 12 + \left(\frac{1}{2}\right)^2 = 20\frac{3}{4} - 1\frac{1}{2} \times 12 + \frac{1}{4}$

$= 20\frac{3}{4} - \frac{3}{2} \times 12 + \frac{1}{4}$

$= 20\frac{3}{4} - \frac{36}{2} + \frac{1}{4}$

$= 20\frac{3}{4} - 18 + \frac{1}{4}$

$= 2\frac{3}{4} + \frac{1}{4}$

$= 2\frac{4}{4} = 3$

62. $\dfrac{\frac{1}{2}+\frac{1}{4}+\frac{1}{3}+\frac{1}{5}}{4} = \dfrac{\frac{30}{60}+\frac{15}{60}+\frac{20}{60}+\frac{12}{60}}{4}$

$= \dfrac{\frac{77}{60}}{4}$

$= \frac{77}{60} \cdot \frac{1}{4}$

$= \frac{77}{240}$

63. $\dfrac{\frac{1}{3}}{\frac{5}{8}-1} = \dfrac{\frac{1}{3}}{\frac{5}{8}-\frac{8}{8}} = \dfrac{\frac{1}{3}}{-\frac{3}{8}} = \frac{1}{3} \div \left(-\frac{3}{8}\right) = \frac{1}{3} \cdot \left(-\frac{8}{3}\right) =$

$-\frac{1 \cdot 8}{3 \cdot 3} = -\frac{8}{9}$

64. $\dfrac{-\frac{x}{7}}{\frac{3}{14}} = -\frac{x}{7} \div \frac{3}{14} = -\frac{x}{7} \cdot \frac{14}{3} = -\frac{x \cdot 14}{7 \cdot 3} =$

$-\frac{x \cdot 2 \cdot \not{7}}{\not{7} \cdot 3} = -\frac{2x}{3}$

65. $\frac{1}{4} + \frac{2}{5} \div 5^2 = \frac{1}{4} + \frac{2}{5} \div 25$

$= \frac{1}{4} + \frac{2}{5} \cdot \frac{1}{25}$

$= \frac{1}{4} + \frac{2}{125}$

$= \frac{1}{4} \cdot \frac{125}{125} + \frac{2}{125} \cdot \frac{4}{4}$

$= \frac{125}{4 \cdot 125} + \frac{8}{4 \cdot 125}$

$= \frac{133}{500}$

Answer A is correct.

66. $x + \frac{2}{3} = 5$

$x + \frac{2}{3} - \frac{2}{3} = 5 - \frac{2}{3}$

$x + 0 = 5 \cdot \frac{3}{3} - \frac{2}{3}$

$x = \frac{15}{3} - \frac{2}{3} = \frac{15-2}{3}$

$x = \frac{13}{3}$, or $4\frac{1}{3}$

The solution is $\frac{13}{3}$, or $4\frac{1}{3}$. Answer D is correct.

67. $\frac{1}{100} + \frac{1}{150} + \frac{1}{200} = \frac{1}{100} \cdot \frac{6}{6} + \frac{1}{150} \cdot \frac{4}{4} + \frac{1}{200} \cdot \frac{3}{3}$

$= \frac{6}{600} + \frac{4}{600} + \frac{3}{600}$

$= \frac{13}{600}$

Thus, $\frac{1}{r} = \frac{13}{600}$, so r is the reciprocal of $\frac{13}{600}$, or $\frac{600}{13}$, or $46\frac{2}{13}$.

68. Since the largest fraction we can form is $\frac{6}{3}$, or 2, and $3\frac{1}{4} - 2 = \frac{5}{4}$, we know that both fractions must be greater than 1. By trial, we find true equation $\frac{6}{3} + \frac{5}{4} = 3\frac{1}{4}$.

69. a) $\frac{7}{\square}$ is greater than 1 when the denominator is less than the numerator. Thus, for this fraction the largest integer denominator possible is 6.

b) $\frac{11}{\square}$ is greater than 1 when the denominator is less than the numerator. Thus, for this fraction the largest integer denominator possible is 10.

c) $\frac{\square}{-27}$ is greater than 1 when the numerator is greater than the denominator. Thus, for this fraction the largest integer numerator possible is -28.

d) $\dfrac{\square}{-\frac{1}{2}}$ is greater than 1 when the numerator is greater than the denominator. Thus, for this fraction the largest integer numerator possible is -1.

Chapter 4 Discussion and Writing Exercises

1. No; if the sum of the fractional parts of the mixed numerals is $\frac{n}{n}$, then the sum of the mixed numerals is an integer. For example, $1\frac{1}{5} + 6\frac{4}{5} = 7\frac{5}{5} = 8$.

2. A wheel makes $33\frac{1}{3}$ revolutions per minute. It rotates for $4\frac{1}{2}$ min. How many revolutions does it make?

3. The student is multiplying the whole numbers to get the whole number portion of the answer and multiplying fractions to get the fractional part of the answer. The student should have converted each mixed numeral to fractional notation, multiplied, simplified, and then converted back to a mixed numeral. The correct answer is $4\frac{6}{7}$.

4. It might be necessary to find the least common denominator before adding or subtracting. The least common denominator is the least common multiple of the denominators.

5. Suppose that a room has dimensions $15\frac{3}{4}$ ft by $28\frac{5}{8}$ ft. The equation $2 \cdot 15\frac{3}{4} + 2 \cdot 28\frac{5}{8} = 88\frac{3}{4}$ gives the perimeter of the room, in feet. Answers may vary.

6. The products $5 \cdot 3$ and $5 \cdot \frac{2}{7}$ should be added rather than multiplied together. The student could also have converted $3\frac{2}{7}$ to fractional notation, multiplied, simplified, and converted back to a mixed numeral. The correct answer is $16\frac{3}{7}$.

Chapter 4 Test

1. $12 = 2 \cdot 2 \cdot 3 = 2^2 \cdot 3$
$16 = 2 \cdot 2 \cdot 2 \cdot 2 = 2^4$

We form the LCM using the greatest power of each factor. The LCM is $2^4 \cdot 3$, or 48.

2. $\frac{1}{2} + \frac{5}{2} = \frac{1+5}{2} = \frac{6}{2} = 3$

3.
$$\begin{aligned} & -\frac{7}{8} + \frac{2}{3} \\ &= \frac{-7}{8} + \frac{2}{3} \quad \text{8 and 3 have no common factors, so the LCD is } 8 \cdot 3, \text{ or } 24. \\ &= \frac{-7}{8} \cdot \frac{3}{3} + \frac{2}{3} \cdot \frac{8}{8} \\ &= \frac{-21}{24} + \frac{16}{24} \\ &= \frac{-5}{24} \end{aligned}$$

4. $\frac{5}{t} - \frac{3}{t} = \frac{5-3}{t} = \frac{2}{t}$

5. The LCM of 6 and 4 is 12.

$$\begin{aligned} \frac{5}{6} - \frac{3}{4} &= \frac{5}{6} \cdot \frac{2}{2} - \frac{3}{4} \cdot \frac{3}{3} \\ &= \frac{10}{12} - \frac{9}{12} = \frac{1}{12} \end{aligned}$$

6. The LCM of 8 and 24 is 24.

$$\begin{aligned} \frac{5}{8} - \frac{17}{24} &= \frac{5}{8} \cdot \frac{3}{3} - \frac{17}{24} \\ &= \frac{15}{24} - \frac{17}{24} = \frac{15-17}{24} \\ &= \frac{-2}{24} = \frac{-1 \cdot \cancel{2}}{\cancel{2} \cdot 12} \\ &= \frac{-1}{12}, \text{ or } -\frac{1}{12} \end{aligned}$$

7.
$$\begin{aligned} x + \frac{2}{3} &= \frac{11}{12} \\ x + \frac{2}{3} - \frac{2}{3} &= \frac{11}{12} - \frac{2}{3} \quad \text{Subtracting } \frac{2}{3} \text{ on both sides} \\ x + 0 &= \frac{11}{12} - \frac{2}{3} \cdot \frac{4}{4} \quad \text{The LCD is 12.} \\ x &= \frac{11}{12} - \frac{8}{12} = \frac{3}{12} \\ x &= \frac{3 \cdot 1}{3 \cdot 4} = \frac{3}{3} \cdot \frac{1}{4} \\ x &= \frac{1}{4} \end{aligned}$$

8.
$$\begin{aligned} -5x - 3 &= 9 \\ -5x - 3 + 3 &= 9 + 3 \\ -5x &= 12 \\ \frac{-5x}{-5} &= \frac{12}{-5} \\ x &= \frac{12}{-5} \end{aligned}$$

The solution is $\frac{12}{-5}$, or $\frac{-12}{5}$, or $-\frac{12}{5}$.

9. The LCM of the denominators is 12.

$$\begin{aligned} \frac{3}{4} &= \frac{1}{2} + \frac{5}{3}x \\ 12 \cdot \frac{3}{4} &= 12\left(\frac{1}{2} + \frac{5}{3}x\right) \\ \frac{36}{4} &= 12 \cdot \frac{1}{2} + \frac{12 \cdot 5}{3}x \\ 9 &= \frac{12}{2} + \frac{60}{3}x \\ 9 &= 6 + 20x \\ 9 - 6 &= 6 + 20x - 6 \\ 3 &= 20x \\ \frac{3}{20} &= \frac{20x}{20} \\ \frac{3}{20} &= x \end{aligned}$$

The solution is $\frac{3}{20}$.

10. The LCD is 175.

$$\frac{6}{7} \cdot \frac{25}{25} = \frac{150}{175}$$
$$\frac{21}{25} \cdot \frac{7}{7} = \frac{147}{175}$$

Since $150 > 147$, it follows that $\frac{150}{175} > \frac{147}{175}$, so $\frac{6}{7} > \frac{21}{25}$.

11. $3\frac{1}{2} = \frac{7}{2}$ $\quad (3 \cdot 2 = 6,\ 6 + 1 = 7)$

12. $-9\frac{3}{8} = -\left(9 + \frac{3}{8}\right) = -\frac{75}{8}$

$(9 \cdot 8 = 72;\ 72 + 3 = 75;$ include the negative sign.)

13. First consider $\frac{74}{9}$.

$$\begin{array}{r} 8 \\ 9\overline{)\,74} \\ \underline{72} \\ 2 \end{array} \qquad \frac{74}{9} = 8\frac{2}{9}$$

Since $\frac{74}{9} = 8\frac{2}{9}$, we have $-\frac{74}{9} = -8\frac{2}{9}$.

14.

$$\begin{array}{r} 162 \\ 11\overline{)\,1789} \\ \underline{11} \\ 68 \\ \underline{66} \\ 29 \\ \underline{22} \\ 7 \end{array}$$

The answer is $162\frac{7}{11}$.

15.

$$\begin{array}{r} 6\frac{2}{5} \\ \underline{+7\frac{4}{5}} \\ 13\frac{6}{5} \end{array} = 13 + \frac{6}{5}$$

$$= 13 + 1\frac{1}{5}$$

$$= 14\frac{1}{5}$$

16. The LCD is 12.

$$\begin{array}{rcl} 3\boxed{\frac{1}{4} \cdot \frac{3}{3}} & = & 3\frac{3}{12} \\ +9\boxed{\frac{1}{6} \cdot \frac{2}{2}} & = & +9\frac{2}{12} \\ \hline & & 12\frac{5}{12} \end{array}$$

17. The LCD is 24.

$$\begin{array}{rcccl} 10\boxed{\frac{1}{6} \cdot \frac{4}{4}} & = & 10\frac{4}{24} & = & 9\frac{28}{24} \\ -5\boxed{\frac{7}{8} \cdot \frac{3}{3}} & = & -5\frac{21}{24} & = & -5\frac{21}{24} \\ \hline & & \uparrow & & 4\frac{7}{24} \end{array}$$

(Since $\frac{4}{24}$ is smaller than $\frac{21}{24}$, we cannot subtract until we borrow: $10\frac{4}{24} = 9 + \frac{24}{24} + \frac{4}{24} = 9 + \frac{28}{24} = 9\frac{28}{24}$.)

18. $14 + \left(-5\frac{3}{7}\right) = 13\frac{7}{7} + \left(-5\frac{3}{7}\right) = 8\frac{4}{7}$

19. $3\frac{4}{5} - 9\frac{1}{2} = 3\frac{4}{5} + \left(-9\frac{1}{2}\right)$

Since $-9\frac{1}{2}$ has the larger absolute value, the answer will be negative. We find the difference in absolute values.

$$\begin{array}{rcccl} 9\boxed{\frac{1}{2} \cdot \frac{5}{5}} & = & 9\frac{5}{10} & = & 8\frac{15}{10} \\ -3\boxed{\frac{4}{5} \cdot \frac{2}{2}} & = & -3\frac{8}{10} & = & -3\frac{8}{10} \\ \hline & & & & 5\frac{7}{10} \end{array}$$

Thus, $3\frac{4}{5} - 9\frac{1}{2} = -5\frac{7}{10}$.

20.

$$\begin{aligned} \frac{3}{8}x - \frac{1}{2}x &= \frac{3}{8}x - \frac{1}{2} \cdot \frac{4}{4} \cdot x \\ &= \frac{3}{8}x - \frac{4}{8}x \\ &= -\frac{1}{8}x \end{aligned}$$

21.

$$\begin{aligned} 5\frac{2}{11}a - 3\frac{1}{5}a &= \left(5\frac{2}{11} - 3\frac{1}{5}\right)a \\ &= \left(5\frac{10}{55} - 3\frac{11}{55}\right)a \\ &= \left(4\frac{65}{55} - 3\frac{11}{55}\right)a \\ &= 1\frac{54}{55}a \end{aligned}$$

22. $9 \cdot 4\frac{1}{3} = 9 \cdot \frac{13}{3} = \frac{9 \cdot 13}{3} = \frac{3 \cdot 3 \cdot 13}{3 \cdot 1} = \frac{3}{3} \cdot \frac{3 \cdot 13}{1} = 39$

23. $6\frac{3}{4} \cdot \left(-2\frac{2}{3}\right) = \frac{27}{4} \cdot \left(-\frac{8}{3}\right) = -\frac{27 \cdot 8}{4 \cdot 3} = -\frac{3 \cdot 9 \cdot 2 \cdot 4}{4 \cdot 3 \cdot 1} =$
$-\frac{9 \cdot 2}{1} \cdot \frac{3 \cdot 4}{3 \cdot 4} = -18$

24. $33 \div 5\frac{1}{2} = 33 \div \frac{11}{2} = 33 \cdot \frac{2}{11} = \frac{33 \cdot 2}{11} = \frac{3 \cdot 11 \cdot 2}{11 \cdot 1} =$
$\frac{11}{11} \cdot \frac{3 \cdot 2}{1} = 6$

25. $2\frac{1}{3} \div 1\frac{1}{6} = \frac{7}{3} \div \frac{7}{6} = \frac{7}{3} \cdot \frac{6}{7} = \frac{7 \cdot 6}{3 \cdot 7} = \frac{7 \cdot 2 \cdot 3}{3 \cdot 7 \cdot 1} =$
$\frac{7 \cdot 3}{7 \cdot 3} \cdot \frac{2}{1} = 2$

26. $\frac{2}{3}ab = \frac{2}{3} \cdot 7 \cdot 4\frac{1}{5} = \frac{2 \cdot 7}{3} \cdot 4\frac{1}{5} = \frac{2 \cdot 7}{3} \cdot \frac{21}{5} = \frac{2 \cdot 7 \cdot 21}{3 \cdot 5} =$
$\frac{2 \cdot 7 \cdot \cancel{3} \cdot 7}{\cancel{3} \cdot 5} = \frac{98}{5}$, or $19\frac{3}{5}$

27. $$\begin{aligned} 4+mn &= 4+7\frac{2}{5}\cdot 3\frac{1}{4} \\ &= 4+\frac{37}{5}\cdot\frac{13}{4} \\ &= 4+\frac{481}{20} \\ &= 4+24\frac{1}{20} \\ &= 28\frac{1}{20} \end{aligned}$$

28. ***Familiarize***. Let w = Rezazadeh's body weight, in kilograms.

Translate.

$$\underbrace{\text{Weight lifted}}\ \text{is}\ 2\frac{1}{2}\ \text{times}\ \underbrace{\text{body weight}}$$
$$263 = 2\frac{1}{2} \cdot w$$

Solve. We will divide by $2\frac{1}{2}$ on both sides of the equation.

$$\begin{aligned} 263 &= 2\frac{1}{2}w \\ 263\div 2\frac{1}{2} &= w \\ 263\div\frac{5}{2} &= w \\ 263\cdot\frac{2}{5} &= w \\ \frac{526}{5} &= w \\ 105\frac{1}{5} &= w \\ 105 &\approx w \end{aligned}$$

Check. Since $2\frac{1}{2}\cdot 105=\frac{5}{2}\cdot 105=\frac{525}{2}=262\frac{1}{2}\approx 263$, the answer checks.

State. Rezazadeh weighs about 105 kg.

29. ***Familiarize***. Let b = the number of books in the order.

Translate.

$$\underbrace{\text{Weight per book}}\ \text{times}\ \underbrace{\text{Number of books}}\ \text{is}\ \underbrace{\text{Total weight}}$$
$$2\frac{3}{4} \cdot b = 220$$

Solve

$$\begin{aligned} 2\frac{3}{4}\cdot b &= 220 \\ \frac{11}{4}\cdot b &= 220 \\ \frac{4}{11}\cdot\frac{11}{4}\cdot b &= \frac{4}{11}\cdot 220 \\ b &= \frac{4\cdot 220}{11}=\frac{4\cdot \cancel{11}\cdot 20}{\cancel{11}\cdot 1} \\ b &= 80 \end{aligned}$$

Check. Since $2\frac{3}{4}\cdot 80=\frac{11}{4}\cdot 80=\frac{880}{4}=220$, the answer checks.

State. There are 80 books in the order.

30. ***Familiarize***. We add the three lengths across the top to find a and the three lengths across the bottom to find b.

Translate.

$$a = 1\frac{1}{8}+\frac{3}{4}+1\frac{1}{8}$$
$$b = \frac{3}{4}+3+\frac{3}{4}$$

Solve. We carry out the additions.

$$a = 1\frac{1}{8}+\frac{6}{8}+1\frac{1}{8}=2\frac{8}{8}=2+1=3$$
$$b = \frac{3}{4}+3+\frac{3}{4}=3\frac{6}{4}=3+1\frac{2}{4}=3+1\frac{1}{2}=4\frac{1}{2}$$

Check. We can repeat the calculations. The answer checks.

State. a) The short length a across the top is 3 in.

b) The length b across the bottom is $4\frac{1}{2}$ in.

31. ***Familiarize***. Let t = the number of inches by which $\frac{3}{4}$ in. exceeds the actual thickness of the plywood.

Translate.

$$\underbrace{\text{Actual thickness}}\ \text{plus}\ \underbrace{\text{Excess thickness}}\ \text{is}\ \underbrace{\frac{3}{4}\ \text{in.}}$$
$$\frac{11}{16} + t = \frac{3}{4}$$

Solve. We will subtract $\frac{11}{16}$ on both sides of the equation.

$$\begin{aligned} \frac{11}{16}+t &= \frac{3}{4} \\ \frac{11}{16}+t-\frac{11}{16} &= \frac{3}{4}-\frac{11}{16} \\ t &= \frac{3}{4}\cdot\frac{4}{4}-\frac{11}{16} \\ t &= \frac{12}{16}-\frac{11}{16} \\ t &= \frac{1}{16} \end{aligned}$$

Check. Since $\frac{11}{16}+\frac{1}{16}=\frac{12}{16}=\frac{3}{4}$, the answer checks.

State. A $\frac{3}{4}$-in. piece of plywood is actually $\frac{1}{16}$ in. thinner than its name indicates.

32. We add the heights and divide by the number of addends.

$$\frac{6\frac{5}{12}+5\frac{11}{12}+6\frac{7}{12}}{3}=\frac{17\frac{23}{12}}{3}=\frac{17+1\frac{11}{12}}{3}=$$
$$\frac{18\frac{11}{12}}{3}=\frac{227}{12}\div 3=\frac{227}{12}\cdot\frac{1}{3}=\frac{227}{36}=6\frac{11}{36}$$

The women's average height is $6\frac{11}{36}$ ft.

33. $\frac{2}{3}+1\frac{1}{3}\cdot 2\frac{1}{8}=\frac{2}{3}+\frac{4}{3}\cdot\frac{17}{8}=\frac{2}{3}+\frac{4\cdot 17}{3\cdot 8}=\frac{2}{3}+\frac{4\cdot 17}{3\cdot 2\cdot 4}=$
$\frac{2}{3}+\frac{4}{4}\cdot\frac{17}{3\cdot 2}=\frac{2}{3}+\frac{17}{6}=\frac{2}{3}\cdot\frac{2}{2}+\frac{17}{6}=\frac{4}{6}+\frac{17}{6}=\frac{21}{6}=$
$3\frac{3}{6}=3\frac{1}{2}$

34. $-1\frac{1}{2}-\frac{1}{2}\left(\frac{1}{2}\div\frac{1}{4}\right)+\left(\frac{1}{2}\right)^2=-1\frac{1}{2}-\frac{1}{2}\left(\frac{1}{2}\div\frac{1}{4}\right)+\frac{1}{4}=$
$-1\frac{1}{2}-\frac{1}{2}\left(\frac{1}{2}\cdot\frac{4}{1}\right)+\frac{1}{4}=-1\frac{1}{2}-\frac{1}{2}\left(\frac{4}{2}\right)+\frac{1}{4}=$
$-1\frac{1}{2}-\frac{4}{4}+\frac{1}{4}=-1\frac{1}{2}-1+\frac{1}{4}=-2\frac{1}{2}+\frac{1}{4}=-\frac{5}{2}+\frac{1}{4}=$
$-\frac{10}{4}+\frac{1}{4}=-\frac{9}{4}$, or $-2\frac{1}{4}$

35. $\frac{\frac{1}{3}-\frac{7}{9}}{\frac{1}{2}+\frac{1}{6}}=\frac{\frac{3}{9}-\frac{7}{9}}{\frac{3}{6}+\frac{1}{6}}=\frac{-\frac{4}{9}}{\frac{4}{6}}=\frac{-\frac{4}{9}}{\frac{2}{3}}=-\frac{4}{9}\div\frac{2}{3}=-\frac{4}{9}\cdot\frac{3}{2}=$
$-\frac{4\cdot 3}{9\cdot 2}=-\frac{\not{2}\cdot 2\cdot\not{3}}{\not{3}\cdot 3\cdot\not{2}}=-\frac{2}{3}$

36. a) Find the prime factorization of each number.

$12=2\cdot 2\cdot 3$
$36=2\cdot 2\cdot 3\cdot 3$
$60=2\cdot 2\cdot 3\cdot 5$

b) Create a product by writing factors that appear in the factorizations of 12, 36, and 60, using each factor the greatest number of times it appears in any one factorization.

The LCM is $2\cdot 2\cdot 3\cdot 3\cdot 5$, or 180.

Answer D is correct.

37. a) We find some common multiples of 8 and 6.

Multiples of 8: 8, 16, 24, 32, 40, 48, 56, 64, 72, ...
Multiples of 6: 6, 12, 18, 24, 30, 36, 42, 48, 54, 60, 66, 72, ...

Some common multiples are 24, 48, and 72. These are some class sizes for which study groups of 8 students or of 6 students can be organized with no students left out.

b) The smallest such class size is the least common multiple, 24.

38. ***Familiarize***. First compare $\frac{1}{7}$ mi and $\frac{1}{8}$ mi. The LCD is 56.

$\frac{1}{7}=\frac{1}{7}\cdot\frac{8}{8}=\frac{8}{56}$

$\frac{1}{8}=\frac{1}{8}\cdot\frac{7}{7}=\frac{7}{56}$

Since $8>7$, then $\frac{8}{56}>\frac{7}{56}$ so $\frac{1}{7}>\frac{1}{8}$.

This tells us that Rebecca walks farther than Trent.

Next we will find how much farther Rebecca walks on each lap and then multiply by 17 to find how much farther she walks in 17 laps. Let d represent how much farther Rebecca walks on each lap, in miles.

Translate. An equation that fits this situation is

$\frac{1}{8}+d=\frac{1}{7}$, or $\frac{7}{56}+d=\frac{8}{56}$

Solve.

$\frac{7}{56}+d=\frac{8}{56}$

$\frac{7}{56}+d-\frac{7}{56}=\frac{8}{56}-\frac{7}{56}$

$d=\frac{1}{56}$

Now we multiply: $17\cdot\frac{1}{56}=\frac{17}{56}$.

Check. We can think of the problem in a different way. In 17 laps Rebecca walks $17\cdot\frac{1}{7}$, or $\frac{17}{7}$ mi, and Trent walks $17\cdot\frac{1}{8}$, or $\frac{17}{8}$ mi. Then $\frac{17}{7}-\frac{17}{8}=\frac{17}{7}\cdot\frac{8}{8}-\frac{17}{8}\cdot\frac{7}{7}=\frac{136}{56}-\frac{119}{56}=\frac{17}{56}$, so Rebecca walks $\frac{17}{56}$ mi farther and our answer checks.

State. Rebecca walks $\frac{17}{56}$ mi farther than Trent.

Chapter 5

Decimal Notation

Exercise Set 5.1

1. 486.34

a) Write a word name for the whole number. — Four hundred eighty-six

b) Write "and" for the decimal point. — Four hundred eighty-six and

c) Write a word name for the number to the right of the decimal point, followed by the place value of the last digit. — Four hundred eighty-six and thirty-four hundredths

A word name for 486.34 is four hundred eighty-six and thirty-four hundredths.

3. 0.146

The whole number (the number to the left of the decimal point) is zero, so we write only a word name for the number to the right of the decimal point, followed by the place value of the last digit.

A word name for 0.146 is one hundred forty-six thousandths.

5. A word name for 249.89 is two hundred forty-nine and eighty-nine hundredths.

7. A word name for 3.7854 is three and seven thousand, eight hundred fifty-four ten-thousandths.

9. A word name for 5.4 is five and four tenths.

11. Write "and 95 cents" as "and $\frac{95}{100}$ dollars." A word name for \$524.95 is five hundred twenty-four and $\frac{95}{100}$ dollars.

13. Write "and 72 cents" as "and $\frac{72}{100}$ dollars." A word name for \$36.72 is thirty-six and $\frac{72}{100}$ dollars.

15. 7.3   7.3. $\frac{73}{10}$

1 place   1 zero

$$7.3 = \frac{73}{10}$$

To write 7.3 as a mixed numeral, we rewrite the whole number part and express the rest in fraction form.

$$7.3 = 7\frac{3}{10}$$

17. 21.67   21.67. $\frac{2167}{100}$

2 places   2 zeros

$$21.67 = \frac{2167}{100}$$

19. −2.703   −2.703. $\frac{-2073}{1000}$

3 places   3 zeros

$$-2.703 = \frac{-2073}{1000}, \text{ or } -\frac{2703}{1000}$$

To write −2.703 as a mixed numeral, we rewrite the whole number part and express the rest in fraction form.

$$-2.703 = -2\frac{703}{1000}$$

21. 0.0109   0.0109. $\frac{109}{10,000}$

4 places   4 zeros

$$0.0109 = \frac{109}{10,000}$$

Since the whole number part of 0.0109 is zero, we cannot express this number as a mixed numeral.

23. −4.0003   −4.0003. $\frac{-40,003}{10,000}$

4 places   4 zeros

$$-4.0003 = -\frac{40,003}{10,000}$$

To write −4.0003 as a mixed numeral, we rewrite the whole number part and express the rest in fraction form.

$$-4.0003 = -4\frac{3}{10,000}$$

25. −0.0207   −0.0207. $-\frac{207}{10,000}$

4 places   4 zeros

$$-0.0207 = -\frac{207}{10,000}$$

Since the whole number part of −0.0207 is zero, we cannot express this number as a mixed numeral.

27. 70.00105   70.00105. $\frac{7,000,105}{100,000}$

5 places   5 zeros

$$70.00105 = -\frac{7,000,105}{100,000}$$

To write 70.00105 as a mixed numeral, we rewrite the

whole number part and express the rest in fraction form.

$$70.00105 = 70\frac{105}{100,000}$$

29. $\frac{3}{10}$ 0.3.

1 zero Move 1 place.

$\frac{3}{10} = 0.3$

31. $-\frac{59}{100}$ −0.59.

2 zeros Move 2 places.

$-\frac{59}{100} = -0.59$

33. $\frac{3798}{1000}$ 3.798.

3 zeros Move 3 places.

$\frac{3798}{1000} = 3.798$

35. $\frac{78}{10,000}$ 0.0078.

4 zeros Move 4 places.

$\frac{78}{10,000} = 0.0078$

37. $\frac{-18}{100,000}$ −0.00018.

5 zeros Move 5 places.

$\frac{-18}{100,000} = -0.00018$

39. $\frac{486,197}{1,000,000}$ 0.486197.

6 zeros Move 6 places.

$\frac{486,197}{1,000,000} = 0.486197$

41. $7\frac{13}{1000} = 7 + \frac{13}{1000} = 7$ and $\frac{13}{1000} = 7.013$

43. $-8\frac{431}{1000}$

First consider $8\frac{431}{1000}$:

$8\frac{431}{1000} = 8 + \frac{431}{1000} = 8$ and $\frac{431}{1000} = 8.431$

Since $8\frac{431}{1000} = 8.431$, we have $-8\frac{431}{1000} = -8.431$.

45. $2\frac{1739}{10,000} = 2 + \frac{1739}{10,000} = 2$ and $\frac{1739}{10,000} = 2.1739$

47. $8\frac{953,073}{1,000,000} = 8 + \frac{953,073}{1,000,000} =$

8 and $\frac{953,073}{1,000,000} = 8.953073$

49. To compare two positive numbers in decimal notation, start at the left and compare corresponding digits moving from left to right. When two digits differ, the number with the larger digit is the larger of the two numbers.

0.06

Different; 5 is larger than 0.

0.58

Thus, 0.58 is larger.

51. 0.403

Starting at the left, these digits are the first to differ; 1 is larger than 0.

0.410

Thus, 0.410 is larger.

53. To compare two negative numbers in decimal notation, start at the left and compare corresponding digits moving from left to right. When two digits differ, the number with the smaller digit is the larger of the two numbers.

−5.046

Starting at the left, these digits are the first to differ, and 3 is smaller than 6.

−5.043

Thus, −5.043 is larger.

55. 234.07

Starting at the left, these digits are the first to differ, and 5 is larger than 4.

235.07

Thus, 235.07 is larger.

57. $\frac{7}{100} = 0.07$ so we compare 0.007 and 0.07.

0.007

Starting at the left, these digits are the first to differ, and 7 is larger than 0.

0.07

Thus, 0.07 or $\frac{7}{100}$ is larger.

59. −0.872

Starting at the left, these digits are the first to differ, and 2 is smaller than 3.

−0.873

Thus, −0.872 is larger.

61.

0.2[3] Hundredths digit is 4 or lower.
Round down.

0.2

63.

−0.3[7]2 Hundredths digit is 5 or higher.
Round from −0.372 to −0.4.

−0.4

65. $2.\underline{9}\boxed{5}1$ Hundredths digit is 5 or higher. Round up.

$\downarrow$

3.0

(When we make the tenths digit a 10, we carry 1 to the ones place.)

67. $-327.\underline{2}\boxed{3}47$ Hundredths digit is 4 or lower. Round from -327.2347 to -327.2.

$\downarrow$

-327.2

69. $0.8\underline{9}\boxed{3}$ Thousandths digit is 4 or lower. Round down.

$\downarrow$

0.89

71. $-0.6\underline{6}\boxed{6}6$ Thousandths digit is 5 or higher. Round from -0.6666 to -0.67.

$\downarrow$

-0.67

73. $0.9\underline{9}\boxed{5}2$ Thousandths digit is 5 or higher. Round up.

$\downarrow$

1.00

(When we make the hundredths digit a 10, we carry 1 to the tenths place. This then requires us to carry 1 to the ones place.)

75. $-0.0\underline{3}\boxed{4}88$ Thousandths digit is 4 or lower. Round from -0.03488 to -0.03.

$\downarrow$

-0.03

77. $0.57\underline{2}\boxed{4}$ Ten-thousandths digit is 4 or lower. Round down.

$\downarrow$

0.572

79. $17.00\underline{1}\boxed{5}$ Ten-thousandths digit is 5 or higher. Round up.

$\downarrow$

17.002

81. $-20.20\underline{2}\boxed{0}2$ Ten-thousandths digit is 4 or lower. Round -20.20202 to -20.202.

$\downarrow$

-20.202

83. $9.98\underline{4}\boxed{8}$ Ten-thousandths digit is 5 or higher. Round up.

$\downarrow$

9.985

85. $809.\underline{4}\boxed{7}321$ Hundredths digit is 5 or higher. Round up.

$\downarrow$

809.5

87. $809.4\underline{7}\boxed{3}21$ Thousandths digit is 4 or lower. Round down.

$\downarrow$

809.47

89.
$$\begin{array}{r} {\scriptstyle 1\,1} \\ 6\,8\,1 \\ +\,1\,4\,9 \\ \hline 8\,3\,0 \end{array}$$

91.
$$\begin{array}{r} {\scriptstyle 1\ 16} \\ \not{2}\,\not{6}\,7 \\ -\quad 8\,5 \\ \hline 1\,8\,2 \end{array}$$

93. $\frac{37}{55} - \frac{49}{55} = \frac{37-49}{55} = \frac{-12}{55}$, or $-\frac{12}{55}$

95.
$$\begin{array}{r} {\scriptstyle 18} \\ {\scriptstyle 3\ \not{8}\ 9\ 13} \\ 3\,4,\not{9}\,\not{0}\,\not{3} \\ -\quad 1\,9\,4\,5 \\ \hline 3\,2,9\,5\,8 \end{array}$$

97. All of the numbers except -1.09 have 3 decimal places, so we add a zero to -1.09 so that each number has 3 decimal places. Then we start at the left and compare corresponding digits moving from left to right. Keep in mind that, when we compare two negative numbers and when digits differ, the number with the smaller digit is the larger of the two numbers. The numbers, listed from smallest to largest, are -1.09, -1.009, -0.989, -0.898, and -0.098.

99. $6.7834\underline{6}\boxed{123}$ ←Drop all decimal places past the fifth place.

The answer is 6.78346.

101. $99.9999\underline{9}\boxed{9999}$ ←Drop all decimal places past the fifth place.

The answer is 99.99999.

103. From the graph we see that the years for which the vertical bars lie above $+1.2°$ are 1998 and 2006.

105. From the graph we see that the last year for which the corresponding vertical bar lies below $0.0°$ is 1996.

Exercise Set 5.2

1.
$$\begin{array}{r} {\scriptstyle 1} \\ 4\,2\,6.2\,5 \\ +\quad 3\,8.1\,2 \\ \hline 4\,6\,4.3\,7 \end{array}$$

Add hundredths.
Add tenths.
Write a decimal point in the answer.
Add ones.
Add tens.
Add hundreds.

3.
$$\begin{array}{r} {\scriptstyle 1\,1} \\ 6\,5\,9.4\,0\,3 \\ +\ \ 9\,1\,6.6\,1\,2 \\ \hline 1\,5\,7\,6.0\,1\,5 \end{array}$$

Add thousandths.
Add hundredths.
Add tenths.
Write a decimal point in the answer.
Add ones.
Add tens.
Add hundreds.

5.
$$\begin{array}{r} {\scriptstyle 1}{\scriptstyle 1} \\ 9.1\,0\,4 \\ +\,1\,2\,3.4\,5\,6 \\ \hline 1\,3\,2.5\,6\,0 \end{array}$$

7. Line up the decimal points.

$$\begin{array}{r} {}^{1} \\ 2.006 \\ +\,5.817 \\ \hline 7.823 \end{array}$$

9. Line up the decimal points.

$$\begin{array}{r} 20.7 \\ +\,30.0124 \\ \hline 50.7124 \end{array}$$

11. We write 9 as 9.00. Line up the decimal points.

$$\begin{array}{r} 1.06 \\ +\ \ 9.00 \\ \hline 10.06 \end{array}$$

13. Line up the decimal points.

$$\begin{array}{rl} {}^{1} & \\ 0.340 & \text{Writing an extra zero} \\ 3.500 & \text{Writing 2 extra zeros} \\ 0.127 & \\ +\,768.000 & \text{Writing in the decimal point} \\ \hline & \text{and 3 extra zeros} \\ 771.967 & \text{Adding} \end{array}$$

15.

$$\begin{array}{rl} {}^{1}\ \ {}^{1\,1} & \\ 17.0000 & \text{Writing in the decimal point.} \\ 3.2400 & \text{You may find it helpful to} \\ 0.2560 & \text{write extra zeros.} \\ +\ \ 0.3689 & \\ \hline 20.8649 & \end{array}$$

17.

$$\begin{array}{r} {}^{1\,2\,1}\ \ \ {}^{1} \\ 2.7030 \\ 78.3300 \\ 28.0009 \\ +\,118.4341 \\ \hline 227.4680 \end{array}$$

19.

$$\begin{array}{rl} 47.596 & \text{Subtract thousandths.} \\ -\ \ 6.215 & \text{Subtract hundredths.} \\ \hline 41.381 & \text{Subtract tenths.} \\ & \text{Write a decimal point in the answer.} \\ & \text{Subtract ones.} \\ & \text{Subtract tens.} \end{array}$$

21.

$$\begin{array}{rl} {}^{4\ 11\ 2\ 11} & \text{Borrow tenths to subtract hundredths.} \\ \not{5}\,\not{1}.\not{3}\,\not{1} & \text{Subtract hundredths.} \\ -\ \ 2.29 & \text{Subtract tenths.} \\ \hline 49.02 & \text{Write a decimal point in the answer.} \\ & \text{Borrow tens to subtract ones.} \\ & \text{Subtract ones.} \\ & \text{Subtract tens.} \end{array}$$

23.

$$\begin{array}{rl} {}^{5\ 9\ 10} & \\ 3.\not{6}\,\not{0}\,\not{0} & \text{Writing 2 extra zeros} \\ -\,0.036 & \\ \hline 3.564 & \end{array}$$

25.

$$\begin{array}{r} {}^{11} \\ {}^{8\ \not{1}\ 13} \\ \not{9}\,\not{2}.\not{3}\,41 \\ -\ \ 6.42 \\ \hline 85.921 \end{array}$$

27.

$$\begin{array}{r} {}^{2\ 9\ 10\ 6\ 14} \\ \not{3}.\not{0}\,\not{0}\,\not{7}\,\not{4} \\ -\,1.3408 \\ \hline 1.6666 \end{array}$$

29.

$$\begin{array}{rl} {}^{6\ 9\ 10} & \\ 6.0\not{7}\,\not{0}\,\not{0} & \text{Writing 2 extra zeros} \\ -\,2.0078 & \\ \hline 4.0622 & \end{array}$$

31. Line up the decimal points. Write an extra zero if desired.

$$\begin{array}{r} {}^{11\ 13} \\ {}^{2\ 9\ \not{1}\ \not{3}\ 10} \\ \not{3}\,\not{0}.\not{2}\,\not{4}\,\not{0} \\ -\ \ 0.241 \\ \hline 29.999 \end{array}$$

33.

$$\begin{array}{r} {}^{3\ 10} \\ 3\not{4}.\not{0}\,7 \\ -\,30.7 \\ \hline 3.37 \end{array}$$

35.

$$\begin{array}{r} {}^{4\ 10} \\ 8.4\not{5}\,\not{0} \\ -\,7.405 \\ \hline 1.045 \end{array}$$

37.

$$\begin{array}{r} {}^{5\ 10} \\ \not{6}.\not{0}\,03 \\ -\,2.3 \\ \hline 3.703 \end{array}$$

39.

$$\begin{array}{rl} {}^{1\ 9\ 9\ 9\ 10} & \\ \not{2}.\not{0}\,\not{0}\,\not{0}\,\not{0} & \text{Writing in the decimal point} \\ -\,1.0908 & \text{and 4 extra zeros} \\ \hline 0.9092 & \text{Subtracting} \end{array}$$

41.

$$\begin{array}{r} {}^{13} \\ {}^{1\ \not{3}\ 9\ 10} \\ 6\not{2}\,\not{4}.\not{0}\,\not{0} \\ -\ \ 18.79 \\ \hline 605.21 \end{array}$$

43.

$$\begin{array}{r} 57.803 \\ -\ \ 4.6 \\ \hline 53.203 \end{array}$$

45.

$$\begin{array}{r} {}^{6\ 10} \\ 263.\not{7}\,\not{0} \\ -\,102.08 \\ \hline 161.62 \end{array}$$

47.

$$\begin{array}{r} {}^{4\ 9\ 9\ 10} \\ 4\not{5}.\not{0}\,\not{0}\,\not{0} \\ -\ \ 0.999 \\ \hline 44.001 \end{array}$$

49. $-5.02 + 1.73$ A positive and a negative number

a) $|-5.02| = 5.02$, $|1.73| = 1.73$, and $|-5.02| > |-1.73|$, so the answer is negative.

b)

$$\begin{array}{rl} {}^{4\ 9\ 12} & \\ \not{5}.\not{0}\,\not{2} & \text{Find the difference in} \\ -\,1.73 & \text{the absolute values.} \\ \hline 3.29 & \end{array}$$

c) $-5.02 + 1.73 = -3.29$

51. $12.9 - 15.4 = 12.9 + (-15.4)$
We add the opposite of 15.4. We have a positive and a negative number.

a) $|12.9| = 12.9$, $|-15.4| = 15.4$, and $|15.4| > |12.9|$, so the answer is negative.

b) $\begin{array}{r} \overset{4\ 14}{1\ \not{5}.\not{4}} \\ -\ 1\ 2.9 \\ \hline 2.5 \end{array}$ Finding the difference in the absolute values

c) $12.9 - 15.4 = -2.5$

53. $-2.9 + (-4.3)$ Two negative numbers

a) $\begin{array}{r} \overset{1}{2}.9 \\ +\ 4.3 \\ \hline 7.2 \end{array}$ Adding the absolute values

b) $-2.9 + (-4.3) = -7.2$ The sum of two negative numbers is negative.

55. $-4.301 + 7.68$ A negative and a positive number

a) $|-4.301| = 4.301$, $|7.68| = 7.68$, and $|7.68| > |-4.301|$, so the answer is positive.

b) $\begin{array}{r} 7.6\ \overset{7}{\not{8}}\ \overset{10}{\not{0}} \\ -\ 4.3\ 0\ 1 \\ \hline 3.3\ 7\ 9 \end{array}$ Finding the difference in the absolute values

c) $-4.301 + 7.68 = 3.379$

57. $-12.9 - 3.7$
$= -12.9 + (-3.7)$ Adding the opposite of 3.7
$= -16.6$ The sum of two negatives is negative.

59. $-2.1 - (-4.6)$
$= -2.1 + 4.6$ Adding the opposite of -4.6
$= 2.5$ Subtracting absolute values. Since 4.6 has the larger absolute value, the answer is positive.

61. $14.301 + (-17.82)$
$= -3.519$ Subtracting absolute values. Since -17.82 has the larger absolute value, the answer is negative.

63. $7.201 - (-2.4)$
$= 7.201 + 2.4$ Adding the opposite of -2.4
$= 9.601$ Adding

65. $96.9 + (-21.4)$
$= 75.5$ Subtracting absolute values. Since 96.9 has the larger absolute value, the answer is positive.

67. $-3 - (-12.7)$
$= -3 + 12.7$ Adding the opposite of -12.7
$= 9.7$ Subtracting absolute values. Since 12.7 has the larger absolute value, the answer is positive.

69. $-4.9 - 5.392$
$= -4.9 + (-5.392)$ Adding the opposite of 5.392
$= -10.292$ The sum of two negatives is negative.

71. $14.7 - 15$
$= 14.7 + (-15)$ Adding the opposite of 15
$= -0.3$ Subtracting absolute values. Since -15 has the larger absolute value, the answer is negative.

73. $1.8x + 3.9x$
$= (1.8 + 3.9)x$ Using the distributive law
$= 5.7x$ Adding

75. $17.59a - 12.73a$
$= (17.59 - 12.73)a$
$= 4.86a$

77. $15.2t + 7.9 + 5.9t$
$= 15.2t + 5.9t + 7.9$ Using the commutative law
$= (15.2 + 5.9)t + 7.9$ Using the distributive law
$= 21.1t + 7.9$

79. $5.217x - 8.134x$
$= (5.217 - 8.134)x$ Using the distributive law
$= (5.217 + (-8.134))x$ Adding the opposite of 8.134
$= -2.917x$ Subtracting absolute values. The coefficient is negative since -8.134 has the larger absolute value.

81. $4.906y - 7.1 + 3.2y$
$= 4.906y + 3.2y - 7.1$
$= (4.906 + 3.2)y - 7.1$
$= 8.106y - 7.1$

83. $4.8x + 1.9y - 5.7x + 1.2y$
$= 4.8x + 1.9y + (-5.7x) + 1.2y$ Rewriting as addition
$= 4.8x + (-5.7x) + 1.9y + 1.2y$ Using the commutative law
$= (4.8 + (-5.7))x + (1.9 + 1.2)y$
$= -0.9x + 3.1y$

85. $4.9 - 3.9t - 6 - 4.5t$
$= 4.9 + (-3.9t) + (-6) + (-4.5t)$
$= 4.9 + (-6) + (-3.9t) + (-4.5t)$
$= [4.9 + (-6)] + [-3.9 + (-4.5)]t$
$= -1.1 + (-8.4t)$
$= -1.1 - 8.4t$

87. $\frac{3}{5} \cdot \frac{4}{7} = \frac{3 \cdot 4}{5 \cdot 7} = \frac{12}{35}$

89. $\frac{3}{10} \cdot \frac{21}{100} = \frac{3 \cdot 21}{10 \cdot 100} = \frac{63}{1000}$

91. $5 - 3x^2$
$= 5 - 3 \cdot (-2)^2$ Substituting -2 for x
$= 5 - 3 \cdot 4$
$= 5 - 12$
$= -7$

93. $-3.928 - 4.39a + 7.4b - 8.073 + 2.0001a - 9.931b - 9.8799a + 12.897b$
$= -3.928 - 8.073 - 4.39a + 2.0001a - 9.8799a + 7.4b - 9.931b + 12.897b$
$= -12.001 - 12.2698a + 10.366b$

95. $39.123a - 42.458b - 72.457a + 31.462b - 59.491 + 37.927a$
$= 39.123a - 72.457a + 37.927a - 42.458b + 31.462b - 59.491$
$= 4.593a - 10.996b - 59.491$

97. First "undo" the incorrect subtraction by adding 349.2 to the incorrect answer:

$-836.9 + 349.2 = -487.7$

Now add 349.2 to -487.7:

$-487.7 + 349.2 = -138.5$

The correct answer is -138.5.

99.
$$\begin{array}{r} {}^{3}\,{}^{13} \\ 9\,3.\,a\,\not{4}\,\not{3} \\ -8\,7.\,9\,6\,9 \\ \hline 5.\,2\,7\,4 \end{array}$$

We need to borrow 1 tenth in order to subtract hundredths. Then when we subtract 9 tenths from $(a-1)$ tenths we get 2. Since $11 - 9 = 2$, we know that we had to borrow a one in order to subtract tenths and this was added to 1 tenth, or $a - 1$. Then if $a - 1 = 1$, we know that $a = 2$.

Exercise Set 5.3

1.
$$\begin{array}{rl} 6.\,8 & \text{(1 decimal place)} \\ \times \quad 7 & \text{(0 decimal places)} \\ \hline 4\,7.\,6 & \text{(1 decimal place)} \end{array}$$

3.
$$\begin{array}{rl} 0.\,8\,4 & \text{(2 decimal places)} \\ \times \quad 8 & \text{(0 decimal places)} \\ \hline 6.\,7\,2 & \text{(2 decimal places)} \end{array}$$

5.
$$\begin{array}{rl} 6.\,3 & \text{(1 decimal place)} \\ \times\, 0.\,0\,4 & \text{(2 decimal places)} \\ \hline 0.\,2\,5\,2 & \text{(3 decimal places)} \end{array}$$

7.
$$\begin{array}{rl} 8\,7 & \text{(0 decimal places)} \\ \times\, 0.0\,0\,6 & \text{(3 decimal places)} \\ \hline 0.5\,2\,2 & \text{(3 decimal places)} \end{array}$$

9. $1\underline{0} \times 42.63$ 42.6.3

1 zero   Move 1 place to the right.

$10 \times 42.63 = 426.3$

11. $-1000 \times 783.686852 = -(1000 \times 783.686852)$

$-(1\underline{000} \times 783.686852)$ $-783.686.852$

3 zeros   Move 3 places to the right.

$-1000 \times 783.686852 = -783{,}686.852$

13. $-7.8 \times 1\underline{00}$ $-7.80.$

2 zeros   Move 2 places to the right.

$-7.8 \times 100 = -780$

15. $0.\underline{1} \times 79.18$ 7.9.18

1 decimal place   Move 1 place to the left.

$0.1 \times 79.18 = 7.918$

17. $0.\underline{001} \times 97.68$ 0.097.68

3 decimal places   Move 3 places to the left.

$0.001 \times 97.68 = 0.09768$

19. $28.7 \times (-0.01) = -(28.7 \times 0.01)$

$-(28.7 \times 0.\underline{01})$ $-0.28.7$

2 decimal places   Move 2 places to the left.

$28.7 \times (-0.01) = -0.287$

21.
$$\begin{array}{rl} 2.\,7\,3 & \text{(2 decimal places)} \\ \times \quad 1\,6 & \text{(0 decimal places)} \\ \hline 1\,6\,3\,8 & \\ 2\,7\,3\,0 & \\ \hline 4\,3.\,6\,8 & \text{(2 decimal places)} \end{array}$$

23.
$$\begin{array}{rl} 0.\,9\,8\,4 & \text{(3 decimal places)} \\ \times 0.\,0\,3\,1 & \text{(3 decimal places)} \\ \hline 9\,8\,4 & \\ 2\,9\,5\,2\,0 & \\ \hline 0.0\,3\,0\,5\,0\,4 & \text{(6 decimal places)} \end{array}$$

25. We multiply the absolute values.
$$\begin{array}{rl} 3\,7.\,4 & \text{(1 decimal place)} \\ \times \quad 2.\,4 & \text{(1 decimal place)} \\ \hline 1\,4\,9\,6 & \\ 7\,4\,8\,0 & \\ \hline 8\,9.\,7\,6 & \text{(2 decimal places)} \end{array}$$

Since the product of two negative numbers is positive, the answer is 89.76.

27. We multiply the absolute values.
$$\begin{array}{rl} 7\,4\,9 & \text{(0 decimal places)} \\ \times \quad 0.\,4\,3 & \text{(2 decimal places)} \\ \hline 2\,2\,4\,7 & \\ 2\,9\,9\,6\,0 & \\ \hline 3\,2\,2.\,0\,7 & \text{(2 decimal places)} \end{array}$$

Since the product of a positive number and a negative number is negative, the answer is -322.07.

29.
$$\begin{array}{rl} 0.\,8\,7 & \text{(2 decimal places)} \\ \times \quad 6\,4 & \text{(0 decimal places)} \\ \hline 3\,4\,8 & \\ 5\,2\,2\,0 & \\ \hline 5\,5.\,6\,8 & \text{(2 decimal places)} \end{array}$$

31.

```
        4 6. 5 0    (2 decimal places)
    ×       7 5     (0 decimal places)
      2 3 2 5 0
    3 2 5 5 0 0
    3 4 8 7. 5 0    (2 decimal places)
```

Since the last decimal place is 0, we could also write this answer as 3487.5.

33. We multiply the absolute values.

```
      0. 2 3 1     (3 decimal places)
    ×       0. 5   (1 decimal place)
    0. 1 1 5 5     (4 decimal places)
```

Since the product of two negative numbers is positive, the answer is 0.1155.

35. $9.42 \times (-1000) = -(9.42 \times 1000)$

$-(9.42 \times 1\underline{000})$ $-9.420.$

3 zeros Move 3 places to the right.

$9.42 \times (-1000) = -9420$

37. $-95.3 \times (-0.0001) = 95.3 \times 0.0001$

$95.3 \times 0.\underline{0001}$ $0.0095.3$

4 decimal places Move 4 places to the left.

$-95.3 \times (-0.0001) = 0.00953$

39. Move 2 places to the right.

\$57.06.¢

Change from \$ sign in front to ¢ sign at end.

\$57.06 = 5706¢

41. Move 2 places to the right.

\$0.95.¢

Change from \$ sign in front to ¢ sign at end.

\$0.95 = 95¢

43. Move 2 places to the right.

\$0.01.¢

Change from \$ sign in front to ¢ sign at end.

\$0.01 = 1¢

45. Move 2 places to the left.

\$0.72.¢

Change from ¢ sign at end to \$ sign in front.

72¢ = \$0.72

47. Move 2 places to the left.

\$0.02.¢

Change from ¢ sign at end to \$ sign in front.

2¢ = \$0.02

49. Move 2 places to the left.

\$63.99.¢

Change from ¢ sign at end to \$ sign in front.

6399¢ = \$63.99

51. $11.1 \text{ million} = 11.1 \times 1,\underbrace{000,000}_{6 \text{ zeros}}$

11.100000.

Move 6 places to the right.

$11.1 \text{ million} = 11,100,000$

53. $152.7 \text{ billion} = 152.7 \times 1,\underbrace{000,000,000}_{9 \text{ zeros}}$

152.700000000.

Move 9 places to the right.

$152.7 \text{ billion} = 152,700,000,000$

55. $3.156 \text{ billion} = 3.156 \times 1,\underbrace{000,000,000}_{9 \text{ zeros}}$

3.156000000.

Move 9 places to the right.

$3.156 \text{ billion} = 3,156,000,000$

57.

$$\begin{aligned} & P + Prt \\ &= 10,000 + 10,000(0.04)(2.5) && \text{Substituting} \\ &= 10,000 + 400(2.5) && \text{Multiplying and dividing} \\ &= 10,000 + 1000 && \text{in order from left to right} \\ &= 11,000 && \text{Adding} \end{aligned}$$

59.

$$\begin{aligned} & vt + 0.5at^2 \\ &= 10(1.5) + 0.5(9.8)(1.5)^2 \\ &= 10(1.5) + 4.9(1.5)(1.5) \\ &= 10(1.5) + 4.9(2.25) && \text{Squaring first} \\ &= 15 + 11.025 \\ &= 26.025 \end{aligned}$$

61. a)

$$\begin{aligned} P &= 2l + 2w \\ &= 2(12.5) + 2(9.5) \\ &= 25 + 19 \\ &= 44 \end{aligned}$$

The perimeter is 44 ft.

b)

$$\begin{aligned} A &= l \cdot w \\ &= (12.5)(9.5) \\ &= 118.75 \end{aligned}$$

The area is 118.75 ft^2.

63. a) $P = 2l + 2w$
$= 2(10.5) + 2(8.4)$
$= 21 + 16.8$
$= 37.8$

The perimeter is 37.8 m.

b) $A = l \cdot w$
$= (10.5)(8.4)$
$= 88.2$

The area is 88.2 m^2.

65. In 2015, $t = 2015 - 2000 = 15$. Substitute 15 for t.

$$0.0375t + 2.2 = 0.0375(15) + 2.2$$
$$= 0.5625 + 2.2$$
$$= 2.7625$$

In 2015 there will be 2.7625 million, or 2,762,500 nurses in the United States.

67. $-162 \div 6 = -27$

(The signs are different, so the quotient is negative.)

69. $-1035 \div (-15) = 69$

(The signs are the same, so the quotient is positive.)

71. $-525 \div 25 = -21$

(The signs are different, so the quotient is negative.)

73.
```
       1 2 5 7
 1 8 ) 2 2, 6 2 6
       1 8
         4 6
         3 6
         1 0 2
           9 0
           1 2 6
           1 2 6
               0
```

The answer is 1257.

75. (1 trillion) $\cdot$ (1 billion)

$$= 1,\underbrace{000,000,000,000}_{12 \text{ zeros}} \times 1,\underbrace{000,000,000}_{9 \text{ zeros}}$$
$$= 1,\underbrace{000,000,000,000,000,000,000}_{21 \text{ zeros}}$$
$$= 10^{21} = 1 \text{ sextillion}$$

77. (1 trillion) $\cdot$ (1 trillion)

$$= 1,\underbrace{000,000,000,000}_{12 \text{ zeros}} \times 1,\underbrace{000,000,000,000}_{12 \text{ zeros}}$$
$$= 1,\underbrace{000,000,000,000,000,000,000,000}_{24 \text{ zeros}}$$
$$= 10^{24} = 1 \text{ septillion}$$

79. For the British number 6.6 billion, we have:

6.6 billion
$= 6.6 \times 1,000,000 \times 1,000,000$
$= 6.6 \times 1,000,000,000,000$
$= 6,600,000,000,000$ Moving the decimal point 12 places to the right

81. Use a calculator.

$d + vt + at^2$
$= 79.2 + 3.029(7.355) + 4.9(7.355)^2$
$= 79.2 + 3.029(7.355) + 4.9(54.096025)$
$= 79.2 + 22.278295 + 265.0705225$
$= 101.478295 + 265.0705225$
$= 366.5488175$

83. Use a calculator.

$$0.5(b_1 + b_2)h = 0.5(9.7 \text{ cm} + 13.4 \text{ cm})(6.32 \text{ cm})$$
$$= 0.5(23.1 \text{ cm})(6.32 \text{ cm})$$
$$= 72.996 \text{ cm}^2$$

85. The period from April 20 to May 20 consists of 30 days, so the "customer charge" is 30 × \$0.374, or \$11.22. The "energy charge" for the first 250 kilowatt-hours is 250 × \$0.1174, or \$29.35.

We subtract to find the number of kilowatt-hours in excess of 250: 480 − 250 = 230. Then the "energy charge" for the 230 kilowatt-hours in excess of 250 kilowatt-hours is 230 × \$0.09079, or \$20.88 (rounding to the nearest cent).

Finally, we add to find the total bill:

\$11.22 + \$29.35 + \$20.88 = \$61.45.

Exercise Set 5.4

1.
```
      2. 9 9
  2 ) 5. 9 8
      4
      1 9
      1 8
        1 8
        1 8
          0
```

Divide as though dividing whole numbers. Place the decimal point directly above the decimal point in the dividend.

3.
```
     2 3. 7 8
 4 ) 9 5. 1 2
     8
     1 5
     1 2
       3 1
       2 8
         3 2
         3 2
           0
```

Divide as though dividing whole numbers. Place the decimal point directly above the decimal point in the dividend.

5.
```
          7. 0 8
  1 2 ) 8 4. 9 6
        8 4
            9 6
            9 6
              0
```

7.
```
          1.2
  1 5 ) 1 8.0
        1 5
          3 0    Write an extra 0.
          3 0
            0
```

9. We first consider $5.4 \div 6$.

```
   0. 9
6 ⟌5. 4
   5 4
   ---
     0
```

Since a positive number divided by a negative number is negative, the answer is -0.9.

11. We first find $30 \div 0.005$.

```
            6 0 0 0.
0.0 0 5∧ ⟌3 0.0 0 0∧
          3 0
          ---
            0
```

Multiply the divisor by 1000 (move the decimal point 3 places). Multiply the same way in the dividend (move 3 places). Then divide.

Since a negative number divided by a positive number is negative, the answer is -6000.

13.

```
        1 4 0.
0.0 6∧ ⟌8.4 0∧
        6
        2 4
        2 4
        ---
          0
```

Multiply the divisor by 100 (move the decimal point 2 places). Multiply the same way in the dividend (move 2 places). Then divide.

15.

```
         4 0.
2. 6∧ ⟌1 0 4.0∧
       1 0 4
       -----
           0
```

Put a decimal point at the end of the whole number. Multiply the divisor by 10 (move the decimal point 1 place). Multiply the same way in the dividend (move 1 place), adding an extra 0. Then divide.

17. We first consider $1.8 \div 12$.

```
     0.1 5
1 2 ⟌1.8 0
     1 2
     ---
       6 0
       6 0
       ---
         0
```

Divide as though dividing whole numbers. Place the decimal point directly above the decimal point in the dividend.

Since a positive number divided by a negative number is negative, the answer is -0.15.

19.

```
            3.2
8.5∧ ⟌2 7.2∧0
      2 5 5
      -----
        1 7 0    Write an extra 0.
        1 7 0
        -----
            0
```

21. We first find $31.59 \div 8.1$.

```
           3.9
8.1∧ ⟌3 1.5∧9
      2 4 3
      -----
        7 2 9
        7 2 9
        -----
            0
```

Since a negative number divided by a positive number is negative, the answer is -3.9.

23. We first consider $5 \div 8$.

```
   0 .6 2 5
8 ⟌5 .0 0 0
   4 8
   ---
     2 0     Write an extra 0.
     1 6
     ---
       4 0   Write an extra 0.
       4 0
       ---
         0
```

Since a negative number divided by a negative number is positive, the answer is 0.625.

25.

```
              0.2 6
0.4 7∧ ⟌0. 1 2∧2 2
             9 4
             ---
             2 8 2
             2 8 2
             -----
                 0
```

27.

```
                2.3 4
0.0 3 2∧ ⟌0. 0 7 4∧8 8
                6 4
                ---
                1 0 8
                  9 6
                -----
                  1 2 8
                  1 2 8
                  -----
                      0
```

29. We first consider $24.969 \div 82$.

```
     0.3 0 4 5
8 2 ⟌2 4.9 6 9 0
     2 4 6 0 0
     ---------
         3 6 9
         3 2 8
         -----
           4 1 0   Write an extra 0.
           4 1 0
           -----
               0
```

Since a negative number divided by a positive number is negative, the answer is -0.3045.

31. $\dfrac{213.4567}{100}$ 2.13.4567 (↑ arrow)

2 zeros   Move 2 places to the left.

$$\frac{213.4567}{100} = 2.134567$$

33. $\dfrac{-23.59}{10}$ $-2.3.59$ (↑ arrow)

1 zero   Move 1 place to the left.

$$\frac{-23.59}{10} = -2.359$$

35. $\dfrac{1.0237}{0.001}$ 1.023.7 (↑ arrow)

3 decimal places   Move 3 places to the right.

$$\frac{1.0237}{0.001} = 1023.7$$

37. $\dfrac{-92.36}{0.0\underline{1}}$ -92.36 (decimal point moved 2 places to the right)

2 decimal places Move 2 places to the right.

$\dfrac{92.36}{-0.01} = \dfrac{-92.36}{0.01} = -9236$

39. $\dfrac{0.8172}{1\underline{0}}$ 0.0.8172 (decimal point moved 1 place to the left)

1 zero Move 1 place to the left.

$\dfrac{0.8172}{10} = 0.08172$

41. $\dfrac{0.97}{0.\underline{1}}$ 0.9.7 (decimal point moved 1 place to the right)

1 decimal place Move 1 place to the right.

$\dfrac{0.97}{0.1} = 9.7$

43. $\dfrac{52.7}{-1000} = \dfrac{-52.7}{1000}$

$\dfrac{-52.7}{1\underline{000}}$ −0.052.7 (decimal point moved 3 places to the left)

3 zeros Move 3 places to the left.

$\dfrac{52.7}{-1000} = \dfrac{-52.7}{1000} = -0.0527$

45. $\dfrac{75.3}{-0.001} = -\dfrac{75.3}{0.001}$

$-\dfrac{75.3}{0.\underline{001}}$ −75.300. (decimal point moved 3 places to the right)

3 decimal places Move 3 places to the right.

$\dfrac{75.3}{-0.001} = -75,300$

47. $\dfrac{-75.3}{1\underline{000}}$ −0.075.3 (decimal point moved 3 places to the left)

3 zeros Move 3 places to the left.

$\dfrac{-75.3}{1000} = -0.0753$

49. $14 \times (82.6 + 67.9)$

$= 14 \times (150.5)$ Doing the calculation inside the parentheses

$= 2107$ Multiplying

51. $0.003 + 3.03 \div (-0.01) = 0.003 - 303$ Dividing first

$= -302.997$ Subtracting

53. $(4.9 - 18.6) \times 13$

$= -13.7 \times 13$ Doing the calculation inside the parentheses

$= -178.1$ Multiplying

55. $210.3 - 4.24 \times 1.01$

$= 210.3 - 4.2824$ Multiplying

$= 206.0176$ Subtracting

57. $0.04 \times 0.1 \div 0.4 \times 50$

$= 0.004 \div 0.4 \times 50$ Doing the first multiplication

$= 0.01 \times 50$ Dividing

$= 0.5$ Multiplying

59. $12 \div (-0.03) - 12 \times 0.03^2$

$= 12 \div (-0.03) - 12 \times 0.0009$ Evaluating the exponential expression

$= -400 - 0.0108$ Dividing and multiplying in order from left to right

$= -400.0108$ Subtracting

61. $(4 - 2.5)^2 \div 100 + 0.1 \times 6.5$

$= (1.5)^2 \div 100 + 0.1 \times 6.5$ Doing the calculation inside the parentheses

$= 2.25 \div 100 + 0.1 \times 6.5$ Evaluating the exponential expression

$= 0.0225 + 0.65$ Dividing and multiplying in order from left to right

$= 0.6725$ Adding

63. $6 \times 0.9 - 0.1 \div 4 + 0.2^3$

$= 6 \times 0.9 - 0.1 \div 4 + 0.008$ Evaluating the exponential expression

$= 5.4 - 0.025 + 0.008$ Multiplying and dividing in order from left to right

$= 5.383$ Subtracting and adding in order from left to right

65. $12^2 \div (12 + 2.4) - [(2 - 2.4) \div 0.8]$

$= 12^2 \div (12 + 2.4) - [-0.4 \div 0.8]$ Doing the calculations in the innermost parentheses first

$= 12^2 \div 14.4 - [-0.5]$ Doing the calculations inside the parentheses

$= 12^2 \div 14.4 + 0.5$ Simplifying

$= 144 \div 14.4 + 0.5$ Evaluating the exponential expression

$= 10 + 0.5$ Dividing

$= 10.5$ Adding

67. We add the amounts and divide by the number of addends, 5.

$$\frac{5.4 + 5.2 + 5.0 + 5.1 + 5.0}{5}$$

$$= \frac{25.7}{5} = 5.14$$

The average number of stays was 5.14 million stays per year.

69. We add the amounts and divide by the number of addends, 8.

$$\frac{29.0 + 29.5 + 29.5 + 30.3 + 30.1 + 31.2 + 31.2 + 32.6}{8}$$

$$= \frac{243.4}{8} = 30.425$$

The average number of miles per gallon was 30.425 mpg.

71. We add the amounts in the "Length, in miles" row of the table and divide by the number of addends, 5.

$$\frac{35.5+34.4+33.5+31.3+21.5}{5}$$
$$=\frac{156.2}{5}=31.24$$

The average length of the tunnels is 31.24 mi.

73. $\frac{33}{44}=\frac{3\cdot 11}{4\cdot 11}=\frac{3}{4}\cdot\frac{11}{11}=\frac{3}{4}$

75. $-\frac{27}{18}=-\frac{3\cdot 9}{2\cdot 9}=-\frac{3}{2}\cdot\frac{9}{9}=-\frac{3}{2}$

77. $\frac{9a}{27}=\frac{9\cdot a}{9\cdot 3}=\frac{9}{9}\cdot\frac{a}{3}=\frac{a}{3}$

79. $\frac{4r}{20r}=\frac{4\cdot r\cdot 1}{5\cdot 4\cdot r}=\frac{4\cdot r}{4\cdot r}\cdot\frac{1}{5}=\frac{1}{5}$

81. Use a calculator.

$$7.434\div(-1.2)\times 9.5+1.47^2$$
$$=7.434\div(-1.2)\times 9.5+2.1609$$
Evaluating the exponential expression
$$=-6.195\times 9.5+2.1609$$
$$=-58.8525+2.1609$$
Multiplying and dividing in order from left to right
$$=-56.6916$$
Adding

83. Use a calculator.

$$9.0534-2.041^2\times 0.731\div 1.043^2$$
$$=9.0534-4.165681\times 0.731\div 1.087849$$
Evaluating the exponential expressions
$$=9.0534-3.045112811\div 1.087849$$
$$=9.0534-2.799205415$$
Multiplying and dividing in order from left to right
$$=6.254194585$$
Subtracting

85.
$$439.57\times 0.01\div 1000\cdot x=4.3957$$
$$4.3957\div 1000\cdot x=4.3957$$
$$0.0043957\cdot x=4.3957$$
$$x=\frac{4.3957}{0.0043957}$$
$$x=1000$$

The solution is 1000.

87.
$$0.0329\div 0.001\times 10^4\div x=3290$$
$$0.0329\div 0.001\times 10,000\div x=3290$$
$$32.9\times 10,000\div x=3290$$
$$329,000\div x=3290$$

We need to divide 329,000 by a number that moves the decimal point 2 places to the left. Thus, we need to divide by 100. The solution is 100.

89. We divide. Note that 78.2 million $=78.2\times 1,000,000=78,200,000$.

```
                68
1,150,000 ) 78,200,000
            69 000 000
             9 200 000
             9 200 000
                     0
```

The game received 68 rating points.

91. The period from August 20 to September 20 consists of 31 days.

The "customer charge" is $31\times \$0.374=\$11.59$ (rounded to the nearest cent). The "energy charge" for the first 250 kilowatt-hours is $250\times \$0.1174=\$29.35$.

Subtract to find the "energy charge" for the kilowatt-hours in excess of 250:

$\$59.10-\$11.59-\$29.35=\$18.16$

Divide to find the number of kilowatt-hours in excess of 250:

$18.16\div \$0.09079=200$ (rounded to the nearest hour)

The total number of kilowatt-hours of electricity used is $250+200=450$ kwh.

Chapter 5 Mid-Chapter Review

1. In the number 308.00567, the number 6 names the ten-thousandths place. The statement is false.

2. The statement is true. See page 305 in the text.

3. Since $-2.3<-2.2$, we know that -2.3 is to the left of -2.2 on the number line. The given statement is true.

4.
$$P(1+r)=5000(1+0.045)$$
$$=5000(1.045)$$
$$=5225$$

5.
$$5.6+4.3\times(6.5-0.25)^2=5.6+4.3\times(6.25)^2$$
$$=5.6+4.3\times 39.0625$$
$$=5.6+167.96875$$
$$=173.56875$$

6. A word name for 9.69 is nine and sixty-nine hundredths.

7.
$$1.05\text{ million}=1.05\times 1\text{ million}$$
$$=1.05\times 1,000,000$$
$$=1,050,000$$

8. 4.$\underline{53}$ 4.53. (Move 2 places.) $\frac{453}{1\underline{00}}$

2 places   Move 2 places.   2 zeros

$$4.53=\frac{453}{100}$$
$$4.53=4+0.53=4+\frac{53}{100}=4\frac{53}{100}$$

9. 0.$\underline{287}$ 0.287. (Move 3 places.) $\frac{287}{1\underline{000}}$

3 places   Move 3 places.   3 zeros

$$0.287=\frac{287}{1000}$$

10. 0.07

0.13

Starting at the left, these digits are the first to differ; 1 is larger than 0.

Thus, 0.13 is larger.

11. -5.2

-5.09

Starting at the left, these digits are the first to differ; 0 is smaller than 2.

Thus, -5.09 is larger.

12. $\frac{7}{10}$ 0.7.

1 zero Move 1 place.

$\frac{7}{10} = 0.7$

13. $\frac{639}{100}$ 6.39.

2 zeros Move 2 places.

$\frac{639}{100} = 6.39$

14. First consider $35\frac{67}{100}$.

$35\frac{67}{100} = 35 + \frac{67}{100} = 35$ and $\frac{67}{100} = 35.67$

We have $35\frac{67}{100} = 35.67$, so $-35\frac{67}{100} = -35.67$.

15. $8\frac{2}{1000} = 8 + \frac{2}{1000} = 8$ and $\frac{2}{1000} = 8.002$

16. 28.461[5] Ten-thousandths digit is 5 or higher. Round up.

28.462

17. 28.46[1]5 Thousandths digit is 4 or lower. Round down.

28.46

18. 28.4[6]15 Hundredths digit is 5 or higher. Round up.

28.5

19. 28.[4]615 Tenths digit is 4 or lower. Round down.

28

20.

$$\begin{array}{r} \scriptstyle 1\,1\quad 1 \\ 47.638 \\ +\ \ 2.457 \\ \hline 50.095 \end{array}$$

21.

$$\begin{array}{rl} \scriptstyle 1\,1\,1\,1\,1 & \\ 15.600 & \text{Writing 2 extra zeros} \\ 234.729 & \\ 3.080 & \text{Writing an extra zero} \\ +\ 961.453 & \\ \hline 1214.862 & \end{array}$$

22. $-10.5 + 0.27$

The difference in absolute values is $10.5 - 0.27 = 10.23$. Since the negative number has the larger absolute value, the answer is negative.

$-10.5 + 0.27 = -10.23$

23. Line up the decimal points.

$$\begin{array}{rl} \scriptstyle 1 & \\ 16.00 & \text{Writing in the decimal point and 2 extra zeros} \\ 0.34 & \\ +\ \ 1.90 & \text{Writing an extra zero} \\ \hline 18.24 & \end{array}$$

24.

$$\begin{array}{r} \scriptstyle 11 \\ \scriptstyle 2\ \ 1\ 11\ 4\ 17 \\ \not{3}\,\not{2}\,\not{1}.\not{5}\,\not{7} \\ -\ \ 49.38 \\ \hline 272.19 \end{array}$$

25.

$$\begin{array}{rl} \scriptstyle 5\ 9\ 10 & \\ 5.\not{6}\,\not{0}\,\not{0} & \text{Writing 2 extra zeros} \\ -\ 0.007 & \\ \hline 5.593 & \end{array}$$

26. Line up the decimal points. Write an extra zero if desired.

$$\begin{array}{r} \scriptstyle 13\ 12 \\ \scriptstyle 2\ \not{3}\ \ \not{2}\ 10 \\ \not{3}\,\not{4}.\not{3}\,\not{0} \\ -\ 18.75 \\ \hline 15.55 \end{array}$$

27. $-6.9 - 13 = -6.9 + (-13) = -19.9$

28.

$$\begin{array}{rl} 4.6 & \text{(1 decimal place)} \\ \times\ 0.9 & \text{(1 decimal place)} \\ \hline 4.14 & \text{(2 decimal places)} \end{array}$$

29.

$$\begin{array}{rl} 15.3 & \text{(1 decimal place)} \\ \times\ 6.07 & \text{(2 decimal places)} \\ \hline 1071 & \\ 91800 & \\ \hline 92.871 & \text{(3 decimal places)} \end{array}$$

30. 100×81.236 81.23.6

2 zeros Move 2 places to the right.

$100 \times 81.236 = 8123.6$

31. $0.1 \times (-0.483)$ $-.0.483$

1 decimal place Move 1 place to the left.

$0.1 \times (-0.483) = -0.0483$

32. First we find $20.24 \div 4$.

$$\begin{array}{r} 5.0\,6 \\ 4\,\overline{)\,2\,0.2\,4} \\ \underline{2\,0} \\ 2\,4 \\ \underline{2\,4} \\ 0 \end{array}$$

A negative number divided by a negative number is positive, so the answer is 5.06.

33.
$$\begin{array}{r} 3\,.2 \\ 6.8_{\wedge}\overline{)\,2\,1.7_{\wedge}6} \\ \underline{2\,0\,4} \\ 1\,3\,\,6 \\ \underline{1\,3\,\,6} \\ 0 \end{array}$$

34. $\dfrac{76.3}{0.\underline{1}}$ $\qquad$ 76.3.

1 decimal place $\qquad$ Move 1 place to the right.

$76.3 \div 0.1 = 763$

35. $\dfrac{914.036}{1\underline{000}}$ $\qquad$ 0.914.036

3 zeros $\qquad$ Move 3 places to the left.

$914.036 \div 1000 = 0.914036$

36. Move 2 places to the right.

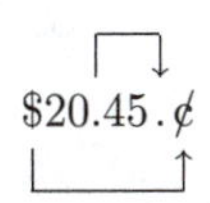

Change from \$ sign in front to ¢ sign at end.

\$20.45=2045¢

37. Move 2 places to the left.

\$1.47.¢

Change from ¢ sign at end to \$ sign in front.

147¢ = \$1.47

38.
$$\begin{aligned} &3.08x - 7.1 - 4.3x \\ &= (3.08 - 4.3)x - 7.1 \\ &= -1.22x - 7.1 \end{aligned}$$

39.
$$\begin{aligned} 2(l + w) &= 2(1.3 + 0.8) \\ &= 2(2.1) \\ &= 4.2 \end{aligned}$$

40.
$$\begin{aligned} &6.594 + 0.5318 \div 0.01 \\ &= 6.594 + 53.18 \qquad \text{Dividing} \\ &= 59.774 \qquad \text{Adding} \end{aligned}$$

41.
$$\begin{aligned} &7.3 \times 4.6 - 0.8 \div 3.2 \\ &= 33.58 - 0.25 \qquad \text{Multiplying and dividing} \\ &= 33.33 \qquad \text{Subtracting} \end{aligned}$$

42. The student probably rounded over successively from the thousandths place as follows: $236.448 \approx 236.45 \approx 236.5 \approx 237$. The student should have considered only the tenths place and rounded down.

43. The decimal points were not lined up before the subtraction was carried out.

44. $10 \div 0.2 = \dfrac{10}{0.2} = \dfrac{10}{0.2} \cdot \dfrac{10}{10} = \dfrac{100}{2}$, or $100 \div 2$

45. $0.247 \div 0.1 = \dfrac{247}{1000} \div \dfrac{1}{10} = \dfrac{247}{1000} \cdot \dfrac{10}{1} = \dfrac{247 \cdot 10}{10 \cdot 100} = \dfrac{247}{100} = 2.47 \neq 0.0247$;

$0.247 \div 10 = \dfrac{247}{1000} \div 10 = \dfrac{247}{1000} \cdot \dfrac{1}{10} = \dfrac{247}{10,000} = 0.0247 \neq 2.47$

Exercise Set 5.5

1. Since $\dfrac{3}{8}$ means $3 \div 8$ we have:

$$\begin{array}{r} 0.\,3\,7\,5 \\ 8\,\overline{)\,3.\,0\,0\,0} \\ \underline{2\,4} \\ 6\,0 \\ \underline{5\,6} \\ 4\,0 \\ \underline{4\,0} \\ 0 \end{array}$$

$\dfrac{3}{8} = 0.375$

3. Since $\dfrac{-1}{2}$ is negative, we divide 1 by 2 and make the results negative.

$$\begin{array}{r} 0.\,5 \\ 2\,\overline{)\,1.\,0} \\ \underline{1\,0} \\ 0 \end{array}$$

Thus, $\dfrac{-1}{2} = -0.5$.

5. Since $\dfrac{3}{25}$ means $3 \div 25$, we have:

$$\begin{array}{r} 0.\,1\,2 \\ 2\,5\,\overline{)\,3.\,0\,0} \\ \underline{2\,5} \\ 5\,0 \\ \underline{5\,0} \\ 0 \end{array}$$

$\dfrac{3}{25} = 0.12$

7. Since $\dfrac{9}{40}$ means $9 \div 40$, we have:

$$\begin{array}{rl} & 0.225 \\ 40 & \overline{)9.000} \\ & \underline{80} \\ & 100 \\ & \underline{80} \\ & 200 \\ & \underline{200} \\ & 0 \end{array}$$

$\dfrac{9}{40} = 0.225$

9. Since $\dfrac{13}{25}$ means $13 \div 25$, we have:

$$\begin{array}{rl} & 0.52 \\ 25 & \overline{)13.00} \\ & \underline{125} \\ & 50 \\ & \underline{50} \\ & 0 \end{array}$$

$\dfrac{13}{25} = 0.52$

11. Since $\dfrac{-17}{20}$ is negative, we divide 17 by 20 and make the result negative.

$$\begin{array}{rl} & 0.85 \\ 20 & \overline{)17.00} \\ & \underline{160} \\ & 100 \\ & \underline{100} \\ & 0 \end{array}$$

Thus, $\dfrac{-17}{20} = -0.85$.

13. Since $-\dfrac{9}{16}$ is negative, we divide 9 by 16 and make the result negative.

$$\begin{array}{rl} & 0.5625 \\ 16 & \overline{)9.0000} \\ & \underline{80} \\ & 100 \\ & \underline{96} \\ & 40 \\ & \underline{32} \\ & 80 \\ & \underline{80} \\ & 0 \end{array}$$

Thus, $-\dfrac{9}{16} = -0.5625$.

15. Since $\dfrac{7}{5}$ means $7 \div 5$, we have:

$$\begin{array}{rl} & 1.4 \\ 5 & \overline{)7.0} \\ & \underline{5} \\ & 20 \\ & \underline{20} \\ & 0 \end{array}$$

$\dfrac{7}{5} = 1.4$

17. Since $\dfrac{28}{25}$ means $28 \div 25$, we have:

$$\begin{array}{rl} & 1.12 \\ 25 & \overline{)28.00} \\ & \underline{25} \\ & 30 \\ & \underline{25} \\ & 50 \\ & \underline{50} \\ & 0 \end{array}$$

$\dfrac{28}{25} = 1.12$

19. Since $\dfrac{11}{-8}$ is negative, we divide 11 by 8 and make the result negative.

$$\begin{array}{rl} & 1.375 \\ 8 & \overline{)11.000} \\ & \underline{8} \\ & 30 \\ & \underline{24} \\ & 60 \\ & \underline{56} \\ & 40 \\ & \underline{40} \\ & 0 \end{array}$$

Thus, $\dfrac{11}{-8} = -1.375$.

21. Since $-\dfrac{39}{40}$ is negative, we divide 39 by 40 and make the result negative.

$$\begin{array}{rl} & 0.975 \\ 40 & \overline{)39.000} \\ & \underline{360} \\ & 300 \\ & \underline{280} \\ & 200 \\ & \underline{200} \\ & 0 \end{array}$$

Thus, $-\dfrac{39}{40} = -0.975$.

23. Since $\dfrac{121}{200}$ means $121 \div 200$, we have:

$$\begin{array}{rl} & 0.605 \\ 200 & \overline{)121.000} \\ & \underline{1200} \\ & 1000 \\ & \underline{1000} \\ & 0 \end{array}$$

$\dfrac{121}{200} = 0.605$

25. Since $\frac{8}{15}$ means $8 \div 15$, we have:

$$\begin{array}{r} 0.5\,3\,3 \\ 15\overline{)8.0\,0\,0} \\ \underline{7\,5} \\ 5\,0 \\ \underline{4\,5} \\ 5\,0 \\ \underline{4\,5} \\ 5 \end{array}$$

Since 5 keeps reappearing as a remainder, the digits repeat and

$\frac{8}{15} = 0.533\ldots$ or $0.5\overline{3}$.

27. Since $\frac{1}{3}$ means $1 \div 3$, we have:

$$\begin{array}{r} 0.3\,3\,3 \\ 3\overline{)1.0\,0\,0} \\ \underline{9} \\ 1\,0 \\ \underline{9} \\ 1\,0 \\ \underline{9} \\ 1 \end{array}$$

Since 1 keeps reappearing as a remainder, the digits repeat and

$\frac{1}{3} = 0.333\ldots$ or $0.\overline{3}$.

29. Since $\frac{-4}{3}$ is negative, we divide by 4 and 3 and make the result negative.

$$\begin{array}{r} 1.3\,3 \\ 3\overline{)4.0\,0} \\ \underline{3} \\ 1\,0 \\ \underline{9} \\ 1\,0 \\ \underline{9} \\ 1 \end{array}$$

Since 1 keeps reappearing as a remainder, the digits repeat and

$\frac{4}{3} = 1.333\ldots$ or $1.\overline{3}$.

Thus, $\frac{-4}{3} = -1.\overline{3}$.

31. Since $\frac{7}{6}$ means $7 \div 6$, we have:

$$\begin{array}{r} 1.1\,6\,6 \\ 6\overline{)7.0\,0\,0} \\ \underline{6} \\ 1\,0 \\ \underline{6} \\ 4\,0 \\ \underline{3\,6} \\ 4\,0 \\ \underline{3\,6} \\ 4 \end{array}$$

Since 4 keeps reappearing as a remainder, the digits repeat and

$\frac{7}{6} = 1.166\ldots$ or $1.1\overline{6}$.

33. Since $-\frac{14}{11}$ is negative, we divide 14 by 11 and make the result negative.

$$\begin{array}{r} 1.2\,7\,2\,7 \\ 11\overline{)14.0\,0\,0\,0} \\ \underline{1\,1} \\ 3\,0 \\ \underline{2\,2} \\ 8\,0 \\ \underline{7\,7} \\ 3\,0 \\ \underline{2\,2} \\ 8\,0 \\ \underline{7\,7} \\ 3 \end{array}$$

Since 3 and 8 keep reappearing as remainders, the sequence of digits "27" repeats in the quotient and $\frac{14}{11} = 1.2727\ldots$, or $1.\overline{27}$.

Thus, $-\frac{14}{11} = -1.\overline{27}$.

35. Since $-\frac{5}{12}$ is negative, we divide 5 by 12 and make the result negative.

$$\begin{array}{r} 0.4\,1\,6\,6 \\ 12\overline{)5.0\,0\,0\,0} \\ \underline{4\,8} \\ 2\,0 \\ \underline{1\,2} \\ 8\,0 \\ \underline{7\,2} \\ 8\,0 \\ \underline{7\,2} \\ 8 \end{array}$$

Since 8 keeps reappearing as a remainder, the digits repeat and

$\frac{5}{12} = 0.4166\ldots$, or $0.41\overline{6}$.

Thus, $\frac{-5}{12} = -0.41\overline{6}$.

37. Since $\frac{127}{500}$ means $127 \div 500$, we have:

$$\begin{array}{r} 0.2\,5\,4 \\ 500\overline{)127.0\,0\,0} \\ \underline{1\,0\,0\,0} \\ 2\,7\,0\,0 \\ \underline{2\,5\,0\,0} \\ 2\,0\,0\,0 \\ \underline{2\,0\,0\,0} \\ 0 \end{array}$$

$\frac{127}{500} = 0.254$

39. Since $\frac{4}{33}$ means $4 \div 33$, we have:

$$\begin{array}{r}
0.1212 \\
33 \overline{)4.0000} \\
\underline{3\,3} \\
70 \\
\underline{66} \\
40 \\
\underline{33} \\
70 \\
\underline{66} \\
4
\end{array}$$

Since 7 and 4 keep reappearing as remainders, the sequence of digits "12" repeats in the quotient and

$\frac{4}{33} = 0.1212\ldots$, or $0.\overline{12}$.

41. Since $-\frac{12}{55}$ is negative, we divide 12 by 55 and make the result negative.

$$\begin{array}{r}
0.21818 \\
55 \overline{)12.00000} \\
\underline{11\,0} \\
1\,00 \\
\underline{55} \\
450 \\
\underline{440} \\
100 \\
\underline{55} \\
450 \\
\underline{440} \\
10
\end{array}$$

Since 10 and 45 keep reappearing as remainders, the sequence of digits "18" repeats in the quotient and $\frac{12}{55} = 0.21818\ldots$, or $0.2\overline{18}$.

Thus, $\frac{-12}{55} = -0.2\overline{18}$.

43. Since $\frac{4}{7}$ means $4 \div 7$, we have:

$$\begin{array}{r}
0.571428 \\
7 \overline{)4.000000} \\
\underline{3\,5} \\
50 \\
\underline{49} \\
10 \\
\underline{7} \\
30 \\
\underline{28} \\
20 \\
\underline{14} \\
60 \\
\underline{56} \\
4
\end{array}$$

Since we have already divided 7 into 4, the sequence of digits "571428" repeats in the quotient and

$\frac{4}{7} = 0.571428571428\ldots$, or $0.\overline{571428}$.

45. In Example 4 we see that $\frac{4}{11} = 0.\overline{36}$.

Round 0.3[6]36 . . . to the nearest tenth.
Hundredths digit is 6 or more.
0.4 Round up.

Round 0.36[3]6 . . . to the nearest hundredth.
Thousandths digit is 4 or less.
0.36 Round down.

Round 0.363[6] . . . to the nearest thousandth.
Ten-thousandths digit is 5 or more.
0.364 Round up.

47. First we find decimal notation for $-\frac{5}{3}$.

$$\begin{array}{r}
1.66 \\
3 \overline{)5.00} \\
\underline{3} \\
2\,0 \\
\underline{1\,8} \\
20 \\
\underline{18} \\
2
\end{array}$$

$\frac{5}{3} = 1.\overline{6}$, so $-\frac{5}{3} = -1.\overline{6}$

Round −1.6[6]66 . . . to the nearest tenth.
Hundredths digit is 5 or more.
−1.7 Round to −1.7.

Round −1.66[6]6 . . . to the nearest hundredth.
Thousandths digit is 5 or more.
−1.67 Round to −1.67.

Round −1.666[6] . . . to the nearest thousandth.
Ten-thousandths digit is 5 or more.
−1.667 Round to −1.667.

49. First we find decimal notation for $\frac{-8}{17}$.

$$\begin{array}{r}
0.470588 \\
17 \overline{)8.000000} \\
\underline{6\,8} \\
1\,20 \\
\underline{1\,19} \\
100 \\
\underline{85} \\
150 \\
\underline{136} \\
140 \\
\underline{136} \\
4
\end{array}$$

The digits repeat eventually but we have enough decimal places now to be able to round as instructed.

We have $\frac{8}{17} \approx 0.47059$, so $\frac{-8}{17} \approx -0.47059$.

Round $-0.4\boxed{7}059\ldots$ to the nearest tenth.
Hundredths digit is 5 or more.
-0.5 Round to -0.5.

Round $-0.47\boxed{0}59\ldots$ to the nearest hundredth.
Thousandths digit is 4 or less.
-0.47 Round to -0.47.

Round $-0.470\boxed{5}9\ldots$ to the nearest thousandth.
Ten-thousandths digit is 5 or more.
-0.471 Round to -0.471.

51. First find decimal notation for $\frac{7}{12}$.

$$\begin{array}{r} 0.5833 \\ 12\overline{)7.0000} \\ \underline{60} \\ 100 \\ \underline{96} \\ 40 \\ \underline{36} \\ 40 \\ \underline{36} \\ 4 \end{array}$$

$\frac{7}{12} = 0.58\overline{3}$

Round $0.5\boxed{8}33\ldots$ to the nearest tenth.
Hundredths digit is 5 or more.
0.6 Round up.

Round $0.58\boxed{3}3\ldots$ to the nearest hundredth.
Thousandths digit is 4 or less.
0.58 Round down.

Round $0.583\boxed{3}\ldots$ to the nearest thousandth.
Ten-thousandths digit is 4 or less.
0.583 Round down.

53. First find decimal notation for $\frac{29}{-150}$.

$$\begin{array}{r} 0.1933 \\ 150\overline{)29.0000} \\ \underline{150} \\ 1400 \\ \underline{1350} \\ 500 \\ \underline{450} \\ 500 \\ \underline{450} \\ 50 \end{array}$$

We have $\frac{29}{150} = 0.19\overline{3}$, so $\frac{29}{-150} = -0.19\overline{3}$.

Round $-0.1\boxed{9}33\ldots$ to the nearest tenth.
Hundredths digit is 5 or more.
-0.2 Round to -0.2.

Round $-0.19\boxed{3}3\ldots$ to the nearest hundredth.
Thousandths digit is 3 or less.
-0.19 Round to -0.19.

Round $-0.193\boxed{3}\ldots$ to the nearest thousandth.
Ten-thousandths digit is 3 or less.
-0.193 Round to -0.193.

55. First find decimal notation for $\frac{7}{-9}$.

$$\begin{array}{r} 0.77 \\ 9\overline{)7.00} \\ \underline{63} \\ 70 \\ \underline{63} \\ 7 \end{array}$$

We have $\frac{7}{9} = 0.\overline{7}$, so $\frac{7}{-9} = -0.\overline{7}$.

Round $-0.7\boxed{7}77\ldots$ to the nearest tenth.
Hundredths digit is 5 or more.
-0.8 Round to -0.8.

Round $-0.77\boxed{7}7\ldots$ to the nearest hundredth.
Thousandths digit is 5 or more.
-0.78 Round to -0.78.

Round $-0.777\boxed{7}\ldots$ to the nearest thousandth.
Ten-thousandths digit is 5 or more.
-0.778 Round to -0.778.

57. Round $0.1\boxed{8}18\ldots$ to the nearest tenth.
Hundredths digit is 5 or higher.
0.2 Round up.

Round $0.18\boxed{1}8\ldots$ to the nearest hundredth.
Thousandths digit is 4 or lower.
0.18 Round down.

Round $0.181\boxed{8}\ldots$ to the nearest thousandth.
Ten-thousandths digit is 5 or higher.
0.182 Round up.

59. Round $0.2\boxed{7}77\ldots$ to the nearest tenth.
Hundredths digit is 5 or higher.
0.3 Round up.

Round $0.27\boxed{7}7\ldots$ to the nearest hundredth.
Thousandths digit is 5 or higher.
0.28 Round up.

Round $0.277\boxed{7}\ldots$ to the nearest thousandth.
Ten-thousandths digit is 5 or higher.
0.278 Round up.

61. Note that there are 3 women and 4 men, so there are $3+4$, or 7, people.

(a) $\dfrac{\text{Women}}{\text{Number of people}} = \dfrac{3}{7} = 0.\overline{428571} \approx 0.429$

(b) $\dfrac{\text{Women}}{\text{Men}} = \dfrac{3}{4} = 0.75$

(c) $\dfrac{\text{Men}}{\text{Number of people}} = \dfrac{4}{7} = 0.\overline{571428} \approx 0.571$

(d) $\dfrac{\text{Men}}{\text{Women}} = \dfrac{4}{3} = 1.\overline{3} \approx 1.333$

63. $\dfrac{\text{Miles driven}}{\text{Gasoline used}} = \dfrac{285}{18} = 15.833\ldots \approx 15.8$

The gasoline mileage was about 15.8 miles per gallon.

65. $\dfrac{\text{Miles driven}}{\text{Gasoline used}} = \dfrac{324.8}{18.2} \approx 17.8$

The gasoline mileage was about 17.8 miles per gallon.

67. We will use the second method discussed in the text.

$$\frac{7}{8} \times 12.64 = \frac{7}{8} \times \frac{1264}{100} = \frac{7 \cdot 1264}{8 \cdot 100}$$
$$= \frac{7 \cdot 2 \cdot 2 \cdot 2 \cdot 2 \cdot 79}{2 \cdot 2 \cdot 2 \cdot 2 \cdot 2 \cdot 5 \cdot 5}$$
$$= \frac{2 \cdot 2 \cdot 2 \cdot 2}{2 \cdot 2 \cdot 2 \cdot 2} \cdot \frac{7 \cdot 79}{2 \cdot 5 \cdot 5}$$
$$= 1 \cdot \frac{7 \cdot 79}{2 \cdot 5 \cdot 5}$$
$$= \frac{7 \cdot 79}{2 \cdot 5 \cdot 5} = \frac{553}{50}, \text{ or } 11.06$$

69. $6.84 \div 2\frac{1}{2} = 6.84 \div 2.5$ Writing $2\frac{1}{2}$ using decimal notation

$= 2.736$ Dividing

71. We will use the third method discussed in the text.

$$\frac{47}{9}(-79.95) = \frac{47}{9} \cdot \left(-\frac{7995}{100}\right)$$
$$= -\frac{47 \cdot 7995}{9 \cdot 100}$$
$$= -\frac{47 \cdot \not{3} \cdot \not{5} \cdot 533}{\not{3} \cdot 3 \cdot \not{5} \cdot 20}$$
$$= -\frac{25,051}{60}$$
$$= -417.51\overline{6}$$

73. $\frac{1}{2} - 0.5 = 0.5 - 0.5$ Writing $\frac{1}{2}$ using decimal notation

$= 0$

75. We will use the first method discussed in the text.

$$\left(\frac{1}{6}\right)0.0765 + \left(\frac{3}{4}\right)0.1124 = \frac{1}{6} \times \frac{0.0765}{1} + \frac{3}{4} \times \frac{0.1124}{1}$$
$$= \frac{0.0765}{6} + \frac{3 \times 0.1124}{4}$$
$$= \frac{0.0765}{6} + \frac{0.3372}{4}$$
$$= 0.01275 + 0.0843$$
$$= 0.09705$$

77. We will use the third method discussed in the text.

$$\frac{3}{4} \times 2.56 - \frac{7}{8} \times 3.94$$
$$= \frac{3}{4} \times \frac{256}{100} - \frac{7}{8} \times \frac{394}{100}$$
$$= \frac{768}{400} - \frac{2758}{800}$$
$$= \frac{768}{400} \cdot \frac{2}{2} - \frac{2758}{800}$$
$$= \frac{1536}{800} - \frac{2758}{800}$$
$$= \frac{-1222}{800} = -\frac{1222}{800}$$
$$= -\frac{2 \cdot 611}{2 \cdot 400} = \frac{2}{2} \cdot \left(-\frac{611}{400}\right)$$
$$= -\frac{611}{400}, \text{ or } -1.5275$$

79. We will use the second method discussed in the text.

$$5.2 \times 1\frac{7}{8} \div 0.4 = 5.2 \times 1.875 \div 0.4$$
$$= 9.75 \div 0.4$$
$$= 24.375$$

81. ***Familiarize.*** We draw a picture and recall that the formula for the area A of a triangle with base b and height h is $A = \frac{1}{2} \times b \times h$.

1.8 m

1.2 m

Translate. We substitute 1.2 for b and 1.8 for h.

$$A = \frac{1}{2} \times b \times h = \frac{1}{2} \times 1.2 \times 1.8$$

Solve. We carry out the computation.

$A = \frac{1}{2} \times 1.2 \times 1.8$

$= \frac{1.2}{2} \times 1.8$ Multiplying $\frac{1}{2}$ and 1.2

$= 0.6 \times 1.8$ Dividing

$= 1.08$ Multiplying

Check. We repeat the calculations using a different method.

$\frac{1}{2} \times 1.2 \times 1.8 = 0.5 \times (1.2 \times 1.8) = 0.5 \times 2.16 = 1.08$

Our answer checks.

State. The area of the shawl is 1.08 m^2.

83. ***Familiarize.*** We draw a picture and recall that the formula for the area A of a triangle with base b and height h is $A = \frac{1}{2} \times b \times h$.

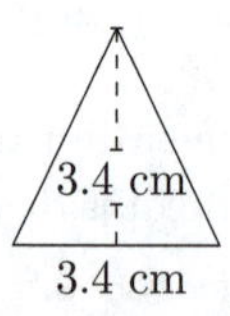

Translate. We substitute 3.4 for b and 3.4 for h.

$$A = \frac{1}{2} \times b \times h = \frac{1}{2} \times 3.4 \times 3.4$$

Solve. We carry out the computation.

$$\begin{aligned} A &= \frac{1}{2} \times 3.4 \times 3.4 \\ &= \frac{3.4}{2} \times 3.4 && \text{Multiplying } \frac{1}{2} \text{ and } 3.4 \\ &= 1.7 \times 3.4 && \text{Dividing} \\ &= 5.78 && \text{Multiplying} \end{aligned}$$

Check. We repeat the calculations using a different method.

$$\frac{1}{2} \times 3.4 \times 3.4 = 0.5 \times (3.4 \times 3.4) = 0.5 \times 11.56 = 5.78$$

Our answer checks.

State. The area of the stamp is 5.78 cm^2.

85. ***Familiarize***. First combine the lengths of the two 19.5-in. segments: 19.5 in. + 19.5 in. = 39 in. Now we can think of the area as the sum of the areas of two triangles, one with base 39 in. and height 11.25 in. and the other with base 39 in. and height 29.31 in.

Translate. We use the formula $A = \frac{1}{2}bh$ twice and add.

$$A = \frac{1}{2} \times 39 \times 11.25 + \frac{1}{2} \times 39 \times 29.31$$

Solve. We carry out the computation.

$$\begin{aligned} A &= \frac{1}{2} \times 39 \times 11.25 + \frac{1}{2} \times 39 \times 29.31 \\ &= \frac{39}{2} \times 11.25 + \frac{39}{2} \times 29.31 && \text{Multiplying } \frac{1}{2} \text{ and 39 twice} \\ &= 19.5 \times 11.25 + 19.5 \times 29.31 && \text{Dividing} \\ &= 219.375 + 571.545 \\ &= 790.92 \end{aligned}$$

Check. We repeat the calculation using a different method.

$$\begin{aligned} &\frac{1}{2} \times 39 \times 11.25 + \frac{1}{2} \times 39 \times 29.31 \\ &= 0.5 \times (39 \times 11.25) + 0.5 \times (39 \times 29.31) \\ &= 0.5 \times 438.75 + 0.5 \times 1143.09 \\ &= 219.375 + 571.545 \\ &= 790.92 \end{aligned}$$

The answer checks.

State. The area of the kite is 790.92 in^2.

87. 3 5 $\underline{7}$ $\boxed{2}$

The ones digit is 4 or less so we round down to 3570.

89. 7 8, $\underline{9}$ $\boxed{5}$ 1

The tens digit is 5 or more so we round up to 79,000.

91. $\frac{n}{1} = n$, for any integer n.

Thus, $\frac{95}{-1} = \frac{-95}{1} = -95$.

93.
$$\begin{aligned} &9 - 4 + 2 \div (-1) \cdot 6 \\ &= 9 - 4 - 2 \cdot 6 && \text{Multiplying and dividing in order from left to right} \\ &= 9 - 4 - 12 \\ &= 5 - 12 && \text{Adding and subtracting in order from left to right} \\ &= -7 \end{aligned}$$

95. Using a calculator we find that $\frac{1}{7} = 1 \div 7 = 0.\overline{142857}$.

97. Using a calculator we find that $\frac{3}{7} = 3 \div 7 = 0.\overline{428571}$.

99. Using a calculator we find that $\frac{5}{7} = 5 \div 7 = 0.\overline{714285}$.

101. Using a calculator we find that $\frac{1}{9} = 1 \div 9 = 0.\overline{1}$.

103. Using a calculator we find that $\frac{1}{999} = 0.\overline{001}$.

105. We substitute $\frac{22}{7}$ for π and 2.1 for r.

$$\begin{aligned} A &= \pi r^2 \\ &= \frac{22}{7}(2.1)^2 \\ &= \frac{22}{7}(4.41) \\ &= \frac{22 \times 4.41}{7} \\ &= \frac{97.02}{7} \\ &= 13.86 \text{ cm}^2 \end{aligned}$$

107. We substitute 3.14 for π and $\frac{3}{4}$ for r.

$$\begin{aligned} A &= \pi r^2 \\ &= 3.14\left(\frac{3}{4}\right)^2 \\ &= 3.14\left(\frac{9}{16}\right) \\ &= \frac{3.14 \times 9}{16} \\ &= \frac{28.26}{16} \\ &= 1.76625 \text{ ft}^2 \end{aligned}$$

When the calculation is done using the π key on a calculator, the result is 1.767145868 ft^2.

Exercise Set 5.6

1. We are estimating the sum

$$\$279.89 + \$149.99.$$

We round both numbers to the nearest ten. The estimate is

$$\$280 + \$150 = \$430.$$

Answer (d) is correct.

3. We are estimating the difference

$$\$279.89 - \$149.99.$$

We round both numbers to the nearest ten. The estimate is

$$\$280 - \$150 = \$130.$$

Answer (c) is correct.

5. We are estimating the product

$$6 \times \$79.99.$$

We round \$79.99 to the nearest ten. The estimate is

$$6 \times \$80 = \$480.$$

Answer (a) is correct.

7. We are estimating the quotient

$$\$830 \div \$79.99.$$

We round \$830 to the nearest hundred and \$79.99 to the nearest ten. The estimate is

$$\$800 \div \$80 = 10.$$

Answer (c) is correct.

9. This is about $0.0 + 1.3 + 0.3$, so the answer is about 1.6.

11. This is about $6 + 0 + 0$, so the answer is about 6.

13. This is about $52 + 1 + 7$, so the answer is about 60.

15. This is about $2.7 - 0.4$, so the answer is about 2.3.

17. This is about $200 - 20$, so the answer is about 180.

19. This is about 50×8, rounding 49 to the nearest ten and 7.89 to the nearest one, so the answer is about 400. Answer (a) is correct.

21. This is about 100×0.08, rounding 98.4 to the nearest ten and 0.083 to the nearest hundredth, so the answer is about 8. Answer (c) is correct.

23. This is about $4 \div 4$, so the answer is about 1. Answer (b) is correct.

25. This is about $75 \div 25$, so the answer is about 3. Answer (b) is correct.

27. We estimate the quotient $1760 \div 8.625$.

$1800 \div 9 = 200$

We estimate that 200 posts will be needed. Answers may vary depending on how the rounding was done.

29. We estimate the product $\$1.89 \times 12$.

$\$2 \cdot 12 = \$24$

We estimate the cost to be \$24. Answers may vary depending on how the rounding was done.

31. The decimal $0.57\overline{3}$ is an example of a repeating decimal.

33. The sentence $5(3+8) = 5\cdot3+5\cdot8$ illustrates the distributive law.

35. The number 1 is the multiplicative identity.

37. The least common denominator of two or more fractions is the least common multiple of their denominators.

39. We round each factor to the nearest ten. The estimate is $180 \times 60 = 10,800$. The estimate is close to the result given, so the decimal point was placed correctly.

41. We round each number on the left to the nearest one. The estimate is $19 - 1 \times 4 = 19 - 4 = 15$. The estimate is not close to the result given, so the decimal point was not placed correctly.

43. a) Observe that $2^{13} = 8192 \approx 8000$, $156,876.8 \approx 160,000$, and $8000 \times 20 = 160,000$. Thus, we want to find the product of 2^{13} and a number that is approximately 20. Since $0.37 + 18.78 = 19.15 \approx 20$, we add inside the parentheses and then multiply:

$$(0.37 + 18.78) \times 2^{13} = 156,876.8$$

We can use a calculator to confirm this result.

b) Observe that $312.84 \approx 6 \cdot 50$. We start by multiplying 6.4 and 51.2, getting 327.68. Then we can use a calculator to find that if we add 2.56 to this product and then subtract 17.4, we have the desired result. Thus,we have

$$2.56 + 6.4 \times 51.2 - 17.4 = 312.84.$$

Exercise Set 5.7

1. $5x = 27$

$\frac{5x}{5} = \frac{27}{5}$ Dividing both sides by 5

$x = 5.4$

Check: $5x = 27$

$5(5.4) \ ? \ 27$

$27 \mid 27$ TRUE

The solution is 5.4.

3. $x + 15.7 = 3.1$

$x + 15.7 - 15.7 = 3.1 - 15.7$ Adding -15.7 to (or subtracting 15.7 from) both sides

$x = -12.6$

Check: $x + 15.7 = 3.1$

$-12.6 + 15.7 \ ? \ 3.1$

$3.1 \mid 3.1$ TRUE

The solution is -12.6.

5. $5x - 8 = 22$

$5x - 8 + 8 = 22 + 8$ Adding 8 to both sides

$5x = 30$

$\frac{5x}{5} = \frac{30}{5}$ Dividing both sides by 5

$x = 6$

The solution is 6.

7.
$$\begin{aligned} 6.9x - 8.4 &= 4.02 \\ 6.9x - 8.4 + 8.4 &= 4.02 + 8.4 \quad \text{Adding 8.4 to both sides} \\ 6.9x &= 12.42 \\ \frac{6.9x}{6.9} &= \frac{12.42}{6.9} \quad \text{Dividing both sides by 6.9} \\ x &= 1.8 \end{aligned}$$

The solution is 1.8.

9.
$$\begin{aligned} 21.6 + 4.1t &= 6.43 \\ 21.6 + 4.1t - 21.6 &= 6.43 - 21.6 \quad \text{Subtracting 21.6 from both sides} \\ 4.1t &= -15.17 \\ \frac{4.1t}{4.1} &= \frac{-15.17}{4.1} \quad \text{Dividing both sides by 4.1} \\ t &= -3.7 \end{aligned}$$

The solution is -3.7.

11.
$$\begin{aligned} -26.25 &= 7.5x + 9 \\ -26.25 - 9 &= 7.5x + 9 - 9 \\ -35.25 &= 7.5x \\ -4.7 &= x \end{aligned}$$

The solution is -4.7.

13.
$$\begin{aligned} -4.2x + 3.04 &= -4.1 \\ -4.2x + 3.04 - 3.04 &= -4.1 - 3.04 \\ -4.2x &= -7.14 \\ \frac{-4.2x}{-4.2} &= \frac{-7.14}{-4.2} \\ x &= 1.7 \end{aligned}$$

The solution is 1.7.

15.
$$\begin{aligned} -3.05 &= 7.24 - 3.5t \\ -3.05 - 7.24 &= 7.24 - 3.5t - 7.24 \\ -10.29 &= -3.5t \\ \frac{-10.29}{-3.5} &= \frac{-3.5t}{-3.5} \\ 2.94 &= t \end{aligned}$$

The solution is 2.94.

17.
$$\begin{aligned} 3 - 1.2y &= -2.4 \\ 3 - 1.2y - 3 &= -2.4 - 3 \\ -1.2y &= -5.4 \\ \frac{-1.2y}{-1.2} &= \frac{-5.4}{-1.2} \\ y &= 4.5 \end{aligned}$$

The solution is 4.5.

19.
$$\begin{aligned} 9x - 2 &= 5x + 34 \\ 9x - 2 + 2 &= 5x + 34 + 2 \quad \text{Adding 2 to both sides} \\ 9x &= 5x + 36 \\ 9x - 5x &= 5x + 36 - 5x \quad \text{Subtracting } 5x \text{ from both sides} \\ 4x &= 36 \\ \frac{4x}{4} &= \frac{36}{4} \quad \text{Dividing both sides by 4} \\ x &= 9 \end{aligned}$$

Check:
$$\begin{array}{r|l} \multicolumn{2}{c}{9x - 2 = 5x + 34} \\ \hline 9 \cdot 9 - 2 \ ? & 5 \cdot 9 + 34 \\ 81 - 2 & 45 + 34 \\ 79 & 79 \end{array} \quad \text{TRUE}$$

The solution is 9.

21.
$$\begin{aligned} 2x + 6 &= 7x - 10 \\ 2x + 6 - 6 &= 7x - 10 - 6 \quad \text{Subtracting 6 from both sides} \\ 2x &= 7x - 16 \\ 2x - 7x &= 7x - 16 - 7x \quad \text{Subtracting } 7x \text{ from both sides} \\ -5x &= -16 \\ \frac{-5x}{-5} &= \frac{-16}{-5} \quad \text{Dividing both sides by } -5 \\ x &= 3.2 \end{aligned}$$

Check:
$$\begin{array}{r|l} \multicolumn{2}{c}{2x + 6 = 7x - 10} \\ \hline 2(3.2) + 6 \ ? & 7(3.2) - 10 \\ 6.4 + 6 & 22.4 - 10 \\ 12.4 & 12.4 \end{array} \quad \text{TRUE}$$

The solution is 3.2.

23.
$$\begin{aligned} 5y - 3 &= 4 + 9y \\ 5y - 3 + 3 &= 4 + 9y + 3 \\ 5y &= 9y + 7 \\ 5y - 9y &= 9y + 7 - 9y \\ -4y &= 7 \\ \frac{-4y}{-4} &= \frac{7}{-4} \\ y &= -1.75 \end{aligned}$$

The solution is -1.75.

25.
$$\begin{aligned} 5.9x + 67 &= 7.6x + 16 \\ 5.9x + 67 - 16 &= 7.6x + 16 - 16 \\ 5.9x + 51 &= 7.6x \\ 5.9x + 51 - 5.9x &= 7.6x - 5.9x \\ 51 &= 1.7x \\ \frac{51}{1.7} &= \frac{1.7x}{1.7} \\ 30 &= x \end{aligned}$$

The solution is 30.

27.
$$\begin{aligned} 7.8a + 2 &= 2.4a + 19.28 \\ 7.8a + 2 - 2 &= 2.4a + 19.28 - 2 \\ 7.8a &= 2.4a + 17.28 \\ 7.8a - 2.4a &= 2.4a + 17.28 - 2.4a \\ 5.4a &= 17.28 \\ \frac{5.4a}{5.4} &= \frac{17.28}{5.4} \\ a &= 3.2 \end{aligned}$$

The solution is 3.2

29.
$$\begin{aligned}
6(x+2) &= 4x+30 \\
6x+12 &= 4x+30 \quad \text{Using the distributive law} \\
6x+12-12 &= 4x+30-12 \\
6x &= 4x+18 \\
6x-4x &= 4x+18-4x \\
2x &= 18 \\
\frac{2x}{2} &= \frac{18}{2} \\
x &= 9
\end{aligned}$$

Check: $6(x+2) = 4x+30$

$$\begin{array}{r|l}
6(9+2) \; ? & 4\cdot 9+30 \\
6(11) & 36+30 \\
66 & 66 \qquad \text{TRUE}
\end{array}$$

The solution is 9.

31.
$$\begin{aligned}
5(x+3) &= 15x-6 \\
5x+15 &= 15x-6 \quad \text{Using the distributive law} \\
5x+15-15 &= 15x-6-15 \\
5x &= 15x-21 \\
5x-15x &= 15x-21-15x \\
-10x &= -21 \\
\frac{-10x}{-10} &= \frac{-21}{-10} \\
x &= 2.1
\end{aligned}$$

Check: $5(x+3) = 15x-6$

$$\begin{array}{r|l}
5(2.1+3) \; ? & 15(2.1)-6 \\
5(5.1) & 31.5-6 \\
25.5 & 25.5 \qquad \text{TRUE}
\end{array}$$

The solution is 2.1.

33.
$$\begin{aligned}
7a-9 &= 15(a-3) \\
7a-9 &= 15a-45 \quad \text{Using the distributive law} \\
7a-9+9 &= 15a-45+9 \\
7a &= 15a-36 \\
7a-15a &= 15a-36-15a \\
-8a &= -36 \\
\frac{-8a}{-8} &= \frac{-36}{-8} \\
a &= 4.5
\end{aligned}$$

The solution is 4.5.

35.
$$\begin{aligned}
1.5(y-6) &= 1.3-y \\
1.5y-9 &= 1.3-y \\
1.5y-9+9 &= 1.3-y+9 \\
1.5y &= 10.3-y \\
1.5y+y &= 10.3-y+y \\
2.5y &= 10.3 \\
\frac{2.5y}{2.5} &= \frac{10.3}{2.5} \\
y &= 4.12
\end{aligned}$$

The solution is 4.12.

37.
$$\begin{aligned}
2.9(x+8.1) &= 7.8x-3.95 \\
2.9x+23.49 &= 7.8x-3.95 \\
2.9x+23.49-23.49 &= 7.8x-3.95-23.49 \\
2.9x &= 7.8x-27.44 \\
2.9x-7.8x &= 7.8x-27.44-7.8x \\
-4.9x &= -27.44 \\
\frac{-4.9x}{-4.9} &= \frac{-27.44}{-4.9} \\
x &= 5.6
\end{aligned}$$

The solution is 5.6.

39.
$$\begin{aligned}
-6.21-4.3t &= 9.8(t+2.1) \\
-6.21-4.3t &= 9.8t+20.58 \\
-6.21-4.3t+6.21 &= 9.8t+20.58+6.21 \\
-4.3t &= 9.8t+26.79 \\
-4.3t-9.8t &= 26.79 \\
-14.1t &= 26.79 \\
\frac{-14.1t}{-14.1} &= \frac{26.79}{-14.1} \\
t &= -1.9
\end{aligned}$$

The solution is -1.9.

41.
$$\begin{aligned}
4(x-2)-9 &= 2x+9 \\
4x-8-9 &= 2x+9 \\
4x-17 &= 2x+9 \\
4x-17+17 &= 2x+9+17 \\
4x &= 2x+26 \\
4x-2x &= 2x+26-2x \\
2x &= 26 \\
\frac{2x}{2} &= \frac{26}{2} \\
x &= 13
\end{aligned}$$

The solution is 13.

43.
$$\begin{aligned}
2(4y-1.8)+0.4 &= 8(2y-0.4) \\
8y-3.6+0.4 &= 16y-3.2 \\
8y-3.2 &= 16y-3.2 \\
8y-3.2+3.2 &= 16y-3.2+3.2 \\
8y &= 16y \\
8y-16y &= 16y-16y \\
-8y &= 0 \\
\frac{-8y}{-8} &= \frac{0}{-8} \\
y &= 0
\end{aligned}$$

The solution is 0.

45.
$$\begin{aligned}
43(7-2x)+34 &= 50(x-4.1)+744\\
301-86x+34 &= 50x-205+744\\
-86x+335 &= 50x+539\\
-86x+335-335 &= 50x+539-335\\
-86x &= 50x+204\\
-86x-50x &= 50x+204-50x\\
-136x &= 204\\
\frac{-136x}{-136} &= \frac{204}{-136}\\
x &= -1.5
\end{aligned}$$

The solution is -1.5.

47. We use the formula $A=\frac{1}{2}\cdot b\cdot h$ and substitute 7 m for b and 4 m for h.

$$\begin{aligned}
A &= \frac{1}{2}\cdot b\cdot h\\
&= \frac{1}{2}\cdot 7\text{ m}\cdot 4\text{ m}\\
&= \frac{7\cdot 4}{2}\text{ m}^2\\
&= 14\text{ m}^2
\end{aligned}$$

49. We use the formula $A=\frac{1}{2}\cdot b\cdot h$ and substitute 5 in. for b and 5 in. for h.

$$\begin{aligned}
A &= \frac{1}{2}\cdot b\cdot h\\
&= \frac{1}{2}\cdot 5\text{ in.}\cdot 5\text{ in.}\\
&= \frac{5\cdot 5}{2}\text{ in}^2\\
&= \frac{25}{2}\text{ in}^2,\text{ or }12.5\text{ in}^2
\end{aligned}$$

51. The area of the figure is the sum of the areas of two triangles, each with base 5 ft and height 1 ft. Then we have

$$\begin{aligned}
A &= \frac{1}{2}\cdot b\cdot h+\frac{1}{2}\cdot b\cdot h\\
&= \frac{1}{2}\cdot 5\text{ ft}\cdot 1\text{ ft}+\frac{1}{2}\cdot 5\text{ ft}\cdot 1\text{ ft}\\
&= \frac{5\cdot 1}{2}\text{ ft}^2+\frac{5\cdot 1}{2}\text{ ft}^2\\
&= \frac{5}{2}\text{ ft}^2+\frac{5}{2}\text{ ft}^2\\
&= 5\text{ ft}^2
\end{aligned}$$

53.
$$\begin{aligned}
\frac{3}{25}-\frac{7}{10} &= \frac{3}{25}\cdot\frac{2}{2}-\frac{7}{10}\cdot\frac{5}{5} \quad \text{The LCM is 50.}\\
&= \frac{6}{50}-\frac{35}{50}\\
&= -\frac{29}{50}
\end{aligned}$$

55. We add in order from left to right.

$-17+24+(-9)=7+(-9)=-2$

57.
$$\begin{aligned}
7.035(4.91x-8.21)+17.401 &=\\
&\quad 23.902x-7.372815\\
34.54185x-57.75735+17.401 &=\\
&\quad 23.902x-7.372815\\
34.54185x-40.35635 &=\\
&\quad 23.902x-7.372815\\
34.54185x-40.35635-23.902x &=\\
&\quad 23.902x-7.372815-23.902x\\
10.63985x-40.35635 &= -7.372815\\
10.63985x-40.35635+40.35635 &=\\
&\quad -7.372815+40.35635\\
10.63985x &= 32.983535\\
\frac{10.63985x}{10.63985} &= \frac{32.983535}{10.63985}\\
x &= 3.1
\end{aligned}$$

The solution is 3.1.

59.
$$\begin{aligned}
5(x-4.2)+3[2x-5(x+7)] &=\\
&\quad 39+2(7.5-6x)+3x\\
5(x-4.2)+3[2x-5x-35] &=\\
&\quad 39+2(7.5-6x)+3x\\
5(x-4.2)+3[-3x-35] &= 39+2(7.5-6x)+3x\\
5x-21-9x-105 &= 39+15-12x+3x\\
-4x-126 &= 54-9x\\
-4x-126+9x &= 54-9x+9x\\
5x-126 &= 54\\
5x-126+126 &= 54+126\\
5x &= 180\\
\frac{5x}{5} &= \frac{180}{5}\\
x &= 36
\end{aligned}$$

The solution is 36.

61.
$$\begin{aligned}
3.5(4.8x-2.9)+4.5 &= 9.4x-3.4(x-1.9)\\
16.8x-10.15+4.5 &= 9.4x-3.4x+6.46\\
16.8x-5.65 &= 6x+6.46\\
16.8x-5.65-6x &= 6x+6.46-6x\\
10.8x-5.65 &= 6.46\\
10.8x-5.65+5.65 &= 6.46+5.65\\
10.8x &= 12.11\\
\frac{10.8x}{10.8} &= \frac{12.11}{10.8}\\
x &\approx 1.1212963
\end{aligned}$$

The solution is approximately 1.1212963.

Exercise Set 5.8

1. Let $a=$ Ron's age; $a+5$, or $5+a$

3. $b+6$, or $6+b$

5. $c-9$

7. Let $n=$ the number; $n-16$

9. Let $s=$ Nate's speed; $8s$

11. $\dfrac{x}{17}$; or $x \div 17$

13. Let x = the number; $\dfrac{1}{2}x + 20$, or $20 + \dfrac{1}{2}x$

15. Let x = the number; $4x - 20$

17. Let l = the length and w = the width; $l + w$, or $w + l$

19. Let r = the rate and t = the time; $rt + 10$, or $10 + rt$

21. Let n = the number; $10n + n$, or $n + 10n$

23. Let x and y = the numbers; $5(x - y)$

25. ***Familiarize.*** Let t = the amount *Titanic* took in during its lifetime, in billions of dollars.

Translate.

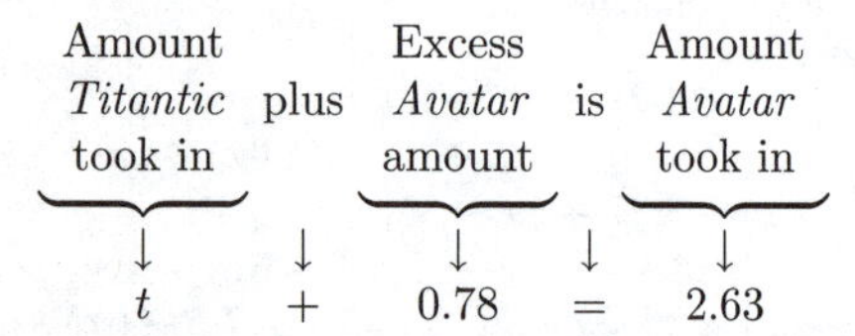

Solve. We subtract 0.78 from both sides of the equation.

$$t + 0.78 = 2.63$$
$$t + 0.78 - 0.78 = 2.63 - 0.78$$
$$t = 1.85$$

Check. We can add 1.85 and 0.78 to get 2.63. The answer checks.

State. *Titanic* took in \$1.85 billion during its lifetime.

27. ***Familiarize.*** We let a = the amount by which the amount of damage from Hurricane Katrina exceeded the amount of damage from Hurricane Andrew, in billions of dollars.

Translate.

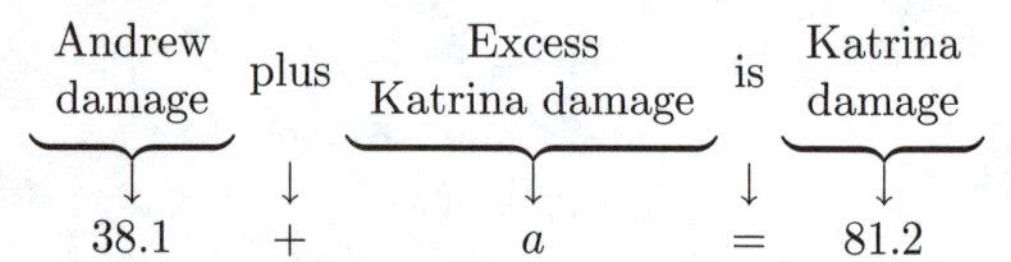

Solve. We subtract 38.1 on both sides of the equation.

$$38.1 + a = 81.2$$
$$38.1 + a - 38.1 = 81.2 - 38.1$$
$$a = 43.1$$

Check. We can check by adding 43.1 to 38.1 to get 81.2. The answer checks.

State. Hurricane Katrina was \$43.1 billion more costly than Hurricane Andrew.

29. ***Familiarize.*** Repeated addition fits this situation. We let c = the cost of 20.4 gal of gasoline, in dollars.

Translate.

Cost per gallon times Number of gallons is Total cost

$$2.249 \cdot 20.4 = c$$

Solve. We carry out the multiplication.

$$\begin{array}{r} 2.249 \\ \times \quad 20.4 \\ \hline 8996 \\ 449800 \\ \hline 45.8796 \end{array}$$

Thus, $c = 45.8796$.

Check. We obtain a partial check by rounding and estimating:

$$2.249 \times 20.4 \approx 2.25 \times 20 = 45 \approx 45.8796.$$

State. We round \$45.8796 to the nearest cent and find that the cost of the gasoline is \$45.88.

31. ***Familiarize.*** Let t = the cost of drinking tap water for a year.

Translate.

Cost of drinking tap water plus Excess cost of bottled water is Cost of drinking bottled water

$$t + 918.31 = 918.82$$

Solve. We subtract 918.31 on both sides of the equation.

$$t + 918.31 = 918.82$$
$$t + 918.31 - 918.31 = 918.82 - 918.31$$
$$t = 0.51$$

Check. We add 0.51 and 918.31 to get 918.82. The answer checks.

State. It costs \$0.51 to drink tap water for a year.

33. ***Familiarize.*** We visualize the situation. Let w = each winner's share.

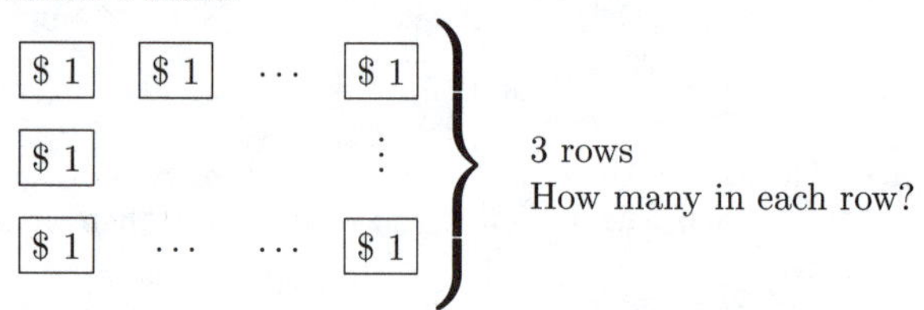

Translate.

Total prize ÷ Number of winners = Each winner's share

$$193{,}000{,}000 \div 3 = w$$

Solve. We carry out the division, obtaining $64{,}333{,}333.\overline{3}$. Rounding to the nearest cent, or hundredth, we get $w = 64{,}333{,}333.33$.

Check. We can repeat the calculation. The answer checks.

State. Each winner's share is \$64,333,333.33.

35. ***Familiarize.*** Let A = the area, in sq cm, and P = the perimeter, in cm.

Translate. We use the formulas $A = l \cdot w$ and $P = l + w + l + w$ and substitute 3.25 for l and 2.5 for w.

$$A = l \cdot w = (3.25) \cdot (2.5)$$
$$P = l + w + l + w = 3.25 + 2.5 + 3.25 + 2.5$$

Solve. To find the area we carry out the multiplication.

$$\begin{array}{r} 3.25 \\ \times\ 2.5 \\ \hline 1625 \\ 6500 \\ \hline 8.125 \end{array}$$

Thus, $A = 8.125$

To find the perimeter we carry out the addition.

$$\begin{array}{r} 3.25 \\ 2.5 \\ 3.25 \\ +\ 2.5 \\ \hline 11.50 \end{array}$$

Then $P = 11.5$.

Check. We can obtain partial checks by estimating.

$(3.25) \times (2.5) \approx 3 \times 3 \approx 9 \approx 8.125$

$3.25 + 2.5 + 3.25 + 2.5 \approx 3 + 3 + 3 + 3 = 12 \approx 11.5$

The answers check.

State. The area of the stamp is 8.125 sq cm, and the perimeter is 11.5 cm.

37. ***Familiarize.*** We visualize the situation. We let m = the odometer reading at the end of the trip.

22,456.8 mi	234.7 mi
m	

Translate. We are combining amounts.

Reading before trip	plus	Miles driven	is	Reading at end of trip
↓	↓	↓	↓	↓
$22,456.8$	$+$	234.7	$=$	m

Solve. To solve the equation we carry out the addition.

$$\begin{array}{r} \scriptsize{1\ 1} \\ 22,456.8 \\ +\ \ \ 234.7 \\ \hline 22,691.5 \end{array}$$

Thus, $m = 22,691.5$.

Check. We can check by repeating the addition. We can also check by rounding:

$22,456.8 + 234.7 \approx 22,460 + 230 = 22,690 \approx 22,691.5$

State. The odometer reading at the end of the trip was 22,691.5.

39. ***Familiarize.*** This is a two-step problem. First, we find the number of miles that have been driven between fillups. This is a "how-much-more" situation. We let n = the number of miles driven.

Translate and Solve.

First odometer reading	plus	Number of miles driven	is	Second odometer reading
↓	↓	↓	↓	↓
$26,342.8$	$+$	n	$=$	$26,736.7$

To solve the equation we subtract 26,342.8 on both sides.

$n = 26,736.7 - 26,342.8$

$n = 393.9$

$$\begin{array}{r} 26,736.7 \\ -\ 26,342.8 \\ \hline 393.9 \end{array}$$

Second, we divide the total number of miles driven by the number of gallons. This gives us m = the number of miles per gallon.

$393.9 \div 19.5 = m$

To find the number m, we divide.

$$\begin{array}{r} 20.2 \\ 19.5_\wedge \overline{)\,393.9_\wedge 0} \\ 3900 \\ \hline 390 \\ 390 \\ \hline 0 \end{array}$$

Thus, $m = 20.2$.

Check. To check, we first multiply the number of miles per gallon times the number of gallons:

$19.5 \times 20.2 = 393.9$

Then we add 393.9 to 26,342.8:

$26,342.8 + 393.9 = 26,736.7$

The number 20.2 checks.

State. The van gets 20.2 miles per gallon.

41. ***Familiarize.*** This is a two-step problem. First we find the total cost of the purchase. Then we find the amount of change Andrew received. Let t = the total cost of the purchase, in dollars.

Translate and Solve.

Purchase price	plus	Sales tax	is	Total cost
↓	↓	↓	↓	↓
23.99	$+$	1.68	$=$	t

To solve the equation we carry out the addition.

$$\begin{array}{r} \scriptsize{1\ 1} \\ 23.99 \\ +\ \ 1.68 \\ \hline 25.67 \end{array}$$

Thus, $t = 25.67$.

Now we find the amount of the change.

We visualize the situation. We let c = the amount of change.

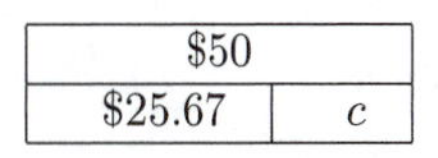

Amount paid	minus	Amount of purchase	is	Amount of change
↓	↓	↓	↓	↓
$50	$-$	$25.67	$=$	c

To solve the equation we carry out the subtraction.

$$\begin{array}{r} \overset{4}{\cancel{5}}\overset{9}{\cancel{0}}.\overset{9}{\cancel{0}}\overset{10}{\cancel{0}} \\ -\ 2\ 5.\ 6\ 7 \\ \hline 2\ 4.\ 3\ 3 \end{array}$$

Thus, $c = \$24.33$.

Check. We check by adding \$24.33 to \$25.67 to get \$50. This checks.

State. The change was \$24.33.

43. ***Familiarize***. Let $a =$ the amount of insulin consumed in a week, in cubic centimeters. Note that 38 units of insulin corresponds to 0.38 cc. (See Example 4.)

Translate.

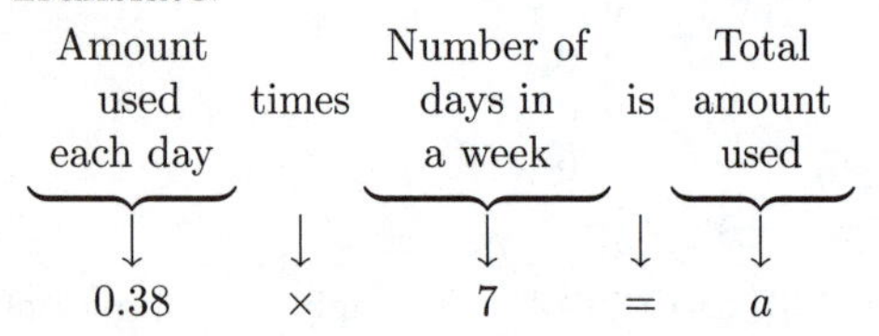

Solve. We carry out the multiplication.

$$\begin{array}{r} 0.\ 3\ 8 \\ \times\quad 7 \\ \hline 2.\ 6\ 6 \end{array}$$

Thus, $a = 2.66$.

Check. We can approximate the product.

$0.38 \times 7 \approx 0.4 \times 7 = 2.8 \approx 2.66$

The answer checks.

State. Phil consumes 2.66 cc of insulin in a week.

45. ***Familiarize***. This is a two-step problem. First we find the number of eggs in 20 dozen (1 dozen = 12). We let n represent this number.

Translate and Solve. We think (Number of dozens) $\cdot$ (Number in a dozen) = (Number of eggs).

$20 \cdot 12 = n$

$240 = n$

Second, we find the cost c of one egg. We think (Total cost) $\div$ (Number of eggs) = (Cost of one egg).

$\$25.80 \div 240 = c$

We carry out the division.

$$\begin{array}{r} 0.1\,0\,7\,5 \\ 2\,4\,0\,\overline{)\,2\,5.8\,0\,0\,0} \\ 2\,4\,0\,0\quad\;\; \\ \hline 1\,8\,0\,0\;\; \\ 1\,6\,8\,0\;\; \\ \hline 1\,2\,0\,0 \\ 1\,2\,0\,0 \\ \hline 0 \end{array}$$

Thus, $c = 0.1075 \approx 0.108$ (rounded to the nearest tenth of a cent).

Check. We repeat the calculations.

State. Each egg cost about \$0.108, or 10.8¢.

47. ***Familiarize***. This is a multi-step problem. We find the total area of the poster and the area devoted to the painting and then we subtract to find the area not devoted to the painting.

Translate and Solve. First we use the formula $A = l \times w$ to find the total area of the poster.

$A = l \times w$

$= 27.4 \text{ in.} \times 19.3 \text{ in.}$

$= 528.82 \text{ in}^2$

$$\begin{array}{r} 2\ 7.\ 4 \\ \times\ 1\ 9.\ 3 \\ \hline 8\ 2\ 2 \\ 2\ 4\ 6\ 6\ 0 \\ 2\ 7\ 4\ 0\ 0 \\ \hline 5\ 2\ 8.\ 8\ 2 \end{array}$$

Next we use the formula $A = l \times w$ again to find the area of the picture.

$A = l \times w$

$= 22.2 \text{ in.} \times 15.7 \text{ in.}$

$= 348.54 \text{ in}^2$

$$\begin{array}{r} 2\ 2.\ 2 \\ \times\ 1\ 5.\ 7 \\ \hline 1\ 5\ 5\ 4 \\ 1\ 1\ 1\ 0\ 0 \\ 2\ 2\ 2\ 0\ 0 \\ \hline 3\ 4\ 8.\ 5\ 4 \end{array}$$

Let $a =$ the area not devoted to the painting. We subtract to find this area.

$a = 528.82 \text{ in}^2 - 348.54 \text{ in}^2$

$= 180.28 \text{ in}^2$

$$\begin{array}{r} \overset{4}{\cancel{5}}\ \overset{12}{\cancel{2}}\ 8.\ \overset{7}{\cancel{8}}\ \overset{12}{\cancel{2}} \\ -\ 3\ 4\ 8.\ 5\ 4 \\ \hline 1\ 8\ 0.\ 2\ 8 \end{array}$$

Check. We repeat the calculations. The answer checks.

State. The area not devoted to the painting is 180.28 in^2.

49. ***Familiarize***. Let $n =$ the area available for notes. The total area available is twice the area of one side of the card. Recall that the formula for the area of a rectangle with length l and width w is $A = l \times w$.

Translate.

Total area	is	twice	Area of one side
↓	↓	↓	↓
n	$=$	$2\times$	12.7×7.6

Solve. We find the product.

$n = 2 \times 12.7 \times 7.6$

$= 25.4 \times 7.6$

$= 193.04$

$$\begin{array}{r} 2\ 5.\ 4 \\ \times\quad 7.\ 6 \\ \hline 1\ 5\ 2\ 4 \\ 1\ 7\ 7\ 8\ 0 \\ \hline 1\ 9\ 3.\ 0\ 4 \end{array}$$

Check. We can approximate the product.

$2 \times 12.7 \times 7.6 \approx 2 \times 13 \times 7.5 = 195 \approx 193.04$

The answer checks.

State. The area available for notes is 193.04 cm^2.

51. ***Familiarize***. This is a multistep problem. First we find the sum s of the two 0.8 cm segments. Then we use this length to find d.

Translate and Solve.

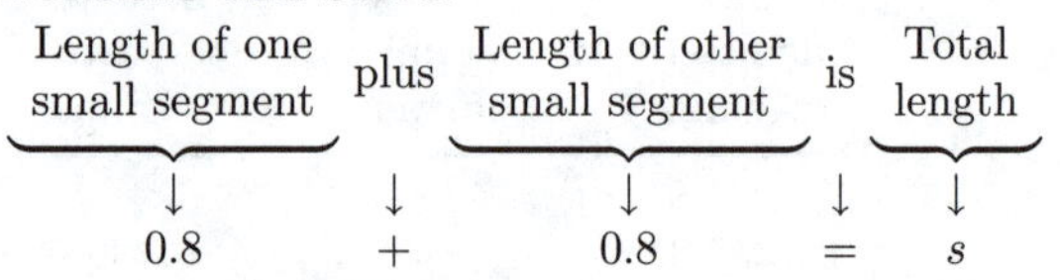

To solve we carry out the addition.

$$\begin{array}{r} \overset{1}{0.8} \\ +\,0.8 \\ \hline 1.6 \end{array}$$

Thus, $s = 1.6$.

Now we find d.

$$\underbrace{\text{Total length of smaller segments}}_{\downarrow\; 1.6} \;\; \underset{+}{\text{plus}} \;\; \underbrace{\text{length of } d}_{\downarrow\; d} \;\; \underset{=}{\text{is}} \;\; \underset{3.91}{\text{3.91 cm}}$$

To solve we subtract 1.6 on both sides of the equation.

$d = 3.91 - 1.6$
$d = 2.31$

$$\begin{array}{r} 3.91 \\ -\,1.60 \\ \hline 2.31 \end{array}$$

Check. We repeat the calculations.

State. The length d is 2.31 cm.

53. ***Familiarize.*** We make and label a drawing. The question deals with a rectangle and a circle, so we also list the relevant area formulas. We let $d =$ the amount of decking needed.

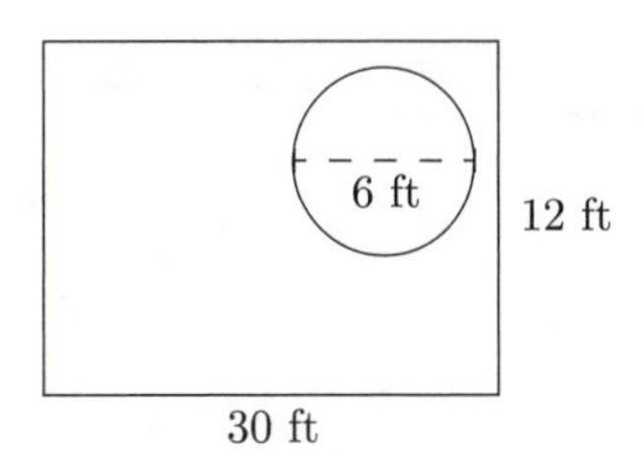

Area of a rectangle with length l and width w:
$A = l \times w$

Area of a circle with radius r: $A = \pi r^2$, where $\pi \approx 3.14$

Translate. We subtract the area of the circle from the area of the rectangle. Recall that a circle's radius is half of its diameter.

$$\underbrace{\text{Area of rectangle}}_{\downarrow\; 30 \times 12} \;\; \underset{-}{\text{minus}} \;\; \underbrace{\text{Area of circle}}_{\downarrow\; 3.14\left(\frac{6}{2}\right)^2} \;\; \underset{=}{\text{is}} \;\; \underbrace{\text{Area covered by decking}}_{\downarrow\; d}$$

Solve. We carry out the computations.

$$30 \times 12 - 3.14\left(\frac{6}{2}\right)^2 = d$$
$$30 \times 12 - 3.14(3)^2 = d$$
$$30 \times 12 - 3.14 \times 9 + d$$
$$360 - 28.26 = d$$
$$331.74 = d$$

Check. We can repeat the calculations. Also note that 331.74 is less than the area of the yard but more than the area of the flower garden. This agrees with the impression given by our drawing.

State. The amount of decking needed is 331.74 ft^2.

55. ***Familiarize.*** Let $m =$ the number of gigabytes Nikki paid for. We will express 3 cents as \$0.03.

Translate.

$$\underbrace{\text{Access fee}}_{\downarrow\; 8} \;\; \underset{+}{\text{plus}} \;\; \underset{0.03}{\$0.03} \;\; \underset{\cdot}{\text{times}} \;\; \underbrace{\text{Number of gigabytes}}_{\downarrow\; m} \;\; \underset{=}{\text{is}} \;\; \underbrace{\text{Total bill}}_{\downarrow\; 11.75}$$

Solve.

$$8 + 0.03 \cdot m = 11.75$$
$$8 + 0.03 \cdot m - 8 = 11.75 - 8$$
$$0.03 \cdot m = 3.75$$
$$\frac{0.03 \cdot m}{0.03} = \frac{3.75}{0.03}$$
$$m = 125$$

Check. $\$0.03 \cdot 125 = \$3.75$ and $\$3.75 + \$8 = \$11.75$, so the answer checks.

State. Nikki paid for 125 gigabytes.

57. ***Familiarize.*** This is a multistep problem. First, we find the cost of the cheese. We let $c =$ the cost of the cheese.

Translate and Solve.

$$\underbrace{\text{Number of pounds}}_{\downarrow\; 6} \;\; \underset{\cdot}{\text{times}} \;\; \underbrace{\text{Price per pound}}_{\downarrow\; \$4.79} \;\; \underset{=}{\text{is}} \;\; \underbrace{\text{Cost of cheese}}_{\downarrow\; c}$$

To solve the equation we carry out the multiplication.

$$\begin{array}{r} \$4.79 \\ \times\quad 6 \\ \hline \$28.74 \end{array}$$

Thus, $c = \$28.74$.

Next, we subtract to find how much money m is left to purchase seltzer.

$m = \$40 - \$28.74$
$m = \$11.26$

$$\begin{array}{r} \overset{3\ 9\ 9\ 10}{\cancel{4}\cancel{0}.\cancel{0}\cancel{0}} \\ -\,28.74 \\ \hline 11.26 \end{array}$$

Finally, we divide the amount of money left over by the cost of a bottle of seltzer to find how many bottles can be purchased. We let $b =$ the number of bottles of seltzer that can be purchased.

$\$11.26 \div \$0.64 = b$

To find b we carry out the division.

$$\begin{array}{r} 17. \\ 0.64_\wedge \overline{)\,11.26_\wedge} \\ 640 \\ \hline 486 \\ 448 \\ \hline 38 \end{array}$$

We stop dividing at this point, because Frank cannot purchase a fraction of a bottle. Thus, $b = 17$ (rounded to the nearest 1).

Check. The cost of the seltzer is $17 \cdot \$0.64$ or \$10.88. The cost of the cheese is $6 \cdot \$4.79$, or \$28.74. Frank has spent a total of \$10.88 + \$28.74, or \$39.62. Frank has \$40 − \$39.62, or \$0.38 left over. This is not enough to purchase another bottle of seltzer, so our answer checks.

State. Frank should buy 17 bottles of seltzer.

59. ***Familiarize.*** Let l = the number of hours Ben worked doing lawn maintenance. Then $l + 6$ = the number of hours he worked as a waiter.

Translate.

$$\underbrace{\text{Waiter hours}}_{l+6} \quad \underset{+}{\text{plus}} \quad \underbrace{\text{Lawn Maintenance hours}}_{l} \quad \underset{=}{\text{is}} \quad \underbrace{\text{Total hours}}_{27}$$

Solve.

$$\begin{aligned} l+6+l &= 27 \\ 2l+6 &= 27 \\ 2l+6-6 &= 27-6 \\ 2l &= 21 \\ \frac{2l}{2} &= \frac{21}{2} \\ l &= 10.5 \end{aligned}$$

Then $l + 6 = 10.5 + 6 = 16.5$.

Check. 16.5 hr is 6 hr more than 10.5 hr, and $16.5+10.5 = 27$. The answer checks.

State. Ben worked 16.5 hr as a waiter and 10.5 hr doing lawn maintenance.

61. ***Familiarize.*** Let p = the number of protected acres, in millions. Then $p + 0.4$ = the number of unprotected acres.

Translate.

$$\underbrace{\text{Protected acres}}_{p} \quad \underset{+}{\text{plus}} \quad \underbrace{\text{Unprotected acres}}_{p+0.4} \quad \underset{=}{\text{is}} \quad \underbrace{\text{Total area}}_{5.4}$$

Solve.

$$\begin{aligned} p+p+0.4 &= 5.4 \\ 2p+0.4 &= 5.4 \\ 2p+0.4-0.4 &= 5.4-0.4 \\ 2p &= 5 \\ \frac{2p}{2} &= \frac{5}{2} \\ p &= 2.5 \end{aligned}$$

Then $p + 0.4 = 2.5 + 0.4 = 2.9$.

Check. 2.9 million acres is 0.4 million acres more than 2.5 million acres, and 2.5 million acres + 2.9 million acres = 5.4 million acres. The answer checks.

State. The habitat has 2.5 million protected acres and 2.9 million unprotected acres.

63. ***Familiarize.*** Let n = the number of non-spam messages sent each day, in billions. Then $5n$ = the number of spam messages.

Translate.

$$\underbrace{\text{Non-spam messages}}_{n} \quad \underset{+}{\text{plus}} \quad \underbrace{\text{Spam messages}}_{5n} \quad \underset{=}{\text{is}} \quad \underbrace{\text{Total messages}}_{210}$$

Solve.

$$\begin{aligned} n+5n &= 210 \\ 6n &= 210 \\ \frac{6n}{6} &= \frac{210}{6} \\ n &= 35 \end{aligned}$$

Then $5n = 5 \cdot 35 = 175$.

Check. 175 billion is five times 35 billion, and 35 billion + 175 billion = 210 billion. The answer checks.

State. Each day 35 billion non-spam messages and 175 billion spam messages are sent.

65. ***Familiarize.*** Let f = the amount spent for food. Then $3f$ = the amount spent for lodging.

Translate.

$$\underbrace{\text{Amount spent for food}}_{f} \quad \underset{+}{\text{plus}} \quad \underbrace{\text{Amount spent for lodging}}_{3f} \quad \underset{=}{\text{is}} \quad \underbrace{\text{Total spending}}_{261.20}$$

Solve.

$$\begin{aligned} f+3f &= 261.20 \\ 4f &= 261.20 \\ \frac{4f}{4} &= \frac{261.20}{4} \\ f &= 65.30 \end{aligned}$$

Then $3f = 3 \cdot 65.30 = 195.90$.

Check. \$195.90 is 3 times \$65.30, and \$65.30 + \$195.90 = \$261.20. The answer checks.

State. Emily spent \$65.30 for food and \$195.90 for lodging.

67. $\frac{0}{n} = 0$, for any integer n that is not 0.

Thus, $\frac{0}{-13} = 0$.

69. $\frac{n}{n} = 1$, for any integer n that is not 0.

Thus, $\frac{-76}{-76} = 1$.

71. $$\begin{aligned} \frac{8}{11} - \frac{4}{3} &= \frac{8}{11}\cdot\frac{3}{3} - \frac{4}{3}\cdot\frac{11}{11} \quad \text{The LCM is 33.} \\ &= \frac{24}{33} - \frac{44}{33} \\ &= \frac{-20}{33}, \text{ or } -\frac{20}{33} \end{aligned}$$

73.
$$\begin{aligned} 4x-7&=9x+13\\ 4x-7+7&=9x+13+7\\ 4x&=9x+20\\ 4x-9x&=9x+20-9x\\ -5x&=20\\ \frac{-5x}{-5}&=\frac{20}{-5}\\ x&=-4 \end{aligned}$$

The solution is -4.

75. First we subtract to find the yearly decreases, in millions.

$24.3-23.1=1.2$;

$23.1-22.8=0.3$;

$22.8-22.3=0.5$.

Now we find the average yearly decrease.

$$\frac{1.2+0.3+0.5}{3}=\frac{2}{3}, \text{ or } 0.\overline{6}$$

The average yearly decrease was $\frac{2}{3}$ million, or $0.\overline{6}$ million.

77. ***Familiarize.*** Let $p=$ the highest price per mile that Lindsey can afford

Translate.

Daily rate	plus	Price per mile	times	Number of miles	is	Total cost
↓	↓	↓	↓	↓	↓	↓
18.90	+	p	·	190	=	55

Solve.

$$\begin{aligned} 18.90+p\cdot 190&=55\\ 18.90+p\cdot 190-18.90&=55-18.90\\ p\cdot 190&=36.10\\ \frac{p\cdot 190}{190}&=\frac{36.10}{190}\\ p&=0.19 \end{aligned}$$

Check. At \$0.19 per mile, the mileage charge for 190 mi will be $\$0.19\cdot 190$, or \$36.10, and $\$18.90+\$36.10=\$55$. The answer checks.

State. Lindsey can afford to pay at most \$0.19 per mile.

79. We must make some assumptions. First we assume that the figures are nested squares formed by connecting the midpoints of consecutive sides of the next larger square. Next assume that the shaded area is the same as the area of the innermost square. (It appears that if we folded the shaded area into the innermost square, it would exactly fill the square.) Finally assume that the length of a side of the innermost square is 5 cm. (If we project the vertices of the innermost square onto the corresponding sides of the largest square, it appears that the distance between each projection and the nearest vertex of the largest square is one-fourth the length of a side of the largest square. Thus, the distance between projections on each side of the largest square is $\frac{1}{2}\cdot 10$ cm, or 5 cm and, hence, the length of a side of the innermost square is 5 cm.) Then the area of the innermost square is 5 cm · 5 cm, or 25 cm^2, so the shaded area is 25 cm^2.

Chapter 5 Concept Reinforcement

1. One thousand billions $=1000\times 1{,}000{,}000{,}000$
$=1{,}000{,}000{,}000{,}000$
$=$ one trillion

The statement is true.

2. The number of decimal places in the product of two number is the *sum* of the number of decimal places in the factors. The given statement is false.

3. Dividing a number by 0.1, 0.01, 0.001, and so on is equivalent to multiplying the number by 10, 100, 1000, and so on, so the quotient is larger than the dividend. The given statement is true.

4. A terminating decimal occurs only when the denominator of a fraction has only 2's or 5's, or both, as factors. The given statement is false.

5. An estimate found by rounding to the nearest ten uses numbers that are closer to the actual numbers in the calculation than one found by rounding to the nearest hundred, so an estimate found by rounding to the nearest ten is more accurate. The given statement is true.

Chapter 5 Important Concepts

1. 0.03   0.03. $\frac{3}{100}$

2 places   Move 2 places.   2 zeros

$0.03=\frac{3}{100}$

2. $\frac{817}{10}$ 81.7.

1 zero   Move 1 place.

$\frac{817}{10}=81.7$

3. $42\frac{159}{1000}=42+\frac{159}{1000}=42$ and $\frac{159}{1000}=42.159$

4. 153.34[6]   Thousandths digit is 5 or higher. Round up.

↓

153.35

5. Line up the decimal points. Add an extra zero if desired.

$$\begin{array}{r} \scriptstyle 1\\ 5.540\\ +\,33.071\\ \hline 38.611 \end{array}$$

6. Line up the decimal points. Add an extra zero if desired.

$$\begin{array}{r} 2\,2\,1.0\,4\,0 \\ -\quad 1\,3.1\,9\,2 \\ \hline 2\,0\,7.8\,4\,8 \end{array}$$

7.

$$\begin{array}{rl} 5.\,4\,6 & \text{(2 decimal places)} \\ \times \quad 3.\,5 & \text{(1 decimal place)} \\ \hline 2\,7\,3\,0 & \\ 1\,6\,3\,8\,0 & \\ \hline 1\,9.\,1\,1\,0 & \end{array}$$

8. $17.6 \times 0.\underline{01}$ — 0.17.6

2 decimal places — Move 2 places to the left.

$17.6 \times 0.01 = 0.176$

9. $1\underline{000} \times 60.437$ — 60.437.

3 zeros — Move 3 places to the right.

$1000 \times 60.437 = 60,437$

10.

$$\begin{array}{r} 7.4 \\ 3.6_\wedge \overline{)\,2\,6.6_\wedge 4} \\ 2\,5\,2 \\ \hline 1\,4\,4 \\ 1\,4\,4 \\ \hline 0 \end{array}$$

11. $\frac{4.7}{1\underline{00}}$ — 0.04.7

2 zeros — Move 2 places to the left.

$\frac{4.7}{100} = 0.047$

12. $\frac{156.9}{0.\underline{01}}$ — 156.90.

2 decimal places — Move 2 places to the right.

$\frac{156.9}{0.01} = 15,690$

Chapter 5 Review Exercises

1. 6.59 million $= 6.59 \times 1,\underbrace{000,000}_{6\text{ zeros}}$

6.590000.

Move 6 places to the right.

6.59 million $= 6,590,000$

2. 3.1 billion $= 3.1 \times 1,\underbrace{000,000,000}_{9\text{ zeros}}$

3.100000000.

Move 9 places to the right.

3.1 billion $= 3,100,000,000$

3. A word name for 3.47 is three and forty-seven hundredths.

4. A word name for 0.031 is thirty-one thousandths.

5. A word name for 27.0001 is twenty-seven and one ten thousandth.

6. A word name for 0.9 is nine tenths.

7. $0.\underline{09}$ — 0.09. — $\frac{9}{1\underline{00}}$

2 places — Move 2 places. — 2 zeros

$0.09 = \frac{9}{100}$

8. $-4.\underline{561}$ — $-4.561.$ — $-\frac{4561}{1\underline{000}}$

3 places — Move 3 places. — 3 zeros

$-4.561 = -\frac{4561}{1000}$, and $-4.561 = -4\frac{561}{1000}$.

9. $-0.\underline{089}$ — $-0.089.$ — $-\frac{89}{1\underline{000}}$

3 places — Move 3 places. — 3 zeros

$-0.089 = -\frac{89}{1000}$

10. $3.\underline{0227}$ — 3.0227. — $\frac{30,227}{1\underline{0,000}}$

4 places — Move 4 places. — 4 zeros

$3.0227 = \frac{30,227}{10,000}$, and $3.0227 = 3\frac{227}{10,000}$

11. $\frac{34}{1\underline{000}}$ — 0.034.

3 zeros — Move 3 places.

$\frac{34}{1000} = 0.034$

12. $\frac{42,603}{1\underline{0,000}}$ — 4.2603.

4 zeros — Move 4 places.

$\frac{42,603}{10,000} = 4.2603$

13. $27\frac{91}{100} = 27 + \frac{91}{100} = 27$ and $\frac{91}{100} = 27.91$

14. First we consider $867\frac{6}{1000}$.

$867\frac{6}{1000} = 867 + \frac{6}{1000} = 867$ and $\frac{6}{1000} = 867.006$

$867\frac{6}{1000} = 867.006$, so $-867\frac{6}{1000} = -867.006$.

15. 0.034

Starting at the left, these digits are the first to differ; 3 is larger than 1.

0.0185

Thus, 0.034 is larger.

16. -0.91

Starting at the left, these digits are the first to differ; 1 is smaller than 9.

-0.19

Thus, -0.19 is larger.

17. 0.741

Starting at the left, these digits are the first to differ; 7 is larger than 6.

0.6943

Thus, 0.741 is larger.

18. 1.038

Starting at the left, these digits are the first to differ, and 4 is larger than 3.

1.041

Thus, 1.041 is larger.

19. $17.4\boxed{2}87$ Hundredths digit is 4 or lower. Round down.

$\downarrow$

17.4

20. $17.42\boxed{8}7$ Thousandths digit is 5 or higher. Round up.

$\downarrow$

17.43

21. $17.428\boxed{7}$ Ten-thousandths digit is 5 or higher. Round up.

$\downarrow$

17.429

22. $17.\boxed{4}287$ Tenths digit is 4 or lower. Round down.

$\downarrow$

17

23.
$$\begin{array}{r} \scriptstyle 1 \quad\quad \\ 236.231 \\ 263.4 \\ +\quad 0.198 \\ \hline 499.829 \end{array}$$

24.
$$\begin{array}{r} \scriptstyle 13 \quad\; \\ \scriptstyle 2\ 17\ 5\ \not{3}\ 15 \\ \not{3}\,\not{7}.\not{6}\,\not{4}\,\not{5} \\ -\quad 8.4\,9\,7 \\ \hline 29.1\,4\,8 \end{array}$$

25.
$$\begin{array}{r} \scriptstyle 1\ 1\quad \\ 219.3 \\ 2.8 \\ +\quad 7.0 \\ \hline 229.1 \end{array}$$

26.
$$\begin{array}{r} \scriptstyle 13\ 14\quad 10\quad\; \\ \scriptstyle 6\ \not{3}\ \ \not{4}\ 9\ \not{0}\ 10 \\ \not{7}\,\not{4}\,\not{5}.\not{0}\,\not{1}\,\not{0}\,9 \\ -\quad 59.95\,9\,0 \\ \hline 685.05\,1\,9 \end{array}$$

27. $-37.8+(-19.5)$

Add the absolute values: $37.8+19.5=57.3$

Make the answer negative: $-37.8+(-19.5)=-57.3$

28. $-7.52-(-9.89)=-7.52+9.89=2.37$

29.
$$\begin{array}{r} 48 \\ \times\ 0.27 \\ \hline 336 \\ 960 \\ \hline 12.96 \end{array}$$

30. $-3.7(0.29)=-1.073$

31.
$$\begin{array}{r} 24.68 \\ \times\ 1000 \\ \hline \end{array}$$

The number 1000 has 3 zeros so we move the decimal point in 24.68 three places to the right. The product is 24,680.

32.
$$\begin{array}{r} 3.2 \\ 25\,\overline{)\,80.0} \\ 75 \\ \hline 50 \\ 50 \\ \hline 0 \end{array}$$

33. First we consider $11.52 \div 7.2$.

$$\begin{array}{r} 1.6 \\ 7.2_\wedge\,\overline{)\,11.5_\wedge 2} \\ 720 \\ \hline 432 \\ 432 \\ \hline 0 \end{array}$$

Since we have a positive number divided by a negative number, the answer is -1.6.

34. $\dfrac{276.3}{1000}$

The number 1000 has 3 zeros, so we move the decimal point in the numerator 3 places to the left.

$\dfrac{276.3}{1000}=0.2763$

35.
$$\begin{aligned} &3.7x-5.2y-1.5x-3.9y \\ &=3.7x+(-5.2y)+(-1.5x)+(-3.9y) \\ &=3.7x+(-1.5x)+(-5.2y)+(-3.9y) \\ &=[3.7+(-1.5)]x+[-5.2+(-3.9)]y \\ &=2.2x-9.1y \end{aligned}$$

36.
$$\begin{aligned} &7.94-3.89a+4.63+1.05a \\ &=7.94+(-3.89a)+4.63+1.05a \\ &=-3.89a+1.05a+7.94+4.63 \\ &=(-3.89+1.05)a+(7.94+4.63) \\ &=-2.84a+12.57 \end{aligned}$$

37.
$$\begin{aligned} P-Prt &= 1000-1000(0.05)(1.5) \\ &= 1000-50(1.5) \\ &= 1000-75 \\ &= 925 \end{aligned}$$

38.
$$\begin{aligned} & 9-3.2(-1.5)+5.2^2 \\ &= 9-3.2(-1.5)+5.2(5.2) \\ &= 9-3.2(-1.5)+27.04 \\ &= 9+4.8+27.04 \\ &= 13.8+27.04 \\ &= 40.84 \end{aligned}$$

39. Move 2 places to the left.

\$15.49.¢

Change from ¢ sign at end to \$ sign in front.

1549¢ = \$15.49

40. Round

2 4 8. 2 7 [2] 7 ... to the nearest hundredth.

Thousandths digit is 4 or less.

2 4 8. 2 7 Round down.

41. $\frac{13}{5} = \frac{13}{5} \cdot \frac{2}{2} = \frac{26}{10} = 2.6$

42. $\frac{32}{25} = \frac{32}{25} \cdot \frac{4}{4} = \frac{128}{100} = 1.28$

43.
$$\begin{array}{r} 3.25 \\ 4\overline{)13.00} \\ \underline{12} \\ 1\,0 \\ \underline{8} \\ 20 \\ \underline{20} \\ 0 \end{array}$$

$\frac{13}{4} = 3.25$

44. Since $-\frac{7}{6}$ is negative, we divide 7 by 6 and make the result negative.

$$\begin{array}{r} 1.166 \\ 6\overline{)7.000} \\ \underline{6} \\ 1\,0 \\ \underline{6} \\ 40 \\ \underline{36} \\ 40 \\ \underline{36} \\ 4 \end{array}$$

Since 4 keeps reappearing as a remainder, the digits repeat and

$\frac{7}{6} = 1.166\ldots$, or $1.1\overline{6}$.

Thus, $-\frac{7}{6} = -1.1\overline{6}$.

45. $\frac{4}{15} \times 79.05 = \frac{4}{15} \times \frac{79.05}{1} = \frac{4 \times 79.05}{15 \times 1} = \frac{316.2}{15} = 21.08$

46.
$$\begin{aligned} t-4.3 &= -7.5 \\ t-4.3+4.3 &= -7.5+4.3 \\ t &= -3.2 \end{aligned}$$

The solution is -3.2.

47.
$$\begin{aligned} 4.1x+5.6 &= -6.7 \\ 4.1x+5.6-5.6 &= -6.7-5.6 \\ 4.1x &= -12.3 \\ \frac{4.1x}{4.1} &= \frac{-12.3}{4.1} \\ x &= -3 \end{aligned}$$

The solution is -3.

48.
$$\begin{aligned} 6x-11 &= 8x+4 \\ 6x-11+11 &= 8x+4+11 \\ 6x &= 8x+15 \\ 6x-8x &= 8x+15-8x \\ -2x &= 15 \\ \frac{-2x}{-2} &= \frac{15}{-2} \\ x &= -7.5 \end{aligned}$$

The solution is -7.5.

49.
$$\begin{aligned} 3(x+2) &= 5x-7 \\ 3x+6 &= 5x-7 \\ 3x+6-6 &= 5x-7-6 \\ 3x &= 5x-13 \\ 3x-5x &= 5x-13-5x \\ -2x &= -13 \\ \frac{-2x}{-2} &= \frac{-13}{-2} \\ x &= 6.5 \end{aligned}$$

The solution is 6.5.

50. ***Familiarize.*** Let t = the number by which the number of telephone poles for every 100 people in the U.S. exceeds the number in Canada.

Translate. We have a "how much more" situation.

Number of poles in Canada	plus	How many more poles	is	Number of poles in U.S.
↓	↓	↓	↓	↓
40.65	+	t	=	51.81

Solve.

$$\begin{aligned} 40.65+t &= 51.81 \\ 40.65+t-40.65 &= 51.81-40.65 \\ t &= 11.16 \end{aligned}$$

Check. Since $40.65+11.16=51.81$, the answer checks.

State. There are 11.16 more telephone poles for every 100 people in the U.S. than in Canada.

51. ***Familiarize.*** Let $h =$ Stacia's hourly wage.

Translate.

Hourly wage	times	Number of hours worked	is	Total earnings
↓	↓	↓	↓	↓
h	$\cdot$	40	$=$	620.74

Solve.

$$h \cdot 40 = 620.74$$
$$\frac{h \cdot 40}{40} = \frac{620.74}{40}$$
$$h \approx 15.52$$

Check. $40 \cdot \$15.52 = \$620.80 \approx \$620.74$, so the answer checks. (Remember, we rounded the solution of the equation.)

State. Stacia earns \$15.52 per hour.

52. ***Familiarize.*** We let $a =$ the area of grass in the yard. Recall that the area of a rectangle with length l and width w is $A = l \times w$ and the area of a circle with radius r is $A = \pi r^2$, where $\pi \approx 3.14$.

Translate. We subtract the area of the base of the fountain from the area of the yard. Recall that a circle's radius is half of its diameter, or width.

Area of yard	minus	Area of fountain	is	Area to be seeded
↓	↓	↓	↓	↓
20×15	$-$	$3.14\left(\frac{8}{2}\right)^2$	$=$	a

Solve. We carry out the computations.

$$20 \times 15 - 3.14\left(\frac{8}{2}\right)^2 = a$$
$$20 \times 15 - 3.14(4)^2 = a$$
$$20 \times 15 - 3.14(16) = a$$
$$300 - 50.24 = a$$
$$249.76 = a$$

Check. We recheck the calculations. Our answer checks.

State. The area of grass in the yard is 249.76 ft^2.

53. ***Familiarize.*** Let $a =$ the amount left in the account after the purchase was made.

Translate. We write a subtraction sentence.

$$a = 6274.35 - 485.79$$

Solve. We carry out the subtraction.

$$\begin{array}{r} 6\,2\,7\,4.\,3\,5 \\ -\quad 4\,8\,5.\,7\,9 \\ \hline 5\,7\,8\,8.\,5\,6 \end{array}$$

Thus, $a = 5788.56$.

Check. $\$5788.56 + \$485.79 = \$6274.35$, so the answer checks.

State. There is \$5788.56 left in the account.

54. ***Familiarize.*** Let $t =$ the number of transactions processed the first month. We will express 21¢ as \$0.21.

Translate.

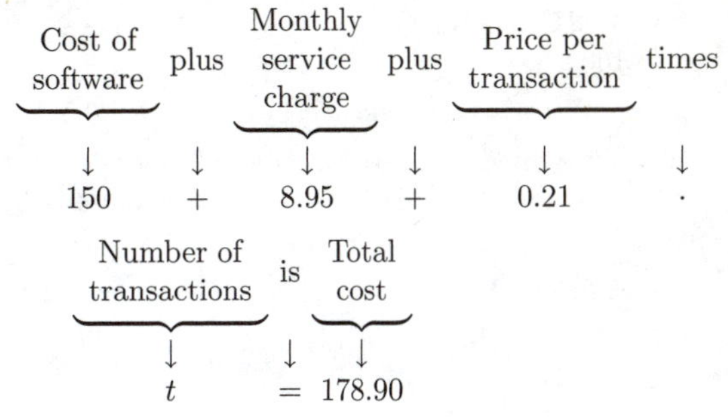

Solve.

$$150 + 8.95 + 0.21 \cdot t = 178.90$$
$$158.95 + 0.21 \cdot t = 178.90$$
$$158.95 + 0.21t - 158.95 = 178.90 - 158.95$$
$$0.21t = 19.95$$
$$\frac{0.21t}{0.21} = \frac{19.95}{0.21}$$
$$t = 95$$

Check. $\$0.21(95) = \$19.95$, and $\$150 + \$8.95 + \$19.95 = \$178.90$. The answer checks.

State. 95 transactions were processed the first month.

55. ***Familiarize.*** Let $g =$ the number of pounds of glass Lisa processed. Then $g + 130 =$ the number of pounds of newspaper she processed.

Translate.

Pounds of glass	plus	Pounds of newspaper	is	Total weight
↓	↓	↓	↓	↓
g	$+$	$g + 130$	$=$	261.4

Solve.

$$g + g + 130 = 261.4$$
$$2g + 130 = 261.4$$
$$2g + 130 - 130 = 261.4 - 130$$
$$2g = 131.4$$
$$g = 65.7$$

Then $g + 130 = 65.7 + 130 = 195.7$.

Check. 195.7 lb is 130 lb more than 65.7 lb, and 65.7 lb + 195.7 lb = 261.4 lb. The answer checks.

State. Lisa processed 65.7 lb of glass and 195.7 lb of newspaper.

56. ***Familiarize.*** This is a two-step problem. First, we find the number of miles that have been driven between fillups. This is a "how-much-more" situation. We let $n =$ the number of miles driven.

Translate and Solve.

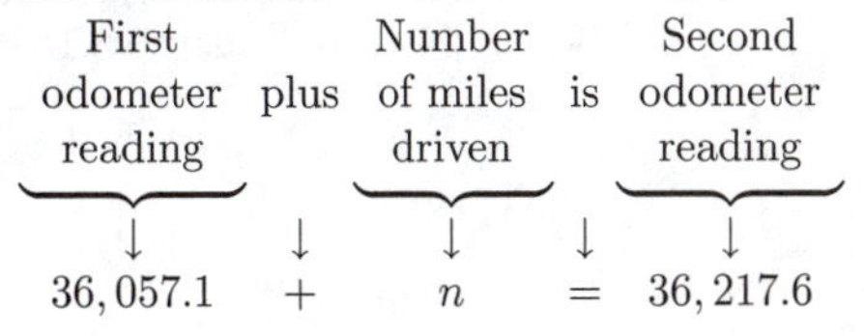

To solve the equation we subtract 36,057.1 on both sides.

$n = 36,217.6 - 36,057.1$
$n = 160.5$

$$\begin{array}{r} 3\,6,\,2\,1\,7.6 \\ -\;3\,6,\,0\,5\,7.1 \\ \hline 1\,6\,0.5 \end{array}$$

Second, we divide the total number of miles driven by the number of gallons. This gives us m = the number of miles per gallon.

$160.5 \div 11.1 = m$

To find the number m, we divide.

$$\begin{array}{r} 1\,4.4\,5 \\ 1\,1.1_\wedge \overline{)\,1\,6\,0.\,5_\wedge 0\,0} \\ 1\,1\,1\,0 \\ \hline 4\,9\,5 \\ 4\,4\,4 \\ \hline 5\,1\,0 \\ 4\,4\,4 \\ \hline 6\,6\,0 \\ 5\,5\,5 \\ \hline 1\,0\,5 \end{array}$$

Thus, $m \approx 14.5$.

Check. To check, we first multiply the number of miles per gallon times the number of gallons:

$11.1 \times 14.5 = 160.95$

Then we add 160.95 to 36,057.1:

$36,057.1 + 160.95 = 36,218.05 \approx 36,217.6$

The number 14.5 checks.

State. Inge gets 14.5 miles per gallon.

57. a) ***Familiarize.*** Let s = the total consumption of seafood per person, in pounds, for the six given years.

Translate. We add the six amounts shown in the graph in the text.

$s = 16.3 + 16.5 + 16.1 + 16.5 + 16.7 + 16.0$

Solve. We carry out the addition.

$$\begin{array}{r} {}^{3}\,{}^{2} \\ 1\,6.\,3 \\ 1\,6.\,5 \\ 1\,6.\,1 \\ 1\,6.\,5 \\ 1\,6.\,7 \\ +\,1\,6.\,0 \\ \hline 9\,8.\,1 \end{array}$$

Check. We repeat the calculation. The answer checks.

State. The total consumption of seafood per person for the six given years was 98.1 lb.

b) We add the amounts and divide by the number of addends. From part (a) we know that the sum of the six numbers is 98.1, so we have $98.1 \div 6$:

$$\begin{array}{r} 1\,6.\,3\,5 \\ 6\,\overline{)\,9\,8.\,1\,0} \\ 6 \\ \hline 3\,8 \\ 3\,6 \\ \hline 2\,1 \\ 1\,8 \\ \hline 3\,0 \\ 3\,0 \\ \hline 0 \end{array}$$

We find that the average seafood consumption per person was 16.35 lb.

58. ***Familiarize.*** Let d = the number of miles that an out-of-towner can travel for \$15.23. We will express 95¢ as \$0.95.

Translate.

Initial charge	plus	\$0.95	times	Distance traveled	is	Fare
↓	↓	↓	↓	↓	↓	↓
7.25	+	0.95	·	d	=	15.23

Solve.

$$7.25 + 0.95 \cdot d = 15.23$$
$$7.25 + 0.95 \cdot d - 7.25 = 15.23 - 7.25$$
$$0.95 \cdot d = 7.98$$
$$\frac{0.95 \cdot d}{0.95} = \frac{7.98}{0.95}$$
$$d = 8.4$$

Check. $\$0.95 \cdot 8.4 = \$7.98$ and $\$7.98 + \$7.25 = \$15.23$, so the answer checks.

State. An out-of-towner can travel 8.4 mi for \$15.23.

59. ***Familiarize.*** Let c = the cost per serving.

Translate.

Cost	divided by	Number of servings	is	Cost per serving
↓	↓	↓	↓	↓
5.99	÷	4.5	=	c

Solve. We carry out the division.

$$\begin{array}{r} 1.3\,3\,1 \\ 4.5_\wedge \overline{)\,5.9_\wedge 9\,0\,0} \\ 4\,5\,0 \\ \hline 1\,4\,9 \\ 1\,3\,5 \\ \hline 1\,4\,0 \\ 1\,3\,5 \\ \hline 5\,0 \\ 4\,5 \\ \hline 5 \end{array}$$

Rounding to the nearest cent, we have $c \approx 1.33$.

Check. We find the cost of 4.5 servings at \$1.33 per serving.

$4.5 \cdot \$1.33 = \$5.985 \approx \$5.99$

The answer checks

State. The ham costs about \$1.33 per serving.

60. ***Familiarize***. We will find the perimeter P of the room to determine how much crown molding is needed, and we will find the area A of the floor to determine how many square feet of bamboo tiles are needed. We will use the formulas $P = 2l + 2w$ and $A = l \cdot w$.

Translate.

$$P = 2 \cdot 14.5 \text{ ft} + 2 \cdot 16.25 \text{ ft}$$
$$A = 16.25 \text{ ft} \cdot 14.5 \text{ ft}$$

Solve. We carry out the calculations.

$$\begin{aligned} P &= 2 \cdot 14.5 \text{ ft} + 2 \cdot 16.25 \text{ ft} \\ &= 29 \text{ ft} + 32.5 \text{ ft} \\ &= 61.5 \text{ ft} \end{aligned}$$
$$A = 16.25 \text{ ft} \cdot 14.5 \text{ ft} = 235.625 \text{ ft}^2$$

Check. We can repeat the calculations. The answer checks.

State. 61.5 ft of crown molding and 235.625 ft^2 of bamboo tiles are needed.

61. $82.304 \div 17.287 \approx 80 \div 20 = 4$

Answer B is correct.

62. Let p = the price; we have $2p - 15$.

Answer A is correct.

63. a) By trial we find the following true sentence.

$2.56 - 6.4 + 51.2 - 17.4 + 89.7 = 119.66$

b) By trial we find the following true sentence.

$(11.12 - 0.29)3^4 = 877.23$

64. First we find decimal notation for each fraction. Then we compare these numbers in decimal notation.

$$-\frac{2}{3} = -0.\overline{6}, \; -\frac{15}{19} \approx -0.789474, \; -\frac{11}{13} = -0.\overline{846153},$$
$$\frac{-5}{7} = -0.\overline{714285}, \; \frac{-13}{15} = -0.8\overline{6}, \; \frac{-17}{20} = -0.85$$

Arranging these numbers from smallest to largest and writing them in fraction notation, we have

$$\frac{-13}{15}, \frac{-17}{20}, -\frac{11}{13}, -\frac{15}{19}, \frac{-5}{7}, -\frac{2}{3}$$

65. ***Familiarize***. Let m = the number of miles Quentin drove the car in 2010. Then $m - 10,000$ = the number of miles in excess of 10,000. At \$225 per month, the leasing cost for 1 year, or 12 months, is $12 \cdot \$225$. We will express 20 cents as \$0.20.

Translate.

Leasing cost	plus	\$0.20	times	Miles over 10,000	is	Total bill
↓	↓	↓	↓	↓	↓	↓
$12 \cdot 225$	$+$	0.20	$\cdot$	$(m - 10,000)$	$=$	5952

Solve.

$$\begin{aligned} 12 \cdot 225 + 0.20 \cdot (m - 10,000) &= 5952 \\ 2700 + 0.20 \cdot (m - 10,000) &= 5952 \\ 2700 + 0.20m - 2000 &= 5952 \\ 700 + 0.20m &= 5952 \\ 700 + 0.20m - 700 &= 5952 - 700 \\ 0.20m &= 5252 \\ \frac{0.20m}{0.20} &= \frac{5252}{0.20} \\ m &= 26,260 \end{aligned}$$

Check. If Quentin drives 26,260 mi, then he drives $26,260 - 10,000$, or 16,260 mi, in excess of 10,000. The charge for the excess miles is $\$0.20 \cdot 16,260$, or \$3252. The leasing fee is $12 \cdot \$225$, or \$2700, and $\$2700 + \$3252 = \$5952$, so the answer checks.

State. Quentin drove the car 26,260 mi in 2010.

66. Area of Sicilian pizza $= 17 \cdot 20 = 340 \text{ in}^2$

Area of round pie $\approx 3.14 \cdot \left(\frac{18}{2}\right)^2 = 254.34 \text{ in}^2$

Cost per in^2 of Sicilian pie $= \frac{\$15}{340 \text{ in}^2} \approx \frac{\$0.044}{\text{in}^2}$, or $\frac{4.4¢}{\text{in}^2}$

Cost per in^2 of round pie $= \frac{\$14}{254.34 \text{ in}^2} \approx \frac{\$0.055}{\text{in}^2}$, or $\frac{5.5¢}{\text{in}^2}$

The Sicilian pizza is a better buy.

Chapter 5 Discussion and Writing Exercises

1. Count the number of decimal places. Move the decimal point that many places to the right and write the result over a denominator of 1 followed by that many zeros.

2. $346.708 \times 0.1 = \frac{346,708}{1000} \times \frac{1}{10} = \frac{346,708}{10,000} =$

$34.6708 \neq 3467.08$

3. When the denominator of a fraction is a multiple of 10, long division is not the fastest way to convert the fraction to decimal notation. Many times when the denominator is a factor of some multiple of 10 this is also the case. The latter situation occurs when the denominator has only 2's or 5's or both as factors.

4. Multiply by 1 to get a denominator that is a power of 10:

$$\frac{44}{125} = \frac{44}{125} \cdot \frac{8}{8} = \frac{352}{1000} = 0.352.$$

We can also divide to find that $\frac{44}{125} = 0.352$.

Chapter 5 Test

1. 18.4 million
$= 18.4 \times 1$ million
$= 18.4 \times 1,000,000$ 6 zeros
$= 18,400,000$ Moving the decimal point 6 places to the right

2. 13.1 billion
$= 13.1 \times 1$ billion
$= 13.1 \times 1,000,000,000$ 9 zeros
$= 13,100,000,000$ Moving the decimal point 9 places to the right

3. 2.34

a) Write a word name for the whole number. — Two

b) Write "and" for the decimal point. — Two and

c) Write a word name for the number to the right of the decimal point, followed by the place value of the last digit. — Two and thirty-four hundredths

A word name for 2.34 is two and thirty-four hundredths.

4. 105.0005

a) Write a word name for the whole number. — One hundred five

b) Write "and" for the decimal point. — One hundred five and

c) Write a word name for the number to the right of the decimal point, followed by the place value of the last digit. — One hundred five and five ten-thousandths

A word name for 105.0005 is one hundred five and five ten-thousandths.

5. $-0.\underline{91}$ $-0.91.$ $-\dfrac{91}{100}$

2 places   Move 2 places.   2 zeros

$-0.91 = -\dfrac{91}{100}$

6. $2.\underline{769}$ $2.769.$ $\dfrac{2769}{1000}$

3 places   Move 3 places.   3 zeros

$2.769 = \dfrac{2769}{1000}$, or $2\dfrac{769}{1000}$

7. $\dfrac{74}{1000}$ $0.074.$

3 zeros   Move 3 places.

$\dfrac{74}{1000} = 0.074$

8. $-\dfrac{37,047}{10,000}$ $-3.7047.$

4 zeros   Move 4 places.

$-\dfrac{37,047}{10,000} = -3.7047$

9. $756\dfrac{9}{100} = 756 + \dfrac{9}{100} = 756$ and $\dfrac{9}{100} = 756.09$

10. $91\dfrac{703}{1000} = 91 + \dfrac{703}{1000} = 91$ and $\dfrac{703}{1000} = 91.703$

11. To compare two positive numbers in decimal notation, start at the left and compare corresponding digits moving from left to right. When two digits differ, the number with the larger digit is the larger of the two numbers.

0.07
0.162 — Different; 1 is larger than 0.

Thus, 0.162 is larger.

12. 0.078
0.06 — Starting at the left, these digits are the first to differ; 7 is larger than 6.

Thus, 0.078 is larger.

13. To compare two negative numbers in decimal notation, start at the left and compare corresponding digits moving from left to right. When two digits differ, the number with the smaller digit is the larger of the two numbers.

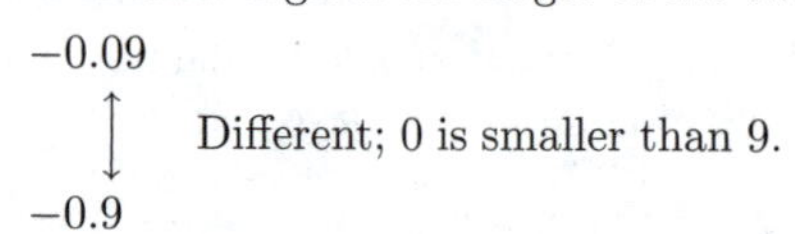

-0.09
-0.9 — Different; 0 is smaller than 9.

Thus, -0.09 is larger.

14. 5.[6]783 → 6   Tenths digit is 5 or higher. Round up.

15. 5.67[8]3 → 5.68   Thousandths digit is 5 or higher. Round up.

16. 5.678[3] → 5.678   Ten-thousandths digit is 4 or lower. Round down.

17. 5.6[7]83 — Hundredths digit is 5 or higher. Round up.

$5.6\boxed{7}83 \to 5.7$

18.

$$\begin{array}{r} \overset{1}{4}02.3 \\ 2.81 \\ +\ 0.109 \\ \hline 405.219 \end{array}$$

19.

$$\begin{array}{rl} 0.125 & \text{(3 decimal places)} \\ \times\ 0.24 & \text{(2 decimal places)} \\ \hline 500 & \\ 2500 & \\ \hline 0.03000 & \text{(5 decimal places)} \end{array}$$

20. $0.\underline{001} \times 213.45$ — $0.213.45$

3 decimal places — Move 3 places to the left.

$0.001 \times 213.45 = 0.21345$

21.

$$\begin{array}{r} 52.091 \\ -\ 7.345 \\ \hline 44.746 \end{array}$$

(borrowing: 4 11 10 8 11)

22.

$$\begin{array}{r} 342.9 \\ 8.1 \\ +\ 5.37 \\ \hline 356.37 \end{array}$$

23. $-9.5 + 7.3$

The absolute values are 9.5 and 7.3. The difference is 2.2. The negative number has the larger absolute value, so the answer is negative.

$-9.5 + 7.3 = -2.2$

24. We write extra zeros.

$$\begin{array}{r} 2.0000 \\ -0.0054 \\ \hline 1.9946 \end{array}$$

(borrowing: 1 9 9 9 10)

25. $1\underline{000} \times 73.962$ — $73.962.$

3 zeros — Move 3 places to the right.

$1000 \times 73.962 = 73,962$

26.

$$\begin{array}{r} 4.75 \\ 4\overline{)19.00} \\ 16 \\ \hline 30 \\ 28 \\ \hline 20 \\ 20 \\ \hline 0 \end{array}$$

27.

$$\begin{array}{r} 30.4 \\ 3.3_{\wedge}\overline{)100.3_{\wedge}2} \\ 9900 \\ \hline 132 \\ 132 \\ \hline 0 \end{array}$$

28. $\dfrac{-346.82}{1\underline{000}}$ — $-0.346.82$

3 zeros — Move 3 places to the left.

$\dfrac{-346.82}{1000} = -0.34682$

29. $\dfrac{346.82}{0.\underline{01}}$ — $346.82.$

2 decimal places — Move 2 places to the right.

$\dfrac{346.82}{0.01} = 34,682$

30. Move 2 places to the right.

$179.82.¢

Change from $ sign in front to ¢ sign at end.

$179.82 = 17,982¢

31.

$$\begin{aligned} & 4.1x + 5.2 - 3.9y + 5.7x - 9.8 \\ &= 4.1x + 5.2 + (-3.9y) + 5.7x + (-9.8) \\ &= 4.1x + 5.7x + (-3.9y) + 5.2 + (-9.8) \\ &= (4.1 + 5.7)x + (-3.9y) + (5.2 + (-9.8)) \\ &= 9.8x - 3.9y - 4.6 \end{aligned}$$

32.

$$\begin{aligned} 2l + 4w + 2h &= 2 \cdot 2.4 + 4 \cdot 1.3 + 2 \cdot 0.8 \\ &= 4.8 + 5.2 + 1.6 \\ &= 10.0 + 1.6 \\ &= 11.6 \end{aligned}$$

33.

$$\begin{aligned} 20 \div 5(-2)^2 - 8.4 &= 20 \div 5 \cdot 4 - 8.4 \\ &= 4 \cdot 4 - 8.4 \\ &= 16 - 8.4 \\ &= 7.6 \end{aligned}$$

34. $\dfrac{8}{5} = \dfrac{8}{5} \cdot \dfrac{2}{2} = \dfrac{16}{10} = 1.6$

35. $\dfrac{21}{4} = \dfrac{21}{4} \cdot \dfrac{25}{25} = \dfrac{525}{100} = 5.25$

36. First consider $\frac{7}{16}$.

$$\begin{array}{r} 0.4375 \\ 16\overline{)7.0000} \\ \underline{64} \\ 60 \\ \underline{48} \\ 120 \\ \underline{112} \\ 80 \\ \underline{80} \\ 0 \end{array}$$

Since $\frac{7}{16} = 0.4375$, we have $-\frac{7}{16} = -0.4375$.

37.
$$\begin{array}{r} 1.55 \\ 9\overline{)14.00} \\ \underline{9} \\ 5\,0 \\ \underline{4\,5} \\ 50 \\ \underline{45} \\ 5 \end{array}$$

Since 5 keeps reappearing as a remainder, the digit 5 repeats and

$$\frac{14}{9} = 1.55\ldots = 1.\overline{5}.$$

38. $1.55\boxed{5}5\ldots$ Thousandths digit is 5 or higher.

1.56 Round up.

39.
$$\begin{aligned} 3 \div (-0.3) \cdot 2 - 1.5^2 &= 3 \div (-0.3) \cdot 2 - 2.25 \\ &= -10 \cdot 2 - 2.25 \\ &= -20 - 2.25 \\ &= -22.25 \end{aligned}$$

40.
$$\begin{aligned} &(8 - 1.23) \div 4 + 5.6 \times 0.02 \\ &= 6.77 \div 4 + 5.6 \times 0.02 \\ &= 1.6925 + 0.112 \\ &= 1.8045 \end{aligned}$$

41.
$$\begin{aligned} &\frac{3}{8} \times 45.6 - \frac{1}{5} \times 36.9 \\ &= \frac{3 \times 45.6}{8} - \frac{36.9}{5} \\ &= \frac{136.8}{8} - \frac{36.9}{5} \\ &= 17.1 - 7.38 \\ &= 9.72 \end{aligned}$$

42.
$$\begin{aligned} 17y - 3.12 &= -58.2 \\ 17y - 3.12 + 3.12 &= -58.2 + 3.12 \\ 17y &= -55.08 \\ \frac{17y}{17} &= \frac{-55.08}{17} \\ y &= -3.24 \end{aligned}$$

The solution is -3.24.

43.
$$\begin{aligned} 9t - 4 &= 6t + 26 \\ 9t - 4 + 4 &= 6t + 26 + 4 \\ 9t &= 6t + 30 \\ 9t - 6t &= 6t + 30 - 6t \\ 3t &= 30 \\ \frac{3t}{3} &= \frac{30}{3} \\ t &= 10 \end{aligned}$$

The solution is 10.

44.
$$\begin{aligned} 4 + 2(x - 3) &= 7x - 9 \\ 4 + 2x - 6 &= 7x - 9 \\ 2x - 2 &= 7x - 9 \\ 2x - 2 + 2 &= 7x - 9 + 2 \\ 2x &= 7x - 7 \\ 2x - 7x &= 7x - 7 - 7x \\ -5x &= -7 \\ \frac{-5x}{-5} &= \frac{-7}{-5} \\ x &= 1.4 \end{aligned}$$

The solution is 1.4.

45. ***Familiarize***. This is a multistep problem. First we will find the number of minutes in excess of 1400. Then we will find the total cost of these minutes and, finally, we will find the total cell phone bill. Let m = the number of minutes in excess of 1400.

Translate and Solve.

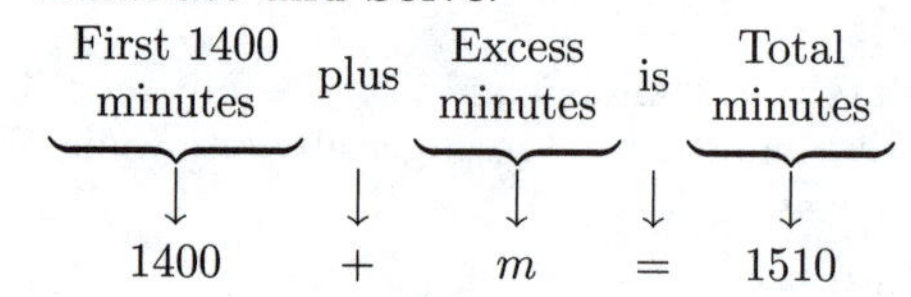

To solve the equation, we subtract 1400 on both sides.

$m = 1510 - 1400 = 110$

Next we multiply by \$0.40 to find the cost c of the 110 excess minutes.

$c = \$0.40 \cdot 110 = \$44$

Finally, we add the cost of the first 1400 minutes and the cost of the excess minutes to find the total charge, t.

$t = \$89.99 + \$44 = \$133.99$

Check. We can repeat the calculations. The answer checks.

State. The charge was \$133.99.

46. ***Familiarize***. This is a two-step problem. First we will find the number of miles that are driven between fillups. Then we find the gas mileage. Let n = the number of miles driven between fillups.

Translate and Solve.

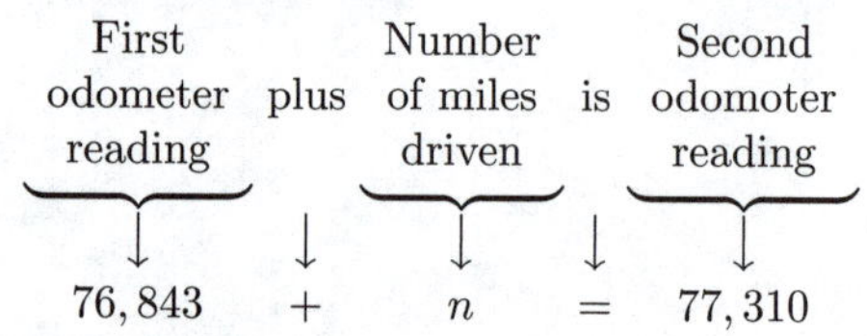

To solve the equation, we subtract 76,843 on both sides.

$n = 77,310 - 76,843 = 467$

Now let $m =$ the number of miles driven per gallon.

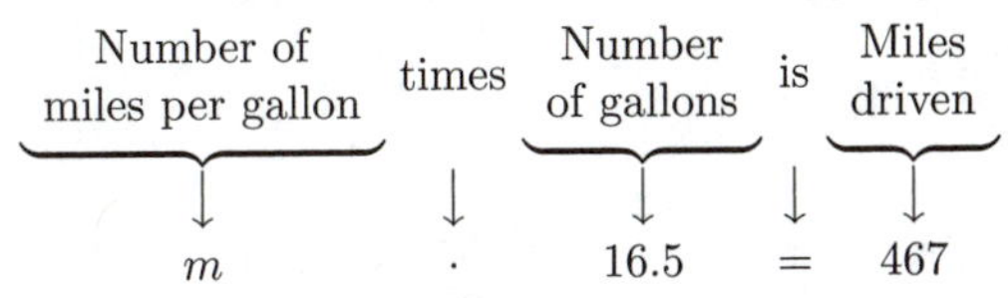

We divide by 16.5 on both sides to find m.

$m = 467 \div 16.5$

$m = 28.\overline{30}$

$m \approx 28.3$ Rounding to the nearest tenth

Check. First we multiply the number of miles per gallon by the number of gallons to find the number of miles driven:

$16.5 \cdot 28.3 = 466.95 \approx 467$

Then we add 467 mi to the first odometer reading:

$76,843 + 467 = 77,310$

This is the second odometer reading, so the answer checks.

State. The gas mileage is about 28.3 miles per gallon.

47. ***Familiarize.*** Let $b =$ the balance after the purchases are made.

Translate. We subtract the amounts of the three purchases from the original balance:

$b = 820 - 123.89 - 56.68 - 46.98$

Solve. We carry out the calculations.

$$\begin{aligned} b &= 820 - 123.89 - 56.68 - 46.98 \\ &= 696.11 - 56.68 - 46.98 \\ &= 639.43 - 46.98 \\ &= 592.45 \end{aligned}$$

Check. We can find the total amount of the purchases and then subtract to find the new balance.

$\$123.89 + \$56.68 + \$46.98 = \$227.55$

$\$820 - \$227.55 = \$592.45$

The answer checks.

State. After the purchases were made, the balance was $592.45.

48. ***Familiarize.*** Let $c =$ the total cost of the copy paper.

Translate.

Cost per case	times	Number of cases	is	Total cost
↓	↓	↓	↓	↓
41.99	·	7	=	c

Solve. We carry out the multiplication.

$$\begin{array}{rl} 4\,1.9\,9 & \text{(2 decimal places)} \\ \times \quad\; 7 & \\ \hline 2\,9\,3.9\,3 & \text{(2 decimal places)} \end{array}$$

Thus, $c = 293.93$.

Check. We can obtain a partial check by rounding and estimating:

$41.99 \times 7 \approx 40 \times 7 = 280 \approx 293.93$

State. The total cost of the copy paper is $293.93.

49. We add the numbers and divide by the number of addends.

$$\frac{3.45 + 2.75 + 2.75 + 2.70 + 2.67}{5} = \frac{14.32}{5} = 2.864$$

The average number of passengers is 2.864 million.

50. $20 \div 2.749 \approx 20 \div 3 = 6.\overline{6} \approx 7$

Answer C is correct.

51. a) The product of two numbers greater than 0 and less than 1 is always less than 1.

b) The product of two numbers greater than 1 is never less than 1.

c) The product of a number greater than 1 and a number less than 1 is sometimes equal to 1.

d) The product of a number greater than 1 and a number less than 1 is sometimes equal to 0.

52. ***Familiarize.*** This is a two-step problem. First we will find the cost of membership for six months without the coupon. Let $c =$ this cost. Then we will find how much Allise will save if she uses the coupon.

Translate and Solve.

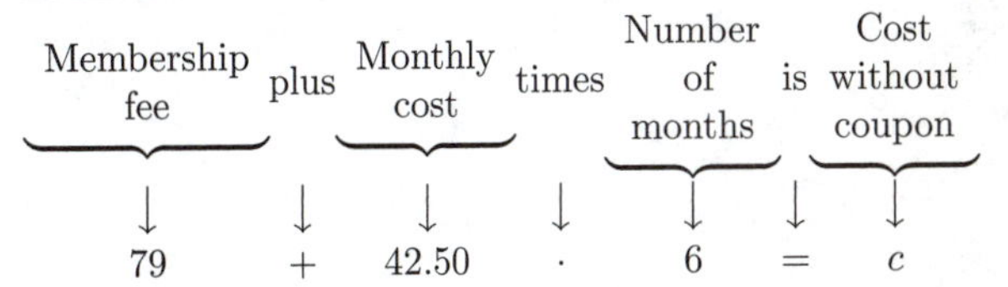

To solve the equation, we carry out the calculation.

$c = 79 + 42.50 \cdot 6 = 79 + 255 = 334$

Now let $s =$ the coupon savings.

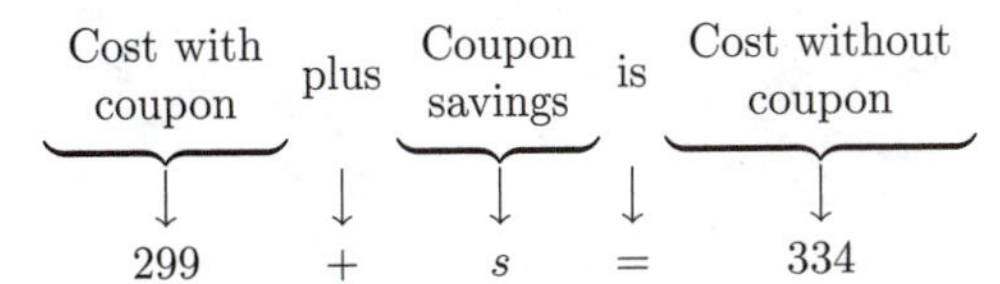

We subtract 299 on both sides.

$s = 334 - 299 = 35$

Check. We repeat the calculations. The answer checks.

State. Allise will save $35 if she uses the coupon.

53. The cost to drive roundtrip is $\$1.35 \cdot 600$, or $810.

a) Since $\$229 < \$810$, it is more economical for an individual to fly.

b) The airfare for a couple is $2 \cdot \$229$, or $458. Since $\$458 < \$810$, it is more economical for a couple to fly.

c) The airfare for a family of 4 is $4 \cdot \$229$, or $916. Since $\$916 > \$810$, it is more economical for a family of 4 to drive.

Chapter 6

Percent Notation

Exercise Set 6.1

1. The ratio of 178 to 572 is $\frac{178}{572}$.

3. The ratio of $8\frac{3}{4}$ to $9\frac{5}{6}$ is $\frac{8\frac{3}{4}}{9\frac{5}{6}}$.

5. The ratio of the time of the current trip to the time of the trip on the space plane is $\frac{21}{4}$.

The ratio of the time of the trip on the space plane to the time of the current trip is $\frac{4}{21}$.

7. The ratio of 4 to 6 is $\frac{4}{6} = \frac{2 \cdot 2}{2 \cdot 3} = \frac{2}{2} \cdot \frac{2}{3} = \frac{2}{3}$.

9. The ratio of 2.8 to 3.6 is $\frac{2.8}{3.6} = \frac{2.8}{3.6} \cdot \frac{10}{10} = \frac{28}{36} = \frac{4 \cdot 7}{4 \cdot 9} = \frac{4}{4} \cdot \frac{7}{9} = \frac{7}{9}$.

11. The ratio of length to width is $\frac{478}{213}$.

The ratio of width to length is $\frac{213}{478}$.

13. $\frac{120 \text{ km}}{3 \text{ hr}} = 40 \frac{\text{km}}{\text{hr}}$

15. $\frac{217 \text{ mi}}{29 \text{ sec}} \approx 7.48 \frac{\text{mi}}{\text{sec}}$

17. $\frac{624 \text{ mi}}{19.5 \text{ gal}} = 32 \text{ mpg}$

19. $\frac{32{,}796 \text{ people}}{0.75 \text{ sq mi}} = 43{,}728 \text{ people/sq mi}$

21. $\frac{186{,}000 \text{ mi}}{1 \text{ sec}} = 186{,}000 \frac{\text{mi}}{\text{sec}}$

23. $\frac{623 \text{ gal}}{1000 \text{ sq ft}} = 0.623 \text{ gal/ft}^2$

25. $\frac{310 \text{ km}}{2.5 \text{ hr}} = 124 \frac{\text{km}}{\text{hr}}$

27. We can use cross products:

$5 \cdot 9 = 45$ (5, 7 / 6, 9) $6 \cdot 7 = 42$

Since the cross products are not the same, $45 \neq 42$, we know that the numbers are not proportional.

29. We can use cross products:

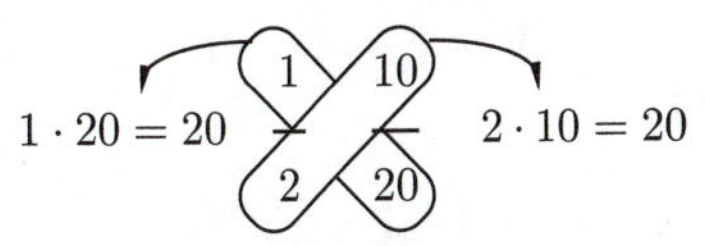

Since the cross products are the same, $20 = 20$, we know that $\frac{1}{2} = \frac{10}{20}$, so the numbers are proportional.

31. We can use cross products:

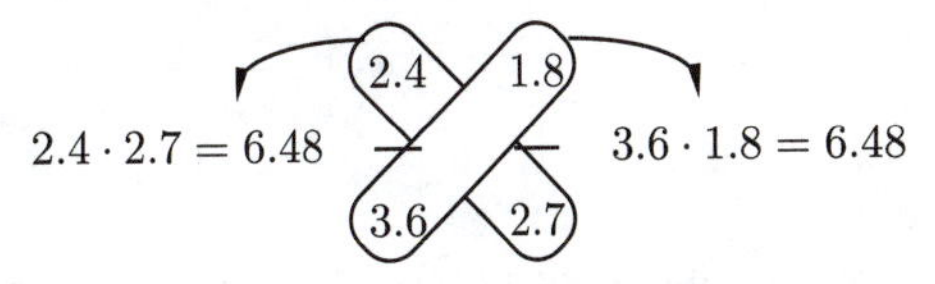

Since the cross products are the same, $6.48 = 6.48$, we know that $\frac{2.4}{3.6} = \frac{1.8}{2.7}$, so the numbers are proportional.

33. We can use cross products:

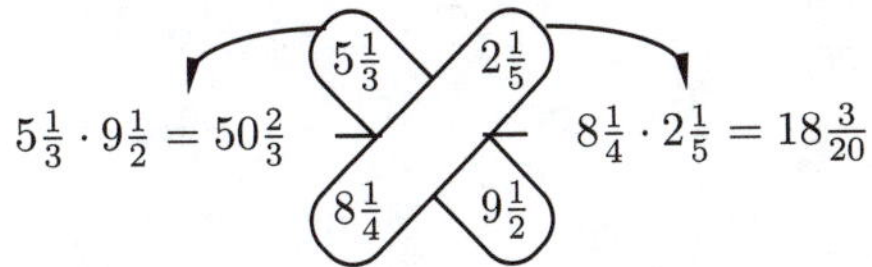

Since the cross products are not the same, $50\frac{2}{3} \neq 18\frac{3}{20}$, we know that the numbers are not proportional.

35.
$$\begin{aligned}
\frac{x}{8} &= \frac{9}{6} \\
6 \cdot x &= 8 \cdot 9 && \text{Equating cross products} \\
\frac{6 \cdot x}{6} &= \frac{8 \cdot 9}{6} && \text{Dividing by 6} \\
x &= \frac{8 \cdot 9}{6} \\
x &= \frac{72}{6} && \text{Multiplying} \\
x &= 12 && \text{Dividing}
\end{aligned}$$

37.
$$\begin{aligned}
\frac{2}{5} &= \frac{8}{n} \\
2 \cdot n &= 5 \cdot 8 \\
\frac{2 \cdot n}{2} &= \frac{5 \cdot 8}{2} \\
n &= \frac{5 \cdot 8}{2} \\
n &= \frac{40}{2} \\
n &= 20
\end{aligned}$$

39. $\dfrac{16}{12} = \dfrac{24}{x}$

$16 \cdot x = 12 \cdot 24$

$\dfrac{16 \cdot x}{16} = \dfrac{12 \cdot 24}{16}$

$x = \dfrac{12 \cdot 24}{16}$

$x = \dfrac{288}{16}$

$x = 18$

41. $\dfrac{12}{9} = \dfrac{x}{7}$

$12 \cdot 7 = 9 \cdot x$

$\dfrac{12 \cdot 7}{9} = \dfrac{9 \cdot x}{9}$

$\dfrac{12 \cdot 7}{9} = x$

$\dfrac{84}{9} = x$

$\dfrac{28}{3} = x$ Simplifying

$9\dfrac{1}{3} = x$ Writing a mixed numeral

43. $\dfrac{1.2}{4} = \dfrac{x}{9}$

$1.2 \cdot 9 = 4 \cdot x$

$\dfrac{1.2 \cdot 9}{4} = \dfrac{4 \cdot x}{4}$

$\dfrac{1.2 \cdot 9}{4} = x$

$\dfrac{10.8}{4} = x$

$2.7 = x$

45. $\dfrac{8}{2.4} = \dfrac{6}{y}$

$8 \cdot y = 2.4 \cdot 6$

$\dfrac{8 \cdot y}{8} = \dfrac{2.4 \cdot 6}{8}$

$y = \dfrac{2.4 \cdot 6}{8}$

$y = \dfrac{14.4}{8}$

$y = 1.8$

47. $\dfrac{\frac{y}{\frac{3}{5}}}{} = \dfrac{\frac{7}{12}}{\frac{14}{15}}$

$y \cdot \dfrac{14}{15} = \dfrac{3}{5} \cdot \dfrac{7}{12}$

$y = \dfrac{15}{14} \cdot \dfrac{3}{5} \cdot \dfrac{7}{12}$ Dividing by $\dfrac{14}{15}$

$y = \dfrac{15 \cdot 3 \cdot 7}{14 \cdot 5 \cdot 12} = \dfrac{\cancel{3} \cdot \cancel{5} \cdot 3 \cdot \cancel{7}}{2 \cdot \cancel{7} \cdot \cancel{5} \cdot \cancel{3} \cdot 4}$

$y = \dfrac{3}{8}$

49. $\dfrac{x}{1\frac{3}{5}} = \dfrac{2}{15}$

$x \cdot 15 = 1\dfrac{3}{5} \cdot 2$

$x \cdot 15 = \dfrac{8}{5} \cdot 2$

$x = \dfrac{1}{15} \cdot \dfrac{8}{5} \cdot 2$ Dividing by 15

$x = \dfrac{16}{75}$

51. $\dfrac{0.5}{n} = \dfrac{2.5}{3.5}$

$0.5 \cdot 3.5 = n \cdot 2.5$

$\dfrac{0.5 \cdot 3.5}{2.5} = \dfrac{n \cdot 2.5}{2.5}$

$\dfrac{0.5 \cdot 3.5}{2.5} = n$

$\dfrac{1.75}{2.5} = n$

$0.7 = n$

53. $\dfrac{\frac{1}{5}}{\frac{1}{10}} = \dfrac{\frac{1}{10}}{x}$

$\dfrac{1}{5} \cdot x = \dfrac{1}{10} \cdot \dfrac{1}{10}$

$x = \dfrac{5}{1} \cdot \dfrac{1}{10} \cdot \dfrac{1}{10}$ Dividing by $\dfrac{1}{5}$

$x = \dfrac{5 \cdot 1 \cdot 1}{1 \cdot 10 \cdot 10} = \dfrac{\cancel{5} \cdot 1}{2 \cdot \cancel{5} \cdot 10}$

$x = \dfrac{1}{20}$

55. ***Familiarize.*** Let $d =$ the number of defective bulbs in a lot of 2500.

Translate. We translate to a proportion.

$$\begin{array}{rcl}\text{Defective bulbs} \rightarrow & \dfrac{7}{100} = \dfrac{d}{2500} & \leftarrow \text{Defective bulbs} \\ \text{Bulbs in lot} \rightarrow & & \leftarrow \text{Bulbs in lot}\end{array}$$

Solve.

$7 \cdot 2500 = 100 \cdot d$

$\dfrac{7 \cdot 2500}{100} = d$

$\dfrac{7 \cdot 25 \cdot 100}{100} = d$

$7 \cdot 25 = d$

$175 = d$

Check. We substitute in the proportion and check cross products.

$$\frac{7}{100} = \frac{175}{2500}$$

$7 \cdot 2500 = 17,500$; $100 \cdot 175 = 17,500$

The cross products are the same, so the answer checks.

State. There will be 175 defective bulbs in a lot of 2500.

57. ***Familiarize.*** Let D = the number of deer in the game preserve.

Translate. We translate to a proportion.

$$\begin{array}{ccc} \text{Deer tagged originally} & \rightarrow 318 \quad 56 \leftarrow & \text{Tagged deer caught later} \\ \text{Deer in game preserve} & \rightarrow D \quad 168 \leftarrow & \text{Deer caught later} \end{array}$$

$$\frac{318}{D} = \frac{56}{168}$$

Solve.

$$318 \cdot 168 = 56 \cdot D$$
$$\frac{318 \cdot 168}{56} = D$$
$$954 = D$$

Check. We substitute in the proportion and check cross products.

$$\frac{318}{954} = \frac{56}{168};\ 318 \cdot 168 = 53,424;\ 954 \cdot 56 = 53,424$$

Since the cross products are the same, the answer checks.

State. We estimate that there are 954 deer in the game preserve.

59. ***Familiarize.*** Let z = the number of pounds of zinc in the alloy.

Translate. We translate to a proportion.

$$\begin{array}{ccc} \text{Zinc} \rightarrow & 3 \quad\quad z & \leftarrow \text{Zinc} \\ \text{Copper} \rightarrow & 13 \quad 520 & \leftarrow \text{Copper} \end{array}$$

$$\frac{3}{13} = \frac{z}{520}$$

Solve.

$$3 \cdot 520 = 13 \cdot z$$
$$\frac{3 \cdot 520}{13} = z$$
$$\frac{3 \cdot 13 \cdot 40}{13} = z$$
$$3 \cdot 40 = z$$
$$120 = z$$

Check. We substitute in the proportion and check cross products.

$$\frac{3}{13} = \frac{120}{520}$$

$$3 \cdot 520 = 1560;\ 13 \cdot 120 = 1560$$

The cross products are the same, so the answer checks.

State. There are 120 lb of zinc in the alloy.

61. ***Familiarize.*** Let s = the number of ounces of grass seed needed for 5000 ft^2 of lawn.

Translate. We translate to a proportion.

$$\begin{array}{ccc} \text{Seed} \rightarrow & 60 \quad\quad s & \leftarrow \text{Seed} \\ \text{Area} \rightarrow & 3000 \quad 5000 & \leftarrow \text{Area} \end{array}$$

$$\frac{60}{3000} = \frac{s}{5000}$$

Solve.

$$60 \cdot 5000 = 3000 \cdot s$$
$$\frac{60 \cdot 5000}{3000} = s$$
$$100 = s$$

Check. We find the number of ounces of seed needed for 1 ft^2 of lawn and then multiply this number by 5000:

$$60 \div 3000 = 0.02 \text{ and } 5000(0.02) = 100$$

The answer checks.

State. 100 oz of grass seed would be needed to seed 5000 ft^2 of lawn.

63. ***Familiarize.*** Let a = the number of Americans who would be considered overweight or obese in 2015, in millions.

Translate. We translate to a proportion.

$$\begin{array}{ccc} \text{Overweight/obese} \rightarrow & 66 \quad\quad a & \leftarrow \text{Overweight/obese} \\ \text{Total Americans} \rightarrow & 100 \quad 322 & \leftarrow \text{Total Americans} \end{array}$$

$$\frac{66}{100} = \frac{a}{322}$$

Solve.

$$66 \cdot 322 = 100 \cdot a \quad \text{Equating cross products}$$
$$\frac{66 \cdot 322}{100} = a$$
$$212.52 = a$$

Check. We substitute in the proportion and check cross products.

$$\frac{66}{100} = \frac{212.52}{322}$$

$$66 \cdot 322 = 21,252;\ 100 \cdot 212.52 = 21,252$$

The cross products are the same, so the answer checks.

State. At the given rate, 212.52 million, or 212,520,000 Americans would be overweight or obese in 2015.

65. a) ***Familiarize.*** Let g = the number of gallons of gasoline needed to drive 2690 mi.

Translate. We translate to a proportion.

$$\begin{array}{ccc} \text{Gallons} \rightarrow & 15.5 \quad\quad g & \leftarrow \text{Gallons} \\ \text{Miles} \rightarrow & 341 \quad 2690 & \leftarrow \text{Miles} \end{array}$$

$$\frac{15.5}{341} = \frac{g}{2690}$$

Solve.

$$15.5 \cdot 2690 = 341 \cdot g \quad \text{Equating cross products}$$
$$\frac{15.5 \cdot 2690}{341} = g$$
$$122 \approx g$$

Check. We find how far the car can be driven on 1 gallon of gasoline and then divide to find the number of gallons required for a 2690-mi trip.

$$341 \div 15.5 = 22 \text{ and } 2690 \div 22 \approx 122$$

The answer checks.

State. It will take about 122 gal of gasoline to drive 2690 mi.

b) ***Familiarize.*** Let d = the number of miles the car can be driven on 140 gal of gasoline.

Translate. We translate to a proportion.

$$\begin{array}{ccc} \text{Gallons} \rightarrow & 15.5 \quad 140 & \leftarrow \text{Gallons} \\ \text{Miles} \rightarrow & 341 \quad\quad d & \leftarrow \text{Miles} \end{array}$$

$$\frac{15.5}{341} = \frac{140}{d}$$

Solve.

$$15.5 \cdot d = 341 \cdot 140 \quad \text{Equating cross products}$$
$$d = \frac{341 \cdot 140}{15.5}$$
$$d = 3080$$

Check. From the check in part (a) we know that the car can be driven 22 mi on 1 gal of gasoline. We multiply to find how far it can be driven on 140 gal.

$140 \cdot 22 = 3080$

The answer checks.

State. The car can be driven 3080 mi on 140 gal of gasoline.

67. ***Familiarize***. Let $c =$ the number of calories in 6 cups of cereal.

Translate. We translate to a proportion, keeping the number of calories in the numerators.

$$\begin{array}{l} \text{Calories} \rightarrow \\ \text{Cups} \rightarrow \end{array} \frac{110}{3/4} = \frac{c}{6} \begin{array}{l} \leftarrow \text{Calories} \\ \leftarrow \text{Cups} \end{array}$$

Solve. We solve the proportion.

$$110 \cdot 6 = \frac{3}{4} \cdot c \quad \text{Equating cross products}$$

$$\frac{110 \cdot 6}{3/4} = \frac{\frac{3}{4} \cdot c}{3/4}$$

$$\frac{110 \cdot 6}{3/4} = c$$

$$110 \cdot 6 \cdot \frac{4}{3} = c$$

$$880 = c$$

Check. We substitute into the proportion and check cross products.

$$\frac{110}{3/4} = \frac{880}{6}$$

$$110 \cdot 6 = 660; \frac{3}{4} \cdot 880 = 660$$

The cross products are the same, so the answer checks.

State. There are 880 calories in 6 cups of cereal.

69. ***Familiarize***. Let $d =$ the actual distance between the cities.

Translate. We translate to a proportion.

$$\begin{array}{l} \text{Map distance} \rightarrow \\ \text{Actual distance} \rightarrow \end{array} \frac{1}{16.6} = \frac{3.5}{d} \begin{array}{l} \leftarrow \text{Map distance} \\ \leftarrow \text{Actual distance} \end{array}$$

Solve.

$$1 \cdot d = 16.6 \cdot 3.5$$

$$d = 58.1$$

Check. We use a different approach. Since 1 in. represents 16.6 mi, we multiply 16.6 by 3.5:

$3.5(16.6) = 58.1$

The answer checks.

State. The cities are 58.1 mi apart.

71. ***Familiarize***. Let $p =$ the number of gallons of paint Helen should buy.

Translate. We translate to a proportion.

$$\begin{array}{l} \text{Area} \rightarrow \\ \text{Paint} \rightarrow \end{array} \frac{950}{2} = \frac{30,000}{p} \begin{array}{l} \leftarrow \text{Area} \\ \leftarrow \text{Paint} \end{array}$$

Solve.

$$950 \cdot p = 2 \cdot 30,000$$

$$p = \frac{2 \cdot 30,000}{950}$$

$$p = \frac{2 \cdot 50 \cdot 600}{19 \cdot 50}$$

$$p = \frac{2 \cdot 600}{19}$$

$$p = \frac{1200}{19}, \text{ or } 63\frac{3}{19}$$

Check. We find the area covered by 1 gal of paint and then divide to find the number of gallons needed to paint a 30,000-ft^2 wall.

$$950 \div 2 = 475 \text{ and } 30,000 \div 475 = 63\frac{3}{19}$$

The answer checks.

State. Since Helen is buying paint in one gallon cans, she will have to buy 64 cans of paint.

73. ***Familiarize***. Let $t =$ the number of drive-in movie theaters in 1958.

Translate.

Number of theaters in 2007	plus	Excess number of theaters in 1958	is	Number of theaters in 1958
↓	↓	↓	↓	↓
405	+	3658	=	t

Solve. We carry out the addition.

$$\begin{array}{r} {}^{1}\ {}^{1}\quad\ \ \\ 4\,0\,5 \\ +\,3\,6\,5\,8 \\ \hline 4\,0\,6\,3 \end{array}$$

Thus, $4063 = t$.

Check. We can repeat the calculation. We can also do a partial check by rounding to the nearest hundred.

$405 + 3658 \approx 400 + 3700 = 4100 \approx 4063.$

Since 4100 is close to 4063, the answer is reasonable.

State. There were 4063 drive-in movie theaters in 1958.

75. ***Familiarize***. Let $p =$ the price of the home in Austin.

Translate. We translate to a proportion.

$$\begin{array}{l} \text{Price in Austin} \rightarrow \\ \text{Price in Denver} \rightarrow \end{array} \frac{189,000}{437,850} = \frac{p}{350,000} \begin{array}{l} \leftarrow \text{Price in Austin} \\ \leftarrow \text{Price in Denver} \end{array}$$

Solve.

$$189,000 \cdot 350,000 = 437,850 \cdot p$$

$$\frac{189,000 \cdot 350,000}{437,850} = p$$

$$151,079 \approx p \quad \text{Rounding to the nearest one}$$

Check. We substitute in the proportion and check cross products.

$$\frac{189,000}{437,850} = \frac{151,079}{350,000}$$

$$189,000 \cdot 350,000 = 66,150,000,000;$$

$$437,850 \cdot 151,079 \approx 66,150,000,000$$

The cross products are approximately the same. (Remember that we rounded the value of p.) The answer checks.

State. The price of the home in Austin would be about \$151,000 (rounded to the nearest thousand).

Exercise Set 6.2

1. $90\% = \frac{90}{100}$ A ratio of 90 to 100

$90\% = 90 \times \frac{1}{100}$ Replacing % with $\times \frac{1}{100}$

$90\% = 90 \times 0.01$ Replacing % with $\times 0.01$

3. $12.5\% = \frac{12.5}{100}$ A ratio of 12.5 to 100

$12.5\% = 12.5 \times \frac{1}{100}$ Replacing % with $\times \frac{1}{100}$

$12.5\% = 12.5 \times 0.01$ Replacing % with $\times 0.01$

5. 67%

a) Replace the percent symbol with ×0.01.

67×0.01

b) Move the decimal point two places to the left.

0.67.

Thus, 67% = 0.67.

7. 45.6%

a) Replace the percent symbol with ×0.01.

45.6×0.01

b) Move the decimal point two places to the left.

0.45.6

Thus, 45.6% = 0.456.

9. 59.01%

a) Replace the percent symbol with ×0.01.

59.01×0.01

b) Move the decimal point two places to the left.

0.59.01

Thus, 59.01% = 0.5901.

11. 10%

a) Replace the percent symbol with ×0.01.

10×0.01

b) Move the decimal point two places to the left.

0.10.

Thus, 10% = 0.1.

13. 1%

a) Replace the percent symbol with ×0.01.

1×0.01

b) Move the decimal point two places to the left.

0.01.

Thus, 1% = 0.01.

15. 200%

a) Replace the percent symbol with ×0.01.

200×0.01

b) Move the decimal point two places to the left.

2.00.

Thus, 200% = 2.

17. 0.1%

a) Replace the percent symbol with ×0.01.

0.1×0.01

b) Move the decimal point two places to the left.

0.00.1

Thus, 0.1% = 0.001.

19. 0.09%

a) Replace the percent symbol with ×0.01.

0.09×0.01

b) Move the decimal point two places to the left.

0.00.09

Thus, 0.09% = 0.0009.

21. 0.18%

a) Replace the percent symbol with ×0.01.

0.18×0.01

b) Move the decimal point two places to the left.

0.00.18

Thus, 0.18% = 0.0018.

23. 23.19%

a) Replace the percent symbol with ×0.01.

23.19×0.01

b) Move the decimal point two places to the left.

0.23.19

Thus, 23.19% = 0.2319.

25. $14\frac{7}{8}\%$

a) Convert $14\frac{7}{8}$ to decimal notation and replace the percent symbol with $\times 0.01$.

14.875×0.01

b) Move the decimal point two places to the left.

0.14.875

Thus, $14\frac{7}{8}\% = 0.14875$.

27. $56\frac{1}{2}\%$

a) Convert $56\frac{1}{2}$ to decimal notation and replace the percent symbol with $\times 0.01$.

56.5×0.01

b) Move the decimal point two places to the left.

0.56.5

Thus, $56\frac{1}{2}\% = 0.565$.

29. 97%

a) Replace the percent symbol with $\times 0.01$.

97×0.01

b) Move the decimal point two places to the left.

0.97.

Thus, 97% = 0.97.

58%

a) Replace the percent symbol with $\times 0.01$.

58×0.01

b) Move the decimal point two places to the left.

0.58.

Thus, 58% = 0.58.

31. 7%

a) Replace the percent symbol with $\times 0.01$.

7×0.01

b) Move the decimal point two places to the left.

0.07.

Thus, 7% = 0.07.

8%

a) Replace the percent symbol with $\times 0.01$.

8×0.01

b) Move the decimal point two places to the left.

0.08.

Thus, 8% = 0.08.

33. 54.8%

a) Replace the percent symbol with $\times 0.01$.

54.8×0.01

b) Move the decimal point two places to the left.

0.54.8

Thus, 54.8% = 0.548.

35. 0.47

a) Move the decimal point two places to the right.

0.47.

b) Write a percent symbol: 47%

Thus, 0.47 = 47%.

37. 0.03

a) Move the decimal point two places to the right.

0.03.

b) Write a percent symbol: 3%

Thus, 0.03 = 3%.

39. 8.7

a) Move the decimal point two places to the right.

8.70.

b) Write a percent symbol: 870%

Thus, 8.7 = 870%.

41. 0.334

a) Move the decimal point two places to the right.

0.33.4

b) Write a percent symbol: 33.4%

Thus, 0.334 = 33.4%.

43. 0.75

a) Move the decimal point two places to the right.

0.75.

b) Write a percent symbol: 75%

Thus, 0.75 = 75%.

45. 0.4

a) Move the decimal point two places to the right.

0.40.

b) Write a percent symbol: 40%

Thus, 0.4 = 40%.

47. 0.006

a) Move the decimal point two places to the right.

0.00.6

b) Write a percent symbol: 0.6%

Thus, 0.006 = 0.6%.

49. 0.017

a) Move the decimal point two places to the right.

0.01.7

b) Write a percent symbol: 1.7%

Thus, 0.017 = 1.7%.

51. 0.2718

a) Move the decimal point two places to the right.

0.27.18

b) Write a percent symbol: 27.18%

Thus, 0.2718 = 27.18%.

53. 0.0239

a) Move the decimal point two places to the right.

0.02.39

b) Write a percent symbol: 2.39%

Thus, 0.0239 = 2.39%.

55. 0.27

a) Move the decimal point two places to the right.

0.27.

b) Write a percent symbol: 27%

Thus, 0.27 = 27%.

57. 0.057

a) Move the decimal point two places to the right.

0.05.7

b) Write a percent symbol: 5.7%

Thus, 0.057 = 5.7%.

0.176

a) Move the decimal point two places to the right.

0.17.6

b) Write a percent symbol: 17.6%

Thus, 0.176 = 17.6%.

59. 0.906

a) Move the decimal point two places to the right.

0.90.6

b) Write a percent symbol: 90.6%

Thus, 0.906 = 90.6%.

0.88

a) Move the decimal point two places to the right.

0.88.

b) Write a percent symbol: 88%

Thus, 0.88 = 88%.

61. 64%

a) Replace the percent symbol with $\times 0.01$.

64×0.01

b) Move the decimal point two places to the left.

0.64.

Thus, 64% = 0.64.

30%

a) Replace the percent symbol with $\times 0.01$.

30×0.01

b) Move the decimal point two places to the left.

0.30.

Thus, 30% = 0.3.

4%

a) Replace the percent symbol with $\times 0.01$.

4×0.01

b) Move the decimal point two places to the left.

0.04.

Thus, 4% = 0.04.

2%

a) Replace the percent symbol with $\times 0.01$.

2×0.01

b) Move the decimal point two places to the left.

0.02.

Thus, 2% = 0.02.

63. To convert $\dfrac{100}{3}$ to a mixed numeral, we divide.

$$\begin{array}{r} 33 \\ 3\overline{)100} \\ \underline{90} \\ 10 \\ \underline{9} \\ 1 \end{array} \qquad \frac{100}{3} = 33\frac{1}{3}$$

65. To convert $\dfrac{75}{8}$ to a mixed numeral, we divide.

$$\begin{array}{r} 9 \\ 8\overline{)75} \\ \underline{72} \\ 3 \end{array} \qquad \frac{75}{8} = 9\frac{3}{8}$$

67. To convert $\dfrac{567}{98}$ to a mixed numeral, we divide.

$$\begin{array}{r} 5 \\ 98\overline{)567} \\ \underline{490} \\ 77 \end{array} \qquad \frac{567}{98} = 5\frac{77}{98} = 5\frac{11}{14}$$

69. To convert $\dfrac{2}{3}$ to decimal notation, we divide.

$$\begin{array}{r} 0.66 \\ 3\overline{)2.00} \\ \underline{18} \\ 20 \\ \underline{18} \\ 2 \end{array}$$

Since 2 keeps reappearing as a remainder, the digits repeat and

$$\frac{2}{3} = 0.66\ldots \quad \text{or} \quad 0.\overline{6}.$$

71. To convert $\dfrac{5}{6}$ to decimal notation, we divide.

$$\begin{array}{r} 0.83 \\ 6\overline{)5.00} \\ \underline{48} \\ 20 \\ \underline{18} \\ 2 \end{array}$$

Since 2 keeps reappearing as a remainder, the digits repeat and

$$\frac{5}{6} = 0.833\ldots \quad \text{or} \quad 0.8\overline{3}.$$

73. To convert $\dfrac{8}{3}$ to decimal notation, we divide.

$$\begin{array}{r} 2.66 \\ 3\overline{)8.00} \\ \underline{6} \\ 20 \\ \underline{18} \\ 20 \end{array}$$

Since 2 keeps reappearing as a remainder, the digits repeat and

$$\frac{8}{3} = 2.66\ldots \text{ or } 2.\overline{6}.$$

75. $\dfrac{1}{2} = \dfrac{1}{2} \cdot \dfrac{50}{50} = \dfrac{50}{100} = 50\%$

77. $\dfrac{7}{10} = \dfrac{7}{10} \cdot \dfrac{10}{10} = \dfrac{70}{100} = 70\%$

79. One of the five equal-sized portions of the figure is shaded, so $\dfrac{1}{5}$ is shaded. We find percent notation for $\dfrac{1}{5}$.

$$\frac{1}{5} = \frac{1}{5} \cdot \frac{20}{20} = \frac{20}{100} = 20\%$$

Thus, 20% is shaded.

Exercise Set 6.3

1. We use the definition of percent as a ratio.

$$\frac{41}{100} = 41\%$$

3. We use the definition of percent as a ratio.

$$\frac{5}{100} = 5\%$$

5. We multiply by 1 to get 100 in the denominator.

$$\frac{2}{10} = \frac{2}{10} \cdot \frac{10}{10} = \frac{20}{100} = 20\%$$

7. We multiply by 1 to get 100 in the denominator.

$$\frac{3}{10} = \frac{3}{10} \cdot \frac{10}{10} = \frac{30}{100} = 30\%$$

9. $\dfrac{1}{2} = \dfrac{1}{2} \cdot \dfrac{50}{50} = \dfrac{50}{100} = 50\%$

11. Find decimal notation by division.

$$\begin{array}{r} 0.875 \\ 8\overline{)7.000} \\ \underline{64} \\ 60 \\ \underline{56} \\ 40 \\ \underline{40} \\ 0 \end{array}$$

$$\frac{7}{8} = 0.875$$

Convert to percent notation.

0.87.5

$$\frac{7}{8} = 87.5\%, \text{or } 87\frac{1}{2}\%$$

13. $\dfrac{4}{5} = \dfrac{4}{5} \cdot \dfrac{20}{20} = \dfrac{80}{100} = 80\%$

15. First find decimal notation by division.

$$\begin{array}{r} 0.666 \\ 3\overline{)2.000} \\ \underline{18} \\ 20 \\ \underline{18} \\ 20 \\ \underline{18} \\ 2 \end{array}$$

We get a repeating decimal: $\frac{2}{3} = 0.66\overline{6}$

Convert to percent notation.

$0.66.\overline{6}$

$\frac{2}{3} = 66.\overline{6}\%$, or $66\frac{2}{3}\%$

17.
$$\begin{array}{r} 0.1\,6\,6\\ 6\,\overline{)1.0\,0\,0}\\ \underline{6}\\ 4\,0\\ \underline{3\,6}\\ 4\,0\\ \underline{3\,6}\\ 4\end{array}$$

We get a repeating decimal: $\frac{1}{6} = 0.16\overline{6}$

Convert to percent notation.

$0.16.\overline{6}$

$\frac{1}{6} = 16.\overline{6}\%$, or $16\frac{2}{3}\%$

19.
$$\begin{array}{r} 0.1\,8\,7\,5\\ 1\,6\,\overline{)3.0\,0\,0\,0}\\ \underline{1\,6}\\ 1\,4\,0\\ \underline{1\,2\,8}\\ 1\,2\,0\\ \underline{1\,1\,2}\\ 8\,0\\ \underline{8\,0}\\ 0\end{array}$$

$\frac{3}{16} = 0.1875$

Convert to percent notation.

$0.18.75$

$\frac{3}{16} = 18.75\%$, or $18\frac{3}{4}\%$

21.
$$\begin{array}{r} 0.8\,1\,2\,5\\ 1\,6\,\overline{)1\,3.0\,0\,0\,0}\\ \underline{1\,2\,8}\\ 2\,0\\ \underline{1\,6}\\ 4\,0\\ \underline{3\,2}\\ 8\,0\\ \underline{8\,0}\\ 0\end{array}$$

$\frac{13}{16} = 0.8125$

Convert to percent notation.

$0.81.25$

$\frac{13}{16} = 81.25\%$, or $81\frac{1}{4}\%$

23. $\frac{4}{25} = \frac{4}{25}\cdot\frac{4}{4} = \frac{16}{100} = 16\%$

25. $\frac{1}{20} = \frac{1}{20}\cdot\frac{5}{5} = \frac{5}{100} = 5\%$

27. $\frac{17}{50} = \frac{17}{50}\cdot\frac{2}{2} = \frac{34}{100} = 34\%$

29. $\frac{2}{25} = \frac{2}{25}\cdot\frac{4}{4} = \frac{8}{100} = 8\%$

$\frac{59}{100} = 59\%$

31. $\frac{11}{50} = \frac{11}{50}\cdot\frac{2}{2} = \frac{22}{100} = 22\%$

33. $\frac{3}{25} = \frac{3}{25}\cdot\frac{4}{4} = \frac{12}{100} = 12\%$

35. $\frac{3}{20} = \frac{3}{20}\cdot\frac{5}{5} = \frac{15}{100} = 15\%$

37. $85\% = \frac{85}{100}$ Definition of percent

$= \frac{5\cdot 17}{5\cdot 20}$

$= \frac{5}{5}\cdot\frac{17}{20}$ } Simplifying

$= \frac{17}{20}$

39. $62.5\% = \frac{62.5}{100}$ Definition of percent

$= \frac{62.5}{100}\cdot\frac{10}{10}$ Multiplying by 1 to eliminate the decimal point in the numerator

$= \frac{625}{1000}$

$= \frac{5\cdot 125}{8\cdot 125}$

$= \frac{5}{8}\cdot\frac{125}{125}$ } Simplifying

$= \frac{5}{8}$

41. $33\frac{1}{3}\% = \frac{100}{3}\%$ Converting from mixed numeral to fraction notation

$= \frac{100}{3}\times\frac{1}{100}$ Definition of percent

$= \frac{100\cdot 1}{3\cdot 100}$ Multiplying

$= \frac{1}{3}\cdot\frac{100}{100}$ } Simplifying

$= \frac{1}{3}$

43. $16.\overline{6}\% = 16\frac{2}{3}\%$ $\quad (16.\overline{6} = 16\frac{2}{3})$

$= \frac{50}{3}\%$ Converting from mixed numeral to fractional notation

$= \frac{50}{3} \times \frac{1}{100}$ Definition of percent

$= \frac{50 \cdot 1}{3 \cdot 50 \cdot 2}$ Multiplying

$= \frac{1}{2 \cdot 3} \cdot \frac{50}{50}$
$= \frac{1}{6}$ } Simplifying

45. $7.25\% = \frac{7.25}{100} = \frac{7.25}{100} \cdot \frac{100}{100}$
$= \frac{725}{10,000} = \frac{29 \cdot 25}{400 \cdot 25} = \frac{29}{400} \cdot \frac{25}{25}$
$= \frac{29}{400}$

47. $0.8\% = \frac{0.8}{100} = \frac{0.8}{100} \cdot \frac{10}{10}$
$= \frac{8}{1000} = \frac{1 \cdot 8}{125 \cdot 8} = \frac{1}{125} \cdot \frac{8}{8}$
$= \frac{1}{125}$

49. $25\frac{3}{8}\% = \frac{203}{8}\%$
$= \frac{203}{8} \times \frac{1}{100}$ Definition of percent
$= \frac{203}{800}$

51. $78\frac{2}{9}\% = \frac{704}{9}\%$
$= \frac{704}{9} \times \frac{1}{100}$ Definition of percent
$= \frac{4 \cdot 176 \cdot 1}{9 \cdot 4 \cdot 25}$
$= \frac{4}{4} \cdot \frac{176 \cdot 1}{9 \cdot 25}$
$= \frac{176}{225}$

53. $64\frac{7}{11}\% = \frac{711}{11}\%$
$= \frac{711}{11} \times \frac{1}{100}$
$= \frac{711}{1100}$

55. $150\% = \frac{150}{100} = \frac{3 \cdot 50}{2 \cdot 50} = \frac{3}{2} \cdot \frac{50}{50} = \frac{3}{2}$

57. $0.0325\% = \frac{0.0325}{100} = \frac{0.0325}{100} \cdot \frac{10,000}{10,000} = \frac{325}{1,000,000} = \frac{25 \cdot 13}{25 \cdot 40,000} = \frac{25}{25} \cdot \frac{13}{40,000} = \frac{13}{40,000}$

59. Note that $33.\overline{3}\% = 33\frac{1}{3}\%$ and proceed as in Exercise 41; $33.\overline{3}\% = \frac{1}{3}$.

61. $6\% = \frac{6}{100}$
$= \frac{2 \cdot 3}{2 \cdot 50} = \frac{2}{2} \cdot \frac{3}{50}$
$= \frac{3}{50}$

63. $12\% = \frac{12}{100}$
$= \frac{4 \cdot 3}{4 \cdot 25} = \frac{4}{4} \cdot \frac{3}{25}$
$= \frac{3}{25}$

65. $75\% = \frac{75}{100}$
$= \frac{25 \cdot 3}{25 \cdot 4} = \frac{25}{25} \cdot \frac{3}{4}$
$= \frac{3}{4}$

67. $15\% = \frac{15}{100}$
$= \frac{5 \cdot 3}{5 \cdot 20} = \frac{5}{5} \cdot \frac{3}{20}$
$= \frac{3}{20}$

69. $20.9\% = \frac{20.9}{100}$
$= \frac{20.9}{100} \cdot \frac{10}{10}$
$= \frac{209}{1000}$

71. $\frac{1}{8} = 1 \div 8$

$$\begin{array}{r} 0.1\,2\,5 \\ 8\,\overline{)\,1.0\,0\,0} \\ \underline{8} \\ 2\,0 \\ \underline{1\,6} \\ 4\,0 \\ \underline{4\,0} \\ 0 \end{array}$$

$\mathbf{\frac{1}{8} = 0.125 = 12\frac{1}{2}\%}$**, or 12.5%**

$\frac{1}{6} = 1 \div 6$

$$\begin{array}{r} 0.1\,6\,6 \\ 6\,\overline{)\,1.0\,0\,0} \\ \underline{6} \\ 4\,0 \\ \underline{3\,6} \\ 4\,0 \\ \underline{3\,6} \\ 4 \end{array}$$

We get a repeating decimal: $0.1\overline{6}$

$0.16.\overline{6}$ $\quad 0.1\overline{6} = 16.\overline{6}\%$

$\frac{\mathbf{1}}{\mathbf{6}} = \mathbf{0.1\overline{6} = 16.\overline{6}\%}$, **or** $\mathbf{16\frac{2}{3}\%}$

$20\% = \frac{20}{100} = \frac{1}{5} \cdot \frac{20}{20} = \frac{1}{5}$

$0.20.$ $\quad 20\% = 0.2$

$\frac{\mathbf{1}}{\mathbf{5}} = \mathbf{0.2 = 20\%}$

$0.25.$ $\quad 0.25 = 25\%$

$25\% = \frac{25}{100} = \frac{1}{4} \cdot \frac{25}{25} = \frac{1}{4}$

$\frac{\mathbf{1}}{\mathbf{4}} = \mathbf{0.25 = 25\%}$

$33\frac{1}{3}\% = \frac{100}{3}\% = \frac{100}{3} \times \frac{1}{100} = \frac{100}{300} = \frac{1}{3} \cdot \frac{100}{100} = \frac{1}{3}$

$0.33.\overline{3}$ $\quad 33.\overline{3}\% = 0.33\overline{3}$, or $0.\overline{3}$

$\frac{\mathbf{1}}{\mathbf{3}} = \mathbf{0.\overline{3} = 33\frac{1}{3}\%}$, **or** $\mathbf{33.\overline{3}\%}$

$37.5\% = \frac{37.5}{100} = \frac{37.5}{100} \cdot \frac{10}{10} = \frac{375}{1000} = \frac{3}{8} \cdot \frac{125}{125} = \frac{3}{8}$

$0.37.5$ $\quad 37.5\% = 0.375$

$\frac{\mathbf{3}}{\mathbf{8}} = \mathbf{0.375 = 37\frac{1}{2}\%}$, **or** $\mathbf{37.5\%}$

$40\% = \frac{40}{100} = \frac{2}{5} \cdot \frac{20}{20} = \frac{2}{5}$

$0.40.$ $\quad 40\% = 0.4$

$\frac{\mathbf{2}}{\mathbf{5}} = \mathbf{0.4 = 40\%}$

$\frac{1}{2} = \frac{1}{2} \cdot \frac{5}{5} = \frac{5}{10} = 0.5$

$\frac{1}{2} = \frac{1}{2} \cdot \frac{50}{50} = \frac{50}{100} = 5\%$

$\frac{\mathbf{1}}{\mathbf{2}} = \mathbf{0.5 = 50\%}$

73. $0.50.$ $\quad 0.5 = 50\%$

$50\% = \frac{50}{100} = \frac{1}{2} \cdot \frac{50}{50} = \frac{1}{2}$

$\frac{\mathbf{1}}{\mathbf{2}} = \mathbf{0.5 = 50\%}$

$\frac{1}{3} = 1 \div 3$

$$\begin{array}{r} 0.3 \\ 3\,\overline{)\,1.0} \\ \underline{9} \\ 1 \end{array}$$

We get a repeating decimal: $0.\overline{3}$

$0.33.\overline{3}$ $\quad 0.\overline{3} = 33.\overline{3}\%$

$\frac{\mathbf{1}}{\mathbf{3}} = \mathbf{0.\overline{3} = 33.\overline{3}\%}$, **or** $\mathbf{33\frac{1}{3}\%}$

$25\% = \frac{25}{100} = \frac{25}{25} \cdot \frac{1}{4} = \frac{1}{4}$

$0.25.$ $\quad 25\% = 0.25$

$\frac{\mathbf{1}}{\mathbf{4}} = \mathbf{0.25 = 25\%}$

$16\frac{2}{3}\% = \frac{50}{3}\% = \frac{50}{3} \times \frac{1}{100} = \frac{50 \cdot 1}{3 \cdot 2 \cdot 50} = \frac{50}{50} \cdot \frac{1}{6} = \frac{1}{6}$

$\frac{1}{6} = 1 \div 6$

$$\begin{array}{r} 0.1\,6 \\ 6\,\overline{)\,1.0\,0} \\ \underline{6} \\ 4\,0 \\ \underline{3\,6} \\ 4 \end{array}$$

We get a repeating decimal: $0.1\overline{6}$

$\frac{\mathbf{1}}{\mathbf{6}} = \mathbf{0.1\overline{6} = 16\frac{2}{3}\%}$, **or** $\mathbf{16.\overline{6}\%}$

$0.12.5$ $\quad 0.125 = 12.5\%$

$12.5\% = \frac{12.5}{100} = \frac{12.5}{100} \cdot \frac{10}{10} = \frac{125}{1000} = \frac{125}{125} \cdot \frac{1}{8} = \frac{1}{8}$

$\frac{\mathbf{1}}{\mathbf{8}} = \mathbf{0.125 = 12.5\%}$, **or** $\mathbf{12\frac{1}{2}\%}$

$\frac{3}{4} = \frac{3}{4}\cdot\frac{25}{25} = \frac{75}{100} = 75\%$

0.75. ↑└┘ $75\% = 0.75$

$\mathbf{\frac{3}{4} = 0.75 = 75\%}$

$0.8\overline{3} = 0.83.\overline{3}$ $0.8\overline{3} = 83.\overline{3}\%$
└┘↑

$83.\overline{3}\% = 83\frac{1}{3}\% = \frac{250}{3}\% = \frac{250}{3}\times\frac{1}{100} = \frac{5\cdot 50}{3\cdot 2\cdot 50} = \frac{5}{6}\cdot\frac{50}{50} = \frac{5}{6}$

$\mathbf{\frac{5}{6} = 0.8\overline{3} = 83.\overline{3}\%}$**, or** $\mathbf{83\frac{1}{3}\%}$

$\frac{3}{8} = 3 \div 8$

$$\begin{array}{r} 0.375 \\ 8\,\overline{)\,3.000} \\ \underline{2\,4} \\ 60 \\ \underline{56} \\ 40 \\ \underline{40} \\ 0 \end{array}$$

$\frac{3}{8} = 0.375$

0.37.5 $0.375 = 37.5\%$
└┘↑

$\mathbf{\frac{3}{8} = 0.375 = 37.5\%}$**, or** $\mathbf{37\frac{1}{2}\%}$

75. $13\cdot x = 910$

$\frac{13\cdot x}{13} = \frac{910}{13}$

$x = 70$

77. $0.05\times b = 20$

$\frac{0.05\times b}{0.05} = \frac{20}{0.05}$

$b = 400$

79. $\frac{24}{37} = \frac{15}{x}$

$24\cdot x = 37\cdot 15$ Equating cross products

$x = \frac{37\cdot 15}{24}$

$x = 23.125$

81. $\frac{9}{10} = \frac{x}{5}$

$9\cdot 5 = 10\cdot x$

$\frac{9\cdot 5}{10} = x$

$\frac{45}{10} = x$

$\frac{9}{2} = x$, or

$4.5 = x$

83.
$$\begin{array}{r} 33 \\ 3\,\overline{)\,100} \\ \underline{9} \\ 10 \\ \underline{9} \\ 1 \end{array}$$

$\frac{100}{3} = 33\frac{1}{3}$

85.
$$\begin{array}{r} 83 \\ 3\,\overline{)\,250} \\ \underline{24} \\ 10 \\ \underline{9} \\ 1 \end{array}$$

$\frac{250}{3} = 83\frac{1}{3}$

87.
$$\begin{array}{r} 43 \\ 8\,\overline{)\,345} \\ \underline{32} \\ 25 \\ \underline{24} \\ 1 \end{array}$$

$\frac{345}{8} = 43\frac{1}{8}$

89.
$$\begin{array}{r} 18 \\ 4\,\overline{)\,75} \\ \underline{4} \\ 35 \\ \underline{32} \\ 3 \end{array}$$

$\frac{75}{4} = 18\frac{3}{4}$

91. $1\frac{1}{17} = \frac{18}{17}$ $(1\cdot 17 = 17,\ 17 + 1 = 18)$

93. $101\frac{1}{2} = \frac{203}{2}$ $(101\cdot 2 = 202,\ 202 + 1 = 203)$

95. Use a calculator.

$\frac{41}{369} = 0.11.\overline{1} = 11.\overline{1}\%$
└┘↑

97. $2.5\overline{74631} = 2.57.\overline{46317} = 257.\overline{46317}\%$
└┘↑

99. $\frac{14}{9}\% = \frac{14}{9}\times\frac{1}{100} = \frac{2\cdot 7\cdot 1}{9\cdot 2\cdot 50} = \frac{2}{2}\cdot\frac{7}{450} = \frac{7}{450}$

To find decimal notation for $\frac{7}{450}$ we divide.

$$\begin{array}{r} 0.0155 \\ 450\,\overline{)\,7.0000} \\ \underline{450} \\ 2500 \\ \underline{2250} \\ 2500 \\ \underline{2250} \\ 250 \end{array}$$

We get a repeating decimal: $\frac{14}{9}\% = 0.01\overline{5}$

101. $\frac{729}{7}\% = \frac{729}{7} \times \frac{1}{100} = \frac{729}{700}$

To find decimal notation for $\frac{729}{700}$ we divide.

$$\begin{array}{r} 1.04142857 \\ 700\overline{)729.00000000} \\ \underline{700} \\ 2900 \\ \underline{2800} \\ 1000 \\ \underline{700} \\ 3000 \\ \underline{2800} \\ 2000 \\ \underline{1400} \\ 6000 \\ \underline{5600} \\ 4000 \\ \underline{3500} \\ 5000 \\ \underline{4900} \\ 100 \end{array}$$

We get a repeating decimal: $\frac{729}{7}\% = 1.04\overline{142857}$.

103. We will express each number in decimal notation.

$16\frac{1}{6}\% = 0.161\overline{6}$

1.6

$\frac{1}{6}\% = 0.001\overline{6}$

$\frac{1}{2} = 0.5$

0.2

$1.6\% = 0.016$

$1\frac{1}{6}\% = 0.011\overline{6}$

$0.5\% = 0.005$

$\frac{2}{7}\% = 0.00\overline{285714}$

$0.\overline{54} = 0.5454\overline{54}$

Arranging the numbers from smallest to largest, we have $\frac{1}{6}\%$, $\frac{2}{7}\%$, 0.5%, $1\frac{1}{6}\%$, 1.6%, $16\frac{1}{6}\%$, 0.2, $\frac{1}{2}$, $0.\overline{54}$, 1.6.

Exercise Set 6.4

1. What is 32% of 78?

$$\begin{array}{ccccc} \downarrow & \downarrow & \downarrow & \downarrow & \downarrow \\ a & = & 32\% & \times & 78 \end{array}$$

3. 89 is what percent of 99?

$$\begin{array}{ccccc} \downarrow & \downarrow & \downarrow & \downarrow & \downarrow \\ 89 & = & p & \times & 99 \end{array}$$

5. 13 is 25% of what?

$$\begin{array}{ccccc} \downarrow & \downarrow & \downarrow & \downarrow & \downarrow \\ 13 & = & 25\% & \times & b \end{array}$$

7. What is 85% of 276?

Translate: $a = 85\% \cdot 276$

Solve: The letter is by itself. To solve the equation we convert 85% to decimal notation and multiply.

$$\begin{array}{r} 276 \\ \times\ 0.85 \\ \hline 1380 \\ 22080 \\ \hline a = 234.60 \end{array} \qquad (85\% = 0.85)$$

234.6 is 85% of 276. The answer is 234.6.

9. 150% of 30 is what?

Translate: $150\% \times 30 = a$

Solve: Convert 150% to decimal notation and multiply.

$$\begin{array}{r} 30 \\ \times\ 1.5 \\ \hline 150 \\ 300 \\ \hline a = 45.0 \end{array} \qquad (150\% = 1.5)$$

150% of 30 is 45. The answer is 45.

11. What is 6% of \$300?

Translate: $a = 6\% \cdot \$300$

Solve: Convert 6% to decimal notation and multiply.

$$\begin{array}{r} \$\ 300 \\ \times\ 0.06 \\ \hline a = \$\ 18.00 \end{array} \qquad (6\% = 0.06)$$

\$18 is 6% of \$300. The answer is \$18.

13. 3.8% of 50 is what?

Translate: $3.8\% \cdot 50 = a$

Solve: Convert 3.8% to decimal notation and multiply.

$$\begin{array}{r} 50 \\ \times\ 0.038 \\ \hline 400 \\ 1500 \\ \hline a = 1.900 \end{array} \qquad (3.8\% = 0.038)$$

3.8% of 50 is 1.9. The answer is 1.9.

15. \$39 is what percent of \$50?

Translate: $39 = n \times 50$

Solve: To solve the equation we divide on both sides by 50 and convert the answer to percent notation.

$$n \cdot 50 = 39$$

$$\frac{n \cdot 50}{50} = \frac{39}{50}$$

$$n = 0.78 = 78\%$$

\$39 is 78% of \$50. The answer is 78%.

17. 20 is what percent of 10?

Translate: $20 = n \times 10$

Solve: To solve the equation we divide on both sides by 10 and convert the answer to percent notation.

$$n \cdot 10 = 20$$
$$\frac{n \cdot 10}{10} = \frac{20}{10}$$
$$n = 2 = 200\%$$

20 is 200% of 10. The answer is 200%.

19. What percent of $300 is $150?

Translate: $n \times 300 = 150$

Solve: $n \cdot 300 = 150$

$$\frac{n \cdot 300}{300} = \frac{150}{300}$$
$$n = 0.5 = 50\%$$

50% of $300 is $150. The answer is 50%.

21. What percent of 80 is 100?

Translate: $n \times 80 = 100$

Solve: $n \cdot 80 = 100$

$$\frac{n \cdot 80}{80} = \frac{100}{80}$$
$$n = 1.25 = 125\%$$

125% of 80 is 100. The answer is 125%.

23. 20 is 50% of what?

Translate: $20 = 50\% \times b$

Solve: To solve the equation we divide on both sides by 50%:

$$\frac{20}{50\%} = \frac{50\% \times b}{50\%}$$
$$\frac{20}{0.5} = b \quad (50\% = 0.5)$$
$$40 = b$$

```
          4 0 .
0. 5^)2 0. 0^
      2 0
        0
        0
        0
```

20 is 50% of 40. The answer is 40.

25. 40% of what is $16?

Translate: $40\% \times b = 16$

Solve: To solve the equation we divide on both sides by 40%:

$$\frac{40\% \times b}{40\%} = \frac{16}{40\%}$$
$$b = \frac{16}{0.4} \quad (40\% = 0.4)$$
$$b = 40$$

```
          4 0 .
0. 4^)1 6. 0^
      1 6
        0
        0
        0
```

40% of $40 is $16. The answer is $40.

27. 56.32 is 64% of what?

Translate: $56.32 = 64\% \times b$

Solve: $\frac{56.32}{64\%} = \frac{64\% \times b}{64\%}$

$$\frac{56.32}{0.64} = b$$
$$88 = b$$

```
              8 8 .
0. 6 4^)5 6. 3 2^
        5 1 2
          5 1 2
          5 1 2
              0
```

56.32 is 64% of 88. The answer is 88.

29. 70% of what is 14?

Translate: $70\% \times b = 14$

Solve: $\frac{70\% \times b}{70\%} = \frac{14}{70\%}$

$$b = \frac{14}{0.7}$$
$$b = 20$$

```
          2 0 .
0. 7^)1 4. 0^
      1 4
        0
        0
        0
```

70% of 20 is 14. The answer is 20.

31. What is $62\frac{1}{2}\%$ of 10?

Translate: $a = 62\frac{1}{2}\% \times 10$

Solve: $a = 0.625 \times 10 \quad (62\frac{1}{2}\% = 0.625)$

$a = 6.25$ Multiplying

6.25 is $62\frac{1}{2}\%$ of 10. The answer is 6.25.

33. What is 8.3% of $10,200?

Translate: $a = 8.3\% \times 10,200$

Solve: $a = 8.3\% \times 10,200$

$a = 0.083 \times 10,200 \quad (8.3\% = 0.083)$

$a = 846.6$ Multiplying

$846.60 is 8.3% of $10,200. The answer is $846.60.

35. 2.5% of what is 30.4?

Translate: $2.5\% \times b = 30.4$

Solve: $\frac{2.5\% \times b}{2.5\%} = \frac{30.4}{2.5\%}$

$$b = \frac{30.4}{0.025}$$
$$b = 1216$$

```
                 1 2 1 6 .
0. 0 2 5^)3 0. 4 0 0^
          2 5
            5 4
            5 0
              4 0
              2 5
              1 5 0
              1 5 0
                  0
```

2.5% of 1216 is 30.4. The answer is 1216.

37. $0.\underline{09} = \dfrac{9}{1\underline{00}}$

2 decimal places — 2 zeros

39. $0.\underline{875} = \dfrac{875}{1\underline{000}}$

3 decimal places — 3 zeros

$$\frac{875}{1000} = \frac{7 \cdot 125}{8 \cdot 125} = \frac{7}{8} \cdot \frac{125}{125} = \frac{7}{8}$$

Thus, $0.875 = \dfrac{875}{1000}$, or $\dfrac{7}{8}$.

41. $0.\underline{9375} = \dfrac{9375}{1\underline{0,000}}$

4 decimal places — 4 zeros

$$\frac{9375}{10,000} = \frac{15 \cdot 625}{16 \cdot 625} = \frac{15}{16} \cdot \frac{625}{625} = \frac{15}{16}$$

Thus, $0.9375 = \dfrac{9375}{10,000}$, or $\dfrac{15}{16}$

43. $\dfrac{89}{1\underline{00}}$ — $0.89.$

2 zeros — Move 2 places

$$\frac{89}{100} = 0.89$$

45. $\dfrac{3}{1\underline{0}}$ — $0.3.$

1 zero — Move 1 place

$$\frac{3}{10} = 0.3$$

47. Estimate: Round 7.75% to 8% and $10,880 to $11,000. Then translate:

What is 8% of $11,000?

$a = 8\% \times 11,000$

We convert 8% to decimal notation and multiply.

$$\begin{array}{r} 11,000 \\ \times \quad 0.08 \\ \hline 880.00 \end{array} \qquad (8\% = 0.08)$$

$880 is about 7.75% of $10,880. (Answers may vary.)

Calculate: First we translate.

What is 7.75% of $10,880?

$a = 7.75\% \times 10,880$

Use a calculator to multiply:

$$0.0775 \times 10,880 = 843.2$$

$843.20 is 7.75% of $10,880.

49. Estimate: Round $2496 to $2500 and 24% to 25%. Then translate:

$2500 is 25% of what?

$2500 = 25\% \times b$

We convert 25% to decimal notation and divide.

$$\frac{2500}{0.25} = \frac{0.25 \times b}{0.25}$$

$$10,000 = b$$

$2496 is 24% of about $10,000. (Answers may vary.)

Calculate: First we translate.

$2496 is 24% of what?

$2496 = 0.24 \times b$

Use a calculator to divide:

$$\frac{2496}{0.24} = 10,400$$

$2496 is 24% of $10,400.

51. We translate:

40% of $18\frac{3}{4}$% of $25,000 is what?

$40\% \times 18\frac{3}{4}\% \times 25,000 = a$

We convert 40% and $18\frac{3}{4}$% to decimal notation and multiply.

$$0.4 \times 0.1875 \times 25,000 = a$$

$$\begin{array}{r} 0.1875 \\ \times \quad 0.4 \\ \hline 0.07500 \end{array}$$

$$\begin{array}{r} 25,000 \\ \times \quad 0.075 \\ \hline 125000 \\ 1750000 \\ \hline 1875.000 \end{array}$$

40% of $18\frac{3}{4}$% of $25,000 is $1875.

Exercise Set 6.5

1. What is 37% of 74?

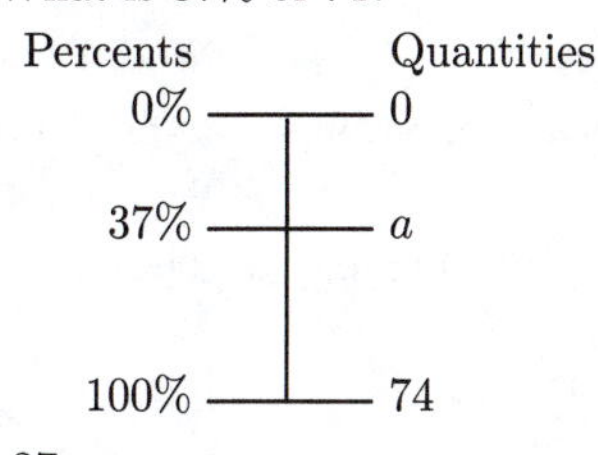

$$\frac{37}{100} = \frac{a}{74}$$

3. 4.3 is what percent of 5.9?

Percents — Quantities

0% — 0

N% — 4.3

100% — 5.9

$$\frac{N}{100} = \frac{4.3}{5.9}$$

5. 14 is 25% of what?

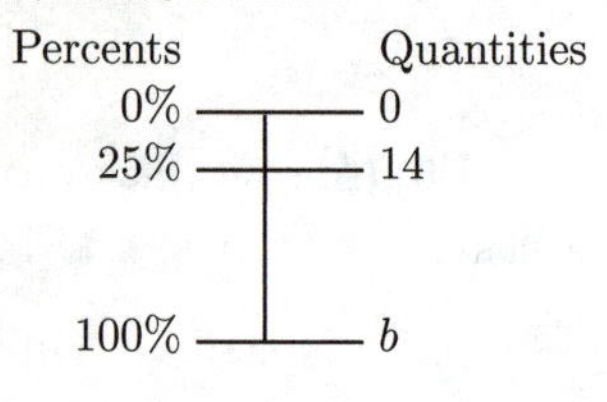

$$\frac{25}{100} = \frac{14}{b}$$

7. What is 76% of 90?

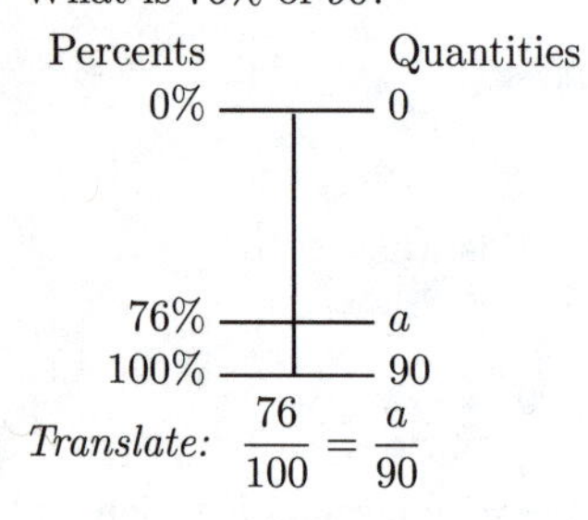

Translate: $\dfrac{76}{100} = \dfrac{a}{90}$

Solve: $76 \cdot 90 = 100 \cdot a$ Equating cross-products

$\dfrac{76 \cdot 90}{100} = \dfrac{100 \cdot a}{100}$ Dividing by 100

$\dfrac{6840}{100} = a$

$68.4 = a$ Simplifying

68.4 is 76% of 90. The answer is 68.4.

9. 70% of 660 is what?

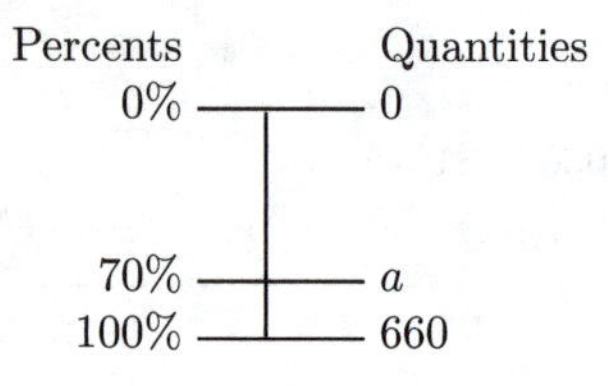

Translate: $\dfrac{70}{100} = \dfrac{a}{660}$

Solve: $70 \cdot 660 = 100 \cdot a$ Equating cross-products

$\dfrac{70 \cdot 660}{100} = \dfrac{100 \cdot a}{100}$ Dividing by 100

$\dfrac{46,200}{100} = a$

$462 = a$ Simplifying

70% of 660 is 462. The answer is 462.

11. What is 4% of 1000?

Percents Quantities
0% — 0
4% — a
100% — 1000

Translate: $\dfrac{4}{100} = \dfrac{a}{1000}$

Solve: $4 \cdot 1000 = 100 \cdot a$

$\dfrac{4 \cdot 1000}{100} = \dfrac{100 \cdot a}{100}$

$\dfrac{4000}{100} = a$

$40 = a$

40 is 4% of 1000. The answer is 40.

13. 4.8% of 60 is what?

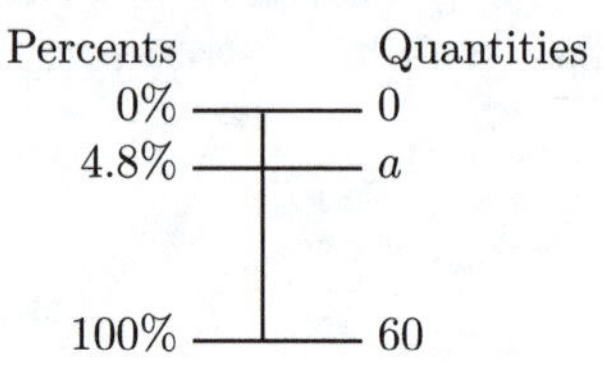

Translate: $\dfrac{4.8}{100} = \dfrac{a}{60}$

Solve: $4.8 \cdot 60 = 100 \cdot a$

$\dfrac{4.8 \cdot 60}{100} = \dfrac{100 \cdot a}{100}$

$\dfrac{288}{100} = a$

$2.88 = a$

4.8% of 60 is 2.88. The answer is 2.88.

15. \$24 is what percent of \$96?

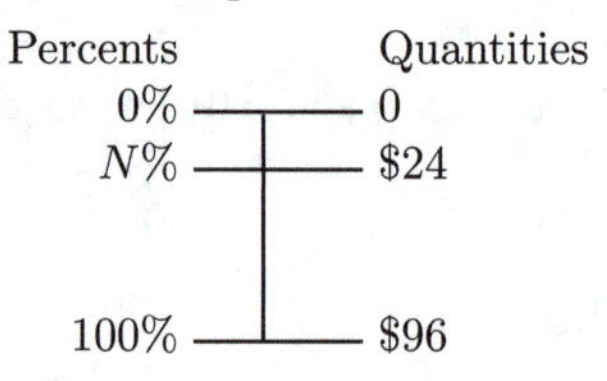

Translate: $\dfrac{N}{100} = \dfrac{24}{96}$

Solve: $96 \cdot N = 100 \cdot 24$

$\dfrac{96N}{96} = \dfrac{100 \cdot 24}{96}$

$N = \dfrac{100 \cdot 24}{96}$

$N = 25$

\$24 is 25% of \$96. The answer is 25%.

17. 102 is what percent of 100?

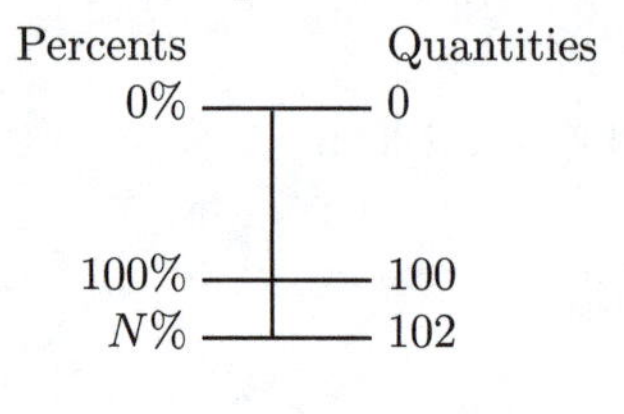

Translate: $\dfrac{N}{100} = \dfrac{102}{100}$

Solve: $100 \cdot N = 100 \cdot 102$

$$\frac{100 \cdot N}{100} = \frac{100 \cdot 102}{100}$$

$$N = \frac{100 \cdot 102}{100}$$

$$N = 102$$

102 is 102% of 100. The answer is 102%.

19. What percent of \$480 is \$120?

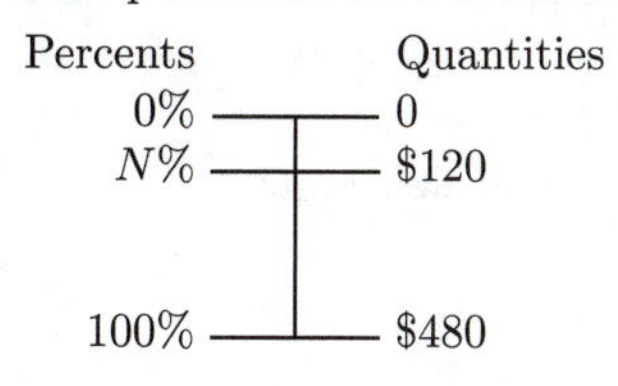

Translate: $\dfrac{N}{100} = \dfrac{120}{480}$

Solve: $480 \cdot N = 100 \cdot 120$

$$\frac{480 \cdot N}{480} = \frac{100 \cdot 120}{480}$$

$$N = \frac{100 \cdot 120}{480}$$

$$N = 25$$

25% of \$480 is \$120. The answer is 25%.

21. What percent of 160 is 150?

Percents — Quantities
0% — 0
N% — 150
100% — 160

Translate: $\dfrac{N}{100} = \dfrac{150}{160}$

Solve: $160 \cdot N = 100 \cdot 150$

$$\frac{160 \cdot N}{160} = \frac{100 \cdot 150}{160}$$

$$N = \frac{100 \cdot 150}{160}$$

$$N = 93.75$$

93.75% of 160 is 150. The answer is 93.75%, or $93\frac{3}{4}$%.

23. \$18 is 25% of what?

Percents — Quantities
0% — 0
25% — \$18
100% — b

Translate: $\dfrac{25}{100} = \dfrac{18}{b}$

Solve: $25 \cdot b = 100 \cdot 18$

$$\frac{25 \cdot b}{b} = \frac{100 \cdot 18}{25}$$

$$b = \frac{100 \cdot 18}{25}$$

$$b = 72$$

\$18 is 25% of \$72. The answer is \$72.

25. 60% of what is \$54.

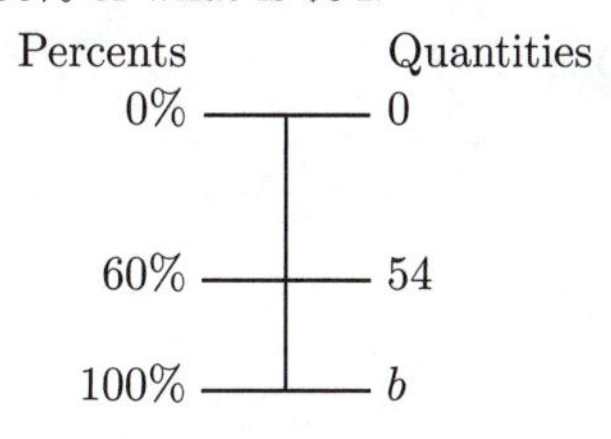

Translate: $\dfrac{60}{100} = \dfrac{54}{b}$

Solve: $60 \cdot b = 100 \cdot 54$

$$\frac{60 \cdot b}{b} = \frac{100 \cdot 54}{60}$$

$$b = \frac{100 \cdot 54}{60}$$

$$b = 90$$

60% of 90 is 54. The answer is 90.

27. 65.12 is 74% of what?

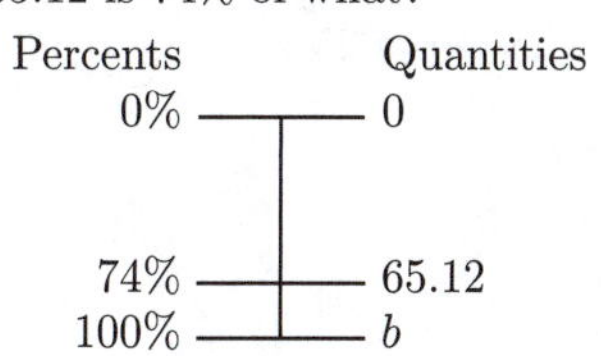

Translate: $\dfrac{74}{100} = \dfrac{65.12}{b}$

Solve: $74 \cdot b = 100 \cdot 65.12$

$$\frac{74 \cdot b}{74} = \frac{100 \cdot 65.12}{74}$$

$$b = \frac{100 \cdot 65.12}{74}$$

$$b = 88$$

65.12 is 74% of 88. The answer is 88.

29. 80% of what is 16?

Percents — Quantities
0% — 0
80% — 16
100% — b

Translate: $\dfrac{80}{100} = \dfrac{16}{b}$

Solve: $80 \cdot b = 100 \cdot 16$

$$\frac{80 \cdot b}{80} = \frac{100 \cdot 16}{80}$$

$$b = \frac{100 \cdot 16}{80}$$

$$b = 20$$

80% of 20 is 16. The answer is 20.

31. What is $62\frac{1}{2}\%$ of 40?

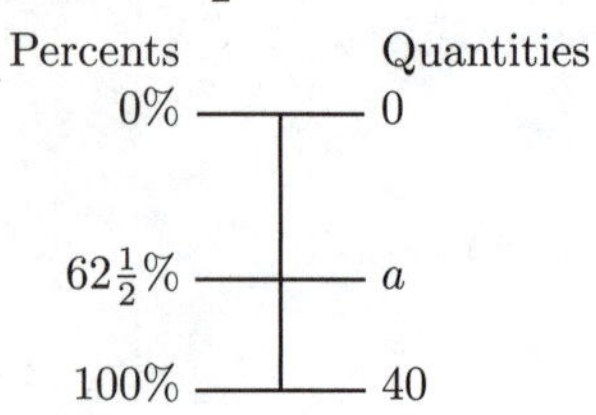

Translate: $\frac{62\frac{1}{2}}{100} = \frac{a}{40}$

Solve: $62\frac{1}{2} \cdot 40 = 100 \cdot a$

$$\frac{125}{2} \cdot \frac{40}{1} = 100 \cdot a$$

$$2500 = 100 \cdot a$$

$$\frac{2500}{100} = \frac{100 \cdot a}{100}$$

$$25 = a$$

25 is $62\frac{1}{2}\%$ of 40. The answer is 25.

33. What is 9.4% of \$8300?

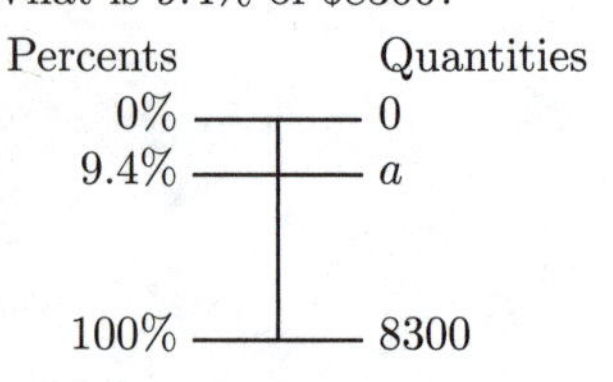

Translate: $\frac{9.4}{100} = \frac{a}{8300}$

Solve: $9.4 \cdot 8300 = 100 \cdot a$

$$\frac{9.4 \cdot 8300}{100} = \frac{100 \cdot a}{100}$$

$$\frac{78,020}{100} = a$$

$$780.2 = a$$

\$780.20 is 9.4% of \$8300. The answer is \$780.20.

35. 80.8 is $40\frac{2}{5}\%$ of what?

Percents — Quantities

0% — 0

$40\frac{2}{5}\%$ — 80.8

100% — b

Translate: $\frac{40\frac{2}{5}}{100} = \frac{80.8}{b}$

Solve: $40\frac{2}{5} \cdot b = 100 \cdot 80.8$

$$40.4 \cdot b = 100 \cdot 80.8 \qquad \left(40\frac{2}{5} = 40.4\right)$$

$$\frac{40.4 \cdot b}{40.4} = \frac{100 \cdot 80.8}{40.4}$$

$$b = \frac{100 \cdot 80.8}{40.4}$$

$$b = 200$$

80.8 is $40\frac{2}{5}\%$ of 200. The answer is 200.

37. $\frac{x}{188} = \frac{2}{47}$

$$47 \cdot x = 188 \cdot 2$$

$$x = \frac{188 \cdot 2}{47}$$

$$x = \frac{4 \cdot 47 \cdot 2}{47}$$

$$x = 8$$

39. $\frac{4}{7} = \frac{x}{14}$

$$4 \cdot 14 = 7 \cdot x$$

$$\frac{4 \cdot 14}{7} = x$$

$$\frac{4 \cdot 2 \cdot 7}{7} = x$$

$$8 = x$$

41. $\frac{5000}{t} = \frac{3000}{60}$

$$5000 \cdot 60 = 3000 \cdot t$$

$$\frac{5000 \cdot 60}{3000} = t$$

$$\frac{5 \cdot 1000 \cdot 3 \cdot 20}{3 \cdot 1000} = t$$

$$100 = t$$

43. $\frac{x}{1.2} = \frac{36.2}{5.4}$

$$5.4 \cdot x = 1.2(36.2)$$

$$x = \frac{1.2(36.2)}{5.4}$$

$$x = 8.0\overline{4}$$

45. ***Familiarize***. Let q = the number of quarts of liquid ingredients the recipe calls for.

Translate.

$$\underbrace{\text{Butter-milk}}_{\frac{1}{2}} \ \underset{+}{\text{plus}} \ \underbrace{\text{Skim milk}}_{\frac{1}{3}} \ \underset{+}{\text{plus}} \ \underbrace{\text{Oil}}_{\frac{1}{16}} \ \underset{=}{\text{is}} \ \underbrace{\text{Total liquid ingredients}}_{q}$$

Solve. We carry out the addition. The LCM of the denominators is 48, so the LCD is 48.

$$\frac{1}{2}\cdot\frac{24}{24}+\frac{1}{3}\cdot\frac{16}{16}+\frac{1}{16}\cdot\frac{3}{3}=q$$
$$\frac{24}{48}+\frac{16}{48}+\frac{3}{48}=q$$
$$\frac{43}{48}=q$$

Check. We repeat the calculation. The answer checks.

State. The recipe calls for $\frac{43}{48}$ qt of liquid ingredients.

47. Estimate: Round 8.85% to 9%, and \$12,640 to \$12,600.

What is 9% of \$12,600?

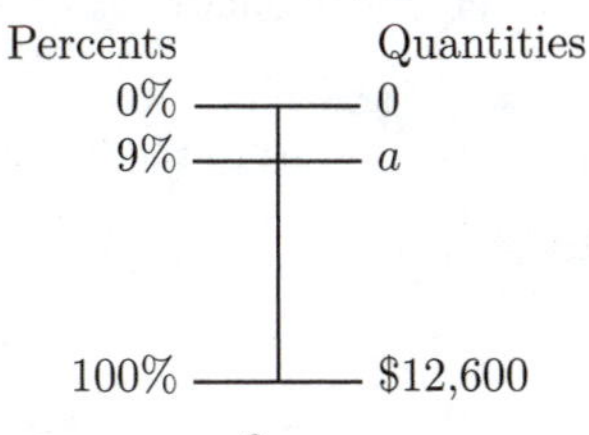

Translate: $\frac{9}{100}=\frac{a}{12,600}$

Solve: $9\cdot 12,600=100\cdot a$

$$\frac{9\cdot 12,600}{100}=\frac{100\cdot a}{100}$$
$$\frac{113,400}{100}=a$$
$$1134=a$$

\$1134 is about 8.85% of \$12,640. (Answers may vary.)

Calculate:

What is 8.85% of \$12,640?

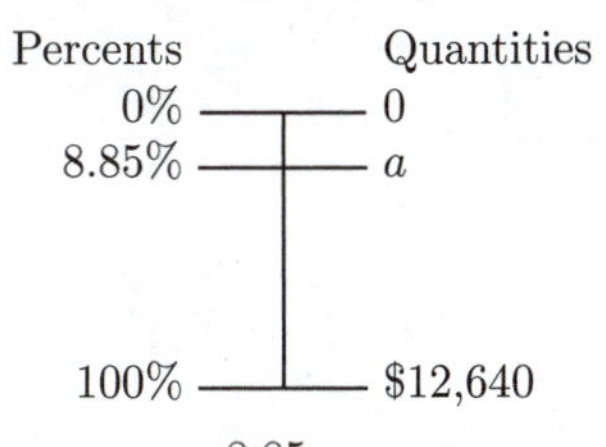

Translate: $\frac{8.85}{100}=\frac{a}{12,640}$

Solve: $8.85\cdot 12,640=100\cdot a$

$$\frac{8.85\cdot 12,640}{100}=\frac{100\cdot a}{100}$$
$$\frac{111,864}{100}=a \quad \text{Use a calculator to multiply and divide.}$$
$$1118.64=a$$

\$1118.64 is 8.85% of \$12,640.

Chapter 6 Mid-Chapter Review

1. If $\frac{x}{t}=\frac{y}{s}$, then $x\cdot s=t\cdot y$. Thus, the given statement is false.

2. The statement is true. See page 403 in the text.

3. The statement is false. The symbol % is equivalent to $\times 0.01$.

4. We begin by writing decimal notation for each number.

$\frac{1}{10}=0.1$

$1\%=0.01$

$0.1\%=0.001$

$10\%=0.1$

$\frac{1}{100}=0.01$

We see that the smallest number is 0.001, or 0.1%, so the given statement is true.

5. $\frac{1}{2}\%=\frac{1}{2}\cdot\frac{1}{100}=\frac{1}{200}$

6. $\frac{80}{1000}=\frac{8}{100}=8\%$

7. $5.5\%=\frac{5.5}{100}=\frac{55}{1000}=\frac{11}{200}$

8. $0.375=\frac{375}{1000}=\frac{37.5}{100}=37.5\%$

9.
$$15=p\times 80$$
$$\frac{15}{80}=\frac{p\times 80}{80}$$
$$\frac{15}{80}=p$$
$$0.1875=p$$
$$18.75\%=p$$

10.
$$\frac{x}{4}=\frac{3}{6}$$
$$x\cdot 6=4\cdot 3$$
$$\frac{x\cdot 6}{6}=\frac{4\cdot 3}{6}$$
$$x=\frac{4\cdot 3}{6}$$
$$x=2$$

11. $\frac{25}{75}=\frac{25\cdot 1}{3\cdot 25}=\frac{25}{25}\cdot\frac{1}{3}=\frac{1}{3}$

12. $\frac{2.4}{8.4}=\frac{2.4}{8.4}\cdot\frac{10}{10}=\frac{24}{84}=\frac{2\cdot 12}{7\cdot 12}=\frac{2}{7}\cdot\frac{12}{12}=\frac{2}{7}$

13. $\frac{146\text{ km}}{3\text{ hr}}=\frac{146}{3}\cdot\frac{\text{km}}{\text{hr}}\approx 48.67\text{ km/hr}$

14. $\frac{243\text{ mi}}{4\text{ hr}}=\frac{243}{4}\frac{\text{mi}}{\text{hr}}=60.75\text{ mi/hr, or }60.75\text{ mph}$

15. $$\frac{x}{24} = \frac{30}{18}$$
$$x \cdot 18 = 24 \cdot 30$$
$$\frac{x \cdot 18}{18} = \frac{24 \cdot 30}{18}$$
$$x = \frac{24 \cdot 30}{18}$$
$$x = \frac{720}{18}$$
$$x = 40$$

16. $$\frac{12}{y} = \frac{20}{15}$$
$$12 \cdot 15 = y \cdot 20$$
$$\frac{12 \cdot 15}{20} = \frac{y \cdot 20}{20}$$
$$\frac{12 \cdot 15}{20} = y$$
$$\frac{180}{20} = y$$
$$9 = y$$

17. $$\frac{0.24}{0.02} = \frac{y}{0.36}$$
$$0.24 \cdot 0.36 = 0.02 \cdot y$$
$$\frac{0.24 \cdot 0.36}{0.02} = \frac{0.02 \cdot y}{0.02}$$
$$\frac{0.24 \cdot 0.36}{0.02} = y$$
$$\frac{0.0864}{0.02} = y$$
$$4.32 = y$$

18. $$\frac{\frac{1}{4}}{x} = \frac{\frac{1}{8}}{\frac{1}{4}}$$
$$\frac{1}{4} \cdot \frac{1}{4} = x \cdot \frac{1}{8}$$
$$\frac{1}{4} \cdot \frac{1}{4} \cdot \frac{8}{1} = x \quad \text{Dividing by } \frac{1}{8}$$
$$\frac{8}{16} = x$$
$$\frac{1}{2} = x$$

19. $\dfrac{\$5.99}{12 \text{ oz}} = \dfrac{599¢}{12 \text{ oz}} \approx 49.917¢/\text{oz}$

20. $\dfrac{\$2.09}{18 \text{ oz}} = \dfrac{209¢}{18 \text{ oz}} \approx 11.611¢/\text{oz}$

21. 28%

a) Replace the percent symbol with $\times 0.01$.

28×0.01

b) Move the decimal point two places to the left.

0.28.
↑⌴

Thus, $28\% = 0.28$.

22. 0.15%

a) Replace the percent symbol with $\times 0.01$.

0.15×0.01

b) Move the decimal point two places to the left.

0.00.15
↑⌴

Thus, $0.15\% = 0.0015$.

23. $5\frac{3}{8}\% = 5.375\%$

a) Replace the percent symbol with $\times 0.01$.

5.375×0.01

b) Move the decimal point two places to the left.

0.05.375
↑⌴

Thus, $5\frac{3}{8}\% = 0.05375$.

24. 240%

a) Replace the percent symbol with $\times 0.01$.

240×0.01

b) Move the decimal point two places to the left.

2.40.
↑⌴

Thus, $240\% = 2.4$.

25. 0.71

a) Move the decimal point two places to the right.

0.71.
⌴↑

b) Write a percent symbol: 71%

Thus, $0.71 = 71\%$.

26. We use the definition of percent as a ratio.

$$\frac{9}{100} = 9\%$$

27. 0.3891

a) Move the decimal point two places to the right.

0.38.91
⌴↑

b) Write a percent symbol: 38.91%

Thus, $0.3891 = 38.91\%$.

28. Find decimal notation by division.

$$\begin{array}{r} 0.1875 \\ 16 \overline{)3.0000} \\ \underline{16} \\ 140 \\ \underline{128} \\ 120 \\ \underline{112} \\ 80 \\ \underline{80} \\ 0 \end{array}$$

$\frac{3}{16} = 0.1875$

Convert to percent notation.

0.18.75

$\frac{3}{16} = 18.75\%$, or $18\frac{3}{4}\%$

29. 0.005

a) Move the decimal point two places to the right.

0.00.5

b) Write a percent symbol: 0.5%

Thus, 0.005 = 0.5%.

30. $\frac{37}{50} = \frac{37}{50} \cdot \frac{2}{2} = \frac{74}{100} = 74\%$

31. $6 = 6 \cdot \frac{100}{100} = \frac{600}{100} = 600\%$

32. Find decimal notation by division.

$$\begin{array}{r} 0.833 \\ 6\overline{)5.000} \\ \underline{48} \\ 20 \\ \underline{18} \\ 20 \\ \underline{18} \\ 2 \end{array}$$

We get a repeating decimal: $\frac{5}{6} = 0.83\overline{3}$

Convert to percent notation.

$0.83.\overline{3}$

$\frac{5}{6} = 83.\overline{3}\%$, or $83\frac{1}{3}\%$

33. $85\% = \frac{85}{100} = \frac{5 \cdot 17}{5 \cdot 20} = \frac{5}{5} \cdot \frac{17}{20} = \frac{17}{20}$

34.
$$\begin{aligned} 0.048\% &= \frac{0.048}{100} \\ &= \frac{0.048}{100} \cdot \frac{1000}{1000} \\ &= \frac{48}{100,000} = \frac{16 \cdot 3}{16 \cdot 6250} \\ &= \frac{16}{16} \cdot \frac{3}{6250} = \frac{3}{6250} \end{aligned}$$

35.
$$\begin{aligned} 22\frac{3}{4}\% = 22.75\% &= \frac{22.75}{100} \\ &= \frac{22.75}{100} \cdot \frac{100}{100} = \frac{2275}{10,000} \\ &= \frac{25 \cdot 91}{25 \cdot 400} = \frac{25}{25} \cdot \frac{91}{400} \\ &= \frac{91}{400} \end{aligned}$$

36.
$$\begin{aligned} 16.\overline{6} = 16\frac{2}{3}\% &= \frac{50}{3}\% \\ &= \frac{50}{3} \cdot \frac{1}{100} = \frac{50}{300} \\ &= \frac{50 \cdot 1}{50 \cdot 6} = \frac{50}{50} \cdot \frac{1}{6} \\ &= \frac{1}{6} \end{aligned}$$

37. Five of the eight equal parts are shaded, so the shaded area is $\frac{5}{8}$ of the figure. We convert $\frac{5}{8}$ to decimal notation and then to percent notation.

$$\begin{array}{r} 0.625 \\ 8\overline{)5.000} \\ \underline{48} \\ 20 \\ \underline{16} \\ 40 \\ \underline{40} \\ 0 \end{array}$$

$\frac{5}{8} = 0.625 = 62.5\%$, or $62\frac{1}{2}\%$

38. Nine of the twenty equal parts are shaded, so $\frac{9}{20}$ of the area is shaded.

$\frac{9}{20} = \frac{9}{20} \cdot \frac{5}{5} = \frac{45}{100} = 45\%$

39. 25% of what is 14.5?

Translate: $25\% \times b = 14.5$

Solve:
$$\begin{aligned} \frac{25\% \times b}{25\%} &= \frac{14.5}{25\%} \\ b &= \frac{14.5}{0.25} \\ b &= 58 \end{aligned}$$

25% of 58 is 14.5. The answer is 58.

40. 220 is what percent of 1320?

Translate: $220 = p \times 1320$

Solve:
$$\begin{aligned} \frac{220}{1320} &= \frac{p \times 1320}{1320} \\ 0.16\overline{6} &= p \\ 16.\overline{6}\% &= p \end{aligned}$$

220 is $16.\overline{6}\%$ of 1320. The answer is $16.\overline{6}\%$, or $16\frac{2}{3}\%$.

41. What is 3.2% of 80,000?

Translate: $a = 3.2\% \cdot 80,000$

Solve: Convert 3.2% to decimal notation and multiply.

$$\begin{array}{r} 80,000 \\ \times \quad 0.032 \\ \hline 160000 \\ 2400000 \\ \hline a = 2560.000 \end{array}$$

2560 is 3.2% of 80,000. The answer is 2560.

42. \$17.50 is 35% of what?

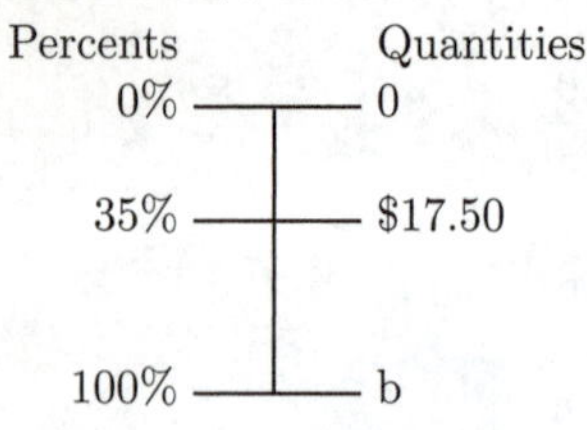

Translate: $\dfrac{35}{100} = \dfrac{\$17.50}{b}$

Solve: $35 \cdot b = 100 \cdot \$17.50$

$$\frac{35 \cdot b}{35} = \frac{100 \cdot \$17.50}{35}$$

$$b = \frac{\$1750}{35}$$

$$b = \$50$$

\$17.50 is 35% of \$50. The answer is \$50,.

43. We begin by writing the numbers in decimal notation.

$\frac{1}{2}\% = 0.5\% = 0.005$

$5\% = 0.05$

0.275

$\frac{13}{100} = 0.13$

$1\% = 0.01$

$0.1\% = 0.001$

$0.05\% = 0.0005$

$\frac{3}{10} = 0.3$

$\frac{7}{20} = \frac{7}{20} \cdot \frac{5}{5} = \frac{35}{100} = 0.35$

$10\% = 0.1$

Arranging the numbers from smallest to largest, we have 0.05%, 0.1%, $\frac{1}{2}\%$, 1%, 5%, 10%, $\frac{13}{100}$, 0.275, $\frac{3}{10}$, $\frac{7}{20}$.

44. \$102,000 is what percent of \$3.6 million?

Translate: $102,000 = p \times 3,600,000$

Solve: $\dfrac{102,000}{3,600,000} = \dfrac{p \times 3,600,000}{3,600,000}$

$$\frac{102,000}{3,600,000} = p$$

$$0.028\overline{3} = p$$

$$2.8\overline{3}\% = p, \text{ or}$$

$$2\frac{5}{6}\% = p$$

Answer B is correct.

45. Some will say that the conversion will be done most accurately by first finding decimal notation. Others will say that it is more efficient to become familiar with some or all of the fraction and percent equivalents that appear inside the back cover and to make the conversion by going directly from fraction notation to percent notation.

46. Since $40\% \div 10 = 4\%$, we can divide 36.8 by 10, obtaining 3.68. Since $400\% = 40\% \times 10$, we can multiply 36.8 by 10, obtaining 368.

47. The student's approach is not necessarily a bad one. However, when we use the approach of equating cross-products we eliminate the need to find the least common denominator.

48. They all represent the same number.

Exercise Set 6.6

1. ***Familiarize.*** Let s and j represent the number of foreign students from South Korea and Japan, respectively.

Translate. We translate to percent equations.

$$\underbrace{\text{What number}}_{\downarrow\; s} \;\underset{=}{\text{is}}\; \underset{10.70\%}{10.70\%} \;\underset{\cdot}{\text{of}}\; \underset{583,000}{583,000?}$$

$$\underbrace{\text{What number}}_{\downarrow\; j} \;\underset{=}{\text{is}}\; \underset{6.05\%}{6.05\%} \;\underset{\cdot}{\text{of}}\; \underset{583,000}{583,000?}$$

Solve. We convert the percent notation to decimal notation and multiply.

$$s = 0.1070 \cdot 583,000 = 62,381$$

$$j = 0.0605 \cdot 583,000 = 35,271.5 \approx 35,272$$

Check. We can repeat the calculations. We can also do a partial check by estimating:

$10.70\% \cdot 583,000 \approx 10\% \cdot 600,000 = 60,000;$

$0.0605 \cdot 583,000 \approx 6\% \cdot 600,000 = 36,000.$

Since 60,000 is close to 62,381 and 36,000 is close to 35,272, our answer seems reasonable.

State. There were 62,381 students from South Korea and 35,272 students from Japan.

3. ***Familiarize.*** We note that the amount of the raise can be found and then added to the old salary. A drawing helps us visualize the situation.

\$43,200	\$?
100%	8%

We let $x =$ the new salary.

Translate. We translate to a percent equation.

$$\underset{x}{\text{What}} \;\underset{=}{\text{is}}\; \underbrace{\text{the old salary}}_{43,200} \;\underset{+}{\text{plus}}\; \underset{8\%}{8\%} \;\underset{\times}{\text{of}}\; \underbrace{\text{the old salary?}}_{43,200}$$

Solve. We convert 8% to a decimal and simplify.

$$\begin{aligned} x &= 43,200 + 0.08 \times 43,200 \\ &= 43,200 + 3456 \qquad \text{The raise is \$3456.} \\ &= 46,656 \end{aligned}$$

Check. To check, we note that the new salary is 100% of the old salary plus 8% of the old salary, or 108% of the old salary. Since $1.08 \times 43,200 = 46,656$, our answer checks.

State. The new salary is \$46,656.

5. ***Familiarize.*** Let a = the number of items on the test.

Translate. We translate to a proportion.

$$\frac{85}{100}=\frac{119}{a}$$

Solve.

$$\frac{85}{100}=\frac{119}{a}$$
$$85\cdot a=100\cdot 119$$
$$\frac{85\cdot a}{85}=\frac{100\cdot 119}{85}$$
$$a=140$$

Check. We can repeat the calculation. Also note that $\frac{119}{140}=85\%$. The answer checks.

State. There were 140 items on the test.

7. ***Familiarize.*** Let f = the total number of acres of farm land in the United States.

Translate. We translate to a percent equation.

47,000,000 is 5% of what number?

$$47,000,000 = 5\% \cdot f$$

Solve.

$$47,000,000=0.05\cdot f \quad (5\%=0.05)$$
$$\frac{47,000,000}{0.05}=\frac{0.05\cdot f}{0.05} \quad \text{Dividing by 0.05}$$
$$940,000,000=f$$

Check. We find 5% of 940,000,000: $0.05\cdot 940,000,000 = 47,000,000$. Since this is the number of acres of farm land in Kansas, the answer checks.

State. There are 940,000,000 acres of farm land in the United States.

9. ***Familiarize.*** Since the car depreciates 25% in the first year, its value after the first year is 100% − 25%, or 75%, of the original value. To find the decrease in value, we ask:

$27,300 is 75% of what?

Let b = the original cost.

Translate. We translate to an equation.

$27,300 is 75% of what?

$$\$27,300 = 75\% \times b$$

Solve.

$$27,300=75\%\times b$$
$$\frac{27,300}{75\%}=\frac{75\%\times b}{75\%}$$
$$\frac{27,300}{0.75}=b$$
$$36,400=b$$

Check. We find 25% of 36,400 and then subtract this amount from 36,400:

$$0.25\times 36,400=9100 \text{ and}$$
$$36,400-9100=27,300$$

The answer checks.

State. The original cost was $36,400.

11. ***Familiarize.*** First we find the number of items Pedro got correct. Let b represent this number.

Translate. We translate to a percent equation.

What number is 93% of 80?

$$b = 93\% \cdot 80$$

Solve. We convert 93% to decimal notation and multiply.

$$b=0.93\cdot 80=74.4$$

We subtract to find the number of items Pedro got incorrect:

$$80-74.4=5.6$$

Check. We can repeat the calculation. Also note that $93\%\cdot 80\approx 90\%\cdot 80=72\approx 74.4$. The answer checks.

State. Pedro got 74.4 items correct and 5.6 items incorrect.

13. ***Familiarize.*** Let x and y represent the number of people under the age of 15 in Egypt and in the United States, respectively.

Translate. We translate to proportions.

$$\frac{32.6}{100}=\frac{x}{75,449,000} \text{ and } \frac{20.4}{100}=\frac{y}{305,468,000}$$

Solve.

$$\frac{32.6}{100}=\frac{x}{75,449,000}$$
$$32.6\cdot 75,449,000=100\cdot x$$
$$\frac{32.6\cdot 75,449,000}{100}=\frac{100\cdot x}{100}$$
$$24,596,374=x$$

$$\frac{20.4}{100}=\frac{y}{305,468,000}$$
$$20.4\cdot 305,468,000=100\cdot y$$
$$\frac{20.4\cdot 305,468,000}{100}=\frac{100\cdot y}{100}$$
$$62,315,472=y$$

Check. We can repeat the calculations. Also note that $\frac{24,596,374}{75,449,000}\approx\frac{25,000,000}{75,000,000}=33\frac{1}{3}\%\approx 32.4\%$ and $\frac{62,315,472}{305,468,000}\approx\frac{60,000,000}{300,000,000}=20\%\approx 20.4\%$. The answer checks.

State. In Egypt, 24,596,374 people are under the age of 15. There are 62,315,472 people in this age group in the United States.

15. ***Familiarize.*** Let p = the percent of patients waiting for transplants who are waiting for a liver transplant.

Translate. We write a percent equation.

16,737 is what percent of 96,749?

$$16,737 = p \cdot 96,749$$

Solve. We divide on both sides by 96,749 and express the result in percent notation.

$$16,737 = p \cdot 96,749$$
$$\frac{16,737}{96,749} = \frac{p \cdot 96,749}{96,749}$$
$$0.173 \approx p$$
$$17.3\% = p$$

Check. We find 17.3% of 96,749:

$17.3\% \cdot 96,749 = 0.173 \cdot 96,749 = 16,737.577 \approx 16,737.$ The answer checks.

State. About 17.3% of patients waiting for a transplant are waiting for a liver transplant.

17. ***Familiarize.*** Let $t =$ the total cost of the meal, including the tip.

Translate.

$$\underbrace{\text{Total cost}}_{\downarrow\; t} \;\underset{=}{\text{is}}\; \underbrace{\text{Food cost}}_{\downarrow\; 195} \;\underset{+}{\text{plus}}\; \underset{18\%}{18\%} \;\underset{\cdot}{\text{of}}\; \underbrace{\text{Food cost}}_{\downarrow\; 195}$$

Solve. We convert 18% to decimal notation and simplify.

$$t = 195 + 0.18 \cdot 195$$
$$t = 195 + 35.10 \quad \text{The tip is \$35.10.}$$
$$t = 230.10$$

Check. To check, note that the total cost of the meal is 100% of the cost of the food plus 18% of the cost of the food, or 118% of the cost of the food. Since $1.18 \times \$195 = \$230.10$, the answer checks.

State. The total cost of the meal is $230.10.

19. ***Familiarize.*** Let $f =$ the number of fast-food cooks in the United States.

Translate. We translate to a percent equation.

$$392,850 \text{ is } 64.2\% \text{ of } \underbrace{\text{what number}}?$$
$$392,850 = 64.2\% \times f$$

Solve.

$$392,850 = 64.2\% \times f$$
$$392,850 = 0.642 \times f \quad (64.2\% = 0.642)$$
$$\frac{392,850}{0.642} = \frac{0.642 \times f}{0.642} \quad \text{Dividing by 0.642}$$
$$612,000 \approx f$$

Check. We find 64.2% of 612,000:

$64.2\% \cdot 612,000 = 0.642 \times 612,000 = 392,904.$ Since 392,904 is close to 392,850, the answer checks. (Remember, we rounded to get 612,000.)

State. There are about 612,000 fast-food cooks in the United States.

21. ***Familiarize.*** First we find the amount of the solution that is alcohol. We let $a =$ this amount, in mL.

Translate. We translate to a percent equation.

$$\text{What is } 8\% \text{ of } 540?$$
$$a = 8\% \times 540$$

Solve. We convert 8% to decimal notation and multiply.

$$8\% \times 540 = 0.08 \times 540 = 43.2$$

Now we find the amount that is water. We let $w =$ this amount, in mL.

$$\underbrace{\text{Total amount}}_{\downarrow\; 540} \;\underset{-}{\text{minus}}\; \underbrace{\text{Amount of alcohol}}_{\downarrow\; 43.2} \;\underset{=}{\text{is}}\; \underbrace{\text{Amount of water}}_{\downarrow\; w}$$

To solve the equation we carry out the subtraction.

$$w = 540 - 43.2 = 496.8$$

Check. We can repeat the calculations. Also, observe that, since 8% of the solution is alcohol, 92% is water. Because 92% of $540 = 0.92 \times 540 = 496.8$, our answer checks.

State. The solution contains 43.2 mL of alcohol and 496.8 mL of water.

23. ***Familiarize.*** Let f, a, n, and m represent the percent of people in active military service in the Air Force, Army, Navy, and Marines, respectively.

Translate. We translate to proportions.

$$\frac{f}{100} = \frac{349,000}{1,385,000}; \quad \frac{a}{100} = \frac{505,000}{1,385,000};$$
$$\frac{n}{100} = \frac{350,000}{1,385,000}; \quad \frac{m}{100} = \frac{180,000}{1,385,000}$$

Solve. We solve each proportion, beginning by equating cross products.

$$f \cdot 1,385,000 = 100 \cdot 349,000$$
$$\frac{f \cdot 1,385,000}{1,385,000} = \frac{100 \cdot 349,000}{1,385,000}$$
$$f \approx 25.2$$

$$a \cdot 1,385,000 = 100 \cdot 505,000$$
$$\frac{a \cdot 1,385,000}{1,385,000} = \frac{100 \cdot 505,000}{1,385,000}$$
$$a \approx 36.5$$

$$n \cdot 1,385,000 = 100 \cdot 350,000$$
$$\frac{n \cdot 1,385,000}{1,385,000} = \frac{100 \cdot 350,000}{1,385,000}$$
$$n \approx 25.3$$

$$m \cdot 1,385,000 = 100 \cdot 180,000$$
$$\frac{m \cdot 1,385,000}{1,385,000} = \frac{100 \cdot 180,000}{1,385,000}$$
$$m \approx 13.0$$

Check. We can repeat the calculations. Also note that the sum of the percents is $25.2\% + 36.5\% + 25.3\% + 13.0\%$, or 100%, so the answer seems reasonable.

State. The percent of the people in active military service in the United States are as follows: Air Force: 25.2%, Army: 36.5%, Navy: 25.3%, Marines: 13.0%.

25. ***Familiarize.*** Use the drawing in the text to visualize the situation. Note that the increase in the amount was \$42.

Let $n =$ the percent of increase.

Translate. We translate to a percent equation.

$$\underbrace{\$42 \text{ is what percent of } \$800?}$$

$$42 = n \times 800$$

Solve. We divide by 800 on both sides and convert the result to percent notation.

$$42 = n \times 800$$
$$\frac{42}{800} = \frac{n \times 800}{800}$$
$$0.05 = n$$
$$5\% = n$$

Check. Find 5% of 840: $5\% \times 840 = 0.05 \times 840 = 42$. Since this is the amount of the increase, the answer checks.

State. The percent of increase was 5%.

27. ***Familiarize.*** We use the drawing in the text to visualize the situation. Note that the weight loss is 24 lb.

We let $n =$ the percent of decrease.

Translate. We translate to a percent equation.

\$24 is what percent of \$160?

$$24 = n \times 160$$

Solve. To solve the equation, we divide on both sides by 160 and convert the result to percent notation.

$$\frac{24}{160} = \frac{n \times 160}{160}$$
$$0.15 = n$$
$$15\% = n$$

Check. We find 15% of 160: $15\% \times 160 = 0.15 \times 160 = 24$. Since this is the amount of weight lost, the answer checks.

State. The percent of decrease was 15%.

29. ***Familiarize.*** First we find the amount of decrease.

$$\begin{array}{r} \$\ 23.43 \\ -\ 15.31 \\ \hline \$\ \ 8.12 \end{array}$$

Let $N =$ the percent of decrease.

Translate. We translate to a proportion.

$$\frac{N}{100} = \frac{8.12}{23.43}$$

Solve.

$$\frac{N}{100} = \frac{8.12}{23.43}$$
$$N \cdot 23.43 = 100 \cdot 8.12$$
$$\frac{N \cdot 23.43}{23.43} = \frac{100 \cdot 8.12}{23.43}$$
$$N \approx 34.7$$

Check. We can repeat the calculations. Also note that 34.7% of $\$23.43 \approx 30\% \cdot \$25 = \$7.50 \approx \$8.12$. The answer checks.

State. The percent of decrease is about 34.7%.

31. ***Familiarize.*** First we find the amount of increase, in billions of rides.

$$\begin{array}{r} 2.8 \\ -\ 2.1 \\ \hline 0.7 \end{array}$$

Let $p =$ the percent of increase.

Translate. We translate to a percent equation.

0.7 is what percent of 2.1?

$$0.7 = p \times 2.1$$

Solve. We divide on both sides by 2.1 and convert the result to percent notation.

$$0.7 = p \times 2.1$$
$$\frac{0.7}{2.1} = \frac{p \times 2.1}{2.1}$$
$$0.\overline{3} = p$$
$$33.\overline{3}\% = p, \text{ or}$$
$$33\frac{1}{3}\% = p$$

Check. We find $33\frac{1}{3}\%$ of 2.1: $33\frac{1}{3}\% \times 2.1 = \frac{1}{3} \times 2.1 = 0.7$. This is the amount of the increase, so the answer checks.

State. The percent of increase was $33\frac{1}{3}\%$.

33. ***Familiarize.*** First we find the amount of overdraft fees in 2001, in billions of dollars.

$$\begin{array}{r} 45.6 \\ -\ 15.1 \\ \hline 30.5 \end{array}$$

Let $p =$ the percent of increase.

Translate. We translate to a percent equation.

15.1 is what percent of 30.5?

$$15.1 = p \times 30.5$$

Solve. We divide on both sides by 30.5 and convert the result to percent notation.

$$15.1 = p \times 30.5$$
$$\frac{15.1}{30.5} = \frac{p \times 30.5}{30.5}$$
$$0.495 \approx p$$
$$49.5\% \approx p$$

Check. We find 49.5% of 30.5:

$$49.5\% \times 30.5 = 0.495 \times 30.5 = 15.0975 \approx 15.1.$$

This is the amount of the increase, so the answer checks.

State. The percent of increase is about 49.5%.

35. ***Familiarize.*** First we find the amount of decrease.

$$\begin{array}{r} 114,469 \\ -\ \ 46,866 \\ \hline 67,603 \end{array}$$

Let $N =$ the percent of decrease.

Translate. We translate to a proportion.

$$\frac{N}{100} = \frac{67,603}{114,469}$$

Solve.

$$\frac{N}{100} = \frac{67,603}{114,469}$$
$$N \cdot 114,469 = 100 \cdot 67,603$$
$$\frac{N \cdot 114,469}{114,469} = \frac{100 \cdot 67,603}{114,469}$$
$$N \approx 59$$

Check. If the percent of decrease is 59%, then the 2008 number is 100% − 59%, or 41% of the 2007 number. We find 41% of 114,469: $41\% \cdot 114,469 = 0.41 \cdot 114,469 \approx 46,932 \approx 46,866$. The answer checks.

State. The percent of decrease is about 59%.

37. ***Familiarize.*** First we find the amount of decrease.

$$\begin{array}{r} 1\,7\,3,7\,9\,4 \\ -\ 1\,5\,7,2\,8\,4 \\ \hline 1\,6,5\,1\,0 \end{array}$$

Let p = the percent of decrease.

Translate. We translate to a percent equation.

$$\underbrace{16,510}\ \text{is}\ \underbrace{\text{what percent}}\ \text{of}\ 173,794?$$
$$16,510 = p \cdot 173,794$$

Solve. We divide on both sides by 173,794 and convert the result to percent notation.

$$16,510 = p \cdot 173,794$$
$$\frac{16,510}{173,794} = \frac{p \cdot 173,794}{173,794}$$
$$0.095 \approx p$$
$$9.5\% \approx p$$

Check. We find 9.5% of 173,794: $9.5\% \times 173,794 = 0.095 \times 173,794 = 16,510.43 \approx 16,510$. Since we get approximately the amount of decrease, the answer checks.

State. The percent of decrease is about 9.5%.

39. ***Familiarize.*** This is a multistep problem. First we find the area of a cross-section of a finished board and of a rough board using the formula $A = l \cdot w$. Then we find the amount of wood removed in planing and drying and finally we find the percent of wood removed. Let f = the area of a cross-section of a finished board and let r = the area of a cross-section of a rough board.

Translate. We find the areas.

$$f = 3\frac{1}{2} \cdot 1\frac{1}{2}$$
$$r = 4 \cdot 2$$

Solve. We carry out the multiplications.

$$f = 3\frac{1}{2} \cdot 1\frac{1}{2} = \frac{7}{2} \cdot \frac{3}{2} = \frac{21}{4}$$
$$r = 4 \cdot 2 = 8$$

Now we subtract to find the amount of wood removed in planing and drying.

$$8 - \frac{21}{4} = \frac{32}{4} - \frac{21}{4} = \frac{11}{4}$$

Finally we find p, the percent of wood removed in planing and drying.

$$\frac{11}{4}\ \text{is}\ \underbrace{\text{what percent}}\ \text{of}\ 8?$$
$$\frac{11}{4} = p \cdot 8$$

We solve the equation.

$$\frac{11}{4} = p \cdot 8$$
$$\frac{1}{8} \cdot \frac{11}{4} = p$$
$$\frac{11}{32} = p$$
$$0.34375 = p$$
$$34.375\% = p,\ \text{or}$$
$$34\frac{3}{8}\% = p$$

Check. We repeat the calculations. The answer checks.

State. 34.375%, or $34\frac{3}{8}\%$, of the wood is removed in planing and drying.

41. ***Familiarize.*** First we subtract to find the amount of change.

$$\begin{array}{r} 6\,2\,3,9\,0\,8 \\ -\,6\,0\,8,8\,2\,7 \\ \hline 1\,5,0\,8\,1 \end{array}$$

Now let N = the percent of change.

Translate. We translate to a proportion.

$$\frac{N}{100} = \frac{15,081}{608,827}$$

Solve.

$$\frac{N}{100} = \frac{15,081}{608,827}$$
$$N \cdot 608,827 = 100 \cdot 15,081$$
$$N = \frac{100 \cdot 15,081}{608,827}$$
$$N \approx 2.5$$

Check. We can repeat the calculations. Also note that $102.5\% \cdot 608,827 \approx 624,028 \approx 623,908$. The answer checks.

State. The population of Vermont increased by 15,081. This was a 2.5% increase.

43. ***Familiarize.*** First we subtract to find the population in 2000.

$$\begin{array}{r} 6,1\,6\,6,3\,1\,8 \\ -\,1,0\,3\,5,6\,8\,6 \\ \hline 5,1\,3\,0,6\,3\,2 \end{array}$$

Now let p = the percent of change.

Translate. We translate to an equation.

$$1,035,686\ \text{is}\ \underbrace{\text{what percent}}\ \text{of}\ 5,130,632?$$
$$1,035,686 = p \cdot 5,130,632$$

Solve.

$$1{,}035{,}686 = p \cdot 5{,}130{,}632$$
$$\frac{1{,}035{,}686}{5{,}130{,}632} = p$$
$$0.202 \approx p$$
$$20.2\% \approx p$$

Check. We can repeat the calculations. Also note that $120.2\% \cdot 5{,}130{,}632 \approx 6{,}167{,}020 \approx 6{,}166{,}318$. The answer checks.

State. The population of Arizona was 5,130,632 in 2000. The population had increased by about 20.2% in 2006.

45. *Familiarize*. First we add to find the population in 2006.

$$\begin{array}{r} 1{,}293{,}953 \\ +\ 172{,}512 \\ \hline 1{,}466{,}465 \end{array}$$

Now let $N =$ the percent of change.

Translate. We translate to a proportion.

$$\frac{N}{100} = \frac{172{,}512}{1{,}293{,}953}$$

Solve.

$$\frac{N}{100} = \frac{172{,}512}{1{,}293{,}953}$$
$$N \cdot 1{,}293{,}953 = 100 \cdot 172{,}512$$
$$N = \frac{100 \cdot 172{,}512}{1{,}293{,}953}$$
$$N \approx 13.3$$

Check. We can repeat the calculations. Also note that $113.2\% \cdot 1{,}293{,}953 \approx 1{,}464{,}755 \approx 1{,}466{,}465$. The answer checks.

State. The population of Idaho in 2006 was 1,466,465. The population had increased by about 13.2% in 2006.

47. *Familiarize*. First we find the amount of decrease.

$$\begin{array}{r} 642{,}200 \\ -635{,}867 \\ \hline 6333 \end{array}$$

Let $p =$ the percent of decrease.

Translate. We translate to an equation.

6333 is what percent of 642, 200?

$$6333 = p \cdot 642{,}200$$

Solve.

$$6333 = p \cdot 642{,}200$$
$$\frac{6333}{642{,}200} = \frac{p \cdot 642{,}200}{642{,}200}$$
$$0.0010 \approx p$$
$$1.0\% \approx p$$

Check. We can repeat the calculations. Also note that the population in 2006 was $100\% - 1\%$, or 99%, of the population in 2000. We find 99% of 642,200: $0.99 \times 642{,}200 = 635{,}778 \approx 635{,}867$. The answer checks.

State. The percent of decrease was about 1.0%.

49. $\frac{25}{11} = 25 \div 11$

$$\begin{array}{r} 2.27 \\ 11\overline{)25.00} \\ 22 \\ \hline 30 \\ 22 \\ \hline 80 \\ 77 \\ \hline 3 \end{array}$$

Since the remainders begin to repeat, we have a repeating decimal.

$$\frac{25}{11} = 2.\overline{27}$$

51. $\frac{27}{8} = 27 \div 8$

$$\begin{array}{r} 3.375 \\ 8\overline{)27.000} \\ 24 \\ \hline 30 \\ 24 \\ \hline 60 \\ 56 \\ \hline 40 \\ 40 \\ \hline 0 \end{array}$$

$$\frac{27}{8} = 3.375$$

We could also do this conversion as follows:

$$\frac{27}{8} = \frac{27}{8} \cdot \frac{125}{125} = \frac{3375}{1000} = 3.375$$

53. $\frac{23}{25} = \frac{23}{25} \cdot \frac{4}{4} = \frac{92}{100} = 0.92$

55. $\frac{14}{32} = 14 \div 32$

$$\begin{array}{r} 0.4375 \\ 32\overline{)14.0000} \\ 128 \\ \hline 120 \\ 96 \\ \hline 240 \\ 224 \\ \hline 160 \\ 160 \\ \hline 0 \end{array}$$

$$\frac{14}{32} = 0.4375$$

(Note that we could have simplified the fraction first, getting $\frac{7}{16}$ and then found the quotient $7 \div 16$.)

57. Since 10,000 has 4 zeros, we move the decimal point in the number in the numerator 4 places to the left.

$$\frac{34{,}809}{10{,}000} = 3.4809$$

59. ***Familiarize.*** Let $c =$ the cost of the dinner before the tip and without the coupon.

Translate.

$$\underbrace{\text{Cost of food}}_{\downarrow\; c}\ \underset{\downarrow\;+}{\text{plus}}\ \underset{\downarrow\;20\%}{20\%}\ \underset{\downarrow\;\cdot}{\text{of}}\ \underbrace{\text{Cost of food}}_{\downarrow\; c}\ \underset{\downarrow\;-}{\text{minus}}\ \underset{\downarrow\;10}{\$10}\ \underset{\downarrow\;=}{\text{is}}\ \underset{\downarrow\;40.40}{\$40.40}$$

Solve.

$$c + 20\% \cdot c - 10 = 40.40$$
$$1 \cdot c + 0.2 \cdot c - 10 = 40.40$$
$$1.2 \cdot c - 10 = 40.40$$
$$1.2 \cdot c - 10 + 10 = 40.40 + 10$$
$$1.2 \cdot c = 50.40$$
$$\frac{1.2 \cdot c}{1.2} = \frac{50.40}{1.2}$$
$$c = 42$$

Check. 20% of \$42 is $0.2 \times \$42 = \$8.40$; $\$42 + \$8.40 = \$50.40$ and $\$50.40 - \$10 = \$40.40$. The answer checks.

State. Before the tip and without the coupon, the meal would cost \$42.

Exercise Set 6.7

1. The sales tax on an item costing \$239 is

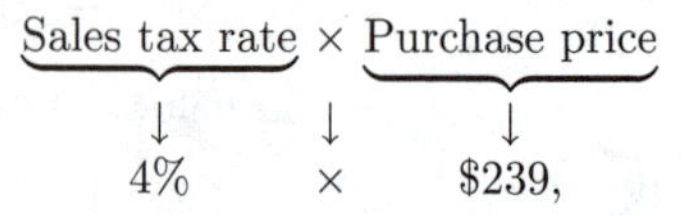

$$\underbrace{\text{Sales tax rate}}_{\downarrow\; 4\%} \underset{\downarrow\;\times}{\times} \underbrace{\text{Purchase price}}_{\downarrow\; \$239,}$$

or 0.04×239, or 9.56. Thus the tax is \$9.56.

3. The sales tax on an item costing \$29.50 is

$$\underbrace{\text{Sales tax rate}}_{\downarrow\; 5.5\%} \underset{\downarrow\;\times}{\times} \underbrace{\text{Purchase price}}_{\downarrow\; \$29.50,}$$

or 0.055×29.50, or 1.62 (rounded to the nearest cent). Thus the tax is \$1.62.

5. a) We first find the cost of the pillows. It is

$4 \times \$39.95 = \$159.80.$

b) The sales tax on items costing \$159.80 is

$$\underbrace{\text{Sales tax rate}}_{\downarrow\; 7.25\%} \underset{\downarrow\;\times}{\times} \underbrace{\text{Purchase price}}_{\downarrow\; \$159.80,}$$

or 0.0725×159.80, or about 11.59. Thus the tax is \$11.59.

c) The total price is given by the purchase price plus the sales tax:

$\$159.80 + \$11.59 = \$171.39.$

To check, note that the total price is the purchase price plus 7.25% of the purchase price. Thus the total price is 107.25% of the purchase price. Since $1.0725 \times \$159.80 = \$171.3855 \approx \$171.39$, we have a check. The total price is \$171.39.

7. *Rephrase:* $\underbrace{\text{Sales tax}}$ is $\underbrace{\text{what percent}}$ of $\underbrace{\text{purchase price?}}$

Translate: $30 = r \times 750$

To solve the equation, we divide on both sides by 750.

$$\frac{30}{750} = \frac{r \times 750}{750}$$
$$0.04 = r$$
$$4\% = r$$

The sales tax rate is 4%.

9. *Rephrase:* $\underbrace{\text{Sales tax}}$ is 2% of what?

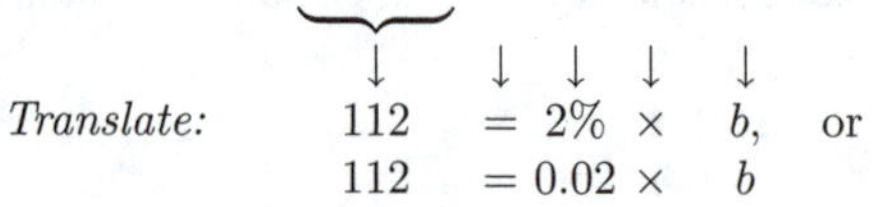

Translate: $112 = 2\% \times b$, or
$112 = 0.02 \times b$

To solve the equation, we divide on both sides by 0.02.

$$\frac{112}{0.02} = \frac{0.02 \times b}{0.02}$$
$$5600 = b$$

The purchase price is \$5600.

11. a) We first find the cost of the chocolates. It is

$6 \times \$17.95 = \$107.70.$

b) The total tax rate is the city tax rate plus the state tax rate, or $4.375\% + 4\% = 8.375\%$. The sales tax paid on items costing \$107.70 is

$$\underbrace{\text{Sales tax rate}}_{\downarrow\; 8.375\%} \underset{\downarrow\;\times}{\times} \underbrace{\text{Purchase price}}_{\downarrow\; \$107.70,}$$

or $0.08375 \times \$107.70$, or about \$9.02. Thus the tax is \$9.02.

c) The total price is given by the purchase price plus the sales tax:

$\$107.70 + \$9.02 = \$116.72.$

To check, note that the total price is the purchase price plus 8.375% of the purchase price. Thus the total price is 108.375% of the purchase price. Since $1.08375 \times 107.70 \approx 116.72$, we have a check. The total amount paid for the 6 boxes of chocolates is \$116.72.

13. a) We first find the cost of the ceiling fans. It is

$3 \times \$84.49 = \$253.47.$

b) The total tax rate is the county tax rate plus the state tax rate, or $2.5\% + 6.5\% = 9\%$. The sales tax paid on items costing \$253.47 is

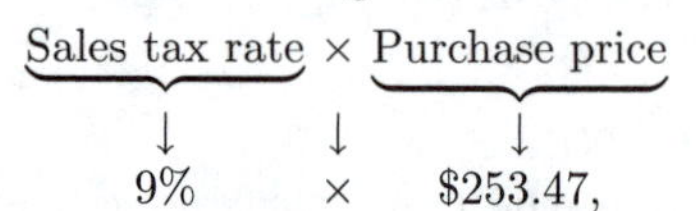

$$\underbrace{\text{Sales tax rate}}_{\downarrow\; 9\%} \underset{\downarrow\;\times}{\times} \underbrace{\text{Purchase price}}_{\downarrow\; \$253.47,}$$

or $0.09 \times \$253.47$, or about \$22.81. Thus the tax is \$22.81.

c) The total price is given by the purchase price plus the sales tax:

$\$253.47 + \$22.81 = \$276.28.$

To check, note that the total price is the purchase price plus 9% of the purchase price. Thus the total price is 109% of the purchase price. Since $1.09 \times 253.47 \approx 276.28$, we have a check. The total amount paid for the 3 ceiling fans is \$276.28.

15. a) We first find the cost of the basketballs. It is

$$6 \times \$29.95 = \$179.70.$$

b) The total tax rate is the city tax rate plus the county tax rate plus the state tax rate, or $1\% + 3\% + 4\% = 8\%$. The sales tax paid on items costing \$179.70 is

$$\underbrace{\text{Sales tax rate}}_{\downarrow\ 8\%} \ \underset{\downarrow}{\times}\ \underbrace{\text{Purchase price}}_{\downarrow\ \$179.70,}$$

or $0.08 \times \$179.70$, or about \$14.38. Thus the tax is \$14.38.

c) The total price is given by the purchase price plus the sales tax:

$$\$179.70 + \$14.38 = \$194.08.$$

To check, note that the total price is the purchase price plus 8% of the purchase price. Thus the total price is 108% of the purchase price. Since $1.08 \times 179.70 \approx 194.08$, we have a check. The total amount paid for the 6 basketballs is \$194.08.

17. Commission = Commission rate × Sales

$$C = 21\% \times 12{,}500$$

This tells us what to do. We multiply.

$$\begin{array}{r} 12,500 \\ \times \quad 0.21 \\ \hline 12500 \\ 250000 \\ \hline 2625.00 \end{array} \qquad (21\% = 0.21)$$

The commission is \$2625.

19. Commission = Commission rate × Sales

$$408 = r \times 3400$$

To solve this equation we divide on both sides by 3400:

$$\frac{408}{3400} = \frac{r \times 3400}{3400}$$
$$0.12 = r$$
$$12\% = r$$

The commission rate is 12%.

21. Commission = Commission rate × S

$$\$552 = 40\% \times S$$

To solve this equation we divide on both sides by 0.4.

$$\frac{552}{0.4} = \frac{0.4 \times S}{0.4}$$
$$1380 = S$$

\$1380 worth of clothing was sold.

23. Commission = Commission rate × Sales

$$1147.50 = r \times 7650$$

To solve this equation we divide on both sides by 7650.

$$\frac{1147.50}{7650} = \frac{r \times 7650}{7650}$$
$$0.15 = r$$
$$15\% = r$$

The commission rate is 15%.

25. First we find the commission on the first \$1000 of sales.

Commission = Commission rate × Sales

$$C = 4\% \times 1000$$

This tells us what to do. We multiply.

$$\begin{array}{r} 1000 \\ \times \ 0.04 \\ \hline 40.00 \end{array}$$

The commission on the first \$1000 of sales is \$40.

Next we subtract to find the amount of sales over \$1000.

$$\$5500 - \$1000 = \$4500$$

Sabrina had \$4500 in sales over \$1000.

Then we find the commission on the sales over \$1000.

Commission = Commission rate × Sales

$$C = 7\% \times 4500$$

This tells us what to do. We multiply.

$$\begin{array}{r} 4500 \\ \times \ 0.07 \\ \hline 315.00 \end{array}$$

The commission on the sales over \$1000 is \$315.

Finally we add to find the total commission.

$$\$40 + \$315 = \$355$$

The total commission is \$355.

27. Discount = Rate of discount × Original price

$$D = 10\% \times \$300$$

Convert 10% to decimal notation and multiply.

$$\begin{array}{r} 300 \\ \times \quad 0.1 \\ \hline 30.0 \end{array} \qquad (10\% = 0.10 = 0.1)$$

The discount is \$30.

Sale price = Original price − Discount

$$S = 300 - 30$$

We subtract:

$$\begin{array}{r} 300 \\ -\ 30 \\ \hline 270 \end{array}$$

To check, note that the sale price is 90% of the original price: $0.9 \times 300 = 270$.

The sale price is \$270.

29. Discount = Rate of discount × Original price

$$12.50 = 10\% \times M$$

To solve the equation we divide on both sides by 0.1.

$$\frac{12.50}{0.1} = \frac{0.1 \times M}{0.1}$$
$$125 = M$$

The original price is \$125.

Sale price = Original price − Discount
$S = 125.00 - 12.50$

We subtract:
$$\begin{array}{r} 125.00 \\ -\ 12.50 \\ \hline 112.50 \end{array}$$

To check, note that the sale price is 90% of the original price: $0.9 \times 125 = 112.50$.

The sale price is $112.50.

31. Discount = Rate of discount × Original price
$240 = r \times 600$

To solve the equation we divide on both sides by 600.
$$\frac{240}{600} = \frac{r \times 600}{600}$$
We can simplify by removing a factor of 1:
$$r = \frac{240}{600} = \frac{2}{5} \cdot \frac{120}{120} = \frac{2}{5} = 0.4 = 40\%$$
The rate of discount is 40%.

Sale price = Original price − Discount
$S = 600 - 240$

We subtract:
$$\begin{array}{r} 600 \\ -\ 240 \\ \hline 360 \end{array}$$

To check, note that a 40% discount rate means that 60% of the original price is paid. Since $\frac{360}{600} = 0.6$, or 60%, we have a check.

The sale price is $360.

33. Discount = Original price − Sale price
$D = 3999 - 3150$

We subtract:
$$\begin{array}{r} 3999 \\ -\ 3150 \\ \hline 849 \end{array}$$

The discount is $849.

Discount = Rate of discount × Original price
$849 = R \times 3999$

To solve the equation we divide on both sides by 3999.
$$\frac{849}{3999} = \frac{R \times 3999}{3999}$$
$$0.212 \approx R$$
$$21.2\% \approx R$$

To check, note that a discount rate of 21.2% means that 78.8% of the original price is paid: $0.788 \times 3999 = 3151.212 \approx 3150$. Since that is the sale price, the answer checks.

The rate of discount is 21.2%.

35. $\frac{5}{9} = 5 \div 9$

$$\begin{array}{r} 0.55 \\ 9\overline{)5.00} \\ 45 \\ \hline 50 \\ 45 \\ \hline 5 \end{array}$$

We get a repeating decimal.
$$\frac{5}{9} = 0.\overline{5}$$

37. $\frac{11}{12} = 11 \div 12$

$$\begin{array}{r} 0.9166 \\ 12\overline{)11.0000} \\ 108 \\ \hline 20 \\ 12 \\ \hline 80 \\ 72 \\ \hline 80 \\ 72 \\ \hline 8 \end{array}$$

We get a repeating decimal.
$$\frac{11}{12} = 0.91\overline{6}$$

39. $\frac{15}{7} = 15 \div 7$

$$\begin{array}{r} 2.142857 \\ 7\overline{)15.000000} \\ 14 \\ \hline 10 \\ 7 \\ \hline 30 \\ 28 \\ \hline 20 \\ 14 \\ \hline 60 \\ 56 \\ \hline 40 \\ 35 \\ \hline 50 \\ 49 \\ \hline 1 \end{array}$$

We get a repeating decimal.
$$\frac{15}{7} = 2.\overline{142857}$$

41.
$$\begin{aligned} 4.03 \text{ trillion} &= 4.03 \times 1 \text{ trillion} \\ &= 4.03 \times 1{,}000{,}000{,}000{,}000 \\ &= 4{,}030{,}000{,}000{,}000 \end{aligned}$$

43.
$$\begin{aligned} 42.7 \text{ million} &= 42.7 \times 1 \text{ million} \\ &= 42.7 \times 1{,}000{,}000 \\ &= 42{,}700{,}000 \end{aligned}$$

45. First we find the commission on the first $5000 in sales.

Commission = Commission rate × Sales
$C = 10\% \times 5000$

Using a calculator we find that $0.1 \times 5000 = 500$, so the commission on the first $5000 in sales was $500. We subtract to find the additional commission:

$2405 − $500 = $1905

Now we find the amount of sales required to earn $1905 at a commission rate of 15%.

Commission = Commission rate × Sales
$1905 = 15\% \times S$

Using a calculator to divide 1905 by 15%, or 0.15, we get 12,700.

Finally we add to find the total sales:

$$\$5000 + \$12,700 = \$17,700$$

Exercise Set 6.8

1. $I = P \cdot r \cdot t$
$= \$200 \times 4\% \times 1$
$= \$200 \times 0.04$
$= \$8$

3. $I = P \cdot r \cdot t$
$= \$4300 \times 10.56\% \times \dfrac{1}{4}$
$= \dfrac{\$4300 \times 0.1056}{4}$
$= \$113.52$

5. $I = P \cdot r \cdot t$
$= \$20,000 \times 4\frac{5}{8}\% \times 1$
$= \$20,000 \times 0.04625$
$= \$925$

7. $I = P \cdot r \cdot t$
$= \$50,000 \times 5\frac{3}{8}\% \times \dfrac{1}{4}$
$= \dfrac{\$50,000 \times 0.05375}{4}$
$\approx \$671.88$

9. a) We express 60 days as a fractional part of a year and find the interest.

$I = P \cdot r \cdot t$
$= \$10,000 \times 9\% \times \dfrac{60}{365}$
$= \$10,000 \times 0.09 \times \dfrac{60}{365}$
$\approx \$147.95$ Using a calculator

The interest due for 60 days is \$147.95.

b) The total amount that must be paid after 60 days is the principal plus the interest.

$$10,000 + 147.95 = 10,147.95$$

The total amount due is \$10,147.95.

11. a) We express 90 days as a fractional part of a year and find the interest.

$I = P \cdot r \cdot t$
$= \$6500 \times 5\frac{1}{4}\% \times \dfrac{90}{365}$
$= \$6500 \times 0.0525 \times \dfrac{90}{365}$
$\approx \$84.14$ Using a calculator

The interest due for 90 days is \$84.14.

b) The total amount that must be paid after 90 days is the principal plus the interest.

$$6500 + 84.14 = 6584.14$$

The total amount due is \$6584.14.

13. a) We express 30 days as a fractional part of a year and find the interest.

$I = P \cdot r \cdot t$
$= \$5600 \times 10\% \times \dfrac{30}{365}$
$= \$5600 \times 0.1 \times \dfrac{30}{365}$
$\approx \$46.03$ Using a calculator

The interest due for 30 days is \$46.03.

b) The total amount that must be paid after 30 days is the principal plus the interest.

$$5600 + 46.03 = 5646.03$$

The total amount due is \$5646.03.

15. a) After 1 year, the account will contain 105% of \$400.

$1.05 \times \$400 = \$420$

$$\begin{array}{r} 4\,0\,0 \\ \times\ \ 1.0\,5 \\ \hline 2\,0\,0\,0 \\ 4\,0\,0\,0\,0 \\ \hline 4\,2\,0.0\,0 \end{array}$$

b) At the end of the second year, the account will contain 1.05% of \$420.

$1.05 \times \$420 = \$441$

$$\begin{array}{r} 4\,2\,0 \\ \times\ \ 1.0\,5 \\ \hline 2\,1\,0\,0 \\ 4\,2\,0\,0\,0 \\ \hline 4\,4\,1.0\,0 \end{array}$$

The amount in the account after 2 years is \$441.

(Note that we could have used the formula $A = P \cdot \left(1 + \dfrac{r}{n}\right)^{n \cdot t}$, substituting \$400 for P, 5% for r, 1 for n, and 2 for t.)

17. We use the compound interest formula, substituting \$2000 for P, 8.8% for r, 1 for n, and 4 for t.

$A = P \cdot \left(1 + \dfrac{r}{n}\right)^{n \cdot t}$
$= \$2000 \cdot \left(1 + \dfrac{8.8\%}{1}\right)^{1 \cdot 4}$
$= \$2000 \cdot (1 + 0.088)^4$
$= \$2000 \cdot (1.088)^4$
$\approx \$2802.50$

The amount in the account after 4 years is \$2802.50.

19. We use the compound interest formula, substituting \$4300 for P, 10.56% for r, 1 for n, and 6 for t.

$A = P \cdot \left(1 + \dfrac{r}{n}\right)^{n \cdot t}$
$= \$4300 \cdot \left(1 + \dfrac{10.56\%}{1}\right)^{1 \cdot 6}$
$= \$4300 \cdot (1 + 0.1056)^6$
$= \$4300 \cdot (1.1056)^6$
$\approx \$7853.38$

The amount in the account after 6 years is \$7853.38.

21. We use the compound interest formula, substituting \$20,000 for P, $6\frac{5}{8}\%$ for r, 1 for n, and 25 for t.

$$\begin{aligned} A &= P \cdot \left(1 + \frac{r}{n}\right)^{n \cdot t} \\ &= \$20,000 \cdot \left(1 + \frac{6\frac{5}{8}\%}{1}\right)^{1 \cdot 25} \\ &= \$20,000 \cdot (1 + 0.06625)^{25} \\ &= \$20,000 \cdot (1.06625)^{25} \\ &\approx \$99,427.40 \end{aligned}$$

The amount in the account after 25 years is \$99,427.40.

23. We use the compound interest formula, substituting \$4000 for P, 6% for r, 2 for n, and 1 for t.

$$\begin{aligned} A &= P \cdot \left(1 + \frac{r}{n}\right)^{n \cdot t} \\ &= \$4000 \cdot \left(1 + \frac{6\%}{2}\right)^{2 \cdot 1} \\ &= \$4000 \cdot \left(1 + \frac{0.06}{2}\right)^{2} \\ &= \$4000 \cdot (1.03)^{2} \\ &= \$4243.60 \end{aligned}$$

The amount in the account after 1 year is \$4243.60.

25. We use the compound interest formula, substituting \$20,000 for P, 8.8% for r, 2 for n, and 4 for t.

$$\begin{aligned} A &= P \cdot \left(1 + \frac{r}{n}\right)^{n \cdot t} \\ &= \$20,000 \cdot \left(1 + \frac{8.8\%}{2}\right)^{2 \cdot 4} \\ &= \$20,000 \cdot \left(1 + \frac{0.088}{2}\right)^{8} \\ &= \$20,000 \cdot (1.044)^{8} \\ &\approx \$28,225.00 \end{aligned}$$

The amount in the account after 4 years is \$28,225.00.

27. We use the compound interest formula, substituting \$5000 for P, 10.56% for r, 2 for n, and 6 for t.

$$\begin{aligned} A &= P \cdot \left(1 + \frac{r}{n}\right)^{n \cdot t} \\ &= \$5000 \cdot \left(1 + \frac{10.56\%}{2}\right)^{2 \cdot 6} \\ &= \$5000 \cdot \left(1 + \frac{0.1056}{2}\right)^{12} \\ &= \$5000 \cdot (1.0528)^{12} \\ &\approx \$9270.87 \end{aligned}$$

The amount in the account after 6 years is \$9270.87.

29. We use the compound interest formula, substituting \$20,000 for P, $7\frac{5}{8}\%$ for r, 2 for n, and 25 for t.

$$\begin{aligned} A &= P \cdot \left(1 + \frac{r}{n}\right)^{n \cdot t} \\ &= \$20,000 \cdot \left(1 + \frac{7\frac{5}{8}\%}{2}\right)^{2 \cdot 25} \\ &= \$20,000 \cdot \left(1 + \frac{0.07625}{2}\right)^{50} \\ &= \$20,000 \cdot (1.038125)^{50} \\ &\approx \$129,871.09 \end{aligned}$$

The amount in the account after 25 years is \$129,871.09.

31. We use the compound interest formula, substituting \$4000 for P, 6% for r, 12 for n, and $\frac{5}{12}$ for t.

$$\begin{aligned} A &= P \cdot \left(1 + \frac{r}{n}\right)^{n \cdot t} \\ &= \$4000 \cdot \left(1 + \frac{6\%}{12}\right)^{12 \cdot \frac{5}{12}} \\ &= \$4000 \cdot \left(1 + \frac{0.06}{12}\right)^{5} \\ &= \$4000 \cdot (1.005)^{5} \\ &\approx \$4101.01 \end{aligned}$$

The amount in the account after 5 months is \$4101.01.

33. We use the compound interest formula, substituting \$1200 for P, 10% for r, 4 for n, and 1 for t.

$$\begin{aligned} A &= P \cdot \left(1 + \frac{r}{n}\right)^{n \cdot t} \\ &= \$1200 \cdot \left(1 + \frac{10\%}{4}\right)^{4 \cdot 1} \\ &= \$1200 \cdot \left(1 + \frac{0.1}{4}\right)^{4} \\ &= \$1200 \cdot (1.025)^{4} \\ &\approx \$1324.58 \end{aligned}$$

The amount in the account after 1 year is \$1324.58.

35. First we find the amount of interest on \$1278.56 at 19.6% for one month.

$$\begin{aligned} I &= P \cdot r \cdot t \\ &= \$1278.56 \times 0.196 \times \frac{1}{12} \\ &\approx \$20.88 \end{aligned}$$

We subtract to find the amount applied to decrease the principal.

$$\$25.57 - \$20.88 = \$4.69$$

We also subtract to find the balance after the payment.

$$\$1278.56 - \$4.69 = \$1273.87$$

37. a) We multiply the balance by 2%:

$$0.02 \times \$4876.54 = \$97.5308.$$

Antonio's minimum payment, rounded to the nearest dollar, is \$98.

b) We find the amount of interest on \$4876.54 at 21.3% for one month.

$$I = P \cdot r \cdot t$$
$$= \$4876.54 \times 0.213 \times \frac{1}{12}$$
$$\approx \$86.56$$

We subtract to find the amount applied to decrease the principal in the first payment.

$$\$98 - \$86.56 = \$11.44$$

The principal is decreased by \$11.44 with the first payment.

c) We find the amount of interest on \$4876.54 at 12.6% for one month.

$$I = P \cdot r \cdot t$$
$$= \$4876.54 \times 0.126 \times \frac{1}{12}$$
$$\approx \$51.20$$

We subtract to find the amount applied to decrease the principal in the first payment.

$$\$98 - \$51.20 = \$46.80.$$

The principal is decreased by \$46.80 with the first payment.

d) With the 12.6% rate the principal was decreased by \$46.80 − \$11.44, or \$35.36 more than at the 21.3% rate. This also means that the interest at 12.6% is \$35.36 less than at 21.3%.

39. If the product of two numbers is 1, they are reciprocals of each other.

41. The number 0 is the additive identity.

43. The distance around an object is its perimeter.

45. A natural number that has exactly two different factors, only itself and 1, is called a prime number.

47. For a principle P invested at 9% compounded monthly, to find the amount in the account at the end of 1 year we would multiply P by $(1 + 0.09/12)^{12}$. Since $(1 + 0.09/12)^{12} = 1.0075^{12} \approx 1.0938$, the effective yield is approximately 9.38%.

Chapter 6 Concept Reinforcement

1. The statement is true. See Example 5 on page 258 in the text.

2. Find cross products. For $\frac{a}{b} = \frac{c}{d}$, we have $ad = bc$. For $\frac{c}{a} = \frac{d}{b}$, we have $cb = ad$, or $bc = ad$, or $ad = bc$. The proportions have the same cross products, we see that $\frac{a}{b} = \frac{c}{d}$ can be written as $\frac{c}{a} = \frac{d}{b}$. The given statement is true.

3. The statement is true.

4. We begin by writing each number in decimal notation.

$$0.5\% = 0.005$$
$$\frac{5}{1000} = 0.005$$
$$\frac{1}{2}\% = 0.5\% = 0.005$$
$$\frac{1}{5} = 0.2$$
$$0.\overline{1}$$

We see that $0.2 > 0.\overline{1}$, so $0.\overline{1}$ is not the largest number. The given statement is false.

5. Principal A grows to the amount

$$A\left(1 + \frac{0.04}{4}\right)^{4\cdot 2} = A(1 + 0.01)^8 = A(1.01)^8.$$

Principal B grows to the amount

$$B\left(1 + \frac{0.02}{2}\right)^{2\cdot 4} = B(1 + 0.01)^8 = B(1.01)^8.$$

Since $A = B$, the interest from the investments is the same. The given statement is true.

Chapter 6 Important Concepts

1. The ratio is $\frac{17}{3}$.

2. $\frac{\$120}{16\text{ hr}} = \frac{120}{16}\frac{\$}{\text{hr}} = \$7.5/\text{hr}$, or \$7.50/hr

3. We can use cross products.

$$7 \cdot 27 = 189 \qquad \frac{7}{9} \;\; \frac{21}{27} \qquad 9 \cdot 21 = 189$$

Since the cross products are the same, 189 = 189, we know that the numbers are proportional.

4.

$$\frac{9}{x} = \frac{8}{3}$$
$$9 \cdot 3 = x \cdot 8 \quad \text{Equating cross products}$$
$$\frac{9 \cdot 3}{8} = \frac{x \cdot 8}{8}$$
$$\frac{9 \cdot 3}{8} = x$$
$$\frac{27}{8} = x$$

5. ***Familiarize***. Let d = the distance between the cities in reality, in miles.

Translate. We translate to a proportion.

$$\begin{array}{rcl} \text{Map distance} \rightarrow & \dfrac{\frac{1}{2}}{50} = \dfrac{1\frac{3}{4}}{d} & \leftarrow \text{Map distance} \\ \text{Actual distance} \rightarrow & & \leftarrow \text{Actual distance} \end{array}$$

Solve.

$$\frac{1}{2}\cdot d = 50\cdot 1\frac{3}{4}$$

$$\frac{1}{2}\cdot d = 50\cdot \frac{7}{4}$$

$$d = \frac{2}{1}\cdot\frac{50}{1}\cdot\frac{7}{4} \quad \text{Dividing by } \frac{1}{2}$$

$$d = \frac{\cancel{2}\cdot\cancel{2}\cdot 25\cdot 7}{1\cdot 1\cdot\cancel{2}\cdot\cancel{2}}$$

$$d = 175$$

Check. We substitute in the proportion and check cross products.

$$\frac{\frac{1}{2}}{50} = \frac{1\frac{3}{4}}{175}$$

$$\frac{1}{2}\cdot 175 = \frac{175}{2};\ 50\cdot 1\frac{3}{4} = 50\cdot\frac{7}{4} = \frac{350}{4} = \frac{175}{2}$$

The cross products are the same so the answer checks.

State. The cities are 175 mi apart in reality.

6. $62\frac{5}{8}\% = 62.625\% = 62.625\times 0.01 = 0.62625$

7. First we divide to find decimal notation.

```
      0.6 3 6
1 1 ) 7.0 0 0
      6 6
        4 0
        3 3
          7 0
          6 6
            4
```

We get a repeating decimal, $0.6\overline{36}$.

$$\frac{7}{11} = 0.6\overline{36} = 63.\overline{63}\%, \text{ or } 63\frac{7}{11}\%$$

8. $6.8\% = \frac{6.8}{100} = \frac{6.8}{100}\cdot\frac{10}{10} = \frac{68}{1000} = \frac{4\cdot 17}{4\cdot 250} =$

$$\frac{4}{4}\cdot\frac{17}{250} = \frac{17}{250}$$

9. We translate to a percent equation.

$$12 = p\cdot 288$$

$$\frac{12}{288} = \frac{p\cdot 288}{288}$$

$$\frac{12}{288} = p$$

$$0.041\overline{6} = p$$

$$4.1\overline{6}\% = p, \text{ or}$$

$$4\frac{1}{6}\% = p$$

Thus, 12 is $4.1\overline{6}\%$, or $4\frac{1}{6}\%$, of 288.

10. We translate to a proportion.

$$\frac{3}{100} = \frac{300}{b}$$

$$3\cdot b = 100\cdot 300 \quad \text{Equating cross products}$$

$$\frac{3\cdot b}{3} = \frac{100\cdot 300}{3}$$

$$b = \frac{100\cdot 300}{3}$$

$$b = 10,000$$

Thus, 3% of 10,000 is 300.

11. ***Familiarize.*** First we find the amount of increase.

```
  4 6. 2 0
- 4 0. 0 7
  --------
    6. 1 3
```

Let $p =$ the percent of increase.

Translate.

6.13 is what percent of 40.07?

$$6.13 = p \times 40.07$$

Solve.

$$6.13 = p\times 40.07$$

$$\frac{6.13}{40.07} = \frac{p\times 40.07}{40.07}$$

$$\frac{6.13}{40.07} = p$$

$$0.153 \approx p$$

$$15.3\% = p$$

Check. If the percent of increase is 15.3%, then the 2008 cost was the 2007 cost plus 15.3% of the 2007 cost, or 115.3% of the 2007 cost. Since 115.3% of \$40.07 is $1.153\times$ \$40.07 = \$46.20071 ≈ \$46.20, the answer checks.

State. The percent of increase was about 15.3%.

12. Sales tax is what percent of purchase price

$$1102.20 = r \times 18,370$$

$$\frac{1102.20}{18,370} = \frac{r\times 18,370}{18,370}$$

$$\frac{1102.20}{18,370} = r$$

$$0.06 = r$$

$$6\% = r$$

The sales tax rate is 6%.

13. Commission = Commission rate × Sales

$$12,950 = 7\% \times S$$

To solve this equation we divide on both sides by 0.07:

$$\frac{12,950}{0.07} = \frac{0.07\times S}{0.07}$$

$$185,000 = S$$

The home sold for \$185,000.

14. We express 60 days as a fractional part of a year.

$$\begin{aligned} I &= P \cdot r \cdot t \\ &= \$2500 \times 5\frac{1}{2}\% \times \frac{60}{365} \\ &= \$2500 \times 0.055 \times \frac{60}{365} \\ &\approx \$22.60 \end{aligned}$$

The interest is \$22.60.

The total amount due is \$2500 + \$22.60 = \$2522.60.

15.
$$\begin{aligned} A &= P\left(1+\frac{r}{n}\right)^{n\cdot t} \\ &= \$6000\left(1+\frac{4\frac{3}{4}\%}{4}\right)^{4\cdot 2} \\ &= \$6000\left(1+\frac{0.0475}{4}\right)^{8} \\ &= \$6000(1.011875)^8 \\ &= \$6594.26 \end{aligned}$$

Chapter 6 Review Exercises

1. The ratio of 47 to 84 is $\frac{47}{84}$.

2. The ratio of 46 to 1.27 is $\frac{46}{1.27}$.

3. The ratio of 83 to 100 is $\frac{83}{100}$.

4. The ratio of 0.72 to 197 is $\frac{0.72}{197}$.

5. a) The ratio of 12,480 to 16,640 is $\frac{12,480}{16,640}$.

We can simplify this ratio as follows:

$$\frac{12,480}{16,640} = \frac{3\cdot 5\cdot 8\cdot 8\cdot 13}{4\cdot 5\cdot 8\cdot 8\cdot 13} = \frac{3}{4}\cdot\frac{5\cdot 8\cdot 8\cdot 13}{5\cdot 8\cdot 8\cdot 13} = \frac{3}{4}$$

b) The total of both kinds of fish sold is 12,480 lb + 16,640 lb, or 29,120 lb. Then the ratio of salmon sold to the total amount of both kinds of fish sold is $\frac{16,640}{29,120}$.

We can simplify this ratio as follows:

$$\frac{16,640}{29,120} = \frac{4\cdot 5\cdot 8\cdot 8\cdot 13}{5\cdot 7\cdot 8\cdot 8\cdot 13} = \frac{4}{7}\cdot\frac{5\cdot 8\cdot 8\cdot 13}{5\cdot 8\cdot 8\cdot 13} = \frac{4}{7}$$

6. $\frac{9}{12} = \frac{3\cdot 3}{3\cdot 4} = \frac{3}{3}\cdot\frac{3}{4} = \frac{3}{4}$

7. $\frac{3.6}{6.4} = \frac{3.6}{6.4}\cdot\frac{10}{10} = \frac{36}{64} = \frac{4\cdot 9}{4\cdot 16} = \frac{4}{4}\cdot\frac{9}{16} = \frac{9}{16}$

8. $\frac{377 \text{ mi}}{14.5 \text{ gal}} = 26\ \frac{\text{mi}}{\text{gal}}$, or 26 mpg

9. $\frac{472,500 \text{ revolutions}}{75 \text{ min}} = 6300\ \frac{\text{revolutions}}{\text{min}}$, or 6300 rpm

10. $\frac{319 \text{ gal}}{500 \text{ ft}^2} = 0.638 \text{ gal/ft}^2$

11. We can use cross products:

$9 \cdot 60 = 540$ $\frac{9}{15}$ ✕ $\frac{36}{60}$ $15 \cdot 36 = 540$

Since the cross products are the same, 540 = 540, we know that the numbers are proportional.

12. We can use cross products:

$24 \cdot 46.25 = 1110$ $\frac{24}{37}$ ✕ $\frac{40}{46.25}$ $37 \cdot 40 = 1480$

Since the cross products are not the same, $1110 \neq 1480$, we know that the numbers are not proportional.

13.
$$\begin{aligned} \frac{8}{9} &= \frac{x}{36} \\ 8\cdot 36 &= 9\cdot x \quad \text{Equating cross products} \\ \frac{8\cdot 36}{9} &= \frac{9\cdot x}{9} \\ \frac{288}{9} &= x \\ 32 &= x \end{aligned}$$

14.
$$\begin{aligned} \frac{6}{x} &= \frac{48}{56} \\ 6\cdot 56 &= x\cdot 48 \\ \frac{6\cdot 56}{48} &= \frac{x\cdot 48}{48} \\ \frac{336}{48} &= x \\ 7 &= x \end{aligned}$$

15.
$$\begin{aligned} \frac{120}{\frac{3}{7}} &= \frac{7}{x} \\ 120\cdot x &= \frac{3}{7}\cdot 7 \\ 120\cdot x &= 3 \\ \frac{120\cdot x}{120} &= \frac{3}{120} \\ x &= \frac{1}{40} \end{aligned}$$

16.
$$\begin{aligned} \frac{4.5}{120} &= \frac{0.9}{x} \\ 4.5\cdot x &= 120\cdot 0.9 \\ \frac{4.5\cdot x}{4.5} &= \frac{120\cdot 0.9}{4.5} \\ x &= \frac{108}{4.5} \\ x &= 24 \end{aligned}$$

17. ***Familiarize.*** Let d = the number of defective circuits in a lot of 585.

Translate. We translate to a proportion.

$$\begin{array}{lcl} \text{Defective} \rightarrow & \dfrac{3}{65} = \dfrac{d}{585} & \leftarrow \text{Defective} \\ \text{Total circuits} \rightarrow & & \leftarrow \text{Total circuits} \end{array}$$

Solve. We solve the proportion.

$$\frac{3}{65} = \frac{d}{585}$$
$$3 \cdot 585 = 65 \cdot d$$
$$\frac{3 \cdot 585}{65} = d$$
$$27 = d$$

Check. We substitute in the proportion and check cross products.

$$\frac{3}{65} = \frac{27}{585}$$
$$3 \cdot 585 = 1755;\ 65 \cdot 27 = 1755$$

The cross products are the same, so the answer checks.

State. It would be expected that 27 defective circuits would occur in a lot of 585 circuits.

18. a) ***Familiarize***. Let $c =$ the number of Canadian dollars equivalent to 250 U.S. dollars.

Translate. We translate to a proportion.

$$\begin{array}{lcl} \text{U.S. dollars} \rightarrow & \dfrac{1}{1.068} = \dfrac{250}{c} & \leftarrow \text{U.S. dollars} \\ \text{Canadian dollars} \rightarrow & & \leftarrow \text{Canadian dollars} \end{array}$$

Solve.

$1 \cdot c = 1.068 \cdot 250$ Equating cross products

$c = 267$

Check. We substitute in the proportion and check cross products.

$$\frac{1}{1.068} = \frac{250}{267}$$
$$1 \cdot 267 = 267;\ 1.068 \cdot 250 = 267$$

The cross products are the same, so the answer checks.

State. 250 U.S. dollars would be worth 267 Canadian dollars.

b) ***Familiarize***. Let $c =$ the cost of the sweatshirt in U.S. dollars.

Translate. We translate to a proportion.

$$\begin{array}{lcl} \text{U.S. dollars} \rightarrow & \dfrac{1}{1.068} = \dfrac{c}{50} & \leftarrow \text{U.S. dollars} \\ \text{Canadian dollars} \rightarrow & & \leftarrow \text{Canadian dollars} \end{array}$$

Solve.

$1 \cdot 50 = 1.068 \cdot c$ Equating cross products

$$\frac{1 \cdot 50}{1.068} = c$$
$$46.82 \approx c$$

Check. We substitute in the proportion and check cross products.

$$\frac{1}{1.068} = \frac{46.82}{50}$$
$$1 \cdot 50 = 50;\ 1.068 \cdot 46.82 \approx 50.004$$

The cross products are about the same. Remember that we rounded the value of c. The answer checks.

State. The sweatshirt cost \$46.82 in U.S. dollars.

19. ***Familiarize***. Let $d =$ the number of miles the train will travel in 13 hr.

Translate. We translate to a proportion.

$$\begin{array}{lcl} \text{Miles} \rightarrow & \dfrac{448}{7} = \dfrac{d}{13} & \leftarrow \text{Miles} \\ \text{Hours} \rightarrow & & \leftarrow \text{Hours} \end{array}$$

Solve.

$448 \cdot 13 = 7 \cdot d$ Equating cross products

$$\frac{448 \cdot 13}{7} = \frac{7 \cdot d}{7}$$
$$832 = d$$

Check. We find how far the train travels in 1 hr and then multiply by 13:

$$448 \div 7 = 64 \text{ and } 64 \cdot 13 = 832$$

The answer checks.

State. The train will travel 832 mi in 13 hr.

20. ***Familiarize***. Let $a =$ the number of acres required to produce 97.2 bushels of tomatoes.

Translate. We translate to a proportion.

$$\begin{array}{lc} \text{Acres} \rightarrow & \dfrac{15}{54} = \dfrac{a}{97.2} \\ \text{Bushels} \rightarrow & \end{array}$$

Solve.

$15 \cdot 97.2 = 54 \cdot a$ Equating cross products

$$\frac{15 \cdot 97.2}{54} = \frac{54 \cdot a}{54}$$
$$27 = a$$

Check. We substitute in the proportion and check cross products.

$$\frac{15}{54} = \frac{27}{97.2}$$
$$15 \cdot 97.2 = 1458;\ 54 \cdot 27 = 1458$$

The answer checks.

State. 27 acres are required to produce 97.2 bushels of tomatoes.

21. ***Familiarize***. Let $g =$ the number of pounds of trash produced in Austin, Texas in one day.

Translate. We translate to a proportion.

$$\begin{array}{lcl} \text{Trash} \rightarrow & \dfrac{23}{5} = \dfrac{g}{743,074} & \leftarrow \text{Trash} \\ \text{People} \rightarrow & & \leftarrow \text{People} \end{array}$$

Solve.

$23 \cdot 743,074 = 5 \cdot g$ Equating cross products

$$\frac{23 \cdot 743,074}{5} = \frac{5 \cdot g}{5}$$
$$3,418,140 \approx g$$

Check. We can divide to find the amount of garbage produced by one person and then multiply to find the amount produced by 743,074 people.

$23 \div 5 = 4.6$ and $4.6 \cdot 743,074 = 3,418,140.4 \approx 3,418,140$. The answer checks.

State. About 3,418,140 lb of garbage is produced in Austin, Texas in one day.

22. ***Familiarize.*** Let w = the number of inches of water to which $4\frac{1}{2}$ ft of snow melts.

Translate. We translate to a proportion.

$$\begin{array}{l} \text{Snow} \rightarrow \\ \text{Water} \rightarrow \end{array} \frac{1\frac{1}{2}}{2} = \frac{4\frac{1}{2}}{w} \begin{array}{l} \leftarrow \text{Snow} \\ \leftarrow \text{Water} \end{array}$$

Solve.

$1\frac{1}{2} \cdot w = 2 \cdot 4\frac{1}{2}$ Equating cross products

$\frac{3}{2} \cdot w = 2 \cdot \frac{9}{2}$

$\frac{3}{2} \cdot w = 9$

$w = 9 \div \frac{3}{2}$ Dividing by $\frac{3}{2}$ on both sides

$w = 9 \cdot \frac{2}{3}$

$w = \frac{9 \cdot 2}{3}$

$w = 6$

Check. We substitute in the proportion and check cross products.

$$\frac{1\frac{1}{2}}{2} = \frac{4\frac{1}{2}}{6}$$

$1\frac{1}{2} \cdot 6 = \frac{3}{2} \cdot 6 = \frac{3 \cdot 6}{2} = 9$; $2 \cdot 4\frac{1}{2} = 2 \cdot \frac{9}{2} = \frac{2 \cdot 9}{2} = 9$

The cross products are the same, so the answer checks.

State. $4\frac{1}{2}$ ft of snow will melt to 6 in. of water.

23. ***Familiarize.*** Let l = the number of lawyers we would expect to find in Chicago.

Translate. We translate to a proportion.

$$\begin{array}{l} \text{Lawyers} \rightarrow \\ \text{Population} \rightarrow \end{array} \frac{4.8}{1000} = \frac{l}{2,842,518} \begin{array}{l} \leftarrow \text{Lawyers} \\ \leftarrow \text{Population} \end{array}$$

Solve.

$4.8 \cdot 2,842,518 = 1000 \cdot l$ Equating cross products

$\frac{4.8 \cdot 2,842,518}{1000} = \frac{1000 \cdot l}{1000}$

$13,644 \approx l$

Check. We substitute in the proportion and check cross products.

$$\frac{4.8}{1000} = \frac{13,644}{2,842,518}$$

$4.8 \cdot 2,842,518 = 13,644,086.4$;

$1000 \cdot 13,644 = 13,644,000 \approx 13,644,086.4$

The answer checks.

State. We would expect that there would be about 13,644 lawyers in Chicago.

24. $4\% = 4 \times 0.01 = 0.04$

$14.4\% = 14.4 \times 0.01 = 0.144$

25. $62.1\% = 62.1 \times 0.01 = 0.621$

$84.2\% = 84.2 \times 0.01 = 0.842$

26. Move the decimal point two places to the right and write a percent symbol.

$1.7 = 170\%$

27. Move the decimal point two places to the right and write a percent symbol.

$0.065 = 6.5\%$

28. First we divide to find decimal notation.

$$\begin{array}{r} 0.375 \\ 8\overline{)3.000} \\ \underline{2\,4} \\ 60 \\ \underline{56} \\ 40 \\ \underline{40} \\ 0 \end{array}$$

$\frac{3}{8} = 0.375$

Now convert 0.375 to percent notation by moving the decimal point two places to the right and writing a percent symbol.

$\frac{3}{8} = 37.5\%$

29. First we divide to find decimal notation.

$$\begin{array}{r} 0.333 \\ 3\overline{)1.000} \\ \underline{9} \\ 10 \\ \underline{9} \\ 10 \\ \underline{9} \\ 1 \end{array}$$

We get a repeating decimal: $\frac{1}{3} = 0.33\overline{3}$. We convert $0.33\overline{3}$ to percent notation by moving the decimal point two places to the right and writing a percent symbol.

$\frac{1}{3} = 33.\overline{3}\%$, or $33\frac{1}{3}\%$

30. $24\% = \frac{24}{100} = \frac{4 \cdot 6}{4 \cdot 25} = \frac{4}{4} \cdot \frac{6}{25} = \frac{6}{25}$

31. $6.3\% = \frac{6.3}{100} = \frac{6.3}{100} \cdot \frac{10}{10} = \frac{63}{1000}$

32. *Translate.* $30.6 = p \times 90$

Solve. We divide by 90 on both sides and convert to percent notation.

$30.6 = p \times 90$

$\frac{30.6}{90} = \frac{p \times 90}{90}$

$0.34 = p$

$34\% = p$

30.6 is 34% of 90.

33. *Translate.* $63 = 84\% \times b$

Solve. We divide by 84% on both sides.

$$63 = 84\% \times b$$
$$\frac{63}{84\%} = \frac{84\% \times b}{84\%}$$
$$\frac{63}{0.84} = b$$
$$75 = b$$

63 is 84% of 75.

34. *Translate.* $a = 38\frac{1}{2}\% \times 168$

Solve. Convert $38\frac{1}{2}\%$ to decimal notation and multiply.

$$\begin{array}{r} 1\,6\,8 \\ \times\ 0.\,3\,8\,5 \\ \hline 8\,4\,0 \\ 1\,3\,4\,4\,0 \\ 5\,0\,4\,0\,0 \\ \hline 6\,4.\,6\,8\,0 \end{array}$$

64.68 is $38\frac{1}{2}\%$ of 168.

35. 24 percent of what is 16.8?

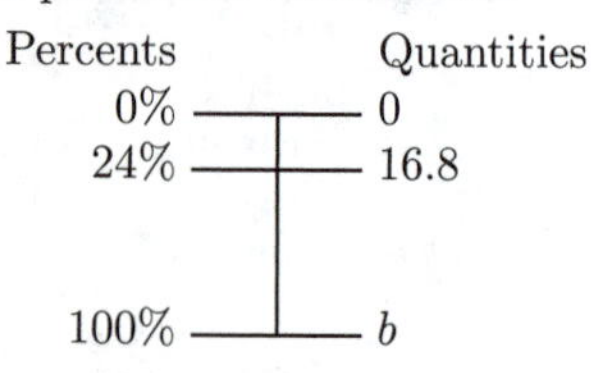

Translate: $\frac{24}{100} = \frac{16.8}{b}$

Solve: $24 \cdot b = 100 \cdot 16.8$

$$\frac{24 \cdot b}{24} = \frac{100 \cdot 16.8}{24}$$
$$b = \frac{100 \cdot 16.8}{24}$$
$$b = 70$$

24% of 70 is 16.8. The answer is 16.8.

36. 42 is what percent of 30?

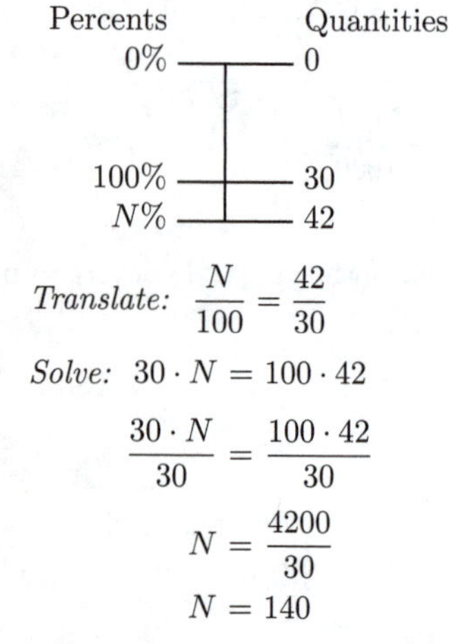

Translate: $\frac{N}{100} = \frac{42}{30}$

Solve: $30 \cdot N = 100 \cdot 42$

$$\frac{30 \cdot N}{30} = \frac{100 \cdot 42}{30}$$
$$N = \frac{4200}{30}$$
$$N = 140$$

42 is 140% of 30. The answer is 140%.

37. What is 10.5% of 84?

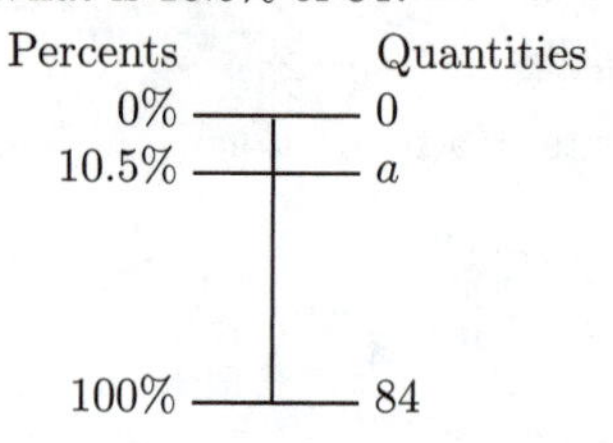

Translate: $\frac{10.5}{100} = \frac{a}{84}$

Solve: $10.5 \cdot 84 = 100 \cdot a$

$$\frac{10.5 \cdot 84}{100} = \frac{100 \cdot a}{100}$$
$$\frac{882}{100} = a$$
$$8.82 = a$$

8.82 is 10.5% of 84. The answer is 8.82.

38. ***Familiarize.*** Let c = the number of students who would choose chocolate as their favorite ice cream and b = the number who would choose butter pecan.

Translate. We translate to two equations.

$\underbrace{\text{What number}}$ is 8.9% of 2000?
$\downarrow \quad \downarrow \quad \downarrow \quad \downarrow \quad \downarrow$
$c = 8.9\% \cdot 2000$

$\underbrace{\text{What number}}$ is 4.2% of 2000?
$\downarrow \quad \downarrow \quad \downarrow \quad \downarrow \quad \downarrow$
$b = 4.2\% \cdot 2000$

Solve. We convert percent notation to decimal notation and multiply.

$c = 0.089 \cdot 2000 = 178$

$b = 0.042 \cdot 2000 = 84$

Check. We can repeat the calculation. We can also do partial checks by estimating.

$8.9\% \cdot 2000 \approx 9\% \cdot 2000 = 180;$

$4.2\% \cdot 2000 \approx 4\% \cdot 2000 = 80$

Since 180 is close to 178 and 80 is close to 84, our answers seem reasonable.

State. 178 students would choose chocolate as their favorite ice cream and 84 would choose butter pecan.

39. ***Familiarize.*** Let p = the percent of people in the U.S. who take at least one kind of prescription drug per day.

Translate. We translate to a proportion.

$$\frac{p}{100} = \frac{140.3}{305}$$

Solve. We equate cross products.

$$p \cdot 305 = 100 \cdot 140.3$$
$$\frac{p \cdot 305}{305} = \frac{100 \cdot 140.3}{305}$$
$$p = 0.46$$
$$p = 46\%$$

Check. $46\% \cdot 305$ million $= 0.46 \cdot 305$ million $= 140.3$ million. The answer checks.

State. In the U.S. 46% of the people take at least one kind of prescription drug per day.

40. ***Familiarize***. Let $w =$ the total output of water from the body per day.

Translate.

200 mL is 8% of what number?

$$200 = 8\% \cdot w$$

Solve.

$$200 = 8\% \cdot w$$
$$200 = 0.08 \cdot w$$
$$\frac{200}{0.08} = \frac{0.08 \cdot w}{0.08}$$
$$2500 = w$$

Check. $8\% \cdot 2500 = 0.08 \cdot 2500 = 200$, so the answer checks.

State. The total output of water from the body is 2500 mL per day.

41. ***Familiarize***. First we subtract to find the amount of the increase.

$$\begin{array}{r} \overset{7\ 14}{\not 8\ \not 4} \\ -\ 7\ 5 \\ \hline 9 \end{array}$$

Now let $p =$ the percent of increase.

Translate. We translate to a proportion.

$$\frac{p}{100} = \frac{9}{75}$$

Solve. We equate cross products.

$$p \cdot 75 = 100 \cdot 9$$
$$\frac{p \cdot 75}{75} = \frac{100 \cdot 9}{75}$$
$$p = 12$$

Check. $12\% \cdot 75 = 0.12 \cdot 75 = 9$, the amount of the increase, so the answer checks.

State. Jason's score increased 12%.

42. ***Familiarize***. Let $s =$ the new score. Note that the new score is the original score plus 15% of the original score.

New score is Original score plus 15% of Original score

$$s = 80 + 15\% \cdot 80$$

Solve. We convert 15% to decimal notation and carry out the computation.

$$s = 80 + 0.15 \cdot 80 = 80 + 12 = 92$$

Check. We repeat the calculation. The answer checks.

State. Jenny's new score was 92.

43. The meals tax is

Meal tax rate × Cost of meal

$$7\frac{1}{2}\% \times \$320,$$

or $0.075 \times \$320$, or \$24.

44. Sales tax is what percent of purchase price?

$$\$453.60 = r \times 7560$$

To solve the equation, we divide on both sides by 7560.

$$\frac{\$453.60}{7560} = \frac{r \times 7560}{7560}$$
$$0.06 = r$$
$$6\% = r$$

The sales tax rate is 6%.

45. Commission = Commission rate × Sales

$$753.50 = r \times 6850$$

To solve this equation, we divide on both sides by 6850.

$$\frac{753.50}{6850} = \frac{r \times 6850}{6850}$$
$$0.11 = r$$
$$11\% = r$$

The commission rate is 11%.

46. Discount = Rate of discount × Original price

$$D = 12\% \times \$350$$

Convert 12% to decimal notation and multiply.

$$\begin{array}{r} 3\ 5\ 0 \\ \times\ 0.\ 1\ 2 \\ \hline 7\ 0\ 0 \\ 3\ 5\ 0\ 0 \\ \hline 4\ 2.\ 0\ 0 \end{array}$$

The discount is \$42.

Sale price = Original price − Discount

$$S = \$350 - \$42$$

We subtract:

$$\begin{array}{r} \overset{4\ 10}{3\ \not 5\ \not 0} \\ -\ \ 4\ 2 \\ \hline 3\ 0\ 8 \end{array}$$

The sale price is \$308.

47. First we find the discount.

$$\$305 - \$262.30 = \$42.70$$

Now we find the rate of discount.

Discount = Rate of discount × Original price

$$42.70 = r \times 305$$

$$\frac{42.70}{305} = \frac{r \times 305}{305}$$
$$\frac{42.70}{305} = r$$
$$0.14 = r$$
$$14\% = r$$

The rate of discount is 14%.

48. Commission = Commission rate × Sales
$C = 7\% \times 42{,}000$

We convert 7% to decimal notation and multiply.

$$\begin{array}{r} 42,000 \\ \times \quad 0.07 \\ \hline 2940.00 \end{array}$$

The commission is \$2940.

49. First we subtract to find the discount.

$\$82 - \$67 = \$15$

Discount = Rate of discount × Original price
$15 = r \times 82$

We divide on both sides by 82.

$$\frac{15}{82} = \frac{r \times 82}{82}$$

$0.183 \approx r$

$18.3\% \approx r$

The rate of discount is about 18.3%.

50. $I = P \cdot r \cdot t$

$= \$1800 \times 6\% \times \frac{1}{3}$

$= \$1800 \times 0.06 \times \frac{1}{3}$

$= \$36$

51. a) $I = P \cdot r \cdot t$

$= \$24,000 \times 10\% \times \frac{60}{365}$

$= \$24,000 \times 0.1 \times \frac{60}{365}$

$\approx \$394.52$

b) $\$24,000 + \$394.52 = \$24,394.52$

52. $A = P \cdot \left(1 + \frac{r}{n}\right)^{n \cdot t}$

$= \$7500 \cdot \left(1 + \frac{4\%}{12}\right)^{12 \cdot \frac{1}{4}}$

$= \$7500 \cdot \left(1 + \frac{0.04}{12}\right)^{3}$

$\approx \$7575.25$

53. $A = P \cdot \left(1 + \frac{r}{n}\right)^{n \cdot t}$

$= \$8000 \cdot \left(1 + \frac{9\%}{1}\right)^{1 \cdot 2}$

$= \$8000 \cdot (1 + 0.09)^2$

$= \$8000 \cdot (1.09)^2$

$= \$9504.80$

54. a) 2% of $\$6428.74 = 0.02 \times \$6428.74 \approx \$129$

b) $I = P \cdot r \cdot t$

$= \$6428.74 \times 0.187 \times \frac{1}{12}$

$\approx \$100.18$

The amount of interest is \$100.18.

$\$129 - \$100.18 = \$28.82$, so the principal is reduced by \$28.82.

c) $I = P \cdot r \cdot t$

$= \$6428.74 \times 0.132 \times \frac{1}{12}$

$\approx \$70.72$

The amount of interest is \$70.72.

$\$129 - \$70.72 = \$58.28$, so the principal is reduced by \$58.28 with the lower interest rate.

d) With the 13.2% rate the principal was decreased by \$58.28 − \$28.82, or \$29.46, more than at the 18.7% rate. This also means that the interest at 13.2% is \$29.46 less than at 18.7%.

55. ***Familiarize***. Let p = the price of 5 dozen eggs.

Translate. We translate to a proportion.

$$\begin{array}{l} \text{Eggs} \rightarrow \\ \text{Price} \rightarrow \end{array} \frac{3}{5.04} = \frac{5}{p} \begin{array}{l} \leftarrow \text{Eggs} \\ \leftarrow \text{Price} \end{array}$$

Solve. We solve the proportion.

$$\frac{3}{5.04} = \frac{5}{p}$$

$3 \cdot p = 5.04 \cdot 5$

$$\frac{3 \cdot p}{3} = \frac{5.04 \cdot 5}{3}$$

$p = 8.4$

Check. We substitute in the proportion and check cross products.

$$\frac{3}{5.04} = \frac{5}{8.4}$$

$3 \cdot 8.4 = 25.2;\ 5.04 \cdot 5 = 25.2$

The cross products are the same, so the answer checks.

State. 5 dozen eggs would cost \$8.40.

Answer C is correct.

56. $A = 10,500\left(1 + \frac{6\%}{2}\right)^{2 \cdot 1\frac{1}{2}}$

$= 10,500\left(1 + \frac{0.06}{2}\right)^{2 \cdot \frac{3}{2}}$

$= 10,500(1 + 0.03)^3$

$= 10,500(1.03)^3$

$\approx 11,473.63$

The amount in the account is \$11,473.63. Answer C is correct.

57. Let S = the original salary. After a 3% raise, the salary becomes $103\% \cdot S$, or $1.03S$. After a 6% raise, the new salary is $1.06\% \cdot 1.03S$, or $1.06(1.03S)$. Finally, after a 9% raise, the salary is $109\% \cdot 1.06(1.03S)$, or $1.09(1.06)(1.03S)$. Multiplying, we get $1.09(1.06)(1.03S) = 1.190062S$. This is equivalent to $119.0062\% \cdot S$, so the original salary has increased about 19%.

58. First we divide to find how many gallons of finishing paint are needed.

$4950 \div 450 = 11$ gal

Next we write and solve a proportion to find how many gallons of primer are needed. Let $p =$ the amount of primer needed.

$$\begin{matrix}\text{Finishing paint} \rightarrow \\ \text{Primer} \rightarrow\end{matrix} \frac{2}{3} = \frac{11}{p} \begin{matrix}\leftarrow \text{Finishing paint} \\ \leftarrow \text{Primer}\end{matrix}$$

$$2 \cdot p = 3 \cdot 11$$
$$p = \frac{3 \cdot 11}{2}$$
$$p = \frac{33}{2}, \text{ or } 16.5$$

Thus, 11 gal of finishing paint and 16.5 gal of primer should be purchased.

Chapter 6 Discussion and Writing Exercises

1. A 40% discount is better. When successive discounts are taken, each is based on the previous discounted price rather than on the original price. A 20% discount followed by a 22% discount is the same as a 37.6% discount off the original price.

2. In terms of cost, a low faculty-to-student ratio is less expensive than a high faculty-to-student ratio. In terms of quality of education and student satisfaction, a high faculty-to-student ratio is more desirable. A college president must balance the cost and quality issues.

3. No; the 10% discount was based on the original price rather than on the sale price.

4. Let S = the original salary. After both raises have been given, the two situations yield the same salary: $1.05 \cdot 1.1S = 1.1 \cdot 1.05S$. However, the first situation is better for the wage earner, because $1.1S$ is earned the first year when a 10% raise is given while in the second situation $1.05S$ is earned that year.

5. For a number n, 40% of 50% of n is $0.4(0.5n)$, or $0.2n$, or 20% of n. Thus, taking 40% of 50% of a number is the same as taking 20% of the number.

6. The interest due on the 30 day loan will be \$41.10 while that due on the 60 day loan will be \$131.51. This could be an argument in favor of the 30 day loan. On the other hand the 60 day loan puts twice as much cash at the firm's disposal for twice as long as the 30 day loan. This could be an argument in favor of the 60 day loan.

Chapter 6 Test

1. The ratio of 85 to 97 is $\frac{85}{97}$.

2. The ratio of 0.34 to 124 is $\frac{0.34}{124}$.

3. $\frac{18}{20} = \frac{2 \cdot 9}{2 \cdot 10} = \frac{2}{2} \cdot \frac{9}{10} = \frac{9}{10}$

4. $\frac{0.75}{0.96} = \frac{0.75}{0.96} \cdot \frac{100}{100}$ Clearing the decimals

$$= \frac{75}{96}$$
$$= \frac{3 \cdot 25}{3 \cdot 33} = \frac{3}{3} \cdot \frac{25}{33}$$
$$= \frac{25}{33}$$

5. $\frac{16 \text{ servings}}{12 \text{ lb}} = \frac{16}{12} \frac{\text{servings}}{\text{lb}} = \frac{4}{3}$ servings/lb, or $1\frac{1}{3}$ servings/lb

6. $\frac{464 \text{ mi}}{14.5 \text{ gal}} = \frac{464}{14.5} \frac{\text{mi}}{\text{gal}} = 32$ mpg

7. We can use cross products:

$7 \cdot 72 = 504$ $\frac{7}{8}$ ✕ $\frac{63}{72}$ $8 \cdot 63 = 504$

Since the cross products are the same, $504 = 504$, we know that $\frac{7}{8} = \frac{63}{72}$, so the numbers are proportional.

8. We can use cross products:

$1.3 \cdot 15.2 = 19.76$ $\frac{1.3}{3.4}$ ✕ $\frac{5.6}{15.2}$ $3.4 \cdot 5.6 = 19.04$

Since the cross products are not the same, $19.76 \neq 19.04$, we know that $\frac{1.3}{3.4} \neq \frac{5.6}{15.2}$, so the numbers are not proportional.

9.
$$\frac{68}{y} = \frac{17}{25}$$
$$68 \cdot 25 = y \cdot 17 \quad \text{Equating cross products}$$
$$\frac{68 \cdot 25}{17} = \frac{y \cdot 17}{17}$$
$$\frac{4 \cdot \cancel{17} \cdot 25}{\cancel{17} \cdot 1} = y$$
$$100 = y$$

10.
$$\frac{150}{2.5} = \frac{x}{6}$$
$$150 \cdot 6 = 2.5 \cdot x \quad \text{Equating cross products}$$
$$\frac{150 \cdot 6}{2.5} = \frac{2.5 \cdot x}{2.5}$$
$$\frac{900}{2.5} = x$$
$$360 = x$$

11. ***Familiarize.*** Let d = the actual distance between the cities.

Translate. We translate to a proportion.

$$\begin{matrix}\text{Map distance} \rightarrow \\ \text{Actual distance} \rightarrow\end{matrix} \frac{3}{225} = \frac{7}{d} \begin{matrix}\leftarrow \text{Map distance} \\ \leftarrow \text{Actual distance}\end{matrix}$$

Solve.

$$3 \cdot d = 225 \cdot 7 \quad \text{Equating cross products}$$
$$\frac{3 \cdot d}{3} = \frac{225 \cdot 7}{3}$$
$$d = 525 \quad \text{Multiplying and dividing}$$

Check. We substitute in the proportion and check cross products.

$$\frac{3}{225} = \frac{7}{525}$$
$$3 \cdot 525 = 1575;\ 225 \cdot 7 = 1575$$

The cross products are the same, so the answer checks.

State. The cities are 525 mi apart.

12. a) ***Familiarize.*** Let $c =$ the value of 450 U.S. dollars in Hong Kong dollars.

Translate. We translate to a proportion.

$$\begin{array}{r} \text{U.S. dollars} \rightarrow \\ \text{Hong Kong dollars} \rightarrow \end{array} \frac{1}{7.781} = \frac{450}{c} \begin{array}{l} \leftarrow \text{U.S. dollars} \\ \leftarrow \text{Hong Kong dollars} \end{array}$$

Solve.

$$1 \cdot c = 7.781 \cdot 450 \quad \text{Equating cross products}$$
$$c = 3501.45$$

Check. We substitute in the proportion and check cross products.

$$\frac{1}{7.781} = \frac{450}{3501.45}$$
$$1 \cdot 3501.45 = 3501.45;\ 7.781 \cdot 450 = 3501.45$$

The cross products are the same, so the answer checks.

State. 450 U.S. dollars would be worth 3501.45 Hong Kong dollars.

b) ***Familiarize.*** Let $d =$ the price of the DVD player in U.S. dollars.

Translate. We translate to a proportion.

$$\begin{array}{r} \text{U.S. dollars} \rightarrow \\ \text{Hong Kong dollars} \rightarrow \end{array} \frac{1}{7.781} = \frac{d}{795} \begin{array}{l} \leftarrow \text{U.S. dollars} \\ \leftarrow \text{Hong Kong dollars} \end{array}$$

Solve.

$$1 \cdot 795 = 7.781 \cdot d \quad \text{Equating cross products}$$
$$\frac{1 \cdot 795}{7.781} = \frac{7.781 \cdot d}{7.781}$$
$$102.17 \approx d$$

Check. We use a different approach. Since 1 U.S. dollar is worth 7.781 Hong Kong dollars, we multiply 102.17 by 7.781:

$$102.17(7.781) \approx 795.$$

This is the price in Hong Kong dollars, so the answer checks.

State. The DVD player would cost $102.17 in U.S. dollars.

13. ***Familiarize.*** Let $c =$ the cost of a turkey dinner for 14 people.

Translate. We translate to a proportion.

$$\begin{array}{r} \text{People} \rightarrow \\ \text{Cost} \rightarrow \end{array} \frac{8}{33.81} = \frac{14}{c} \begin{array}{l} \leftarrow \text{People} \\ \leftarrow \text{Cost} \end{array}$$

Solve.

$$8 \cdot c = 33.81 \cdot 14$$
$$c = \frac{33.81 \cdot 14}{8}$$
$$c \approx 59.17$$

Check. We substitute in the proportion and check cross products.

$$\frac{8}{33.81} = \frac{14}{59.17}$$
$$8 \cdot 59.17 = 473.36;\ 33.81 \cdot 14 = 473.34$$

Since $473.36 \approx 473.34$, the answer checks.

State. It would cost about $59.17 to serve a turkey dinner for 14 people.

14. ***Familiarize.*** Let $m =$ the number of minutes the watch will lose in 24 hr.

Translate. We translate to a proportion.

$$\begin{array}{r} \text{Minutes lost} \rightarrow \\ \text{Hours} \rightarrow \end{array} \frac{2}{10} = \frac{m}{24} \begin{array}{l} \leftarrow \text{Minutes lost} \\ \leftarrow \text{Hours} \end{array}$$

Solve.

$$2 \cdot 24 = 10 \cdot m \quad \text{Equating cross products}$$
$$\frac{2 \cdot 24}{10} = \frac{10 \cdot m}{10}$$
$$4.8 = m \quad \text{Multiplying and dividing}$$

Check. We substitute in the proportion and check cross products.

$$\frac{2}{10} = \frac{4.8}{24}$$
$$2 \cdot 24 = 48;\ 10 \cdot 4.8 = 48$$

The cross products are the same, so the answer checks.

State. The watch will lose 4.8 min in 24 hr.

15. 14.7%

a) Replace the percent symbol with $\times 0.01$.

14.7×0.01

b) Move the decimal point two places to the left.

0.14.7 (decimal point moved two places left)

Thus, $14.7\% = 0.147$.

16. 0.38

a) Move the decimal point two places to the right.

0.38. (decimal point moved two places right)

b) Write a percent symbol: 38%

Thus, $0.38 = 38\%$.

17.

$$\begin{array}{r} 1.375 \\ 8\overline{)11.000} \\ \underline{8} \\ 3\,0 \\ \underline{2\,4} \\ 6\,0 \\ \underline{5\,6} \\ 4\,0 \\ \underline{4\,0} \\ 0 \end{array}$$

$\frac{11}{8} = 1.375$

Convert to percent notation.

1.37.5

$\frac{11}{8} = 137.5\%$, or $137\frac{1}{2}\%$

18. $65\% = \frac{65}{100}$ Definition of percent

$= \frac{5 \cdot 13}{5 \cdot 20}$

$= \frac{5}{5} \cdot \frac{13}{20}$ Simplifying

$= \frac{13}{20}$

19. Translate: What is 40% of 55?

$a = 40\% \cdot 55$

Solve: We convert 40% to decimal notation and multiply.

$a = 40\% \cdot 55$

$= 0.4 \cdot 55 = 22$

The answer is 22.

20. What percent of 80 is 65?

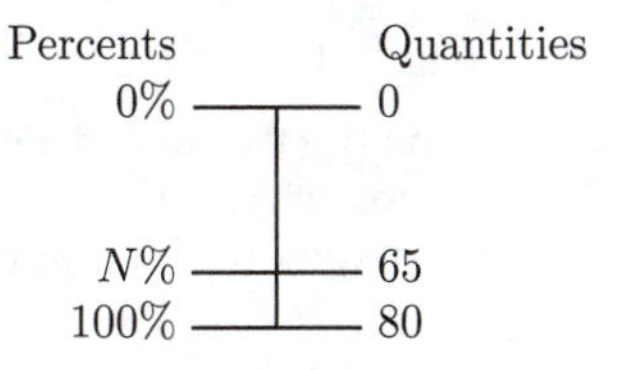

Translate: $\frac{N}{100} = \frac{65}{80}$

Solve: $80 \cdot N = 100 \cdot 65$

$\frac{80 \cdot N}{80} = \frac{100 \cdot 65}{80}$

$N = \frac{6500}{80}$

$N = 81.25$

The answer is 81.25%.

21. ***Familiarize.*** Let k = the number of kidney transplants, l = the number of liver transplants, and h = the number of heart transplants in 2006.

Translate. We translate to three equations.

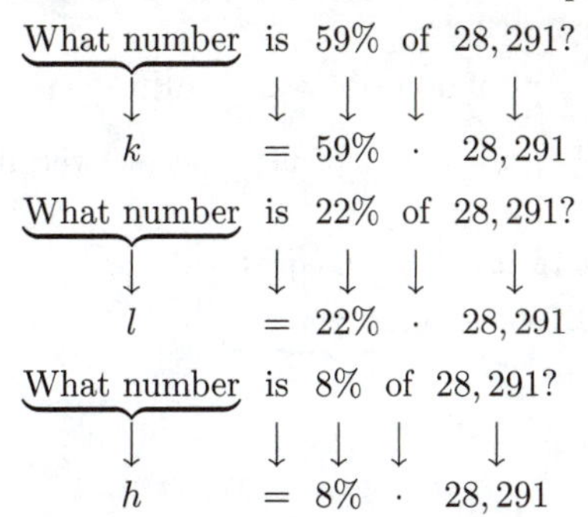

What number is 59% of 28,291?

$k = 59\% \cdot 28,291$

What number is 22% of 28,291?

$l = 22\% \cdot 28,291$

What number is 8% of 28,291?

$h = 8\% \cdot 28,291$

Solve. To solve each equation we convert percent notation to decimal notation and multiply.

$k = 59\% \cdot 28,291 = 0.59 \cdot 28,291 \approx 16,692$

$l = 22\% \cdot 28,291 = 0.22 \cdot 28,291 \approx 6224$

$h = 8\% \cdot 28,291 = 0.08 \cdot 28,291 \approx 2263$

Check. We repeat the calculations. The answers check.

State. In 2006, there were 16,692 kidney transplants, 6224 liver transplants, and 2263 heart transplants.

22. ***Familiarize.*** Let b = the number of at-bats.

Translate. We translate to a proportion. We are asking "175 is 28.64% of what?"

$\frac{28.64}{100} = \frac{175}{b}$

Solve.

$28.64 \cdot b = 100 \cdot 175$ Equating cross products

$\frac{28.64 \cdot b}{28.64} = \frac{100 \cdot 175}{28.64}$

$b \approx 611$

Check. We can repeat the calculation. Also note that $\frac{175}{611} \approx \frac{150}{600} = \frac{1}{4} = 25\% \approx 28.64\%$. The answer checks.

State. Garrett Atkins had about 611 at-bats.

23. ***Familiarize.*** We first find the amount of decrease.

$$\begin{array}{r} 2\,0,\,6\,7\,9 \\ -\,1\,9,\,2\,9\,2 \\ \hline 1\,3\,8\,7 \end{array}$$

Let p = the percent of decrease.

Translate. We translate to an equation.

1387 is what percent of 20,679?

$1387 = p \cdot 20,679$

Solve.

$1387 = p \cdot 20,679$

$\frac{1387}{20,679} = \frac{p \cdot 20,679}{20,679}$

$0.067 \approx p$

$6.7\% \approx p$

Check. Note that $6.7\% \approx 7\%$. With a decrease of approximately 7%, the number of adoptions in 2007 should be about $100\% - 7\%$, or 93%, of the number of adoptions in 2006. Since $93\% \cdot 20,679 = 0.93 \cdot 20,679 \approx 19,231 \approx 19,292$, the answer checks.

State. The percent of decrease was about 6.7%.

24. ***Familiarize***. Let $p =$ the percent of people who live in Asia.

Translate. We translate to an equation.

$$\begin{array}{ccccc} 4,002,000,000 & \text{is} & \underbrace{\text{what percent}} & \text{of} & 6,603,000,000 \\ \downarrow & \downarrow & \downarrow & \downarrow & \downarrow \\ 4,002,000,000 & = & p & \cdot & 6,603,000,000 \end{array}$$

Solve.

$$4,002,000,000 = p \cdot 6,603,000,000$$
$$\frac{4,002,000,000}{6,603,000,000} = \frac{p \cdot 6,603,000,000}{6,603,000,000}$$
$$0.606 \approx p$$
$$60.6\% \approx p$$

Check. We find 60.6% of 6,603,000,000:

$$60.6\% \cdot 6,603,000,000 = 0.606 \cdot 6,603,000,000 \approx 4,001,418,000 \approx 4,002,000,000$$

The answer checks.

State. About 60.6% of people living in the world today live in Asia.

25. The sales tax on an item costing $560 is

$$\begin{array}{ccc} \underbrace{\text{Sales tax rate}} & \times & \underbrace{\text{Purchase price}} \\ \downarrow & \downarrow & \downarrow \\ 4.5\% & \times & \$560, \end{array}$$

or $0.045 \times \$560$, or \$25.20. Thus the tax is \$25.20.

The total price is given by the purchase price plus the sales tax:

$$\$560 + \$25.20 = \$585.20$$

26.

$$\begin{array}{lcll} \text{Commission} & = & \text{Commission rate} & \times \text{ Sales} \\ C & = & 15\% & \times\ 4200 \\ C & = & 0.15 & \times\ 4200 \\ C & = & 630 & \end{array}$$

The commission is \$630.

27.

$$\begin{array}{lcccc} \text{Discount} & = & \text{Rate of discount} & \times & \text{Marked price} \\ D & = & 20\% & \times & \$200 \end{array}$$

Convert 20% to decimal notation and multiply.

$$\begin{array}{r} 2\,0\,0 \\ \times\ \ 0.\,2 \\ \hline 4\,0.\,0 \end{array} \qquad (20\% = 0.20 = 0.2)$$

The discount is \$40.

$$\begin{array}{lcccc} \text{Sale price} & = & \text{Marked price} & - & \text{Discount} \\ S & = & 200 & - & 40 \end{array}$$

We subtract:

$$\begin{array}{r} 2\,0\,0 \\ -\ \ 4\,0 \\ \hline 1\,6\,0 \end{array}$$

To check, note that the sale price is 80% of the marked price: $0.8 \times 200 = 160$.

The sale price is \$160.

28.

$$\begin{aligned} I = P \cdot r \cdot t &= \$120 \times 7.1\% \times 1 \\ &= \$120 \times 0.071 \times 1 \\ &= \$8.52 \end{aligned}$$

29.

$$\begin{aligned} I = P \cdot r \cdot t &= \$5200 \times 6\% \times \frac{1}{2} \\ &= \$5200 \times 0.06 \times \frac{1}{2} \\ &= \$312 \times \frac{1}{2} \\ &= \$156 \end{aligned}$$

The interest earned is \$156. The amount in the account is the principal plus the interest: $\$5200 + \$156 = \$5356$.

30.

$$\begin{aligned} A &= P \cdot \left(1 + \frac{r}{n}\right)^{n \cdot t} \\ &= \$1000 \cdot \left(1 + \frac{5\frac{3}{8}\%}{1}\right)^{1 \cdot 2} \\ &= \$1000 \cdot \left(1 + \frac{0.05375}{1}\right)^{2} \\ &= \$1000(1.05375)^2 \\ &\approx \$1110.39 \end{aligned}$$

31. **Plumber**: We add to find the number of jobs in 2016: $705,000 + 52,000 = 757,000$, so we project that there will be 757,000 jobs for plumbers in 2016. We solve an equation to find the percent of increase, p.

$$\begin{array}{ccccc} 52,000 & \text{is} & \underbrace{\text{what percent}} & \text{of} & 705,000? \\ \downarrow & \downarrow & \downarrow & \downarrow & \downarrow \\ 52,000 & = & p & \cdot & 705,000 \end{array}$$

Solve.

$$\frac{52,000}{705,000} = \frac{p \cdot 705,000}{705,000}$$
$$0.074 \approx p$$
$$7.4\% \approx p$$

The percent of increase is about 7.4%.

Veterinary Assistant: We subtract to find the change: $100,000 - 71,000 = 29,000$, so we project that the change will be 29,000. We solve an equation to find the percent of increase, p.

$$\begin{array}{ccccc} 29,000 & \text{is} & \underbrace{\text{what percent}} & \text{of} & 71,000? \\ \downarrow & \downarrow & \downarrow & \downarrow & \downarrow \\ 29,000 & = & p & \cdot & 71,000 \end{array}$$

Solve.

$$\frac{29,000}{71,000} = \frac{p \cdot 71,000}{71,000}$$
$$0.408 \approx p$$
$$40.8\% \approx p$$

The percent of increase is about 40.8%.

Motorcycle Repair Technician: We subtract to find the number of jobs in 2006; $24,000 - 3000 = 21,000$, so there were 21,000 jobs for motorcycle repair technicians in 2006. We solve an equation to find the percent of increase, p.

$$\begin{array}{ccccc} 3000 & \text{is} & \underbrace{\text{what percent}} & \text{of} & 21,000? \\ \downarrow & \downarrow & \downarrow & \downarrow & \downarrow \\ 3000 & = & p & \cdot & 21,000 \end{array}$$

Solve.

$$\frac{3000}{21,000} = \frac{p \cdot 21,000}{21,000}$$
$$0.143 \approx p$$
$$14.3\% \approx p$$

The percent of increase is about 14.3%.

Fitness Professional: Let n = the number of jobs in 2006. If the number of jobs increases by 26.8%, then the number of jobs in 2016 represents 100% of n plus 26.8% of n, or 126.8% of n, or $1.268n$. We solve an equation to find n.

$$\begin{array}{ccc} \underbrace{\text{The number of jobs in 2006 increased by 26.8\%}} & \text{is} & \underbrace{298{,}000} \\ \downarrow & \downarrow & \downarrow \\ 1.268 \cdot n & = & 298,000 \end{array}$$

Solve.

$$\frac{1.268 \cdot n}{1.268} = \frac{298,000}{1.268}$$
$$n \approx 235,000$$

There were about 235,000 jobs for fitness professionals in 2006.

We subtract to find the change; $298,000 - 235,000 = 63,000$, so we project that the change will be 63,000 jobs.

32. $A = P \cdot \left(1 + \frac{r}{n}\right)^{n \cdot t}$

$= \$10,000 \cdot \left(1 + \frac{4.9\%}{12}\right)^{12 \cdot 3}$

$= \$10,000 \cdot \left(1 + \frac{0.049}{12}\right)^{36}$

$\approx \$11,580.07$

33. Discount = Original price − Sale price

$D = 349.99 - 299.99$

We subtract:

$$\begin{array}{r} 349.99 \\ -\ 299.99 \\ \hline 50.00 \end{array}$$

The discount is \$50.

Discount = Rate of discount × Original price

$50 = R \times 349.99$

To solve the equation we divide on both sides by 349.99.

$$\frac{50}{349.99} = \frac{R \times 349.99}{349.99}$$
$$0.143 \approx R$$
$$14.3\% \approx R$$

To check, note that a discount rate of 14.3% means that 85.7% of the original price is paid: $0.857 \times 349.99 \approx 299.94 \approx 299.99$. Since that is the sale price, the answer checks.

The rate of discount is about 14.3%.

34. We first use the formula $I = P \cdot r \cdot t$ to find the amount of interest paid.

$$I = P \cdot r \cdot t = \$2704.27 \cdot 0.163 \cdot \frac{1}{12} \approx \$36.73.$$

Then the amount by which the principal is reduced is

$\$54 - \$36.73 = \$17.27.$

Finally, we find that the principal after the payment is

$\$2704.27 - \$17.27 = \$2687.$

35. 0.75% of what number is 300?

Translate: $300 = 0.75\% \times b$

Solve: We divide on both sides by 0.75%:

$$\frac{300}{0.75\%} = \frac{0.75\% \times b}{0.75\%}$$
$$\frac{300}{0.0075} = b \quad (0.75\% = 0.0075)$$
$$40,000 = b$$

300 is 0.75% of 40,000. The correct answer is B.

36. $\dfrac{4\frac{1}{2}\text{ mi}}{1\frac{1}{2}\text{hr}} = \dfrac{4\frac{1}{2}}{1\frac{1}{2}}\dfrac{\text{mi}}{\text{hr}} = \dfrac{\frac{9}{2}}{\frac{3}{2}}\dfrac{\text{mi}}{\text{hr}} = \dfrac{9}{2} \cdot \dfrac{2\text{ mi}}{3\text{ hr}} = \dfrac{9 \cdot 2\text{ mi}}{2 \cdot 3\text{ hr}} =$

$\dfrac{3 \cdot 3 \cdot 2\text{ mi}}{2 \cdot 3 \cdot 1\text{ hr}} = \dfrac{3 \cdot 2}{3 \cdot 2} \cdot \dfrac{3\text{ mi}}{1\text{ hr}} = 3\dfrac{\text{mi}}{\text{hr}}$, or 3 mph

The correct answer is C.

37. ***Familiarize***. Let p = the price for which a realtor would have to sell the house in order for Juan and Marie to receive \$180,000 from the sale. The realtor's commission would be $7.5\% \cdot p$, or $0.075 \cdot p$, and Juan and Marie would receive 100% of p − 7.5% of p, or 92.5% of p, or $0.925 \cdot p$.

Translate.

$$\begin{array}{ccc} \underbrace{\text{Amount Juan and Marie receive}} & \text{is} & \$180,000 \\ \downarrow & \downarrow & \downarrow \\ 0.925 \cdot p & = & 180,000 \end{array}$$

Solve.

$$\frac{0.925 \cdot p}{0.925} = \frac{180,000}{0.925}$$
$$p \approx 194,600 \quad \text{Rounding to the nearest hundred}$$

Check. 7.5% of $\$194,600 = 0.075 \cdot \$194,600 = \$14,595$ and $\$194,600 - \$14,595 = \$180,005 \approx \$180,000$. The answer checks.

State. A realtor would need to sell the house for about \$194,600.

38. First we find the commission.

Commission = Commission rate × Sales

$C = 16\% \times \$15,000$

$C = 0.16 \times \$15,000$

$C = \$2400$

Now we find the amount in the account after 6 months.

$$\begin{aligned}
A &= P \cdot \left(1 + \frac{r}{n}\right)^{n \cdot t} \\
&= \$2400 \cdot \left(1 + \frac{12\%}{4}\right)^{4 \cdot \frac{1}{2}} \\
&= \$2400 \cdot \left(1 + \frac{0.12}{4}\right)^{2} \\
&= \$2400 \cdot (1 + 0.03)^2 \\
&= \$2400 \cdot (1.03)^2 \\
&= \$2400(1.0609) \\
&= \$2546.16
\end{aligned}$$

Chapter 7

Data, Graphs, and Statistics

Exercise Set 7.1

1. To find the average, we first add the numbers. Then divide by the number of addends.

$$\frac{2+2+17+30+90+110+52+21+5}{9} = \frac{329}{9} = 36.\overline{5}$$

The average is $36.\overline{5}$.

To find the median, first list the numbers in order from smallest to largest. Then locate the middle number.

$$2, 2, 5, 17, 21, 30, 52, 90, 110$$
$$\uparrow$$
Middle number

The median is 21.

Find the mode:

The number that occurs most often is 2. The mode is 2.

3. To find the average, add the numbers. Then divide by the number of addends.

$$\frac{17+19+29+18+14+29}{6} = \frac{126}{6} = 21$$

The average is 21.

To find the median, first list the numbers in order from smallest to largest. Then locate the middle number.

$$14, 17, 18, 19, 29, 29$$
$$\uparrow$$
Middle number

The median is halfway between 18 and 19. It is the average of the two middle numbers:

$$\frac{18+19}{2} = \frac{37}{2} = 18.5$$

Find the mode:

The number that occurs most often is 29. The mode is 29.

5. To find the average, add the numbers. Then divide by the number of addends.

$$\frac{5+37+20+20+35+5+25}{7} = \frac{147}{7} = 21$$

The average is 21.

To find the median, first list the numbers in order from smallest to largest. Then locate the middle number.

$$5, 5, 20, 20, 25, 35, 37$$
$$\uparrow$$
Middle number

The median is 20.

Find the mode:

There are two numbers that occur most often, 5 and 20. Thus the modes are 5 and 20.

7. Find the average:

$$\frac{4.3+7.4+1.2+5.7+8.3}{5} = \frac{26.9}{5} = 5.38$$

The average is 5.38.

Find the median:

$$1.2, 4.3, 5.7, 7.4, 8.3$$
$$\uparrow$$
Middle number

The median is 5.7.

Find the mode:

All the numbers are equally represented. No mode exists.

9. Find the average:

$$\frac{234+228+234+229+234+278}{6} = \frac{1437}{6} = 239.5$$

The average is 239.5.

Find the median:

$$228, 229, 234, 234, 234, 278$$
$$\uparrow$$
Middle number

The median is halfway between 234 and 234. Although it seems clear that this is 234, we can compute it as follows:

$$\frac{234+234}{2} = \frac{468}{2} = 234$$

The median is 234.

Find the mode:

The number that occurs most often is 234. The mode is 234.

11. We divide the total number of miles, 253, by the number of gallons, 11.

$$\frac{253}{11} = 23$$

The average was 23 miles per gallon.

13. To find the GPA we first add the grade point values for each hour taken. This is done by first multiplying the grade point value by the number of hours in the course and then adding as follows:

$$\begin{array}{ll} \text{B} & 3.0 \cdot 4 = 12 \\ \text{A} & 4.0 \cdot 5 = 20 \\ \text{D} & 1.0 \cdot 3 = \ \ 3 \\ \text{C} & 2.0 \cdot 4 = \ \ \underline{8} \\ & \qquad\quad \ 43 \text{ (Total)} \end{array}$$

The total number of hours taken is

$4 + 5 + 3 + 4$, or 16.

We divide 43 by 16 and round to the nearest tenth.

$$\frac{43}{16} = 2.6875 \approx 2.7$$

The student's grade point average is 2.7.

15. Find the average:

$$\frac{\$3.99 + \$4.49 + \$4.99 + \$3.99 + \$3.49}{5} = \frac{\$20.95}{5} = \$4.19$$

Find the median:

$$\$3.49, \$3.99, \$3.99, \$4.49, \$4.99$$

↑ Middle number

The median is \$3.99.

Find the mode:

The number that occurs most often is \$3.99. The mode is \$3.99.

17. We can find the total of the five scores needed as follows:

$$80 + 80 + 80 + 80 + 80 = 400.$$

The total of the scores on the first four tests is

$$80 + 74 + 81 + 75 = 310.$$

Thus Rich needs to get at least

$$400 - 310, \text{ or } 90$$

to get a B. We can check this as follows:

$$\frac{80 + 74 + 81 + 75 + 90}{5} = \frac{400}{5} = 80.$$

19. We can find the total number of days needed as follows:

$$266 + 266 + 266 + 266 = 1064.$$

The total number of days for Marta's first three pregnancies is

$$270 + 259 + 272 = 801.$$

Thus, Marta's fourth pregnancy must last

$$1064 - 801 = 263 \text{ days}$$

in order to equal the worldwide average.

We can check this as follows:

$$\frac{270 + 259 + 272 + 263}{4} = \frac{1064}{4} = 266.$$

21. Compare the averages of the two sets of data.

Bulb A: Average $= (983 + 964 + 1214 + 1417 + 1211 + 1521 + 1084 + 1075 + 892 + 1423 + 949 + 1322)/12 = 1171.25$

Bulb B: Average $= (979 + 1083 + 1344 + 984 + 1445 + 975 + 1492 + 1325 + 1283 + 1325 + 1352 + 1432)/12 \approx 1251.58$

Since the average life of Bulb A is 1171.25 hr and of Bulb B is about 1251.58 hr, Bulb B is better.

23.

```
     1 2. 8 6   (2 decimal places)
   × 1 7. 5     (1 decimal place)
     6 4 3 0
   9 0 0 2 0
 1 2 8 6 0 0
 2 2 5. 0 5 0   (3 decimal places)
```

25.
$$\frac{4}{5} \cdot \frac{3}{28} = \frac{4 \cdot 3}{5 \cdot 28} = \frac{4 \cdot 3}{5 \cdot 4 \cdot 7} = \frac{4}{4} \cdot \frac{3}{5 \cdot 7} = \frac{3}{35}$$

27. First we divide to find the decimal notation.

```
       1. 1 8 7 5
 1 6 ) 1 9. 0 0 0 0
       1 6
         3 0
         1 6
         1 4 0
         1 2 8
           1 2 0
           1 1 2
               8 0
               8 0
                 0
```

Then we move the decimal point two places to the right and write a percent symbol.

$$\frac{19}{16} = 1.1875 = 118.75\%$$

29. First we divide to find the decimal notation.

```
         0. 5 1 2
 1 2 5 ) 6 4. 0 0 0
         6 2 5
           1 5 0
           1 2 5
             2 5 0
             2 5 0
                 0
```

Then we move the decimal point two places to the right and write a percent symbol.

$$\frac{64}{125} = 0.512 = 51.2\%$$

31. Since a is the middle number, it is the median, 30. Now find the average.

$$\frac{18 + 21 + 24 + 30 + 36 + 37 + b}{7} = 32$$
$$\frac{166 + b}{7} = 32$$
$$7\left(\frac{166 + b}{7}\right) = 7 \cdot 32$$
$$166 + b = 224$$
$$166 + b - 166 = 224 - 166$$
$$b = 58$$

Thus we have $a = 30$ and $b = 58$.

33. Amy's second offer: $\dfrac{\$3600 + \$3200}{2} = \$3400$

Jim's second offer: $\dfrac{\$3400 + \$3600}{2} = \$3500$

Amy's third offer: $\dfrac{\$3500 + \$3400}{2} = \$3450$

Jim's third offer: $\dfrac{\$3500 + \$3450}{2} = \$3475$

Amy will pay $3475 for the car.

Exercise Set 7.2

1. Go down the Product column to Franklin Farms Portabella Fresh. Then go across to the column headed Calories and read the entry, 100. There are 100 calories in a Franklin Farms Portabella Fresh veggie burger.

3. Go down the column headed Calories. For each entry that is less than 110, go across to the Product column and read the entry. The veggie burgers with less than 110 calories are the Boca All American Flame Grilled Meatless, the Franklin Farms Portabella Fresh, and the Gardenburger Portabella. (Note that we also knew from Exercise 1 that the Franklin Farms Portabella Fresh veggie burger has less than 110 calories.)

5. Find the average fiber content:

$$\frac{3+3+4+3+5+3+3+4}{8} = \frac{28}{8} = 3.5$$

The average fiber content is 3.5 g.

Find the median:

$$3, 3, 3, 3, 3, 4, 4, 5$$

↑

Middle number

The median is halfway between 3 and 3. It is the average of the two numbers.

$$\frac{3+3}{2} = \frac{6}{2} = 3$$

The median fiber content is 3 g.

Find the mode:

The number that occurs most often is 3. The mode is 3 g.

7. To find the most expensive veggie burger, find the largest number in the Cost column. It is 1.48. Go across to the Product column and read the entry. We see that the Lightlife Meatless Light veggie burger is the most expensive at $1.48 per patty.

To find the least expensive veggie burger, find the smallest number in the Cost column. It is 0.96. Go across to the Product column and read the entry. We see that the Boca All American Flame Grilled Meatless veggie burger is the least expensive at $0.96 per patty.

9. To find which veggie burger contains the most fat, find the largest number in the Fat column. It is 6.0. Go across to the Product column and read the entry. We see that the Veggie Patch Garlic Portabella veggie burger contains the most fat, 6.0 g per patty.

To find which veggie burger contains the least fat, find the smallest number in the Fat column. It is 1.5 and it occurs twice. Go across from each of these entries to the Product column and read the entries there. We see that the Franklin Farms Portabella Fresh and the Lightlife Meatless Light veggie burgers contain the least fat, 1.5 g per patty.

11. Go down the column headed Actual Temperature (°F) to 80°. Then go across to the Relative Humidity column headed 60%. The entry is 92, so the apparent temperature is 92°F.

13. Go down the column headed Actual Temperature (°F) to 85°. Then go across the Relative Humidity column headed 90%. The entry is 108, so the apparent temperature is 108°F.

15. The number 100 appears in the columns headed Apparent Temperature (°F) 3 times, so there are 3 temperature-humidity combinations that give an apparent temperature of 100°. They are 85°and 60% humidity, 90°and 40% humidity, and 100°and 10% humidity.

17. Go down the Relative Humidity column headed 50% and find all the entries greater than 100. The last 4 entries are greater than 100. Then go across to the column headed Actual Temperature (°F) and read the temperatures that correspond to these entries. At 50% humidity, the actual temperatures 90° and higher give an apparent temperature above 100°.

19. Go down the column headed Actual Temperature (°F) to 95°. Then read across to locate the entries greater than 100. All of the entries except the first two are greater than 100. Go up from each entry to find the corresponding relative humidity. At an actual temperature of 95°, relative humidities of 30% and higher give an apparent temperature above 100°.

21. Go down the column headed Actual Temperature (°F) to 85°, then across to 94, and up to find that the corresponding relative humidity is 40%. Similarly, go down to 85°, across to 108, and up to 90%. At an actual temperature of 85°, difference in humidities required to raise the apparent temperature from 94° to 108° is

$$90\% - 40\%, \text{ or } 50\%.$$

23. Go down the Planet column to Jupiter. Then go across to the column headed Average Distance from Sun (in miles) and read the entry, 483,612,200. The average distance from the sun to Jupiter is 483,612,200 miles.

25. Go down the column headed Time of Revolution in Earth Time (in years) to 164.78. Then go across the Planet column. The entry there is Neptune, so Neptune has a time of revolution of 164.78 days.

27. All of the entries in the column headed Average Distance from Sun (in miles) are greater than 1,000,000. Thus, all of the planets have an average distance from the sun that is greater than 1,000,000 mi.

29. Go down the Planet column to Earth and then across to the Diameter (in miles) column to find that the diameter of Earth is 7926 mi. Similarly, find that the diameter of Jupiter is 88,846 mi. Then divide:

$$\frac{88,846}{7926} \approx 11$$

It would take about 11 Earth diameters to equal one Jupiter diameter.

31. Find the average of all the numbers in the column headed Diameter (in miles):

$(3031 + 7520 + 7926 + 4221 + 88,846 + 74,898 + 31,763 + 31,329)/8 = 31,191.75.$

The average of the diameters of the planets is 31,191.75 mi.

To find the median of the diameters of the planets we first list the diameters in order from smallest to largest:

3031, 4221, 7520, 7926, 31,329, 31,763, 74,898, 88,846.

The median is the average of the two middle numbers, 7926 and 31,329:

$$\frac{7926 + 31,329}{2} = 19,627.5.$$

The median of the diameters is 19,627.5 mi.

Since no number appears more than once in the Diameter (in miles) column, there is no mode.

33. The white rhino is represented by the greatest number of symbols, so this species has the greatest number of rhinos.

35. From the graph we see that there are about 12.5×300, or 3750, black rhinos and 8×300, or 2400, Indian rhinos. Thus there are about $3750 - 2400$, or 1350, more black rhinos than Indian rhinos.

37. From Exercise 35, we know that there are about 3750 black rhinos and 2400 Indian rhinos. Now we find the number of rhinos in the other three species.

White rhinos: $48.5 \times 300 = 14,550$

Javan rhino: $\frac{1}{6} \times 300 = 50$

Sumatran rhino: $\frac{1}{10} \times 300 = 30$

We find the average of the five numbers.

$$\frac{3750 + 2400 + 14,550 + 50 + 30}{5} = \frac{20,780}{5} = 4156$$

The average number of rhinos in these five species is about 4156 rhinos. (Answers may vary depending on how partial symbols are interpreted.)

39. Cabinets: 50% of $\$26,888 = 0.5(\$26,888) = \$13,444$

Countertops: 15% of $\$26,888 = 0.15(\$26,888) = \$4033.20$

Appliances: 8% of $\$26,888 = 0.08(\$26,888) = \$2151.04$

Fixtures: 3% of $\$26,888 = 0.03(\$26,888) = \$806.64$

Chapter 7 Mid-Chapter Review

1. The given statement is true. See pages 473 and 477 in the text.

2. Consider the set of data 3, 3, 3. The average is

$$\frac{3+3+3}{3} = \frac{9}{3} = 3$$

The median is 3 and the mode is also 3. Thus, it is true that it is possible for the average, the median, and the mode of a set of data to be the same number.

3. The given statement is false. If there is an even number of items in a set of data, the median is the average of the two middle numbers.

4. $\frac{60 + 45 + 115 + 15 + 35}{5} = \frac{270}{5} = 54$

5. We first arrange the numbers from smallest to largest.

2.1, 4.8, 6.3, 8.7, 11.3, 14.5

The median is the average of the two middle numbers, 6.3 and 8.7.

$$\frac{6.3 + 8.7}{2} = \frac{15}{2} = 7.5.$$

The median is 7.5.

6. Find the average:

$$\frac{56 + 29 + 45 + 240 + 175 + 7 + 29}{7} = \frac{581}{7} = 83$$

To find the median, we first arrange the numbers from smallest to largest.

7, 29, 29, 45, 56, 175, 240

The median is the middle number, 45.

Find the mode:

The number that occurs most often is 29. The mode is 29.

7. Find the average:

$$\frac{2.12 + 18.42 + 9.37 + 43.89}{4} = \frac{73.8}{4} = 18.45$$

To find the median, we first arrange the numbers from smallest to largest.

2.12, 9.37, 18.42, 43.89

The median is the average of the two middle numbers, 9.37 and 18.42.

$$\frac{9.37 + 18.42}{2} = \frac{27.79}{2} = 13.895$$

Find the mode:

Each number occurs the same number of times, so no mode exists.

8. Find the average:

$$\frac{\frac{5}{10} + \frac{1}{10} + \frac{7}{10} + \frac{9}{10} + \frac{3}{10}}{5} = \frac{\frac{25}{10}}{5} = \frac{25}{10} \cdot \frac{1}{5} = \frac{25 \cdot 1}{10 \cdot 5} =$$

$$\frac{5 \cdot 5 \cdot 1}{2 \cdot 5 \cdot 5} = \frac{5 \cdot 5}{5 \cdot 5} \cdot \frac{1}{2} = \frac{1}{2}$$

To find the median, we first arrange the numbers from smallest to largest.

$$\frac{1}{10}, \frac{3}{10}, \frac{5}{10}, \frac{7}{10}, \frac{9}{10}$$

The median is the middle number, $\frac{5}{10}$.

Find the mode:

Each number occurs the same number of times, so no mode exists.

9. Find the average:

$$\frac{160+102+102+116+160+116}{6} = \frac{756}{6} = 126$$

To find the median, we first arrange the numbers from smallest to largest.

102, 102, 116, 116, 160, 160

The median is the average of the two middle numbers, 116 and 116.

$$\frac{116+116}{2} = \frac{232}{2} = 116$$

Find the mode:

Each number occurs the same number of times, so no mode exists.

10. Find the average:

$$\frac{\$4.96+\$5.24+\$4.96+\$10.05+\$5.24}{5} = \frac{\$30.45}{5} = \$6.09$$

To find the median, we first arrange the numbers from smallest to largest.

\$4.96, \$4.96, \$5.24, \$5.24, \$10.05

The median is the middle number, \$5.24.

Find the mode:

The numbers \$4.96 and \$5.24 each occur two times while \$10.05 occurs only one time. Thus, the modes are \$4.96 and \$5.24.

11. Find the average:

$$\frac{\frac{1}{2}+\frac{3}{4}+\frac{7}{8}+\frac{5}{4}}{4} = \frac{\frac{4}{8}+\frac{6}{8}+\frac{7}{8}+\frac{10}{8}}{4} = \frac{\frac{27}{8}}{4} = \frac{27}{8}\cdot\frac{1}{4} = \frac{27}{32}$$

To find the median, we first arrange the numbers from smallest to largest. We will write the fractions with a common denominator as we did above.

$$\frac{4}{8}, \frac{6}{8}, \frac{7}{8}, \frac{10}{8}$$

The median is the average of the two middle numbers, $\frac{6}{8}$ and $\frac{7}{8}$.

$$\frac{\frac{6}{8}+\frac{7}{8}}{2} = \frac{\frac{13}{8}}{2} = \frac{13}{8}\cdot\frac{1}{2} = \frac{13}{16}$$

Find the mode:

Each number occurs the same number of times, so no mode exists.

12. Find the average:

$$\frac{2+5+7+7+8+5+5+7+8}{9} = \frac{54}{9} = 6$$

To find the median, we first arrange the numbers from smallest to largest.

2, 5, 5, 5, 7, 7, 7, 8, 8

The median is the middle number, 7.

Find the mode: There are two numbers that occur most often, 5 and 7. Thus the modes are 5 and 7.

13. Find the average:

$$\frac{38.2+38.2+38.2+38.2}{4} = \frac{152.8}{4} = 38.2$$

Find the median:

The numbers are arranged from smallest to largest. The median is the average of the two middle numbers, 38.2 and 38.2.

$$\frac{38.2+38.2}{2} = \frac{76.4}{2} = 38.2$$

Find the mode:

Each number occurs the same number of times, so no mode exists.

14. Find Breyer's ice cream in the Product column and go across to the Size column. We see that the old package contained 56 oz and the new package contains 48 oz. The difference is

56 oz − 48 oz, or 8 oz.

15. Find Hellman's mayonnaise in the Product column and go across to the Percent Smaller column. We see that the new package is 6% smaller than the old package.

16. Find the largest number in the Percent Smaller column. It is 15. Go across to the Product column and read the entry. We see that Hershey's Special Dark chocolate bar has the greatest percent of decrease.

17. Find Tropicana orange juice in the Product column and go across to the Size column. We see that the old package contained 96 oz and the new package contains 89 oz. The difference is

96 oz − 89 oz = 7 oz.

18. Find the smallest number in the Percent Smaller column. It is 5. Go across to the Product column and read the entry. We see that Nabisco Chips Ahoy cookies have the smallest percent of decrease.

19. The United States is represented by 18 symbols. Thus, the gun ownership rate is 18×5, or 90 guns per 100 citizens.

20. From Exercise 19 we know that the gun ownership rate in the United States is 90 guns per 100 citizens. Canada is represented by 6 symbols, so the gun ownership rate is 6×5, or 30 guns per 100 citizens. We find that the gun ownership in the United States is $90 - 30$, or 60 more guns per 100 citizens than in Canada.

21. Switzerland is represented by 9 symbols, so the gun ownership rate is 9×5, or 45 guns per 100 citizens.

Finland is represented by 11 symbols, so the gun ownership rate is 11×5, or 55 guns per 100 citizens.

22. The country with the smallest number of symbols is India. Of the countries represented, India has the lowest gun ownership rate.

23. From our work in the preceding exercises we know the following gun ownership rates:

Canada: 30 guns per 100 citizens

Finland: 55 guns per 100 citizens

Switzerland: 45 guns per 100 citizens

United States: 90 guns per 100 citizens

We find the rates for the other two countries.

India: $1 \times 5 = 5$ guns per 100 citizens

Yemen: $12 \times 5 = 60$ guns per 100 citizens

We find the average:

$$\frac{30+55+45+90+5+60}{6} = \frac{285}{6} = 47.5$$

The average gun ownership rate for the six countries is 47.5 guns per 100 citizens.

24. We list the rates from smallest to largest.

5, 30, 45, 55, 60, 90

The median is the average of the two middle numbers, 45 and 55.

$$\frac{45+55}{2} = \frac{100}{2} = 50 \text{ guns per 100 citizens}$$

25. Yes; if the trip took $1\frac{1}{2}$ hr then the average speed would be 20 mph. ($30 \text{ mi} \div 1\frac{1}{2} \text{ hr} = 30 \text{ mi} \div \frac{3}{2} \text{ hr} = 30 \text{ mi} \cdot \frac{2}{3} \text{ hr} = 20$ mph) But the driver could have driven at a speed of 75 mph for a brief period during that time.

26. Answers may vary. Some would ask for the average salary since it is a center point that places equal emphasis on all the salaries in the firm. Some would ask for the median salary since it is a center point that deemphasizes the extremely high and extremely low salaries. Some would ask for the mode of the salaries since it might indicate the salary you are most likely to earn.

Exercise Set 7.3

1. Go across from 17 on the vertical scale to a bar whose shading begins at this level. Then go down to the horizontal scale and read the variety. We see that the miniature tall bearded iris has a minimum height of 17 in.

3. From the top of each blossom go across to the vertical scale and read the corresponding height. The maximum heights are 9 in., 26 in., 34 in., 27 in., 15 in., and 28 in. We find the average

$$\frac{9+26+34+27+15+28}{6} = \frac{139}{6} \approx 23.2$$

The average of the maximum heights is about 23.2 in.

5. Find the bar representing the border bearded iris. From the top and the bottom of the bar, go across to the vertical scale and read the corresponding heights. We see that this iris ranges in height from 16 in. to 26 in.

7. Find the shortest bar and blossom and read the corresponding variety from the horizontal scale. We see that the tall bearded iris has the smallest range in heights.

9. The tallest bar and blossom represent the tall bearded iris. The top of this bar corresponds to a height of 34 in. This is the height of the tallest iris with the largest maximum height.

The shortest bar and blossom represent the miniature dwarf bearded iris. The top of this bar corresponds to a height of 9 in.

We subtract to find the difference in heights:

35 in. − 9 in. = 25 in.

11. Move to the right along the bar representing 1 cup of hot cocoa with skim milk. We read that there are about 185 calories in the cup of cocoa.

13. The longest bar is for 1 slice of chocolate cake with fudge frosting. Thus, it has the highest caloric content.

15. We locate 460 calories at the bottom of the graph and then go up until we reach a bar that ends at approximately 460 calories. Now go across to the left and read the dessert, 1 cup of premium chocolate ice cream.

17. From the graph we see that 1 cup of hot cocoa made with whole milk has about 310 calories and 1 cup of hot cocoa made with skim milk has about 190 calories. We subtract to find the difference:

$310 - 190 = 120$

The cocoa made with whole milk has about 120 more calories than the cocoa made with skim milk. (Answers may vary.)

19. Find the instances for which the bar representing men's degrees is taller than the corresponding bar representing women's degrees and then read the years they represent on the horizontal scale. We see that more bachelor's degrees were conferred on men than on women in 1950 and 1970.

21. Find 2000 on the horizontal scale, go across to the vertical scale from the top of each bar, and read the corresponding numbers. We see that about 530,000 men and 705,000 women received bachelor's degrees in 2000. The difference is

$705,000 - 530,000$, or 175,000.

Thus, about 175,000 more bachelor's degrees were conferred on women than on men in 2000.

23. First write the seven items in the City column of the table in equally-spaced intervals on the vertical scale and title the scale "City." Next, scale the horizontal axis. We see that the average costs range from \$32 to \$107. We will start the horizontal scale at 30, indicating the gap from 0 to 30 with a jagged line, and extend it by 10's through 110. Label this scale "Average daily cost of adult day care." Finally, draw horizontal bars to show the average costs.

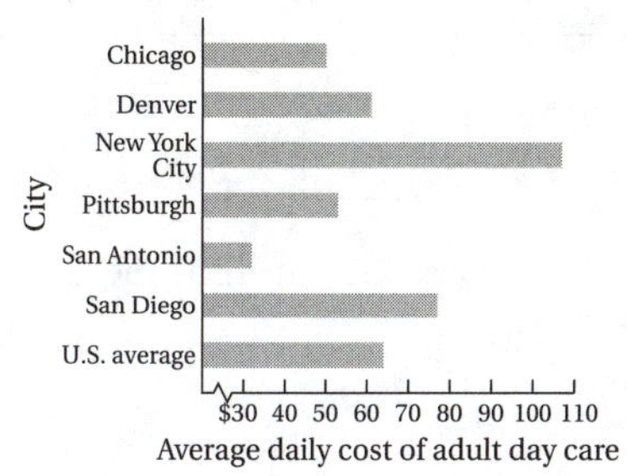

25. From the table we see that the average daily cost of adult day car is less than \$75 in Chicago, Denver, Pittsburgh, and San Antonio. We could also read this from the bar graph by finding the cities associated with the bars whose lengths do not extend to 75 on the horizontal scale.

27. From the table we see that the average daily cost of adult day care in New York City is \$107 and the national average cost is \$64. The difference is \$107 − \$64, or \$43. Thus, the average daily cost of adult day care in New York City is \$43 higher than the national average cost.

29. From the table we see that New York has the greatest commuting time. We could also find this by observing that the longest bar in the graph drawn in Exercise 28 corresponds to New York.

31. We list the times from smallest to largest

21.6, 24.7, 25.9, 28.1, 33.1, 39.0

The median is the average of the two middle terms, 25.9 and 28.1.

$$\frac{25.9 + 28.1}{2} = \frac{54}{2} = 27$$

The median commuting time is 27 min.

33. Find 2004 on the horizontal scale, go up to the graph, and then go left to the vertical scale and read the value. We estimate that about 3.2 million international tourists visited Beijing in 2004.

Find 2007 on the horizontal scale, go up to the graph, and then go left to the vertical scale and read the value. We estimate that about 4.3 million international tourists visited Beijing in 2007.

35. We look for segments of the graph that slant down from left to right. The only such segment goes from the point associated with 2007 to the point associated with 2008. Thus, the number of international tourists to Beijing decreased between 2007 and 2008.

37. The highest point on the graph corresponds to the interval labeled 9-11 A.M. on the horizontal scale. This is the time interval in which most bank crimes occurred.

39. We see that 1900 bank crimes occurred between 9 A.M. and 11 A.M. while 1600 bank crimes occurred between 3 P.M. and 6 P.M. Thus, 1900 − 1600, or 300, more bank crimes occurred between 9 A.M. and 11 A.M. than between 3 P.M. and 6 P.M.

41. First indicate the years on the horizontal scale and label it "Year." The years range from 1980 to 2030 and increase by 10's. We could start the vertical scale at 0, but the graph will be more compact if we start at a higher number. The years lived beyond age 65 range from 14 to 17.5 so we choose to label the vertical scale from 13 to 18. We use a jagged line to indicate that we are not starting at 0. Label the vertical scale "Average number of years men are estimated to live beyond 65." Next, at the appropriate level above each year on the horizontal scale, mark the corresponding number of years. Finally, draw line segments connecting the points.

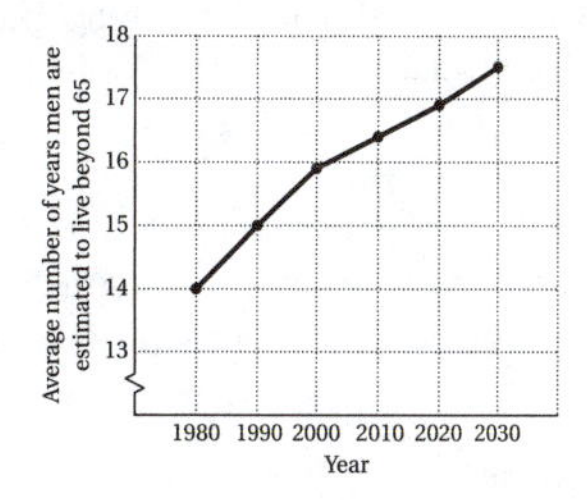

43. First we subtract to find the amount of the increase:

$$17.5 - 14 = 3.5$$

Let $p =$ the percent of increase. We write and solve an equation to find p.

$$3.5 = p \cdot 14$$
$$\frac{3.5}{14} = p$$
$$0.25 = p$$
$$25\% = p$$

Longevity is estimated to increase 25% between 1980 and 2030.

45. First we subtract to find the amount of increase:

$$17.5 - 15.9 = 1.6$$

Let $p =$ the percent of increase. We write and solve an equation to find p.

$$1.6 = p \cdot 15.9$$
$$\frac{1.6}{15.9} = p$$
$$0.101 \approx p$$
$$10.1\% \approx p$$

Longevity is estimated to increase about 10.1% between 2000 and 2030.

47. To find the average of a set of numbers, add the numbers and then divide by the number of items of data.

49. If an item has a regular price of \$100 and is on sale for 25% off, \$100 is called the marked price, 25% the rate of discount, 25% of \$100, or \$25, the discount and \$100 − \$25, or \$75, the sale price.

51. The decimal $0.\overline{1518}$ is an example of a repeating decimal.

53. The decimal 0.125 is an example of a terminating decimal.

Exercise Set 7.4

1. We see from the graph that 11% of foreign students were from South Korea.

3. We see from the graph that 15% of foreign students were from India. Find 15% of 625,000:

   $0.15 \times 625,000 = 93,750$ students

5. The section of the graph labeled 6% represents the foreign students from Japan.

7. From the graph we see that 18% of those who participate in outdoor activities look forward most to forgetting about work.

9. We see from the graph that 38% of the participants in the survey listed exercise as their main reason for looking forward to outdoor recreation. We find 38% of the 1027 participants:

   $0.38 \times 1027 \approx 390$ people

11. We first draw a line from the center of the circle to any tick mark. From that tick mark we count off 57 tick marks to graph 57% and label the wedge "College." We continue in this manner with the remaining items in the table. The graph is shown below.

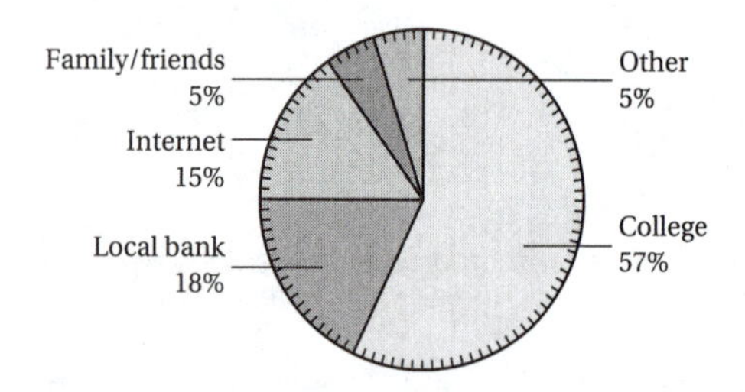

13. We first draw a line from the center of the circle to any tick mark. From that tick mark we count off 5 tick marks to graph 5% and label this wedge "12-17." We continue in this manner with the remaining items in the table. The graph is shown below.

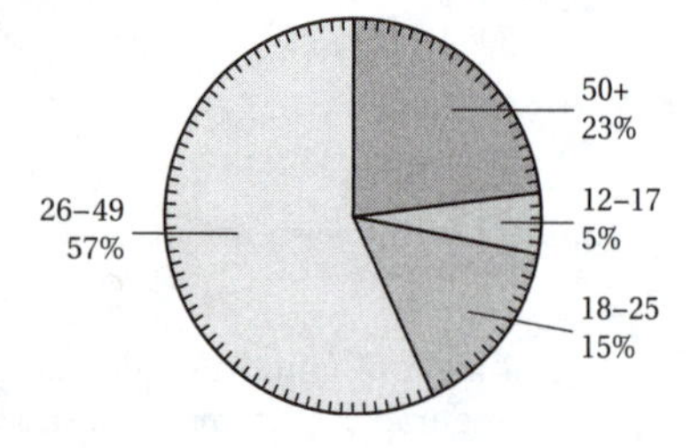

Chapter 7 Concept Reinforcement

1. The statement is false. To find the average of a set of numbers, add the numbers and then *divide* by the number of items of data.

2. The statement is true. See page 477 in the text.

3. The statement is true. See page 477 in the text.

Chapter 7 Important Concepts

1. Find the average:

$$\frac{8+13+1+4+8+7+15}{7} = \frac{56}{7} = 8$$

   To find the median we first arrange the numbers from smallest to largest.

   1, 4, 7, 8, 8, 13, 15

   The median is the middle number, 8.

   Find the mode:

   The number 8 occurs most often. It is the mode.

2. Find the largest number in the Cost column, \$0.54, and then read the corresponding entry in the Product column. We see that Quaker Organic Maple & Brown Sugar oatmeal has the highest cost per packet or serving.

3. Find Kashi oatmeal in the Product column and then read the corresponding number in the Sugar column, 12. We see that Kashi oatmeal has 12 g of sugar per packet or serving.

4. Read the names of the stadiums corresponding to the bars that do not extend to \$100 million. We see that the building costs of Arrowhead Stadium and Candlestick Park were less than \$100 million.

5. The bar representing Invesco Field extends to about \$360 million while the bar representing Bank of America Stadium extends to about \$250 million. Thus, Invesco Field cost about

   \$360 million – \$250 million, or \$110 million,

   more to build than Bank of America Stadium.

6. The smallest wedge represents the 80 and older age group. This is the group with the fewest people.

7. We see from the graph that 27% of the population is under 20 years old.

Chapter 7 Review Exercises

1. $\frac{26+34+43+51}{4} = \frac{154}{4} = 38.5$

2. $\frac{11+14+17+18+7}{5} = \frac{67}{5} = 13.4$

3. $\frac{0.2+1.7+1.9+2.4}{4} = \frac{6.2}{4} = 1.55$

4. $\frac{700+2700+3000+900+1900}{5} = \frac{9200}{5} = 1840$

5. $\frac{\$2+\$14+\$17+\$17+\$21+\$29}{6} = \frac{\$100}{6} = \$16.\overline{6}$

6. $\frac{20+190+280+470+470+500}{6} = \frac{1930}{6} = 321.\overline{6}$

7. We can find the total of the four scores needed as follows:

$$90 + 90 + 90 + 90 = 360.$$

The total of the scores on the first three tests is

$$94 + 78 + 92 = 264.$$

Thus the student needs to get at least

$$360 - 264 = 96$$

to get an A. We can check this as follows:

$$\frac{90 + 78 + 92 + 96}{4} = \frac{360}{4} = 90.$$

8. We divide the number of miles, 532, by the number of gallons, 19.

$\frac{532}{19} = 28$

The average was 28 miles per gallon.

9. To find the GPA we first add the grade point values for each hour taken. This is done by first multiplying the grade point value by the number of hours in the course and then adding as follows:

A	$4.0 \cdot 5 =$	20
B	$3.0 \cdot 3 =$	9
C	$2.0 \cdot 4 =$	8
B	$3.0 \cdot 3 =$	9
B	$3.0 \cdot 1 =$	3
		49 (Total)

The total number of hours taken is

$5 + 3 + 4 + 3 + 1$, or 16.

We divide 49 by 16 and round to the nearest tenth.

$\frac{49}{16} = 3.0625 \approx 3.1$

The student's grade point average is 3.1.

10. $26, 34, 43, 51$

↑ Middle number

The median is halfway between 34 and 43. It is the average of the two middle numbers.

$\frac{34 + 43}{2} = \frac{77}{2} = 38.5$

The median is 38.5.

11. $7, 11, 14, 17, 18$

↑ Middle number

The median is 14.

12. $0.2, 1.7, 1.9, 2.4$

↑ Middle number

The median is halfway between 1.7 and 1.9. It is the average of the two middle numbers.

$\frac{1.7 + 1.9}{2} = \frac{3.6}{2} = 1.8$

The median is 1.8.

13. $700, 900, 1900, 2700, 3000$

↑ Middle number

The median is 1900.

14. We arrange the numbers from smallest to largest.

\$2, \$14, \$17, \$17, \$21, \$29

↑ Middle number

The median is halfway between \$17 and \$17. Although it seems clear that this is \$17, we can compute it as follows:

$\frac{\$17 + \$17}{2} = \frac{\$34}{2} = \$17$

The median is \$17.

15. We arrange the numbers from smallest to largest.

$20, 190, 280, 470, 470, 500$

↑ Middle number

The median is halfway between 280 and 470. It is the average of the two middle numbers.

$\frac{280 + 470}{2} = \frac{750}{2} = 375$

The median is 375.

16. Find the average:

$$\frac{\$360 + \$192 + \$240 + \$216 + \$420 + \$132}{6} = \frac{\$1560}{6} = \$260$$

To find the median we first arrange the numbers from smallest to largest.

\$132, \$192, \$216, \$240, \$360, \$420

The median is the average of the two middle numbers, \$216 and \$240.

$\frac{\$216 + \$240}{2} = \frac{\$456}{2} = \$228$

17. The number that occurs most often is 26, so 26 is the mode.

18. The numbers that occur most often are 11 and 17. They are the modes.

19. The number that occurs most often is 0.2, so 0.2 is the mode.

20. The numbers that occur most often are 700 and 800. They are the modes.

21. The number that occurs most often is \$17, so \$17 is the mode.

22. The number that occurs most often is 20, so 20 is the mode.

23. Battery A:

$(38.9 + 39.3 + 40.4 + 53.1 + 41.7 + 38.0 + 36.8 + 47.7 +$

$48.1 + 38.2 + 46.9 + 47.4) \div 12 = \dfrac{516.5}{12} \approx 43.04$

Battery B:

$(39.3 + 38.6 + 38.8 + 37.4 + 47.6 + 37.9 + 46.9 + 37.8 +$

$38.1 + 47.9 + 50.1 + 38.2) \div 12 = \dfrac{498.6}{12} \approx 41.55$

Because the average time for Battery A is longer, it is the better battery.

24. Go down the UPS Package column to 5 and then go across to the Next Day Air Delivery column and read the cost, \$30.37.

25. Go down the UPS Package column to 8 and then go across to the UPS Ground column and read the cost, \$10.85.

26. First go down the UPS Package column to 5 and then go across to the Next Day Air Saver Delivery column and read the cost, \$26.61. The amount saved is the difference in the costs, \$30.37 – \$26.61, or \$3.76.

27. The 25-29 years age group is represented by 18 symbols, so there are 18 PGA champions in this age group.

28. The 30-34 years age group is represented by the most symbols, so this is the age group with the most PGA champions.

29. We see from the graph that the 30-34 years age group has 39 PGA champions while the 35-39 years age group has 20 champions. Thus, the 30-34 years age group has 39-20, or 19, more PGA champions than the 35-39 years age group.

30. We find the years associated with the bars that do not extend to 100. They are 2004, 2007, and 2008.

31. From the top of the bar for 2008, go to the vertical scale and read that the runoff was 60%.

32. Reading from the graph as described in Exercise 31, we see that the runoff in 2005 was about 105% and it was about 53% in 2007. Thus the difference in the runoffs was 105% – 53%, or 52%.

33. The only bar that extends beyond 170 on the vertical scale is associated with 2006. This is the year in which the runoff was over 170%.

34. The lowest point on the graph corresponds to 2001. This is the year in which Phil Mickelson had his lowest score.

35. The highest point on the graph corresponds to 2007. This is the year in which Phil Mickelson got his highest score.

36. The points on the graph that lie below 280 correspond to 2001, 2004, and 2009. These are the years in which Phil Mickelson scored less than 280.

37. Phil Mickelson's total score was 279 in the years that correspond to 279 on the graph. They are 2004 and 2009.

38. From the graph we see that the highest and lowest scores were 299 and 275, respectively. Thus, the highest score was 299 – 275, or 24, higher than the lowest score.

39. We read the scores from the graph and arrange them from smallest to largest.

275, 279, 279, 280, 281, 283, 285, 286, 299

The median is the middle score, 281.

40. We find the average of the scores in Exercise 39.

$$\frac{275 + 279 + 279 + 280 + 281 + 283 + 285 + 286 + 299}{9} =$$

$$\frac{2547}{9} \approx 283$$

41. From the graph we see that the Army is 36% of the Armed Forces.

42. From the graph we see that the Marines represent 13% of the Armed Forces.

43. From the graph we see that 25% of the Armed Forces were in the Air Force. We find 25% of 1,400,000:

$$0.25 \times 1,400,000 = 350,000$$

44. The Army and Navy represent 36% and 26% of the Armed Forces, respectively. Thus, the Army represents 36% – 26%, or 10% more of the Armed Forces than the Navy does.

45. 6, 9, 6, 8, 8, 5, 10, 5, 9, 10

Each number occurs the same number of times, so no mode exists. Answer D is correct.

46. $\dfrac{\frac{1}{2}+\frac{1}{3}+\frac{1}{4}+\frac{1}{5}}{4} = \dfrac{\frac{30}{60}+\frac{20}{60}+\frac{15}{60}+\frac{12}{60}}{4} = \dfrac{\frac{77}{60}}{4} =$

$\dfrac{77}{60} \cdot \dfrac{1}{4} = \dfrac{77}{240}$

Answer A is correct.

47. On the horizontal scale in eight equally spaced intervals indicate the years. Label this scale "Year." Then label the vertical scale "Cost of first-class postage (in cents)." The smallest cost is 32¢ and the largest is 44¢, so we start the vertical scale at 0 and extend it to 50¢, labeling it by 10's. Finally, draw vertical bars above the years to show the cost of the postage.

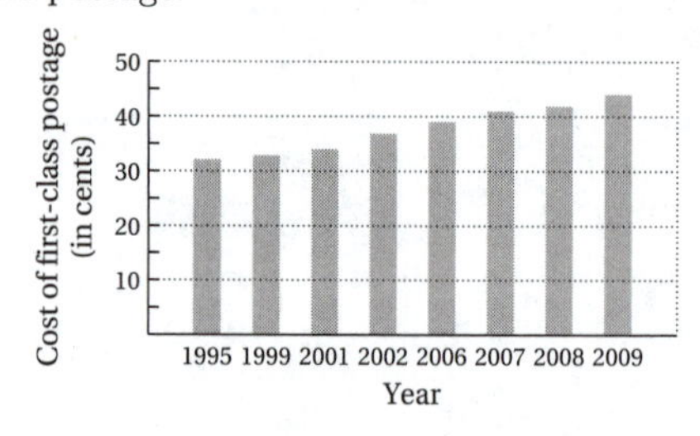

48. Prepare horizontal and vertical scales as described in Exercise 47. Then, at the appropriate level above each year, mark the corresponding postage. Finally, draw line segments connecting the points.

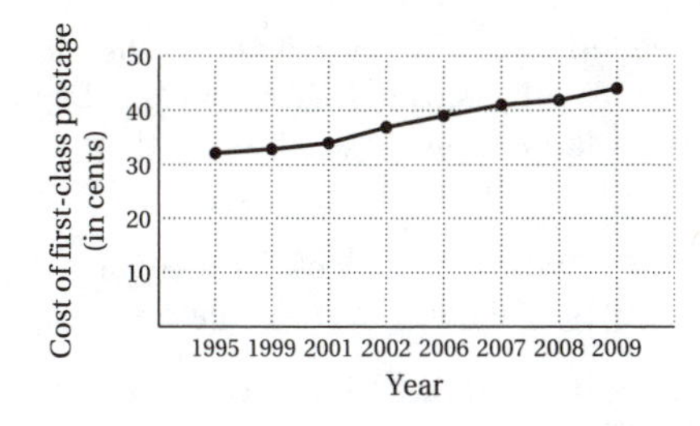

49. Using a circle with 100 equally-spaced tick marks, first draw a line from the center to any tick mark. Then count off 8 tick marks to graph 8% and label the wedge "20-24 years." Continue in this manner with the remaining age groups. The graph is shown below.

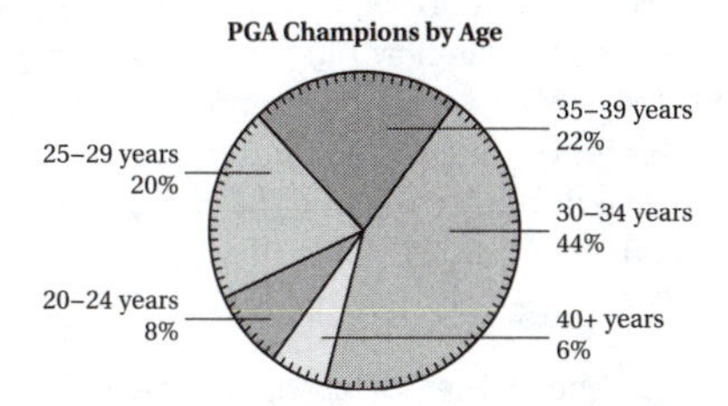

50. a is the middle number and the median is 316, so $a = 316$.

The average is 326 so the data must add to $326 + 326 + 326 + 326 + 326 + 326 + 326$, or 2282.

The sum of the known data items, including a, is $298 + 301 + 305 + 316 + 323 + 390$, or 1933.

We subtract to find b:

$$b = 2282 - 1933 = 349$$

Chapter 7 Discussion and Writing Exercises

1. The equation could represent a person's average income during a 4-yr period. Answers may vary.

2. Bar graphs that show change over time, such as the one in Exercise 47 in the Review Exercises, can be successfully converted to line graphs. Other bar graphs, such as the one in Example 1 in Section 7.3, cannot be successfully converted to line graphs.

3. We can use circle graphs to visualize how the number of items in various categories compare in size.

4. A bar graph is convenient for showing comparisons. A line graph is convenient for showing a change over time as well as to indicate patterns or trends. The choice of which to use to graph a particular set of data would probably depend on the type of data analysis desired.

5. The average, the median, and the mode are "center points" that characterize a set of data. You might use the average to find a center point that is midway between the extreme values of the data. The median is a center point that is in the middle of all the data. That is, there are as many values less than the median as there are values greater than the median. The mode is a center point that represents the value or values that occur most frequently.

6. Circle graphs are similar to bar graphs in that both allow us to tell at a glance how items in various categories compare in size. They differ in that circle graphs show percents whereas bar graphs show actual numbers of items in a given category.

Chapter 7 Test

1. We add the numbers and then divide by the number of items of data.

$$\frac{45 + 49 + 52 + 52}{4} = \frac{198}{4} = 49.5$$

2. We add the numbers and then divide by the number of items of data.

$$\frac{1 + 1 + 3 + 5 + 3}{5} = \frac{13}{5} = 2.6$$

3. We add the numbers and then divide by the number of items of data.

$$\frac{3 + 17 + 17 + 18 + 18 + 20}{6} = \frac{93}{6} = 15.5$$

4. 45, 49, 52, 53

Find the median: There is an even number of numbers. The median is the average of the two middle numbers:

$$\frac{49 + 52}{2} = \frac{101}{2} = 50.5$$

Find the mode: Each number occurs only once, so there is no mode.

5. Find the median: First we rearrange the numbers from the smallest to largest.

$$1, 1, 3, 3, 5$$

↑

Middle number

The median is 3.

Find the mode: There are two numbers that occur most often, 1 and 3. They are the modes.

6. 3, 17, 17, 18, 18, 20

Find the median: There is an even number of numbers. The median is the average of the two middle numbers:

$$\frac{17 + 18}{2} = \frac{35}{2} = 17.5$$

Find the mode: There are two numbers that occur most often, 17 and 18. They are the modes.

7. We divide the number of miles by the number of gallons.

$$\frac{462}{14} = 33 \text{ mpg}$$

8. The total of the four scores needed is

$$70 + 70 + 70 + 70 = 4 \cdot 70, \text{ or } 280.$$

The total of the scores on the first three tests is

$$68 + 71 + 65 = 204.$$

Thus the student needs to get at least

$$280 - 204, \text{ or } 76$$

on the fourth test.

9. To find the GPA we first add the grade point values for each class taken. This is done by first multiplying the grade point value by the number of hours in the course and then adding as follows:

$$\begin{array}{ll} \text{B} & 3.0 \cdot 3 = \ 9 \\ \text{A} & 4.0 \cdot 3 = 12 \\ \text{C} & 2.0 \cdot 4 = \ 8 \\ \text{B} & 3.0 \cdot 3 = \ 9 \\ \text{B} & 3.0 \cdot 2 = \ \underline{6} \\ & \qquad\quad \ 44 \ \text{(Total)} \end{array}$$

The total number of hours taken is

$3 + 3 + 4 + 3 + 2$, or 15.

We divide 44 by 15 and round to the nearest tenth.

$\frac{44}{15} = 2.9\overline{3} \approx 2.9$

The grade point average is 2.9.

10. We find the average of each set of ratings.

Pecan:

$\frac{9 + 10 + 8 + 10 + 9 + 7 + 6 + 9 + 10 + 7 + 8 + 8}{12} =$

$\frac{101}{12} \approx 8.417$

Hazelnut:

$\frac{10 + 6 + 8 + 9 + 10 + 10 + 8 + 7 + 6 + 9 + 10 + 8}{12} =$

$\frac{101}{12} \approx 8.417$

Since the averages are equal, the chocolate bars are of equal quality.

11. Go down the column in the first table labeled "Height" to the entry "6 ft 1 in." Then go to the right and read the entry in the column headed "Medium Frame." We see that the desirable weight is 179 lb.

12. Locate the number 120 in the second table and observe that it is in the column headed "Medium Frame." Then go to the left and observe that the corresponding entry in the "Height" column is 5 ft 3 in. Thus a 5 ft 3 in. woman with a medium frame has a desirable weight of 120 lb.

13. From Exercise 12, we know that the desirable weight for a 5 ft 3 in. woman with a medium frame is 120 lb. To find the desirable weight for a 5 ft 3 in. woman with a small frame, first locate 5 ft 3 in. in the "Height" column. Then go to the right and read the entry in the "Small Frame" column. We see that the desirable weight is 111 lb. We subtract to find the difference: 120 lb − 111 lb = 9 lb. Thus, the desirable weight for a 5 ft 3 in. woman with a medium frame is 9 lb more than for a woman of the same height with a small frame.

14. To find the desirable weight for a 6 ft 3 in. man with a large frame, first locate 6 ft 3 in. in the "Height" column. Then go to the right and read the entry in the "Large Frame" column, 204 lb.

To find the desirable weight for a 6 ft 3 in. man with a small frame, locate 6 ft 3 in. in the "Height" column and then go to the right and read the entry in the "Small Frame" column, 172 lb. We subtract to find the difference: 204 lb− 172 lb = 32 lb. Thus, the desirable weight for a 6 ft 3 in. man with a large frame is 32 lb more than for a man of the same height with a small frame.

15. Since $1300 \div 100 = 13$, we look for a country represented by 13 symbols. We find that it is Spain.

16. Since $1500 \div 100 = 15$, we look for countries represented by more than 15 symbols. Those countries are Norway and the United States.

17. Canada is represented by 9 symbols, so $9 \cdot 100$, or 900 lb, of waste is generated per person per year in Canada.

18. The United States is represented by 17 symbols, so $17 \cdot 100$, or 1700 lb, of waste is generated per person per year in the United States. Mexico is represented by 7 symbols, so $7 \cdot 100$, or 700 lb, of waste is generated per person per year in Mexico. We subtract to find the difference: 1700 lb − 700 lb = 1000 lb. Thus, in the United States, 1000 lb more of waste is generated per person per year than in Mexico.

19. First indicate the names of the animals in eight equally spaced intervals on the horizontal scale. Title this scale "Animal." Now note that the lowest speed is 9 mph and the highest is 70 mph. We start the vertical scaling at 0 and label the marks on the scale by 10's from 0 to 80. Title this scale "Maximum speed (in miles per hour)." Finally, draw vertical bars above the names of the animals to show the speeds.

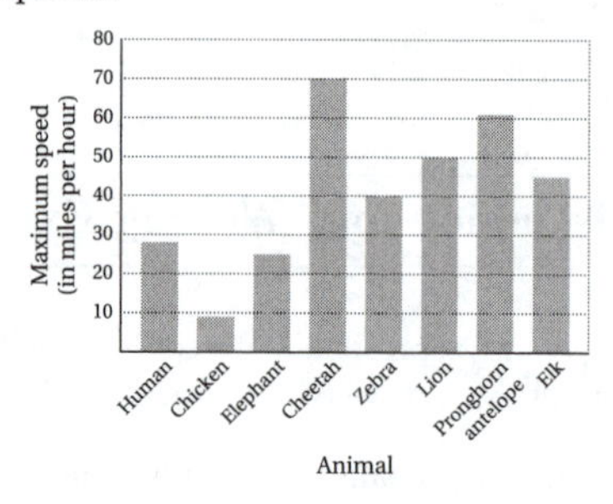

20. From the table or the bar graph, we see that the slowest speed is 9 mph and the fastest is 70 mph. Then the fastest speed exceeds the slowest by

$70 - 9$, or 61 mph.

21. The fastest human's maximum speed is 28 mph and a zebra's maximum speed is 40 mph. Thus a human cannot outrun a zebra because a zebra can run $40 - 28$, or 12 mph, faster than a human.

22. We add the speeds and then divide by the number of speeds.

$\frac{28 + 9 + 25 + 70 + 40 + 50 + 61 + 45}{8} = \frac{328}{8} = 41$ mph

23. First we write the numbers from smallest to largest.

9, 25, 28, 40, 45, 50, 61, 70

There is an even number of numbers. The median is the average of the two middle numbers.

$\frac{40 + 45}{2} = \frac{85}{2} = 42.5$ mph

24. Find 2010 on the bottom scale and move up from there to the line. The line is labeled 53% at that point, so 53% of food dollars will be spent away from home in 2010.

25. Find 1985 halfway between 1980 and 1990 on the bottom scale and move up from that point to the line. Then go straight across to the left and find that about 41% of food dollars were spent away from home in 1985.

26. Locate 30% on the vertical scale. Then move to the right to the line. Look down to the bottom scale and observe that the year 1967 corresponds to this point.

27. Locate 50% on the vertical scale. Then move to the right to the line. Look down to the bottom scale and observe that the year 2006 corresponds to this point.

28. We will make a vertical bar graph. First indicate the days in seven equally-spaced intervals on the horizontal scale and title this scale "Day." Now note that the number of books ranges from 160 to 420. We start the vertical scaling with 150, using a jagged line to indicate that we are not showing the range of numbers from 0 to 150, and label the marks by 50's up to 450. Title this scale "Number of Books." Finally, draw vertical bars to show the number of books checked out each day.

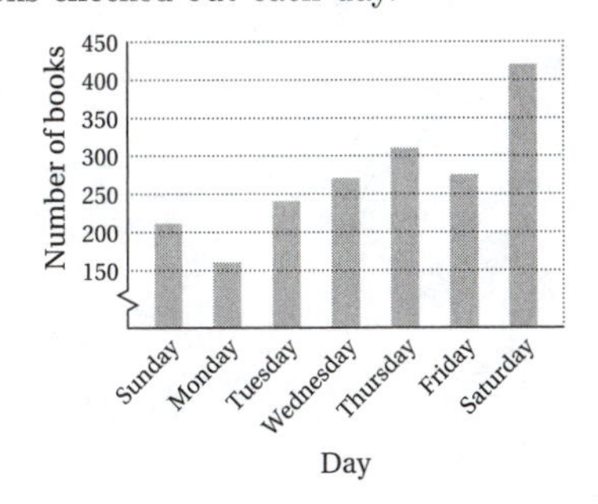

29. First indicate the days on the horizontal scale and title this scale "Day." We scale the vertical axis by 50's from 150 to 450, using a jagged line to indicate that we are not showing the range of numbers from 0 to 150. Next mark the number of books checked out above the days. Then draw line segments connecting adjacent points.

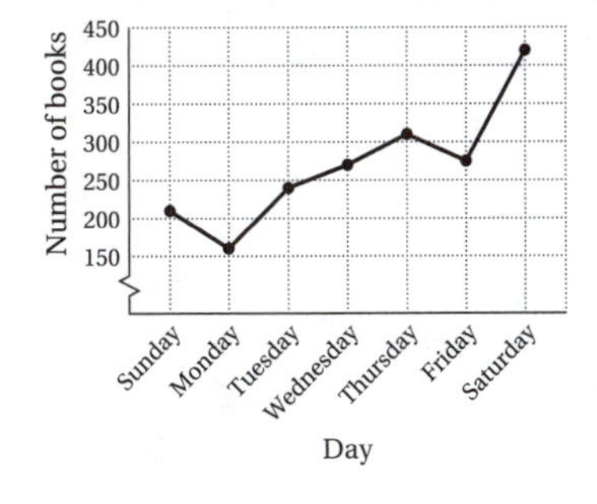

30. Using a circle with 100 equally spaced tick marks, we first draw a line from the center to any tick mark. From that tick mark, count off 23 tick marks and draw another line to graph 23%. Label this wedge "Meat, poultry, fish, and eggs" and "23%." Continue in this manner with the other types of food categories.

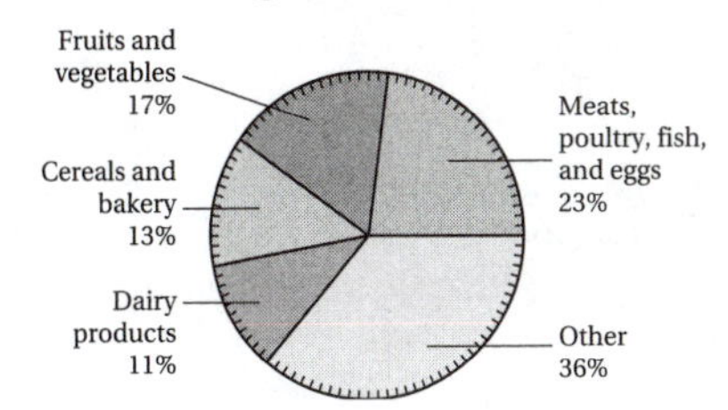

31. 13% of \$664 $= 0.13(\$664) = \$86.32$

Answer C is correct.

32. a is the middle number in the ordered set of data, so a is the median, 74.

Since the mean, or average, is 82, the total of the seven numbers is $7 \cdot 82$, or 574.

The total of the known numbers is

$$69 + 71 + 73 + 74 + 78 + 98, \text{ or } 463.$$

Then $b = 574 - 463 = 111$.

Chapter 8

Geometry

Exercise Set 8.1

1. The segment consists of the endpoints G and H and all points between them.

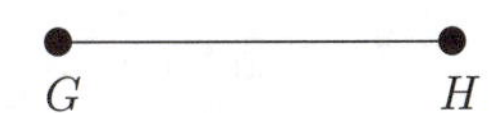

It can be named $\overline{GH}$ or $\overline{HG}$.

3. The ray with endpoint Q extends forever in the direction of point D.

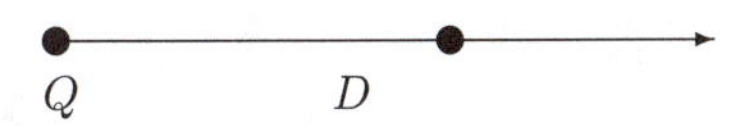

In naming a ray, the endpoint is always given first. This ray is named $\overrightarrow{QD}$.

5.

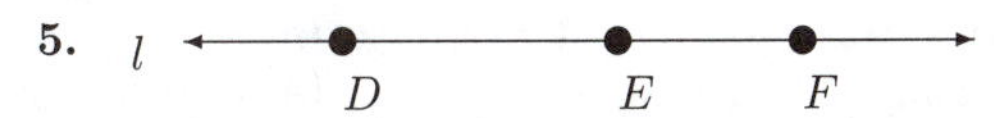

The line can be named with the small letter l, or it can be named by any two points on it. This line can be named

l, $\overleftrightarrow{DE}$, $\overleftrightarrow{ED}$, $\overleftrightarrow{DF}$, $\overleftrightarrow{FD}$, $\overleftrightarrow{EF}$, or $\overleftrightarrow{FE}$.

7. The angle can be named in five different ways:

angle GHI, angle IHG, $\angle\, GHI$, $\angle\, IHG$, or $\angle\, H$.

9. Place the $\triangle$ of the protractor at the vertex of the angle, and line up one of the sides at 0°. We choose the horizontal side. Since 0° is on the inside scale, we check where the other side of the angle crosses the inside scale. It crosses at 10°. Thus, the measure of the angle is 10°.

11. Place the $\triangle$ of the protractor at the vertex of the angle, point B. Line up one of the sides at 0°. We choose the side that contains point A. Since 0° is on the outside scale, we check where the other side crosses the outside scale. It crosses at 180°. Thus, the measure of the angle is 180°.

13. Place the $\triangle$ of the protractor at the vertex of the angle, and line up one of the sides at 0°. We choose the horizontal side. Since 0° is on the inside scale, we check where the other side crosses the inside scale. It crosses at 130°. Thus, the measure of the angle is 130°.

15. Using a protractor, we find that the measure of the angle in Exercise 7 is 148°. Since its measure is greater than 90°and less than 180°, it is an obtuse angle.

17. The measure of the angle in Exercise 9 is 10°. Since its measure is greater than 0°and less than 90°, it is an acute angle.

19. The measure of the angle in Exercise 11 is 180°. It is a straight angle.

21. The measure of the angle in Exercise 13 is 130°. Since its measure is greater than 90°and less than 180°, it is an obtuse angle.

23. The measure of the angle in Margin Exercise 12 is 30°. Since its measure is greater than 0°and less than 90°, it is an acute angle.

25. The measure of the angle in Margin Exercise 14 is 126°. Since its measure is greater than 90°and less than 180°, it is an obtuse angle.

27. Using a protractor, we find that the lines do not intersect to form a right angle. They are not perpendicular.

29. Using a protractor, we find that the lines intersect to form a right angle. They are perpendicular.

31. All the sides are of different lengths. The triangle is a scalene triangle.

One angle is an obtuse angle. The triangle is an obtuse triangle.

33. All the sides are of different lengths. The triangle is a scalene triangle.

One angle is a right angle. The triangle is a right triangle.

35. All the sides are the same length. The triangle is an equilateral triangle.

All three angles are acute. The triangle is an acute triangle.

37. All the sides are of different lengths. The triangle is a scalene triangle.

One angle is an obtuse angle. The triangle is an obtuse triangle.

39. The polygon has 4 sides. It is a quadrilateral.

41. The polygon has 5 sides. It is a pentagon.

43. The polygon has 3 sides. It is a triangle.

45. The polygon has 5 sides. It is a pentagon.

47. The polygon has 6 sides. It is a hexagon.

49. If a polygon has n sides, the sum of its angle measures is $(n-2)\cdot 180°$. A decagon has 10 sides. Substituting 10 for n in the formula, we get

$$\begin{aligned}(n-2)\cdot 180° &= (10-2)\cdot 180°\\ &= 8\cdot 180°\\ &= 1440°.\end{aligned}$$

51. If a polygon has n sides, the sum of its angle measures is $(n-2)\cdot 180°$. A heptagon has 7 sides. Substituting 7 for n in the formula, we get

$$\begin{aligned}(n-2)\cdot 180° &= (7-2)\cdot 180°\\ &= 5\cdot 180°\\ &= 900°.\end{aligned}$$

53. If a polygon has n sides, the sum of its angle measures is $(n-2)\cdot 180°$. To find the sum of the angle measures for a 14-sided polygon, substitute 14 for n in the formula.

$$\begin{aligned}(n-2)\cdot 180° &= (14-2)\cdot 180°\\ &= 12\cdot 180°\\ &= 2160°\end{aligned}$$

55. If a polygon has n sides, the sum of its angle measures is $(n-2)\cdot 180°$. To find the sum of the angle measures for a 20-sided polygon, substitute 20 for n in the formula.

$$\begin{aligned}(n-2)\cdot 180° &= (20-2)\cdot 180°\\ &= 18\cdot 180°\\ &= 3240°\end{aligned}$$

57.
$$\begin{aligned}m(\angle A)+m(\angle B)+m(\angle C) &= 180°\\ 42°+92°+x &= 180°\\ 134°+x &= 180°\\ x &= 180°-134°\\ x &= 46°\end{aligned}$$

59.
$$\begin{aligned}31°+29°+x &= 180°\\ 60°+x &= 180°\\ x &= 180°-60°\\ x &= 120°\end{aligned}$$

61.
$$\begin{aligned}m(\angle R)+m(\angle S)+m(\angle T) &= 180°\\ x+58°+79° &= 180°\\ x+137° &= 180°\\ x &= 180°-137°\\ x &= 43°\end{aligned}$$

63.
$$\begin{aligned}I &= P\cdot r\cdot t\\ &= \$2000\cdot 8\%\cdot 1\\ &= \$2000\cdot 0.08\cdot 1\\ &= \$160\end{aligned}$$

65.
$$\begin{aligned}I &= P\cdot r\cdot t\\ &= \$4000\cdot 7.4\%\cdot \frac{1}{2}\\ &= \$4000\cdot 0.074\cdot \frac{1}{2}\\ &= \$148\end{aligned}$$

67.
$$\begin{aligned}A &= P\cdot\left(1+\frac{r}{n}\right)^{n\cdot t}\\ &= \$25,000\cdot\left(1+\frac{6\%}{2}\right)^{2\cdot 5}\\ &= \$25,000\cdot\left(1+\frac{0.06}{2}\right)^{10}\\ &= \$25,000(1.03)^{10}\\ &\approx \$33,597.91\end{aligned}$$

69.
$$\begin{aligned}A &= P\cdot\left(1+\frac{r}{n}\right)^{n\cdot t}\\ &= \$150,000\cdot\left(1+\frac{7.4\%}{2}\right)^{2\cdot 20}\\ &= \$150,000\cdot\left(1+\frac{0.074}{2}\right)^{40}\\ &= \$150,000(1.037)^{40}\\ &\approx \$641,566.26\end{aligned}$$

71. $\angle ACB$ and $\angle ACD$ are complementary angles. Since $m\angle ACD = 40°$ and $90°-40° = 50°$, we have $m\angle ACB = 50°$.

Now consider triangle ABC. We know that the sum of the measures of the angles is 180°. Then

$$\begin{aligned}m\angle ABC+m\angle BCA+m\angle CAB &= 180°\\ 50°+90°+m\angle CAB &= 180°\\ 140°+m\angle CAB &= 180°\\ m\angle CAB &= 180°-140°\\ m\angle CAB &= 40°,\end{aligned}$$

so $m\angle CAB = 40°$.

To find $m\angle EBC$ we first find $m\angle CEB$. We note that $\angle DEC$ and $\angle CEB$ are supplementary angles. Since $m\angle DEC = 100°$ and $180°-100° = 80°$, we have $m\angle CEB = 80°$. Now consider triangle BCE. We know that the sum of the measures of the angles is 180°. Note that $\angle ACB$ can also be named $\angle BCE$. Then

$$\begin{aligned}m\angle BCE+m\angle CEB+m\angle EBC &= 180°\\ 50°+80°+m\angle EBC &= 180°\\ 130°+m\angle EBC &= 180°\\ m\angle EBC &= 180°-130°\\ m\angle EBC &= 50°,\end{aligned}$$

so $m\angle EBC = 50°$.

$\angle EBA$ and $\angle EBC$ are complementary angles. Since $m\angle EBC = 50°$ and $90°-50° = 40°$, we have $m\angle EBA = 40°$.

Now consider triangle ABE. We know that the sum of the measures of the angles is 180°. Then

$$\begin{aligned}m\angle CAB+m\angle EBA+m\angle AEB &= 180°\\ 40°+40°+m\angle AEB &= 180°\\ 80°+m\angle AEB &= 180°\\ m\angle AEB &= 180°-80°\\ m\angle AEB &= 100°,\end{aligned}$$

so $m\angle AEB = 100°$.

To find $m\angle ADB$ we first find $m\angle EDC$. Consider triangle CDE. We know that the sum of the measures of the angles is 180°. Then

$$\begin{aligned}m\angle DEC+m\angle ECD+m\angle EDC &= 180°\\ 100°+40°+m\angle EDC &= 180°\\ 140°+m\angle EDC &= 180°\\ m\angle EDC &= 180°-140°\\ m\angle EDC &= 40°,\end{aligned}$$

so $m\angle EDC = 40°$. We now note that $\angle ADB$ and $\angle EDC$ are complementary angles. Since $m\angle EDC = 40°$ and $90°-40° = 50°$, we have $m\angle ADB = 50°$.

Exercise Set 8.2

1.
$$\begin{aligned}\text{Perimeter} &= 4\text{ mm} + 6\text{ mm} + 7\text{ mm}\\ &= (4+6+7)\text{ mm}\\ &= 17\text{ mm}\end{aligned}$$

3.
$$\begin{aligned}\text{Perimeter} &= 3.5\text{ in.} + 3.5\text{ in.} + 4.25\text{ in.} +\\ &\qquad 0.5\text{ in.} + 3.5\text{ in.}\\ &= (3.5+3.5+4.25+0.5+3.5)\text{ in.}\\ &= 15.25\text{ in.}\end{aligned}$$

5. $P = 2\cdot(l+w)$ Perimeter of a rectangle
$P = 2\cdot(5.6\text{ km} + 3.4\text{ km})$
$P = 2\cdot(9\text{ km})$
$P = 18\text{ km}$

7. $P = 2\cdot(l+w)$ Perimeter of a rectangle
$P = 2\cdot(5\text{ ft} + 10\text{ ft})$
$P = 2\cdot(15\text{ ft})$
$P = 30\text{ ft}$

9. $P = 2\cdot(l+w)$ Perimeter of a rectangle
$P = 2\cdot\left(3\frac{1}{2}\text{ yd} + 4\frac{1}{2}\text{ yd}\right)$
$P = 2\cdot(8\text{ yd})$
$P = 16\text{ yd}$

11. $P = 4\cdot s$ Perimeter of a square
$P = 4\cdot 22\text{ ft}$
$P = 88\text{ ft}$

13. $P = 4\cdot s$ Perimeter of a square
$P = 4\cdot 45.5\text{ mm}$
$P = 182\text{ mm}$

15. ***Familiarize.*** We let P = the perimeter. We also make a drawing.

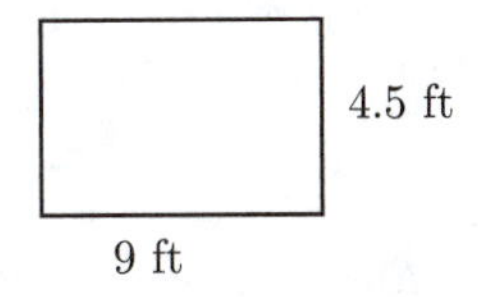

Translate. The perimeter of the billiard table is given by

$$P = 2\cdot(l+w) = 2\cdot(9\text{ ft} + 4.5\text{ ft}).$$

Solve. We calculate the perimeter.

$$P = 2\cdot(9\text{ ft} + 4.5\text{ ft}) = 2\cdot(13.5\text{ ft}) = 27\text{ ft}$$

Check. Repeat the calculations. The answer checks.

State. The perimeter of the table is 27 ft.

17. ***Familiarize.*** We make a drawing and let P = the perimeter.

30.5 cm

30.5 cm

Translate. The perimeter of the square is given by

$$P = 4\cdot s = 4\cdot(30.5\text{ cm}).$$

Solve. We do the calculation.

$$P = 4\cdot(30.5\text{ cm}) = 122\text{ cm}.$$

Check. Repeat the calculation. The answer checks.

State. The perimeter of the tile is 122 cm.

19. ***Familiarize.*** We make a drawing and let P = the perimeter. We will express 2 ft 8 in. as 32 in. and 4 ft 6 in. as 54 in.

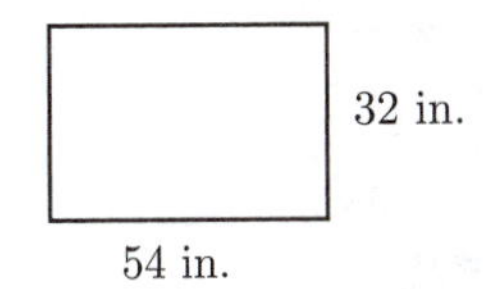

Translate. The perimeter of the backboard is given by

$$P = 2\cdot(l+w) = 2\cdot(54\text{ in.} + 32\text{ in.})$$

Solve. We calculate the perimeter.

$$P = 2\cdot(54\text{ in.} + 32\text{ in.}) = 2\cdot(86\text{ in.}) = 172\text{ in.}$$

Check. Repeat the calculation. The answer checks.

State. The perimeter of the backboard is 172 in., or 14 ft 4 in.

21. ***Familiarize.*** We label the missing lengths on the drawing and let P = the perimeter.

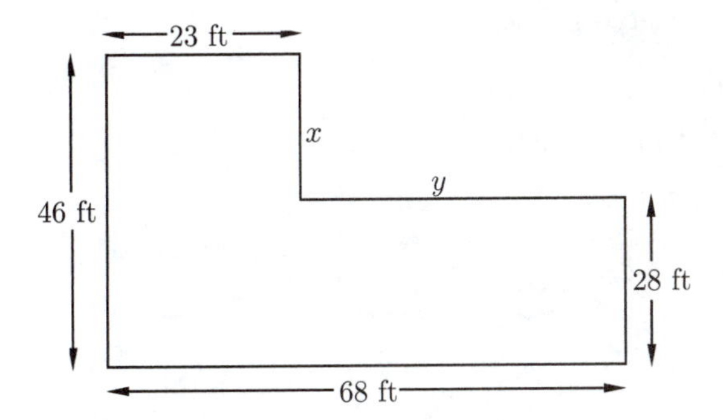

Translate. First we find the missing lengths x and y.

$$\begin{array}{ccccc} \underbrace{28\text{ ft}} & \text{plus} & \underbrace{\text{how many more ft}} & \text{is} & \underbrace{46\text{ ft}} \\ \downarrow & \downarrow & \downarrow & \downarrow & \downarrow \\ 28 & + & x & = & 46 \end{array}$$

$$\begin{array}{ccccc} \underbrace{23\text{ ft}} & \text{plus} & \underbrace{\text{how many more ft}} & \text{is} & \underbrace{68\text{ ft}} \\ \downarrow & \downarrow & \downarrow & \downarrow & \downarrow \\ 23 & + & y & = & 68 \end{array}$$

Solve. We solve for x and y.

$$\begin{aligned}28 + x &= 46\\ x &= 46 - 28\\ x &= 18\end{aligned} \qquad \begin{aligned}23 + y &= 68\\ y &= 68 - 23\\ y &= 45\end{aligned}$$

a) To find the perimeter we add the lengths of the sides of the house.

$$\begin{aligned} P &= 23 \text{ ft} + 18 \text{ ft} + 45 \text{ ft} + 28 \text{ ft} + 68 \text{ ft} + 46 \text{ ft} \\ &= (23 + 18 + 45 + 28 + 68 + 46) \text{ ft} \\ &= 228 \text{ ft} \end{aligned}$$

b) Next we find t, the total cost of the gutter.

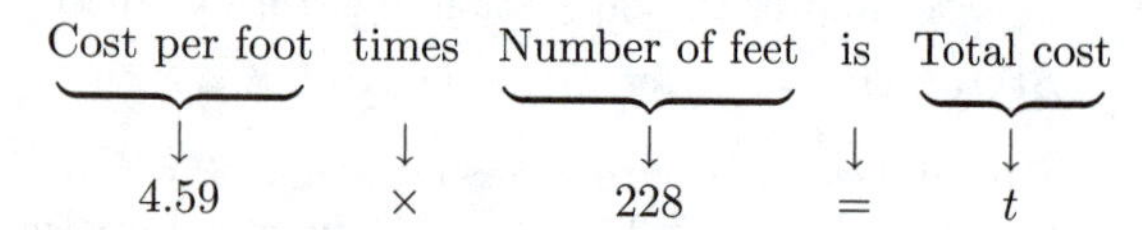

We carry out the multiplication.

```
        2 2 8
      × 4 .5 9
      2 0 5 2
    1 1 4 0 0
    9 1 2 0 0
  1 0 4 6 .5 2
```

Thus, $t = 1046.52$.

Check. We can repeat the calculations. The answer checks.

State. (a) The perimeter of the house is 228 ft. (b) The total cost of the gutter is \$1046.52.

23. $\text{Interest} = P \cdot r \cdot t$

$$\begin{aligned} &= \$600 \times 6.4\% \times \frac{1}{2} \\ &= \frac{\$600 \times 0.064}{2} \\ &= \$19.20 \end{aligned}$$

The interest is \$19.20.

25. $10^3 = 10 \cdot 10 \cdot 10 = 1000$

27. $15^2 = 15 \cdot 15 = 225$

29. $7^2 = 7 \cdot 7 = 49$

31. *Rephrase:* Sales tax is what percent of purchase price?

Translate: $878 = r \times 17,560$

To solve the equation we divide on both sides by 17,560.

$$\frac{878}{17,560} = \frac{r \times 17,560}{17,560}$$

$$0.05 = r$$

$$5\% = r$$

The sales tax rate is 5%.

33. Excluding the amount of ribbon required for the bow, the ribbon needed is equivalent to the perimeters of two rectangles, one measuring 8 in. by 4 in. and the other measuring 7 in. by 4 in.

For the 8 in. by 4 in. rectangle,

$$P = 2 \cdot (8 \text{ in.} + 4 \text{ in.}) = 2 \cdot (12 \text{ in.}) = 24 \text{ in.}$$

For the 7 in. by 4 in. rectangle,

$$P = 2 \cdot (7 \text{ in.} + 4 \text{ in.}) = 2 \cdot (11 \text{ in.}) = 22 \text{ in.}$$

Then, including the bow, the amount of ribbon required is

$$18 \text{ in.} + 24 \text{ in.} + 22 \text{ in.} = 64 \text{ in.}$$

Exercise Set 8.3

1. $A = l \cdot w$ Area of a rectangular region

$A = (5 \text{ km}) \cdot (3 \text{ km})$

$A = 5 \cdot 3 \cdot \text{ km} \cdot \text{ km}$

$A = 15 \text{ km}^2$

3. $A = l \cdot w$ Area of a rectangular region

$A = (2 \text{ in.}) \cdot (0.7 \text{ in.})$

$A = 2 \cdot 0.7 \cdot \text{ in.} \cdot \text{ in.}$

$A = 1.4 \text{ in}^2$

5. $A = s \cdot s$ Area of a square

$A = \left(2\frac{1}{2} \text{ yd}\right) \cdot \left(2\frac{1}{2} \text{ yd}\right)$

$A = \left(\frac{5}{2} \text{ yd}\right) \cdot \left(\frac{5}{2} \text{ yd}\right)$

$A = \frac{5}{2} \cdot \frac{5}{2} \cdot \text{ yd} \cdot \text{ yd}$

$A = \frac{25}{4} \text{ yd}^2, \text{ or } 6\frac{1}{4} \text{ yd}^2$

7. $A = s \cdot s$ Area of a square

$A = (90 \text{ ft}) \cdot (90 \text{ ft})$

$A = 90 \cdot 90 \cdot \text{ ft} \cdot \text{ ft}$

$A = 8100 \text{ ft}^2$

9. $A = l \cdot w$ Area of a rectangular region

$A = (10 \text{ ft}) \cdot (5 \text{ ft})$

$A = 10 \cdot 5 \cdot \text{ ft} \cdot \text{ ft}$

$A = 50 \text{ ft}^2$

11. $A = l \cdot w$ Area of a rectangular region

$A = (34.67 \text{ cm}) \cdot (4.9 \text{ cm})$

$A = 34.67 \cdot 4.9 \cdot \text{ cm} \cdot \text{ cm}$

$A = 169.883 \text{ cm}^2$

13. $A = l \cdot w$ Area of a rectangular region

$A = \left(4\frac{2}{3} \text{ in.}\right) \cdot \left(8\frac{5}{6} \text{ in.}\right)$

$A = \left(\frac{14}{3} \text{ in.}\right) \cdot \left(\frac{53}{6} \text{ in.}\right)$

$A = \frac{14}{3} \cdot \frac{53}{6} \cdot \text{ in.} \cdot \text{ in.}$

$A = \frac{2 \cdot 7 \cdot 53}{3 \cdot 2 \cdot 3} \text{ in}^2$

$A = \frac{2}{2} \cdot \frac{7 \cdot 53}{3 \cdot 3} \text{ in}^2$

$A = \frac{371}{9} \text{ in}^2, \text{ or } 41\frac{2}{9} \text{ in}^2$

15. $A = s \cdot s$ Area of a square

$A = (22 \text{ ft}) \cdot (22 \text{ ft})$

$A = 22 \cdot 22 \cdot \text{ ft} \cdot \text{ ft}$

$A = 484 \text{ ft}^2$

17. $A = s \cdot s$ Area of a square
$A = (56.9 \text{ km}) \cdot (56.9 \text{ km})$
$A = 56.9 \cdot 56.9 \cdot \text{ km} \cdot \text{ km}$
$A = 3237.61 \text{ km}^2$

19. $A = s \cdot s$ Area of a square
$A = \left(5\frac{3}{8} \text{ yd}\right) \cdot \left(5\frac{3}{8} \text{ yd}\right)$
$A = \left(\frac{43}{8} \text{ yd}\right) \cdot \left(\frac{43}{8} \text{ yd}\right)$
$A = \frac{43}{8} \cdot \frac{43}{8} \cdot \text{ yd} \cdot \text{ yd}$
$A = \frac{1849}{64} \text{ yd}^2$, or $28\frac{57}{64} \text{ yd}^2$

21. $A = b \cdot h$ Area of a parallelogram
$A = 8 \text{ cm} \cdot 4 \text{ cm}$ Substituting 8 cm for b and 4 cm for h
$A = 32 \text{ cm}^2$

23. $A = \frac{1}{2} \cdot b \cdot h$ Area of a triangle
$A = \frac{1}{2} \cdot 15 \text{ in.} \cdot 8 \text{ in.}$ Substituting 15 in. for b and 8 in. for h
$A = 60 \text{ in}^2$

25. $A = \frac{1}{2} \cdot h \cdot (a + b)$ Area of a trapezoid
$A = \frac{1}{2} \cdot 8 \text{ ft} \cdot (6 + 20) \text{ ft}$ Substituting 8 ft for h, 6 ft for a, and 20 ft for b
$A = \frac{8 \cdot 26}{2} \text{ ft}^2$
$A = 104 \text{ ft}^2$

27. $A = \frac{1}{2} \cdot h \cdot (a + b)$ Area of a trapezoid
$A = \frac{1}{2} \cdot 7 \text{ in.} \cdot (4.5 + 8.5) \text{ in.}$ Substituting 7 in. for h, 4.5 in. for a, and 8.5 in. for b
$A = \frac{7 \cdot 13}{2} \text{ in}^2$
$A = \frac{91}{2} \text{ in}^2$
$A = 45.5 \text{ in}^2$

29. $A = b \cdot h$ Area of a parallelogram
$A = 2.3 \text{ cm} \cdot 3.5 \text{ cm}$ Substituting 2.3 cm for b and 3.5 cm for h
$A = 8.05 \text{ cm}^2$

31. $A = \frac{1}{2} \cdot h \cdot (a + b)$ Area of a trapezoid
$A = \frac{1}{2} \cdot 18 \text{ cm} \cdot (9 + 24) \text{ cm}$ Substituting 18 cm for h, 9 cm for a, and 24 cm for b
$A = \frac{18 \cdot 33}{2} \text{ cm}^2$
$A = 297 \text{ cm}^2$

33. $A = \frac{1}{2} \cdot b \cdot h$ Area of a triangle
$A = \frac{1}{2} \cdot 4 \text{ m} \cdot 3.5 \text{ m}$ Substituting 4 m for b and 3.5 m for h
$A = \frac{4 \cdot 3.5}{2} \text{ m}^2$
$A = 7 \text{ m}^2$

35. ***Familiarize.*** We draw a picture.

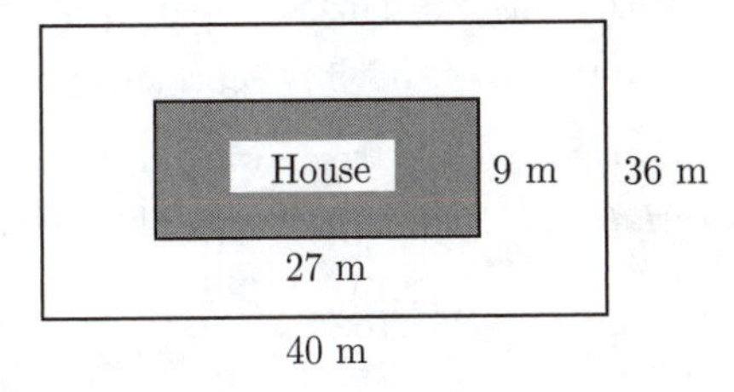

Translate. We let $A =$ the area left over.

Area left over	is	Area of lot	minus	Area of house
↓	↓	↓	↓	↓
A	$=$	$(40 \text{ m}) \cdot (36 \text{ m})$	$-$	$(27 \text{ m}) \cdot (9 \text{ m})$

Solve. The area of the lot is

$$(40 \text{ m}) \cdot (36 \text{ m}) = 40 \cdot 36 \cdot \text{ m} \cdot \text{ m} = 1440 \text{ m}^2.$$

The area of the house is

$$(27 \text{ m}) \cdot (9 \text{ m}) = 27 \cdot 9 \cdot \text{ m} \cdot \text{ m} = 243 \text{ m}^2.$$

The area left over is

$$A = 1440 \text{ m}^2 - 243 \text{ m}^2 = 1197 \text{ m}^2.$$

Check. Repeat the calculations. The answer checks.

State. The area left over for the lawn is 1197 m^2.

37. a) First find the area of the entire yard, including the basketball court:

$$A = l \cdot w = \left(110\frac{2}{3} \text{ ft}\right) \cdot (80 \text{ ft})$$
$$= \left(\frac{332}{3} \text{ ft}\right) \cdot (80 \text{ ft})$$
$$= \frac{26,560}{3} \text{ ft}^2$$
$$= 8853\frac{1}{3} \text{ ft}^2$$

Now find the area of the basketball court:

$A = s \cdot s = \left(19\frac{1}{2} \text{ ft}\right) \cdot \left(19\frac{1}{2} \text{ ft}\right) =$
$\frac{39}{2} \cdot \frac{39}{2} \text{ ft}^2 = \frac{1521}{4} \text{ ft}^2 = 380\frac{1}{4} \text{ ft}^2$

Finally, subtract to find the area of the lawn:

$8853\frac{1}{3} \text{ ft}^2 - 380\frac{1}{4} \text{ ft}^2 = 8853\frac{4}{12} \text{ ft}^2 - 380\frac{3}{12} \text{ ft}^2 =$
$8473\frac{1}{12} \text{ ft}^2 \approx 8473 \text{ ft}^2$

b) Let $c =$ the cost of mowing the lawn. We translate to an equation.

$$\underbrace{\text{The cost of mowing}}_{c} \ \underset{=}{\text{is}} \ \underset{0.012}{\$0.012} \ \underset{\cdot}{\text{times}} \ \underbrace{\text{the area of the lawn.}}_{8473}$$

We multiply to solve the equation.

$c = 0.012 \cdot 8473 \approx \$102$

The total cost of the mowing is about \$102.

39. ***Familiarize.*** We use the drawing in the text.

Translate. We let $A =$ the area of the sidewalk, in square feet.

$$\underbrace{\text{Area of sidewalk}}_{A} \ \underset{=}{\text{is}} \ \underbrace{\text{Total area}}_{(113.4 \text{ ft})\times(75.4 \text{ ft})} \ \underset{-}{\text{minus}} \ \underbrace{\text{Area of building}}_{(110 \text{ ft})\times(72 \text{ ft})}$$

Solve. The total area is

$$(113.4 \text{ ft})\times(75.4 \text{ ft}) = 113.4\times 75.4\times \text{ ft}\times \text{ ft} = 8550.36 \text{ ft}^2.$$

The area of the building is

$$(110 \text{ ft}) \times (72 \text{ ft}) = 110 \times 72 \times \text{ ft} \times \text{ ft} = 7920 \text{ ft}^2.$$

The area of the sidewalk is

$$A = 8550.36 \text{ ft}^2 - 7920 \text{ ft}^2 = 630.36 \text{ ft}^2.$$

Check. Repeat the calculations. The answer checks.

State. The area of the sidewalk is 630.36 ft^2.

41. ***Familiarize.*** The dimensions are as follows:

Two walls are 15 ft by 8 ft.

Two walls are 20 ft by 8 ft.

The ceiling is 15 ft by 20 ft.

The total area of the walls and ceiling is the total area of the rectangles described above less the area of the windows and the door.

Translate. a) We let $A =$ the total area of the walls and ceiling. The total area of the two 15 ft by 8 ft walls is

$$2 \cdot (15 \text{ ft}) \cdot (8 \text{ ft}) = 2 \cdot 15 \cdot 8 \cdot \text{ ft} \cdot \text{ ft} = 240 \text{ ft}^2$$

The total area of the two 20 ft by 8 ft walls is

$$2 \cdot (20 \text{ ft}) \cdot (8 \text{ ft}) = 2 \cdot 20 \cdot 8 \cdot \text{ ft} \cdot \text{ ft} = 320 \text{ ft}^2$$

The area of the ceiling is

$$(15 \text{ ft}) \cdot (20 \text{ ft}) = 15 \cdot 20 \cdot \text{ ft} \cdot \text{ ft} = 300 \text{ ft}^2$$

The area of the two windows is

$$2 \cdot (3 \text{ ft}) \cdot (4 \text{ ft}) = 2 \cdot 3 \cdot 4 \cdot \text{ ft} \cdot \text{ ft} = 24 \text{ ft}^2$$

The area of the door is

$$\left(2\tfrac{1}{2} \text{ ft}\right) \cdot \left(6\tfrac{1}{2} \text{ ft}\right) = \left(\frac{5}{2} \text{ ft}\right) \cdot \left(\frac{13}{2} \text{ ft}\right)$$
$$= \frac{5}{2} \cdot \frac{13}{2} \cdot \text{ ft} \cdot \text{ ft}$$
$$= \frac{65}{4} \text{ ft}^2, \text{ or } 16\frac{1}{4} \text{ ft}^2$$

Thus

$$A = 240 \text{ ft}^2 + 320 \text{ ft}^2 + 300 \text{ ft}^2 - 24 \text{ ft}^2 - 16\frac{1}{4} \text{ ft}^2$$
$$= 819\frac{3}{4} \text{ ft}^2, \text{ or } 819.75 \text{ ft}^2$$

b) We divide to find how many gallons of paint are needed.

$$819.75 \div 360.625 \approx 2.27$$

It will be necessary to buy 3 gallons of paint in order to have the required 2.27 gallons.

c) We multiply to find the cost of the paint.

$$3 \times \$24.95 = \$74.85$$

Check. We repeat the calculations. The answer checks.

State. (a) The total area of the walls and ceiling is 819.75 ft^2. (b) 3 gallons of paint are needed. (c) It will cost \$74.85 to paint the room.

43.

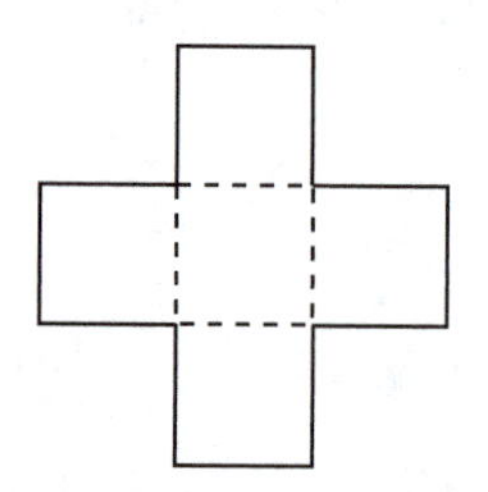

Each side is 4 cm.

The region is composed of 5 squares, each with sides of length 4 cm. The area is

$$A = 5 \cdot (s \cdot s) = 5 \cdot (4 \text{ cm} \cdot 4 \text{ cm}) = 5 \cdot 4 \cdot 4 \text{ cm} \cdot \text{ cm} = 80 \text{ cm}^2$$

45. ***Familiarize.*** We look for the kinds of figures whose areas we can calculate using area formulas that we already know.

Translate. The shaded region consists of a square region with a triangular region removed from it. The sides of the square are 30 cm, and the triangle has base 30 cm and height 15 cm. We find the area of the square using the formula $A = s \cdot s$, and the area of the triangle using $A = \frac{1}{2} \cdot b \cdot h$. Then we subtract.

Solve. Area of the square: $A = 30 \text{ cm} \cdot 30 \text{ cm} = 900 \text{ cm}^2$.

Area of the triangle: $A = \frac{1}{2} \cdot 30 \text{ cm} \cdot 15 \text{ cm} = 225 \text{ cm}^2$.

Area of the shaded region: $A = 900 \text{ cm}^2 - 225 \text{ cm}^2 = 675 \text{ cm}^2$.

Check. We repeat the calculations. The answer checks.

State. The area of the shaded region is 675 cm^2.

47. ***Familiarize.*** We have one large triangle with height and base each 6 cm. We also have 6 small triangles, each with height and base 1 cm.

Translate. We will find the area of each type of triangle using the formula $A = \frac{1}{2} \cdot b \cdot h$. Next we will multiply the area of the smaller triangle by 6. And, finally, we will add this product to the area of the larger triangle to find the total area.

Solve.

For the large triangle: $A = \frac{1}{2} \cdot 6 \text{ cm} \cdot 6 \text{ cm} = 18 \text{ cm}^2$

For one small triangle: $A = \frac{1}{2} \cdot 1 \text{ cm} \cdot 1 \text{ cm} = \frac{1}{2} \text{ cm}^2$

Find the area of the 6 small triangles: $6 \cdot \frac{1}{2} \text{ cm}^2 = 3 \text{ cm}^2$

Add to find the total area: $18 \text{ cm}^2 + 3 \text{ cm}^2 = 21 \text{ cm}^2$

Check. We repeat the calculations.

State. The area of the shaded region is 21 cm^2.

49. ***Familiarize.*** The sail consists of a triangle with base 9 ft and height 12 ft as well as three rectangles, one that measures $\frac{1}{2}$ ft by 9 ft, one that measures $\frac{1}{2}$ ft by 12 ft, and one that measures $\frac{1}{2}$ ft by 15 ft. The piece of sailcloth from which the sail is cut is a rectangle that measures 18 ft by 12 ft.

Translate. We will use the formula $A = \frac{1}{2} \cdot b \cdot h$ to find the area of the triangle and the formula $A = l \cdot w$ to find the area of the rectangles. Then we will add to find the area of the sail and, finally, subtract to find how much fabric is left over.

Solve.

Area of the triangle: $A = \frac{1}{2} \cdot 9 \text{ ft} \cdot 12 \text{ ft} = 54 \text{ ft}^2$

Area of the $\frac{1}{2}$ ft by 9 ft rectangle: $A = \frac{1}{2} \text{ ft} \cdot 9 \text{ ft} = \frac{9}{2} \text{ ft}^2$

Area of the $\frac{1}{2}$ ft by 12 ft rectangle:

$A = \frac{1}{2} \text{ ft} \cdot 12 \text{ ft} = 6 \text{ ft}^2$

Area of the $\frac{1}{2}$ ft by 15 ft rectangle:

$A = \frac{1}{2} \text{ ft} \cdot 15 \text{ ft} = \frac{15}{2} \text{ ft}^2$

Area of the 18 ft by 12 ft rectangle:
$A = 18 \text{ ft} \cdot 12 \text{ ft} = 216 \text{ ft}^2$

Total area of the sail:

$54 \text{ ft}^2 + \frac{9}{2} \text{ ft}^2 + 6 \text{ ft}^2 + \frac{15}{2} \text{ ft}^2 = 72 \text{ ft}^2$

Area of fabric left over: $216 \text{ ft}^2 - 72 \text{ ft}^2 = 144 \text{ ft}^2$

Check. We repeat the calculations. The answer checks.

State. The area of the fabric left over is 144 ft^2.

51. A number is divisible by 8 if the number named by the last <u>three</u> digits is divisible by 8.

53. Two lines are <u>perpendicular</u> if they intersect to form a right angle.

55. An <u>angle</u> is a set of points consisting of two rays.

57. The <u>perimeter</u> of a polygon is the sum of the lengths of its sides.

59.

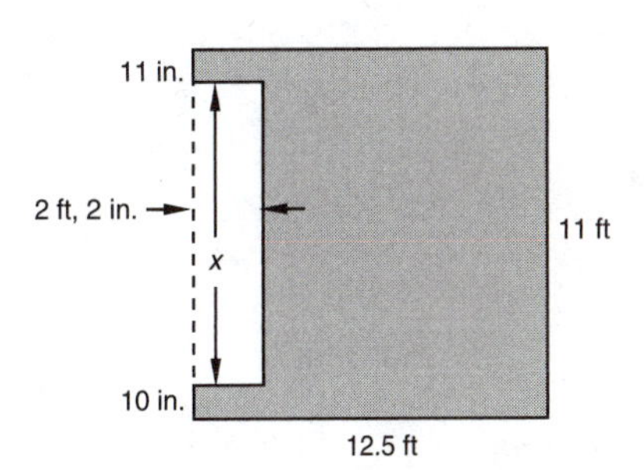

2 ft $= 2 \times 1$ ft $= 2 \times 12$ in. $= 24$ in., so 2 ft, 2 in. $= 2$ ft $+$ 2 in. $= 24$ in. $+ 2$ in. $= 26$ in.

11 ft $= 11 \times 1$ ft $= 11 \times 12$ in. $= 132$ in.

12.5 ft $= 12.5 \times 1$ ft $= 12.5 \times 12$ in. $= 150$ in.

We solve an equation to find x, in inches:

$$\begin{aligned} 11 + x + 10 &= 132 \\ 21 + x &= 132 \\ 21 + x - 21 &= 132 - 21 \\ x &= 111 \end{aligned}$$

Then the area of the shaded region is the area of a 150 in. by 132 in. rectangle less the area of a 111 in. by 26 in. rectangle.

$$\begin{aligned} A &= (150 \text{ in.}) \cdot (132 \text{ in.}) - (111 \text{ in.}) \cdot (26 \text{ in.}) \\ A &= 19,800 \text{ in}^2 - 2886 \text{ in}^2 \\ A &= 16,914 \text{ in}^2 \end{aligned}$$

Exercise Set 8.4

1. $d = 2 \cdot r$

$d = 2 \cdot 7 \text{ cm} = 14 \text{ cm}$

$C = 2 \cdot \pi \cdot r$

$C \approx 2 \cdot \frac{22}{7} \cdot 7 \text{ cm} = \frac{2 \cdot 22 \cdot 7}{7} \text{ cm} = 44 \text{ cm}$

$A = \pi \cdot r \cdot r$

$A \approx \frac{22}{7} \cdot 7 \text{ cm} \cdot 7 \text{ cm} = \frac{22}{7} \cdot 49 \text{ cm}^2 = 154 \text{ cm}^2$

3. $d = 2 \cdot r$

$d = 2 \cdot \frac{3}{4} \text{ in.} = \frac{6}{4} \text{ in.} = \frac{3}{2} \text{ in., or } 1\frac{1}{2} \text{ in.}$

$C = 2 \cdot \pi \cdot r$

$C \approx 2 \cdot \frac{22}{7} \cdot \frac{3}{4} \text{ in.} = \frac{2 \cdot 22 \cdot 3}{7 \cdot 4} \text{ in.} = \frac{132}{28} \text{ in.} = \frac{33}{7} \text{ in.,}$ or $4\frac{5}{7}$ in.

$A = \pi \cdot r \cdot r$

$A \approx \frac{22}{7} \cdot \frac{3}{4} \text{ in.} \cdot \frac{3}{4} \text{ in.} = \frac{22 \cdot 3 \cdot 3}{7 \cdot 4 \cdot 4} \text{ in}^2 = \frac{99}{56} \text{ in}^2, \text{ or } 1\frac{43}{56} \text{ in}^2$

5. $r = \frac{d}{2}$

$r = \frac{32 \text{ ft}}{2} = 16 \text{ ft}$

$C = \pi \cdot d$

$C \approx 3.14 \cdot 32 \text{ ft} = 100.48 \text{ ft}$

$A = \pi \cdot r \cdot r$

$A \approx 3.14 \cdot 16 \text{ ft} \cdot 16 \text{ ft} \quad (r = \frac{d}{2}; r = \frac{32 \text{ ft}}{2} = 16 \text{ ft})$

$A = 3.14 \cdot 256 \text{ ft}^2$

$A = 803.84 \text{ ft}^2$

7. $r = \frac{d}{2}$

$r = \frac{1.4 \text{ cm}}{2} = 0.7 \text{ cm}$

$C = \pi \cdot d$

$C \approx 3.14 \cdot 1.4 \text{ cm} = 4.396 \text{ cm}$

$A = \pi \cdot r \cdot r$

$A \approx 3.14 \cdot 0.7 \text{ cm} \cdot 0.7 \text{ cm}$

$(r = \frac{d}{2}; r = \frac{1.4 \text{ cm}}{2} = 0.7 \text{ cm})$

$A = 3.14 \cdot 0.49 \text{ cm}^2 = 1.5386 \text{ cm}^2$

9. $d = 2 \cdot r$

$d = 2 \cdot 6.37 \text{ ft} = 12.74 \text{ ft}$

$C = 2 \cdot \pi \cdot r$

$C \approx 2 \cdot 3.14 \cdot 6.37 \text{ ft} \approx 40 \text{ ft}$

$A = \pi \cdot r \cdot r$

$A \approx 3.14 \cdot 6.37 \text{ ft} \cdot 6.37 \text{ ft} \approx 127.41 \text{ ft}^2$

The areas of the nets described in Example 10 in Section 8.3 are about 80.74 ft^2 and 107.65 ft^2. The differences in the areas are

$127.41 \text{ ft}^2 - 80.74 \text{ ft}^2$, or 46.67 ft^2

and

$127.41 \text{ ft}^2 - 107.65 \text{ ft}^2$, or 19.76 ft^2.

Thus, the Adventure II net is 46.67 ft^2 larger than the medium net and 19.76 ft^2 larger than the large net.

11. Area of circular pizza ($r = \frac{12 \text{ in.}}{2} = 6$ in.):

$\pi \cdot r \cdot r \approx 3.14 \cdot 6 \text{ in.} \cdot 6 \text{ in.} = 113.04 \text{ in}^2$

Area of square pizza:

$s \cdot s = 12 \text{ in.} \cdot 12 \text{ in.} = 144 \text{ in}^2$

Difference in areas:

$144 \text{ in}^2 - 113.04 \text{ in}^2 = 30.96 \text{ in}^2$

The square pizza is 30.96 in^2 larger.

13. $C = \pi \cdot d$

$C \approx 3.14 \cdot 7926.41 \approx 24,889 \text{ mi}$

15. Maximum circumference of the barrel:

$C = \pi \cdot d$

$C \approx \frac{22}{7} \cdot 2\frac{3}{4} = \frac{22}{7} \cdot \frac{11}{4} = \frac{\not{2} \cdot 11 \cdot 11}{7 \cdot \not{2} \cdot 2} = \frac{121}{14}$, or $8\frac{9}{14}$ in.

Minimum circumference of the handle:

$C = \pi \cdot d$

$C \approx \frac{22}{7} \cdot \frac{16}{19} = \frac{352}{133}$, or $2\frac{86}{133}$ in.

17. Find the area of the larger circle (pool plus walk). Its diameter is 1 yd + 20 yd + 1 yd, or 22 yd. Thus its radius is $\frac{22}{2}$ yd, or 11 yd.

$A = \pi \cdot r \cdot r$

$A \approx 3.14 \cdot 11 \text{ yd} \cdot 11 \text{ yd} = 379.94 \text{ yd}^2$

Find the area of the pool. Its diameter is 20 yd. Thus its radius is $\frac{20}{2}$ yd, or 10 yd.

$A = \pi \cdot r \cdot r$

$A \approx 3.14 \cdot 10 \text{ yd} \cdot 10 \text{ yd} = 314 \text{ yd}^2$

We subtract to find the area of the walk:

$A = 379.94 \text{ yd}^2 - 314 \text{ yd}^2$

$A = 65.94 \text{ yd}^2$

The area of the walk is 65.94 yd^2.

19. The perimeter consists of the circumferences of three semicircles, each with diameter 8 ft, and one side of a square of length 8 ft. We first find the circumference of one semicircle. This is one-half the circumference of a circle with diameter 8 ft:

$\frac{1}{2} \cdot \pi \cdot d \approx \frac{1}{2} \cdot 3.14 \cdot 8 \text{ ft} = 12.56 \text{ ft}$

Then we multiply by 3:

$3 \cdot (12.56 \text{ ft}) = 37.68 \text{ ft}$

Finally we add the circumferences of the semicircles and the length of the side of the square:

$37.68 \text{ ft} + 8 \text{ ft} = 45.68 \text{ ft}$

The perimeter is 45.68 ft.

21. The perimeter consists of three-fourths of the circumference of a circle with radius 4 yd and two sides of a square with sides of length 4 yd. We first find three-fourths of the circumference of the circle:

$\frac{3}{4} \cdot 2 \cdot \pi \cdot r \approx 0.75 \cdot 2 \cdot 3.14 \cdot 4 \text{ yd} = 18.84 \text{ yd}$

Then we add this length to the lengths of two sides of the square:

$18.84 \text{ yd} + 4 \text{ yd} + 4 \text{ yd} = 26.84 \text{ yd}$

The perimeter is 26.84 yd.

23. The perimeter consists of three-fourths of the perimeter of a square with side of length 10 yd and the circumference of a semicircle with diameter 10 yd. First we find three-fourths of the perimeter of the square:

$$\frac{3}{4} \cdot 4 \cdot s = \frac{3}{4} \cdot 4 \cdot 10 \text{ yd} = 30 \text{ yd}$$

Then we find one-half of the circumference of a circle with diameter 10 yd:

$$\frac{1}{2} \cdot \pi \cdot d \approx \frac{1}{2} \cdot 3.14 \cdot 10 \text{ yd} = 15.7 \text{ yd}$$

Then we add:

$$30 \text{ yd} + 15.7 \text{ yd} = 45.7 \text{ yd}$$

The perimeter is 45.7 yd.

25. The shaded region consists of a circle of radius 8 m, with two circles each of diameter 8 m, removed. First we find the area of the large circle:

$$A = \pi \cdot r \cdot r \approx 3.14 \cdot 8 \text{ m} \cdot 8 \text{ m} = 200.96 \text{ m}^2$$

Then we find the area of one of the small circles:
The radius is $\frac{8 \text{ m}}{2} = 4$ m.

$$A = \pi \cdot r \cdot r \approx 3.14 \cdot 4 \text{ m} \cdot 4 \text{ m} = 50.24 \text{ m}^2$$

We multiply this area by 2 to find the area of the two small circles:

$$2 \cdot 50.24 \text{ m}^2 = 100.48 \text{ m}^2$$

Finally we subtract to find the area of the shaded region:

$$200.96 \text{ m}^2 - 100.48 \text{ m}^2 = 100.48 \text{ m}^2$$

The area of the shaded region is 100.48 m^2.

27. The shaded region consists of one-half of a circle with diameter 2.8 cm and a triangle with base 2.8 cm and height 2.8 cm. First we find the area of the semicircle. The radius is $\frac{2.8 \text{ cm}}{2} = 1.4$ cm.

$$A = \frac{1}{2} \cdot \pi \cdot r \cdot r \approx \frac{1}{2} \cdot 3.14 \cdot 1.4 \text{ cm} \cdot 1.4 \text{ cm} = 3.0772 \text{ cm}^2$$

Then we find the area of the triangle:

$$A = \frac{1}{2} \cdot b \cdot h = \frac{1}{2} \cdot 2.8 \text{ cm} \cdot 2.8 \text{ cm} = 3.92 \text{ cm}^2$$

Finally we add to find the area of the shaded region:

$$3.0772 \text{ cm}^2 + 3.92 \text{ cm}^2 = 6.9972 \text{ cm}^2$$

The area of the shaded region is 6.9972 cm^2.

29. The shaded area consists of a rectangle of dimensions 11.4 in. by 14.6 in., with the area of two semicircles, each of diameter 11.4 in., removed. This is equivalent to removing one circle with diameter 11.4 in. from the rectangle. First we find the area of the rectangle:

$$l \cdot w = (11.4 \text{ in.}) \cdot (14.6 \text{ in.}) = 166.44 \text{ in}^2$$

Then we find the area of the circle. The radius is $\frac{11.4 \text{ in.}}{2} = 5.7$ in.

$$\pi \cdot r \cdot r \approx 3.14 \cdot 5.7 \text{ in.} \cdot 5.7 \text{ in.} = 102.0186 \text{ in}^2$$

Finally we subtract to find the area of the shaded region:

$$166.44 \text{ in}^2 - 102.0186 \text{ in}^2 = 64.4214 \text{ in}^2$$

31. $2^4 = 2 \cdot 2 \cdot 2 \cdot 2 = 16$

33. We divide.

$$\begin{array}{r} 0.375 \\ 8\overline{)3.000} \\ \underline{24} \\ 60 \\ \underline{56} \\ 40 \\ \underline{40} \\ 0 \end{array}$$

$\frac{3}{8} = 0.375 = 37.5\%$, or $37\frac{1}{2}\%$

35. ***Familiarize***. Let $w =$ the weight of the brain, in pounds, for a 200-lb person.

Translate.

What is 2.5% of 200 lb?
↓ ↓ ↓ ↓ ↓
$w = 2.5\% \times 200$

Solve. We convert 2.5% to decimal notation and multiply.

$w = 0.025 \times 200 = 5$

Check. We repeat the calculation. The answer checks.

State. For a 200-lb person, the brain weighs 5 lb.

37. $A = l \cdot w = 580.8 \text{ ft} \cdot 75 \text{ ft} = 43,560 \text{ ft}^2$

$P = 2 \cdot (l + w) = 2 \cdot (580.8 \text{ ft} + 75 \text{ ft}) = 2 \cdot (655.8 \text{ ft}) = 1311.6 \text{ ft}$

Number of rolls of fencing needed:

$1311.6 \text{ ft} \div 330 \text{ ft} \approx 3.97$

Thus, 4 rolls of fencing must be purchased.

Cost of fencing: $4 \cdot \$149.99 = \$599.96$

39. $A = \pi \cdot r \cdot r \approx 3.14 \cdot 117.83 \text{ ft} \cdot 117.83 \text{ ft} = 43,595.47395 \text{ ft}^2$

$C = 2 \cdot \pi \cdot r \approx 2 \cdot 3.14 \cdot 117.83 \text{ ft} = 739.9724 \text{ ft}$

Number of rolls of fencing needed:

$739.9724 \text{ ft} \div 330 \text{ ft} \approx 2.24$

Thus, 3 rolls of fencing must be purchased.

Cost of fencing: $3 \cdot \$149.99 = \$449.97$

41. $A = l \cdot w = 242 \text{ ft} \cdot 180 \text{ ft} = 43,560 \text{ ft}^2$

$P = 2 \cdot (l + w) = 2 \cdot (242 \text{ ft} + 180 \text{ ft}) = 2 \cdot (422 \text{ ft}) = 844 \text{ ft}$

Number of rolls of fencing needed:

$844 \text{ ft} \div 330 \text{ ft} \approx 2.58$

Thus, 3 rolls of fencing must be purchased.

Cost of fencing: $3 \cdot \$149.99 = \$449.97$

Chapter 8 Mid-Chapter Review

1. The statement is true. Each area is $8 \text{ cm} \cdot 5 \text{ cm}$, or 40 cm^2.

2. We find the area of the square by squaring 4 in. We find the area of the circle by multiplying the square of 4 by π, so the area of the square is less than the area of the circle. The given statement is true.

3. The perimeter of the rectangle is $2\cdot(6 \text{ ft}+3 \text{ ft})=2\cdot(9 \text{ ft})=18$ ft. The circumference of the circle is $2 \cdot \pi \cdot 3 \text{ ft} \approx 2 \cdot 3.14 \cdot 3 \text{ ft} = 18.84$ ft. The perimeter of the rectangle is less than the circumference of the circle, so the given statement is false.

4. $C = \pi \cdot d$, so $C/d = \pi$. The given statement is true.

5.
$$A = \frac{1}{2} \cdot b \cdot h$$
$$A = \frac{1}{2} \cdot 12 \text{ cm} \cdot 8 \text{ cm}$$
$$A = \frac{12 \cdot 8}{2} \text{ cm}^2$$
$$A = \frac{96}{2} \text{ cm}^2 = 48 \text{ cm}^2$$

6.
$$C = \pi \cdot d$$
$$C \approx 3.14 \cdot 10.2 \text{ in.}$$
$$C = 32.028 \text{ in.}$$

$$r = \frac{d}{2} = \frac{10.2 \text{ in.}}{2} = 5.1$$
$$A = \pi \cdot r \cdot r$$
$$A \approx 3.14 \cdot 5.1 \text{ in.} \cdot 5.1 \text{ in.}$$
$$A = 81.6714 \text{ in}^2$$

7.
$$\begin{aligned}(n-2)\cdot 180^\circ &= (19-2)\cdot 180^\circ\\ &= 17 \cdot 180^\circ\\ &= 3060^\circ\end{aligned}$$

8.
$$\begin{aligned}27^\circ + 138^\circ + m\angle x &= 180^\circ\\ 165^\circ + m\angle x &= 180^\circ\\ m\angle x &= 180^\circ - 165^\circ = 15^\circ\end{aligned}$$

9. The polygon has 6 sides. It is a hexagon.

10. All the sides of the triangle have different lengths.

The triangle is scalene. The triangle has a 90° angle. It is a right triangle.

11. Two sides of the triangle are the same length, and the third has a different length. The triangle is isosceles.

One angle of the triangle is obtuse. It is an obtuse triangle.

12. All the sides of the triangle are the same length. The triangle is equilateral.

All three angles are acute. It is an acute triangle.

13.
$$\begin{aligned}\text{Perimeter} &= 23 \text{ mm} + 8 \text{ mm} + 10 \text{ mm} + 7 \text{ mm} + \\ &\quad 13 \text{ mm} + 15 \text{ mm}\\ &= (23+8+10+7+13+15) \text{ mm}\\ &= 76 \text{ mm}\end{aligned}$$

14.
$$\begin{aligned}P &= 4 \cdot s\\ &= 4 \cdot 12\tfrac{2}{3} \text{ ft} = 4 \cdot \frac{38}{3} \text{ ft}\\ &= \frac{152}{3} \text{ ft, or } 50\tfrac{2}{3} \text{ ft}\end{aligned}$$

$$\begin{aligned}A &= s \cdot s\\ &= 12\tfrac{2}{3} \text{ ft} \cdot 12\tfrac{2}{3} \text{ ft}\\ &= \frac{38}{3} \text{ ft} \cdot \frac{38}{3} \text{ ft} = \frac{38}{3} \cdot \frac{38}{3} \cdot \text{ ft} \cdot \text{ ft}\\ &= \frac{1444}{9} \text{ ft}^2, \text{ or } 160\tfrac{4}{9} \text{ ft}^2\end{aligned}$$

15.
$$A = b \cdot h$$
$$A = 40 \text{ in.} \cdot 20 \text{ in.}$$
$$A = 800 \text{ in}^2$$

16.
$$A = \frac{1}{2} \cdot b \cdot h$$
$$A = \frac{1}{2} \cdot \frac{3}{4} \text{ yd} \cdot 1\tfrac{1}{2} \text{ yd} = \frac{1}{2} \cdot \frac{3}{4} \text{ yd} \cdot \frac{3}{2} \text{ yd}$$
$$A = \frac{3 \cdot 3}{2 \cdot 4 \cdot 2} \text{ yd}^2 = \frac{9}{16} \text{ yd}^2$$

17.
$$\begin{aligned}A &= \frac{1}{2} \cdot h \cdot (a+b)\\ A &= \frac{1}{2} \cdot 6 \text{ km} \cdot (13 \text{ km} + 9 \text{ km})\\ &= \frac{6 \cdot 22}{2} \text{ km}^2\\ &= 66 \text{ km}^2\end{aligned}$$

18.
$$C = 2 \cdot \pi \cdot r$$
$$C \approx 2 \cdot 3.14 \cdot 7 \text{ in.} = 43.96 \text{ in.}$$

$$A = \pi \cdot r \cdot r$$
$$A \approx 3.14 \cdot 7 \text{ in.} \cdot 7 \text{ in.} = 153.86 \text{ in}^2$$

19.
$$C = \pi \cdot d$$
$$C \approx 3.14 \cdot 8.6 \text{ cm} = 27.004 \text{ cm}$$

$$r = \frac{d}{2} = \frac{8.6 \text{ cm}}{2} = 4.3 \text{ cm}$$
$$A = \pi \cdot r \cdot r$$
$$A \approx 3.14 \cdot 4.3 \text{ cm} \cdot 4.3 \text{ cm} = 58.0586 \text{ cm}^2$$

20. Area of a circle with radius 4 ft:
$$A = \pi \cdot 4 \text{ ft} \cdot 4 \text{ ft} = 16 \cdot \pi \text{ ft}^2$$
Area of a square with side 4 ft:
$$A = 4 \text{ ft} \cdot 4 \text{ ft} = 16 \text{ ft}^2$$
Circumference of a circle with radius 4 ft:
$$C = 2 \cdot \pi \cdot 4 \text{ ft} = 8 \cdot \pi \text{ ft}$$

Area of a rectangle with length 8 ft and width 4 ft:

$A = 8 \text{ ft} \cdot 4 \text{ ft} = 32 \text{ ft}^2$

Area of a triangle with base 4 ft and height 8 ft:

$A = \frac{1}{2} \cdot 4 \text{ ft} \cdot 8 \text{ ft} = 16 \text{ ft}^2$

Perimeter of a square with side 4 ft:

$P = 4 \cdot 4 \text{ ft} = 16 \text{ ft}$

Perimeter of a rectangle with length 8 ft and width 4 ft:

$P = 2 \cdot (8 \text{ ft} + 4 \text{ ft}) = 24 \text{ ft}$

21. The area of a 16-in.-diameter pizza is approximately $3.14 \cdot 8$ in. $\cdot$ 8in., or 200.96 in^2. At \$16.25, its unit price is $\frac{\$16.25}{200.96 \text{ in}^2}$, or about \$0.08/in^2. The area of a 10-in.-diameter pizza is approximately $3.14 \cdot 5$ in. $\cdot$ 5 in., or 78.5 in^2. At \$7.85, its unit price is $\frac{\$7.85}{78.5 \text{ in}^2}$, or \$0.10/in^2. Since the 16-in.-diameter pizza has the lower unit price, it is a better buy.

22. No; let l and w represent the length and width of the smaller rectangle. Then $3 \cdot l$ and $3 \cdot w$ represent the length and width of the larger rectangle. The area of the first rectangle is $l \cdot w$, but the area of the second is $3 \cdot l \cdot 3 \cdot w = 3 \cdot 3 \cdot l \cdot w = 9 \cdot l \cdot w$, or 9 times the area of the smaller rectangle.

23. Yes; let s = the length of a side of the larger square. Then $\frac{1}{2} \cdot s$ = the length of a side of the smaller square. The perimeter of the larger square is $4 \cdot s$, and the perimeter of the smaller square is $4 \cdot \frac{1}{2} \cdot s$, or $\frac{1}{2} \cdot (4 \cdot s)$, or $\frac{1}{2}$ the perimeter of the larger square.

24. For a rectangle with length l and width w,

$$P = l + w + l + w$$
$$P = (l + w) + (l + w)$$
$$P = 2 \cdot (l + w).$$

We also have

$$P = l + w + l + w$$
$$P = (l + l) + (w + w)$$
$$P = 2 \cdot l + 2 \cdot w.$$

25. See page 536 of the text.

26. No; let r = the radius of the smaller circle. Then its area is $\pi \cdot r \cdot r$, or πr^2. The radius of the larger circle is $2r$, and its area is $\pi \cdot 2r \cdot 2r$, or $4\pi r^2$, or $4 \cdot \pi r^2$. Thus, the area of the larger circle is 4 times the area of the smaller circle.

Exercise Set 8.5

1. $V = l \cdot w \cdot h$

$V = 12 \text{ cm} \cdot 8 \text{ cm} \cdot 8 \text{ cm}$

$V = 12 \cdot 64 \text{ cm}^3$

$V = 768 \text{ cm}^3$

$$\begin{aligned} SA &= 2lw + 2lh + 2wh \\ &= 2 \cdot 12 \text{ cm} \cdot 8 \text{ cm} + 2 \cdot 12 \text{ cm} \cdot 8 \text{ cm} + \\ &\quad 2 \cdot 8 \text{ cm} \cdot 8 \text{ cm} \\ &= 192 \text{ cm}^2 + 192 \text{ cm}^2 + 128 \text{ cm}^2 \\ &= 512 \text{ cm}^2 \end{aligned}$$

3. $V = l \cdot w \cdot h$

$V = 7.5 \text{ in.} \cdot 2 \text{ in.} \cdot 3 \text{ in.}$

$V = 7.5 \cdot 6 \text{ in}^3$

$V = 45 \text{ in}^3$

$$\begin{aligned} SA &= 2lw + 2lh + 2wh \\ &= 2 \cdot 7.5 \text{ in.} \cdot 2 \text{ in.} + 2 \cdot 7.5 \text{ in.} \cdot 3 \text{ in.} + \\ &\quad 2 \cdot 2 \text{ in.} \cdot 3 \text{ in.} \\ &= 30 \text{ in}^2 + 45 \text{ in}^2 + 12 \text{ in}^2 \\ &= 87 \text{ in}^2 \end{aligned}$$

5. $V = l \cdot w \cdot h$

$V = 10 \text{ m} \cdot 5 \text{ m} \cdot 1.5 \text{ m}$

$V = 10 \cdot 7.5 \text{ m}^3$

$V = 75 \text{ m}^3$

$$\begin{aligned} SA &= 2lw + 2lh + 2wh \\ &= 2 \cdot 10 \text{ m} \cdot 5 \text{ m} + 2 \cdot 10 \text{ m} \cdot 1.5 \text{ m} + \\ &\quad 2 \cdot 5 \text{ m} \cdot 1.5 \text{ m} \\ &= 100 \text{ m}^2 + 30 \text{ m}^2 + 15 \text{ m}^2 \\ &= 145 \text{ m}^2 \end{aligned}$$

7. $V = l \cdot w \cdot h$

$V = 6\frac{1}{2} \text{ yd} \cdot 5\frac{1}{2} \text{ yd} \cdot 10 \text{ yd}$

$V = \frac{13}{2} \cdot \frac{11}{2} \cdot 10 \text{ yd}^3$

$V = \frac{715}{2} \text{ yd}^3$

$V = 357\frac{1}{2} \text{ yd}^3$

$$\begin{aligned} SA &= 2lw + 2lh + 2wh \\ &= 2 \cdot 6\tfrac{1}{2} \text{ yd} \cdot 5\tfrac{1}{2} \text{ yd} + 2 \cdot 6\tfrac{1}{2} \text{ yd} \cdot 10 \text{ yd} + \\ &\quad 2 \cdot 5\tfrac{1}{2} \text{ yd} \cdot 10 \text{ yd} \\ &= 2 \cdot \frac{13}{2} \cdot \frac{11}{2} \text{ yd}^2 + 2 \cdot \frac{13}{2} \cdot 10 \text{ yd}^2 + \\ &\quad 2 \cdot \frac{11}{2} \cdot 10 \text{ yd}^2 \\ &= \frac{143}{2} \text{ yd}^2 + 130 \text{ yd}^2 + 110 \text{ yd}^2 \\ &= 311\tfrac{1}{2} \text{ yd}^2 \end{aligned}$$

9. $$\begin{aligned} V &= Bh = \pi \cdot r^2 \cdot h \\ &\approx 3.14 \times 8 \text{ in.} \times 8 \text{ in.} \times 4 \text{ in.} \\ &= 803.84 \text{ in}^3 \end{aligned}$$

11. $$\begin{aligned} V &= Bh = \pi \cdot r^2 \cdot h \\ &\approx 3.14 \times 5 \text{ cm} \times 5 \text{ cm} \times 4.5 \text{ cm} \\ &= 353.25 \text{ cm}^3 \end{aligned}$$

13. $$\begin{aligned} V &= Bh = \pi \cdot r^2 \cdot h \\ &\approx \frac{22}{7} \times 210 \text{ yd} \times 210 \text{ yd} \times 300 \text{ yd} \\ &= 41{,}580{,}000 \text{ yd}^3 \end{aligned}$$

15. $$\begin{aligned} V &= \frac{4}{3} \cdot \pi \cdot r^3 \\ &\approx \frac{4}{3} \times 3.14 \times (100 \text{ in.})^3 \\ &= \frac{4 \times 3.14 \times 1{,}000{,}000 \text{ in}^3}{3} \\ &\approx 4{,}186{,}666.67 \text{ in}^3 \end{aligned}$$

17. $$\begin{aligned} V &= \frac{4}{3} \cdot \pi \cdot r^3 \\ &\approx \frac{4}{3} \times 3.14 \times (3.1 \text{ m})^3 \\ &= \frac{4 \times 3.14 \times 29.791 \text{ m}^3}{3} \\ &\approx 124.72 \text{ m}^3 \end{aligned}$$

19. $$\begin{aligned} V &= \frac{4}{3} \cdot \pi \cdot r^3 \\ &\approx \frac{4}{3} \times \frac{22}{7} \times \left(7\frac{3}{4} \text{ ft}\right)^3 \\ &= \frac{4}{3} \times \frac{22}{7} \times \left(\frac{31}{4} \text{ ft}\right)^3 \\ &= \frac{4 \times 22 \times 29{,}791 \text{ ft}^3}{3 \times 7 \times 64} \\ &\approx 1950\frac{101}{168} \text{ ft}^3 \end{aligned}$$

21. $$\begin{aligned} V &= \frac{1}{3} \cdot \pi \cdot r^2 \cdot h \\ &\approx \frac{1}{3} \times 3.14 \times 33 \text{ ft} \times 33 \text{ ft} \times 100 \text{ ft} \\ &\approx 113{,}982 \text{ ft}^3 \end{aligned}$$

23. $$\begin{aligned} V &= \frac{1}{3} \cdot \pi \cdot r^2 \cdot h \\ &\approx \frac{1}{3} \times \frac{22}{7} \times 1.4 \text{ cm} \times 1.4 \text{ cm} \times 12 \text{ cm} \\ &\approx 24.64 \text{ cm}^3 \end{aligned}$$

25. $$\begin{aligned} V &= \frac{1}{3} \cdot \pi \cdot r^2 \cdot h \\ &\approx \frac{1}{3} \times \frac{22}{7} \times \frac{3}{4} \text{ yd} \times \frac{3}{4} \text{ yd} \times \frac{7}{5} \text{ yd} \\ &= \frac{33}{40} \text{ yd}^3 \end{aligned}$$

27. We must find the radius of the base in order to use the formula for the volume of a circular cylinder.

$$\begin{aligned} r &= \frac{d}{2} = \frac{12 \text{ cm}}{2} = 6 \text{ cm} \\ V &= Bh = \pi \cdot r^2 \cdot h \\ &\approx 3.14 \times 6 \text{ cm} \times 6 \text{ cm} \times 42 \text{ cm} \\ &\approx 4747.68 \text{ cm}^3 \end{aligned}$$

29. We first find the radius of the ball in order to use the formula for the volume of a sphere.

$$\begin{aligned} r &= \frac{d}{2} = \frac{12 \text{ ft}}{2} = 6 \text{ ft} \\ V &= \frac{4}{3} \cdot \pi \cdot r^3 \\ &\approx \frac{4}{3} \times 3.14 \times (6 \text{ ft})^3 \\ &\approx 904 \text{ ft}^3 \end{aligned}$$

31. First we find the radius of the ball:

$$r = \frac{d}{2} = \frac{6.5 \text{ cm}}{2} = 3.25 \text{ cm}$$

Then we find the volume, using the formula for the volume of a sphere.

$$\begin{aligned} V &= \frac{4}{3} \cdot \pi \cdot r^3 \\ &\approx \frac{4}{3} \cdot 3.14 \cdot (3.25 \text{ cm})^3 \\ &\approx 143.72 \text{ cm}^3 \end{aligned}$$

33. First we find the radius of the earth:

$$\frac{3980 \text{ mi}}{2} = 1990 \text{ mi}$$

Then we find the volume, using the formula for the volume of a sphere.

$$\begin{aligned} V &= \frac{4}{3} \cdot \pi \cdot r^3 \\ &\approx \frac{4}{3} \cdot 3.14 \cdot (1990 \text{ mi})^3 \\ &\approx 32{,}993{,}440{,}000 \text{ mi}^3 \end{aligned}$$

35. First we find the radius of the base.

$$\begin{aligned} r &= \frac{d}{2} = \frac{4.875 \text{ in.}}{2} = 2.4375 \text{ in.} \\ V &= \frac{1}{3} \cdot \pi \cdot r^2 \cdot h \\ &\approx \frac{1}{3} \times 3.14 \times 2.4375 \text{ in.} \times 2.4375 \text{ in.} \times 12.5 \text{ in.} \\ &\approx 77.7 \text{ in}^3 \end{aligned}$$

37. $$\begin{aligned} V &= Bh = \pi \cdot r^2 \cdot h \\ &\approx \frac{22}{7} \cdot 14 \text{ m} \cdot 14 \text{ m} \cdot 100 \text{ m} \\ &= 61{,}600 \text{ m}^3 \end{aligned}$$

39. A cube is a rectangular solid.

$$\begin{aligned} V &= l \cdot w \cdot h \\ &= 18 \text{ yd} \cdot 18 \text{ yd} \cdot 18 \text{ yd} \\ &= 5832 \text{ yd}^3 \end{aligned}$$

41. First we find the radius of the can.

$$r = \frac{d}{2} = \frac{6.5 \text{ cm}}{2} = 3.25 \text{ cm}$$

The height of the can is the length of the diameters of 3 tennis balls.

$$h = 3(6.5 \text{ cm}) = 19.5 \text{ cm}$$

Now we find the volume.

$$V = Bh = \pi \cdot r^2 \cdot h$$
$$\approx 3.14 \times 3.25 \text{ cm} \times 3.25 \text{ cm} \times 19.5 \text{ cm}$$
$$\approx 646.74 \text{ cm}^3$$

43. $24\% = \frac{24}{100} = \frac{4 \cdot 6}{4 \cdot 25} = \frac{4}{4} \cdot \frac{6}{25} = \frac{6}{25}$

45. $12.75\% = \frac{12.75}{100} = \frac{12.75}{100} \cdot \frac{100}{100}$
$= \frac{1275}{10,000} = \frac{25 \cdot 51}{25 \cdot 400} = \frac{25}{25} \cdot \frac{51}{400}$
$= \frac{51}{400}$

47. $35\% = \frac{35}{100} = \frac{5 \cdot 7}{5 \cdot 20} = \frac{5}{5} \cdot \frac{7}{20} = \frac{7}{20}$

49. $37\frac{1}{2}\% = 37.5\% = \frac{37.5}{100}$
$= \frac{37.5}{100} \cdot \frac{10}{10} = \frac{375}{1000}$
$= \frac{125 \cdot 3}{125 \cdot 8} = \frac{125}{125} \cdot \frac{3}{8}$
$= \frac{3}{8}$

51. $83.\overline{3}\% = 83\frac{1}{3}\% = \frac{250}{3}\%$
$= \frac{250}{3} \times \frac{1}{100} = \frac{250}{300}$
$= \frac{50 \cdot 5}{50 \cdot 6} = \frac{50}{50} \cdot \frac{5}{6}$
$= \frac{5}{6}$

53. $\frac{1}{2}\% = 0.5\% = \frac{0.5}{100}$
$= \frac{0.5}{100} \cdot \frac{10}{10} = \frac{5}{1000}$
$= \frac{5 \cdot 1}{5 \cdot 200} = \frac{5}{5} \cdot \frac{1}{200}$
$= \frac{1}{200}$

55. ***Familiarize***. Let m = the number of cubic miles by which the volume of water in Lake Michigan exceeds the volume of water in Lake Erie.

Translate.

Water volume in Lake Erie	plus	Excess amount in Lake Michigan	is	Water volume in Lake Michigan
↓	↓	↓	↓	↓
116	+	m	=	1180

Solve.

$$116 + m = 1180$$
$$116 + m - 116 = 1180 - 116$$
$$m = 1064$$

Check. Since $116 + 1064 = 1180$, the answer checks.

State. The volume of water in Lake Michigan is 1064 mi^3 greater than the volume of water in Lake Erie.

57. $\frac{483 + 279 + 195 + 62 + 283}{5} = \frac{1302}{5} = 260.4$ ft

59. $850 \text{ mi}^3 = 850 \times 1 \text{ mi} \times 1 \text{ mi} \times 1 \text{ mi}$
$\approx 850 \times 1.609 \text{ km} \times 1.609 \text{ km} \times 1.609 \text{ km}$
$\approx 3540.68 \text{ km}^3$

61. The length of a side of the cube is the length of a diameter of the sphere, 1 m.

$$V = l \cdot w \cdot h$$
$$= 1 \text{ m} \cdot 1 \text{ m} \cdot 1 \text{ m} = 1 \text{ m}^3$$

The radius of the sphere is $\frac{1 \text{ m}}{2}$, or 0.5 m.

$$V = \frac{4}{3} \cdot \pi \cdot r^3$$
$$\approx \frac{4}{3} \cdot 3.14 \cdot (0.5 \text{ m})^3 \approx 0.523 \text{ m}^3$$

$1 \text{ m}^3 - 0.523 \text{ m}^3 = 0.477 \text{ m}^3$, so there is 0.477 m^3 more volume in the cube.

Exercise Set 8.6

1. Two angles are complementary if the sum of their measures is 90°.

$$90° - 11° = 79°.$$

The measure of a complement is 79°.

3. Two angles are complementary if the sum of their measures is 90°.

$$90° - 67° = 23°.$$

The measure of a complement is 23°.

5. Two angles are complementary if the sum of their measures is 90°.

$$90° - 58° = 32°.$$

The measure of a complement is 32°.

7. Two angles are complementary if the sum of their measures is 90°.

$$90° - 29° = 61°.$$

The measure of a complement is 61°.

9. Two angles are supplementary if the sum of their measures is 180°.

$$180° - 3° = 177°.$$

The measure of a supplement is 177°.

11. Two angles are supplementary if the sum of their measures is 180°.

$$180° - 139° = 41°.$$

The measure of a supplement is 41°.

13. Two angles are supplementary if the sum of their measures is 180°.

$$180° - 85° = 95°.$$

The measure of a supplement is 95°.

15. Two angles are supplementary if the sum of their measures is 180°.

$$180° - 102° = 78°.$$

The measure of a supplement is 78°.

17. The segments have different lengths. They are not congruent.

19. $m\angle G = m\angle R$, so $\angle G \cong \angle R$.

21. Since $\angle 2$ and $\angle 5$ are vertical angles, $m\angle 2 = 67°$. Likewise, $\angle 1$ and $\angle 4$ are vertical angles, so $m\angle 4 = 80°$.

$$\begin{aligned} m\angle 1 + m\angle 2 + m\angle 3 &= 180° \\ 80° + 67° + m\angle 3 &= 180° \quad \text{Substituting} \\ 147° + m\angle 3 &= 180° \\ m\angle 3 &= 180° - 147° \\ m\angle 3 &= 33° \end{aligned}$$

Since $\angle 3$ and $\angle 6$ are vertical angles, $m\angle 6 = 33°$.

23. a) The pairs of corresponding angles are

$\angle 1$ and $\angle 3$,

$\angle 2$ and $\angle 4$,

$\angle 8$ and $\angle 6$,

$\angle 7$ and $\angle 5$.

b) The interior angles are $\angle 2$, $\angle 3$, $\angle 6$, and $\angle 7$.

c) The pairs of alternate interior angles are

$\angle 2$ and $\angle 6$,

$\angle 3$ and $\angle 7$.

25. $\angle 4$ and $\angle 6$ are vertical angles, so $m\angle 6 = 125°$.

$\angle 4$ and $\angle 2$ are corresponding angles. By Property 1, $m\angle 2 = 125°$.

$\angle 6$ and $\angle 8$ are corresponding angles. By Property 1, $m\angle 8 = 125°$.

$\angle 2$ and $\angle 3$ are interior angles on the same side of the transversal. Using Property 4 and $m\angle 2 = 125°$, $m\angle 3 = 55°$.

$\angle 6$ and $\angle 7$ are interior angles on the same side of the transversal. Using Property 4 and $m\angle 6 = 125°$, $m\angle 7 = 55°$.

$\angle 3$ and $\angle 5$ are vertical angles, so $m\angle 5 = 55°$.

$\angle 7$ and $\angle 1$ are vertical angles, so $m\angle 1 = 55°$.

27. Considering the transversal $\overleftrightarrow{BC}$, $\angle ABE$ and $\angle DCE$ are alternate interior angles. By Property 2, $\angle ABE \cong \angle DCE$. Then $m\angle ABE = m\angle DCE = 95°$.

Considering the transversal $\overleftrightarrow{AD}$, $\angle BAE$ and $\angle CDE$ are alternate interior angles. By Property 2, $\angle BAE \cong \angle CDE$. We cannot determine the measure of these angles.

$\angle AEB$ and $\angle DEC$ are vertical angles, so $\angle AEB \cong \angle DEC$. We cannot determine the measure of these angles.

$\angle BED$ and $\angle AEC$ are also vertical angles, so $\angle BED \cong \angle AEC$. We cannot determine their measures.

29. Considering the transversal $\overleftrightarrow{CE}$, $\angle AEC$ and $\angle DCE$ are alternate interior angles. By Property 2, $\angle AEC \cong \angle DCE$. Then $m\angle AEC = m\angle DCE = 50°$.

Considering the transversal $\overleftrightarrow{DE}$, $\angle BED$ and $\angle EDC$ are alternate interior angles. By Property 2, $\angle BED \cong \angle EDC$. Then $m\angle BED = m\angle EDC = 41°$.

31.
$$\begin{aligned} 6 \times 1\frac{7}{8} &= 6 \times \frac{15}{8} \\ &= \frac{6 \times 15}{8} \\ &= \frac{2 \times 3 \times 15}{2 \times 4} \\ &= \frac{2}{2} \times \frac{3 \times 15}{4} \\ &= \frac{45}{4}, \text{ or } 11\frac{1}{4} \end{aligned}$$

33.
$$\begin{aligned} 8\frac{3}{7} \times 14 &= \frac{59}{7} \times 14 \\ &= \frac{59 \times 14}{7} \\ &= \frac{59 \times 2 \times 7}{7 \times 1} \\ &= \frac{7}{7} \times \frac{59 \times 2}{1} \\ &= 118 \end{aligned}$$

Exercise Set 8.7

1. The notation tells us the way in which the vertices of the two triangles are matched.

$\triangle ABC \cong \triangle RST$

$\triangle ABC \cong \triangle RST$ means

$\angle A \cong \angle R$	and	$\overline{AB} \cong \overline{RS}$
$\angle B \cong \angle S$		$\overline{AC} \cong \overline{RT}$
$\angle C \cong \angle T$		$\overline{BC} \cong \overline{ST}$

3. The notation tells us the way in which the vertices of the two triangles are matched.

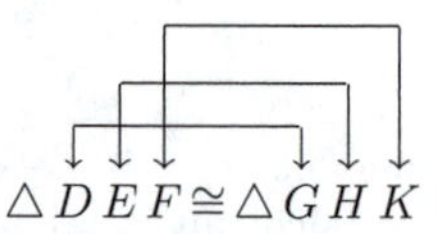

$\triangle DEF \cong \triangle GHK$ means

$\angle D \cong \angle G$	and	$\overline{DE} \cong \overline{GH}$
$\angle E \cong \angle H$		$\overline{DF} \cong \overline{GK}$
$\angle F \cong \angle K$		$\overline{EF} \cong \overline{HK}$

5. The notation tells us the way in which the vertices of the two triangles are matched.

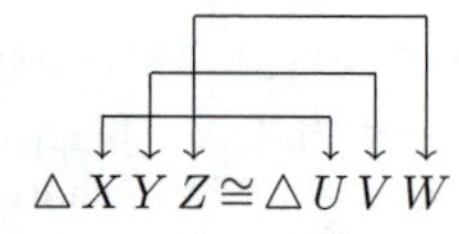

$\triangle XYZ \cong \triangle UVW$ means

$\angle X \cong \angle U$	and	$\overline{XY} \cong \overline{UV}$
$\angle Y \cong \angle V$		$\overline{XZ} \cong \overline{UW}$
$\angle Z \cong \angle W$		$\overline{YZ} \cong \overline{VW}$

7. The notation tells us the way in which the vertices of the two triangles are matched.

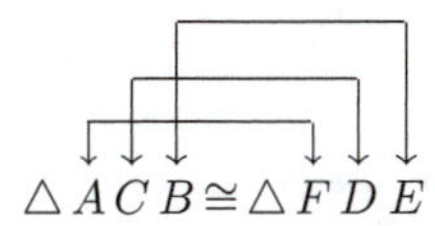

$\triangle ACB \cong \triangle FDE$ means

$\angle A \cong \angle F$	and	$\overline{AC} \cong \overline{FD}$
$\angle C \cong \angle D$		$\overline{AB} \cong \overline{FE}$
$\angle B \cong \angle E$		$\overline{CB} \cong \overline{DE}$

9. The notation tells us the way in which the vertices of the two triangles are matched.

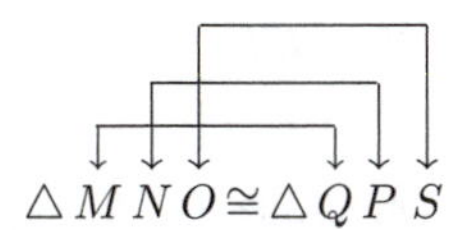

$\triangle MNO \cong \triangle QPS$ means

$\angle M \cong \angle Q$	and	$\overline{MN} \cong \overline{QP}$
$\angle N \cong \angle P$		$\overline{MO} \cong \overline{QS}$
$\angle O \cong \angle S$		$\overline{NO} \cong \overline{PS}$

11. We cannot determine from the information given that two sides of one triangle and the included angle are congruent to two sides and the included angle of the other triangle. Therefore, we cannot use the SAS Property.

13. Two sides of one triangle and the included angle are congruent to two sides and the included angle of the other triangle. They are congruent by the SAS Property.

15. Two sides of one triangle and the included angle are congruent to two sides and the included angle of the other triangle. They are congruent by the SAS Property.

17. We cannot determine from the information given that three sides of one triangle are congruent to three sides of the other triangle. Therefore, we cannot use the SSS Property.

19. Three sides of one triangle are congruent to three sides of the other triangle. They are congruent by the SSS Property.

21. Three sides of one triangle are congruent to three sides of the other triangle. They are congruent by the SSS Property.

23. Two angles and the included side of one triangle are congruent to two angles and the included side of the other triangle. They are congruent by the ASA Property.

25. Two angles and the included side of one triangle are congruent to two angles and the included side of the other triangle. They are congruent by the ASA Property.

27. The vertical angles are congruent so two angles and the included side of one triangle are congruent to two angles and the included side of the other triangle. They are congruent by the ASA Property.

29. Two angles and the included side of one triangle are congruent to two angles and the included side of the other triangle. They are congruent by the ASA Property.

31. Two sides of one triangle and the included angle are congruent to two sides and the included angle of the other triangle. They are congruent by the SAS Property.

33. Three sides of one triangle are congruent to three sides of the other triangle. In addition, two sides of one triangle and the included angle are congruent to two sides and the included angle of the other triangle. Therefore, we can use either the SSS Property or the SAS Property to show that they are congruent.

35. Since R is the midpoint of $\overline{PT}$, $\overline{PR} \cong \overline{TR}$.

Since R is the midpoint of $\overline{QS}$, $\overline{RQ} \cong \overline{RS}$.

$\angle PRQ$ and $\angle TRS$ are vertical angles, so $\angle PRQ \cong \angle TRS$.

Two sides and the included angle of $\triangle PRQ$ are congruent to two sides and the included angle of $\triangle TRS$, so $\triangle PRQ \cong \triangle TRS$ by the SAS Property.

37. Since $GL \perp KM$, $m\angle GLK = m\angle GLM = 90°$. Then $\angle GLK \cong \angle GLM$.

Since L is the midpoint of $\overline{KM}$, $\overline{KL} \cong \overline{LM}$.

$\overline{GL} \cong \overline{GL}$.

Two sides and the included angle of $\triangle KLG$ are congruent to two sides and the included angle of $\triangle MLG$, so $\triangle KLG \cong \triangle MLG$ by the SAS Property.

39. The information given tells us that $\overline{AE} \cong \overline{CB}$ and $\overline{AB} \cong \overline{CD}$.

Since B is the midpoint of $\overline{ED}$, $\overline{EB} \cong \overline{BD}$.

Three sides of $\triangle AEB$ are congruent to three sides of $\triangle CDB$, so $\triangle AEB \cong \triangle CDB$ by the SSS Property.

41. The information given tells us that $\overline{HK} \cong \overline{KJ}$ and $\overline{GK} \cong \overline{LK}$.

Since $\overline{GK} \perp \overline{LJ}$, $m\angle HKL = m\angle GKJ = 90°$.

Then $\angle HKL \cong \angle GKJ$.

Two sides and the included angle of $\triangle LKH$ are congruent to two sides and the included angle of $\triangle GKJ$, so $\triangle LKH \cong \triangle GKJ$ by the SAS Property. This means that the remaining corresponding parts of the two triangles are congruent. That is, $\angle HLK \cong \angle JGK$, $\angle LHK \cong \angle GJK$, and $\overline{LH} \cong \overline{GJ}$.

43. Two angles and the included side of $\triangle PED$ are congruent to two angles and the included side of $\triangle PFG$, so $\triangle PED \cong \triangle PFG$ by the ASA Property. Then corresponding parts of the two triangles are congruent, so $\overline{EP} \cong \overline{FP}$. Therefore, P is the midpoint of $\overline{EF}$.

45. $\angle A$ and $\angle C$ are opposite angles, so $m\angle A = 70°$ by Property 2.

$\angle C$ and $\angle B$ are consecutive angles, so the are supplementary by Property 4. Then

$$m\angle B = 180° - m\angle C$$
$$m\angle B = 180° - 70°$$
$$m\angle B = 110°.$$

$\angle B$ and $\angle D$ are opposite angles, so $m\angle D = 110°$ by Property 2.

47. $\angle M$ and $\angle K$ are opposite angles, so $m\angle M = 71°$ by Property 2.

$\angle K$ and $\angle L$ are consecutive angles, so they are supplementary by Property 4. Then

$$m\angle L = 180° - m\angle K$$
$$m\angle L = 180° - 71°$$
$$m\angle L = 109°.$$

$\angle J$ and $\angle L$ are opposite angles, so $m\angle J = 109°$ by Property 2.

49. $\overline{ON}$ and $\overline{TU}$ are opposite sides of the parallelogram. So are $\overline{OT}$ and $\overline{NU}$. The opposite sides of a parallelogram are congruent (Property 3), so $TU = 9$ and $NU = 15$.

51. $\overline{JM}$ and $\overline{KL}$ are opposite sides of the parallelogram. So are $\overline{JK}$ and $\overline{ML}$. The opposite sides of a parallelogram are congruent (Property 3). Then $KL = 3\frac{1}{2}$ and $JK + LM = 22 - 3\frac{1}{2} - 3\frac{1}{2} = 15$. Thus, $JK = LM = \frac{1}{2} \cdot 15 = 7\frac{1}{2}$.

53. The diagonals of a parallelogram bisect each other (Property 5). Then

$$AC = 2 \cdot AB = 2 \cdot 14 = 28,$$
$$ED = 2 \cdot BD = 2 \cdot 19 = 38.$$

55. 0.452   0.45.2   Move the decimal point 2 places to the right.

Write a % symbol: 45.2%

0.452 = 45.2%

57. We multiply by 1 to get 100 in the denominator.

$$\frac{11}{20} = \frac{11}{20} \cdot \frac{5}{5} = \frac{55}{100} = 55\%$$

59. The ratio of the amount spent in Florida to the total amount spent is $\frac{2.7}{13.1}$. This can also be expressed as follows:

$$\frac{2.7}{13.1} = \frac{2.7}{13.1} \cdot \frac{10}{10} = \frac{27}{131}$$

The ratio of the total amount spent to the amount spent in Florida is $\frac{13.1}{2.7}$, or $\frac{131}{27}$.

61.

$$\begin{array}{r} 1.75 \\ 12\overline{)21.00} \\ \underline{12} \\ 90 \\ \underline{84} \\ 60 \\ \underline{60} \\ 0 \end{array}$$

The answer is 1.75.

63. To divide by 100, move the decimal point 2 places to the left.

23.4   .23.4

$23.4 \div 100 = 0.234$

65.

3.14	(2 decimal places)
× 4.41	(2 decimal places)
314	
12560	
125600	
13.8474	(4 decimal places)

Round

13.8 4 [7] 4   to the nearest hundredth.

Thousandths digit is 5 or higher.

13.85   Round up.

Exercise Set 8.8

1. Vertex R is matched with vertex A, vertex S is matched with vertex B, and vertex T is matched with vertex C. Then

$$\begin{array}{lcl} \overline{RS} \longleftrightarrow \overline{AB} & \text{and} & \angle R \longleftrightarrow \angle A \\ \overline{ST} \longleftrightarrow \overline{BC} & & \angle S \longleftrightarrow \angle B \\ \overline{TR} \longleftrightarrow \overline{CA} & & \angle T \longleftrightarrow \angle C \end{array}$$

3. Vertex C is matched with vertex W, vertex B is matched with vertex J, and vertex S is matched with vertex Z. Then

$$\begin{array}{lcl} \overline{CB} \longleftrightarrow \overline{WJ} & \text{and} & \angle C \longleftrightarrow \angle W \\ \overline{BS} \longleftrightarrow \overline{JZ} & & \angle B \longleftrightarrow \angle J \\ \overline{SC} \longleftrightarrow \overline{ZW} & & \angle S \longleftrightarrow \angle Z \end{array}$$

5. The notation tells us the way in which the vertices are matched.

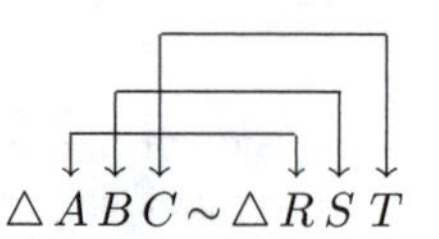

$\triangle ABC \sim \triangle RST$ means

$$\angle A \cong \angle R$$
$$\angle B \cong \angle S \quad \text{and} \quad \frac{AB}{RS} = \frac{AC}{RT} = \frac{BC}{ST}.$$
$$\angle C \cong \angle T$$

7. The notation tells us the way in which the vertices are matched.

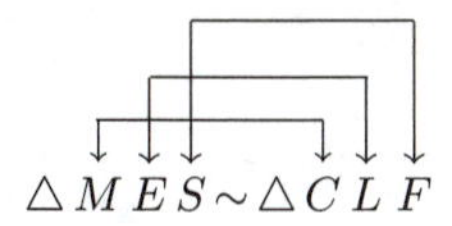

$\triangle MES \sim \triangle CLF$ means

$$\angle M \cong \angle C$$
$$\angle E \cong \angle L \quad \text{and} \quad \frac{ME}{CL} = \frac{MS}{CF} = \frac{ES}{LF}.$$
$$\angle S \cong \angle F$$

9. If we match P with N, S with D, and Q with M, the corresponding angles will be congruent. That is, $\triangle PSQ \sim \triangle NDM$. Then

$$\frac{PS}{ND} = \frac{PQ}{NM} = \frac{SQ}{DM}.$$

11. If we match T with G, A with F, and W with C, the corresponding angles will be congruent. That is, $\triangle TAW \sim \triangle GFC$. Then

$$\frac{TA}{GF} = \frac{TW}{GC} = \frac{AW}{FC}.$$

13. Since $\triangle ABC \sim \triangle PQR$, the corresponding sides are proportional. Then

$$\frac{3}{6} = \frac{4}{PR} \quad \text{and} \quad \frac{3}{6} = \frac{5}{QR}$$
$$3(PR) = 6 \cdot 4 \qquad 3(QR) = 6 \cdot 5$$
$$3(PR) = 24 \qquad 3(QR) = 30$$
$$PR = 8 \qquad QR = 10$$

15. Recall that if a transversal intersects two parallel lines, then the alternate interior angles are congruent. Thus,

$$\angle A \cong \angle B \text{ and } \angle D \cong \angle C.$$

Since $\angle AED$ and $\angle CEB$ are vertical angles, they are congruent. Thus,

$$\angle AED \cong \angle CEB.$$

Then $\triangle AED \sim \triangle CEB$, and the lengths of the corresponding sides are proportional.

$$\frac{AD}{CB} = \frac{ED}{EC}$$
$$\frac{7}{21} = \frac{6}{EC}$$
$$7 \cdot EC = 126$$
$$EC = 18$$

17. If we use the sun's rays to represent the third side of a triangle in a drawing of the situation, we see that we have similar triangles. We let $h =$ the height of the tree.

Sun's rays 4 ft 3 ft Sun's rays h 27 ft

The ratio of h to 4 is the same as the ratio of 27 to 3. We have the proportion

$$\frac{h}{4} = \frac{27}{3}.$$

Solve: $3 \cdot h = 4 \cdot 27$

$$h = \frac{4 \cdot 27}{3}$$
$$h = 36$$

The tree is 36 ft tall.

19. Since the ratio of d to 25 ft is the same as the ratio of 40 ft to 10 ft, we have the proportion

$$\frac{d}{25} = \frac{40}{10}.$$

Solve: $10 \cdot d = 25 \cdot 40$

$$d = \frac{25 \cdot 40}{10}$$
$$d = 100$$

The distance across the river is 100 ft.

21. $2\frac{4}{5} \times 10\frac{1}{2} = \frac{14}{5} \times \frac{21}{2} = \frac{14 \times 21}{5 \times 2} =$
$\frac{2 \times 7 \times 21}{5 \times 2} = \frac{\not{2} \times 7 \times 21}{5 \times \not{2}} = \frac{147}{5} = 29\frac{2}{5}$

23. $8 \times 9\frac{3}{4} = \frac{8}{1} \times \frac{39}{4} = \frac{8 \times 39}{1 \times 4} = \frac{2 \times 4 \times 39}{1 \times 4} =$
$\frac{2 \times \not{4} \times 39}{1 \times \not{4}} = \frac{78}{1} = 78$

Chapter 8 Concept Reinforcement

1. True; the sum of the measures of the acute angles of a right triangle is $180° - 90°$, or $90°$.

2. False; two angles are supplementary if the sum of their measures is $180°$.

3. Using a calculator, we find that $\pi \approx 3.14159$ and $\frac{22}{7} \approx 3.14288$. Thus, π is greater than 3.14 but less than $\frac{22}{7}$. The given statement is false.

4. True; the volume of a sphere with diameter 6 ft, or radius 3 ft, is $\frac{4}{3} \cdot \pi \cdot (3\text{ ft})^3 \approx \frac{4}{3} \times 3.14 \times (3\text{ ft})^3 = 113.04\text{ ft}^3$.

The volume of a rectangular solid that measures 6 ft by 6 ft by 6 ft is $6\text{ ft} \times 6\text{ ft} \times 6\text{ ft} = 216\text{ ft}^3$.

5. True; the measure of an obtuse angle is greater than $90°$ while the measure of an acute angle is less than $90°$.

Chapter 8 Important Concepts

1. The angle can be named as angle WAQ, angle QAW, $\angle WAQ$, $\angle QAW$, or $\angle A$.

Using a protractor, we find that the measure of the angle is $26°$.

2. (a) The measure of the angle is 180°. It is a straight angle.

(b) The measure of the angle is greater than 0°and less than 90°. It is an acute angle.

(c) The measure of the angle is greater than 90°and less than 180°. It is an obtuse angle.

(d) The measure of the angle is 90°. It is a right angle.

3. (a) Two sides of the triangle are the same length, and the third has a different length. The triangle is isosceles.

The triangle has a 90°angle. It is a right triangle.

(b) All the sides of the triangle have the same length. The triangle is equilateral.

All three angles are acute. It is an acute triangle.

(c) All the sides of the triangle have different lengths. The triangle is scalene.

All three angles are acute. It is an acute triangle.

(d) All the sides of the triangle have different lengths. The triangle is scalene.

One angle of the triangle is obtuse. It is an obtuse triangle.

4.
$$\begin{aligned} x + 21^\circ + 72^\circ &= 180^\circ \\ x + 93^\circ &= 180^\circ \\ x &= 180^\circ - 93^\circ \\ x &= 87^\circ \end{aligned}$$

5.
$$\begin{aligned} (n-2)\cdot 180^\circ &= (9-2)\cdot 180^\circ \\ &= 7 \cdot 180^\circ \\ &= 1260^\circ \end{aligned}$$

6.
$$\begin{aligned} P &= 2\cdot(l+w) \\ &= 2\cdot(8.2 \text{ ft} + 5.7 \text{ ft}) \\ &= 2\cdot(13.9 \text{ ft}) = 27.8 \text{ ft} \end{aligned}$$

$$\begin{aligned} A &= l \cdot w \\ &= 8.2 \text{ ft} \cdot 5.7 \text{ ft} \\ &= 8.2 \cdot 5.7 \cdot \text{ ft} \cdot \text{ ft} = 46.74 \text{ ft}^2 \end{aligned}$$

7.
$$\begin{aligned} A &= b \cdot h \\ &= 6.2 \text{ m} \cdot 2.5 \text{ m} \\ &= 6.2 \cdot 2.5 \cdot \text{ m} \cdot \text{ m} = 15.5 \text{ m}^2 \end{aligned}$$

8.
$$\begin{aligned} A &= \frac{1}{2}\cdot b \cdot h \\ &= \frac{1}{2} \times 3.5 \text{ ft} \cdot 5 \text{ ft} \\ &= \frac{3.5 \cdot 5}{2} \text{ ft}^2 = 8.75 \text{ ft}^2 \end{aligned}$$

9.
$$\begin{aligned} A &= \frac{1}{2}\cdot h \cdot (a+b) \\ &= \frac{1}{2} \times 8 \text{ m} \times (5 \text{ m} + 15 \text{ m}) \\ &= \frac{1}{2} \times 8 \text{ m} \times (20 \text{ m}) \\ &= \frac{8 \times 20}{2} \text{ m}^2 = 80 \text{ m}^2 \end{aligned}$$

10.
$$\begin{aligned} C &= 2\cdot\pi\cdot r \\ &\approx 2 \cdot 3.14 \cdot 6 \text{ in.} = 37.68 \text{ in.} \end{aligned}$$

11.
$$\begin{aligned} A &= \pi \cdot r \cdot r \\ &\approx \frac{22}{7} \cdot 14 \text{ cm} \cdot 14 \text{ cm} \\ &= \frac{22 \cdot 14 \cdot 14}{7} \text{ cm}^2 \\ &= 616 \text{ cm}^2 \end{aligned}$$

12.
$$\begin{aligned} V &= l \cdot w \cdot h \\ &= 18.1 \text{ m} \times 15 \text{ m} \times 6.2 \text{ m} \\ &= 1683.3 \text{ m}^3 \end{aligned}$$

13.
$$\begin{aligned} V &= \pi \cdot r^2 \cdot h \\ &\approx \frac{22}{7} \times 1\frac{1}{3} \text{ ft} \times 1\frac{1}{3} \text{ ft} \times 5\frac{2}{5} \text{ ft} \\ &= \frac{22}{7} \times \frac{4}{3} \text{ ft} \times \frac{4}{3} \text{ ft} \times \frac{27}{5} \text{ ft} \\ &= \frac{22 \times 4 \times 4 \times 27}{7 \times 3 \times 3 \times 5} \text{ ft}^3 \\ &= \frac{1056}{35} \text{ ft}^3, \text{ or } 30\frac{6}{35} \text{ ft}^3 \end{aligned}$$

14.
$$\begin{aligned} V &= \frac{4}{3} \cdot \pi \cdot r^3 \\ &\approx \frac{4}{3} \times 3.14 \times (7.4 \text{ cm})^3 \\ &= 1696.537813 \text{ cm}^3 \end{aligned}$$

15.
$$\begin{aligned} V &= \frac{1}{3} \cdot \pi \cdot r^2 \cdot h \\ &\approx \frac{1}{3} \times 3.14 \times 2.25 \text{ ft} \times 2.25 \text{ ft} \times 5 \text{ ft} \\ &= 26.49375 \text{ ft}^3 \end{aligned}$$

16. Measure of the complement: $90^\circ - 38^\circ = 52^\circ$

Measure of the supplement: $180^\circ - 38^\circ = 142^\circ$

17. $\angle 8$ and $\angle 11$ are vertical angles, so $m\angle 11 = 65^\circ$.

$\angle 12$ and $\angle 9$ are vertical angles, so $m\angle 9 = 55^\circ$.

$$\begin{aligned} m\angle 9 + m\angle 10 + m\angle 11 &= 180^\circ \\ 55^\circ + m\angle 10 + 65^\circ &= 180^\circ \\ 120^\circ + m\angle 10 &= 180^\circ \\ m\angle 10 &= 180^\circ - 120^\circ \\ m\angle 10 &= 60^\circ \end{aligned}$$

$\angle 10$ and $\angle 7$ are vertical angles, so $m\angle 7 = 60^\circ$.

18. $\angle 5$ and $\angle 4$ are vertical angles, so $m\angle 4 = 105^\circ$.

$\angle 4$ and $\angle 8$ are corresponding angles, so $m\angle 8 = 105^\circ$.

$\angle 8$ and $\angle 1$ are vertical angles, so $m\angle 1 = 105^\circ$.

$\angle 5$ and $\angle 6$ are supplementary angles, so

$$m\angle 6 = 180^\circ - 105^\circ = 75^\circ.$$

$\angle 6$ and $\angle 3$ are vertical angles, so $m\angle 3 = 75^\circ$.

$\angle 6$ and $\angle 2$ are corresponding angles, so $m\angle 2 = 75^\circ$.

$\angle 2$ and $\angle 7$ are vertical angles, so $m\angle 7 = 75^\circ$.

19. a) Two sides and the included angle of one triangle are congruent to two sides and the included angle of the other triangle, so the triangles are congruent by the SAS property.

b) Two angles and the included side of one triangle are congruent to two angles and the included side of the other triangle, so the triangles are congruent by the ASA property.

20. Since $AD = 45.5$, then $BC = 45.5$

$$\text{Perimeter} = AB + BC + DC + AD$$
$$237 = AB + 45.5 + DC + 45.5$$
$$237 = AB + DC + 91$$
$$146 = AB + DC$$

Since $AB = DC$, we know that $AB = \frac{146}{2} = 73 = DC$.

$$m\angle A = m\angle C = 30°$$
$$m\angle A + m\angle B = 180°$$
$$30° + m\angle B = 180°$$
$$m\angle B = 180° - 30° = 150°$$
$$m\angle D = m\angle B = 150°$$

21. The corresponding sides of the congruent triangles are proportional.

$$\frac{75}{ZA} = \frac{175}{35}$$
$$75 \cdot 35 = (ZA)175$$
$$2625 = (ZA)175$$
$$15 = ZA$$

$$\frac{150}{AT} = \frac{175}{35}$$
$$150 \cdot 35 = (AT)175$$
$$5250 = (AT)175$$
$$30 = AT$$

Chapter 8 Review Exercises

1. Place the △ of the protractor at the vertex of the angle, and line up one of the sides at 0°. We choose the nearly horizontal side. Since 0° is on the inside scale, we check where the other side of the angle crosses the inside scale. It crosses at 54°. Thus, the measure of the angle is 54°.

2. Place the △ of the protractor at the vertex of the angle, point B. Line up one of the sides at 0°. We choose the side that contains point P. Since 0° is on the outside scale, we check where the other side crosses the outside scale. It crosses at 180°. Thus, the measure of the angle is 180°.

3. Place the △ of the protractor at the vertex of the angle, and line up one of the sides at 0°. We choose the horizontal side. Since 0° is on the inside scale, we check where the other side crosses the inside scale. It crosses at 140°. Thus, the measure of the angle is 140°.

4. Place the △ of the protractor at the vertex of the angle, and line up one of the sides at 0°. We choose the horizontal side. Since 0° is on the inside scale, we check where the other side crosses the inside scale. It crosses at 90°. Thus, the measure of the angle is 90°.

5. The measure of the angle in Exercise 1 is 54°. Since its measure is greater than 0°and less than 90°, it is an acute angle.

6. The measure of the angle in Exercise 2 is 180°. It is a straight angle.

7. The measure of the angle in Exercise 3 is 140°. Since its measure is greater than 90°and less than 180°, it is an obtuse angle.

8. The measure of the angle in Exercise 4 is 90°. It is a right angle.

9.
$$30° + 90° + x = 180°$$
$$120° + x = 180°$$
$$x = 180° - 120°$$
$$x = 60°$$

10. All the sides are of different lengths. The triangle is a scalene triangle.

11. One angle is a right angle. The triangle is a right triangle.

12.
$$(n-2) \cdot 180°$$
$$= (6-2) \cdot 180° \quad \text{Subtituting 6 for } n$$
$$= 4 \cdot 180°$$
$$= 720°$$

13.
$$\text{Perimeter} = 5 \text{ ft} + 7 \text{ ft} + 4 \text{ ft} + 4 \text{ ft} + 3 \text{ ft}$$
$$= (5 + 7 + 4 + 4 + 3) \text{ ft}$$
$$= 23 \text{ ft}$$

14.
$$\text{Perimeter} = 0.5 \text{ m} + 1.9 \text{ m} + 1.2 \text{ m} + 0.8 \text{ m}$$
$$= (0.5 + 1.9 + 1.2 + 0.8) \text{ m}$$
$$= 4.4 \text{ m}$$

15.
$$P = 2 \cdot l + 2 \cdot w$$
$$P = 2 \cdot 78 \text{ ft} + 2 \cdot 36 \text{ ft}$$
$$P = 156 \text{ ft} + 72 \text{ ft}$$
$$P = 228 \text{ ft}$$

$$A = l \cdot w$$
$$A = 78 \text{ ft} \cdot 36 \text{ ft}$$
$$A = 2808 \text{ ft}^2$$

16.
$$P = 4 \cdot s$$
$$P = 4 \cdot 9 \text{ ft}$$
$$P = 36 \text{ ft}$$

$$A = s \cdot s$$
$$A = 9 \text{ ft} \cdot 9 \text{ ft}$$
$$A = 81 \text{ ft}^2$$

17. $P = 2 \cdot (l + w)$

$P = 2 \cdot (7 \text{ cm} + 1.8 \text{ cm})$

$P = 2 \cdot (8.8 \text{ cm})$

$P = 17.6 \text{ cm}$

$A = l \cdot w$

$A = 7 \text{ cm} \cdot 1.8 \text{ cm}$

$A = 12.6 \text{ cm}^2$

18. $A = b \cdot h$

$A = 12 \text{ cm} \cdot 5 \text{ cm}$

$A = 60 \text{ cm}^2$

19. $A = \frac{1}{2} \cdot h \cdot (a + b)$

$A = \frac{1}{2} \cdot 5 \text{ mm} \cdot (4 + 10) \text{ mm}$

$A = \frac{5 \cdot 14}{2} \text{ mm}^2$

$A = 35 \text{ mm}^2$

20. $A = \frac{1}{2} \cdot b \cdot h$

$A = \frac{1}{2} \cdot 15 \text{ m} \cdot 3 \text{ m}$

$A = \frac{15 \cdot 3}{2} \text{ m}^2$

$A = 22.5 \text{ m}^2$

21. $A = \frac{1}{2} \cdot b \cdot h$

$A = \frac{1}{2} \cdot 11.4 \text{ yd} \cdot 5.2 \text{ yd}$

$A = \frac{11.4 \cdot 5.2}{2} \text{ yd}^2$

$A = 29.64 \text{ yd}^2$

22. $A = \frac{1}{2} \cdot h \cdot (a + b)$

$A = \frac{1}{2} \cdot 8 \text{ m} \cdot (5 + 17) \text{ m}$

$A = \frac{8 \cdot 22}{2} \text{ m}^2$

$A = 88 \text{ m}^2$

23. $A = b \cdot h$

$A = 21\frac{5}{6} \text{ in.} \cdot 6\frac{2}{3} \text{ in.}$

$A = \frac{131}{6} \cdot \frac{20}{3} \text{ in}^2$

$A = \frac{131 \cdot 20}{6 \cdot 3} \text{ in}^2$

$A = \frac{1310}{9} \text{ in}^2$

$A = 145\frac{5}{9} \text{ in}^2$

24. ***Familiarize***. The seeded area is the total area of the house and the seeded area less the area of the house. From the drawing in the text we see that the total area is the area of a rectangle with length 70 ft and width 25 ft+7 ft, or 32 ft. The length of the rectangular house is 70 ft − 7 ft − 7 ft, or 56 ft, and its width is 25 ft. We let A = the seeded area.

Translate.

$$\underbrace{\text{Seeded area}}_{A} \quad \underbrace{\text{is}}_{=} \quad \underbrace{\text{Total area}}_{70 \text{ ft} \cdot 32 \text{ ft}} \quad \underbrace{\text{minus}}_{-} \quad \underbrace{\text{Area of house}}_{56 \text{ ft} \cdot 25 \text{ ft}}$$

Solve.

$A = 70 \text{ ft} \cdot 32 \text{ ft} - 56 \text{ ft} \cdot 25 \text{ ft}$

$A = 2240 \text{ ft}^2 - 1400 \text{ ft}^2$

$A = 840 \text{ ft}^2$

Check. We can repeat the calculations. The answer checks.

State. The seeded area is 840 ft^2.

25. $r = \frac{d}{2} = \frac{16 \text{ m}}{2} = 8 \text{ m}$

26. $r = \frac{d}{2} = \frac{\frac{28}{11} \text{ in.}}{2} = \frac{28}{11} \text{ in.} \cdot \frac{1}{2}$

$= \frac{28}{11 \cdot 2} \text{ in.} = \frac{\cancel{2} \cdot 14}{11 \cdot \cancel{2}} \text{ in.}$

$= \frac{14}{11} \text{ in., or } 1\frac{3}{11} \text{ in.}$

27. $d = 2 \cdot r = 2 \cdot 7 \text{ ft} = 14 \text{ ft}$

28. $d = 2 \cdot r = 2 \cdot 10 \text{ cm} = 20 \text{ cm}$

29. $C = \pi \cdot d$

$C \approx 3.14 \cdot 16 \text{ m}$

$= 50.24 \text{ m}$

30. $C = \pi \cdot d$

$C \approx \frac{22}{7} \cdot \frac{28}{11} \text{ in.}$

$= \frac{22 \cdot 28}{7 \cdot 11} \text{ in.} = \frac{2 \cdot \cancel{11} \cdot 4 \cdot \cancel{7}}{\cancel{7} \cdot \cancel{11} \cdot 1} \text{ in.}$

$= 8 \text{ in.}$

31. In Exercise 25 we found that the radius of the circle is 8 m.

$A = \pi \cdot r \cdot r$

$A \approx 3.14 \cdot 8 \text{ m} \cdot 8 \text{ m}$

$A = 200.96 \text{ m}^2$

32. In Exercise 26 we found that the radius of the circle is $\frac{14}{11}$ in.

$A = \pi \cdot r \cdot r$

$A \approx \frac{22}{7} \cdot \frac{14}{11} \text{ in.} \cdot \frac{14}{11} \text{ in.}$

$A = \frac{22 \cdot 14 \cdot 14}{7 \cdot 11 \cdot 11} \text{ in}^2 = \frac{2 \cdot \cancel{11} \cdot 2 \cdot \cancel{7} \cdot 14}{\cancel{7} \cdot \cancel{11} \cdot 11}$

$A = \frac{56}{11} \text{ in}^2, \text{ or } 5\frac{1}{11} \text{ in}^2$

33. The shaded area is the area of a circle with radius of 21 ft less the area of a circle with a diameter of 21 ft. The radius of the smaller circle is $\frac{21 \text{ ft}}{2}$, or 10.5 ft.

$$A = \pi \cdot 21 \text{ ft} \cdot 21 \text{ ft} - \pi \cdot 10.5 \text{ ft} \cdot 10.5 \text{ ft}$$
$$A \approx 3.14 \cdot 21 \text{ ft} \cdot 21 \text{ ft} - 3.14 \cdot 10.5 \text{ ft} \cdot 10.5 \text{ ft}$$
$$A = 1384.74 \text{ ft}^2 - 346.185 \text{ ft}^2$$
$$A = 1038.555 \text{ ft}^2$$

34. The window is composed of half of a circle with radius 2 ft and of a rectangle with length 5 ft and width twice the radius of the half circle, or $2 \cdot 2$ ft, or 4 ft. To find the area of the window we add one-half the area of a circle with radius 2 ft and the area of a rectangle with length 5 ft and width 4 ft.

$$A = \frac{1}{2} \cdot \pi \cdot 2 \text{ ft} \cdot 2 \text{ ft} + 5 \text{ ft} \cdot 4 \text{ ft}$$
$$\approx \frac{1}{2} \cdot 3.14 \cdot 2 \text{ ft} \cdot 2 \text{ ft} + 5 \text{ ft} \cdot 4 \text{ ft}$$
$$= \frac{3.14 \cdot 2 \cdot \not{2}}{\not{2}} \text{ ft}^2 + 20 \text{ ft}^2$$
$$= 6.28 \text{ ft}^2 + 20 \text{ ft}^2$$
$$= 26.28 \text{ ft}^2$$

The perimeter is composed of one-half the circumference of a circle with radius 2 ft along with the lengths of three sides of the rectangle.

$$P = \frac{1}{2} \cdot 2 \cdot \pi \cdot 2 \text{ ft} + 5 \text{ ft} + 4 \text{ ft} + 5 \text{ ft}$$
$$\approx \frac{1}{2} \cdot 2 \cdot 3.14 \cdot 2 \text{ ft} + 5 \text{ ft} + 4 \text{ ft} + 5 \text{ ft}$$
$$= \frac{\not{2} \cdot 3.14 \cdot 2 \text{ ft}}{\not{2}} + 5 \text{ ft} + 4 \text{ ft} + 5 \text{ ft}$$
$$= 6.28 \text{ ft} + 5 \text{ ft} + 4 \text{ ft} + 5 \text{ ft}$$
$$= 20.28 \text{ ft}$$

35.
$$V = l \cdot w \cdot h$$
$$V = 12 \text{ yd} \cdot 3 \text{ yd} \cdot 2.6 \text{ yd}$$
$$V = 36 \cdot 2.6 \text{ yd}^3$$
$$V = 93.6 \text{ yd}^3$$
$$SA = 2lw + 2lh + 2wh$$
$$= 2 \cdot 12 \text{ yd} \cdot 3 \text{ yd} + 2 \cdot 12 \text{ yd} \cdot 2.6 \text{ yd} + 2 \cdot 3 \text{ yd} \cdot 2.6 \text{ yd}$$
$$= 72 \text{ yd}^2 + 62.4 \text{ yd}^2 + 15.6 \text{ yd}^2$$
$$= 150 \text{ yd}^2$$

36.
$$V = l \cdot w \cdot h$$
$$V = 4.6 \text{ cm} \cdot 3 \text{ cm} \cdot 14 \text{ cm}$$
$$V = 13.8 \cdot 14 \text{ cm}^3$$
$$V = 193.2 \text{ cm}^3$$
$$SA = 2lw + 2lh + 2wh$$
$$= 2 \cdot 4.6 \text{ cm} \cdot 3 \text{ cm} + 2 \cdot 4.6 \text{ cm} \cdot 14 \text{ cm} + 2 \cdot 3 \text{ cm} \cdot 14 \text{ cm}$$
$$= 27.6 \text{ cm}^2 + 128.8 \text{ cm}^2 + 84 \text{ cm}^2$$
$$= 240.4 \text{ cm}^2$$

37.
$$r = \frac{20 \text{ ft}}{2} = 10 \text{ ft}$$
$$V = B \cdot h = \pi \cdot r^2 \cdot h$$
$$\approx 3.14 \times 10 \text{ ft} \times 10 \text{ ft} \times 100 \text{ ft}$$
$$= 31,400 \text{ ft}^3$$

38.
$$V = \frac{4}{3} \cdot \pi \cdot r^3$$
$$\approx \frac{4}{3} \times 3.14 \times (2 \text{ cm})^3$$
$$= \frac{4 \times 3.14 \times 8 \text{ cm}^3}{3}$$
$$= 33.49\overline{3} \text{ cm}^3$$

39.
$$V = \frac{1}{3} \cdot \pi \cdot r^2 \cdot h$$
$$\approx \frac{1}{3} \times 3.14 \times 1 \text{ in.} \times 1 \text{ in.} \times 4.5 \text{ in.}$$
$$= 4.71 \text{ in}^3$$

40.
$$V = B \cdot h = \pi \cdot r^2 \cdot h$$
$$\approx 3.14 \times 5 \text{ cm} \times 5 \text{ cm} \times 12 \text{ cm}$$
$$= 942 \text{ cm}^3$$

41. Two angles are complementary if the sum of their measures is 90°.

$$90° - 41° = 49°.$$

The measure of a complement of $\angle BAC$ is 49°.

42. $90° - 82° = 9°$

43. $90° - 5° = 85°$

44. $180° - 33° = 147°$

45. $180° - 133° = 47°$

46. $\angle 1$ and $\angle 4$ are vertical angles, so $m\angle 4 = 38°$. Likewise, $\angle 5$ and $\angle 2$ and vertical angles, so $m\angle 2 = 105°$.

$$m\angle 1 + m\angle 2 + m\angle 3 = 180°$$
$$38° + 105° + m\angle 3 = 180°$$
$$143° + m\angle 3 = 180°$$
$$m\angle 3 = 37°$$

Since $\angle 3$ and $\angle 6$ are vertical angles, $m\angle 6 = 37°$.

47. a) The pairs of corresponding angles are

$\angle 1$ and $\angle 5$,
$\angle 2$ and $\angle 6$,
$\angle 3$ and $\angle 7$,
$\angle 4$ and $\angle 8$.

b) The interior angles are $\angle 4$, $\angle 2$, $\angle 5$, and $\angle 7$.

c) The pairs of alternate interior angles are

$\angle 2$ and $\angle 5$,
$\angle 4$ and $\angle 7$.

48. $\angle 4$ and $\angle 6$ are vertical angles, so $m\angle 6 = 135°$.

$\angle 4$ and $\angle 2$ are corresponding angles. By Property 1, $m\angle 2 = 135°$.

$\angle 6$ and $\angle 8$ are corresponding angles. By Property 1, $m\angle 8 = 135°$.

$\angle 2$ and $\angle 3$ are interior angles on the same side of the transversal. Using Property 4 and $m\angle 2 = 135°$, $m\angle 3 = 45°$.

$\angle 6$ and $\angle 7$ are interior angles on the same side of the transversal. Using Property 4 and $m\angle 6 = 135°$, $m\angle 7 = 45°$.

$\angle 3$ and $\angle 5$ are vertical angles, so $m\angle 5 = 45°$.

$\angle 7$ and $\angle 1$ are vertical angles, so $m\angle 1 = 45°$.

49. The notation tells us the way in which the vertices of the two triangles are matched.

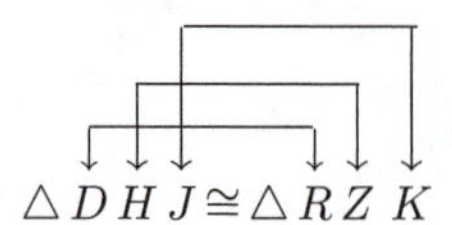

$\triangle DHJ \cong \triangle RZK$ means

$$\begin{array}{lll} \angle D \cong \angle R & \text{and} & \overline{DH} \cong \overline{RZ} \\ \angle H \cong \angle Z & & \overline{DJ} \cong \overline{RK} \\ \angle J \cong \angle K & & \overline{HJ} \cong \overline{ZK}. \end{array}$$

50. $\triangle ABC \cong \triangle GDF$

The notation tells us the way in which the vertices of the two triangles are matched.

$\triangle ABC \cong \triangle GDF$

$$\begin{array}{lll} \angle A \cong \angle G & \text{and} & \overline{AB} \cong \overline{GD} \\ \angle B \cong \angle D & & \overline{AC} \cong \overline{GF} \\ \angle C \cong \angle F & & \overline{BC} \cong \overline{DF}. \end{array}$$

51. Two angles and the included side of one triangle are congruent to two angles and the included side of the other triangle. They are congruent by the ASA Property.

52. Three sides of one triangle are congruent to three sides of the other triangle. They are congruent by the SSS Property.

53. Since we know only that three angles of one triangle are congruent to three angles of the other triangle, none of the properties can be used to show that the triangles are congruent.

54. Since J is the midpoint of $\overline{JK}$, $IJ \cong KJ$.

$\angle HJI \cong \angle LJK$ because they are vertical angles.

$\angle HIJ \cong \angle LKJ$ because they are opposite interior angles.

Two angles and the included side of $\triangle JIH$ are congruent to two angles and the included side of $\triangle JKL$, so $\triangle JIH \cong \triangle JKL$ by the ASA Property.

55. $\angle A$ and $\angle C$ are opposite angles, so $m\angle A = 63°$ by Property 2.

$\angle C$ and $\angle B$ are consecutive angles, so they are supplementary by Property 4. Then

$$\begin{aligned} m\angle B &= 180° - m\angle C \\ m\angle B &= 180° - 63° \\ m\angle B &= 117°. \end{aligned}$$

$\angle B$ and $\angle D$ are opposite angles, so $m\angle D = 117°$ by Property 2.

The opposite sides of a parallelogram are congruent (Property 3), so $CD = 13$ and $BC = 23$.

56. The notation tells us the way in which the vertices are matched.

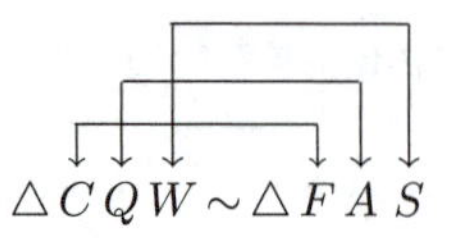

$\triangle CQW \sim \triangle FAS$ means

$$\begin{array}{l} \angle C \cong \angle F \\ \angle Q \cong \angle A \quad \text{and} \quad \dfrac{CQ}{FA} = \dfrac{CW}{FS} = \dfrac{QW}{AS}. \\ \angle W \cong \angle S \end{array}$$

57. Since $\triangle NMO \sim \triangle STR$, the corresponding sides are proportional. Thus

$$\begin{aligned} \frac{10}{15} &= \frac{MO}{21} \\ 10 \cdot 21 &= 15(MO) \\ 210 &= 15(MO) \\ 14 &= MO. \end{aligned}$$

58. $180° - 20\frac{3}{4}° = 179\frac{4}{4}° - 20\frac{3}{4}° = 159\frac{1}{4}°$

Answer B is correct.

59. $r = \dfrac{d}{2} = \dfrac{\frac{7}{9}\text{ in.}}{2} = \dfrac{7}{9}\text{ in.} \times \dfrac{1}{2} = \dfrac{7}{18}\text{ in.}$

$$\begin{aligned} A &= \pi \cdot r \cdot r \\ &\approx \frac{22}{7} \cdot \frac{7}{18}\text{ in.} \cdot \frac{7}{18}\text{ in.} \\ &= \frac{22 \cdot 7 \cdot 7}{7 \cdot 18 \cdot 18} \cdot \text{in.} \cdot \text{in.} \\ &= \frac{77}{162}\text{ in}^2 \end{aligned}$$

Answer B is correct.

60. ***Familiarize***. Let $s =$ the length of a side of the square, in feet. When the square is cut in half the resulting rectangle has length s and width $s/2$.

Translate.

$$\underbrace{\text{Perimeter of rectangle}}_{\downarrow\; 2 \cdot s + 2 \cdot \frac{s}{2}} \quad \underset{=}{\overset{\text{is}}{\downarrow}} \quad \underbrace{30\text{ ft.}}_{\downarrow\; 30}$$

Solve.

$$2 \cdot s + 2 \cdot \frac{s}{2} = 30$$
$$2 \cdot s + s = 30$$
$$3 \cdot s = 30$$
$$s = 10$$

If $s = 10$, then the area of the square is 10 ft · 10 ft, or 100 ft^2.

Check. If $s = 10$, then $s/2 = 10/2 = 5$ and the perimeter of a rectangle with length 10 ft and width 5 ft is $2 \cdot 10$ ft + $2 \cdot 5$ ft = 20 ft + 10 ft = 30 ft. We can also recheck the calculation for the area of the square. The answer checks.

State. The area of the square is 100 ft^2.

61. The area A of the shaded region is the area of a square with sides 2.8 m less the areas of the four small squares cut out at each corner. Each of the small squares has sides of 1.8 mm, or 0.0018 m.

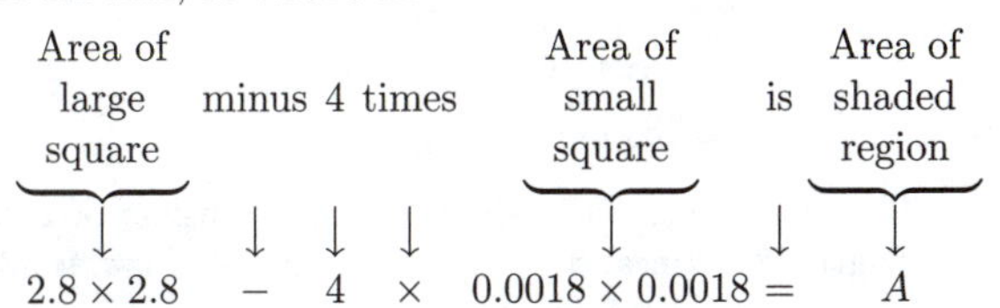

We carry out the calculations.

$7.84 - 4 \times 0.00000324 = 7.84 - 0.00001296 = 7.83998704$

The area of the shaded region is 7.83998704 m^2.

62. The shaded region consists of one large triangle with base 84 mm, or 8.4 cm, and height 100 mm, or 10 cm, and 8 small triangles, each with height 1.25 cm and base 1.05 cm. Let A = the area of the shaded region.

Area of shaded region | is | Area of large triangle | plus | 8 | times | Area of small triangle

$$A \quad = \quad \frac{1}{2} \cdot 8.4 \cdot 10 \quad + \quad 8 \quad \cdot \quad \frac{1}{2} \cdot 1.25 \cdot 1.05$$

We carry out the computations.

$$A = \frac{1}{2} \cdot 8.4 \cdot 10 + 8 \cdot \frac{1}{2} \cdot 1.25 \cdot 1.05$$
$$A = 42 + 5.25$$
$$A = 47.25$$

The area of the shaded region is 47.25 cm^2.

Chapter 8 Discussion and Writing Exercises

1. Add 90° to the measure of the angle's complement.

2. This could be done using the technique in Example 8 in Section 8.5. We could also approximate the volume with the volume of a similarly-shaped rectangular solid. Another method is to break the egg and measure the capacity of its contents.

3. Linear measure is one-dimensional, area is two-dimensional, and volume is three-dimensional.

4. Divide the figure into 3 triangles.

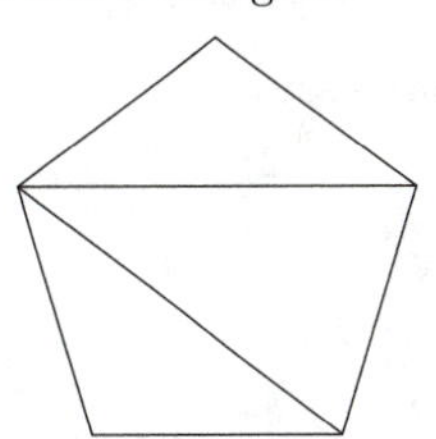

The sum of the measures of the angles of each triangle is 180°, so the sum of the measures of the angles of the figure is $3 \cdot 180°$, or 540°.

5. The volume of the cone is half the volume of the dome. It can be argued that a cone-cap is more energy-efficient since there is less air under it to be heated and cooled.

6. Volume of two spheres, each with radius r: $2\left(\frac{4}{3}\pi r^3\right) = \frac{8}{3}\pi r^3$; volume of one sphere with radius $2r$: $\frac{4}{3}\pi(2r)^3 = \frac{32}{3}\pi r^3$. The volume of the sphere with radius $2r$ is four times the volume of the two spheres, each with radius r: $\frac{32}{3}\pi r^3 = 4 \cdot \frac{8}{3}\pi r^3$.

Chapter 8 Test

1. Using a protractor, we find that the measure of the angle is 90°.

2. Using a protractor, we find that the measure of the angle is 35°.

3. Using a protractor, we find that the measure of the angle is 180°.

4. Using a protractor, we find that the measure of the angle is 113°.

5. The measure of the angle in Exercise 1 is 90°. It is a right angle.

6. The measure of the angle in Exercise 2 is 35°. Since its measure is greater than 0°and less than 90°, it is an acute angle.

7. The measure of the angle in Exercise 3 is 180°. It is a straight angle.

8. The measure of the angle in Exercise 4 is 113°. Since its measure is greater than 90°and less than 180°, it is an obtuse angle.

9.
$$m(\angle A) + m(\angle H) + m(\angle F) = 180°$$
$$35° + 110° + x = 180°$$
$$145° + x = 180°$$
$$x = 180° - 145°$$
$$x = 35°$$

10. From the labels on the triangle, we see that two sides are the same length. By measuring we find that the third side is a different length. Thus, this is an isosceles triangle.

11. One angle is an obtuse angle, so this is an obtuse triangle.

12. A pentagon has 5 sides.

$(n-2)\cdot 180° = (5-2)\cdot 180° = 3\cdot 180° = 540°$

13.
$$\begin{aligned} P &= 2\cdot(l+w) \\ &= 2\cdot(9.4 \text{ cm} + 7.01 \text{ cm}) \\ &= 2\cdot(16.41 \text{ cm}) \\ &= 32.82 \text{ cm} \end{aligned}$$
$$\begin{aligned} A &= l\cdot w \\ &= (9.4 \text{ cm})\cdot(7.01 \text{ cm}) \\ &= 9.4\cdot 7.01\cdot \text{cm}\cdot\text{cm} \\ &= 65.894 \text{ cm}^2 \end{aligned}$$

14.
$$\begin{aligned} P &= 4\cdot s \\ &= 4\cdot 4\tfrac{7}{8} \text{ in.} \\ &= 4\cdot\frac{39}{8} \text{ in.} \\ &= \frac{4\cdot 39}{8} \text{ in.} \\ &= \frac{\cancel{4}\cdot 39}{2\cdot\cancel{4}} \text{ in.} \\ &= \frac{39}{2} \text{ in., or } 19\tfrac{1}{2} \text{ in.} \end{aligned}$$
$$\begin{aligned} A &= s\cdot s \\ &= \left(4\tfrac{7}{8} \text{ in.}\right)\cdot\left(4\tfrac{7}{8} \text{ in.}\right) \\ &= 4\tfrac{7}{8}\cdot 4\tfrac{7}{8}\cdot \text{in.}\cdot\text{in.} \\ &= \frac{39}{8}\cdot\frac{39}{8}\cdot \text{in}^2 \\ &= \frac{1521}{64} \text{ in}^2, \text{ or } 23\tfrac{49}{64} \text{ in}^2 \end{aligned}$$

15.
$$\begin{aligned} A &= b\cdot h \\ &= 10 \text{ cm}\cdot 2.5 \text{ cm} \\ &= 25 \text{ cm}^2 \end{aligned}$$

16.
$$\begin{aligned} A &= \frac{1}{2}\cdot b\cdot h \\ &= \frac{1}{2}\cdot 8 \text{ m}\cdot 3 \text{ m} \\ &= \frac{8\cdot 3}{2} \text{ m}^2 \\ &= 12 \text{ m}^2 \end{aligned}$$

17.
$$\begin{aligned} A &= \frac{1}{2}\cdot h\cdot(a+b) \\ &= \frac{1}{2}\cdot 3 \text{ ft}\cdot(8 \text{ ft} + 4 \text{ ft}) \\ &= \frac{1}{2}\cdot 3 \text{ ft}\cdot 12 \text{ ft} \\ &= \frac{3\cdot 12}{2} \text{ ft}^2 \\ &= 18 \text{ ft}^2 \end{aligned}$$

18. $d = 2\cdot r = 2\cdot\frac{1}{8} \text{ in.} = \frac{1}{4} \text{ in.}$

19. $r = \frac{d}{2} = \frac{18 \text{ cm}}{2} = 9 \text{ cm}$

20.
$$\begin{aligned} C &= 2\cdot\pi\cdot r \\ &\approx 2\cdot\frac{22}{7}\cdot\frac{1}{8} \text{ in.} \\ &= \frac{2\cdot 22\cdot 1}{7\cdot 8} \text{ in.} \\ &= \frac{\cancel{2}\cdot\cancel{2}\cdot 11\cdot 1}{7\cdot\cancel{2}\cdot\cancel{2}\cdot 2} \text{ in.} \\ &= \frac{11}{14} \text{ in.} \end{aligned}$$

21. In Exercise 19 we found that the radius of the circle is 9 cm.
$$\begin{aligned} A &= \pi\cdot r\cdot r \\ &\approx 3.14\cdot 9 \text{ cm}\cdot 9 \text{ cm} \\ &= 3.14\cdot 81 \text{ cm}^2 \\ &= 254.34 \text{ cm}^2 \end{aligned}$$

22. The perimeter of the shaded region consists of 2 sides of length 18.6 km and the circumferences of two semicircles with diameter 9.0 km. Note that the sum of the circumferences of the two semicircles is the same as the circumference of one circle with diameter 9.0 km.

The total length of the 2 sides of length 18.6 km is

$2\cdot 18.6 \text{ km} = 37.2 \text{ km}.$

Next we find the perimeter, or circumference, of the circle.
$$\begin{aligned} C &= \pi\cdot d \\ &\approx 3.14\cdot 9.0 \text{ km} \\ &= 28.26 \text{ km} \end{aligned}$$

Finally we add to find the perimeter of the shaded region.

$37.2 \text{ km} + 28.26 \text{ km} = 65.46 \text{ km}$

The shaded region is the area of a rectangle that is 18.6 km by 9.0 km less the area of two semicircles, each with diameter 9.0 km. Note that the two semicircles have the same area as one circle with diameter 9.0 km.

First we find the area of the rectangle.
$$\begin{aligned} A &= l\cdot w \\ &= 18.6 \text{ km}\cdot 9.0 \text{ km} \\ &= 167.4 \text{ km}^2 \end{aligned}$$

Now find the area of the circle. The radius is $\frac{9.0 \text{ km}}{2}$, or 4.5 km.
$$\begin{aligned} A &= \pi\cdot r\cdot r \\ &\approx 3.14\cdot 4.5 \text{ km}\cdot 4.5 \text{ km} \\ &= 3.14\cdot 20.25 \text{ km}^2 \\ &= 63.585 \text{ km}^2 \end{aligned}$$

Finally, we subtract to find the area of the shaded region.

$167.4 \text{ km}^2 - 63.585 \text{ km}^2 = 103.815 \text{ km}^2$

23. $V = l \cdot w \cdot h$

$= 4 \text{ cm} \cdot 2 \text{ cm} \cdot 10.5 \text{ cm}$

$= 8 \cdot 10.5 \text{ cm}^3$

$= 84 \text{ cm}^3$

$SA = 2lw + 2lh + 2wh$

$= 2 \cdot 4\text{ cm} \cdot 2\text{ cm} + 2 \cdot 4 \text{ cm} \cdot 10.5 \text{ cm} + 2 \cdot 2 \text{ cm} \cdot 10.5 \text{ cm}$

$= 16 \text{ cm}^2 + 84 \text{ cm}^2 + 42 \text{ cm}^2$

$= 142 \text{ cm}^2$

24. $V = l \cdot w \cdot h$

$= 10\frac{1}{2} \text{ in.} \cdot 8 \text{ in.} \cdot 5 \text{ in.}$

$= \frac{21}{2} \text{ in.} \cdot 8 \text{ in.} \cdot 5 \text{ in.}$

$= \frac{21 \cdot 8 \cdot 5}{2} \text{ in}^3$

$= \frac{21 \cdot \cancel{2} \cdot 4 \cdot 5}{\cancel{2} \cdot 1} \text{ in}^3$

$= 420 \text{ in}^3$

25. $V = \pi \cdot r^2 \cdot h$

$\approx 3.14 \times 5 \text{ ft} \times 5 \text{ ft} \times 15 \text{ ft}$

$= 1177.5 \text{ ft}^3$

26. $r = \frac{d}{2} = \frac{20 \text{ yd}}{2} = 10 \text{ yd}$

$V = \frac{4}{3} \cdot \pi \cdot r^3$

$\approx \frac{4}{3} \times 3.14 \times (10 \text{ yd})^3$

$= 4186.\overline{6} \text{ yd}^3$

27. $V = \frac{1}{3}\pi \cdot r^2 \cdot h$

$\approx \frac{1}{3} \times 3.14 \times 3 \text{ cm} \times 3 \text{ cm} \times 12 \text{ cm}$

$= 113.04 \text{ cm}^3$

28. $\angle CAD = 65°$

$90° - 65° = 25°$, so the measure of a complement is 25°.

$180 - 65° = 115°$, so the measure of a supplement is 115°.

29. $\angle 1$ and $\angle 4$ are vertical angles, so $m\angle 4 = 62°$. Likewise, $\angle 5$ and $\angle 2$ are vertical angles, so $m\angle 2 = 110°$.

$m\angle 1 + m\angle 2 + m\angle 3 = 180°$

$62° + 110° + m\angle 3 = 180°$

$172° + m\angle 3 = 180°$

$m\angle 3 = 8°$

Since $\angle 3$ and $\angle 6$ are vertical angles, $m\angle 6 = 8°$.

30. $\angle 4$ and $\angle 6$ are vertical angles, so $m\angle 6 = 120°$.

$\angle 4$ and $\angle 2$ are corresponding angles. By Property 1, $m\angle 2 = 120°$.

$\angle 6$ and $\angle 8$ are corresponding angles. By Property 1, $m\angle 8 = 120°$.

$\angle 2$ and $\angle 3$ are interior angles on the same side of the transversal. Using Property 4 and $m\angle 2 = 120°$, $m\angle 3 = 60°$.

$\angle 6$ and $\angle 7$ are interior angles on the same side of the transversal. Using Property 4 and $m\angle 6 = 120°$, $m\angle 7 = 60°$.

$\angle 3$ and $\angle 5$ are vertical angles, so $m\angle 5 = 60°$.

$\angle 7$ and $\angle 1$ are vertical angles, so $m\angle 1 = 60°$.

31. The notation tells us the way in which the vertices of the two triangles are matched.

$\triangle CWS \cong \triangle ATZ$

$\triangle CWS \sim \triangle ATZ$ means

$\angle C \cong \angle A$ and $\overline{CW} \cong \overline{AT}$
$\angle W \cong \angle T$ $\overline{CS} \cong \overline{AZ}$
$\angle S \cong \angle Z$ $\overline{WS} \cong \overline{TZ}$.

32. Two sides of one triangle and the included angle are congruent to two sides and the included angle of the other triangle. They are congruent by the SAS Property.

33. Since we know only that three angles of one triangle are congruent to three angles of the other triangle, none of the properties can be used to show that the triangles are congruent.

34. Two angles and the included side of one triangle are congruent to two angles and the included side of the other triangle. They are congruent by the ASA Property.

35. We know only that two sides of one triangle as well as an angle that is not the included angle are congruent to the corresponding sides and angle of the other triangle. Thus, none of the properties can be used to show that the triangles are congruent.

36. $\angle E$ and $\angle G$ are opposite angles, so $m\angle G = 105°$ by Property 2.

$\angle D$ and $\angle E$ are consecutive angles, so they are supplementary by Property 4. Then

$m\angle D = 180° - m\angle E$

$m\angle D = 180° - 105°$

$m\angle D = 75°$

$\angle D$ and $\angle F$ are opposite angles, so $m\angle F = 75°$ by Property 2.

The opposite sides of a parallelogram are congruent (Property 3), so $EF = 11$. Similarly, $DE = GF$. Let $l =$ the length of DE. Then $l =$ the length of GF also. The perimeter of the parallelogram is 62, so we have:

$11 + l + 11 + l = 62$

$22 + 2 \cdot l = 62$

$2 \cdot l = 40$

$l = 20$

Then $DE = GF = 20$.

37. The diagonals of a parallelogram bisect each other (Property 5). Then

$$LJ = 2 \cdot JN = 2 \cdot 3.2 = 6.4,$$
$$KM = 2 \cdot KN = 2 \cdot 3 = 6.$$

38. The notation tells us the way in which the vertices are matched.

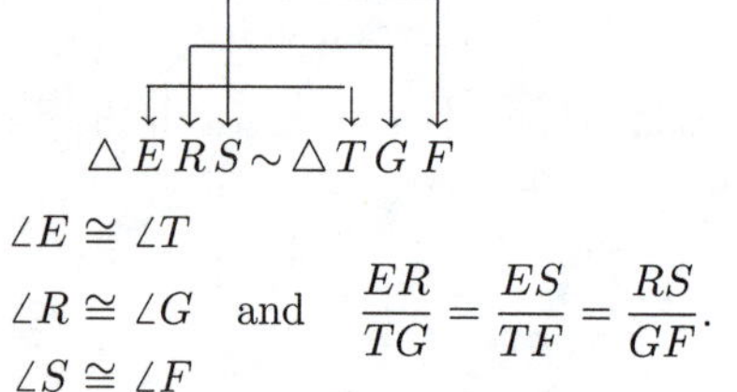

$\angle E \cong \angle T$

$\angle R \cong \angle G$ and $\dfrac{ER}{TG} = \dfrac{ES}{TF} = \dfrac{RS}{GF}$.

$\angle S \cong \angle F$

39. The corresponding sides of the triangle are proportional.

$$\frac{24}{8} = \frac{EK}{6} \qquad \frac{24}{8} = \frac{ZK}{9}$$
$$24 \cdot 6 = 8(EK) \qquad 24 \cdot 9 = 8(ZK)$$
$$144 = 8(EK) \qquad 216 = 8(ZK)$$
$$18 = EK \qquad 27 = ZK$$

40. $r = \dfrac{d}{2} = \dfrac{42 \text{ cm}}{2} = 21 \text{ cm}$

$$V = \frac{4}{3} \cdot \pi \cdot r^3$$
$$\approx \frac{4}{3} \cdot \frac{22}{7} \cdot (21 \text{ cm})^3$$
$$= 38,808 \text{ cm}^3$$

Answer D is correct.

41. First we convert 3 in. to feet.

$$3 \text{ in.} = 3 \text{ in.} \cdot \frac{1 \text{ ft}}{12 \text{ in.}}$$
$$= \frac{3}{12} \cdot \frac{\text{in.}}{\text{in.}} \cdot 1 \text{ ft}$$
$$= \frac{1}{4} \text{ ft}$$

Now we find the area of the rectangle.

$$A = l \cdot w$$
$$= 8 \text{ ft} \cdot \frac{1}{4} \text{ ft}$$
$$= \frac{8}{4} \text{ ft}^2$$
$$= 2 \text{ ft}^2$$

42. We convert both units of measure to feet. From Exercise 41 we know that 3 in. $= \dfrac{1}{4}$ ft. We also have

$$5 \text{ yd} = 5 \times 1 \text{ yd}$$
$$= 5 \times 3 \text{ ft}$$
$$= 15 \text{ ft}.$$

Now we find the area.

$$A = \frac{1}{2} \cdot b \cdot h$$
$$= \frac{1}{2} \cdot 15 \text{ ft} \cdot \frac{1}{4} \text{ ft}$$
$$= \frac{15}{2 \cdot 4} \text{ ft}^2$$
$$= \frac{15}{8} \text{ ft}^2, \text{ or } 1.875 \text{ ft}^2$$

43. First we convert 2.6 in. and 3 in. to feet.

$$2.6 \text{ in.} = 2.6 \text{ in.} \cdot \frac{1 \text{ ft}}{12 \text{ in.}}$$
$$= \frac{2.6}{12} \cdot \frac{\text{in.}}{\text{in.}} \cdot 1 \text{ ft}$$
$$= \frac{2.6}{12} \text{ ft}$$

From Exercise 41 we know that 3 in. $= \dfrac{1}{4}$ ft. Now we find the volume.

$$V = l \cdot w \cdot h$$
$$= 12 \text{ ft} \cdot \frac{1}{4} \text{ ft} \cdot \frac{2.6}{12} \text{ ft}$$
$$= \frac{12 \cdot 2.6}{4 \cdot 12} \text{ ft}^3$$
$$= 0.65 \text{ ft}^3$$

44. First we convert 1 in. to feet.

$$1 \text{ in.} = 1 \text{ in.} \cdot \frac{1 \text{ ft}}{12 \text{ in.}}$$
$$= \frac{1}{12} \cdot \frac{\text{in.}}{\text{in.}} \cdot 1 \text{ ft}$$
$$= \frac{1}{12} \text{ ft}$$

Now we find the volume.

$$V = \frac{1}{3}\pi \cdot r^2 \cdot h$$
$$\approx \frac{1}{3} \cdot 3.14 \cdot \frac{1}{12} \text{ ft} \cdot \frac{1}{12} \text{ ft} \cdot 4.5 \text{ ft}$$
$$= \frac{3.14 \cdot 4.5}{3 \cdot 12 \cdot 12} \text{ ft}^3$$
$$\approx 0.033 \text{ ft}^3$$

45. First we find the radius of the cylinder.

$$r = \frac{d}{2} = \frac{\frac{3}{4} \text{ in.}}{2} = \frac{3}{4} \text{ in.} \cdot \frac{1}{2} = \frac{3}{8} \text{ in.}$$

Now we convert $\dfrac{3}{8}$ in. to feet.

$$\frac{3}{8} \text{ in.} = \frac{3}{8} \text{ in.} \cdot \frac{1 \text{ ft}}{12 \text{ in.}}$$
$$= \frac{3}{8 \cdot 12} \cdot \frac{\text{in.}}{\text{in.}} \cdot 1 \text{ ft}$$
$$= \frac{1}{32} \text{ ft}$$

Finally, we find the volume.

$$\begin{aligned} V &= B \cdot h = \pi \cdot r^2 \cdot h \\ &\approx 3.14 \cdot \frac{1}{32} \text{ ft} \cdot \frac{1}{32} \text{ ft} \cdot 18 \text{ ft} \\ &= \frac{3.14 \cdot 18}{32 \cdot 32} \text{ ft}^3 \\ &\approx 0.055 \text{ ft}^3 \end{aligned}$$

Chapter 9

Introduction to Real Numbers and Algebraic Expressions

Exercise Set 9.1

1. Substitute 56 for x: $56 - 24 = 32$, so it takes Erin 32 min to get to work if it takes George 56 min.

Substitute 93 for x: $93 - 24 = 69$, so it takes Erin 69 min to get to work if it takes George 93 min.

Substitute 105 for x: $105 - 24 = 81$, so it takes Erin 81 min to get to work if it takes George 105 min.

3. Substitute 45 m for b and 86 m for h, and carry out the multiplication:

$$\begin{aligned} A = \frac{1}{2}bh &= \frac{1}{2}(45\text{ m})(86\text{ m}) \\ &= \frac{1}{2}(45)(86)(\text{m})(\text{m}) \\ &= 1935\text{ m}^2 \end{aligned}$$

5. Substitute 65 for r and 4 for t, and carry out the multiplication:

$$d = rt = 65 \cdot 4 = 260\text{ mi}$$

7. We substitute 6 ft for l and 4 ft for w in the formula for the area of a rectangle.

$$\begin{aligned} A = lw &= (6\text{ ft})(4\text{ ft}) \\ &= (6)(4)(\text{ft})(\text{ft}) \\ &= 24\text{ ft}^2 \end{aligned}$$

9. $8x = 8 \cdot 7 = 56$

11. $\frac{a}{b} = \frac{24}{3} = 8$

13. $\frac{3p}{q} = \frac{3 \cdot 2}{6} = \frac{6}{6} = 1$

15. $\frac{x+y}{5} = \frac{10+20}{5} = \frac{30}{5} = 6$

17. $\frac{x-y}{8} = \frac{20-4}{8} = \frac{16}{8} = 2$

19. $b + 7$, or $7 + b$

21. $c - 12$

23. $4 + q$, or $q + 4$

25. $a + b$, or $b + a$

27. $x \div y$, or $\frac{x}{y}$, or x/y, or $x \cdot \frac{1}{y}$

29. $x + w$, or $w + x$

31. $n - m$

33. $x + y$, or $y + x$

35. $2z$

37. $3m$

39. $4a + 6$, or $6 + 4a$

41. $xy - 8$

43. $2t - 5$

45. $3n + 11$, or $11 + 3n$

47. $4x + 3y$, or $3y + 4x$

49. Let s represent your salary. Then we have $89\%s$, or $0.89s$.

51. A 5% increase in s is represented by $0.05s$, so we have $s + 0.05s$.

53. The distance traveled is the product of the speed and the time. Thus, Danielle traveled $65t$ miles.

55. $\$50 - x$

57. $\$8.50n$

59. We use a factor tree.

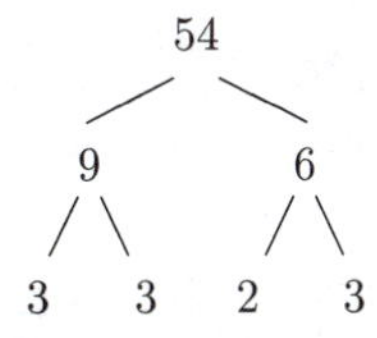

The prime factorization is $2 \cdot 3 \cdot 3 \cdot 3$.

61. We use the list of primes. The first prime that is a factor of 108 is 2.

$$108 = 2 \cdot 54$$

We keep dividing by 2 until it is no longer possible to do so.

$$108 = 2 \cdot 2 \cdot 27$$

Now we do the same thing for the next prime, 3.

$$108 = 2 \cdot 2 \cdot 3 \cdot 3 \cdot 3$$

This is the prime factorization of 108.

63. We use the list of primes. The first prime number that is a factor of 1023 is 3.

$$1023 = 3 \cdot 341$$

We continue through the list of prime numbers until we have

$1023 = 3 \cdot 11 \cdot 31.$

Since 3, 11, and 31 are prime numbers, the prime factorization of 1023 is $3 \cdot 11 \cdot 31$.

65. $6 = 2 \cdot 3$

$24 = 2 \cdot 2 \cdot 2 \cdot 3$

$32 = 2 \cdot 2 \cdot 2 \cdot 2 \cdot 2$

The LCM is $2 \cdot 2 \cdot 2 \cdot 2 \cdot 2 \cdot 3$, or 96.

67. $16 = 2 \cdot 2 \cdot 2 \cdot 2$

$24 = 2 \cdot 2 \cdot 2 \cdot 3$

$32 = 2 \cdot 2 \cdot 2 \cdot 2 \cdot 2$

The LCM is $2 \cdot 2 \cdot 2 \cdot 2 \cdot 2 \cdot 3$, or 96.

69. $$\frac{a-2b+c}{4b-a} = \frac{20-2\cdot 10+5}{4\cdot 10-20} = \frac{20-20+5}{40-20} = \frac{0+5}{20} = \frac{5}{20} = \frac{\cancel{5}\cdot 1}{\cancel{5}\cdot 4} = \frac{1}{4}$$

71. $$\frac{12-c}{c+12b} = \frac{12-12}{12+12\cdot 1} = \frac{0}{12+12} = \frac{0}{24} = 0$$

Exercise Set 9.2

1. The integer -282 corresponds to 282 ft below sea level.

3. The integer 24 corresponds to 24° above zero; the integer -2 corresponds to 2° below zero.

5. The integer 3,600,000,000 corresponds to a temperature of 3,600,000,000°F; the integer -460 corresponds to a temperature of 460°F below zero.

7. The integer -34 describes the situation from the Alley Cats' point of view. The integer 34 describes the situation from the Strikers' point of view.

9. The number $\frac{10}{3}$ can be named $3\frac{1}{3}$, or $3.\overline{3}$. The graph is $\frac{1}{3}$ of the way from 3 to 4.

$\frac{10}{3}$

−6 −5 −4 −3 −2 −1 0 1 2 3 4 5 6

11. The graph of -5.2 is $\frac{2}{10}$ of the way from -5 to -6.

−5.2

−6 −5 −4 −3 −2 −1 0 1 2 3 4 5 6

13. The graph of $-4\frac{2}{5}$ is $\frac{2}{5}$ of the way from -4 to -5.

$-4\frac{2}{5}$

−6 −5 −4 −3 −2 −1 0 1 2 3 4 5 6

15. We first find decimal notation for $\frac{7}{8}$. Since $\frac{7}{8}$ means $7 \div 8$, we divide.

$$\begin{array}{r} 0.875 \\ 8\overline{)7.000} \\ \underline{64} \\ 60 \\ \underline{56} \\ 40 \\ \underline{40} \\ 0 \end{array}$$

Thus $\frac{7}{8} = 0.875$, so $-\frac{7}{8} = -0.875$.

17. $\frac{5}{6}$ means $5 \div 6$, so we divide.

$$\begin{array}{r} 0.833\ldots \\ 6\overline{)5.000} \\ \underline{48} \\ 20 \\ \underline{18} \\ 20 \\ \underline{18} \\ 2 \end{array}$$

We have $\frac{5}{6} = 0.8\overline{3}$.

19. First we find decimal notation for $\frac{7}{6}$. Since $\frac{7}{6}$ means $7 \div 6$, we divide.

$$\begin{array}{r} 1.166\ldots \\ 6\overline{)7.000} \\ \underline{6} \\ 10 \\ \underline{6} \\ 40 \\ \underline{36} \\ 40 \\ \underline{36} \\ 4 \end{array}$$

Thus $\frac{7}{6} = 1.1\overline{6}$, so $-\frac{7}{6} = -1.1\overline{6}$.

21. $\frac{2}{3}$ means $2 \div 3$, so we divide.

$$\begin{array}{r} 0.666\ldots \\ 3\overline{)2.000} \\ \underline{18} \\ 20 \\ \underline{18} \\ 20 \\ \underline{18} \\ 2 \end{array}$$

We have $\frac{2}{3} = 0.\overline{6}$.

23. $\frac{1}{10}$ means $1 \div 10$, so we divide.

$$\begin{array}{r} 0.1 \\ 10\overline{)1.0} \\ \underline{10} \\ 0 \end{array}$$

We have $\frac{1}{10} = 0.1$

25. We first find decimal notation for $\frac{1}{2}$. Since $\frac{1}{2}$ means $1 \div 2$, we divide.

$$\begin{array}{r} 0.5 \\ 2\overline{)1.0} \\ \underline{1\,0} \\ 0 \end{array}$$

Thus $\frac{1}{2} = 0.5$, so $-\frac{1}{2} = -0.5$

27. $\frac{4}{25}$ means $4 \div 25$, so we divide.

$$\begin{array}{r} 0.1\,6 \\ 2\,5\overline{)4.0\,0} \\ \underline{2\,5} \\ 1\,5\,0 \\ \underline{1\,5\,0} \\ 0 \end{array}$$

We have $\frac{4}{25} = 0.16$.

29. Since 8 is to the right of 0, we have $8 > 0$.

31. Since -8 is to the left of 3, we have $-8 < 3$.

33. Since -8 is to the left of 8, we have $-8 < 8$.

35. Since -8 is to the left of -5, we have $-8 < -5$.

37. Since -5 is to the right of -11, we have $-5 > -11$.

39. Since -6 is to the left of -5, we have $-6 < -5$.

41. Since 2.14 is to the right of 1.24, we have $2.14 > 1.24$.

43. Since -14.5 is to the left of 0.011, we have $-14.5 < 0.011$.

45. Since -12.88 is to the left of -6.45, we have $-12.88 < -6.45$.

47. $-\frac{1}{2} = -\frac{1}{2} \cdot \frac{3}{3} = -\frac{3}{6}$

$-\frac{2}{3} = -\frac{2}{3} \cdot \frac{2}{2} = -\frac{4}{6}$

Since $-\frac{3}{6}$ is to the right of $-\frac{4}{6}$, then $-\frac{1}{2}$ is to the right of $-\frac{2}{3}$, and we have $-\frac{1}{2} > -\frac{2}{3}$.

49. Since $-\frac{2}{3}$ is to the left of $\frac{1}{3}$, we have $-\frac{2}{3} < \frac{1}{3}$.

51. Convert to decimal notation $\frac{5}{12} = 0.4166\ldots$ and $\frac{11}{25} = 0.44$. Since $0.4166\ldots$ is to the left of 0.44, $\frac{5}{12} < \frac{11}{25}$.

53. $x < -6$ has the same meaning as $-6 > x$.

55. $y \geq -10$ has the same meaning as $-10 \leq y$.

57. $-5 \leq -6$ is false since neither $-5 < -6$ nor $-5 = -6$ is true.

59. $4 \geq 4$ is true since $4 = 4$ is true.

61. $-3 \geq -11$ is true since $-3 > -11$ is true.

63. $0 \geq 8$ is false since neither $0 > 8$ nor $0 = 8$ is true.

65. The distance of -3 from 0 is 3, so $|-3| = 3$.

67. The distance of 10 from 0 is 10, so $|10| = 10$.

69. The distance of 0 from 0 is 0, so $|0| = 0$.

71. The distance of -30.4 from 0 is 30.4, so $|-30.4| = 30.4$.

73. The distance of $-\frac{2}{3}$ from 0 is $\frac{2}{3}$, so $\left|-\frac{2}{3}\right| = \frac{2}{3}$.

75. The distance of $\frac{0}{4}$ from 0 is $\frac{0}{4}$, or 0, so $\left|\frac{0}{4}\right| = 0$.

77. The distance of -2.65 from 0 is 2.65, so $|-2.65| = 2.65$.

79. The distance of $-7\frac{4}{5}$ from 0 is $7\frac{4}{5}$, so $\left|-7\frac{4}{5}\right| = 7\frac{4}{5}$.

81. 63% 0.63.

Move the decimal point 2 places to the left.

63% = 0.63

83. 110% 1.10.

Move the decimal point 2 places to the left.

110% = 1.1

85. $\frac{13}{25} = 0.52 = 52\%$

87. From Exercise 17 we know that $\frac{5}{6} = 0.8\overline{3}$, or $0.83\overline{3}$, so $\frac{5}{6} = 83.\overline{3}\%$, or $83\frac{1}{3}\%$.

89. $\frac{2}{3} = 0.\overline{6}$, $-\frac{1}{7} = -0.\overline{142857}$, $\frac{1}{3} = 0.\overline{3}$,

$-\frac{2}{7} = -0.\overline{285714}$, $-\frac{2}{3} = -0.\overline{6}$, $\frac{2}{5} = 0.4$,

$-\frac{1}{3} = -0.\overline{3}$, $-\frac{2}{5} = -0.4$, $\frac{9}{8} = 1.125$

Listing from least to greatest, we have

$-\frac{2}{3}, -\frac{2}{5}, -\frac{1}{3}, -\frac{2}{7}, -\frac{1}{7}, \frac{1}{3}, \frac{2}{5}, \frac{2}{3}, \frac{9}{8}$.

91. Note that $7^1 = 7$; $|-6| = 6$; $|3| = 3$; $1^7 = 1$; $\frac{14}{4} = \frac{7}{2}$, or $3\frac{1}{2}$; and $\frac{-67}{8} = -8\frac{3}{8}$.

Listing from least to greatest, we have

$-100, -8\frac{7}{8}, -8\frac{5}{8}, -\frac{67}{8}, -5, 0, 1^7, |3|, \frac{14}{4}, 4, |-6|, 7^1$.

93. $0.\overline{9} = 3(0.\overline{3}) = 3 \cdot \frac{1}{3} = \frac{1}{1}$

Exercise Set 9.3

1. $2+(-9)$ The absolute values are 2 and 9. The difference is $9-2$, or 7. The negative number has the larger absolute value, so the answer is negative. $2+(-9)=-7$

3. $-11+5$ The absolute values are 11 and 5. The difference is $11-5$, or 6. The negative number has the larger absolute value, so the answer is negative. $-11+5=-6$

5. $-8+8$ A negative and a positive number. The numbers have the same absolute value. The sum is 0. $-8+8=0$

7. $-3+(-5)$ Two negatives. Add the absolute values, getting 8. Make the answer negative. $-3+(-5)=-8$

9. $-7+0$ One number is 0. The answer is the other number. $-7+0=-7$

11. $0+(-27)$ One number is 0. The answer is the other number. $0+(-27)=-27$

13. $17+(-17)$ A negative and a positive number. The numbers have the same absolute value. The sum is 0. $17+(-17)=0$

15. $-17+(-25)$ Two negatives. Add the absolute values, getting 42. Make the answer negative. $-17+(-25)=-42$

17. $18+(-18)$ A positive and a negative number. The numbers have the same absolute value. The sum is 0. $18+(-18)=0$

19. $-28+28$ A negative and a positive number. The numbers have the same absolute value. The sum is 0. $-28+28=0$

21. $8+(-5)$ The absolute values are 8 and 5. The difference is $8-5$, or 3. The positive number has the larger absolute value, so the answer is positive. $8+(-5)=3$

23. $-4+(-5)$ Two negatives. Add the absolute values, getting 9. Make the answer negative. $-4+(-5)=-9$

25. $13+(-6)$ The absolute values are 13 and 6. The difference is $13-6$, or 7. The positive number has the larger absolute value, so the answer is positive. $13+(-6)=7$

27. $-25+25$ A negative and a positive number. The numbers have the same absolute value. The sum is 0. $-25+25=0$

29. $53+(-18)$ The absolute values are 53 and 18. The difference is $53-18$, or 35. The positive number has the larger absolute value, so the answer is positive. $53+(-18)=35$

31. $-8.5+4.7$ The absolute values are 8.5 and 4.7. The difference is $8.5-4.7$, or 3.8. The negative number has the larger absolute value, so the answer is negative. $-8.5+4.7=-3.8$

33. $-2.8+(-5.3)$ Two negatives. Add the absolute values, getting 8.1. Make the answer negative. $-2.8+(-5.3)=-8.1$

35. $-\frac{3}{5}+\frac{2}{5}$ The absolute values are $\frac{3}{5}$ and $\frac{2}{5}$. The difference is $\frac{3}{5}-\frac{2}{5}$, or $\frac{1}{5}$. The negative number has the larger absolute value, so the answer is negative. $-\frac{3}{5}+\frac{2}{5}=-\frac{1}{5}$

37. $-\frac{2}{9}+\left(-\frac{5}{9}\right)$ Two negatives. Add the absolute values, getting $\frac{7}{9}$. Make the answer negative. $-\frac{2}{9}+\left(-\frac{5}{9}\right)=-\frac{7}{9}$

39. $-\frac{5}{8}+\frac{1}{4}$ The absolute values are $\frac{5}{8}$ and $\frac{1}{4}$. The difference is $\frac{5}{8}-\frac{2}{8}$, or $\frac{3}{8}$. The negative number has the larger absolute value, so the answer is negative. $-\frac{5}{8}+\frac{1}{4}=-\frac{3}{8}$

41. $-\frac{5}{8}+\left(-\frac{1}{6}\right)$ Two negatives. Add the absolute values, getting $\frac{15}{24}+\frac{4}{24}$, or $\frac{19}{24}$. Make the answer negative.

$-\frac{5}{8}+\left(-\frac{1}{6}\right)=-\frac{19}{24}$

43. $-\frac{3}{8}+\frac{5}{12}$ The absolute values are $\frac{3}{8}$ and $\frac{5}{12}$. The difference is $\frac{10}{24}-\frac{9}{24}$, or $\frac{1}{24}$. The positive number has the larger absolute value, so the answer is positive. $-\frac{3}{8}+\frac{5}{12}=\frac{1}{24}$

45. $-\frac{1}{6}+\frac{7}{10}$ The absolute values are $\frac{1}{6}$ and $\frac{7}{10}$. The difference is $\frac{21}{30}-\frac{5}{30}=\frac{16}{30}=\frac{\cancel{2}\cdot 8}{\cancel{2}\cdot 15}=\frac{8}{15}$. The positive number has the larger absolute value, so the answer is positive. $-\frac{1}{6}+\frac{7}{10}=\frac{8}{15}$

47. $\frac{7}{15}+\left(-\frac{1}{9}\right)$ The absolute values are $\frac{7}{15}$ and $\frac{1}{9}$. The difference is $\frac{21}{45}-\frac{5}{45}=\frac{16}{45}$. The positive number has the larger absolute value, so the answer is positive. $\frac{7}{15}+\left(-\frac{1}{9}\right)=\frac{16}{45}$

49. $76+(-15)+(-18)+(-6)$

a) Add the negative numbers: $-15+(-18)+(-6)=-39$

b) Add the results: $76+(-39)=37$

51. $-44+\left(-\frac{3}{8}\right)+95+\left(-\frac{5}{8}\right)$

a) Add the negative numbers: $-44+\left(-\frac{3}{8}\right)+\left(-\frac{5}{8}\right)=-45$

b) Add the results: $-45+95=50$

53. We add from left to right.

$$\begin{aligned} & 98+(-54)+113+(-998)+44+(-612) \\ =\ & 44+113+(-998)+44+(-612) \\ =\ & 157+(-998)+44+(-612) \\ =\ & -841+44+(-612) \\ =\ & -797+(-612) \\ =\ & -1409 \end{aligned}$$

55. The additive inverse of 24 is -24 because $24 + (-24) = 0$.

57. The additive inverse of -26.9 is 26.9 because $-26.9 + 26.9 = 0$.

59. If $x = 8$, then $-x = -8$. (The opposite of 8 is -8.)

61. If $x = -\frac{13}{8}$ then $-x = -\left(-\frac{13}{8}\right) = \frac{13}{8}$. (The opposite of $-\frac{13}{8}$ is $\frac{13}{8}$.)

63. If $x = -43$ then $-(-x) = -(-(-43)) = -43$. (The opposite of the opposite of -43 is -43.)

65. If $x = \frac{4}{3}$ then $-(-x) = -\left(-\frac{4}{3}\right) = \frac{4}{3}$. (The opposite of the opposite of $\frac{4}{3}$ is $\frac{4}{3}$.)

67. $-(-24) = 24$ (The opposite of -24 is 24.)

69. $-\left(-\frac{3}{8}\right) = \frac{3}{8}$ (The opposite of $-\frac{3}{8}$ is $\frac{3}{8}$.)

71. Let E = the elevation of Mauna Kea above sea level.

$$\underbrace{\text{Elevation above sea level}}_{E} \ \underbrace{\text{is}}_{=} \ \underbrace{\text{total height}}_{33,480} \ \underbrace{\text{plus}}_{+} \ \underbrace{\text{elevation below sea level.}}_{(-19,684)}$$

We carry out the addition.

$$E = 33,480 + (-19,684) = 13,796$$

The elevation of Mauna Kea is 13,796 ft above sea level.

73. Let T = the final temperature. We will express the rise in temperature as a positive number and a decrease in the temperature as a negative number.

$$\underbrace{\text{Final temperature}}_{T} \ \underbrace{\text{is}}_{=} \ \underbrace{\text{original temperature}}_{32} \ \underbrace{\text{plus}}_{+}$$

$$\underbrace{\text{rise in temperature}}_{15} \ \underbrace{\text{plus}}_{+} \ \underbrace{\text{decrease in temperature.}}_{(-50)}$$

We add from left to right.

$$\begin{aligned} T &= 32 + 15 + (-50) \\ &= 47 + (-50) \\ &= -3 \end{aligned}$$

The final temperature was $-3°$F.

75. Let S = the sum of the profits and losses. We add the five numbers in the bar graph to find S.

$S = \$10,500 + (-\$16,600) + (-\$12,800) + (-\$9600) + \$8200 = -\$20,300$

The sum of the profits and losses is $-\$20,300$.

77. Let B = the new balance in the account at the end of August. We will express the payments as positive numbers and the original balance and the amount of the new charge as negative numbers.

$$\underbrace{\text{New balance}}_{B} \ \underbrace{\text{is}}_{=} \ \underbrace{\text{original balance}}_{-470} \ \underbrace{\text{plus}}_{+} \ \underbrace{\text{amount of first payment}}_{45} \ \underbrace{\text{plus}}_{+}$$

$$\underbrace{\text{amount of new charge}}_{(-160)} \ \underbrace{\text{plus}}_{+} \ \underbrace{\text{amount of second payment.}}_{500}$$

We add from left to right.

$$\begin{aligned} B &= -470 + 45 + (-160) + 500 \\ &= -425 + (-160) + 500 \\ &= -585 + 500 \\ &= -85 \end{aligned}$$

The balance in the account is $-\$85$. Lyle owes \$85.

79. 71.3% 0.71.3

Move the decimal point two places to the left.

71.3% = 0.713

81. $\frac{1}{8} = 0.125 = 12.5\%$

83. $\frac{2}{3} \div \frac{5}{12} = \frac{2}{3} \cdot \frac{12}{5} = \frac{2 \cdot 12}{3 \cdot 5} = \frac{2 \cdot \cancel{3} \cdot 4}{\cancel{3} \cdot 5} = \frac{8}{5}$

85. When x is positive, the opposite of x, $-x$, is negative, so $-x$ is negative for all positive numbers x.

87. If a is positive, $-a$ is negative. Thus $-a + b$, the sum of two negative numbers, is negative. The correct answer is B.

Exercise Set 9.4

1. $2 - 9 = 2 + (-9) = -7$

3. $-8 - (-2) = -8 + 2 = -6$

5. $-11 - (-11) = -11 + 11 = 0$

7. $12 - 16 = 12 + (-16) = -4$

9. $20 - 27 = 20 + (-27) = -7$

11. $-9 - (-3) = -9 + 3 = -6$

13. $-40 - (-40) = -40 + 40 = 0$

15. $7 - (-7) = 7 + 7 = 14$

17. $8 - (-3) = 8 + 3 = 11$

19. $-6 - 8 = -6 + (-8) = -14$

21. $-4 - (-9) = -4 + 9 = 5$

23. $-6-(-5)=-6+5=-1$

25. $8-(-10)=8+10=18$

27. $-5-(-2)=-5+2=-3$

29. $-7-14=-7+(-14)=-21$

31. $0-(-5)=0+5=5$

33. $-8-0=-8+0=-8$

35. $7-(-5)=7+5=12$

37. $2-25=2+(-25)=-23$

39. $-42-26=-42+(-26)=-68$

41. $-71-2=-71+(-2)=-73$

43. $24-(-92)=24+92=116$

45. $-50-(-50)=-50+50=0$

47. $-\frac{3}{8}-\frac{5}{8}=-\frac{3}{8}+\left(-\frac{5}{8}\right)=-\frac{8}{8}=-1$

49. $\frac{3}{4}-\frac{2}{3}=\frac{3}{4}+\left(-\frac{2}{3}\right)=\frac{9}{12}+\left(-\frac{8}{12}\right)=\frac{1}{12}$

51. $-\frac{3}{4}-\frac{2}{3}=-\frac{3}{4}+\left(-\frac{2}{3}\right)=-\frac{9}{12}+\left(-\frac{8}{12}\right)=-\frac{17}{12}$

53. $-\frac{5}{8}-\left(-\frac{3}{4}\right)=-\frac{5}{8}+\frac{3}{4}=-\frac{5}{8}+\frac{6}{8}=\frac{1}{8}$

55. $6.1-(-13.8)=6.1+13.8=19.9$

57. $-2.7-5.9=-2.7+(-5.9)=-8.6$

59. $0.99-1=0.99+(-1)=-0.01$

61. $-79-114=-79+(-114)=-193$

63. $0-(-500)=0+500=500$

65. $-2.8-0=-2.8+0=-2.8$

67. $7-10.53=7+(-10.53)=-3.53$

69. $\frac{1}{6}-\frac{2}{3}=\frac{1}{6}+\left(-\frac{2}{3}\right)=\frac{1}{6}+\left(-\frac{4}{6}\right)=-\frac{3}{6}$, or $-\frac{1}{2}$

71. $-\frac{4}{7}-\left(-\frac{10}{7}\right)=-\frac{4}{7}+\frac{10}{7}=\frac{6}{7}$

73. $-\frac{7}{10}-\frac{10}{15}=-\frac{7}{10}+\left(-\frac{10}{15}\right)=-\frac{21}{30}+\left(-\frac{20}{30}\right)=-\frac{41}{30}$

75. $\frac{1}{5}-\frac{1}{3}=\frac{1}{5}+\left(-\frac{1}{3}\right)=\frac{3}{15}+\left(-\frac{5}{15}\right)=-\frac{2}{15}$

77. $\frac{5}{12}-\frac{7}{16}=\frac{5}{12}+\left(-\frac{7}{16}\right)=\frac{20}{48}+\left(-\frac{21}{48}\right)=-\frac{1}{48}$

79. $-\frac{2}{15}-\frac{7}{12}=-\frac{2}{15}+\left(-\frac{7}{12}\right)=-\frac{8}{60}+\left(-\frac{35}{60}\right)=-\frac{43}{60}$

81. $18-(-15)-3-(-5)+2=18+15+(-3)+5+2=37$

83. $-31+(-28)-(-14)-17=(-31)+(-28)+14+(-17)=-62$

85. $-34-28+(-33)-44=(-34)+(-28)+(-33)+(-44)=-139$

87. $-93-(-84)-41-(-56)=(-93)+84+(-41)+56=6$

89. $-5.4-(-30.9)+30.8+40.2-(-12)=$
$-5.4+30.9+30.8+40.2+12=108.5$

91. $-\frac{7}{12}+\frac{3}{4}-\left(-\frac{5}{8}\right)-\frac{13}{24}=-\frac{7}{12}+\frac{3}{4}+\frac{5}{8}+\left(-\frac{13}{24}\right)=$
$-\frac{28}{48}+\frac{36}{48}+\frac{30}{48}+\left(-\frac{26}{48}\right)=\frac{12}{48}=\frac{\cancel{12}\cdot 1}{4\cdot\cancel{12}}=\frac{1}{4}$

93. Let $D=$ the difference in elevation.

Difference in elevation	is	larger depth	minus	smaller depth.
↓	↓	↓	↓	↓
D	$=$	$10,924$	$-$	8605

We carry out the subtraction.

$D=10,924-8605=2319$

The difference in elevation is 2319 m.

95. Let $A=$ the amount owed.

Amount owed	is	amount of charge	minus	amount of return.
↓	↓	↓	↓	↓
A	$=$	476.89	$-$	128.95

We subtract.

$A=476.89-128.95=347.94$

Claire owes $347.94.

97. Let $D=$ the difference in the elevations.

Difference in elevations	is	higher elevation	minus	lower elevation.
↓	↓	↓	↓	↓
D	$=$	5672	$-$	(-4)

We carry out the subtraction.

$D=5672-(-4)=5672+4=5676$

The difference in the elevations is 5676 ft.

99. Let $D=$ the difference in the elevations.

Difference in elevations	is	higher elevation	minus	lower elevation.
↓	↓	↓	↓	↓
D	$=$	-131	$-$	(-512)

We carry out the subtraction.

$D=-131-(-512)=-131+512=381$

Lake Assal is 381 ft lower than the Valdes Peninsula.

101. Let T = the difference in the temperatures.

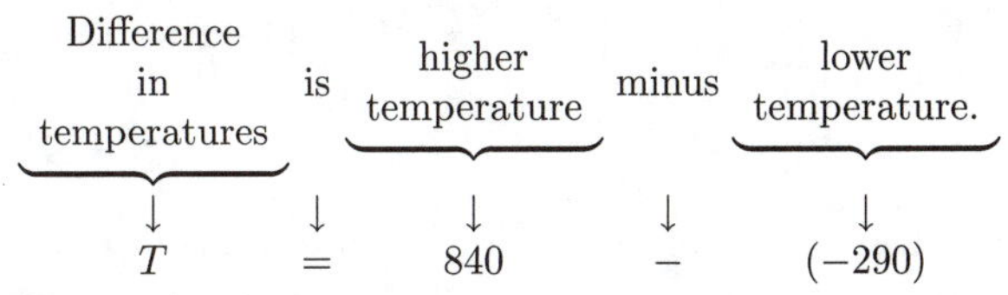

We carry out the subtraction.

$T = 840 - (-290) = 840 + 290 = 1130$

The difference between the temperatures is 1130°F.

103. $256 \div 64 \div 2^3 + 100 = 256 \div 64 \div 8 + 100$
$= 4 \div 8 + 100$
$= \frac{1}{2} + 100$
$= 100\frac{1}{2}$, or 100.5

105. $2^5 \div 4 + 20 \div 2^2 = 32 \div 4 + 20 \div 4 = 8 + 5 = 13$

107. $\frac{1}{8} + \frac{7}{12} + \frac{5}{24} = \frac{3}{24} + \frac{14}{24} + \frac{5}{24} = \frac{22}{24} = \frac{\not{2} \cdot 11}{\not{2} \cdot 12} = \frac{11}{12}$

109. False. $3 - 0 = 3, 0 - 3 = -3, 3 - 0 \neq 0 - 3$

111. True

113. True by definition of opposites.

Chapter 9 Mid-Chapter Review

1. The statement is true. See page 623 in the text.

2. The statement is false. If $a > b$, then a lies to the *right* of b on the number line.

3. The statement is true. See page 628 in the text.

4. The statement is false. We translate 7 less than y as $y - 7$. The expression $7 - y$ represents y less than 7.

5. $-x = -(-4) = 4;$
   $-(-x) = -(-(-4)) = -(4) = -4$

6. $5 - 13 = 5 + (-13) = -8$

7. $-6 - 7 = -6 + (-7) = -13$

8. $\frac{3m}{n} = \frac{3 \cdot 8}{6} = \frac{3 \cdot 2 \cdot 4}{3 \cdot 2 \cdot 1} = \frac{3 \cdot 2}{3 \cdot 2} \cdot \frac{4}{1} = 4$

9. $\frac{a+b}{2} = \frac{5+17}{2} = \frac{22}{2} = 11$

10. Let y represent the number. We have $3y$.

11. Let n represent the number. We have $n - 5$.

12. The integer 450 corresponds to a \$450 deposit. The integer −79 corresponds to writing a check for \$79.

13. −3.5 is halfway between −4 and −3 on the number line.

−3.5

−6 −5 −4 −3 −2 −1 0 1 2 3 4 5 6

14. We first find decimal notation for $\frac{4}{5}$. Since $\frac{4}{5}$ means $4 \div 5$, we divide.

$$\begin{array}{r} 0.8 \\ 5\overline{)4.0} \\ \underline{4\,0} \\ 0 \end{array}$$

Thus $\frac{4}{5} = 0.8$, so $-\frac{4}{5} = -0.8$.

15. $\frac{7}{3}$ means $7 \div 3$, so we divide.

$$\begin{array}{r} 2.3\,3 \\ 3\overline{)7.0\,0} \\ \underline{6} \\ 1\,0 \\ \underline{9} \\ 1\,0 \\ \underline{9} \\ 1 \end{array}$$

We have $\frac{7}{3} = 2.\overline{3}$.

16. Since −5 is to the left of −3, we have $-5 < -3$.

17. Since −9.9 is to the right of −10.1, we have $-9.9 > -10.1$.

18. $-8 \geq -5$ is false because neither $-8 > -5$ nor $-8 = -5$ is true.

19. $-4 \leq -4$ is true because $-4 = -4$ is true.

20. $5 > y$ has the same meaning as $y < 5$.

21. $t \leq -3$ has the same meaning as $-3 \geq t$.

22. The distance of 15.6 from 0 is 15.6, so $|15.6| = 15.6$.

23. The distance of −18 from 0 is 18, so $|-18| = 18$.

24. The distance of 0 from 0 is 0, so $|0| = 0$.

25. The distance of $-\frac{12}{5}$ from 0 is $\frac{12}{5}$, so $\left|-\frac{12}{5}\right| = \frac{12}{5}$.

26. The opposite of −5.6 is 5.6 because $-5.6 + 5.6 = 0$.

27. The opposite of $\frac{7}{4}$ is $-\frac{7}{4}$ because $\frac{7}{4} + \left(-\frac{7}{4}\right) = 0$.

28. The opposite of 0 is 0 because $0 + 0 = 0$.

29. The opposite of −49 is 49 because $-49 + 49 = 0$.

30. $-x = -(-19) = 19$

31. $-(-x) = -(-2.3) = 2.3$

32. $7 + (-9)$ The absolute values are 7 and 9. The difference is $9 - 7$, or 2. The negative number has the larger absolute value, so the answer is negative. $7 + (-9) = -2$

33. $-\frac{3}{8} + \frac{1}{4}$ The absolute values are $\frac{3}{8}$ and $\frac{1}{4}$. The difference is $\frac{3}{8} - \frac{1}{4}$, or $\frac{3}{8} - \frac{2}{8}$, or $\frac{1}{8}$. The negative number has the larger absolute value, so the answer is negative.
$-\frac{3}{8} + \frac{1}{4} = -\frac{1}{8}$

34. $3.6 + (-3.6)$ A negative and a positive number. The numbers have the same absolute value. The sum is 0. $3.6 + (-3.6) = 0$

35. $-8+(-9)$ Two negative numbers. Add the absolute values, getting 17. Make the answer negative.

$-8 + (-9) = -17$

36. $\frac{2}{3} + \left(-\frac{9}{8}\right)$ The absolute values are $\frac{2}{3}$ and $\frac{9}{8}$. The difference is $\frac{9}{8} - \frac{2}{3}$, or $\frac{27}{24} - \frac{16}{24}$, or $\frac{11}{24}$. The negative number has the larger absolute value, so the answer is negative.

$\frac{2}{3} + \left(-\frac{9}{8}\right) = -\frac{11}{24}$

37. $-4.2 + (-3.9)$ Two negative numbers. Add the absolute values, getting 8.1. Make the answer negative.

$-4.2 + (-3.9) = -8.1$

38. $-14+5$ The absolute values are 14 and 5. The difference is $14-5$, or 9. The negative number has the larger absolute value, so the answer is negative. $-14 + 5 = -9$

39. $19+(-21)$ The absolute values are 19 and 21. The difference is $21 - 19$, or 2. The negative number has the larger absolute value, so the answer is negative.

$19 + (-21) = -2$

40. $-4.1 - 6.3 = -4.1 + (-6.3) = -10.4$

41. $5 - (-11) = 5 + 11 = 16$

42. $-\frac{1}{4} - \left(-\frac{3}{5}\right) = -\frac{1}{4} + \frac{3}{5} = -\frac{5}{20} + \frac{12}{20} = \frac{7}{20}$

43. $12 - 24 = 12 + (-24) = -12$

44. $-8 - (-4) = -8 + 4 = -4$

45. $-\frac{1}{2} - \frac{5}{6} = -\frac{1}{2} + \left(-\frac{5}{6}\right) = -\frac{3}{6} + \left(-\frac{5}{6}\right) = -\frac{8}{6} = -\frac{4}{3}$

46. $12.3 - 14.1 = 12.3 + (-14.1) = -1.8$

47. $6 - (-7) = 6 + 7 = 13$

48.
$$\begin{aligned} 16 - (-9) - 20 - (-4) &= 16 + 9 + (-20) + 4 \\ &= 25 + (-20) + 4 \\ &= 5 + 4 \\ &= 9 \end{aligned}$$

49.
$$\begin{aligned} -4 + (-10) - (-3) - 12 &= -4 + (-10) + 3 + (-12) \\ &= -14 + 3 + (-12) \\ &= -11 + (-12) \\ &= -23 \end{aligned}$$

50.
$$\begin{aligned} 17 - (-25) + 15 - (-18) &= 17 + 25 + 15 + 18 \\ &= 42 + 15 + 18 \\ &= 57 + 18 \\ &= 75 \end{aligned}$$

51.
$$\begin{aligned} -9 + (-3) + 16 - (-10) &= -9 + (-3) + 16 + 10 \\ &= -12 + 16 + 10 \\ &= 4 + 10 \\ &= 14 \end{aligned}$$

52. Let T = the difference in the temperatures.

Difference in temperature	is	higher temperature	minus	lower temperature.
↓	↓	↓	↓	↓
T	$=$	25	$-$	(-8)

We carry out the subtraction.

$T = 25 - (-8) = 25 + 8 = 33$

The difference in temperatures is 33°C.

53. Let P = the sum of the gains and losses.

$P = 56.12 + (-1.18) + 1.22 + (-1.36) = 54.80$

The value of the stock at the end of the day was $54.80.

54. Answers may vary. Three examples are $\frac{6}{13}$, -23.8, and $\frac{43}{5}$. These are rational numbers because they can be named in the form $\frac{a}{b}$, where a and b are integers and b is not 0. They are not integers, however, because they are not whole numbers or the opposites of whole numbers.

55. Answers may vary. Three examples are π, $-\sqrt{7}$, and 0.31311311131111.... Rational numbers can be named as described in Exercise 54 above. Real numbers that are not rational are irrational. Decimal notation for rational numbers either terminates or repeats. Decimal notation for irrational numbers neither terminates nor repeats.

56. Answers may vary. Think of adding two negative numbers on the number line. We start to the left of 0 and move to the left, so the result is also to the left of 0, or negative.

57. Yes; consider $m - (-n)$ where both m and n are positive. (Note that, if n is positive, $-n$ is negative.) Then

$m - (-n) = m + n$. Now $m + n$, the sum of two positive numbers, is positive.

Exercise Set 9.5

1. -8

3. -48

5. -24

7. -72

9. 16

11. 42

13. -120

15. -238

17. 1200

19. 98

21. -72

23. -12.4

25. 30

27. 21.7

29. $\frac{2}{3}\cdot\left(-\frac{3}{5}\right)=-\left(\frac{2\cdot 3}{3\cdot 5}\right)=-\left(\frac{2}{5}\cdot\frac{3}{3}\right)=-\frac{2}{5}$

31. $-\frac{3}{8}\cdot\left(-\frac{2}{9}\right)=\frac{3\cdot 2}{8\cdot 9}=\frac{3\cdot 2\cdot 1}{4\cdot 2\cdot 3\cdot 3}=\frac{3\cdot 2}{3\cdot 2}\cdot\frac{1}{4\cdot 3}=\frac{1}{12}$

33. -17.01

35. $-\frac{5}{9}\cdot\frac{3}{4}=-\left(\frac{5\cdot 3}{9\cdot 4}\right)=-\frac{5\cdot 3}{3\cdot 3\cdot 4}=-\frac{5}{3\cdot 4}\cdot\frac{3}{3}=-\frac{5}{12}$

37. $7\cdot(-4)\cdot(-3)\cdot 5=7\cdot 12\cdot 5=7\cdot 60=420$

39. $-\frac{2}{3}\cdot\frac{1}{2}\cdot\left(-\frac{6}{7}\right)=-\frac{2}{6}\cdot\left(-\frac{6}{7}\right)=\frac{2\cdot 6}{7\cdot 6}=\frac{2}{7}\cdot\frac{6}{6}=\frac{2}{7}$

41. $-3\cdot(-4)\cdot(-5)=12\cdot(-5)=-60$

43. $-2\cdot(-5)\cdot(-3)\cdot(-5)=10\cdot 15=150$

45. $-\frac{2}{45}$

47. $-7\cdot(-21)\cdot 13=147\cdot 13=1911$

49. $-4\cdot(-1.8)\cdot 7=(7.2)\cdot 7=50.4$

51. $-\frac{1}{9}\cdot\left(-\frac{2}{3}\right)\cdot\left(\frac{5}{7}\right)=\frac{2}{27}\cdot\frac{5}{7}=\frac{10}{189}$

53. $4\cdot(-4)\cdot(-5)\cdot(-12)=-16\cdot(60)=-960$

55. $0.07\cdot(-7)\cdot 6\cdot(-6)=0.07\cdot 6\cdot(-7)\cdot(-6)=0.42\cdot(42)=17.64$

57. $\left(-\frac{5}{6}\right)\left(\frac{1}{8}\right)\left(-\frac{3}{7}\right)\left(-\frac{1}{7}\right)=\left(-\frac{5}{48}\right)\left(\frac{3}{49}\right)=-\frac{5\cdot 3}{16\cdot 3\cdot 49}=$
$-\frac{5}{16\cdot 49}\cdot\frac{3}{3}=-\frac{5}{784}$

59. 0, The product of 0 and any real number is 0.

61. $(-8)(-9)(-10)=72(-10)=-720$

63. $(-6)(-7)(-8)(-9)(-10)=42\cdot 72\cdot(-10)=3024\cdot(-10)=$
$-30,240$

65.
$$\begin{aligned}&(-1)^{12}\\&=(-1)(-1)(-1)(-1)(-1)(-1)(-1)(-1)(-1)(-1)(-1)(-1)\\&=1\cdot 1\cdot 1\cdot 1\cdot 1\cdot 1=1\end{aligned}$$

67. For $x=4$:
$$(-x)^2=(-4)^2=16$$
$$-x^2=-(4)^2=-(16)=-16$$
For $x=-4$:
$$(-x)^2=[-(-4)]^2=[4]^2=16$$
$$-x^2=-(-4)^2=-(16)=-16$$

69.
$$\begin{aligned}(-3x)^2&=(-3\cdot 7)^2 &&\text{Substituting}\\&=(-21)^2 &&\text{Multiplying inside the parentheses}\\&=(-21)(-21) &&\text{Evaluating the power}\\&=441\end{aligned}$$
$$\begin{aligned}-3x^2&=-3(7)^2 &&\text{Substituting}\\&=-3\cdot 49 &&\text{Evaluating the power}\\&=-147\end{aligned}$$

71. When $x=2$:
$$\begin{aligned}5x^2&=5(2)^2 &&\text{Substituting}\\&=5\cdot 4 &&\text{Evaluating the power}\\&=20\end{aligned}$$
When $x=-2$:
$$\begin{aligned}5x^2&=5(-2)^2 &&\text{Substituting}\\&=5\cdot 4 &&\text{Evaluating the power}\\&=20\end{aligned}$$

73. When $x=1$, $-2x^3=-2\cdot 1^3=-2\cdot 1=-2$.
When $x=-1$, $-2x^3=-2(-1)^3=-2(-1)=2$.

75. Let w = the total weight change. Since Dave's weight decreases 2 lb each week for 10 weeks we have
$$w=10\cdot(-2)=-20.$$
Thus, the total weight change is -20 lb.

77. This is a multistep problem. First we find the number of degrees the temperature dropped. Since it dropped 3°C each minute for 18 minutes we have a drop d given by
$$d=18\cdot(-3)=-54.$$
Now let T = the temperature at 10:18 AM.
$$T=0+(-54)=-54$$
The temperature was -54°C at 10:18 AM.

79. This is a multistep problem. First we find the total decrease in price. Since it decreased \$1.38 each hour for 8 hours we have a decrease in price d given by
$$d=8(-\$1.38)=-\$11.04.$$
Now let P = the price of the stock after 8 hours.
$$P=\$23.75+(-\$11.04)=\$12.71$$
After 8 hours the price of the stock was \$12.71.

81. This is a multistep problem. First we find the total distance the diver rises. Since the diver rises 7 meters each minute for 9 minutes, the total distance d the diver rises is given by
$$d=9\cdot 7=63.$$
Now let E = the diver's elevation after 9 minutes.
$$E=-95+63=-32$$
The diver's elevation is -32 m, or 32 m below the surface.

83. This is a multistep problem. First we find the total drop in temperature. Since it dropped 6°F per hour for 4 hr, the total drop in temperature d is given by
$$d=4(-6°\text{F})=-24°\text{F}.$$
Now let T = the temperature at the end of the 4-hr period.
$$T=62°\text{F}+(-24°\text{F})=38°\text{F}$$
At the end of the 4-hr period the temperature was 38°F.

85. $36 = 2 \cdot 2 \cdot 3 \cdot 3$
$60 = 2 \cdot 2 \cdot 3 \cdot 5$
$\text{LCM} = 2 \cdot 2 \cdot 3 \cdot 3 \cdot 5$, or 180

87. $\frac{26}{39} = \frac{2 \cdot \cancel{13}}{3 \cdot \cancel{13}} = \frac{2}{3}$

89. $\frac{264}{484} = \frac{\cancel{2} \cdot \cancel{2} \cdot 2 \cdot 3 \cdot \cancel{11}}{\cancel{2} \cdot \cancel{2} \cdot \cancel{11} \cdot 11} = \frac{6}{11}$

91. $\frac{275}{800} = \frac{\cancel{25} \cdot 11}{\cancel{25} \cdot 32} = \frac{11}{32}$

93. $\frac{11}{264} = \frac{\cancel{11} \cdot 1}{\cancel{11} \cdot 24} = \frac{1}{24}$

95. If a is positive and b is negative, then ab is negative and thus $-ab$ is positive. The correct answer is A.

97. To locate $2x$, start at 0 and measure off two adjacent lengths of x to the right of 0.

To locate $3x$, start at 0 and measure off three adjacent lengths of x to the right of 0.

To locate $2y$, start at 0 and measure off two adjacent lengths of y to the right of 0.

To locate $-x$, start at 0 and measure off the length x to the left of 0.

To locate $-y$, start at 0 and measure off the length y to the left of 0.

To locate $x + y$, start at 0 and measure off the length x to the right of 0 followed by the length y immediately to the right of x. (We could also measure off y followed by x.)

To locate $x - y$, start at 0 and measure off the length x to the right of 0. Then, from that point, measure off the length y going to the left.

To locate $x - 2y$, first locate $x - y$ as described above. Then, from that point, measure off another length y going to the left.

The graph is shown in the answer section in the text.

Exercise Set 9.6

1. $48 \div (-6) = -8$ Check: $-8(-6) = 48$

3. $\frac{28}{-2} = -14$ Check: $-14(-2) = 28$

5. $\frac{-24}{8} = -3$ Check: $-3 \cdot 8 = -24$

7. $\frac{-36}{-12} = 3$ Check: $3(-12) = -36$

9. $\frac{-72}{9} = -8$ Check: $-8 \cdot 9 = -72$

11. $-100 \div (-50) = 2$ Check: $2(-50) = -100$

13. $-108 \div 9 = -12$ Check: $9(-12) = -108$

15. $\frac{200}{-25} = -8$ Check: $-8(-25) = 200$

17. Not defined

19. $\frac{0}{-2.6} = 0$ Check: $0(-2.6) = 0$

21. The reciprocal of $\frac{15}{7}$ is $\frac{7}{15}$ because $\frac{15}{7} \cdot \frac{7}{15} = 1$.

23. The reciprocal of $-\frac{47}{13}$ is $-\frac{13}{47}$ because $\left(-\frac{47}{13}\right) \cdot \left(-\frac{13}{47}\right) = 1$.

25. The reciprocal of 13 is $\frac{1}{13}$ because $13 \cdot \frac{1}{13} = 1$.

27. The reciprocal of -32 is $-\frac{1}{32}$ because $-32 \cdot \left(-\frac{1}{32}\right) = 1$.

29. The reciprocal of $-\frac{1}{7.1}$ is -7.1 because $\left(-\frac{1}{7.1}\right)(-7.1) = 1$.

31. The reciprocal of $\frac{1}{9}$ is $\frac{9}{1}$, or 9, because $\frac{1}{9} \cdot 9 = 1$.

33. The reciprocal of $\frac{1}{4y}$ is $4y$ because $\frac{1}{4y} \cdot 4y = 1$.

35. The reciprocal of $\frac{2a}{3b}$ is $\frac{3b}{2a}$ because $\frac{2a}{3b} \cdot \frac{3b}{2a} = 1$.

37. $4 \cdot \frac{1}{17}$

39. $8 \cdot \left(-\frac{1}{13}\right)$

41. $13.9 \cdot \left(-\frac{1}{1.5}\right)$

43. $\frac{2}{3} \cdot \left(-\frac{5}{4}\right)$

45. $x \cdot y$

47. $(3x + 4)\left(\frac{1}{5}\right)$

49. $\frac{3}{4} \div \left(-\frac{2}{3}\right) = \frac{3}{4} \cdot \left(-\frac{3}{2}\right) = -\frac{9}{8}$

51. $-\frac{5}{4} \div \left(-\frac{3}{4}\right) = -\frac{5}{4} \cdot \left(-\frac{4}{3}\right) = \frac{20}{12} = \frac{5 \cdot 4}{3 \cdot 4} = \frac{5}{3}$

53. $-\frac{2}{7} \div \left(-\frac{4}{9}\right) = -\frac{2}{7} \cdot \left(-\frac{9}{4}\right) = \frac{18}{28} = \frac{9 \cdot 2}{14 \cdot 2} = \frac{9}{14}$

55. $-\frac{3}{8} \div \left(-\frac{8}{3}\right) = -\frac{3}{8} \cdot \left(-\frac{3}{8}\right) = \frac{9}{64}$

57. $-\frac{5}{6} \div \frac{2}{3} = -\frac{5}{6} \cdot \frac{3}{2} = -\frac{15}{12} = -\frac{3 \cdot 5}{3 \cdot 4} = -\frac{5}{4}$

59. $-\frac{9}{4} \div \frac{5}{12} = -\frac{9}{4} \cdot \frac{12}{5} = -\frac{108}{20} = -\frac{4 \cdot 27}{4 \cdot 5} = -\frac{27}{5}$

61. $\frac{-11}{-13} = \frac{11}{13}$ The opposite of a number divided by the opposite of another number is the quotient of the two numbers.

63. $-6.6 \div 3.3 = -2$ Do the long division. Make the answer negative.

65. $\frac{48.6}{-3} = -16.2$ Do the long division. Make the answer negative.

67. $-\dfrac{12.5}{5} = -2.5$ Do the long division. Make the answer negative.

69. $11.25 \div (-9) = -1.25$ Do the long division. Make the answer negative.

71. $\dfrac{-9}{17-17} = \dfrac{-9}{0}$ Division by 0 is not defined.

73. $\dfrac{4}{17} \approx 0.235 = 23.5\%$

75. $\dfrac{-116}{3527} \approx -0.033 = 3.3\%$

77. $2^3 - 5 \cdot 3 + 8 \cdot 10 \div 2$

$= 8 - 5 \cdot 3 + 8 \cdot 10 \div 2$ Evaluating the power
$= 8 - 15 + 80 \div 2$ Multiplying and dividing
$= 8 - 15 + 40$ in order from left to right
$= -7 + 40$ Adding and subtracting
$= 33$ in order from left to right

79. $1000 \div 100 \div 10$

$= 10 \div 10$ Dividing in order from
$= 1$ left to right

81. $\dfrac{264}{468} = \dfrac{4 \cdot 66}{4 \cdot 117} = \dfrac{\not{4} \cdot \not{3} \cdot 22}{\not{4} \cdot \not{3} \cdot 39} = \dfrac{22}{39}$

83. $\dfrac{7}{8} = 0.875 = 87.5\%$

85. $\dfrac{12}{25} \div \dfrac{32}{75} = \dfrac{12}{25} \cdot \dfrac{75}{32}$

$= \dfrac{12 \cdot 75}{25 \cdot 32}$

$= \dfrac{3 \cdot \not{4} \cdot 3 \cdot \not{25}}{\not{25} \cdot \not{4} \cdot 8}$

$= \dfrac{9}{8}$

87. The reciprocal of -10.5 is $\dfrac{1}{-10.5}$.

The reciprocal of $\dfrac{1}{-10.5} = -10.5$.

We see that the reciprocal of the reciprocal is the original number.

89. $-a$ is positive and b is negative, so $\dfrac{-a}{b}$ is the quotient of a positive and a negative number and, thus, is negative.

91. a is negative and $-b$ is positive, so $\dfrac{a}{-b}$ is the quotient of a negative number and a positive number and, thus, is negative. Then $-\left(\dfrac{a}{-b}\right)$ is the opposite of a negative number and, thus, is positive.

93. $-a$ and $-b$ are both positive, so $\dfrac{-a}{-b}$ is the quotient of two positive numbers and, thus, is positive. Then $-\left(\dfrac{-a}{-b}\right)$ is the opposite of a positive number and, thus, is negative.

Exercise Set 9.7

1. Note that $5y = 5 \cdot y$. We multiply by 1, using y/y as an equivalent expression for 1:

$$\frac{3}{5} = \frac{3}{5} \cdot 1 = \frac{3}{5} \cdot \frac{y}{y} = \frac{3y}{5y}$$

3. Note that $15x = 3 \cdot 5x$. We multiply by 1, using $5x/5x$ as an equivalent expression for 1:

$$\frac{2}{3} = \frac{2}{3} \cdot 1 = \frac{2}{3} \cdot \frac{5x}{5x} = \frac{10x}{15x}$$

5. Note that $x^2 = x \cdot x$. We multiply by 1, using x/x as an equivalent expression for 1.

$$\frac{2}{x} = \frac{2}{x} \cdot 1 = \frac{2}{x} \cdot \frac{x}{x} = \frac{2x}{x^2}$$

7. $-\dfrac{24a}{16a} = -\dfrac{3 \cdot 8a}{2 \cdot 8a}$

$= -\dfrac{3}{2} \cdot \dfrac{8a}{8a}$

$= -\dfrac{3}{2} \cdot 1$ $\left(\dfrac{8a}{8a} = 1\right)$

$= -\dfrac{3}{2}$ Identity property of 1

9. $-\dfrac{42ab}{36ab} = -\dfrac{7 \cdot 6ab}{6 \cdot 6ab}$

$= -\dfrac{7}{6} \cdot \dfrac{6ab}{6ab}$

$= -\dfrac{7}{6} \cdot 1$ $\left(\dfrac{6ab}{6ab} = 1\right)$

$= -\dfrac{7}{6}$ Identity property of 1

11. $\dfrac{20st}{15t} = \dfrac{4s \cdot 5t}{3 \cdot 5t}$

$= \dfrac{4s}{3} \cdot \dfrac{5t}{5t}$

$= \dfrac{4s}{3} \cdot 1$ $\left(\dfrac{5t}{5t} = 1\right)$

$= \dfrac{4s}{3}$ Identity property of 1

13. $8 + y$, commutative law of addition

15. nm, commutative law of multiplication

17. $xy + 9$, commutative law of addition
$9 + yx$, commutative law of multiplication

19. $c + ab$, commutative law of addition
$ba + c$, commutative law of multiplication

21. $(a + b) + 2$, associative law of addition

23. $8(xy)$, associative law of multiplication

25. $a + (b + 3)$, associative law of addition

27. $(3a)b$, associative law of multiplication

29. a) $(a+b)+2=a+(b+2)$, associative law of addition
b) $(a+b)+2=(b+a)+2$, commutative law of addition
c) $(a+b)+2=(b+a)+2$ Using the commutative law first,
$=b+(a+2)$ then the associative law
There are other correct answers.

31. a) $5+(v+w)=(5+v)+w$, associative law of addition
b) $5+(v+w)=5+(w+v)$, commutative law of addition
c) $5+(v+w)=5+(w+v)$ Using the commutative law first,
$=(5+w)+v$ then the associative law
There are other correct answers.

33. a) $(xy)3=x(y3)$, associative law of multiplication
b) $(xy)3=(yx)3$, commutative law of multiplication
c) $(xy)3=(yx)3$ Using the commutative law first,
$=y(x3)$ then the associative law
There are other correct answers.

35. a) $7(ab)=(7a)b$
b) $7(ab)=(7a)b=b(7a)$
c) $7(ab)=7(ba)=(7b)a$
There are other correct answers.

37. $2(b+5)=2\cdot b+2\cdot 5=2b+10$

39. $7(1+t)=7\cdot 1+7\cdot t=7+7t$

41. $6(5x+2)=6\cdot 5x+6\cdot 2=30x+12$

43. $7(x+4+6y)=7\cdot x+7\cdot 4+7\cdot 6y=7x+28+42y$

45. $7(x-3)=7\cdot x-7\cdot 3=7x-21$

47. $-3(x-7)=-3\cdot x-(-3)\cdot 7=-3x-(-21)=-3x+21$

49. $\frac{2}{3}(b-6)=\frac{2}{3}\cdot b-\frac{2}{3}\cdot 6=\frac{2}{3}b-4$

51. $7.3(x-2)=7.3\cdot x-7.3\cdot 2=7.3x-14.6$

53. $-\frac{3}{5}(x-y+10)=-\frac{3}{5}\cdot x-\left(-\frac{3}{5}\right)\cdot y+\left(-\frac{3}{5}\right)\cdot 10=$
$-\frac{3}{5}x-\left(-\frac{3}{5}y\right)+(-6)=-\frac{3}{5}x+\frac{3}{5}y-6$

55. $-9(-5x-6y+8)=-9(-5x)-(-9)6y+(-9)8$
$=45x-(-54y)+(-72)=45x+54y-72$

57. $-4(x-3y-2z)=-4\cdot x-(-4)3y-(-4)2z$
$=-4x-(-12y)-(-8z)=-4x+12y+8z$

59. $3.1(-1.2x+3.2y-1.1)=3.1(-1.2x)+(3.1)3.2y-3.1(1.1)$
$=-3.72x+9.92y-3.41$

61. $4x+3z$ Parts are separated by plus signs. The terms are $4x$ and $3z$.

63. $7x+8y-9z=7x+8y+(-9z)$ Separating parts with plus signs
The terms are $7x$, $8y$, and $-9z$.

65. $2x+4=2\cdot x+2\cdot 2=2(x+2)$

67. $30+5y=5\cdot 6+5\cdot y=5(6+y)$

69. $14x+21y=7\cdot 2x+7\cdot 3y=7(2x+3y)$

71. $14t-7=7\cdot 2t-7\cdot 1=7(2t-1)$

73. $8x-24=8\cdot x-8\cdot 3=8(x-3)$

75. $18a-24b=6\cdot 3a-6\cdot 4b=6(3a-4b)$

77. $-4y+32=-4\cdot y-4(-8)=-4(y-8)$
We could also factor this expression as follows:
$-4y+32=4(-y)+4\cdot 8=4(-y+8)$

79. $5x+10+15y=5\cdot x+5\cdot 2+5\cdot 3y=5(x+2+3y)$

81. $16m-32n+8=8\cdot 2m-8\cdot 4n+8\cdot 1=8(2m-4n+1)$

83. $12a+4b-24=4\cdot 3a+4\cdot b-4\cdot 6=4(3a+b-6)$

85. $8x+10y-22=2\cdot 4x+2\cdot 5y-2\cdot 11=2(4x+5y-11)$

87. $ax-a=a\cdot x-a\cdot 1=a(x-1)$

89. $ax-ay-az=a\cdot x-a\cdot y-a\cdot z=a(x-y-z)$

91. $-18x+12y+6=-6\cdot 3x-6(-2y)-6(-1)=-6(3x-2y-1)$
We could also factor this expression as follows:
$-18x+12y+6=6(-3x)+6\cdot 2y+6\cdot 1=6(-3x+2y+1)$

93. $\frac{2}{3}x-\frac{5}{3}y+\frac{1}{3}=\frac{1}{3}\cdot 2x-\frac{1}{3}\cdot 5y+\frac{1}{3}\cdot 1=$
$\frac{1}{3}(2x-5y+1)$

95. $36x-6y+18z=6\cdot 6x-6\cdot y+6\cdot 3z=6(6x-y+3z)$

97. $9a+10a=(9+10)a=19a$

99. $10a-a=10a-1\cdot a=(10-1)a=9a$

101. $2x+9z+6x=2x+6x+9z=(2+6)x+9z=8x+9z$

103. $7x+6y^2+9y^2=7x+(6+9)y^2=7x+15y^2$

105. $41a+90-60a-2=41a-60a+90-2$
$=(41-60)a+(90-2)$
$=-19a+88$

107. $23+5t+7y-t-y-27$
$=23-27+5t-1\cdot t+7y-1\cdot y$
$=(23-27)+(5-1)t+(7-1)y$
$=-4+4t+6y$, or $4t+6y-4$

109. $\frac{1}{2}b+\frac{1}{2}b=\left(\frac{1}{2}+\frac{1}{2}\right)b=1b=b$

111. $2y+\frac{1}{4}y+y=2y+\frac{1}{4}y+1\cdot y=\left(2+\frac{1}{4}+1\right)y=3\frac{1}{4}y$, or $\frac{13}{4}y$

113. $11x-3x=(11-3)x=8x$

115. $6n - n = (6-1)n = 5n$

117. $y - 17y = (1-17)y = -16y$

119.
$$\begin{aligned} & -8 + 11a - 5b + 6a - 7b + 7 \\ &= 11a + 6a - 5b - 7b - 8 + 7 \\ &= (11+6)a + (-5-7)b + (-8+7) \\ &= 17a - 12b - 1 \end{aligned}$$

121. $9x + 2y - 5x = (9-5)x + 2y = 4x + 2y$

123. $11x + 2y - 4x - y = (11-4)x + (2-1)y = 7x + y$

125. $2.7x + 2.3y - 1.9x - 1.8y = (2.7-1.9)x + (2.3-1.8)y = 0.8x + 0.5y$

127.
$$\begin{aligned} & \frac{13}{2}a + \frac{9}{5}b - \frac{2}{3}a - \frac{3}{10}b - 42 \\ &= \left(\frac{13}{2} - \frac{2}{3}\right)a + \left(\frac{9}{5} - \frac{3}{10}\right)b - 42 \\ &= \left(\frac{39}{6} - \frac{4}{6}\right)a + \left(\frac{18}{10} - \frac{3}{10}\right)b - 42 \\ &= \frac{35}{6}a + \frac{15}{10}b - 42 \\ &= \frac{35}{6}a + \frac{3}{2}b - 42 \end{aligned}$$

129. $16 = 2 \cdot 2 \cdot 2 \cdot 2$

$18 = 2 \cdot 3 \cdot 3$

The LCM is $2 \cdot 2 \cdot 2 \cdot 2 \cdot 3 \cdot 3$, or 144.

131. $16 = 2 \cdot 2 \cdot 2 \cdot 2$

$18 = 2 \cdot 3 \cdot 3$

$24 = 2 \cdot 2 \cdot 2 \cdot 3$

The LCM is $2 \cdot 2 \cdot 2 \cdot 2 \cdot 3 \cdot 3$, or 144.

133. $16 = 2 \cdot 2 \cdot 2 \cdot 2$

$32 = 2 \cdot 2 \cdot 2 \cdot 2 \cdot 2$

The LCM is $2 \cdot 2 \cdot 2 \cdot 2 \cdot 2$, or 32.

135. $15 = 3 \cdot 5$

$45 = 3 \cdot 3 \cdot 5$

$90 = 2 \cdot 3 \cdot 3 \cdot 5$

The LCM is $2 \cdot 3 \cdot 3 \cdot 5$, or 90.

137.
$$\begin{aligned} \frac{11}{12} + \frac{15}{16} &= \frac{11}{12} \cdot \frac{4}{4} + \frac{15}{16} \cdot \frac{3}{3} \quad \text{LCD is 48} \\ &= \frac{44}{48} + \frac{45}{48} \\ &= \frac{89}{48} \end{aligned}$$

139. $\frac{1}{8} - \frac{1}{3} = \frac{1}{8} + \left(-\frac{1}{3}\right) = \frac{3}{24} + \left(-\frac{8}{24}\right) = -\frac{5}{24}$

141. No; for any replacement other than 5 the two expressions do not have the same value. For example, let $t = 2$. Then $3 \cdot 2 + 5 = 6 + 5 = 11$, but $3 \cdot 5 + 2 = 15 + 2 = 17$.

143. Yes; commutative law of addition

145.
$$\begin{aligned} & q + qr + qrs + qrst && \text{There are no like terms.} \\ &= q \cdot 1 + q \cdot r + q \cdot rs + q \cdot rst \\ &= q(1 + r + rs + rst) && \text{Factoring} \end{aligned}$$

Exercise Set 9.8

1. $-(2x+7) = -2x - 7$ Changing the sign of each term

3. $-(8-x) = -8 + x$ Changing the sign of each term

5. $-4a + 3b - 7c$

7. $-6x + 8y - 5$

9. $-3x + 5y + 6$

11. $8x + 6y + 43$

13.
$$\begin{aligned} 9x - (4x+3) &= 9x - 4x - 3 && \text{Removing parentheses by changing the sign of every term} \\ &= 5x - 3 && \text{Collecting like terms} \end{aligned}$$

15. $2a - (5a - 9) = 2a - 5a + 9 = -3a + 9$

17. $2x + 7x - (4x + 6) = 2x + 7x - 4x - 6 = 5x - 6$

19. $2x - 4y - 3(7x - 2y) = 2x - 4y - 21x + 6y = -19x + 2y$

21.
$$\begin{aligned} & 15x - y - 5(3x - 2y + 5z) \\ &= 15x - y - 15x + 10y - 25z && \text{Multiplying each term in parentheses by } -5 \\ &= 9y - 25z \end{aligned}$$

23. $(3x + 2y) - 2(5x - 4y) = 3x + 2y - 10x + 8y = -7x + 10y$

25.
$$\begin{aligned} & (12a - 3b + 5c) - 5(-5a + 4b - 6c) \\ &= 12a - 3b + 5c + 25a - 20b + 30c \\ &= 37a - 23b + 35c \end{aligned}$$

27.
$$\begin{aligned} 9 - 2(5-4) &= 9 - 2 \cdot 1 && \text{Computing } 5 - 4 \\ &= 9 - 2 && \text{Computing } 2 \cdot 1 \\ &= 7 \end{aligned}$$

29. $8[7 - 6(4-2)] = 8[7 - 6(2)] = 8[7 - 12] = 8[-5] = -40$

31.
$$\begin{aligned} & [4(9-6) + 11] - [14 - (6+4)] \\ &= [4(3) + 11] - [14 - 10] \\ &= [12 + 11] - [14 - 10] \\ &= 23 - 4 \\ &= 19 \end{aligned}$$

33.
$$\begin{aligned} & [10(x+3) - 4] + [2(x-1) + 6] \\ &= [10x + 30 - 4] + [2x - 2 + 6] \\ &= [10x + 26] + [2x + 4] \\ &= 10x + 26 + 2x + 4 \\ &= 12x + 30 \end{aligned}$$

35.
$$\begin{aligned} & [7(x+5) - 19] - [4(x-6) + 10] \\ &= [7x + 35 - 19] - [4x - 24 + 10] \\ &= [7x + 16] - [4x - 14] \\ &= 7x + 16 - 4x + 14 \\ &= 3x + 30 \end{aligned}$$

37.
$$\begin{aligned} & 3\{[7(x-2) + 4] - [2(2x-5) + 6]\} \\ &= 3\{[7x - 14 + 4] - [4x - 10 + 6]\} \\ &= 3\{[7x - 10] - [4x - 4]\} \\ &= 3\{7x - 10 - 4x + 4\} \\ &= 3\{3x - 6\} \\ &= 9x - 18 \end{aligned}$$

39. $4\{[5(x-3)+2]-3[2(x+5)-9]\}$
$=4\{[5x-15+2]-3[2x+10-9]\}$
$=4\{[5x-13]-3[2x+1]\}$
$=4\{5x-13-6x-3\}$
$=4\{-x-16\}$
$=-4x-64$

41. $8-2\cdot3-9=8-6-9$ Multiplying
$=2-9$ Doing all additions and subtractions in order from left to right
$=-7$

43. $(8-2\cdot3)-9=(8-6)-9$ Multiplying inside the parentheses
$=2-9$ Subtracting inside the parentheses
$=-7$

45. $[(-24)\div(-3)]\div\left(-\frac{1}{2}\right)=8\div\left(-\frac{1}{2}\right)=8\cdot(-2)=-16$

47. $16\cdot(-24)+50=-384+50=-334$

49. $2^4+2^3-10=16+8-10=24-10=14$

51. $5^3+26\cdot71-(16+25\cdot3)=5^3+26\cdot71-(16+75)=$
$5^3+26\cdot71-91=125+26\cdot71-91=125+1846-91=$
$1971-91=1880$

53. $4\cdot5-2\cdot6+4=20-12+4=8+4=12$

55. $4^3/8=64/8=8$

57. $8(-7)+6(-5)=-56-30=-86$

59. $19-5(-3)+3=19+15+3=34+3=37$

61. $9\div(-3)+16\div8=-3+2=-1$

63. $-4^2+6=-16+6=-10$

65. $-8^2-3=-64-3=-67$

67. $12-20^3=12-8000=-7988$

69. $2\cdot10^3-5000=2\cdot1000-5000=2000-5000=-3000$

71. $6[9-(3-4)]=6[9-(-1)]=6[9+1]=6[10]=60$

73. $-1000\div(-100)\div10=10\div10=1$

75. $8-(7-9)=8-(-2)=8+2=10$

77. $\frac{10-6^2}{9^2+3^2}=\frac{10-36}{81+9}=\frac{-26}{90}=-\frac{13}{45}$

79. $\frac{3(6-7)-5\cdot4}{6\cdot7-8(4-1)}=\frac{3(-1)-5\cdot4}{42-8\cdot3}=\frac{-3-20}{42-24}=-\frac{23}{18}$

81. $\frac{|2^3-3^2|+|12\cdot5|}{-32\div(-16)\div(-4)}=\frac{|8-9|+|12\cdot5|}{-32\div(-16)\div(-4)}=$
$\frac{|-1|+|60|}{2\div(-4)}=\frac{1+60}{-\frac{1}{2}}=\frac{61}{-\frac{1}{2}}=61(-2)=-122$

83. The set of integers is
$\{\ldots,-5,-4,-3,-2,-1,0,1,2,3,\ldots\}$.

85. The commutative law of addition says that $a+b=b+a$ for any real numbers a and b.

87. The associative law of addition says that $a+(b+c)=(a+b)+c$ for any real numbers a, b, and c.

89. Two numbers whose product is 1 are called multiplicative inverses of each other.

91. $6y+2x-3a+c=6y-(-2x)-3a-(-c)=6y-(-2x+3a-c)$

93. $6m+3n-5m+4b=6m-(-3n)-5m-(-4b)=$
$6m-(-3n+5m-4b)$

95. $\{x-[f-(f-x)]+[x-f]\}-3x$
$=\{x-[f-f+x]+[x-f]\}-3x$
$=\{x-[x]+[x-f]\}-3x$
$=\{x-x+x-f\}-3x=x-f-3x=-2x-f$

97. a) $x^2+3=7^2+3=49+3=52;$
$x^2+3=(-7)^2+3=49+3=52;$
$x^2+3=(-5.013)^2+3=25.130169+3=28.130169$

b) $1-x^2=1-5^2=1-25=-24;$
$1-x^2=1-(-5)^2=1-25=-24;$
$1-x^2=1-(-10.455)^2=1-109.307025=$
-108.307025

99. $\frac{-15+20+50+(-82)+(-7)+(-2)}{6}=\frac{-36}{6}=-6$

Chapter 9 Concept Reinforcement

1. True; the set of integers is composed of the set of whole numbers and the set of opposites of the natural numbers.

2. True; see page 652 in the text.

3. False; the product of a number and its multiplicative inverse is 1.

4. False; $a<b$ has the same meaning as $b>a$.

Chapter 9 Important Concepts

1. $2a+b=2(-1)+16=-2+16=14$

2. Since -6 is to the left of -3, we have $-6<-3$.

3. The number $-\frac{5}{4}$ is negative, so we make it positive.
$$\left|-\frac{5}{4}\right|=\frac{5}{4}$$

4. $-5.6+(-2.9)$ Two negative numbers. Add the absolute values, getting 8.5. Make the answer negative.
$-5.6+(-2.9)=-8.5$

5. $7-9=7+(-9)=-2$

6. $-8(-7)=56$

7. $-48\div6=-8$

8. $-\frac{3}{4} \div \left(-\frac{5}{3}\right) = -\frac{3}{4} \cdot \left(-\frac{3}{5}\right) = \frac{9}{20}$

9. $\frac{45y}{27y} = \frac{5 \cdot 9y}{3 \cdot 9y} = \frac{5}{3} \cdot \frac{9y}{9y} = \frac{5}{3} \cdot 1 = \frac{5}{3}$

10. $5(x + 3y - 4z) = 5 \cdot x + 5 \cdot 3y - 5 \cdot 4z$
$= 5x + 15y - 20z$

11. $27x + 9y - 36z = 9 \cdot 3x + 9 \cdot y - 9 \cdot 4z = 9(3x + y - 4z)$

12. $6a - 4b - a + 2b = 6a - a - 4b + 2b$
$= (6 - 1)a + (-4 + 2)b$
$= 5a - 2b$

13. $8a - b - (4a + 3b) = 8a - b - 4a - 3b = 4a - 4b$

14. $75 \div (-15) + 24 \div 8$
$= -5 + 3$ Dividing
$= -2$ Adding

14. $-11 \geq -3$ is false since neither $-11 > -3$ nor $-11 = -3$ is true.

15. The opposite of 3.8 is -3.8 because $3.8 + (-3.8) = 0$.

16. The opposite of $-\frac{3}{4}$ is $\frac{3}{4}$ because $-\frac{3}{4} + \frac{3}{4} = 0$.

17. The reciprocal of $\frac{3}{8}$ is $\frac{8}{3}$ because $\frac{3}{8} \cdot \frac{8}{3} = 1$.

18. The reciprocal of -7 is $-\frac{1}{7}$ because $-7 \cdot \left(-\frac{1}{7}\right) = 1$.

19. If $x = -34$, then $-x = -(-34) = 34$.

20. If $x = 5$, then $-(-x) = -(-5) = 5$.

21. $4 + (-7)$
The absolute values are 4 and 7. The difference is $7 - 4$, or 3. The negative number has the larger absolute value, so the answer is negative. $4 + (-7) = -3$

22. $6 + (-9) + (-8) + 7$
a) Add the negative numbers: $-9 + (-8) = -17$
b) Add the positive numbers: $6 + 7 = 13$
c) Add the results: $-17 + 13 = -4$

23. $-3.8 + 5.1 + (-12) + (-4.3) + 10$
a) Add the negative numbers: $-3.8 + (-12) + (-4.3) = -20.1$
b) Add the positive numbers: $5.1 + 10 = 15.1$
c) Add the results: $-20.1 + 15.1 = -5$

24. $-3 - (-7) + 7 - 10 = -3 + 7 + 7 + (-10)$
$= 4 + 7 + (-10)$
$= 11 + (-10)$
$= 1$

25. $-\frac{9}{10} - \frac{1}{2} = -\frac{9}{10} - \frac{5}{10} = -\frac{9}{10} + \left(-\frac{5}{10}\right) = -\frac{14}{10} = -\frac{7 \cdot 2}{5 \cdot 2} = -\frac{7}{5} \cdot \frac{2}{2} = -\frac{7}{5}$

26. $-3.8 - 4.1 = -3.8 + (-4.1) = -7.9$

27. $-9 \cdot (-6) = 54$

28. $-2.7(3.4) = -9.18$

29. $\frac{2}{3} \cdot \left(-\frac{3}{7}\right) = -\left(\frac{2 \cdot 3}{3 \cdot 7}\right) = -\left(\frac{2}{7} \cdot \frac{3}{3}\right) = -\frac{2}{7}$

30. $3 \cdot (-7) \cdot (-2) \cdot (-5) = -21 \cdot 10 = -210$

31. $35 \div (-5) = -7$ Check: $-7 \cdot (-5) = 35$

32. $-5.1 \div 1.7 = -3$ Check: $-3 \cdot (1.7) = -5.1$

33. $-\frac{3}{11} \div -\frac{4}{11} = -\frac{3}{11} \cdot \left(-\frac{11}{4}\right) = \frac{3 \cdot 11}{11 \cdot 4} = \frac{3}{4} \cdot \frac{11}{11} = \frac{3}{4}$

34. $(-3.4 - 12.2) - 8(-7) = -15.6 - 8(-7)$
$= -15.6 + 56$
$= 40.4$

Chapter 9 Review Exercises

1. Substitute 17 for x and 5 for y and carry out the computation.
$\frac{x - y}{3} = \frac{17 - 5}{3} = \frac{12}{3} = 4$

2. $19\%x$, $0.19x$

3. The integer -45 corresponds to a debt of \$45; the integer 72 corresponds to having \$72 in a savings account.

4. The distance of -38 from 0 is 38, so $|-38| = 38$.

5. The distance of 126 from 0 is 126, so $|126| = 126$.

6. The graph of -2.5 is halfway between -3 and -2.

−2.5
−6 −5 −4 −3 −2 −1 0 1 2 3 4 5 6

7. The graph of $\frac{8}{9}$ is $\frac{8}{9}$ of the way from 0 to 1.

8/9
−6 −5 −4 −3 −2 −1 0 1 2 3 4 5 6

8. Since -3 is to the left of 10, we have $-3 < 10$.

9. Since -1 is to the right of -6, we have $-1 > -6$.

10 Since 0.126 is to the right of -12.6, we have $0.126 > -12.6$.

11. $-\frac{2}{3} = -\frac{2}{3} \cdot \frac{10}{10} = -\frac{20}{30}$
$-\frac{1}{10} = -\frac{1}{10} \cdot \frac{3}{3} = -\frac{3}{30}$
Since $-\frac{20}{30}$ is to the left of $-\frac{3}{30}$, then $-\frac{2}{3}$ is to the left of $-\frac{1}{10}$ and we have $-\frac{2}{3} < -\frac{1}{10}$.

12. $x > -3$ has the same meaning as $-3 < x$.

13. $-9 \leq 11$ is true since $-9 < 11$ is true.

35. $$\frac{-12(-3)-2^3-(-9)(-10)}{3\cdot 10+1} = \frac{-12(-3)-8-(-9)(-10)}{30+1} = \frac{36-8-90}{31} = \frac{28-90}{31} = \frac{-62}{31} = -2$$

36. $$-16 \div 4 - 30 \div (-5) = -4-(-6) = -4+6 = 2$$

37. $$\frac{-4[7-(10-13)]}{|-2(8)-4|} = \frac{-4[7-(-3)]}{|-16-4|} = \frac{-4[7+3]}{|-20|} = \frac{-4[10]}{20} = \frac{-40}{20} = -2$$

38. Let t = the total gain or loss. We represent the gains as positive numbers and the loss as a negative number. We add the gains and the loss to find t.

$$t = 5+(-12)+15 = -7+15 = 8$$

There is a total gain of 8 yd.

39. Let a = Kaleb's total assets after he borrows \$300.

Total assets	is	Initial assets	minus	Amount of loan
↓	↓	↓	↓	↓
a	$=$	170	$-$	300

We carry out the subtraction.

$$a = 170-300 = -130$$

Kaleb's total assets were $-\$130$.

40. First we multiply to find the total drop d in the price:

$$d = 8(-\$1.63) = -\$13.04$$

Now we add this number to the opening price to find the price p after 8 hr:

$$p = \$17.68 + (-\$13.04) = \$4.64$$

After 8 hr the price of the stock was \$4.64 per share.

41. Yuri spent the \$68 in his account plus an additional \$64.65, so he spent a total of \$68 + \$64.65, or \$132.65, on seven equally-priced DVDs. Then each DVD cost $\frac{\$132.65}{7}$, or \$18.95.

42. $5(3x-7) = 5\cdot 3x - 5\cdot 7 = 15x-35$

43. $-2(4x-5) = -2\cdot 4x - (-2)(5) = -8x-(-10) = -8x+10$

44. $10(0.4x+1.5) = 10\cdot 0.4x + 10\cdot 1.5 = 4x+15$

45. $-8(3-6x) = -8\cdot 3 - (-8)(6x) = -24-(-48x) = -24+48x$

46. $2x-14 = 2\cdot x - 2\cdot 7 = 2(x-7)$

47. $-6x+6 = -6\cdot x - 6(-1) = -6(x-1)$

The expression can also be factored as follows:

$-6x+6 = 6(-x)+6\cdot 1 = 6(-x+1)$

48. $5x+10 = 5\cdot x + 5\cdot 2 = 5(x+2)$

49. $-3x+12y-12 = -3\cdot x - 3(-4y) - 3\cdot 4 = -3(x-4y+4)$

We could also factor this expression as follows:

$-3x+12y-12 = 3(-x)+3\cdot 4y+3(-4) = 3(-x+4y-4)$

50. $$11a+2b-4a-5b = 11a-4a+2b-5b = (11-4)a+(2-5)b = 7a-3b$$

51. $$7x-3y-9x+8y = 7x-9x-3y+8y = (7-9)x+(-3+8)y = -2x+5y$$

52. $$6x+3y-x-4y = 6x-x+3y-4y = (6-1)x+(3-4)y = 5x-y$$

53. $$-3a+9b+2a-b = -3a+2a+9b-b = (-3+2)a+(9-1)b = -a+8b$$

54. $2a-(5a-9) = 2a-5a+9 = -3a+9$

55. $3(b+7)-5b = 3b+21-5b = -2b+21$

56. $3[11-3(4-1)] = 3[11-3\cdot 3] = 3[11-9] = 3\cdot 2 = 6$

57. $2[6(y-4)+7] = 2[6y-24+7] = 2[6y-17] = 12y-34$

58. $$[8(x+4)-10]-[3(x-2)+4] = [8x+32-10]-[3x-6+4] = 8x+22-[3x-2] = 8x+22-3x+2 = 5x+24$$

59. $$5\{[6(x-1)+7]-[3(3x-4)+8]\} = 5\{[6x-6+7]-[9x-12+8]\} = 5\{6x+1-[9x-4]\} = 5\{6x+1-9x+4\} = 5\{-3x+5\} = -15x+25$$

60. $18x-6y+30 = 6\cdot 3x - 6\cdot y + 6\cdot 5 = 6(3x-y+5)$

Answer D is correct.

61. $mn+5 = nm+5$ by the commutative law of multiplication.

$mn+5 = 5+mn$ by the commutative law of addition.

$mn+5 = 5+mn = 5+nm$ by the commutative laws of addition and multiplication.

Thus we can eliminate answers A, C, and D. It is not possible to write $mn+5$ as $5n+m$ using the commutative and associative laws, so answer B is correct.

62. $$-\left|\frac{7}{8}-\left(-\frac{1}{2}\right)-\frac{3}{4}\right| = -\left|\frac{7}{8}+\frac{1}{2}-\frac{3}{4}\right|$$
$$= -\left|\frac{7}{8}+\frac{4}{8}-\frac{6}{8}\right|$$
$$= -\left|\frac{11}{8}-\frac{6}{8}\right|$$
$$= -\left|\frac{5}{8}\right|$$
$$= -\frac{5}{8}$$

63. $$(|2.7-3|+3^2-|-3|)\div(-3)$$
$$= (|2.7-3|+9-|-3|)\div(-3)$$
$$= (|-0.3|+9-|-3|)\div(-3)$$
$$= (0.3+9-3)\div(-3)$$
$$= (9.3-3)\div(-3)$$
$$= 6.3\div(-3)$$
$$= -2.1$$

64. $$\underbrace{2000-1990}_{10} + \underbrace{1980-1970}_{10} + \ldots + \underbrace{20-10}_{10}$$

Counting by 10's from 10 through 2000 gives us 2000/10, or 200, numbers in the expression. There are 200/2, or 100, pairs of numbers in the expression. Each pair is a difference that is equivalent to 10. Thus, the expression is equal to $100 \cdot 10$, or 1000.

65. Note that the sum of the lengths of the three horizontal segments at the top of the figure, two of which are not labeled and one of which is labeled b, is equivalent to the length of the horizontal segment at the bottom of the figure, a. Then the perimeter is $a+b+b+a+a+a$, or $4a+2b$.

Chapter 9 Discussion and Writing Exercises

1. The sum of each pair of opposites such as -50 and 50, -49 and 49, and so on is 0. The sum of these sums and the remaining integer 0 is 0.

2. The product of an even number of negative numbers is positive, so $(-7)^8$ is positive. The product of an odd number of negative numbers is negative, so $(-7)^{11}$ is negative.

3. Consider $\frac{a}{b} = q$ where a and b are both negative integers. Then $q \cdot b = a$, so q must be a positive number.

4. Consider $\frac{a}{b} = q$ where a is a negative integer and b is a positive integer. Then $q \cdot b = a$, so q must be a negative number.

5. We use the distributive law when we collect like terms even though we might not always write this step.

6. Jake expects the calculator to multiply 2 and 3 first and then divide 18 by that product. This procedure does not follow the rules for order of operations.

Chapter 9 Test

1. Substitute 10 for x and 5 for y and carry out the computations.
$$\frac{3x}{y} = \frac{3\cdot 10}{5} = \frac{30}{5} = 6$$

2. Using x for "some number," we have $x-9$.

3. Since -3 is the right of -8 on the number line, we have $-3 > -8$.

4. Since $-\frac{1}{2}$ is to the left of $-\frac{1}{8}$ on the number line, we have $-\frac{1}{2} < -\frac{1}{8}$.

5. Since -0.78 is to the right of -0.87 on the number line, we have $-0.78 > -0.87$.

6. $-2 > x$ has the same meaning as $x < -2$.

7. $-13 \le -3$ is true since $-13 < -3$ is true.

8. The distance of -7 from 0 is 7, so $|-7| = 7$.

9. The distance of $\frac{9}{4}$ from 0 is $\frac{9}{4}$, so $\left|\frac{9}{4}\right| = \frac{9}{4}$.

10. The distance of -2.7 from 0 is 2.7, so $|-2.7| = 2.7$.

11. The opposite of $\frac{2}{3}$ is $-\frac{2}{3}$ because $\frac{2}{3}+\left(-\frac{2}{3}\right) = 0$.

12. The opposite of -1.4 is 1.4 because $-1.4+1.4 = 0$.

13. The reciprocal of -2 is $-\frac{1}{2}$ because $-2\left(-\frac{1}{2}\right) = 1$.

14. The reciprocal of $\frac{4}{7}$ is $\frac{7}{4}$ because $\frac{4}{7}\cdot\frac{7}{4} = 1$.

15. If $x = -8$, then $-x = -(-8) = 8$.

16. $3.1-(-4.7) = 3.1+4.7 = 7.8$

17. $$-8+4+(-7)+3 = -4+(-7)+3$$
$$= -11+3$$
$$= -8$$

18. $$-\frac{1}{5}+\frac{3}{8} = -\frac{1}{5}\cdot\frac{8}{8}+\frac{3}{8}\cdot\frac{5}{5}$$
$$= -\frac{8}{40}+\frac{15}{40}$$
$$= \frac{7}{40}$$

19. $2-(-8) = 2+8 = 10$

20. $3.2-5.7 = 3.2+(-5.7) = -2.5$

21. $\frac{1}{8}-\left(-\frac{3}{4}\right)=\frac{1}{8}+\frac{3}{4}$
$=\frac{1}{8}+\frac{3}{4}\cdot\frac{2}{2}$
$=\frac{1}{8}+\frac{6}{8}$
$=\frac{7}{8}$

22. $4\cdot(-12)=-48$

23. $-\frac{1}{2}\cdot\left(-\frac{3}{8}\right)=\frac{3}{16}$

24. $-45\div 5=-9$ Check: $-9\cdot 5=-45$

25. $-\frac{3}{5}\div\left(-\frac{4}{5}\right)=-\frac{3}{5}\cdot\left(-\frac{5}{4}\right)=\frac{3\cdot 5}{5\cdot 4}=\frac{3\cdot \cancel{5}}{\cancel{5}\cdot 4}=\frac{3}{4}$

26. $4.864\div(-0.5)=-9.728$

27. $-2(16)-|2(-8)-5^3|=-2(16)-|2(-8)-125|$
$=-2(16)-|-16-125|$
$=-2(16)-|-141|$
$=-2(16)-141$
$=-32-141$
$=-173$

28. $-20\div(-5)+36\div(-4)=4+(-9)=-5$

29. Let $P=$ the number of points by which the market has changed over the five week period.

Total change = Week 1 change + Week 2 change + Week 3 change + Week 4 change + Week 5 change

$P = -13 + (-16) + 36 + (-11) + 19$

We carry out the computation.

$P=-13+(-16)+36+(-11)+19$
$=-29+36+(-11)+19$
$=7+(-11)+19$
$=-4+19$
$=15$

The market rose 15 points.

30. Let $D=$ the difference in the temperatures.

Difference in temperature is Higher temperature minus Lower temperature

$D = -67 - (-81)$

We carry out the subtraction.

$D=-67-(-81)=-67+81=14$

The average high temperature is 14°F higher than the average low temperature.

31. First we multiply to find the total decrease d in the population.

$d=6\cdot 420=2520$

The population decreased by 2520 over the six year period.

Now we subtract to find the new population p.

$18,600-2520=16,080$

After 6 yr the population was 16,080.

32. First we subtract to find the total drop in temperature t.

$t=16°\text{C}-(-17°\text{C})=16°\text{C}+17°\text{C}=33°\text{C}$

Then we divide to find by how many degrees d the temperature dropped each minute in the 44 minutes from 11:08 A.M. to 11:52 A.M.

$d=33\div 44=0.75$

The temperature changed about $-0.75°$C each minute.

33. $3(6-x)=3\cdot 6-3\cdot x=18-3x$

34. $-5(y-1)=-5\cdot y-(-5)(1)=-5y-(-5)=-5y+5$

35. $12-22x=2\cdot 6-2\cdot 11x=2(6-11x)$

36. $7x+21+14y=7\cdot x+7\cdot 3+7\cdot 2y=7(x+3+2y)$

37. $6+7-4-(-3)=6+7+(-4)+3$
$=13+(-4)+3$
$=9+3$
$=12$

38. $5x-(3x-7)=5x-3x+7=2x+7$

39. $4(2a-3b)+a-7=8a-12b+a-7$
$=9a-12b-7$

40. $4\{3[5(y-3)+9]+2(y+8)\}$
$=4\{3[5y-15+9]+2y+16\}$
$=4\{3[5y-6]+2y+16\}$
$=4\{15y-18+2y+16\}$
$=4\{17y-2\}$
$=68y-8$

41. $256\div(-16)\div 4=-16\div 4=-4$

42. $2^3-10[4-(-2+18)3]=2^3-10[4-(16)3]$
$=2^3-10[4-48]$
$=2^3-10[-44]$
$=8-10[-44]$
$=8+440$
$=448$

43. Choices A, C, and D are true because $-5=-5$ is true. Choice B is not true because -5 is not to the left of -5 on the number line. The correct answer is B.

44. $|-27-3(4)|-|-36|+|-12|$
$= |-27-12|-|-36|+|-12|$
$= |-39|-|-36|+|-12|$
$= 39-36+12$
$= 3+12$
$= 15$

45. $a-\{3a-[4a-(2a-4a)]\}$
$= a-\{3a-[4a-(-2a)]\}$
$= a-\{3a-[4a+2a]\}$
$= a-\{3a-6a\}$
$= a-\{-3a\}$
$= a+3a$
$= 4a$

46. The perimeter is equivalent to the perimeter of a square with sides x along with four additional segments of length y. We have $x+x+x+x+y+y+y+y=4x+4y$.

Chapter 10

Solving Equations and Inequalities

Exercise Set 10.1

1. $\begin{array}{c|c} x+17 = 32 & \\ \hline 15+17 \ ? \ 32 & \\ 32 & \end{array}$ Writing the equation; Substituting 15 for x; TRUE

Since the left-hand and right-hand sides are the same, 15 is a solution of the equation.

3. $\begin{array}{c|c} x-7 = 12 & \\ \hline 21-7 \ ? \ 12 & \\ 14 & \end{array}$ Writing the equation; Substituting 21 for x; FALSE

Since the left-hand and right-hand sides are not the same, 21 is not a solution of the equation.

5. $\begin{array}{c|c} 6x = 54 & \\ \hline 6(-7) \ ? \ 54 & \\ -42 & \end{array}$ Writing the equation; Substituting; FALSE

-7 is not a solution of the equation.

7. $\begin{array}{c|c} \frac{x}{6} = 5 & \\ \hline \frac{30}{6} \ ? \ 5 & \\ 5 & \end{array}$ Writing the equation; Substituting; TRUE

5 is a solution of the equation.

9. $\begin{array}{c|c} 5x+7 = 107 & \\ \hline 5 \cdot 20+7 \ ? \ 107 & \\ 100+7 & \\ 107 & \end{array}$ Substituting; TRUE

20 is a solution of the equation.

11. $\begin{array}{c|c} 7(y-1) = 63 & \\ \hline 7(-10-1) \ ? \ 63 & \\ 7(-11) & \\ -77 & \end{array}$ Substituting; FALSE

-10 is not a solution of the equation.

13. $x+2 = 6$

$x+2-2 = 6-2$ Subtracting 2 on both sides

$x = 4$ Simplifying

Check: $\begin{array}{c|c} x+2 = 6 & \\ \hline 4+2 \ ? \ 6 & \\ 6 & \end{array}$ TRUE

The solution is 4.

15. $x+15 = -5$

$x+15-15 = -5-15$ Subtracting 15 on both sides

$x = -20$

Check: $\begin{array}{c|c} x+15 = -5 & \\ \hline -20+15 \ ? \ -5 & \\ -5 & \end{array}$ TRUE

The solution is -20.

17. $x+6 = -8$

$x+6-6 = -8-6$

$x = -14$

Check: $\begin{array}{c|c} x+6 = -8 & \\ \hline -14+6 \ ? \ -8 & \\ -8 & \end{array}$ TRUE

The solution is -14.

19. $x+16 = -2$

$x+16-16 = -2-16$

$x = -18$

Check: $\begin{array}{c|c} x+16 = -2 & \\ \hline -18+16 \ ? \ -2 & \\ -2 & \end{array}$ TRUE

The solution is -18.

21. $x-9 = 6$

$x-9+9 = 6+9$

$x = 15$

Check: $\begin{array}{c|c} x-9 = 6 & \\ \hline 15-9 \ ? \ 6 & \\ 6 & \end{array}$ TRUE

The solution is 15.

23. $x-7 = -21$

$x-7+7 = -21+7$

$x = -14$

Check: $\begin{array}{c|c} x-7 = -21 & \\ \hline -14-7 \ ? \ -21 & \\ -21 & \end{array}$ TRUE

The solution is -14.

25. $5+t = 7$

$-5+5+t = -5+7$

$t = 2$

Check: $\begin{array}{c|c} 5+t = 7 & \\ \hline 5+2 \ ? \ 7 & \\ 7 & \end{array}$ TRUE

The solution is 2.

27.
$$-7+y=13$$
$$7+(-7)+y=7+13$$
$$y=20$$

Check: $-7+y=13$

$-7+20 \ ? \ 13$

$13 \mid 13$ TRUE

The solution is 20.

29.
$$-3+t=-9$$
$$3+(-3)+t=3+(-9)$$
$$t=-6$$

Check: $-3+t=-9$

$-3+(-6) \ ? \ -9$

$-9 \mid -9$ TRUE

The solution is -6.

31.
$$x+\frac{1}{2}=7$$
$$x+\frac{1}{2}-\frac{1}{2}=7-\frac{1}{2}$$
$$x=6\frac{1}{2}$$

Check: $x+\frac{1}{2}=7$

$6\frac{1}{2}+\frac{1}{2} \ ? \ 7$

$7 \mid 7$ TRUE

The solution is $6\frac{1}{2}$.

33.
$$12=a-7.9$$
$$12+7.9=a-7.9+7.9$$
$$19.9=a$$

Check: $12=a-7.9$

$12 \ ? \ 19.9-7.9$

$12 \mid 12$ TRUE

The solution is 19.9.

35.
$$r+\frac{1}{3}=\frac{8}{3}$$
$$r+\frac{1}{3}-\frac{1}{3}=\frac{8}{3}-\frac{1}{3}$$
$$r=\frac{7}{3}$$

Check: $r+\frac{1}{3}=\frac{8}{3}$

$\frac{7}{3}+\frac{1}{3} \ ? \ \frac{8}{3}$

$\frac{8}{3} \mid \frac{8}{3}$ TRUE

The solution is $\frac{7}{3}$.

37.
$$m+\frac{5}{6}=-\frac{11}{12}$$
$$m+\frac{5}{6}-\frac{5}{6}=-\frac{11}{12}-\frac{5}{6}$$
$$m=-\frac{11}{12}-\frac{5}{6}\cdot\frac{2}{2}$$
$$m=-\frac{11}{12}-\frac{10}{12}$$
$$m=-\frac{21}{12}=-\frac{\cancel{3}\cdot 7}{\cancel{3}\cdot 4}$$
$$m=-\frac{7}{4}$$

Check: $m+\frac{5}{6}=-\frac{11}{12}$

$-\frac{7}{4}+\frac{5}{6} \ ? \ -\frac{11}{12}$

$-\frac{21}{12}+\frac{10}{12}$

$-\frac{11}{12} \mid -\frac{11}{12}$ TRUE

The solution is $-\frac{7}{4}$.

39.
$$x-\frac{5}{6}=\frac{7}{8}$$
$$x-\frac{5}{6}+\frac{5}{6}=\frac{7}{8}+\frac{5}{6}$$
$$x=\frac{7}{8}\cdot\frac{3}{3}+\frac{5}{6}\cdot\frac{4}{4}$$
$$x=\frac{21}{24}+\frac{20}{24}$$
$$x=\frac{41}{24}$$

Check: $x-\frac{5}{6}=\frac{7}{8}$

$\frac{41}{24}-\frac{5}{6} \ ? \ \frac{7}{8}$

$\frac{41}{24}-\frac{20}{24} \mid \frac{21}{24}$

$\frac{21}{24}$ TRUE

The solution is $\frac{41}{24}$.

41. $$-\frac{1}{5}+z=-\frac{1}{4}$$
$$\frac{1}{5}-\frac{1}{5}+z=\frac{1}{5}-\frac{1}{4}$$
$$z=\frac{1}{5}\cdot\frac{4}{4}-\frac{1}{4}\cdot\frac{5}{5}$$
$$z=\frac{4}{20}-\frac{5}{20}$$
$$z=-\frac{1}{20}$$

Check: $-\frac{1}{5}+z=-\frac{1}{4}$

$$-\frac{1}{5}+\left(-\frac{1}{20}\right) \ ? \ -\frac{1}{4}$$
$$-\frac{4}{20}+\left(-\frac{1}{20}\right) \ \Big| \ -\frac{5}{20}$$
$$-\frac{5}{20} \qquad \text{TRUE}$$

The solution is $-\frac{1}{20}$.

43. $$x+2.3=7.4$$
$$x+2.3-2.3=7.4-2.3$$
$$x=5.1$$

Check: $x+2.3=7.4$

$$5.1+2.3 \ ? \ 7.4$$
$$7.4 \ \Big| \qquad \text{TRUE}$$

The solution is 5.1.

45. $$7.6=x-4.8$$
$$7.6+4.8=x-4.8+4.8$$
$$12.4=x$$

Check: $7.6=x-4.8$

$$7.6 \ ? \ 12.4-4.8$$
$$\Big| \ 7.6 \qquad \text{TRUE}$$

The solution is 12.4.

47. $$-9.7=-4.7+y$$
$$4.7+(-9.7)=4.7+(-4.7)+y$$
$$-5=y$$

Check: $-9.7=-4.7+y$

$$-9.7 \ ? \ -4.7+(-5)$$
$$\Big| \ -9.7 \qquad \text{TRUE}$$

The solution is -5.

49. $$5\frac{1}{6}+x=7$$
$$-5\frac{1}{6}+5\frac{1}{6}+x=-5\frac{1}{6}+7$$
$$x=-\frac{31}{6}+\frac{42}{6}$$
$$x=\frac{11}{6}, \text{ or } 1\frac{5}{6}$$

Check: $5\frac{1}{6}+x=7$

$$5\frac{1}{6}+1\frac{5}{6} \ ? \ 7$$
$$7 \ \Big| \qquad \text{TRUE}$$

The solution is $\frac{11}{6}$, or $1\frac{5}{6}$.

51. $$q+\frac{1}{3}=-\frac{1}{7}$$
$$q+\frac{1}{3}-\frac{1}{3}=-\frac{1}{7}-\frac{1}{3}$$
$$q=-\frac{1}{7}\cdot\frac{3}{3}-\frac{1}{3}\cdot\frac{7}{7}$$
$$q=-\frac{3}{21}-\frac{7}{21}$$
$$q=-\frac{10}{21}$$

Check: $q+\frac{1}{3}=-\frac{1}{7}$

$$-\frac{10}{21}+\frac{1}{3} \ ? \ -\frac{1}{7}$$
$$-\frac{10}{21}+\frac{7}{21} \ \Big| \ -\frac{3}{21}$$
$$-\frac{3}{21} \qquad \text{TRUE}$$

The solution is $-\frac{10}{21}$.

53. $-3+(-8)$ Two negative numbers. We add the absolute values, getting 11, and make the answer negative.

$-3+(-8)=-11$

55. $-\frac{2}{3}\cdot\frac{5}{8}=-\frac{2\cdot 5}{3\cdot 8}=-\frac{\cancel{2}\cdot 5}{3\cdot\cancel{2}\cdot 4}=-\frac{5}{12}$

57. $\frac{2}{3}\div\left(-\frac{4}{9}\right)=\frac{2}{3}\cdot\left(-\frac{9}{4}\right)=-\frac{2\cdot 9}{3\cdot 4}=-\frac{\cancel{2}\cdot\cancel{3}\cdot 3}{\cancel{3}\cdot\cancel{2}\cdot 2}=-\frac{3}{2}$

59. $$-\frac{2}{3}-\left(-\frac{5}{8}\right)=-\frac{2}{3}+\frac{5}{8}$$
$$=-\frac{2}{3}\cdot\frac{8}{8}+\frac{5}{8}\cdot\frac{3}{3}$$
$$=-\frac{16}{24}+\frac{15}{24}$$
$$=-\frac{1}{24}$$

61. The translation is \$83 $- x$.

63. $$-356.788=-699.034+t$$
$$699.034+(-356.788)=699.034+(-699.034)+t$$
$$342.246=t$$

The solution is 342.246.

65.
$$x+\frac{4}{5}=-\frac{2}{3}-\frac{4}{15}$$
$$x+\frac{4}{5}=-\frac{2}{3}\cdot\frac{5}{5}-\frac{4}{15} \quad \text{Adding on the right side}$$
$$x+\frac{4}{5}=-\frac{10}{15}-\frac{4}{15}$$
$$x+\frac{4}{5}=-\frac{14}{15}$$
$$x+\frac{4}{5}-\frac{4}{5}=-\frac{14}{15}-\frac{4}{5}$$
$$x=-\frac{14}{15}-\frac{4}{5}\cdot\frac{3}{3}$$
$$x=-\frac{14}{15}-\frac{12}{15}$$
$$x=-\frac{26}{15}$$

The solution is $-\frac{26}{15}$.

67.
$$16+x-22=-16$$
$$x-6=-16 \quad \text{Adding on the left side}$$
$$x-6+6=-16+6$$
$$x=-10$$

The solution is -10.

69.
$$x+3=3+x$$
$$x+3-3=3+x-3$$
$$x=x$$

$x=x$ is true for all real numbers. Thus the solution is all real numbers.

71.
$$-\frac{3}{2}+x=-\frac{5}{17}-\frac{3}{2}$$
$$\frac{3}{2}-\frac{3}{2}+x=\frac{3}{2}-\frac{5}{17}-\frac{3}{2}$$
$$x=\left(\frac{3}{2}-\frac{3}{2}\right)-\frac{5}{17}$$
$$x=-\frac{5}{17}$$

The solution is $-\frac{5}{17}$.

73.
$$|x|+6=19$$
$$|x|+6-6=19-6$$
$$|x|=13$$

x represents a number whose distance from 0 is 13. Thus $x=-13$ or $x=13$.

The solutions are -13 and 13.

Exercise Set 10.2

1.
$$6x=36$$
$$\frac{6x}{6}=\frac{36}{6} \quad \text{Dividing by 6 on both sides}$$
$$1\cdot x=6 \quad \text{Simplifying}$$
$$x=6 \quad \text{Identity property of 1}$$

Check: $6x=36$

$6\cdot 6\ ?\ 36$

$36 \mid 36$ TRUE

The solution is 6.

3.
$$5y=45$$
$$\frac{5y}{5}=\frac{45}{5} \quad \text{Dividing by 5 on both sides}$$
$$1\cdot y=9 \quad \text{Simplifying}$$
$$y=9 \quad \text{Identity property of 1}$$

Check: $5y=45$

$5\cdot 9\ ?\ 45$

$45 \mid 45$ TRUE

The solution is 9.

5.
$$84=7x$$
$$\frac{84}{7}=\frac{7x}{7} \quad \text{Dividing by 7 on both sides}$$
$$12=1\cdot x$$
$$12=x$$

Check: $84=7x$

$84\ ?\ 7\cdot 12$

$84 \mid 84$ TRUE

The solution is 12.

7.
$$-x=40$$
$$-1\cdot x=40$$
$$\frac{-1\cdot x}{-1}=\frac{40}{-1}$$
$$1\cdot x=-40$$
$$x=-40$$

Check: $-x=40$

$-(-40)\ ?\ 40$

$40 \mid 40$ TRUE

The solution is -40.

9.
$$-1=-z$$
$$-1=-1\cdot z$$
$$\frac{-1}{-1}=\frac{-1\cdot z}{-1}$$
$$1=1\cdot z$$
$$1=z$$

Check: $-1=-z$

$-1\ ?\ -(1)$

$-1 \mid -1$ TRUE

The solution is 1.

11.
$$7x=-49$$
$$\frac{7x}{7}=\frac{-49}{7}$$
$$1\cdot x=-7$$
$$x=-7$$

Check: $7x=-49$

$7(-7)\ ?\ -49$

$-49 \mid -49$ TRUE

The solution is -7.

13. $-12x = 72$

$$\frac{-12x}{-12} = \frac{72}{-12}$$

$$1 \cdot x = -6$$

$$x = -6$$

Check: $-12x = 72$

$-12(-6) \;?\; 72$

72 | TRUE

The solution is -6.

15. $-21x = -126$

$$\frac{-21x}{-21} = \frac{-126}{-21}$$

$$1 \cdot x = 6$$

$$x = 6$$

Check: $-21x = -126$

$-21 \cdot 6 \;?\; -126$

-126 | TRUE

The solution is 6.

17. $\frac{t}{7} = -9$

$$7 \cdot \frac{1}{7}t = 7 \cdot (-9)$$

$$1 \cdot t = -63$$

$$t = -63$$

Check: $\frac{t}{7} = -9$

$\frac{-63}{7} \;?\; -9$

-9 | TRUE

The solution is -63.

19. $\frac{n}{-6} = 8$

$$-6\left(\frac{n}{-6}\right) = -6 \cdot 8$$

$$1 \cdot n = -48$$

$$n = -48$$

Check: $\frac{n}{-6} = 8$

$\frac{-48}{-6} \;?\; 8$

8 | TRUE

The solution is -48.

21. $\frac{3}{4}x = 27$

$$\frac{4}{3} \cdot \frac{3}{4}x = \frac{4}{3} \cdot 27$$

$$1 \cdot x = \frac{4 \cdot \not{3} \cdot 3 \cdot 3}{\not{3} \cdot 1}$$

$$x = 36$$

Check: $\frac{3}{4}x = 27$

$\frac{3}{4} \cdot 36 \;?\; 27$

27 | TRUE

The solution is 36.

23. $-\frac{2}{3}x = 6$

$$-\frac{3}{2}\left(-\frac{2}{3}x\right) = -\frac{3}{2} \cdot 6$$

$$1 \cdot x = -\frac{3 \cdot \not{2} \cdot 3}{\not{2} \cdot 1}$$

$$x = -9$$

Check: $-\frac{2}{3}x = 6$

$-\frac{2}{3}(-9) \;?\; 6$

6 | TRUE

The solution is -9.

25. $\frac{-t}{3} = 7$

$$3 \cdot \frac{1}{3} \cdot (-t) = 3 \cdot 7$$

$$-t = 21$$

$$-1 \cdot (-1 \cdot t) = -1 \cdot 21$$

$$1 \cdot t = -21$$

$$t = -21$$

Check: $\frac{-t}{3} = 7$

$\frac{-(-21)}{3} \;?\; 7$

$\frac{21}{3}$

7 | TRUE

The solution is -21.

27. $-\frac{m}{3} = \frac{1}{5}$

$$-\frac{1}{3} \cdot m = \frac{1}{5}$$

$$-3 \cdot \left(-\frac{1}{3} \cdot m\right) = -3 \cdot \frac{1}{5}$$

$$m = -\frac{3}{5}$$

Check: $-\dfrac{m}{3} = \dfrac{1}{5}$

$$-\dfrac{-\frac{3}{5}}{3} \;?\; \dfrac{1}{5}$$
$$-\left(-\dfrac{3}{5} \div 3\right)$$
$$-\left(-\dfrac{3}{5} \cdot \dfrac{1}{3}\right)$$
$$-\left(-\dfrac{1}{5}\right)$$
$$\dfrac{1}{5} \qquad \text{TRUE}$$

The solution is $-\dfrac{3}{5}$.

29.
$$-\dfrac{3}{5}r = \dfrac{9}{10}$$
$$-\dfrac{5}{3} \cdot \left(-\dfrac{3}{5}r\right) = -\dfrac{5}{3} \cdot \dfrac{9}{10}$$
$$1 \cdot r = -\dfrac{\cancel{5} \cdot \cancel{3} \cdot 3}{\cancel{3} \cdot \cancel{5} \cdot 2}$$
$$r = -\dfrac{3}{2}$$

Check: $-\dfrac{3}{5}r = \dfrac{9}{10}$

$$-\dfrac{3}{5} \cdot \left(-\dfrac{3}{2}\right) \;?\; \dfrac{9}{10}$$
$$\dfrac{9}{10} \qquad \text{TRUE}$$

The solution is $-\dfrac{3}{2}$.

31.
$$-\dfrac{3}{2}r = -\dfrac{27}{4}$$
$$-\dfrac{2}{3} \cdot \left(-\dfrac{3}{2}r\right) = -\dfrac{2}{3} \cdot \left(-\dfrac{27}{4}\right)$$
$$1 \cdot r = \dfrac{\cancel{2} \cdot \cancel{3} \cdot 3 \cdot 3}{\cancel{3} \cdot \cancel{2} \cdot 2}$$
$$r = \dfrac{9}{2}$$

Check: $-\dfrac{3}{2}r = -\dfrac{27}{4}$

$$-\dfrac{3}{2} \cdot \dfrac{9}{2} \;?\; -\dfrac{27}{4}$$
$$-\dfrac{27}{4} \qquad \text{TRUE}$$

The solution is $\dfrac{9}{2}$.

33.
$$6.3x = 44.1$$
$$\dfrac{6.3x}{6.3} = \dfrac{44.1}{6.3}$$
$$1 \cdot x = 7$$
$$x = 7$$

Check: $6.3x = 44.1$

$$6.3 \cdot 7 \;?\; 44.1$$
$$44.1 \qquad \text{TRUE}$$

The solution is 7.

35.
$$-3.1y = 21.7$$
$$\dfrac{-3.1y}{-3.1} = \dfrac{21.7}{-3.1}$$
$$1 \cdot y = -7$$
$$y = -7$$

Check: $-3.1y = 21.7$

$$-3.1(-7) \;?\; 21.7$$
$$21.7 \qquad \text{TRUE}$$

The solution is -7.

37.
$$38.7m = 309.6$$
$$\dfrac{38.7m}{38.7} = \dfrac{309.6}{38.7}$$
$$1 \cdot m = 8$$
$$m = 8$$

Check: $38.7m = 309.6$

$$38.7 \cdot 8 \;?\; 309.6$$
$$309.6 \qquad \text{TRUE}$$

The solution is 8.

39.
$$-\dfrac{2}{3}y = -10.6$$
$$-\dfrac{3}{2} \cdot (-\dfrac{2}{3}y) = -\dfrac{3}{2} \cdot (-10.6)$$
$$1 \cdot y = \dfrac{31.8}{2}$$
$$y = 15.9$$

Check: $-\dfrac{2}{3}y = -10.6$

$$-\dfrac{2}{3} \cdot (15.9) \;?\; -10.6$$
$$-\dfrac{31.8}{3}$$
$$-10.6 \qquad \text{TRUE}$$

The solution is 15.9.

41.
$$\dfrac{-x}{5} = 10$$
$$5 \cdot \dfrac{-x}{5} = 5 \cdot 10$$
$$-x = 50$$
$$-1 \cdot (-x) = -1 \cdot 50$$
$$x = -50$$

Check: $\frac{-x}{5} = 10$

$$\begin{array}{c|c} \frac{-(-50)}{5} & ?\ 10 \\ \frac{50}{5} & \\ 10 & \text{TRUE} \end{array}$$

The solution is -50.

43.
$$-\frac{t}{2} = 7$$
$$2 \cdot \left(-\frac{t}{2}\right) = 2 \cdot 7$$
$$-t = 14$$
$$-1 \cdot (-t) = -1 \cdot 14$$
$$t = -14$$

Check: $-\frac{t}{2} = 7$

$$\begin{array}{c|c} -\frac{-14}{2} & ?\ 7 \\ -(-7) & \\ 7 & \text{TRUE} \end{array}$$

The solution is -14.

45. $3x + 4x = (3 + 4)x = 7x$

47. $-4x + 11 - 6x + 18x = (-4 - 6 + 18)x + 11 = 8x + 11$

49. $3x - (4 + 2x) = 3x - 4 - 2x = x - 4$

51. $8y - 6(3y + 7) = 8y - 18y - 42 = -10y - 42$

53. The translation is $8r$ miles.

55.
$$-0.2344m = 2028.732$$
$$\frac{-0.2344m}{-0.2344} = \frac{2028.732}{-0.2344}$$
$$1 \cdot m = -8655$$
$$m = -8655$$

The solution is -8655.

57. For all x, $0 \cdot x = 0$. There is no solution to $0 \cdot x = 9$.

59.
$$2|x| = -12$$
$$\frac{2|x|}{2} = \frac{-12}{2}$$
$$1 \cdot |x| = -6$$
$$|x| = -6$$

Absolute value cannot be negative. The equation has no solution.

61.
$$3x = \frac{b}{a}$$
$$\frac{1}{3} \cdot 3x = \frac{1}{3} \cdot \frac{b}{a}$$
$$x = \frac{b}{3a}$$

The solution is $\frac{b}{3a}$.

63.
$$\frac{a}{b}x = 4$$
$$\frac{b}{a} \cdot \frac{a}{b}x = \frac{b}{a} \cdot 4$$
$$x = \frac{4b}{a}$$

The solution is $\frac{4b}{a}$.

Exercise Set 10.3

1.

$5x + 6 = 31$	
$5x + 6 - 6 = 31 - 6$	Subtracting 6 on both sides
$5x = 25$	Simplifying
$\frac{5x}{5} = \frac{25}{5}$	Dividing by 5 on both sides
$x = 5$	Simplifying

Check: $5x + 6 = 31$

$$\begin{array}{c|c} 5 \cdot 5 + 6 & ?\ 31 \\ 25 + 6 & \\ 31 & \text{TRUE} \end{array}$$

The solution is 5.

3.

$8x + 4 = 68$	
$8x + 4 - 4 = 68 - 4$	Subtracting 4 on both sides
$8x = 64$	Simplifying
$\frac{8x}{8} = \frac{64}{8}$	Dividing by 8 on both sides
$x = 8$	Simplifying

Check: $8x + 4 = 68$

$$\begin{array}{c|c} 8 \cdot 8 + 4 & ?\ 68 \\ 64 + 4 & \\ 68 & \text{TRUE} \end{array}$$

The solution is 8.

5.

$4x - 6 = 34$	
$4x - 6 + 6 = 34 + 6$	Adding 6 on both sides
$4x = 40$	
$\frac{4x}{4} = \frac{40}{4}$	Dividing by 4 on both sides
$x = 10$	

Check: $4x - 6 = 34$

$$\begin{array}{c|c} 4 \cdot 10 - 6 & ?\ 34 \\ 40 - 6 & \\ 34 & \text{TRUE} \end{array}$$

The solution is 10.

7.
$$3x - 9 = 33$$
$$3x - 9 + 9 = 33 + 9$$
$$3x = 42$$
$$\frac{3x}{3} = \frac{42}{3}$$
$$x = 14$$

Check: $\begin{array}{c|c} 3x-9 = 33 & \\ \hline 3\cdot 14-9 \ ? & 33 \\ 42-9 & \\ 33 & \end{array}$ TRUE

The solution is 14.

9. $$\begin{aligned} 7x+2 &= -54 \\ 7x+2-2 &= -54-2 \\ 7x &= -56 \\ \frac{7x}{7} &= \frac{-56}{7} \\ x &= -8 \end{aligned}$$

Check: $\begin{array}{c|c} 7x+2 = -54 & \\ \hline 7(-8)+2 \ ? & -54 \\ -56+2 & \\ -54 & \end{array}$ TRUE

The solution is -8.

11. $$\begin{aligned} -45 &= 6y+3 \\ -45-3 &= 6y+3-3 \\ -48 &= 6y \\ \frac{-48}{6} &= \frac{6y}{6} \\ -8 &= y \end{aligned}$$

Check: $\begin{array}{c|c} -45 = 6y+3 & \\ \hline -45 \ ? & 6(-8)+3 \\ & -48+3 \\ & -45 \end{array}$ TRUE

The solution is -8.

13. $$\begin{aligned} -4x+7 &= 35 \\ -4x+7-7 &= 35-7 \\ -4x &= 28 \\ \frac{-4x}{-4} &= \frac{28}{-4} \\ x &= -7 \end{aligned}$$

Check: $\begin{array}{c|c} -4x+7 = 35 & \\ \hline -4(-7)+7 \ ? & 35 \\ 28+7 & \\ 35 & \end{array}$ TRUE

The solution is -7.

15. $$\begin{aligned} \frac{5}{4}x-18 &= -3 \\ \frac{5}{4}x-18+18 &= -3+18 \\ \frac{5}{4}x &= 15 \\ \frac{4}{5}\cdot\frac{5}{4}x &= \frac{4}{5}\cdot 15 \\ x &= 12 \end{aligned}$$

Check: $\begin{array}{c|c} \frac{5}{4}x-18 = -3 & \\ \hline \frac{5}{4}\cdot 12-18 \ ? & -3 \\ 15-18 & \\ -3 & \end{array}$ TRUE

The solution is 12.

17. $$\begin{aligned} 5x+7x &= 72 \\ 12x &= 72 && \text{Collecting like terms} \\ \frac{12x}{12} &= \frac{72}{12} && \text{Dividing by 12 on both sides} \\ x &= 6 \end{aligned}$$

Check: $\begin{array}{c|c} 5x+7x = 72 & \\ \hline 5\cdot 6+7\cdot 6 \ ? & 72 \\ 30+42 & \\ 72 & \end{array}$ TRUE

The solution is 6.

19. $$\begin{aligned} 8x+7x &= 60 \\ 15x &= 60 && \text{Collecting like terms} \\ \frac{15x}{15} &= \frac{60}{15} && \text{Dividing by 15 on both sides} \\ x &= 4 \end{aligned}$$

Check: $\begin{array}{c|c} 8x+7x = 60 & \\ \hline 8\cdot 4+7\cdot 4 \ ? & 60 \\ 32+28 & \\ 60 & \end{array}$ TRUE

The solution is 4.

21. $$\begin{aligned} 4x+3x &= 42 \\ 7x &= 42 \\ \frac{7x}{7} &= \frac{42}{7} \\ x &= 6 \end{aligned}$$

Check: $\begin{array}{c|c} 4x+3x = 42 & \\ \hline 4\cdot 6+3\cdot 6 \ ? & 42 \\ 24+18 & \\ 42 & \end{array}$ TRUE

The solution is 6.

23. $$\begin{aligned} -6y-3y &= 27 \\ -9y &= 27 \\ \frac{-9y}{-9} &= \frac{27}{-9} \\ y &= -3 \end{aligned}$$

Check: $\begin{array}{c|c} -6y-3y = 27 & \\ \hline -6(-3)-3(-3) \ ? & 27 \\ 18+9 & \\ 27 & \end{array}$ TRUE

The solution is -3.

25. $-7y - 8y = -15$

$$-15y = -15$$
$$\frac{-15y}{-15} = \frac{-15}{-15}$$
$$y = 1$$

Check:
$$\begin{array}{c|l} -7y - 8y & = -15 \\ \hline -7 \cdot 1 - 8 \cdot 1 & ?\ -15 \\ -7 - 8 & \\ -15 & \qquad \text{TRUE} \end{array}$$

The solution is 1.

27. $x + \frac{1}{3}x = 8$

$$\left(1 + \frac{1}{3}\right)x = 8$$
$$\frac{4}{3}x = 8$$
$$\frac{3}{4} \cdot \frac{4}{3}x = \frac{3}{4} \cdot 8$$
$$x = 6$$

Check:
$$\begin{array}{c|l} x + \frac{1}{3}x & = 8 \\ \hline 6 + \frac{1}{3} \cdot 6 & ?\ 8 \\ 6 + 2 & \\ 8 & \qquad \text{TRUE} \end{array}$$

The solution is 6.

29. $10.2y - 7.3y = -58$

$$2.9y = -58$$
$$\frac{2.9y}{2.9} = \frac{-58}{2.9}$$
$$y = -20$$

Check:
$$\begin{array}{c|l} 10.2y - 7.3y & = -58 \\ \hline 10.2(-20) - 7.3(-20) & ?\ -58 \\ -204 + 146 & \\ -58 & \qquad \text{TRUE} \end{array}$$

The solution is -20.

31. $8y - 35 = 3y$

$8y = 3y + 35$ Adding 35 and simplifying

$8y - 3y = 35$ Subtracting $3y$ and simplifying

$5y = 35$ Collecting like terms

$\frac{5y}{5} = \frac{35}{5}$ Dividing by 5

$y = 7$

Check:
$$\begin{array}{c|l} 8y - 35 & = 3y \\ \hline 8 \cdot 7 - 35 & ?\ 3 \cdot 7 \\ 56 - 35 & 21 \\ 21 & \qquad \text{TRUE} \end{array}$$

The solution is 7.

33. $8x - 1 = 23 - 4x$

$8x + 4x = 23 + 1$ Adding 1 and $4x$ and simplifying

$12x = 24$ Collecting like terms

$\frac{12x}{12} = \frac{24}{12}$ Dividing by 12

$x = 2$

Check:
$$\begin{array}{c|l} 8x - 1 & = 23 - 4x \\ \hline 8 \cdot 2 - 1 & ?\ 23 - 4 \cdot 2 \\ 16 - 1 & 23 - 8 \\ 15 & 15 \qquad \text{TRUE} \end{array}$$

The solution is 2.

35. $2x - 1 = 4 + x$

$2x - x = 4 + 1$ Adding 1 and $-x$

$x = 5$ Collecting like terms

Check:
$$\begin{array}{c|l} 2x - 1 & = 4 + x \\ \hline 2 \cdot 5 - 1 & ?\ 4 + 5 \\ 10 - 1 & 9 \\ 9 & \qquad \text{TRUE} \end{array}$$

The solution is 5.

37. $6x + 3 = 2x + 11$

$$6x - 2x = 11 - 3$$
$$4x = 8$$
$$\frac{4x}{4} = \frac{8}{4}$$
$$x = 2$$

Check:
$$\begin{array}{c|l} 6x + 3 & = 2x + 11 \\ \hline 6 \cdot 2 + 3 & ?\ 2 \cdot 2 + 11 \\ 12 + 3 & 4 + 11 \\ 15 & 15 \qquad \text{TRUE} \end{array}$$

The solution is 2.

39. $5 - 2x = 3x - 7x + 25$

$$5 - 2x = -4x + 25$$
$$4x - 2x = 25 - 5$$
$$2x = 20$$
$$\frac{2x}{2} = \frac{20}{2}$$
$$x = 10$$

Check:
$$\begin{array}{c|l} 5 - 2x & = 3x - 7x + 25 \\ \hline 5 - 2 \cdot 10 & ?\ 3 \cdot 10 - 7 \cdot 10 + 25 \\ 5 - 20 & 30 - 70 + 25 \\ -15 & -40 + 25 \\ & -15 \qquad \text{TRUE} \end{array}$$

The solution is 10.

41. $4 + 3x - 6 = 3x + 2 - x$

$3x - 2 = 2x + 2$ Collecting like terms on each side

$3x - 2x = 2 + 2$

$x = 4$

Check: $4+3x-6=3x+2-x$

$4+3\cdot4-6$? $3\cdot4+2-4$	
$4+12-6$	$12+2-4$
$16-6$	$14-4$
10	10 TRUE

The solution is 4.

43. $4y-4+y+24=6y+20-4y$

$$5y+20=2y+20$$
$$5y-2y=20-20$$
$$3y=0$$
$$y=0$$

Check: $4y-4+y+24=6y+20-4y$

$4\cdot0-4+0+24$? $6\cdot0+20-4\cdot0$	
$0-4+0+24$	$0+20-0$
20	20 TRUE

The solution is 0.

45. $\frac{7}{2}x+\frac{1}{2}x=3x+\frac{3}{2}+\frac{5}{2}x$

The least common multiple of all the denominators is 2. We multiply by 2 on both sides.

$$2\left(\frac{7}{2}x+\frac{1}{2}x\right)=2\left(3x+\frac{3}{2}+\frac{5}{2}x\right)$$
$$2\cdot\frac{7}{2}x+2\cdot\frac{1}{2}x=2\cdot3x+2\cdot\frac{3}{2}+2\cdot\frac{5}{2}x$$
$$7x+x=6x+3+5x$$
$$8x=11x+3$$
$$8x-11x=3$$
$$-3x=3$$
$$\frac{-3x}{-3}=\frac{3}{-3}$$
$$x=-1$$

Check: $\frac{7}{2}x+\frac{1}{2}x=3x+\frac{3}{2}+\frac{5}{2}x$

$\frac{7}{2}(-1)+\frac{1}{2}(-1)$? $3(-1)+\frac{3}{2}+\frac{5}{2}(-1)$	
$-\frac{7}{2}-\frac{1}{2}$	$-3+\frac{3}{2}-\frac{5}{2}$
-4	$-\frac{8}{2}$
	-4 TRUE

The solution is -1.

47. $\frac{2}{3}+\frac{1}{4}t=\frac{1}{3}$

The least common multiple of all the denominators is 12. We multiply by 12 on both sides.

$$12\left(\frac{2}{3}+\frac{1}{4}t\right)=12\cdot\frac{1}{3}$$
$$12\cdot\frac{2}{3}+12\cdot\frac{1}{4}t=12\cdot\frac{1}{3}$$
$$8+3t=4$$
$$3t=4-8$$
$$3t=-4$$
$$\frac{3t}{3}=\frac{-4}{3}$$
$$t=-\frac{4}{3}$$

Check: $\frac{2}{3}+\frac{1}{4}t=\frac{1}{3}$

$\frac{2}{3}+\frac{1}{4}\left(-\frac{4}{3}\right)$? $\frac{1}{3}$	
$\frac{2}{3}-\frac{1}{3}$	
$\frac{1}{3}$	TRUE

The solution is $-\frac{4}{3}$.

49. $\frac{2}{3}+3y=5y-\frac{2}{15}$, LCM is 15

$$15\left(\frac{2}{3}+3y\right)=15\left(5y-\frac{2}{15}\right)$$
$$15\cdot\frac{2}{3}+15\cdot3y=15\cdot5y-15\cdot\frac{2}{15}$$
$$10+45y=75y-2$$
$$10+2=75y-45y$$
$$12=30y$$
$$\frac{12}{30}=\frac{30y}{30}$$
$$\frac{2}{5}=y$$

Check: $\frac{2}{3}+3y=5y-\frac{2}{15}$

$\frac{2}{3}+3\cdot\frac{2}{5}$? $5\cdot\frac{2}{5}-\frac{2}{15}$	
$\frac{2}{3}+\frac{6}{5}$	$2-\frac{2}{15}$
$\frac{10}{15}+\frac{18}{15}$	$\frac{30}{15}-\frac{2}{15}$
$\frac{28}{15}$	$\frac{28}{15}$ TRUE

The solution is $\frac{2}{5}$.

51.
$$\frac{5}{3}+\frac{2}{3}x=\frac{25}{12}+\frac{5}{4}x+\frac{3}{4},\quad \text{LCM is 12}$$
$$12\left(\frac{5}{3}+\frac{2}{3}x\right)=12\left(\frac{25}{12}+\frac{5}{4}x+\frac{3}{4}\right)$$
$$12\cdot\frac{5}{3}+12\cdot\frac{2}{3}x=12\cdot\frac{25}{12}+12\cdot\frac{5}{4}x+12\cdot\frac{3}{4}$$
$$20+8x=25+15x+9$$
$$20+8x=15x+34$$
$$20-34=15x-8x$$
$$-14=7x$$
$$\frac{-14}{7}=\frac{7x}{7}$$
$$-2=x$$

Check: $\frac{5}{3}+\frac{2}{3}x=\frac{25}{12}+\frac{5}{4}x+\frac{3}{4}$

$\frac{5}{3}+\frac{2}{3}(-2)$?	$\frac{25}{12}+\frac{5}{4}(-2)+\frac{3}{4}$
$\frac{5}{3}-\frac{4}{3}$	$\frac{25}{12}-\frac{5}{2}+\frac{3}{4}$
$\frac{1}{3}$	$\frac{25}{12}-\frac{30}{12}+\frac{9}{12}$
	$\frac{4}{12}$
	$\frac{1}{3}$ TRUE

The solution is -2.

53.
$$2.1x+45.2=3.2-8.4x$$
Greatest number of decimal places is 1
$$10(2.1x+45.2)=10(3.2-8.4x)$$
Multiplying by 10 to clear decimals
$$10(2.1x)+10(45.2)=10(3.2)-10(8.4x)$$
$$21x+452=32-84x$$
$$21x+84x=32-452$$
$$105x=-420$$
$$\frac{105x}{105}=\frac{-420}{105}$$
$$x=-4$$

Check: $2.1x+45.2=3.2-8.4x$

$2.1(-4)+45.2$?	$3.2-8.4(-4)$
$-8.4+45.2$	$3.2+33.6$
36.8	36.8 TRUE

The solution is -4.

55.
$$1.03-0.62x=0.71-0.22x$$
Greatest number of decimal places is 2
$$100(1.03-0.62x)=100(0.71-0.22x)$$
Multiplying by 100 to clear decimals
$$100(1.03)-100(0.62x)=100(0.71)-100(0.22x)$$
$$103-62x=71-22x$$
$$32=40x$$
$$\frac{32}{40}=\frac{40x}{40}$$
$$\frac{4}{5}=x,\text{ or}$$
$$0.8=x$$

Check: $1.03-0.62x=0.71-0.22x$

$1.03-0.62(0.8)$?	$0.71-0.22(0.8)$
$1.03-0.496$	$0.71-0.176$
0.534	0.534 TRUE

The solution is $\frac{4}{5}$, or 0.8.

57.
$$\frac{2}{7}x-\frac{1}{2}x=\frac{3}{4}x+1,\text{ LCM is 28}$$
$$28\left(\frac{2}{7}x-\frac{1}{2}x\right)=28\left(\frac{3}{4}x+1\right)$$
$$28\cdot\frac{2}{7}x-28\cdot\frac{1}{2}x=28\cdot\frac{3}{4}x+28\cdot 1$$
$$8x-14x=21x+28$$
$$-6x=21x+28$$
$$-6x-21x=28$$
$$-27x=28$$
$$x=-\frac{28}{27}$$

Check: $\frac{2}{7}x-\frac{1}{2}x=\frac{3}{4}x+1$

$\frac{2}{7}\left(-\frac{28}{27}\right)-\frac{1}{2}\left(-\frac{28}{27}\right)$?	$\frac{3}{4}\left(-\frac{28}{27}\right)+1$
$-\frac{8}{27}+\frac{14}{27}$	$-\frac{21}{27}+1$
$\frac{6}{27}$	$\frac{6}{27}$ TRUE

The solution is $-\frac{28}{27}$.

59.
$3(2y-3)=27$
$6y-9=27$ Using a distributive law
$6y=27+9$ Adding 9
$6y=36$
$y=6$ Dividing by 6

Check: $3(2y-3)=27$

$3(2\cdot 6-3)$?	27
$3(12-3)$	
$3\cdot 9$	
27	TRUE

The solution is 6.

61.
$$40 = 5(3x+2)$$
$$40 = 15x + 10 \quad \text{Using a distributive law}$$
$$40 - 10 = 15x$$
$$30 = 15x$$
$$2 = x$$

Check: $40 = 5(3x+2)$

40	?	$5(3\cdot 2+2)$	
		$5(6+2)$	
		$5\cdot 8$	
		40	TRUE

The solution is 2.

63.
$$-23 + y = y + 25$$
$$-y - 23 + y = -y + y + 25$$
$$-23 = 25 \quad \text{FALSE}$$

The equation has no solution.

65.
$$-23 + x = x - 23$$
$$-x - 23 + x = -x + x - 23$$
$$-23 = -23 \quad \text{TRUE}$$

All real numbers are solutions.

67.
$$2(3+4m) - 9 = 45$$
$$6 + 8m - 9 = 45$$
$$8m - 3 = 45 \quad \text{Collecting like terms}$$
$$8m = 45 + 3$$
$$8m = 48$$
$$m = 6$$

Check: $2(3+4m) - 9 = 45$

$2(3+4\cdot 6) - 9$	?	45	
$2(3+24) - 9$			
$2\cdot 27 - 9$			
$54 - 9$			
45			TRUE

The solution is 6.

69.
$$5r - (2r+8) = 16$$
$$5r - 2r - 8 = 16$$
$$3r - 8 = 16 \quad \text{Collecting like terms}$$
$$3r = 16 + 8$$
$$3r = 24$$
$$r = 8$$

Check: $5r - (2r+8) = 16$

$5\cdot 8 - (2\cdot 8+8)$	?	16	
$40 - (16+8)$			
$40 - 24$			
16			TRUE

The solution is 8.

71.
$$6 - 2(3x-1) = 2$$
$$6 - 6x + 2 = 2$$
$$8 - 6x = 2$$
$$8 - 2 = 6x$$
$$6 = 6x$$
$$1 = x$$

Check: $6 - 2(3x-1) = 2$

$6 - 2(3\cdot 1 - 1)$	?	2	
$6 - 2(3-1)$			
$6 - 2\cdot 2$			
$6 - 4$			
2			TRUE

The solution is 1.

73.
$$5x + 5 - 7x = 15 - 12x + 10x - 10$$
$$-2x + 5 = 5 - 2x \quad \text{Collecting like terms}$$
$$2x - 2x + 5 = 2x + 5 - 2x \quad \text{Adding } 2x$$
$$5 = 5 \quad \text{TRUE}$$

All real numbers are solutions.

75.
$$22x - 5 - 15x + 3 = 10x - 4 - 3x + 11$$
$$7x - 2 = 7x + 7 \quad \text{Collecting like terms}$$
$$-7x + 7x - 2 = -7x + 7x + 7$$
$$-2 = 7 \quad \text{FALSE}$$

The equation has no solution.

77.
$$5(d+4) = 7(d-2)$$
$$5d + 20 = 7d - 14$$
$$20 + 14 = 7d - 5d$$
$$34 = 2d$$
$$17 = d$$

Check: $5(d+4) = 7(d-2)$

$5(17+4)$	?	$7(17-2)$	
$5\cdot 21$		$7\cdot 15$	
105		105	TRUE

The solution is 17.

79.
$$8(2t+1) = 4(7t+7)$$
$$16t + 8 = 28t + 28$$
$$16t - 28t = 28 - 8$$
$$-12t = 20$$
$$t = -\frac{20}{12}$$
$$t = -\frac{5}{3}$$

Check:
$$
\begin{array}{r|l}
\multicolumn{2}{c}{8(2t+1) = 4(7t+7)} \\
\hline
8\left(2\left(-\frac{5}{3}\right)+1\right) \ ? & 4\left(7\left(-\frac{5}{3}\right)+7\right) \\
8\left(-\frac{10}{3}+1\right) & 4\left(-\frac{35}{3}+7\right) \\
8\left(-\frac{7}{3}\right) & 4\left(-\frac{14}{3}\right) \\
-\frac{56}{3} & -\frac{56}{3} \quad \text{TRUE}
\end{array}
$$

The solution is $-\frac{5}{3}$.

81.
$$
\begin{aligned}
3(r-6)+2 &= 4(r+2)-21 \\
3r-18+2 &= 4r+8-21 \\
3r-16 &= 4r-13 \\
13-16 &= 4r-3r \\
-3 &= r
\end{aligned}
$$

Check:
$$
\begin{array}{r|l}
\multicolumn{2}{c}{3(r-6)+2 = 4(r+2)-21} \\
\hline
3(-3-6)+2 \ ? & 4(-3+2)-21 \\
3(-9)+2 & 4(-1)-21 \\
-27+2 & -4-21 \\
-25 & -25 \quad \text{TRUE}
\end{array}
$$

The solution is -3.

83.
$$
\begin{aligned}
19-(2x+3) &= 2(x+3)+x \\
19-2x-3 &= 2x+6+x \\
16-2x &= 3x+6 \\
16-6 &= 3x+2x \\
10 &= 5x \\
2 &= x
\end{aligned}
$$

Check:
$$
\begin{array}{r|l}
\multicolumn{2}{c}{19-(2x+3) = 2(x+3)+x} \\
\hline
19-(2\cdot 2+3) \ ? & 2(2+3)+2 \\
19-(4+3) & 2\cdot 5+2 \\
19-7 & 10+2 \\
12 & 12 \quad \text{TRUE}
\end{array}
$$

The solution is 2.

85.
$$
\begin{aligned}
2[4-2(3-x)]-1 &= 4[2(4x-3)+7]-25 \\
2[4-6+2x]-1 &= 4[8x-6+7]-25 \\
2[-2+2x]-1 &= 4[8x+1]-25 \\
-4+4x-1 &= 32x+4-25 \\
4x-5 &= 32x-21 \\
-5+21 &= 32x-4x \\
16 &= 28x \\
\frac{16}{28} &= x \\
\frac{4}{7} &= x
\end{aligned}
$$

The check is left to the student.

The solution is $\frac{4}{7}$.

87.
$$
\begin{aligned}
11-4(x+1)-3 &= 11+2(4-2x)-16 \\
11-4x-4-3 &= 11+8-4x-16 \\
4-4x &= 3-4x \\
4x+4-4x &= 4x+3-4x \\
4 &= 3 \qquad \text{FALSE}
\end{aligned}
$$

The equation has no solution.

89.
$$
\begin{aligned}
22x-1-12x &= 5(2x-1)+4 \\
22x-1-12x &= 10x-5+4 \\
10x-1 &= 10x-1 \\
-10x+10x-1 &= -10x+10x-1 \\
-1 &= -1 \qquad \text{TRUE}
\end{aligned}
$$

All real numbers are solutions.

91.
$$
\begin{aligned}
0.7(3x+6) &= 1.1-(x+2) \\
2.1x+4.2 &= 1.1-x-2 \\
10(2.1x+4.2) &= 10(1.1-x-2) \quad \text{Clearing decimals} \\
21x+42 &= 11-10x-20 \\
21x+42 &= -10x-9 \\
21x+10x &= -9-42 \\
31x &= -51 \\
x &= -\frac{51}{31}
\end{aligned}
$$

The check is left to the student.

The solution is $-\frac{51}{31}$.

93. Do the long division. The answer is negative.

$$
\begin{array}{r}
6.5 \\
3.4_\wedge \overline{)\,22.1_\wedge 0} \\
\underline{20\,4} \\
1\,7\,0 \\
\underline{1\,7\,0} \\
0
\end{array}
$$

$-22.1 \div 3.4 = -6.5$

95. $7x-21-14y = 7\cdot x-7\cdot 3-7\cdot 2y = 7(x-3-2y)$

97.
$$
\begin{aligned}
-3+2(-5)^2(-3)-7 &= -3+2(25)(-3)-7 \\
&= -3+50(-3)-7 \\
&= -3-150-7 \\
&= -153-7 \\
&= -160
\end{aligned}
$$

99. $23(2x-4)-15(10-3x) = 46x-92-150+45x = 91x-242$

101. First we multiply to remove the parentheses.

$$\frac{2}{3}\left(\frac{7}{8}-4x\right)-\frac{5}{8}=\frac{3}{8}$$
$$\frac{7}{12}-\frac{8}{3}x-\frac{5}{8}=\frac{3}{8},\ \text{LCM is 24}$$
$$24\left(\frac{7}{12}-\frac{8}{3}x-\frac{5}{8}\right)=24\cdot\frac{3}{8}$$
$$24\cdot\frac{7}{12}-24\cdot\frac{8}{3}x-24\cdot\frac{5}{8}=9$$
$$14-64x-15=9$$
$$-1-64x=9$$
$$-64x=10$$
$$x=-\frac{10}{64}$$
$$x=-\frac{5}{32}$$

The solution is $-\frac{5}{32}$.

103.
$$\frac{4-3x}{7}=\frac{2+5x}{49}-\frac{x}{14}$$
$$98\left(\frac{4-3x}{7}\right)=98\left(\frac{2+5x}{49}-\frac{x}{14}\right),\ \text{LCM is 98}$$
$$\frac{98(4-3x)}{7}=98\left(\frac{2+5x}{49}\right)-98\cdot\frac{x}{14}$$
$$14(4-3x)=2(2+5x)-7x$$
$$56-42x=4+10x-7x$$
$$56-42x=4+3x$$
$$56-42x+42x=4+3x+42x$$
$$56=4+45x$$
$$56-4=4+45x-4$$
$$52=45x$$
$$\frac{52}{45}=x$$

The solution is $\frac{52}{45}$.

Exercise Set 10.4

1. a) We substitute 1900 for a and calculate B.

$$B=30a=30\cdot 1900=57,000$$

The minimum furnace output is 57,000 Btu's.

b) $B=30a$

$$\frac{B}{30}=\frac{30a}{30}\quad\text{Dividing by 30}$$
$$\frac{B}{30}=a$$

3. a) We substitute 8 for t and calculate M.

$$M=\frac{1}{5}\cdot 8=\frac{8}{5},\ \text{or } 1.6$$

The storm is 1.6 miles away.

b) $M=\frac{1}{5}t$

$$5\cdot M=5\cdot\frac{1}{5}t$$
$$5M=t$$

5. a) We substitute 21,345 for n and calculate f.

$$f=\frac{21,345}{15}=1423$$

There are 1423 full-time equivalent students.

b) $f=\frac{n}{15}$

$$15\cdot f=15\cdot\frac{n}{15}$$
$$15f=n$$

7. We substitute 84 for c and 8 for w and calculate D.

$$D=\frac{c}{w}=\frac{84}{8}=10.5$$

The calorie density is 10.5 calories per oz.

9. We substitute 7 for n and calculate N.

$$N=n^2-n=7^2-7=49-7=42$$

42 games are played.

11.
$$y=5x$$
$$\frac{y}{5}=\frac{5x}{5}$$
$$\frac{y}{5}=x$$

13.
$$a=bc$$
$$\frac{a}{b}=\frac{bc}{b}$$
$$\frac{a}{b}=c$$

15.
$$n=m+11$$
$$n-11=m+11-11$$
$$n-11=m$$

17.
$$y=x-\frac{3}{5}$$
$$y+\frac{3}{5}=x-\frac{3}{5}+\frac{3}{5}$$
$$y+\frac{3}{5}=x$$

19.
$$y=13+x$$
$$y-13=13+x-13$$
$$y-13=x$$

21.
$$y=x+b$$
$$y-b=x+b-b$$
$$y-b=x$$

23.
$$y=5-x$$
$$y-5=5-x-5$$
$$y-5=-x$$
$$-1\cdot(y-5)=-1\cdot(-x)$$
$$-y+5=x,\ \text{or}$$
$$5-y=x$$

25.
$$\begin{aligned} y &= a - x \\ y - a &= a - x - a \\ y - a &= -x \\ -1\cdot(y-a) &= -1\cdot(-x) \\ -y + a &= x, \text{ or} \\ a - y &= x \end{aligned}$$

27.
$$\begin{aligned} 8y &= 5x \\ \frac{8y}{8} &= \frac{5x}{8} \\ y &= \frac{5x}{8}, \text{ or } \frac{5}{8}x \end{aligned}$$

29.
$$\begin{aligned} By &= Ax \\ \frac{By}{A} &= \frac{Ax}{A} \\ \frac{By}{A} &= x \end{aligned}$$

31.
$$\begin{aligned} W &= mt + b \\ W - b &= mt + b - b \\ W - b &= mt \\ \frac{W-b}{m} &= \frac{mt}{m} \\ \frac{W-b}{m} &= t \end{aligned}$$

33.
$$\begin{aligned} y &= bx + c \\ y - c &= bx + c - c \\ y - c &= bx \\ \frac{y-c}{b} &= \frac{bx}{b} \\ \frac{y-c}{b} &= x \end{aligned}$$

35.
$$\begin{aligned} A &= bh \\ \frac{A}{b} &= \frac{bh}{b} \quad \text{Dividing by } b \\ \frac{A}{b} &= h \end{aligned}$$

37.
$$\begin{aligned} P &= 2l + 2w \\ P - 2l &= 2l + 2w - 2l \quad \text{Subtracting } 2l \\ P - 2l &= 2w \\ \frac{P-2l}{2} &= \frac{2w}{2} \quad \text{Dividing by 2} \\ \frac{P-2l}{2} &= w, \text{ or} \\ \frac{1}{2}P - l &= w \end{aligned}$$

39.
$$\begin{aligned} A &= \frac{a+b}{2} \\ 2A &= a + b \quad \text{Multiplying by 2} \\ 2A - b &= a \quad \text{Subtracting } b \end{aligned}$$

41.
$$\begin{aligned} A &= \frac{a+b+c}{3} \\ 3A &= a + b + c \quad \text{Multiplying by 3} \\ 3A - a - c &= b \quad \text{Subtracting } a \text{ and } c \end{aligned}$$

43.
$$\begin{aligned} A &= at + b \\ A - b &= at \quad \text{Subtracting } b \\ \frac{A-b}{a} &= t \quad \text{Dividing by } a \end{aligned}$$

45.
$$\begin{aligned} Ax + By &= c \\ Ax &= c - By \quad \text{Subtracting } By \\ \frac{Ax}{A} &= \frac{c-By}{A} \quad \text{Dividing by } A \\ x &= \frac{c-By}{A} \end{aligned}$$

47.
$$\begin{aligned} F &= ma \\ \frac{F}{m} &= \frac{ma}{m} \quad \text{Dividing by } m \\ \frac{F}{m} &= a \end{aligned}$$

49.
$$\begin{aligned} E &= mc^2 \\ \frac{E}{m} &= \frac{mc^2}{m} \quad \text{Dividing by } m \\ \frac{E}{m} &= c^2 \end{aligned}$$

51.
$$\begin{aligned} v &= \frac{3k}{t} \\ tv &= t\cdot\frac{3k}{t} \quad \text{Multiplying by } t \\ tv &= 3k \\ \frac{tv}{v} &= \frac{3k}{v} \quad \text{Dividing by } v \\ t &= \frac{3k}{v} \end{aligned}$$

53. We divide:

```
       0.9 2
 2 5 ) 2 3.0 0
       2 2 5
       -----
           5 0
           5 0
           ---
             0
```

Decimal notation for $\frac{23}{25}$ is 0.92.

55. $0.082 + (-9.407) = -9.325$

57. $-45.8 - (-32.6) = -45.8 + 32.6 = -13.2$

59. 3.1% 0.03.1

Move the decimal point 2 places to the left.

$3.1\% = 0.031$

61.
$$\begin{aligned} -\frac{2}{3} + \frac{5}{6} &= -\frac{2}{3}\cdot\frac{2}{2} + \frac{5}{6} \\ &= -\frac{4}{6} + \frac{5}{6} \\ &= \frac{1}{6} \end{aligned}$$

63. a) We substitute 120 for w, 67 for h, and 23 for a and calculate K.

$$K = 917 + 6(w + h - a)$$
$$K = 917 + 6(120 + 67 - 23)$$
$$K = 917 + 6(164)$$
$$K = 917 + 984$$
$$K = 1901 \text{ calories}$$

b) Solve for a:

$$K = 917 + 6(w + h - a)$$
$$K = 917 + 6w + 6h - 6a$$
$$K + 6a = 917 + 6w + 6h$$
$$6a = 917 + 6w + 6h - K$$
$$a = \frac{917 + 6w + 6h - K}{6}$$

Solve for h:

$$K = 917 + 6(w + h - a)$$
$$K = 917 + 6w + 6h - 6a$$
$$K - 917 - 6w + 6a = 6h$$
$$\frac{K - 917 - 6w + 6a}{6} = h$$

Solve for w:

$$K = 917 + 6(w + h - a)$$
$$K = 917 + 6w + 6h - 6a$$
$$K - 917 - 6h + 6a = 6w$$
$$\frac{K - 917 - 6h + 6a}{6} = w$$

65.

$$H = \frac{2}{a - b}$$
$$(a - b)H = (a - b)\left(\frac{2}{a - b}\right)$$
$$Ha - Hb = 2$$
$$Ha - Hb - Ha = 2 - Ha$$
$$-Hb = 2 - Ha$$
$$-1(-Hb) = -1(2 - Ha)$$
$$Hb = -2 + Ha$$
$$\frac{Hb}{H} = \frac{-2 + Ha}{H}$$
$$b = \frac{-2 + Ha}{H}, \text{ or } \frac{Ha - 2}{H}, \text{ or } a - \frac{2}{H}$$

$$H = \frac{2}{a - b}$$
$$(a - b)H = (a - b) \cdot \frac{2}{a - b}$$
$$Ha - Hb = 2$$
$$Ha - Hb + Hb = 2 + Hb$$
$$Ha = 2 + Hb$$
$$\frac{Ha}{H} = \frac{2 + Hb}{H}$$
$$a = \frac{2 + Hb}{H}, \text{ or } \frac{2}{H} + b$$

67. $A = lw$

When l and w both double, we have

$$2l \cdot 2w = 4lw = 4A,$$

so A quadruples.

69. $A = \frac{1}{2}bh$

When b increases by 4 units we have

$$\frac{1}{2}(b + 4)h = \frac{1}{2}bh + 2h = A + 2h,$$

so A increases by $2h$ units.

Chapter 10 Mid-Chapter Review

1. The solution of $3 - x = 4x$ is $\frac{3}{5}$; the solution of $5x = -3$ is $-\frac{3}{5}$. The equations have different solutions, so they are not equivalent. The given statement is false.

2. True; see page 699 in the text.

3. True; see page 704 in the text.

4. False; see page 716 in the text.

5.

$$x + 5 = -3$$
$$x + 5 - 5 = -3 - 5$$
$$x + 0 = -8$$
$$x = -8$$

6.

$$-6x = 42$$
$$\frac{-6x}{-6} = \frac{42}{-6}$$
$$1 \cdot x = -7$$
$$x = -7$$

7.

$$5y + z = t$$
$$5y + z - z = t - z$$
$$5y = t - z$$
$$\frac{5y}{5} = \frac{t - z}{5}$$
$$y = \frac{t - z}{5}$$

8.

$$x + 5 = 11$$
$$x + 5 - 5 = 11 - 5$$
$$x = 6$$

The solution is 6.

9.

$$x + 9 = -3$$
$$x + 9 - 9 = -3 - 9$$
$$x = -12$$

The solution is -12.

10.

$$8 = t + 1$$
$$8 - 1 = t + 1 - 1$$
$$7 = t$$

The solution is 7.

11. $$\begin{aligned} -7 &= y+3 \\ -7-3 &= y+3-3 \\ -10 &= y \end{aligned}$$
The solution is -10.

12. $$\begin{aligned} x-6 &= 14 \\ x-6+6 &= 14+6 \\ x &= 20 \end{aligned}$$
The solution is 20.

13. $$\begin{aligned} y-7 &= -2 \\ y-7+7 &= -2+7 \\ y &= 5 \end{aligned}$$
The solution is 5.

14. $$\begin{aligned} -\frac{3}{2}+z &= -\frac{3}{4} \\ -\frac{3}{2}+z+\frac{3}{2} &= -\frac{3}{4}+\frac{3}{2} \\ z &= -\frac{3}{4}+\frac{6}{4} \\ z &= \frac{3}{4} \end{aligned}$$
The solution is $\frac{3}{4}$.

15. $$\begin{aligned} -3.3 &= -1.9+t \\ -3.3+1.9 &= -1.9+t+1.9 \\ -1.4 &= t \end{aligned}$$
The solution is -1.4.

16. $$\begin{aligned} 7x &= 42 \\ \frac{7x}{7} &= \frac{42}{7} \\ x &= 6 \end{aligned}$$
The solution is 6.

17. $$\begin{aligned} 17 &= -t \\ 17 &= -1\cdot t \\ \frac{17}{-1} &= \frac{-1\cdot t}{-1} \\ -17 &= t \end{aligned}$$
The solution is -17.

18. $$\begin{aligned} 6x &= -54 \\ \frac{6x}{6} &= \frac{-54}{6} \\ x &= -9 \end{aligned}$$
The solution is -9.

19. $$\begin{aligned} -5y &= -85 \\ \frac{-5y}{-5} &= \frac{-85}{-5} \\ y &= 17 \end{aligned}$$
The solution is 17.

20. $$\begin{aligned} \frac{x}{7} &= 3 \\ \frac{1}{7}\cdot x &= 3 \\ 7\cdot\frac{1}{7}x &= 7\cdot 3 \\ x &= 21 \end{aligned}$$
The solution is 21.

21. $$\begin{aligned} \frac{2}{3}x &= 12 \\ \frac{3}{2}\cdot\frac{2}{3}x &= \frac{3}{2}\cdot 12 \\ x &= \frac{3\cdot\not{2}\cdot 6}{\not{2}\cdot 1} \\ x &= 18 \end{aligned}$$
The solution is 18.

22. $$\begin{aligned} -\frac{t}{5} &= 3 \\ -\frac{1}{5}\cdot t &= 3 \\ -5\left(-\frac{1}{5}\cdot t\right) &= -5\cdot 3 \\ t &= -15 \end{aligned}$$
The solution is -15.

23. $$\begin{aligned} \frac{3}{4}x &= -\frac{9}{8} \\ \frac{4}{3}\cdot\frac{3}{4}x &= \frac{4}{3}\left(-\frac{9}{8}\right) \\ x &= -\frac{\not{4}\cdot\not{3}\cdot 3}{\not{3}\cdot 2\cdot\not{4}} \\ x &= -\frac{3}{2} \end{aligned}$$
The solution is $-\frac{3}{2}$.

24. $$\begin{aligned} 3x+2 &= 5 \\ 3x+2-2 &= 5-2 \\ 3x &= 3 \\ \frac{3x}{3} &= \frac{3}{3} \\ x &= 1 \end{aligned}$$
The solution is 1.

25. $$\begin{aligned} 5x+4 &= -11 \\ 5x+4-4 &= -11-4 \\ 5x &= -15 \\ \frac{5x}{5} &= \frac{-15}{5} \\ x &= -3 \end{aligned}$$
The solution is -3.

26.
$$\begin{aligned} 6x-7&=2\\ 6x-7+7&=2+7\\ 6x&=9\\ \frac{6x}{6}&=\frac{9}{6}\\ x&=\frac{9}{6}=\frac{3\cdot \cancel{3}}{2\cdot \cancel{3}}\\ x&=\frac{3}{2} \end{aligned}$$

The solution is $\frac{3}{2}$.

27.
$$\begin{aligned} -4x-9&=-5\\ -4x-9+9&=-5+9\\ -4x&=4\\ \frac{-4x}{-4}&=\frac{4}{-4}\\ x&=-1 \end{aligned}$$

The solution is -1.

28.
$$\begin{aligned} 6x+5x&=33\\ 11x&=33\\ \frac{11x}{11}&=\frac{33}{11}\\ x&=3 \end{aligned}$$

The solution is 3.

29.
$$\begin{aligned} -3y-4y&=49\\ -7y&=49\\ \frac{-7y}{-7}&=\frac{49}{-7}\\ y&=-7 \end{aligned}$$

The solution is -7.

30.
$$\begin{aligned} 3x-4&=12-x\\ 3x-4+x&=12-x+x\\ 4x-4&=12\\ 4x-4+4&=12+4\\ 4x&=16\\ \frac{4x}{4}&=\frac{16}{4}\\ x&=4 \end{aligned}$$

The solution is 4.

31.
$$\begin{aligned} 5-6x&=9-8x\\ 5-6x+8x&=9-8x+8x\\ 5+2x&=9\\ 5+2x-5&=9-5\\ 2x&=4\\ \frac{2x}{2}&=\frac{4}{2}\\ x&=2 \end{aligned}$$

The solution is 2.

32.
$$\begin{aligned} 4y-\frac{3}{2}&=\frac{3}{4}+2y\\ 4\left(4y-\frac{3}{2}\right)&=4\left(\frac{3}{4}+2y\right) \quad \text{Clearing fractions}\\ 4\cdot 4y-4\cdot\frac{3}{2}&=4\cdot\frac{3}{4}+4\cdot 2y\\ 16y-6&=3+8y\\ 16y-6-8y&=3+8y-8y\\ 8y-6&=3\\ 8y-6+6&=3+6\\ 8y&=9\\ \frac{8y}{8}&=\frac{9}{8} \end{aligned}$$

The solution is $\frac{9}{8}$.

33.
$$\begin{aligned} \frac{4}{5}+\frac{1}{6}t&=\frac{1}{10}\\ 30\left(\frac{4}{5}+\frac{1}{6}t\right)&=30\cdot\frac{1}{10} \quad \text{Clearing fractions}\\ 30\cdot\frac{4}{5}+30\cdot\frac{1}{6}t&=\frac{30}{10}\\ 24+5t&=3\\ 24+5t-24&=3-24\\ 5t&=-21\\ \frac{5t}{5}&=\frac{-21}{5}\\ t&=-\frac{21}{5} \end{aligned}$$

The solution is $-\frac{21}{5}$.

34.
$$\begin{aligned} 0.21n-1.05&=2.1-0.14n\\ 100(0.21n-1.05)&=100(2.1-0.14n) \quad \text{Clearing decimals}\\ 100(0.21n)-100(1.05)&=100(2.1)-100(0.14n)\\ 21n-105&=210-14n\\ 21n-105+14n&=210-14n+14n\\ 35n-105&=210\\ 35n-105+105&=210+105\\ 35n&=315\\ \frac{35n}{35}&=\frac{315}{35}\\ n&=9 \end{aligned}$$

The solution is 9.

35.
$$\begin{aligned} 5(3y-1)&=-35\\ 15y-5&=-35\\ 15y-5+5&=-35+5\\ 15y&=-30\\ \frac{15y}{15}&=\frac{-30}{15}\\ y&=-2 \end{aligned}$$

The solution is -2.

36. $$\begin{aligned} 7-2(5x+3)&=1\\ 7-10x-6&=1\\ 1-10x&=1\\ 1-10x-1&=1-1\\ -10x&=0\\ \frac{-10x}{-10}&=\frac{0}{-10}\\ x&=0 \end{aligned}$$

The solution is 0.

37. $$\begin{aligned} -8+t&=t-8\\ -8+t-t&=t-8-t\\ -8&=-8 \end{aligned}$$

We have an equation that is true for all real numbers. Thus, all real numbers are solutions.

38. $$\begin{aligned} z+12&=-12+z\\ z+12-z&=-12+z-z\\ 12&=-12 \end{aligned}$$

We have a false equation. There are no solutions.

39. $$\begin{aligned} 4(3x+2)&=5(2x-1)\\ 12x+8&=10x-5\\ 12x+8-10x&=10x-5-10x\\ 2x+8&=-5\\ 2x+8-8&=-5-8\\ 2x&=-13\\ \frac{2x}{2}&=\frac{-13}{2}\\ x&=-\frac{13}{2} \end{aligned}$$

The solution is $-\frac{13}{2}$.

40. $$\begin{aligned} 8x-6-2x&=3(2x-4)+6\\ 6x-6&=6x-12+6\\ 6x-6&=6x-6\\ 6x-6-6x&=6x-6-6x\\ -6&=-6 \end{aligned}$$

We have an equation that is true for all real numbers. Thus, all real numbers are solutions.

41. $$\begin{aligned} A&=4b\\ \frac{A}{4}&=\frac{4b}{4}\\ \frac{A}{4}&=b \end{aligned}$$

42. $$\begin{aligned} y&=x-1.5\\ y+1.5&=x-1.5+1.5\\ y+1.5&=x \end{aligned}$$

43. $$\begin{aligned} n&=s-m\\ n-s&=s-m-s\\ n-s&=-m\\ -1(n-s)&=-1(-m)\\ -n+s&=m, \text{ or}\\ s-n&=m \end{aligned}$$

44. $$\begin{aligned} 4t&=9w\\ \frac{4t}{4}&=\frac{9w}{4}\\ t&=\frac{9w}{4} \end{aligned}$$

45. $$\begin{aligned} B&=at-c\\ B+c&=at-c+c\\ B+c&=at\\ \frac{B+c}{a}&=\frac{at}{a}\\ \frac{B+c}{a}&=t \end{aligned}$$

46. $$\begin{aligned} M&=\frac{x+y+z}{2}\\ 2\cdot M&=2\left(\frac{x+y+z}{2}\right)\\ 2M&=x+y+z\\ 2M-x-z&=x+y+z-x-z\\ 2M-x-z&=y \end{aligned}$$

47. Equivalent expressions have the same value for all possible replacements for the variable(s). Equivalent equations have the same solution(s).

48. The equations are not equivalent because they do not have the same solutions. Although 5 is a solution of both equations, -5 is a solution of $x^2 = 25$ but not of $x = 5$.

49. For an equation $x + a = b$, add the opposite of a (or subtract a) on both sides of the equation.

50. It appears that the student added $\frac{1}{3}$ on the right side of the equation rather than subtracting $\frac{1}{3}$.

51. For an equation $ax = b$, multiply by $1/a$ (or divide by a) on both sides of the equation.

52. Answers may vary. A walker who knows how far and how long she walks each day wants to know her average speed each day.

Exercise Set 10.5

1. ***Translate.***

$$\underbrace{\text{What percent}}_{\downarrow\; p} \;\text{of}\; \underbrace{180}_{\downarrow} \;\text{is}\; \underbrace{36?}_{\downarrow}$$

$$p \cdot 180 = 36$$

Solve. We divide by 36 on both sides and convert the answer to percent notation.

$$p \cdot 180 = 36$$
$$\frac{p \cdot 180}{180} = \frac{36}{180}$$
$$p = 0.2$$
$$p = 20\%$$

Thus, 36 is 20% of 180. The answer is 20%.

3. ***Translate.***

45 is 30% of what?

$$45 = 30\% \cdot b$$

Solve. We solve the equation.

$$45 = 30\% \cdot b$$
$$45 = 0.3b \quad \text{Converting to decimal notation}$$
$$\frac{45}{0.3} = \frac{0.3b}{0.3}$$
$$150 = b$$

Thus, 45 is 30% of 150. The answer is 150.

5. ***Translate.***

What is 65% of 840?

$$a = 65\% \cdot 840$$

Solve. We convert 65% to decimal notation and multiply.

$$a = 65\% \cdot 840$$
$$a = 0.65 \times 840$$
$$a = 546$$

Thus, 546 is 65% of 840. The answer is 546.

7. ***Translate.***

30 is what percent of 125?

$$30 = p \cdot 125$$

Solve. We solve the equation.

$$30 = p \cdot 125$$
$$\frac{30}{125} = \frac{p \cdot 125}{125}$$
$$0.24 = p$$
$$24\% = p$$

Thus, 30 is 24% of 125. The answer is 24%.

9. ***Translate.***

12% of what number is 0.3?

$$12\% \cdot b = 0.3$$

Solve. We solve the equation.

$$12\% \cdot b = 0.3$$
$$0.12b = 0.3 \quad \text{Converting to decimal notation}$$
$$\frac{0.12b}{0.12} = \frac{0.3}{0.12}$$
$$b = 2.5$$

Thus, 12% of 2.5 is 0.3. The answer is 2.5.

11. ***Translate.***

2 is what percent of 40?

$$2 = p \cdot 40$$

Solve. We divide by 40 on both sides and convert the answer to percent notation.

$$2 = p \cdot 40$$
$$\frac{2}{40} = \frac{p \cdot 40}{40}$$
$$0.05 = p$$
$$5\% = p$$

Thus, 2 is 5% of 40. The answer is 5%.

13. ***Translate.***

What percent of 68 is 17?

$$p \cdot 68 = 17$$

Solve. We divide by 68 on both sides and then convert to percent notation.

$$p \cdot 68 = 17$$
$$p = \frac{17}{68}$$
$$p = 0.25 = 25\%$$

The answer is 25%.

15. ***Translate.***

What is 35% of 240?

$$a = 35\% \cdot 240$$

Solve. We convert 35% to decimal notation and multiply.

$$a = 35\% \cdot 240$$
$$a = 0.35 \cdot 240$$
$$a = 84$$

The answer is 84.

17. ***Translate.***

What percent of 125 is 30?

$$p \cdot 125 = 30$$

Solve. We divide by 125 on both sides and then convert to percent notation.

$$p \cdot 125 = 30$$
$$p = \frac{30}{125}$$
$$p = 0.24 = 24\%$$

The answer is 24%.

19. ***Translate.***

What percent of 300 is 48?

↓ ↓ ↓ ↓ ↓

$p \cdot 300 = 48$

Solve. We divide by 300 on both sides and then convert to percent notation.

$$p \cdot 300 = 48$$
$$p = \frac{48}{300}$$
$$p = 0.16 = 16\%$$

The answer is 16%.

21. ***Translate.***

14 is 30% of what number?

↓ ↓ ↓ ↓ ↓

$14 = 30\% \cdot b$

Solve. We solve the equation.

$$14 = 0.3b \quad (30\% = 0.3)$$
$$\frac{14}{0.3} = b$$
$$46.\overline{6} = b$$

The answer is $46.\overline{6}$, or $46\frac{2}{3}$, or $\frac{140}{3}$.

23. ***Translate.***

What is 2% of 40?

↓ ↓ ↓ ↓ ↓

$a = 2\% \cdot 40$

Solve. We convert 2% to decimal notation and multiply.

$$a = 2\% \cdot 40$$
$$a = 0.02 \cdot 40$$
$$a = 0.8$$

The answer is 0.8.

25. ***Translate.***

0.8 is 16% of what number?

↓ ↓ ↓ ↓ ↓

$0.8 = 16\% \cdot b$

Solve. We solve the equation.

$$0.8 = 0.16b \quad (16\% = 0.16)$$
$$\frac{0.8}{0.16} = b$$
$$5 = b$$

The answer is 5.

27. ***Translate.***

54 is 135% of what number?

↓ ↓ ↓ ↓ ↓

$54 = 135\% \cdot b$

Solve. We solve the equation.

$$54 = 1.35b \quad (135\% = 1.35)$$
$$\frac{54}{1.35} = b$$
$$40 = b$$

The answer is 40.

29. First we reword and translate.

What is 39% of \$41.2?

↓ ↓ ↓ ↓ ↓

$a = 39\% \cdot 41.2$

Solve. We convert 39% to decimal notation and multiply.

$$a = 39\% \cdot 41.2 = 0.39 \cdot 41.2 \approx 16.1$$

About \$16.1 billion was spent on pet food.

31. First we reword and translate.

What is 5% of \$41.2?

↓ ↓ ↓ ↓ ↓

$a = 5\% \cdot 41.2$

Solve. We convert 5% to decimal notation and multiply.

$$a = 5\% \cdot 41.2 = 0.05 \cdot 41.2 \approx 2.1$$

About \$2.1 billion was spent purchasing pets.

33. First we reword and translate.

25.1 is what percent of 209?

↓ ↓ ↓ ↓ ↓

$25.1 = p \cdot 209$

Solve. We divide by 209 on both sides and convert to percent notation.

$$25.1 = p \cdot 209$$
$$\frac{25.1}{209} = p$$
$$0.12 \approx p$$
$$12\% \approx p$$

About 12% of total TV sales are projected to be smart TV sales.

35. First we reword and translate.

What is 58.8% of \$4.5?

↓ ↓ ↓ ↓ ↓

$a = 58.8\% \cdot 4.5$

Solve. We convert 58.8% to decimal notation and multiply.

$$a = 58.8\% \cdot 4.5 = 0.588 \cdot 4.5 = 2.646$$

In 2008, \$2.646 billion was given as graduation gifts.

37. First we reword and translate.

What is 6% of \$6500?

↓ ↓ ↓ ↓ ↓

$a = 6\% \cdot 6500$

Solve. We convert 6% to decimal notation and multiply.

$$a = 6\% \cdot 6500 = 0.06 \cdot 6500 = 390$$

Sarah will pay \$390 in interest.

39. a) First we reword and translate.

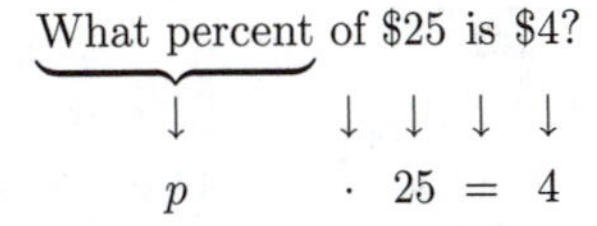

Solve. We divide by 25 on both sides and convert to percent notation.

$$p \cdot 25 = 4$$
$$\frac{p \cdot 25}{25} = \frac{4}{25}$$
$$p = 0.16$$
$$p = 16\%$$

The tip was 16% of the cost of the meal.

b) We add to find the total cost of the meal, including tip:

$\$25 + \$4 = \$29$

41. a) First we reword and translate.

$$\text{What is } 15\% \text{ of } \$25?$$
$$\downarrow \quad \downarrow \quad \downarrow \quad \downarrow \quad \downarrow$$
$$a = 15\% \cdot 25$$

Solve. We convert 15% to decimal notation and multiply.

$$a = 15\% \cdot 25$$
$$a = 0.15 \times 25$$
$$a = 3.75$$

The tip was \$3.75.

b) We add to find the total cost of the meal, including tip:

$\$25 + \$3.75 = \$28.75$

43. a) First we reword and translate.

$$15\% \text{ of what is } \$4.50?$$
$$\downarrow \quad \downarrow \quad \downarrow \quad \downarrow \quad \downarrow$$
$$15\% \cdot b = 4.50$$

Solve. We solve the equation.

$$15\% \cdot b = 4.50$$
$$0.15 \cdot b = 4.50$$
$$\frac{0.15 \cdot b}{0.15} = \frac{4.50}{0.15}$$
$$b = 30$$

The cost of the meal before the tip was \$30.

b) We add to find the total cost of the meal, including tip:

$\$30 + \$4.50 = \$34.50$

45. First we reword and translate.

$$15.1\% \text{ of what is } 12,959?$$
$$\downarrow \quad \downarrow \quad \downarrow \quad \downarrow \quad \downarrow$$
$$15.1\% \cdot b = 12,959$$

Solve. We solve the equation.

$$15.1\% \cdot b = 12,959$$
$$0.151 \cdot b = 12,959$$
$$\frac{0.151 \cdot b}{0.151} = \frac{12,959}{0.151}$$
$$b \approx 85,821$$

The total acreage of Portland is about 85,821 acres.

47. We subtract to find the increase.

$2304 - 1879 = 425$

The increase was 425 ft^2.

Now we find the percent of increase.

$$425 \text{ is } \underbrace{\text{what percent}} \text{ of } 1879?$$
$$\downarrow \; \downarrow \qquad \downarrow \qquad \downarrow \; \downarrow$$
$$425 = p \cdot 1879$$

We divide by 1879 on both sides and then convert to percent notation.

$$425 = p \cdot 1879$$
$$\frac{425}{1879} = p$$
$$0.226 \approx p$$
$$22.6\% \approx p$$

The percent of increase was about 22.6%.

49. We subtract to find the increase.

$36 - 4 = 32$

The increase is 32 billion gal.

Now we find the percent of increase.

$$32 \text{ is } \underbrace{\text{what percent}} \text{ of } 4?$$
$$\downarrow \; \downarrow \qquad \downarrow \qquad \downarrow \; \downarrow$$
$$32 = p \cdot 4$$

We divide by 4 on both sides and convert to percent notation.

$$32 = p \cdot 4$$
$$\frac{32}{4} = p$$
$$8 = p$$
$$800\% \approx p$$

The percent of increase is 800%.

51. We subtract to find the decrease.

$50 - 45 = 5$

The decrease is 5 thousand jobs.

Now we find the percent of decrease.

$$5 \text{ is } \underbrace{\text{what percent}} \text{ of } 50?$$
$$\downarrow \; \downarrow \qquad \downarrow \qquad \downarrow \; \downarrow$$
$$5 = p \cdot 50$$

We divide by 50 on both sides and convert to percent notation.

$$5 = p \cdot 50$$
$$\frac{5}{50} = p$$
$$0.1 = p$$
$$10\% = p$$

The percent of decrease is 10%.

53. We subtract to find the increase.

$127 - 52 = 75$

The increase was 75 partnerships.

Now we find the percent of increase.

75 is what percent of 52?

$$75 = p \cdot 52$$

We divide by 52 on both sides and convert to percent notation.

$$75 = p \cdot 52$$
$$\frac{75}{52} = p$$
$$1.44 \approx p$$
$$144\% \approx p$$

The percent of increase was about 144%.

55.
```
          1 8 1.5 2
0.0 5^ ⟌ 9.0 7^6 0
          5
          4 0
          4 0
              7
              5
              2 6
              2 5
                1 0
                1 0
                  0
```

The answer is 181.52.

57.
```
    1 1 1
    1. 0 8 9 0
   1 0. 8 9 0 0
 +  0. 1 0 8 9
   1 2. 0 8 7 9
```

59.
$$-5a + 3c - 2(c - 3a)$$
$$= -5a + 3c - 2 \cdot c - 2(-3a)$$
$$= -5a + 3c - 2c + 6a$$
$$= (-5 + 6)a + (3 - 2)c$$
$$= 1 \cdot a + 1 \cdot c$$
$$= a + c$$

61. $-6.5 + 2.6 = -3.9$ The absolute values are 6.5 and 2.6. The difference is 3.9. The negative number has the larger absolute value, so the answer is negative, -3.9.

63. To simplify the calculation $18 - 24 \div 3 - 48 \div (-4)$, do all the division calculations first, and then the subtraction calculations.

65. Since 6 ft = 6×1 ft = 6×12 in. = 72 in., we can express 6 ft 4 in. as 72 in. + 4 in., or 76 in.

Translate. We reword the problem.

96.1% of what is 76 in.?

$$96.1\% \cdot b = 76$$

Solve. We solve the equation.

$$96.1\% \cdot b = 76$$
$$0.961 \cdot b = 76$$
$$\frac{0.961 \cdot b}{0.961} = \frac{76}{0.961}$$
$$b \approx 79$$

Note that 79 in. = 72 in. + 7 in. = 6 ft 7 in.

Jaraan's final adult height will be about 6 ft 7 in.

Exercise Set 10.6

1. ***Familiarize.*** Let p = Florida's manatee population in 2006.

Translate.

2006 population less 296 is 2007 population.

$$p - 296 = 2817$$

Solve. We solve the equation.

$$p - 296 = 2817$$
$$p - 296 + 296 = 2817 + 296$$
$$p = 3113$$

Check. 296 less than 3113 is $3113 - 296$, or 2817. This is the population in 2007, so the answer checks.

State. In 2006, Florida's manatee population was 3113 manatees.

3. ***Familiarize.*** Using the labels on the drawing in the text, we let x = the length of the shorter piece, in inches, and $3x$ = the length of the longer piece, in inches.

Translate. We reword the problem.

The length of the shorter piece plus the length of the longer piece is 240 ft.

$$x + 3x = 240$$

Solve. We solve the equation.

$$x + 3x = 240$$
$$4x = 240 \quad \text{Collecting like terms}$$
$$\frac{4x}{4} = \frac{240}{4}$$
$$x = 60$$

If x is 60, then $3x = 3 \cdot 60$, or 180.

Check. 180 is three times 60, and $60 + 180 = 240$. The answer checks.

State. The lengths of the pieces are 60 in. and 180 in.

5. ***Familiarize.*** Let c = the average cost of movie tickets for a family of four in 1993.

Translate.

\$11.76 more than cost in 1993 is cost in 2008.

$$11.76 + c = 28.32$$

Solve. We solve the equation.

$$11.76 + c = 28.32$$
$$11.76 + c - 11.76 = 28.32 - 11.76$$
$$c = 16.56$$

Check. \$11.76 more than \$16.56 is \$11.76 + \$16.56, or \$28.32, the cost in 2008. The answer checks.

State. The average cost of movie tickets for a family of four was \$16.56 in 1993.

7. ***Familiarize***. Let d = the musher's distance from Nome, in miles. Then $2d$ = the distance from Anchorage, in miles. This is the number of miles the musher has completed. The sum of the two distances is the length of the race, 1049 miles.

Translate.

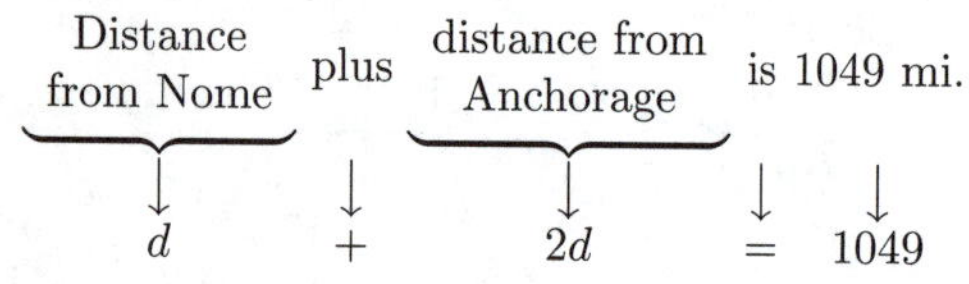

$$d + 2d = 1049$$

Solve. We solve the equation.

$$d + 2d = 1049$$
$$3d = 1049 \quad \text{Collecting like terms}$$
$$\frac{3d}{3} = \frac{1049}{3}$$
$$d = \frac{1049}{3}$$

If $d = \frac{1049}{3}$, then $2d = 2 \cdot \frac{1049}{3} = \frac{2098}{3} = 699\frac{1}{3}$.

Check. $\frac{2098}{3}$ is twice $\frac{1049}{3}$, and $\frac{1049}{3} + \frac{2098}{3} = \frac{3147}{3} = 1049$. The result checks.

State. The musher has traveled $699\frac{1}{3}$ miles.

9. ***Familiarize.*** Let x = the smaller number and $x+1$ = the larger number.

Translate. We reword the problem.

$$\underbrace{\text{First number}}_{x} + \underbrace{\text{second number}}_{(x+1)} \text{ is } \underbrace{2409}_{= 2409}$$

Solve. We solve the equation.

$$x + (x + 1) = 2409$$
$$2x + 1 = 2409 \quad \text{Collecting like terms}$$
$$2x + 1 - 1 = 2409 - 1 \quad \text{Subtracting 1}$$
$$2x = 2408$$
$$\frac{2x}{2} = \frac{2408}{2} \quad \text{Dividing by 2}$$
$$x = 1204$$

If x is 1204, then $x + 1$ is 1205.

Check. 1204 and 1205 are consecutive integers, and their sum is 2409. The answer checks.

State. The apartment numbers are 1204 and 1205.

11. ***Familiarize.*** Let a = the first number. Then $a + 1$ = the second number, and $a + 2$ = the third number.

Translate. We reword the problem.

$$\underbrace{\text{First number}}_{a} + \underbrace{\text{second number}}_{(a+1)} + \underbrace{\text{third number}}_{(a+2)} \text{ is } 126$$
$$a + (a+1) + (a+2) = 126$$

Solve. We solve the equation.

$$a + (a + 1) + (a + 2) = 126$$
$$3a + 3 = 126 \quad \text{Collecting like terms}$$
$$3a + 3 - 3 = 126 - 3$$
$$3a = 123$$
$$\frac{3a}{3} = \frac{123}{3}$$
$$a = 41$$

If a is 41, then $a + 1$ is 42 and $a + 2$ is 43.

Check. 41, 42, and 43 are consecutive integers, and their sum is 126. The answer checks.

State. The numbers are 41, 42, and 43.

13. ***Familiarize.*** Let x = the first odd integer. Then $x + 2$ = the next odd integer and $(x + 2) + 2$, or $x + 4$ = the third odd integer.

Translate. We reword the problem.

$$\underbrace{\text{First odd integer}}_{x} + \underbrace{\text{second odd integer}}_{(x+2)} + \underbrace{\text{third odd integer}}_{(x+4)} \text{ is } \underbrace{189}_{= 189}$$

Solve. We solve the equation.

$$x + (x + 2) + (x + 4) = 189$$
$$3x + 6 = 189 \quad \text{Collecting like terms}$$
$$3x + 6 - 6 = 189 - 6$$
$$3x = 183$$
$$\frac{3x}{3} = \frac{183}{3}$$
$$x = 61$$

If x is 61, then $x + 2$ is 63 and $x + 4$ is 65.

Check. 61, 63, and 65 are consecutive odd integers, and their sum is 189. The answer checks.

State. The integers are 61, 63, and 65.

15. ***Familiarize***. Using the labels on the drawing in the text, we let w = the width and $3w + 6$ = the length. The perimeter P of a rectangle is given by the formula $2l+2w = P$, where l = the length and w = the width.

Translate. Substitute $3w + 6$ for l and 124 for P:

$$2l + 2w = P$$
$$2(3w + 6) + 2w = 124$$

Solve. We solve the equation.

$$\begin{aligned} 2(3w+6)+2w &= 124 \\ 6w+12+2w &= 124 \\ 8w+12 &= 124 \\ 8w+12-12 &= 124-12 \\ 8w &= 112 \\ \frac{8w}{8} &= \frac{112}{8} \\ w &= 14 \end{aligned}$$

The possible dimensions are $w = 14$ ft and $l = 3w + 6 = 3(14) + 6$, or 48 ft.

Check. The length, 48 ft, is 6 ft more than three times the width, 14 ft. The perimeter is $2(48 \text{ ft}) + 2(14 \text{ ft}) = 96 \text{ ft} + 28 \text{ ft} = 124$ ft. The answer checks.

State. The width is 14 ft, and the length is 48 ft.

17. ***Familiarize***. Let $p =$ the regular price of the shoes. At 15% off, Amy paid 85% of the regular price.

Translate.

$$\underbrace{\$63.75}\ \text{is}\ \underbrace{85\%}\ \text{of}\ \underbrace{\text{the regular price.}}$$

$$63.75 = 0.85 \cdot p$$

Solve. We solve the equation.

$$\begin{aligned} 63.75 &= 0.85p \\ \frac{63.75}{0.08} &= p \qquad \text{Dividing both sides by 0.85} \\ 75 &= p \end{aligned}$$

Check. 85% of \$75, or 0.85(\$75), is \$63.75. The answer checks.

State. The regular price was \$75.

19. ***Familiarize***. Let $b =$ the price of the jacket itself. When the sales tax rate is 5%, the tax paid on the jacket is 5% of b, or $0.05b$.

Translate.

$$\underbrace{\text{Price of jacket}}\ \text{plus}\ \underbrace{\text{sales tax}}\ \text{is}\ \$89.25.$$

$$b + 0.05b = 89.25$$

Solve. We solve the equation.

$$\begin{aligned} b + 0.05b &= 89.25 \\ 1.05b &= 89.25 \\ b &= \frac{89.25}{1.05} \\ b &= 85 \end{aligned}$$

Check. 5% of \$85, or 0.05(\$85), is \$4.25 and \$85 + \$4.25 is \$89.25, the total cost. The answer checks.

State. The jacket itself cost \$85.

21. ***Familiarize***. Let $n =$ the number of visits required for a total parking cost of \$27.00. The parking cost for each $1\frac{1}{2}$ hour visit is \$1.50 for the first hour plus \$1.00 for part of a second hour, or \$2.50. Then the total parking cost for n visits is $2.50n$ dollars.

Translate. We reword the problem.

$$\underbrace{\text{Total parking cost}}\ \text{is}\ \$27.00.$$

$$2.50n = 27.00$$

Solve. We solve the equation.

$$\begin{aligned} 2.5n &= 27 \\ 10(2.5n) &= 10(27) \qquad \text{Clearing the decimal} \\ 25n &= 270 \\ \frac{25n}{25} &= \frac{270}{25} \\ n &= 10.8 \end{aligned}$$

If the total parking cost is \$27.00 for 10.8 visits, then the cost will be more than \$27.00 for 11 or more visits.

Check. The parking cost for 10 visits is \$2.50(10), or \$25, and the parking cost for 11 visits is \$2.50(11), or \$27.50. Since 11 is the smallest number for which the parking cost exceeds \$27.00, the answer checks.

State. The minimum number of weekly visits for which it is worthwhile to buy a parking pass is 11.

23. ***Familiarize.*** Let $x =$ the measure of the first angle. Then $3x =$ the measure of the second angle, and $x + 40 =$ the measure of the third angle. Recall that the sum of measures of the angles of a triangle is 180°.

Translate.

$$\underbrace{\text{Measure of first angle}} + \underbrace{\text{measure of second angle}} + \underbrace{\text{measure of third angle}}\ \text{is}\ 180.$$

$$x + 3x + (x+40) = 180$$

Solve. We solve the equation.

$$\begin{aligned} x + 3x + (x+40) &= 180 \\ 5x + 40 &= 180 \\ 5x + 40 - 40 &= 180 - 40 \\ 5x &= 140 \\ \frac{5x}{5} &= \frac{140}{5} \\ x &= 28 \end{aligned}$$

Possible answers for the angle measures are as follows:

First angle: $x = 28°$

Second angle: $3x = 3(28) = 84°$

Third angle: $x + 40 = 28 + 40 = 68°$

Check. Consider 28°, 84°, and 68°. The second angle is three times the first, and the third is 40° more than the first. The sum, 28° + 84° + 68°, is 180°. These numbers check.

State. The measures of the angles are 28°, 84°, and 68°.

25. ***Familiarize.*** Using the labels on the drawing in the text, we let $x =$ the measure of the first angle, $x + 5 =$ the measure of the second angle, and $3x + 10 =$ the measure of the third angle. Recall that the sum of measures of the angles of a triangle is 180°.

Translate.

$$\underbrace{\text{Measure of first angle}} + \underbrace{\text{measure of second angle}} + \underbrace{\text{measure of third angle}} \text{ is } 180.$$

$$x + (x+5) + (3x+10) = 180$$

Solve. We solve the equation.

$$\begin{aligned} x+(x+5)+(3x+10) &= 180 \\ 5x+15 &= 180 \\ 5x+15-15 &= 180-15 \\ 5x &= 165 \\ \frac{5x}{5} &= \frac{165}{5} \\ x &= 33 \end{aligned}$$

Possible answers for the angle measures are as follows:

First angle: $x = 33°$

Second angle: $x+5 = 33+5 = 38°$

Third angle: $3x+10 = 3(33)+10 = 109°$

Check. The second angle is 5° more than the first, and the third is 10° more than 3 times the first. The sum, 33° + 38° + 109°, is 180°. The numbers check.

State. The measures of the angles are 33°, 38°, and 109°.

27. ***Familiarize.*** Let a = the amount Sarah invested. The investment grew by 28% of a, or $0.28a$.

Translate.

$$\underbrace{\text{Amount invested}} \text{ plus } \underbrace{\text{amount of growth}} \text{ is } \$448.$$

$$a + 0.28a = 448$$

Solve. We solve the equation.

$$\begin{aligned} a+0.28a &= 448 \\ 1.28a &= 448 \\ a &= 350 \end{aligned}$$

Check. 28% of \$350 is 0.28(\$350), or \$98, and \$350 + \$98 = \$448. The answer checks.

State. Sarah invested \$350.

29. ***Familiarize.*** Let b = the balance on the credit card at the beginning of the month. The balance grew by 2% of b, or $0.02b$.

Translate.

$$\underbrace{\text{Original balance}} \text{ plus } \underbrace{\text{amount of growth}} \text{ is } \$870.$$

$$b + 0.02b = 870$$

Solve. We solve the equation.

$$\begin{aligned} b+0.02b &= 870 \\ 1.02b &= 870 \\ b &\approx \$852.94 \end{aligned}$$

Check. 2% of \$852.94 is 0.02(\$852.94), or \$17.06, and \$852.94 + \$17.06 = \$870. The answer checks.

State. The balance at the beginning of the month was \$852.94.

31. ***Familiarize.*** The total cost is the initial charge plus the mileage charge. Let d = the distance, in miles, that Courtney can travel for \$12. The mileage charge is the cost per mile times the number of miles traveled or $0.75d$.

Translate.

$$\underbrace{\text{Initial charge}} \text{ plus } \underbrace{\text{mileage charge}} \text{ is } \$12.$$

$$3 + 0.75d = 12$$

Solve. We solve the equation.

$$\begin{aligned} 3+0.75d &= 12 \\ 0.75d &= 9 \\ d &= 12 \end{aligned}$$

Check. A 12-mi taxi ride from the airport would cost \$3 + 12(\$0.75), or \$3 + \$9, or \$12. The answer checks.

State. Courtney can travel 12 mi from the airport for \$12.

33. ***Familiarize.*** Let c = the cost of the meal before the tip. We know that the cost of the meal before the tip plus the tip, 15% of the cost, is the total cost, \$41.40.

Translate.

$$\underbrace{\text{Cost of meal}} \text{ plus tip is } \$41.40$$

$$c + 15\%c = 41.40$$

Solve. We solve the equation.

$$\begin{aligned} c+15\%c &= 41.40 \\ c+0.15c &= 41.40 \\ 1c+0.15c &= 41.40 \\ 1.15c &= 41.40 \\ \frac{1.15c}{1.15} &= \frac{41.40}{1.15} \\ c &= 36 \end{aligned}$$

Check. We find 15% of \$36 and add it to \$36:

$15\% \times \$36 = 0.15 \times \$36 = \$5.40$ and $\$36 + \$5.40 = \$41.40$.

The answer checks.

State. The cost of the meal before the tip was added was \$36.

35. ***Familiarize.*** Tom paid a total of $3 \cdot \$34$, or \$102, for the three ties. Let t = the price of one of the ties. Then $2t$ = the price of another and we are told that the remaining tie cost \$27.

Translate.

$$\underbrace{\text{Total cost of the ties}} \text{ is } \$102.$$

$$t + 2t + 27 = 102$$

Solve. We solve the equation.

$$\begin{aligned} t+2t+27 &= 102 \\ 3t+27 &= 102 \\ 3t+27-27 &= 102-27 \\ 3t &= 75 \\ \frac{3t}{3} &= \frac{75}{3} \\ t &= 25 \end{aligned}$$

If $t = 25$, then $2t = 2 \cdot 25 = 50$.

Check. The \$50 tie costs twice as much as the \$25 tie, and the total cost of the ties is \$25 + \$50 + \$27, or \$102. The answer checks.

State. One tie cost \$25 and another cost \$50.

37. ***Familiarize.*** Let $y =$ the number.

Translate. We reword the problem.

$$\underbrace{\text{Two times a number}}_{\downarrow \; 2\cdot y} \quad \underset{+}{\text{plus}} \quad \underset{16}{16} \quad \underset{=}{\text{is}} \quad \underbrace{\frac{2}{3}\text{ of the number}}_{\downarrow \; \frac{2}{3}\cdot y}$$

Solve. We solve the equation.

$$\begin{aligned} 2y + 16 &= \frac{2}{3}y \\ 3(2y+16) &= 3 \cdot \frac{2}{3}y && \text{Clearing the fraction} \\ 6y + 48 &= 2y \\ 48 &= -4y && \text{Subtracting 6y} \\ -12 &= y && \text{Dividing by } -8 \end{aligned}$$

Check. We double -12 and get -24. Adding 16, we get -8. Also, $\frac{2}{3}(-12) = -8$. The answer checks.

State. The number is -12.

39. $$\begin{aligned} -\frac{4}{5} - \frac{3}{8} &= -\frac{4}{5} + \left(-\frac{3}{8}\right) \\ &= -\frac{32}{40} + \left(-\frac{15}{40}\right) \\ &= -\frac{47}{40} \end{aligned}$$

41. $$\begin{aligned} -\frac{4}{5} \cdot \frac{3}{8} &= -\frac{4 \cdot 3}{5 \cdot 8} \\ &= -\frac{4 \cdot 3}{5 \cdot 2 \cdot 4} \\ &= -\frac{\not{4} \cdot 3}{5 \cdot 2 \cdot \not{4}} \\ &= -\frac{3}{10} \end{aligned}$$

43. $\frac{1}{10} \div \left(-\frac{1}{100}\right) = \frac{1}{10} \cdot \left(-\frac{100}{1}\right) = -\frac{1 \cdot 100}{10 \cdot 1} =$

$-\frac{\not{1} \cdot \not{10} \cdot 10}{\not{10} \cdot \not{1} \cdot 1} = -\frac{10}{1} = -10$

45. $-25.6(-16) = 409.6$

47. $-25.6 + (-16) = -41.6$

49. ***Familiarize.*** Let $a =$ the original number of apples. Then $\frac{1}{3}a$, $\frac{1}{4}a$, $\frac{1}{8}a$, and $\frac{1}{5}a$ are given to four people, respectively. The fifth and sixth people get 10 apples and 1 apple, respectively.

Translate. We reword the problem.

$$\underbrace{\text{The total number of apples}}_{\downarrow \; \frac{1}{3}a + \frac{1}{4}a + \frac{1}{8}a + \frac{1}{5}a + 10 + 1} \quad \underset{=}{\text{is}} \quad \underbrace{a}_{a}$$

Solve. We solve the equation.

$$\begin{aligned} \frac{1}{3}a + \frac{1}{4}a + \frac{1}{8}a + \frac{1}{5}a + 10 + 1 &= a, \text{ LCD is 120} \\ 120\left(\frac{1}{3}a + \frac{1}{4}a + \frac{1}{8}a + \frac{1}{5}a + 11\right) &= 120 \cdot a \\ 40a + 30a + 15a + 24a + 1320 &= 120a \\ 109a + 1320 &= 120a \\ 1320 &= 11a \\ 120 &= a \end{aligned}$$

Check. If the original number of apples was 120, then the first four people got $\frac{1}{3} \cdot 120$, $\frac{1}{4} \cdot 120$, $\frac{1}{8} \cdot 120$, and $\frac{1}{5} \cdot 120$, or 40, 30, 15, and 24 apples, respectively. Adding all the apples we get 40 + 30 + 15 + 24 + 10 + 1, or 120. The result checks.

State. There were originally 120 apples in the basket.

51. Divide the largest triangle into three triangles, each with a vertex at the center of the circle and with height x as shown.

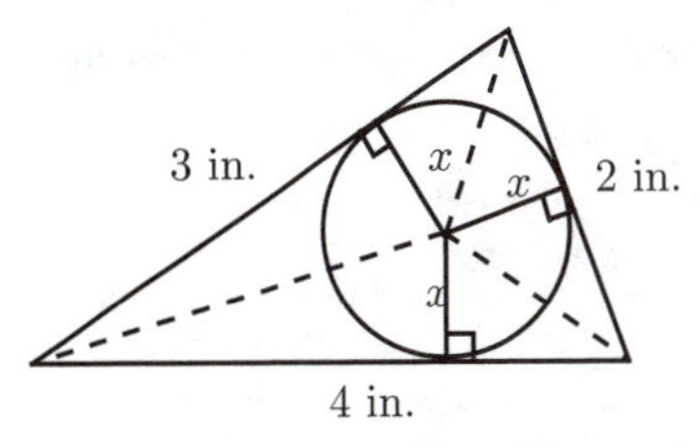

Then the sum of the areas of the three smaller triangles is the area of the original triangle. We have:

$$\begin{aligned} \frac{1}{2} \cdot 3x + \frac{1}{2} \cdot 2x + \frac{1}{2} \cdot 4x &= 2.9047 \\ 2\left(\frac{1}{2} \cdot 3x + \frac{1}{2} \cdot 2x + \frac{1}{2} \cdot 4x\right) &= 2(2.9047) \\ 3x + 2x + 4x &= 5.8094 \\ 9x &= 5.8094 \\ x &\approx 0.65 \end{aligned}$$

Thus, x is about 0.65 in.

53. ***Familiarize.*** Let $p =$ the price of the gasoline as registered on the pump. Then the sales tax will be $9\%p$.

Translate. We reword the problem.

$$\underbrace{\text{Price on pump}}_{\downarrow \; p} \quad \underset{+}{\text{plus}} \quad \underbrace{\text{sales tax}}_{\downarrow \; 9\%p} \quad \underset{=}{\text{is}} \quad \underbrace{\$10}_{\downarrow \; 10}$$

Solve. We solve the equation.

$$p + 9\%p = 10$$
$$1p + 0.09p = 10$$
$$1.09p = 10$$
$$\frac{1.09p}{1.09} = \frac{10}{1.09}$$
$$p \approx 9.17$$

Check. We find 9% of \$9.17 and add it to \$9.17:

$$9\% \times \$9.17 = 0.09 \times \$9.17 \approx \$0.83$$

Then \$9.17 + \$0.83 = \$10, so \$9.17 checks.

State. The attendant should have filled the tank until the pump read \$9.17, not \$9.10.

Exercise Set 10.7

1. $x > -4$

a) Since $4 > -4$ is true, 4 is a solution.

b) Since $0 > -4$ is true, 0 is a solution.

c) Since $-4 > -4$ is false, -4 is not a solution.

d) Since $6 > -4$ is true, 6 is a solution.

e) Since $5.6 > -4$ is true, 5.6 is a solution.

3. $x \geq 6.8$

a) Since $-6 \geq 6.8$ is false, -6 is not a solution.

b) Since $0 \geq 6.8$ is false, 0 is not a solution.

c) Since $6 \geq 6.8$ is false, 6 is not a solution.

d) Since $8 \geq 6.8$ is true, 8 is a solution.

e) Since $-3\frac{1}{2} \geq 6.8$ is false, $-3\frac{1}{2}$ is not a solution.

5. The solutions of $x > 4$ are those numbers greater than 4. They are shown on the graph by shading all points to the right of 4. The open circle at 4 indicates that 4 is not part of the graph.

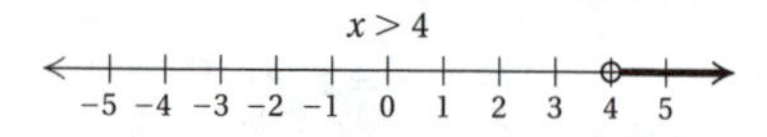

7. The solutions of $t < -3$ are those numbers less than -3. They are shown on the graph by shading all points to the left of -3. The open circle at -3 indicates that -3 is not part of the graph.

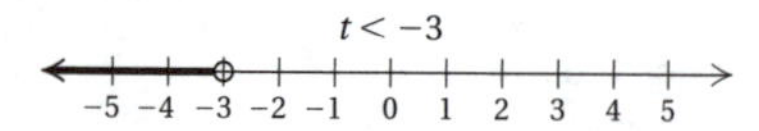

9. The solutions of $m \geq -1$ are are shown by shading the point for -1 and all points to the right of -1. The closed circle at -1 indicates that -1 is part of the graph.

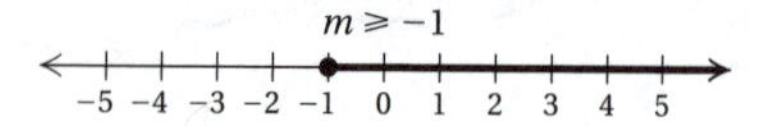

11. In order to be a solution of the inequality $-3 < x \leq 4$, a number must be a solution of both $-3 < x$ and $x \leq 4$. The solution set is graphed as follows:

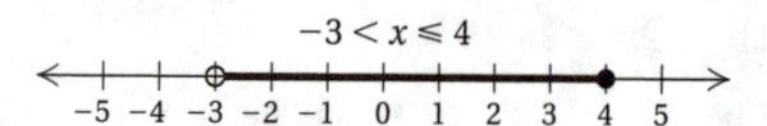

The open circle at -3 means that -3 is not part of the graph. The closed circle at 4 means that 4 is part of the graph.

13. In order to be a solution of the inequality $0 < x < 3$, a number must be a solution of both $0 < x$ and $x < 3$. The solution set is graphed as follows:

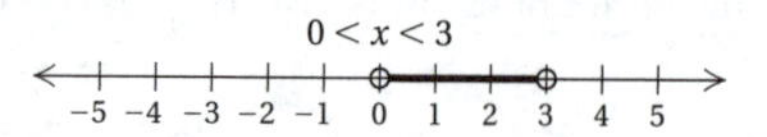

The open circles at 0 and at 3 mean that 0 and 3 are not part of the graph.

15.
$$\begin{aligned} x + 7 &> 2 \\ x + 7 - 7 &> 2 - 7 \quad \text{Subtracting 7} \\ x &> -5 \quad \text{Simplifying} \end{aligned}$$

The solution set is $\{x|x > -5\}$.

The graph is as follows:

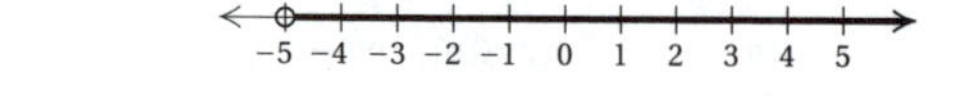

17.
$$\begin{aligned} x + 8 &\leq -10 \\ x + 8 - 8 &\leq -10 - 8 \quad \text{Subtracting 8} \\ x &\leq -18 \quad \text{Simplifying} \end{aligned}$$

The solution set is $\{x|x \leq -18\}$.

The graph is as follows:

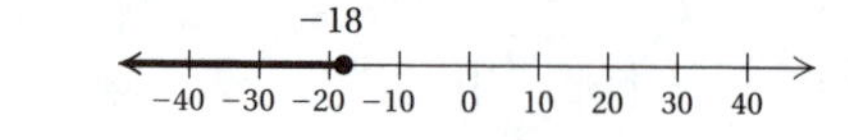

19.
$$\begin{aligned} y - 7 &> -12 \\ y - 7 + 7 &> -12 + 7 \quad \text{Adding 7} \\ y &> -5 \quad \text{Simplifying} \end{aligned}$$

The solution set is $\{y|y > -5\}$.

21.
$$\begin{aligned} 2x + 3 &> x + 5 \\ 2x + 3 - 3 &> x + 5 - 3 \quad \text{Subtracting 3} \\ 2x &> x + 2 \quad \text{Simplifying} \\ 2x - x &> x + 2 - x \quad \text{Subtracting } x \\ x &> 2 \quad \text{Simplifying} \end{aligned}$$

The solution set is $\{x|x > 2\}$.

23.
$$\begin{aligned} 3x + 9 &\leq 2x + 6 \\ 3x + 9 - 9 &\leq 2x + 6 - 9 \quad \text{Subtracting 9} \\ 3x &\leq 2x - 3 \quad \text{Simplifying} \\ 3x - 2x &\leq 2x - 3 - 2x \quad \text{Subtracting } 2x \\ x &\leq -3 \quad \text{Simplifying} \end{aligned}$$

The solution set is $\{x|x \leq -3\}$.

25.
$$\begin{aligned} 5x - 6 &< 4x - 2 \\ 5x - 6 + 6 &< 4x - 2 + 6 \\ 5x &< 4x + 4 \\ 5x - 4x &< 4x + 4 - 4x \\ x &< 4 \end{aligned}$$

The solution set is $\{x|x < 4\}$.

27.
$$\begin{aligned} -9+t &> 5 \\ -9+t+9 &> 5+9 \\ t &> 14 \end{aligned}$$

The solution set is $\{t|t > 14\}$.

29.
$$\begin{aligned} y+\frac{1}{4} &\le \frac{1}{2} \\ y+\frac{1}{4}-\frac{1}{4} &\le \frac{1}{2}-\frac{1}{4} \\ y &\le \frac{2}{4}-\frac{1}{4} && \text{Obtaining a common denominator} \\ y &\le \frac{1}{4} \end{aligned}$$

The solution set is $\left\{y\middle|y \le \frac{1}{4}\right\}$.

31.
$$\begin{aligned} x-\frac{1}{3} &> \frac{1}{4} \\ x-\frac{1}{3}+\frac{1}{3} &> \frac{1}{4}+\frac{1}{3} \\ x &> \frac{3}{12}+\frac{4}{12} && \text{Obtaining a common denominator} \\ x &> \frac{7}{12} \end{aligned}$$

The solution set is $\left\{x\middle|x > \frac{7}{12}\right\}$.

33.
$$\begin{aligned} 5x &< 35 \\ \frac{5x}{5} &< \frac{35}{5} && \text{Dividing by 5} \\ x &< 7 \end{aligned}$$

The solution set is $\{x|x < 7\}$. The graph is as follows:

−8 −6 −4 −2 0 2 4 6 8

35.
$$\begin{aligned} -12x &> -36 \\ \frac{-12x}{-12} &< \frac{-36}{-12} && \text{Dividing by } -12 \end{aligned}$$

↑ The symbol has to be reversed.

$$x < 3 \qquad \text{Simplifying}$$

The solution set is $\{x|x < 3\}$. The graph is as follows:

−5 −4 −3 −2 −1 0 1 2 3 4 5

37.
$$\begin{aligned} 5y &\ge -2 \\ \frac{5y}{5} &\ge \frac{-2}{5} && \text{Dividing by 5} \\ y &\ge -\frac{2}{5} \end{aligned}$$

The solution set is $\left\{y\middle|y \ge -\frac{2}{5}\right\}$.

39.
$$\begin{aligned} -2x &\le 12 \\ \frac{-2x}{-2} &\ge \frac{12}{-2} && \text{Dividing by } -2 \end{aligned}$$

↑ The symbol has to be reversed.

$$x \ge -6 \qquad \text{Simplifying}$$

The solution set is $\{x|x \ge -6\}$.

41.
$$\begin{aligned} -4y &\ge -16 \\ \frac{-4y}{-4} &\le \frac{-16}{-4} && \text{Dividing by } -4 \end{aligned}$$

↑ The symbol has to be reversed.

$$y \le 4 \qquad \text{Simplifying}$$

The solution set is $\{y|y \le 4\}$.

43.
$$\begin{aligned} -3x &< -17 \\ \frac{-3x}{-3} &> \frac{-17}{-3} && \text{Dividing by } -3 \end{aligned}$$

↑ The symbol has to be reversed.

$$x > \frac{17}{3} \qquad \text{Simplifying}$$

The solution set is $\left\{x\middle|x > \frac{17}{3}\right\}$.

45.
$$\begin{aligned} -2y &> \frac{1}{7} \\ -\frac{1}{2}\cdot(-2y) &< -\frac{1}{2}\cdot\frac{1}{7} \end{aligned}$$

↑ The symbol has to be reversed.

$$y < -\frac{1}{14}$$

The solution set is $\left\{y\middle|y < -\frac{1}{14}\right\}$.

47.
$$\begin{aligned} -\frac{6}{5} &\le -4x \\ -\frac{1}{4}\cdot\left(-\frac{6}{5}\right) &\ge -\frac{1}{4}\cdot(-4x) \\ \frac{6}{20} &\ge x \\ \frac{3}{10} &\ge x, \text{ or } x \le \frac{3}{10} \end{aligned}$$

The solution set is $\left\{x\middle|\frac{3}{10} \ge x\right\}$, or $\left\{x\middle|x \le \frac{3}{10}\right\}$.

49.
$$\begin{aligned} 4+3x &< 28 \\ -4+4+3x &< -4+28 && \text{Adding } -4 \\ 3x &< 24 && \text{Simplifying} \\ \frac{3x}{3} &< \frac{24}{3} && \text{Dividing by 3} \\ x &< 8 \end{aligned}$$

The solution set is $\{x|x < 8\}$.

51.
$$\begin{aligned} 3x-5 &\le 13 \\ 3x-5+5 &\le 13+5 && \text{Adding 5} \\ 3x &\le 18 \\ \frac{3x}{3} &\le \frac{18}{3} && \text{Dividing by 3} \\ x &\le 6 \end{aligned}$$

The solution set is $\{x|x \le 6\}$.

53.
$$\begin{aligned} 13x-7 &< -46 \\ 13x-7+7 &< -46+7 \\ 13x &< -39 \\ \frac{13x}{13} &< \frac{-39}{13} \\ x &< -3 \end{aligned}$$

The solution set is $\{x|x < -3\}$.

55.
$$30 > 3 - 9x$$
$$30 - 3 > 3 - 9x - 3 \quad \text{Subtracting 3}$$
$$27 > -9x$$
$$\frac{27}{-9} < \frac{-9x}{-9} \quad \text{Dividing by } -9$$
↑ The symbol has to be reversed.
$$-3 < x$$
The solution set is $\{x| -3 < x\}$, or $\{x|x > -3\}$.

57.
$$4x + 2 - 3x \le 9$$
$$x + 2 \le 9 \quad \text{Collecting like terms}$$
$$x + 2 - 2 \le 9 - 2$$
$$x \le 7$$
The solution set is $\{x|x \le 7\}$.

59.
$$-3 < 8x + 7 - 7x$$
$$-3 < x + 7 \quad \text{Collecting like terms}$$
$$-3 - 7 < x + 7 - 7$$
$$-10 < x$$
The solution set is $\{x| -10 < x\}$, or $\{x|x > -10\}$.

61.
$$6 - 4y > 4 - 3y$$
$$6 - 4y + 4y > 4 - 3y + 4y \quad \text{Adding } 4y$$
$$6 > 4 + y$$
$$-4 + 6 > -4 + 4 + y \quad \text{Adding } -4$$
$$2 > y, \text{ or } y < 2$$
The solution set is $\{y|2 > y\}$, or $\{y|y < 2\}$.

63.
$$5 - 9y \le 2 - 8y$$
$$5 - 9y + 9y \le 2 - 8y + 9y$$
$$5 \le 2 + y$$
$$-2 + 5 \le -2 + 2 + y$$
$$3 \le y, \text{ or } y \ge 3$$
The solution set is $\{y|3 \le y\}$, or $\{y|y \ge 3\}$.

65.
$$19 - 7y - 3y < 39$$
$$19 - 10y < 39 \quad \text{Collecting like terms}$$
$$-19 + 19 - 10y < -19 + 39$$
$$-10y < 20$$
$$\frac{-10y}{-10} > \frac{20}{-10}$$
↑ The symbol has to be reversed.
$$y > -2$$
The solution set is $\{y|y > -2\}$.

67.
$$0.9x + 19.3 > 5.3 - 2.6x$$
$$10(0.9x + 19.3) > 10(5.3 - 2.6x) \quad \text{Multiplying by 10 to clear decimals}$$
$$9x + 193 > 53 - 26x$$
$$35x > -140 \quad \text{Adding } 26x \text{ and subtracting 193}$$
$$x > -4 \quad \text{Dividing by 35}$$
The solution set is $\{x|x > -4\}$.

69.
$$\frac{x}{3} - 2 \le 1$$
$$3\left(\frac{x}{3} - 2\right) \le 3 \cdot 1 \quad \text{Multiplying by 3 to to clear the fraction}$$
$$x - 6 \le 3 \quad \text{Simplifying}$$
$$x \le 9 \quad \text{Adding 6}$$
The solution set is $\{x|x \le 9\}$.

71.
$$\frac{y}{5} + 1 \le \frac{2}{5}$$
$$5\left(\frac{y}{5} + 1\right) \le 5 \cdot \frac{2}{5} \quad \text{Clearing fractions}$$
$$y + 5 \le 2$$
$$y \le -3 \quad \text{Subtracting 5}$$
The solution set is $\{y|y \le -3\}$.

73.
$$3(2y - 3) < 27$$
$$6y - 9 < 27 \quad \text{Removing parentheses}$$
$$6y < 36 \quad \text{Adding 9}$$
$$y < 6 \quad \text{Dividing by 6}$$
The solution set is $\{y|y < 6\}$.

75.
$$2(3 + 4m) - 9 \ge 45$$
$$6 + 8m - 9 \ge 45 \quad \text{Removing parentheses}$$
$$8m - 3 \ge 45 \quad \text{Collecting like terms}$$
$$8m \ge 48 \quad \text{Adding 3}$$
$$m \ge 6 \quad \text{Dividing by 8}$$
The solution set is $\{m|m \ge 6\}$.

77.
$$8(2t + 1) > 4(7t + 7)$$
$$16t + 8 > 28t + 28$$
$$16t - 28t > 28 - 8$$
$$-12t > 20$$
$$t < -\frac{20}{12} \quad \text{Dividing by } -12 \text{ and reversing the symbol}$$
$$t < -\frac{5}{3}$$
The solution set is $\left\{t\middle|t < -\frac{5}{3}\right\}$.

79.
$$3(r - 6) + 2 < 4(r + 2) - 21$$
$$3r - 18 + 2 < 4r + 8 - 21$$
$$3r - 16 < 4r - 13$$
$$-16 + 13 < 4r - 3r$$
$$-3 < r, \text{ or } r > -3$$
The solution set is $\{r|r > -3\}$.

81.
$$0.8(3x + 6) \ge 1.1 - (x + 2)$$
$$2.4x + 4.8 \ge 1.1 - x - 2$$
$$10(2.4x + 4.8) \ge 10(1.1 - x - 2) \quad \text{Clearing decimals}$$
$$24x + 48 \ge 11 - 10x - 20$$
$$24x + 48 \ge -10x - 9 \quad \text{Collecting like terms}$$
$$24x + 10x \ge -9 - 48$$
$$34x \ge -57$$
$$x \ge -\frac{57}{34}$$
The solution set is $\left\{x\middle|x \ge -\frac{57}{34}\right\}$.

83. $\frac{5}{3}+\frac{2}{3}x<\frac{25}{12}+\frac{5}{4}x+\frac{3}{4}$

The number 12 is the least common multiple of all the denominators. We multiply by 12 on both sides.

$$12\left(\frac{5}{3}+\frac{2}{3}x\right)<12\left(\frac{25}{12}+\frac{5}{4}x+\frac{3}{4}\right)$$
$$12\cdot\frac{5}{3}+12\cdot\frac{2}{3}x<12\cdot\frac{25}{12}+12\cdot\frac{5}{4}x+12\cdot\frac{3}{4}$$
$$20+8x<25+15x+9$$
$$20+8x<34+15x$$
$$20-34<15x-8x$$
$$-14<7x$$
$$-2<x,\text{ or } x>-2$$

The solution set is $\{x|x>-2\}$.

85. $-56+(-18)$ Two negative numbers. Add the absolute values and make the answer negative.

$-56+(-18)=-74$

87. $-\frac{3}{4}+\frac{1}{8}$ One negative and one positive number. Find the difference of the absolute values. Then make the answer negative, since the negative number has the larger absolute value.

$-\frac{3}{4}+\frac{1}{8}=-\frac{6}{8}+\frac{1}{8}=-\frac{5}{8}$

89. $-56-(-18)=-56+18=-38$

91. $-2.3-7.1=-2.3+(-7.1)=-9.4$

93.
$$\begin{aligned}5-3^2+(8-2)^2\cdot 4&=5-3^2+6^2\cdot 4\\&=5-9+36\cdot 4\\&=5-9+144\\&=-4+144\\&=140\end{aligned}$$

95. $5(2x-4)-3(4x+1)=10x-20-12x-3=$ $-2x-23$

97. $|x|<3$

a) Since $|0|=0$ and $0<3$ is true, 0 is a solution.

b) Since $|-2|=2$ and $2<3$ is true, -2 is a solution.

c) Since $|-3|=3$ and $3<3$ is false, -3 is not a solution.

d) Since $|4|=4$ and $4<3$ is false, 4 is not a solution.

e) Since $|3|=3$ and $3<3$ is false, 3 is not a solution.

f) Since $|1.7|=1.7$ and $1.7<3$ is true, 1.7 is a solution.

g) Since $|-2.8|=2.8$ and $2.8<3$ is true, -2.8 is a solution.

99.
$$x+3<3+x$$
$$x-x<3-3\quad\text{Subtracting } x \text{ and } 3$$
$$0<0$$

We get a false inequality. There is no solution.

Exercise Set 10.8

1. $n\ge 7$

3. $w>2$ kg

5. 90 mph $<s<$ 110 mph

7. $w\le 20$ hr

9. $c\ge \$1.50$

11. $x>8$

13. $y\le -4$

15. $n\ge 1300$

17. $w\le 500$ L

19. $3x+2<13$, or $2+3x<13$

21. ***Familiarize***. Let s represent the score on the fourth test.

Translate.

$$\underbrace{\text{The average score}}\ \underbrace{\text{is at least}}\ 80.$$
$$\frac{82+76+78+s}{4}\quad\ge\quad 80$$

Solve.

$$\frac{82+76+78+s}{4}\ge 80$$
$$4\left(\frac{82+76+78+s}{4}\right)\ge 4\cdot 80$$
$$82+76+78+s\ge 320$$
$$236+s\ge 320$$
$$s\ge 84$$

Check. As a partial check we show that the average is at least 80 when the fourth test score is 84.

$$\frac{82+76+78+84}{4}=\frac{320}{4}=80$$

State. James will get at least a B if the score on the fourth test is at least 84. The solution set is $\{s|s\ge 84\}$.

23. ***Familiarize***. We use the formula for converting Celsius temperatures to Fahrenheit temperatures, $F=\frac{9}{5}C+32$.

Translate.

$$\underbrace{\text{Fahrenheit temperature}}\ \underbrace{\text{is less than}}\ 1945.4.$$
$$\frac{9}{5}C+32\quad\le\quad 1945.4$$

Solve.

$$\frac{9}{5}C+32<1945.4$$
$$\frac{9}{5}C<1913.4$$
$$\frac{5}{9}\cdot\frac{9}{5}C<\frac{5}{9}(1913.4)$$
$$C<1063$$

Check. As a partial check we can show that the Fahrenheit temperature is less than 1945.4°for a Celsius temperature less than 1063°and is greater than 1945.4°for a Celsius temperature greater than 1063°.

$$F = \frac{9}{5} \cdot 1062 + 32 = 1943.6 < 1945.4$$

$$F = \frac{9}{5} \cdot 1064 + 32 = 1947.2 > 1945.4$$

State. Gold stays solid for temperatures less than 1063°C. The solution set is $\{C|C < 1063°\}$.

25. ***Familiarize.*** $R = -0.075t + 3.85$

In the formula R represents the world record and t represents the years since 1930. When $t = 0$ (1930), the record was $-0.075 \cdot 0 + 3.85$, or 3.85 minutes. When $t = 2$ (1932), the record was $-0.075(2) + 3.85$, or 3.7 minutes. For what values of t will $-0.075t + 3.85$ be less than 3.5?

Translate. The record is to be less than 3.5. We have the inequality

$$R < 3.5.$$

To find the t values which satisfy this condition we substitute $-0.075t + 3.85$ for R.

$$-0.075t + 3.85 < 3.5$$

Solve.

$$\begin{aligned} -0.075t + 3.85 &< 3.5 \\ -0.075t &< 3.5 - 3.85 \\ -0.075t &< -0.35 \\ t &> \frac{-0.35}{-0.075} \\ t &> 4\frac{2}{3} \end{aligned}$$

Check. With inequalities it is impossible to check each solution. But we can check to see if the solution set we obtained seems reasonable.

When $t = 4\frac{1}{2}$, $R = -0.075(4.5) + 3.85$, or 3.5125.

When $t = 4\frac{2}{3}$, $R = -0.075\left(\frac{14}{3}\right) + 3.85$, or 3.5.

When $t = 4\frac{3}{4}$, $R = -0.075(4.75) + 3.85$, or 3.49375.

Since $r = 3.5$ when $t = 4\frac{2}{3}$ and R decreases as t increases, R will be less than 3.5 when t is greater than $4\frac{2}{3}$.

State. The world record will be less than 3.5 minutes more than $4\frac{2}{3}$ years after 1930. If we let $Y =$ the year, then the solution set is $\{Y|Y \geq 1935\}$.

27. ***Familiarize.*** As in the drawing in the text, we let $L =$ the length of the envelope. Recall that the area of a rectangle is the product of the length and the width.

Translate.

$$\underbrace{\text{Length}}_{\downarrow \atop L} \quad \underset{\downarrow \atop \cdot}{\text{times}} \quad \underbrace{\text{width}}_{\downarrow \atop 3\frac{1}{2}} \quad \underbrace{\text{is at least}}_{\downarrow \atop \geq} \quad \underbrace{17\tfrac{1}{2}\text{ in}^2}_{\downarrow \atop 17\frac{1}{2}}$$

Solve.

$$\begin{aligned} L \cdot 3\frac{1}{2} &\geq 17\frac{1}{2} \\ L \cdot \frac{7}{2} &\geq \frac{35}{2} \\ L \cdot \frac{7}{2} \cdot \frac{2}{7} &\geq \frac{35}{2} \cdot \frac{2}{7} \\ L &\geq 5 \end{aligned}$$

The solution set is $\{L|L \geq 5\}$.

Check. We can obtain a partial check by substituting a number greater than or equal to 5 in the inequality. For example, when $L = 6$:

$$L \cdot 3\frac{1}{2} = 6 \cdot 3\frac{1}{2} = 6 \cdot \frac{7}{2} = 21 \geq 17\frac{1}{2}$$

The result appears to be correct.

State. Lengths of 5 in. or more will satisfy the constraints. The solution set is $\{L|L \geq 5 \text{ in.}\}$.

29. ***Familiarize.*** Let $c =$ the number of copies Myra has made. The total cost of the copies is the setup fee of \$5 plus \$4 times the number of copies, or $\$4 \cdot c$.

Translate.

$$\underbrace{\text{Setup fee}}_{\downarrow \atop 5} \quad \underset{\downarrow \atop +}{\text{plus}} \quad \underbrace{\text{copying cost}}_{\downarrow \atop 4c} \quad \underbrace{\text{cannot exceed}}_{\downarrow \atop \leq} \quad \underset{\downarrow \atop 65}{\$65.}$$

Solve. We solve the inequality.

$$\begin{aligned} 5 + 4c &\leq 65 \\ 4c &\leq 60 \\ c &\leq 15 \end{aligned}$$

Check. As a partial check, we show that Myra can have 15 copies made and not exceed her \$65 budget.

$$\$5 + \$4 \cdot 15 = 5 + 60 = \$65$$

State. Myra can have 15 or fewer copies made and stay within her budget.

31. ***Familiarize.*** Let m represent the length of a telephone call, in minutes.

Translate.

$$\underbrace{\text{\$0.75 charge}}_{\downarrow \atop 0.75} \quad \underset{\downarrow \atop +}{\text{plus}} \quad \underbrace{\text{charge for time used}}_{\downarrow \atop 0.45m} \quad \underbrace{\text{is at least}}_{\downarrow \atop \geq} \quad \underset{\downarrow \atop 3}{\$3.00.}$$

Solve. We solve the inequality.

$$\begin{aligned} 0.75 + 0.45m &\geq 3 \\ 0.45m &\geq 2.25 \\ m &\geq 5 \end{aligned}$$

Check. As a partial check, we can show that if a call lasts 5 minutes it costs at least \$3.00:

$$\$0.75 + \$0.45(5) = \$0.75 + \$2.25 = \$3.00.$$

State. Simon's calls last at least 5 minutes each.

33. ***Familiarize.*** Let c = the number of courses for which Angelica registers. Her total tuition is the \$35 registration fee plus \$375 times the number of courses for which she registers, or $\$375 \cdot c$.

Translate.

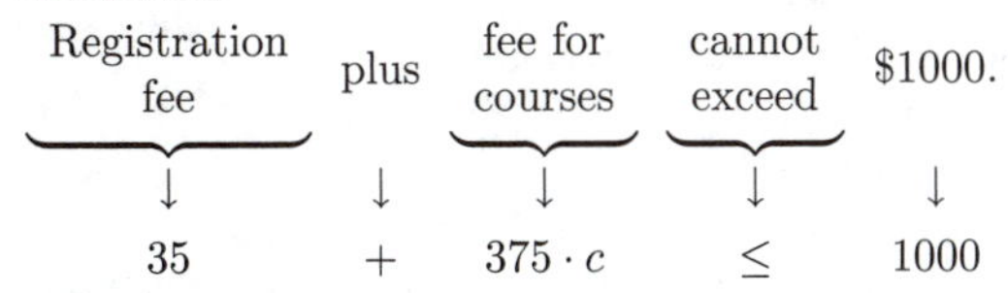

Solve. We solve the inequality.

$$35 + 375c \leq 1000$$
$$375c \leq 965$$
$$c \leq 2.57\overline{3}$$

Check. Although the solution set of the inequality is all numbers less than or equal to $2.57\overline{3}$, since c represents the number of courses for which Angelica registers, we round down to 2. If she registers for 2 courses, her tuition is $\$35 + \$375 \cdot 2$, or \$785 which does not exceed \$1000. If she registers for 3 courses, her tuition is $\$35 + \$375 \cdot 3$, or \$1160 which exceeds \$1000.

State. Angelica can register for at most 2 courses.

35. ***Familiarize.*** Let s = the number of servings of fruits or vegetables Dale eats on Saturday.

Translate.

Average number of fruit or vegetable servings ⟶ $\dfrac{4+6+7+4+6+4+s}{7}$

is at least ⟶ $\geq$

5. ⟶ 5

Solve. We first multiply by 7 to clear the fraction.

$$7\left(\frac{4+6+7+4+6+4+s}{7}\right) \geq 7 \cdot 5$$
$$4+6+7+4+6+4+s \geq 35$$
$$31 + s \geq 35$$
$$s \geq 4$$

Check. As a partial check, we show that Dale can eat 4 servings of fruits or vegetables on Saturday and average at least 5 servings per day for the week:

$$\frac{4+6+7+4+6+4+4}{7} = \frac{35}{7} = 5$$

State. Dale should eat at least 4 servings of fruits or vegetables on Saturday.

37. ***Familiarize.*** We first make a drawing. We let l represent the length, in feet.

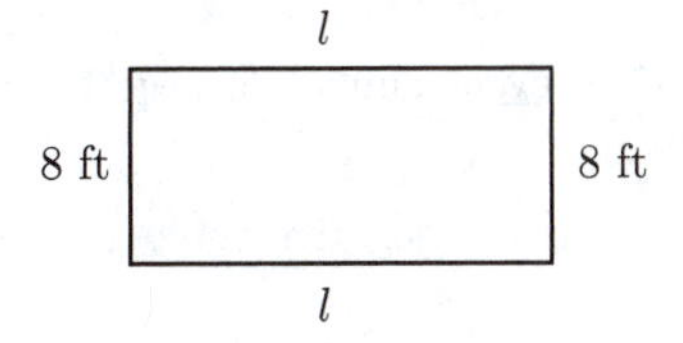

The perimeter is $P = 2l + 2w$, or $2l + 2 \cdot 8$, or $2l + 16$.

Translate. We translate to 2 inequalities.

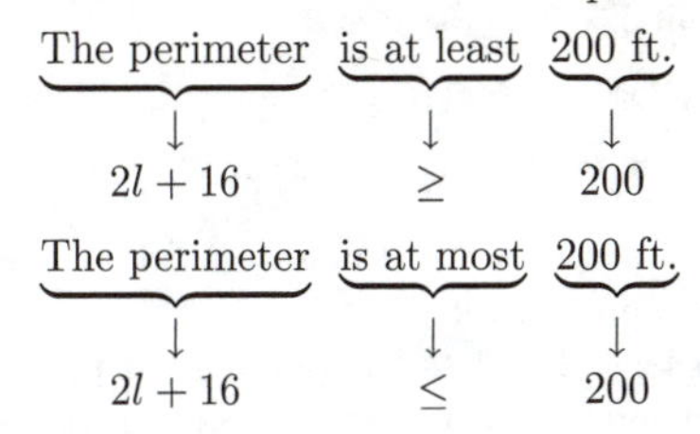

Solve. We solve each inequality.

$$2l + 16 \geq 200 \qquad 2l + 16 \leq 200$$
$$2l \geq 184 \qquad 2l \leq 184$$
$$l \geq 92 \qquad l \leq 92$$

Check. We check to see if the solutions seem reasonable.

When $l = 91$ ft, $P = 2 \cdot 91 + 16$, or 198 ft.

When $l = 92$ ft, $P = 2 \cdot 92 + 16$, or 200 ft.

When $l = 93$ ft, $P = 2 \cdot 93 + 16$, or 202 ft.

From these calculations, it appears that the solutions are correct.

State. Lengths greater than or equal to 92 ft will make the perimeter at least 200 ft. Lengths less than or equal to 92 ft will make the perimeter at most 200 ft.

39. ***Familiarize.*** Using the label on the drawing in the text, we let L represent the length.

The area is the length times the width, or $4L$.

Translate.

Area ⟶ $4L$

is less than ⟶ $<$

86 cm². ⟶ 86

Solve.

$$4L < 86$$
$$L < 21.5$$

Check. We check to see if the solution seems reasonable.

When $L = 22$, the area is $22 \cdot 4$, or 88 cm^2.

When $L = 21.5$, the area is $21.5(4)$, or 86 cm^2.

When $L = 21$, the area is $21 \cdot 4$, or 84 cm^2.

From these calculations, it would appear that the solution is correct.

State. The area will be less than 86 cm^2 for lengths less than 21.5 cm.

41. ***Familiarize.*** Let v = the blue book value of the car. Since the car was repaired, we know that \$8500 does not exceed $0.8v$ or, in other words, $0.8v$ is at least \$8500.

Translate.

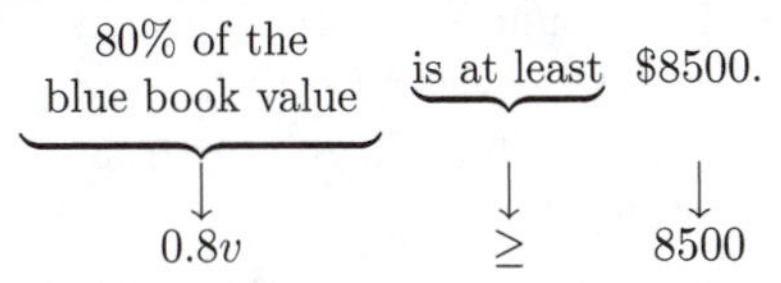

Solve.

$$0.8v \geq 8500$$
$$v \geq \frac{8500}{0.8}$$
$$v \geq 10,625$$

Check. As a partial check, we show that 80% of $10,625 is at least $8500:

$0.8(\$10,625) = \$8500$

State. The blue book value of the car was at least $10,625.

43. ***Familiarize***. Let r = the amount of fat in a serving of the regular peanut butter, in grams. If reduced fat peanut butter has at least 25% less fat than regular peanut butter, then it has at most 75% as much fat as the regular peanut butter.

Translate.

12 g of fat	is at most	75%	of	the amount of fat in regular peanut butter.
↓	↓	↓	↓	↓
12	$\leq$	0.75	$\cdot$	r

Solve.

$12 \leq 0.75r$

$16 \leq r$

Check. As a partial check, we show that 12 g of fat does not exceed 75% of 16 g of fat:

$0.75(16) = 12$

State. Regular peanut butter contains at least 16 g of fat per serving.

45. ***Familiarize.*** Let w = the number of weeks after July 1. After w weeks the water level has dropped $\frac{2}{3}w$ ft.

Translate.

Original depth	minus	drop in water level	does not exceed	21 ft.
↓	↓	↓	↓	↓
25	$-$	$\frac{2}{3}w$	$\leq$	21

Solve. We solve the inequality.

$$25 - \frac{2}{3}w \leq 21$$
$$-\frac{2}{3}w \leq -4$$
$$w \geq -\frac{3}{2}(-4)$$
$$w \geq 6$$

Check. As a partial check we show that the water level is 21 ft 6 weeks after July 1.

$$25 - \frac{2}{3} \cdot 6 = 25 - 4 = 21 \text{ ft}$$

Since the water level continues to drop during the weeks after July 1, the answer seems reasonable.

State. The water level will not exceed 21 ft for dates at least 6 weeks after July 1.

47. ***Familiarize***. Let h = the height of the triangle, in ft. Recall that the formula for the area of a triangle with base b and height h is $A = \frac{1}{2}bh$.

Translate.

Area	is at least	3 ft².
↓	↓	↓
$\frac{1}{2}\left(1\frac{1}{2}\right)h$	$\geq$	3

Solve. We solve the inequality.

$$\frac{1}{2}\left(1\frac{1}{2}\right)h \geq 3$$
$$\frac{1}{2} \cdot \frac{3}{2} \cdot h \geq 3$$
$$\frac{3}{4}h \geq 3$$
$$h \geq \frac{4}{3} \cdot 3$$
$$h \geq 3$$

Check. As a partial check, we show that the area of the triangle is 3 ft^2 when the height is 4 ft.

$$\frac{1}{2}\left(1\frac{1}{2}\right)(4) = \frac{1}{2} \cdot \frac{3}{2} \cdot \frac{4}{1} = 3$$

State. The height should be at least 4 ft.

49. ***Familiarize***. The average number of calls per week is the sum of the calls for the three weeks divided by the number of weeks, 3. We let c represent the number of calls made during the third week.

Translate. The average of the three weeks is given by

$$\frac{17 + 22 + c}{3}.$$

Since the average must be at least 20, this means that it must be greater than or equal to 20. Thus, we can translate the problem to the inequality

$$\frac{17 + 22 + c}{3} \geq 20.$$

Solve. We first multiply by 3 to clear the fraction.

$$3\left(\frac{17 + 22 + c}{3}\right) \geq 3 \cdot 20$$
$$17 + 22 + c \geq 60$$
$$39 + c \geq 60$$
$$c \geq 21$$

Check. Suppose c is a number greater than or equal to 21. Then by adding 17 and 22 on both sides of the inequality we get

$$17 + 22 + c \geq 17 + 22 + 21$$
$$17 + 22 + c \geq 60$$

so

$$\frac{17 + 22 + c}{3} \geq \frac{60}{3}, \text{ or } 20.$$

State. 21 calls or more will maintain an average of at least 20 for the three-week period.

51. The product of an even number of negative numbers is always positive.

53. The additive inverse of a negative number is always positive.

55. Equations with the same solutions are called equivalent equations.

57. The multiplication principle for inequalities asserts that when we multiply or divide by a negative number on both sides of an inequality, the direction of the inequality symbol is reversed.

59. ***Familiarize***. We use the formula $F = \frac{9}{5}C + 32$.

Translate. We are interested in temperatures such that $5° < F < 15°$. Substituting for F, we have:

$$5 < \frac{9}{5}C + 32 < 15$$

Solve.

$$\begin{aligned} 5 &< \frac{9}{5}C + 32 < 15 \\ 5 \cdot 5 &< 5\left(\frac{9}{5}C + 32\right) < 5 \cdot 15 \\ 25 &< 9C + 160 < 75 \\ -135 &< 9C < -85 \\ -15 &< C < -9\frac{4}{9} \end{aligned}$$

Check. The check is left to the student.

State. Green ski wax works best for temperatures between $-15°$C and $-9\frac{4}{9}°$C.

61. ***Familiarize***. Let f = the fat content of a serving of regular tortilla chips, in grams. A product that contains 60% less fat than another product has 40% of the fat content of that product. If Reduced Fat Tortilla Pops cannot be labeled lowfat, then they contain at least 3 g of fat.

Translate.

40%	of	the fat content of regular tortilla chips	is at least	3 grams of fat
↓	↓	↓	↓	↓
0.4	·	f	$\geq$	3

Solve.

$$\begin{aligned} 0.4f &\geq 3 \\ f &\geq 7.5 \end{aligned}$$

Check. As a partial check, we show that 40% of 7.5 g is not less than 3 g.

$$0.4(7.5) = 3$$

State. A serving of regular tortilla chips contains at least 7.5 g of fat.

Chapter 10 Concept Reinforcement

1. True; see page 716 in the text.

2. True; for any number n, $n \geq n$ is true because $n = n$ is true.

3. False; the solution set of $2x - 7 \leq 11$ is $\{x|x \leq 9\}$; the solution set of $x < 2$ is $\{x|x < 2\}$. The inequalities do not have the same solution set, so they are not equivalent.

4. True; if $x > y$, then $-1 \cdot x < -1 \cdot y$ (reversing the inequality symbol), or $-x < -y$.

Chapter 10 Important Concepts

1.

$$\begin{aligned} 4(x-3) &= 6(x+2) \\ 4x - 12 &= 6x + 12 \\ 4x - 12 - 6x &= 6x + 12 - 6x \\ -2x - 12 &= 12 \\ -2x - 12 + 12 &= 12 + 12 \\ -2x &= 24 \\ \frac{-2x}{-2} &= \frac{24}{-2} \\ x &= -12 \end{aligned}$$

The solution is -12.

2.

$$\begin{aligned} 4 + 3y - 7 &= 3 + 3(y-2) \\ 4 + 3y - 7 &= 3 + 3y - 6 \\ 3y - 3 &= -3 + 3y \\ 3y - 3 - 3y &= -3 + 3y - 3y \\ -3 &= -3 \end{aligned}$$

Every real number is a solution of the equation $-3 = -3$, so all real numbers are solutions of the original equation.

3.

$$\begin{aligned} 4(x-3) + 7 &= -5 + 4x + 10 \\ 4x - 12 + 7 &= -5 + 4x + 10 \\ 4x - 5 &= 5 + 4x \\ 4x - 5 - 4x &= 5 + 4x - 4x \\ -5 &= 5 \end{aligned}$$

We get a false equation, so the original equation has no solution.

4.

$$\begin{aligned} A &= \frac{1}{2}bh \\ 2 \cdot A &= 2 \cdot \frac{1}{2}bh \\ 2A &= bh \\ \frac{2A}{h} &= \frac{bh}{h} \\ \frac{2A}{h} &= b \end{aligned}$$

5. Graph: $x > 1$

The solutions of $x > 1$ are all numbers greater than 1. We shade all points to the right of 1 and use an open circle at 1 to indicate that 1 is not part of the graph.

$x > 1$

−6 −5 −4 −3 −2 −1 0 1 2 3 4 5 6

6. Graph: $x \leq -1$

The solutions of $x \leq -1$ are all numbers less than or equal to -1. We shade all points to the left of -1 and use a closed circle at -1 to indicate that -1 is part of the graph.

$x \leq -1$

−1 0

7.
$$\begin{aligned} 6y+5 &> 3y-7 \\ 6y+5-3y &> 3y-7-3y \\ 3y+5 &> -7 \\ 3y+5-5 &> -7-5 \\ 3y &> -12 \\ \frac{3y}{3} &> \frac{-12}{3} \\ y &> -4 \end{aligned}$$
The solution set is $\{y|y > -4\}$.

Chapter 10 Review Exercises

1.
$$\begin{aligned} x+5 &= -17 \\ x+5-5 &= -17-5 \\ x &= -22 \end{aligned}$$
The solution is -22.

2.
$$\begin{aligned} n-7 &= -6 \\ n-7+7 &= -6+7 \\ n &= 1 \end{aligned}$$
The solution is 1.

3.
$$\begin{aligned} x-11 &= 14 \\ x-11+11 &= 14+11 \\ x &= 25 \end{aligned}$$
The solution is 25.

4.
$$\begin{aligned} y-0.9 &= 9.09 \\ y-0.9+0.9 &= 9.09+0.9 \\ y &= 9.99 \end{aligned}$$
The solution is 9.99.

5.
$$\begin{aligned} -\frac{2}{3}x &= -\frac{1}{6} \\ -\frac{3}{2}\cdot\left(-\frac{2}{3}x\right) &= -\frac{3}{2}\cdot\left(-\frac{1}{6}\right) \\ 1\cdot x &= \frac{\cancel{3}\cdot 1}{2\cdot 2\cdot \cancel{3}} \\ x &= \frac{1}{4} \end{aligned}$$
The solution is $\frac{1}{4}$.

6.
$$\begin{aligned} -8x &= -56 \\ \frac{-8x}{-8} &= \frac{-56}{-8} \\ x &= 7 \end{aligned}$$
The solution is 7.

7.
$$\begin{aligned} -\frac{x}{4} &= 48 \\ 4\cdot\frac{1}{4}\cdot(-x) &= 4\cdot 48 \\ -x &= 192 \\ -1\cdot(-1\cdot x) &= -1\cdot 192 \\ x &= -192 \end{aligned}$$
The solution is -192.

8.
$$\begin{aligned} 15x &= -35 \\ \frac{15x}{15} &= \frac{-35}{15} \\ x &= -\frac{\cancel{5}\cdot 7}{3\cdot\cancel{5}} \\ x &= -\frac{7}{3} \end{aligned}$$
The solution is $-\frac{7}{3}$.

9.
$$\begin{aligned} \frac{4}{5}y &= -\frac{3}{16} \\ \frac{5}{4}\cdot\frac{4}{5}y &= \frac{5}{4}\cdot\left(-\frac{3}{16}\right) \\ y &= -\frac{15}{64} \end{aligned}$$
The solution is $-\frac{15}{64}$.

10.
$$\begin{aligned} 5-x &= 13 \\ 5-x-5 &= 13-5 \\ -x &= 8 \\ -1\cdot(-1\cdot x) &= -1\cdot 8 \\ x &= -8 \end{aligned}$$
The solution is -8.

11.
$$\begin{aligned} \frac{1}{4}x-\frac{5}{8} &= \frac{3}{8} \\ \frac{1}{4}x-\frac{5}{8}+\frac{5}{8} &= \frac{3}{8}+\frac{5}{8} \\ \frac{1}{4}x &= 1 \\ 4\cdot\frac{1}{4}x &= 4\cdot 1 \\ x &= 4 \end{aligned}$$
The solution is 4.

12.
$$\begin{aligned} 5t+9 &= 3t-1 \\ 5t+9-3t &= 3t-1-3t \\ 2t+9 &= -1 \\ 2t+9-9 &= -1-9 \\ 2t &= -10 \\ \frac{2t}{2} &= \frac{-10}{2} \\ t &= -5 \end{aligned}$$
The solution is -5.

13.
$$\begin{aligned} 7x-6 &= 25x \\ 7x-6-7x &= 25x-7x \\ -6 &= 18x \\ \frac{-6}{18} &= \frac{18x}{18} \\ -\frac{\cancel{6}\cdot 1}{3\cdot\cancel{6}} &= x \\ -\frac{1}{3} &= x \end{aligned}$$
The solution is $-\frac{1}{3}$.

14.
$$\begin{aligned}14y &= 23y - 17 - 10\\ 14y &= 23y - 27 \quad \text{Collecting like terms}\\ 14y - 23y &= 23y - 27 - 23y\\ -9y &= -27\\ \frac{-9y}{-9} &= \frac{-27}{-9}\\ y &= 3\end{aligned}$$
The solution is 3.

15.
$$\begin{aligned}0.22y - 0.6 &= 0.12y + 3 - 0.8y\\ 0.22y - 0.6 &= -0.68y + 3 \quad \text{Collecting like terms}\\ 0.22y - 0.6 + 0.68y &= -0.68y + 3 + 0.68y\\ 0.9y - 0.6 &= 3\\ 0.9y - 0.6 + 0.6 &= 3 + 0.6\\ 0.9y &= 3.6\\ \frac{0.9y}{0.9} &= \frac{3.6}{0.9}\\ y &= 4\end{aligned}$$
The solution is 4.

16.
$$\begin{aligned}\frac{1}{4}x - \frac{1}{8}x &= 3 - \frac{1}{16}x\\ \frac{2}{8}x - \frac{1}{8}x &= 3 - \frac{1}{16}x\\ \frac{1}{8}x &= 3 - \frac{1}{16}x\\ \frac{1}{8}x + \frac{1}{16}x &= 3 - \frac{1}{16}x + \frac{1}{16}x\\ \frac{2}{16}x + \frac{1}{16}x &= 3\\ \frac{3}{16}x &= 3\\ \frac{16}{3} \cdot \frac{3}{16}x &= \frac{16}{3} \cdot 3\\ x &= \frac{16 \cdot \cancel{3}}{\cancel{3} \cdot 1}\\ x &= 16\end{aligned}$$
The solution is 16.

17.
$$\begin{aligned}14y + 17 + 7y &= 9 + 21y + 8\\ 21y + 17 &= 21y + 17\\ 21y + 17 - 21y &= 21y + 17 - 21y\\ 17 &= 17 \qquad \text{TRUE}\end{aligned}$$
All real numbers are solutions.

18.
$$\begin{aligned}4(x + 3) &= 36\\ 4x + 12 &= 36\\ 4x + 12 - 12 &= 36 - 12\\ 4x &= 24\\ \frac{4x}{4} &= \frac{24}{4}\\ x &= 6\end{aligned}$$
The solution is 6.

19.
$$\begin{aligned}3(5x - 7) &= -66\\ 15x - 21 &= -66\\ 15x - 21 + 21 &= -66 + 21\\ 15x &= -45\\ \frac{15x}{15} &= \frac{-45}{15}\\ x &= -3\end{aligned}$$
The solution is -3.

20.
$$\begin{aligned}8(x - 2) - 5(x + 4) &= 20 + x\\ 8x - 16 - 5x - 20 &= 20 + x\\ 3x - 36 &= 20 + x\\ 3x - 36 - x &= 20 + x - x\\ 2x - 36 &= 20\\ 2x - 36 + 36 &= 20 + 36\\ 2x &= 56\\ \frac{2x}{2} &= \frac{56}{2}\\ x &= 28\end{aligned}$$
The solution is 28.

21.
$$\begin{aligned}-5x + 3(x + 8) &= 16\\ -5x + 3x + 24 &= 16\\ -2x + 24 &= 16\\ -2x + 24 - 24 &= 16 - 24\\ -2x &= -8\\ \frac{-2x}{-2} &= \frac{-8}{-2}\\ x &= 4\end{aligned}$$
The solution is 4.

22.
$$\begin{aligned}6(x - 2) - 16 &= 3(2x - 5) + 11\\ 6x - 12 - 16 &= 6x - 15 + 11\\ 6x - 28 &= 6x - 4\\ 6x - 28 - 6x &= 6x - 4 - 6x\\ -28 &= -4 \qquad \text{False}\end{aligned}$$
There is no solution.

23. Since $-3 \le 4$ is true, -3 is a solution.

24. Since $7 \le 4$ is false, 7 is not a solution.

25. Since $4 \le 4$ is true, 4 is a solution.

26.
$$\begin{aligned}y + \frac{2}{3} &\ge \frac{1}{6}\\ y + \frac{2}{3} - \frac{2}{3} &\ge \frac{1}{6} - \frac{2}{3}\\ y &\ge \frac{1}{6} - \frac{4}{6}\\ y &\ge -\frac{3}{6}\\ y &\ge -\frac{1}{2}\end{aligned}$$
The solution set is $\left\{y \middle| y \ge -\frac{1}{2}\right\}$.

27.
$$9x \geq 63$$
$$\frac{9x}{9} \geq \frac{63}{9}$$
$$x \geq 7$$
The solution set is $\{x|x \geq 7\}$.

28.
$$2 + 6y > 14$$
$$2 + 6y - 2 > 14 - 2$$
$$6y > 12$$
$$\frac{6y}{6} > \frac{12}{6}$$
$$y > 2$$
The solution set is $\{y|y > 2\}$.

29.
$$7 - 3y \geq 27 + 2y$$
$$7 - 3y - 2y \geq 27 + 2y - 2y$$
$$7 - 5y \geq 27$$
$$7 - 5y - 7 \geq 27 - 7$$
$$-5y \geq 20$$
$$\frac{-5y}{-5} \leq \frac{20}{-5} \quad \text{Reversing the inequality symbol}$$
$$y \leq -4$$
The solution set is $\{y|y \leq -4\}$.

30.
$$3x + 5 < 2x - 6$$
$$3x + 5 - 2x < 2x - 6 - 2x$$
$$x + 5 < -6$$
$$x + 5 - 5 < -6 - 5$$
$$x < -11$$
The solution set is $\{x|x < -11\}$.

31.
$$-4y < 28$$
$$\frac{-4y}{-4} > \frac{28}{-4} \quad \text{Reversing the inequality symbol}$$
$$y > -7$$
The solution set is $\{y|y > -7\}$.

32.
$$4 - 8x < 13 + 3x$$
$$4 - 8x - 3x < 13 + 3x - 3x$$
$$4 - 11x < 13$$
$$4 - 11x - 4 < 13 - 4$$
$$-11x < 9$$
$$\frac{-11x}{-11} > \frac{9}{-11} \quad \text{Reversing the inequality symbol}$$
$$x > -\frac{9}{11}$$
The solution set is $\left\{x\middle|x > -\frac{9}{11}\right\}$.

33.
$$-4x \leq \frac{1}{3}$$
$$-\frac{1}{4} \cdot (-4x) \geq -\frac{1}{4} \cdot \frac{1}{3} \quad \text{Reversing the inequality symbol}$$
$$x \geq -\frac{1}{12}$$
The solution set is $\left\{x\middle|x \geq -\frac{1}{12}\right\}$.

34.
$$4x - 6 < x + 3$$
$$4x - 6 - x < x + 3 - x$$
$$3x - 6 < 3$$
$$3x - 6 + 6 < 3 + 6$$
$$3x < 9$$
$$\frac{3x}{3} < \frac{9}{3}$$
$$x < 3$$
The solution set is $\{x|x < 3\}$. The graph is as follows:

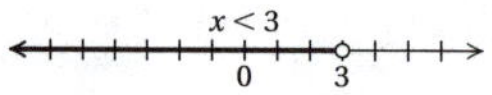

$x < 3$

0 3

35. In order to be a solution of $-2 < x \leq 5$, a number must be a solution of both $-2 < x$ and $x \leq 5$. The solution set is graphed as follows:

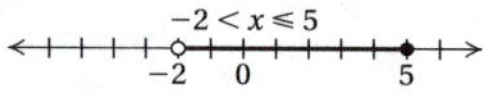

$-2 < x \leq 5$

−2 0 5

36. The solutions of $y > 0$ are those numbers greater than 0. The graph is as follows:

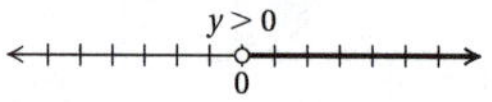

$y > 0$

0

37.
$$C = \pi d$$
$$\frac{C}{\pi} = \frac{\pi d}{\pi}$$
$$\frac{C}{\pi} = d$$

38.
$$V = \frac{1}{3}Bh$$
$$3 \cdot V = 3 \cdot \frac{1}{3}Bh$$
$$3V = Bh$$
$$\frac{3V}{h} = \frac{Bh}{h}$$
$$\frac{3V}{h} = B$$

39.
$$A = \frac{a+b}{2}$$
$$2 \cdot A = 2 \cdot \left(\frac{a+b}{2}\right)$$
$$2A = a + b$$
$$2A - b = a + b - b$$
$$2A - b = a$$

40.

$$y = mx + b$$
$$y - b = mx + b - b$$
$$y - b = mx$$
$$\frac{y-b}{m} = \frac{mx}{m}$$
$$\frac{y-b}{m} = x$$

41. ***Familiarize.*** Let $w =$ the width, in miles. Then $w + 90 =$ the length. Recall that the perimeter P of a rectangle with length l and width w is given by $P = 2l + 2w$.

Translate. Substitute 1280 for P and $w + 90$ for l in the formula above.

$$P = 2l + 2w$$
$$1280 = 2(w + 90) + 2w$$

Solve. We solve the equation.

$$1280 = 2(w + 90) + 2w$$
$$1280 = 2w + 180 + 2w$$
$$1280 = 4w + 180$$
$$1280 - 180 = 4w + 180 - 180$$
$$1100 = 4w$$
$$\frac{1100}{4} = \frac{4w}{4}$$
$$275 = w$$

If $w = 275$, then $w + 90 = 275 + 90 = 365$.

Check. The length, 365 mi, is 90 mi more than the width, 275 mi. The perimeter is $2 \cdot 365$ mi $+ 2 \cdot 275$ mi $= 730$ mi $+$ 550 mi $= 1280$ mi. The answer checks.

State. The length is 365 mi, and the width is 275 mi.

42. ***Familiarize.*** Let $x =$ the number on the first marker. Then $x + 1 =$ the number on the second marker.

Translate.

$$\underbrace{\text{First number}}_{x} \quad \underbrace{\text{plus}}_{+} \quad \underbrace{\text{second number}}_{(x+1)} \quad \underbrace{\text{is}}_{=} \quad \underbrace{691.}_{691}$$

Solve. We solve the equation.

$$x + (x + 1) = 691$$
$$2x + 1 = 691$$
$$2x + 1 - 1 = 691 - 1$$
$$2x = 690$$
$$\frac{2x}{2} = \frac{690}{2}$$
$$x = 345$$

If $x = 345$, then $x + 1 = 345 + 1 = 346$.

Check. 345 and 346 are consecutive integers and $345 + 346 = 691$. The answer checks.

State. The numbers on the markers are 345 and 346.

43. ***Familiarize.*** Let $c =$ the cost of the entertainment center in February.

Translate.

$$\underbrace{\text{Cost in February}}_{c} \quad \underbrace{\text{plus}}_{+} \quad \underbrace{\$332}_{332} \quad \underbrace{\text{is}}_{=} \quad \underbrace{\text{Cost in June}}_{2449}$$

Solve. We solve the equation.

$$c + 332 = 2449$$
$$c + 332 - 332 = 2449 - 332$$
$$c = 2117$$

Check. $\$2117 + \$332 = \$2449$, so the answer checks.

State. The entertainment center cost $2117 in February.

44. ***Familiarize.*** Let $a =$ the number of subscriptions Ty sold.

Translate.

$$\underbrace{\text{Commission per subscription}}_{4} \quad \underbrace{\text{times}}_{\times} \quad \underbrace{\text{number sold}}_{a} \quad \underbrace{\text{is}}_{=} \quad \underbrace{\text{Total commission}}_{108}$$

Solve. We solve the equation.

$$4 \cdot a = 108$$
$$\frac{4 \cdot a}{4} = \frac{108}{4}$$
$$a = 27$$

Check. $\$4 \cdot 27 = \$108$, so the answer checks.

State. Ty sold 27 magazine subscriptions.

45. ***Familiarize.*** Let $x =$ the measure of the first angle. Then $x + 50 =$ the measure of the second angle, and $2x - 10 =$ the measure of the third angle. Recall that the sum of measures of the angles of a triangle is 180°.

Translate.

$$\underbrace{\text{Measure of first angle}}_{x} \quad \underbrace{+}_{+} \quad \underbrace{\text{measure of second angle}}_{(x+50)} \quad \underbrace{+}_{+} \quad \underbrace{\text{measure of third angle}}_{(2x-10)} \quad \underbrace{\text{is}}_{=} \quad \underbrace{180^\circ.}_{180}$$

Solve. We solve the equation.

$$x + (x + 50) + (2x - 10) = 180$$
$$4x + 40 = 180$$
$$4x + 40 - 40 = 180 - 40$$
$$4x = 140$$
$$\frac{4x}{4} = \frac{140}{4}$$
$$x = 35$$

If $x = 35$, then $x + 50 = 35 + 50 = 85$ and $2x - 10 = 2 \cdot 35 - 10 = 70 - 10 = 60$.

Check. The measure of the second angle is 50° more than the measure of the first angle, and the measure of the third angle is 10° less than twice the measure of the first angle. The sum of the measure is $35^\circ + 85^\circ + 60^\circ = 180^\circ$. The answer checks.

State. The measures of the angles are 35°, 85°, and 60°.

46. *Translate.*

$$\underbrace{\text{What number}}_{a} \text{ is } 20\% \text{ of } 75?$$

$$a = 20\% \cdot 75$$

Solve. We convert 20% to decimal notation and multiply.

$$a = 20\% \cdot 75$$
$$a = 0.2 \cdot 75$$
$$a = 15$$

Thus, 15 is 20% of 75.

47. *Translate.*

$$15 \text{ is } \underbrace{\text{what percent}}_{p} \text{ of } 80?$$

$$15 = p \cdot 80$$

Solve. We solve the equation.

$$15 = p \cdot 80$$
$$\frac{15}{80} = \frac{p \cdot 80}{80}$$
$$0.1875 = p$$
$$18.75\% = p$$

Thus, 15 is 18.75% of 80.

48. *Translate.*

$$18 \text{ is } 3\% \text{ of } \underbrace{\text{what number}}_{b}?$$

$$18 = 3\% \cdot b$$

Solve. We solve the equation.

$$18 = 3\% \cdot b$$
$$18 = 0.03 \cdot b$$
$$\frac{18}{0.03} = \frac{0.03 \cdot b}{0.03}$$
$$600 = b$$

Thus, 18 is 3% of 600.

49. We subtract to find the increase, in millions.

$$1.636 - 1.388 = 0.248$$

The increase is 0.248 million.

Now we find the percent of increase.

$$0.248 \text{ is } \underbrace{\text{what percent}}_{p} \text{ of } 1.388?$$

$$0.248 = p \cdot 1.388$$

We divide by 1.388 on both sides and then convert to percent notation.

$$0.248 = p \cdot 1.388$$
$$\frac{0.248}{1.388} = \frac{p \cdot 1.388}{1.388}$$
$$0.18 \approx p$$
$$18\% \approx p$$

The percent of increase is about 18%.

50. ***Familiarize.*** Let $p =$ the price before the reduction.

Translate.

$$\underbrace{\text{Price before reduction}}_{p} \text{ minus } 30\% \text{ of price is } \$154.$$

$$p - 30\% \cdot p = 154$$

Solve. We solve the equation.

$$p - 30\% \cdot p = 154$$
$$p - 0.3p = 154$$
$$0.7p = 154$$
$$\frac{0.7p}{0.7} = \frac{154}{0.7}$$
$$p = 220$$

Check. 30% of \$220 is $0.3 \cdot \$220 = \$66$ and $\$220 - \$66 = \$154$, so the answer checks.

State. The price before the reduction was \$220.

51. ***Familiarize.*** Let $s =$ the previous salary.

Translate.

$$\underbrace{\text{Previous salary}}_{s} \text{ plus } 15\% \text{ of } \underbrace{\text{previous salary}}_{s} \text{ is } \$61,410.$$

$$s + 15\% \cdot s = 61,410$$

Solve. We solve the equation.

$$s + 15\% \cdot s = 61,410$$
$$s + 0.15s = 61,410$$
$$1.15s = 61,410$$
$$\frac{1.15s}{1.15} = \frac{61,410}{1.15}$$
$$s = 53,400$$

Check. 15% of $\$53,400 = 0.15 \cdot \$53,400 = \$8010$ and $\$53,400 + \$8010 = \$61,410$, so the answer checks.

State. The previous salary was \$53,400.

52. ***Familiarize.*** Let $a =$ the amount the organization actually owes. This is the price of the supplies without sales tax added. Then the incorrect amount is $a + 5\%$ of a, or $a + 0.05a$, or $1.05a$.

Translate.

$$\underbrace{\text{Incorrect amount}}_{1.05a} \text{ is } \$145.90.$$

$$1.05a = 145.90$$

Solve. We solve the equation.

$$1.05a = 145.90$$
$$\frac{1.05a}{1.05} = \frac{145.90}{1.05}$$
$$a \approx 138.95$$

Check. 5% of \$138.95 is $0.05 \cdot \$138.95 \approx \$6.95$, and $\$138.95 + \$6.95 = \$145.90$, so the answer checks.

State. The organization actually owes \$138.95.

53. ***Familiarize.*** Let s represent the score on the next test.

Translate.

$$\underbrace{\text{The average score}}_{\frac{71+75+82+86+s}{5}} \ \underbrace{\text{is at least}}_{\geq} \ 80.$$

$$\frac{71 + 75 + 82 + 86 + s}{5} \geq 80$$

Solve.

$$\frac{71+75+82+86+s}{5} \geq 80$$

$$5\left(\frac{71+75+82+86+s}{5}\right) \geq 5 \cdot 80$$

$$71+75+82+86+s \geq 400$$

$$314+s \geq 400$$

$$s \geq 86$$

Check. As a partial check we show that the average is at least 80 when the next test score is 86.

$$\frac{71+75+82+86+86}{5} = \frac{400}{5} = 80$$

State. The lowest grade Jacinda can get on the next test and have an average test score of 80 is 86.

54. ***Familiarize.*** Let w represent the width of the rectangle, in cm. The perimeter is given by $P = 2l+2w$, or $2\cdot 43+2w$, or $86+2w$.

Translate.

$$\underbrace{\text{The perimeter}}_{86+2w} \quad \underbrace{\text{is greater than}}_{>} \quad \underbrace{120\text{ cm}}_{120}.$$

Solve.

$$86+2w > 120$$

$$2w > 34$$

$$w > 17$$

Check. We check to see if the solution seems reasonable.

When $w = 16$ cm, $P = 2\cdot 43 + 2\cdot 16$, or 118 cm.

When $w = 17$ cm, $P = 2\cdot 43 + 2\cdot 17$, or 120 cm.

When $w = 18$ cm, $P = 2\cdot 43 + 2\cdot 18$, or 122 cm.

It appears that the solution is correct.

State. The solution set is $\{w|w > 17 \text{ cm}\}$.

55.

$$4(3x-5)+6 = 8+x$$

$$12x-20+6 = 8+x$$

$$12x-14 = 8+x$$

$$12x-14-x = 8+x-x$$

$$11x-14 = 8$$

$$11x-14+14 = 8+14$$

$$11x = 22$$

$$\frac{11x}{11} = \frac{22}{11}$$

$$x = 2$$

The solution is 2. This is between 1 and 5, so the correct answer is C.

56.

$$3x+4y = P$$

$$3x+4y-3x = P-3x$$

$$4y = P-3x$$

$$\frac{4y}{4} = \frac{P-3x}{4}$$

$$y = \frac{P-3x}{4}$$

Answer A is correct.

57.

$$2|x|+4 = 50$$

$$2|x| = 46$$

$$|x| = 23$$

The solutions are the numbers whose distance from 0 is 23. Those numbers are -23 and 23.

58. $|3x| = 60$

The solutions are the values of x for which the distance of $3\cdot x$ from 0 is 60. Then we have:

$$3x = -60 \quad or \quad 3x = 60$$

$$x = -20 \quad or \quad x = 20$$

The solutions are -20 and 20.

59.

$$y = 2a - ab + 3$$

$$y-3 = 2a-ab$$

$$y-3 = a(2-b)$$

$$\frac{y-3}{2-b} = a$$

Chapter 10 Discussion and Writing Exercises

1. The end result is the same either way. If s is the original salary, the new salary after a 5% raise followed by an 8% raise is $1.08(1.05s)$. If the raises occur in the opposite order, the new salary is $1.05(1.08s)$. By the commutative and associate laws of multiplication, we see that these are equal. However, it would be better to receive the 8% raise first, because this increase yields a higher salary initially than a 5% raise.

2. No; Erin paid 75% of the original price and was offered credit for 125% of this amount, not to be used on sale items. Now 125% of 75% is 93.75%, so Erin would have a credit of 93.75% of the original price. Since this credit can be applied only to nonsale items, she has less purchasing power than if the amount she paid were refunded and she could spend it on sale items.

3. The inequalities are equivalent by the multiplication principle for inequalities. If we multiply both sides of one inequality by -1, the other inequality results.

4. For any pair of numbers, their relative position on the number line is reversed when both are multiplied by the same negative number. For example, -3 is to the left of 5 on the number line ($-3 < 5$), but 12 is to the right of -20. That is, $-3(-4) > 5(-4)$.

5. Answers may vary. Fran is more than 3 years older than Todd.

6. Let n represent "a number." Then "five more than a number" translates to $n+5$, or $5+n$, and "five is more than a number" translates to $5 > n$.

Chapter 10 Test

1.
$$\begin{aligned} x+7 &= 15 \\ x+7-7 &= 15-7 \\ x &= 8 \end{aligned}$$
The solution is 8.

2.
$$\begin{aligned} t-9 &= 17 \\ t-9+9 &= 17+9 \\ t &= 26 \end{aligned}$$
The solution is 26.

3.
$$\begin{aligned} 3x &= -18 \\ \frac{3x}{3} &= \frac{-18}{3} \\ x &= -6 \end{aligned}$$
The solution is -6.

4.
$$\begin{aligned} -\frac{4}{7}x &= -28 \\ -\frac{7}{4}\cdot\left(-\frac{4}{7}x\right) &= -\frac{7}{4}\cdot(-28) \\ x &= \frac{7\cdot \cancel{4}\cdot 7}{\cancel{4}\cdot 1} \\ x &= 49 \end{aligned}$$
The solution is 49.

5.
$$\begin{aligned} 3t+7 &= 2t-5 \\ 3t+7-2t &= 2t-5-2t \\ t+7 &= -5 \\ t+7-7 &= -5-7 \\ t &= -12 \end{aligned}$$
The solution is -12.

6.
$$\begin{aligned} \frac{1}{2}x-\frac{3}{5} &= \frac{2}{5} \\ \frac{1}{2}x-\frac{3}{5}+\frac{3}{5} &= \frac{2}{5}+\frac{3}{5} \\ \frac{1}{2}x &= 1 \\ 2\cdot\frac{1}{2}x &= 2\cdot 1 \\ x &= 2 \end{aligned}$$
The solution is 2.

7.
$$\begin{aligned} 8-y &= 16 \\ 8-y-8 &= 16-8 \\ -y &= 8 \\ -1\cdot(-1\cdot y) &= -1\cdot 8 \\ y &= -8 \end{aligned}$$
The solution is -8.

8.
$$\begin{aligned} -\frac{2}{5}+x &= -\frac{3}{4} \\ -\frac{2}{5}+x+\frac{2}{5} &= -\frac{3}{4}+\frac{2}{5} \\ x &= -\frac{15}{20}+\frac{8}{20} \\ x &= -\frac{7}{20} \end{aligned}$$
The solution is $-\frac{7}{20}$.

9.
$$\begin{aligned} 3(x+2) &= 27 \\ 3x+6 &= 27 \\ 3x+6-6 &= 27-6 \\ 3x &= 21 \\ \frac{3x}{3} &= \frac{21}{3} \\ x &= 7 \end{aligned}$$
The solution is 7.

10.
$$\begin{aligned} -3x-6(x-4) &= 9 \\ -3x-6x+24 &= 9 \\ -9x+24 &= 9 \\ -9x+24-24 &= 9-24 \\ -9x &= -15 \\ \frac{-9x}{-9} &= \frac{-15}{-9} \\ x &= \frac{\cancel{3}\cdot 5}{\cancel{3}\cdot 3} \\ x &= \frac{5}{3} \end{aligned}$$
The solution is $\frac{5}{3}$.

11. We multiply by 10 to clear the decimals.
$$\begin{aligned} 0.4p+0.2 &= 4.2p-7.8-0.6p \\ 10(0.4p+0.2) &= 10(4.2p-7.8-0.6p) \\ 4p+2 &= 42p-78-6p \\ 4p+2 &= 36p-78 \\ 4p+2-36p &= 36p-78-36p \\ -32p+2 &= -78 \\ -32p+2-2 &= -78-2 \\ -32p &= -80 \\ \frac{-32p}{-32} &= \frac{-80}{-32} \\ p &= \frac{5\cdot \cancel{16}}{2\cdot \cancel{16}} \\ p &= \frac{5}{2} \end{aligned}$$
The solution is $\frac{5}{2}$.

12. $4(3x-1)+11 = 2(6x+5)-8$
$$12x-4+11 = 12x+10-8$$
$$12x+7 = 12x+2$$
$$12x+7-12x = 12x+2-12x$$
$$7 = 2 \qquad \text{FALSE}$$
There are no solutions.

13. $-2+7x+6 = 5x+4+2x$
$$7x+4 = 7x+4$$
$$7x+4-7x = 7x+4-7x$$
$$4 = 4 \qquad \text{TRUE}$$
All real numbers are solutions.

14. $x+6 \le 2$
$$x+6-6 \le 2-6$$
$$x \le -4$$
The solution set is $\{x|x \le -4\}$.

15. $14x+9 > 13x-4$
$$14x+9-13x > 13x-4-13x$$
$$x+9 > -4$$
$$x+9-9 > -4-9$$
$$x > -13$$
The solution set is $\{x|x > -13\}$.

16. $12x \le 60$
$$\frac{12x}{12} \le \frac{60}{12}$$
$$x \le 5$$
The solution set is $\{x|x \le 5\}$.

17. $-2y \ge 26$
$$\frac{-2y}{-2} \le \frac{26}{-2} \quad \text{Reversing the inequality symbol}$$
$$y \le -13$$
The solution set is $\{y|y \le -13\}$.

18. $-4y \le -32$
$$\frac{-4y}{-4} \ge \frac{-32}{-4} \quad \text{Reversing the inequality symbol}$$
$$y \ge 8$$
The solution set is $\{y|y \ge 8\}$.

19. $-5x \ge \frac{1}{4}$
$$-\frac{1}{5}\cdot(-5x) \le -\frac{1}{5}\cdot\frac{1}{4} \quad \text{Reversing the inequality symbol}$$
$$x \le -\frac{1}{20}$$
The solution set is $\left\{x\middle|x \le -\frac{1}{20}\right\}$.

20. $4-6x > 40$
$$4-6x-4 > 40-4$$
$$-6x > 36$$
$$\frac{-6x}{-6} < \frac{36}{-6} \quad \text{Reversing the inequality symbol}$$
$$x < -6$$
The solution set is $\{x|x < -6\}$.

21. $5-9x \ge 19+5x$
$$5-9x-5x \ge 19+5x-5x$$
$$5-14x \ge 19$$
$$5-14x-5 \ge 19-5$$
$$-14x \ge 14$$
$$\frac{-14x}{-14} \le \frac{14}{-14} \quad \text{Reversing the inequality symbol}$$
$$x \le -1$$
The solution set is $\{x|x \le -1\}$.

22. The solutions of $y \le 9$ are shown by shading the point for 9 and all points to the left of 9. The closed circle at 9 indicates that 9 is part of the graph.

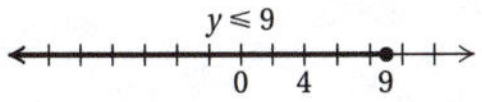

$y \le 9$ (number line: 0, 4, 9)

23. $6x-3 < x+2$
$$6x-3-x < x+2-x$$
$$5x-3 < 2$$
$$5x-3+3 < 2+3$$
$$5x < 5$$
$$\frac{5x}{5} < \frac{5}{5}$$
$$x < 1$$
The solution set is $\{x|x < 1\}$. The graph is as follows:

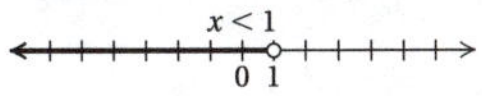

$x < 1$ (number line: 0, 1)

24. In order to be a solution of the inequality $-2 \le x \le 2$, a number must be a solution of both $-2 \le x$ and $x \le 2$. The solution set is graphed as follows:

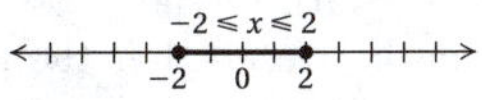

$-2 \le x \le 2$ (number line: −2, 0, 2)

25. *Translate.*

What number is 24% of 75?
$$a = 24\% \cdot 75$$

Solve. We convert 24% to decimal notation and multiply.
$$a = 24\% \cdot 75$$
$$a = 0.24 \cdot 75$$
$$a = 18$$
Thus, 18 is 24% of 75.

26. *Translate.*

$$\underbrace{15.84 \text{ is } \overbrace{\text{what percent}} \text{ of } 96?}$$

$$15.84 = p \cdot 96$$

Solve.

$$15.84 = p \cdot 96$$
$$\frac{15.84}{96} = \frac{p \cdot 96}{96}$$
$$0.165 = p$$
$$16.5\% = p$$

Thus, 15.84 is 16.5% of 96.

27. *Translate.*

800 is 2% of what number?

$$800 = 2\% \cdot b$$

Solve.

$$800 = 2\% \cdot b$$
$$800 = 0.02 \cdot b$$
$$\frac{800}{0.02} = \frac{0.02 \cdot b}{0.02}$$
$$40,000 = b$$

Thus, 800 is 2% of 40,000.

28. We subtract to find the increase.

$$83,000 - 66,000 = 17,000$$

Now we find the percent of increase.

17,000 is what percent of 66,000?

$$17,000 = p \cdot 66,000$$

We divide by 66,000 on both sides and then convert to percent notation.

$$17,000 = p \cdot 66,000$$
$$\frac{17,000}{66,000} = \frac{p \cdot 66,000}{66,000}$$
$$0.258 \approx p$$
$$25.8\% \approx p$$

The percent of increase is about 25.8%.

29. ***Familiarize.*** Let w = the width of the photograph, in cm. Then $w+4$ = the length. Recall that the perimeter P of a rectangle with length l and width w is given by $P = 2l + 2w$.

Translate. We substitute 36 for P and $w+4$ for l in the formula above.

$$P = 2l + 2w$$
$$36 = 2(w+4) + 2w$$

Solve. We solve the equation.

$$36 = 2(w+4) + 2w$$
$$36 = 2w + 8 + 2w$$
$$36 = 4w + 8$$
$$36 - 8 = 4w + 8 - 8$$
$$28 = 4w$$
$$\frac{28}{4} = \frac{4w}{4}$$
$$7 = w$$

If $w = 7$, then $w + 4 = 7 + 4 = 11$.

Check. The length, 11 cm, is 4 cm more than the width, 7 cm. The perimeter is $2 \cdot 11$ cm $+ 2 \cdot 7$ cm $= 22$ cm $+ 14$ cm $= 36$ cm. The answer checks.

State. The width is 7 cm, and the length is 11 cm.

30. ***Familiarize.*** Let c = the amount of contributions to all charities, in billions of dollars.

Translate.

Contribution to religious charities is 33% of contributions to all charities

$$102.3 = 33\% \cdot c$$

Solve. We write 33% in decimal notation and solve the equation.

$$102.3 = 0.33 \cdot c$$
$$\frac{102.3}{0.33} = \frac{0.33 \cdot c}{0.33}$$
$$310 = c$$

Check. We find 33% of \$310 billion.

$$0.33(\$310 \text{ billion}) = \$102.3 \text{ billion}$$

The answer checks.

State. In 2007, \$310 billion was given to charities.

31. ***Familiarize.*** Let x = the first integer. Then $x+1$ = the second and $x+2$ = the third.

Translate.

First integer plus second integer plus third integer is 7530.

$$x + (x+1) + (x+2) = 7530$$

Solve.

$$x + (x+1) + (x+2) = 7530$$
$$3x + 3 = 7530$$
$$3x + 3 - 3 = 7530 - 3$$
$$3x = 7527$$
$$\frac{3x}{3} = \frac{7527}{3}$$
$$x = 2509$$

If $x = 2509$, then $x + 1 = 2510$ and $x + 2 = 2511$.

Check. The numbers 2509, 2510, and 2511 are consecutive integers and $2509+2510+2511 = 7530$. The answer checks.

State. The integers are 2509, 2510, and 2511.

32. ***Familiarize.*** Let $x =$ the amount originally invested. Using the formula for simple interest, $I = Prt$, the interest earned in one year will be $x \cdot 5\% \cdot 1$, or $5\%x$.

Translate.

$$\underbrace{\text{Amount invested}} \quad \text{plus} \quad \text{interest} \quad \text{is} \quad \underbrace{\text{amount after 1 year.}}$$
$$x \quad + \quad 5\%x \quad = \quad 924$$

Solve. We solve the equation.

$$x + 5\%x = 924$$
$$x + 0.05x = 924$$
$$1.05x = 924$$
$$\frac{1.05x}{1.05} = \frac{924}{1.05}$$
$$x = 880$$

Check. 5% of \$880 is $0.05 \cdot \$880 = \$44$ and $\$880 + \$44 = \$924$, so the answer checks.

State. \$880 was originally invested.

33. ***Familiarize.*** Using the labels on the drawing in the text, we let $x =$ the length of the shorter piece, in meters, and $x + 2 =$ the length of the longer piece.

Translate.

$$\underbrace{\text{Length of shorter piece}} \quad \text{plus} \quad \underbrace{\text{length of longer piece}} \quad \text{is} \quad \underbrace{8\text{ m}}.$$
$$x \quad + \quad (x+2) \quad = \quad 8$$

Solve. We solve the equation.

$$x + (x + 2) = 8$$
$$2x + 2 = 8$$
$$2x + 2 - 2 = 8 - 2$$
$$2x = 6$$
$$\frac{2x}{2} = \frac{6}{2}$$
$$x = 3$$

If $x = 3$, then $x + 2 = 3 + 2 = 5$.

Check. One piece is 2 m longer than the other and the sum of the lengths is 3 m+5 m, or 8 m. The answer checks.

State. The lengths of the pieces are 3 m and 5 m.

34. ***Familiarize.*** Let $l =$ the length of the rectangle, in yd. The perimeter is given by $P = 2l + 2w$, or $2l + 2 \cdot 96$, or $2l + 192$.

Translate.

$$\underbrace{\text{The perimeter}} \quad \underbrace{\text{is at least}} \quad \underbrace{540\text{ yd}}.$$
$$2l + 192 \quad \geq \quad 540$$

Solve.

$$2l + 192 \geq 540$$
$$2l \geq 348$$
$$l \geq 174$$

Check. We check to see if the solution seems reasonable.

When $l = 174$ yd, $P = 2 \cdot 174 + 2 \cdot 96$, or 540 yd.

When $l = 175$ yd, $P = 2 \cdot 175 + 2 \cdot 96$, or 542 yd.

It appears that the solution is correct.

State. For lengths that are at least 174 yd, the perimeter will be at least 540 yd. The solution set can be expressed as $\{l|l \geq 174 \text{ yd}\}$.

35. ***Familiarize.*** Let $s =$ the amount Jason spends in the sixth month.

Translate.

$$\underbrace{\text{Average spending}} \quad \underbrace{\text{is no more than}} \quad \$95.$$
$$\frac{98 + 89 + 110 + 85 + 83 + s}{6} \quad \leq \quad 95$$

Solve.

$$\frac{98 + 89 + 110 + 85 + 83 + s}{6} \leq 95$$
$$6\left(\frac{98 + 89 + 110 + 85 + 83 + s}{6}\right) \leq 6 \cdot 95$$
$$98 + 89 + 110 + 85 + 83 + s \leq 570$$
$$465 + s \leq 570$$
$$s \leq 105$$

Check. As a partial check we show that the average spending is \$95 when Jason spends \$105 in the sixth month.

$$\frac{98 + 89 + 110 + 85 + 83 + 105}{6} = \frac{570}{6} = 95$$

State. Jason can spend no more than \$105 in the sixth month. The solution set can be expressed as $\{s|s \leq \$105\}$.

36. ***Familiarize.*** Let $c =$ the number of copies made. For 3 months, the rental charge is $3 \cdot \$225$, or \$675. Expressing 1.2¢ as \$0.012, the charge for the copies is given by $\$0.012 \cdot c$.

Translate.

$$\underbrace{\text{Rental charge}} \quad \text{plus} \quad \underbrace{\text{copy charge}} \quad \underbrace{\text{is no more than}} \quad \$2400.$$
$$675 \quad + \quad 0.012c \quad \leq \quad 2400$$

Solve.

$$675 + 0.012c \leq 2400$$
$$0.012c \leq 1725$$
$$c \leq 143,750$$

Check. We check to see if the solution seems reasonable.

When $c = 143,749$, the total cost is $\$675 + \$0.012(143,749)$, or about \$2399.99.

When $c = 143,750$, the total cost is $\$675 + \$0.012(143,750)$, or about \$2400.

It appears that the solution is correct.

State. No more than 143,750 copies can be made. The solution set can be expressed as $\{c|c \leq 143,750\}$.

37.
$$\begin{aligned} A &= 2\pi rh \\ \frac{A}{2\pi h} &= \frac{2\pi rh}{2\pi h} \\ \frac{A}{2\pi h} &= r \end{aligned}$$

38.
$$\begin{aligned} y &= 8x + b \\ y - b &= 8x + b - b \\ y - b &= 8x \\ \frac{y-b}{8} &= \frac{8x}{8} \\ \frac{y-b}{8} &= x \end{aligned}$$

39. We subtract to find the increase, in millions.

$70.3 - 40.4 = 29.9$

Now we find the percent of increase.

$$\underbrace{29.9}\ \text{is}\ \underbrace{\text{what percent}}\ \text{of}\ 40.4?$$
$$29.9 = p \cdot 40.4$$

We divide by 40.4 on both sides and then convert to percent notation.

$$\begin{aligned} 29.9 &= p \cdot 40.4 \\ \frac{29.9}{40.4} &= \frac{p \cdot 40.4}{40.4} \\ 0.74 &\approx p \\ 74\% &\approx p \end{aligned}$$

The percent of increase is about 74%. Answer D is correct.

40.
$$\begin{aligned} c &= \frac{1}{a-d} \\ (a-d) \cdot c &= a - d \cdot \left(\frac{1}{a-d}\right) \\ ac - dc &= 1 \\ ac - dc - ac &= 1 - ac \\ -dc &= 1 - ac \\ \frac{-dc}{-c} &= \frac{1-ac}{-c} \\ d &= \frac{1-ac}{-c} \end{aligned}$$

Since $\frac{1-ac}{-c} = \frac{-1}{-1} \cdot \frac{1-ac}{-c} = \frac{-1(1-ac)}{-1(-c)} = \frac{-1+ac}{c}$, or $\frac{ac-1}{c}$, we can also express the result as $d = \frac{ac-1}{c}$.

41.
$$\begin{aligned} 3|w| - 8 &= 37 \\ 3|w| &= 45 \\ |w| &= 15 \end{aligned}$$

The solutions are the numbers whose distance from 0 is 15. They are -15 and 15.

42. ***Familiarize***. Let t = the number of tickets given away.

Translate. We add the number of tickets given to the five people.

$$\frac{1}{3}t + \frac{1}{4}t + \frac{1}{5}t + 8 + 5 = t$$

Solve.

$$\begin{aligned} \frac{1}{3}t + \frac{1}{4}t + \frac{1}{5}t + 8 + 5 &= t \\ \frac{20}{60}t + \frac{15}{60}t + \frac{12}{60}t + 8 + 5 &= t \\ \frac{47}{60}t + 13 &= t \\ 13 &= t - \frac{47}{60}t \\ 13 &= \frac{60}{60}t - \frac{47}{60}t \\ 13 &= \frac{13}{60}t \\ \frac{60}{13} \cdot 13 &= \frac{60}{13} \cdot \frac{13}{60}t \\ 60 &= t \end{aligned}$$

Check. $\frac{1}{3} \cdot 60 = 20$, $\frac{1}{4} \cdot 60 = 15$, $\frac{1}{5} \cdot 60 = 12$; then $20 + 15 + 12 + 8 + 5 = 60$. The answer checks.

State. 60 tickets were given away.

Chapter 11

Graphs of Linear Equations

Exercise Set 11.1

1. $(2,5)$ is 2 units right and 5 units up.

$(-1,3)$ is 1 unit left and 3 units up.

$(3,-2)$ is 3 units right and 2 units down.

$(-2,-4)$ is 2 units left and 4 units down.

$(0,4)$ is 0 units left or right and 4 units up.

$(0,-5)$ is 0 units left or right and 5 units down.

$(5,0)$ is 5 units right and 0 units up or down.

$(-5,0)$ is 5 units left and 0 units up or down.

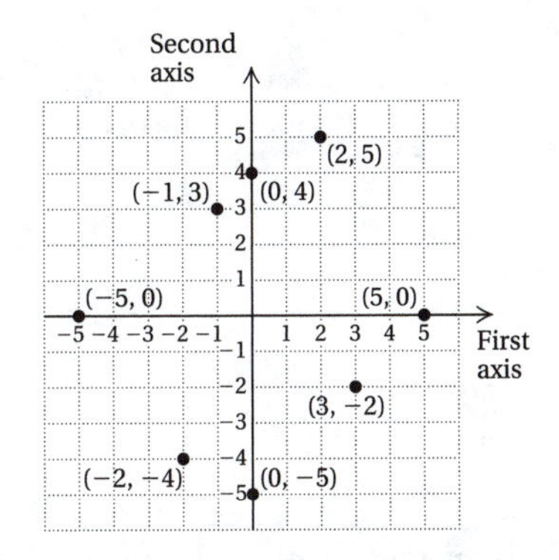

3. Since the first coordinate is negative and the second coordinate positive, the point $(-5,3)$ is located in quadrant II.

5. Since the first coordinate is positive and the second coordinate negative, the point $(100,-1)$ is in quadrant IV.

7. Since both coordinates are negative, the point $(-6,-29)$ is in quadrant III.

9. Since one of the coordinates is 0, the point $(3.8,0)$ lies on an axis. It is not in a quadrant.

11. Since the first coordinate is negative and the second coordinate is positive, the point $\left(-\frac{1}{3},\frac{15}{7}\right)$ is in quadrant II.

13. Since the first coordinate is positive and the second coordinate is negative, the point $\left(12\frac{7}{8},-1\frac{1}{2}\right)$ is in quadrant IV.

15.

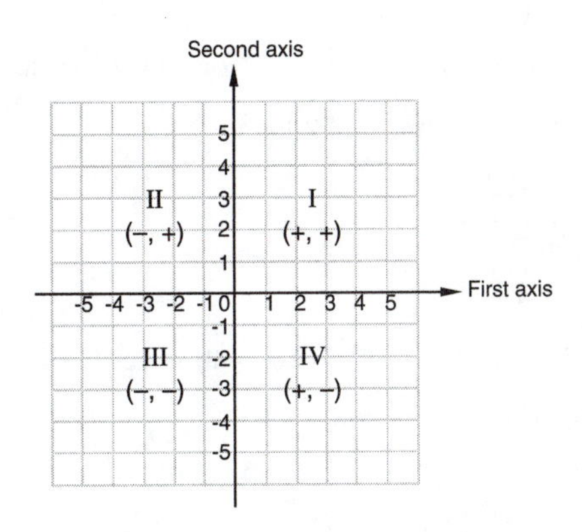

If the first coordinate is negative and the second coordinate is positive, the point is in quadrant II.

17. See the figure in Exercise 15.

If the first coordinate is positive, then the point must be in either quadrant I or quadrant IV.

19. If the first and second coordinates are equal, they must either be both positive or both negative. The point must be in either quadrant I (both positive) or quadrant III (both negative).

21.

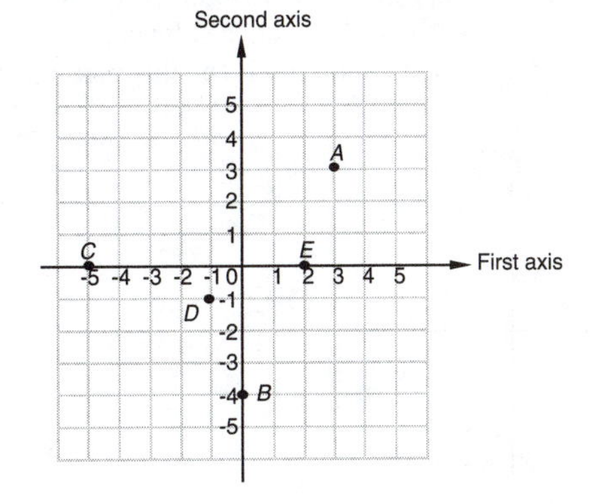

Point A is 3 units right and 3 units up. The coordinates of A are $(3,3)$.

Point B is 0 units left or right and 4 units down. The coordinates of B are $(0,-4)$.

Point C is 5 units left and 0 units up or down. The coordinates of C are $(-5,0)$.

Point D is 1 unit left and 1 unit down. The coordinates of D are $(-1,-1)$.

Point E is 2 units right and 0 units up or down. The coordinates of E are $(2,0)$.

23. We substitute 2 for x and 9 for y (alphabetical order of variables).

$$\begin{array}{c|l} y = 3x - 1 & \\ \hline 9 \ ? & 3 \cdot 2 - 1 \\ & 6 - 1 \\ & 5 \qquad \text{FALSE} \end{array}$$

Since $9 = 5$ is false, the pair $(2, 9)$ is not a solution.

25. We substitute 4 for x and 2 for y.

$$\begin{array}{c|l} 2x + 3y = 12 & \\ \hline 2 \cdot 4 + 3 \cdot 2 & ? \ 12 \\ 8 + 6 & \\ 14 & \qquad \text{FALSE} \end{array}$$

Since $14 = 12$ is false, the pair (4,2) is not a solution.

27. We substitute 3 for a and -1 for b.

$$\begin{array}{c|l} 3a - 4b = 13 & \\ \hline 3 \cdot 3 - 4(-1) & ? \ 13 \\ 9 + 4 & \\ 13 & \qquad \text{TRUE} \end{array}$$

Since $13 = 13$ is true, the pair $(3, -1)$ is a solution.

29. To show that a pair is a solution, we substitute, replacing x with the first coordinate and y with the second coordinate in each pair.

$$\begin{array}{c|l} y = x - 5 & \\ \hline -1 \ ? & 4 - 5 \\ & -1 \qquad \text{TRUE} \end{array} \qquad \begin{array}{c|l} y = x - 5 & \\ \hline -4 \ ? & 1 - 5 \\ & -4 \qquad \text{TRUE} \end{array}$$

In each case the substitution results in a true equation. Thus, $(4, -1)$ and $(1, -4)$ are both solutions of $y = x - 5$. We graph these points and sketch the line passing through them.

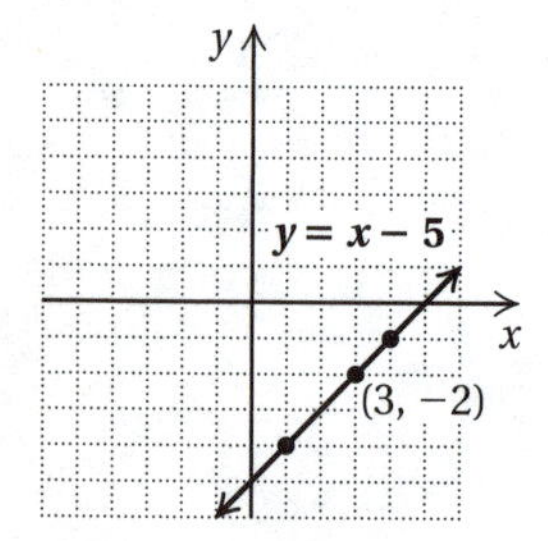

The line appears to pass through $(3, -2)$ also. We check to determine if $(3, -2)$ is a solution of $y = x - 5$.

$$\begin{array}{c|l} y = x - 5 & \\ \hline -2 \ ? & 3 - 5 \\ & -2 \qquad \text{TRUE} \end{array}$$

Thus, $(3, -2)$ is another solution. There are other correct answers, including $(-1, -6)$, $(2, -3)$, $(0, -5)$, $(5, 0)$, and $(6, 1)$.

31. To show that a pair is a solution, we substitute, replacing x with the first coordinate and y with the second coordinate in each pair.

$$\begin{array}{c|l} y = \frac{1}{2}x + 3 & \\ \hline 5 \ ? & \frac{1}{2} \cdot 4 + 3 \\ & 2 + 3 \\ & 5 \qquad \text{TRUE} \end{array} \qquad \begin{array}{c|l} y = \frac{1}{2}x + 3 & \\ \hline 2 \ ? & \frac{1}{2}(-2) + 3 \\ & -1 + 3 \\ & 2 \qquad \text{TRUE} \end{array}$$

In each case the substitution results in a true equation. Thus, $(4, 5)$ and $(-2, 2)$ are both solutions of $y = \frac{1}{2}x + 3$. We graph these points and sketch the line passing through them.

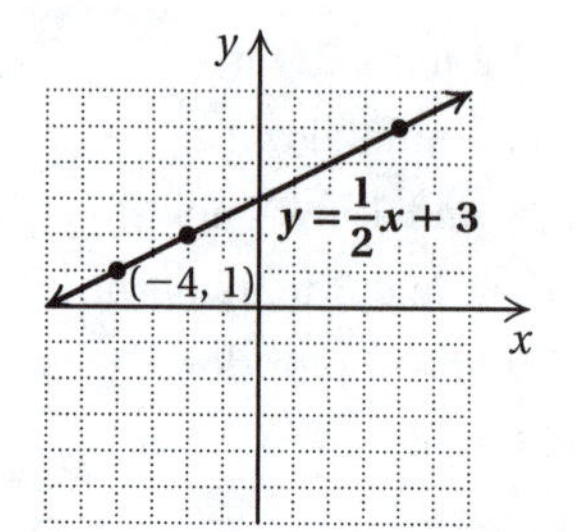

The line appears to pass through $(-4, 1)$ also. We check to determine if $(-4, 1)$ is a solution of $y = \frac{1}{2}x + 3$.

$$\begin{array}{c|l} y = \frac{1}{2}x + 3 & \\ \hline 1 \ ? & \frac{1}{2}(-4) + 3 \\ & -2 + 3 \\ & 1 \qquad \text{TRUE} \end{array}$$

Thus, $(-4, 1)$ is another solution. There are other correct answers, including $(-6, 0)$, $(0, 3)$, $(2, 4)$, and $(6, 6)$.

33. To show that a pair is a solution, we substitute, replacing x with the first coordinate and y with the second coordinate in each pair.

$$\begin{array}{c|l} 4x - 2y = 10 & \\ \hline 4 \cdot 0 - 2(-5) & ? \ 10 \\ 10 & \qquad \text{TRUE} \end{array}$$

$$\begin{array}{c|l} 4x - 2y = 10 & \\ \hline 4 \cdot 4 - 2 \cdot 3 & ? \ 10 \\ 16 - 6 & \\ 10 & \qquad \text{TRUE} \end{array}$$

In each case the substitution results in a true equation. Thus, $(0, -5)$ and $(4, 3)$ are both solutions of $4x - 2y = 10$. We graph these points and sketch the line passing through them.

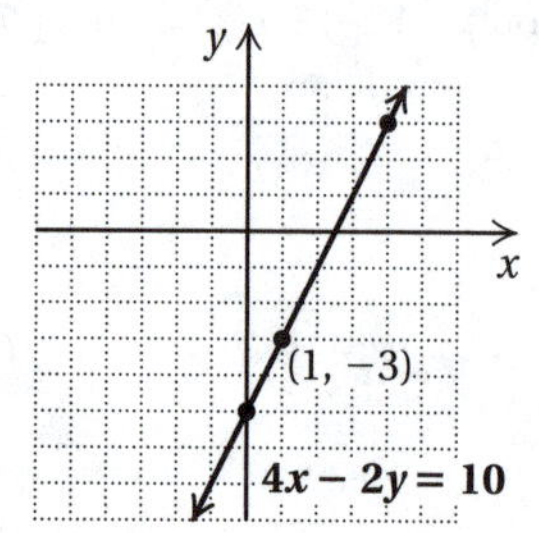

The line appears to pass through $(1, -3)$ also. We check to determine if $(1, -3)$ is a solution of $4x - 2y = 10$.

$$\begin{array}{c|c} \multicolumn{2}{c}{4x - 2y = 10} \\ \hline 4 \cdot 1 - 2(-3) \ ? \ 10 & \\ 4 + 6 & \\ 10 & \quad \text{TRUE} \end{array}$$

Thus, $(1, -3)$ is another solution. There are other correct answers, including $(2, -1)$, $(3, 1)$, and $(5, 5)$.

35. $y = x + 1$

The equation is in the form $y = mx + b$. The y-intercept is $(0, 1)$. We find five other pairs.

When $x = -2$, $y = -2 + 1 = -1$.

When $x = -1$, $y = -1 + 1 = 0$.

When $x = 1$, $y = 1 + 1 = 2$.

When $x = 2$, $y = 2 + 1 = 3$.

When $x = 3$, $y = 3 + 1 = 4$.

x	y
-2	-1
-1	0
0	1
1	2
2	3
3	4

Plot these points, draw the line they determine, and label the graph $y = x + 1$.

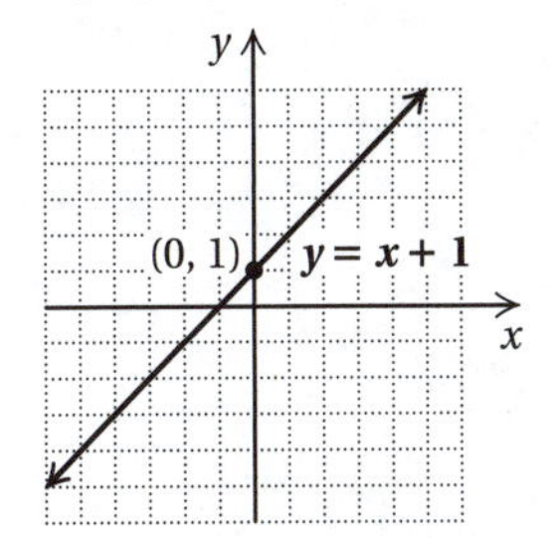

37. $y = x$

The equation is equivalent to $y = x + 0$. The y-intercept is $(0, 0)$. We find five other points.

When $x = -2$, $y = -2$.

When $x = -1$, $y = -1$.

When $x = 1$, $y = 1$.

When $x = 2$, $y = 2$.

When $x = 3$, $y = 3$.

x	y
-2	-2
-1	-1
0	0
1	1
2	2
3	3

Plot these points, draw the line they determine, and label the graph $y = x$.

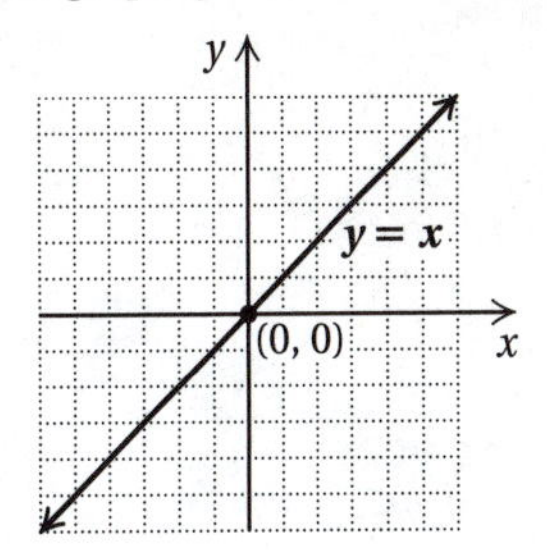

39. $y = \frac{1}{2}x$

The equation is equivalent to $y = \frac{1}{2}x + 0$. The y-intercept is $(0, 0)$. We find two other points.

When $x = -2$, $y = \frac{1}{2}(-2) = -1$.

When $x = 4$, $y = \frac{1}{2} \cdot 4 = 2$.

x	y
-2	-1
0	0
4	2

Plot these points, draw the line they determine, and label the graph $y = \frac{1}{2}x$.

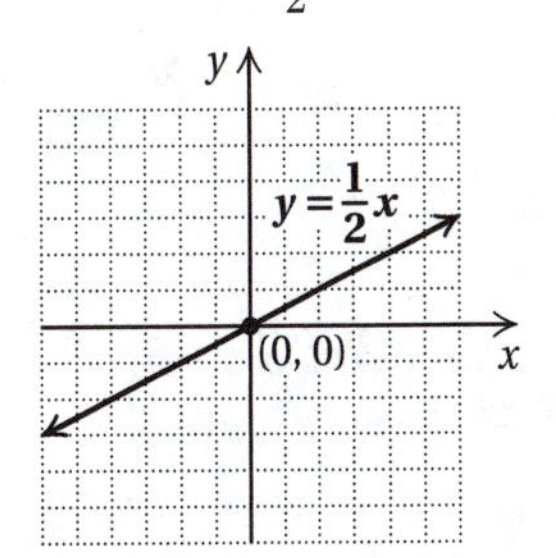

41. $y = x - 3$

The equation is equivalent to $y = x + (-3)$. The y-intercept is $(0, -3)$. We find two other points.

When $x = -2$, $y = -2 - 3 = -5$.

When $x = 4$, $y = 4 - 3 = 1$.

x	y
-2	-5
0	-3
4	1

Plot these points, draw the line they determine, and label the graph $y = x - 3$.

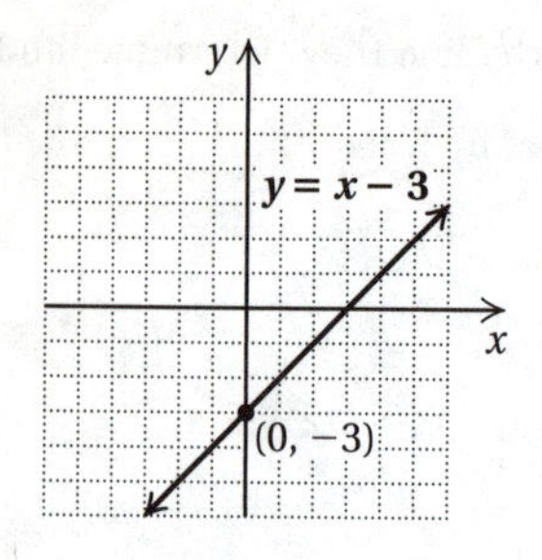

43. $y = 3x - 2 = 3x + (-2)$

The y-intercept is $(0, -2)$. We find two other points.

When $x = -1$, $y = 3(-1) - 2 = -3 - 2 = -5$.

When $x = 2$, $y = 3 \cdot 2 - 2 = 6 - 2 = 4$.

x	y
−1	−5
0	−2
2	4

Plot these points, draw the line they determine, and label the graph $y = 3x - 2$.

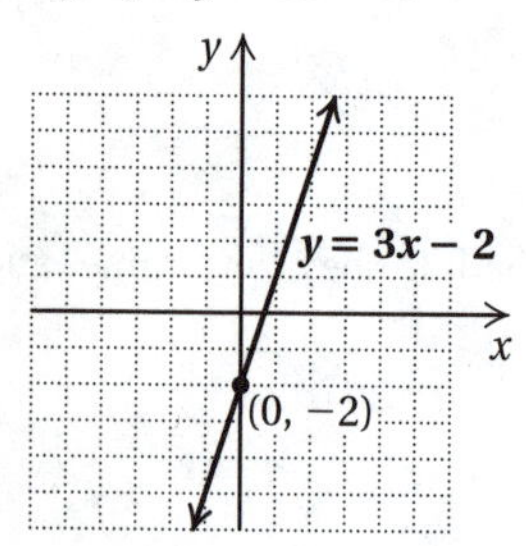

45. $y = \frac{1}{2}x + 1$

The y-intercept is $(0, 1)$. We find two other points using multiples of 2 for x to avoid fractions.

When $x = -4$, $y = \frac{1}{2}(-4) + 1 = -2 + 1 = -1$.

When $x = 4$, $y = \frac{1}{2} \cdot 4 + 1 = 2 + 1 = 3$.

x	y
−4	−1
0	1
4	3

Plot these points, draw the line they determine, and label the graph $y = \frac{1}{2}x + 1$.

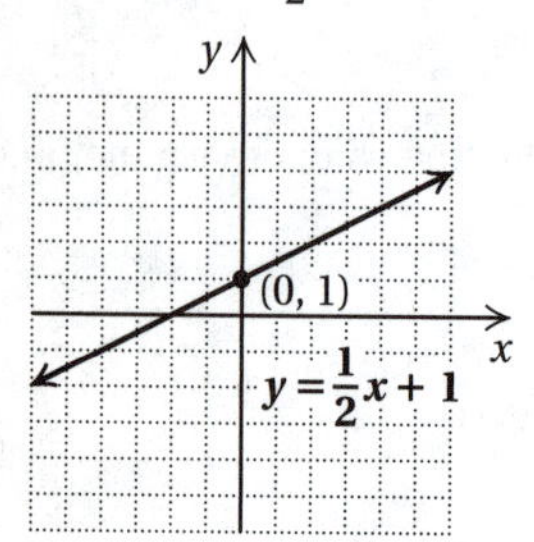

47. $x + y = -5$

$y = -x - 5$

$y = -x + (-5)$

The y-intercept is $(0, -5)$. We find two other points.

When $x = -4$, $y = -(-4) - 5 = 4 - 5 = -1$.

When $x = -1$, $y = -(-1) - 5 = 1 - 5 = -4$.

x	y
−4	−1
0	−5
−1	−4

Plot these points, draw the line they determine, and label the graph $x + y = -5$.

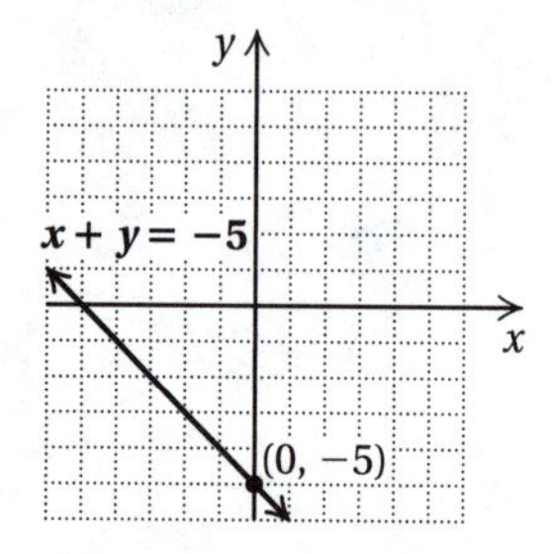

49. $y = \frac{5}{3}x - 2 = \frac{5}{3}x + (-2)$

The y-intercept is $(0, -2)$. We find two other points using multiples of 3 for x to avoid fractions.

When $x = -3$, $y = \frac{5}{3}(-3) - 2 = -5 - 2 = -7$.

When $x = 3$, $y = \frac{5}{3} \cdot 3 - 2 = 5 - 2 = 3$.

x	y
−3	−7
0	−2
3	3

Plot these points, draw the line they determine, and label the graph $y = \frac{5}{3}x - 2$.

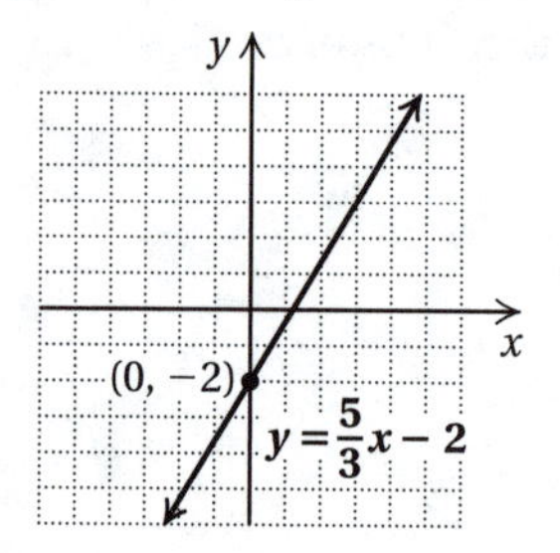

51. $x + 2y = 8$

$2y = -x + 8$

$y = -\frac{1}{2}x + 4$

The y-intercept is $(0, 4)$. We find two other points using multiples of 2 for x to avoid fractions.

When $x = -2$, $y = -\frac{1}{2}(-2) + 4 = 1 + 4 = 5$.

When $x = 4$, $y = -\frac{1}{2} \cdot 4 + 4 = -2 + 4 = 2$.

x	y
-2	5
0	4
4	2

Plot these points, draw the line they determine, and label the graph $x + 2y = 8$.

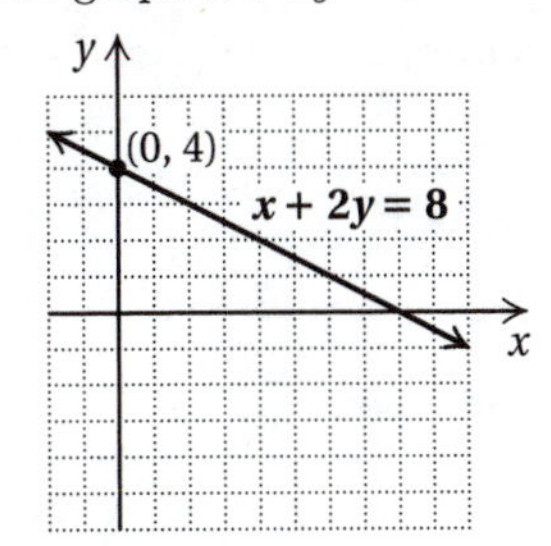

53. $y = \frac{3}{2}x + 1$

The y-intercept is $(0, 1)$. We find two other points using multiples of 2 for x to avoid fractions.

When $x = -4$, $y = \frac{3}{2}(-4) + 1 = -6 + 1 = -5$.

When $x = 2$, $y = \frac{3}{2} \cdot 2 + 1 = 3 + 1 = 4$.

x	y
-4	-5
0	1
2	4

Plot these points, draw the line they determine, and label the graph $y = \frac{3}{2}x + 1$.

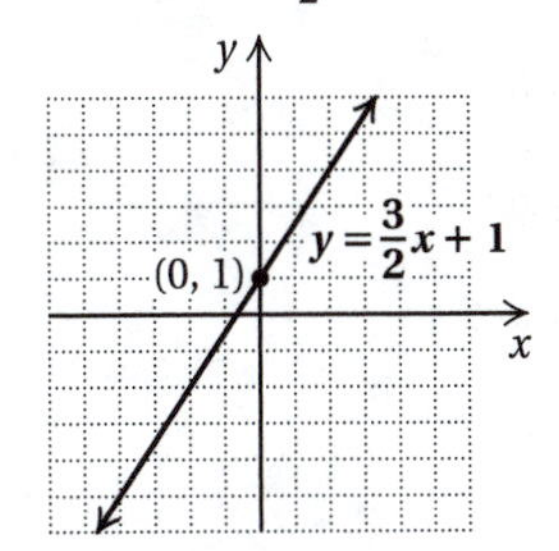

55.
$$8x - 2y = -10$$
$$-2y = -8x - 10$$
$$y = 4x + 5$$

The y-intercept is $(0, 5)$. We find two other points.

When $x = -2$, $y = 4(-2) + 5 = -8 + 5 = -3$.

When $x = -1$, $y = 4(-1) + 5 = -4 + 5 = 1$.

x	y
-2	-3
-1	1
0	5

Plot these points, draw the line they determine, and label the graph $8x - 2y = -10$.

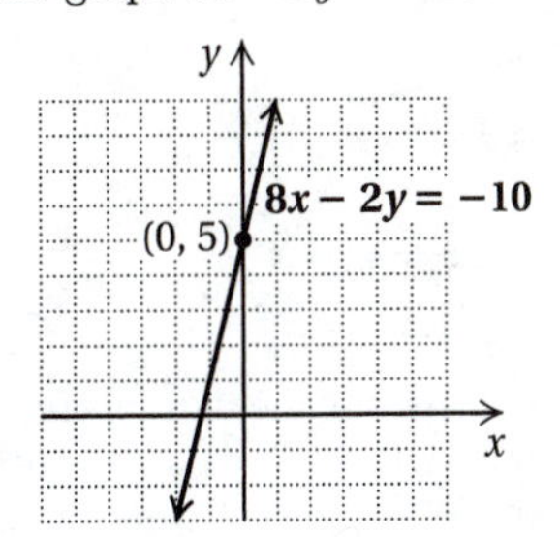

57.
$$8y + 2x = -4$$
$$8y = -2x - 4$$
$$y = -\frac{1}{4}x - \frac{1}{2}$$
$$y = -\frac{1}{4}x + \left(-\frac{1}{2}\right)$$

The y-intercept is $\left(0, -\frac{1}{2}\right)$. We find two other points.

When $x = -2$, $y = -\frac{1}{4}(-2) - \frac{1}{2} = \frac{1}{2} - \frac{1}{2} = 0$.

When $x = 2$, $y = -\frac{1}{4} \cdot 2 - \frac{1}{2} = -\frac{1}{2} - \frac{1}{2} = -1$.

x	y
-2	0
0	$-\frac{1}{2}$
2	-1

Plot these points, draw the line they determine, and label the graph $8y + 2x = -4$.

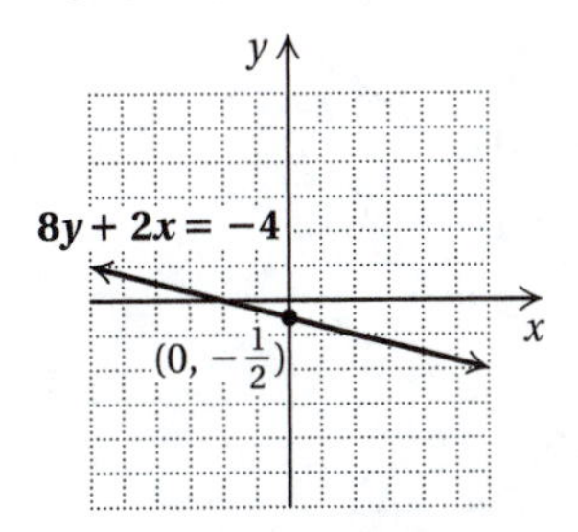

59. a) We substitute 0, 5 $(2007 - 2002 = 5)$, and 8 $(2010 - 2002 = 8)$ for t and then calculate R.

If $t = 0$, then $R = -1698 \cdot 0 + 52,620 = \$52,620$.

If $t = 5$, then $R = -1698 \cdot 5 + 52,620 =$ $-8490 + 52,620 = \$44,130$.

If $t = 8$, then $R = -1698 \cdot 8 + 52,620 =$ $-13,584 + 52,620 = \$39,036$.

b) We plot the three ordered pairs we found in part (a). Note that negative t- and R-values have no meaning in this problem.

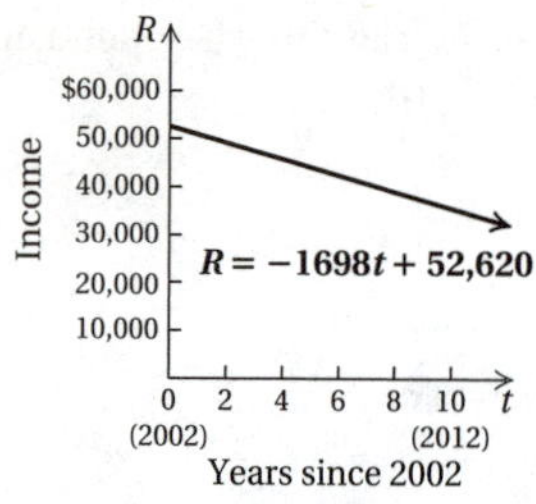

To use the graph to estimate the median income in 2005, we must determine which R-value is paired with $t = 3(2005 - 2002 = 3)$. We locate 3 on the t-axis, go up to the graph, and then find the value on the R-axis that corresponds to this point. It appears that the median income in 2005 was about \$47,500.

c) Substitute 37,338 for R and then solve for t.

$$R = -1698t + 52,620$$
$$37,338 = -1698t + 52,620$$
$$-15,282 = -1698t$$
$$9 = t$$

About 9 yr after 2002, or in 2011, the median income will be \$37,338.

61. a) When $d = 1$, $W = 1.8(1) + 16.44 = 1.8 + 16.44 = 18.24$ gal.

In 2010, $d = 2010 - 2000 = 10$. When $d = 10$, $N = 1.8(10) + 16.44 = 18 + 16.44 = 34.44$ gal.

In 2015, $d = 2015 - 2000 = 15$. When $d = 15$, $N = 1.8(15) + 16.44 = 27 + 16.44 = 43.44$ gal.

b) Plot the three ordered pairs we found in part (a). Note that negative d- and W-values have no meaning in this problem.

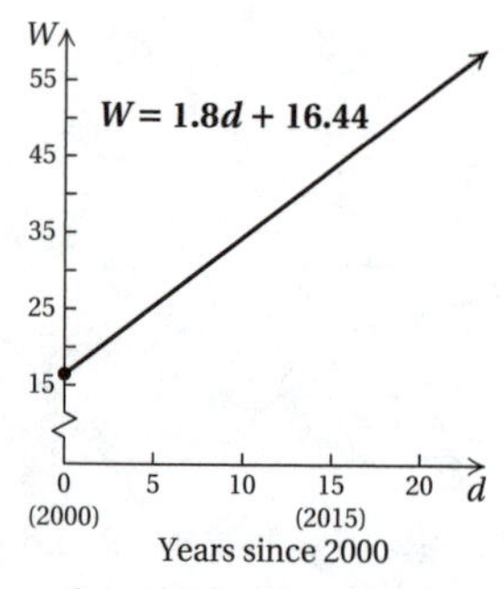

To use the graph to estimate what bottled water consumption was in 2008 we must determine which W-value is paired with 2008, or with $d = 8$. We locate 8 on the d-axis, go up to the graph, and then find the value on the W-axis that corresponds to that point. It appears that bottled water consumption was about 31 gallons in 2008.

c) Substitute 36 for W and then solve for d.

$$W = 1.8d + 16.44$$
$$36 = 1.8d + 16.44$$
$$19.56 = 1.8d$$
$$11 \approx d$$

About 11 yr after 2000, or in 2011, bottled water consumption will be 36 gal.

63. The distance of -12 from 0 is 12, so $|-12| = 12$.

65. The distance of 0 from 0 is 0, so $|0| = 0$.

67. The distance of -3.4 from 0 is 3.4, so $|-3.4| = 3.4$.

69. The distance of $\frac{2}{3}$ from 0 is $\frac{2}{3}$, so $\left|\frac{2}{3}\right| = \frac{2}{3}$.

71. First we reword and translate.

What is 24% of 200?
$\downarrow \quad \downarrow \quad \downarrow \quad \downarrow \quad \downarrow$
$a \quad = 24\% \cdot 200$

To solve the equation, we convert 24% to decimal notation and multiply.

$$a = 24\% \cdot 200$$
$$a = 0.24 \cdot 200$$
$$a = 48$$

48 patients will be age 75 and older.

73.

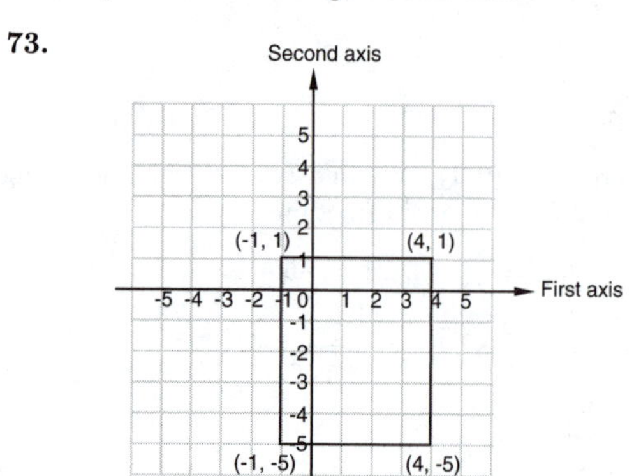

The coordinates of the fourth vertex are $(-1, -5)$.

75. Answers may vary.

We select eight points such that the sum of the coordinates for each point is 6.

$(-1, 7)$	$-1 + 7 = 6$
$(0, 6)$	$0 + 6 = 6$
$(1, 5)$	$1 + 5 = 6$
$(2, 4)$	$2 + 4 = 6$
$(3, 3)$	$3 + 3 = 6$
$(4, 2)$	$4 + 2 = 6$
$(5, 1)$	$5 + 1 = 6$
$(6, 0)$	$6 + 0 = 6$

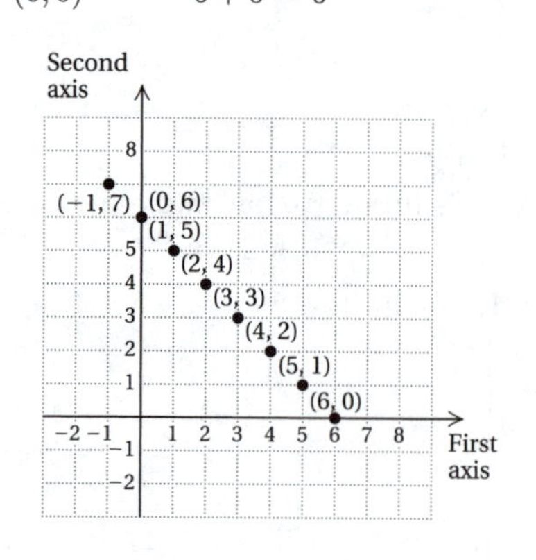

77.

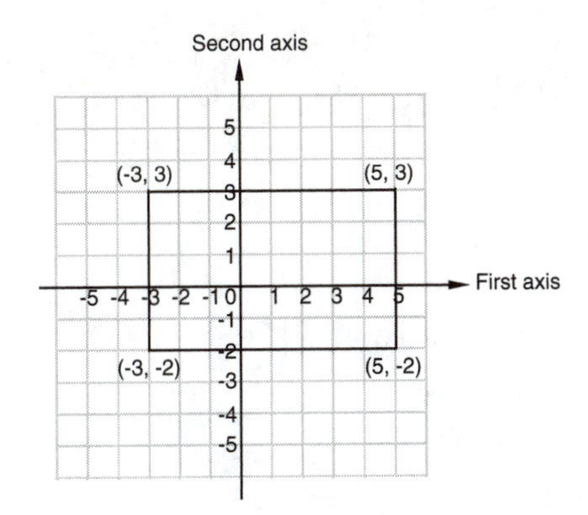

The length is 8 linear units, and the width is 5 linear units.

$P = 2l + 2w$

$P = 2 \cdot 8 + 2 \cdot 5 = 16 + 10 = 26$ linear units

Exercise Set 11.2

1. (a) The graph crosses the y-axis at $(0, 5)$, so the y-intercept is $(0, 5)$.

(b) The graph crosses the x- axis at $(2, 0)$, so the x-intercept is $(2, 0)$.

3. (a) The graph crosses the y-axis at $(0, -4)$, so the y-intercept is $(0, -4)$.

(b) The graph crosses the x-axis at $(3, 0)$, so the x-intercept is $(3, 0)$.

5. $3x + 5y = 15$

(a) To find the y-intercept, let $x = 0$. This is the same as covering up the x-term and then solving.

$$5y = 15$$
$$y = 3$$

The y-intercept is $(0, 3)$.

(b) To find the x-intercept, let $y = 0$. This is the same as covering up the y-term and then solving.

$$3x = 15$$
$$x = 5$$

The x-intercept is $(5, 0)$.

7. $7x - 2y = 28$

(a) To find the y-intercept, let $x = 0$. This is the same as covering up the x-term and then solving.

$$-2y = 28$$
$$y = -14$$

The $y-$intercept is $(0, -14)$.

(b) To find the x-intercept, let $y = 0$. This is the same as covering up the y-term and then solving.

$$7x = 28$$
$$x = 4$$

The x-intercept is $(4, 0)$.

9. $-4x + 3y = 10$

(a) To find the y-intercept, let $x = 0$. This is the same as covering up the x-term and then solving.

$$3y = 10$$
$$y = \frac{10}{3}$$

The y-intercept is $\left(0, \frac{10}{3}\right)$.

(b) To find the x-intercept, let $y = 0$. This is the same as covering up the y-term and then solving.

$$-4x = 10$$
$$x = -\frac{5}{2}$$

The x-intercept is $\left(-\frac{5}{2}, 0\right)$.

11. $6x - 3 = 9y$

$6x - 9y = 3$ Writing the equation in the form $Ax + By = C$

(a) To find the y-intercept, let $x = 0$. This is the same as covering up the x-term and then solving.

$$-9y = 3$$
$$y = -\frac{1}{3}$$

The y-intercept is $\left(0, -\frac{1}{3}\right)$.

(b) To find the x-intercept, let $y = 0$. This is the same as covering up the y-term and then solving.

$$6x = 3$$
$$x = \frac{1}{2}$$

The x-intercept is $\left(\frac{1}{2}, 0\right)$.

13. $x + 3y = 6$

To find the x-intercept, let $y = 0$. Then solve for x.

$$x + 3y = 6$$
$$x + 3 \cdot 0 = 6$$
$$x = 6$$

Thus, $(6, 0)$ is the x-intercept.

To find the y-intercept, let $x = 0$. Then solve for y.

$$x + 3y = 6$$
$$0 + 3y = 6$$
$$3y = 6$$
$$y = 2$$

Thus, $(0, 2)$ is the y-intercept.

Plot these points and draw the line.

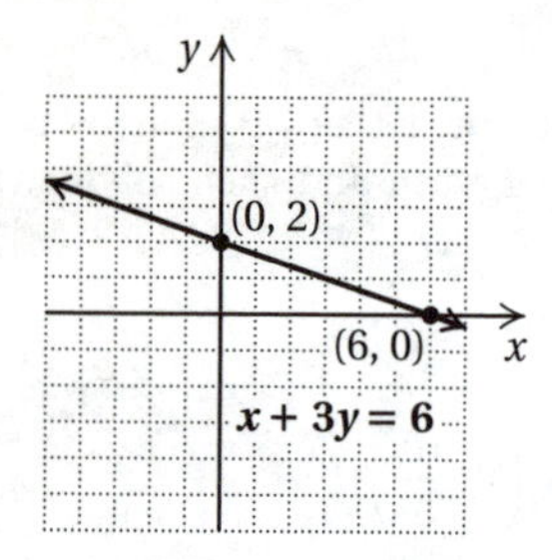

A third point should be used as a check. We substitute any value for x and solve for y.

We let $x = 3$. Then

$$x + 3y = 6$$
$$3 + 3y = 6$$
$$3y = 3$$
$$y = 1$$

The point $(3, 1)$ is on the graph, so the graph is probably correct.

15. $-x + 2y = 4$

To find the x-intercept, let $y = 0$. Then solve for x.

$$-x + 2y = 4$$
$$-x + 2 \cdot 0 = 4$$
$$-x = 4$$
$$x = -4$$

Thus, $(-4, 0)$ is the x-intercept.

To find the y-intercept, let $x = 0$. Then solve for y.

$$-x + 2y = 4$$
$$-0 + 2y = 4$$
$$2y = 4$$
$$y = 2$$

Thus, $(0, 2)$ is the y-intercept.

Plot these points and draw the line.

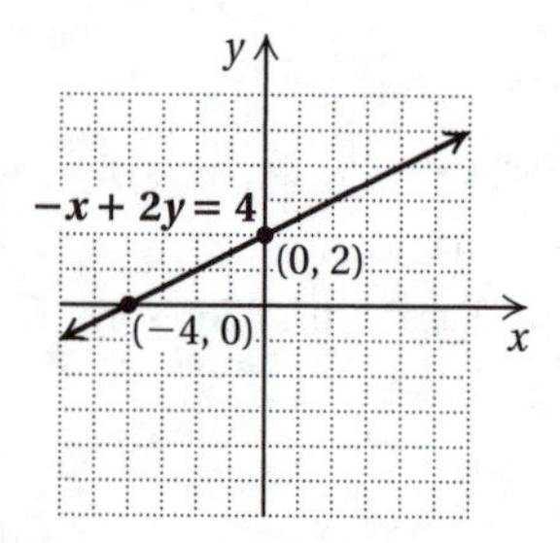

A third point should be used as a check. We substitute any value for x and solve for y.

We let $x = 4$. Then

$$-x + 2y = 4$$
$$-4 + 2y = 4$$
$$2y = 8$$
$$y = 4$$

The point $(4, 4)$ is on the graph, so the graph is probably correct.

17. $3x + y = 6$

To find the x-intercept, let $y = 0$. Then solve for x.

$$3x + y = 6$$
$$3x + 0 = 6$$
$$3x = 6$$
$$x = 2$$

Thus, $(2, 0)$ is the x-intercept.

To find the y-intercept, let $x = 0$. Then solve for y.

$$3x + y = 6$$
$$3 \cdot 0 + y = 6$$
$$y = 6$$

Thus, $(0, 6)$ is the y-intercept.

Plot these points and draw the line.

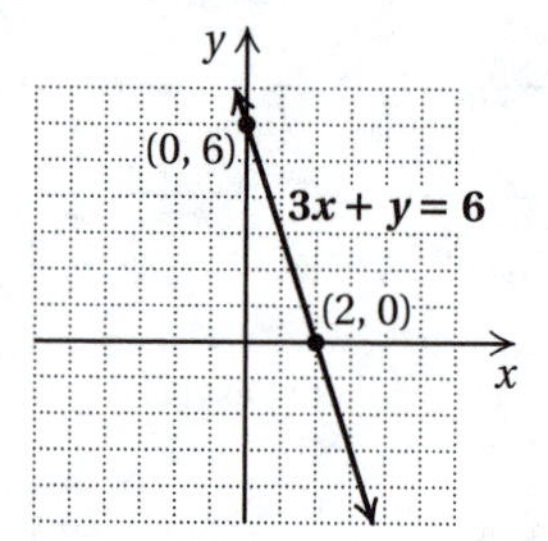

A third point should be used as a check. We substitute any value for x and solve for y.

We let $x = 1$. Then

$$3x + y = 6$$
$$3 \cdot 1 + y = 6$$
$$3 + y = 6$$
$$y = 3$$

The point $(1, 3)$ is on the graph, so the graph is probably correct.

19. $2y - 2 = 6x$

To find the x-intercept, let $y = 0$. Then solve for x.

$$2y - 2 = 6x$$
$$2 \cdot 0 - 2 = 6x$$
$$-2 = 6x$$
$$-\frac{1}{3} = x$$

Thus, $\left(-\frac{1}{3}, 0\right)$ is the x-intercept.

To find the y-intercept, let $x = 0$. Then solve for y.

$$2y - 2 = 6x$$
$$2y - 2 = 6 \cdot 0$$
$$2y - 2 = 0$$
$$2y = 2$$
$$y = 1$$

Thus, $(0, 1)$ is the y-intercept.

It is helpful to plot another point since the intercepts are so close together. This point can also serve as a check.

We let $x = 1$. Then

$$2y - 2 = 6x$$
$$2y - 2 = 6 \cdot 1$$
$$2y - 2 = 6$$
$$2y = 8$$
$$y = 4$$

Plot the point $(1, 4)$ and the intercepts and draw the line.

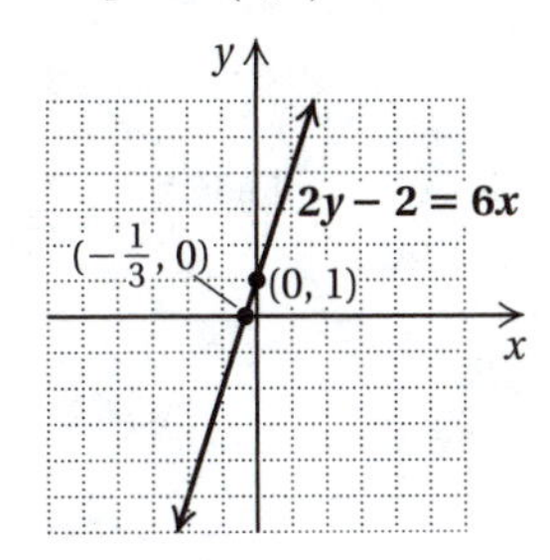

21. $3x - 9 = 3y$

To find the x-intercept, let $y = 0$. Then solve for x.

$$3x - 9 = 3y$$
$$3x - 9 = 3 \cdot 0$$
$$3x - 9 = 0$$
$$3x = 9$$
$$x = 3$$

Thus, $(3, 0)$ is the x-intercept.

To find the y-intercept, let $x = 0$. Then solve for y.

$$3x - 9 = 3y$$
$$3 \cdot 0 - 9 = 3y$$
$$-9 = 3y$$
$$-3 = y$$

Thus, $(0, -3)$ is the y-intercept.

Plot these points and draw the line.

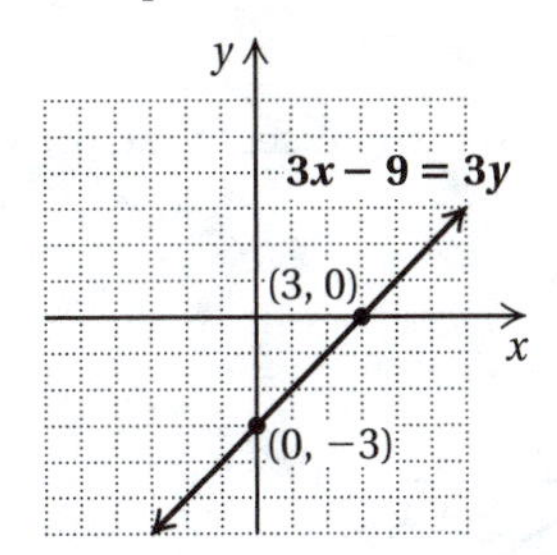

A third point should be used as a check. We substitute any value for x and solve for y.

We let $x = 1$. Then

$$3x - 9 = 3y$$
$$3 \cdot 1 - 9 = 3y$$
$$3 - 9 = 3y$$
$$-6 = 3y$$
$$-2 = y$$

The point $(1, -2)$ is on the graph, so the graph is probably correct.

23. $2x - 3y = 6$

To find the x-intercept, let $y = 0$. Then solve for x.

$$2x - 3y = 6$$
$$2x - 3 \cdot 0 = 6$$
$$2x = 6$$
$$x = 3$$

Thus, $(3, 0)$ is the x-intercept.

To find the y-intercept, let $x = 0$. Then solve for y.

$$2x - 3y = 6$$
$$2 \cdot 0 - 3y = 6$$
$$-3y = 6$$
$$y = -2$$

Thus, $(0, -2)$ is the y-intercept.

Plot these points and draw the line.

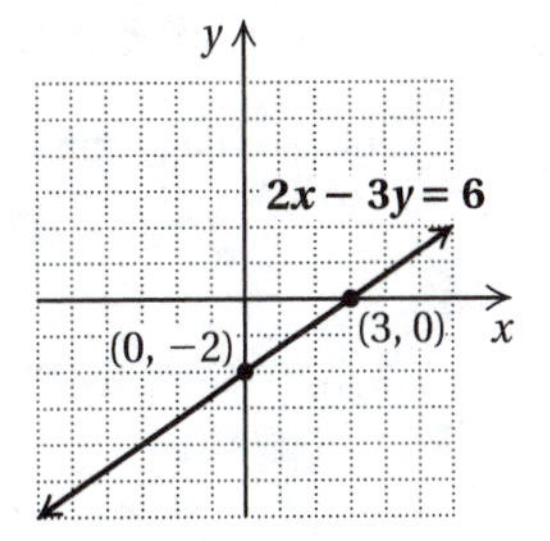

A third point should be used as a check. We substitute any value for x and solve for y.

We let $x = -3$.

$$2x - 3y = 6$$
$$2(-3) - 3y = 6$$
$$-6 - 3y = 6$$
$$-3y = 12$$
$$y = -4$$

The point $(-3, -4)$ is on the graph, so the graph is probably correct.

25. $4x + 5y = 20$

To find the x-intercept, let $y = 0$. Then solve for x.

$$4x + 5y = 20$$
$$4x + 5 \cdot 0 = 20$$
$$4x = 20$$
$$x = 5$$

Thus, $(5, 0)$ is the x-intercept.

To find the y-intercept, let $x = 0$. Then solve for y.

$$4x + 5y = 20$$
$$4 \cdot 0 + 5y = 20$$
$$5y = 20$$
$$y = 4$$

Thus, $(0, 4)$ is the y-intercept.

Plot these points and draw the graph.

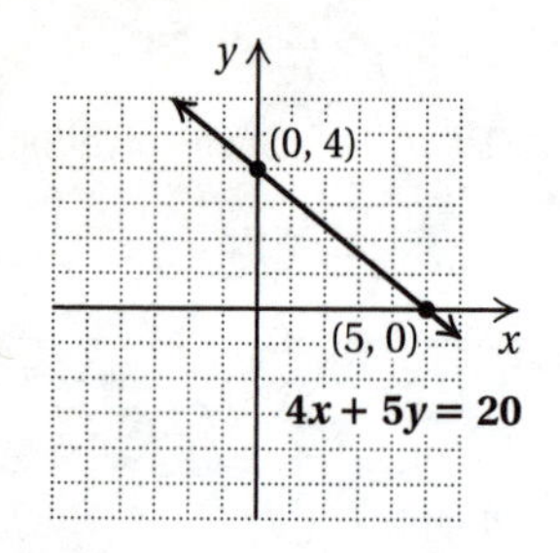

A third point should be used as a check. We substitute any value for x and solve for y.

We let $x = 4$. Then

$$4x + 5y = 20$$
$$4 \cdot 4 + 5y = 20$$
$$16 + 5y = 20$$
$$5y = 4$$
$$y = \frac{4}{5}$$

The point $\left(4, \frac{4}{5}\right)$ is on the graph, so the graph is probably correct.

27. $2x + 3y = 8$

To find the x-intercept, let $y = 0$. Then solve for x.

$$2x + 3y = 8$$
$$2x + 3 \cdot 0 = 8$$
$$2x = 8$$
$$x = 4$$

Thus, $(4, 0)$ is the x-intercept.

To find the y-intercept, let $x = 0$. Then solve for y.

$$2x + 3y = 8$$
$$2 \cdot 0 + 3y = 8$$
$$3y = 8$$
$$y = \frac{8}{3}$$

Thus, $\left(0, \frac{8}{3}\right)$ is the y-intercept.

Plot these points and draw the graph.

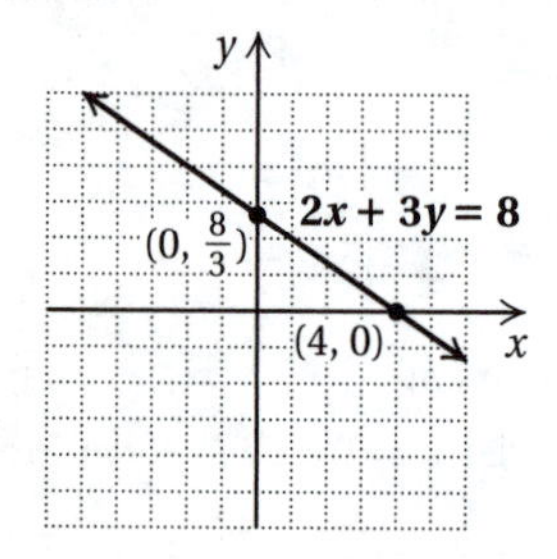

A third point should be used as a check.

We let $x = 1$. Then

$$2x + 3y = 8$$
$$2 \cdot 1 + 3y = 8$$
$$2 + 3y = 8$$
$$3y = 6$$
$$y = 2$$

The point $(1, 2)$ is on the graph, so the graph is probably correct.

29. $3x + 4y = 5$

To find the x-intercept, let $y = 0$. Then solve for x.

$$3x + 4y = 5$$
$$3x + 4 \cdot 0 = 5$$
$$3x = 5$$
$$x = \frac{5}{3}$$

Thus, $\left(\frac{5}{3}, 0\right)$ is the x-intercept.

To find the y-intercept, let $x = 0$. Then solve for y.

$$3x + 4y = 5$$
$$3 \cdot 0 + 4y = 5$$
$$4y = 5$$
$$y = \frac{5}{4}$$

Thus, $\left(0, \frac{5}{4}\right)$ is the y-intercept.

It is helpful to plot another point since the intercepts are so close together. This point can also serve as a check.

We let $x = 3$. Then

$$3x + 4y = 5$$
$$3 \cdot 3 + 4y = 5$$
$$9 + 4y = 5$$
$$4y = -4$$
$$y = -1$$

Plot the point $(3, -1)$ and the intercepts and draw the line.

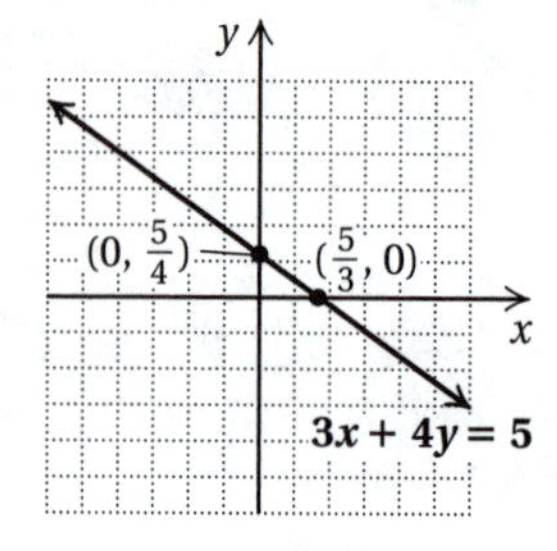

31. $3x - 2 = y$

To find the x-intercept, let $y = 0$. Then solve for x.

$$3x - 2 = y$$
$$3x - 2 = 0$$
$$3x = 2$$
$$x = \frac{2}{3}$$

Thus, $\left(\frac{2}{3}, 0\right)$ is the x-intercept.

To find the y-intercept, let $x = 0$. Then solve for y.

$$3x - 2 = y$$
$$3 \cdot 0 - 2 = y$$
$$-2 = y$$

Thus, $(0, -2)$ is the y-intercept.

Plot these points and draw the line.

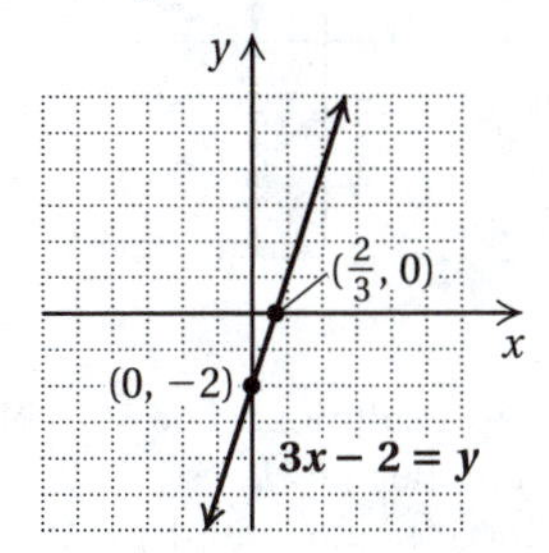

A third point should be used as a check.

We let $x = 2$. Then

$$3x - 2 = y$$
$$3 \cdot 2 - 2 = y$$
$$6 - 2 = y$$
$$4 = y$$

The point $(2, 4)$ is on the graph, so the graph is probably correct.

33. $6x - 2y = 12$

To find the x-intercept, let $y = 0$. Then solve for x.

$$6x - 2y = 12$$
$$6x - 2 \cdot 0 = 12$$
$$6x = 12$$
$$x = 2$$

Thus, $(2, 0)$ is the x-intercept.

To find the y-intercept, let $x = 0$. Then solve for y.

$$6x - 2y = 12$$
$$6 \cdot 0 - 2y = 12$$
$$-2y = 12$$
$$y = -6$$

Thus, $(0, -6)$ is the y-intercept.

Plot these points and draw the line.

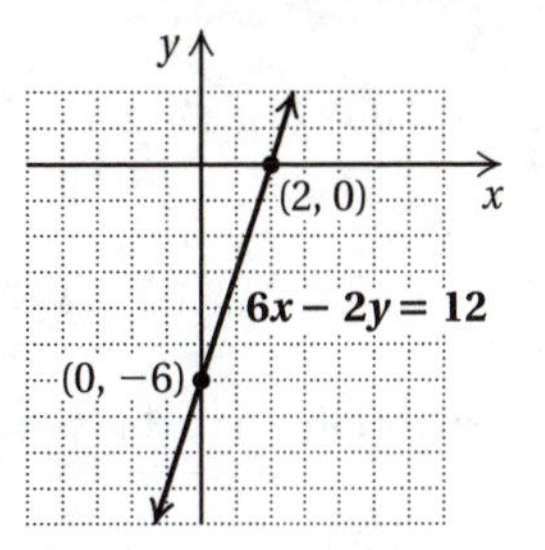

We use a third point as a check.

We let $x = 1$. Then

$$6x - 2y = 12$$
$$6 \cdot 1 - 2y = 12$$
$$6 - 2y = 12$$
$$-2y = 6$$
$$y = -3$$

The point $(1, -3)$ is on the graph, so the graph is probably correct.

35. $y = -3 - 3x$

To find the x-intercept, let $y = 0$. Then solve for x.

$$y = -3 - 3x$$
$$0 = -3 - 3x$$
$$3x = -3$$
$$x = -1$$

Thus, $(-1, 0)$ is the x-intercept.

To find the y-intercept, let $x = 0$. Then solve for y.

$$y = -3 - 3x$$
$$y = -3 - 3 \cdot 0$$
$$y = -3$$

Thus, $(0, -3)$ is the y-intercept.

Plot these points and draw the graph.

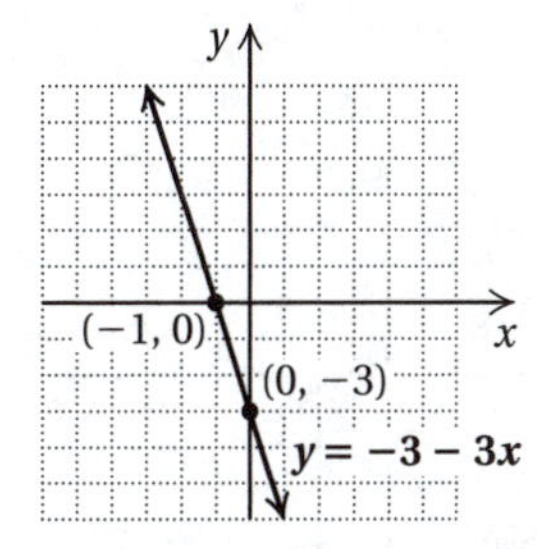

We use a third point as a check.

We let $x = -2$. Then

$$y = -3 - 3x$$
$$y = -3 - 3 \cdot (-2)$$
$$y = -3 + 6$$
$$y = 3$$

The point $(-2, 3)$ is on the graph, so the graph is probably correct.

37. $y - 3x = 0$

To find the x-intercept, let $y = 0$. Then solve for x.

$$0 - 3x = 0$$
$$-3x = 0$$
$$x = 0$$

Thus, $(0, 0)$ is the x-intercept. Note that this is also the y-intercept.

In order to graph the line, we will find a second point.

When $x = 1$, $y - 3 \cdot 1 = 0$
$$y - 3 = 0$$
$$y = 3$$

Plot the points and draw the graph.

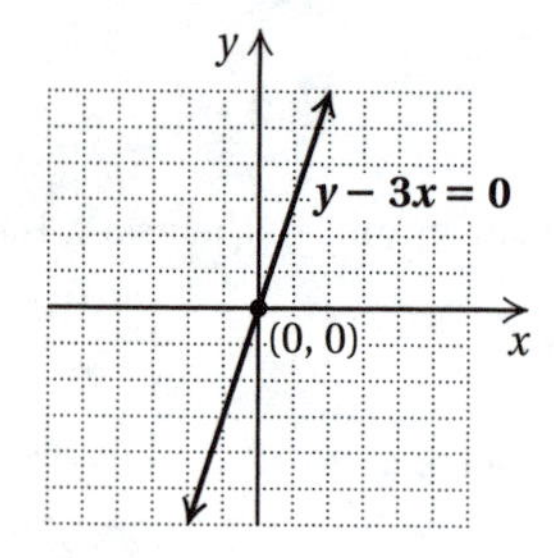

We use a third point as a check.

We let $x = -1$. Then

$$y - 3(-1) = 0$$
$$y + 3 = 0$$
$$y = -3$$

The point $(-1, -3)$ is on the graph, so the graph is probably correct.

39. $x = -2$

Any ordered pair $(-2, y)$ is a solution. The variable x must be -2, but y can be any number we choose. A few solutions are listed below. Plot these points and draw the line.

x	y
-2	-2
-2	0
-2	4

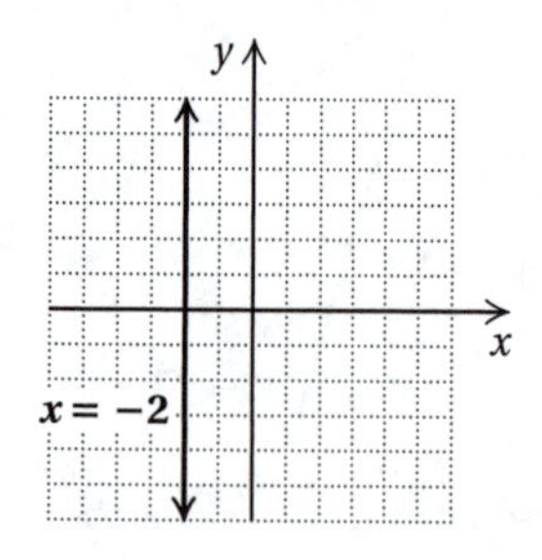

41. $y = 2$

Any ordered pair $(x, 2)$ is a solution. The variable y must be 2, but x can be any number we choose. A few solutions are listed below. Plot these points and draw the line.

x	y
-3	2
0	2
2	2

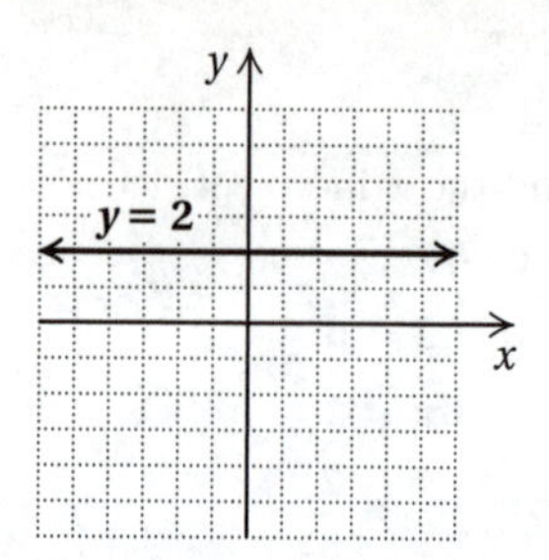

43. $x = 2$

Any ordered pair $(2, y)$ is a solution. The variable x must be 2, but y can be any number we choose. A few solutions are listed below. Plot these points and draw the line.

x	y
2	-1
2	4
2	5

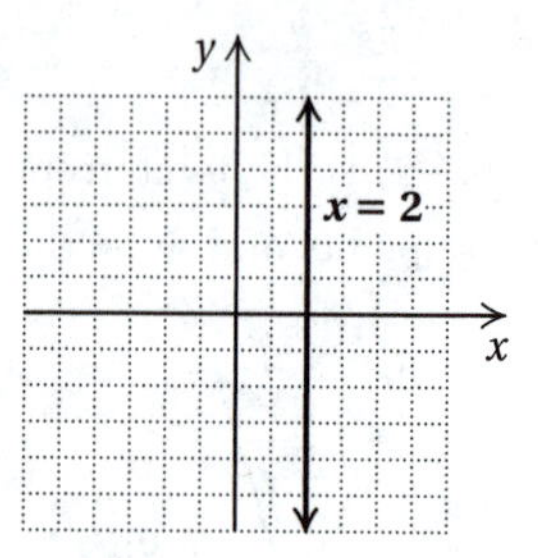

45. $y = 0$

Any ordered pair $(x, 0)$ is a solution. The variable y must be 0, but x can be any number we choose. A few solutions are listed below. Plot these points and draw the line.

x	y
-5	0
-1	0
3	0

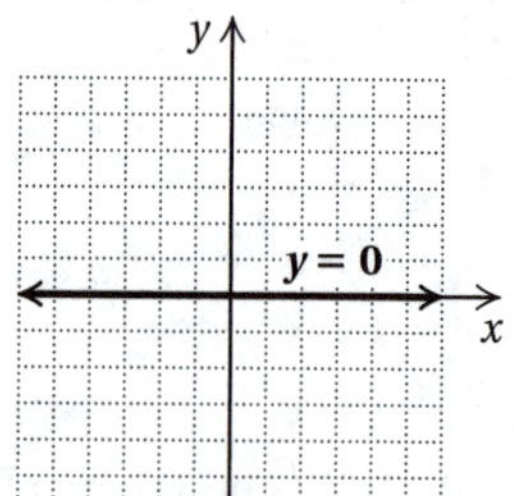

47. $x = \frac{3}{2}$

Any ordered pair $\left(\frac{3}{2}, y\right)$ is a solution. The variable x must be $\frac{3}{2}$, but y can be any number we choose. A few solutions are listed below. Plot these points and draw the line.

x	y
$\frac{3}{2}$	-2
$\frac{3}{2}$	0
$\frac{3}{2}$	4

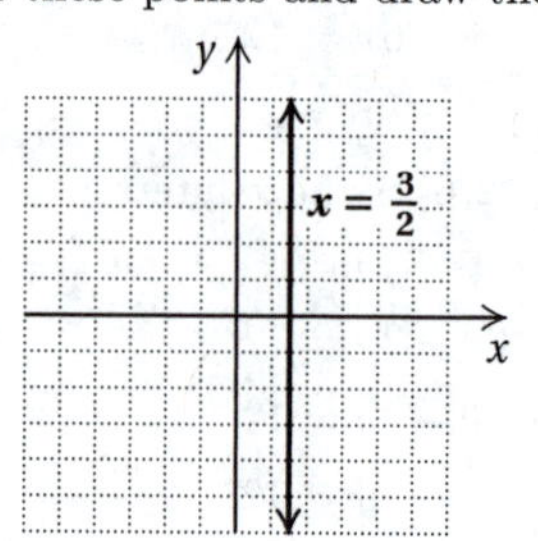

49. $3y = -5$

$y = -\frac{5}{3}$ Solving for y

Any ordered pair $\left(x, -\frac{5}{3}\right)$ is a solution. A few solutions are listed below. Plot these points and draw the line.

x	y
-3	$-\frac{5}{3}$
0	$-\frac{5}{3}$
2	$-\frac{5}{3}$

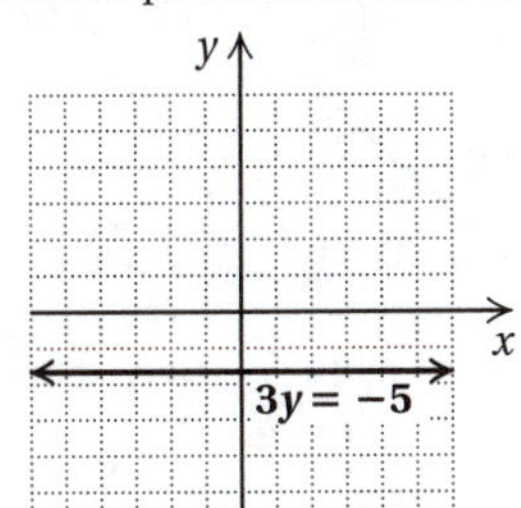

51. $4x + 3 = 0$

$4x = -3$

$x = -\frac{3}{4}$ Solving for x

Any ordered pair $\left(-\frac{3}{4}, y\right)$ is a solution. A few solutions are listed below. Plot these points and draw the line.

x	y
$-\frac{3}{4}$	-2
$-\frac{3}{4}$	0
$-\frac{3}{4}$	3

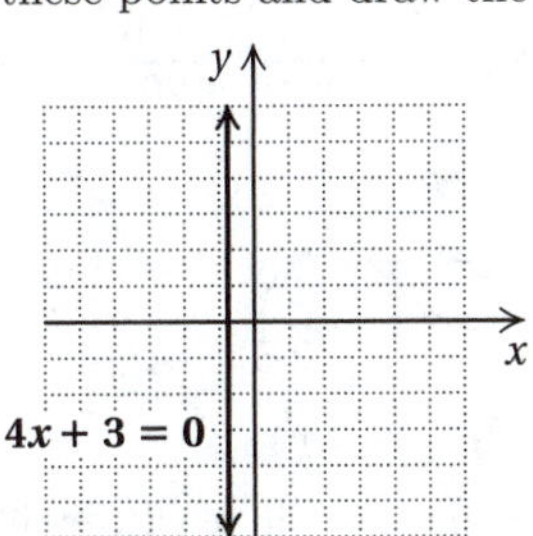

53. $48 - 3y = 0$

$-3y = -48$

$y = 16$ Solving for y

Any ordered pair $(x, 16)$ is a solution. A few solutions are listed below. Plot these points and draw the line.

x	y
-4	16
0	16
2	16

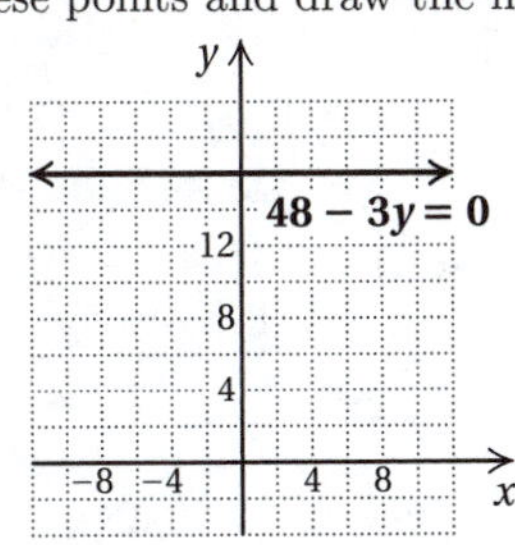

55. Note that every point on the horizontal line passing through $(0, -1)$ has -1 as the y-coordinate. Thus, the equation of the line is $y = -1$.

57. Note that every point on the vertical line passing through $(4, 0)$ has 4 as the x-coordinate. Thus, the equation of the line is $x = 4$.

59. $-1.6x < 64$

$\frac{-1.6x}{-1.6} > \frac{64}{-1.6}$ Dividing by -1.6 and reversing the inequality symbol

$x > -40$

The solution set is $\{x|x > -40\}$.

61. $x + (x - 1) < (x + 2) - (x + 1)$

$2x - 1 < x + 2 - x - 1$

$2x - 1 < 1$

$2x < 2$

$x < 1$

The solution set is $\{x|x < 1\}$.

63. $\frac{2x}{7} - 4 \leq -2$

$\frac{2x}{7} \leq 2$

$\frac{7}{2} \cdot \frac{2x}{7} \leq \frac{7}{2} \cdot 2$

$x \leq 7$

The solution set is $\{x|x \leq 7\}$.

65. First we reword and translate.

What is 8.7% of 1,027,974?

$\downarrow \quad \downarrow \quad \downarrow \quad \downarrow \qquad \downarrow$

$a = 8.7\% \times 1,027,974$

To solve the equation, we convert 8.7% to decimal notation and multiply.

$a = 8.7\% \cdot 1,027,974$

$a = 0.087 \times 1,027,974$

$a \approx 89,434$

About 89,434 residents of Detroit are foreign-born.

67. A line parallel to the x-axis has an equation of the form $y = b$. Since the y-coordinate of one point on the line is -4, then $b = -4$ and the equation is $y = -4$.

69. Substitute -4 for x and 0 for y.

$3(-4) + k = 5 \cdot 0$

$-12 + k = 0$

$k = 12$

Exercise Set 11.3

1. We consider (x_1, y_1) to be $(-3, 5)$ and (x_2, y_2) to be $(4, 2)$.

$$m = \frac{y_2 - y_1}{x_2 - x_1} = \frac{2 - 5}{4 - (-3)} = \frac{-3}{7} = -\frac{3}{7}$$

3. We can choose any two points. We consider (x_1, y_1) to be $(-3, -1)$ and (x_2, y_2) to be $(0, 1)$.

$$m = \frac{y_2 - y_1}{x_2 - x_1} = \frac{1 - (-1)}{0 - (-3)} = \frac{2}{3}$$

5. We can choose any two points. We consider (x_1, y_1) to be $(-4, -2)$ and (x_2, y_2) to be $(4, 4)$.

$$m = \frac{y_2 - y_1}{x_2 - x_1} = \frac{4 - (-2)}{4 - (-4)} = \frac{6}{8} = \frac{3}{4}$$

7. We consider (x_1, y_1) to be $(-4, -2)$ and (x_2, y_2) to be $(3, -2)$.

$$m = \frac{y_2 - y_1}{x_2 - x_1} = \frac{-2-(-2)}{3-(-4)} = \frac{0}{7} = 0$$

9. We plot $(-2, 4)$ and $(3, 0)$ and draw the line containing these points.

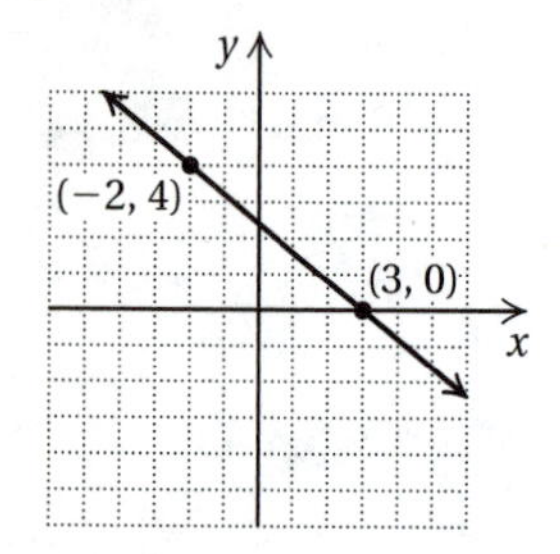

To find the slope, consider (x_1, y_1) to be $(-2, 4)$ and (x_2, y_2) to be $(3, 0)$.

$$m = \frac{y_2 - y_1}{x_2 - x_1} = \frac{0-4}{3-(-2)} = \frac{-4}{5} = -\frac{4}{5}$$

11. We plot $(-4, 0)$ and $(-5, -3)$ and draw the line containing these points.

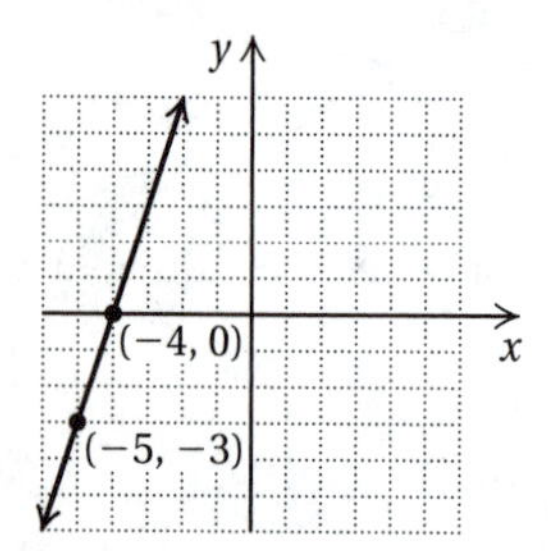

To find the slope, consider (x_1, y_1) to be $(-4, 0)$ and (x_2, y_2) to be $(-5, -3)$.

$$m = \frac{y_2 - y_1}{x_2 - x_1} = \frac{-3-0}{-5-(-4)} = \frac{-3}{-1} = 3$$

13. We plot $(-4, 1)$ and $(2, -3)$ and draw the line containing these points.

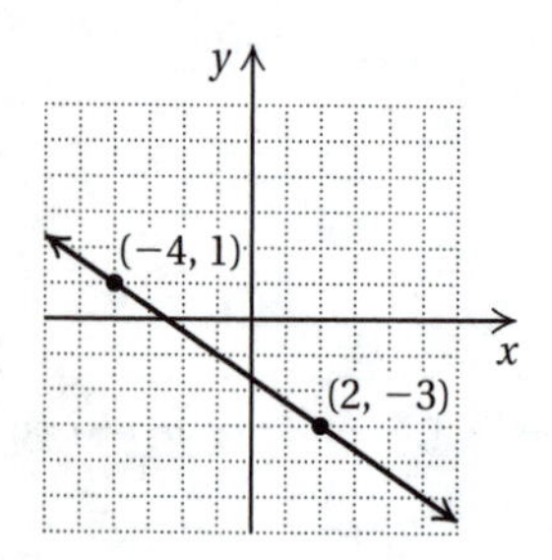

To find the slope, consider (x_1, y_1) to be $(-4, 1)$ and (x_2, y_2) to be $(2, -3)$.

$$m = \frac{y_2 - y_1}{x_2 - x_1} = \frac{-3-1}{2-(-4)} = \frac{-4}{6} = -\frac{2}{3}$$

15. We plot $(5, 3)$ and $(-3, -4)$ and draw the line containing these points.

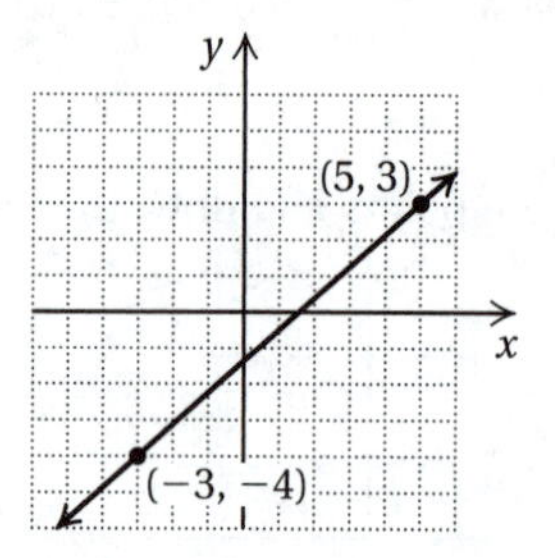

To find the slope, consider (x_1, y_1) to be $(5, 3)$ and (x_2, y_2) to be $(-3, -4)$.

$$m = \frac{y_2 - y_1}{x_2 - x_1} = \frac{-4-3}{-3-5} = \frac{-7}{-8} = \frac{7}{8}$$

17. $m = \dfrac{\frac{1}{2} - \frac{3}{2}}{2-5} = \dfrac{-2}{-3} = \dfrac{2}{3}$

19. $m = \dfrac{-2-3}{4-4} = \dfrac{-5}{0}$

Since division by 0 is not defined, the slope is not defined.

21. $m = \dfrac{-3-7}{15-(-11)} = \dfrac{-10}{26} = -\dfrac{5}{13}$

23. $m = \dfrac{\frac{3}{11} - \frac{3}{11}}{\frac{5}{4} - \left(-\frac{1}{2}\right)} = \dfrac{0}{\frac{7}{4}} = 0$

25. $y = -10x = -10x + 0$

The equation can be written in the form $y = mx + b$, where $m = -10$. Thus, the slope is -10.

27. $y = 3.78x - 4$

The equation is in the form $y = mx + b$, where $m = 3.78$. Thus, the slope is 3.78.

29. We solve for y, obtaining an equation of the form $y = mx + b$.

$$\begin{aligned} 3x - y &= 4 \\ -y &= -3x + 4 \\ -1(-y) &= -1(-3x + 4) \\ y &= 3x - 4 \end{aligned}$$

The slope is 3.

31. We solve for y, obtaining an equation of the form $y = mx + b$.

$$\begin{aligned} x + 5y &= 10 \\ 5y &= -x + 10 \\ y &= \frac{1}{5}(-x + 10) \\ y &= -\frac{1}{5}x + 2 \end{aligned}$$

The slope is $-\dfrac{1}{5}$.

33. We solve for y, obtaining an equation of the form $y = mx + b$.

$$3x + 2y = 6$$
$$2y = -3x + 6$$
$$y = \frac{1}{2}(-3x + 6)$$
$$y = -\frac{3}{2}x + 3$$

The slope is $-\frac{3}{2}$.

35. The graph of $x = \frac{2}{15}$ is a vertical line, so the slope is not defined.

37. $y = 2 - x = -1 \cdot x + 2$

The equation can be written in the form $y = mx+b$, where $m = -1$. Thus, the slope is -1.

39. We solve for y, obtaining an equation of the form $y = mx + b$.

$$9x = 3y + 5$$
$$9x - 5 = 3y$$
$$\frac{1}{3}(9x - 5) = y$$
$$3x - \frac{5}{3} = y$$

The slope is 3.

41. We solve for y, obtaining an equation of the form $y = mx + b$.

$$5x - 4y + 12 = 0$$
$$5x + 12 = 4y$$
$$\frac{1}{4}(5x + 12) = y$$
$$\frac{5}{4}x + 3 = y$$

The slope is $\frac{5}{4}$.

43. $y = 4$

The equation can be thought of as $y = 0 \cdot x + 4$, so the slope is 0.

45.

$$x = \frac{3}{4}y - 2$$
$$x + 2 = \frac{3}{4}y$$
$$\frac{4}{3}(x + 2) = \frac{4}{3} \cdot \frac{3}{4}y$$
$$\frac{4}{3}x + \frac{8}{3} = y$$

The slope is $\frac{4}{3}$.

47.

$$\frac{2}{3}y = -\frac{7}{4}x$$
$$\frac{3}{2} \cdot \frac{2}{3}y = \frac{3}{2}\left(-\frac{7}{4}x\right)$$
$$y = -\frac{21}{8}x$$

The slope is $-\frac{21}{8}$.

49. $m = \frac{\text{rise}}{\text{run}} = \frac{2.4}{8.2} = \frac{2.4}{8.2} \cdot \frac{10}{10} = \frac{24}{82}$

$= \frac{\cancel{2} \cdot 12}{\cancel{2} \cdot 41} = \frac{12}{41}$

51. $m = \frac{\text{rise}}{\text{run}} = \frac{56}{258} = \frac{\cancel{2} \cdot 28}{\cancel{2} \cdot 129} = \frac{28}{129}$

53. Grade $= \frac{8\frac{1}{2}}{280} = \frac{8.5}{280} \approx 0.030 = 3.0\%$

The grade meets the rapid-transit rail standards.

55. Rate of change $= m = \frac{932,000,000 - 945,000,000}{2006 - 2000}$

$= \frac{-13,000,000}{6}$

$\approx -2,170,000$ acres per year

57. Rate of change $= m = \frac{2,495,529 - 1,998,257}{2006 - 2000}$

$= \frac{497,272}{6}$

$\approx 82,900$ people per year

59. We use the data points (2004, 137,000) and (2006, 175,000).

Rate of change $= m = \frac{175,000 - 137,000}{2006 - 2004}$

$= \frac{38,000}{2}$

$\approx 19,000$ tons per year

61. $16\% = \frac{16}{100} = \frac{\cancel{4} \cdot 4}{\cancel{4} \cdot 25} = \frac{4}{25}$

63. $37.5\% = \frac{37.5}{100} = \frac{37.5}{100} \cdot \frac{10}{10} = \frac{375}{1000} = \frac{3 \cdot \cancel{125}}{8 \cdot \cancel{125}} = \frac{3}{8}$

65. ***Translate.***

What is 15% of \$23.80?
↓ ↓ ↓ ↓ ↓
$a = 15\% \cdot 23.80$

Solve. We convert to decimal notation and multiply.

$a = 15\% \cdot 23.80 = 0.15 \cdot 23.80 = 3.57$

The answer is \$3.57.

67. ***Familiarize.*** Let $p =$ the percent of the cost of the meal represented by the tip.

Translate. We reword the problem.

\$8.50 is what percent of \$42.50?
↓ ↓ ↓ ↓ ↓
$8.50 = p \cdot 42.50$

Solve. We solve the equation.

$$8.50 = p \cdot 42.50$$
$$0.2 = p$$
$$20\% = p$$

Check. We can find 20% of 42.50.

$$20\% \cdot 42.50 = 0.2 \cdot 42.50 = 8.50$$

The answer checks.

State. The tip was 20% of the cost of the meal.

69. ***Familiarize.*** Let $c =$ the cost of the meal before the tip was added. Then the tip is $15\% \cdot c$.

Translate. We reword the problem.

$$\underbrace{\text{Cost of meal}}_{\downarrow\; c} \;\; \underset{\downarrow\;+}{\text{plus}} \;\; \underset{\downarrow\;15\%\cdot c}{\text{tip}} \;\; \underset{\downarrow\;=}{\text{is}} \;\; \underbrace{\text{total cost}}_{\downarrow\;51.92}$$

Solve. We solve the equation.

$$c + 15\% \cdot c = 51.92$$
$$1 \cdot c + 0.15c = 51.92$$
$$1.15c = 51.92$$
$$c \approx 45.15$$

Check. We can find 15% of 45.15 and then add this to 45.15.

$15\% \cdot 45.15 = 0.15 \cdot 45.15 \approx 6.77$ and $45.15 + 6.77 = 51.92$

The answer checks.

State. Before the tip the meal cost $45.15.

71. Note that the sum of the coordinates of each point on the graph is 5. Thus, we have $x + y = 5$, or $y = -x + 5$.

73. Note that each y-coordinate is 2 more than the corresponding x-coordinate. Thus, we have $y = x + 2$.

Exercise Set 11.4

1. $y = -4x - 9$

The equation is already in the form $y = mx + b$. The slope is -4 and the y-intercept is $(0, -9)$.

3. $y = 1.8x$

We can think of $y = 1.8x$ as $y = 1.8x + 0$. The slope is 1.8 and the y-intercept is $(0, 0)$.

5. We solve for y.

$$-8x - 7y = 21$$
$$-7y = 8x + 21$$
$$y = -\frac{1}{7}(8x + 21)$$
$$y = -\frac{8}{7}x - 3$$

The slope is $-\frac{8}{7}$ and the y-intercept is $(0, -3)$.

7. We solve for y.

$$4x = 9y + 7$$
$$4x - 7 = 9y$$
$$\frac{1}{9}(4x - 7) = y$$
$$\frac{4}{9}x - \frac{7}{9} = y$$

The slope is $\frac{4}{9}$ and the y-intercept is $\left(0, -\frac{7}{9}\right)$.

9. We solve for y.

$$-6x = 4y + 2$$
$$-6x - 2 = 4y$$
$$\frac{1}{4}(-6x - 2) = y$$
$$-\frac{3}{2}x - \frac{1}{2} = y$$

The slope is $-\frac{3}{2}$ and the y-intercept is $\left(0, -\frac{1}{2}\right)$.

11. $y = -17$

We can think of $y = -17$ as $y = 0x - 17$. The slope is 0 and the y-intercept is $(0, -17)$.

13. We substitute -7 for m and -13 for b in the equation $y = mx + b$.

$$y = -7x - 13$$

15. We substitute 1.01 for m and -2.6 for b in the equation $y = mx + b$.

$$y = 1.01x - 2.6$$

17. We substitute 0 for m and -5 for b in the equation $y = mx + b$.

$$y = 0 \cdot x - 5, \text{ or } y = -5$$

19. We know the slope is -2, so the equation is $y = -2x + b$. Using the point $(-3, 0)$, we substitute -3 for x and 0 for y in $y = -2x + b$. Then we solve for b.

$$y = -2x + b$$
$$0 = -2(-3) + b$$
$$0 = 6 + b$$
$$-6 = b$$

Thus, we have the equation $y = -2x - 6$.

21. We know the slope is $\frac{3}{4}$, so the equation is $y = \frac{3}{4}x + b$. Using the point $(2, 4)$, we substitute 2 for x and 4 for y in $y = \frac{3}{4}x + b$. Then we solve for b.

$$y = \frac{3}{4}x + b$$
$$4 = \frac{3}{4} \cdot 2 + b$$
$$4 = \frac{3}{2} + b$$
$$\frac{5}{2} = b$$

Thus, we have the equation $y = \frac{3}{4}x + \frac{5}{2}$.

23. We know the slope is 1, so the equation is $y = 1 \cdot x + b$, or $y = x + b$. Using the point $(2, -6)$, we substitute 2 for x and -6 for y in $y = x + b$. Then we solve for y.

$$y = x + b$$
$$-6 = 2 + b$$
$$-8 = b$$

Thus, we have the equation $y = x - 8$.

25. We substitute -3 for m and 3 for b in the equation $y = mx + b$.

$$y = -3x + 3$$

27. $(12, 16)$ and $(1, 5)$

First we find the slope.

$$m = \frac{16 - 5}{12 - 1} = \frac{11}{11} = 1$$

Thus, $y = 1 \cdot x + b$, or $y = x + b$. We can use either point to find b. We choose $(1, 5)$. Substitute 1 for x and 5 for y in $y = x + b$.

$$y = x + b$$
$$5 = 1 + b$$
$$4 = b$$

Thus, the equation is $y = x + 4$.

29. $(0, 4)$ and $(4, 2)$

First we find the slope.

$$m = \frac{4 - 2}{0 - 4} = \frac{2}{-4} = -\frac{1}{2}$$

Thus, $y = -\frac{1}{2}x + b$. One of the given points is the y-intercept $(0, 4)$. Thus, we substitute 4 for b in $y = -\frac{1}{2}x + b$. The equation is $y = -\frac{1}{2}x + 4$.

31. $(3, 2)$ and $(1, 5)$

First we find the slope.

$$m = \frac{2 - 5}{3 - 1} = \frac{-3}{2} = -\frac{3}{2}$$

Thus, $y = -\frac{3}{2}x + b$. We can use either point to find b. We choose $(3, 2)$. Substitute 3 for x and 2 for y in $y = -\frac{3}{2}x + b$.

$$y = -\frac{3}{2}x + b$$
$$2 = -\frac{3}{2} \cdot 3 + b$$
$$2 = -\frac{9}{2} + b$$
$$\frac{13}{2} = b$$

Thus, the equation is $y = -\frac{3}{2}x + \frac{13}{2}$.

33. $\left(4, -\frac{2}{5}\right)$ and $\left(4, \frac{2}{5}\right)$

First we find the slope.

$$m = \frac{-\frac{2}{5} - \frac{2}{5}}{4 - 4} = \frac{-\frac{4}{5}}{0}$$

The slope is not defined. Thus, we have a vertical line. Since the first coordinates of the given points are 4, the equation of the line is $x = 4$.

35. $(-4, 5)$ and $(-2, -3)$

First we find the slope.

$$m = \frac{5 - (-3)}{-4 - (-2)} = \frac{8}{-2} = -4$$

Thus, $y = -4x + b$. We can use either point to find b. We choose $(-4, 5)$. Substitute -4 for x and 5 for y in $y = -4x + b$.

$$y = -4x + b$$
$$5 = -4(-4) + b$$
$$5 = 16 + b$$
$$-11 = b$$

Thus, the equation is $y = -4x - 11$.

37. $\left(-2, \frac{1}{4}\right)$ and $\left(3, \frac{1}{4}\right)$

First we find the slope.

$$m = \frac{\frac{1}{4} - \frac{1}{4}}{-2 - 3} = \frac{0}{-5} = 0$$

We could observe here that, because the slope is 0, we have a horizontal line. Both y– coordinates are $\frac{1}{4}$, so the equation is $y = \frac{1}{4}$.

We could also use one of the points to find that the y-intercept is $\frac{1}{4}$ and then write the equation, $y = 0 \cdot x + \frac{1}{4}$, or $y = \frac{1}{4}$.

39. a) First we find the slope.

$$m = \frac{66.54 - 44.50}{7 - 0} = \frac{22.04}{7} = \frac{22.04}{7} \cdot \frac{100}{100}$$
$$= \frac{2204}{700} = \frac{\not{4} \cdot 551}{\not{4} \cdot 175} = \frac{551}{175}$$

We see from the graph that the y-intercept is $(0, 44.50)$, so the equation of the line is $H = \frac{551}{175}x + 44.50$. Since $\frac{551}{175} \approx 3.15$, we could also write the equation as $H = 3.15x + 44.50$.

b) The rate of change is the slope, $\$\frac{551}{175}$ per year, or about \$3.15 per year.

c) In 2012, $t = 2012 - 2001$, or 11.

$H = 3.15(11) + 44.50 = 34.65 + 44.50 = \$79.15$.

41. $3x - 4(9 - x) = 17$
$3x - 36 + 4x = 17$
$7x - 36 = 17$
$7x = 53$
$x = \frac{53}{7}$
The solution is $\frac{53}{7}$.

43. $4(a - 3) + 6 = 21 - \frac{1}{2}a$
$4a - 12 + 6 = 21 - \frac{1}{2}a$
$4a - 6 = 21 - \frac{1}{2}a$
$2(4a - 6) = 2\left(21 - \frac{1}{2}a\right)$
$8a - 12 = 42 - a$
$9a - 12 = 42$
$9a = 54$
$a = 6$
The solution is 6.

45. $40(2x - 7) = 50(4 - 6x)$
$80x - 280 = 200 - 300x$
$380x - 280 = 200$
$380x = 480$
$x = \frac{480}{380}$
$x = \frac{24}{19}$
The solution is $\frac{24}{19}$.

47. $3x - 9x + 21x - 15x = 6x - 12 - 24x + 18$
$0 = -18x + 6$
$18x = 6$
$x = \frac{6}{18}$
$x = \frac{1}{3}$
The solution is $\frac{1}{3}$.

49. $3(x - 9x) + 21(x - 15x) = 6(x - 12) - 24(x + 18)$
$3(-8x) + 21(-14x) = 6x - 72 - 24x - 432$
$-24x - 294x = -18x - 504$
$-318x = -18x - 504$
$-300x = -504$
$x = \frac{504}{300} = \frac{\not{2} \cdot \not{6} \cdot 6 \cdot 7}{\not{2} \cdot \not{6} \cdot 5 \cdot 5}$
$x = \frac{42}{25}$
The solution is $\frac{42}{25}$.

51. First find the slope of $3x - y + 4 = 0$.
$3x - y + 4 = 0$
$3x + 4 = y$
The slope is 3.
Thus, $y = 3x + b$. Using the point $(2, -3)$, we substitute 2 for x and -3 for y in $y = 3x + b$. Then we solve for b.
$y = 3x + b$
$-3 = 3 \cdot 2 + b$
$-3 = 6 + b$
$-9 = b$
Thus, the equation is $y = 3x - 9$.

53. First find the slope of $3x - 2y = 8$.
$3x - 2y = 8$
$-2y = -3x + 8$
$y = \frac{3}{2}x - 4$
The slope is $\frac{3}{2}$.
Then find the y-intercept of $2y + 3x = -4$.
$2y + 3x = -4$
$2y = -3x - 4$
$y = -\frac{3}{2}x - 2$
The y-intercept is $(0, -2)$.
Finally, write the equation of the line with slope $\frac{3}{2}$ and y-intercept $(0, -2)$.
$y = mx + b$
$y = \frac{3}{2}x + (-2)$
$y = \frac{3}{2}x - 2$

Chapter 11 Mid-Chapter Review

1. $\left|-\frac{3}{4}\right| = \frac{3}{4}$; $\left|-\frac{5}{2}\right| = \frac{5}{2}$; since $\frac{5}{2} > \frac{3}{4}$, a slope of $-\frac{5}{2}$ is steeper than a slope of $-\frac{3}{4}$. Thus, the given statement is false.

2. True; $\frac{d - b}{c - a}$ is $\frac{\text{rise}}{\text{run}}$, or slope.

3. True; when $x = 0$, we have $y = \frac{C}{B}$.

4. False; in quadrant IV, the first coordinate is positive and the second coordinate is negative.

5. a) The y-intercept is $(0, -3)$.
b) The x-intercept is $(-3, 0)$.
c) The slope is $\frac{-3 - 0}{0 - (-3)} = \frac{-3}{3} = -1$.
d) The equation of the line in $y = mx + b$ form is $y = -1 \cdot x + (-3)$, or $y = -x - 3$.

6. a) The x-intercept is $(c, 0)$.

b) The y-intercept is $(0, d)$.

c) The slope is $\frac{d-0}{0-c} = \frac{d}{-c} = -\frac{d}{c}$.

d) The equation of the line in $y = mx + b$ form is $y = -\frac{d}{c}x + d$.

7.
$$\begin{array}{c|c} -2q - 7p & 19 \\ \hline -2(-5) - 7\cdot 8 \ ? & 19 \\ 10 - 56 & \\ -46 & \end{array} \quad \text{FALSE}$$

$(8, -5)$ is not a solution.

8.
$$\begin{array}{c|c} 6y & -3x + 1 \\ \hline 6\cdot\frac{2}{3} \ ? & -3(-1) + 1 \\ 4 & 3 + 1 \\ & 4 \end{array} \quad \text{TRUE}$$

$\left(-1, \frac{2}{3}\right)$ is a solution.

9. To find the x-intercept, let $y = 0$. Then solve for x.

$$-3x + 2y = 18$$
$$-3x + 2\cdot 0 = 18$$
$$-3x = 18$$
$$x = -6$$

The x-intercept is $(-6, 0)$.

To find the y-intercept, let $x = 0$. Then solve for y.

$$-3x + 2y = 18$$
$$-3\cdot 0 + 2y = 18$$
$$2y = 18$$
$$y = 9$$

The y-intercept is $(0, 9)$.

10. To find the x-intercept, let $y = 0$. Then solve for x.

$$x - \frac{1}{2} = 10y$$
$$x - \frac{1}{2} = 10\cdot 0$$
$$x - \frac{1}{2} = 0$$
$$x = \frac{1}{2}$$

The x-intercept is $\left(\frac{1}{2}, 0\right)$.

To find the y-intercept, let $x = 0$. Then solve for y.

$$x - \frac{1}{2} = 10y$$
$$0 - \frac{1}{2} = 10y$$
$$-\frac{1}{2} = 10y$$
$$-\frac{1}{20} = y$$

The y-intercept is $\left(0, -\frac{1}{20}\right)$.

11. Graph: $-2x + y = -3$.

$$-2x + y = -3$$
$$y = 2x - 3$$

The y-intercept is $(0, -3)$.

We find the x-intercept.

$$-2x + 0 = -3$$
$$-2x = -3$$
$$x = \frac{3}{2}$$

The x-intercept is $\left(\frac{3}{2}, 0\right)$.

We find a third point as a check. Let $x = 1$.

$$y = 2\cdot 1 - 3 = 2 - 3 = -1$$

A third point is $(1, -1)$. We plot this point and the intercepts and draw the line.

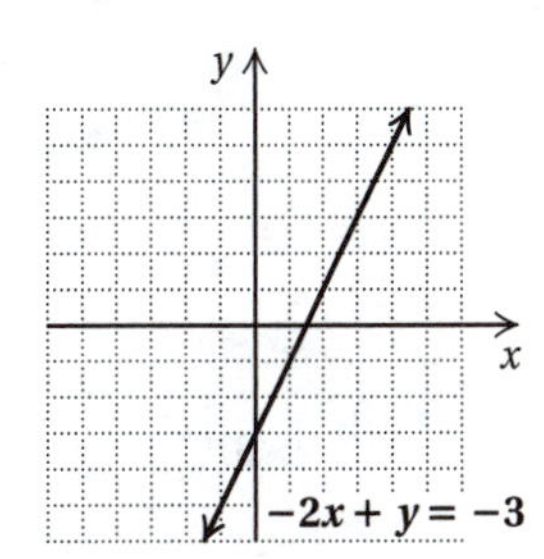

12. Graph: $y = -\frac{3}{2}$.

This is an equation of the form $y = b$ with $b = -\frac{3}{2}$. Its graph is a horizontal line with y-intercept $(0, b)$, or $\left(0, -\frac{3}{2}\right)$.

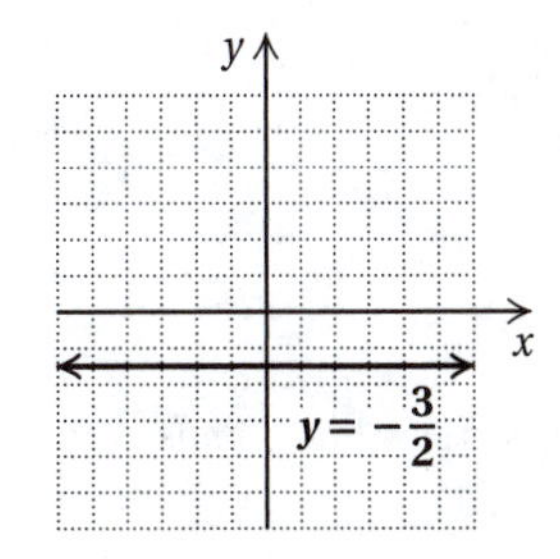

13. Graph: $y = -x + 4$.

The y-intercept is $(0, 4)$.

We find the x-intercept.

$$0 = -x + 4$$
$$x = 4$$

The x-intercept is $(4, 0)$.

We find a third point as a check. Let $x = 3$.

$$y = -3 + 4 = 1$$

A third point is $(3, 1)$. We plot this point and the intercepts and draw the line.

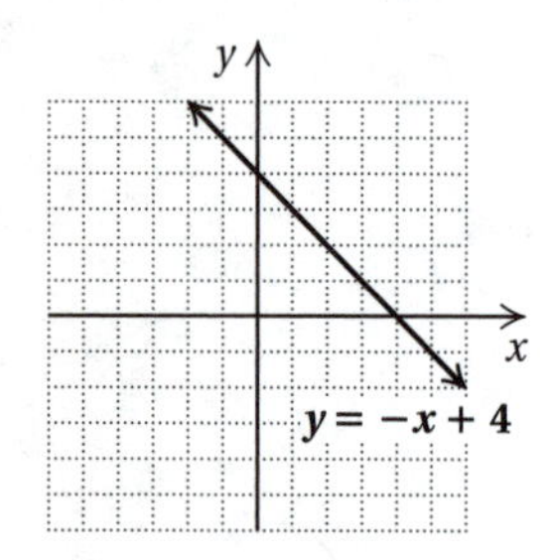

14. Graph: $x = 0$.

This is an equation of the form $x = a$ with $a = 0$. Its graph is a vertical line with x-intercept $(0, a)$, or $(0, 0)$.

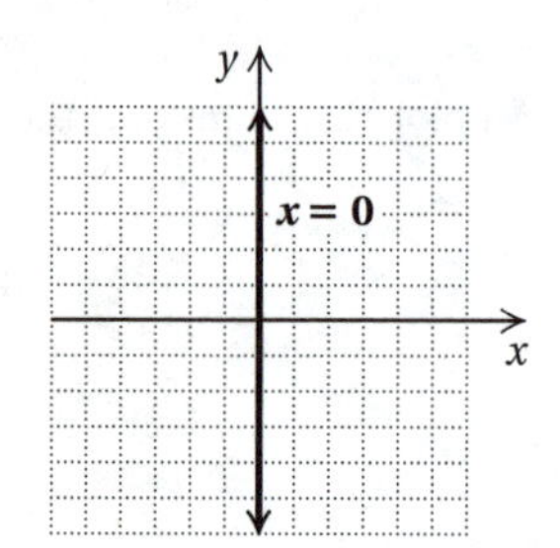

15. $m = \dfrac{4-(-6)}{-2-\frac{1}{4}} = \dfrac{10}{-2\frac{1}{4}} = \dfrac{10}{-\frac{9}{4}} = 10\left(-\dfrac{4}{9}\right) = -\dfrac{40}{9}$

16. $m = \dfrac{-3-3}{6-(-6)} = \dfrac{-6}{12} = -\dfrac{1}{2}$

17. $y = 0.728$, or $y = 0 \cdot x + 0.728$

The slope is 0.

18.
$$13x - y = -5$$
$$13x = y - 5$$
$$13x + 5 = y$$

The slope is 13.

19.
$$12x + 7 = 0$$
$$12x = -7$$
$$x = -\frac{7}{12}$$

This is the equation of a vertical line. Thus, the slope is not defined.

20. We use the data points (2000, 4,468,976) and (2006, 4,287,768).

Rate of change $= m = \dfrac{4,287,768 - 4,468,976}{2006 - 2000} = \dfrac{-181,208}{6} \approx -30,200$ people per year

21. $y = -1$, or $y = 0 \cdot x - 1$

The slope is 0 and the y-intercept is $(0, -1)$. Choice D is correct.

22. $x = 1$

This is the equation of a vertical line with x-intercept $(1, 0)$. The slope of a vertical line is not defined. Thus, choice C is correct.

23. $y = -x - 1$

The slope is -1 and the y-intercept is $(0, -1)$. Choice B is correct.

24. $y = x - 1$

The slope is 1 so either A or E is the correct choice. We find the x-intercept.

$$0 = x - 1$$
$$1 = x$$

The x-intercept is $(1, 0)$. Thus, choice E is correct.

25. $y = x + 1$

The slope is 1 so either A or E is the correct choice. We find the x-intercept.

$$0 = x + 1$$
$$-1 = x$$

The x-intercept is $(-1, 0)$. Thus choice A is correct.

26. The slope is -3 so the equation is $y = -3x + b$. Using the point $\left(-\dfrac{1}{3}, 3\right)$, we substitute $-\dfrac{1}{3}$ for x and 3 for y in $y = -3x + b$. Then we solve for b.

$$y = -3x + b$$
$$3 = -3 \cdot -\frac{1}{3} + b$$
$$3 = 1 + b$$
$$2 = b$$

The equation is $y = -3x + 2$.

27. $\left(\dfrac{1}{2}, 6\right)$ and $\left(\dfrac{1}{2}, -6\right)$

First we find the slope.

$$m = \frac{6-(-6)}{\frac{1}{2}-\frac{1}{2}} = \frac{12}{0}$$

The slope is not defined. Thus, we have a vertical line with x-coordinate $\dfrac{1}{2}$. The equation is $x = \dfrac{1}{2}$.

28. $(3, -4)$ and $(-7, -2)$

First we find the slope.

$$m = \frac{-2-(-4)}{-7-3} = \frac{2}{-10} = -\frac{1}{5}$$

Thus $y = -\frac{1}{5}x + b$. We use the point $(3, -4)$ to find b.

$$y = -\frac{1}{5}x + b$$
$$-4 = -\frac{1}{5} \cdot 3 + b$$
$$-4 = -\frac{3}{5} + b$$
$$-\frac{20}{5} + \frac{3}{5} = b$$
$$-\frac{17}{5} = b$$

The equation is $y = -\frac{1}{5}x - \frac{17}{5}$.

29. $(3, -4)$ and $(2, -4)$

First we find the slope.

$$m = \frac{-4 - (-4)}{3 - 2} = \frac{0}{1} = 0$$

The slope is 0, so we have a horizontal line with y-coordinate -4. The equation is $y = -4$.

30. No; an equation $x = a$, $a \neq 0$, does not have a y-intercept.

31. Most would probably say that the second equation would be easier to graph because it has been solved for y. This makes it more efficient to find the y-value that corresponds to a given x-value.

32. $A = 0$. If the line is horizontal, then regardless of the value of x, the value of y remains constant. Thus, Ax must be 0 and, hence, $A = 0$.

33. Any ordered pair $(7, y)$ is a solution of $x = 7$. Thus, all points on the graph are 7 units to the right of the y-axis, so they lie on a vertical line.

Exercise Set 11.5

1. Slope $\frac{2}{5}$; y-intercept $(0, 1)$

We plot $(0, 1)$ and from there move up 2 units and right 5 units. This locates the point $(5, 3)$. We plot $(5, 3)$ and draw a line passing through $(0, 1)$ and $(5, 3)$.

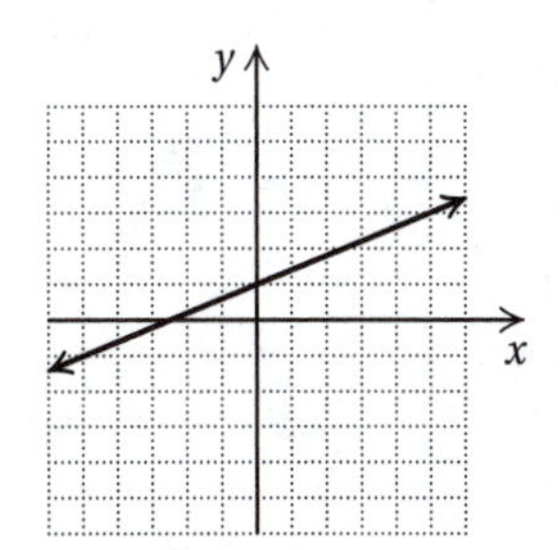

3. Slope $\frac{5}{3}$; y-intercept $(0, -2)$

We plot $(0, -2)$ and from there move up 5 units and right 3 units. This locates the point $(3, 3)$. We plot $(3, 3)$ and draw a line passing through $(0, -2)$ and $(3, 3)$.

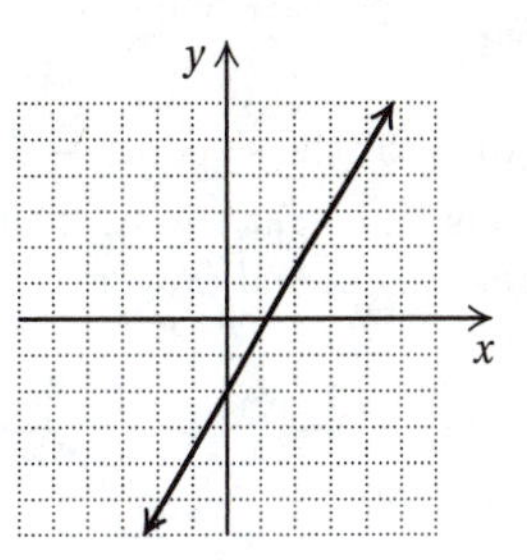

5. Slope $-\frac{3}{4}$; y-intercept $(0, 5)$

We plot $(0, 5)$. We can think of the slope as $\frac{-3}{4}$, so from $(0, 5)$ we move down 3 units and right 4 units. This locates the point $(4, 2)$. We plot $(4, 2)$ and draw a line passing through $(0, 5)$ and $(4, 2)$.

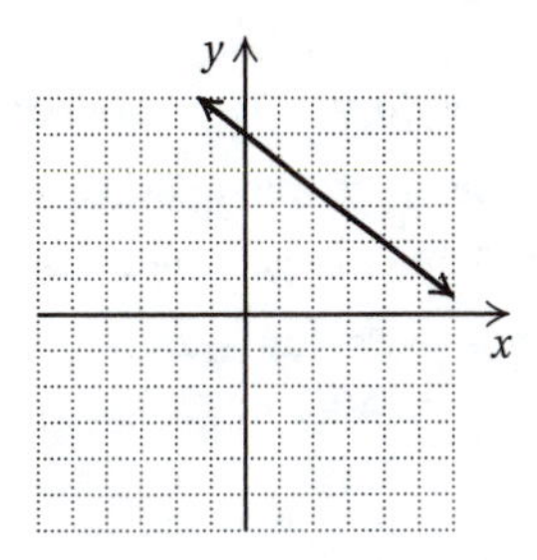

7. Slope $-\frac{1}{2}$; y-intercept $(0, 3)$

We plot $(0, 3)$. We can think of the slope as $\frac{-1}{2}$, so from $(0, 3)$ we move down 1 unit and right 2 units. This locates the point $(2, 2)$. We plot $(2, 2)$ and draw a line passing through $(0, 3)$ and $(2, 2)$

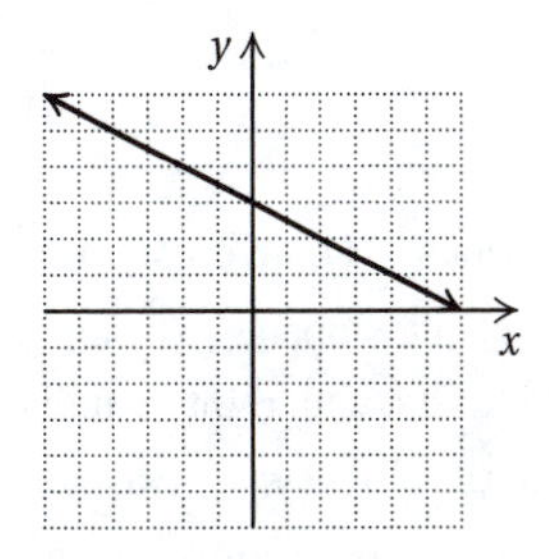

9. Slope 2; y-intercept $(0, -4)$

We plot $(0, -4)$. We can think of the slope as $\frac{2}{1}$, so from $(0, -4)$ we move up 2 units and right 1 unit. This locates the point $(1, -2)$. We plot $(1, -2)$ and draw a line passing through $(0, -4)$ and $(1, -2)$.

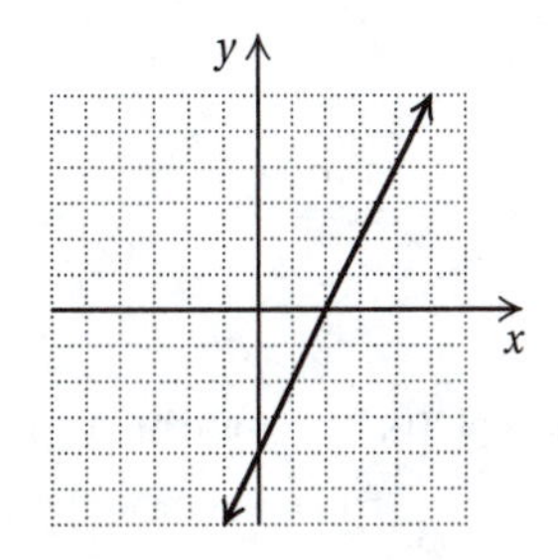

11. Slope -3; y-intercept $(0,2)$

We plot $(0,2)$. We can think of the slope as $\frac{-3}{1}$, so from $(0,2)$ we move down 3 units and right 1 unit. This locates the point $(1,-1)$. We plot $(1,-1)$ and draw a line passing through $(0,2)$ and $(1,-1)$.

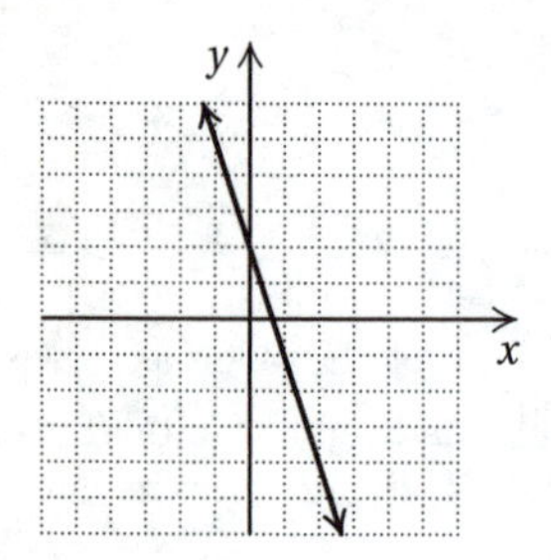

13. $y=\frac{3}{5}x+2$

First we plot the y-intercept $(0,2)$. We can start at the y-intercept and use the slope, $\frac{3}{5}$, to find another point. We move up 3 units and right 5 units to get a new point $(5,5)$. Thinking of the slope as $\frac{-3}{-5}$ we can start at $(0,2)$ and move down 3 units and left 5 units to get another point $(-5,-1)$.

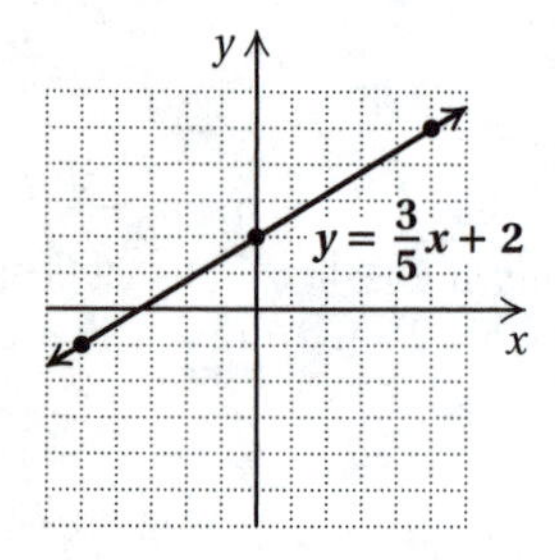

15. $y=-\frac{3}{5}x+1$

First we plot the y-intercept $(0,1)$. We can start at the y-intercept and, thinking of the slope as $\frac{-3}{5}$, find another point by moving down 3 units and right 5 units to the point $(5,-2)$. Thinking of the slope as $\frac{3}{-5}$ we can start at $(0,1)$ and move up 3 units and left 5 units to get another point $(-5,4)$.

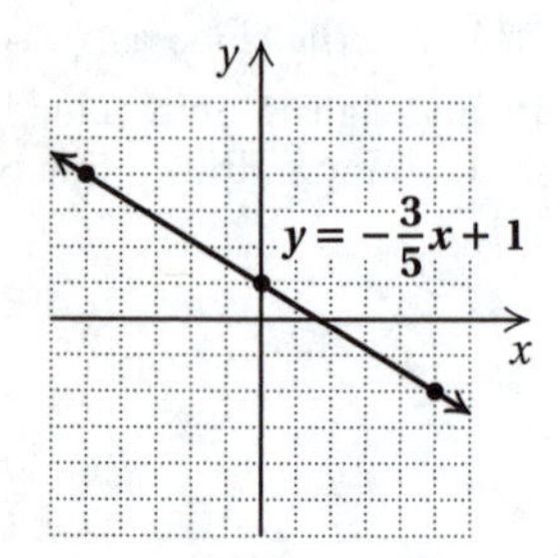

17. $y=\frac{5}{3}x+3$

First we plot the y-intercept $(0,3)$. We can start at the y-intercept and use the slope, $\frac{5}{3}$, to find another point. We move up 5 units and right 3 units to get a new point $(3,8)$. Thinking of the slope as $\frac{-5}{-3}$ we can start at $(0,3)$ and move down 5 units and left 3 units to get another point $(-3,-2)$.

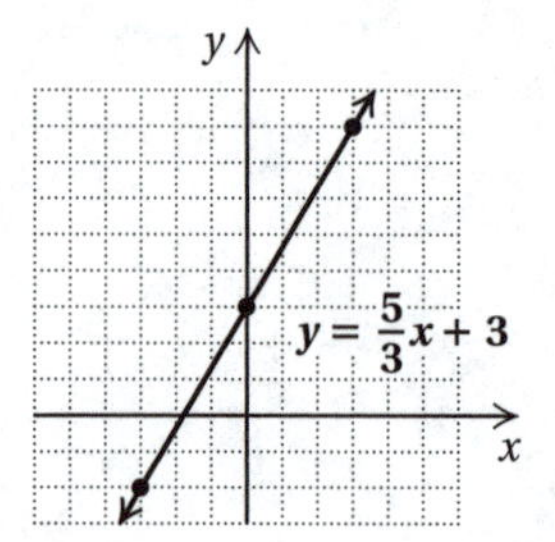

19. $y=-\frac{3}{2}x-2$

First we plot the y-intercept $(0,-2)$. We can start at the y-intercept and, thinking of the slope as $\frac{-3}{2}$, find another point by moving down 3 units and right 2 units to the point $(2,-5)$. Thinking of the slope as $\frac{3}{-2}$ we can start at $(0,-2)$ and move up 3 units and left 2 units to get another point $(-2,1)$.

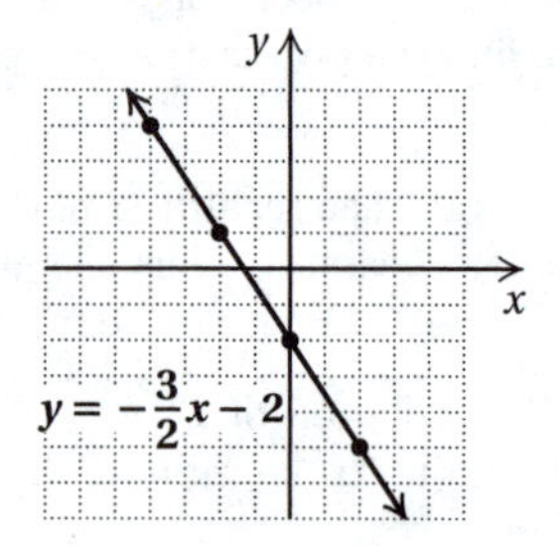

21. We first rewrite the equation in slope-intercept form.

$$2x+y=1$$
$$y=-2x+1$$

Now we plot the y-intercept $(0,1)$. We can start at the y-intercept and, thinking of the slope as $\frac{-2}{1}$, find another point by moving down 2 units and right 1 unit to the point $(1,-1)$. In a similar manner, we can move from the point $(1,-1)$ to find a third point $(2,-3)$.

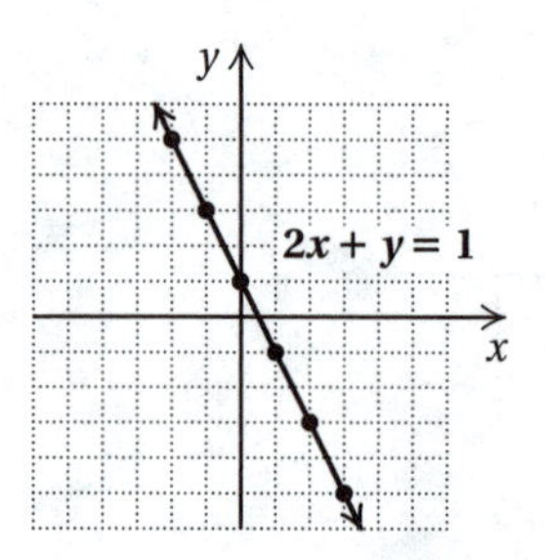

23. We first rewrite the equation in slope-intercept form.

$$3x-y=4$$
$$-y=-3x+4$$
$$y=3x-4 \quad \text{Multiplying by } -1$$

Now we plot the y-intercept $(0,-4)$. We can start at the y-intercept and, thinking of the slope as $\frac{3}{1}$, find another point by moving up 3 units and right 1 unit to the point $(1,-1)$. In a similar manner, we can move from the point $(1,-1)$ to find a third point $(2,2)$.

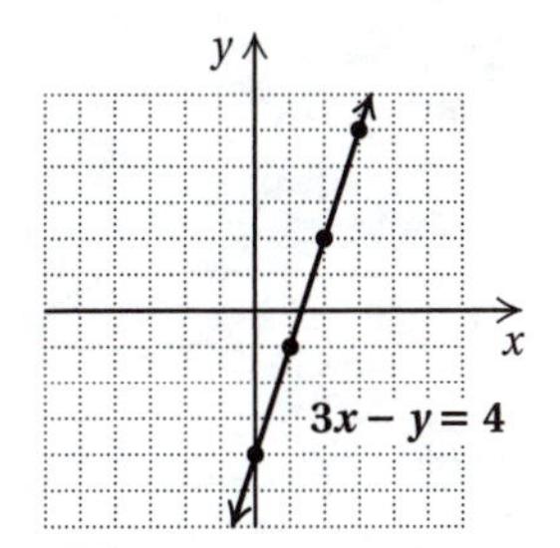

25. We first rewrite the equation in slope-intercept form.

$$2x+3y=9$$
$$3y=-2x+9$$
$$y=\frac{1}{3}(-2x+9)$$
$$y=-\frac{2}{3}x+3$$

Now we plot the y-intercept $(0,3)$. We can start at the y-intercept and, thinking of the slope as $\frac{-2}{3}$, find another point by moving down 2 units and right 3 units to the point $(3,1)$. Thinking of the slope as $\frac{2}{-3}$ we can start at $(0,3)$ and move up 2 units and left 3 units to get another point $(-3,5)$.

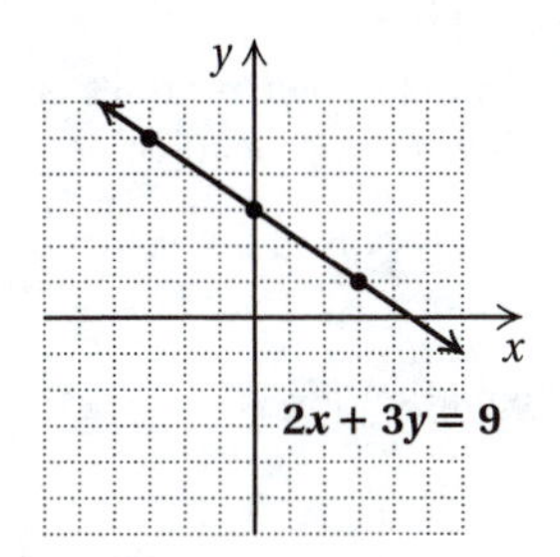

27. We first rewrite the equation in slope-intercept form.

$$x-4y=12$$
$$-4y=-x+12$$
$$y=-\frac{1}{4}(-x+12)$$
$$y=\frac{1}{4}x-3$$

Now we plot the y-intercept $(0,-3)$. We can start at the y-intercept and use the slope, $\frac{1}{4}$, to find another point. We move up 1 unit and right 4 units to the point $(4,-2)$. Thinking of the slope as $\frac{-1}{-4}$ we can start at $(0,-3)$ and move down 1 unit and left 4 units to get another point $(-4,-4)$.

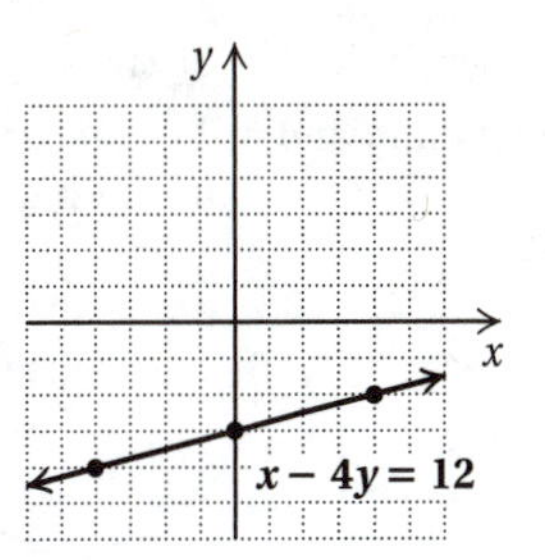

29. We first rewrite the equation in slope-intercept form.

$$x+2y=6$$
$$2y=-x+6$$
$$y=\frac{1}{2}(-x+6)$$
$$y=-\frac{1}{2}x+3$$

Now we plot the y-intercept $(0,3)$. We can start at the y-intercept and, thinking of the slope as $\frac{-1}{2}$, find another point by moving down 1 unit and right 2 units to the point $(2,2)$. Thinking of the slope as $\frac{1}{-2}$ we can start at $(0,3)$ and move up 1 unit and left 2 units to get another point $(-2,4)$.

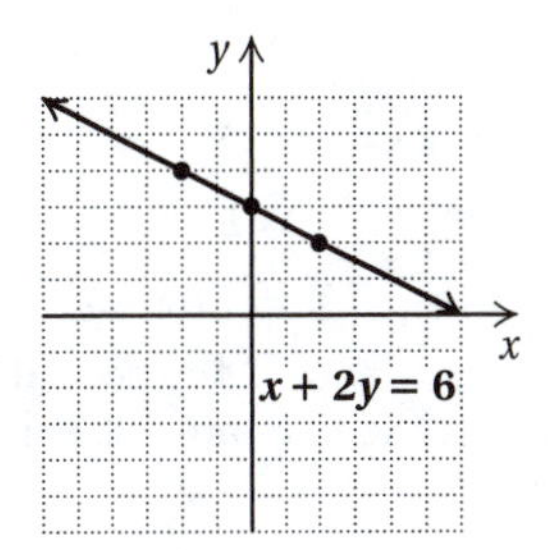

31. $m=\frac{y_2-y_1}{x_2-x_1}=\frac{7-(-6)}{8-(-2)}=\frac{13}{10}$

33. $m=\frac{y_2-y_1}{x_2-x_1}=\frac{4.6-(-2.3)}{14.5-4.5}=\frac{6.9}{10}=\frac{69}{100}$, or 0.69

35. $m=\frac{y_2-y_1}{x_2-x_1}=\frac{-6-(-6)}{8-(-2)}=\frac{0}{10}=0$

37. $m=\frac{y_2-y_1}{x_2-x_1}=\frac{-4-(-1)}{11-11}=\frac{-3}{0}$

Since division by 0 is not defined, the slope is not defined.

39. Rate of change $=\frac{\text{Change in number of transplants}}{\text{Change in years}}=$

$\frac{17,094-9358}{2006-1990}=\frac{7736}{16}\approx 484$

Kidney transplants are increasing at a rate of about 484 per year. The slope of the line is 484.

41. For residents in excess of 2, the rate of change is 1.5 ft^3 per person. For 1-2 people, the number of residents in excess of 2 is 0, so the x-intercept is $(0,16)$. Then the equation is $y=1.5x+16$.

43. First we plot $(-3,1)$. Then, thinking of the slope as $\frac{2}{1}$, from $(-3,1)$ we move up 2 units and right 1 unit to locate the point $(-2,3)$. We plot $(-2,3)$ and draw a line passing through $(-3,1)$ and $(-2,3)$.

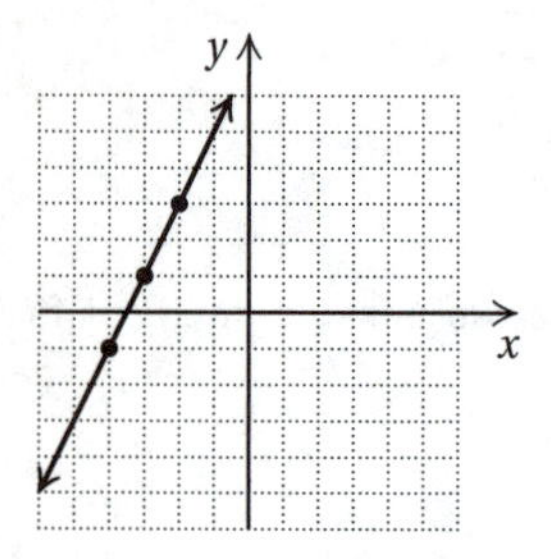

Exercise Set 11.6

1. 1. The first equation is already solved for y:

$$y = x + 4$$

2. We solve the second equation for y:

$$y - x = -3$$
$$y = x - 3$$

The slope of each line is 1. The y-intercepts, $(0,4)$ and $(0,-3)$, are different. The lines are parallel.

3. We solve each equation for y:

1. $y + 3 = 6x$
$$y = 6x - 3$$

2. $-6x - y = 2$
$$-y = 6x + 2$$
$$y = -6x - 2$$

The slope of the first line is 6 and of the second is -6. Since the slopes are different, the lines are not parallel.

5. We solve each equation for y:

1. $10y+32x = 16.4$
$$10y = -32x + 16.4$$
$$y = -3.2x + 1.64$$

2. $y+3.5 = 0.3125x$
$$y = 0.3125x - 3.5$$

The slope of the first line is -3.2 and of the second is 0.3125. Since the slopes are different, the lines are not parallel.

7. 1. The first equation is already solved for y:

$$y = 2x + 7$$

2. We solve the second equation for y:

$$5y + 10x = 20$$
$$5y = -10x + 20$$
$$y = -2x + 4$$

The slope of the first line is 2 and of the second is -2. Since the slopes are different, the lines are not parallel.

9. We solve each equation for y:

1. $3x - y = -9$
$$3x + 9 = y$$

2. $2y - 6x = -2$
$$2y = 6x - 2$$
$$y = 3x - 1$$

The slope of each line is 3. The y-intercepts, $(0,9)$ and $(0,-1)$ are different. The lines are parallel.

11. $x = 3$,

$x = 4$

These are vertical lines with equations of the form $x = p$ and $x = q$, where $p \neq q$. Thus, they are parallel.

13. 1. The first equation is already solved for y:

$$y = -4x + 3$$

2. We solve the second equation for y:

$$4y + x = -1$$
$$4y = -x - 1$$
$$y = -\frac{1}{4}x - \frac{1}{4}$$

The slopes are -4 and $-\frac{1}{4}$. Their product is $-4\left(-\frac{1}{4}\right) = 1$. Since the product of the slopes is not -1, the lines are not perpendicular.

15. We solve each equation for y:

1. $x + y = 6$
$$y = -x + 6$$

2. $4y - 4x = 12$
$$4y = 4x + 12$$
$$y = x + 3$$

The slopes are -1 and 1. Their product is $-1 \cdot 1 = -1$. The lines are perpendicular.

17. 1. The first equation is already solved for y:

$$y = -0.3125x + 11$$

2. We solve the second equation for y:

$$y - 3.2x = -14$$
$$y = 3.2x - 14$$

The slopes are -0.3125 and 3.2. Their product is $-0.3125(3.2) = -1$. The lines are perpendicular.

19. 1. The first equation is already solved for y:

$$y = -x + 8$$

2. We solve the second equation for y:

$$x - y = -1$$
$$x + 1 = y$$

The slopes are -1 and 1. Their product is $-1 \cdot 1 = -1$. The lines are perpendicular.

21. We solve each equation for y:

1. $\frac{3}{8}x - \frac{y}{2} = 1$
$$8\left(\frac{3}{8}x - \frac{y}{2}\right) = 8 \cdot 1$$
$$8 \cdot \frac{3}{8}x - 8 \cdot \frac{y}{2} = 8$$
$$3x - 4y = 8$$
$$-4y = -3x + 8$$
$$y = \frac{3}{4}x - 2$$

2. $\frac{4}{3}x - y + 1 = 0$

$\frac{4}{3}x + 1 = y$

The slopes are $\frac{3}{4}$ and $\frac{4}{3}$. Their product is $\frac{3}{4}\left(\frac{4}{3}\right) = 1$. Since the product of the slopes is not -1, the lines are not perpendicular.

23. $x = 0,$

$y = -2$

The first line is vertical and the second is horizontal, so the lines are perpendicular.

25. We solve each equation for y:

1. $3y + 21 = 2x$
$3y = 2x - 21$
$y = \frac{2}{3}x - 7$

2. $3y = 2x + 24$
$y = \frac{2}{3}x + 8$

The slope of each line is $\frac{2}{3}$. The y-intercepts, $(0, -7)$ and $(0, 8)$, are different. The lines are parallel.

27. We solve each equation for y:

1. $3y = 2x - 21$
$y = \frac{2}{3}x - 7$

2. $2y - 16 = 3x$
$2y = 3x + 16$
$y = \frac{3}{2}x + 8$

The slopes, $\frac{2}{3}$ and $\frac{3}{2}$, are different so the lines are not parallel. The product of the slopes is $\frac{2}{3} \cdot \frac{3}{2} = 1 \neq -1$, so the lines are not perpendicular. Thus, the lines are neither parallel nor perpendicular.

29. Equations with the same solutions are called equivalent equations.

31. The multiplication principle for equations asserts that when we multiply or divide by the same non-zero number on both sides of an equation, we get equivalent equations.

33. Vertical lines are graphs of equations of the type $x = a$.

35. The x-intercept of a line, if it exists, indicates where the line crosses the x-axis.

37. First we find the slope of the given line:

$y - 3x = 4$

$y = 3x + 4$

The slope is 3.

Then we use the slope-intercept equation to write the equation of a line with slope 3 and y-intercept $(0, 6)$:

$y = mx + b$

$y = 3x + 6$ Substituting 3 for m and 6 for b

39. First we find the slope of the given line:

$3y - x = 0$

$3y = x$

$y = \frac{1}{3}x$

The slope is $\frac{1}{3}$.

We can find the slope of the line perpendicular to the given line by taking the reciprocal of $\frac{1}{3}$ and changing the sign. We get -3.

Then we use the slope-intercept equation to write the equation of a line with slope -3 and y-intercept $(0, 2)$:

$y = mx + b$

$y = -3x + 2$ Substituting -3 for m and 2 for b

41. First we find the slope of the given line:

$4x - 8y = 12$

$-8y = -4x + 12$

$y = \frac{1}{2}x - \frac{3}{2}$

The slope is $\frac{1}{2}$, so the equation is $y = \frac{1}{2}x + b$. Substitute -2 for x and 0 for y and solve for b.

$y = \frac{1}{2}x + b$

$0 = \frac{1}{2}(-2) + b$

$0 = -1 + b$

$1 = b$

Thus, the equation is $y = \frac{1}{2}x + 1$.

43. We find the slope of each line:

1. $4y = kx - 6$
$y = \frac{k}{4}x - \frac{3}{2}$

2. $5x + 20y = 12$
$20y = -5x + 12$
$y = -\frac{1}{4}x + \frac{3}{5}$

The slopes are $\frac{k}{4}$ and $-\frac{1}{4}$. If the lines are perpendicular, the product of their slopes is -1.

$\frac{k}{4}\left(-\frac{1}{4}\right) = -1$

$-\frac{k}{16} = -1$

$k = 16$

45. First we find the equation of A, a line containing the points $(1, -1)$ and $(4, 3)$:

The slope is $\frac{3 - (-1)}{4 - 1} = \frac{4}{3}$, so the equation is $y = \frac{4}{3}x + b$. Use either point to find b. We choose $(1, -1)$.

$$y = \frac{4}{3}x + b$$
$$-1 = \frac{4}{3} \cdot 1 + b$$
$$-1 = \frac{4}{3} + b$$
$$-\frac{7}{3} = b$$

Thus, the equation of line A is $y = \frac{4}{3}x - \frac{7}{3}$.

The slope of A is $\frac{4}{3}$. Since A and B are perpendicular we find the slope of B by taking the reciprocal of $\frac{4}{3}$ and changing the sign. We get $-\frac{3}{4}$, so the equation is $y = -\frac{3}{4}x + b$. We use the point $(1, -1)$ to find b.

$$y = -\frac{3}{4}x + b$$
$$-1 = -\frac{3}{4} \cdot 1 + b$$
$$-1 = -\frac{3}{4} + b$$
$$-\frac{1}{4} = b$$

Thus, the equation of line B is $y = -\frac{3}{4}x - \frac{1}{4}$.

Exercise Set 11.7

1. We use alphabetical order to replace x by -3 and y by -5.

$$\begin{array}{c|c} -x - 3y < 18 & \\ \hline -(-3) - 3(-5) \; ? \; 18 & \\ 3 + 15 & \\ 18 & \text{FALSE} \end{array}$$

Since $18 < 18$ is false, $(-3, -5)$ is not a solution.

3. We use alphabetical order to replace x by 1 and y by -10.

$$\begin{array}{c|c} 7y - 9x \leq -3 & \\ \hline 7(-10) - 9 \cdot 1 \; ? \; -3 & \\ -70 - 9 & \\ -79 & \text{TRUE} \end{array}$$

Since $-79 \leq -3$ is true, $(1, -10)$ is a solution.

5. Graph $x > 2y$.

First graph the line $x = 2y$, or $y = \frac{1}{2}x$. Two points on the line are $(0, 0)$ and $(4, 2)$. We draw a dashed line since the inequality symbol is $>$. Then we pick a test point that is not on the line. We try $(-2, 1)$.

$$\begin{array}{c|c} x > 2y & \\ \hline -2 \; ? \; 2 \cdot 1 & \\ & 2 \quad \text{FALSE} \end{array}$$

We see that $(-2, 1)$ is not a solution of the inequality, so we shade the points in the region that does not contain $(-2, 1)$.

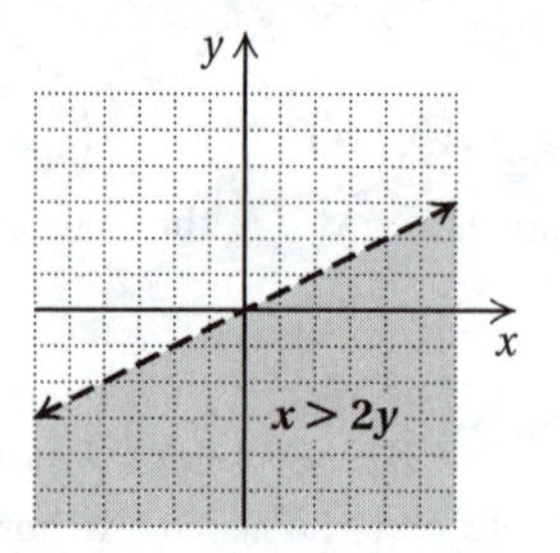

7. Graph $y \leq x - 3$.

First graph the line $y = x - 3$. The intercepts are $(0, -3)$ and $(3, 0)$. We draw a solid line since the inequality symbol is $\leq$. Then we pick a test point that is not on the line. We try $(0, 0)$.

$$\begin{array}{c|c} y \leq x - 3 & \\ \hline 0 \; ? \; 0 - 3 & \\ & -3 \quad \text{FALSE} \end{array}$$

We see that $(0, 0)$ is not a solution of the inequality, so we shade the region that does not contain $(0, 0)$.

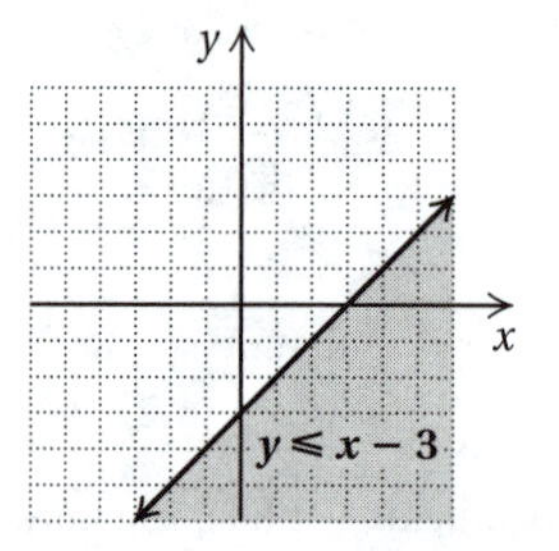

9. Graph $y < x + 1$.

First graph the line $y = x + 1$. The intercepts are $(0, 1)$ and $(-1, 0)$. We draw a dashed line since the inequality symbol is $<$. Then we pick a test point that is not on the line. We try $(0, 0)$.

$$\begin{array}{c|c} y < x + 1 & \\ \hline 0 \; ? \; 0 + 1 & \\ & 1 \quad \text{TRUE} \end{array}$$

Since $(0, 0)$ is a solution of the inequality, we shade the region that contains $(0, 0)$.

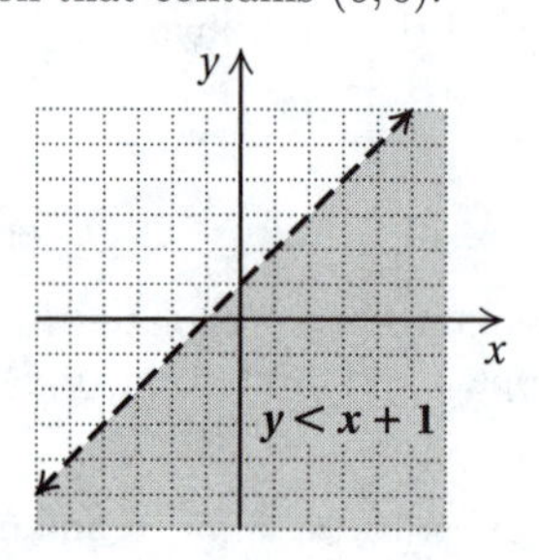

11. Graph $y \geq x - 2$.

First graph the line $y = x - 2$. The intercepts are $(0, -2)$ and $(2, 0)$. We draw a solid line since the inequality symbol is $\geq$. Then we test the point $(0,0)$.

$$\begin{array}{c|c} y \geq x-2 & \\ \hline 0 \ ?\ 0-2 & \\ & -2 \quad \text{TRUE} \end{array}$$

Since $(0,0)$ is a solution of the inequality, we shade the region containing $(0,0)$.

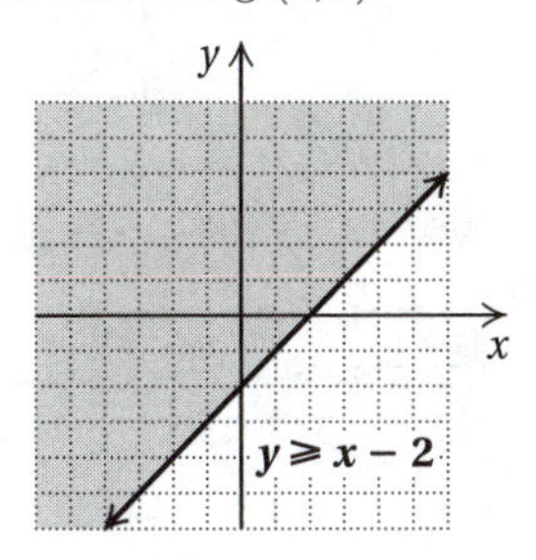

13. Graph $y \leq 2x-1$.

First graph the line $y=2x-1$. The intercepts are $(0,-1)$ and $\left(\frac{1}{2},0\right)$. We draw a solid line since the inequality symbol is $\leq$. Then we test the point $(0,0)$.

$$\begin{array}{c|c} y \leq 2x-1 & \\ \hline 0 \ ?\ 2\cdot 0-1 & \\ & -1 \quad \text{FALSE} \end{array}$$

Since $(0,0)$ is not a solution of the inequality, we shade the region that does not contain $(0,0)$.

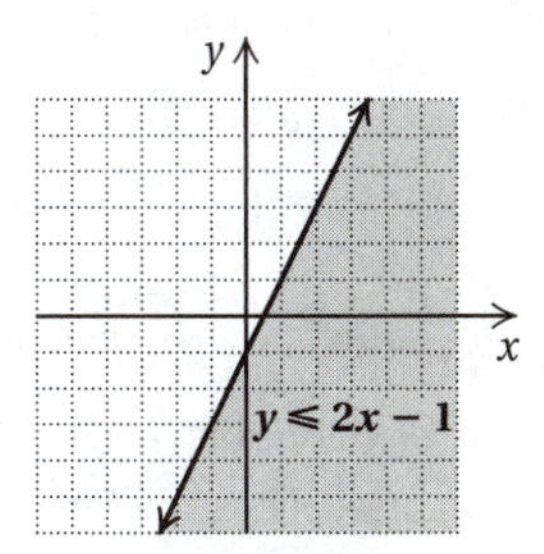

15. Graph $x+y \leq 3$.

First graph the line $x+y=3$. The intercepts are $(0,3)$ and $(3,0)$. We draw a solid line since the inequality symbol is $\leq$. Then we test the point $(0,0)$.

$$\begin{array}{c|c} x+y \leq 3 & \\ \hline 0+0 \ ?\ 3 & \\ 0 & \text{TRUE} \end{array}$$

Since $(0,0)$ is a solution of the inequality, we shade the region that contains $(0,0)$.

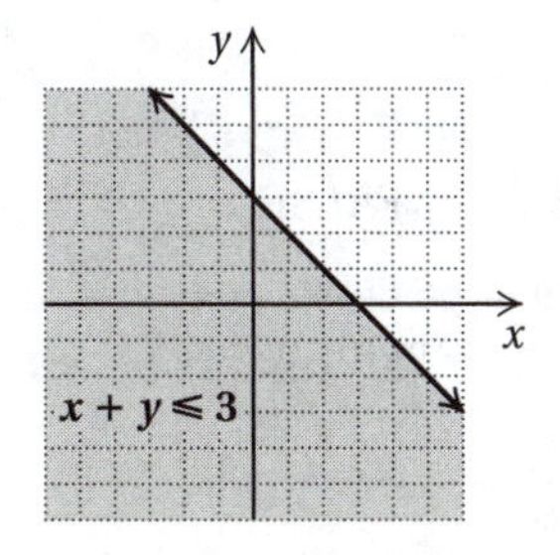

17. Graph $x-y>7$.

First graph the line $x-y=7$. The intercepts are $(0,-7)$ and $(7,0)$. We draw a dashed line since the inequality symbol is $>$. Then we test the point $(0,0)$.

$$\begin{array}{c|c} x-y > 7 & \\ \hline 0-0 \ ?\ 7 & \\ 0 & \text{FALSE} \end{array}$$

Since $(0,0)$ is not a solution of the inequality, we shade the region that does not contain $(0,0)$.

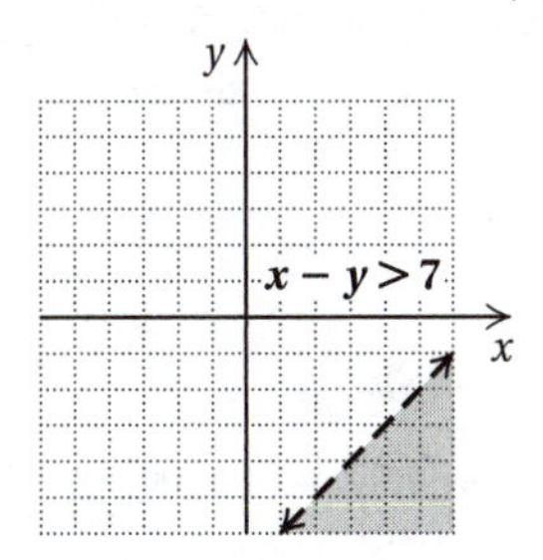

19. Graph $2x+3y \leq 12$.

First graph the line $2x+3y=12$. The intercepts are $(0,4)$ and $(6,0)$. We draw a solid line since the inequality symbol is $\leq$. Then we test the point $(0,0)$.

$$\begin{array}{c|c} 2x+3y \leq 12 & \\ \hline 2\cdot 0+3\cdot 0 \ ?\ 12 & \\ 0 & \text{TRUE} \end{array}$$

Since $(0,0)$ is a solution of the inequality, we shade the region containing $(0,0)$.

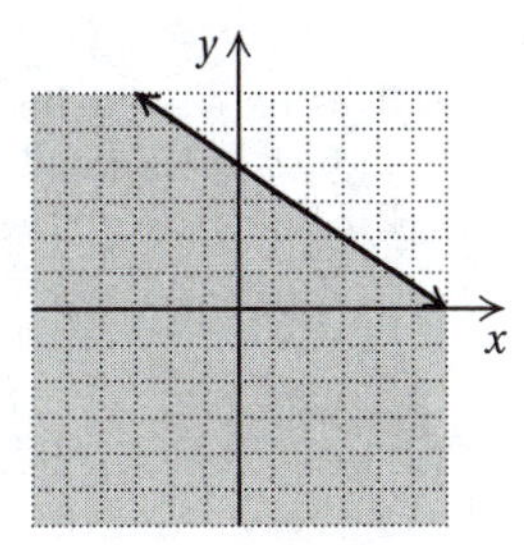

2x + 3y ≤ 12

21. Graph $y \geq 1-2x$.

First graph the line $y=1-2x$. The intercepts are $(0,1)$ and $\left(\frac{1}{2},0\right)$. We draw a solid line since the inequality symbol is $\geq$. Then we test the point $(0,0)$.

$$\begin{array}{c|c} y \geq 1-2x & \\ \hline 0 \ ?\ 1-2\cdot 0 & \\ & 1 \quad \text{FALSE} \end{array}$$

Since $(0,0)$ is not a solution of the inequality, we shade the region that does not contain $(0,0)$.

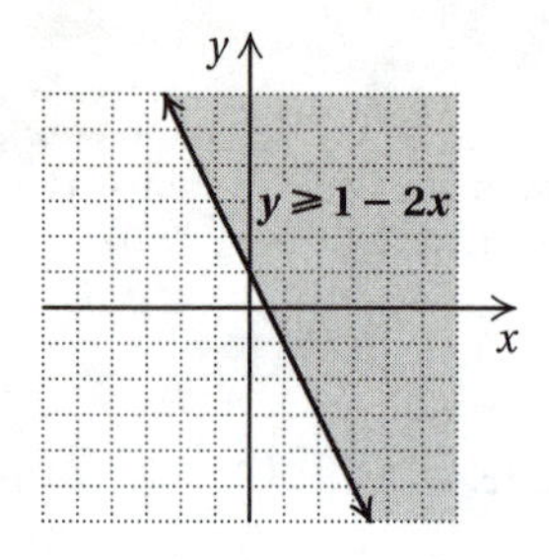

23. Graph $2x - 3y > 6$.

First graph the line $2x-3y=6$. The intercepts are $(0,-2)$ and $(3,0)$. We draw a dashed line since the inequality symbol is $>$. Then we test the point $(0,0)$.

$$\begin{array}{c|c} 2x-3y > 6 & \\ \hline 2\cdot 0 - 3\cdot 0 \ ?\ 6 & \\ 0 & \text{FALSE} \end{array}$$

Since $(0,0)$ is not a solution of the inequality, we shade the region that does not contain $(0,0)$.

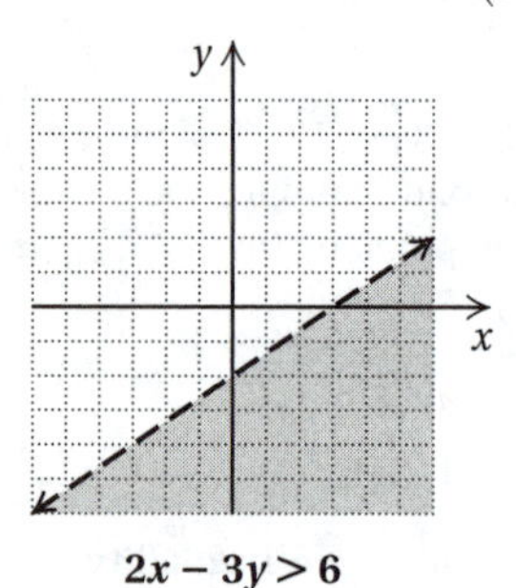

25. Graph $y \leq 3$.

First graph the line $y = 3$ using a solid line since the inequality symbol is $\leq$. Then pick a test point that is not on the line. We choose $(1,-2)$. We can write the inequality as $0x + y \leq 3$.

$$\begin{array}{c|c} 0x + y \leq 3 & \\ \hline 0\cdot 1 + (-2) \ ?\ 3 & \\ -2 & \text{TRUE} \end{array}$$

Since $(1,-2)$ is a solution of the inequality, we shade the region containing $(1,-2)$.

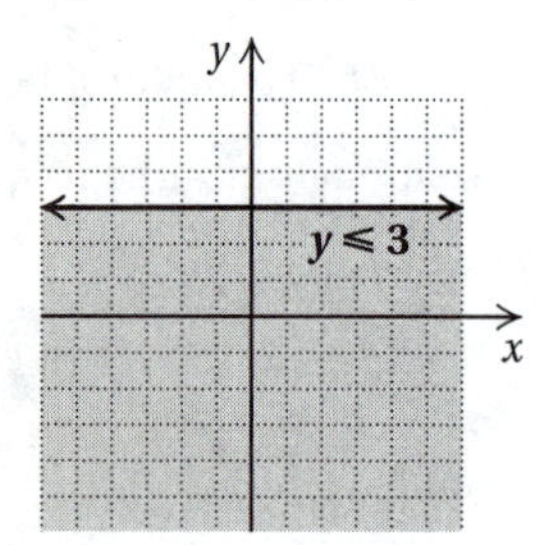

27. Graph $x \geq -1$.

Graph the line $x = -1$ using a solid line since the inequality symbol is $\geq$. Then pick a test point that is not on the line. We choose $(2,3)$. We can write the inequality as $x + 0y \geq -1$.

$$\begin{array}{c|c} x + 0y \geq -1 & \\ \hline 2 + 0\cdot 3 \ ?\ -1 & \\ 2 & \text{TRUE} \end{array}$$

Since $(2,3)$ is a solution of the inequality, we shade the region containing $(2,3)$.

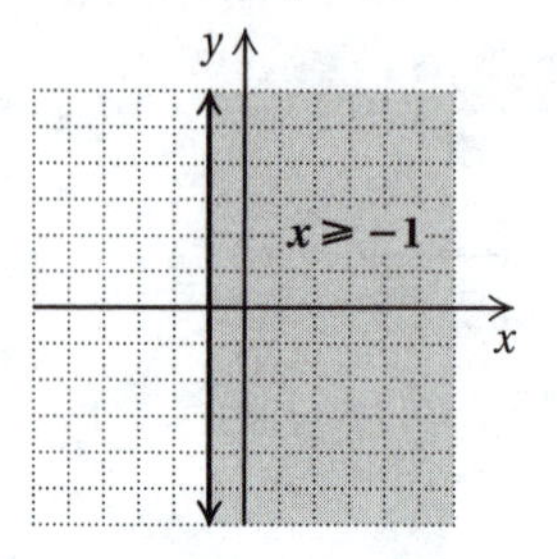

29. First we solve each equation for y:

1. $5y + 50 = 4x$
   $5y = 4x - 50$
   $y = \frac{4}{5}x - 10$

2. $5y = 4x + 15$
   $y = \frac{4}{5}x + 3$

The slope of each line is $\frac{4}{5}$. The y-intercepts, $(0,-10)$ and $(0,3)$, are different. The lines are parallel.

31. First we solve each equation for y:

1. $5y + 50 = 4x$
   $5y = 4x - 50$
   $y = \frac{4}{5}x - 10$

2. $4y = 5x + 12$
   $y = \frac{5}{4}x + 3$

The slope, $\frac{4}{5}$ and $\frac{5}{4}$, are different, so the lines are not parallel. The product of the slopes is $\frac{4}{5}\cdot\frac{5}{4} = 1 \neq -1$, so the lines are not perpendicular. Thus, the lines are neither parallel nor perpendicular.

33. The c children weigh $35c$ kg, and the a adults weigh $75a$ kg. Together, the children and adults weigh $35c + 75a$ kg. When this total is more than 1000 kg the elevator is overloaded, so we have $35c + 75a > 1000$. (Of course, c and a would also have to be nonnegative, but we will not deal with nonnegativity constraints here.)

To graph $35c+75a > 1000$, we first graph $35c+75a = 1000$ using a dashed line. Two points on the line are $(4,20)$ and $(11,5)$. (We are using alphabetical order of variables.) Then we test the point $(0,0)$.

$$\begin{array}{c|c} 35c + 75a > 1000 & \\ \hline 35\cdot 0 + 75\cdot 0 \ ?\ 1000 & \\ 0 & \text{FALSE} \end{array}$$

Since $(0,0)$ is not a solution of the inequality, we shade the region that does not contain $(0,0)$.

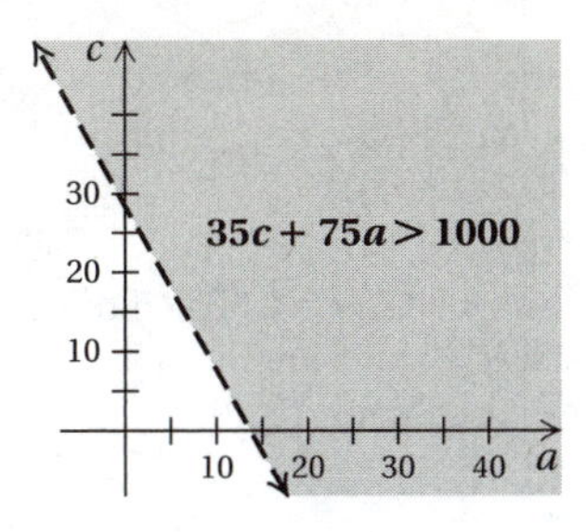

Chapter 11 Concept Reinforcement

1. True; $(0,0)$ is a solution of $y = mx$, so the x- and y-intercepts are both $(0,0)$.

2. False; parallel lines have different y-intercepts.

3. True; the product of the slopes is $m\left(-\frac{1}{m}\right) = -1$.

4. False; $0 > 0$ is false.

5. True; see page 785 in the text.

6. False; when $y = 0$, $x = \frac{C}{A}$, not $\frac{A}{C}$.

7. False; the slope of the line that passes through $(0,t)$ and $(-t,0)$ is $\frac{t-0}{0-(-t)} = \frac{t}{t} = 1$.

Chapter 11 Important Concepts

1. Point F is 2 units right and 4 units up. Its coordinates are $(2,4)$.

   Point G is 2 units left and 0 units up or down. Its coordinates are $(-2,0)$.

   Point H is 3 units left and 5 units down. Its coordinates are $(-3,-5)$.

2. $$x + 2y = 8$$
   $$2y = -x + 8$$
   $$y = -\frac{1}{2}x + 4$$

   The y-intercept is $(0,4)$. We find two other points.

   When $x = 2$, $y = -\frac{1}{2}\cdot 2 + 4 = -1 + 4 = 3$.

   When $x = -2$, $y = -\frac{1}{2}(-2) + 4 = 1 + 4 = 5$.

   We plot $(0,4)$, $(2,3)$, and $(-2,5)$ and draw the line.

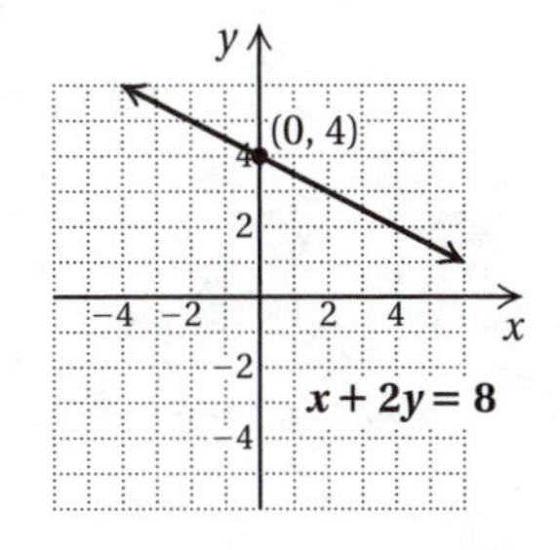

3. To find the x-intercept, we let $y = 0$ and solve for x.

   $$y - 2x = -4$$
   $$0 - 2x = -4$$
   $$-2x = -4$$
   $$x = 2$$

   The x-intercept is $(2,0)$.

   To find the y-intercept, we let $x = 0$ and solve for y.

   $$y - 2x = -4$$
   $$y - 2\cdot 0 = -4$$
   $$y = -4$$

   The y-intercept is $(0,-4)$.

   We find a third point as a check. Let $x = 4$.

   $$y - 2\cdot 4 = -4$$
   $$y - 8 = -4$$
   $$y = 4$$

   A third point is $(4,4)$. We plot this point and the intercepts and draw the graph.

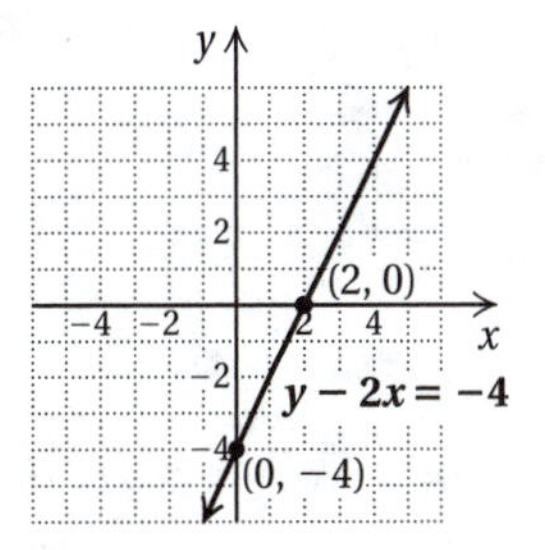

4. Graph: $y = -\frac{5}{2}$.

   The graph is a horizontal line with y-intercept $\left(0, -\frac{5}{2}\right)$.

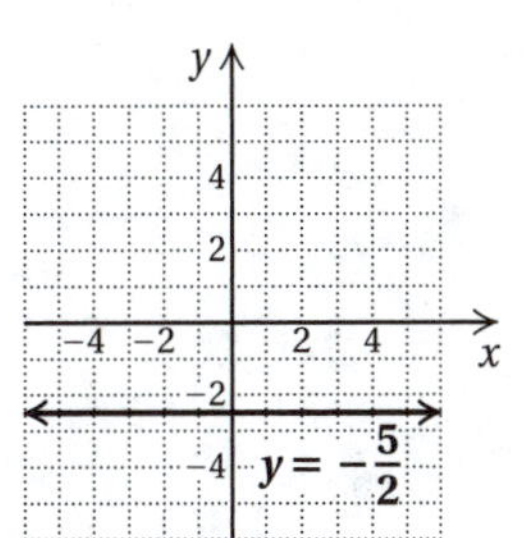

5. Graph: $x = 2$.

   The graph is a vertical line with x-intercept $(2,0)$.

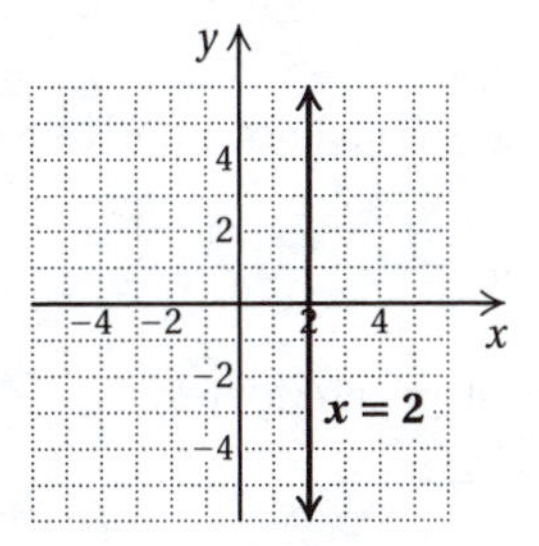

6. $m = \frac{20-14}{-8-(-8)} = \frac{6}{0}$

   The slope is not defined.

7. $m = \dfrac{20-(-1)}{16-2} = \dfrac{21}{14} = \dfrac{3}{2}$

8. $m = \dfrac{2.8-2.8}{0.5-1.5} = \dfrac{0}{-1} = 0$

9. $x = 0.25$ is the equation of a vertical line. The slope is not defined.

10.
$$7y + 14x = -28$$
$$7y = -14x - 28$$
$$y = -2x - 4$$
The slope is -2.

11. $y = -5$, or $y = 0 \cdot x - 5$

The slope is 0.

12. The slope is 6, so the equation is $y = 6x + b$. Use the point $(-1, 1)$ to find b.
$$y = 6x + b$$
$$1 = 6(-1) + b$$
$$1 = -6 + b$$
$$7 = b$$
The equation is $y = 6x + 7$.

13. First we find the slope.
$$m = \frac{-2-(-3)}{1-7} = \frac{1}{-6} = -\frac{1}{6}$$
The equation is $y = -\frac{1}{6}x + b$. We use one of the points to find b.
$$y = -\frac{1}{6}x + b$$
$$-2 = -\frac{1}{6} \cdot 1 + b \qquad \text{Using } (1, -2)$$
$$-2 = -\frac{1}{6} + b$$
$$-\frac{12}{6} + \frac{1}{6} = b$$
$$-\frac{11}{6} = b$$
The equation is $y = -\frac{1}{6}x - \frac{11}{6}$.

14. Write the equations in slope-intercept form.

1. $4y = -x - 12$ $\qquad$ 2. $y - 4x = \frac{1}{2}$

$y = -\frac{1}{4}x - 3$ $\qquad$ $y = 4x + \frac{1}{2}$

The slopes are different, so the lines are not parallel.

The product of the slopes is $-\frac{1}{4} \cdot 4$, or -1, so the lines are perpendicular.

15. Write the equations in slope-intercept form.

1. $2y - x = -4$ $\qquad$ 2. $x - 2y = -12$

$2y = x - 4$ $\qquad$ $-2y = -x - 12$

$y = \frac{1}{2}x - 2$ $\qquad$ $y = \frac{1}{2}x + 6$

The slopes are same, $\frac{1}{2}$. The y-intercepts, $(0, -2)$ and $(0, 6)$, are different. The lines are parallel.

16. Graph: $y - 3x \leq -3$.

We first graph the line $y - 3x = -3$. The intercepts are $(1, 0$ and $(0, -3)$. We draw a solid line because the inequality symbol is $\leq$. We test the point $(0, 0)$.

$$\begin{array}{c|c} y - 3x \leq -3 & \\ \hline 0 - 3 \cdot 0 \ ? \ -3 & \\ 0 & \text{FALSE} \end{array}$$

We see that $(0, 0)$ is not a solution, so we shade the half-plane that does not contain $(0, 0)$.

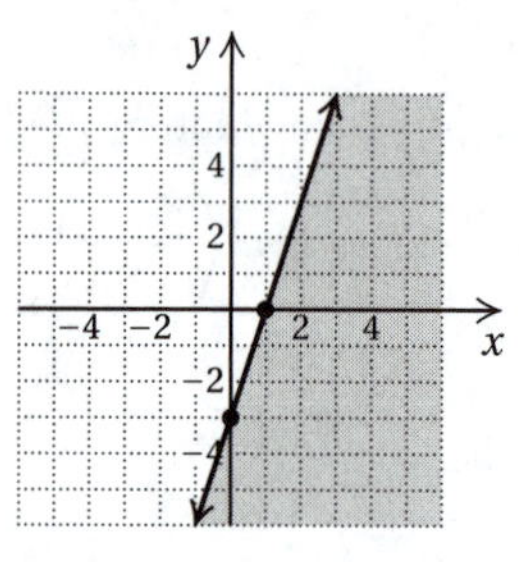

$y - 3x \leq -3$

Chapter 11 Review Exercises

1. Point A is 5 units left and 1 unit down. The coordinates of A are $(-5, -1)$.

2. Point B is 2 units left and 5 units up. The coordinates of B are $(-2, 5)$.

3. Point C is 3 units right and 0 units up or down. The coordinates of C are $(3, 0)$.

4. $(2, 5)$ is 2 units right and 5 units up. See the graph following Exercise 6 below.

5. $(0, -3)$ is 0 units right or left and 3 units down. See the graph following Exercise 6 below.

6. $(-4, -2)$ is 4 units left and 2 units down. See the graph below.

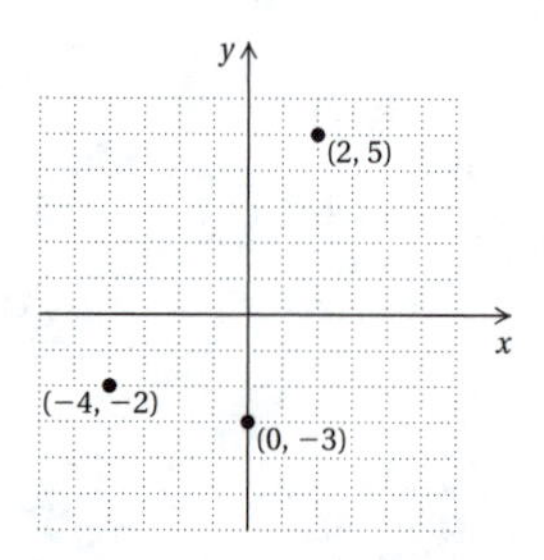

7. Since the first coordinate is positive and the second coordinate is negative, the point $(3, -8)$ is in quadrant IV.

8. Since both coordinates are negative, the point $(-20, -14)$ is in quadrant III.

9. Since both coordinates are positive, the point $(4.9, 1.3)$ is in quadrant I.

10. We substitute 2 for x and -6 for y.

$$\begin{array}{c|l} 2y - x = 10 & \\ \hline 2(-6) - 2 \ ? \ 10 & \\ -12 - 2 & \\ -14 & \text{FALSE} \end{array}$$

Since $-14 = 10$ is false, the pair $(2, -6)$ is not a solution.

11. We substitute 0 for x and 5 for y.

$$\begin{array}{c|l} 2y - x = 10 & \\ \hline 2 \cdot 5 - 0 \ ? \ 10 & \\ 10 - 0 & \\ 10 & \text{TRUE} \end{array}$$

Since $10 = 10$ is true, the pair $(0, 5)$ is a solution.

12. To show that a pair is a solution, we substitute, replacing x with the first coordinate and y with the second coordinate in each pair.

$$\begin{array}{c|l} 2x - y = 3 & \\ \hline 2 \cdot 0 - (-3) \ ? \ 3 & \\ 0 + 3 & \\ 3 & \text{TRUE} \end{array}$$

$$\begin{array}{c|l} 2x - y = 3 & \\ \hline 2 \cdot 2 - 1 \ ? \ 3 & \\ 4 - 1 & \\ 3 & \text{TRUE} \end{array}$$

In each case the substitution results in a true equation. Thus, $(0, -3)$ and $(2, 1)$ are both solutions of $2x - y = 3$. We graph these points and sketch the line passing through them.

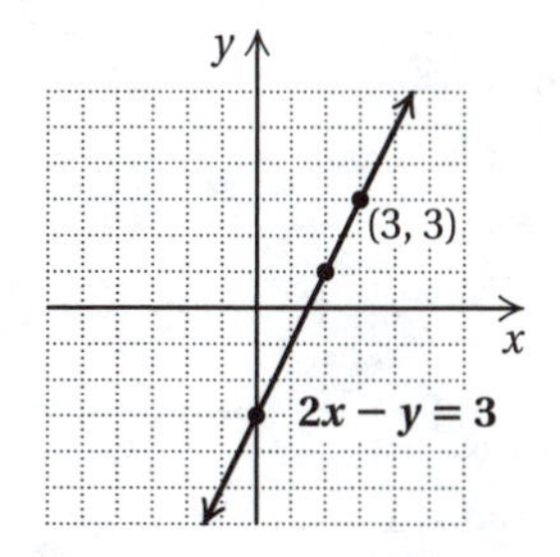

The line appears to pass through $(3, 3)$ also. We check to determine if $(3, 3)$ is a solution of $2x - y = 3$.

$$\begin{array}{c|l} 2x - y = 3 & \\ \hline 2 \cdot 3 - 3 \ ? \ 3 & \\ 6 - 3 & \\ 3 & \text{TRUE} \end{array}$$

Thus, $(3, 3)$ is another solution. There are other correct answers, including $(-1, -5)$ and $(4, 5)$.

13. $y = 2x - 5$

The y-intercept is $(0, -5)$. We find two other points.

When $x = 2$, $y = 2 \cdot 2 - 5 = 4 - 5 = -1$.

When $x = 4$, $y = 2 \cdot 4 - 5 = 8 - 5 = 3$.

x	y
0	-5
2	-1
4	3

Plot these points, draw the line they determine, and label the graph $y = 2x - 5$.

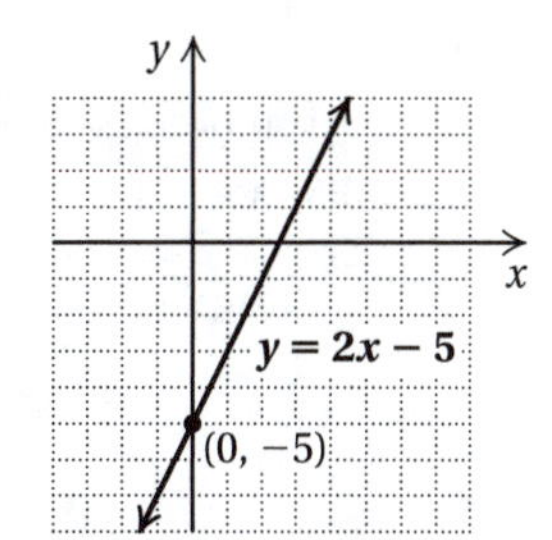

14. $y = -\frac{3}{4}x$

The equation is equivalent to $y = -\frac{3}{4}x + 0$. The y-intercept is $(0, 0)$. We find two other points.

When $x = -4$, $y = -\frac{3}{4}(-4) = 3$.

When $x = 4$, $y = -\frac{3}{4} \cdot 4 = -3$.

x	y
-4	3
0	0
4	-3

Plot these points, draw the line they determine, and label the graph $y = -\frac{3}{4}x$.

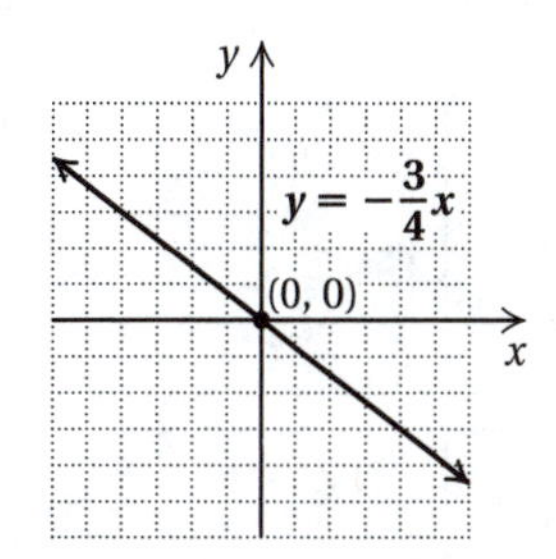

15. $y = -x + 4$

The y-intercept is $(0, 4)$. We find two other points.

When $x = -1$, $y = -(-1) + 4 = 1 + 4 = 5$.

When $x = 4$, $y = -4 + 4 = 0$.

x	y
-1	5
0	4
4	0

Plot these points, draw the line they determine, and label the graph $y = -x + 4$.

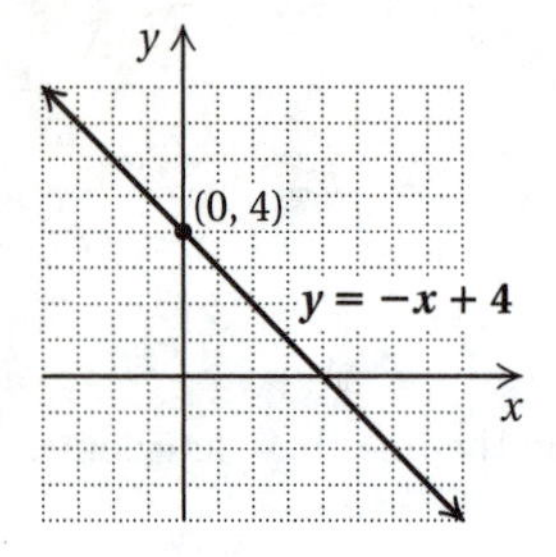

16. $y = 3 - 4x$, or $y = -4x + 3$

The y-intercept is $(0, 3)$. We find two other points.

When $x = 1$, $y = -4 \cdot 1 + 3 = -4 + 3 = -1$.

When $x = 2$, $y = -4 \cdot 2 + 3 = -8 + 3 = -5$.

x	y
0	3
1	-1
2	-5

Plot these points, draw the line they determine, and label the graph $y = 3 - 4x$.

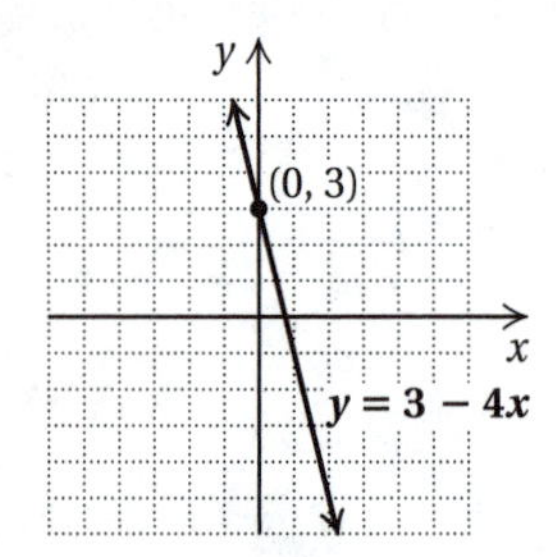

17. $y = 3$

Any ordered pair $(x, 3)$ is a solution. The variable y must be 3, but x can be any number we choose. A few solutions are listed below. Plot these points and draw the line.

x	y
-3	3
0	3
2	3

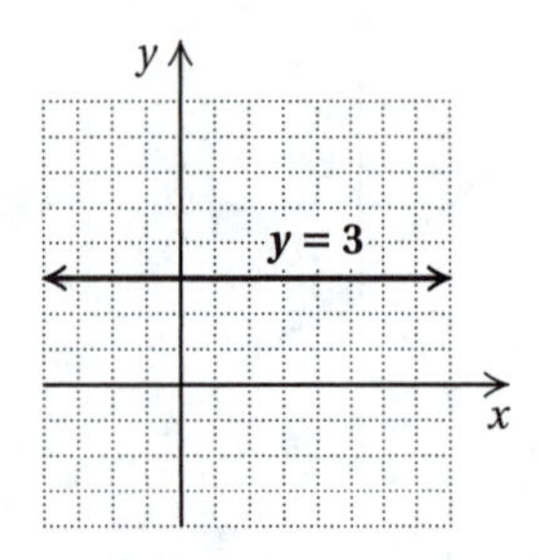

18.
$$5x - 4 = 0$$
$$5x = 4$$
$$x = \frac{4}{5} \quad \text{Solving for } x$$

Any ordered pair $\left(\frac{4}{5}, y\right)$ is a solution. A few solutions are listed below. Plot these points and draw the graph.

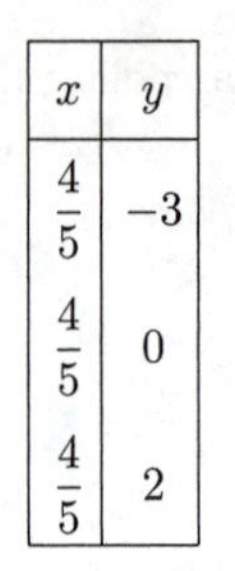

x	y
$\frac{4}{5}$	-3
$\frac{4}{5}$	0
$\frac{4}{5}$	2

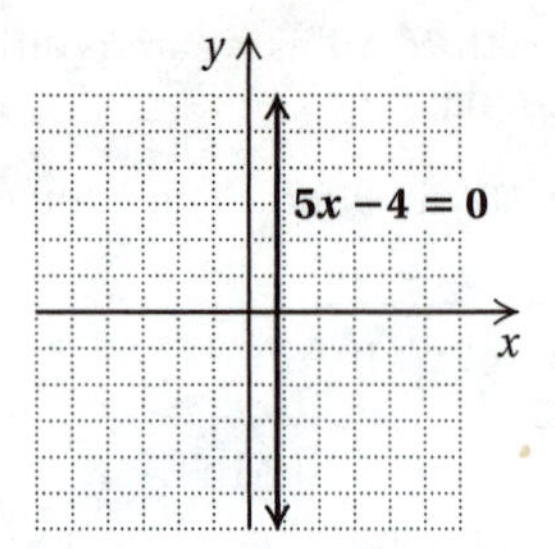

19. $x - 2y = 6$

To find the x-intercept, let $y = 0$. Then solve for x.

$$x - 2y = 6$$
$$x - 2 \cdot 0 = 6$$
$$x = 6$$

Thus, $(6, 0)$ is the x-intercept.

To find the y-intercept, let $x = 0$. Then solve for y.

$$x - 2y = 6$$
$$0 - 2y = 6$$
$$-2y = 6$$
$$y = -3$$

Thus, $(0, -3)$ is the y-intercept.

Plot these points and draw the graph.

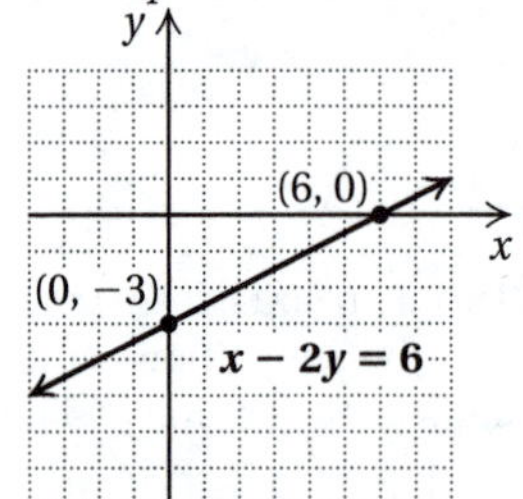

We use a third point as a check.

We let $x = -2$. Then

$$x - 2y = 6$$
$$-2 - 2y = 6$$
$$-2y = 8$$
$$y = -4.$$

The point $(-2, -4)$ is on the graph, so the graph is probably correct.

20. $5x - 2y = 10$

To find the x-intercept, let $y = 0$. Then solve for x.

$$5x - 2y = 10$$
$$5x - 2 \cdot 0 = 10$$
$$5x = 10$$
$$x = 2$$

Thus, $(2, 0)$ is the x-intercept.

To find the y-intercept, let $x = 0$. Then solve for y.

$$5x - 2y = 10$$
$$5 \cdot 0 - 2y = 10$$
$$-2y = 10$$
$$y = -5$$

Thus, $(0, -5)$ is the y-intercept.

Plot these points and draw the graph.

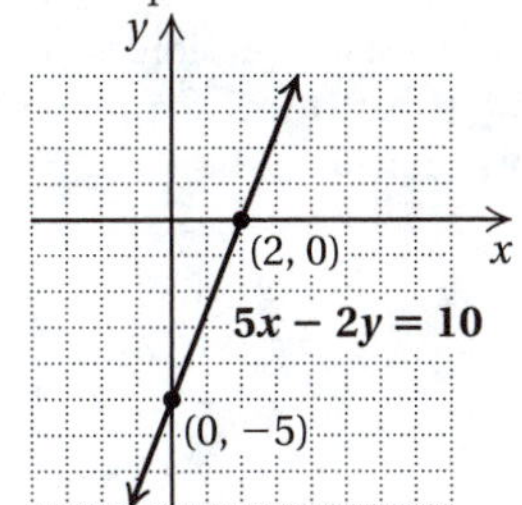

We use a third point as a check.

We let $x = 4$. Then

$$5x - 2y = 10$$
$$5 \cdot 4 - 2y = 10$$
$$20 - 2y = 10$$
$$-2y = -10$$
$$y = 5.$$

The point $(4, 5)$ is on the graph, so the graph is probably correct.

21. a) When $n = 1$, $S = \frac{3}{2} \cdot 1 + 13 = \frac{3}{2} + 13 = 1\frac{1}{2} + 13 = 14\frac{1}{2}$ ft^3.

When $n = 2$, $S = \frac{3}{2} \cdot 2 + 13 = 3 + 13 = 16$ ft^3.

When $n = 5$, $S = \frac{3}{2} \cdot 5 + 13 = \frac{15}{2} + 13 = 7\frac{1}{2} + 13 = 20\frac{1}{2}$ ft^3.

When $n = 10$, $S = \frac{3}{2} \cdot 10 + 13 = 15 + 13 = 28$ ft^3.

b) We plot the points found in part (a): $\left(1, 14\frac{1}{2}\right)$, $(2, 16)$, $\left(5, 20\frac{1}{2}\right)$ and $(10, 28)$. Then we draw the graph.

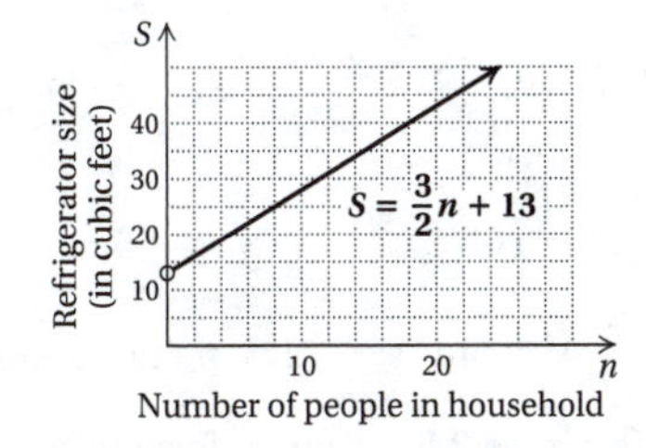

From the graph, it appears that an x-value of 4 corresponds to the S-value of about 20, so the recommended size is about 20.

c) We substitute 22 for S and solve for n.

$$22 = \frac{3}{2}n + 13$$
$$9 = \frac{3}{2}n$$
$$\frac{2}{3} \cdot 9 = \frac{2}{3} \cdot \frac{3}{2}n$$
$$6 = n$$

A 22-ft^3 refrigerator is recommended for a household of 6 residents.

22. 5:30 P.M. is 2.5 hr after 3:00 P.M. In this time the number of driveways plowed was $13 - 7$, or 6.

a) Rate of change $= \frac{6 \text{ driveways}}{2.5 \text{ hr}} = 2.4$ driveways per hour

b) $2.5 \text{ hr} = 2.5 \times 1 \text{ hr} = 2.5 \times 60 \text{ min} = 150 \text{ min}$

Rate of change $= \frac{150 \text{ min}}{6 \text{ driveways}} = 25$ minutes per driveway

23. We will use the points (11:00 A.M., 6 manicures) and (1:00 P.M., 14 manicures) to find the rate of change. Note that 1:00 P.M. is 2 hr after 11:00 A.M.

Rate of change $= \frac{14 \text{ manicures} - 6 \text{ manicures}}{2 \text{ hr}} = \frac{8 \text{ manicures}}{2 \text{ hr}} = 4$ manicures per hour

24. We can choose any two points. We consider (x_1, y_1) to be $(-3, 1)$ and (x_2, y_2) to be $(3, 3)$.

$$m = \frac{y_2 - y_1}{x_2 - x_1} = \frac{3 - 1}{3 - (-3)} = \frac{2}{6} = \frac{1}{3}$$

25. We can choose any two points. We consider (x_1, y_1) to be $(3, 1)$ and (x_2, y_2) to be $(-3, 3)$.

$$m = \frac{3 - 1}{-3 - 3} = \frac{2}{-6} = -\frac{1}{3}$$

26. We plot $(-5, -2)$ and $(5, 4)$ and draw the line containing those points.

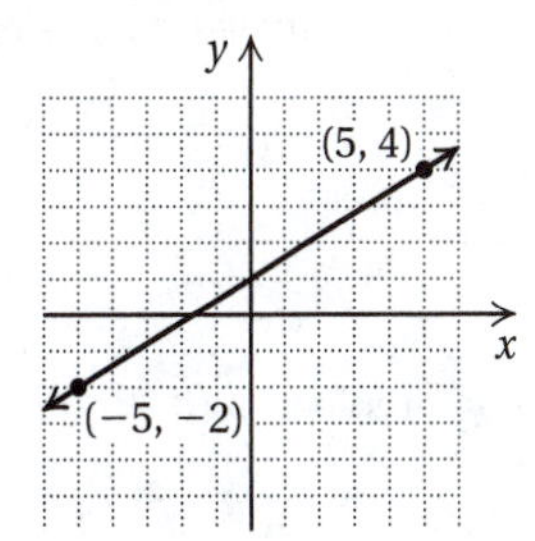

To find the slope, consider (x_1, y_1) to be $(-5, -2)$ and (x_2, y_2) to be $(5, 4)$.

$$m = \frac{y_2 - y_1}{x_2 - x_1} = \frac{4 - (-2)}{5 - (-5)} = \frac{6}{10} = \frac{3}{5}$$

27. We plot $(-5, 5)$ and $(4, -4)$ and draw the line containing those points.

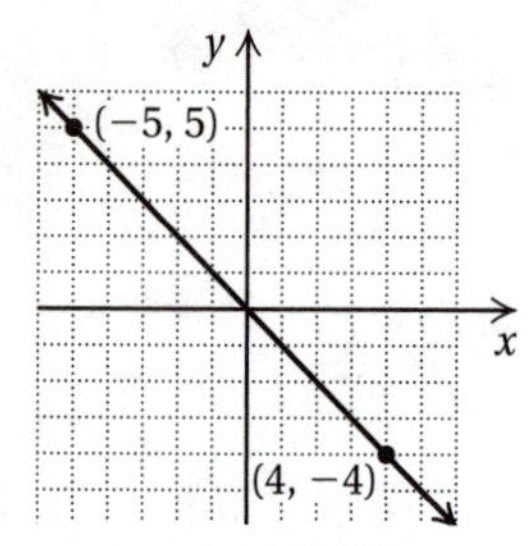

To find the slope, consider (x_1, y_1) to be $(4, -4)$ and (x_2, y_2) to be $(-5, 5)$.

$$m = \frac{y_2 - y_1}{x_2 - x_1} = \frac{5 - (-4)}{-5 - 4} = \frac{9}{-9} = -1$$

28. Grade $= \dfrac{315}{4500} = 0.07 = 7\%$

29. $y = -\dfrac{5}{8}x - 3$

The equation is in the form $y = mx + b$, where $m = -\dfrac{5}{8}$. Thus, the slope is $-\dfrac{5}{8}$.

30. We solve for y, obtaining an equation of the form $y = mx + b$.

$$\begin{aligned} 2x - 4y &= 8 \\ -4y &= -2x + 8 \\ -\frac{1}{4}(-4y) &= -\frac{1}{4}(-2x + 8) \\ y &= \frac{1}{2}x - 2 \end{aligned}$$

The slope is $\dfrac{1}{2}$.

31. The graph of $x = -2$ is a vertical line, so the slope is not defined.

32. $y = 9$, or $y = 0 \cdot x + 9$

The slope is 0.

33. $y = -9x + 46$

The equation is in the form $y = mx + b$. The slope is -9 and the y-intercept is $(0, 46)$.

34. We solve for y.

$$\begin{aligned} x + y &= 9 \\ y &= -x + 9 \end{aligned}$$

The slope is -1 and the y-intercept is $(0, 9)$.

35. We solve for y.

$$\begin{aligned} 3x - 5y &= 4 \\ -5y &= -3x + 4 \\ -\frac{1}{5}(-5y) &= -\frac{1}{5}(-3x + 4) \\ y &= \frac{3}{5}x - \frac{4}{5} \end{aligned}$$

The slope is $\dfrac{3}{5}$ and the y-intercept is $\left(0, -\dfrac{4}{5}\right)$.

36. We substitute -2.8 for m and 19 for b in the equation $y = mx + b$.

$$y = -2.8x + 19$$

37. We substitute $\dfrac{5}{8}$ for m and $-\dfrac{7}{8}$ for b in the equation $y = mx + b$.

$$y = \frac{5}{8}x - \frac{7}{8}$$

38. We know the slope is 3, so the equation is $y = 3x + b$. Using the point $(1, 2)$, we substitute 1 for x and 2 for y in $y = 3x + b$. Then we solve for b.

$$\begin{aligned} y &= 3x + b \\ 2 &= 3 \cdot 1 + b \\ 2 &= 3 + b \\ -1 &= b \end{aligned}$$

Thus, we have the equation $y = 3x - 1$.

39. We know the slope is $\dfrac{2}{3}$, so the equation is $y = \dfrac{2}{3}x + b$. Using the point $(-2, -5)$, we substitute -2 for x and -5 for y in $y = \dfrac{2}{3}x + b$. Then we solve for b.

$$\begin{aligned} y &= \frac{2}{3}x + b \\ -5 &= \frac{2}{3}(-2) + b \\ -5 &= -\frac{4}{3} + b \\ -\frac{11}{3} &= b \end{aligned}$$

Thus, we have the equation $y = \dfrac{2}{3}x - \dfrac{11}{3}$.

40. The slope is -2 and the y-intercept is $(0, -4)$, so we have the equation $y = -2x - 4$.

41. First we find the slope.

$$m = \frac{1 - 7}{-1 - 5} = \frac{-6}{-6} = 1$$

Thus, $y = 1 \cdot x + b$, or $y = x + b$. We can use either point to find b. We choose $(5, 7)$. Substitute 5 for x and 7 for y in $y = x + b$.

$$\begin{aligned} y &= x + b \\ 7 &= 5 + b \\ 2 &= b \end{aligned}$$

Thus, the equation is $y = x + 2$.

42. First we find the slope.

$$m = \frac{-3 - 0}{-4 - 2} = \frac{-3}{-6} = \frac{1}{2}$$

Thus, $y = \dfrac{1}{2}x + b$. We can use either point to find b. We choose $(2, 0)$. Substitute 2 for x and 0 for y in $y = \dfrac{1}{2}x + b$.

$$y = \frac{1}{2}x + b$$
$$0 = \frac{1}{2} \cdot 2 + b$$
$$0 = 1 + b$$
$$-1 = b$$

Thus, the equation is $y = \frac{1}{2}x - 1$.

43. a) First we find the slope.

$$m = \frac{3515 - 2316}{10 - 0} = \frac{1199}{10} = 119.9$$

The y-intercept is $(0, 2316)$, so we have the equation $y = 119.9x + 2316$.

b) The rate of change is the slope, 119.9 prescriptions per year.

c) In 2006, $t = 2006 - 1997 = 9$.

$y = 119.9x + 2316 = 119.9(9) + 2316 =$
$1079.1 + 2316 = 3395.1$ million or
$3,395,100,000$ prescriptions.

44. Slope -1, y-intercept $(0, 4)$

We plot $(0, 4)$. We can think of the slope as $\frac{-1}{1}$, so from $(0, 4)$ we move down 1 unit and right 1 unit. This locates the point $(1, 3)$. We plot $(1, 3)$ and draw a line passing through $(0, 4)$ and $(1, 3)$.

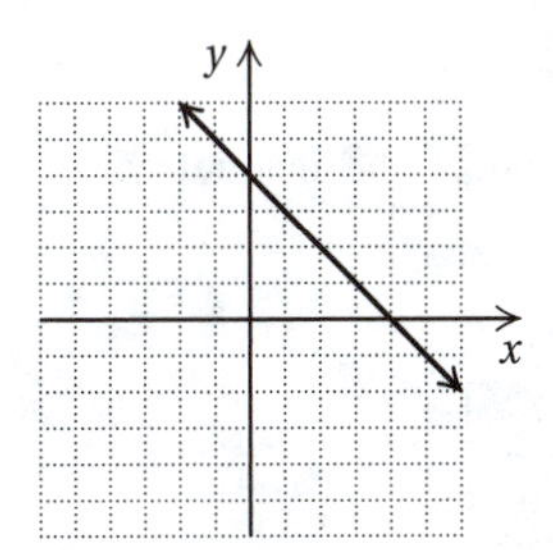

45. Slope $\frac{5}{3}$, y-intercept $(0, -3)$.

Plot $(0, -3)$ and from there move up 5 units and right 3 units. This locates the point $(3, 2)$. We plot $(3, 2)$ and draw a line passing through $(0, -3)$ and $(3, 2)$.

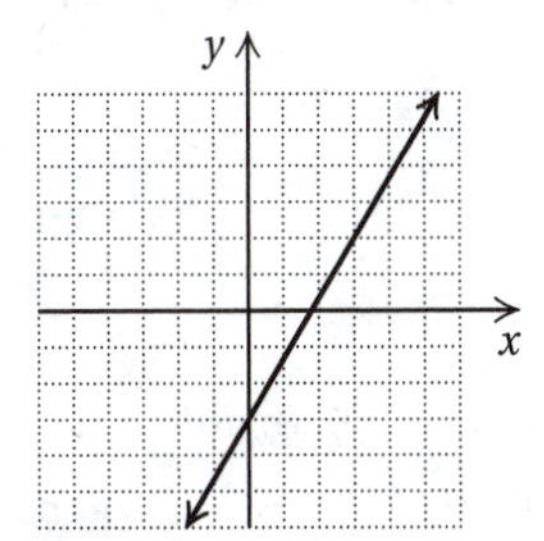

46. $y = -\frac{3}{5}x + 2$

First we plot the y-intercept $(0, 2)$. We can start at the y-intercept and, thinking of the slope as $\frac{-3}{5}$, find another point by moving down 3 units and right 5 units to the point $(5, -1)$. Thinking of the slope as $\frac{3}{-5}$ we can start at $(0, 2)$ and move up 3 units and left 5 units to get another point $(-5, 5)$.

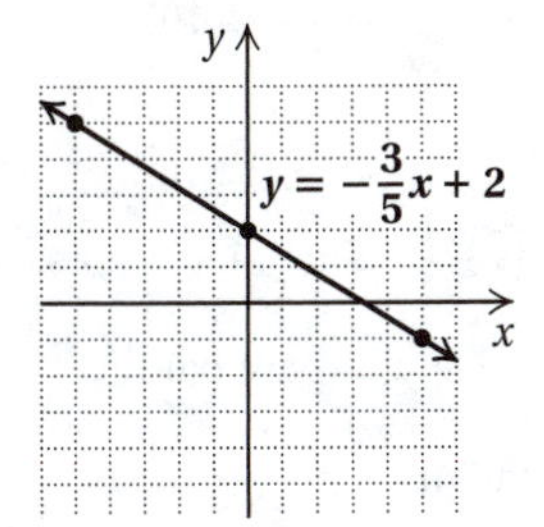

47. First we rewrite the equation in slope-intercept form.

$$2y - 3x = 6$$
$$2y = 3x + 6$$
$$y = \frac{1}{2}(3x + 6)$$
$$y = \frac{3}{2}x + 3$$

Now we plot the y-intercept $(0, 3)$. We can start at the y-intercept and use the slope, $\frac{3}{2}$, to find another point. We move up 3 units and right 2 units to the point $(2, 6)$. Thinking of the slope as $\frac{-3}{-2}$ we can start at $(0, 3)$ and move down 3 units and left 2 units to get another point, $(-2, 0)$.

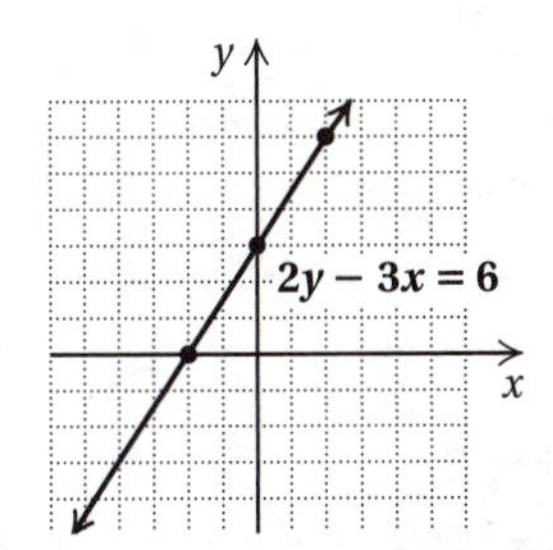

48. First we solve each equation for y:

1. $4x + y = 6$ → $y = -4x + 6$
2. $4x + y = 8$ → $y = -4x + 8$

The slope of each line is -4. The y-intercepts, $(0, 6)$ and $(0, 8)$, are different. The lines are parallel.

49. We solve the first equation for y.

$$2x + y = 10$$
$$y = -2x + 10$$

The second equation is already solved for y.

$$y = \frac{1}{2}x - 4$$

The slopes, -2 and $\frac{1}{2}$, are not the same so the lines are not parallel. The product of the slopes is $-2 \cdot \frac{1}{2} = -1$, so the lines are perpendicular.

50. First we solve each equation for y:

1. $x + 4y = 8$
$$4y = -x + 8$$
$$y = \frac{1}{4}(-x + 8)$$
$$y = -\frac{1}{4}x + 2$$

2. $x = -4y - 10$
$$x + 10 = -4y$$
$$-\frac{1}{4}(x + 10) = y$$
$$-\frac{1}{4}x - \frac{5}{2} = y$$

The slope of each line is $-\frac{1}{4}$. The y-intercepts, $(0, 2)$ and $\left(0, -\frac{5}{2}\right)$, are different. The lines are parallel.

51. First we solve each equation for y:

1. $3x - y = 6$
$$-y = -3x + 6$$
$$y = -1(-3x + 6)$$
$$y = 3x - 6$$

2. $3x + y = 8$
$$y = -3x + 8$$

The slopes, 3 and -3, are not the same so the lines are not parallel. The product of the slopes is $3(-3) = -9 \neq -1$, so the lines are not perpendicular. Thus, the lines are neither parallel nor perpendicular.

52.
$$\begin{array}{c|c} x - 2y > 1 & \\ \hline 0 - 2 \cdot 0 \;?\; 1 & \\ 0 & \text{FALSE} \end{array}$$

Since $0 > 1$ is false, $(0, 0)$ is not a solution.

53.
$$\begin{array}{c|c} x - 2y > 1 & \\ \hline 1 - 2 \cdot 3 \;?\; 1 & \\ 1 - 6 & \\ -5 & \text{FALSE} \end{array}$$

Since $-5 > 1$ is false, $(1, 3)$ is not a solution.

54.
$$\begin{array}{c|c} x - 2y > 1 & \\ \hline 4 - 2(-1) \;?\; 1 & \\ 4 + 2 & \\ 6 & \text{TRUE} \end{array}$$

Since $6 > 1$ is true, $(4, -1)$ is a solution.

55. Graph $x < y$.

First graph the line $x = y$, or $y = x$. Two points on the line are $(0, 0)$ and $(3, 3)$. We draw a dashed line since the inequality symbol is $<$. Then we pick a test point that is not on the line. We try $(1, 2)$.

$$\begin{array}{c} x < y \\ \hline 1 \;?\; 2 \text{ TRUE} \end{array}$$

We see that $(1, 2)$ is a solution of the inequality, so we shade the region that contains $(1, 2)$.

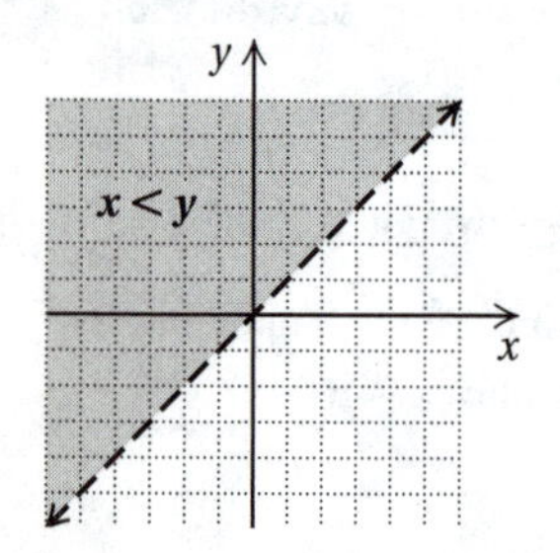

56. Graph $x + 2y \geq 4$.

First graph the line $x + 2y = 4$. The intercepts are $(0, 2)$ and $(4, 0)$. We draw a solid line since the inequality symbol is $\geq$. Then we test the point $(0, 0)$.

$$\begin{array}{c|c} x + 2y \geq 4 & \\ \hline 0 + 2 \cdot 0 \;?\; 1 & \\ 0 & \text{FALSE} \end{array}$$

Since $(0, 0)$ is not a solution of the inequality, we shade the region that does not contain $(0, 0)$.

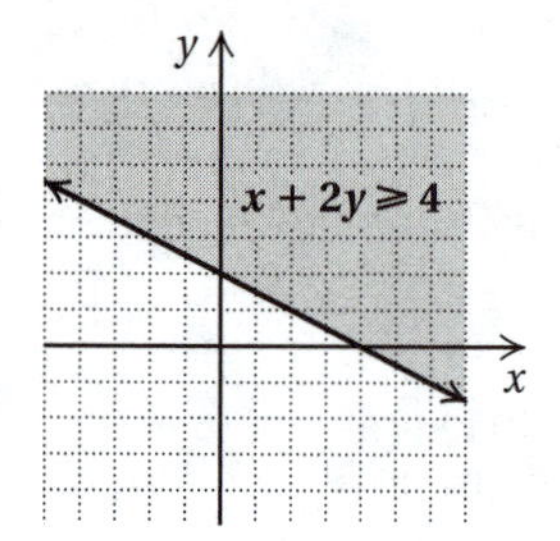

57. Graph $x > -2$.

Graph the line $x = -2$ using a dashed line since the inequality symbol is $>$. Then pick a test point that is not on the line. We choose $(0, 0)$. We can write the inequality as $x + 0y > -2$.

$$\begin{array}{c|c} x + 0y > -2 & \\ \hline 0 + 0 \cdot 0 \;?\; -2 & \\ 0 & \text{TRUE} \end{array}$$

Since $(0, 0)$ is a solution of the inequality, we shade the region containing $(0, 0)$.

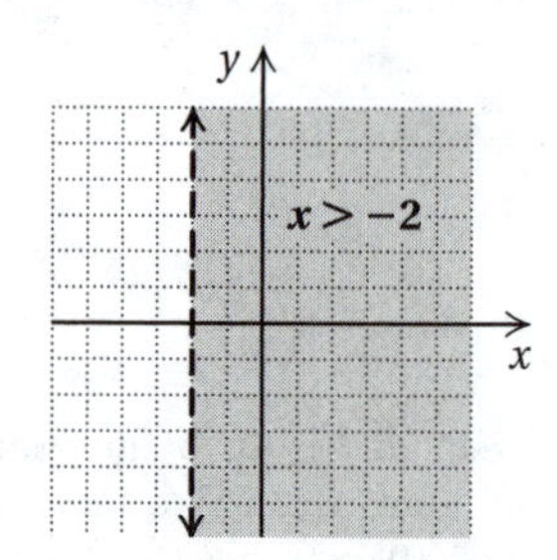

58. We write the equations in slope-intercept form.

1. $-x + \frac{1}{2}y = -2$
$$\frac{1}{2}y = x - 2$$
$$y = 2x - 4$$

2. $2y + x - 8 = 0$
$$2y = -x + 8$$
$$y = -\frac{1}{2}x + 4$$

The slopes are different, so the lines are not parallel. The product of the slopes is $2\left(-\frac{1}{2}\right) = -1$, so the lines are perpendicular. Answer D is correct.

59. The slope is $-\frac{8}{3}$, so the equation is $y = -\frac{8}{3}x + b$.

We use the point $(-3, 8)$ to find b.

$$y = -\frac{8}{3}x + b$$
$$8 = -\frac{8}{3}(-3) + b$$
$$8 = 8 + b$$
$$0 = b$$

The equation is $y = -\frac{8}{3}x + 0$, or $y = -\frac{8}{3}x$. Answer C is correct.

60. We plot the given points. We see that the fourth vertex is $(-2, -3)$.

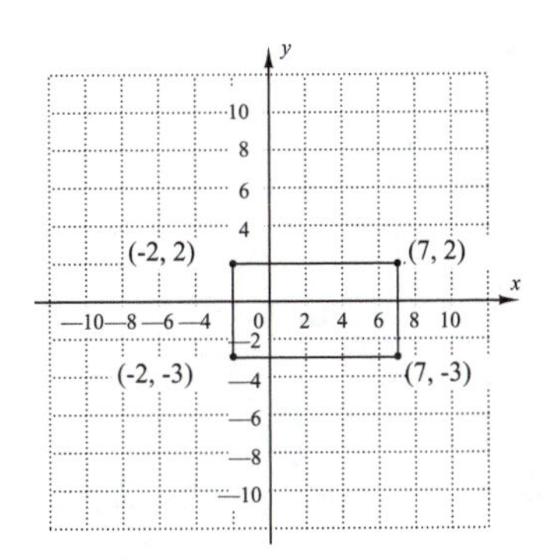

The length of the rectangle is 9 units and the width is 5 units.

$$A = l \cdot w$$
$$A = (9 \text{ units})(5 \text{ units}) = 45 \text{ square units}$$

$$P = 2l + 2w$$
$$P = 2 \cdot 9 \text{ units} + 2 \cdot 5 \text{ units}$$
$$P = 18 \text{ units} + 10 \text{ units} = 28 \text{ units}$$

61. a) From Telluride to Station St. Sophia, the ascent is 10,550 ft − 8725 = 1825 ft. From Station St. Sophia to Mountain View, the descent is 10,550 ft − 9500 ft = 1050 ft. The total ascent and descent is 1825 ft + 1050 ft = 2875 ft. The time that elapses from 11:55 A.M. to 12:07 P.M. is 12 minutes Then the average rate of ascent and descent is

$$\frac{2875 \text{ ft}}{12 \text{ min}} = 239.58\overline{3} \text{ feet per minute.}$$

b) The average rate of ascent and descent in minutes per foot is

$$\frac{12 \text{ min}}{2875 \text{ ft}} \approx 0.004 \text{ minutes per foot.}$$

Chapter 11 Discussion and Writing Exercises

1. If one equation represents a vertical line (that is, is of the form $x = a$) and the other represents a horizontal line (that is, is of the form $y = b$), then the graphs are perpendicular. If neither line is of one of the forms above, then solve each equation for y in order to determine the slope of each. Then, if the product of the slopes is -1, the graphs are perpendicular.

2. If $b > 0$, then the y-intercept of $y = mx + b$ is on the positive y-axis and the graph of $y = mx + b$ lies "above" the origin. Using $(0, 0)$ as a test point, we have the false inequality $0 > b$ so the region above $y = mx + b$ is shaded.

If $b = 0$, the line $y = mx + b$ or $y = mx$, passes through the origin. Testing a point above the line, such as $(1, m + 1)$, we have the true inequality $m + 1 > m$ so the region above the line is shaded.

If $b < 0$, then the y-intercept of $y = mx + b$ is on the negative y-axis and the graph of $y = mx + b$ lies "below" the origin. Using $(0, 0)$ as a test point we get the true inequality $0 > b$ so the region above $y = mx + b$ is shaded.

Thus, we see that in any case the graph of any inequality of the form $y > mx + b$ is always shaded above the line $y = mx + b$.

3. The y-intercept is the point at which the graph crosses the y-axis. Since a point on the y-axis is neither left nor right of the origin, the first or x-coordinate of the point is 0.

4. The graph of $x < 1$ on a number line consists of the points in the set $\{x|x < 1\}$.

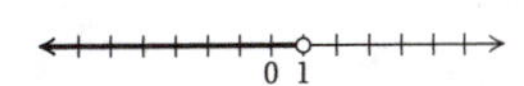

The graph of $x < 1$ on a plane consists of the points, or ordered pairs, in the set $\{(x, y)|x + 0 \cdot y < 1\}$. This is the set of ordered pairs with first coordinate less than 1.

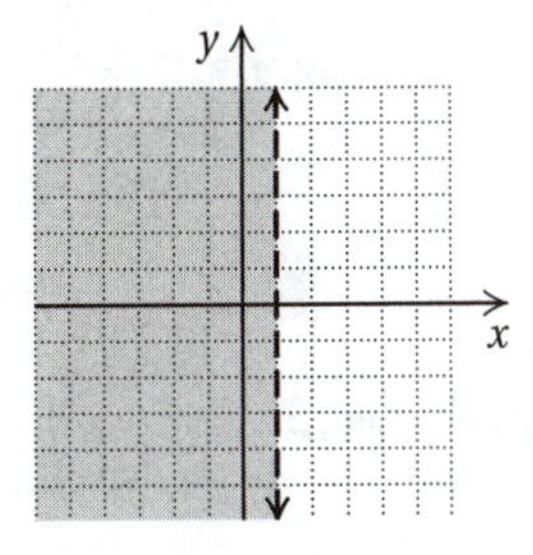

5. First plot the y-intercept, $(0, 2458)$. Then, thinking of the slope as $\frac{37}{100}$, plot a second point on the line by moving up 37 units and right 100 units from the y-intercept and plot a third point by moving down 37 units and left 100 units. Finally, draw a line through the three points.

6. If the equations are of the form $x = p$ and $x = q$, where $p \neq q$, then the graphs are parallel vertical lines. If neither equation is of the form $x = p$, then solve each for y in order to determine the slope and y-intercept of each. If the slopes are the same and the y-intercepts are different, the lines are parallel.

Chapter 11 Test

1. Since the first coordinate is negative and the second coordinate is positive, the point $\left(-\frac{1}{2}, 7\right)$ is in quadrant II.

2. Since both coordinates are negative, the point $(-5, -6)$ is in quadrant III.

3. Point A is 5 units left and 1 unit up. The coordinates of A are $(-5, 1)$.

4. Point B is 0 units left or right and 4 units down. The coordinates of B are $(0, -4)$.

5. $$\begin{array}{c|c} y - 2x & 5 \\ \hline -3 - 2(-4) \ ? & 5 \\ -3 + 8 & \\ 5 & \end{array} \quad \text{TRUE}$$

$$\begin{array}{c|c} y - 2x & 5 \\ \hline 3 - 2(-1) \ ? & 5 \\ 3 + 2 & \\ 5 & \end{array} \quad \text{TRUE}$$

In each case we get a true equation, so $(-4, -3)$ and $(-1, 3)$ are solutions of $y - 2x = 5$. We plot these points and draw the line passing through them.

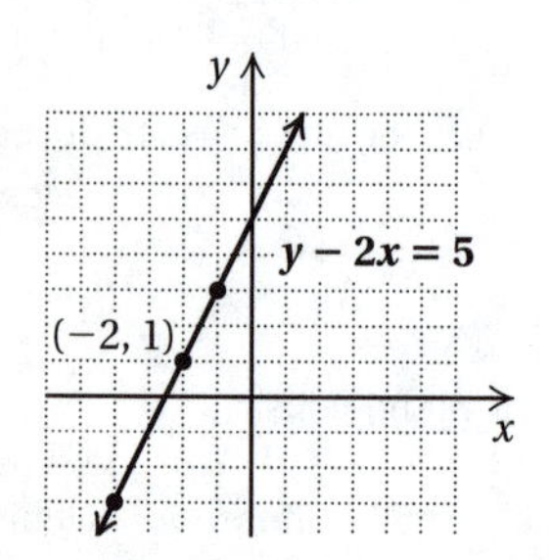

The line appears to pass through $(-2, 1)$. We check to determine if $(-2, 1)$ is a solution of $y - 2x = 5$.

$$\begin{array}{c|c} y - 2x & 5 \\ \hline 1 - 2(-2) \ ? & 5 \\ 1 + 4 & \\ 5 & \end{array} \quad \text{TRUE}$$

Thus, $(-2, 1)$ is another solution. There are other correct answers, including $(-5, -5)$, $(-3, -1)$, and $(0, 5)$.

6. $y = 2x - 1$

The y-intercept is $(0, -1)$. We find two other points.

When $x = -2$, $y = 2(-2) - 1 = -4 - 1 = -5$.

When $x = 3$, $y = 2 \cdot 3 - 1 = 6 - 1 = 5$.

x	y
-2	-5
0	-1
3	5

Plot these points, draw the line they determine, and label the graph $y = 2x - 1$.

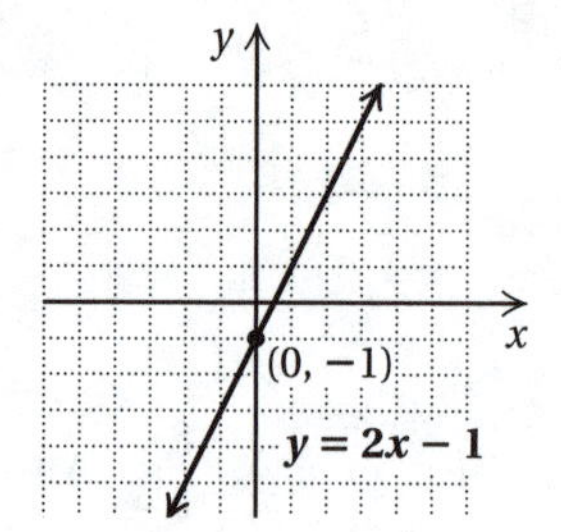

7. $y = -\frac{3}{2}x$, or $y = -\frac{3}{2}x + 0$

The y-intercept is $(0, 0)$. We find two other points.

When $x = -2$, $y = -\frac{3}{2}(-2) = 3$.

When $x = 2$, $y = -\frac{3}{2} \cdot 2 = -3$.

x	y
-2	3
0	0
2	-3

Plot these points, draw the line they determine, and label the graph $y = -\frac{3}{2}x$.

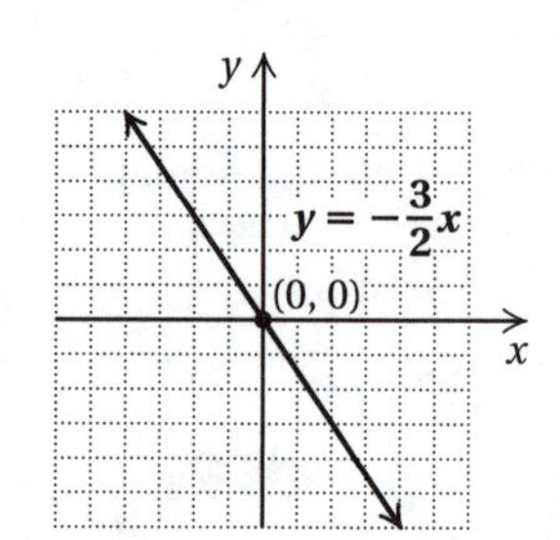

8. $2x + 8 = 0$

$2x = -8$

$x = -4$ Solving for x

Any ordered pair $(-4, y)$ is a solution. A few solutions are listed below. Plot these points and draw the graph.

x	y
-4	-3
-4	0
-4	2

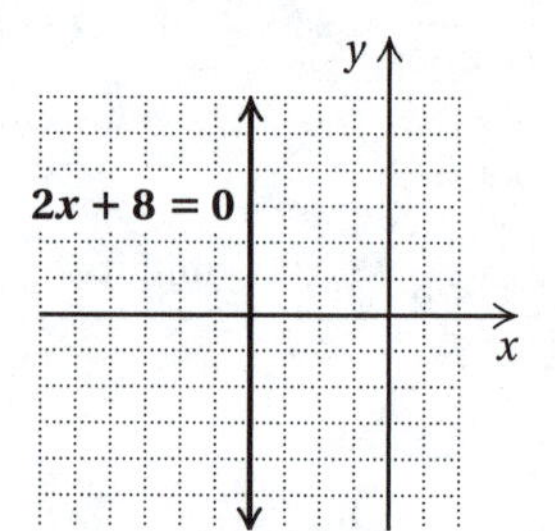

9. $y = 5$

Any ordered pair $(x, 5)$ is a solution. A few solutions are listed below. Plot these points and draw the graph.

x	y
-2	5
0	5
3	5

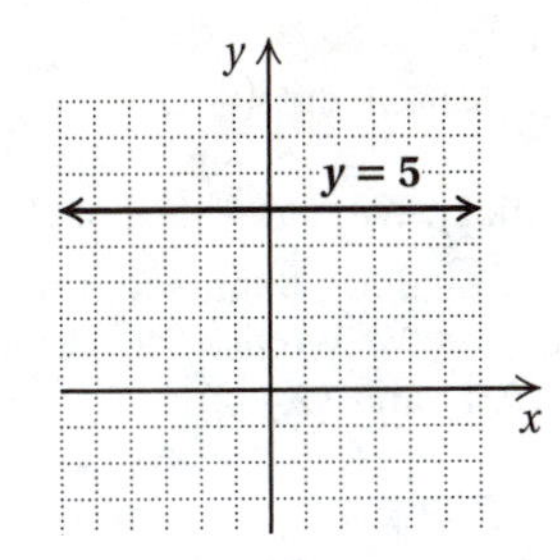

10. $2x - 4y = -8$

To find the x-intercept, let $y = 0$. Then solve for x.

$$2x - 4y = -8$$
$$2x - 4 \cdot 0 = -8$$
$$2x = -8$$
$$x = -4$$

Thus, $(-4, 0)$ is the x-intercept.

To find the y-intercept, let $x = 0$. Then solve for y.

$$2x - 4y = -8$$
$$2 \cdot 0 - 4y = -8$$
$$-4y = -8$$
$$y = 2$$

Thus, $(0, 2)$ is the y-intercept.

Plot these points and draw the graph.

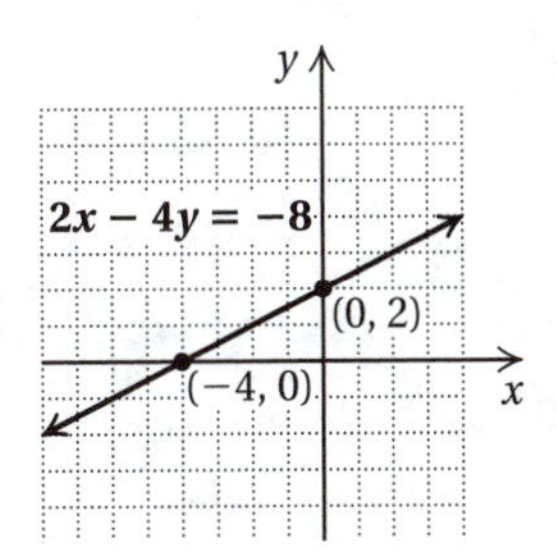

We use a third point as a check.

We let $x = 4$. Then

$$2x - 4y = -8$$
$$2 \cdot 4 - 4y = -8$$
$$8 - 4y = -8$$
$$-4y = -16$$
$$y = 4$$

The point $(4, 4)$ is on the graph, so the graph is probably correct.

11. $2x - y = 3$

To find the x-intercept, let $y = 0$. Then solve for x.

$$2x - y = 3$$
$$2x - 0 = 3$$
$$2x = 3$$
$$x = \frac{3}{2}$$

Thus, $\left(\frac{3}{2}, 0\right)$ is the x-intercept.

To find the y-intercept, let $x = 0$. Then solve for y.

$$2x - y = 3$$
$$2 \cdot 0 - y = 3$$
$$-y = 3$$
$$y = -3$$

Thus, $(0, -3)$ is the y-intercept.

Plot these points and draw the graph.

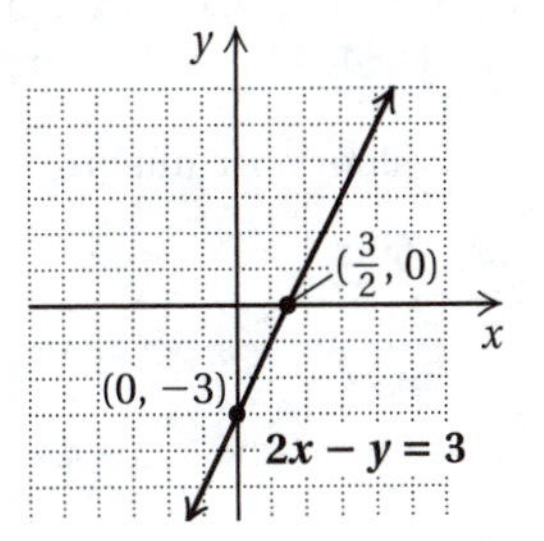

We use a third point as a check.

We let $x = 3$. Then

$$2x - y = 3$$
$$2 \cdot 3 - y = 3$$
$$6 - y = 3$$
$$-y = -3$$
$$y = 3$$

The point $(3, 3)$ is on the line, so the graph is probably correct.

12. a) In 1990, $n = 0$, and $T = 0.7(0) + 7.8 = 7.8$, so the cost of tuition was \$7.8 thousand, or \$7800.

In 1996, $n = 1996 - 1990$, or 6, and $T = 0.7(6) + 7.8 = 4.2 + 7.8 = 12$, so the cost of tuition was \$12 thousand, or \$12,000.

In 2005, $n = 2005 - 1990 = 15$, and $T = 0.7(15) + 7.8 = 10.5 + 7.8 = 18.3$, so the cost of tuition was \$18.3 thousand, or \$18,300.

In 2010, $n = 2010 - 1990 = 20$, and $T = 0.7(20) + 7.8 = 14 + 7.8 = 21.8$, so the cost of tuition was \$21.8 thousand, or \$21,800.

b) We plot the points found in part (a), $(0, 7.8)$, $(6, 12)$, $(15, 18.3)$, and $(20, 21.8)$. Then we draw the line passing through these points.

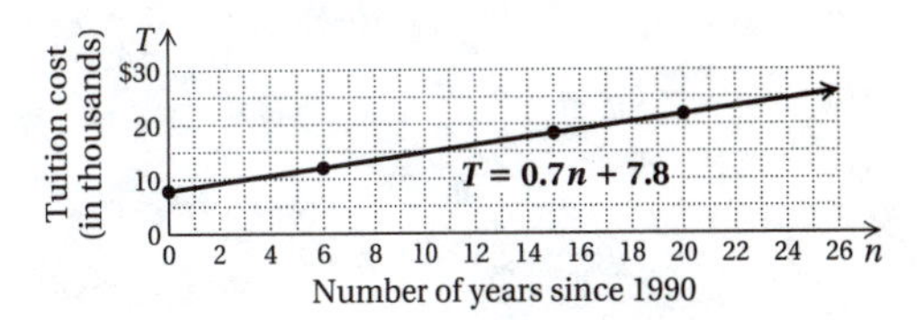

In 2015, $n = 2015 - 1990 = 25$. From the graph it appears that an n-value of 25 corresponds to the T-value of about 25.3, so we estimate that the cost of tuition will be \$25.3 thousand, or \$25,300, in 2015.

c) We substitute 28.8 for T and solve for n.

$$T = 0.7n + 7.8$$
$$28.8 = 0.7n + 7.8$$
$$21 = 0.7n$$
$$30 = n$$

We predict that a tuition cost of \$28,800 will occur 30 years after 1990, or in 2020.

13. The time that elapses from 2:38 to 2:40 is 2 minutes. In that time the elevator travels 34 − 5, or 29 floors.

a) $\text{Rate} = \dfrac{29 \text{ floors}}{2 \text{ minutes}} = 14.5$ floors per minute

b) 2 min = 2 × 1 min = 2 × 60 sec = 120 sec

$\text{Rate} = \dfrac{120 \text{ seconds}}{29 \text{ floors}} = \dfrac{120}{29}$ seconds per floor $=$

$4\frac{4}{29}$ seconds per floor

14. The time that elapses from 1:00 P.M. to 5:00 P.M. is 4 hours.

$\text{Rate} = \dfrac{450 \text{ miles} - 100 \text{ miles}}{4 \text{ hours}} = \dfrac{350 \text{ miles}}{4 \text{ hours}} =$

87.5 miles per hour

15. We can choose any two points. We consider (x_1, y_1) to be $(2, 4)$ and (x_2, y_2) to be $(5, -2)$.

$$m = \frac{y_2 - y_1}{x_2 - x_1} = \frac{-2 - 4}{5 - 2} = \frac{-6}{3} = -2$$

16. We plot $(-3, 1)$ and $(5, 4)$ and draw the line containing these points.

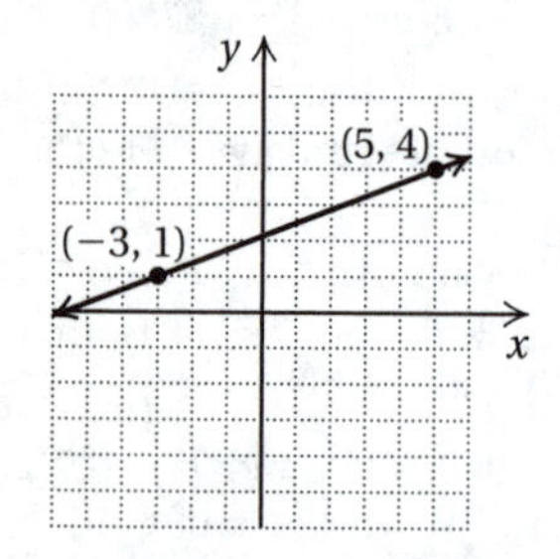

To find the slope, consider (x_1, y_1) to be $(5, 4)$ and (x_2, y_2) to be $(-3, 1)$.

$$m = \frac{y_2 - y_1}{x_2 - x_1} = \frac{1 - 4}{-3 - 5} = \frac{-3}{-8} = \frac{3}{8}$$

17. a) We solve for y.

$$2x - 5y = 10$$
$$-5y = -2x + 10$$
$$y = -\frac{1}{5}(-2x + 10)$$
$$y = \frac{2}{5}x - 2$$

The slope is $\frac{2}{5}$.

b) The graph of $x = -2$ is a vertical line, so the slope is not defined.

18. $\text{Slope} = \dfrac{-54}{1080} = -\dfrac{1}{20}$

19. Slope $-\frac{3}{2}$, y-intercept $(0, 1)$

We plot $(0, 1)$. We can think of the slope as $\frac{-3}{2}$, so from $(0, 1)$ we move down 3 units and right 2 units. This locates the point $(2, -2)$. We plot $(2, -2)$ and draw a line passing through $(0, 1)$ and $(2, -2)$.

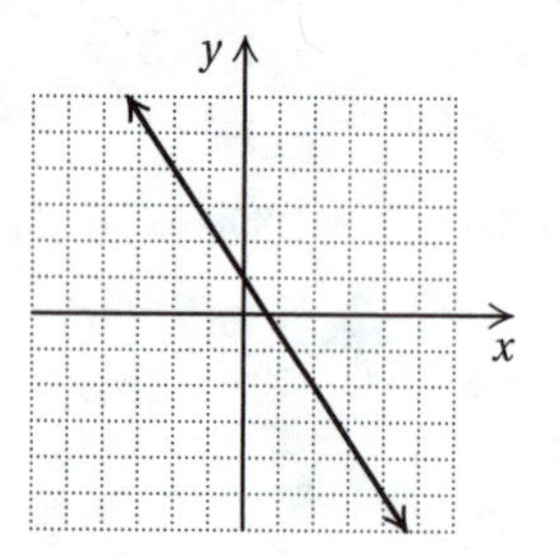

20. $y = 2x - 3$

The slope is 2 and the y-intercept is $(0, -3)$. We plot $(0, -3)$. We can start at the y-intercept and, thinking of the slope as $\frac{2}{1}$, find another point by moving up 2 units and right 1 unit to the point $(1, -1)$. Thinking of the slope as $\frac{-2}{-1}$ we can start at $(0, -3)$ and move down 2 units and left 1 unit to the point $(-1, -5)$.

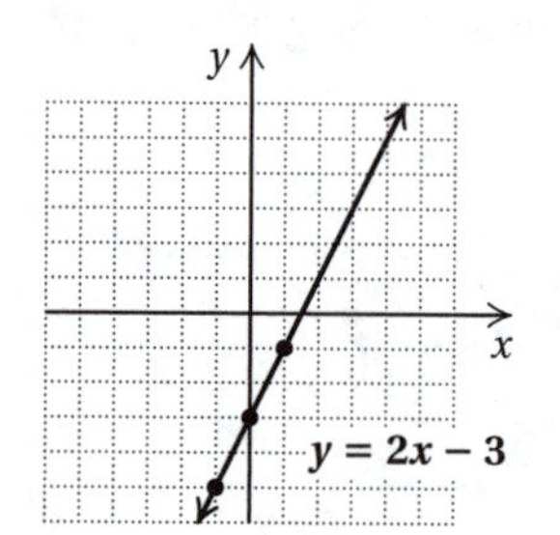

21. $y = 2x - \frac{1}{4}$

The equation is in the form $y = mx + b$. The slope is 2 and the y-intercept is $\left(0, -\frac{1}{4}\right)$.

22. We solve for y.

$$-4x + 3y = -6$$
$$3y = 4x - 6$$
$$y = \frac{1}{3}(4x - 6)$$
$$y = \frac{4}{3}x - 2$$

The slope is $\frac{4}{3}$ and the y-intercept is $(0, -2)$.

23. We substitute 1.8 for m and −7 for b in the equation $y = mx + b$.

$$y = 1.8x - 7$$

24. We substitute $-\frac{3}{8}$ for m and $-\frac{1}{8}$ for b in the equation $y = mx + b$.

$$y = -\frac{3}{8}x - \frac{1}{8}$$

25. We know the slope is 1, so the equation is $y = 1 \cdot x + b$, or $y = x + b$. Using the point $(3, 5)$, we substitute 3 for x and 5 for y in $y = x + b$. Then we solve for b.

$$y = x + b$$
$$5 = 3 + b$$
$$2 = b$$

Thus, the equation is $y = x + 2$.

26. We know the slope is -3, so the equation is $y = -3x + b$. Using the point $(-2, 0)$, we substitute -2 for x and 0 for y in $y = -3x + b$. Then we solve for b.

$$y = -3x + b$$
$$0 = -3(-2) + b$$
$$0 = 6 + b$$
$$-6 = b$$

Thus, the equation is $y = -3x - 6$.

27. First we find the slope.

$$m = \frac{-2-1}{2-1} = \frac{-3}{1} = -3$$

Thus, $y = -3x + b$. We can use either point to find b. We choose $(1, 1)$. Substitute 1 for both x and y in $y = -3x+b$.

$$y = -3x + b$$
$$1 = -3 \cdot 1 + b$$
$$1 = -3 + b$$
$$4 = b$$

Thus, the equation is $y = -3x + 4$.

28. First we find the slope.

$$m = \frac{-1-(-3)}{4-(-4)} = \frac{2}{8} = \frac{1}{4}$$

Thus, $y = \frac{1}{4}x + b$. We can use either point to find b. We choose $(4, -1)$. Substitute 4 for x and -1 for y in $y = \frac{1}{4}x + b$.

$$y = \frac{1}{4}x + b$$
$$-1 = \frac{1}{4} \cdot 4 + b$$
$$-1 = 1 + b$$
$$-2 = b$$

Thus, the equation is $y = \frac{1}{4}x - 2$.

29. a) First we find the slope.

$$m = \frac{90.9 - 84.7}{6 - 0} = \frac{6.2}{6} \approx 1.03$$

The slope is 1.03 and the y-intercept is $(0, 84.7)$, so the equation is $y = 1.03x + 84.7$.

b) The rate of change is the slope, 1.03 billion eggs per year.

c) In 2012, $x = 2012 - 2000 = 12$.

$$y = 1.03x + 84.7$$
$$y = 1.03(12) + 84.7$$
$$y = 12.36 + 84.7$$
$$y = 97.06$$

We estimate the number of eggs produced in 2012 to be 97.06 billion.

30. First we solve each equation for y:

1. $2x + y = 8$, $\quad y = -2x + 8$

2. $2x + y = 4$, $\quad y = -2x + 4$

The slope of each line is -2. The y-intercepts, $(0, 8)$ and $(0, 4)$ are different. The lines are parallel.

31. We solve the first equation for y.

$$2x + 5y = 2$$
$$5y = -2x + 2$$
$$y = \frac{1}{5}(-2x + 2)$$
$$y = -\frac{2}{5}x + \frac{2}{5}$$

The second equation is in the form $y = mx + b$:

$$y = 2x + 4$$

The slopes, $-\frac{2}{5}$ and 2, are not the same, so the lines are not parallel. The product of the slopes is $-\frac{2}{5} \cdot 2 = -\frac{4}{5} \neq -1$, so the lines are not perpendicular. Thus, the lines are neither parallel nor perpendicular.

32. First we solve each equation for y:

1. $x + 2y = 8$
$$2y = -x + 8$$
$$y = \frac{1}{2}(-x + 8)$$
$$y = -\frac{1}{2}x + 4$$

2. $-2x + y = 8$
$$y = 2x + 8$$

The slopes, $-\frac{1}{2}$ and 2, are not the same, so the lines are not parallel. The product of the slope is $-\frac{1}{2} \cdot 2 = -1$, so the lines are perpendicular.

33.

$3y - 2x < -2$	
$3 \cdot 0 - 2 \cdot 0$	? -2
0	FALSE

Since $0 < -2$ is false, $(0, 0)$ is not a solution.

34.

$3y - 2x < -2$	
$3(-10) - 2(-4)$	? -2
$-30 + 8$	
-22	TRUE

Since $-22 < -2$ is true, $(-4, -10)$ is a solution.

35. Graph $y > x - 1$.

First graph the line $y = x - 1$. Two points on the line are $(0, -1)$ and $(4, 3)$. We draw a dashed line since the inequality symbol is $>$. Then we test a point that is not on the line. We try $(0,0)$.

$$\begin{array}{c|c} y > x-1 & \\ \hline 0 \; ? \; 0-1 & \\ & -1 \quad \text{TRUE} \end{array}$$

Since $(0,0)$ is a solution of the inequality, we shade the region containing $(0,0)$.

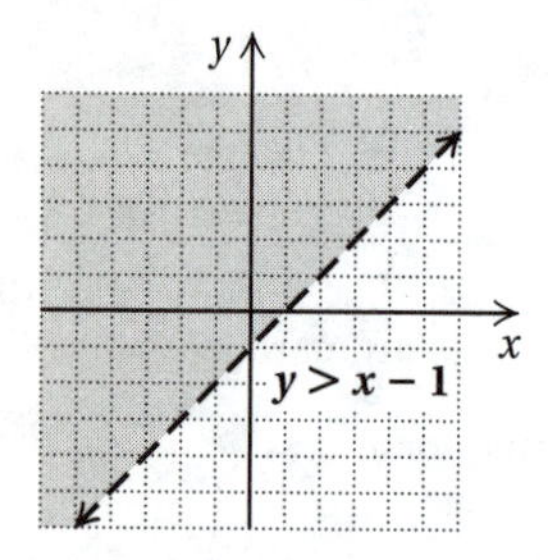

36. Graph $2x - y \leq 4$.

First graph the line $2x - y = 4$. The intercepts are $(0,-4)$ and $(2,0)$. We draw a solid line since the inequality symbol is $\leq$. Then we test a point that is not on the line. We try $(0,0)$.

$$\begin{array}{c|c} 2x - y \leq 4 & \\ \hline 2\cdot 0 - 0 \; ? \; 4 & \\ 0 & \quad \text{TRUE} \end{array}$$

Since $(0,0)$ is a solution of the inequality, we shade the region containing $(0,0)$.

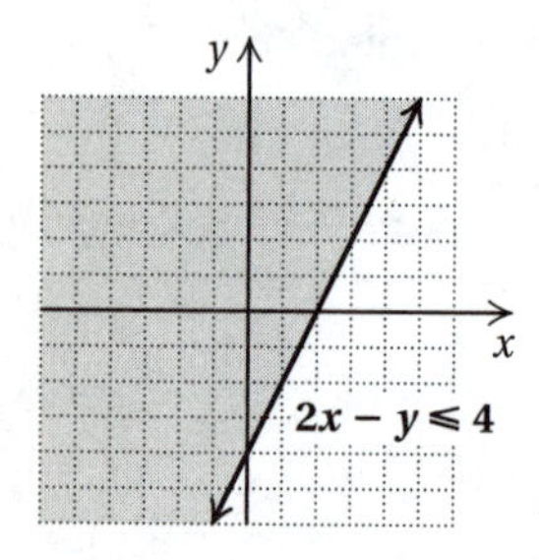

37. First we solve each equation for y.

1. $15x + 21y = 7$

$$21y = -15x + 7$$
$$y = \frac{1}{21}(-15x+7)$$
$$y = -\frac{5}{7}x + \frac{1}{3}$$

2. $35y + 14 = -25x$

$$35y = -25x - 14$$
$$y = \frac{1}{35}(-25x - 14)$$
$$y = -\frac{5}{7}x - \frac{2}{5}$$

The lines have the same slope, $-\frac{5}{7}$. The y-intercepts, $\left(0, \frac{1}{3}\right)$ and $\left(0, -\frac{2}{5}\right)$ are different. Thus, the lines are parallel. Answer A is correct.

38. We plot the given points. We see that the other vertices of the square are $(-3,4)$ and $(2,-1)$.

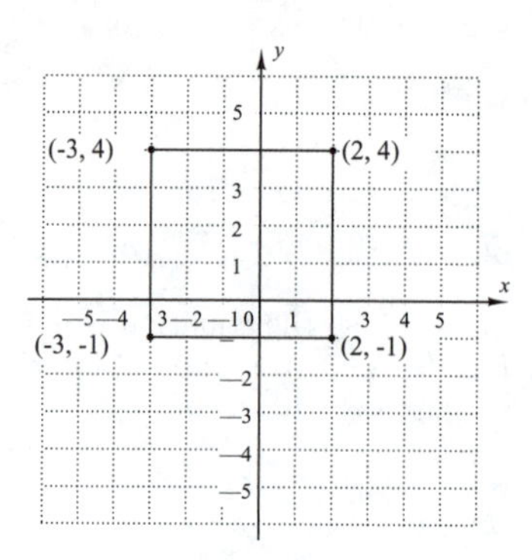

The length of the sides of the square is 5 units.

$$A = s^2 = (5 \text{ units})^2 = 25 \text{ square units}$$
$$P = 4s = 4 \cdot 5 \text{ units} = 20 \text{ linear units}$$

39. First solve each equation for y.

$$3x + 7y = 14$$
$$7y = -3x + 14$$
$$y = \frac{1}{7}(-3x + 14)$$
$$y = -\frac{3}{7}x + 2$$

$$ky - 7x = -3$$
$$ky = 7x - 3$$
$$y = \frac{1}{k}(7x - 3)$$
$$y = \frac{7}{k}x - \frac{3}{k}$$

If the lines are perpendicular, the product of their slopes is -1.

$$-\frac{3}{7} \cdot \frac{7}{k} = -1$$
$$-\frac{3 \cdot \not{7}}{\not{7} \cdot k} = -1$$
$$-\frac{3}{k} = -1$$
$$k\left(-\frac{3}{k}\right) = k(-1)$$
$$-3 = -k$$
$$3 = k$$

Chapter 12

Polynomials: Operations

Exercise Set 12.1

1. 3^4 means $3 \cdot 3 \cdot 3 \cdot 3$.

3. $(-1.1)^5$ means $(-1.1)(-1.1)(-1.1)(-1.1)(-1.1)$.

5. $\left(\frac{2}{3}\right)^4$ means $\left(\frac{2}{3}\right)\left(\frac{2}{3}\right)\left(\frac{2}{3}\right)\left(\frac{2}{3}\right)$.

7. $(7p)^2$ means $(7p)(7p)$.

9. $8k^3$ means $8 \cdot k \cdot k \cdot k$.

11. $-6y^4$ means $-6 \cdot y \cdot y \cdot y \cdot y$.

13. $a^0 = 1,\ a \neq 0$

15. $b^1 = b$

17. $\left(\frac{2}{3}\right)^0 = 1$

19. $(-7.03)^1 = -7.03$

21. $8.38^0 = 1$

23. $(ab)^1 = ab$

25. $ab^0 = a \cdot b^0 = a \cdot 1 = a$

27. $m^3 = 3^3 = 3 \cdot 3 \cdot 3 = 27$

29. $p^1 = 19^1 = 19$

31. $-x^4 = -(-3)^4 = -(-3)(-3)(-3)(-3) = -81$

33. $x^4 = 4^4 = 4 \cdot 4 \cdot 4 \cdot 4 = 256$

35. $y^2 - 7 = 10^2 - 7$
$= 100 - 7$ Evaluating the power
$= 93$ Subtracting

37. $161 - b^2 = 161 - 5^2$
$= 161 - 25$ Evaluating the power
$= 136$ Subtracting

39. $x^1 + 3 = 7^1 + 3$
$= 7 + 3 \quad (7^1 = 7)$
$= 10$

$x^0 + 3 = 7^0 + 3$
$= 1 + 3 \quad (7^0 = 1)$
$= 4$

41. $A = \pi r^2 \approx 3.14 \times (34\text{ ft})^2$
$\approx 3.14 \times 1156\text{ ft}^2$ Evaluating the power
$\approx 3629.84\text{ ft}^2$

43. $3^{-2} = \frac{1}{3^2} = \frac{1}{9}$

45. $10^{-3} = \frac{1}{10^3} = \frac{1}{1000}$

47. $7^{-3} = \frac{1}{7^3} = \frac{1}{343}$

49. $a^{-3} = \frac{1}{a^3}$

51. $\frac{1}{8^{-2}} = 8^2 = 64$

53. $\frac{1}{y^{-4}} = y^4$

55. $\frac{1}{z^{-n}} = z^n$

57. $\frac{1}{4^3} = 4^{-3}$

59. $\frac{1}{x^3} = x^{-3}$

61. $\frac{1}{a^5} = a^{-5}$

63. $2^4 \cdot 2^3 = 2^{4+3} = 2^7$

65. $8^5 \cdot 8^9 = 8^{5+9} = 8^{14}$

67. $x^4 \cdot x^3 = x^{4+3} = x^7$

69. $9^{17} \cdot 9^{21} = 9^{17+21} = 9^{38}$

71. $(3y)^4(3y)^8 = (3y)^{4+8} = (3y)^{12}$

73. $(7y)^1(7y)^{16} = (7y)^{1+16} = (7y)^{17}$

75. $3^{-5} \cdot 3^8 = 3^{-5+8} = 3^3$

77. $x^{-2} \cdot x = x^{-2+1} = x^{-1} = \frac{1}{x}$

79. $x^{14} \cdot x^3 = x^{14+3} = x^{17}$

81. $x^{-7} \cdot x^{-6} = x^{-7+(-6)} = x^{-13} = \frac{1}{x^{13}}$

83. $a^{11} \cdot a^{-3} \cdot a^{-18} = a^{11+(-3)+(-18)} = a^{-10} = \frac{1}{a^{10}}$

85. $t^8 \cdot t^{-8} = t^{8+(-8)} = t^0 = 1$

87. $\frac{7^5}{7^2} = 7^{5-2} = 7^3$

89. $\frac{8^{12}}{8^6} = 8^{12-6} = 8^6$

91. $\frac{y^9}{y^5} = y^{9-5} = y^4$

93. $\dfrac{16^2}{16^8} = 16^{2-8} = 16^{-6} = \dfrac{1}{16^6}$

95. $\dfrac{m^6}{m^{12}} = m^{6-12} = m^{-6} = \dfrac{1}{m^6}$

97. $\dfrac{(8x)^6}{(8x)^{10}} = (8x)^{6-10} = (8x)^{-4} = \dfrac{1}{(8x)^4}$

99. $\dfrac{(2y)^9}{(2y)^9} = (2y)^{9-9} = (2y)^0 = 1$

101. $\dfrac{x}{x^{-1}} = x^{1-(-1)} = x^2$

103. $\dfrac{x^7}{x^{-2}} = x^{7-(-2)} = x^9$

105. $\dfrac{z^{-6}}{z^{-2}} = z^{-6-(-2)} = z^{-4} = \dfrac{1}{z^4}$

107. $\dfrac{x^{-5}}{x^{-8}} = x^{-5-(-8)} = x^3$

109. $\dfrac{m^{-9}}{m^{-9}} = m^{-9-(-9)} = m^0 = 1$

111. $5^2 = 5 \cdot 5 = 25$

$5^{-2} = \dfrac{1}{5^2} = \dfrac{1}{25}$

$\left(\dfrac{1}{5}\right)^2 = \dfrac{1}{5} \cdot \dfrac{1}{5} = \dfrac{1}{25}$

$\left(\dfrac{1}{5}\right)^{-2} = \dfrac{1}{\left(\dfrac{1}{5}\right)^2} = \dfrac{1}{\dfrac{1}{25}} = 1 \cdot \dfrac{25}{1} = 25$

$-5^2 = -(5)(5) = -25$

$(-5)^2 = (-5)(-5) = 25$

$-\left(-\dfrac{1}{5}\right)^2 = -\left(-\dfrac{1}{5}\right)\left(-\dfrac{1}{5}\right) = -\dfrac{1}{25}$

$\left(-\dfrac{1}{5}\right)^{-2} = \dfrac{1}{\left(-\dfrac{1}{5}\right)^2} = \dfrac{1}{\dfrac{1}{25}} = 1 \cdot \dfrac{25}{1} = 25$

113. ***Familiarize.*** Let x = the length of the shorter piece. Then $2x$ = the length of the longer piece.

Translate.

Length of shorter piece	plus	length of longer piece	is	12 in.
↓	↓	↓	↓	↓
x	$+$	$2x$	$=$	12

Solve.

$$\begin{aligned} x + 2x &= 12 \\ 3x &= 12 \\ \frac{3x}{3} &= \frac{12}{3} \\ x &= 4 \end{aligned}$$

If $x = 4$, $2x = 2 \cdot 4 = 8$.

Check. The longer piece, 8 in., is twice as long as the shorter piece, 4 in. Also, 4 in. + 8 in. = 12 in., the total length of the sandwich. The answer checks.

State. The lengths of the pieces are 4 in. and 8 in.

115. ***Familiarize.*** Let w = the width. Then $w + 15$ = the length. We draw a picture.

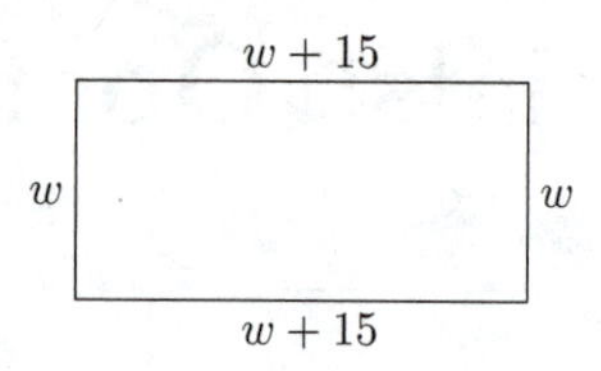

We will use the fact that the perimeter is 640 ft to find w (the width). Then we can find $w + 15$ (the length) and multiply the length and the width to find the area.

Translate.

Width+Width+ Length + Length =Perimeter

$w \quad + \quad w \quad +(w+15)+(w+15) = \quad 640$

Solve.

$$\begin{aligned} w + w + (w+15) + (w+15) &= 640 \\ 4w + 30 &= 640 \\ 4w &= 610 \\ w &= 152.5 \end{aligned}$$

If the width is 152.5, then the length is 152.5+15, or 167.5. The area is (167.5)(152.5), or 25,543.75 ft^2.

Check. The length, 167.5 ft, is 15 ft greater than the width, 152.5 ft. The perimeter is 152.5 + 152.5 + 167.5 + 167.5, or 640 ft. We should also recheck the computation we used to find the area. The answer checks.

State. The area is 25,543.75 ft^2.

117.
$$\begin{aligned} -6(2-x) + 10(5x-7) &= 10 \\ -12 + 6x + 50x - 70 &= 10 \\ 56x - 82 &= 10 \quad \text{Collecting like terms} \\ 56x - 82 + 82 &= 10 + 82 \quad \text{Adding 82} \\ 56x &= 92 \\ \frac{56x}{56} &= \frac{92}{56} \quad \text{Dividing by 56} \\ x &= \frac{23}{14} \end{aligned}$$

The solution is $\dfrac{23}{14}$.

119. $4x - 12 + 24y = 4 \cdot x - 4 \cdot 3 + 4 \cdot 6y = 4(x - 3 + 6y)$

121. Let $y_1 = (x+1)^2$ and $y_2 = x^2+1$. A graph of the equations or a table of values shows that $(x+1)^2 = x^2+1$ is not correct.

123. Let $y_1 = (5x)^0$ and $y_2 = 5x^0$. A graph of the equations or a table of values shows that $(5x)^0 = 5x^0$ is not correct.

125. $(y^{2x})(y^{3x}) = y^{2x+3x} = y^{5x}$

127. $\dfrac{a^{6t}(a^{7t})}{a^{9t}} = \dfrac{a^{6t+7t}}{a^{9t}} = \dfrac{a^{13t}}{a^{9t}} = a^{13t-9t} = a^{4t}$

129. $\dfrac{(0.8)^5}{(0.8)^3(0.8)^2} = \dfrac{(0.8)^5}{(0.8)^{3+2}} = \dfrac{(0.8)^5}{(0.8)^5} = 1$

131. Since the bases are the same, the expression with the larger exponent is larger. Thus, $3^5 > 3^4$.

133. Since the exponents are the same, the expression with the larger base is larger. Thus, $4^3 < 5^3$.

135. $\dfrac{1}{-z^4} = \dfrac{1}{-(-10)^4} = \dfrac{1}{-(-10)(-10)(-10)(-10)} =$
$\dfrac{1}{-10,000} = -\dfrac{1}{10,000}$

Exercise Set 12.2

1. $(2^3)^2 = 2^{3\cdot 2} = 2^6$

3. $(5^2)^{-3} = 5^{2(-3)} = 5^{-6} = \dfrac{1}{5^6}$

5. $(x^{-3})^{-4} = x^{(-3)(-4)} = x^{12}$

7. $(a^{-2})^9 = a^{-2\cdot 9} = a^{-18} = \dfrac{1}{a^{18}}$

9. $(t^{-3})^{-6} = t^{(-3)(-6)} = t^{18}$

11. $(t^4)^{-3} = t^{4(-3)} = t^{-12} = \dfrac{1}{t^{12}}$

13. $(x^{-2})^{-4} = x^{-2(-4)} = x^8$

15. $(ab)^3 = a^3b^3$ Raising each factor to the third power

17. $(ab)^{-3} = a^{-3}b^{-3} = \dfrac{1}{a^3b^3}$

19. $(mn^2)^{-3} = m^{-3}(n^2)^{-3} = m^{-3}n^{2(-3)} =$
$m^{-3}n^{-6} = \dfrac{1}{m^3n^6}$

21. $(4x^3)^2 = 4^2(x^3)^2$ Raising each factor to the second power
$= 16x^6$

23. $(3x^{-4})^2 = 3^2(x^{-4})^2 = 3^2x^{-4\cdot 2} = 9x^{-8} = \dfrac{9}{x^8}$

25. $(x^4y^5)^{-3} = (x^4)^{-3}(y^5)^{-3} = x^{4(-3)}y^{5(-3)} =$
$x^{-12}y^{-15} = \dfrac{1}{x^{12}y^{15}}$

27. $(x^{-6}y^{-2})^{-4} = (x^{-6})^{-4}(y^{-2})^{-4} = x^{(-6)(-4)}y^{(-2)(-4)} =$
$x^{24}y^8$

29. $(a^{-2}b^7)^{-5} = (a^{-2})^{-5}(b^7)^{-5} = a^{10}b^{-35} = \dfrac{a^{10}}{b^{35}}$

31. $(5r^{-4}t^3)^2 = 5^2(r^{-4})^2(t^3)^2 = 25r^{-4\cdot 2}t^{3\cdot 2} =$
$25r^{-8}t^6 = \dfrac{25t^6}{r^8}$

33. $(a^{-5}b^7c^{-2})^3 = (a^{-5})^3(b^7)^3(c^{-2})^3 =$
$a^{-5\cdot 3}b^{7\cdot 3}c^{-2\cdot 3} = a^{-15}b^{21}c^{-6} = \dfrac{b^{21}}{a^{15}c^6}$

35. $(3x^3y^{-8}z^{-3})^2 = 3^2(x^3)^2(y^{-8})^2(z^{-3})^2 =$
$9x^6y^{-16}z^{-6} = \dfrac{9x^6}{y^{16}z^6}$

37. $(-4x^3y^{-2})^2 = (-4)^2(x^3)^2(y^{-2})^2 = 16x^6y^{-4} = \dfrac{16x^6}{y^4}$

39. $(-a^{-3}b^{-2})^{-4} = (-1\cdot a^{-3}b^{-2})^{-4} =$
$(-1)^{-4}(a^{-3})^{-4}(b^{-2})^{-4} = \dfrac{1}{(-1)^4}\cdot a^{12}b^8 = \dfrac{a^{12}b^8}{1} = a^{12}b^8$

41. $\left(\dfrac{y^3}{2}\right)^2 = \dfrac{(y^3)^2}{2^2} = \dfrac{y^6}{4}$

43. $\left(\dfrac{a^2}{b^3}\right)^4 = \dfrac{(a^2)^4}{(b^3)^4} = \dfrac{a^8}{b^{12}}$

45. $\left(\dfrac{y^2}{2}\right)^{-3} = \dfrac{(y^2)^{-3}}{2^{-3}} = \dfrac{y^{-6}}{2^{-3}} = \dfrac{\frac{1}{y^6}}{\frac{1}{2^3}} = \dfrac{1}{y^6}\cdot\dfrac{2^3}{1} = \dfrac{8}{y^6}$

47. $\left(\dfrac{7}{x^{-3}}\right)^2 = \dfrac{7^2}{(x^{-3})^2} = \dfrac{49}{x^{-6}} = 49x^6$

49. $\left(\dfrac{x^2y}{z}\right)^3 = \dfrac{(x^2)^3y^3}{z^3} = \dfrac{x^6y^3}{z^3}$

51. $\left(\dfrac{a^2b}{cd^3}\right)^{-2} = \dfrac{(a^2)^{-2}b^{-2}}{c^{-2}(d^3)^{-2}} = \dfrac{a^{-4}b^{-2}}{c^{-2}d^{-6}} = \dfrac{\frac{1}{a^4}\cdot\frac{1}{b^2}}{\frac{1}{c^2}\cdot\frac{1}{d^6}} = \dfrac{\frac{1}{a^4b^2}}{\frac{1}{c^2d^6}} =$
$\dfrac{1}{a^4b^2}\cdot\dfrac{c^2d^6}{1} = \dfrac{c^2d^6}{a^4b^2}$

53. 2.8,000,000,000.
↑ 10 places
Large number, so the exponent is positive.
$28,000,000,000 = 2.8\times 10^{10}$

55. 9.07,000,000,000,000,000.
↑ 17 places
Large number, so the exponent is positive.
$907,000,000,000,000,000 = 9.07\times 10^{17}$

57. 0.000003.04
↑ 6 places
Small number, so the exponent is negative.
$0.00000304 = 3.04\times 10^{-6}$

59. 0.00000001.8
↑ 8 places
Small number, so the exponent is negative.
$0.000000018 = 1.8\times 10^{-8}$

61. 1.00,000,000,000.
↑ 11 places
Large number, so the exponent is positive.
$100,000,000,000 = 1.0\times 10^{11} = 10^{11}$

63. 419,854,000

4.19,854,000.

↑ 8 places

Large number, so the exponent is positive.

$419,854,000 = 4.19854 \times 10^8$

65. 2,400,000,000

2.400,000,000.

↑ 9 places

Large number, so the exponent is positive.

$2,400,000,000 = 2.4 \times 10^9$

67. 8.74×10^7

Positive exponent, so the answer is a large number.

8.7400000.

↑ 7 places

$8.74 \times 10^7 = 87,400,000$

69. 5.704×10^{-8}

Negative exponent, so the answer is a small number.

0.00000005.704

↑ 8 places

$5.704 \times 10^{-8} = 0.00000005704$

71. $10^7 = 1 \times 10^7$

Positive exponent, so the answer is a large number.

1.0000000.

↑ 7 places

$10^7 = 10,000,000$

73. $10^{-5} = 1 \times 10^{-5}$

Negative exponent, so the answer is a small number.

0.00001.

↑ 5 places

$10^{-5} = 0.00001$

75. $$\begin{aligned}(3 \times 10^4)(2 \times 10^5) &= (3 \cdot 2) \times (10^4 \cdot 10^5)\\ &= 6 \times 10^9\end{aligned}$$

77. $$\begin{aligned}(5.2 \times 10^5)(6.5 \times 10^{-2}) &= (5.2 \cdot 6.5) \times (10^5 \cdot 10^{-2})\\ &= 33.8 \times 10^3\end{aligned}$$

The answer at this stage is 33.8×10^3 but this is not scientific notation since 33.8 is not a number between 1 and 10. We convert 33.8 to scientific notation and simplify.

$33.8 \times 10^3 = (3.38 \times 10^1) \times 10^3 = 3.38 \times (10^1 \times 10^3) = 3.38 \times 10^4$

The answer is 3.38×10^4.

79. $$\begin{aligned}(9.9 \times 10^{-6})(8.23 \times 10^{-8}) &= (9.9 \cdot 8.23) \times (10^{-6} \cdot 10^{-8})\\ &= 81.477 \times 10^{-14}\end{aligned}$$

The answer at this stage is 81.477×10^{-14}. We convert 81.477 to scientific notation and simplify.

$81.477 \times 10^{-14} = (8.1477 \times 10^1) \times 10^{-14} =$

$8.1477 \times (10^1 \times 10^{-14}) = 8.1477 \times 10^{-13}$.

The answer is 8.1477×10^{-13}.

81. $$\begin{aligned}\frac{8.5 \times 10^8}{3.4 \times 10^{-5}} &= \frac{8.5}{3.4} \times \frac{10^8}{10^{-5}}\\ &= 2.5 \times 10^{8-(-5)}\\ &= 2.5 \times 10^{13}\end{aligned}$$

83. $$\begin{aligned}(3.0 \times 10^6) \div (6.0 \times 10^9) &= \frac{3.0 \times 10^6}{6.0 \times 10^9}\\ &= \frac{3.0}{6.0} \times \frac{10^6}{10^9}\\ &= 0.5 \times 10^{6-9}\\ &= 0.5 \times 10^{-3}\end{aligned}$$

The answer at this stage is 0.5×10^{-3}. We convert 0.5 to scientific notation and simplify.

$0.5 \times 10^{-3} = (5.0 \times 10^{-1}) \times 10^{-3} = 5.0 \times (10^{-1} \times 10^{-3}) = 5.0 \times 10^{-4}$

85. $$\begin{aligned}\frac{7.5 \times 10^{-9}}{2.5 \times 10^{12}} &= \frac{7.5}{2.5} \times \frac{10^{-9}}{10^{12}}\\ &= 3.0 \times 10^{-9-12}\\ &= 3.0 \times 10^{-21}\end{aligned}$$

87. There are 60 seconds in one minute and 60 minutes in one hour, so there are 60(60), or 3600 seconds in one hour. There are 24 hours in one day and 365 days in one year, so there are 3600(24)(365), or 31,536,000 seconds in one year.

$$\begin{aligned}&4,200,000 \times 31,536,000\\ &= (4.2 \times 10^6) \times (3.1536 \times 10^7)\\ &= (4.2 \times 3.1536) \times (10^6 \times 10^7)\\ &\approx 13.25 \times 10^{13}\\ &\approx (1.325 \times 10) \times 10^{13}\\ &\approx 1.325 \times (10 \times 10^{13})\\ &\approx 1.325 \times 10^{14}\end{aligned}$$

About 1.325×10^{14} cubic feet of water is discharged from the Amazon River in 1 yr.

89. $$\begin{aligned}\frac{1.908 \times 10^{24}}{6 \times 10^{21}} &= \frac{1.908}{6} \times \frac{10^{24}}{10^{21}}\\ &= 0.318 \times 10^3\\ &= (3.18 \times 10^{-1}) \times 10^3\\ &= 3.18 \times (10^{-1} \times 10^3)\\ &= 3.18 \times 10^2\end{aligned}$$

The mass of Jupiter is 3.18×10^2 times the mass of Earth.

91. $$\begin{aligned}10 \text{ billion trillion} &= 1 \times 10 \times 10^9 \times 10^{12}\\ &= 1 \times 10^{22}\end{aligned}$$

There are 1×10^{22} stars in the known universe.

93. We divide the mass of the sun by the mass of earth.

$$\begin{aligned}\frac{1.998 \times 10^{27}}{6 \times 10^{21}} &= 0.333 \times 10^6\\ &= (3.33 \times 10^{-1}) \times 10^6\\ &= 3.33 \times 10^5\end{aligned}$$

The mass of the sun is 3.33×10^5 times the mass of Earth.

95. First we divide the distance from the earth to the moon by 3 days to find the number of miles per day the space vehicle travels. Note that $240,000 = 2.4 \times 10^5$.

$$\frac{2.4 \times 10^5}{3} = 0.8 \times 10^5 = 8 \times 10^4$$

The space vehicle travels 8×10^4 miles per day. Now divide the distance from the earth to Mars by 8×10^4 to find how long it will take the space vehicle to reach Mars. Note that $35,000,000 = 3.5 \times 10^7$.

$$\frac{3.5 \times 10^7}{8 \times 10^4} = 0.4375 \times 10^3 = 4.375 \times 10^2$$

It takes 4.375×10^2 days for the space vehicle to travel from the earth to Mars.

97. $9x - 36 = 9 \cdot x - 9 \cdot 4 = 9(x - 4)$

99. $3s + 3t + 24 = 3 \cdot s + 3 \cdot t + 3 \cdot 8 = 3(s + t + 8)$

101.
$$\begin{aligned}
2x - 4 - 5x + 8 &= x - 3 \\
-3x + 4 &= x - 3 && \text{Collecting like terms} \\
-3x + 4 - 4 &= x - 3 - 4 && \text{Subtracting 4} \\
-3x &= x - 7 \\
-3x - x &= x - 7 - x && \text{Subtracting } x \\
-4x &= -7 \\
\frac{-4x}{-4} &= \frac{-7}{-4} && \text{Dividing by } -4 \\
x &= \frac{7}{4}
\end{aligned}$$

The solution is $\frac{7}{4}$.

103.
$$\begin{aligned}
8(2x + 3) - 2(x - 5) &= 10 \\
16x + 24 - 2x + 10 &= 10 && \text{Removing parentheses} \\
14x + 34 &= 10 && \text{Collecting like terms} \\
14x + 34 - 34 &= 10 - 34 && \text{Subtracting 34} \\
14x &= -24 \\
\frac{14x}{14} &= \frac{-24}{14} && \text{Dividing by 14} \\
x &= -\frac{12}{7} && \text{Simplifying}
\end{aligned}$$

The solution is $-\frac{12}{7}$.

105. $y = x - 5$

The equation is equivalent to $y = x + (-5)$. The y-intercept is $(0, -5)$. We find two other points.

When $x = 2$, $y = 2 - 5 = -3$.

When $x = 4$, $y = 4 - 5 = -1$.

x	y
0	−5
2	−3
4	−1

Plot these points, draw the line they determine, and label the graph $y = x - 5$.

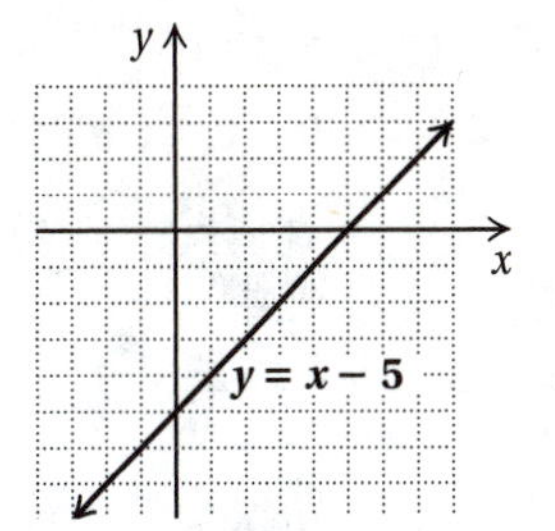

107.
$$\begin{aligned}
\frac{(5.2 \times 10^6)(6.1 \times 10^{-11})}{1.28 \times 10^{-3}} &= \frac{(5.2 \cdot 6.1)}{1.28} \times \frac{(10^6 \cdot 10^{-11})}{10^{-3}} \\
&= 24.78125 \times 10^{-2} \\
&= (2.478125 \times 10^1) \times 10^{-2} \\
&= 2.478125 \times 10^{-1}
\end{aligned}$$

109. $\frac{(5^{12})^2}{5^{25}} = \frac{5^{24}}{5^{25}} = 5^{24-25} = 5^{-1} = \frac{1}{5}$

111. $\frac{(3^5)^4}{3^5 \cdot 3^4} = \frac{3^{5 \cdot 4}}{3^{5+4}} = \frac{3^{20}}{3^9} = 3^{20-9} = 3^{11}$

113. $\frac{49^{18}}{7^{35}} = \frac{(7^2)^{18}}{7^{35}} = \frac{7^{36}}{7^{35}} = 7$

115. $\frac{(0.4)^5}{[(0.4)^3]^2} = \frac{(0.4)^5}{(0.4)^6} = (0.4)^{-1} = \frac{1}{0.4}$, or 2.5

117. False; let $x = 2$, $y = 3$, $m = 4$, and $n = 2$:

$2^4 \cdot 3^2 = 16 \cdot 9 = 144$, but

$(2 \cdot 3)^{4 \cdot 2} = 6^8 = 1,679,616$

119. False; let $x = 5$, $y = 3$, and $m = 2$:

$(5 - 3)^2 = 2^2 = 4$, but

$5^2 - 3^2 = 25 - 9 = 16$

121. True; $(-x)^{2m} = (-1 \cdot x)^{2m} = (-1)^{2m} \cdot x^{2m} = [(-1)^2]^m \cdot x^{2m} = 1^m \cdot x^{2m} = x^{2m}$

Exercise Set 12.3

1. $-5x + 2 = -5 \cdot 4 + 2 = -20 + 2 = -18$;

$-5x + 2 = -5(-1) + 2 = 5 + 2 = 7$

3. $2x^2 - 5x + 7 = 2 \cdot 4^2 - 5 \cdot 4 + 7 = 2 \cdot 16 - 20 + 7 = 32 - 20 + 7 = 19$;

$2x^2 - 5x + 7 = 2(-1)^2 - 5(-1) + 7 = 2 \cdot 1 + 5 + 7 = 2 + 5 + 7 = 14$

5. $x^3 - 5x^2 + x = 4^3 - 5 \cdot 4^2 + 4 = 64 - 5 \cdot 16 + 4 = 64 - 80 + 4 = -12$;

$x^3 - 5x^2 + x = (-1)^3 - 5(-1)^2 + (-1) = -1 - 5 \cdot 1 - 1 = -1 - 5 - 1 = -7$

7. $\frac{1}{3}x + 5 = \frac{1}{3}(-2) + 5 = -\frac{2}{3} + 5 = -\frac{2}{3} + \frac{15}{3} = \frac{13}{3}$;

$\frac{1}{3}x + 5 = \frac{1}{3} \cdot 0 + 5 = 0 + 5 = 5$

9. $x^2 - 2x + 1 = (-2)^2 - 2(-2) + 1 = 4 + 4 + 1 = 9;$

$x^2 - 2x + 1 = 0^2 - 2 \cdot 0 + 1 = 0 - 0 + 1 = 1$

11. $-3x^3 + 7x^2 - 3x - 2 = -3(-2)^3 + 7(-2)^2 - 3(-2) - 2 = -3(-8) + 7(4) - 3(-2) - 2 = 24 + 28 + 6 - 2 = 56;$

$-3x^3 + 7x^2 - 3x - 2 = -3 \cdot 0^3 + 7 \cdot 0^2 - 3 \cdot 0 - 2 = -3 \cdot 0 + 7 \cdot 0 - 0 - 2 = 0 + 0 - 0 - 2 = -2$

13. We evaluate the polynomial for $t = 10$:

$S = 11.12t^2 = 11.12(10)^2 = 11.12(100) = 1112$

The skydiver has fallen approximately 1112 ft.

15. We evaluate the polynomial for $x = 75$:

$$\begin{aligned} R = 280x - 0.4x^2 &= 280(75) - 0.4(75)^2 \\ &= 280(75) - 0.4(5625) \\ &= 21,000 - 2250 \\ &= 18,750 \end{aligned}$$

The total revenue from the sale of 75 TVs is $18,750.

We evaluate the polynomial for $x = 100$:

$$\begin{aligned} R = 280x - 0.4x^2 &= 280(100) - 0.4(100)^2 \\ &= 280(100) - 0.4(10,000) \\ &= 28,000 - 4000 \\ &= 24,000 \end{aligned}$$

The total revenue from the sale of 100 TVs is $24,000.

17. Locate -3 on the x-axis. Then move vertically to the graph and horizontally to the y-axis. It appears that the y-value that is paired with -3 is -4. Thus, the value of $y = 5 - x^2$ is -4 when $x = -3$.

Locate -1 on the x-axis. Then move vertically to the graph and horizontally to the y-axis. It appears that the y-value that is paired with -1 is 4. Thus, the value of $y = 5 - x^2$ is 4 when $x = -1$.

Locate 0 on the x-axis. Then move vertically to the graph. We arrive at a point on the y-axis with the y-value 5. Thus, the value of $5 - x^2$ is 5 when $x = 0$.

Locate 1.5 on the x-axis. Then move vertically to the graph and horizontally to the y-axis. It appears that the y-value that is paired with 1.5 is 2.75. Thus, the value of $y = 5 - x^2$ is 2.75 when $x = 1.5$.

Locate 2 on the x-axis. Then move vertically to the graph and horizontally to the y-axis. It appears that the y-value that is paired with 2 is 1. Thus, the value of $y = 5 - x^2$ is 1 when $x = 2$.

19. a) In 2010, $t = 0$.

$E = 158.68(0) + 2728.4 = 2728.4$.

The consumption of electricity in 2010 is 2728.4 billion kilowatt-hours.

In 2015, $t = 2015 - 2010 = 5$.

$E = 158.68(5) + 2728.4 = 793.4 + 2728.4 = 3521.8$

The consumption of electricity in 2015 will be 3521.8 billion kilowatt-hours.

In 2020, $t = 2020 - 2010 = 10$.

$E = 158.68(10) + 2728.4 = 1586.8 + 2728.4 = 4315.2$

The consumption of electricity in 2020 will be 4315.2 billion kilowatt-hours.

In 2025, $t = 2025 - 2010 = 15$.

$E = 158.68(15) + 2728.4 = 2380.2 + 2728.4 = 5108.6$

The consumption of electricity in 2025 will be 5108.6 billion kilowatt-hours.

In 2030, $t = 2030 - 2010 = 20$.

$E = 158.68(20) + 2728.4 = 3173.6 + 2728.4 = 5902$

The consumption of electricity in 2030 will be 5902 billion kilowatt-hours.

b) It appears that the points $(0, 2728.4)$, $(5, 3521.8)$, $(10, 4315.2)$, $(15, 5108.6)$, and $(20, 5902)$ are on the graph, so the results check.

21. Locate 10 on the horizontal axis. From there move vertically to the graph and then horizontally to the M-axis. This locates an M-value of about 9. Thus, about 9 words were memorized in 10 minutes.

23. Locate 8 on the horizontal axis. From there move vertically to the graph and then horizontally to the M-axis. This locates an M-value of about 6. Thus, the value of $-0.001t^3 + 0.1t^2$ for $t = 8$ is approximately 6.

25. Locate 13 on the horizontal axis. It is halfway between 12 and 14. From there move vertically to the graph and then horizontally to the M-axis. This locates an M-value of about 15. Thus, the value of $-0.001t^3 + 0.1t^2$ when t is 13 is approximately 15.

27. $2 - 3x + x^2 = 2 + (-3x) + x^2$

The terms are 2, $-3x$, and x^2.

29. $-2x^4 + \frac{1}{3}x^3 - x + 3 = -2x^4 + \frac{1}{3}x^3 + (-x) + 3$

The terms are $-2x^4$, $\frac{1}{3}x^3$, $-x$, and 3.

31. $5x^3 + 6x^2 - 3x^2$

Like terms: $6x^2$ and $-3x^2$ Same variable and exponent

33. $2x^4 + 5x - 7x - 3x^4$

Like terms: $2x^4$ and $-3x^4$ Same variable and
Like terms: $5x$ and $-7x$ exponent

35. $3x^5 - 7x + 8 + 14x^5 - 2x - 9$

Like terms: $3x^5$ and $14x^5$
Like terms: $-7x$ and $-2x$
Like terms: 8 and -9 Constant terms are like terms.

37. $-3x + 6$

The coefficient of $-3x$, the first term, is -3.

The coefficient of 6, the second term, is 6.

39. $5x^2 + \frac{3}{4}x + 3$

The coefficient of $5x^2$, the first term, is 5.

The coefficient of $\frac{3}{4}x$, the second term, is $\frac{3}{4}$.

The coefficient of 3, the third term, is 3.

41. $-5x^4 + 6x^3 - 2.7x^2 + 8x - 2$

The coefficient of $-5x^4$, the first term, is -5.

The coefficient of $6x^3$, the second term, is 6.

The coefficient of $-2.7x^2$, the third term, is -2.7.

The coefficient of $8x$, the fourth term, is 8.

The coefficient of -2, the fifth term, is -2.

43. $2x - 5x = (2-5)x = -3x$

45. $x - 9x = 1x - 9x = (1-9)x = -8x$

47. $5x^3 + 6x^3 + 4 = (5+6)x^3 + 4 = 11x^3 + 4$

49. $5x^3 + 6x - 4x^3 - 7x = (5-4)x^3 + (6-7)x =$

$1x^3 + (-1)x = x^3 - x$

51. $6b^5 + 3b^2 - 2b^5 - 3b^2 = (6-2)b^5 + (3-3)b^2 =$

$4b^5 + 0b^2 = 4b^5$

53. $\frac{1}{4}x^5 - 5 + \frac{1}{2}x^5 - 2x - 37 =$

$\left(\frac{1}{4} + \frac{1}{2}\right)x^5 - 2x + (-5 - 37) = \frac{3}{4}x^5 - 2x - 42$

55. $6x^2 + 2x^4 - 2x^2 - x^4 - 4x^2 =$

$6x^2 + 2x^4 - 2x^2 - 1x^4 - 4x^2 =$

$(6 - 2 - 4)x^2 + (2-1)x^4 = 0x^2 + 1x^4 =$

$0 + x^4 = x^4$

57. $\frac{1}{4}x^3 - x^2 - \frac{1}{6}x^2 + \frac{3}{8}x^3 + \frac{5}{16}x^3 =$

$\frac{1}{4}x^3 - 1x^2 - \frac{1}{6}x^2 + \frac{3}{8}x^3 + \frac{5}{16}x^3 =$

$\left(\frac{1}{4} + \frac{3}{8} + \frac{5}{16}\right)x^3 + \left(-1 - \frac{1}{6}\right)x^2 =$

$\left(\frac{4}{16} + \frac{6}{16} + \frac{5}{16}\right)x^3 + \left(-\frac{6}{6} - \frac{1}{6}\right)x^2 = \frac{15}{16}x^3 - \frac{7}{6}x^2$

59. $x^5 + x + 6x^3 + 1 + 2x^2 = x^5 + 6x^3 + 2x^2 + x + 1$

61. $5y^3 + 15y^9 + y - y^2 + 7y^8 =$

$15y^9 + 7y^8 + 5y^3 - y^2 + y$

63. $3x^4 - 5x^6 - 2x^4 + 6x^6 = x^4 + x^6 = x^6 + x^4$

65. $-2x + 4x^3 - 7x + 9x^3 + 8 = -9x + 13x^3 + 8 =$

$13x^3 - 9x + 8$

67. $3x + 3x + 3x - x^2 - 4x^2 = 9x - 5x^2 = -5x^2 + 9x$

69. $-x + \frac{3}{4} + 15x^4 - x - \frac{1}{2} - 3x^4 = -2x + \frac{1}{4} + 12x^4 =$

$12x^4 - 2x + \frac{1}{4}$

71. $2x - 4 = 2x^1 - 4x^0$

The degree of $2x$ is 1.

The degree of -4 is 0.

The degree of the polynomial is 1, the largest exponent.

73. $3x^2 - 5x + 2 = 3x^2 - 5x^1 + 2x^0$

The degree of $3x^2$ is 2.

The degree of $-5x$ is 1.

The degree of 2 is 0.

The degree of the polynomial is 2, the largest exponent.

75. $-7x^3 + 6x^2 + \frac{3}{5}x + 7 = -7x^3 + 6x^2 + \frac{3}{5}x^1 + 7x^0$

The degree of $-7x^3$ is 3.

The degree of $6x^2$ is 2.

The degree of $\frac{3}{5}x$ is 1.

The degree of 7 is 0.

The degree of the polynomial is 3, the largest exponent.

77. $x^2 - 3x + x^6 - 9x^4 = x^2 - 3x^1 + x^6 - 9x^4$

The degree of x^2 is 2.

The degree of $-3x$ is 1.

The degree of x^6 is 6.

The degree of $-9x^4$ is 4.

The degree of the polynomial is 6, the largest exponent.

79. See the answer section in the text.

81. In the polynomial $x^3 - 27$, there are no x^2 or x terms. The x^2 term (or second-degree term) and the x term (or first-degree term) are missing.

83. In the polynomial $x^4 - x$, there are no x^3, x^2, or x^0 terms. The x^3 term (or third-degree term), the x^2 term (or second-degree term), and the x^0 term (or zero-degree term) are missing.

85. No terms are missing in the polynomial

$2x^3 - 5x^2 + x - 3$.

87. $x^3 - 27 = x^3 + 0x^2 + 0x - 27$

$x^3 - 27 = x^3 \qquad\qquad\qquad - 27$

89. $x^4 - x = x^4 + 0x^3 + 0x^2 - x + 0x^0$

$x^4 - x = x^4 \qquad\qquad\qquad - x$

91. There are no missing terms.

93. The polynomial $x^2 - 10x + 25$ is a *trinomial* because it has just three terms.

95. The polynomial $x^3 - 7x^2 + 2x - 4$ is *none of these* because it has more than three terms.

97. The polynomial $4x^2 - 25$ is a *binomial* because it has just two terms.

99. The polynomial $40x$ is a *monomial* because it has just one term.

101. ***Familiarize.*** Let a = the number of apples the campers had to begin with. Then the first camper ate $\frac{1}{3}a$ apples and $a - \frac{1}{3}a$, or $\frac{2}{3}a$, apples were left. The second camper ate $\frac{1}{3}\left(\frac{2}{3}a\right)$, or $\frac{2}{9}a$, apples, and $\frac{2}{3}a - \frac{2}{9}a$, or $\frac{4}{9}a$, apples were left. The third camper ate $\frac{1}{3}\left(\frac{4}{9}a\right)$, or $\frac{4}{27}a$, apples, and $\frac{4}{9}a - \frac{4}{27}a$, or $\frac{8}{27}a$, apples were left.

Translate. We write an equation for the number of apples left after the third camper eats.

$$\underbrace{\text{Number of apples left}}_{\frac{8}{27}a} \; \underbrace{\text{is}}_{=} \; \underbrace{8.}_{8}$$

Solve. We solve the equation.

$$\frac{8}{27}a = 8$$
$$a = \frac{27}{8} \cdot 8$$
$$a = 27$$

Check. If the campers begin with 27 apples, then the first camper eats $\frac{1}{3} \cdot 27$, or 9, and $27 - 9$, or 18, are left. The second camper then eats $\frac{1}{3} \cdot 18$, or 6 apples and $18 - 6$, or 12, are left. Finally, the third camper eats $\frac{1}{3} \cdot 12$, or 4 apples and $12 - 4$, or 8, are left. The answer checks.

State. The campers had 27 apples to begin with.

103. $\frac{1}{8} - \frac{5}{6} = \frac{1}{8} + \left(-\frac{5}{6}\right)$, LCM is 24

$$= \frac{1}{8} \cdot \frac{3}{3} + \left(-\frac{5}{6}\right)\left(\frac{4}{4}\right)$$
$$= \frac{3}{24} + \left(-\frac{20}{24}\right)$$
$$= -\frac{17}{24}$$

105. $5.6 - 8.2 = 5.6 + (-8.2) = -2.6$

107.

$C = ab - r$	
$C + r = ab$	Adding r
$\frac{C+r}{a} = \frac{ab}{a}$	Dividing by a
$\frac{C+r}{a} = b$	Simplifying

109. $3x - 15y + 63 = 3 \cdot x - 3 \cdot 5y + 3 \cdot 21 = 3(x - 5y + 21)$

111.

$$(3x^2)^3 + 4x^2 \cdot 4x^4 - x^4(2x)^2 + [(2x)^2]^3 - 100x^2(x^2)^2$$
$$= 27x^6 + 4x^2 \cdot 4x^4 - x^4 \cdot 4x^2 + (2x)^6 - 100x^2 \cdot x^4$$
$$= 27x^6 + 16x^6 - 4x^6 + 64x^6 - 100x^6$$
$$= 3x^6$$

113. $(5m^5)^2 = 5^2m^{5\cdot2} = 25m^{10}$

The degree is 10.

115. Graph $y = 5 - x^2$. Then use VALUE from the CALC menu to find the y-values that correspond to $x = -3$, $x = -1$, $x = 0$, $x = 1.5$, and $x = 2$. As before, we find that these values are -4, 4, 5, 2.75, and 1, respectively.

117. Graph $y = -0.00006x^3 + 0.006x^2 - 0.1x + 1.9$. Then use VALUE from the CALC menu to find the y-values that correspond to $x = 20$ and $x = 40$. As before, we find that these values are 1.82 and 3.66, respectively. These results represent 1,820,000 and 3,660,000 hearing-impaired Americans.

Exercise Set 12.4

1. $(3x + 2) + (-4x + 3) = (3 - 4)x + (2 + 3) = -x + 5$

3. $(-6x + 2) + \left(x^2 + \frac{1}{2}x - 3\right) = x^2 + \left(-6 + \frac{1}{2}\right)x + (2 - 3) =$

$x^2 + \left(-\frac{12}{2} + \frac{1}{2}\right)x + (2 - 3) = x^2 - \frac{11}{2}x - 1$

5. $(x^2 - 9) + (x^2 + 9) = (1 + 1)x^2 + (-9 + 9) = 2x^2$

7. $(3x^2 - 5x + 10) + (2x^2 + 8x - 40) =$

$(3 + 2)x^2 + (-5 + 8)x + (10 - 40) = 5x^2 + 3x - 30$

9. $(1.2x^3 + 4.5x^2 - 3.8x) + (-3.4x^3 - 4.7x^2 + 23) =$

$(1.2 - 3.4)x^3 + (4.5 - 4.7)x^2 - 3.8x + 23 =$

$-2.2x^3 - 0.2x^2 - 3.8x + 23$

11. $(1 + 4x + 6x^2 + 7x^3) + (5 - 4x + 6x^2 - 7x^3) =$

$(1 + 5) + (4 - 4)x + (6 + 6)x^2 + (7 - 7)x^3 =$

$6 + 0x + 12x^2 + 0x^3 = 6 + 12x^2$, or $12x^2 + 6$

13. $\left(\frac{1}{4}x^4 + \frac{2}{3}x^3 + \frac{5}{8}x^2 + 7\right) + \left(-\frac{3}{4}x^4 + \frac{3}{8}x^2 - 7\right) =$

$\left(\frac{1}{4} - \frac{3}{4}\right)x^4 + \frac{2}{3}x^3 + \left(\frac{5}{8} + \frac{3}{8}\right)x^2 + (7 - 7) =$

$-\frac{2}{4}x^4 + \frac{2}{3}x^3 + \frac{8}{8}x^2 + 0 =$

$-\frac{1}{2}x^4 + \frac{2}{3}x^3 + x^2$

15. $(0.02x^5 - 0.2x^3 + x + 0.08) + (-0.01x^5 + x^4 - 0.8x - 0.02) =$

$(0.02 - 0.01)x^5 + x^4 - 0.2x^3 + (1 - 0.8)x + (0.08 - 0.02) =$

$0.01x^5 + x^4 - 0.2x^3 + 0.2x + 0.06$

17. $9x^8 - 7x^4 + 2x^2 + 5) + (8x^7 + 4x^4 - 2x) +$

$(-3x^4 + 6x^2 + 2x - 1) = 9x^8 + 8x^7 + (-7 + 4 - 3)x^4 +$

$(2 + 6)x^2 + (-2 + 2)x + (5 - 1) =$

$9x^8 + 8x^7 - 6x^4 + 8x^2 + 4$

19. Rewrite the problem so the coefficients of like terms have the same number of decimal places.

$$\begin{array}{rrrrr} 0.15x^4 & +\ 0.10x^3 & -\ 0.90x^2 & & \\ & -\ 0.01x^3 & +\ 0.01x^2 & +\ x & \\ 1.25x^4 & & +\ 0.11x^2 & & +\ 0.01 \\ & 0.27x^3 & & & +\ 0.99 \\ -0.35x^4 & & +\ 15.00x^2 & & -\ 0.03 \\ \hline 1.05x^4 & +\ 0.36x^3 & +\ 14.22x^2 & +\ x & +\ 0.97 \end{array}$$

21. We change the sign of the term inside the parentheses.

$-(-5x) = 5x$

23. We change the sign of every term inside the parentheses.

$-(-x^2 + \frac{3}{2}x - 2) = x^2 - \frac{3}{2}x + 2$

25. We change the sign of every term inside the parentheses.

$-(12x^4 - 3x^3 + 3) = -12x^4 + 3x^3 - 3$

27. We change the sign of every term inside parentheses.

$-(3x - 7) = -3x + 7$

29. We change the sign of every term inside parentheses.

$-(4x^2 - 3x + 2) = -4x^2 + 3x - 2$

31. We change the sign of every term inside parentheses.

$-\left(-4x^4 + 6x^2 + \frac{3}{4}x - 8\right) = 4x^4 - 6x^2 - \frac{3}{4}x + 8$

33. $(3x + 2) - (-4x + 3) = 3x + 2 + 4x - 3$ Changing the sign of every term inside parentheses

$= 7x - 1$

35. $(-6x + 2) - (x^2 + x - 3) = -6x + 2 - x^2 - x + 3$

$= -x^2 - 7x + 5$

37. $(x^2 - 9) - (x^2 + 9) = x^2 - 9 - x^2 - 9 = -18$

39. $(6x^4 + 3x^3 - 1) - (4x^2 - 3x + 3)$

$= 6x^4 + 3x^3 - 1 - 4x^2 + 3x - 3$

$= 6x^4 + 3x^3 - 4x^2 + 3x - 4$

41. $(1.2x^3 + 4.5x^2 - 3.8x) - (-3.4x^3 - 4.7x^2 + 23)$

$= 1.2x^3 + 4.5x^2 - 3.8x + 3.4x^3 + 4.7x^2 - 23$

$= 4.6x^3 + 9.2x^2 - 3.8x - 23$

43. $\frac{5}{8}x^3 - \frac{1}{4}x - \frac{1}{3} - \left(-\frac{1}{8}x^3 + \frac{1}{4}x - \frac{1}{3}\right)$

$= \frac{5}{8}x^3 - \frac{1}{4}x - \frac{1}{3} + \frac{1}{8}x^3 - \frac{1}{4}x + \frac{1}{3}$

$= \frac{6}{8}x^3 - \frac{2}{4}x$

$= \frac{3}{4}x^3 - \frac{1}{2}x$

45. $(0.08x^3 - 0.02x^2 + 0.01x) - (0.02x^3 + 0.03x^2 - 1)$

$= 0.08x^3 - 0.02x^2 + 0.01x - 0.02x^3 - 0.03x^2 + 1$

$= 0.06x^3 - 0.05x^2 + 0.01x + 1$

47.

$$\begin{array}{rrr} x^2 & +\ 5x & +\ 6 \\ x^2 & +\ 2x & \\ \hline \end{array}$$

$$\begin{array}{rrrl} x^2 & +\ 5x & +\ 6 & \\ -x^2 & -\ 2x & & \text{Changing signs} \\ \hline & 3x & +\ 6 & \text{Adding} \end{array}$$

49.

$$\begin{array}{rrrrr} 5x^4 & +\ 6x^3 & -\ 9x^2 & & \\ -6x^4 & -\ 6x^3 & & +\ 8x & +\ 9 \\ \hline \end{array}$$

$$\begin{array}{rrrrrl} 5x^4 & +\ 6x^3 & -\ 9x^2 & & & \\ 6x^4 & +\ 6x^3 & & -\ 8x & -\ 9 & \text{Changing signs} \\ \hline 11x^4 & +\ 12x^3 & -\ 9x^2 & -\ 8x & -\ 9 & \text{Adding} \end{array}$$

51.

$$\begin{array}{rrrrrr} x^5 & & & & & -\ 1 \\ x^5 & -\ x^4 & +\ x^3 & -\ x^2 & +\ x & -\ 1 \\ \hline \end{array}$$

$$\begin{array}{rrrrrrl} x^5 & & & & & -\ 1 & \\ -x^5 & +\ x^4 & -\ x^3 & +\ x^2 & -\ x & +\ 1 & \text{Changing signs} \\ \hline & x^4 & -\ x^3 & +\ x^2 & -\ x & & \text{Adding} \end{array}$$

53. We add the lengths of the sides:

$4a + 7 + a + \frac{1}{2}a + 3 + a + 2a + 3a$

$= \left(4 + 1 + \frac{1}{2} + 1 + 2 + 3\right)a + (7 + 3)$

$= 11\frac{1}{2}a + 10$, or $\frac{23}{2}a + 10$

55.

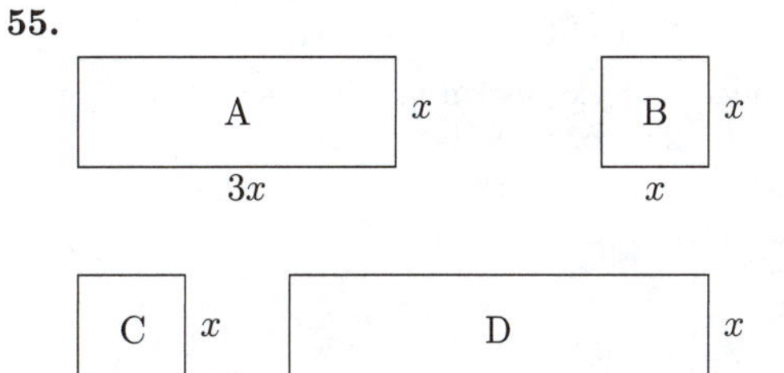

The area of a rectangle is the product of the length and width. The sum of the areas is found as follows:

$$\begin{array}{ccccccccc} & \text{Area of } A & + & \text{Area of } B & + & \text{Area of } C & + & \text{Area of } D \\ = & 3x \cdot x & + & x \cdot x & + & x \cdot x & + & 4 \cdot x \\ = & 3x^2 & + & x^2 & + & x^2 & + & 4x \\ = & 5x^2 & + & 4x & & & & \end{array}$$

A polynomial for the sum of the areas is $5x^2 + 4x$.

57.

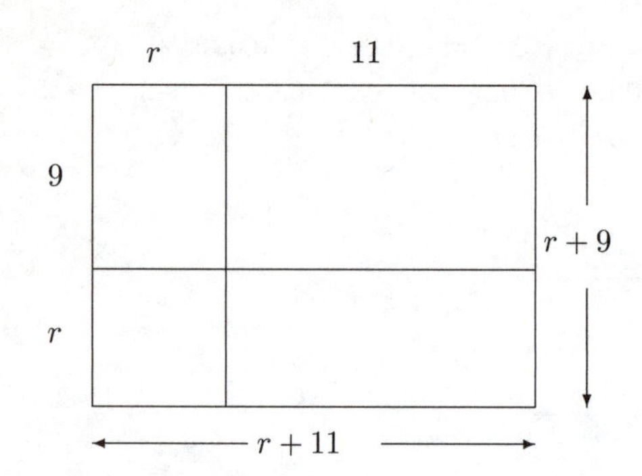

The length and width of the figure can be expressed as $r + 11$ and $r + 9$, respectively. The area of this figure (a rectangle) is the product of the length and width. An algebraic expression for the area is $(r + 11) \cdot (r + 9)$.

The area of the figure can also be found by adding the areas of the four rectangles A, B, C, and D. The area of a rectangle is the product of the length and the width.

	Area of A	+	Area of B	+	Area of C	+	Area of D
=	$9 \cdot r$	+	$11 \cdot 9$	+	$r \cdot r$	+	$11 \cdot r$
=	$9r$	+	99	+	r^2	+	$11r$

A second algebraic expression for the area of the figure is $9r + 99 + r^2 + 11r$, or $r^2 + 20r + 99$.

59.

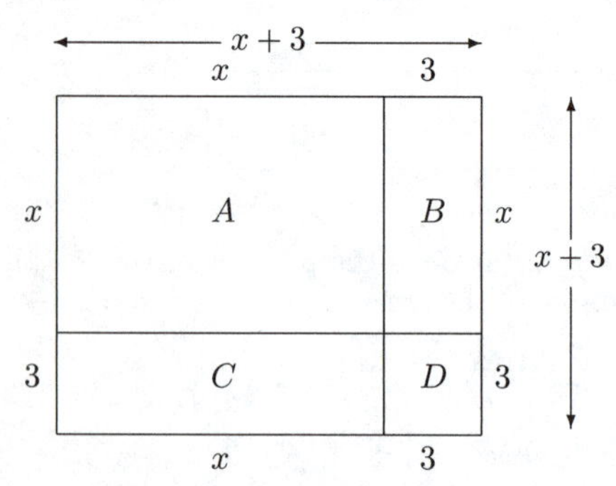

The length and width of the figure can each be expressed as $x + 3$. The area can be expressed as $(x + 3) \cdot (x + 3)$, or $(x + 3)^2$.

Another way to express the area is to find an expression for the sum of the areas of the four rectangles A, B, C, and D. The area of each rectangle is the product of its length and width.

	Area of A	+	Area of B	+	Area of C	+	Area of D
=	$x \cdot x$	+	$3 \cdot x$	+	$3 \cdot x$	+	$3 \cdot 3$
=	x^2	+	$3x$	+	$3x$	+	9

Then a second algebraic expression for the area of the figure is $x^2 + 3x + 3x + 9$, or $x^2 + 6x + 9$.

61.

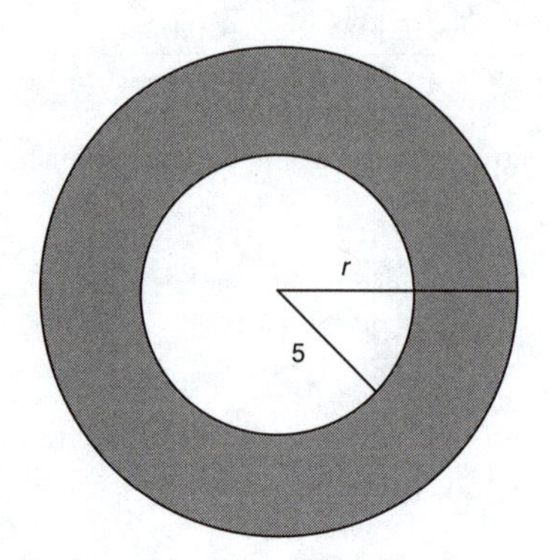

Familiarize. Recall that the area of a circle is the product of π and the square of the radius, r^2.

$$A = \pi r^2$$

Translate.

Area of circle with radius r	$-$	Area of circle with radius 5	$=$	Shaded area
$\pi \cdot r^2$	$-$	$\pi \cdot 5^2$	$=$	Shaded area

Carry out. We simplify the expression.

$$\pi \cdot r^2 - \pi \cdot 5^2 = \pi r^2 - 25\pi$$

Check. We can go over our calculations. We can also assign some value to r, say 7, and carry out the computation in two ways.

Difference of areas: $\pi \cdot 7^2 - \pi \cdot 5^2 = 49\pi - 25\pi = 24\pi$

Substituting in the polynomial: $\pi \cdot 7^2 - 25\pi = 49\pi - 25\pi = 24\pi$

Since the results are the same, our solution is probably correct.

State. A polynomial for the shaded area is $\pi r^2 - 25\pi$.

63. ***Familiarize***. We label the figure with additional information.

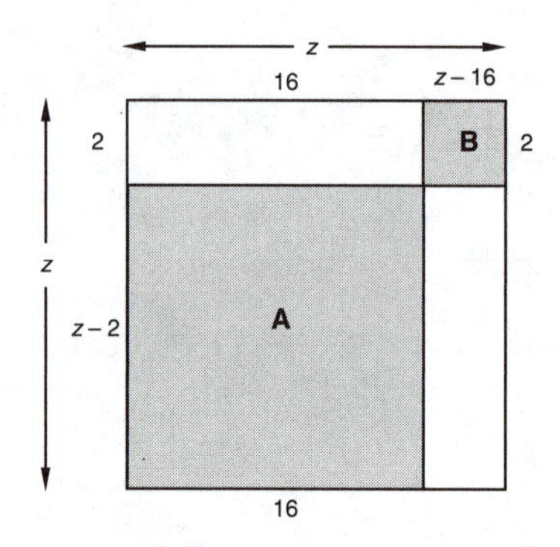

Translate.

Area of shaded sections = Area of A + Area of B

Area of shaded sections = $16(z-2)+2(z-16)$

Carry out. We simplify the expression.

$16(z-2)+2(z-16) = 16z-32+2z-32 = 18z-64$

Check. We can go over the calculations. We can also assign some value to z, say 30, and carry out the computation in two ways.

Sum of areas:

$16 \cdot 28 + 2 \cdot 14 = 448 + 28 = 476$

Substituting in the polynomial:

$18 \cdot 30 - 64 = 540 - 64 = 476$

Since the results are the same, our solution is probably correct.

State. A polynomial for the shaded area is $18z-64$.

65.
$$\begin{aligned} 8x+3x &= 66 \\ 11x &= 66 \quad \text{Collecting like terms} \\ \frac{11x}{11} &= \frac{66}{11} \quad \text{Dividing by 11} \\ x &= 6 \end{aligned}$$

The solution is 6.

67.
$$\begin{aligned} \frac{3}{8}x+\frac{1}{4}-\frac{3}{4}x &= \frac{11}{16}+x, \quad \text{LCM is 16} \\ 16\left(\frac{3}{8}x+\frac{1}{4}-\frac{3}{4}x\right) &= 16\left(\frac{11}{16}+x\right) \quad \text{Clearing fractions} \\ 6x+4-12x &= 11+16x \\ -6x+4 &= 11+16x \quad \text{Collecting like terms} \\ -6x+4-4 &= 11+16x-4 \quad \text{Subtracting 4} \\ -6x &= 7+16x \\ -6x-16x &= 7+16x-16x \quad \text{Subtracting } 16x \\ -22x &= 7 \\ \frac{-22x}{-22} &= \frac{7}{-22} \quad \text{Dividing by } -22 \\ x &= -\frac{7}{22} \end{aligned}$$

The solution is $-\frac{7}{22}$.

69.
$$\begin{aligned} 1.5x-2.7x &= 22-5.6x \\ 10(1.5x-2.7x) &= 10(22-5.6x) \quad \text{Clearing decimals} \\ 15x-27x &= 220-56x \\ -12x &= 220-56x \quad \text{Collecting like terms} \\ 44x &= 220 \quad \text{Adding } 56x \\ x &= \frac{220}{44} \quad \text{Dividing by 44} \\ x &= 5 \quad \text{Simplifying} \end{aligned}$$

The solution is 5.

71.
$$\begin{aligned} 6(y-3)-8 &= 4(y+2)+5 \\ 6y-18-8 &= 4y+8+5 \quad \text{Removing parentheses} \\ 6y-26 &= 4y+13 \quad \text{Collecting like terms} \\ 6y-26+26 &= 4y+13+26 \quad \text{Adding 26} \\ 6y &= 4y+39 \\ 6y-4y &= 4y+39-4y \quad \text{Subtracting } 4y \\ 2y &= 39 \\ \frac{2y}{2} &= \frac{39}{2} \quad \text{Dividing by 2} \\ y &= \frac{39}{2} \end{aligned}$$

The solution is $\frac{39}{2}$.

73.
$$\begin{aligned} 3x-7 &\le 5x+13 \\ -2x-7 &\le 13 \quad \text{Subtracting } 5x \\ -2x &\le 20 \quad \text{Adding 7} \\ x &\ge -10 \quad \text{Dividing by } -2 \text{ and reversing the inequality symbol} \end{aligned}$$

The solution set is $\{x|x \ge -10\}$.

75. ***Familiarize.*** The surface area is $2lw+2lh+2wh$, where l = length, w = width, and h = height of the rectangular solid. Here we have $l=3$, $w=w$, and $h=7$.

Translate. We substitute in the formula above.

$2 \cdot 3 \cdot w + 2 \cdot 3 \cdot 7 + 2 \cdot w \cdot 7$

Carry out. We simplify the expression.

$$\begin{aligned} &2 \cdot 3 \cdot w + 2 \cdot 3 \cdot 7 + 2 \cdot w \cdot 7 \\ &= 6w+42+14w \\ &= 20w+42 \end{aligned}$$

Check. We can go over the calculations. We can also assign some value to w, say 6, and carry out the computation in two ways.

Using the formula: $2 \cdot 3 \cdot 6 + 2 \cdot 3 \cdot 7 + 2 \cdot 6 \cdot 7 = 36+42+84 = 162$

Substituting in the polynomial: $20 \cdot 6 + 42 = 120 + 42 = 162$

Since the results are the same, our solution is probably correct.

State. A polynomial for the surface area is $20w+42$.

77. ***Familiarize.*** The surface area is $2lw+2lh+2wh$, where l = length, w = width, and h = height of the rectangular solid. Here we have $l=x$, $w=x$, and $h=5$.

Translate. We substitute in the formula above.

$2 \cdot x \cdot x + 2 \cdot x \cdot 5 + 2 \cdot x \cdot 5$

Carry out. We simplify the expression.

$$\begin{aligned} &2 \cdot x \cdot x + 2 \cdot x \cdot 5 + 2 \cdot x \cdot 5 \\ &= 2x^2+10x+10x \\ &= 2x^2+20x \end{aligned}$$

Check. We can go over the calculations. We can also assign some value to x, say 3, and carry out the computation in two ways.

Using the formula: $2 \cdot 3 \cdot 3 + 2 \cdot 3 \cdot 5 + 2 \cdot 3 \cdot 5 =$
$18 + 30 + 30 = 78$

Substituting in the polynomial: $2 \cdot 3^2 + 20 \cdot 3 =$
$2 \cdot 9 + 60 = 18 + 60 = 78$

Since the results are the same, our solution is probably correct.

State. A polynomial for the surface area is $2x^2 + 20x$.

79.

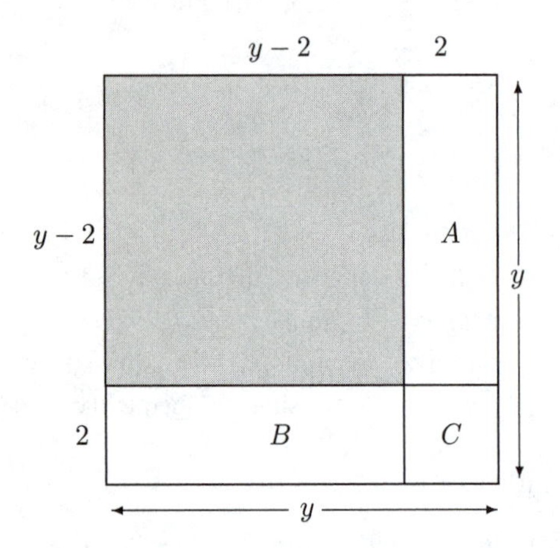

The shaded area is $(y-2)^2$. We find it as follows:

$$\begin{array}{ccccccccc} \text{Shaded area} & = & \text{Area of square} & - & \text{Area of } A & - & \text{Area of } B & - & \text{Area of } C \\ (y-2)^2 & = & y^2 & - & 2(y-2) & - & 2(y-2) & - & 2\cdot 2 \end{array}$$

$(y-2)^2 = y^2 - 2y + 4 - 2y + 4 - 4$

$(y-2)^2 = y^2 - 4y + 4$

81. $(7y^2 - 5y + 6) - (3y^2 + 8y - 12) + (8y^2 - 10y + 3)$
$= 7y^2 - 5y + 6 - 3y^2 - 8y + 12 + 8y^2 - 10y + 3$
$= 12y^2 - 23y + 21$

83. $(-y^4 - 7y^3 + y^2) + (-2y^4 + 5y - 2) - (-6y^3 + y^2)$
$= -y^4 - 7y^3 + y^2 - 2y^4 + 5y - 2 + 6y^3 - y^2$
$= -3y^4 - y^3 + 5y - 2$

Chapter 12 Mid-Chapter Review

1. True; $a^n \cdot a^{-n} = a^0 = 1$.

2. False; $x^2 \cdot x^3 = x^{2+3} = x^5 \neq x^6$.

3. False; the variable is raised to different powers, so these are not like terms.

4. True; $a^0 = 1$ for any nonzero number a.

5. $4w^3 + 6w - 8w^3 - 3w = (4-8)w^3 + (6-3)w$
$= -4w^3 + 3w$

6. $3y^4 - y^2 + 11 - (y^4 - 4y^2 + 5)$
$= 3y^4 - y^2 + 11 - y^4 + 4y^2 - 5$
$= 2y^4 + 3y^2 + 6$

7. $z^1 = z$

8. $4.56^0 = 1$

9. $a^5 = (-2)^5 = -32$

10. $-x^3 = -(-1)^3 = -(-1) = 1$

11. $5^3 \cdot 5^4 = 5^{3+4} = 5^7$

12. $(3a)^2(3a)^7 = (3a)^{2+7} = (3a)^9$

13. $x^{-8} \cdot x^5 = x^{-8+5} = x^{-3} = \dfrac{1}{x^3}$

14. $t^4 \cdot t^{-4} = t^{4+(-4)} = t^0 = 1$

15. $\dfrac{7^8}{7^4} = 7^{8-4} = 7^4$

16. $\dfrac{x}{x^3} = x^{1-3} = x^{-2} = \dfrac{1}{x^2}$

17. $\dfrac{w^5}{w^{-3}} = w^{5-(-3)} = w^{5+3} = w^8$

18. $\dfrac{y^{-6}}{y^{-2}} = y^{-6-(-2)} = y^{-6+2} = y^{-4} = \dfrac{1}{y^4}$

19. $(3^5)^3 = 3^{5\cdot 3} = 3^{15}$

20. $(x^{-3}y^2)^{-6} = x^{-3(-6)}y^{2(-6)} = x^{18}y^{-12} = \dfrac{x^{18}}{y^{12}}$

21. $\left(\dfrac{a^4}{5}\right)^6 = \dfrac{a^{4\cdot 6}}{5^6} = \dfrac{a^{24}}{5^6}$

22. $\left(\dfrac{2y^3}{xz^2}\right)^{-2} = \dfrac{2^{-2}y^{3(-2)}}{x^{-2}(z^2)^{-2}} = \dfrac{2^{-2}y^{-6}}{x^{-2}z^{-4}} = \dfrac{x^2z^4}{4y^6}$

23. 25,430,000

2.5,430,000.
7 places

Large number, so exponent is positive.

$25,430,000 = 2.543 \times 10^7$

24. 0.00012

0.0001.2
4 places

Small number, so the exponent is negative.

$0.00012 = 1.2 \times 10^{-4}$

25. 3.6×10^{-5}

Negative exponent, so the answer is a small number.

0.00003.6
5 places

$3.6 \times 10^{-5} = 0.000036$

26. 1.44×10^8

Positive exponent, so the answer is a large number.

1.44000000.
8 places

$1.44 \times 10^8 = 144,000,000$

27. $(3 \times 10^6)(2 \times 10^{-3}) = (3 \times 2) \times (10^6 \times 10^{-3}) =$
6×10^3

28. $\dfrac{1.2 \times 10^{-4}}{2.4 \times 10^2} = \dfrac{1.2}{2.4} \times \dfrac{10^{-4}}{10^2} = 0.5 \times 10^{-6} =$
$(5 \times 10^{-1}) \times 10^{-6} = 5 \times 10^{-7}$

29. $-3x + 7 = -3(-3) + 7 = 9 + 7 = 16;$
$-3x + 7 = -3 \cdot 2 + 7 = -6 + 7 = 1$

30. $x^3 - 2x + 5 = (-3)^3 - 2(-3) + 5 = -27 + 6 + 5 = -16;$
$x^3 - 2x + 5 = 2^3 - 2 \cdot 2 + 5 = 8 - 4 + 5 = 9$

31. $3x - 2x^5 + x - 5x^2 + 2 = (3 + 1)x - 2x^5 - 5x^2 + 2 =$
$4x - 2x^5 - 5x^2 + 2$, or $-2x^5 - 5x^2 + 4x + 2$

32. $4x^3 - 9x^2 - 2x^3 + x^2 + 8x^6 = (4-2)x^3 + (-9+1)x^2 + 8x^6 =$
$2x^3 - 8x^2 + 8x^6$, or $8x^6 - 2x^3 - 8x^2$

33. The degree of $5x^3$ is 3.
The degree of $-x$, or $-x^1$, is 1.
The degree of 4, or 4^0, is 0.
The degree of the polynomial is 3.

34. The degree of $2x$, or $2x^1$, is 1.
The degree of $-x^4$ is 4.
The degree of $3x^6$ is 6.
The degree of the polynomial is 6.

35. $x - 9$ has two terms. It is a binomial.

36. $x^5 - 2x^3 + 6x^2$ has three terms. It is a trinomial.

37. $(3x^2 - 1) + (5x^2 + 6) = (3 + 5)x^2 + (-1 + 6) = 8x^2 + 5$

38. $(x^3 + 2x - 5) + (4x^3 - 2x^2 - 6) =$
$(1 + 4)x^3 - 2x^2 + 2x + (-5 - 6) =$
$5x^3 - 2x^2 + 2x - 11$

39. $(5x - 8) - (9x + 2) = 5x - 8 - 9x - 2 = -4x - 10$

40.
$$\begin{aligned} &(0.1x^2 - 2.4x + 3.6) - (0.5x^2 + x - 5.4) \\ &= 0.1x^2 - 2.4x + 3.6 - 0.5x^2 - x + 5.4 \\ &= -0.4x^2 - 3.4x + 9 \end{aligned}$$

41. The areas of the rectangles are $3y$, y^2, and $2y^2$. The sum of the areas is $3y + y^2 + 2y^2 = 3y + 3y^2$.

42. Let s = the length of a side of the smaller square. Then $3s$ = the length of a side of the larger square. The area of the smaller square is s^2, and the area of the larger square is $(3s)^2$, or $9s^2$, so the area of the larger square is 9 times the area of the smaller square.

43. Let s = the width of the smaller cube. Then $2s$ = the width of the larger cube. The volume of the smaller cube is s^3, and the volume of the larger cube is $(2s)^3$, or $8s^3$, so the volume of the larger cube is 8 times the volume of the smaller cube.

44. Exponents are added when powers with like bases are multiplied. Exponents are multiplied when a power is raised to a power.

45. $3^{-29} = \dfrac{1}{3^{29}}$ and
$2^{-29} = \dfrac{1}{2^{29}}$. Since $3^{29} > 2^{29}$, we have $\dfrac{1}{3^{29}} < \dfrac{1}{2^{29}}$.

46. It is better to evaluate a polynomial after like terms have been collected, because there are fewer terms to evaluate.

47. Yes; consider the following.
$$(x^2 + 4) + (4x - 7) = x^2 + 4x - 3$$

Exercise Set 12.5

1. $(8x^2)(5) = (8 \cdot 5)x^2 = 40x^2$

3. $(-x^2)(-x) = (-1x^2)(-1x) = (-1)(-1)(x^2 \cdot x) = x^3$

5. $(8x^5)(4x^3) = (8 \cdot 4)(x^5 \cdot x^3) = 32x^8$

7. $(0.1x^6)(0.3x^5) = (0.1)(0.3)(x^6 \cdot x^5) = 0.03x^{11}$

9. $\left(-\dfrac{1}{5}x^3\right)\left(-\dfrac{1}{3}x\right) = \left(-\dfrac{1}{5}\right)\left(-\dfrac{1}{3}\right)(x^3 \cdot x) = \dfrac{1}{15}x^4$

11. $(-4x^2)(0) = 0$ Any number multiplied by 0 is 0.

13. $(3x^2)(-4x^3)(2x^6) = (3)(-4)(2)(x^2 \cdot x^3 \cdot x^6) = -24x^{11}$

15.
$$\begin{aligned} 2x(-x + 5) &= 2x(-x) + 2x(5) \\ &= -2x^2 + 10x \end{aligned}$$

17.
$$\begin{aligned} -5x(x - 1) &= -5x(x) - 5x(-1) \\ &= -5x^2 + 5x \end{aligned}$$

19.
$$\begin{aligned} x^2(x^3 + 1) &= x^2(x^3) + x^2(1) \\ &= x^5 + x^2 \end{aligned}$$

21.
$$\begin{aligned} 3x(2x^2 - 6x + 1) &= 3x(2x^2) + 3x(-6x) + 3x(1) \\ &= 6x^3 - 18x^2 + 3x \end{aligned}$$

23.
$$\begin{aligned} -6x^2(x^2 + x) &= -6x^2(x^2) - 6x^2(x) \\ &= -6x^4 - 6x^3 \end{aligned}$$

25.
$$\begin{aligned} 3y^2(6y^4 + 8y^3) &= 3y^2(6y^4) + 3y^2(8y^3) \\ &= 18y^6 + 24y^5 \end{aligned}$$

27.
$$\begin{aligned} (x + 6)(x + 3) &= (x + 6)x + (x + 6)3 \\ &= x \cdot x + 6 \cdot x + x \cdot 3 + 6 \cdot 3 \\ &= x^2 + 6x + 3x + 18 \\ &= x^2 + 9x + 18 \end{aligned}$$

29.
$$\begin{aligned} (x + 5)(x - 2) &= (x + 5)x + (x + 5)(-2) \\ &= x \cdot x + 5 \cdot x + x(-2) + 5(-2) \\ &= x^2 + 5x - 2x - 10 \\ &= x^2 + 3x - 10 \end{aligned}$$

31.
$$\begin{aligned} (x - 1)(x + 4) &= (x - 1)x + (x - 1)4 \\ &= x \cdot x - 1 \cdot x + x \cdot 4 - 1 \cdot 4 \\ &= x^2 - x + 4x - 4 \\ &= x^2 + 3x - 4 \end{aligned}$$

33. $(x-4)(x-3) = (x-4)x + (x-4)(-3)$
$= x \cdot x - 4 \cdot x + x(-3) - 4(-3)$
$= x^2 - 4x - 3x + 12$
$= x^2 - 7x + 12$

35. $(x+3)(x-3) = (x+3)x + (x+3)(-3)$
$= x \cdot x + 3 \cdot x + x(-3) + 3(-3)$
$= x^2 + 3x - 3x - 9$
$= x^2 - 9$

37. $(x-4)(x+4) = (x-4)x + (x-4)4$
$= x \cdot x - 4 \cdot x + x \cdot 4 - 4 \cdot 4$
$= x^2 - 4x + 4x - 16$
$= x^2 - 16$

39. $(5+3x)(2+5x) = (5+3x)2 + (5+3x)5x$
$= 5 \cdot 2 + 3x \cdot 2 + 5 \cdot 5x + 3x \cdot 5x$
$= 10 + 6x + 25x + 15x^2$
$= 10 + 31x + 15x^2$

41. $(5-x)(5-2x) = (5-x)5 + (5-x)(-2x)$
$= 5 \cdot 5 - x \cdot 5 + 5(-2x) - x(-2x)$
$= 25 - 5x - 10x + 2x^2$
$= 25 - 15x + 2x^2$

43. $(2x+5)(2x+5) = (2x+5)2x + (2x+5)5$
$= 2x \cdot 2x + 5 \cdot 2x + 2x \cdot 5 + 5 \cdot 5$
$= 4x^2 + 10x + 10x + 25$
$= 4x^2 + 20x + 25$

45. $(x-3)(x-3) = (x-3)x + (x-3)(-3)$
$= x \cdot x - 3 \cdot x + x(-3) - 3(-3)$
$= x^2 - 3x - 3x + 9$
$= x^2 - 6x + 9$

47. $\left(x - \frac{5}{2}\right)\left(x + \frac{2}{5}\right) = \left(x - \frac{5}{2}\right)x + \left(x - \frac{5}{2}\right)\frac{2}{5}$
$= x \cdot x - \frac{5}{2} \cdot x + x \cdot \frac{2}{5} - \frac{5}{2} \cdot \frac{2}{5}$
$= x^2 - \frac{5}{2}x + \frac{2}{5}x - 1$
$= x^2 - \frac{25}{10}x + \frac{4}{10}x - 1$
$= x^2 - \frac{21}{10}x - 1$

49. $(x-2.3)(x+4.7) = (x-2.3)x + (x-2.3)4.7$
$= x \cdot x - 2.3 \cdot x + x \cdot 4.7 - 2.3(4.7)$
$= x^2 - 2.3x + 4.7x - 10.81$
$= x^2 + 2.4x - 10.81$

51. The length of the rectangle is $x+6$ and the width is $x+2$, so the area is $(x+6)(x+2)$. If we carry out the multiplication, we have $x^2 + 8x + 12$.

53. The length of the rectangle is $x+6$ and the width is $x+1$, so the area is $(x+6)(x+1)$. If we carry out the multiplication, we have $x^2 + 7x + 6$.

55. Illustrate $x(x+5)$ as the area of a rectangle with width x and length $x+5$.

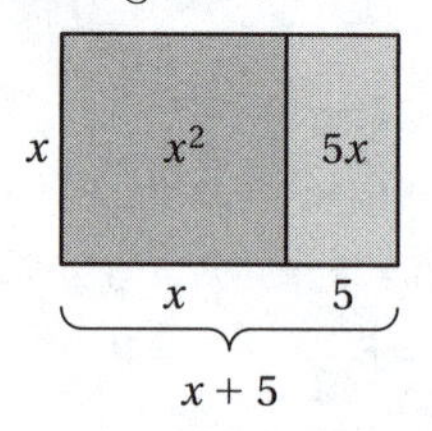

57. Illustrate $(x+1)(x+2)$ as the area of a rectangle with width $x+1$ and length $x+2$.

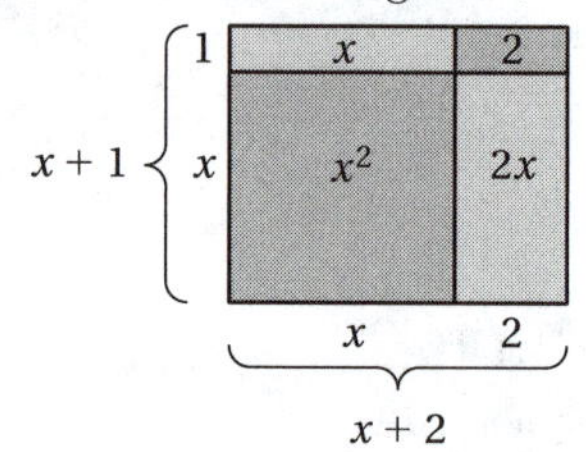

59. Illustrate $(x+5)(x+3)$ as the area of a rectangle with length $x+5$ and width $x+3$.

3 | 3x | 15
x + 3 | x | x² | 5x
x | 5
x + 5

61. $(x^2 + x + 1)(x - 1)$
$= (x^2 + x + 1)x + (x^2 + x + 1)(-1)$
$= x^2 \cdot x + x \cdot x + 1 \cdot x + x^2(-1) + x(-1) + 1(-1)$
$= x^3 + x^2 + x - x^2 - x - 1$
$= x^3 - 1$

63. $(2x+1)(2x^2 + 6x + 1)$
$= 2x(2x^2 + 6x + 1) + 1(2x^2 + 6x + 1)$
$= 2x \cdot 2x^2 + 2x \cdot 6x + 2x \cdot 1 + 1 \cdot 2x^2 + 1 \cdot 6x + 1 \cdot 1$
$= 4x^3 + 12x^2 + 2x + 2x^2 + 6x + 1$
$= 4x^3 + 14x^2 + 8x + 1$

65. $(y^2 - 3)(3y^2 - 6y + 2)$
$= y^2(3y^2 - 6y + 2) - 3(3y^2 - 6y + 2)$
$= y^2 \cdot 3y^2 + y^2(-6y) + y^2 \cdot 2 - 3 \cdot 3y^2 - 3(-6y) - 3 \cdot 2$
$= 3y^4 - 6y^3 + 2y^2 - 9y^2 + 18y - 6$
$= 3y^4 - 6y^3 - 7y^2 + 18y - 6$

67. $(x^3 + x^2)(x^3 + x^2 - x)$
$= x^3(x^3 + x^2 - x) + x^2(x^3 + x^2 - x)$
$= x^3 \cdot x^3 + x^3 \cdot x^2 + x^3(-x) + x^2 \cdot x^3 + x^2 \cdot x^2 + x^2(-x)$
$= x^6 + x^5 - x^4 + x^5 + x^4 - x^3$
$= x^6 + 2x^5 - x^3$

69. $(-5x^3-7x^2+1)(2x^2-x)$
$= (-5x^3-7x^2+1)2x^2+(-5x^3-7x^2+1)(-x)$
$= -5x^3\cdot 2x^2 - 7x^2\cdot 2x^2 + 1\cdot 2x^2 - 5x^3(-x) - 7x^2(-x) + 1(-x)$
$= -10x^5-14x^4+2x^2+5x^4+7x^3-x$
$= -10x^5-9x^4+7x^3+2x^2-x$

71.
$$\begin{array}{rl}
1+x+x^2 & \text{Line up like terms} \\
-1-x+x^2 & \text{in columns} \\
\hline
x^2+x^3+x^4 & \text{Multiplying the top row by } x^2 \\
-x-x^2-x^3 & \text{Multiplying by } -x \\
-1-x-x^2 & \text{Multiplying by } -1 \\
\hline
-1-2x-x^2\qquad+x^4 &
\end{array}$$

73.
$$\begin{array}{rl}
2t^2-t-4 & \\
3t^2+2t-1 & \\
\hline
-2t^2+t+4 & \text{Multiplying by } -1 \\
4t^3-2t^2-8t & \text{Multiplying by } 2t \\
6t^4-3t^3-12t^2 & \text{Multiplying by } 3t^2 \\
\hline
6t^4+t^3-16t^2-7t+4 &
\end{array}$$

75.
$$\begin{array}{rl}
x\quad -x^3\quad +x^5 & \\
-1+x^2\quad+x^4 & \text{Rewriting in ascending order} \\
\hline
x^5-x^7+x^9 & \text{Multiplying by } x^4 \\
x^3-x^5+x^7 & \text{Multiplying by } x^2 \\
-x+x^3-x^5 & \text{Multiplying by } -1 \\
\hline
-x+2x^3-x^5\qquad+x^9 &
\end{array}$$

77.
$$\begin{array}{r}
x^3+x^2+x+1 \\
x-1 \\
\hline
-x^3-x^2-x-1 \\
x^4+x^3+x^2+x \\
\hline
x^4\qquad\qquad-1
\end{array}$$

79. We will multiply horizontally while still aligning like terms.

$(x+1)(x^3+7x^2+5x+4)$

$$\begin{array}{rl}
= x^4+7x^3+5x^2+4x & \text{Multiplying by } x \\
+x^3+7x^2+5x+4 & \text{Multiplying by } 1 \\
\hline
= x^4+8x^3+12x^2+9x+4 &
\end{array}$$

81. We will multiply horizontally while still aligning like terms.

$\left(x-\frac{1}{2}\right)\left(2x^3-4x^2+3x-\frac{2}{5}\right)$

$$\begin{array}{r}
= 2x^4-4x^3+3x^2-\frac{2}{5}x\qquad \\
-x^3+2x^2-\frac{3}{2}x+\frac{1}{5} \\
\hline
2x^4-5x^3+5x^2-\frac{19}{10}x+\frac{1}{5}
\end{array}$$

83. $-\frac{1}{4}-\frac{1}{2}=-\frac{1}{4}-\frac{1}{2}\cdot\frac{2}{2}=-\frac{1}{4}-\frac{2}{4}=-\frac{3}{4}$

85. $(10-2)(10+2)=8\cdot 12=96$

87. $15x-18y+12=3\cdot 5x-3\cdot 6y+3\cdot 4=$
$3(5x-6y+4)$

89. $-9x-45y+15=-3\cdot 3x-3\cdot 15y-3(-5)=$
$-3(3x+15y-5)$

91. $y=\frac{1}{2}x-3$

The equation is equivalent to $y=\frac{1}{2}x+(-3)$. The y-intercept is $(0,-3)$. We find two other points, using multiples of 2 for x to avoid fractions.

When $x=-2$, $y=\frac{1}{2}(-2)-3=-1-3=-4$.

When $x=4$, $y=\frac{1}{2}\cdot 4-3=2-3=-1$.

x	y
0	-3
-2	-4
4	-1

Plot these points, draw the line they determine, and label the graph $y=\frac{1}{2}x-3$.

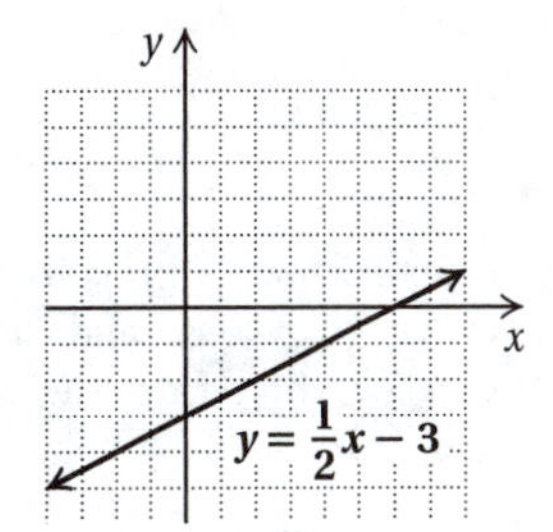

93. The shaded area is the area of the large rectangle, $6y(14y-5)$ less the area of the unshaded rectangle, $3y(3y+5)$. We have:

$6y(14y-5)-3y(3y+5)$
$=84y^2-30y-9y^2-15y$
$=75y^2-45y$

95.

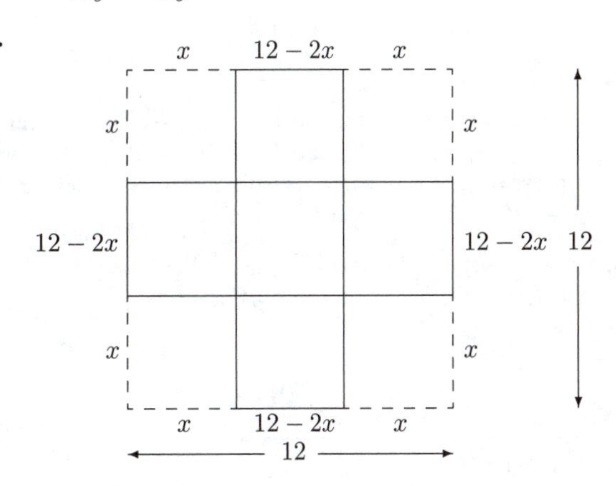

The dimensions, in inches, of the box are $12-2x$ by $12-2x$ by x. The volume is the product of the dimensions (volume = length × width × height):

Volume $=(12-2x)(12-2x)x$
$=(144-48x+4x^2)x$
$=(144x-48x^2+4x^3)\text{ in}^3$, or
$(4x^3-48x^2+144x)\text{ in}^3$

The outside surface area is the sum of the area of the bottom and the areas of the four sides. The dimensions, in inches, of the bottom are $12 - 2x$ by $12 - 2x$, and the dimensions, in inches, of each side are x by $12 - 2x$.

$$\text{Surface area} = \text{Area of bottom} + 4 \cdot \text{Area of each side}$$
$$= (12 - 2x)(12 - 2x) + 4 \cdot x(12 - 2x)$$
$$= 144 - 24x - 24x + 4x^2 + 48x - 8x^2$$
$$= 144 - 48x + 4x^2 + 48x - 8x^2$$
$$= (144 - 4x^2) \text{ in}^2, \text{ or } (-4x^2 + 144) \text{ in}^2$$

97. Let n = the missing number.

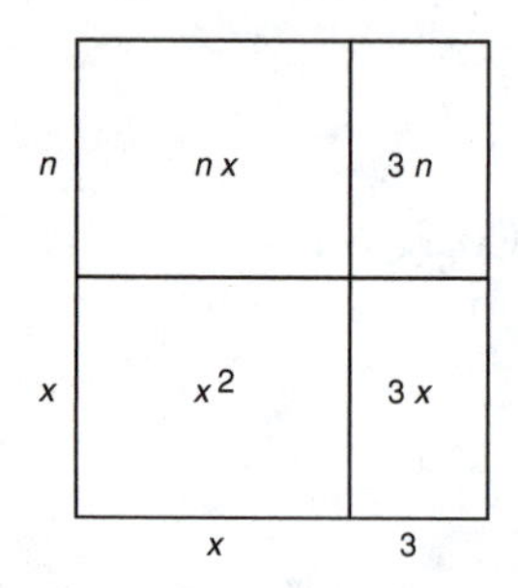

The area of the figure is $x^2+3x+nx+3n$. This is equivalent to $x^2 + 8x + 15$, so we have $3x + nx = 8x$ and $3n = 15$. Solving either equation for n, we find that the missing number is 5.

99. We have a rectangular solid with dimensions x m by x m by $x+2$ m with a rectangular solid piece with dimensions 6 m by 5 m by 7 m cut out of it.

$$\text{Volume} = \text{Volume of large solid} - \text{Volume of small solid}$$
$$= (x \text{ m})(x \text{ m})(x + 2 \text{ m}) - (6 \text{ m})(5 \text{ m})(7 \text{ m})$$
$$= x^2(x + 2) \text{ m}^3 - 210 \text{ m}^3$$
$$= (x^3 + 2x^2 - 210) \text{ m}^3$$

101. $(x - 2)(x - 7) - (x - 7)(x - 2)$

First observe that, by the commutative law of multiplication, $(x - 2)(x - 7)$ and $(x - 7)(x - 2)$ are equivalent expressions. Then when we subtract $(x - 7)(x - 2)$ from $(x - 2)(x - 7)$, the result is 0.

103.
$$(x - a)(x - b) \cdots (x - x)(x - y)(x - z)$$
$$= (x - a)(x - b) \cdots 0 \cdot (x - y)(x - z)$$
$$= 0$$

Exercise Set 12.6

1. $(x + 1)(x^2 + 3)$

F O I L

$$= x \cdot x^2 + x \cdot 3 + 1 \cdot x^2 + 1 \cdot 3$$
$$= x^3 + 3x + x^2 + 3$$

3. $(x^3 + 2)(x + 1)$

F O I L

$$= x^3 \cdot x + x^3 \cdot 1 + 2 \cdot x + 2 \cdot 1$$
$$= x^4 + x^3 + 2x + 2$$

5. $(y + 2)(y - 3)$

F O I L

$$= y \cdot y + y \cdot (-3) + 2 \cdot y + 2 \cdot (-3)$$
$$= y^2 - 3y + 2y - 6$$
$$= y^2 - y - 6$$

7. $(3x + 2)(3x + 2)$

F O I L

$$= 3x \cdot 3x + 3x \cdot 2 + 2 \cdot 3x + 2 \cdot 2$$
$$= 9x^2 + 6x + 6x + 4$$
$$= 9x^2 + 12x + 4$$

9. $(5x - 6)(x + 2)$

F O I L

$$= 5x \cdot x + 5x \cdot 2 + (-6) \cdot x + (-6) \cdot 2$$
$$= 5x^2 + 10x - 6x - 12$$
$$= 5x^2 + 4x - 12$$

11. $(3t - 1)(3t + 1)$

F O I L

$$= 3t \cdot 3t + 3t \cdot 1 + (-1) \cdot 3t + (-1) \cdot 1$$
$$= 9t^2 + 3t - 3t - 1$$
$$= 9t^2 - 1$$

13. $(4x - 2)(x - 1)$

F O I L

$$= 4x \cdot x + 4x \cdot (-1) + (-2) \cdot x + (-2) \cdot (-1)$$
$$= 4x^2 - 4x - 2x + 2$$
$$= 4x^2 - 6x + 2$$

15. $\left(p - \frac{1}{4}\right)\left(p + \frac{1}{4}\right)$

F O I L

$$= p \cdot p + p \cdot \frac{1}{4} + \left(-\frac{1}{4}\right) \cdot p + \left(-\frac{1}{4}\right) \cdot \frac{1}{4}$$
$$= p^2 + \frac{1}{4}p - \frac{1}{4}p - \frac{1}{16}$$
$$= p^2 - \frac{1}{16}$$

17. $(x - 0.1)(x + 0.1)$

F O I L

$$= x \cdot x + x \cdot (0.1) + (-0.1) \cdot x + (-0.1)(0.1)$$
$$= x^2 + 0.1x - 0.1x - 0.01$$
$$= x^2 - 0.01$$

19. $(2x^2 + 6)(x + 1)$

F O I L

$$= 2x^3 + 2x^2 + 6x + 6$$

21. $(-2x + 1)(x + 6)$

F O I L

$$= -2x^2 - 12x + x + 6$$
$$= -2x^2 - 11x + 6$$

23. $(a+7)(a+7)$
F O I L
$= a^2 + 7a + 7a + 49$
$= a^2 + 14a + 49$

25. $(1+2x)(1-3x)$
F O I L
$= 1 - 3x + 2x - 6x^2$
$= 1 - x - 6x^2$

27. $\left(\frac{3}{8}y - \frac{5}{6}\right)\left(\frac{3}{8}y - \frac{5}{6}\right)$
F O I L
$= \frac{9}{64}y^2 - \frac{15}{48}y - \frac{15}{48}y + \frac{25}{36}$
$= \frac{9}{64}y^2 - \frac{30}{48}y + \frac{25}{36}$
$= \frac{9}{64}y^2 - \frac{5}{8}y + \frac{25}{36}$

29. $(x^2+3)(x^3-1)$
F O I L
$= x^5 - x^2 + 3x^3 - 3$

31. $(3x^2-2)(x^4-2)$
F O I L
$= 3x^6 - 6x^2 - 2x^4 + 4$

33. $(2.8x-1.5)(4.7x+9.3)$
F O I L
$= 2.8x(4.7x) + 2.8x(9.3) - 1.5(4.7x) - 1.5(9.3)$
$= 13.16x^2 + 26.04x - 7.05x - 13.95$
$= 13.16x^2 + 18.99x - 13.95$

35. $(3x^5+2)(2x^2+6)$
F O I L
$= 6x^7 + 18x^5 + 4x^2 + 12$

37. $(8x^3+1)(x^3+8)$
F O I L
$= 8x^6 + 64x^3 + x^3 + 8$
$= 8x^6 + 65x^3 + 8$

39. $(4x^2+3)(x-3)$
F O I L
$= 4x^3 - 12x^2 + 3x - 9$

41. $(4y^4+y^2)(y^2+y)$
F O I L
$= 4y^6 + 4y^5 + y^4 + y^3$

43. $(x+4)(x-4)$ Product of sum and difference of two terms
$= x^2 - 4^2$
$= x^2 - 16$

45. $(2x+1)(2x-1)$ Product of sum and difference of two terms
$= (2x)^2 - 1^2$
$= 4x^2 - 1$

47. $(5m-2)(5m+2)$ Product of sum and difference of two terms
$= (5m)^2 - 2^2$
$= 25m^2 - 4$

49. $(2x^2+3)(2x^2-3)$ Product of sum and difference of two terms
$= (2x^2)^2 - 3^2$
$= 4x^4 - 9$

51. $(3x^4-4)(3x^4+4)$
$= (3x^4)^2 - 4^2$
$= 9x^8 - 16$

53. $(x^6-x^2)(x^6+x^2)$
$= (x^6)^2 - (x^2)^2$
$= x^{12} - x^4$

55. $(x^4+3x)(x^4-3x)$
$= (x^4)^2 - (3x)^2$
$= x^8 - 9x^2$

57. $(x^{12}-3)(x^{12}+3)$
$= (x^{12})^2 - 3^2$
$= x^{24} - 9$

59. $(2y^8+3)(2y^8-3)$
$= (2y^8)^2 - 3^2$
$= 4y^{16} - 9$

61. $\left(\frac{5}{8}x - 4.3\right)\left(\frac{5}{8}x + 4.3\right)$
$= \left(\frac{5}{8}x\right)^2 - (4.3)^2$
$= \frac{25}{64}x^2 - 18.49$

63. $(x+2)^2 = x^2 + 2 \cdot x \cdot 2 + 2^2$ Square of a binomial sum
$= x^2 + 4x + 4$

65. $(3x^2+1)$ Square of a binomial sum
$= (3x^2)^2 + 2 \cdot 3x^2 \cdot 1 + 1^2$
$= 9x^4 + 6x^2 + 1$

67. $\left(a - \frac{1}{2}\right)^2$ Square of a binomial sum
$= a^2 - 2 \cdot a \cdot \frac{1}{2} + \left(\frac{1}{2}\right)^2$
$= a^2 - a + \frac{1}{4}$

69. $(3+x)^2 = 3^2 + 2\cdot 3\cdot x + x^2$
$= 9 + 6x + x^2$

71. $(x^2+1)^2 = (x^2)^2 + 2\cdot x^2\cdot 1 + 1^2$
$= x^4 + 2x^2 + 1$

73. $(2-3x^4)^2 = 2^2 - 2\cdot 2\cdot 3x^4 + (3x^4)^2$
$= 4 - 12x^4 + 9x^8$

75. $(5+6t^2)^2 = 5^2 + 2\cdot 5\cdot 6t^2 + (6t^2)^2$
$= 25 + 60t^2 + 36t^4$

77. $\left(x-\frac{5}{8}\right)^2 = x^2 - 2\cdot x\cdot\frac{5}{8} + \left(\frac{5}{8}\right)^2$
$= x^2 - \frac{5}{4}x + \frac{25}{64}$

79. $(3-2x^3)^2 = 3^2 - 2\cdot 3\cdot 2x^3 + (2x^3)^2$
$= 9 - 12x^3 + 4x^6$

81. $4x(x^2+6x-3)$ Product of a monomial and a trinomial
$= 4x\cdot x^2 + 4x\cdot 6x + 4x(-3)$
$= 4x^3 + 24x^2 - 12x$

83. $\left(2x^2-\frac{1}{2}\right)\left(2x^2-\frac{1}{2}\right)$ Square of a binomial difference
$= (2x^2)^2 - 2\cdot 2x^2\cdot\frac{1}{2} + \left(\frac{1}{2}\right)^2$
$= 4x^4 - 2x^2 + \frac{1}{4}$

85. $(-1+3p)(1+3p)$
$= (3p-1)(3p+1)$ Product of the sum and difference of two terms
$= (3p)^2 - 1^2$
$= 9p^2 - 1$

87. $3t^2(5t^3 - t^2 + t)$ Product of a monomial and a trinomial
$= 3t^2\cdot 5t^3 + 3t^2(-t^2) + 3t^2\cdot t$
$= 15t^5 - 3t^4 + 3t^3$

89. $(6x^4+4)^2$ Square of a binomial sum
$= (6x^4)^2 + 2\cdot 6x^4\cdot 4 + 4^2$
$= 36x^8 + 48x^4 + 16$

91. $(3x+2)(4x^2+5)$ Product of two binomials; use FOIL
$= 3x\cdot 4x^2 + 3x\cdot 5 + 2\cdot 4x^2 + 2\cdot 5$
$= 12x^3 + 15x + 8x^2 + 10$

93. $(8-6x^4)^2$ Square of a binomial difference
$= 8^2 - 2\cdot 8\cdot 6x^4 + (6x^4)^2$
$= 64 - 96x^4 + 36x^8$

95.

$$\begin{array}{r} t^2+t+1 \\ t-1 \\ \hline -t^2-t-1 \\ t^3+t^2+t \quad \\ \hline t^3 \qquad\quad -1 \end{array}$$

97. $3^2 + 4^2 = 9 + 16 = 25$
$(3+4)^2 = 7^2 = 49$

99. $9^2 - 5^2 = 81 - 25 = 56$
$(9-5)^2 = 4^2 = 16$

101.

	a	1
1	A	B
a	D	C

We can find the shaded area in two ways.

Method 1: The figure is a square with side $a+1$, so the area is $(a+1)^2 = a^2 + 2a + 1$.

Method 2: We add the areas of A, B, C, and D.
$1\cdot a + 1\cdot 1 + 1\cdot a + a\cdot a = a + 1 + a + a^2 = a^2 + 2a + 1$.

Either way we find that the total shaded area is $a^2 + 2a + 1$.

103.

	t	6
t	A	B
4	D	C

We can find the shaded area in two ways.

Method 1: The figure is a rectangle with dimensions $t+6$ by $t+4$, so the area is $(t+6)(t+4) = t^2 + 4t + 6t + 24 = t^2 + 10t + 24$.

Method 2: We add the areas of A, B, C, and D.
$t\cdot t + t\cdot 6 + 6\cdot 4 + 4\cdot t = t^2 + 6t + 24 + 4t = t^2 + 10t + 24$.

Either way, we find that the total shaded area is $t^2 + 10t + 24$.

105. ***Familiarize***. Let t = the number of watts used by the television set. Then $10t$ = the number of watts used by the lamps, and $40t$ = the number of watts used by the air conditioner.

Translate.

$$\underbrace{\text{Lamp watts}} + \underbrace{\text{Air conditioner watts}} + \underbrace{\text{Television watts}} = \underbrace{\text{Total watts}}$$
$$10t + 40t + t = 2550$$

Solve. We solve the equation.

$$10t + 40t + t = 2550$$
$$51t = 2550$$
$$t = 50$$

The possible solution is:

Television, t: 50 watts

Lamps, $10t$: $10 \cdot 50$, or 500 watts

Air conditioner, $40t$: $40 \cdot 50$, or 2000 watts

Check. The number of watts used by the lamps, 500, is 10 times 50, the number used by the television. The number of watts used by the air conditioner, 2000, is 40 times 50, the number used by the television. Also, $50+500+2000 = 2550$, the total wattage used.

State. The television uses 50 watts, the lamps use 500 watts, and the air conditioner uses 2000 watts.

107. $3(x-2) = 5(2x+7)$

$3x - 6 = 10x + 35$ Removing parentheses

$3x - 6 + 6 = 10x + 35 + 6$ Adding 6

$3x = 10x + 41$

$3x - 10x = 10x + 41 - 10x$ Subtracting $10x$

$-7x = 41$

$\dfrac{-7x}{-7} = \dfrac{41}{-7}$ Dividing by -7

$x = -\dfrac{41}{7}$

The solution is $-\dfrac{41}{7}$.

109. $3x - 2y = 12$

$-2y = -3x + 12$ Subtracting $3x$

$\dfrac{-2y}{-2} = \dfrac{-3x+12}{-2}$ Dividing by -2

$y = \dfrac{3x-12}{2}$, or

$y = \dfrac{3}{2}x - 6$

111. $5x(3x-1)(2x+3)$

$= 5x(6x^2 + 7x - 3)$ Using FOIL

$= 30x^3 + 35x^2 - 15x$

113. $[(a-5)(a+5)]^2$

$= (a^2 - 25)^2$ Finding the product of a sum and difference of same two terms

$= a^4 - 50a^2 + 625$ Squaring a binomial

115. $(3t^4 - 2)^2(3t^4 + 2)^2$

$= [(3t^4 - 2)(3t^4 + 2)]^2$

$= (9t^8 - 4)^2$

$= 81t^{16} - 72t^8 + 16$

117. $(x+2)(x-5) = (x+1)(x-3)$

$x^2 - 5x + 2x - 10 = x^2 - 3x + x - 3$

$x^2 - 3x - 10 = x^2 - 2x - 3$

$-3x - 10 = -2x - 3$ Adding $-x^2$

$-3x + 2x = 10 - 3$ Adding $2x$ and 10

$-x = 7$

$x = -7$

The solution is -7.

119. See the answer section in the text.

121. Enter $y_1 = (x-1)^2$ and $y_2 = x^2 - 2x + 1$. Then compare the graphs or the y_1-and y_2-values in a table. It appears that the graphs are the same and that the y_1-and y_2-values are the same, so $(x-1)^2 = x^2 - 2x + 1$ is correct.

123. Enter $y_1 = (x-3)(x+3)$ and $y_2 = x^2 - 6$. Then compare the graphs or the y_1-and y_2-values in a table. The graphs are not the same nor are the y_1-and y_2-values, so $(x-3)(x+3) = x^2 - 6$ is not correct.

Exercise Set 12.7

1. We replace x by 3 and y by -2.

$x^2 - y^2 + xy = 3^2 - (-2)^2 + 3(-2) = 9 - 4 - 6 = -1$

3. We replace x by 3 and y by -2.

$x^2 - 3y^2 + 2xy = 3^2 - 3(-2)^2 + 2 \cdot 3(-2) =$
$9 - 3 \cdot 4 + 2 \cdot 3(-2) = 9 - 12 - 12 = -15$

5. We replace x by 3, y by -2, and z by -5.

$8xyz = 8 \cdot 3 \cdot (-2) \cdot (-5) = 240$

7. We replace x by 3, y by -2, and z by -5.

$xyz^2 - z = 3(-2)(-5)^2 - (-5) = 3(-2)(25) - (-5) =$
$-150 + 5 = -145$

9. We replace h by 165 and A by 20.

$C = 0.041h - 0.018A - 2.69$

$= 0.041(165) - 0.018(20) - 2.69$

$= 6.765 - 0.36 - 2.69$

$= 6.405 - 2.69$

$= 3.715$

The lung capacity of a 20-year-old person who is 165 cm tall is 3.715 liters.

11. Evaluate the polynomial for $h = 160$, $v = 30$, and $t = 3$.

$h = h_0 + vt - 4.9t^2$

$= 160 + 30 \cdot 3 - 4.9(3)^2$

$= 160 + 90 - 44.1$

$= 205.9$

The ball will be 205.9 m above the ground 3 seconds after it is thrown.

13. Replace h by 4.7, r by 1.2, and π by 3.14.

$$\begin{aligned} S &= 2\pi rh + 2\pi r^2 \\ &\approx 2(3.14)(1.2)(4.7) + 2(3.14)(1.2)^2 \\ &\approx 2(3.14)(1.2)(4.7) + 2(3.14)(1.44) \\ &\approx 35.4192 + 9.0432 \\ &\approx 44.46 \end{aligned}$$

The surface area of the can is about 44.46 in^2.

15. Evaluate the polynomial for $h = 7\frac{1}{2}$, or $\frac{15}{2}$, $r = 1\frac{1}{4}$, or $\frac{5}{4}$, and $\pi \approx 3.14$.

$$\begin{aligned} S &= 2\pi rh + \pi r^2 \\ &\approx 2(3.14)\left(\frac{5}{4}\right)\left(\frac{15}{2}\right) + (3.14)\left(\frac{5}{4}\right)^2 \\ &\approx 2(3.14)\left(\frac{5}{4}\right)\left(\frac{15}{2}\right) + (3.14)\left(\frac{25}{16}\right) \\ &\approx 58.875 + 4.90625 \\ &\approx 63.78125 \end{aligned}$$

The surface area is about 63.78125 in^2.

17. $x^3y - 2xy + 3x^2 - 5$

Term	Coefficient	Degree	
x^3y	1	4	(Think: $x^3y = x^3y^1$)
$-2xy$	-2	2	(Think: $-2xy = -2x^1y^1$)
$3x^2$	3	2	
-5	-5	0	(Think: $-5 = -5x^0$)

The degree of the polynomial is the degree of the term of highest degree. The term of highest degree is x^3y. Its degree is 4. The degree of the polynomial is 4.

19. $17x^2y^3 - 3x^3yz - 7$

Term	Coefficient	Degree	
$17x^2y^3$	17	5	
$-3x^3yz$	-3	5	(Think: $-3x^3yz = -3x^3y^1z^1$)
-7	-7	0	(Think: $-7 = -7x^0$)

The terms of highest degree are $17x^2y^3$ and $-3x^3yz$. Each has degree 5. The degree of the polynomial is 5.

21. $a + b - 2a - 3b = (1-2)a + (1-3)b = -a - 2b$

23. $3x^2y - 2xy^2 + x^2$

There are *no* like terms, so none of the terms can be collected.

25.
$$\begin{aligned} &6au + 3av + 14au + 7av \\ &= (6+14)au + (3+7)av \\ &= 20au + 10av \end{aligned}$$

27.
$$\begin{aligned} &2u^2v - 3uv^2 + 6u^2v - 2uv^2 \\ &= (2+6)u^2v + (-3-2)uv^2 \\ &= 8u^2v - 5uv^2 \end{aligned}$$

29.
$$\begin{aligned} &(2x^2 - xy + y^2) + (-x^2 - 3xy + 2y^2) \\ &= (2-1)x^2 + (-1-3)xy + (1+2)y^2 \\ &= x^2 - 4xy + 3y^2 \end{aligned}$$

31.
$$\begin{aligned} &(r - 2s + 3) + (2r + s) + (s + 4) \\ &= (1+2)r + (-2+1+1)s + (3+4) \\ &= 3r + 0s + 7 \\ &= 3r + 7 \end{aligned}$$

33.
$$\begin{aligned} &(b^3a^2 - 2b^2a^3 + 3ba + 4) + (b^2a^3 - 4b^3a^2 + 2ba - 1) \\ &= (1-4)b^3a^2 + (-2+1)b^2a^3 + (3+2)ba + (4-1) \\ &= -3b^3a^2 - b^2a^3 + 5ba + 3 \end{aligned}$$

35.
$$\begin{aligned} &(a^3 + b^3) - (a^2b - ab^2 + b^3 + a^3) \\ &= a^3 + b^3 - a^2b + ab^2 - b^3 - a^3 \\ &= (1-1)a^3 - a^2b + ab^2 + (1-1)b^3 \\ &= -a^2b + ab^2, \text{ or } ab^2 - a^2b \end{aligned}$$

37.
$$\begin{aligned} &(xy - ab - 8) - (xy - 3ab - 6) \\ &= xy - ab - 8 - xy + 3ab + 6 \\ &= (1-1)xy + (-1+3)ab + (-8+6) \\ &= 2ab - 2 \end{aligned}$$

39.
$$\begin{aligned} &(-2a + 7b - c) - (-3b + 4c - 8d) \\ &= -2a + 7b - c + 3b - 4c + 8d \\ &= -2a + (7+3)b + (-1-4)c + 8d \\ &= -2a + 10b - 5c + 8d \end{aligned}$$

41.
$$\begin{aligned} (3z - u)(2z + 3u) &= \overset{F}{6z^2} + \overset{O}{9zu} - \overset{I}{2uz} - \overset{L}{3u^2} \\ &= 6z^2 + 7zu - 3u^2 \end{aligned}$$

43.
$$\begin{aligned} (a^2b - 2)(a^2b - 5) &= \overset{F}{a^4b^2} - \overset{O}{5a^2b} - \overset{I}{2a^2b} + \overset{L}{10} \\ &= a^4b^2 - 7a^2b + 10 \end{aligned}$$

45.
$$\begin{aligned} (a^3 + bc)(a^3 - bc) &= (a^3)^2 - (bc)^2 \\ &\quad [(A+B)(A-B) = A^2 - B^2] \\ &= a^6 - b^2c^2 \end{aligned}$$

47.
$$\begin{array}{r} y^4x + y^2 + 1 \\ y^2 + 1 \\ \hline y^4x + y^2 + 1 \\ y^6x + y^4 \qquad + y^2 \\ \hline y^6x + y^4 + y^4x + 2y^2 + 1 \end{array}$$

49.
$$\begin{aligned} &(3xy - 1)(4xy + 2) \\ &\quad\ \ F \qquad\ O \qquad I \quad\ L \\ &= 12x^2y^2 + 6xy - 4xy - 2 \\ &= 12x^2y^2 + 2xy - 2 \end{aligned}$$

51.
$$\begin{aligned} &(3 - c^2d^2)(4 + c^2d^2) \\ &\quad\ F \qquad O \qquad\ I \qquad L \\ &= 12 + 3c^2d^2 - 4c^2d^2 - c^4d^4 \\ &= 12 - c^2d^2 - c^4d^4 \end{aligned}$$

53.
$$\begin{aligned} &(m^2 - n^2)(m + n) \\ &\quad\ F \qquad O \qquad I \qquad L \\ &= m^3 + m^2n - mn^2 - n^3 \end{aligned}$$

55.
$$\begin{aligned} &(xy + x^5y^5)(x^4y^4 - xy) \\ &\quad\ F \qquad O \qquad I \qquad L \\ &= x^5y^5 - x^2y^2 + x^9y^9 - x^6y^6 \\ &= x^9y^9 - x^6y^6 + x^5y^5 - x^2y^2 \end{aligned}$$

57.
$$\begin{aligned} &(x + h)^2 \\ &= x^2 + 2xh + h^2 \quad [(A+B)^2 = A^2 + 2AB + B^2] \end{aligned}$$

59. $(3a+2b)^2$

$= (3a)^2 + 2\cdot 3a\cdot 2b + (2b)^2$

$[(A+B)^2 = A^2 + 2AB + B^2]$

$= 9a^2 + 12ab + 4b^2$

61. $(r^3t^2-4)^2$

$= (r^3t^2)^2 - 2\cdot r^3t^2\cdot 4 + 4^2$

$[(A-B)^2 = A^2 - 2AB + B^2]$

$= r^6t^4 - 8r^3t^2 + 16$

63. $(p^4+m^2n^2)^2$

$= (p^4)^2 + 2\cdot p^4\cdot m^2n^2 + (m^2n^2)^2$

$[(A+B)^2 = A^2 + 2AB + B^2]$

$= p^8 + 2p^4m^2n^2 + m^4n^4$

65. $3a(a-2b)^2 = 3a(a^2-4ab+4b^2)$

$= 3a^3 - 12a^2b + 12ab^2$

67. $(m+n-3)^2$

$= [(m+n)-3]^2$

$= (m+n)^2 - 2(m+n)(3) + 3^2$

$= m^2 + 2mn + n^2 - 6m - 6n + 9$

69. $(a+b)(a-b) = a^2 - b^2$

71. $(2a-b)(2a+b) = (2a)^2 - b^2 = 4a^2 - b^2$

73. $(c^2-d)(c^2+d) = (c^2)^2 - d^2$

$= c^4 - d^2$

75. $(ab+cd^2)(ab-cd^2) = (ab)^2 - (cd^2)^2$

$= a^2b^2 - c^2d^4$

77. $(x+y-3)(x+y+3)$

$= [(x+y)-3][(x+y)+3]$

$= (x+y)^2 - 3^2$

$= x^2 + 2xy + y^2 - 9$

79. $[x+y+z][x-(y+z)]$

$= [x+(y+z)][x-(y+z)]$

$= x^2 - (y+z)^2$

$= x^2 - (y^2+2yz+z^2)$

$= x^2 - y^2 - 2yz - z^2$

81. $(a+b+c)(a-b-c)$

$= [a+(b+c)][a-(b+c)]$

$= a^2 - (b+c)^2$

$= a^2 - (b^2+2bc+c^2)$

$= a^2 - b^2 - 2bc - c^2$

83.

$$\begin{array}{llllll}
x^2 & -\ 4y & +\ 2 & & & \\
3x^2 & +\ 5y & -\ 3 & & & \\
\hline
-3x^2 & +\ 12y & -\ 6 & & & \\
 & +\ 10y & & -\ 20y^2 & +\ 5x^2y & \\
6x^2 & & & & -\ 12x^2y & +\ 3x^4 \\
\hline
3x^2 & +\ 22y & -\ 6 & -\ 20y^2 & -\ 7x^2y & +\ 3x^4
\end{array}$$

We could also write the result as $3x^4 - 7x^2y + 3x^2 - 20y^2 + 22y - 6$.

85. The first coordinate is positive and the second coordinate is negative, so $(2,-5)$ is in quadrant IV.

87. Both coordinates are positive, so $(16,23)$ is in quadrant I.

89. $2x = -10$

$x = -5$

Any ordered pair $(-5,y)$ is a solution. The variable x must be -5, but y can be any number we choose. A few solutions are listed below. Plot these points and draw the line.

x	y
-5	-3
-5	0
-5	4

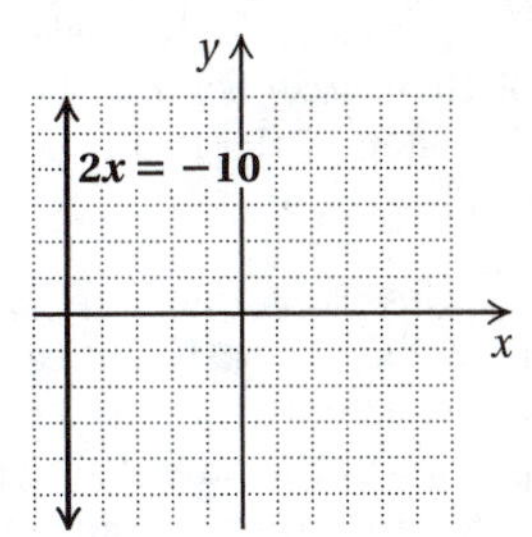

91. $8y - 16 = 0$

$8y = 16$

$y = 2$

Any ordered pair $(x,2)$ is a solution. The variable y must be 2, but x can be any number we choose. A few solutions are listed below. Plot these points and draw the line.

x	y
-4	2
0	2
3	2

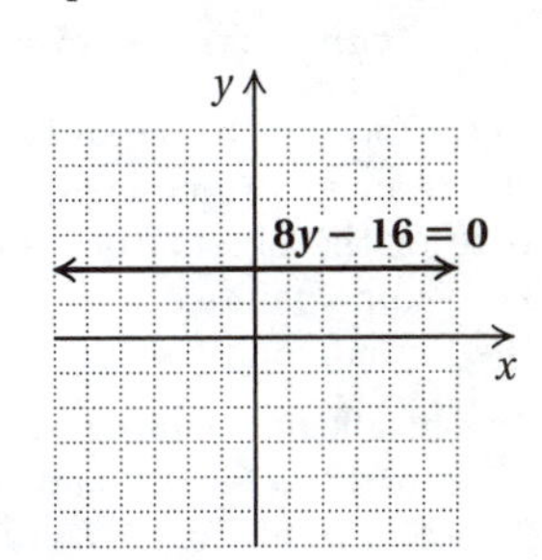

93. It is helpful to add additional labels to the figure.

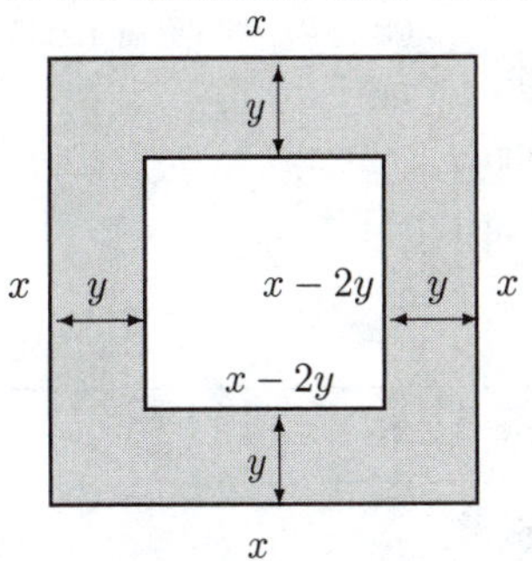

The area of the large square is $x\cdot x$, or x^2. The area of the small square is $(x-2y)(x-2y)$, or $(x-2y)^2$.

Area of shaded region	=	Area of large square	−	Area of small square
Area of shaded region	=	x^2	−	$(x-2y)^2$

$= x^2 - (x^2 - 4xy + 4y^2)$

$= x^2 - x^2 + 4xy - 4y^2$

$= 4xy - 4y^2$

95. It is helpful to add additional labels to the figure.

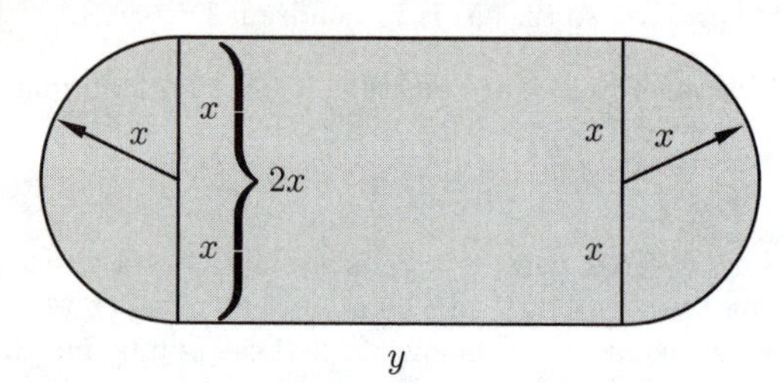

The two semicircles make a circle with radius x. The area of that circle is πx^2. The area of the rectangle is $2x \cdot y$. The sum of the two regions, $\pi x^2 + 2xy$, is the area of the shaded region.

97. The lateral surface area of the outer portion of the solid is the lateral surface area of a right circular cylinder with radius n and height h. The lateral surface area of the inner portion is the lateral surface area of a right circular cylinder with radius m and height h. Recall that the formula for the lateral surface area of a right circular cylinder with radius r and height h is $2\pi rh$.

The surface area of the top is the area of a circle with radius n less the area of a circle with radius m. The surface area of the bottom is the same as the surface area of the top.

Thus, the surface area of the solid is

$$2\pi nh + 2\pi mh + 2\pi n^2 - 2\pi m^2.$$

99. In the formula for the surface area of a silo, $S = 2\pi rh + \pi r^2$, the term πr^2 represents the area of the base. Since the base of the observatory rests on the ground, it will not need to be painted. Thus, we will subtract this term from the formula and find the remaining surface area, $2\pi rh$.

The height of the observatory is 40 ft and its radius is 30/2, or 15 ft, so the surface area is $2\pi rh \approx 2(3.14)(15)(40) \approx 3768\text{ ft}^2$. Since $3768\text{ ft}^2/250\text{ ft}^2 = 15.072$, 16 gallons of paint should be purchased.

101. Substitute \$10,400 for P, 3.5% or 0.035 for r, and 5 for t.

$$\begin{aligned}&P(1+r)^t\\ &= \$10,400(1+0.035)^5\\ &= \$10,400(1.035)^5\\ &\approx \$12,351.94\end{aligned}$$

Exercise Set 12.8

1. $\dfrac{24x^4}{8} = \dfrac{24}{8} \cdot x^4 = 3x^4$

Check: We multiply.

$3x^4 \cdot 8 = 24x^4$

3. $\dfrac{25x^3}{5x^2} = \dfrac{25}{5} \cdot \dfrac{x^3}{x^2} = 5x^{3-2} = 5x$

Check: We multiply.

$5x \cdot 5x^2 = 25x^3$

5. $\dfrac{-54x^{11}}{-3x^8} = \dfrac{-54}{-3} \cdot \dfrac{x^{11}}{x^8} = 18x^{11-8} = 18x^3$

Check: We multiply.

$18x^3(-3x^8) = -54x^{11}$

7. $\dfrac{64a^5b^4}{16a^2b^3} = \dfrac{64}{16} \cdot \dfrac{a^5}{a^2} \cdot \dfrac{b^4}{b^3} = 4a^{5-2}b^{4-3} = 4a^3b$

Check: We multiply.

$(4a^3b)(16a^2b^3) = 64a^5b^4$

9.
$$\begin{aligned}&\frac{24x^4 - 4x^3 + x^2 - 16}{8}\\ &= \frac{24x^4}{8} - \frac{4x^3}{8} + \frac{x^2}{8} - \frac{16}{8}\\ &= 3x^4 - \frac{1}{2}x^3 + \frac{1}{8}x^2 - 2\end{aligned}$$

Check: We multiply.

$$\begin{array}{r}3x^4 - \frac{1}{2}x^3 + \frac{1}{8}x^2 - 2\\ 8\\ \hline 24x^4 - 4x^3 + x^2 - 16\end{array}$$

11.
$$\begin{aligned}&\frac{u - 2u^2 - u^5}{u}\\ &= \frac{u}{u} - \frac{2u^2}{u} - \frac{u^5}{u}\\ &= 1 - 2u - u^4\end{aligned}$$

Check: We multiply.

$$\begin{array}{r}1 - 2u - u^4\\ u\\ \hline u - 2u^2 - u^5\end{array}$$

13.
$$\begin{aligned}&(15t^3 + 24t^2 - 6t) \div (3t)\\ &= \frac{15t^3 + 24t^2 - 6t}{3t}\\ &= \frac{15t^3}{3t} + \frac{24t^2}{3t} - \frac{6t}{3t}\\ &= 5t^2 + 8t - 2\end{aligned}$$

Check: We multiply.

$$\begin{array}{r}5t^2 + 8t - 2\\ 3t\\ \hline 15t^3 + 24t^2 - 6t\end{array}$$

15.
$$\begin{aligned}&(20x^6 - 20x^4 - 5x^2) \div (-5x^2)\\ &= \frac{20x^6 - 20x^4 - 5x^2}{-5x^2}\\ &= \frac{20x^6}{-5x^2} - \frac{20x^4}{-5x^2} - \frac{5x^2}{-5x^2}\\ &= -4x^4 - (-4x^2) - (-1)\\ &= -4x^4 + 4x^2 + 1\end{aligned}$$

Check: We multiply.

$$\begin{array}{r}-4x^4 + 4x^2 + 1\\ -5x^2\\ \hline 20x^6 - 20x^4 - 5x^2\end{array}$$

17. $(24x^5 - 40x^4 + 6x^3) \div (4x^3)$

$= \dfrac{24x^5 - 40x^4 + 6x^3}{4x^3}$

$= \dfrac{24x^5}{4x^3} - \dfrac{40x^4}{4x^3} + \dfrac{6x^3}{4x^3}$

$= 6x^2 - 10x + \dfrac{3}{2}$

Check: We multiply.

$$\begin{array}{r} 6x^2 - 10x + \frac{3}{2} \\ 4x^3 \\ \hline 24x^5 - 40x^4 + 6x^3 \end{array}$$

19. $\dfrac{18x^2 - 5x + 2}{2}$

$= \dfrac{18x^2}{2} - \dfrac{5x}{2} + \dfrac{2}{2}$

$= 9x^2 - \dfrac{5}{2}x + 1$

Check: We multiply.

$$\begin{array}{r} 9x^2 - \frac{5}{2}x + 1 \\ 2 \\ \hline 18x^2 - 5x + 2 \end{array}$$

21. $\dfrac{12x^3 + 26x^2 + 8x}{2x}$

$= \dfrac{12x^3}{2x} + \dfrac{26x^2}{2x} + \dfrac{8x}{2x}$

$= 6x^2 + 13x + 4$

Check: We multiply.

$$\begin{array}{r} 6x^2 + 13x + 4 \\ 2x \\ \hline 12x^3 + 26x^2 + 8x \end{array}$$

23. $\dfrac{9r^2s^2 + 3r^2s - 6rs^2}{3rs}$

$= \dfrac{9r^2s^2}{3rs} + \dfrac{3r^2s}{3rs} - \dfrac{6rs^2}{3rs}$

$= 3rs + r - 2s$

Check: We multiply.

$$\begin{array}{r} 3rs + r - 2s \\ 3rs \\ \hline 9r^2s^2 + 3r^2s - 6rs^2 \end{array}$$

25.

$$\begin{array}{rl} & x + 2 \\ x + 2 \,\big)\!\! & x^2 + 4x + 4 \\ & x^2 + 2x \\ \hline & 2x + 4 \leftarrow (x^2 + 4x) - (x^2 + 2x) \\ & 2x + 4 \\ \hline & 0 \leftarrow (2x + 4) - (2x + 4) \end{array}$$

The answer is $x + 2$.

27.

$$\begin{array}{rl} & x - 5 \\ x - 5 \,\big)\!\! & x^2 - 10x - 25 \\ & x^2 - 5x \\ \hline & -5x - 25 \leftarrow (x^2 - 10x) - (x^2 - 5x) \\ & -5x + 25 \\ \hline & -50 \leftarrow (-5x - 25) - (-5x + 25) \end{array}$$

The answer is $x - 5 + \dfrac{-50}{x - 5}$.

29.

$$\begin{array}{rl} & x - 2 \\ x + 6 \,\big)\!\! & x^2 + 4x - 14 \\ & x^2 + 6x \\ \hline & -2x - 14 \leftarrow (x^2 + 4x) - (x^2 + 6x) \\ & -2x - 12 \\ \hline & -2 \leftarrow (-2x - 14) - (-2x - 12) \end{array}$$

The answer is $x - 2 + \dfrac{-2}{x + 6}$.

31.

$$\begin{array}{rl} & x - 3 \\ x + 3 \,\big)\!\! & x^2 + 0x - 9 \leftarrow \text{Filling in the missing term} \\ & x^2 + 3x \\ \hline & -3x - 9 \leftarrow x^2 - (x^2 + 3x) \\ & -3x - 9 \\ \hline & 0 \leftarrow (-3x - 9) - (-3x - 9) \end{array}$$

The answer is $x - 3$.

33.

$$\begin{array}{rl} & x^4 - x^3 + x^2 - x + 1 \\ x + 1 \,\big)\!\! & x^5 + 0x^4 + 0x^3 + 0x^2 + 0x + 1 \leftarrow \text{Filling in missing terms} \\ & x^5 + x^4 \\ \hline & -x^4 \leftarrow x^5 - (x^5 + x^4) \\ & -x^4 - x^3 \\ \hline & x^3 \leftarrow -x^4 - (-x^4 - x^3) \\ & x^3 + x^2 \\ \hline & -x^2 \leftarrow x^3 - (x^3 + x^2) \\ & -x^2 - x \\ \hline & x + 1 \leftarrow -x^2 - (-x^2 - x) \\ & x + 1 \\ \hline & 0 \leftarrow (x + 1) - (x + 1) \end{array}$$

The answer is $x^4 - x^3 + x^2 - x + 1$.

35.

$$\begin{array}{rl} & 2x^2 - 7x + 4 \\ 4x + 3 \,\big)\!\! & 8x^3 - 22x^2 - 5x + 12 \\ & 8x^3 + 6x^2 \\ \hline & -28x^2 - 5x \leftarrow (8x^3 - 22x^2) - (8x^3 + 6x^2) \\ & -28x^2 - 21x \\ \hline & 16x + 12 \leftarrow (-28x^2 - 5x) - (-28x^2 - 21x) \\ & 16x + 12 \\ \hline & 0 \leftarrow (16x + 12) - (16x + 12) \end{array}$$

The answer is $2x^2 - 7x + 4$.

37.

$$\begin{array}{rl} & x^3 - 6 \\ x^3 - 7 \,\big)\!\! & x^6 - 13x^3 + 42 \\ & x^6 - 7x^3 \\ \hline & -6x^3 + 42 \leftarrow (x^6 - 13x^3) - (x^6 - 7x^3) \\ & -6x^3 + 42 \\ \hline & 0 \leftarrow (-6x^3 + 42) - (-6x^3 + 42) \end{array}$$

The answer is $x^3 - 6$.

39.
$$\begin{array}{r l}
 & t^2+1 \\
t-1 & \overline{)\,t^3-t^2+t-1} \\
 & \underline{t^3-t^2} \quad \leftarrow (t^3-t^2)-(t^3-t^2) \\
 & 0+t-1 \\
 & \underline{t-1} \leftarrow (t-1)-(t-1) \\
 & 0
\end{array}$$

The answer is t^2+1.

41.
$$\begin{array}{r l}
 & y^2-3y+1 \\
y+2 & \overline{)\,y^3-y^2-5y-3} \\
 & \underline{y^3+2y^2} \\
 & -3y^2-5y \\
 & \underline{-3y^2-6y} \\
 & y-3 \\
 & \underline{y+2} \\
 & -5
\end{array}$$

The answer is $y^2-3y+1+\dfrac{-5}{y+2}$.

43.
$$\begin{array}{r l}
 & 3x^2+x+2 \\
5x+1 & \overline{)\,15x^3+8x^2+11x+12} \\
 & \underline{15x^3+3x^2} \\
 & 5x^2+11x \\
 & \underline{5x^2+x} \\
 & 10x+12 \\
 & \underline{10x+2} \\
 & 10
\end{array}$$

The answer is $3x^2+x+2+\dfrac{10}{5x+1}$.

45.
$$\begin{array}{r l}
 & 6y^2-5 \\
2y+7 & \overline{)\,12y^3+42y^2-10y-41} \\
 & \underline{12y^3+42y^2} \\
 & -10y-41 \\
 & \underline{-10y-35} \\
 & -6
\end{array}$$

The answer is $6y^2-5+\dfrac{-6}{2y+7}$.

47. The product rule asserts that when multiplying with exponential notation, if the bases are the same, keep the base and add the exponents.

49. The multiplication principle asserts that when we multiply or divide by the same nonzero number on each side of an equation, we get equivalent equations.

51. A trinomial is a polynomial with three terms, such as $5x^4-7x^2+4$.

53. The absolute value of a number is its distance from zero on a number line.

55.
$$\begin{array}{r l}
 & x^2+5 \\
x^2+4 & \overline{)\,x^4+9x^2+20} \\
 & \underline{x^4+4x^2} \\
 & 5x^2+20 \\
 & \underline{5x^2+20} \\
 & 0
\end{array}$$

The answer is x^2+5.

57.
$$\begin{array}{r l}
 & a+3 \\
5a^2-7a-2 & \overline{)\,5a^3+8a^2-23a-1} \\
 & \underline{5a^3-7a^2-2a} \\
 & 15a^2-21a-1 \\
 & \underline{15a^2-21a-6} \\
 & 5
\end{array}$$

The answer is $a+3+\dfrac{5}{5a^2-7a-2}$.

59. We rewrite the dividend in descending order.
$$\begin{array}{r l}
 & 2x^2+x-3 \\
3x^3-2x-1 & \overline{)\,6x^5+3x^4-13x^3-4x^2+5x+3} \\
 & \underline{6x^5-4x^3-2x^2} \\
 & 3x^4-9x^3-2x^2+5x \\
 & \underline{3x^4-2x^2-x} \\
 & -9x^3+6x+3 \\
 & \underline{-9x^3+6x+3} \\
 & 0
\end{array}$$

The answer is $2x^2+x-3$.

61.
$$\begin{array}{r l}
 & a^5+a^4b+a^3b^2+a^2b^3+ab^4+b^5 \\
a-b & \overline{)\,a^6+0a^5b+0a^4b^2+0a^3b^3+0a^2b^4+0ab^5-b^6} \\
 & \underline{a^6-a^5b} \\
 & a^5b \\
 & \underline{a^5b-a^4b^2} \\
 & a^4b^2 \\
 & \underline{a^4b^2-a^3b^3} \\
 & a^3b^3 \\
 & \underline{a^3b^3-a^2b^4} \\
 & a^2b^4 \\
 & \underline{a^2b^4-ab^5} \\
 & ab^5-b^6 \\
 & \underline{ab^5-b^6} \\
 & 0
\end{array}$$

The answer is $a^5+a^4b+a^3b^2+a^2b^3+ab^4+b^5$.

63.
$$\begin{array}{r l}
 & x+5 \\
x-1 & \overline{)\,x^2+4x+c} \\
 & \underline{x^2-x} \\
 & 5x+c \\
 & \underline{5x-5} \\
 & c+5
\end{array}$$

We set the remainder equal to 0.

$c+5=0$

$c=-5$

Thus, c must be -5.

65.

$$\begin{array}{r l} & c^2x + (-2c+c^2) \\ x-1 \overline{)} & c^2x^2 - 2cx + 1 \\ & \underline{c^2x^2 - c^2x} \\ & (-2c+c^2)x + 1 \\ & \underline{(-2c+c^2)x - (-2c+c^2)} \\ & 1 + (-2c+c^2) \end{array}$$

We set the remainder equal to 0.

$$c^2 - 2c + 1 = 0$$
$$(c-1)^2 = 0$$
$$c = 1$$

Thus, c must be 1.

Chapter 12 Concept Reinforcement

1. True; a trinomial is a polynomial with three terms.
2. False; $(x+y)^2 = x^2 + 2xy + y^2 \neq x^2 + y^2$.
3. False; $(A-B)^2 = A^2 - 2AB + B^2 \neq A^2 - B^2$.
4. True; $(A+B)(A-B) = A^2 - AB + AB - B^2 = A^2 - B^2$.

Chapter 12 Important Concepts

1. $z^5 \cdot z^3 = z^{5+3} = z^8$

2. $\dfrac{a^4b^7}{a^2b} = \dfrac{a^4}{a^2} \cdot \dfrac{b^7}{b} = a^{4-2}b^{7-1} = a^2b^6$

3. $$\begin{aligned} \left(\frac{x^{-4}y^2}{3z^3}\right)^3 &= \frac{(x^{-4}y^2)^3}{(3z^3)^3} \\ &= \frac{(x^{-4})^3(y^2)^3}{3^3(z^3)^3} \\ &= \frac{x^{-4\cdot3}y^{2\cdot3}}{27z^{3\cdot3}} \\ &= \frac{x^{-12}y^6}{27z^9} \\ &= \frac{y^6}{27x^{12}z^9} \end{aligned}$$

4. 763,000

   7.63,000.

   ↑_____| 5 places

   Large number, so the exponent is positive.

   $763{,}000 = 7.63 \times 10^5$

5. 3×10^{-4}

   The exponent is negative so the number is small.

   0.0003.

   ↑____| 4 places

   $3 \times 10^{-4} = 0.0003$

6. $$\begin{aligned} \frac{3.6 \times 10^3}{6.0 \times 10^{-2}} &= \frac{3.6}{6.0} \times \frac{10^3}{10^{-2}} = 0.6 \times 10^5 \\ &= (6 \times 10^{-1}) \times 10^5 = 6 \times 10^4 \end{aligned}$$

7. $$\begin{aligned} 5x^4 - 6x^2 - 3x^4 + 2x^2 - 3 &= (5-3)x^4 + (-6+2)x^2 - 3 \\ &= 2x^4 - 4x^2 - 3 \end{aligned}$$

8. $$\begin{aligned} &(3x^4 - 5x^2 - 4) + (x^3 + 3x^2 + 6) \\ &= 3x^4 + x^3 + (-5+3)x^2 + (-4+6) \\ &= 3x^4 + x^3 - 2x^2 + 2 \end{aligned}$$

9. $(x^4 - 3x^2 + 2)(x^2 - 3)$

$$\begin{array}{r} x^4 - 3x^2 + 2 \\ \underline{x^2 - 3} \\ -3x^4 + 9x^2 - 6 \\ \underline{x^6 - 3x^4 + 2x^2 \quad\quad} \\ x^6 - 6x^4 + 11x^2 - 6 \end{array}$$

10. We use FOIL.

$$\begin{aligned} (y+4)(2y+3) &= y \cdot 2y + y \cdot 3 + 4 \cdot 2y + 4 \cdot 3 \\ &= 2y^2 + 3y + 8y + 12 \\ &= 2y^2 + 11y + 12 \end{aligned}$$

11. $(x+5)(x-5) = x^2 - 5^2 = x^2 - 25$

12. $$\begin{aligned} (3w+4)^2 &= (3w)^2 + 2 \cdot 3w \cdot 4 + 4^2 \\ &= 9w^2 + 24w + 16 \end{aligned}$$

13. $$\begin{aligned} &(a^3b^2 - 5a^2b + 2ab) - (3a^3b^2 - ab^2 + 4ab) \\ &= a^3b^2 - 5a^2b + 2ab - 3a^3b^2 + ab^2 - 4ab \\ &= -2a^3b^2 - 5a^2b + ab^2 - 2ab \end{aligned}$$

14. $$\begin{aligned} (5y^2 - 20y + 8) \div 5 &= \frac{5y^2 - 20y + 8}{5} \\ &= \frac{5y^2}{5} - \frac{20y}{5} + \frac{8}{5} \\ &= y^2 - 4y + \frac{8}{5} \end{aligned}$$

15.

$$\begin{array}{r l} & x - 9 \\ x+5 \overline{)} & x^2 - 4x + 3 \\ & \underline{x^2 + 5x} \\ & -9x + 3 \\ & \underline{-9x - 45} \\ & 48 \end{array}$$

The answer is $x - 9 + \dfrac{48}{x+5}$.

Chapter 12 Review Exercises

1. $7^2 \cdot 7^{-4} = 7^{2+(-4)} = 7^{-2} = \dfrac{1}{7^2}$
2. $y^7 \cdot y^3 \cdot y = y^{7+3+1} = y^{11}$
3. $(3x)^5(3x)^9 = (3x)^{5+9} = (3x)^{14}$
4. $t^8 \cdot t^0 = t^8 \cdot 1 = t^8$, or
   $t^8 \cdot t^0 = t^{8+0} = t^8$
5. $\dfrac{4^5}{4^2} = 4^{5-2} = 4^3$

6. $\dfrac{a^5}{a^8} = a^{5-8} = a^{-3} = \dfrac{1}{a^3}$

7. $\dfrac{(7x)^4}{(7x)^4} = 1$

8. $(3t^4)^2 = 3^2 \cdot (t^4)^2 = 9 \cdot t^{4 \cdot 2} = 9t^8$

9. $(2x^3)^2(-3x)^2 = 2^2 \cdot (x^3)^2(-3)^2x^2 = 4 \cdot x^6 \cdot 9 \cdot x^2 = 36x^8$

10. $\left(\dfrac{2x}{y}\right)^{-3} = \left(\dfrac{y}{2x}\right)^3 = \dfrac{y^3}{2^3 \cdot x^3} = \dfrac{y^3}{8x^3}$

11. $\dfrac{1}{t^5} = t^{-5}$

12. $y^{-4} = \dfrac{1}{y^4}$

13. 0.00003. 28

5 places

Small number, so the exponent is negative.

$0.0000328 = 3.28 \times 10^{-5}$

14. 8.3×10^6

8.300000.

6 places

Positive exponent, so the answer is a large number.

$8.3 \times 10^6 = 8,300,000$

15.
$$\begin{aligned}(3.8 \times 10^4)(5.5 \times 10^{-1}) &= (3.8 \cdot 5.5) \times (10^4 \cdot 10^{-1}) \\ &= 20.9 \times 10^3 \\ &= (2.09 \times 10) \times 10^3 \\ &= 2.09 \times 10^4\end{aligned}$$

16.
$$\begin{aligned}\frac{1.28 \times 10^{-8}}{2.5 \times 10^{-4}} &= \frac{1.28}{2.5} \times \frac{10^{-8}}{10^{-4}} \\ &= 0.512 \times 10^{-4} \\ &= (5.12 \times 10^{-1}) \times 10^{-4} \\ &= 5.12 \times 10^{-5}\end{aligned}$$

17. $46 = 4.6 \times 10$

Also,
$$\begin{aligned}335.8 \text{ million} &= 335.8 \times 10^6 \\ &= (3.358 \times 10^2) \times 10^6 \\ &= 3.358 \times 10^8.\end{aligned}$$

Then we have
$$\begin{aligned}(4.6 \times 10) \times (3.358 \times 10^8) &= (4.6 \times 3.358) \times (10 \times 10^8) \\ &= 15.4468 \times 10^9 \\ &= (1.54468 \times 10) \times 10^9 \\ &= 1.54468 \times 10^{10}\end{aligned}$$

18. $x^2 - 3x + 6 = (-1)^2 - 3(-1) + 6 = 1 + 3 + 6 = 10$

19. $-4y^5 + 7y^2 - 3y - 2 = -4y^5 + 7y^2 + (-3y) + (-2)$

The terms are $-4y^5$, $7y^2$, $-3y$, and -2.

20. In the polynomial x^3+x there are no x^2 or x^0 terms. Thus, the x^2 term (or second-degree term) and the x^0 term (or zero-degree term) are missing.

21. $4x^3 + 6x^2 - 5x + \dfrac{5}{3} = 4x^3 + 6x^2 - 5x^1 + \dfrac{5}{3}x^0$

The degree of $4x^3$ is 3.

The degree of $6x^2$ is 2.

The degree of $-5x$ is 1.

The degree of $\dfrac{5}{3}$ is 0.

The degree of the polynomial is 3, the largest exponent.

22. The polynomial $4x^3 - 1$ is a binomial because it has just two terms.

23. The polynomial $4 - 9t^3 - 7t^4 + 10t^2$ is none of these because it has more than three terms.

24. The polynomial $7y^2$ is a monomial because it has just one term.

25.
$$\begin{aligned}&3x^2 - 2x + 3 - 5x^2 - 1 - x \\ &= (3-5)x^2 + (-2-1)x + (3-1) \\ &= -2x^2 - 3x + 2\end{aligned}$$

26.
$$\begin{aligned}&-x + \frac{1}{2} + 14x^4 - 7x^2 - 1 - 4x^4 \\ &= (14-4)x^4 - 7x^2 - x + \left(\frac{1}{2} - 1\right) \\ &= 10x^4 - 7x^2 - x - \frac{1}{2}\end{aligned}$$

27. $(3x^4-x^3+x-4)+(x^5+7x^3-3x^2-5)+(-5x^4+6x^2-x) = (3-5)x^4+(-1+7)x^3+(1-1)x+(-4-5)+x^5+(-3+6)x^2 = -2x^4+6x^3-9+x^5+3x^2$, or $x^5-2x^4+6x^3+3x^2-9$

28. $(3x^5-4x^4+x^3-3)+(3x^4-5x^3+3x^2)+(-5x^5-5x^2)+(-5x^4+2x^3+5) = (3-5)x^5+(-4+3-5)x^4+(1-5+2)x^3+(-3+5)+(3-5)x^2 = -2x^5-6x^4-2x^3+2-2x^2$, or $-2x^5-6x^4-2x^3-2x^2+2$

29.
$$\begin{aligned}(5x^2 - 4x + 1) - (3x^2 + 1) &= 5x^2 - 4x + 1 - 3x^2 - 1 \\ &= 2x^2 - 4x\end{aligned}$$

30.
$$\begin{aligned}&(3x^5-4x^4+3x^2+3)-(2x^5-4x^4+3x^3+4x^2-5) \\ &= 3x^5 - 4x^4 + 3x^2 + 3 - 2x^5 + 4x^4 - 3x^3 - 4x^2 + 5 \\ &= x^5 - 3x^3 - x^2 + 8\end{aligned}$$

31. $P = 2(w+3) + 2w = 2w + 6 + 2w = 4w + 6$

$A = w(w+3) = w^2 + 3w$

32. Regarding the figure as one large rectangle with length $t+4$ and width $t+3$, we have $(t+4)(t+3)$. We can also add the areas of the four smaller rectangles:

$3 \cdot t + 4 \cdot 3 + 4 \cdot t + t \cdot t$, or $3t + 12 + 4t + t^2$, or $t^2 + 7t + 12$

33.
$$\begin{aligned}\left(x + \frac{2}{3}\right)\left(x + \frac{1}{2}\right) &= x^2 + \frac{1}{2}x + \frac{2}{3}x + \frac{2}{6} \\ &= x^2 + \frac{3}{6}x + \frac{4}{6}x + \frac{1}{3} \\ &= x^2 + \frac{7}{6}x + \frac{1}{3}\end{aligned}$$

34. $(7x+1)^2 = (7x)^2 + 2 \cdot 7x \cdot 1 + 1^2$
$= 49x^2 + 14x + 1$

35.
$$\begin{array}{r} 4x^2 - 5x + 1 \\ 3x - 2 \\ \hline -8x^2 + 10x - 2 \\ 12x^3 - 15x^2 + 3x \\ \hline 12x^3 - 23x^2 + 13x - 2 \end{array}$$

36. $(3x^2+4)(3x^2-4) = (3x^2)^2 - 4^2 = 9x^4 - 16$

37. $5x^4(3x^3 - 8x^2 + 10x + 2)$
$= 5x^4 \cdot 3x^3 - 5x^4 \cdot 8x^2 + 5x^4 \cdot 10x + 5x^4 \cdot 2$
$= 15x^7 - 40x^6 + 50x^5 + 10x^4$

38. $(x+4)(x-7) = x^2 - 7x + 4x - 28 = x^2 - 3x - 28$

39. $(3y^2 - 2y)^2 = (3y^2)^2 - 2 \cdot 3y^2 \cdot 2y + (2y)^2 =$
$9y^4 - 12y^3 + 4y^2$

40. $(2t^2+3)(t^2-7) = 2t^4 - 14t^2 + 3t^2 - 21 =$
$2t^4 - 11t^2 - 21$

41. $2 - 5xy + y^2 - 4xy^3 + x^6$
$= 2 - 5(-1)(2) + 2^2 - 4(-1)(2)^3 + (-1)^6$
$= 2 - 5(-1)(2) + 4 - 4(-1)(8) + 1$
$= 2 + 10 + 4 + 32 + 1$
$= 49$

42. $x^5y - 7xy + 9x^2 - 8$

Term	Coefficient	Degree	
x^5y	1	6	$(x^5y = 1 \cdot x^5y^1)$
$-7xy$	-7	2	$(-7xy = -7x^1y^1)$
$9x^2$	9	2	
-8	-8	0	$(-8 = -8x^0)$

The degree of the polynomial is the degree of the term of highest degree. The term of highest degree is x^5y. Its degree is 6, so the degree of the polynomial is 6.

43. $y + w - 2y + 8w - 5 = (1-2)y + (1+8)w - 5$
$= -y + 9w - 5$

44. $m^6 - 2m^2n + m^2n^2 + n^2m - 6m^3 + m^2n^2 + 7n^2m$
$= m^6 - 2m^2n + (1+1)m^2n^2 + (1+7)n^2m - 6m^3$
$= m^6 - 2m^2n + 2m^2n^2 + 8n^2m - 6m^3$

45. $(5x^2 - 7xy + y^2) + (-6x^2 - 3xy - y^2) + (x^2 + xy - 2y^2)$
$= (5-6+1)x^2 + (-7-3+1)xy + (1-1-2)y^2$
$= -9xy - 2y^2$

46. $(6x^3y^2 - 4x^2y - 6x) - (-5x^3y^2 + 4x^2y + 6x^2 - 6)$
$= 6x^3y^2 - 4x^2y - 6x + 5x^3y^2 - 4x^2y - 6x^2 + 6$
$= (6+5)x^3y^2 + (-4-4)x^2y - 6x - 6x^2 + 6$
$= 11x^3y^2 - 8x^2y - 6x - 6x^2 + 6$

47.
$$\begin{array}{r} p^2 + pq + q^2 \\ p - q \\ \hline -p^2q - pq^2 - q^3 \\ p^3 + p^2q + pq^2 \\ \hline p^3 \qquad\qquad - q^3 \end{array}$$

48. $\left(3a^4 - \frac{1}{3}b^3\right)^2 = (3a^4)^2 - 2 \cdot 3a^4 \cdot \frac{1}{3}b^3 + \left(\frac{1}{3}b^3\right)^2$
$= 9a^8 - 2a^4b^3 + \frac{1}{9}b^6$

49. $\frac{10x^3 - x^2 + 6x}{2x} = \frac{10x^3}{2x} - \frac{x^2}{2x} + \frac{6x}{2x}$
$= 5x^2 - \frac{1}{2}x + 3$

50.
$$\begin{array}{r} 3x^2 - 7x + 4 \\ 2x+3 \overline{\smash{)}\, 6x^3 - 5x^2 - 13x + 13} \\ 6x^3 + 9x^2 \qquad\qquad \\ \hline -14x^2 - 13x \qquad \\ -14x^2 - 21x \qquad \\ \hline 8x + 13 \\ 8x + 12 \\ \hline 1 \end{array}$$

The answer is $3x^2 - 7x + 4 + \frac{1}{2x+3}$.

51. Locate -1 on the x-axis. Then move vertically to the graph and horizontally to the y-axis. It appears that the y-value that is paired with -1 is 0. Thus, the value of $y = 10x^3 - 10x$ is 0 when $x = -1$.

Locate -0.5 on the x-axis. Then move vertically to the graph and horizontally to the y-axis. It appears that the y-value that is paired with -0.5 is about 3.75. Thus, the value of $y = 10x^3 - 10x$ is 3.75 when $x = -0.5$.

Locate 0.5 on the x-axis. Then move vertically to the graph and horizontally to the y-axis. It appears that the y-value that is paired with 0.5 is about -3.75. Thus, the value of $y = 10x^3 - 10x$ is -3.75 when $x = 0.5$.

Locate 1 on the x-axis. Then move vertically to the graph and horizontally to the y-axis. It appears that the y-value that is paired with 1 is 0. Thus, the value of $y = 10x^3 - 10x$ is 0 when $x = 1$.

52. $(2x^2 - 3x + 4) - (x^2 + 2x) = 2x^2 - 3x + 4 - x^2 - 2x = x^2 - 5x + 4$

Answer B is correct.

53. $(x-1)^2 = x^2 - 2 \cdot x \cdot 1 + 1^2 = x^2 - 2x + 1$

Answer D is correct.

54. $A = \frac{1}{2}bh$

$A = \frac{1}{2}(x+y)(x-y) = \frac{1}{2}(x^2 - y^2) = \frac{1}{2}x^2 - \frac{1}{2}y^2$

55. The shaded area is the area of a square with side 20 minus the area of 4 small squares, each with side a.

$A = 20^2 - 4 \cdot a^2 = 400 - 4a^2$

56. $\quad -3x^5 \cdot 3x^3 - x^6(2x)^2 + (3x^4)^2 + (2x^2)^4 - 40x^2(x^3)^2$
$= -3x^5 \cdot 3x^3 - x^6(4x^2) + 9x^8 + 16x^8 - 40x^2(x^6)$
$= -9x^8 - 4x^8 + 9x^2 + 16x^8 - 40x^8$
$= -28x^8$

57.
$$\begin{aligned}(x-7)(x+10) &= (x-4)(x-6)\\ x^2+10x-7x-70 &= x^2-6x-4x+24\\ x^2+3x-70 &= x^2-10x+24\\ 3x-70 &= -10x+24 \quad \text{Subtracting } x^2\\ 13x-70 &= 24 \quad \text{Adding } 10x\\ 13x &= 94 \quad \text{Adding } 70\\ x &= \frac{94}{13}\end{aligned}$$

The solution is $\frac{94}{13}$.

58. Let P represent the other polynomial. Then we have $(x-1)P = x^5 - 1$, or $P = \frac{x^5-1}{x-1}$. We divide to find P.

$$\begin{array}{r} x^4 + x^3 + x^2 + x + 1 \\ x-1 \overline{\left)\, x^5 + 0x^4 + 0x^3 + 0x^2 + 0x - 1\right.} \\ \underline{x^5 - x^4} \qquad\qquad\qquad\qquad \\ x^4 \qquad\qquad\qquad\qquad \\ \underline{x^4 - x^3} \qquad\qquad\qquad \\ x^3 \qquad\qquad\qquad \\ \underline{x^3 - x^2} \qquad\qquad \\ x^2 \qquad\qquad \\ \underline{x^2 - x} \qquad \\ x - 1 \\ \underline{x - 1} \\ 0 \end{array}$$

The other polynomial is $x^4 + x^3 + x^2 + x + 1$.

59. ***Familiarize***. Let w = the width of the garden, in feet. Then $2w$ = the length. From the drawing in the text we see that the width of the garden and the sidewalk together is $w + 4 + 4$, or $w + 8$, and the length of the garden and the sidewalk together is $2w + 4 + 4$, or $2w + 8$.

Translate. The area of the sidewalk is the area of the garden and sidewalk together minus the area of the garden. Recall that the formula for the area of a rectangle is $A = l \cdot w$. Thus, we have

$$1024 = (2w+8)(w+8) - 2w \cdot w.$$

Solve. We solve the equation.

$$\begin{aligned}1024 &= (2w+8)(w+8) - 2w \cdot w\\ 1024 &= 2w^2 + 16w + 8w + 64 - 2w^2\\ 1024 &= 24w + 64\\ 960 &= 24w\\ 40 &= w\end{aligned}$$

If $w = 40$, then $2w = 2 \cdot 40 = 80$.

Check. The dimensions of the garden and the sidewalk together are $2 \cdot 40 + 8$ by $40 + 8$, or 88 by 48. Then the area of the garden and sidewalk together is 88 ft $\cdot$ 48 ft, or 4224 ft^2 and the area of the garden is 80 ft $\cdot$ 40 ft, or 3200 ft^2. Subtracting to find the area of the sidewalk, we get 4224 ft^2 − 3200 ft^2, or 1024 ft^2, so the answer checks.

State. The dimensions of the garden are 80 ft by 40 ft.

Chapter 12 Discussion and Writing Exercises

1. 578.6×10^{-7} is not in scientific notation because 578.6 is larger than 10.

2. When evaluating polynomials it is essential to know the order in which the operations are to be performed.

3. Label the figure as shown.

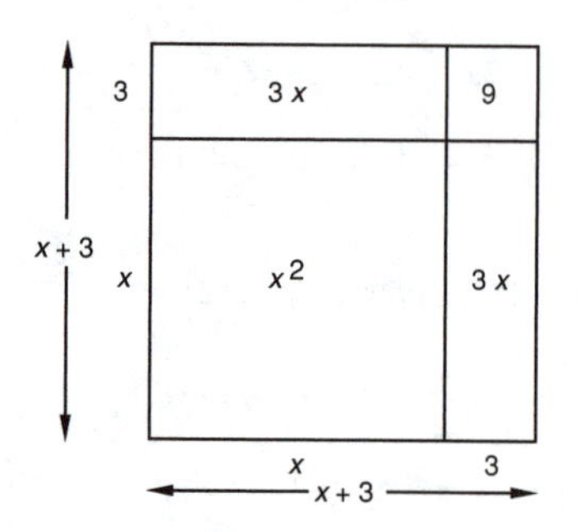

Then we see that the area of the figure is $(x+3)^2$, or $x^2 + 3x + 3x + 9 \neq x^2 + 9$.

4. Emma did not divide *each* term of the polynomial by the divisor. The first term was divided by $3x$, but the second was not. Multiplying Emma's "quotient" by the divisor $3x$ we get $12x^3 - 18x^2 \neq 12x^3 + 6x$. This should convince her that a mistake has been made.

5. Yes; for example, $(x^2+xy+1)+(3x-xy+2) = x^2+3x+3$.

6. Yes; consider $a + b + c + d$. This is a polynomial in 4 variables but it has degree 1.

Chapter 12 Test

1. $6^{-2} \cdot 6^{-3} = 6^{-2+(-3)} = 6^{-5} = \frac{1}{6^5}$

2. $x^6 \cdot x^2 \cdot x = x^{6+2+1} = x^9$

3. $(4a)^3 \cdot (4a)^8 = (4a)^{3+8} = (4a)^{11}$

4. $\frac{3^5}{3^2} = 3^{5-2} = 3^3$

5. $\frac{x^3}{x^8} = x^{3-8} = x^{-5} = \frac{1}{x^5}$

6. $\frac{(2x)^5}{(2x)^5} = 1$

7. $(x^3)^2 = x^{3 \cdot 2} = x^6$

8. $(-3y^2)^3 = (-3)^3(y^2)^3 = -27y^{2\cdot 3} = -27y^6$

9. $(2a^3b)^4 = 2^4(a^3)^4 \cdot b^4 = 16a^{12}b^4$

10. $\left(\frac{ab}{c}\right)^3 = \frac{(ab)^3}{c^3} = \frac{a^3b^3}{c^3}$

11. $(3x^2)^3(-2x^5)^3 = 3^3(x^2)^3(-2)^3(x^5)^3 = 27x^6(-8)x^{15} = -216x^{21}$

12. $3(x^2)^3(-2x^5)^3 = 3x^6(-2)^3(x^5)^3 = 3x^6(-8)x^{15} = -24x^{21}$

13. $2x^2(-3x^2)^4 = 2x^2(-3)^4(x^2)^4 = 2x^2 \cdot 81x^8 = 162x^{10}$

14. $(2x)^2(-3x^2)^4 = 2^2x^2(-3)^4(x^2)^4 = 4x^2 \cdot 81x^8 = 324x^{10}$

15. $5^{-3} = \frac{1}{5^3}$

16. $\frac{1}{y^8} = y^{-8}$

17. 3,900,000,000

3.900,000,000.

↑______| 9 places

Large number, so the exponent is positive.

$3,900,000,000 = 3.9 \times 10^9$

18. 5×10^{-8}

Negative exponent, so the answer is a small number.

0.00000005.

↑______| 8 places

$5 \times 10^{-8} = 0.00000005$

19. $\frac{5.6\times10^6}{3.2\times10^{-11}} = \frac{5.6}{3.2} \times \frac{10^6}{10^{-11}} = 1.75\times10^{6-(-11)} = 1.75\times10^{17}$

20. $(2.4 \times 10^5)(5.4 \times 10^{16}) = (2.4 \cdot 5.4) \times (10^5 \cdot 10^{16}) = 12.96 \times 10^{21} = (1.296 \times 10) \times 10^{21} = 1.296 \times 10^{22}$

21. 600 million $= 600 \times 1$ million $= 600 \times 1,000,000 = 600,000,000 = 6 \times 10^8$

$40,000 = 4 \times 10^4$

We divide:

$\frac{6 \times 10^8}{4 \times 10^4} = 1.5 \times 10^4$

A CD-ROM can hold 1.5×10^4 sound files.

22. $x^5 + 5x - 1 = (-2)^5 + 5(-2) - 1 = -32 - 10 - 1 = -43$

23. $\frac{1}{3}x^5 - x + 7$

The coefficient of $\frac{1}{3}x^5$ is $\frac{1}{3}$.

The coefficient of $-x$, or $-1 \cdot x$, is -1.

The coefficient of 7 is 7.

24. $2x^3 - 4 + 5x + 3x^6$

The degree of $2x^3$ is 3.

The degree of -4, or $-4x^0$, is 0.

The degree of $5x$, or $5x^1$, is 1.

The degree of $3x^6$ is 6.

The degree of the polynomial is 6, the largest exponent.

25. $7 - x$ is a binomial because it has just 2 terms.

26. $4a^2 - 6 + a^2 = (4+1)a^2 - 6 = 5a^2 - 6$

27. $y^2 - 3y - y + \frac{3}{4}y^2 = \left(1 + \frac{3}{4}\right)y^2 + (-3-1)y = \left(\frac{4}{4} + \frac{3}{4}\right)y^2 + (-3-1)y = \frac{7}{4}y^2 - 4y$

28.
$$\begin{aligned} & 3 - x^2 + 2x^3 + 5x^2 - 6x - 2x + x^5 \\ &= 3 + (-1+5)x^2 + 2x^3 + (-6-2)x + x^5 \\ &= 3 + 4x^2 + 2x^3 - 8x + x^5 \\ &= x^5 + 2x^3 + 4x^2 - 8x + 3 \end{aligned}$$

29.
$$\begin{aligned} & (3x^5+5x^3-5x^2-3)+(x^5+x^4-3x^3-3x^2+2x-4) \\ &= (3+1)x^5+x^4+(5-3)x^3+(-5-3)x^2+2x+(-3-4) \\ &= 4x^5 + x^4 + 2x^3 - 8x^2 + 2x - 7 \end{aligned}$$

30.
$$\begin{aligned} & \left(x^4 + \frac{2}{3}x + 5\right) + \left(4x^4 + 5x^2 + \frac{1}{3}x\right) \\ &= (1+4)x^4 + 5x^2 + \left(\frac{2}{3} + \frac{1}{3}\right)x + 5 \\ &= 5x^4 + 5x^2 + x + 5 \end{aligned}$$

31.
$$\begin{aligned} & (2x^4 + x^3 - 8x^2 - 6x - 3) - (6x^4 - 8x^2 + 2x) \\ &= 2x^4 + x^3 - 8x^2 - 6x - 3 - 6x^4 + 8x^2 - 2x \\ &= (2-6)x^4 + x^3 + (-8+8)x^2 + (-6-2)x - 3 \\ &= -4x^4 + x^3 - 8x - 3 \end{aligned}$$

32.
$$\begin{aligned} & (x^3 - 0.4x^2 - 12) - (x^5 + 0.3x^3 + 0.4x^2 + 9) \\ &= x^3 - 0.4x^2 - 12 - x^5 - 0.3x^3 - 0.4x^2 - 9 \\ &= -x^5 + (1-0.3)x^3 + (-0.4-0.4)x^2 + (-12-9) \\ &= -x^5 + 0.7x^3 - 0.8x^2 - 21 \end{aligned}$$

33. $-3x^2(4x^2-3x-5) = -3x^2 \cdot 4x^2 - 3x^2(-3x) - 3x^2(-5) = -12x^4 + 9x^3 + 15x^2$

34. $\left(x - \frac{1}{3}\right)^2 = x^2 - 2 \cdot x \cdot \frac{1}{3} + \left(\frac{1}{3}\right)^2 = x^2 - \frac{2}{3}x + \frac{1}{9}$

35. $(3x+10)(3x-10) = (3x)^2 - 10^2 = 9x^2 - 100$

36. $(3b+5)(b-3) = 3b^2 - 9b + 5b - 15 = 3b^2 - 4b - 15$

37. $(x^6 - 4)(x^8 + 4) = x^{14} + 4x^6 - 4x^8 - 16$, or $x^{14} - 4x^8 + 4x^6 - 16$

38. $(8-y)(6+5y) = 48 + 40y - 6y - 5y^2 = 48 + 34y - 5y^2$

39.
$$\begin{array}{rrrr} & 3x^2 & -\ 5x & -\ 3 \\ & & 2x & +\ 1 \\ \hline & 3x^2 & -\ 5x & -\ 3 \\ 6x^3 & -\ 10x^2 & -\ 6x & \\ \hline 6x^3 & -\ 7x^2 & -\ 11x & -\ 3 \end{array}$$

40. $(5t+2)^2 = (5t)^2 + 2 \cdot 5t \cdot 2 + 2^2 = 25t^2 + 20t + 4$

41.
$$\begin{aligned} & x^3y - y^3 + xy^3 + 8 - 6x^3y - x^2y^2 + 11 \\ &= (1-6)x^3y - y^3 + xy^3 + (8+11) - x^2y^2 \\ &= -5x^3y - y^3 + xy^3 + 19 - x^2y^2 \end{aligned}$$

42. $(8a^2b^2 - ab + b^3) - (-6ab^2 - 7ab - ab^3 + 5b^3)$
$= 8a^2b^2 - ab + b^3 + 6ab^2 + 7ab + ab^3 - 5b^3$
$= 8a^2b^2 + (-1+7)ab + (1-5)b^3 + 6ab^2 + ab^3$
$= 8a^2b^2 + 6ab - 4b^3 + 6ab^2 + ab^3$

43. $(3x^5 - 4y^5)(3x^5 + 4y^5) = (3x^5)^2 - (4y^5)^2 =$
$9x^{10} - 16y^{10}$

44. $(12x^4 + 9x^3 - 15x^2) \div (3x^2)$
$= \dfrac{12x^4 + 9x^3 - 15x^2}{3x^2}$
$= \dfrac{12x^4}{3x^2} + \dfrac{9x^3}{3x^2} - \dfrac{15x^2}{3x^2}$
$= 4x^2 + 3x - 5$

45.
$$\begin{array}{r}
2x^2 - 4x - 2 \\
3x+2 \overline{\big)\, 6x^3 - 8x^2 - 14x + 13} \\
\underline{6x^3 + 4x^2 } \\
-12x^2 - 14x \\
\underline{-12x^2 - 8x } \\
-6x + 13 \\
\underline{-6x - 4} \\
17
\end{array}$$

The answer is $2x^2 - 4x - 2 + \dfrac{17}{3x+2}$.

46. Locate -1 on the x-axis. Then move vertically to the graph and horizontally to the y-axis. It appears that the y-value that is paired with -1 is 3. Thus, the value of $y = x^3 - 5x - 1$ is 3 when $x = -1$.

Locate -0.5 on the x-axis. Then move vertically to the graph and horizontally to the y-axis. It appears that the y-value that is paired with -0.5 is 1.5. Thus, the value of $y = x^3 - 5x - 1$ is 1.5 when $x = -0.5$.

Locate 0.5 on the x-axis. Then move vertically to the graph and horizontally to the y-axis. It appears that the y-value that is paired with 0.5 is -3.5. Thus, the value of $y = x^3 - 5x - 1$ is -3.5 when $x = 0.5$.

Locate 1 on the x-axis. Then move vertically to the graph and horizontally to the y-axis. It appears that the y-value that is paired with 1 is -5. Thus, the value of $y = x^3 - 5x - 1$ is -5 when $x = 1$.

Locate 1.1 on the x-axis. Then move vertically to the graph and horizontally to the y-axis. It appears that the y-value that is paired with 1.1 is -5.25. Thus, the value of $y = x^3 - 5x - 1$ is -5.25 when $x = 1.1$.

47. When we regard the figure as one large rectangle with dimensions $t+2$ by $t+2$, we can express the area as $(t+2)(t+2)$.

Next we will regard the figure as the sum of four smaller rectangles with dimensions t by t, 2 by t, 2 by t, and 2 by 2. The sum of the areas of these rectangles is $t \cdot t + 2 \cdot t + 2 \cdot t + 2 \cdot 2$, or $t^2 + 2t + 2t + 4$, or $t^2 + 4t + 4$.

48. Two sides have dimensions a by 5, two other sides have dimensions a by 9, and the two remaining sides have dimensions 9 by 5. Then the surface area is $2 \cdot a \cdot 5 + 2 \cdot a \cdot 9 + 2 \cdot 9 \cdot 5$, or $10a + 18a + 90$, or $28a + 90$. Answer B is correct.

49. Let l = the length of the box. Then the height is $l-1$ and the width is $l-2$. The volume of the box is length $\times$ width $\times$ height.
$V = l(l-2)(l-1)$
$V = (l^2 - 2l)(l-1)$
$V = l^3 - l^2 - 2l^2 + 2l$
$V = l^3 - 3l^2 + 2l$

50. $(x-5)(x+5) = (x+6)^2$
$x^2 - 25 = x^2 + 12x + 36$
$-25 = 12x + 36$ Subtracting x^2
$-61 = 12x$
$-\dfrac{61}{12} = x$

The solution is $-\dfrac{61}{12}$.

Chapter 13

Polynomials: Factoring

Exercise Set 13.1

1. $x^2 = x^2$

$-6x = -1 \cdot 2 \cdot 3 \cdot x$

The coefficients have no common prime factor. The GCF of the powers of x is x because 1 is the smallest exponent of x. Thus the GCF is x.

3. $3x^4 = 3 \cdot x^4$

$x^2 = x^2$

The coefficients have no common prime factor. The GCF of the powers of x is x^2 because 2 is the smallest exponent of x. Thus the GCF is x^2.

5. $2x^2 = 2 \cdot x^2$

$2x = 2 \cdot x$

$-8 = -1 \cdot 2 \cdot 2 \cdot 2$

Each coefficient has a factor of 2. There are no other common prime factors. The GCF of the powers of x is 0 since -8 has no x-factor. Thus the GCF is 2.

7. $-17x^5y^3 = -1 \cdot 17 \cdot x^5 \cdot y^3$

$34x^3y^2 = 2 \cdot 17 \cdot x^3 \cdot y^2$

$51xy = 3 \cdot 17 \cdot x \cdot y$

Each coefficient has a factor of 17. There are no other common prime factors. The GCF of the powers of x is x because 1 is the smallest exponent of x. Similarly, the GCF of the powers of y is y because 1 is the smallest exponent of y. Thus the GCF is $17xy$.

9. $-x^2 = -1 \cdot x^2$

$-5x = -1 \cdot 5 \cdot x$

$-20x^3 = -1 \cdot 2 \cdot 2 \cdot 5 \cdot x^3$

The coefficients have no common prime factor. (Note that -1 is not a prime number.) The GCF of the powers of x is x because 1 is the smallest exponent of x. Thus the GCF is x.

11. $x^5y^5 = x^5 \cdot y^5$

$x^4y^3 = x^4 \cdot y^3$

$x^3y^3 = x^3 \cdot y^3$

$-x^2y^2 = -1 \cdot x^2 \cdot y^2$

There is no common prime factor. The GCF of the powers of x is x^2 because 2 is the smallest exponent of x. Similarly, the GCF of the powers of y is y^2 because 2 is the smallest exponent of y. Thus the GCF is x^2y^2.

13. $x^2 - 6x = x \cdot x - x \cdot 6$ Factoring each term

$= x(x - 6)$ Factoring out the common factor x

15. $2x^2 + 6x = 2x \cdot x + 2x \cdot 3$ Factoring each term

$= 2x(x + 3)$ Factoring out the common factor $2x$

17. $x^3 + 6x^2 = x^2 \cdot x + x^2 \cdot 6$ Factoring each term

$= x^2(x + 6)$ Factoring out x^2

19. $8x^4 - 24x^2 = 8x^2 \cdot x^2 - 8x^2 \cdot 3$

$= 8x^2(x^2 - 3)$ Factoring out $8x^2$

21. $2x^2 + 2x - 8 = 2 \cdot x^2 + 2 \cdot x - 2 \cdot 4$

$= 2(x^2 + x - 4)$ Factoring out 2

23. $17x^5y^3 + 34x^3y^2 + 51xy$

$= 17xy \cdot x^4y^2 + 17xy \cdot 2x^2y + 17xy \cdot 3$

$= 17xy(x^4y^2 + 2x^2y + 3)$

25. $6x^4 - 10x^3 + 3x^2 = x^2 \cdot 6x^2 - x^2 \cdot 10x + x^2 \cdot 3$

$= x^2(6x^2 - 10x + 3)$

27. $x^5y^5 + x^4y^3 + x^3y^3 - x^2y^2$

$= x^2y^2 \cdot x^3y^3 + x^2y^2 \cdot x^2y + x^2y^2 \cdot xy + x^2y^2(-1)$

$= x^2y^2(x^3y^3 + x^2y + xy - 1)$

29. $2x^7 - 2x^6 - 64x^5 + 4x^3$

$= 2x^3 \cdot x^4 - 2x^3 \cdot x^3 - 2x^3 \cdot 32x^2 + 2x^3 \cdot 2$

$= 2x^3(x^4 - x^3 - 32x^2 + 2)$

31. $1.6x^4 - 2.4x^3 + 3.2x^2 + 6.4x$

$= 0.8x(2x^3) - 0.8x(3x^2) + 0.8x(4x) + 0.8x(8)$

$= 0.8x(2x^3 - 3x^2 + 4x + 8)$

33. $\frac{5}{3}x^6 + \frac{4}{3}x^5 + \frac{1}{3}x^4 + \frac{1}{3}x^3$

$= \frac{1}{3}x^3(5x^3) + \frac{1}{3}x^3(4x^2) + \frac{1}{3}x^3(x) + \frac{1}{3}x^3(1)$

$= \frac{1}{3}x^3(5x^3 + 4x^2 + x + 1)$

35. $x^2(x + 3) + 2(x + 3)$

The binomial $x + 3$ is common to both terms:

$x^2(x + 3) + 2(x + 3) = (x + 3)(x^2 + 2)$

37. $4z^2(3z - 1) + 7(3z - 1)$

The binomial $3z - 1$ is common to both terms.

$4z^2(3z - 1) + 7(3z - 1) = (3z - 1)(4z^2 + 7)$

39. $2x^2(3x + 2) + (3x + 2) = 2x^2(3x + 2) + 1 \cdot (3x + 2)$

$= (3x + 2)(2x^2 + 1)$

41. $5a^3(2a - 7) - (2a - 7)$

$= 5a^3(2a - 7) - 1(2a - 7)$

$= (2a - 7)(5a^3 - 1)$

43. $x^3 + 3x^2 + 2x + 6$
$= (x^3 + 3x^2) + (2x + 6)$
$= x^2(x + 3) + 2(x + 3)$ Factoring each binomial
$= (x + 3)(x^2 + 2)$ Factoring out the common factor $x + 3$

45. $2x^3 + 6x^2 + x + 3$
$= (2x^3 + 6x^2) + (x + 3)$
$= 2x^2(x + 3) + 1(x + 3)$ Factoring each binomial
$= (x + 3)(2x^2 + 1)$

47. $8x^3 - 12x^2 + 6x - 9 = 4x^2(2x - 3) + 3(2x - 3)$
$= (2x - 3)(4x^2 + 3)$

49. $12p^3 - 16p^2 + 3p - 4$
$= 4p^2(3p - 4) + 1(3p - 4)$ Factoring 1 out of the second binomial
$= (3p - 4)(4p^2 + 1)$

51. $5x^3 - 5x^2 - x + 1$
$= (5x^3 - 5x^2) + (-x + 1)$
$= 5x^2(x - 1) - 1(x - 1)$ Check: $-1(x-1) = -x+1$
$= (x - 1)(5x^2 - 1)$

53. $x^3 + 8x^2 - 3x - 24 = x^2(x + 8) - 3(x + 8)$
$= (x + 8)(x^2 - 3)$

55. $2x^3 - 8x^2 - 9x + 36 = 2x^2(x - 4) - 9(x - 4)$
$= (x - 4)(2x^2 - 9)$

57. $-2x < 48$
$x > -24$ Dividing by -2 and reversing the inequality symbol

The solution set is $\{x|x > -24\}$.

59. $\dfrac{-108}{-4} = 27$ (The quotient of two negative numbers is positive.)

61. $(y + 5)(y + 7) = y^2 + 7y + 5y + 35$ Using FOIL
$= y^2 + 12y + 35$

63. $(y + 7)(y - 7) = y^2 - 7^2 = y^2 - 49$
$[(A + B))(A - B) = A^2 - B^2]$

65. $x + y = 4$

To find the x-intercept, let $y = 0$. Then solve for x.

$x + y = 4$
$x + 0 = 4$
$x = 4$

The x-intercept is $(4, 0)$.

To find the y-intercept, let $x = 0$. Then solve for y.

$x + y = 4$
$0 + y = 4$
$y = 4$

The y-intercept is $(0, 4)$.

Plot these points and draw the line.

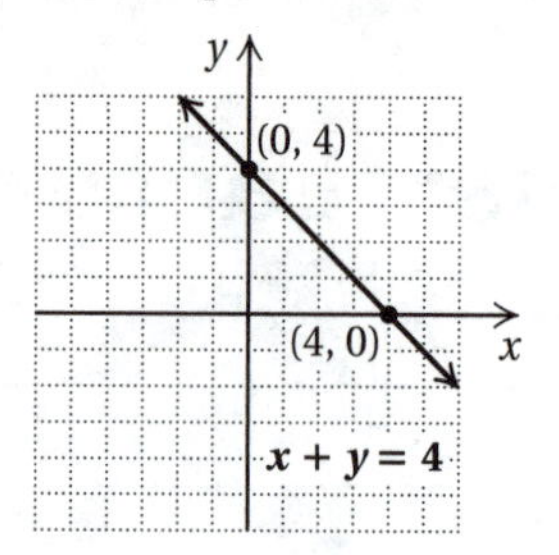

A third point should be used as a check. We substitute any value for x and solve for y. We let $x = 2$. Then

$x + y = 4$
$2 + y = 4$
$y = 2$

The point $(2, 2)$ is on the graph, so the graph is probably correct.

67. $5x - 3y = 15$

To find the x-intercept, let $y = 0$. Then solve for x.

$5x - 3y = 15$
$5x - 3 \cdot 0 = 15$
$5x = 15$
$x = 3$

The x-intercept is $(3, 0)$.

To find the y-intercept, let $x = 0$. Then solve for y.

$5x - 3y = 15$
$5 \cdot 0 - 3y = 15$
$-3y = 15$
$y = -5$

The y-intercept is $(0, -5)$.

Plot these points and draw the line.

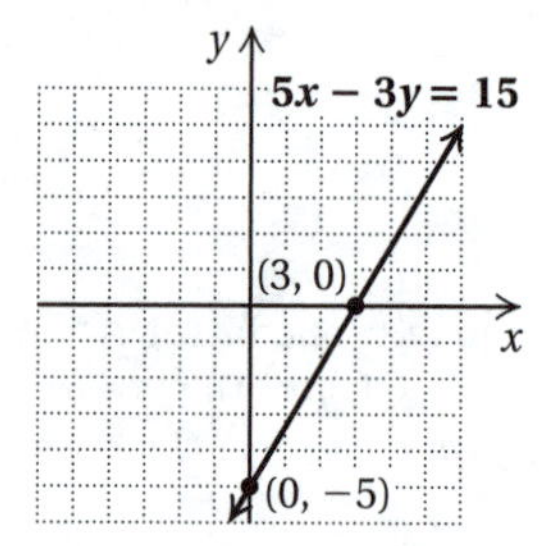

A third point should be used as a check. We substitute any value for x and solve for y. We let $x = 6$. Then

$5x - 3y = 15$
$5 \cdot 6 - 3y = 15$
$30 - 3y = 15$
$-3y = -15$
$y = 5$

The point $(6, 5)$ is on the graph, so the graph is probably correct.

69. $4x^5 + 6x^3 + 6x^2 + 9 = 2x^3(2x^2 + 3) + 3(2x^2 + 3)$
$= (2x^3 + 3)(2x^2 + 3)$

71. $x^{12} + x^7 + x^5 + 1 = x^7(x^5 + 1) + (x^5 + 1)$
$= (x^7 + 1)(x^5 + 1)$

73. $p^3 + p^2 - 3p + 10 = p^2(p + 1) - (3p - 10)$

This polynomial is not factorable using factoring by grouping.

Exercise Set 13.2

1. $x^2 + 8x + 15$

Since the constant term and coefficient of the middle term are both positive, we look for a factorization of 15 in which both factors are positive. Their sum must be 8.

Pairs of factors	Sums of factors
1, 15	16
3, 5	8

The numbers we want are 3 and 5.

$x^2 + 8x + 15 = (x + 3)(x + 5)$.

3. $x^2 + 7x + 12$

Since the constant term is positive and the coefficient of the middle term is positive, we look for a factorization of 12 in which both factors are positive. Their sum must be 7.

Pairs of factors	Sums of factors
1, 12	13
2, 6	8
3, 4	7

The numbers we want are 3 and 4.

$x^2 + 7x + 12 = (x + 3)(x + 4)$.

5. $x^2 - 6x + 9$

Since the constant term is positive and the coefficient of the middle term is negative, we look for a factorization of 9 in which both factors are negative. Their sum must be -6.

Pairs of factors	Sums of factors
$-1, -9$	-10
$-3, -3$	-6

The numbers we want are -3 and -3.

$x^2 - 6x + 9 = (x - 3)(x - 3)$, or $(x - 3)^2$.

7. $x^2 - 5x - 14$

Since the constant term is negative, we look for a factorization of -14 in which one factor is positive and one factor is negative. Their sum must be -5, the coefficient of the middle term.

Pairs of factors	Sums of factors
$-1, 14$	13
$1, -14$	-13
$-2, 7$	5
$2, -7$	-5

The numbers we want are 2 and -7.

$x^2 - 5x - 14 = (x + 2)(x - 7)$.

9. $b^2 + 5b + 4$

Since the constant term is positive and the coefficient of the middle term is positive, we look for a factorization of 4 in which both factors are positive. Their sum must be 5.

Pairs of factors	Sums of factors
1, 4	5
2, 2	4

The numbers we want are 1 and 4.

$b^2 + 5b + 4 = (b + 1)(b + 4)$.

11. $x^2 + \frac{2}{3}x + \frac{1}{9}$

Since the constant term is positive and the coefficient of the middle term is positive, we look for a factorization of $\frac{1}{9}$ in which both factors are positive. Their sum must be $\frac{2}{3}$.

Pairs of factors	Sums of factors
$1, \frac{1}{9}$	$\frac{10}{9}$
$\frac{1}{3}, \frac{1}{3}$	$\frac{2}{3}$

The numbers we want are $\frac{1}{3}$ and $\frac{1}{3}$.

$x^2 + \frac{2}{3}x + \frac{1}{9} = \left(x + \frac{1}{3}\right)\left(x + \frac{1}{3}\right)$, or $\left(x + \frac{1}{3}\right)^2$.

13. $d^2 - 7d + 10$

Since the constant term is positive and the coefficient of the middle term is negative, we look for a factorization of 10 in which both factors are negative. Their sum must be -7.

Pairs of factors	Sums of factors
$-1, -10$	-11
$-2, -5$	-7

The numbers we want are -2 and -5.

$d^2 - 7d + 10 = (d - 2)(d - 5)$.

15. $y^2 - 11y + 10$

Since the constant term is positive and the coefficient of the middle term is negative, we look for a factorization of 10 in which both factors are negative. Their sum must be -11.

Pairs of factors	Sums of factors
$-1, -10$	-11
$-2, -5$	-7

The numbers we want are -1 and -10.

$y^2 - 11y + 10 = (y - 1)(y - 10)$.

17. $x^2 + x + 1$

Since the constant term and the coefficient of the middle term are both positive, we look for a factorization of 1 in which both factors are positive. The sum must be 1. The only possible pair of factors is 1 and 1, but their sum is not 1. Thus, this polynomial is not factorable into binomials. It is prime.

19. $x^2 - 7x - 18$

Since the constant term is negative, we look for a factorization of -18 in which one factor is positive and one factor is negative. Their sum must be -7, the coefficient of the middle term.

Pairs of factors	Sums of factors
$-1, 18$	17
$1, -18$	-17
$-2, 9$	7
$2, -9$	-7
$-3, 6$	3
$3, -6$	-3

The numbers we want are 2 and -9.

$x^2 - 7x - 18 = (x + 2)(x - 9)$.

21. $x^3 - 6x^2 - 16x = x(x^2 - 6x - 16)$

After factoring out the common factor, x, we consider $x^2 - 6x - 16$. Since the constant term is negative, we look for a factorization of -16 in which one factor is positive and one factor is negative. Their sum must be -6, the coefficient of the middle term.

Pairs of factors	Sums of factors
$-1, 16$	15
$1, -16$	-15
$-2, 8$	6
$2, -8$	-6
$-4, 4$	0

The numbers we want are 2 and -8.

Then $x^2 - 6x - 16 = (x + 2)(x - 8)$, so $x^3 - 6x^2 - 16x = x(x + 2)(x - 8)$.

23. $y^3 - 4y^2 - 45y = y(y^2 - 4y - 45)$

After factoring out the common factor, y, we consider $y^2 - 4y - 45$. Since the constant term is negative, we look for a factorization of -45 in which one factor is positive and one factor is negative. Their sum must be -4, the coefficient of the middle term.

Pairs of factors	Sums of factors
$-1, 45$	44
$1, -45$	-44
$-3, 15$	12
$3, -15$	-12
$-5, 9$	4
$5, -9$	-4

The numbers we want are 5 and -9.

Then $y^2 - 4y - 45 = (y + 5)(y - 9)$, so $y^3 - 4y^2 - 45y = y(y + 5)(y - 9)$.

25. $-2x - 99 + x^2 = x^2 - 2x - 99$

Since the constant term is negative, we look for a factorization of -99 in which one factor is positive and one factor is negative. Their sum must be -2, the coefficient of the middle term.

Pairs of factors	Sums of factors
$-1, 99$	98
$1, -99$	-98
$-3, 33$	30
$3, -33$	-30
$-9, 11$	2
$9, -11$	-2

The numbers we want are 9 and -11.

$-2x - 99 + x^2 = (x + 9)(x - 11)$.

27. $c^4 + c^2 - 56$

Consider this trinomial as $(c^2)^2 + c^2 - 56$. We look for numbers p and q such that $c^4 + c^2 - 56 = (c^2 + p)(c^2 + q)$. Since the constant term is negative, we look for a factorization of -56 in which one factor is positive and one factor is negative. Their sum must be 1.

Pairs of factors	Sums of factors
$-1, 56$	55
$1, -56$	-55
$-2, 28$	26
$2, -28$	-26
$-4, 14$	12
$4, -14$	-12
$-7, 8$	1
$7, -8$	-1

The numbers we want are -7 and 8.

$c^4 + c^2 - 56 = (c^2 - 7)(c^2 + 8)$.

29. $a^4 + 2a^2 - 35$

Consider this trinomial as $(a^2)^2 + 2a^2 - 35$. We look for numbers p and q such that $a^4 + 2a^2 - 35 = (a^2 + p)(a^2 + q)$. Since the constant term is negative, we look for a factorization of -35 in which one factor is positive and one factor is negative. Their sum must be 2.

Pairs of factors	Sums of factors
−1, 35	34
1, −35	−34
−5, 7	2
5, −7	−2

The numbers we want are −5 and 7.

$a^4 + 2a^2 - 35 = (a^2 - 5)(a^2 + 7)$.

31. $x^2 + x - 42$

Since the constant term is negative, we look for a factorization of −42 in which one factor is positive and one factor is negative. Their sum must be 1, the coefficient of the middle term.

Pairs of factors	Sums of factors
−1, 42	41
1, −42	−41
−2, 21	19
2, −21	−19
−3, 14	11
3, −14	−11
−6, 7	1
6, −7	−1

The numbers we want are −6 and 7.

$x^2 + x - 42 = (x - 6)(x + 7)$.

33. $7 - 2p + p^2 = p^2 - 2p + 7$

Since the constant term is positive and the coefficient of the middle term is negative, we look for a factorization of 7 in which both factors are negative. The sum must be −2. The only possible pair of factors is −1 and −7, but their sum is not −2. Thus, this polynomial is not factorable into binomials. It is prime.

35. $x^2 + 20x + 100$

We look for two factors, both positive, whose product is 100 and whose sum is 20.

They are 10 and 10. $10 \cdot 10 = 100$ and $10 + 10 = 20$.

$x^2 + 20x + 100 = (x + 10)(x + 10)$, or $(x + 10)^2$.

37. $2z^3 - 2z^2 - 24z = 2z(z^2 - z - 12)$

After factoring out the common factor, we consider $z^2 - z - 12$. Since the constant term is negative, we look for a factorization of −12 in which one factor is positive and one factor is negative. Their sum must be −1, the coefficient of the middle term.

Pairs of factors	Sums of factors
−1, 12	11
1, −12	−11
−2, 6	4
2, −6	−4
−3, 4	1
3, −4	−1

The numbers we want are 3 and −4.

Then $z^2 - z - 12 = (z + 3)(z - 4)$, so $2z^3 - 2z^2 - 24z = 2z(z + 3)(z - 4)$.

39. $3t^4 + 3t^3 + 3t^2 = 3t^2(t^2 + t + 1)$

After factoring out the common factor, we consider t^2+t+1. Since the constant term and the coefficient of the middle term are both positive, we look for a factorization of 1 in which both factors must be positive. Their sum must be 1. Since the only possible factorization of 1 is $1 \cdot 1$ and $1 + 1 \neq 1$, we see that $t^2 + t + 1$ cannot be factored as a product of polynomials. Thus, $3t^4 + 3t^3 + 3t^2 = 3t^2(t^2 + t + 1)$.

41. $x^4 - 21x^3 - 100x^2 = x^2(x^2 - 21x - 100)$

After factoring out the common factor, x^2, we consider $x^2 - 21x - 100$. We look for two factors, one positive and one negative, whose product is −100 and whose sum is −21. They are 4 and −25. $4 \cdot (-25) = -100$ and $4 + (-25) = -21$.

Then $x^2-21x-100 = (x+4)(x-25)$, so $x^4-21x^3-100x^2 = x^2(x + 4)(x - 25)$.

43. $x^2 - 21x - 72$

We look for two factors, one positive and one negative, whose product is −72 and whose sum is −21. They are 3 and −24.

$x^2 - 21x - 72 = (x + 3)(x - 24)$.

45. $x^2 - 25x + 144$

We look for two factors, both negative, whose product is 144 and whose sum is −25. They are −9 and −16.

$x^2 - 25x + 144 = (x - 9)(x - 16)$.

47. $a^2 + a - 132$

We look for two factors, one positive and one negative, whose product is −132 and whose sum is 1. They are −11 and 12.

$a^2 + a - 132 = (a - 11)(a + 12)$.

49. $3t^2 + 6t + 3 = 3(t^2 + 2t + 1)$

After factoring out the common factor, we consider $t^2 + 2t + 1$. We look for two factors, both positive, whose product is 1 and whose sum is 2. They are 1 and 1.

Then $t^2+2t+1 = (t+1)(t+1)$, so $3t^2+6t+3 = 3(t+1)(t+1)$, or $3(t + 1)^2$.

51. $w^4 - 8w^3 + 16w^2 = w^2(w^2 - 8w + 16)$

After factoring out the common factor, we consider $w^2 - 8w + 16$. We look for two factors, both negative, whose product is 16 and whose sum is −8. They are −4 and −4.

Then $w^2-8w+16 = (w-4)(w-4)$, so $w^4-8w^3+16w^2 = w^2(w - 4)(w - 4)$, or $w^2(w - 4)^2$.

53. $30 + 7x - x^2 = -x^2 + 7x + 30 = -1(x^2 - 7x - 30)$

Now we factor $x^2 - 7x - 30$. Since the constant term is negative, we look for a factorization of −30 in which one factor is positive and one factor is negative. Their sum must be −7, the coefficient of the middle term.

Pairs of factors	Sums of factors
$-1,\ 30$	29
$1,\ -30$	-29
$-2,\ 15$	13
$2,\ -15$	-13
$-3,\ 10$	7
$3,\ -10$	-7
$-5,\ 6$	1
$5,\ -6$	-1

The numbers we want are 3 and -10. Then

$x^2 - 7x - 30 = (x+3)(x-10)$, so we have:

$-x^2 + 7x + 30$

$= -1(x+3)(x-10)$

$= (-x-3)(x-10)$ Multiplying $x+3$ by -1

$= (x+3)(-x+10)$ Multiplying $x-10$ by -1

55. $24 - a^2 - 10a = -a^2 - 10a + 24 = -1(a^2 + 10a - 24)$

Now we factor $a^2 + 10a - 24$. Since the constant term is negative, we look for a factorization of -24 in which one factor is positive and one factor is negative. Their sum must be 10, the coefficient of the middle term.

Pairs of factors	Sums of factors
$-1,\ 24$	23
$1,\ -24$	-23
$-2,\ 12$	10
$2,\ -12$	-10
$-3,\ 8$	5
$3,\ -8$	-5
$-4,\ 6$	2
$4,\ -6$	-2

The numbers we want are -2 and 12. Then

$a^2 + 10a - 24 = (a-2)(a+12)$, so we have:

$-a^2 - 10a + 24$

$= -1(a-2)(a+12)$

$= (-a+2)(a+12)$ Multiplying $a-2$ by -1

$= (a-2)(-a-12)$ Multiplying $a+12$ by -1

57. $120 - 23x + x^2 = x^2 - 23x + 120$

We look for two factors, both negative, whose product is 120 and whose sum is -23. They are -8 and -15.

$x^2 - 23x + 120 = (x-8)(x-15)$.

59. First write the polynomial in descending order and factor out -1.

$108 - 3x - x^2 = -x^2 - 3x + 108 = -1(x^2 + 3x - 108)$

Now we factor the polynomial $x^2 + 3x - 108$. We look for two factors, one positive and one negative, whose product is -108 and whose sum is 3. They are -9 and 12.

$x^2 + 3x - 108 = (x-9)(x+12)$

The final answer must include -1 which was factored out above.

$-x^2 - 3x + 108$

$= -1(x-9)(x+12)$

$= (-x+9)(x+12)$ Multiplying $x-9$ by -1

$= (x-9)(-x-12)$ Multiplying $x+12$ by -1

61. $y^2 - 0.2y - 0.08$

We look for two factors, one positive and one negative, whose product is -0.08 and whose sum is-0.2. They are -0.4 and 0.2.

$y^2 - 0.2y - 0.08 = (y-0.4)(y+0.2)$.

63. $p^2 + 3pq - 10q^2 = p^2 + 3qp - 10q^2$

Think of $3q$ as a "coefficient" of p. Then we look for factors of $-10q^2$ whose sum is $3q$. They are $5q$ and $-2q$.

$p^2 + 3pq - 10q^2 = (p+5q)(p-2q)$.

65. $84 - 8t - t^2 = -t^2 - 8t + 84 = -1(t^2 + 8t - 84)$

Now we factor $t^2 + 8t - 84$. We look for two factors, one positive and one negative, whose product is -84 and whose sum is 8. They are 14 and -6.

Then $t^2 + 8t - 84 = (t+14)(t-6)$, so we have:

$-t^2 - 8t + 84$

$= -1(t+14)(t-6)$

$= (-t-14)(t-6)$ Multiplying $t+14$ by -1

$= (t+14)(-t+6)$ Multiplying $t-6$ by -1

67. $m^2 + 5mn + 4n^2 = m^2 + 5nm + 4n^2$

We look for factors of $4n^2$ whose sum is $5n$. They are $4n$ and n.

$m^2 + 5mn + 4n^2 = (m+4n)(m+n)$

69. $s^2 - 2st - 15t^2 = s^2 - 2ts - 15t^2$

We look for factors of $-15t^2$ whose sum is $-2t$. They are $-5t$ and $3t$.

$s^2 - 2st - 15t^2 = (s-5t)(s+3t)$

71. $6a^{10} - 30a^9 - 84a^8 = 6a^8(a^2 - 5a - 14)$

After factoring out the common factor, $6a^8$, we consider $a^2 - 5a - 14$. We look for two factors, one positive and one negative, whose product is -14 and whose sum is -5. They are 2 and -7.

$a^2 - 5a - 14 = (a+2)(a-7)$, so $6a^{10} - 30a^9 - 84a^8 = 6a^8(a+2)(a-7)$.

73. $8x(2x^2 - 6x + 1) = 8x \cdot 2x^2 - 8x \cdot 6x + 8x \cdot 1 = 16x^3 - 48x^2 + 8x$

75. $(7w+6)^2 = (7w)^2 + 2 \cdot 7w \cdot 6 + 6^2 = 49w^2 + 84w + 36$

77. $(4w-11)(4w+11) = (4w)^2 - (11)^2 = 16w^2 - 121$

79. $(3x-5y)(2x+7y) = 3x \cdot 2x + 3x \cdot 7y - 5y \cdot 2x - 5y \cdot 7y = 6x^2 + 21xy - 10xy - 35y^2 = 6x^2 + 11xy - 35y^2$

81. $3x - 8 = 0$

$3x = 8$ Adding 8 on both sides

$x = \frac{8}{3}$ Dividing by 3 on both sides

The solution is $\frac{8}{3}$.

83. ***Familiarize***. Let a = the number of arrests made for counterfeiting in 2007.

Translate.

$$\underbrace{\text{Arrests in 2007}}_{a} \quad \underbrace{\text{plus}}_{+} \quad \underbrace{28\%}_{28\%} \quad \underbrace{\text{of}}_{\cdot} \quad \underbrace{\text{Arrests in 2007}}_{a} \quad \underbrace{\text{is}}_{=} \quad \underbrace{\text{Arrests in 2008.}}_{2231}$$

Solve. We solve the equation.

$a + 28\% \cdot a = 2231$

$1 \cdot a + 0.28a = 2231$

$1.28a = 2231$

$a \approx 1743$

Check. 28% of 1743 is $0.28(1743) \approx 488$ and $1743 + 488 = 2231$, the number of arrests in 2008. The answer checks.

State. About 1743 arrests for counterfeiting were made in 2007.

85. $y^2 + my + 50$

We look for pairs of factors whose product is 50. The sum of each pair is represented by m.

Pairs of factors whose product is −50	Sums of factors
1, 50	51
−1, −50	−51
2, 25	27
−2, −25	−27
5, 10	15
−5, −10	−15

The polynomial $y^2 + my + 50$ can be factored if m is 51, −51, 27, −27, 15, or −15.

87. $x^2 - \frac{1}{2}x - \frac{3}{16}$

We look for two factors, one positive and one negative, whose product is $-\frac{3}{16}$ and whose sum is $-\frac{1}{2}$.

They are $-\frac{3}{4}$ and $\frac{1}{4}$.

$-\frac{3}{4} \cdot \frac{1}{4} = -\frac{3}{16}$ and $-\frac{3}{4} + \frac{1}{4} = -\frac{2}{4} = -\frac{1}{2}$.

$x^2 - \frac{1}{2}x - \frac{3}{16} = \left(x - \frac{3}{4}\right)\left(x + \frac{1}{4}\right)$

89. $x^2 + \frac{30}{7}x - \frac{25}{7}$

We look for two factors, one positive and one negative, whose product is $-\frac{25}{7}$ and whose sum is $\frac{30}{7}$.

They are 5 and $-\frac{5}{7}$.

$5 \cdot \left(-\frac{5}{7}\right) = -\frac{25}{7}$ and $5 + \left(-\frac{5}{7}\right) = \frac{35}{7} + \left(-\frac{5}{7}\right) = \frac{30}{7}$.

$x^2 + \frac{30}{7}x - \frac{25}{7} = (x + 5)\left(x - \frac{5}{7}\right)$

91. $b^{2n} + 7b^n + 10$

Consider this trinomial as $(b^n)^2 + 7b^n + 10$. We look for numbers p and q such that $b^{2n} + 7b^n + 10 = (b^n + p)(b^n + q)$. We find two factors, both positive, whose product is 10 and whose sum is 7. They are 5 and 2.

$b^{2n} + 7b^n + 10 = (b^n + 5)(b^n + 2)$

93. We first label the drawing with additional information.

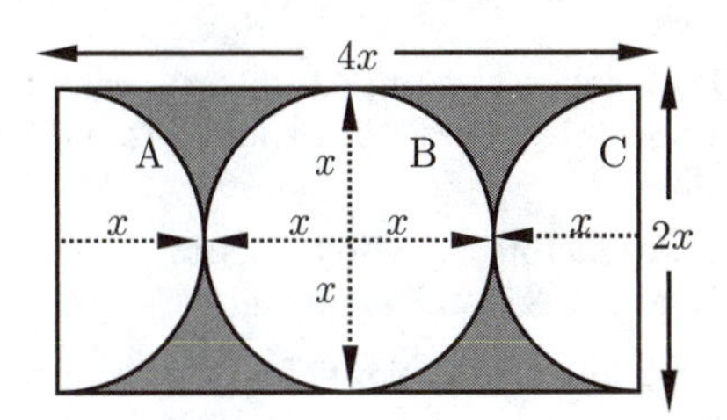

$4x$ represents the length of the rectangle and $2x$ the width. The area of the rectangle is $4x \cdot 2x$, or $8x^2$.

The area of semicircle A is $\frac{1}{2}\pi x^2$.

The area of circle B is πx^2.

The area of semicircle C is $\frac{1}{2}\pi x^2$.

$$\text{Area of shaded region} = \text{Area of rectangle} - \text{Area of } A - \text{Area of } B - \text{Area of } C$$

$$\text{Area of shaded region} = 8x^2 - \frac{1}{2}\pi x^2 - \pi x^2 - \frac{1}{2}\pi x^2$$

$$= 8x^2 - 2\pi x^2$$

$$= 2x^2(4 - \pi)$$

The shaded area can be represented by $2x^2(4 - \pi)$.

Exercise Set 13.3

1. $2x^2 - 7x - 4$

(1) Look for a common factor. There is none (other than 1 or −1).

(2) Factor the first term, $2x^2$. The only possibility is $2x, x$. The desired factorization is of the form:

$(2x+ \quad)(x+ \quad)$

(3) Factor the last term, −4, which is negative. The possibilities are −4, 1 and 4, −1 and 2, −2.

These factors can also be written as 1, −4 and −1, 4 and −2, 2.

(4) Look for combinations of factors from steps (2) and (3) such that the sum of their products is the middle term, $-7x$. We try some possibilities:

$(2x-4)(x+1) = 2x^2 - 2x - 4$

$(2x+4)(x-1) = 2x^2 + 2x - 4$

$(2x+2)(x-2) = 2x^2 - 2x - 4$

$(2x+1)(x-4) = 2x^2 - 7x - 4$

The factorization is $(2x+1)(x-4)$.

3. $5x^2 - x - 18$

(1) There is no common factor (other than 1 or -1).

(2) Factor the first term, $5x^2$. The only possibility is $5x$, x. The desired factorization is of the form:

$$(5x+\quad)(x+\quad)$$

(3) Factor the last term, -18. The possibilities are -18, 1 and 18, -1 and -9, 2 and 9, -2 and -6, 3 and 6, -3.

These factors can also be written as 1, -18 and -1, 18 and 2, -9 and -2, 9 and 3, -6 and -3, 6.

(4) Look for combinations of factors from steps (2) and (3) such that the sum of their products is the middle term, x. We try some possibilities:

$(5x-18)(x+1) = 5x^2 - 13x - 18$

$(5x+18)(x-1) = 5x^2 + 13x - 18$

$(5x+9)(x-2) = 5x^2 - x - 18$

The factorization is $(5x+9)(x-2)$.

5. $6x^2 + 23x + 7$

(1) There is no common factor (other than 1 or -1).

(2) Factor the first term, $6x^2$. The possibilities are $6x$, x and $3x$, $2x$. We have these as possibilities for factorizations:

$$(6x+\quad)(x+\quad) \text{ and } (3x+\quad)(2x+\quad)$$

(3) Factor the last term, 7. The possibilities are 7, 1 and -7, -1.

These factors can also be written as 1, 7 and -1, -7.

(4) Look for combinations of factors from steps (2) and (3) such that the sum of their products is the middle term, $23x$. Since all signs are positive, we need consider only plus signs. We try some possibilities:

$(6x+7)(x+1) = 6x^2 + 13x + 7$

$(3x+7)(2x+1) = 6x^2 + 17x + 7$

$(6x+1)(x+7) = 6x^2 + 43x + 7$

$(3x+1)(2x+7) = 6x^2 + 23x + 7$

The factorization is $(3x+1)(2x+7)$.

7. $3x^2 + 4x + 1$

(1) There is no common factor (other than 1 or -1).

(2) Factor the first term, $3x^2$. The only possibility is $3x$, x. The desired factorization is of the form:

$$(3x+\quad)(x+\quad)$$

(3) Factor the last term, 1. The possibilities are 1, 1 and -1, -1.

(4) Look for combinations of factors from steps (2) and (3) such that the sum of their products is the middle term, $4x$. Since all signs are positive, we need consider only plus signs. There is only one such possibility:

$(3x+1)(x+1) = 3x^2 + 4x + 1$

The factorization is $(3x+1)(x+1)$.

9. $4x^2 + 4x - 15$

(1) There is no common factor (other than 1 or -1).

(2) Factor the first term, $4x^2$. The possibilities are $4x$, x and $2x$, $2x$. We have these as possibilities for factorizations:

$$(4x+\quad)(x+\quad) \text{ and } (2x+\quad)(2x+\quad)$$

(3) Factor the last term, -15. The possibilities are 15, -1 and -15, 1 and 5, -3 and -5, 3.

These factors can also be written as -1, 15 and 1, -15 and -3, 5 and 3, -5.

(4) We try some possibilities:

$(4x+15)(x-1) = 4x^2 + 11x - 15$

$(2x+15)(2x-1) = 4x^2 + 28x - 15$

$(4x-15)(x+1) = 4x^2 - 11x - 15$

$(2x-15)(2x+1) = 4x^2 - 28x - 15$

$(4x+5)(x-3) = 4x^2 - 7x - 15$

$(2x+5)(2x-3) = 4x^2 + 4x - 15$

The factorization is $(2x+5)(2x-3)$.

11. $2x^2 - x - 1$

(1) There is no common factor (other than 1 or -1).

(2) Factor the first term, $2x^2$. The only possibility is $2x$, x. The desired factorization is of the form:

$$(2x+\quad)(x+\quad)$$

(3) Factor the last term, -1. The only possibility is -1, 1.

These factors can also be written as 1, -1.

(4) We try the possibilities:

$(2x-1)(x+1) = 2x^2 + x - 1$

$(2x+1)(x-1) = 2x^2 - x - 1$

The factorization is $(2x+1)(x-1)$.

13. $9x^2 + 18x - 16$

(1) There is no common factor (other than 1 or -1).

(2) Factor the first term, $9x^2$. The possibilities are $9x$, x and $3x$, $3x$. We have these as possibilities for factorizations:

$$(9x+\quad)(x+\quad) \text{ and } (3x+\quad)(3x+\quad)$$

(3) Factor the last term, -16. The possibilities are 16, -1 and -16, 1 and 8, -2 and -8, 2 and 4, -4.

These factors can also be written as -1, 16 and 1, -16 and -2, 8 and 2, -8 and -4, 4.

(4) We try some possibilities:

$(9x+16)(x-1) = 9x^2+7x-16$

$(3x+16)(3x-1) = 9x^2+45x-16$

$(9x-16)(x+1) = 9x^2-7x-16$

$(3x-16)(3x+1) = 9x^2-45x-16$

$(9x+8)(x-2) = 9x^2-10x-16$

$(3x+8)(3x-2) = 9x^2+18x-16$

The factorization is $(3x+8)(3x-2)$.

15. $3x^2-5x-2$

(1) There is no common factor (other than 1 or -1).

(2) Factor the first term, $3x^2$. The only possibility is $3x$, x. The desired factorization is of the form:

$(3x+\quad)(x+\quad)$

(3) Factor the last term, -2. The possibilities are 2, -1 and -2 and 1.

These factors can also be written as -1, 2 and 1, -2.

(4) We try some possibilities:

$(3x+2)(x-1) = 3x^2-x-2$

$(3x-2)(x+1) = 3x^2+x-2$

$(3x-1)(x+2) = 3x^2+5x-2$

$(3x+1)(x-2) = 3x^2-5x-2$

The factorization is $(3x+1)(x-2)$.

17. $12x^2+31x+20$

(1) There is no common factor (other than 1 or -1).

(2) Factor the first term, $12x^2$. The possibilities are $12x$, x and $6x$, $2x$ and $4x$, $3x$. We have these as possibilities for factorizations:

$(12x+\quad)(x+\quad)$ and $(6x+\quad)(2x+\quad)$ and $(4x+\quad)(3x+\quad)$

(3) Factor the last term, 20. Since all signs are positive, we need consider only positive pairs of factors. Those factor pairs are 20, 1 and 10, 2 and 5, 4.

These factors can also be written as 1, 20 and 2, 10 and 4, 5.

(4) We can immediately reject all possibilities in which either factor has a common factor, such as $(12x+20)$ or $(6x+4)$, because we determined at the outset that there are no common factors. We try some of the remaining possibilities:

$(12x+1)(x+20) = 12x^2+241x+20$

$(12x+5)(x+4) = 12x^2+53x+20$

$(6x+1)(2x+20) = 12x^2+122x+20$

$(4x+5)(3x+4) = 12x^2+31x+20$

The factorization is $(4x+5)(3x+4)$.

19. $14x^2+19x-3$

(1) There is no common factor (other than 1 or -1).

(2) Factor the first term, $14x^2$. The possibilities are $14x$, x and $7x$, $2x$. We have these as possibilities for factorizations:

$(14x+\quad)(x+\quad)$ and $(7x+\quad)(2x+\quad)$

(3) Factor the last term, -3. The possibilities are -1, 3 and -3, 1.

These factors can also be written as 3, -1 and 1, -3.

(4) We try some possibilities:

$(14x-1)(x+3) = 14x^2+41x-3$

$(7x-1)(2x+3) = 7x^2+19x-3$

The factorization is $(7x-1)(2x+3)$.

21. $9x^2+18x+8$

(1) There is no common factor (other than 1 or -1).

(2) Factor the first term, $9x^2$. The possibilities are $9x$, x and $3x$, $3x$. We have these as possibilities for factorizations:

$(9x+\quad)(x+\quad)$ and $(3x+\quad)(3x+\quad)$

(3) Factor the last term, 8. Since all signs are positive, we need consider only positive pairs of factors. Those factor pairs are 8, 1 and 4, 2.

These factors can also be written as 1, 8 and 2, 4.

(4) We try some possibilities:

$(9x+8)(x+1) = 9x^2+17x+8$

$(3x+8)(3x+1) = 9x^2+27x+8$

$(9x+4)(x+2) = 9x^2+22x+8$

$(3x+4)(3x+2) = 9x^2+18x+8$

The factorization is $(3x+4)(3x+2)$.

23. $49-42x+9x^2 = 9x^2-42x+49$

(1) There is no common factor (other than 1 or -1).

(2) Factor the first term, $9x^2$. The possibilities are $9x$, x and $3x$, $3x$. We have these as possibilities for factorizations:

$(9x+\quad)(x+\quad)$ and $(3x+\quad)(3x+\quad)$

(3) Factor 49. Since 49 is positive and the middle term is negative, we need consider only negative pairs of factors. Those factor pairs are -49, -1 and -7, -7.

The first pair of factors can also be written as -1, -49.

(4) We try some possibilities:

$(9x-49)(x-1) = 9x^2 - 58x + 49$

$(3x-49)(3x-1) = 9x^2 - 150x + 49$

$(9x-7)(x-7) = 9x^2 - 70x + 49$

$(3x-7)(3x-7) = 9x^2 - 42x + 49$

The factorization is $(3x-7)(3x-7)$, or $(3x-7)^2$. This can also be expressed as follows:

$(3x-7)^2 = (-1)^2(3x-7)^2 = [-1 \cdot (3x-7)]^2 =$
$(-3x+7)^2$, or $(7-3x)^2$

25. $24x^2 + 47x - 2$

(1) There is no common factor (other than 1 or -1).

(2) Factor the first term, $24x^2$. The possibilities are $24x$, x and $12x$, $2x$ and $6x$, $4x$ and $3x$, $8x$. We have these as possibilities for factorizations:

$(24x+\quad)(x+\quad)$ and $(12x+\quad)(2x+\quad)$ and

$(6x+\quad)(4x+\quad)$ and $(3x+\quad)(8x+\quad)$

(3) Factor the last term, -2. The possibilities are 2, -1 and -2, 1.

These factors can also be written as -1, 2 and 1, -2.

(4) We can immediately reject all possibilities in which either factor has a common factor, such as $(24x+2)$ or $(12x-2)$, because we determined at the outset that there are no common factors. We try some of the remaining possibilities:

$(24x-1)(x+2) = 24x^2 + 47x - 2$

The factorization is $(24x-1)(x+2)$.

27. $35x^2 - 57x - 44$

(1) There is no common factor (other than 1 or -1).

(2) Factor the first term, $35x^2$. The possibilities are $35x$, x and $7x$, $5x$. We have these as possibilities for factorizations:

$(35x+\quad)(x+\quad)$ and $(7x+\quad)(5x+\quad)$

(3) Factor the last term, -44. The possibilities are 1, -44 and -1, 44 and 2, -22 and -2, 22 and 4, -11, and -4, 11.

These factors can also be written as -44, 1 and 44, -1 and -22, 2 and 22, -2 and -11, 4 and 11, -4.

(4) We try some possibilities:

$(35x+1)(x-44) = 35x^2 - 1539x - 44$

$(7x+1)(5x-44) = 35x^2 - 303x - 44$

$(35x+2)(x-22) = 35x^2 - 768x - 44$

$(7x+2)(5x-22) = 35x^2 - 144x - 44$

$(35x+4)(x-11) = 35x^2 - 381x - 44$

$(7x+4)(5x-11) = 35x^2 - 57x - 44$

The factorization is $(7x+4)(5x-11)$.

29. $20 + 6x - 2x^2 = -2x^2 + 6x + 20$

We factor out the common factor, -2. Factoring out -2 rather than 2 gives us a positive leading coefficient.

$-2(x^2 - 3x - 10)$

Then we factor the trinomial $x^2 - 3x - 10$. We look for a pair of factors whose product is -10 and whose sum is -3. The numbers are -5 and 2. The factorization of $x^2 - 3x - 10$ is $(x-5)(x+2)$. Then $20 + 6x - 2x^2 = -2(x-5)(x+2)$. If we think of -2 and $-1 \cdot 2$ then we can write other correct factorizations:

$20 + 6x - 2x^2$

$= 2(-x+5)(x+2)$ Multiplying $x-5$ by -1

$= 2(x-5)(-x-2)$ Multlplying $x+2$ by -1

Note that we can also express $2(-x+5)(x+2)$ as $2(5-x)(x+2)$ since $-x+5 = 5-x$ by the commutative law of addition.

31. $12x^2 + 28x - 24$

(1) We factor out the common factor, 4:
$4(3x^2 + 7x - 6)$

Then we factor the trinomial $3x^2 + 7x - 6$.

(2) Factor $3x^2$. The only possibility is $3x$, x. The desired factorization is of the form:

$(3x+\quad)(x+\quad)$

(3) Factor -6. The possibilities are 6, -1 and -6, 1 and 3, -2 and -3, 2.

These factors can also be written as -1, 6 and 1, -6 and -2, 3 and 2, -3.

(4) We can immediately reject all possibilities in which either factor has a common factor, such as $(3x+6)$ or $(3x-3)$, because we factored out the largest common factor at the outset. We try some of the remaining possibilities:

$(3x-1)(x+6) = 3x^2 + 17x - 6$

$(3x-2)(x+3) = 3x^2 + 7x - 6$

The factorization of $3x^2+7x-6$ is $(3x-2)(x+3)$. We must include the common factor in order to get a factorization of the original trinomial.

$12x^2 + 28x - 24 = 4(3x-2)(x+3)$

33. $30x^2 - 24x - 54$

(1) We factor out the common factor, 6:
$6(5x^2 - 4x - 9)$

Then we factor the trinomial $5x^2 - 4x - 9$.

(2) Factor $5x^2$. The only possibility is $5x$, x. The desired factorization is of the form:

$(5x+\quad)(x+\quad)$

(3) Factor -9. The possibilities are 9, -1 and -9, 1 and -3, 3.

These factors can also be written as -1, 9 and 1, -9 and 3, -3.

(4) We try some possibilities:

$(5x+9)(x-1) = 5x^2 + 4x - 9$

$(5x-9)(x+1) = 5x^2 - 4x - 9$

The factorization of $5x^2-4x-9$ is $(5x-9)(x+1)$. We must include the common factor in order to get a factorization of the original trinomial.

$30x^2 - 24x - 54 = 6(5x-9)(x+1)$

35. $4y + 6y^2 - 10 = 6y^2 + 4y - 10$

(1) We factor out the common factor, 2:
$2(3y^2 + 2y - 5)$
Then we factor the trinomial $3y^2 + 2y - 5$.

(2) Factor $3y^2$. The only possibility is $3y$, y. The desired factorization is of the form:

$(3y+\quad)(y+\quad)$

(3) Factor -5. The possibilities are 5, -1 and -5, 1.
These factors can also be written as -1, 5 and 1, -5.

(4) We try some possibilities:

$(3y+5)(y-1) = 3y^2 + 2y - 5$

Then $3y^2 + 2y - 5 = (3y+5)(y-1)$, so $6y^2 + 4y - 10 = 2(3y+5)(y-1)$.

37. $3x^2 - 4x + 1$

(1) There is no common factor (other than 1 or -1).

(2) Factor the first term, $3x^2$. The only possibility is $3x$, x. The desired factorization is of the form:

$(3x+\quad)(x+\quad)$

(3) Factor the last term, 1. Since 1 is positive and the middle term is negative, we need consider only negative factor pairs. The only such pair is -1, -1.

(4) There is only one possibility:

$(3x-1)(x-1) = 3x^2 - 4x + 1$

The factorization is $(3x-1)(x-1)$.

39. $12x^2 - 28x - 24$

(1) We factor out the common factor, 4:
$4(3x^2 - 7x - 6)$
Then we factor the trinomial $3x^2 - 7x - 6$.

(2) Factor $3x^2$. The only possibility is $3x$, x. The desired factorization is of the form:

$(3x+\quad)(x+\quad)$

(3) Factor -6. The possibilities are 6, -1 and -6, 1 and 3, -2 and -3, 2.
These factors can also be written as -1, 6 and 1, -6 and -2, 3 and 2, -3.

(4) We can immediately reject all possibilities in which either factor has a common factor, such as $(3x-6)$ or $(3x+3)$, because we factored out the largest common factor at the outset. We try some of the remaining possibilities:

$(3x-1)(x+6) = 3x^2 + 17x - 6$

$(3x-2)(x+3) = 3x^2 + 7x - 6$

$(3x+2)(x-3) = 3x^2 - 7x - 6$

Then $3x^2 - 7x - 6 = (3x+2)(x-3)$, so $12x^2 - 28x - 24 = 4(3x+2)(x-3)$.

41. $-1 + 2x^2 - x = 2x^2 - x - 1$

(1) There is no common factor (other than 1 or -1).

(2) Factor the first term, $2x^2$. The only possibility is $2x$, x. The desired factorization is of the form:

$(2x+\quad)(x+\quad)$

(3) Factor -1. The only possibility is 1, -1.

(4) We try some possibilities:

$(2x+1)(x-1) = 2x^2 - x - 1$

The factorization is $(2x+1)(x-1)$.

43. $9x^2 - 18x - 16$

(1) There is no common factor (other than 1 or -1).

(2) Factor the first term, $9x^2$. The possibilities are $9x$, x and $3x$, $3x$. We have these as possibilities for factorizations:

$(9x+\quad)(x+\quad)$ and $(3x+\quad)(3x+\quad)$

(3) Factor the last term, -16. The possibilities are 16, -1 and -16, 1 and 8, -2 and -8, 2 and 4, -4.
These factors can also be written as -1, 16 and 1, -16 and -2, 8 and and 2, -8 and -4, 4.

(4) We try some possibilities:

$(9x+16)(x-1) = 9x^2 + 7x - 16$

$(3x+16)(3x-1) = 9x^2 + 45x - 16$

$(9x+8)(x-2) = 9x^2 - 10x - 16$

$(3x+8)(3x-2) = 9x^2 + 18x - 16$

$(3x-8)(3x+2) = 9x^2 - 18x - 16$

The factorization is $(3x-8)(3x+2)$.

45. $15x^2 - 25x - 10$

(1) Factor out the common factor, 5:
$5(3x^2 - 5x - 2)$
Then we factor the trinomial $3x^2 - 5x - 2$. This was done in Exercise 15. We know that
$3x^2 - 5x - 2 = (3x+1)(x-2)$, so
$15x^2 - 25x - 10 = 5(3x+1)(x-2)$.

47. $12p^3 + 31p^2 + 20p$

(1) We factor out the common factor, p:

$p(12p^2 + 31p + 20)$

Then we factor the trinomial $12p^2 + 31p + 20$. This was done in Exercise 17 although the variable is x in that exercise. We know that $12p^2 + 31p + 20 = (3p + 4)(4p + 5)$, so $12p^3 + 31p^2 + 20p = p(3p + 4)(4p + 5)$.

49. $16 + 18x - 9x^2 = -9x^2 + 18x + 16$

$= -1(9x^2 - 18x - 16)$

$= -1(3x - 8)(3x + 2)$ Using the result from Exercise 43

Other correct factorizations are:

$16 + 18x - 9x^2$

$= (-3x + 8)(3x + 2)$ Multiplying $3x - 8$ by -1

$= (3x - 8)(-3x - 2)$ Multiplying $3x + 2$ by -1

We can also express $(-3x+8)(3x+2)$ as $(8-3x)(3x+2)$ since $-3x+8 = 8-3x$ by the commutative law of addition.

51. $-15x^2 + 19x - 6 = -1(15x^2 - 19x + 6)$

Now we factor $15x^2 - 19x + 6$.

(1) There is no common factor (other than 1 or -1).

(2) Factor the first term, $15x^2$. The possibilities are $15x$, x and $5x$, $3x$. We have these as possibilities for factorizations:

$(15x+\quad)(x+\quad)$ and $(5x+\quad)(3x+\quad)$

(3) Factor the last term, 6. The possibilities are 6, 1 and -6, -1 and 3, 2 and -3, -2.

These factors can also be written as 1, 6 and -1, -6 and 2, 3 and -2, -3.

(4) We try some possibilities:

$(15x + 1)(x + 6) = 15x^2 + 91x + 6$

$(5x + 3)(3x + 2) = 15x^2 - 19x + 6$

$(5x - 3)(3x - 2) = 15x^2 - 19x + 6$

The factorization of $15x^2 - 19x + 6$ is $(5x - 3)(3x - 2)$. Then $-15x^2+19x-6 = -1(5x-3)(3x-2)$. Other correct factorizations are:

$-15x^2 + 19x - 6$

$= (-5x + 3)(3x - 2)$ Multiplying $5x-3$ by -1

$= (5x - 3)(-3x + 2)$ Multiplying $3x-2$ by -1

Note that we can also express $(-5x + 3)(3x - 2)$ as $(3-5x)(3x-2)$ since $-5x+3 = 3-5x$ by the commutative law of addition. Similarly, we can express $(5x-3)(-3x+2)$ as $(5x - 3)(2 - 3x)$.

53. $14x^4 + 19x^3 - 3x^2$

(1) Factor out the common factor, x^2: $x^2(14x^2+19x-3)$

Then we factor the trinomial $14x^2 + 19x - 3$. This was done in Exercise 19. We know that $14x^2 + 19x - 3 = (7x - 1)(2x + 3)$, so $14x^4 + 19x^3 - 3x^2 = x^2(7x - 1)(2x + 3)$.

55. $168x^3 - 45x^2 + 3x$

(1) Factor out the common factor, $3x$:

$3x(56x^2 - 15x + 1)$

Then we factor the trinomial $56x^2 - 15x + 1$.

(2) Factor $56x^2$. The possibilities are $56x$, x and $28x$, $2x$ and $14x$, $4x$ and $7x$, $8x$. We have these as possibilities for factorizations:

$(56x+\quad)(x+\quad)$ and $(28x+\quad)(2x+\quad)$ and

$(14x+\quad)(4x+\quad)$ and $(7x+\quad)(8x+\quad)$

(3) Factor 1. Since 1 is positive and the middle term is negative we need consider only the negative factor pair $-1, -1$.

(4) We try some possibilities:

$(56x - 1)(x - 1) = 56x^2 - 57x + 1$

$(28x - 1)(2x - 1) = 56x^2 - 30x + 1$

$(14x - 1)(4x - 1) = 56x^2 - 18x + 1$

$(7x - 1)(8x - 1) = 56x^2 - 15x + 1$

Then $56x^2 - 15x + 1 = (7x - 1)(8x - 1)$, so $168x^3 - 45x^2 + 3x = 3x(7x - 1)(8x - 1)$.

57. $15x^4 - 19x^2 + 6 = 15(x^2)^2 - 19x^2 + 6$

(1) There is no common factor (other than 1 or -1).

(2) Factor the first term, $15x^4$. The possibilities are $15x^2$, x^2 and $5x^2$, $3x^2$. We have these as possibilities for factorizations:

$(15x^2+\quad)(x^2+\quad)$ and $(5x^2+\quad)(3x^2+\quad)$

(3) Factor 6. Since 6 is positive and the middle term is negative, we need consider only negative factor pairs. Those pairs are -6, -1 and -3, -2.

These factors can also be written as -1, -6 and -2, -3.

(4) We can immediately reject all possibilities in which either factor has a common factor, such as $(15x^2-6)$ or $(3x^2 - 3)$, because we determined at the outset that there is no common factor. We try some of the remaining possibilities:

$(15x^2 - 1)(x^2 - 6) = 15x^4 - 91x^2 + 6$

$(15x^2 - 2)(x^2 - 3) = 15x^4 - 47x^2 + 6$

$(5x^2 - 6)(3x^2 - 1) = 15x^4 - 23x^2 + 6$

$(5x^2 - 3)(3x^2 - 2) = 15x^4 - 19x^2 + 6$

The factorization is $(5x^2 - 3)(3x^2 - 2)$.

59. $25t^2 + 80t + 64$

(1) There is no common factor (other than 1 or -1).

(2) Factor the first term, $25t^2$. The possibilities are $25t$, t and $5t$, $5t$. We have these as possibilities for factorizations:

$(25t+\quad)(t+\quad)$ and $(5t+\quad)(5t+\quad)$

(3) Factor the last term, 64. Since all signs are positive, we need consider only positive pairs of factors. Those factor pairs are 64, 1 and 32, 2 and 16, 4 and 8, 8.

These first three pairs can also be written as 1, 64 and 2, 32 and 4, 16.

(4) We try some possibilities:

$$(25t+64)(t+1) = 25t^2+89t+64$$
$$(5t+32)(5t+2) = 25t^2+170t+64$$
$$(25t+16)(t+4) = 25t^2+116t+64$$
$$(5t+8)(5t+8) = 25t^2+80t+64$$

The factorization is $(5t+8)(5t+8)$ or $(5t+8)^2$.

61. $6x^3+4x^2-10x$

(1) Factor out the common factor, $2x$: $2x(3x^2+2x-5)$

Then we factor the trinomial $3x^2+2x-5$. We did this in Exercise 35 (after we factored 2 out of the original trinomial). We know that $3x^2+2x-5 = (3x+5)(x-1)$, so $6x^3+4x^2-10x = 2x(3x+5)(x-1)$.

63. $25x^2+79x+64$

We follow the same procedure as in Exercise 59. None of the possibilities works. Thus, $25x^2+79x+64$ is not factorable. It is prime.

65. $6x^2-19x-5$

(1) There is no common factor (other than 1 or -1).

(2) Factor the first term, $6x^2$. The possibilities are $6x$, x and $3x$, $2x$. We have these as possibilities for factorizations:

$(6x+\quad)(x+\quad)$ and $(3x+\quad)(2x+\quad)$

(3) Factor the last term, -5. The possibilities are -5, 1 and 5, -1.

These factors can also be written as 1, -5 and -1, 5.

(4) We try some possibilities:

$$(6x-5)(x+1) = 6x^2+x-5$$
$$(6x+5)(x-1) = 6x^2-x-5$$
$$(6x+1)(x-5) = 6x^2-29x-5$$
$$(6x-1)(x+5) = 6x^2+29x-5$$
$$(3x-5)(2x+1) = 6x^2-7x-5$$
$$(3x+5)(2x-1) = 6x^2+7x-5$$
$$(3x+1)(2x-5) = 6x^2-13x-5$$
$$(3x-1)(2x+5) = 6x^2+13x-5$$

None of the possibilities works. Thus, $6x^2-19x-5$ is not factorable. It is prime.

67. $12m^2-mn-20n^2$

(1) There is no common factor (other than 1 or -1).

(2) Factor the first term, $12m^2$. The possibilities are $12m$, m and $6m$, $2m$ and $3m$, $4m$. We have these as possibilities for factorizations:

$(12m+\quad)(m+\quad)$ and $(6m+\quad)(2m+\quad)$

and $(3m+\quad)(4m+\quad)$

(3) Factor the last term, $-20n^2$. The possibilities are $20n$, $-n$ and $-20n$, n and $10n$, $-2n$ and $-10n$, $2n$ and $5n$, $-4n$ and $-5n$, $4n$.

These factors can also be written as $-n$, $20n$ and n, $-20n$ and $-2n$, $10n$ and $2n$, $-10n$ and $-4n$, $5n$ and $4n$, $-5n$.

(4) We can immediately reject all possibilities in which either factor has a common factor, such as $(12m+20n)$ or $(4m-2n)$, because we determined at the outset that there is no common factor. We try some of the remaining possibilities:

$$(12m-n)(m+20n) = 12m^2+239mn-20n^2$$
$$(12m+5n)(m-4n) = 12m^2-43mn-20n^2$$
$$(3m-20n)(4m+n) = 12m^2-77mn-20n^2$$
$$(3m-4n)(4m+5n) = 12m^2-mn-20n^2$$

The factorization is $(3m-4n)(4m+5n)$.

69. $6a^2-ab-15b^2$

(1) There is no common factor (other than 1 or -1).

(2) Factor the first term, $6a^2$. The possibilities are $6a$, a and $3a$, $2a$. We have these as possibilities for factorizations:

$(6a+\quad)(a+\quad)$ and $(3a+\quad)(2a+\quad)$

(3) Factor the last term, $-15b^2$. The possibilities are $15b$, $-b$ and $-15b$, b and $5b$, $-3b$ and $-5b$, $3b$.

These factors can also be written as $-b$, $15b$ and b, $-15b$, and $-3b$, $5b$ and $3b$, $-5b$.

(4) We can immediately reject all possibilities in which either factor has a common factor, such as $(6a+15b)$ or $(3a-3b)$, because we determined at the outset that there is no common factor. We try some of the remaining possibilities:

$$(6a-b)(a+15b) = 6a^2+89ab-15b^2$$
$$(3a-b)(2a+15b) = 6a^2+43ab-15b^2$$
$$(6a+5b)(a-3b) = 6a^2-13ab-15b^2$$
$$(3a+5b)(2a-3b) = 6a^2+ab-15b^2$$
$$(3a-5b)(2a+3b) = 6a^2-ab-15b^2$$

The factorization is $(3a-5b)(2a+3b)$.

71. $9a^2 + 18ab + 8b^2$

(1) There is no common factor (other than 1 or -1).

(2) Factor the first term, $9a^2$. The possibilities are $9a$, a and $3a$, $3a$. We have these as possibilities for factorizations:

$$(9a+\quad)(a+\quad) \text{ and } (3a+\quad)(3a+\quad)$$

(3) Factor $8b^2$. Since all signs are positive, we need consider only pairs of factors with positive coefficients. Those factor pairs are $8b$, b and $4b$, $2b$.

These factors can also be written as b, $8b$ and $2b$, $4b$.

(4) We try some possibilities:

$$(9a + 8b)(a + b) = 9a^2 + 17ab + 8b^2$$
$$(3a + 8b)(3a + b) = 9a^2 + 27ab + 8b^2$$
$$(9a + 4b)(a + 2b) = 9a^2 + 22ab + 8b^2$$
$$(3a + 4b)(3a + 2b) = 9a^2 + 18ab + 8b^2$$

The factorization is $(3a + 4b)(3a + 2b)$.

73. $35p^2 + 34pq + 8q^2$

(1) There is no common factor (other than 1 or -1).

(2) Factor the first term, $35p^2$. The possibilities are $35p$, p and $7p$, $5p$. We have these as possibilities for factorizations:

$$(35p+\quad)(p+\quad) \text{ and } (7p+\quad)(5p+\quad)$$

(3) Factor $8q^2$. Since all signs are positive, we need consider only pairs of factors with positive coefficients. Those factor pairs are $8q$, q and $4q$, $2q$.

These factors can also be written as q, $8q$ and $2q$, $4q$.

(4) We try some possibilities:

$$(35p + 8q)(p + q) = 35p^2 + 43pq + 8q^2$$
$$(7p + 8q)(5p + q) = 35p^2 + 47pq + 8q^2$$
$$(35p + 4q)(p + 2q) = 35p^2 + 74pq + 8q^2$$
$$(7p + 4q)(5p + 2q) = 35p^2 + 34pq + 8p^2$$

The factorization is $(7p + 4q)(5p + 2q)$.

75. $18x^2 - 6xy - 24y^2$

(1) Factor out the common factor, 6:

$6(3x^2 - xy - 4y^2)$

Then we factor the trinomial $3x^2 - xy - 4y^2$.

(2) Factor $3x^2$. The only possibility is $3x$, x. The desired factorization is of the form:

$$(3x+\quad)(x+\quad)$$

(3) Factor $-4y^2$. The possibilities are $4y$, $-y$ and $-4y$, y and $2y$, $-2y$.

These factors can also be written as $-y$, $4y$ and y, $-4y$ and $-2y$, $2y$.

(4) We try some possibilities:

$$(3x + 4y)(x - y) = 3x^2 + xy - 4y^2$$
$$(3x - 4y)(x + y) = 3x^2 - xy - 4y^2$$

Then $3x^2 - xy - 4y^2 = (3x - 4y)(x + y)$, so $18x^2 - 6xy - 24y^2 = 6(3x - 4y)(x + y)$.

77.

$$A = pq - 7$$
$$A + 7 = pq \quad \text{Adding 7}$$
$$\frac{A+7}{p} = q \quad \text{Dividing by } p$$

79. $3x + 2y = 6$

$$2y = 6 - 3x \quad \text{Subtracting } 3x$$
$$y = \frac{6 - 3x}{2} \quad \text{Dividing by 2}$$

81. $5 - 4x < -11$

$$-4x < -16 \quad \text{Subtracting 5}$$
$$x > 4 \quad \text{Dividing by } -4 \text{ and reversing the inequality symbol}$$

The solution set is $\{x|x > 4\}$.

83. Graph: $y = \frac{2}{5}x - 1$

Because the equation is in the form $y = mx + b$, we know the y-intercept is $(0, -1)$. We find two other points on the line, substituting multiples of 5 for x to avoid fractions.

When $x = -5$, $y = \frac{2}{5}(-5) - 1 = -2 - 1 = -3$.

When $x = 5$, $y = \frac{2}{5}(5) - 1 = 2 - 1 = 1$.

x	y
0	-1
-5	-3
5	1

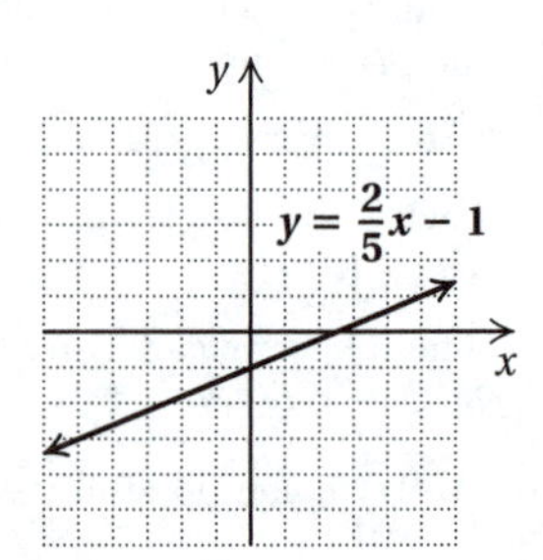

85. $4x - 16y = 64$

To find the x-intercept, let $y = 0$ and solve for x.

$$4x - 16y = 64$$
$$4x - 16 \cdot 0 = 64$$
$$4x = 64$$
$$x = 16$$

The x-intercept is $(16, 0)$.

To find the y-intercept, let $x = 0$ and solve for y.

$$4x - 16y = 64$$
$$4 \cdot 0 - 16y = 64$$
$$-16y = 64$$
$$y = -4$$

The y-intercept is $(0, -4)$.

87. $x - 1.3y = 6.5$

To find the x-intercept, let $y = 0$ and solve for x.

$$x - 1.3y = 6.5$$
$$x - 1.3(0) = 6.5$$
$$x = 6.5$$

The x-intercept is $(6.5, 0)$.

To find the y-intercept, let $x = 0$ and solve for y.

$$x - 1.3y = 6.5$$
$$0 - 1.3y = 6.5$$
$$-1.3y = 6.5$$
$$y = -5$$

The y-intercept is $(0, -5)$.

89. $y = 4 - 5x$

To find the x-intercept, let $y = 0$ and solve for x.

$$y = 4 - 5x$$
$$0 = 4 - 5x$$
$$5x = 4$$
$$x = \frac{4}{5}$$

The x-intercept is $\left(\frac{4}{5}, 0\right)$.

To find the y-intercept, let $x = 0$ and solve for y.

$$y = 4 - 5x$$
$$y = 4 - 5 \cdot 0$$
$$y = 4$$

The y-intercept is $(0, 4)$.

91. $20x^{2n} + 16x^n + 3 = 20(x^n)^2 + 16x^n + 3$

(1) There is no common factor (other than 1 and -1).

(2) Factor the first term, $20x^{2n}$. The possibilities are $20x^n$, x^n and $10x^n$, $2x^n$ and $5x^n$, $4x^n$. We have these as possibilities for factorizations:

$(20x^n+ \quad)(x^n+ \quad)$ and $(10x^n+ \quad)(2x^n+ \quad)$ and $(5x^n+ \quad)(4x^n+ \quad)$

(3) Factor the last term, 3. Since all signs are positive, we need consider only the positive factor pair 3, 1.

(4) We try some possibilities:

$$(20x^n + 3)(x^n + 1) = 20x^{2n} + 23x^n + 3$$
$$(10x^n + 3)(2x^n + 1) = 20x^{2n} + 16x^n + 3$$

The factorization is $(10x^n + 3)(2x^n + 1)$.

93. $3x^{6a} - 2x^{3a} - 1 = 3(x^{3a})^2 - 2x^{3a} - 1$

(1) There is no common factor (other than 1 or -1).

(2) Factor the first term, $3x^{6a}$. The only possibility is $3x^{3a}$, x^{3a}. The desired factorization is of the form:

$$(3x^{3a}+ \quad)(x^{3a}+ \quad)$$

(3) Factor the last term, -1. The only possibility is -1, 1.

(4) We try the possibilities:

$$(3x^{3a} - 1)(x^{3a} + 1) = 3x^{6a} + 2x^{3a} - 1$$
$$(3x^{3a} + 1)(x^{3a} - 1) = 3x^{6a} - 2x^{3a} - 1$$

The factorization is $(3x^{3a} + 1)(x^{3a} - 1)$.

95.-103. Left to the student

Exercise Set 13.4

1. $x^2 + 2x + 7x + 14 = (x^2 + 2x) + (7x + 14)$
$= x(x + 2) + 7(x + 2)$
$= (x + 2)(x + 7)$

3. $x^2 - 4x - x + 4 = (x^2 - 4x) + (-x + 4)$
$= x(x - 4) - 1(x - 4)$
$= (x - 4)(x - 1)$

5. $6x^2 + 4x + 9x + 6 = (6x^2 + 4x) + (9x + 6)$
$= 2x(3x + 2) + 3(3x + 2)$
$= (3x + 2)(2x + 3)$

7. $3x^2 - 4x - 12x + 16 = (3x^2 - 4x) + (-12x + 16)$
$= x(3x - 4) - 4(3x - 4)$
$= (3x - 4)(x - 4)$

9. $35x^2 - 40x + 21x - 24 = (35x^2 - 40x) + (21x - 24)$
$= 5x(7x - 8) + 3(7x - 8)$
$= (7x - 8)(5x + 3)$

11. $4x^2 + 6x - 6x - 9 = (4x^2 + 6x) + (-6x - 9)$
$= 2x(2x + 3) - 3(2x + 3)$
$= (2x + 3)(2x - 3)$

13. $2x^4 + 6x^2 + 5x^2 + 15 = (2x^4 + 6x^2) + (5x^2 + 15)$
$= 2x^2(x^2 + 3) + 5(x^2 + 3)$
$= (x^2 + 3)(2x^2 + 5)$

15. $2x^2 + 7x - 4$

(1) First factor out a common factor, if any. There is none (other than 1 or -1).

(2) Multiply the leading coefficient, 2, and the constant, -4: $2(-4) = -8$.

(3) Look for a factorization of -8 in which the sum of the factors is the coefficient of the middle term, 7.

Pairs of factors	Sums of factors
-1, 8	7
1, -8	-7
-2, 4	2
2, -4	-2

(4) Split the middle term: $7x = -1x + 8x$

(5) Factor by grouping:

$$\begin{aligned}2x^2 + 7x - 4 &= 2x^2 - x + 8x - 4\\ &= (2x^2 - x) + (8x - 4)\\ &= x(2x - 1) + 4(2x - 1)\\ &= (2x - 1)(x + 4)\end{aligned}$$

17. $3x^2 - 4x - 15$

(1) First factor out a common factor, if any. There is none (other than 1 or -1).

(2) Multiply the leading coefficient, 3, and the constant, -15: $3(-15) = -45$.

(3) Look for a factorization of -45 in which the sum of the factors is the coefficient of the middle term, -4.

Pairs of factors	Sums of factors
-1, 45	44
1, -45	-44
-3, 15	12
3, -15	-12
-5, 9	4
5, -9	-4

(4) Split the middle term: $-4x = 5x - 9x$

(5) Factor by grouping:

$$\begin{aligned}3x^2 - 4x - 15 &= 3x^2 + 5x - 9x - 15\\ &= (3x^2 + 5x) + (-9x - 15)\\ &= x(3x + 5) - 3(3x + 5)\\ &= (3x + 5)(x - 3)\end{aligned}$$

19. $6x^2 + 23x + 7$

(1) First factor out a common factor, if any. There is none (other than 1 or -1).

(2) Multiply the leading coefficient, 6, and the constant, 7: $6 \cdot 7 = 42$.

(3) Look for a factorization of 42 in which the sum of the factors is the coefficient of the middle term, 23. We only need to consider positive factors.

Pairs of factors	Sums of factors
1, 42	43
2, 21	23
3, 14	17
6, 7	13

(4) Split the middle term: $23x = 2x + 21x$

(5) Factor by grouping:

$$\begin{aligned}6x^2 + 23x + 7 &= 6x^2 + 2x + 21x + 7\\ &= (6x^2 + 2x) + (21x + 7)\\ &= 2x(3x + 1) + 7(3x + 1)\\ &= (3x + 1)(2x + 7)\end{aligned}$$

21. $3x^2 - 4x + 1$

(1) First factor out a common factor, if any. There is none (other than 1 or -1).

(2) Multiply the leading coefficient, 3, and the constant, 1: $3 \cdot 1 = 3$.

(3) Look for a factorization of 3 in which the sum of the factors is the coefficient of the middle term, -4. The numbers we want are -1 and -3: $-1 \cdot (-3) = 3$ and $-1 + (-3) = -4$.

(4) Split the middle term: $-4x = -1x - 3x$

(5) Factor by grouping:

$$\begin{aligned}3x^2 - 4x + 1 &= 3x^2 - x - 3x + 1\\ &= (3x^2 - x) + (-3x + 1)\\ &= x(3x - 1) - 1(3x - 1)\\ &= (3x - 1)(x - 1)\end{aligned}$$

23. $4x^2 - 4x - 15$

(1) First factor out a common factor, if any. There is none (other than 1 or -1).

(2) Multiply the leading coefficient, 4, and the constant, -15: $4(-15) = -60$.

(3) Look for a factorization of -60 in which the sum of the factors is the coefficient of the middle term, -4.

Pairs of factors	Sums of factors
-1, 60	59
1, -60	-59
-2, 30	28
2, -30	-28
-3, 20	17
3, -20	-17
-4, 15	11
4, -15	-11
-5, 12	7
5, -12	-7
-6, 10	4
6, -10	-4

(4) Split the middle term: $-4x = 6x - 10x$

(5) Factor by grouping:

$$\begin{aligned}4x^2 - 4x - 15 &= 4x^2 + 6x - 10x - 15\\ &= (4x^2 + 6x) + (-10x - 15)\\ &= 2x(2x + 3) - 5(2x + 3)\\ &= (2x + 3)(2x - 5)\end{aligned}$$

25. $2x^2 + x - 1$

(1) First factor out a common factor, if any. There is none (other than 1 or -1).

(2) Multiply the leading coefficient, 2, and the constant, -1: $2(-1) = -2$.

(3) Look for a factorization of -2 in which the sum of the factors is the coefficient of the middle term, 1. The numbers we want are 2 and -1: $2(-1) = -2$ and $2 - 1 = 1$.

(4) Split the middle term: $x = 2x - 1x$

(5) Factor by grouping:

$$\begin{aligned} 2x^2 + x - 1 &= 2x^2 + 2x - x - 1 \\ &= (2x^2 + 2x) + (-x - 1) \\ &= 2x(x+1) - 1(x+1) \\ &= (x+1)(2x-1) \end{aligned}$$

27. $9x^2 - 18x - 16$

(1) First factor out a common factor, if any. There is none (other than 1 or -1).

(2) Multiply the leading coefficient, 9, and the constant, -16: $9(-16) = -144$.

(3) Look for a factorization of -144, so the sum of the factors is the coefficient of the middle term, -18.

Pairs of factors	Sums of factors
$-1,\ 144$	143
$1, -144$	-143
$-2,\ 72$	70
$2, -72$	-70
$-3,\ 48$	45
$3, -48$	-45
$-4,\ 36$	32
$4, -36$	-32
$-6,\ 24$	18
$6, -24$	-18
$-8,\ 18$	10
$8, -18$	-10
$-9,\ 16$	7
$9, -16$	-7
$-12,\ 12$	0

(4) Split the middle term: $-18x = 6x - 24x$

(5) Factor by grouping:

$$\begin{aligned} 9x^2 - 18x - 16 &= 9x^2 + 6x - 24x - 16 \\ &= (9x^2 + 6x) + (-24x - 16) \\ &= 3x(3x+2) - 8(3x+2) \\ &= (3x+2)(3x-8) \end{aligned}$$

29. $3x^2 + 5x - 2$

(1) First factor out a common factor, if any. There is none (other than 1 or -1).

(2) Multiply the leading coefficient, 3, and the constant, -2: $3(-2) = -6$.

(3) Look for a factorization of -6 in which the sum of the factors is the coefficient of the middle term, 5. The numbers we want are -1 and 6: $-1(6) = -6$ and $-1 + 6 = 5$.

(4) Split the middle term: $5x = -1x + 6x$

(5) Factor by grouping:

$$\begin{aligned} 3x^2 + 5x - 2 &= 3x^2 - x + 6x - 2 \\ &= (3x^2 - x) + (6x - 2) \\ &= x(3x-1) + 2(3x-1) \\ &= (3x-1)(x+2) \end{aligned}$$

31. $12x^2 - 31x + 20$

(1) First factor out a common factor, if any. There is none (other than 1 or -1).

(2) Multiply the leading coefficient, 12, and the constant, 20: $12 \cdot 20 = 240$.

(3) Look for a factorization of 240 in which the sum of the factors is the coefficient of the middle term, -31. We only need to consider negative factors.

Pairs of factors	Sums of factors
$-1, -240$	-241
$-2, -120$	-122
$-3,\ -80$	-83
$-4,\ -60$	-64
$-5,\ -48$	-53
$-6,\ -40$	-46
$-8,\ -30$	-38
$-10,\ -24$	-34
$-12,\ -20$	-32
$-15,\ -16$	-31

(4) Split the middle term: $-31x = -15x - 16x$

(5) Factor by grouping:

$$\begin{aligned} 12x^2 - 31x + 20 &= 12x^2 - 15x - 16x + 20 \\ &= (12x^2 - 15x) + (-16x + 20) \\ &= 3x(4x-5) - 4(4x-5) \\ &= (4x-5)(3x-4) \end{aligned}$$

33. $14x^2 - 19x - 3$

(1) First factor out a common factor, if any. There is none (other than 1 or -1).

(2) Multiply the leading coefficient, 14, and the constant, -3: $14(-3) = -42$.

(3) Look for a factorization of -42 so that the sum of the factors is the coefficient of the middle term, -19.

Pairs of factors	Sums of factors
$-1,\ 42$	41
$1, -42$	-41
$-2,\ 21$	19
$2, -21$	-19
$-3,\ 14$	11
$3, -14$	-11
$-6,\ 7$	1
$6,\ -7$	-1

(4) Split the middle term: $-19x = 2x - 21x$

(5) Factor by grouping:

$$\begin{aligned}14x^2 - 19x - 3 &= 14x^2 + 2x - 21x - 3\\ &= (14x^2 + 2x) + (-21x - 3)\\ &= 2x(7x + 1) - 3(7x + 1)\\ &= (7x + 1)(2x - 3)\end{aligned}$$

35. $9x^2 + 18x + 8$

(1) First factor out a common factor, if any. There is none (other than 1 or -1).

(2) Multiply the leading coefficient, 9, and the constant, 8: $9 \cdot 8 = 72$.

(3) Look for a factorization of 72 in which the sum of the factors is the coefficient of the middle term, 18. We only need to consider positive factors.

Pairs of factors	Sums of factors
1, 72	73
2, 36	38
3, 24	27
4, 18	22
6, 12	18
8, 9	17

(4) Split the middle term: $18x = 6x + 12x$

(5) Factor by grouping:

$$\begin{aligned}9x^2 + 18x + 8 &= 9x^2 + 6x + 12x + 8\\ &= (9x^2 + 6x) + (12x + 8)\\ &= 3x(3x + 2) + 4(3x + 2)\\ &= (3x + 2)(3x + 4)\end{aligned}$$

37. $49 - 42x + 9x^2 = 9x^2 - 42x + 49$

(1) First factor out a common factor, if any. There is none (other than 1 or -1).

(2) Multiply the leading coefficient, 9, and the constant, 49: $9 \cdot 49 = 441$.

(3) Look for a factorization of 441 in which the sum of the factors is the coefficient of the middle term, -42. We only need to consider negative factors.

Pairs of factors	Sums of factors
$-1, -441$	-442
$-3, -147$	-150
$-7, -63$	-70
$-9, -49$	-58
$-21, -21$	-42

(4) Split the middle term: $-42x = -21x - 21x$

(5) Factor by grouping:

$$\begin{aligned}9x^2 - 42x + 49 &= 9x^2 - 21x - 21x + 49\\ &= (9x^2 - 21x) + (-21x + 49)\\ &= 3x(3x - 7) - 7(3x - 7)\\ &= (3x - 7)(3x - 7), \text{ or}\\ &\quad (3x - 7)^2\end{aligned}$$

We could also write this as $(7 - 3x)^2$.

39. $24x^2 - 47x - 2$

(1) First factor out a common factor, if any. There is none (other than 1 or -1).

(2) Multiply the leading coefficient, 24, and the constant, -2: $24(-2) = -48$.

(3) Look for a factorization of -48 in which the sum of the factors is the coefficient of the middle term, -47. The numbers we want are -48 and 1: $-48 \cdot 1 = -48$ and $-48 + 1 = -47$.

(4) Split the middle term: $-47x = -48x + 1x$

(5) Factor by grouping:

$$\begin{aligned}24x^2 - 47x - 2 &= 24x^2 - 48x + x - 2\\ &= (24x^2 - 48x) + (x - 2)\\ &= 24x(x - 2) + 1(x - 2)\\ &= (x - 2)(24x + 1)\end{aligned}$$

41. $5 - 9a^2 - 12a = -9a^2 - 12a + 5 = -1(9a^2 + 12a - 5)$

Now we factor $9a^2 + 12a - 5$.

(1) We have already factored out the common factor, -1, to make the leading coefficient positive.

(2) Multiply the leading coefficient, 9, and the constant, -5: $9(-5) = -45$.

(3) Look for a factorization of -45 in which the sum of the factors is the coefficient of the middle term, 12. The numbers we want are 15 and -3: $15(-3) = -45$ and $15 + (-3) = 12$.

(4) Split the middle term: $12a = 15a - 3a$

(5) Factor by grouping:

$$\begin{aligned}9a^2 + 12a - 5 &= 9a^2 + 15a - 3a - 5\\ &= (9a^2 + 15a) + (-3a - 5)\\ &= 3a(3a + 5) - (3a + 5)\\ &= (3a + 5)(3a - 1)\end{aligned}$$

Then we have

$$\begin{aligned}&5 - 9a^2 - 12a\\ &= -1(3a + 5)(3a - 1)\\ &= (3a + 5)(-3a + 1) \quad \text{Multiplying } 3a-1 \text{ by } -1\\ &= (-3a - 5)(3a - 1) \quad \text{Multiplying } 3a+5 \text{ by } -1\end{aligned}$$

Note that we can also express $(3a + 5)(-3a + 1)$ as $(3a + 5)(1 - 3a)$ since $-3a + 1 = 1 - 3a$ by the commutative law of addition.

43. $20 + 6x - 2x^2 = -2x^2 + 6x + 20$

(1) Factor out the common factor -2. We factor out -2 rather than 2 in order to make the leading coefficient of the trinomial factor positive.

$-2x^2 + 6x + 20 = -2(x^2 - 3x - 10)$

To factor $x^2 - 3x - 10$, we look for two factors of -10 whose sum is -3. The numbers we want are -5 and 2. Then $x^2 - 3x - 10 = (x - 5)(x + 2)$, so we have:

$$\begin{aligned} &20 + 6x - 2x^2 \\ &= -2(x-5)(x+2) \\ &= 2(-x+5)(x+2) \quad \text{Multiplying } x-5 \text{ by } -1 \\ &= 2(x-5)(-x-2) \quad \text{Multiplying } x+2 \text{ by } -1 \end{aligned}$$

Note that we can also express $2(-x+5)(x+2)$ as $2(5-x)(x+2)$ since $-x+5 = 5-x$ by the commutative law of addition.

45. $12x^2 + 28x - 24$

(1) Factor out the common factor, 4:

$12x^2 + 28x - 24 = 4(3x^2 + 7x - 6)$

(2) Now we factor the trinomial $3x^2 + 7x - 6$. Multiply the leading coefficient, 3, and the constant, -6: $3(-6) = -18$.

(3) Look for a factorization of -18 in which the sum of the factors is the coefficient of the middle term, 7. The numbers we want are 9 and -2: $9(-2) = -18$ and $9 + (-2) = 7$.

(4) Split the middle term: $7x = 9x - 2x$

(5) Factor by grouping:

$$\begin{aligned} 3x^2 + 7x - 6 &= 3x^2 + 9x - 2x - 6 \\ &= (3x^2 + 9x) + (-2x - 6) \\ &= 3x(x+3) - 2(x+3) \\ &= (x+3)(3x-2) \end{aligned}$$

We must include the common factor to get a factorization of the original trinomial.

$12x^2 + 28x - 24 = 4(x+3)(3x-2)$

47. $30x^2 - 24x - 54$

(1) Factor out the common factor, 6.

$30x^2 - 24x - 54 = 6(5x^2 - 4x - 9)$

(2) Now we factor the trinomial $5x^2 - 4x - 9$. Multiply the leading coefficient, 5, and the constant, -9: $5(-9) = -45$.

(3) Look for a factorization of -45 in which the sum of the factors is the coefficient of the middle term, -4. The numbers we want are -9 and 5: $-9 \cdot 5 = -45$ and $-9 + 5 = -4$.

(4) Split the middle term: $-4x = -9x + 5x$

(5) Factor by grouping:

$$\begin{aligned} 5x^2 - 4x - 9 &= 5x^2 - 9x + 5x - 9 \\ &= (5x^2 - 9x) + (5x - 9) \\ &= x(5x-9) + (5x-9) \\ &= (5x-9)(x+1) \end{aligned}$$

We must include the common factor to get a factorization of the original trinomial.

$30x^2 - 24x - 54 = 6(5x-9)(x+1)$

49. $4y + 6y^2 - 10 = 6y^2 + 4y - 10$

(1) Factor out the common factor, 2.

$6y^2 + 4y - 10 = 2(3y^2 + 2y - 5)$

(2) Now we factor the trinomial $3y^2 + 2y - 5$. Multiply the leading coefficient, 3, and the constant, -5: $3(-5) = -15$.

(3) Look for a factorization of -15 in which the sum of the factors is the coefficient of the middle term, 2. The numbers we want are 5 and -3: $5(-3) = -15$ and $5 + (-3) = 2$.

(4) Split the middle term: $2y = 5y - 3y$

(5) Factor by grouping:

$$\begin{aligned} 3y^2 + 2y - 5 &= 3y^2 + 5y - 3y - 5 \\ &= (3y^2 + 5y) + (-3y - 5) \\ &= y(3y+5) - (3y+5) \\ &= (3y+5)(y-1) \end{aligned}$$

We must include the common factor to get a factorization of the original trinomial.

$4y + 6y^2 - 10 = 2(3y+5)(y-1)$

51. $3x^2 - 4x + 1$

(1) There is no common factor (other than 1 or -1).

(2) Multiply the leading coefficient, 3, and the constant, 1: $3 \cdot 1 = 3$.

(3) Look for a factorization of 3 in which the sum of the factors is the coefficient of the middle term, -4. The numbers we want are -1 and -3: $-1(-3) = 3$ and $-1 + (-3) = -4$.

(4) Split the middle term: $-4x = -1x - 3x$

(5) Factor by grouping:

$$\begin{aligned} 3x^2 - 4x + 1 &= 3x^2 - x - 3x + 1 \\ &= (3x^2 - x) + (-3x + 1) \\ &= x(3x-1) - (3x-1) \\ &= (3x-1)(x-1) \end{aligned}$$

53. $12x^2 - 28x - 24$

(1) Factor out the common factor, 4:

$12x^2 - 28x - 24 = 4(3x^2 - 7x - 6)$

(2) Now we factor the trinomial $3x^2 - 7x - 6$. Multiply the leading coefficient, 3, and the constant, -6: $3(-6) = -18$.

(3) Look for a factorization of -18 in which the sum of the factors is the coefficient of the middle term, -7. The numbers we want are -9 and 2: $-9 \cdot 2 = -18$ and $-9 + 2 = -7$.

(4) Split the middle term: $-7x = -9x + 2x$

(5) Factor by grouping:

$$\begin{aligned} 3x^2 - 7x - 6 &= 3x^2 - 9x + 2x - 6 \\ &= (3x^2 - 9x) + (2x - 6) \\ &= 3x(x - 3) + 2(x - 3) \\ &= (x - 3)(3x + 2) \end{aligned}$$

We must include the common factor to get a factorization of the original trinomial.

$12x^2 - 28x - 24 = 4(x - 3)(3x + 2)$

55. $-1 + 2x^2 - x = 2x^2 - x - 1$

(1) There is no common factor (other than 1 or -1).

(2) Multiply the leading coefficient, 2, and the constant, -1: $2(-1) = -2$.

(3) Look for a factorization of -2 in which the sum of the factors is the coefficient of the middle term, -1. The numbers we want are -2 and 1: $-2 \cdot 1 = -2$ and $-2 + 1 = -1$.

(4) Split the middle term: $-x = -2x + 1x$

(5) Factor by grouping:

$$\begin{aligned} 2x^2 - x - 1 &= 2x^2 - 2x + x - 1 \\ &= (2x^2 - 2x) + (x - 1) \\ &= 2x(x - 1) + (x - 1) \\ &= (x - 1)(2x + 1) \end{aligned}$$

57. $9x^2 + 18x - 16$

(1) There is no common factor (other than 1 or -1).

(2) Multiply the leading coefficient, 9, and the constant, -16: $9(-16) = -144$.

(3) Look for a factorization of -144 in which the sum of the factors is the coefficient of the middle term, 18. The numbers we want are 24 and -6: $24(-6) = -144$ and $24 + (-6) = 18$.

(4) Split the middle term: $18x = 24x - 6x$

(5) Factor by grouping:

$$\begin{aligned} 9x^2 + 18x - 16 &= 9x^2 + 24x - 6x - 16 \\ &= (9x^2 + 24x) + (-6x - 16) \\ &= 3x(3x + 8) - 2(3x + 8) \\ &= (3x + 8)(3x - 2) \end{aligned}$$

59. $15x^2 - 25x - 10$

(1) Factor out the common factor, 5:

$15x^2 - 25x - 10 = 5(3x^2 - 5x - 2)$

(2) Now we factor the trinomial $3x^2 - 5x - 2$. Multiply the leading coefficient, 3, and the constant, -2: $3(-2) = -6$.

(3) Look for a factorization of -6 in which the sum of the factors is the coefficient of the middle term, -5. The numbers we want are -6 and 1: $-6 \cdot 1 = -6$ and $-6 + 1 = -5$.

(4) Split the middle term: $-5x = -6x + 1x$

(5) Factor by grouping:

$$\begin{aligned} 3x^2 - 5x - 2 &= 3x^2 - 6x + x - 2 \\ &= (3x^2 - 6x) + (x - 2) \\ &= 3x(x - 2) + (x - 2) \\ &= (x - 2)(3x + 1) \end{aligned}$$

We must include the common factor to get a factorization of the original trinomial.

$15x^2 - 25x - 10 = 5(x - 2)(3x + 1)$

61. $12p^3 + 31p^2 + 20p$

(1) Factor out the common factor, p:

$12p^3 + 31p^2 + 20p = p(12p^2 + 31p + 20)$

(2) Now we factor the trinomial $12p^2 + 31p + 20$. Multiply the leading coefficient, 12, and the constant, 20: $12 \cdot 20 = 240$.

(3) Look for a factorization of 240 in which the sum of the factors is the coefficient of the middle term, 31. The numbers we want are 15 and 16: $15 \cdot 16 = 240$ and $15 + 16 = 31$.

(4) Split the middle term: $31p = 15p + 16p$

(5) Factor by grouping:

$$\begin{aligned} 12p^2 + 31p + 20 &= 12p^2 + 15p + 16p + 20 \\ &= (12p^2 + 15p) + (16p + 20) \\ &= 3p(4p + 5) + 4(4p + 5) \\ &= (4p + 5)(3p + 4) \end{aligned}$$

We must include the common factor to get a factorization of the original trinomial.

$12p^3 + 31p^2 + 20p = p(4p + 5)(3p + 4)$

63. $4 - x - 5x^2 = -5x^2 - x + 4$

(1) Factor out -1 to make the leading coefficient positive:

$-5x^2 - x + 4 = -1(5x^2 + x - 4)$

(2) Now we factor the trinomial $5x^2 + x - 4$. Multiply the leading coefficient, 5, and the constant, -4: $5(-4) = -20$.

(3) Look for a factorization of -20 in which the sum of the factors is the coefficient of the middle term, 1. The numbers we want are 5 and -4: $5(-4) = -20$ and $5 + (-4) = 1$.

(4) Split the middle term: $x = 5x - 4x$

(5) Factor by grouping:

$$\begin{aligned} 5x^2 + x - 4 &= 5x^2 + 5x - 4x - 4 \\ &= (5x^2 + 5x) + (-4x - 4) \\ &= 5x(x + 1) - 4(x + 1) \\ &= (x + 1)(5x - 4) \end{aligned}$$

We must include the common factor to get a factorization of the original trinomial.

$$4 - x - 5x^2$$
$$= -1(x+1)(5x-4)$$
$$= (x+1)(-5x+4) \quad \text{Multiplying } 5x-4 \text{ by } -1$$
$$= (-x-1)(5x-4) \quad \text{Multiplying } x+1 \text{ by } -1$$

Note that we can also express $(x+1)(-5x+4)$ as $(x+1)(4-5x)$ since $-5x+4 = 4-5x$ by the commutative law of addition.

65. $33t - 15 - 6t^2 = -6t^2 + 33t - 15$

(1) Factor out the common factor, -3. We factor out -3 rather than 3 in order to make the leading coefficient of the trinomial factor positive.

$-6t^2 + 33t - 15 = -3(2t^2 - 11t + 5)$

(2) Now we factor the trinomial $2t^2 - 11t + 5$. Multiply the leading coefficient, 2, and the constant, 5: $2 \cdot 5 = 10$.

(3) Look for a factorization of 10 in which the sum of the factors is the coefficient of the middle term, -11. The numbers we want are -1 and -10: $-1(-10) = 10$ and $-1 + (-10) = -11$.

(4) Split the middle term: $-11t = -1t - 10t$

(5) Factor by grouping:

$$\begin{aligned} 2t^2 - 11t + 5 &= 2t^2 - t - 10t + 5 \\ &= (2t^2 - t) + (-10t + 5) \\ &= t(2t-1) - 5(2t-1) \\ &= (2t-1)(t-5) \end{aligned}$$

We must include the common factor to get a factorization of the original trinomial.

$$33t - 15 - 6t^2$$
$$= -3(2t-1)(t-5)$$
$$= 3(2t-1)(-t+5) \quad \text{Multiplying } t-5 \text{ by } -1$$
$$= 3(-2t+1)(t-5) \quad \text{Multiplying } 2t-1 \text{ by } -1$$

Note that we can also express $3(2t-1)(-t+5)$ as $3(2t-1)(5-t)$ since $-t+5 = 5-t$ by the commutative law of addition. Similarly, we can express $3(-2t+1)(t-5)$ as $3(1-2t)(t-5)$.

67. $14x^4 + 19x^3 - 3x^2$

(1) Factor out the common factor, x^2:

$14x^4 + 19x^3 - 3x^2 = x^2(14x^2 + 19x - 3)$

(2) Now we factor the trinomial $14x^2 + 19x - 3$. Multiply the leading coefficient, 14, and the constant, -3: $14(-3) = -42$.

(3) Look for a factorization of -42 in which the sum of the factors is the coefficient of the middle term, 19. The numbers we want are 21 and -2: $21(-2) = -42$ and $21 + (-2) = 19$.

(4) Split the middle term: $19x = 21x - 2x$

(5) Factor by grouping:

$$\begin{aligned} 14x^2 + 19x - 3 &= 14x^2 + 21x - 2x - 3 \\ &= (14x^2 + 21x) + (-2x - 3) \\ &= 7x(2x+3) - (2x+3) \\ &= (2x+3)(7x-1) \end{aligned}$$

We must include the common factor to get a factorization of the original trinomial.

$14x^4 + 19x^3 - 3x^2 = x^2(2x+3)(7x-1)$

69. $168x^3 - 45x^2 + 3x$

(1) Factor out the common factor, $3x$:

$168x^3 - 45x^2 + 3x = 3x(56x^2 - 15x + 1)$

(2) Now we factor the trinomial $56x^2 - 15x + 1$. Multiply the leading coefficient, 56, and the constant, 1: $56 \cdot 1 = 56$.

(3) Look for a factorization of 56 in which the sum of the factors is the coefficient of the middle term, -15. The numbers we want are -7 and -8: $-7(-8) = 56$ and $-7 + (-8) = -15$.

(4) Split the middle term: $-15x = -7x - 8x$

(5) Factor by grouping:

$$\begin{aligned} 56x^2 - 15x + 1 &= 56x^2 - 7x - 8x + 1 \\ &= (56x^2 - 7x) + (-8x + 1) \\ &= 7x(8x-1) - (8x-1) \\ &= (8x-1)(7x-1) \end{aligned}$$

We must include the common factor to get a factorization of the original trinomial.

$168x^3 - 45x^2 + 3x = 3x(8x-1)(7x-1)$

71. $15x^4 - 19x^2 + 6$

(1) There are no common factors (other than 1 or -1).

(2) Multiply the leading coefficient, 15, and the constant, 6: $15 \cdot 6 = 90$.

(3) Look for a factorization of 90 in which the sum of the factors is the coefficient of the middle term, -19. The numbers we want are -9 and -10: $-9(-10) = 90$ and $-9 + (-10) = -19$.

(4) Split the middle term: $-19x^2 = -9x^2 - 10x^2$

(5) Factor by grouping:

$$\begin{aligned} 15x^4 - 19x^2 + 6 &= 15x^4 - 9x^2 - 10x^2 + 6 \\ &= (15x^4 - 9x^2) + (-10x^2 + 6) \\ &= 3x^2(5x^2 - 3) - 2(5x^2 - 3) \\ &= (5x^2 - 3)(3x^2 - 2) \end{aligned}$$

73. $25t^2 + 80t + 64$

(1) There are no common factors (other than 1 or -1).

(2) Multiply the leading coefficient, 25, and the constant, 64: $25 \cdot 64 = 1600$.

(3) Look for a factorization of 1600 in which the sum of the factors is the coefficient of the middle term, 80. The numbers we want are 40 and 40: $40 \cdot 40 = 1600$ and $40 + 40 = 80$.

(4) Split the middle term: $80t = 40t + 40t$

(5) Factor by grouping:

$$\begin{aligned}25t^2 + 80t + 64 &= 25t^2 + 40t + 40t + 64\\ &= (25t^2 + 40t) + (40t + 64)\\ &= 5t(5t + 8) + 8(5t + 8)\\ &= (5t + 8)(5t + 8), \text{ or}\\ &\quad (5t + 8)^2\end{aligned}$$

75. $6x^3 + 4x^2 - 10x$

(1) Factor out the common factor, $2x$:

$6x^3 + 4x^2 - 10x = 2x(3x^2 + 2x - 5)$

(2) - (5) Now we factor the trinomial $3x^2+2x-5$. We did this in Exercise 49, using the variable y rather than x. We found that $3x^2+2x-5 = (3x+5)(x-1)$. We must include the common factor to get a factorization of the original trinomial.

$6x^3 + 4x^2 - 10x = 2x(3x + 5)(x - 1)$

77. $25x^2 + 79x + 64$

(1) There are no common factors (other than 1 or -1).

(2) Multiply the leading coefficient, 25, and the constant, 64: $25 \cdot 64 = 1600$.

(3) Look for a factorization of 1600 in which the sum of the factors is the coefficient of the middle term, 79. It is not possible to find such a pair of numbers. Thus, $25x^2 + 79x + 64$ cannot be factored into a product of binomial factors. It is prime.

79. $6x^2 - 19x - 5$

(1) There are no common factors (other than 1 or -1).

(2) Multiply the leading coefficient, 6, and the constant, -5: $6(-5) = -30$.

(3) Look for a factorization of -30 in which the sum of the factors is the coefficient of the middle term, -19. There is no such pair of numbers. Thus, $6x^2 - 19x - 5$ cannot be factored into a product of binomial factors. It is prime.

81. $12m^2 - mn - 20n^2$

(1) There are no common factors (other than 1 or -1).

(2) Multiply the leading coefficient, 12, and the constant, -20: $12(-20) = -240$.

(3) Look for a factorization of -240 in which the sum of the factors is the coefficient of the middle term, -1. The numbers we want are 15 and -16: $15(-16) = -240$ and $15 + (-16) = -1$.

(4) Split the middle term: $-mn = 15mn - 16mn$

(5) Factor by grouping:

$$\begin{aligned}&12m^2 - mn - 20n^2\\ &= 12m^2 + 15mn - 16mn - 20n^2\\ &= (12m^2 + 15mn) + (-16mn - 20n^2)\\ &= 3m(4m + 5n) - 4n(4m + 5n)\\ &= (4m + 5n)(3m - 4n)\end{aligned}$$

83. $6a^2 - ab - 15b^2$

(1) There are no common factors (other than 1 or -1).

(2) Multiply the leading coefficient, 6, and the constant, -15: $6(-15) = -90$.

(3) Look for a factorization of -90 in which the sum of the factors is the coefficient of the middle term, -1. The numbers we want are -10 and 9: $-10 \cdot 9 = -90$ and $-10 + 9 = -1$.

(4) Split the middle term: $-ab = -10ab + 9ab$

(5) Factor by grouping:

$$\begin{aligned}6a^2 - ab - 15b^2 &= 6a^2 - 10ab + 9ab - 15b^2\\ &= (6a^2 - 10ab) + (9ab - 15b^2)\\ &= 2a(3a - 5b) + 3b(3a - 5b)\\ &= (3a - 5b)(2a + 3b)\end{aligned}$$

85. $9a^2 - 18ab + 8b^2$

(1) There are no common factors (other than 1 or -1).

(2) Multiply the leading coefficient, 9, and the constant, 8: $9 \cdot 8 = 72$.

(3) Look for a factorization of 72 in which the sum of the factors is the coefficient of the middle term, -18. The numbers we want are -6 and -12: $-6(-12) = 72$ and $-6 + (-12) = -18$.

(4) Split the middle term: $-18ab = -6ab - 12ab$

(5) Factor by grouping:

$$\begin{aligned}9a^2 - 18ab + 8b^2 &= 9a^2 - 6ab - 12ab + 8b^2\\ &= (9a^2 - 6ab) + (-12ab + 8b^2)\\ &= 3a(3a - 2b) - 4b(3a - 2b)\\ &= (3a - 2b)(3a - 4b)\end{aligned}$$

87. $35p^2 + 34pq + 8q^2$

(1) There are no common factors (other than 1 or -1).

(2) Multiply the leading coefficient, 35, and the constant, 8: $35 \cdot 8 = 280$.

(3) Look for a factorization of 280 in which the sum of the factors is the coefficient of the middle term, 34. The numbers we want are 14 and 20: $14 \cdot 20 = 280$ and $14 + 20 = 34$.

(4) Split the middle term: $34pq = 14pq + 20pq$

(5) Factor by grouping:

$$\begin{aligned}35p^2 + 34pq + 8q^2 &= 35p^2 + 14pq + 20pq + 8q^2\\ &= (35p^2 + 14pq) + (20pq + 8q^2)\\ &= 7p(5p + 2q) + 4q(5p + 2q)\\ &= (5p + 2q)(7p + 4q)\end{aligned}$$

89. $18x^2 - 6xy - 24y^2$

(1) Factor out the common factor, 6.

$18x^2 - 6xy - 24y^2 = 6(3x^2 - xy - 4y^2)$

(2) Now we factor the trinomial $3x^2 - xy - 4y^2$. Multiply the leading coefficient, 3, and the constant, -4: $3(-4) = -12$.

(3) Look for a factorization of -12 in which the sum of the factors is the coefficient of the middle term, -1. The numbers we want are -4 and 3: $-4 \cdot 3 = -12$ and $-4 + 3 = -1$.

(4) Split the middle term: $-xy = -4xy + 3xy$

(5) Factor by grouping:

$$\begin{aligned} 3x^2 - xy - 4y^2 &= 3x^2 - 4xy + 3xy - 4y^2 \\ &= (3x^2 - 4xy) + (3xy - 4y^2) \\ &= x(3x - 4y) + y(3x - 4y) \\ &= (3x - 4y)(x + y) \end{aligned}$$

We must include the common factor to get a factorization of the original trinomial.

$18x^2 - 6xy - 24y^2 = 6(3x - 4y)(x + y)$

91. $60x + 18x^2 - 6x^3 = -6x^3 + 18x^2 + 60x$

(1) Factor out the common factor, $-6x$. We factor out $-6x$ rather than $6x$ in order to have a positive leading coefficient in the trinomial factor.

$-6x^3 + 18x^2 + 60x = -6x(x^2 - 3x - 10)$

(2) - (5) We factor $x^2 - 3x - 10$ as we did in Exercise 43, getting the $(x - 5)(x + 2)$. Then we have:

$60x + 18x^2 - 6x^3$

$= -6x(x - 5)(x + 2)$

$= 6x(-x + 5)(x + 2)$

Multiplying $x - 5$ by -1

$= 6x(x - 5)(-x - 2)$

Multiplying $x + 2$ by -1

Note that we can express $6x(-x + 5)(x + 2)$ as $6x(5 - x)(x + 2)$ since $-x + 5 = 5 - x$ by the commutative law of addition.

93. $35x^5 - 57x^4 - 44x^3$

(1) We first factor out the common factor, x^3.

$x^3(35x^2 - 57x - 44)$

(2) Now we factor the trinomial $35x^2 - 57x - 44$. Multiply the leading coefficient, 35, and the constant, -44: $35(-44) = -1540$.

(3) Look for a factorization of -1540 in which the sum of the factors is the coefficient of the middle term, -57.

Pairs of factors	Sums of factors
$7, -220$	-213
$10, -154$	-144
$11, -140$	-129
$14, -110$	-96
$20, -77$	-57

(4) Split the middle term: $-57x = 20x - 77x$

(5) Factor by grouping:

$$\begin{aligned} 35x^2 - 57x - 44 &= 35x^2 + 20x - 77x - 44 \\ &= (35x^2 + 20x) + (-77x - 44) \\ &= 5x(7x + 4) - 11(7x + 4) \\ &= (7x + 4)(5x - 11) \end{aligned}$$

We must include the common factor to get a factorization of the original trinomial.

$35x^5 - 57x^4 - 44x^3 = x^3(7x + 4)(5x - 11)$

95. $-10x > 1000$

$\dfrac{-10x}{-10} < \dfrac{1000}{-10}$ Dividing by -10 and reversing the inequality symbol

$x < -100$

The solution set is $\{x|x < -100\}$.

97. $6 - 3x \geq -18$

$-3x \geq -24$ Subtracting 6

$x \leq 8$ Dividing by -3 and reversing the inequality symbol

The solution set is $\{x|x \leq 8\}$.

99. $\dfrac{1}{2}x - 6x + 10 \leq x - 5x$

$2\left(\dfrac{1}{2}x - 6x + 10\right) \leq 2(x - 5x)$ Multiplying by 2 to clear the fraction

$x - 12x + 20 \leq 2x - 10x$

$-11x + 20 \leq -8x$ Collecting like terms

$20 \leq 3x$ Adding $11x$

$\dfrac{20}{3} \leq x$ Dividing by 3

The solution set is $\left\{x\middle|x \geq \dfrac{20}{3}\right\}$.

101. $3x - 6x + 2(x - 4) > 2(9 - 4x)$

$3x - 6x + 2x - 8 > 18 - 8x$ Removing parentheses

$-x - 8 > 18 - 8x$ Collecting like terms

$7x > 26$ Adding $8x$ and 8

$x > \dfrac{26}{7}$ Dividing by 7

The solution set is $\left\{x\middle|x > \dfrac{26}{7}\right\}$.

103. ***Familiarize***. We will use the formula $C = 2\pi r$, where C is circumference and r is radius, to find the radius in kilometers. Then we will multiply that number by 0.62 to find the radius in miles.

Translate.

$$\underbrace{\text{Circumference}}_{\downarrow} = \underbrace{2 \cdot \pi \cdot \text{radius}}_{\downarrow}$$

$$40{,}000 \approx 2(3.14)r$$

Solve. First we solve the equation.

$$40,000 \approx 2(3.14)r$$
$$40,000 \approx 6.28r$$
$$6369 \approx r$$

Then we multiply to find the radius in miles:

$6369(0.62) \approx 3949$

Check. If $r = 6369$, then $2\pi r = 2(3.14)(6369) \approx 40,000$. We should also recheck the multiplication we did to find the radius in miles. Both values check.

State. The radius of the earth is about 6369 km or 3949 mi. (These values may differ slightly if a different approximation is used for π.)

105. $9x^{10} - 12x^5 + 4$

(a) First factor out a common factor, if any. There is none (other than 1 or -1).

(b) Multiply the leading coefficient, 9, and the constant, 4: $9 \cdot 4 = 36$.

(c) Look for a factorization of 36 in which the sum of the factors is the coefficient of the middle term, -12. The factors we want are -6 and -6.

(d) Split the middle term: $-12x^5 = -6x^5 - 6x^5$

(e) Factor by grouping:

$$\begin{aligned} 9x^{10} - 12x^5 + 4 &= 9x^{10} - 6x^5 - 6x^5 + 4 \\ &= (9x^{10} - 6x^5) + (-6x^5 + 4) \\ &= 3x^5(3x^5 - 2) - 2(3x^5 - 2) \\ &= (3x^5 - 2)(3x^5 - 2), \text{ or} \\ &= (3x^5 - 2)^2 \end{aligned}$$

107. $16x^{10} + 8x^5 + 1$

(a) First factor out a common factor, if any. There is none (other than 1 or -1).

(b) Multiply the leading coefficient, 16, and the constant, 1: $16 \cdot 1 = 16$.

(c) Look for a factorization of 16 in which the sum of the factors is the coefficient of the middle term, 8. The factors we want are 4 and 4.

(d) Split the middle term: $8x^5 = 4x^5 + 4x^5$

(e) Factor by grouping:

$$\begin{aligned} 16x^{10} + 8x^5 + 1 &= 16x^{10} + 4x^5 + 4x^5 + 1 \\ &= (16x^{10} + 4x^5) + (4x^5 + 1) \\ &= 4x^5(4x^5 + 1) + 1(4x^5 + 1) \\ &= (4x^5 + 1)(4x^5 + 1), \text{ or} \\ &= (4x^5 + 1)^2 \end{aligned}$$

109.-117. Left to the student

Chapter 13 Mid-Chapter Review

1. True; the smallest natural number is 1, and the smallest number in the set must be a factor of the GCF.

2. False; look for factors of c whose sum is b.

3. True; a prime polynomial cannot be factored, so it has no common factor other than 1 and -1.

4. False; because the constant term is positive and the middle term is negative, we need to consider only negative pairs of factors of 45.

5. $$\begin{aligned} 10y^3 - 18y^2 + 12y &= 2y \cdot 5y^2 - 2y \cdot 9y + 2y \cdot 6 \\ &= 2y(5y^2 - 9y + 6) \end{aligned}$$

6. $2x^2 - x - 6$

$a \cdot c = 2 \cdot (-6) = -12$

$-x = -4x + 3x$

$$\begin{aligned} 2x^2 - x - 6 &= 2x^2 - 4x + 3x - 6 \\ &= 2x(x - 2) + 3(x - 2) \\ &= (x - 2)(2x + 3) \end{aligned}$$

7. $x^3 = x^3$

$3x = 3 \cdot x$

The coefficients have no common prime factor. The GCF of the powers of x is x because 1 is the smallest exponent of x. Thus the GCF is x.

8. $5x^4 = 5 \cdot x^4$

$x^2 = x^2$

The coefficients have no common prime factor. The GCF of the powers of x is x^2 because 2 is the smallest exponent of x. Thus the GCF is x^2.

9. $6x^5 = 2 \cdot 3 \cdot x^5$

$-12x^3 = -1 \cdot 2 \cdot 2 \cdot 3 \cdot x^3$

Each coefficient has factors of 2 and 3. The GCF of the powers of x is x^3 because 3 is the smallest exponent of x. Thus the GCF is $2 \cdot 3 \cdot x^3$, or $6x^3$.

10. $-8x = -1 \cdot 2 \cdot 2 \cdot 2 \cdot x$

$-12 = -1 \cdot 2 \cdot 2 \cdot 3$

$16x^2 = 2 \cdot 2 \cdot 2 \cdot 2 \cdot x^2$

Each coefficient has two factors of 2. The GCF of the powers of x is x^0 because -12 has no x-factor. Thus the GCF is $2 \cdot 2 \cdot x^0$, or 4.

11. $15x^3y^2 = 3 \cdot 5 \cdot x^3 \cdot y^2$

$5x^2y = 5 \cdot x^2 \cdot y$

$40x^4y^3 = 2 \cdot 2 \cdot 2 \cdot 5 \cdot x^4 \cdot y^3$

Each coefficient has a factor of 5. The GCF of the powers of x is x^2 because 2 is the smallest exponent of x. The GCF of the powers of y is y because 1 is the smallest exponent of y. Thus the GCF is $5x^2y$.

12. $x^2y^4 = x^2 \cdot y^4$

$-x^3y^3 = -1 \cdot x^3 \cdot y^3$

$x^3y^2 = x^3 \cdot y^2$

$x^5y^4 = x^5 \cdot y^4$

There is no common prime factor. The GCFs of the powers of x and y are x^2 and y^2, respectively, because 2 is the smallest exponent of each variable. Thus the GCF is x^2y^2.

13. $x^3 - 8x = x \cdot x^2 - x \cdot 8 = x(x^2 - 8)$

14. $3x^2 + 12x = 3x \cdot x + 3x \cdot 4 = 3x(x + 4)$

15. $2y^2 + 8y - 4 = 2 \cdot y^2 + 2 \cdot 4y - 2 \cdot 2 = 2(y^2 + 4y - 2)$

The trinomial $y^2 + 4y - 2$ cannot be factored, so the factorization is complete.

16. $3t^6 - 5t^4 - 2t^3 = t^3 \cdot 3t^3 - t^3 \cdot 5t - t^3 \cdot 2 = t^3(3t^3 - 5t - 2)$

17. $x^2 + 4x + 3$

Look for a pair of factors of 3 whose sum is 4. The numbers we want are 1 and 3.

$x^2 + 4x + 3 = (x + 1)(x + 3)$

18. $z^2 - 4z + 4$

Look for a pair of factors of 4 whose sum is -4. The numbers we want are -2 and -2.

$z^2 - 4z + 4 = (z - 2)(z - 2)$, or $(z - 2)^2$

19. $x^2 + 4x^2 + 3x + 12 = x^2(x + 4) + 3(x + 4)$
$= (x + 4)(x^2 + 3)$

20. $8y^5 - 48y^3 = 8y^3 \cdot y^2 - 8y^3 \cdot 6 = 8y^3(y^2 - 6)$

21. $6x^3y + 24x^2y^2 - 42xy^3 = 6xy \cdot x^2 + 6xy \cdot 4xy - 6xy \cdot 7y^2 = 6xy(x^2 + 4xy - 7y^2)$

The trinomial $x^2 + 4xy - 7y^2$ cannot be factored, so the factorization is complete.

22. $6 - 11t + 4t^2$, or $4t^2 - 11t + 6$

We will use the FOIL method.

(1) There is no common factor (other than 1 or -1).

(2) We can factor $4t^2$ as

$(4t+ \quad)(t+ \quad)$ or $(2t+ \quad)(2t+ \quad)$.

(3) The constant term is positive and the middle term is negative, so we look for pairs of negative factors of 6. The possibilities are -1, -6 and -2, -3.

The factors can also be written as -6, -1 and -3, -2.

(4) Look for combinations of factors from steps (2) and (3) such that the sum of their products is the middle term, $-11t$.

The factorization is $(4t - 3)(t - 2)$.

23. $z^2 + 4z - 5$

Look for a pair of factors of -5 whose sum is 4. The numbers we want are 5 and -1.

$z^2 + 4z - 5 = (z + 5)(z - 1)$

24. $2z^3 + 8z^3 + 5z + 20 = 2z^2(z + 4) + 5(z + 4)$
$= (z + 4)(2z^2 + 5)$

25. $3p^3 - 2p^2 - 9p + 6 = p^2(3p - 2) - 3(3p - 2)$
$= (3p - 2)(p^2 - 3)$

26. $10x^8 - 25x^6 - 15x^5 + 35x^3 = 5x^3(2x^5 - 5x^3 - 3x^2 + 7)$

27. $2w^3 + 3w^2 - 6w - 9 = w^2(2w + 3) - 3(2w + 3)$
$= (2w + 3)(w^2 - 3)$

28. $4x^4 - 5x^3 + 3x^2 = x^2(4x^2 - 5x + 3)$

The trinomial $4x^2 - 5x + 3$ cannot be factored, so the factorization is complete.

29. $6y^2 + 7y - 10$

We will use the ac-method.

(1) There is no common factor (other than 1 or -1).

(2) Multiply 6 and -10: $6(-10) = -60$.

(3) Look for a factorization of -60 in which the sum of the factors is 7. The numbers we want are 12 and -5.

(4) Split the middle term: $7y = 12y - 5y$.

(5) Factor by grouping.

$$\begin{aligned} 6y^2 + 7y - 10 &= 6y^2 + 12y - 5y - 10 \\ &= 6y(y + 2) - 5(y + 2) \\ &= (y + 2)(6y - 5) \end{aligned}$$

30. $3x^2 - 3x - 18 = 3(x^2 - x - 6)$

Consider $x^2 - x - 6$. Look for a pair of factors of -6 whose sum is -1. The numbers we want are -3 and 2.

$x^2 - x - 6 = (x - 3)(x + 2)$, so $3x^2 - 3x - 18 = 3(x - 3)(x + 2)$.

31. $6x^3 + 4x^2 + 3x + 2 = 2x^2(3x + 2) + (3x + 2)$
$= (3x + 2)(2x^2 + 1)$

32. $15 - 8w + w^2$, or $w^2 - 8w + 15$

Look for a pair of factors of 15 whose sum is -8. The numbers we want are -3 and -5.

$15 - 8w + w^2 = (w - 3)(w - 5)$

33. $8x^3 + 20x^2 + 2x + 5 = 4x^2(2x + 5) + (2x + 5)$
$= (2x + 5)(4x^2 + 1)$

34. $10z^2 - 21z - 10$

We will use the FOIL method.

(1) There is no common factor (other than 1 or -1).

(2) We can factor $10z^2$ as

$(z+ \quad)(10z+ \quad)$ or $(2z+ \quad)(5z+ \quad)$.

(3) Factor the last term, -10. The possibilities are -1, 10 and -10, 1 and -2, 5 and 2, -5.

The factors can also be written as 10, -1 and 1, -10 and 5, -2 and -5, 2.

(4) Look for combinations of factors from steps (2) and (3) such that the sum of their products is the middle term, $-21z$.

The factorization is $(2z - 5)(5z + 2)$.

35. $6x^2 + 7x + 2$

We will use the *ac*-method.

(1) There is no common factor (other than 1 or -1).

(2) Multiply 6 and 2: $6 \cdot 2 = 12$.

(3) Look for a factorization of 12 in which the sum of the factors is 7. The numbers we want are 3 and 4.

(4) Split the middle term: $7x = 3x + 4x$.

(5) Factor by grouping.

$$\begin{aligned} 6x^2 + 7x + 2 &= 6x^2 + 3x + 4x + 2 \\ &= 3x(2x + 1) + 2(2x + 1) \\ &= (2x + 1)(3x + 2) \end{aligned}$$

36. $x^2 - 10xy + 24y^2$

Look for a pair of factors of $24y^2$ whose sum is $-10y$. The factors we want are $-4y$ and $-6y$.

$x^2 - 10xy + 24y^2 = (x - 4y)(x - 6y)$

37. $$\begin{aligned} 6z^3 + 3z^2 + 2z + 1 &= 3z^2(2z + 1) + (2z + 1) \\ &= (2z + 1)(3z^2 + 1) \end{aligned}$$

38. $a^3b^7 + a^4b^5 - a^2b^3 + a^5b^6 = a^2b^3(ab^4 + a^2b^2 - 1 + a^3b^3)$

39. $4y^2 - 7yz - 15z^2$

We will use the FOIL method.

(1) There is no common factor (other than 1 or -1).

(2) We can factor $4y^2$ as

$(4y+ \quad)(y+ \quad)$ or $(2y+ \quad)(2y+ \quad)$.

(3) Factor the last term, $-15z^2$. The possibilities are $-z$, $15z$ and z, $-15z$ and $-3z$, $5z$ and $3z$, $-5z$.

The factors can also be written as $15z$, $-z$ and $-15z$, z and $5z$, $-3z$ and $-5z$, $3z$.

(4) Look for combinations of factors from steps (2) and (3) such that the sum of their products is the middle term, $-7yz$.

The factorization is $(4y + 5z)(y - 3z)$.

40. $3x^3 + 21x^2 + 30x = 3x(x^2 + 7x + 10)$

Consider $x^2 + 7x + 10$. Look for a pair of factors of 10 whose sum is 7. The numbers we want are 2 and 5.

$x^2 + 7x + 10 = (x + 2)(x + 5)$, so $3x^3 + 21x^2 + 30x = 3x(x + 2)(x + 5)$.

41. $$\begin{aligned} x^3 - 3x^2 - 2x + 6 &= x^2(x - 3) - 2(x - 3) \\ &= (x - 3)(x^2 - 2) \end{aligned}$$

42. $9y^2 + 6y + 1$

We will use the *ac*-method.

(1) There is no common factor (other than 1 or -1).

(2) Multiply 9 and 1: $9 \cdot 1 = 9$.

(3) Look for a factorization of 9 in which the sum of the factors is 6. The numbers we want are 3 and 3.

(4) Split the middle term: $6y = 3y + 3y$.

(5) Factor by grouping.

$$\begin{aligned} 9y^2 + 6y + 1 &= 9y^2 + 3y + 3y + 1 \\ &= 3y(3y + 1) + (3y + 1) \\ &= (3y + 1)(3y + 1), \text{ or } (3y + 1)^2 \end{aligned}$$

43. $y^2 + 6y + 8$

Look for a pair of factors of 8 whose sum is 6. The numbers we want are 2 and 4.

$y^2 + 6y + 8 = (y + 2)(y + 4)$

44. $6y^2 + 33y + 45 = 3(2y^2 + 11y + 15)$

We will use the FOIL method to factor $2y^2 + 11y + 15$.

(1) There is no common factor (other than 1 or -1.)

(2) We can factor $2y^2$ as

$(2y+ \quad)(y+ \quad)$.

(3) The constant term and the middle term are both positive, so we look for pairs of positive factors of 15. The possibilities are

1, 15 and 3, 5.

We can also write these factors as

15, 1 and 5, 3.

(4) Look for combinations of factors from steps (2) and (3) such that the sum of their products is the middle term, $11y$.

The factorization of $2y^2+11y+15$ is $(2y+5)(y+3)$, so $6y^2 + 33y + 45 = 3(2y + 5)(y + 3)$.

45. $$\begin{aligned} x^3 - 7x^2 + 4x - 28 &= x^2(x - 7) + 4(x - 7) \\ &= (x - 7)(x^2 + 4) \end{aligned}$$

46. $4 + 3y - y^2 = -y^2 + 3y + 4 = -1(y^2 - 3y - 4)$

To factor $y^2 - 3y - 4$, look for a pair of factors of -4 whose sum is -3. The numbers we want are -4 and 1. Thus $y^2 - 3y - 4 = (y - 4)(y + 1)$. Then we have

$$\begin{aligned} 4 + 3y - y^2 &= -1(y - 4)(y + 1), \text{ or} \\ &\quad (-y + 4)(y + 1), \text{ or} \\ &\quad (y - 4)(-y - 1) \end{aligned}$$

47. $16x^2 - 16x - 60 = 4(4x^2 - 4x - 15)$

We will factor $4x^2 - 4x - 15$ using the *ac*-method.

(1) There is no common factor (other than 1 or -1.)

(2) Multiply 4 and -15: $4(-15) = -60$.

(3) Look for a factorization of -60 in which the sum of the factors is -4. The numbers we want are 6 and -10.

(4) Split the middle term: $-4x = 6x - 10x$.

(5) Factor by grouping.

$$\begin{aligned} 4x^2 - 4x - 15 &= 4x^2 + 6x - 10x - 15 \\ &= 2x(2x + 3) - 5(2x + 3) \\ &= (2x + 3)(2x - 5) \end{aligned}$$

Then $16x^2 - 16x - 60 = 4(2x + 3)(2x - 5)$.

48. $10a^2 - 11ab + 3b^2$

We will use the FOIL method.

(1) There is no common factor (other than 1 or -1).

(2) We can factor $10a^2$ as

$(a+\quad)(10a+\quad)$ or $(2a+\quad)(5a+\quad)$.

(3) The constant term is positive and the middle term is negative, so we look for pairs of negative factors of $3b^2$. The possibilities are $-b$, $3b$ and b, $-3b$.

We can also write these factors as $3b$, $-b$ and $-3b$, b.

(4) Look for combinations of factors from steps (2) and (3) such that the sum of their products is the middle term, $-11ab$.

The factorization is $(2a-b)(5a-3b)$.

49. $6w^3 - 15w^2 - 10w + 25 = 3w^2(2w-5) - 5(2w-5)$
$= (2w-5)(3w^2-5)$

50. $y^3 + 9y^2 + 18y = y(y^2 + 9y + 18)$

To factor $y^2 + 9y + 18$, look for a pair of factors of 18 whose sum is 9. The numbers we want are 3 and 6. Thus $y^2 + 9y + 18 = (y+3)(y+6)$. Then $y^3 + 9y^2 + 18y = y(y+3)(y+6)$.

51. $4x^2 + 11xy + 6y^2$

We will use the ac-method.

(1) There is no common factor (other than 1 or -1).

(2) Multiply 4 and $6y^2$: $4 \cdot 6y^2 = 24y^2$

(3) Look for a factorization of $24y^2$ in which the sum of the factors is $11y$. The numbers we want are $3y$ and $8y$.

(4) Split the middle term: $11xy = 3xy + 8xy$.

(5) Factor by grouping:

$$\begin{aligned}4x^2 + 11xy + 6y^2 &= 4x^2 + 3xy + 8xy + 6y^2\\ &= x(4x+3y) + 2y(4x+3y)\\ &= (4x+3y)(x+2y)\end{aligned}$$

52. $6 - 5z - 6z^2 = -6z^2 - 5z + 6 = -1(6z^2 + 5z - 6)$

We will use the FOIL method to factor $6z^2 + 5z - 6$.

(1) There is no common factor (other than 1 or -1.)

(2) We can factor $6z^2$ as

$(z+\quad)(6z+\quad)$ or $(2z+\quad)(3z+\quad)$.

(3) Factor the last term, -6. The possibilities are -1, 6 and 1, -6 and -2, 3 and 2, -3.

The factors can also be written as 6, -1 and -6, 1 and 3, -2 and -3, 2.

(4) Look for combinations of factors from steps (2) and (3) such that the sum of their products is the middle term, $5z$.

The factorization of $6z^2 + 5z - 6$ is $(2z+3)(3z-2)$, so we have

$$\begin{aligned}6 - 5z - 6z^2 &= -1(2z+3)(3z-2), \text{ or}\\ &(-2z-3)(3z-2), \text{ or}\\ &(2z+3)(-3z+2)\end{aligned}$$

53. $12t^3 + 8t^2 - 9t - 6 = 4t^2(3t+2) - 3(3t+2)$
$= (3t+2)(4t^2-3)$

54. $y^2 + yz - 20z^2$

Look for a pair of factors of $-20z^2$ whose sum is z. The factors we want are $5z$ and $-4z$.

$y^2 + yz - 20z = (y+5z)(y-4z)$

55. $9x^2 - 6xy - 8y^2$

We will use the ac-method.

(1) There is no common factor (other than 1 or -1).

(2) Multiply 9 and $-8y^2$: $9(-8y^2) = -72y^2$.

(3) Look for a factorization of $-72y^2$ in which the sum of the factors is $-6y$. The factors we want and $-12y$ and $6y$.

(4) Split the middle term: $-6xy = -12xy + 6xy$.

(5) Factor by grouping.

$$\begin{aligned}9x^2 - 6xy - 8y^2 &= 9x^2 - 12xy + 6xy - 8y^2\\ &= 3x(3x-4y) + 2y(3x-4y)\\ &= (3x-4y)(3x+2y)\end{aligned}$$

56. $-3 + 8z + 3z^2 = 3z^2 + 8z - 3$

We will use the FOIL method.

(1) There is no common factor (other than 1 or -1).

(2) We can factor $3z^2$ as

$(3z+\quad)(z+\quad)$.

(3) We look for pairs of factors of -3. The possibilities are -1, 3 and 1, -3.

These factors can also be written as 3, -1 and -3, 1.

(4) Look for combinations of factors from steps (2) and (3) such that the sum of their products is the middle term, $8z$.

The factorization is $(3z-1)(z+3)$.

57. $m^2 - 6mn - 16n^2$

Look for a pair of factors of $-16n^2$ whose sum is $-6n$. The numbers we want are $-8n$ and $2n$.

$m^2 - 6mn - 16n^2 = (m-8n)(m+2n)$.

58. $2w^2 - 12w + 18 = 2(w^2 - 6w + 9)$

Consider $w^2 - 6w + 9$. Look for a pair of factors of 9 whose sum is -6. The numbers we want are -3 and -3.

$w^2 - 6w + 9 = (w-3)(w-3)$, so $2w^2 - 12w + 18 = 2(w-3)(w-3)$, or $2(w-3)^2$.

59. $18t^3 - 18t^2 + 4t = 2t(9t^2 - 9t + 2)$

We will use the ac-method to factor $9t^2 - 9t + 2$.

(1) There is no common factor (other than 1 or -1.)

(2) Multiply 9 and 2: $9 \cdot 2 = 18$.

(3) Look for a factorization of 18 in which the sum of the factors is -9. The numbers we want are -6 and -3.

(4) Split the middle term: $-9t = -6t - 3t$.

(5) Factor by grouping.

$$\begin{aligned}9t^2 - 9t + 2 &= 9t^2 - 6t - 3t + 2\\ &= 3t(3t-2) - (3t-2)\\ &= (3t-2)(3t-1)\end{aligned}$$

Then $18t^3 - 18t^2 + 4t = 2t(3t-2)(3t-1)$.

60. $$\begin{aligned}5z^3 + 15z^2 + z + 3 &= 5z^2(z+3) + (z+3)\\ &= (z+3)(5z^2+1)\end{aligned}$$

61. $-14 + 5t + t^2$, or $t^2 + 5t - 14$

Look for a pair of factors of -14 whose sum is 5. The numbers we want are 7 and -2.

$-14 + 5t + t^2 = (t+7)(t-2)$

62. $4t^2 - 20t + 25$

We will use the FOIL method.

(1) There is no common factor (other than 1 or -1).

(2) We can factor $4t^2$ as

$(4t+\quad)(t+\quad)$ or $(2t+\quad)(2t+\quad)$.

(3) Factor the last term, 25. The middle term is negative, so we consider only pairs of negative factors. The possibilities are -1, -25 and -5, -5.

The first pair of factors can also be written as -25, -1.

(4) Look for combinations of factors from steps (2) and (3) such that the sum of their products is the middle term, $-20t$.

The factorization is $(2t-5)(2t-5)$, or $(2t-5)^2$.

63. $t^2 + 4t - 12$

Look for a pair of factors of -12 whose sum is 4. The numbers we want are 6 and -2.

$t^2 + 4t - 12 = (t+6)(t-2)$

64. $12 + 5z - 2z^2 = -1(2z^2 - 5z - 12)$

We use the FOIL method to factor $2z^2 - 5z - 12$.

(1) There is no common factor (other than 1 or -1).

(2) We can factor $2z^2$ as

$(2z+\quad)(z+\quad)$.

(3) Factor the last term, -12. The possibilities are -1, 12 and 1, -12 and -2, 6 and 2, -6 and 3, -4 and -3, 4.

These factors can also be written as 12, -1 and -12, 1 and 6, -2 and -6, 2 and -4, 3 and 4, -3.

(4) Look for combinations of factors from steps (2) and (3) such that the sum of their products is the middle term, $-5z$.

The factorization of $2z^2 - 5z - 12$ is $(2z+3)(z-4)$, so we have

$$\begin{aligned}12 + 5z - 2z^2 &= -1(2z+3)(z-4), \text{ or}\\ &(-2z-3)(z-4), \text{ or}\\ &(2z+3)(-z+4)\end{aligned}$$

65. $12 + 4y - y^2 = -1(y^2 - 4y - 12)$

To factor $y^2 - 4y - 12$, look for a pair of factors of -12 whose sum is -4. The number we want are -6 and 2.

$y^2 - 4y - 12 = (y-6)(y+2)$, so we have

$$\begin{aligned}12 + 4y - y^2 &= -1(y-6)(y+2), \text{ or}\\ &(-y+6)(y+2), \text{ or}\\ &(y-6)(-y-2)\end{aligned}$$

66. Find the product of two binomials. For example, $(ax^2 + b)(cx + d) = acx^3 + adx^2 + bcx + bd$.

67. There is a finite number of pairs of numbers with the correct product, but there are infinitely many pairs with the correct sum.

68. Since both constants are negative, the middle term will be negative so $(x-17)(x-18)$ cannot be a factorization of $x^2 + 35x + 306$.

69. No; both $2x+6$ and $2x+8$ contain a factor of 2, so $2 \cdot 2$, or 4, must be factored out to reach the complete factorization. In other words, the largest common factor is 4 not 2.

Exercise Set 13.5

1. $x^2 - 14x + 49$

(a) We know that x^2 and 49 are squares.

(b) There is no minus sign before either x^2 or 49.

(c) If we multiply the square roots, x and 7, and double the product, we get $2 \cdot x \cdot 7 = 14x$. This is the opposite of the remaining term, $-14x$.

Thus, $x^2 - 14x + 49$ is a trinomial square.

3. $x^2 + 16x - 64$

Both x^2 and 64 are squares, but there is a minus sign before 64. Thus, $x^2 + 16x - 64$ is not a trinomial square.

5. $x^2 - 2x + 4$

(a) Both x^2 and 4 are squares.

(b) There is no minus sign before either x^2 or 4.

(c) If we multiply the square roots, x and 2, and double the product, we get $2 \cdot x \cdot 2 = 4x$. This is neither the remaining term nor its opposite.

Thus, $x^2 - 2x + 4$ is not a trinomial square.

7. $9x^2 - 24x + 16$

(a) Both $9x^2$ and 16 are squares.

(b) There is no minus sign before either $9x^2$ or 16.

(c) If we multiply the square roots, $3x$ and 4, and double the product, we get $2 \cdot 3x \cdot 4 = 24x$. This is the opposite of the remaining term, $-24x$.

Thus, $9x^2 - 24x + 16$ is a trinomial square.

9. $$\begin{aligned}x^2 - 14x + 49 &= \overset{}{x^2} - 2 \cdot \overset{}{x} \cdot \overset{}{7} + 7^2 = (x-7)^2\\ &\quad\uparrow\qquad\ \uparrow\quad\uparrow\quad\uparrow\qquad\uparrow\\ &= A^2 - 2\ \ A\ \ B + B^2 = (A-B)^2\end{aligned}$$

11. $x^2 + 16x + 64 = x^2 + 2 \cdot x \cdot 8 + 8^2 = (x+8)^2$
$\uparrow \quad \uparrow \quad \uparrow \quad \uparrow \quad \uparrow$
$= A^2 + 2 \quad A \quad B + B^2 = (A+B)^2$

13. $x^2 - 2x + 1 = x^2 - 2 \cdot x \cdot 1 + 1^2 = (x-1)^2$

15. $4 + 4x + x^2 = x^2 + 4x + 4$ Changing the order
$= x^2 + 2 \cdot x \cdot 2 + 2^2$
$= (x+2)^2$

17. $y^2 + 12y + 36 = y^2 + 2 \cdot y \cdot 6 + 6^2 = (y+6)^2$

19. $16 + t^2 - 8t = t^2 - 8t + 16$
$= t^2 - 2 \cdot t \cdot 4 + 4^2$
$= (t-4)^2$

21. $q^4 - 6q^2 + 9 = (q^2)^2 - 2 \cdot q^2 \cdot 3 + 3^2 = (q^2 - 3)^2$

23. $49 + 56y + 16y^2 = 16y^2 + 56y + 49$
$= (4y)^2 + 2 \cdot 4y \cdot 7 + 7^2$
$= (4y+7)^2$

25. $2x^2 - 4x + 2 = 2(x^2 - 2x + 1)$
$= 2(x^2 - 2 \cdot x \cdot 1 + 1^2)$
$= 2(x-1)^2$

27. $x^3 - 18x^2 + 81x = x(x^2 - 18x + 81)$
$= x(x^2 - 2 \cdot x \cdot 9 + 9^2)$
$= x(x-9)^2$

29. $12q^2 - 36q + 27 = 3(4q^2 - 12q + 9)$
$= 3[(2q)^2 - 2 \cdot 2q \cdot 3 + 3^2]$
$= 3(2q-3)^2$

31. $49 - 42x + 9x^2 = 7^2 - 2 \cdot 7 \cdot 3x + (3x)^2$
$= (7-3x)^2$, or $(3x-7)^2$

33. $5y^4 + 10y^2 + 5 = 5(y^4 + 2y^2 + 1)$
$= 5[(y^2)^2 + 2 \cdot y^2 \cdot 1 + 1^2]$
$= 5(y^2+1)^2$

35. $1 + 4x^4 + 4x^2 = 1 + 4x^2 + 4x^4$
$= 1^2 + 2 \cdot 1 \cdot 2x^2 + (2x^2)^2$
$= (1+2x^2)^2$

37. $4p^2 + 12pq + 9q^2 = (2p)^2 + 2 \cdot 2p \cdot 3q + (3q)^2$
$= (2p+3q)^2$

39. $a^2 - 6ab + 9b^2 = a^2 - 2 \cdot a \cdot 3b + (3b)^2$
$= (a-3b)^2$

41. $81a^2 - 18ab + b^2 = (9a)^2 - 2 \cdot 9a \cdot b + b^2$
$= (9a-b)^2$

43. $36a^2 + 96ab + 64b^2 = 4(9a^2 + 24ab + 16b^2)$
$= 4[(3a)^2 + 2 \cdot 3a \cdot 4b + (4b)^2]$
$= 4(3a+4b)^2$

45. $x^2 - 4$

(a) The first expression is a square: x^2
The second expression is a square: $4 = 2^2$

(b) The terms have different signs.
$x^2 - 4$ is a difference of squares.

47. $x^2 + 25$
The terms do not have different signs.
$x^2 + 25$ is not a difference of squares.

49. $x^2 - 45$
The number 45 is not a square.
$x^2 - 45$ is not a difference of squares.

51. $-25y^2 + 16x^2$

(a) The first expression is a square: $25y^2 = (5y)^2$
The second expression is a square: $16x^2 = (4x)^2$

(b) The terms have different signs.
$-25y^2 + 16x^2$ is a difference of squares.

53. $y^2 - 4 = y^2 - 2^2 = (y+2)(y-2)$

55. $p^2 - 9 = p^2 - 3^2 = (p+3)(p-3)$

57. $-49 + t^2 = t^2 - 49 = t^2 - 7^2 = (t+7)(t-7)$

59. $a^2 - b^2 = (a+b)(a-b)$

61. $25t^2 - m^2 = (5t)^2 - m^2 = (5t+m)(5t-m)$

63. $100 - k^2 = 10^2 - k^2 = (10+k)(10-k)$

65. $16a^2 - 9 = (4a)^2 - 3^2 = (4a+3)(4a-3)$

67. $4x^2 - 25y^2 = (2x)^2 - (5y)^2 = (2x+5y)(2x-5y)$

69. $8x^2 - 98 = 2(4x^2 - 49) = 2[(2x)^2 - 7^2] =$
$2(2x+7)(2x-7)$

71. $36x - 49x^3 = x(36 - 49x^2) = x[6^2 - (7x)^2] =$
$x(6+7x)(6-7x)$

73. $\frac{1}{16} - 49x^8 = \left(\frac{1}{4}\right)^2 - (7x^4)^2 = \left(\frac{1}{4} + 7x^4\right)\left(\frac{1}{4} - 7x^4\right)$

75. $0.09y^2 - 0.0004 = (0.3y)^2 - (0.02)^2 =$
$(0.3y + 0.02)(0.3y - 0.02)$

77. $49a^4 - 81 = (7a^2)^2 - 9^2 = (7a^2+9)(7a^2-9)$

79. $a^4 - 16$
$= (a^2)^2 - 4^2$
$= (a^2+4)(a^2-4)$ Factoring a difference of squares
$= (a^2+4)(a+2)(a-2)$ Factoring further: $a^2 - 4$ is a difference of squares.

81. $5x^4 - 405$
$5(x^4 - 81)$
$= 5[(x^2)^2 - 9^2]$
$= 5(x^2 + 9)(x^2 - 9)$
$= 5(x^2 + 9)(x + 3)(x - 3)$ Factoring $x^2 - 9$

83. $1 - y^8$
$= 1^2 - (y^4)^2$
$= (1 + y^4)(1 - y^4)$
$= (1 + y^4)(1 + y^2)(1 - y^2)$ Factoring $1 - y^4$
$= (1 + y^4)(1 + y^2)(1 + y)(1 - y)$ Factoring $1 - y^2$

85. $x^{12} - 16$
$= (x^6)^2 - 4^2$
$= (x^6 + 4)(x^6 - 4)$
$= (x^6 + 4)(x^3 + 2)(x^3 - 2)$ Factoring $x^6 - 4$

87. $y^2 - \frac{1}{16} = y^2 - \left(\frac{1}{4}\right)^2$
$= \left(y + \frac{1}{4}\right)\left(y - \frac{1}{4}\right)$

89. $25 - \frac{1}{49}x^2 = 5^2 - \left(\frac{1}{7}x\right)^2$
$= \left(5 + \frac{1}{7}x\right)\left(5 - \frac{1}{7}x\right)$

91. $16m^4 - t^4$
$= (4m^2)^2 - (t^2)^2$
$= (4m^2 + t^2)(4m^2 - t^2)$
$= (4m^2 + t^2)(2m + t)(2m - t)$ Factoring $4m^2 - t^2$

93. $-110 \div 10$ The quotient of a negative number and a positive number is negative.

$-110 \div 10 = -11$

95. $-\frac{2}{3} \div \frac{4}{5} = -\frac{2}{3} \cdot \frac{5}{4} = -\frac{10}{12} = -\frac{2 \cdot 5}{2 \cdot 6} = -\frac{\cancel{2} \cdot 5}{\cancel{2} \cdot 6} = -\frac{5}{6}$

97. $-64 \div (-32)$ The quotient of two negative numbers is a positive number.

$-64 \div (-32) = 2$

99. The shaded region is a square with sides of length $x - y - y$, or $x - 2y$. Its area is $(x - 2y)(x - 2y)$, or $(x - 2y)^2$. Multiplying, we get the polynomial $x^2 - 4xy + 4y^2$.

101. $y^5 \cdot y^7 = y^{5+7} = y^{12}$

103. $y - 6x = 6$

To find the x-intercept, let $y = 0$. Then solve for x.

$y - 6x = 6$
$0 - 6x = 6$
$-6x = 6$
$x = -1$

The x-intercept is $(-1, 0)$.

To find the y-intercept, let $x = 0$. Then solve for y.

$y - 6x = 6$
$y - 6 \cdot 0 = 6$
$y = 6$

The y-intercept is $(0, 6)$.

Plot these points and draw the line.

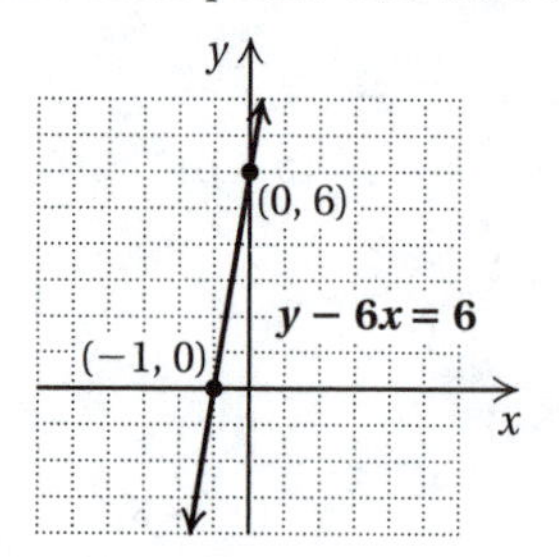

A third point should be used as a check. We substitute any value for x and solve for y. We let $x = -2$. Then

$y - 6x = 6$
$y - 6(-2) = 6$
$y + 12 = 6$
$y = -6$

The point $(-2, -6)$ is on the graph, so the graph is probably correct.

105. $49x^2 - 216$

There is no common factor. Also, $49x^2$ is a square, but 216 is not so this expression is not a difference of squares. It is not factorable. It is prime.

107. $x^2 + 22x + 121 = x^2 + 2 \cdot x \cdot 11 + 11^2$
$= (x + 11)^2$

109. $18x^3 + 12x^2 + 2x = 2x(9x^2 + 6x + 1)$
$= 2x[(3x)^2 + 2 \cdot 3x \cdot 1 + 1^2]$
$= 2x(3x + 1)^2$

111. $x^8 - 2^8$
$= (x^4 + 2^4)(x^4 - 2^4)$
$= (x^4 + 2^4)(x^2 + 2^2)(x^2 - 2^2)$
$= (x^4 + 2^4)(x^2 + 2^2)(x + 2)(x - 2)$, or
$= (x^4 + 16)(x^2 + 4)(x + 2)(x - 2)$

113. $3x^5 - 12x^3 = 3x^3(x^2 - 4) = 3x^3(x + 2)(x - 2)$

115. $18x^3 - \frac{8}{25}x = 2x\left(9x^2 - \frac{4}{25}\right) = 2x\left(3x + \frac{2}{5}\right)\left(3x - \frac{2}{5}\right)$

117. $0.49p - p^3 = p(0.49 - p^2) = p(0.7 + p)(0.7 - p)$

119. $0.64x^2 - 1.21 = (0.8x)^2 - (1.1)^2 = (0.8x + 1.1)(0.8x - 1.1)$

121. $(x+3)^2 - 9 = [(x+3)+3][(x+3)-3] = (x+6)x$, or $x(x+6)$

123. $x^2 - \left(\frac{1}{x}\right)^2 = \left(x + \frac{1}{x}\right)\left(x - \frac{1}{x}\right)$

125. $81 - b^{4k} = 9^2 - (b^{2k})^2$
$= (9 + b^{2k})(9 - b^{2k})$
$= (9 + b^{2k})[3^2 - (b^k)^2]$
$= (9 + b^{2k})(3 + b^k)(3 - b^k)$

127. $9b^{2n} + 12b^n + 4 = (3b^n)^2 + 2 \cdot 3b^n \cdot 2 + 2^2 =$
$(3b^n + 2)^2$

129. $(y + 3)^2 + 2(y + 3) + 1$
$= (y + 3)^2 + 2 \cdot (y + 3) \cdot 1 + 1^2$
$= [(y + 3) + 1]^2$
$= (y + 4)^2$

131. If $cy^2 + 6y + 1$ is the square of a binomial, then $2 \cdot a \cdot 1 = 6$ where $a^2 = c$. Then $a = 3$, so $c = a^2 = 3^2 = 9$. (The polynomial is $9y^2 + 6y + 1$.)

133. Enter $y_1 = x^2 + 9$ and $y_2 = (x + 3)(x + 3)$ and look at a table of values. The y_1-and y_2-values are not the same, so the factorization is not correct.

135. Enter $y_1 = x^2 + 9$ and $y_2 = (x + 3)^2$ and look at a table of values. The y_1-and y_2-values are not the same, so the factorization is not correct.

Exercise Set 13.6

1. $3x^2 - 192 = 3(x^2 - 64)$ 3 is a common factor.
$= 3(x^2 - 8^2)$ Difference of squares
$= 3(x + 8)(x - 8)$

3. $a^2 + 25 - 10a = a^2 - 10a + 25$
$= a^2 - 2 \cdot a \cdot 5 + 5^2$ Trinomial square
$= (a - 5)^2$

5. $2x^2 - 11x + 12$

There is no common factor (other than 1). This polynomial has three terms, but it is not a trinomial square. Multiply the leading coefficient and the constant, 2 and 12: $2 \cdot 12 = 24$. Try to factor 24 so that the sum of the factors is -11. The numbers we want are -3 and -8: $-3(-8) = 24$ and $-3 + (-8) = -11$. Split the middle term and factor by grouping.

$2x^2 - 11x + 12 = 2x^2 - 3x - 8x + 12$
$= (2x^2 - 3x) + (-8x + 12)$
$= x(2x - 3) - 4(2x - 3)$
$= (2x - 3)(x - 4)$

7. $x^3 + 24x^2 + 144x$
$= x(x^2 + 24x + 144)$ x is a common factor.
$= x(x^2 + 2 \cdot x \cdot 12 + 12^2)$ Trinomial square
$= x(x + 12)^2$

9. $x^3 + 3x^2 - 4x - 12$
$= x^2(x + 3) - 4(x + 3)$ Factoring by grouping
$= (x + 3)(x^2 - 4)$
$= (x + 3)(x + 2)(x - 2)$ Factoring the difference of squares

11. $48x^2 - 3 = 3(16x^2 - 1)$ 3 is a common factor.
$= 3[(4x)^2 - 1^2]$ Difference of squares
$= 3(4x + 1)(4x - 1)$

13. $9x^3 + 12x^2 - 45x$
$= 3x(3x^2 + 4x - 15)$ $3x$ is a common factor.
$= 3x(3x - 5)(x + 3)$ Factoring the trinomial

15. $x^2 + 4$ is a *sum* of squares with no common factor. It cannot be factored. It is prime.

17. $x^4 + 7x^2 - 3x^3 - 21x = x(x^3 + 7x - 3x^2 - 21)$
$= x[x(x^2 + 7) - 3(x^2 + 7)]$
$= x[(x^2 + 7)(x - 3)]$
$= x(x^2 + 7)(x - 3)$

19. $x^5 - 14x^4 + 49x^3$
$= x^3(x^2 - 14x + 49)$ x^3 is a common factor.
$= x^3(x^2 - 2 \cdot x \cdot 7 + 7^2)$ Trinomial square
$= x^3(x - 7)^2$

21. $20 - 6x - 2x^2$
$= -2(-10 + 3x + x^2)$ -2 is a common factor.
$= -2(x^2 + 3x - 10)$ Writing in descending order
$= -2(x + 5)(x - 2)$, Using trial and error
or $2(-x - 5)(x - 2)$,
or $2(x + 5)(-x + 2)$

23. $x^2 - 6x + 1$

There is no common factor (other than 1 or -1). This is not a trinomial square, because $-6x \neq 2 \cdot x \cdot 1$ and $-6x \neq -2 \cdot x \cdot 1$. We try factoring using the refined trial and error procedure. We look for two factors of 1 whose sum is -6. There are none. The polynomial cannot be factored. It is prime.

25. $4x^4 - 64$
$= 4(x^4 - 16)$ 4 is a common factor.
$= 4[(x^2)^2 - 4^2]$ Difference of squares
$= 4(x^2 + 4)(x^2 - 4)$ Difference of squares
$= 4(x^2 + 4)(x + 2)(x - 2)$

27. $1 - y^8$ Difference of squares
$= (1 + y^4)(1 - y^4)$ Difference of squares
$= (1 + y^4)(1 + y^2)(1 - y^2)$ Difference of squares
$= (1 + y^4)(1 + y^2)(1 + y)(1 - y)$

29. $x^5 - 4x^4 + 3x^3$
$= x^3(x^2 - 4x + 3)$ x^3 is a common factor.
$= x^3(x - 3)(x - 1)$ Factoring the trinomial using trial and error

31. $\frac{1}{81}x^6 - \frac{8}{27}x^3 + \frac{16}{9}$

$= \frac{1}{9}\left(\frac{1}{9}x^6 - \frac{8}{3}x^3 + 16\right)$ $\frac{1}{9}$ is a common factor.

$= \frac{1}{9}\left[\left(\frac{1}{3}x^3\right)^2 - 2 \cdot \frac{1}{3}x^3 \cdot 4 + 4^2\right]$ Trinomial square

$= \frac{1}{9}\left(\frac{1}{3}x^3 - 4\right)^2$

33. $mx^2 + my^2$

$= m(x^2 + y^2)$ m is a common factor.

The factor with more than one term cannot be factored further, so we have factored completely.

35. $9x^2y^2 - 36xy = 9xy(xy - 4)$

37. $2\pi rh + 2\pi r^2 = 2\pi r(h + r)$

39. $(a + b)(x - 3) + (a + b)(x + 4)$

$= (a + b)[(x - 3) + (x + 4)]$ $(a + b)$ is a common factor.

$= (a + b)(2x + 1)$

41. $(x - 1)(x + 1) - y(x + 1) = (x + 1)(x - 1 - y)$ $(x + 1)$ is a common factor.

43. $n^2 + 2n + np + 2p$

$= n(n + 2) + p(n + 2)$ Factoring by grouping

$= (n + 2)(n + p)$

45. $6q^2 - 3q + 2pq - p$

$= (6q^2 - 3q) + (2pq - p)$

$= 3q(2q - 1) + p(2q - 1)$ Factoring by grouping

$= (2q - 1)(3q + p)$

47. $4b^2 + a^2 - 4ab$

$= a^2 - 4ab + 4b^2$ Rearranging

$= a^2 - 2 \cdot a \cdot 2b + (2b)^2$ Trinomial square

$= (a - 2b)^2$

(Note that if we had rewritten the polynomial as $4b^2 - 4ab + a^2$, we might have written the result as $(2b - a)^2$. The two factorizations are equivalent.)

49. $16x^2 + 24xy + 9y^2$

$= (4x)^2 + 2 \cdot 4x \cdot 3y + (3y)^2$ Trinomial square

$= (4x + 3y)^2$

51. $49m^4 - 112m^2n + 64n^2$

$= (7m^2)^2 - 2 \cdot 7m^2 \cdot 8n + (8n)^2$ Trinomial square

$= (7m^2 - 8n)^2$

53. $y^4 + 10y^2z^2 + 25z^4$

$= (y^2)^2 + 2 \cdot y^2 \cdot 5z^2 + (5z^2)^2$ Trinomial square

$= (y^2 + 5z^2)^2$

55. $\frac{1}{4}a^2 + \frac{1}{3}ab + \frac{1}{9}b^2$

$= \left(\frac{1}{2}a\right)^2 + 2 \cdot \frac{1}{2}a \cdot \frac{1}{3}b + \left(\frac{1}{3}b\right)^2$

$= \left(\frac{1}{2}a + \frac{1}{3}b\right)^2$

57. $a^2 - ab - 2b^2 = (a - 2b)(a + b)$ Using trial and error

59. $2mn - 360n^2 + m^2$

$= m^2 + 2mn - 360n^2$ Rewriting

$= (m + 20n)(m - 18n)$ Using trial and error

61. $m^2n^2 - 4mn - 32 = (mn - 8)(mn + 4)$ Using trial and error

63. $r^5s^2 - 10r^4s + 16r^3$

$= r^3(r^2s^2 - 10rs + 16)$ r^3 is a common factor.

$= r^3(rs - 2)(rs - 8)$ Using trial and error

65. $a^5 + 4a^4b - 5a^3b^2$

$= a^3(a^2 + 4ab - 5b^2)$ a^3 is a common factor.

$= a^3(a + 5b)(a - b)$ Factoring the trinomial

67. $a^2 - \frac{1}{25}b^2$

$= a^2 - \left(\frac{1}{5}b\right)^2$ Difference of squares

$= \left(a + \frac{1}{5}b\right)\left(a - \frac{1}{5}b\right)$

69. $x^2 - y^2 = (x + y)(x - y)$ Difference of squares

71. $16 - p^4q^4$

$= 4^2 - (p^2q^2)^2$ Difference of squares

$= (4 + p^2q^2)(4 - p^2q^2)$ $4 - p^2q^2$ is a difference of squares.

$= (4 + p^2q^2)(2 + pq)(2 - pq)$

73. $1 - 16x^{12}y^{12}$

$= 1^2 - (4x^6y^6)^2$ Difference of squares

$= (1 + 4x^6y^6)(1 - 4x^6y^6)$ $1 - 4x^6y^6$ is a difference of squares.

$= (1 + 4x^6y^6)(1 + 2x^3y^3)(1 - 2x^3y^3)$

75. $q^3 + 8q^2 - q - 8$

$= q^2(q + 8) - (q + 8)$ Factoring by grouping

$= (q + 8)(q^2 - 1)$

$= (q + 8)(q + 1)(q - 1)$ Factoring the difference of squares

77. $6a^3b^3 - a^2b^2 - 2ab$

$= ab(6a^2b^2 - ab - 2)$ ab is a common factor

$= ab(2ab + 1)(3ab - 2)$ Factoring the trinomial

79. $m^4 - 5m^2 + 4$
$= (m^2 - 1)(m^2 - 4)$ Using trial and error
$= (m + 1)(m - 1)(m + 2)(m - 2)$ Factoring two differences of squares

81. $t^4 - 2t^2 + 1$ Trinomial square
$= (t^2 - 1)^2$ Difference of squares
$= [(t + 1)(t - 1)]^2$
$= (t + 1)^2(t - 1)^2$

83. $(-4, 0)$; $m = -3$

The slope is -3, so the equation is $y = -3x + b$. Using the point $(-4, 0)$, we substitute -4 for x and 0 for y in $y = -3x + b$ and then solve for b.

$$y = -3x + b$$
$$0 = -3(-4) + b$$
$$0 = 12 + b$$
$$-12 = b$$

Then the equation is $y = -3x - 12$.

85. $(-4, 5)$; $m = -\frac{2}{3}$

The slope is $-\frac{2}{3}$, so the equation is $y = -\frac{2}{3}x + b$. Using the point $(-4, 5)$, we substitute -4 for x and 5 for y in $y = -\frac{2}{3}x + b$ and then solve for b.

$$y = -\frac{2}{3}x + b$$
$$5 = -\frac{2}{3}(-4) + b$$
$$5 = \frac{8}{3} + b$$
$$\frac{7}{3} = b$$

Then the equation is $y = -\frac{2}{3}x + \frac{7}{3}$.

87. $\frac{7}{5} \div \left(-\frac{11}{10}\right)$
$= \frac{7}{5} \cdot \left(-\frac{10}{11}\right)$ Multiplying by the reciprocal of the divisor
$= -\frac{7 \cdot 10}{5 \cdot 11}$
$= -\frac{7 \cdot 5 \cdot 2}{5 \cdot 11} = -\frac{7 \cdot 2}{11} \cdot \frac{5}{5}$
$= -\frac{14}{11}$

89.
$$A = aX + bX - 7$$
$$A + 7 = aX + bX$$
$$A + 7 = X(a + b)$$
$$\frac{A + 7}{a + b} = X$$

91. $x^3 + 20 - (5x^2 + 4x) = x^3 + 20 - 5x^2 - 4x$
$= x^3 - 5x^2 - 4x + 20$
$= x^2(x - 5) - 4(x - 5)$
$= (x - 5)(x^2 - 4)$
$= (x - 5)(x + 2)(x - 2)$

93. $12.25x^2 - 7x + 1 = (3.5x)^2 - 2 \cdot (3.5x) \cdot 1 + 1^2$
$= (3.5x - 1)^2$

95. $5x^2 + 13x + 7.2$

Multiply the leading coefficient and the constant, 5 and 7.2: $5(7.2) = 36$. Try to factor 36 so that the sum of the factors is 13. The numbers we want are 9 and 4. Split the middle term and factor by grouping:

$$5x^2 + 13x + 7.2 = 5x^2 + 9x + 4x + 7.2$$
$$= (5x^2 + 9x) + (4x + 7.2)$$
$$= 5x(x + 1.8) + 4(x + 1.8)$$
$$= (x + 1.8)(5x + 4)$$

97. $18 + y^3 - 9y - 2y^2$
$= y^3 - 2y^2 - 9y + 18$
$= y^2(y - 2) - 9(y - 2)$
$= (y - 2)(y^2 - 9)$
$= (y - 2)(y + 3)(y - 3)$

99. $x^3 - x^2 - 4x + 4 = x^2(x - 1) - 4(x - 1)$
$= (x - 1)(x^2 - 4)$
$= (x - 1)(x + 2)(x - 2)$

101. $a^3 - 4a^2 - a - 4 = a^2(a - 4) - (a + 4)$

There is no common binomial factor. The polynomial is prime.

103. $y^2(y - 1) - 2y(y - 1) + (y - 1)$
$= (y - 1)(y^2 - 2y + 1)$
$= (y - 1)(y - 1)^2$
$= (y - 1)^3$

105. $(y + 4)^2 + 2x(y + 4) + x^2$
$= (y + 4)^2 + 2 \cdot (y + 4) \cdot x + x^2$ Trinomial square
$= (y + 4 + x)^2$

Exercise Set 13.7

1. $(x + 4)(x + 9) = 0$

$x + 4 = 0$ *or* $x + 9 = 0$ Using the principle of zero products

$x = -4$ *or* $x = -9$ Solving the two equations separately

Check:

For -4

$$\frac{(x + 4)(x + 9) = 0}{(-4 + 4)(-4 + 9) \; ? \; 0}$$
$0 \cdot 5$
0 TRUE

For -9

$$\begin{array}{c|c} (x+4)(x+9) = 0 & \\ \hline (-9+4)(-9+9) \ ? \ 0 & \\ -5 \cdot 0 & \\ 0 & \text{TRUE} \end{array}$$

The solutions are -4 and -9.

3. $(x+3)(x-8) = 0$

$x+3 = 0$ *or* $x-8 = 0$ Using the principle of zero products

$x = -3$ *or* $x = 8$

Check:

For -3

$$\begin{array}{c|c} (x+3)(x-8) = 0 & \\ \hline (-3+3)(-3-8) \ ? \ 0 & \\ 0(-11) & \\ 0 & \text{TRUE} \end{array}$$

For 8

$$\begin{array}{c|c} (x+3)(x-8) = 0 & \\ \hline (8+3)(8-8) \ ? \ 0 & \\ 11 \cdot 0 & \\ 0 & \text{TRUE} \end{array}$$

The solutions are -3 and 8.

5. $(x+12)(x-11) = 0$

$x+12 = 0$ *or* $x-11 = 0$

$x = -12$ *or* $x = 11$

The solutions are -12 and 11.

7. $x(x+3) = 0$

$x = 0$ *or* $x+3 = 0$

$x = 0$ *or* $x = -3$

The solutions are 0 and -3.

9. $0 = y(y+18)$

$y = 0$ *or* $y+18 = 0$

$y = 0$ *or* $y = -18$

The solutions are 0 and -18.

11. $(2x+5)(x+4) = 0$

$2x+5 = 0$ *or* $x+4 = 0$

$2x = -5$ *or* $x = -4$

$x = -\frac{5}{2}$ *or* $x = -4$

The solutions are $-\frac{5}{2}$ and -4.

13. $(5x+1)(4x-12) = 0$

$5x+1 = 0$ *or* $4x-12 = 0$

$5x = -1$ *or* $4x = 12$

$x = -\frac{1}{5}$ *or* $x = 3$

The solutions are $-\frac{1}{5}$ and 3.

15. $(7x-28)(28x-7) = 0$

$7x-28 = 0$ *or* $28x-7 = 0$

$7x = 28$ *or* $28x = 7$

$x = 4$ *or* $x = \frac{7}{28} = \frac{1}{4}$

The solutions are 4 and $\frac{1}{4}$.

17. $2x(3x-2) = 0$

$2x = 0$ *or* $3x-2 = 0$

$x = 0$ *or* $3x = 2$

$x = 0$ *or* $x = \frac{2}{3}$

The solutions are 0 and $\frac{2}{3}$.

19. $\left(\frac{1}{5}+2x\right)\left(\frac{1}{9}-3x\right) = 0$

$\frac{1}{5}+2x = 0$ *or* $\frac{1}{9}-3x = 0$

$2x = -\frac{1}{5}$ *or* $-3x = -\frac{1}{9}$

$x = -\frac{1}{10}$ *or* $x = \frac{1}{27}$

The solutions are $-\frac{1}{10}$ and $\frac{1}{27}$.

21. $(0.3x-0.1)(0.05x+1) = 0$

$0.3x-0.1 = 0$ *or* $0.05x+1 = 0$

$0.3x = 0.1$ *or* $0.05x = -1$

$x = \frac{0.1}{0.3}$ *or* $x = -\frac{1}{0.05}$

$x = \frac{1}{3}$ *or* $x = -20$

The solutions are $\frac{1}{3}$ and -20.

23. $9x(3x-2)(2x-1) = 0$

$9x = 0$ *or* $3x-2 = 0$ *or* $2x-1 = 0$

$x = 0$ *or* $3x = 2$ *or* $2x = 1$

$x = 0$ *or* $x = \frac{2}{3}$ *or* $x = \frac{1}{2}$

The solutions are 0, $\frac{2}{3}$, and $\frac{1}{2}$.

25. $x^2+6x+5 = 0$

$(x+5)(x+1) = 0$ Factoring

$x+5 = 0$ *or* $x+1 = 0$ Using the principle of zero products

$x = -5$ *or* $x = -1$

The solutions are -5 and -1.

27. $x^2+7x-18 = 0$

$(x+9)(x-2) = 0$ Factoring

$x+9 = 0$ *or* $x-2 = 0$ Using the principle of zero products

$x = -9$ *or* $x = 2$

The solutions are -9 and 2.

29. $x^2 - 8x + 15 = 0$
$(x-5)(x-3) = 0$
$x - 5 = 0$ *or* $x - 3 = 0$
$x = 5$ *or* $x = 3$
The solutions are 5 and 3.

31. $x^2 - 8x = 0$
$x(x-8) = 0$
$x = 0$ *or* $x - 8 = 0$
$x = 0$ *or* $x = 8$
The solutions are 0 and 8.

33. $x^2 + 18x = 0$
$x(x+18) = 0$
$x = 0$ *or* $x + 18 = 0$
$x = 0$ *or* $x = -18$
The solutions are 0 and -18.

35. $x^2 = 16$
$x^2 - 16 = 0$ Subtracting 16
$(x-4)(x+4) = 0$
$x - 4 = 0$ *or* $x + 4 = 0$
$x = 4$ *or* $x = -4$
The solutions are 4 and -4.

37. $9x^2 - 4 = 0$
$(3x-2)(3x+2) = 0$
$3x - 2 = 0$ *or* $3x + 2 = 0$
$3x = 2$ *or* $3x = -2$
$x = \frac{2}{3}$ *or* $x = -\frac{2}{3}$
The solutions are $\frac{2}{3}$ and $-\frac{2}{3}$.

39. $0 = 6x + x^2 + 9$
$0 = x^2 + 6x + 9$ Writing in descending order
$0 = (x+3)(x+3)$
$x + 3 = 0$ *or* $x + 3 = 0$
$x = -3$ *or* $x = -3$
There is only one solution, -3.

41. $x^2 + 16 = 8x$
$x^2 - 8x + 16 = 0$ Subtracting $8x$
$(x-4)(x-4) = 0$
$x - 4 = 0$ *or* $x - 4 = 0$
$x = 4$ *or* $x = 4$
There is only one solution, 4.

43. $5x^2 = 6x$
$5x^2 - 6x = 0$
$x(5x-6) = 0$
$x = 0$ *or* $5x - 6 = 0$
$x = 0$ *or* $5x = 6$
$x = 0$ *or* $x = \frac{6}{5}$
The solutions are 0 and $\frac{6}{5}$.

45. $6x^2 - 4x = 10$
$6x^2 - 4x - 10 = 0$
$2(3x^2 - 2x - 5) = 0$
$2(3x-5)(x+1) = 0$
$3x - 5 = 0$ *or* $x + 1 = 0$
$3x = 5$ *or* $x = -1$
$x = \frac{5}{3}$ *or* $x = -1$
The solutions are $\frac{5}{3}$ and -1.

47. $12y^2 - 5y = 2$
$12y^2 - 5y - 2 = 0$
$(4y+1)(3y-2) = 0$
$4y + 1 = 0$ *or* $3y - 2 = 0$
$4y = -1$ *or* $3y = 2$
$y = -\frac{1}{4}$ *or* $y = \frac{2}{3}$
The solutions are $-\frac{1}{4}$ and $\frac{2}{3}$.

49. $t(3t+1) = 2$
$3t^2 + t = 2$ Multiplying on the left
$3t^2 + t - 2 = 0$ Subtracting 2
$(3t-2)(t+1) = 0$
$3t - 2 = 0$ *or* $t + 1 = 0$
$3t = 2$ *or* $t = -1$
$t = \frac{2}{3}$ *or* $t = -1$
The solutions are $\frac{2}{3}$ and -1.

51. $100y^2 = 49$
$100y^2 - 49 = 0$
$(10y+7)(10y-7) = 0$
$10y + 7 = 0$ *or* $10y - 7 = 0$
$10y = -7$ *or* $10y = 7$
$y = -\frac{7}{10}$ *or* $y = \frac{7}{10}$
The solutions are $-\frac{7}{10}$ and $\frac{7}{10}$.

53. $x^2 - 5x = 18 + 2x$

$x^2 - 5x - 18 - 2x = 0$ Subtracting 18 and $2x$

$x^2 - 7x - 18 = 0$

$(x-9)(x+2) = 0$

$x - 9 = 0$ *or* $x + 2 = 0$

$x = 9$ *or* $x = -2$

The solutions are 9 and -2.

55. $10x^2 - 23x + 12 = 0$

$(5x-4)(2x-3) = 0$

$5x - 4 = 0$ *or* $2x - 3 = 0$

$5x = 4$ *or* $2x = 3$

$x = \frac{4}{5}$ *or* $x = \frac{3}{2}$

The solutions are $\frac{4}{5}$ and $\frac{3}{2}$.

57. We let $y = 0$ and solve for x.

$0 = x^2 + 3x - 4$

$0 = (x+4)(x-1)$

$x + 4 = 0$ *or* $x - 1 = 0$

$x = -4$ *or* $x = 1$

The x-intercepts are $(-4, 0)$ and (1,0).

59. We let $y = 0$ and solve for x

$0 = 2x^2 + x - 10$

$0 = (2x+5)(x-2)$

$2x + 5 = 0$ *or* $x - 2 = 0$

$2x = -5$ *or* $x = 2$

$x = -\frac{5}{2}$ *or* $x = 2$

The x-intercepts are $\left(-\frac{5}{2}, 0\right)$ and $(2, 0)$.

61. We let $y = 0$ and solve for x.

$0 = x^2 - 2x - 15$

$0 = (x-5)(x+3)$

$x - 5 = 0$ *or* $x + 3 = 0$

$x = 5$ *or* $x = -3$

The x-intercepts are $(5, 0)$ and $(-3, 0)$.

63. The solutions of the equation are the first coordinates of the x-intercepts of the graph. From the graph we see that the x-intercepts are $(-1, 0)$ and $(4, 0)$, so the solutions of the equation are -1 and 4.

65. The solutions of the equation are the first coordinates of the x-intercepts of the graph. From the graph we see that the x-intercepts are $(-1, 0)$ and $(3, 0)$, so the solutions of the equation are -1 and 3.

67. $(a+b)^2$

69. The two numbers have different signs, so their quotient is negative.

$144 \div -9 = -16$

71. $-\frac{5}{8} \div \frac{3}{16} = -\frac{5}{8} \cdot \frac{16}{3}$

$= -\frac{5 \cdot 16}{8 \cdot 3}$

$= -\frac{5 \cdot \not{8} \cdot 2}{\not{8} \cdot 3}$

$= -\frac{10}{3}$

73. $b(b+9) = 4(5+2b)$

$b^2 + 9b = 20 + 8b$

$b^2 + 9b - 8b - 20 = 0$

$b^2 + b - 20 = 0$

$(b+5)(b-4) = 0$

$b + 5 = 0$ *or* $b - 4 = 0$

$b = -5$ *or* $b = 4$

The solutions are -5 and 4.

75. $(t-3)^2 = 36$

$t^2 - 6t + 9 = 36$

$t^2 - 6t - 27 = 0$

$(t-9)(t+3) = 0$

$t - 9 = 0$ *or* $t + 3 = 0$

$t = 9$ *or* $t = -3$

The solutions are 9 and -3.

77. $x^2 - \frac{1}{64} = 0$

$\left(x - \frac{1}{8}\right)\left(x + \frac{1}{8}\right) = 0$

$x - \frac{1}{8} = 0$ *or* $x + \frac{1}{8} = 0$

$x = \frac{1}{8}$ *or* $x = -\frac{1}{8}$

The solutions are $\frac{1}{8}$ and $-\frac{1}{8}$.

79. $\frac{5}{16}x^2 = 5$

$\frac{5}{16}x^2 - 5 = 0$

$5\left(\frac{1}{16}x^2 - 1\right) = 0$

$5\left(\frac{1}{4}x - 1\right)\left(\frac{1}{4}x + 1\right) = 0$

$\frac{1}{4}x - 1 = 0$ *or* $\frac{1}{4}x + 1 = 0$

$\frac{1}{4}x = 1$ *or* $\frac{1}{4}x = -1$

$x = 4$ *or* $x = -4$

The solutions are 4 and -4.

81. (a) $x = -3$ *or* $x = 4$

$x + 3 = 0$ *or* $x - 4 = 0$

$(x+3)(x-4) = 0$ Principle of zero products

$x^2 - x - 12 = 0$ Multiplying

(b) $x = -3$ *or* $x = -4$

$x + 3 = 0$ *or* $x + 4 = 0$

$(x + 3)(x + 4) = 0$

$x^2 + 7x + 12 = 0$

(c) $x = \frac{1}{2}$ *or* $x = \frac{1}{2}$

$x - \frac{1}{2} = 0$ *or* $x - \frac{1}{2} = 0$

$\left(x - \frac{1}{2}\right)\left(x - \frac{1}{2}\right) = 0$

$x^2 - x + \frac{1}{4} = 0$, or

$4x^2 - 4x + 1 = 0$ Multiplying by 4

(d) $(x - 5)(x + 5) = 0$

$x^2 - 25 = 0$

(e) $(x - 0)(x - 0.1)\left(x - \frac{1}{4}\right) = 0$

$x\left(x - \frac{1}{10}\right)\left(x - \frac{1}{4}\right) = 0$

$x\left(x^2 - \frac{7}{20}x + \frac{1}{40}\right) = 0$

$x^3 - \frac{7}{20}x^2 + \frac{1}{40}x = 0$, or

$40x^3 - 14x^2 + x = 0$ Multiplying by 40

83. 2.33, 6.77

85. 0, 2.74

Exercise Set 13.8

1. ***Familiarize***. We make a drawing. Let w = the width, in inches. Then $3w$ = the length, in inches.

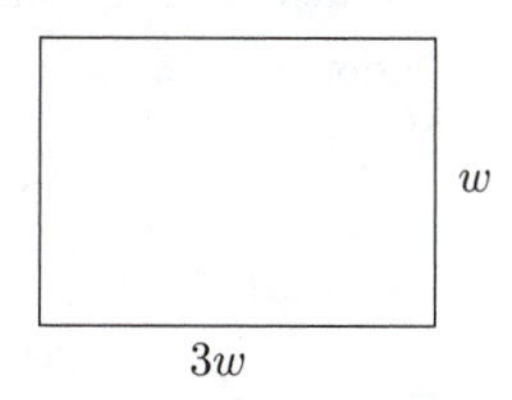

Recall that the area of a rectangle is length times width.

Translate. We reword the problem.

Length times width is $\underbrace{588 \text{ in}^2}$.

$\downarrow \quad \downarrow \quad \downarrow \quad \downarrow \quad \downarrow$

$3w \quad \cdot \quad w \quad = \quad 588$

Solve. We solve the equation.

$3w \cdot w = 588$

$3w^2 = 588$

$3w^2 - 588 = 0$

$w^2 - 196 = 0$

$(w + 14)(w - 14) = 0$

$w + 14 = 0$ *or* $w - 14 = 0$

$w = -14$ *or* $w = 14$

Check. Since the width must be positive, -14 cannot be a solution. If the width is 14 in., then the length is $3 \cdot 14$, or 42 in., and the area is $42 \cdot 14$, or 588 in^2. Thus, 14 checks.

State. The width is 14 in., and the length is 42 in.

3. ***Familiarize***. Let w = the width of the table, in feet. Then $6w$ = the length, in feet. Recall that the area of a rectangle is Length $\cdot$ Width.

Translate.

$\underbrace{\text{The area of the table}}$ is $\underbrace{24 \text{ ft}^2}$.

$\downarrow \quad \downarrow \quad \downarrow$

$6w \cdot w \quad = \quad 24$

Solve. We solve the equation.

$6w \cdot w = 24$

$6w^2 = 24$

$6w^2 - 24 = 0$

$6(w^2 - 4) = 0$

$6(w + 2)(w - 2) = 0$

$w + 2 = 0$ *or* $w - 2 = 0$

$w = -2$ *or* $w = 2$

Check. Since the width must be positive, -2 cannot be a solution. If the width is 2 ft, then the length is $6 \cdot 2$ ft, or 12 ft, and the area is $12 \text{ ft} \cdot 2 \text{ ft} = 24 \text{ ft}^2$. These numbers check.

State. The table is 12 ft long and 2 ft wide.

5. ***Familiarize***. Using the labels shown on the drawing in the text, we let h = the height, in cm, and $h + 10$ = the base, in cm. Recall that the formula for the area of a triangle is $\frac{1}{2} \cdot (\text{base}) \cdot (\text{height})$.

Translate.

$\frac{1}{2}$ times base times height is $\underbrace{28 \text{ cm}^2}$.

$\downarrow \quad \downarrow \quad \downarrow \quad \downarrow \quad \downarrow \quad \downarrow \quad \downarrow$

$\frac{1}{2} \quad \cdot \quad (h + 10) \quad \cdot \quad h \quad = \quad 28$

Solve. We solve the equation.

$\frac{1}{2}(h + 10)h = 28$

$(h + 10)h = 56$ Multiplying by 2

$h^2 + 10h = 56$

$h^2 + 10h - 56 = 0$

$(h + 14)(h - 4) = 0$

$h + 14 = 0$ *or* $h - 4 = 0$

$h = -14$ *or* $h = 4$

Check. Since the height of the triangle must be positive, -14 cannot be a solution. If the height is 4 cm, then the base is $4+10$, or 14 cm, and the area is $\frac{1}{2} \cdot 14 \cdot 4$, or 28 cm^2. Thus, 4 checks.

State. The height of the triangle is 4 cm, and the base is 14 cm.

7. ***Familiarize***. Using the labels shown on the drawing in the text, we let h = the height of the triangle, in meters, and $\frac{1}{2}h$ = the length of the base, in meters. Recall that the formula for the area of a triangle is $\frac{1}{2}\cdot$ (base) $\cdot$ (height).

Translate.

$\frac{1}{2}$ times base times height is $\underbrace{64\text{ m}^2}$.

$$\downarrow \quad \downarrow \quad \downarrow \quad \downarrow \quad \downarrow \quad \downarrow \quad \downarrow$$

$$\frac{1}{2} \quad \cdot \quad \frac{1}{2}h \quad \cdot \quad h \quad = \quad 64$$

Solve. We solve the equation.

$$\frac{1}{2}\cdot\frac{1}{2}h\cdot h = 64$$

$$\frac{1}{4}h^2 = 64$$

$$h^2 = 256 \quad \text{Multiplying by 4}$$

$$h^2 - 256 = 0$$

$$(h+16)(h-16) = 0$$

$$h+16=0 \quad \text{or} \quad h-16=0$$

$$h=-16 \quad \text{or} \quad h=16$$

Check. The height of the triangle cannot be negative, so -16 cannot be a solution. If the height is 16 m, then the length of the base is $\frac{1}{2}\cdot 16$ m, or 8 m, and the area is $\frac{1}{2}\cdot 8\text{ m}\cdot 16\text{ m} = 64\text{ m}^2$. These numbers check.

State. The length of the base is 8 m, and the height is 16 m.

9. ***Familiarize***. Reread Example 4 in Section 12.3.

Translate. Substitute 14 for n.

$$14^2 - 14 = N$$

Solve. We do the computation on the left.

$$14^2 - 14 = N$$

$$196 - 14 = N$$

$$182 = N$$

Check. We can redo the computation, or we can solve the equation $n^2 - n = 182$. The answer checks.

State. 182 games will be played.

11. ***Familiarize***. Reread Example 4 in Section 12.3.

Translate. Substitute 132 for N.

$$n^2 - n = 132$$

Solve.

$$n^2 - n = 132$$

$$n^2 - n - 132 = 0$$

$$(n-12)(n+11) = 0$$

$$n-12=0 \quad \text{or} \quad n+11=0$$

$$n=12 \quad \text{or} \quad n=-11$$

Check. The solutions of the equation are 12 and -11. Since the number of teams cannot be negative, -11 cannot be a solution. But 12 checks since $12^2 - 12 = 144 - 12 = 132$.

State. There are 12 teams in the league.

13. ***Familiarize***. We will use the formula $N = \frac{1}{2}(n^2 - n)$.

Translate. Substitute 100 for n.

$$N = \frac{1}{2}(100^2 - 100)$$

Solve. We do the computation on the right.

$$N = \frac{1}{2}(10{,}000 - 100)$$

$$N = \frac{1}{2}(9900)$$

$$N = 4950$$

Check. We can redo the computation, or we can solve the equation $4950 = \frac{1}{2}(n^2 - n)$. The answer checks.

State. 4950 handshakes are possible.

15. ***Familiarize***. We will use the formula $N = \frac{1}{2}(n^2 - n)$.

Translate. Substitute 300 for N.

$$300 = \frac{1}{2}(n^2 - n)$$

Solve. We solve the equation.

$$2\cdot 300 = 2\cdot\frac{1}{2}(n^2 - n) \quad \text{Multiplying by 2}$$

$$600 = n^2 - n$$

$$0 = n^2 - n - 600$$

$$0 = (n+24)(n-25)$$

$$n+24=0 \quad \text{or} \quad n-25=0$$

$$n=-24 \quad \text{or} \quad n=25$$

Check. The number of people at a meeting cannot be negative, so -24 cannot be a solution. But 25 checks since $\frac{1}{2}(25^2 - 25) = \frac{1}{2}(625 - 25) = \frac{1}{2}\cdot 600 = 300$.

State. There were 25 people at the party.

17. ***Familiarize***. We will use the formula $N = \frac{1}{2}(n^2 - n)$, since "clicks" can be substituted for handshakes.

Translate. Substitute 190 for N.

$$190 = \frac{1}{2}(n^2 - n)$$

Solve.

$$190 = \frac{1}{2}(n^2 - n)$$

$$380 = n^2 - n \quad \text{Multiplying by 2}$$

$$0 = n^2 - n - 380$$

$$0 = (n-20)(n+19)$$

$$n-20=0 \quad \text{or} \quad n+19=0$$

$$n=20 \quad \text{or} \quad n=-19$$

Check. The solutions of the equation are 20 and -19. Since the number of people cannot be negative, -19 cannot be a solution. However, 20 checks since $\frac{1}{2}(20^2 - 20) = \frac{1}{2}(400 - 20) = \frac{1}{2}(380) = 190$.

State. 20 people took part in the toast.

19. ***Familiarize***. The page numbers on facing pages are consecutive integers. Let x = the smaller integer. Then $x+1$ = the larger integer.

Translate. We reword the problem.

$$\underbrace{\text{Smaller integer}}_{x} \;\underbrace{\text{times}}_{\cdot}\; \underbrace{\text{larger integer}}_{(x+1)} \;\underbrace{\text{is}}_{=}\; \underbrace{210.}_{210}$$

Solve. We solve the equation.

$$x(x+1) = 210$$
$$x^2 + x = 210$$
$$x^2 + x - 210 = 0$$
$$(x+15)(x-14) = 0$$
$$x+15 = 0 \quad or \quad x-14 = 0$$
$$x = -15 \quad or \quad x = 14$$

Check. The solutions of the equation are -15 and 14. Since a page number cannot be negative, -15 cannot be a solution of the original problem. We only need to check 14. When $x = 14$, then $x+1 = 15$, and $14 \cdot 15 = 210$. This checks.

State. The page numbers are 14 and 15.

21. ***Familiarize***. Let x = the smaller even integer. Then $x+2$ = the larger even integer.

Translate. We reword the problem.

$$\underbrace{\text{Smaller even integer}}_{x} \;\underbrace{\text{times}}_{\cdot}\; \underbrace{\text{larger even integer}}_{(x+2)} \;\underbrace{\text{is}}_{=}\; \underbrace{168.}_{168}$$

Solve.

$$x(x+2) = 168$$
$$x^2 + 2x = 168$$
$$x^2 + 2x - 168 = 0$$
$$(x+14)(x-12) = 0$$
$$x+14 = 0 \quad or \quad x-12 = 0$$
$$x = -14 \quad or \quad x = 12$$

Check. The solutions of the equation are -14 and 12. When x is -14, then $x+2$ is -12 and $-14(-12) = 168$. The numbers -14 and -12 are consecutive even integers which are solutions of the problem. When x is 12, then $x+2$ is 14 and $12 \cdot 14 = 168$. The numbers 12 and 14 are also consecutive even integers which are solutions of the problem.

State. We have two solutions, each of which consists of a pair of numbers: -14 and -12, and 12 and 14.

23. ***Familiarize***. Let x = the smaller odd integer. Then $x + 2$ = the larger odd integer.

Translate. We reword the problem.

$$\underbrace{\text{Smaller odd integer}}_{x} \;\underbrace{\text{times}}_{\cdot}\; \underbrace{\text{larger odd integer}}_{(x+2)} \;\underbrace{\text{is}}_{=}\; \underbrace{255.}_{255}$$

Solve.

$$x(x+2) = 255$$
$$x^2 + 2x = 255$$
$$x^2 + 2x - 255 = 0$$
$$(x-15)(x+17) = 0$$
$$x-15 = 0 \quad or \quad x+17 = 0$$
$$x = 15 \quad or \quad x = -17$$

Check. The solutions of the equation are 15 and -17. When x is 15, then $x+2$ is 17 and $15 \cdot 17 = 255$. The numbers 15 and 17 are consecutive odd integers which are solutions to the problem. When x is -17, then $x+2$ is -15 and $-17(-15) = 255$. The numbers -17 and -15 are also consecutive odd integers which are solutions to the problem.

State. We have two solutions, each of which consists of a pair of numbers: 15 and 17, and -17 and -15.

25. ***Familiarize***. We make a drawing. Let x = the length of the unknown leg. Then $x+2$ = the length of the hypotenuse.

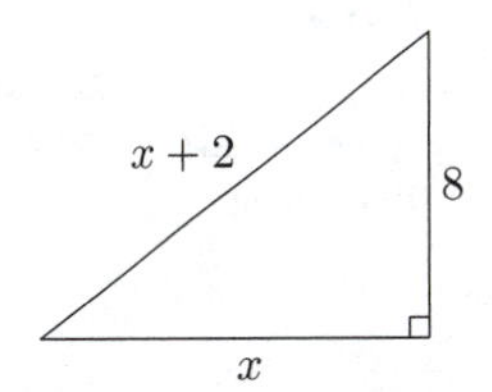

Translate. Use the Pythagorean theorem.

$$a^2 + b^2 = c^2$$
$$8^2 + x^2 = (x+2)^2$$

Solve. We solve the equation.

$$8^2 + x^2 = (x+2)^2$$
$$64 + x^2 = x^2 + 4x + 4$$
$$60 = 4x \qquad \text{Subtracting } x^2 \text{ and } 4$$
$$15 = x$$

Check. When $x = 15$, then $x+2 = 17$ and $8^2 + 15^2 = 17^2$. Thus, 15 and 17 check.

State. The lengths of the hypotenuse and the other leg are 17 ft and 15 ft, respectively.

27. ***Familiarize***. Consider the drawing in the text. We let w = the width of Main Street, in feet.

Translate. Use the Pythagorean theorem.

$$a^2 + b^2 = c^2$$
$$24^2 + w^2 = 40^2$$

Solve. We solve the equation.

$$24^2 + w^2 = 40^2$$
$$576 + w^2 = 1600$$
$$w^2 - 1024 = 0$$
$$(w+32)(w-32) = 0$$
$$w+32 = 0 \quad or \quad w-32 = 0$$
$$w = -32 \quad or \quad w = 32$$

Check. The width of the street cannot be negative, so -32 cannot be a solution. If Main Street is 32 ft wide, we have $24^2 + 32^2 = 576 + 1024 = 1600$, which is 40^2. Thus, 32 ft checks.

State. Main Street is 32 ft wide.

29. ***Familiarize***. Using the labels on the drawing in the text, we let h = the height of a brace, in feet. Note that we have a right triangle with hypotenuse 15 ft and legs of 12 ft and h.

Translate. We use the Pythagorean theorem.

$$a^2 + b^2 = c^2$$

$$12^2 + h^2 = 15^2 \quad \text{Substituting}$$

Solve.

$$12^2 + h^2 = 15^2$$
$$144 + h^2 = 225$$
$$h^2 - 81 = 0$$
$$(h + 9)(h - 9) = 0$$
$$h + 9 = 0 \quad or \quad h - 9 = 0$$
$$h = -9 \quad or \quad h = 9$$

Check. The height of a brace cannot be negative, so -9 cannot be a solution. When $h = 9$, we have $12^2 + 9^2 = 144 + 81 = 225 = 15^2$. This checks.

State. The brace is 9 ft high.

31. ***Familiarize***. We label the drawing. Let x = the length of a side of the dining room, in ft. Then the dining room has dimensions x by x and the kitchen has dimensions x by 10. The entire rectangular space has dimension x by $x + 10$. Recall that we multiply these dimensions to find the area of the rectangle.

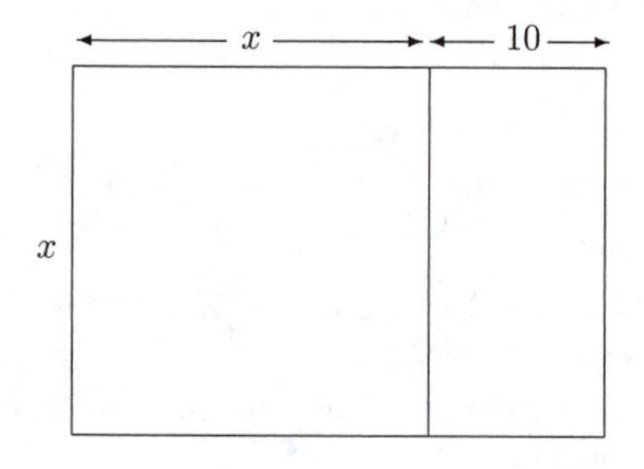

Translate.

The area of the rectangular space is 264 ft^2.

$$x(x + 10) = 264$$

Solve. We solve the equation.

$$x(x + 10) = 264$$
$$x^2 + 10x = 264$$
$$x^2 + 10x - 264 = 0$$
$$(x + 22)(x - 12) = 0$$
$$x + 22 = 0 \quad or \quad x - 12 = 0$$
$$x = -22 \quad or \quad x = 12$$

Check. Since the length of a side of the dining room must be positive, -22 cannot be a solution. If x is 12 ft, then $x+10$ is 22 ft, and the area of the space is $12 \cdot 22$, or 264 ft^2. The number 12 checks.

State. The dining room is 12 ft by 12 ft, and the kitchen is 12 ft by 10 ft.

33. ***Familiarize***. We will use the formula $h = 180t - 16t^2$.

Translate. Substitute 464 for h.

$$464 = 180t - 16t^2$$

Solve. We solve the equation.

$$464 = 180t - 16t^2$$
$$16t^2 - 180t + 464 = 0$$
$$4(4t^2 - 45t + 116) = 0$$
$$4(4t - 29)(t - 4) = 0$$
$$4t - 29 = 0 \quad or \quad t - 4 = 0$$
$$4t = 29 \quad or \quad t = 4$$
$$t = \frac{29}{4} \quad or \quad t = 4$$

Check. The solutions of the equation are $\frac{29}{4}$, or $7\frac{1}{4}$, and 4. Since we want to find how many seconds it takes the rocket to *first* reach a height of 464 ft, we check the smaller number, 4. We substitute 4 for t in the formula.

$$h = 180t - 16t^2$$
$$h = 180 \cdot 4 - 16(4)^2$$
$$h = 180 \cdot 4 - 16 \cdot 16$$
$$h = 720 - 256$$
$$h = 464$$

The answer checks.

State. The rocket will first reach a height of 464 ft after 4 seconds.

35. ***Familiarize***. Let x = the smaller odd positive integer. Then $x + 2$ = the larger odd positive integer.

Translate.

Square of the smaller odd positive integer + Square of the larger odd positive integer is 74

$$x^2 + (x + 2)^2 = 74$$

Solve.

$$x^2 + (x + 2)^2 = 74$$
$$x^2 + x^2 + 4x + 4 = 74$$
$$2x^2 + 4x - 70 = 0$$
$$2(x^2 + 2x - 35) = 0$$
$$2(x + 7)(x - 5) = 0$$
$$x + 7 = 0 \quad or \quad x - 5 = 0$$
$$x = -7 \quad or \quad x = 5$$

Check. The solutions of the equation are -7 and 5. The problem asks for odd positive integers, so -7 cannot be a solution. When x is 5, $x+2$ is 7. The numbers 5 and 7 are consecutive odd positive integers. The sum of their squares, $25+49$, is 74. The numbers check.

State. The integers are 5 and 7.

37. To factor a polynomial is to express it as a product.

39. A factorization of a polynomial is an expression that names that polynomial as a product.

41. The expression $-5x^2+8x-7$ is an example of a trinomial.

43. For the graph of the equation $4x-3y=12$, the pair $(0,-4)$ is known as the y-intercept.

45. ***Familiarize***. First we can use the Pythagorean theorem to find x, in ft. Then the height of the telephone pole is $x+5$.

Translate. We use the Pythagorean theorem.

$$a^2+b^2=c^2$$
$$\left(\frac{1}{2}x+1\right)^2+x^2=34^2$$

Solve. We solve the equation.

$$\left(\frac{1}{2}x+1\right)^2+x^2=34^2$$
$$\frac{1}{4}x^2+x+1+x^2=1156$$
$$x^2+4x+4+4x^2=4624 \quad \text{Multiplying by 4}$$
$$5x^2+4x+4=4624$$
$$5x^2+4x-4620=0$$
$$(5x+154)(x-30)=0$$
$$5x+154=0 \quad or \quad x-30=0$$
$$5x=-154 \quad or \quad x=30$$
$$x=-30.8 \quad or \quad x=30$$

Check. Since the length x must be positive, -30.8 cannot be a solution. If x is 30 ft, then $\frac{1}{2}x+1$ is $\frac{1}{2}\cdot 30+1$, or 16 ft. Since $16^2+30^2=1156=34^2$, the number 30 checks. When x is 30 ft, then $x+5$ is 35 ft.

State. The height of the telephone pole is 35 ft.

47. ***Familiarize***. Using the labels shown on the drawing in the text, we let $x=$ the width of the walk. Then the length and width of the rectangle formed by the pool and walk together are $40+2x$ and $20+2x$, respectively.

Translate.

$$\underbrace{\text{Area}}\ \text{is}\ \underbrace{\text{length}}\ \text{times}\ \underbrace{\text{width.}}$$
$$1500 = (40+2x) \cdot (20+2x)$$

Solve. We solve the equation.

$$1500=(40+2x)(20+2x)$$
$$1500=2(20+x)\cdot 2(10+x) \quad \text{Factoring 2 out of each factor on the right}$$
$$1500=4\cdot(20+x)(10+x)$$
$$375=(20+x)(10+x) \quad \text{Dividing by 4}$$
$$375=200+30x+x^2$$
$$0=x^2+30x-175$$
$$0=(x+35)(x-5)$$
$$x+35=0 \quad or \quad x-5=0$$
$$x=-35 \quad or \quad x=5$$

Check. The solutions of the equation are -35 and 5. Since the width of the walk cannot be negative, -35 is not a solution. When $x=5$, $40+2x=40+2\cdot 5$, or 50 and $20+2x=20+2\cdot 5$, or 30. The total area of the pool and walk is $50\cdot 30$, or 1500 ft^2. This checks.

State. The width of the walk is 5 ft.

49. ***Familiarize***. We make a drawing. Let $w=$ the width of the piece of cardboard. Then $2w=$ the length.

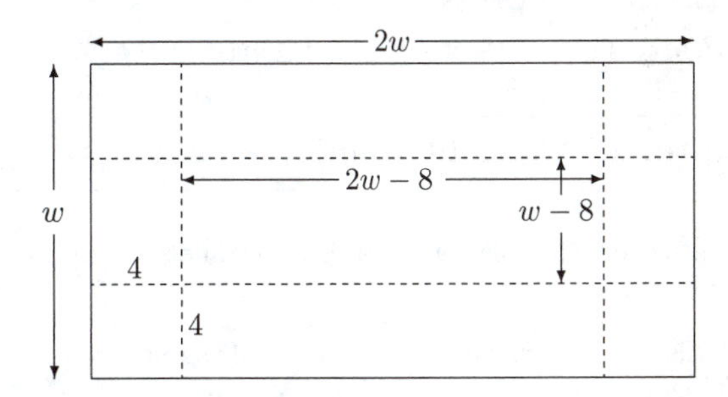

The box will have length $2w-8$, width $w-8$, and height 4. Recall that the formula for volume is $V=$ length $\times$ width $\times$ height.

Translate.

$$\underbrace{\text{The volume}}\ \text{is}\ \underbrace{616\text{cm}^3.}$$
$$(2w-8)(w-8)(4) = 616$$

Solve. We solve the equation.

$$(2w-8)(w-8)(4)=616$$
$$(2w^2-24w+64)(4)=616$$
$$8w^2-96w+256=616$$
$$8w^2-96w-360=0$$
$$8(w^2-12w-45)=0$$
$$w^2-12w-45=0 \quad \text{Dividing by 8}$$
$$(w-15)(w+3)=0$$
$$w-15=0 \quad or \quad w+3=0$$
$$w=15 \quad or \quad w=-3$$

Check. The width cannot be negative, so we only need to check 15. When $w=15$, then $2w=30$ and the dimensions of the box are $30-8$ by $15-8$ by 4, or 22 by 7 by 4. The volume is $22\cdot 7\cdot 4$, or 616.

State. The cardboard is 30 cm by 15 cm.

51. ***Familiarize***. Let $x =$ the length of a side of the base of the box, in inches. Then each of the four sides of the box has dimensions 9 in. by x and the top and bottom each have dimensions x by x.

Translate. We add the areas of the four sides of the box and of the top and bottom.

$$4 \cdot 9 \cdot x + 2 \cdot x^2 = 350$$

Solve. We solve the equation.

$$4 \cdot 9 \cdot x + 2 \cdot x^2 = 350$$
$$36x + 2x^2 = 350$$
$$2x^2 + 36x - 350 = 0$$
$$2(x^2 + 18x - 175) = 0$$
$$x^2 + 18x - 175 = 0 \quad \text{Dividing by 2}$$
$$(x + 25)(x - 7) = 0$$
$$x + 25 = 0 \quad or \quad x - 7 = 0$$
$$x = -25 \quad or \quad x = 7$$

Check. The length of a side cannot be negative, so -25 cannot be a solution. If $x = 7$, then the surface area of the box is $4\cdot 9 \text{ in.} \cdot 7 \text{ in.} + 2\cdot(7 \text{ in.})^2 = 252 \text{ in}^2 + 98 \text{ in}^2 = 350 \text{ in}^2$. The number 7 checks.

State. The length of a side of the base of the box is 7 in.

Chapter 13 Concept Reinforcement

1. False; for example, $w + x + y + z$ cannot be factored by grouping.

2. This is true because the constant and the middle term are both positive.

3. False; $5 \cdot 0 = 0$, but $5 \neq 0$.

4. True; see page 1005 in the text.

Chapter 13 Important Concepts

1. $8x^3y^2 = 2 \cdot 2 \cdot 2 \cdot x^3 \cdot y^2$

$-20xy^3 = -1 \cdot 2 \cdot 2 \cdot 5 \cdot x \cdot y^3$

$32x^2y = 2 \cdot 2 \cdot 2 \cdot 2 \cdot 2 \cdot x^2 \cdot y$

Each coefficient has two factors of 2. The GCFs of the powers of x and of y are x and y, respectively, because 1 is the smallest exponent of both x and y. Thus the GCF is $2 \cdot 2 \cdot x \cdot y$, or $4xy$.

2. $27x^5 - 9x^3 + 18x^2 = 9x^2 \cdot 3x^3 - 9x^2 \cdot x + 9x^2 \cdot 2$
$= 9x^2(3x^3 - x + 2)$

3. $z^3 - 3z^2 + 4z - 12 = z^2(z - 3) + 4(z - 3)$
$= (z - 3)(z^2 + 4)$

4. $x^2 + 6x + 8$

Look for a pair of factors of 8 whose sum is 6. The numbers we want are 2 and 4.

$x^2 + 6x + 8 = (x + 2)(x + 4)$

5. $6z^2 - 21z - 12$

(1) Factor out the largest common factor.

$6z^2 - 21z - 12 = 3(2z^2 - 7z - 4)$

Now we factor $2z^2 - 7z - 4$.

(2) We can factor $2z^2$ as

$(2z+ \quad)(z+ \quad)$.

(3) We can factor -4 as

$-1, 4$ and $1, -4$ and $2, -2$.

These factors can also be written as

$4, -1$ and $-4, 1$ and $-2, 2$.

(4) Look for combinations of factors from steps (2) and (3) such that the sum of their products is the middle term, $-7z$.

The factorization of $2z^2 - 7z - 4$ is $(2z + 1)(z - 4)$, so $6z^2 - 21z - 12 = 3(2z + 1)(z - 4)$.

6. $6y^2 + 7y - 3$

(1) There is no common factor (other than 1 or -1).

(2) Multiply 6 and -3: $6(-3) = -18$.

(3) Look for a factorization of -18 in which the sum of the factors is 7. The numbers we want are 9 and -2.

(4) Split the middle term: $7y = 9y - 2y$.

(5) Factor by grouping.

$6y^2 + 7y - 3 = 6y^2 + 9y - 2y - 3$
$= 3y(2y + 3) - (2y + 3)$
$= (2y + 3)(3y - 1)$

7. $4x^2 + 4x + 1 = (2x)^2 + 2 \cdot 2x \cdot 1 + 1^2 = (2x + 1)^2$

8. $18x^2 - 8 = 2(9x^2 - 4)$
$= 2[(3x)^2 - 2^2]$
$= 2(3x + 2)(3x - 2)$

9.
$$x^2 + 4x = 5$$
$$x^2 + 4x - 5 = 0$$
$$(x + 5)(x - 1) = 0$$
$$x + 5 = 0 \quad or \quad x - 1 = 0$$
$$x = -5 \quad or \quad x = 1$$

The solutions are -5 and 1.

Chapter 13 Review Exercises

1. $-15y^2 = -1 \cdot 3 \cdot 5 \cdot y^2$

$25y^6 = 5 \cdot 5 \cdot y^6$

Each coefficient has a factor of 5. There are no other common prime factors. The GCF of the powers of y is y^2 because 2 is the smallest exponent of y. Thus the GCF is $5y^2$.

2. $12x^3 = 2 \cdot 2 \cdot 3 \cdot x^3$

$-60x^2y = -1 \cdot 2 \cdot 2 \cdot 3 \cdot 5 \cdot x^2 \cdot y$

$36xy = 2 \cdot 2 \cdot 3 \cdot 3 \cdot x \cdot y$

Each coefficient has two factors of 2 and one factor of 3. There are no other common prime factors. The GCF of the powers of x is x because 1 is the smallest exponent of x. The GCF of the powers of y is 1 because $12x^3$ has no y-factor. Thus the GCF is $2 \cdot 2 \cdot 3 \cdot x \cdot 1$, or $12x$.

3. $5 - 20x^6$

$= 5(1 - 4x^6)$ 5 is a common factor.

$= 5(1 - 2x^3)(1 + 2x^3)$ Factoring the difference of squares

4. $x^2 - 3x = x(x - 3)$

5. $9x^2 - 4 = (3x + 2)(3x - 2)$ Factoring a difference of squares

6. $x^2 + 4x - 12$

We look for a pair of factors of -12 whose sum is 4. The numbers we need are 6 and -2.

$x^2 + 4x - 12 = (x + 6)(x - 2)$

7. $x^2 + 14x + 49 = x^2 + 2 \cdot x \cdot 7 + 7^2 = (x + 7)^2$

8. $6x^3 + 12x^2 + 3x = 3x(2x^2 + 4x + 1)$

The trinomial $2x^2 + 4x + 1$ cannot be factored, so the factorization is complete.

9. $x^3 + x^2 + 3x + 3$

$= (x^3 + x^2) + (3x + 3)$

$= x^2(x + 1) + 3(x + 1)$ Factoring by grouping

$= (x + 1)(x^2 + 3)$

10. $6x^2 - 5x + 1$

There is no common factor (other than 1). This polynomial has three terms, but it is not a trinomial square. Multiply the leading coefficient and the constant, 6 and 1: $6 \cdot 1 = 6$. Try to factor 6 so that the sum of the factors is -5. The numbers we want are -2 and -3: $-2(-3) = 6$ and $-2 + (-3) = -5$. Split the middle term and factor by grouping.

$$\begin{aligned} 6x^2 - 5x + 1 &= 6x^2 - 2x - 3x + 1 \\ &= (6x^2 - 2x) + (-3x + 1) \\ &= 2x(3x - 1) - 1(3x - 1) \\ &= (3x - 1)(2x - 1) \end{aligned}$$

11. $x^4 - 81 = (x^2 + 9)(x^2 - 9) = (x^2 + 9)(x + 3)(x - 3)$

12. $9x^3 + 12x^2 - 45x$

$= 3x(3x^2 + 4x - 15)$ $3x$ is a common factor.

$= 3x(3x - 5)(x + 3)$ Using trial and error

13. $2x^2 - 50 = 2(x^2 - 25) = 2(x + 5)(x - 5)$

14.
$$\begin{aligned} x^4 + 4x^3 - 2x - 8 &= (x^4 + 4x^3) + (-2x - 8) \\ &= x^3(x + 4) - 2(x + 4) \\ &= (x + 4)(x^3 - 2) \end{aligned}$$

15. $16x^4 - 1 = (4x^2 + 1)(4x^2 - 1) = (4x^2 + 1)(2x + 1)(2x - 1)$

16. $8x^6 - 32x^5 + 4x^4 = 4x^4(2x^2 - 8x + 1)$

The trinomial $2x^2 - 8x + 1$ cannot be factored, so the factorization is complete.

17.
$$\begin{aligned} 75 + 12x^2 + 60x &= 12x^2 + 60x + 75 \\ &= 3(4x^2 + 20x + 25) \\ &= 3(2x + 5)^2 \end{aligned}$$

18. $x^2 + 9$ is a sum of squares with no common factor, so it is prime.

19. $x^3 - x^2 - 30x = x(x^2 - x - 30) = x(x - 6)(x + 5)$

20. $4x^2 - 25 = (2x + 5)(2x - 5)$

21. $9x^2 + 25 - 30x = 9x^2 - 30x + 25 = (3x - 5)^2$

22.
$$\begin{aligned} 6x^2 - 28x - 48 &= 2(3x^2 - 14x - 24) \\ &= 2(3x + 4)(x - 6) \end{aligned}$$

23. $x^2 - 6x + 9 = (x - 3)^2$

24. $2x^2 - 7x - 4 = (2x + 1)(x - 4)$

25. $18x^2 - 12x + 2 = 2(9x^2 - 6x + 1) = 2(3x - 1)^2$

26. $3x^2 - 27 = 3(x^2 - 9) = 3(x + 3)(x - 3)$

27. $15 - 8x + x^2 = x^2 - 8x + 15 = (x - 3)(x - 5)$

28. $25x^2 - 20x + 4 = (5x - 2)^2$

29.
$$\begin{aligned} 49b^{10} + 4a^8 - 28a^4b^5 &= 49b^{10} - 28a^4b^5 + 4a^8 \\ &= (7b^5)^2 - 2 \cdot 7b^5 \cdot 2a^4 + (2a^4)^2 \\ &= (7b^5 - 2a^4)^2 \end{aligned}$$

30. $x^2y^2 + xy - 12 = (xy + 4)(xy - 3)$

31. $12a^2 + 84ab + 147b^2 = 3(4a^2 + 28ab + 49b^2) =$
$3(2a + 7b)^2$

32.
$$\begin{aligned} m^2 + 5m + mt + 5t &= (m^2 + 5m) + (mt + 5t) \\ &= m(m + 5) + t(m + 5) \\ &= (m + 5)(m + t) \end{aligned}$$

33. $32x^4 - 128y^4z^4 = 32(x^4 - 4y^4z^4) =$
$32(x^2 + 2y^2z^2)(x^2 - 2y^2z^2)$

34. $(x - 1)(x + 3) = 0$

$x - 1 = 0$ *or* $x + 3 = 0$

$x = 1$ *or* $x = -3$

The solutions are 1 and -3.

35. $x^2 + 2x - 35 = 0$

$(x + 7)(x - 5) = 0$

$x + 7 = 0$ *or* $x - 5 = 0$

$x = -7$ *or* $x = 5$

The solutions are -7 and 5.

36.
$$x^2+4x=0$$
$$x(x+4)=0$$
$$x=0 \quad or \quad x+4=0$$
$$x=0 \quad or \quad x=-4$$
The solutions are 0 and -4.

37.
$$3x^2+2=5x$$
$$3x^2-5x+2=0$$
$$(3x-2)(x-1)=0$$
$$3x-2=0 \quad or \quad x-1=0$$
$$3x=2 \quad or \quad x=1$$
$$x=\frac{2}{3} \quad or \quad x=1$$
The solutions are $\frac{2}{3}$ and 1.

38.
$$x^2=64$$
$$x^2-64=0$$
$$(x+8)(x-8)=0$$
$$x+8=0 \quad or \quad x-8=0$$
$$x=-8 \quad or \quad x=8$$
The solutions are -8 and 8.

39.
$$16=x(x-6)$$
$$16=x^2-6x$$
$$0=x^2-6x-16$$
$$0=(x-8)(x+2)$$
$$x-8=0 \quad or \quad x+2=0$$
$$x=8 \quad or \quad x=-2$$
The solutions are 8 and -2.

40. Let $y=0$ and solve for x.
$$0=x^2+9x+20$$
$$0=(x+5)(x+4)$$
$$x+5=0 \quad or \quad x+4=0$$
$$x=-5 \quad or \quad x=-4$$
The x-intercepts are $(-5,0)$ and $(-4,0)$.

41. Let $y=0$ and solve for x.
$$0=2x^2-7x-15$$
$$0=(2x+3)(x-5)$$
$$2x+3=0 \quad or \quad x-5=0$$
$$2x=-3 \quad or \quad x=5$$
$$x=-\frac{3}{2} \quad or \quad x=5$$
The x-intercepts are $\left(-\frac{3}{2},0\right)$ and $(5,0)$.

42. ***Familiarize.*** Let $b=$ the length of the base, in cm. Then $b+1=$ the height.

Translate. We use the formula for the area of a triangle.
$$A=\frac{1}{2}bh$$
$$15=\frac{1}{2}b(b+1) \quad \text{Substituting}$$

Solve.
$$15=\frac{1}{2}b(b+1)$$
$$15=\frac{1}{2}b^2+\frac{1}{2}b$$
$$0=\frac{1}{2}b^2+\frac{1}{2}b-15$$
$$2\cdot 0=2\left(\frac{1}{2}b^2+\frac{1}{2}b-15\right) \quad \text{Clearing fractions}$$
$$0=b^2+b-30$$
$$0=(b+6)(b-5)$$
$$b+6=0 \quad or \quad b-5=0$$
$$b=-6 \quad or \quad b=5$$

Check. The length of the base cannot be negative, so -6 cannot be a solution. If $b=5$, then $b+1=5+1=6$ and the area is $\frac{1}{2}\cdot 5\cdot 6=15\text{ cm}^2$. The number 5 checks.

State. The base is 5 cm, and the height is 6 cm.

43. ***Familiarize.*** Let $x=$ the smaller integer. Then $x+2$ is the other integer.

Translate.

Smaller even integer	times	larger even integer	is	288.
↓	↓	↓	↓	↓
x	$\cdot$	$(x+2)$	$=$	288

Solve.
$$x(x+2)=288$$
$$x^2+2x=288$$
$$x^2+2x-288=0$$
$$(x+18)(x-16)=0$$
$$x+18=0 \quad or \quad x-16=0$$
$$x=-18 \quad or \quad x=16$$

If $x=-18$, then $x+2=-18+2=-16$.

If $x=16$, then $x+2=16+2=18$.

Check. -18 and -16 are consecutive even integers and $-18(-16)=288$. Similarly, 16 and 18 are consecutive even integers and $16\cdot 18=288$. Both pairs of integers check.

State. The integers are -18 and -16 or 16 and 18.

44. ***Familiarize.*** Let $x=$ the smaller integer. Then $x+2$ is the other integer.

Translate.

Smaller odd integer	times	larger odd integer	is	323.
↓	↓	↓	↓	↓
x	$\cdot$	$(x+2)$	$=$	323

Solve.

$$x(x+2) = 323$$
$$x^2 + 2x = 323$$
$$x^2 + 2x - 323 = 0$$
$$(x+19)(x-17) = 0$$
$$x + 19 = 0 \quad \text{or} \quad x - 17 = 0$$
$$x = -19 \quad \text{or} \quad x = 17$$

If $x = -19$, then $x + 2 = -19 + 2 = -17$.

If $x = 17$, then $x + 2 = 17 + 2 = 19$.

Check. -19 and -17 are consecutive odd integers and $-19(-17) = 323$. Similarly, 17 and 19 are consecutive odd integers and $17 \cdot 19 = 323$. Both pairs of integers check.

State. The integers are -19 and -17 or 17 and 19.

45. *Familiarize.* We make a drawing. Let d = the distance from the base of the tree to the point where each cable is attached to the tree. Then $d + 1$ = the distance from the base of the tree to the point on the ground where each cable is anchored. The distances are in feet.

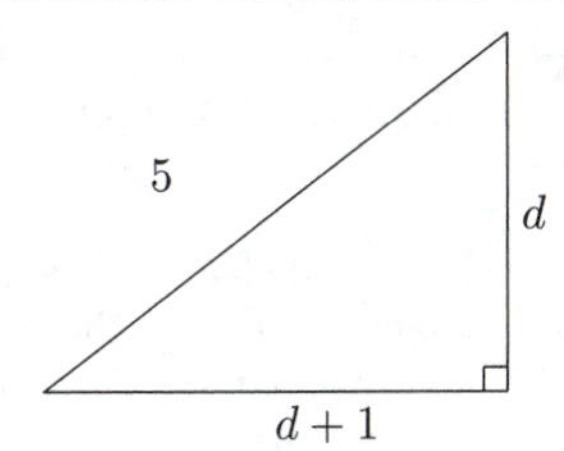

Translate. We use the Pythagorean theorem.

$$a^2 + b^2 = c^2$$
$$(d+1)^2 + d^2 = 5^2 \quad \text{Substituting}$$

Solve.

$$(d+1)^2 + d^2 = 5^2$$
$$d^2 + 2d + 1 + d^2 = 25$$
$$2d^2 + 2d + 1 = 25$$
$$2d^2 + 2d - 24 = 0$$
$$d^2 + d - 12 = 0 \quad \text{Dividing by 2}$$
$$(d+4)(d-3) = 0$$
$$d + 4 = 0 \quad \text{or} \quad d - 3 = 0$$
$$d = -4 \quad \text{or} \quad d = 3$$

Check. A distance cannot be negative, so we check only 3. If $d = 3$, then $d + 1 = 3 + 1 = 4$ and $3^2 + 4^2 = 9 + 16 = 25 = 5^2$. The answer checks.

State. The distance from the base of the tree to the point where each cable is attached to the tree is 3 ft. The distance from the base of the tree to the point on the ground where each cable is anchored is 4 ft.

46. *Familiarize.* Let s = the length of a side of the original square, in km. Then $s + 3$ = the length of a side of the enlarged square.

Translate.

$$\underbrace{\text{Area of enlarged square}}_{(s+3)^2} \ \underset{=}{\text{is}} \ \underbrace{81 \text{ km}^2}_{81}.$$

Solve.

$$(s+3)^2 = 81$$
$$s^2 + 6s + 9 = 81$$
$$s^2 + 6s - 72 = 0$$
$$(s+12)(s-6) = 0$$
$$s + 12 = 0 \quad \text{or} \quad s - 6 = 0$$
$$s = -12 \quad \text{or} \quad s = 6$$

Check. The length of a side of the square cannot be negative, so -12 cannot be a solution. If the length of a side of the original square is 6 km, then the length of a side of the enlarged square is $6+3$, or 9 km. The area of the enlarged square is $(9 \text{ km})^2$, or 81 km^2, so the answer checks.

State. The length of a side of the original square is 6 km.

47. $x^2 - 9x + 8 = (x-1)(x-8)$

Answer B is correct.

48.
$$15x^2 + 5x - 20 = 5(3x^2 + x - 4)$$
$$= 5(3x+4)(x-1)$$

Answer A is correct.

49. *Familiarize.* Let w = the width of the margins, in cm. Then the printed area on each page has dimensions $20 - 2w$ by $15 - 2w$. The area of the margins constitutes one-half the area of each page, so the printed area also constitutes one-half of the area.

Translate.

$$\underbrace{\text{Printed area}}_{(20-2w)(15-2w)} \ \underset{=}{\text{is}} \ \underbrace{\text{one-half}}_{\frac{1}{2}} \ \underset{\cdot}{\text{of}} \ \underbrace{\text{total area.}}_{20 \cdot 15}$$

Solve.

$$(20-2w)(15-2w) = \frac{1}{2} \cdot 20 \cdot 15$$
$$300 - 70w + 4w^2 = 150$$
$$150 - 70w + 4w^2 = 0$$
$$4w^2 - 70w + 150 = 0$$
$$2(2w^2 - 35w + 75) = 0$$
$$2w^2 - 35w + 75 = 0 \quad \text{Dividing by 2}$$
$$(2w-5)(w-15) = 0$$
$$2w - 5 = 0 \quad \text{or} \quad w - 15 = 0$$
$$2w = 5 \quad \text{or} \quad w = 15$$
$$w = 2.5 \quad \text{or} \quad w = 15$$

Check. If $w = 15$, then $20 - 2w$ and $15 - 2w$ are both negative. Since the dimensions of the printed area cannot be negative, 15 cannot be a solution. If $w = 2.5$, then $20 - 2w = 20 - 2(2.5) = 20 - 5 = 15$ and $15 - 2w = 15 - 2(2.5) = 15 - 5 = 10$. Thus the printed area is $15 \cdot 10$, or 150 cm^2. This is one-half of the total area of the page, $20 \cdot 15$, or 300 cm^2. The number 2.5 checks.

State. The width of the margins is 2.5 cm.

50. ***Familiarize.*** Let $n =$ the number.

Translate.

$$\underbrace{\text{The cube of a number}}_{\downarrow\; n^3} \quad \text{is}\;(\downarrow\; =) \quad \text{twice}\;(\downarrow\; 2\cdot) \quad \underbrace{\text{the square of the number.}}_{\downarrow\; n^2}$$

Solve.

$$n^3 = 2n^2$$
$$n^3 - 2n^2 = 0$$
$$n^2(n-2) = 0$$
$$n \cdot n(n-2) = 0$$
$$n = 0 \quad or \quad n = 0 \quad or \quad n - 2 = 0$$
$$n = 0 \quad or \quad n = 0 \quad or \quad n = 2$$

Check. If $n = 0$, then $n^3 = 0^3 = 0$, $n^2 = 0^2 = 0$, and $0 = 2 \cdot 0$. If $n = 2$, then $n^3 = 2^3 = 8$, $n^2 = 2^2 = 4$, and $8 = 2 \cdot 4$. Both numbers check.

State. The number is 0 or 2.

51. ***Familiarize.*** Let $w =$ the width of the original rectangle, in inches. Then $2w =$ the length, in inches. The new length and width are $2w + 20$ and $w - 1$, respectively.

Translate.

$$\underbrace{\text{The new area}}_{\downarrow\; (2w+20)(w-1)} \quad \text{is}\;(\downarrow\; =) \quad 160.\;(\downarrow\; 160)$$

Solve.

$$(2w+20)(w-1) = 160$$
$$2w^2 + 18w - 20 = 160$$
$$2w^2 + 18w - 180 = 0$$
$$2(w^2 + 9w - 90) = 0$$
$$w^2 + 9w - 90 = 0$$
$$(w+15)(w-6) = 0$$
$$w + 15 = 0 \quad or \quad w - 6 = 0$$
$$w = -15 \quad or \quad w = 6$$

Check. The dimensions of the rectangle cannot be negative, so -15 cannot be a solution. If $w = 6$, then $2w = 2 \cdot 6 = 12$, $2w + 20 = 12 + 20 = 32$ and $w - 1 = 6 - 1 = 5$. The area of a rectangle with dimensions 32 by 5 is $32 \cdot 5$, or 160, so the answer checks.

State. The length of the original rectangle is 12 in. and the width is 6 in.

52. $x^2 + 25 = 0$

Since $x^2 + 25$ cannot be factored, the equation has no solution.

53. $(x-2)(x+3)(2x-5) = 0$

$$x - 2 = 0 \quad or \quad x + 3 = 0 \quad or \quad 2x - 5 = 0$$
$$x = 2 \quad or \quad x = -3 \quad or \quad 2x = 5$$
$$x = 2 \quad or \quad x = -3 \quad or \quad x = \frac{5}{2}$$

The solutions are 2, -3, and $\frac{5}{2}$.

54. $(x-3)4x^2 + 3x(x-3) - (x-3)10 = 0$

$$(4x^2 + 3x - 10)(x-3) = 0 \quad \text{Factoring out } x - 3$$
$$(4x-5)(x+2)(x-3) = 0 \quad \text{Factoring } 4x^2 + 3x - 10$$
$$4x - 5 = 0 \quad or \quad x + 2 = 0 \quad or \quad x - 3 = 0$$
$$4x = 5 \quad or \quad x = -2 \quad or \quad x = 3$$
$$x = \frac{5}{4} \quad or \quad x = -2 \quad or \quad x = 3$$

The solutions are $\frac{5}{4}$, -2, and 3.

55. The shaded area is the area of a circle with radius x less the area of a square with a diagonal of length $x + x$, or $2x$. The area of the circle is πx^2. The square can be thought of as two triangles, each with base $2x$ and height x. Then the area of the square is $2 \cdot \frac{1}{2} \cdot 2x \cdot x$, or $2x^2$. We subtract to find the shaded area.

$$\pi x^2 - 2x^2 = x^2(\pi - 2)$$

Chapter 13 Discussion and Writing Exercises

1. Although $x^3 - 8x^2 + 15x$ can be factored as $(x^2 - 5x)(x - 3)$, this is not a complete factorization of the polynomial since $x^2 - 5x = x(x - 5)$. The student should be advised *always* to look for a common factor first.

2. Josh is correct, because answers can easily be checked by multiplying.

3. For $x = -3$:

$$(x-4)^2 = (-3-4)^2 = (-7)^2 = 49$$
$$(4-x)^2 = [4-(-3)]^2 = 7^2 = 49$$

For $x = 1$:

$$(x-4)^2 = (1-4)^2 = (-3)^2 = 9$$
$$(4-x)^2 = (4-1)^2 = 3^2 = 9$$

In general, $(x-4)^2 = [-(-x+4)]^2 = [-(4-x)]^2 = (-1)^2(4-x)^2 = (4-x)^2$.

4. The equation is not in the form $ab = 0$. The correct procedure is:

$$(x-3)(x+4) = 8$$
$$x^2 + x - 12 = 8$$
$$x^2 + x - 20 = 0$$
$$(x+5)(x-4) = 0$$
$$x + 5 = 0 \quad or \quad x - 4 = 0$$
$$x = -5 \quad or \quad x = 4$$

The solutions are -5 and 4.

5. One solution of the equation is 0. Dividing both sides of the equation by x, leaving the solution $x = 3$, is equivalent to dividing by 0.

6. She could use the measuring sticks to draw a right angle as shown below. Then she could use the 3-ft and 4-ft sticks to extend one leg to 7 ft and the 4-ft and 5-ft sticks to extend the other leg to 9 ft.

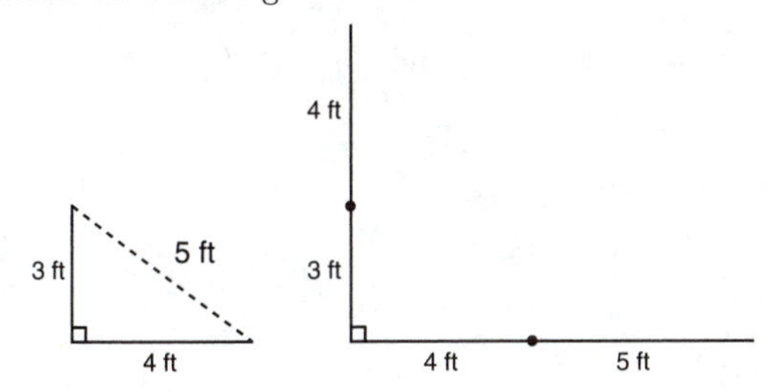

Next she could draw another right angle with either the 7-ft side or the 9-ft side as a side.

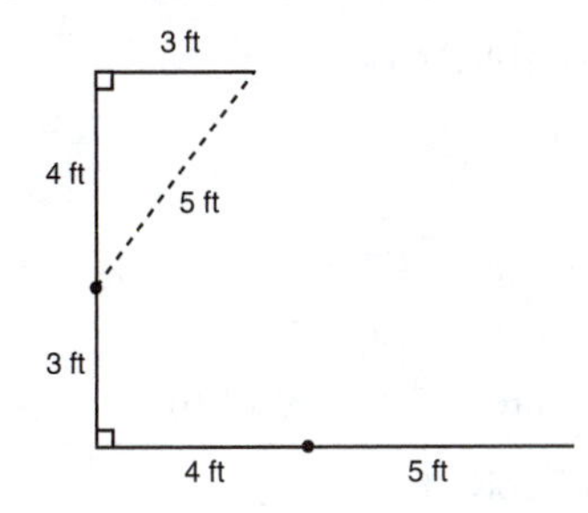

Then she could use the sticks to extend the other side to the appropriate length. Finally she would draw the remaining side of the rectangle.

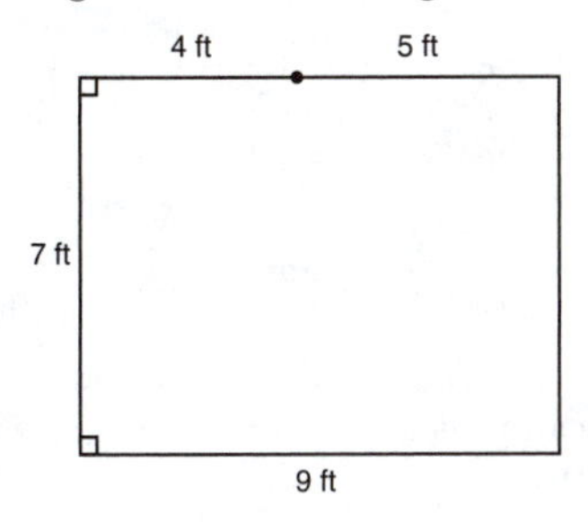

Chapter 13 Test

1. $28x^3 = 2 \cdot 2 \cdot 7 \cdot x^3$

$48x^7 = 2 \cdot 2 \cdot 2 \cdot 2 \cdot 3 \cdot x^7$

The coefficients each have two factors of 2. There are no other common prime factors. The GCF of the powers of x is x^3 because 3 is the smallest exponent of x. Thus the GCF is $2 \cdot 2 \cdot x^3$, or $4x^3$.

2. $x^2 - 7x + 10$

We look for a pair of factors of 10 whose sum is -7. The numbers we need are -2 and -5.

$$x^2 - 7x + 10 = (x-2)(x-5)$$

3. $x^2 + 25 - 10x = x^2 - 10x + 25$

$= x^2 - 2 \cdot x \cdot 5 + 5^2$

$= (x-5)^2$

4. $6y^2 - 8y^3 + 4y^4 = 4y^4 - 8y^3 + 6y^2 =$

$2y^2 \cdot 2y^2 - 2y^2 \cdot 4y + 2y^2 \cdot 3 = 2y^2(2y^2 - 4y + 3)$

Since $2y^2 - 4y + 3$ cannot be factored, the factorization is complete.

5. $x^3 + x^2 + 2x + 2$

$= (x^3 + x^2) + (2x + 2)$

$= x^2(x+1) + 2(x+1)$ Factoring by grouping

$= (x+1)(x^2+2)$

6. $x^2 - 5x = x \cdot x - 5 \cdot x = x(x-5)$

7. $x^3 + 2x^2 - 3x$

$= x(x^2 + 2x - 3)$ x is a common factor.

$= x(x+3)(x-1)$ Factoring the trinomial

8. $28x - 48 + 10x^2$

$= 10x^2 + 28x - 48$

$= 2(5x^2 + 14x - 24)$ 2 is a common factor.

$= 2(5x-6)(x+4)$ Factoring the trinomial

9. $4x^2 - 9 = (2x)^2 - 3^2$ Difference of squares

$= (2x+3)(2x-3)$

10. $x^2 - x - 12$

We look for a pair of factors of -12 whose sum is -1. The numbers we need are -4 and 3.

$$x^2 - x - 12 = (x-4)(x+3)$$

11. $6m^3 + 9m^2 + 3m$

$= 3m(2m^2 + 3m + 1)$ $3m$ is a common factor.

$= 3m(2m+1)(m+1)$ Factoring the trinomial

12. $3w^2 - 75 = 3(w^2 - 25)$ 3 is a common factor.

$= 3(w^2 - 5^2)$ Difference of squares

$= 3(w+5)(w-5)$

13. $60x + 45x^2 + 20$

$= 45x^2 + 60x + 20$

$= 5(9x^2 + 12x + 4)$ 5 is a common factor.

$= 5[(3x)^2 + 2 \cdot 3x \cdot 2 + 2^2]$ Trinomial square

$= 5(3x+2)^2$

14. $3x^4 - 48$

$= 3(x^4 - 16)$ 3 is a common factor.

$= 3[(x^2)^2 - 4^2]$ Difference of squares

$= 3(x^2+4)(x^2-4)$

$= 3(x^2+4)(x^2-2^2)$ Difference of squares

$= 3(x^2+4)(x+2)(x-2)$

15. $49x^2 - 84x + 36$

$= (7x)^2 - 2 \cdot 7x \cdot 6 + 6^2$ Trinomial square

$= (7x-6)^2$

16. $5x^2 - 26x + 5$

There is no common factor (other than 1). This polynomial has 3 terms, but it is not a trinomial square. Using the ac-method we first multiply the leading coefficient and the constant term: $5 \cdot 5 = 25$. Try to factor 25 so that the sum of the factors is -26. The numbers we want are -1 and -25: $-1(-25) = 25$ and $-1 + (-25) = -26$. Split the middle term and factor by grouping:

$$\begin{aligned} 5x^2 - 26x + 5 &= 5x^2 - x - 25x + 5 \\ &= (5x^2 - x) + (-25x + 5) \\ &= x(5x - 1) - 5(5x - 1) \\ &= (5x - 1)(x - 5) \end{aligned}$$

17.
$$\begin{aligned} &x^4 + 2x^3 - 3x - 6 \\ &= (x^4 + 2x^3) + (-3x - 6) \\ &= x^3(x + 2) - 3(x + 2) \quad \text{Factoring by grouping} \\ &= (x + 2)(x^3 - 3) \end{aligned}$$

18.
$$\begin{aligned} &80 - 5x^4 \\ &= 5(16 - x^4) && \text{5 is a common factor.} \\ &= 5[4^2 - (x^2)^2] && \text{Difference of squares} \\ &= 5(4 + x^2)(4 - x^2) \\ &= 5(4 + x^2)(2^2 - x^2) && \text{Difference of squares} \\ &= 5(4 + x^2)(2 + x)(2 - x) \end{aligned}$$

19.
$$\begin{aligned} &6t^3 + 9t^2 - 15t \\ &= 3t(2t^2 + 3t - 5) && 3t \text{ is a common factor.} \\ &= 3t(2t + 5)(t - 1) && \text{Factoring the trinomial} \end{aligned}$$

20.
$$\begin{aligned} x^2 - 3x &= 0 \\ x(x - 3) &= 0 \\ x = 0 \quad &or \quad x - 3 = 0 \\ x = 0 \quad &or \quad x = 3 \end{aligned}$$

The solutions are 0 and 3.

21.
$$\begin{aligned} 2x^2 &= 32 \\ 2x^2 - 32 &= 0 \\ 2(x^2 - 16) &= 0 \\ 2(x + 4)(x - 4) &= 0 \\ x + 4 = 0 \quad &or \quad x - 4 = 0 \\ x = -4 \quad &or \quad x = 4 \end{aligned}$$

The solutions are -4 and 4.

22.
$$\begin{aligned} x^2 - x - 20 &= 0 \\ (x - 5)(x + 4) &= 0 \\ x - 5 = 0 \quad &or \quad x + 4 = 0 \\ x = 5 \quad &or \quad x = -4 \end{aligned}$$

The solutions are 5 and -4.

23.
$$\begin{aligned} 2x^2 + 7x &= 15 \\ 2x^2 + 7x - 15 &= 0 \\ (2x - 3)(x + 5) &= 0 \\ 2x - 3 = 0 \quad &or \quad x + 5 = 0 \\ 2x = 3 \quad &or \quad x = -5 \\ x = \frac{3}{2} \quad &or \quad x = -5 \end{aligned}$$

The solutions are $\frac{3}{2}$ and -5.

24.
$$\begin{aligned} x(x - 3) &= 28 \\ x^2 - 3x &= 28 \\ x^2 - 3x - 28 &= 0 \\ (x - 7)(x + 4) &= 0 \\ x - 7 = 0 \quad &or \quad x + 4 = 0 \\ x = 7 \quad &or \quad x = -4 \end{aligned}$$

The solutions are 7 and -4.

25. We let $y = 0$ and solve for x.

$$\begin{aligned} 0 &= x^2 - 2x - 35 \\ 0 &= (x + 5)(x - 7) \\ x + 5 = 0 \quad &or \quad x - 7 = 0 \\ x = -5 \quad &or \quad x = 7 \end{aligned}$$

The x-intercepts are $(-5, 0)$ and $(7, 0)$.

26. We let $y = 0$ and solve for x.

$$\begin{aligned} 0 &= 3x^2 - 5x + 2 \\ 0 &= (3x - 2)(x - 1) \\ 3x - 2 = 0 \quad &or \quad x - 1 = 0 \\ 3x = 2 \quad &or \quad x = 1 \\ x = \frac{2}{3} \quad &or \quad x = 1 \end{aligned}$$

The x-intercepts are $\left(\frac{2}{3}, 0\right)$ and $(1, 0)$.

27. ***Familiarize***. Let w = the width, in meters. Then $w + 2$ = the length. Recall that the area of a rectangle is (length) $\cdot$ (width).

Translate. We use the formula for the area of a rectangle.

$$48 = (w + 2)w$$

Solve.

$$\begin{aligned} 48 &= (w + 2)w \\ 48 &= w^2 + 2w \\ 0 &= w^2 + 2w - 48 \\ 0 &= (w + 8)(w - 6) \\ w + 8 = 0 \quad &or \quad w - 6 = 0 \\ w = -8 \quad &or \quad w = 6 \end{aligned}$$

Check. The width cannot be negative, so -8 cannot be a solution. If $w = 6$, then $w + 2 = 8$ and the area is $(8 \text{ m}) \cdot (6 \text{ m})$, or 48 m^2. The number 6 checks.

State. The length is 8 m and the width is 6 m.

28. ***Familiarize***. Using the labels on the drawing in the text, we let h = the height of the triangle, in cm, and $2h + 6$ = the base. Recall that the area of a triangle is $\frac{1}{2} \cdot (\text{base}) \cdot (\text{height})$.

Translate. We use the formula for the area of a triangle.

$$28 = \frac{1}{2} \cdot (2h + 6) \cdot h$$

Solve.

$$28 = \frac{1}{2}(2h + 6)h$$
$$28 = h^2 + 3h$$
$$0 = h^2 + 3h - 28$$
$$0 = (h + 7)(h - 4)$$
$$h + 7 = 0 \quad or \quad h - 4 = 0$$
$$h = -7 \quad or \quad h = 4$$

Check. The height cannot be negative, so -7 cannot be a solution. If $h = 4$, then $2h + 6 = 2 \cdot 4 + 6 = 8 + 6 = 14$ and the area is $\frac{1}{2} \cdot (14 \text{ cm}) \cdot (4 \text{ cm})$, or 28 cm^2. The number 4 checks.

State. The height is 4 cm and the base is 14 cm.

29. ***Familiarize***. Using the labels on the drawing in the text, we let $x =$ the distance between the two marked points, in feet. If the corner is a right angle, the lengths 3 ft, 4 ft, and x will satisfy the Pythagorean theorem.

Translate.

$$a^2 + b^2 = c^2$$
$$3^2 + 4^2 = x^2 \quad \text{Substituting}$$

Solve.

$$3^2 + 4^2 = x^2$$
$$9 + 16 = x^2$$
$$25 = x^2$$
$$0 = x^2 - 25$$
$$0 = (x + 5)(x - 5)$$
$$x + 5 = 0 \quad or \quad x - 5 = 0$$
$$x = -5 \quad or \quad x = 5$$

Check. The distance cannot be negative, so -5 cannot be a solution. If $x = 5$, then we have $3^2 + 4^2 = 9 + 16 = 25 = 5^2$, so the answer checks.

State. The distance between the marked points should be 5 ft.

30.

$$2y^4 - 32 = 2(y^4 - 16)$$
$$= 2(y^2 + 4)(y^2 - 4)$$
$$= 2(y^2 + 4)(y + 2)(y - 2)$$

Answer A is correct.

31. ***Familiarize***. Let $w =$ the width of the original rectangle, in meters. Then $5w =$ the length. The new width and length are $w + 2$ and $5w - 3$, respectively.

Translate. We will use the formula for the area of a rectangle, Area = (length) $\cdot$ (width).

$$60 = (5w - 3)(w + 2)$$

Solve.

$$60 = (5w - 3)(w + 2)$$
$$60 = 5w^2 + 7w - 6$$
$$0 = 5w^2 + 7w - 66$$
$$0 = (5w + 22)(w - 3)$$
$$5w + 22 = 0 \quad or \quad w - 3 = 0$$
$$5w = -22 \quad or \quad w = 3$$
$$w = -\frac{22}{5} \quad or \quad w = 3$$

Check. The width cannot be negative, so $-\frac{22}{5}$ cannot be a solution. If $w = 3$, then $5w = 5 \cdot 3 = 15$. The new dimensions are $w + 2$, or $3 + 2$, or 5 and $5w - 3$, or $15 - 3$, or 12, and the new area is $12 \cdot 5$, or 60. The number 3 checks.

State. The original length is 15 m and the original width is 3 m.

32. $(a + 3)^2 - 2(a + 3) - 35$

We can think of $a + 3$ as the variable in this expression. Then we find a pair of factors of -35 whose sum is -2. The numbers we want are -7 and 5.

$(a + 3)^2 - 2(a + 3) - 35 = [(a + 3) - 7][(a + 3) + 5] = (a - 4)(a + 8)$

We could also do this exercise as follows:

$$(a + 3)^2 - 2(a + 3) - 35 = a^2 + 6a + 9 - 2a - 6 - 35$$
$$= a^2 + 4a - 32$$
$$= (a - 4)(a + 8)$$

33.

$$20x(x + 2)(x - 1) = 5x^3 - 24x - 14x^2$$
$$(20x^2 + 40x)(x - 1) = 5x^3 - 24x - 14x^2$$
$$20x^3 + 20x^2 - 40x = 5x^3 - 24x - 14x^2$$
$$15x^3 + 34x^2 - 16x = 0$$
$$x(15x^2 + 34x - 16) = 0$$
$$x(3x + 8)(5x - 2) = 0$$
$$x = 0 \quad or \quad 3x + 8 = 0 \quad or \quad 5x - 2 = 0$$
$$x = 0 \quad or \quad 3x = -8 \quad or \quad 5x = 2$$
$$x = 0 \quad or \quad x = -\frac{8}{3} \quad or \quad x = \frac{2}{5}$$

The solutions are 0, $-\frac{8}{3}$, and $\frac{2}{5}$.

34. $x^2 - y^2 = (x + y)(x - y) = 4 \cdot 6 = 24$, so choice D is correct.

Chapter 14

Rational Expressions and Equations

Exercise Set 14.1

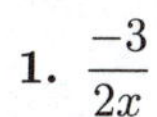

1. $\dfrac{-3}{2x}$

To determine the numbers for which the rational expression is not defined, we set the denominator equal to 0 and solve:

$$2x = 0$$
$$x = 0$$

The expression is not defined for the replacement number 0.

3. $\dfrac{5}{x-8}$

To determine the numbers for which the rational expression is not defined, we set the denominator equal to 0 and solve:

$$x - 8 = 0$$
$$x = 8$$

The expression is not defined for the replacement number 8.

5. $\dfrac{3}{2y+5}$

Set the denominator equal to 0 and solve:

$$2y + 5 = 0$$
$$2y = -5$$
$$y = -\frac{5}{2}$$

The expression is not defined for the replacement number $-\dfrac{5}{2}$.

7. $\dfrac{x^2+11}{x^2-3x-28}$

Set the denominator equal to 0 and solve:

$$x^2 - 3x - 28 = 0$$
$$(x-7)(x+4) = 0$$
$$x - 7 = 0 \quad \text{or} \quad x + 4 = 0$$
$$x = 7 \quad \text{or} \quad x = -4$$

The expression is not defined for the replacement numbers 7 and -4.

9. $\dfrac{m^3-2m}{m^2-25}$

Set the denominator equal to 0 and solve:

$$m^2 - 25 = 0$$
$$(m+5)(m-5) = 0$$
$$m + 5 = 0 \quad \text{or} \quad m - 5 = 0$$
$$m = -5 \quad \text{or} \quad m = 5$$

The expression is not defined for the replacement numbers -5 and 5.

11. $\dfrac{x-4}{3}$

Since the denominator is the constant 3, there are no replacement numbers for which the expression is not defined.

13. $\dfrac{4x}{4x} \cdot \dfrac{3x^2}{5y} = \dfrac{(4x)(3x^2)}{(4x)(5y)}$ Multiplying the numerators and the denominators

15. $\dfrac{2x}{2x} \cdot \dfrac{x-1}{x+4} = \dfrac{2x(x-1)}{2x(x+4)}$ Multiplying the numerators and the denominators

17. $\dfrac{3-x}{4-x} \cdot \dfrac{-1}{-1} = \dfrac{(3-x)(-1)}{(4-x)(-1)}$

19. $\dfrac{y+6}{y+6} \cdot \dfrac{y-7}{y+2} = \dfrac{(y+6)(y-7)}{(y+6)(y+2)}$

21. $\dfrac{8x^3}{32x} = \dfrac{8 \cdot x \cdot x^2}{8 \cdot 4 \cdot x}$ Factoring numerator and denominator

$= \dfrac{8x}{8x} \cdot \dfrac{x^2}{4}$ Factoring the rational expression

$= 1 \cdot \dfrac{x^2}{4}$ $\left(\dfrac{8x}{8x} = 1\right)$

$= \dfrac{x^2}{4}$ We removed a factor of 1.

23. $\dfrac{48p^7q^5}{18p^5q^4} = \dfrac{8 \cdot 6 \cdot p^5 \cdot p^2 \cdot q^4 \cdot q}{6 \cdot 3 \cdot p^5 \cdot q^4}$ Factoring numerator and denominator

$= \dfrac{6p^5q^4}{6p^5q^4} \cdot \dfrac{8p^2q}{3}$ Factoring the rational expression

$= 1 \cdot \dfrac{8p^2q}{3}$ $\left(\dfrac{6p^5q^4}{6p^5q^4} = 1\right)$

$= \dfrac{8p^2q}{3}$ Removing a factor of 1

25. $\dfrac{4x-12}{4x} = \dfrac{4(x-3)}{4 \cdot x}$

$= \dfrac{4}{4} \cdot \dfrac{x-3}{x}$

$= 1 \cdot \dfrac{x-3}{x}$

$= \dfrac{x-3}{x}$

27. $\dfrac{3m^2+3m}{6m^2+9m} = \dfrac{3m(m+1)}{3m(2m+3)}$

$= \dfrac{3m}{3m} \cdot \dfrac{m+1}{2m+3}$

$= 1 \cdot \dfrac{m+1}{2m+3}$

$= \dfrac{m+1}{2m+3}$

29. $\dfrac{a^2-9}{a^2+5a+6} = \dfrac{(a-3)(a+3)}{(a+2)(a+3)}$
$= \dfrac{a-3}{a+2} \cdot \dfrac{a+3}{a+3}$
$= \dfrac{a-3}{a+2} \cdot 1$
$= \dfrac{a-3}{a+2}$

31. $\dfrac{a^2-10a+21}{a^2-11a+28} = \dfrac{(a-7)(a-3)}{(a-7)(a-4)}$
$= \dfrac{a-7}{a-7} \cdot \dfrac{a-3}{a-4}$
$= 1 \cdot \dfrac{a-3}{a-4}$
$= \dfrac{a-3}{a-4}$

33. $\dfrac{x^2-25}{x^2-10x+25} = \dfrac{(x-5)(x+5)}{(x-5)(x-5)}$
$= \dfrac{x-5}{x-5} \cdot \dfrac{x+5}{x-5}$
$= 1 \cdot \dfrac{x+5}{x-5}$
$= \dfrac{x+5}{x-5}$

35. $\dfrac{a^2-1}{a-1} = \dfrac{(a-1)(a+1)}{a-1}$
$= \dfrac{a-1}{a-1} \cdot \dfrac{a+1}{1}$
$= 1 \cdot \dfrac{a+1}{1}$
$= a+1$

37. $\dfrac{x^2+1}{x+1}$ cannot be simplified.

Neither the numerator nor the denominator can be factored.

39. $\dfrac{6x^2-54}{4x^2-36} = \dfrac{2 \cdot 3(x^2-9)}{2 \cdot 2(x^2-9)}$
$= \dfrac{2(x^2-9)}{2(x^2-9)} \cdot \dfrac{3}{2}$
$= 1 \cdot \dfrac{3}{2}$
$= \dfrac{3}{2}$

41. $\dfrac{6t+12}{t^2-t-6} = \dfrac{6(t+2)}{(t-3)(t+2)}$
$= \dfrac{6}{t-3} \cdot \dfrac{t+2}{t+2}$
$= \dfrac{6}{t-3} \cdot 1$
$= \dfrac{6}{t-3}$

43. $\dfrac{2t^2+6t+4}{4t^2-12t-16} = \dfrac{2(t^2+3t+2)}{4(t^2-3t-4)}$
$= \dfrac{2(t+2)(t+1)}{2 \cdot 2(t-4)(t+1)}$
$= \dfrac{2(t+1)}{2(t+1)} \cdot \dfrac{t+2}{2(t-4)}$
$= 1 \cdot \dfrac{t+2}{2(t-4)}$
$= \dfrac{t+2}{2(t-4)}$

45. $\dfrac{t^2-4}{(t+2)^2} = \dfrac{(t-2)(t+2)}{(t+2)(t+2)}$
$= \dfrac{t-2}{t+2} \cdot \dfrac{t+2}{t+2}$
$= \dfrac{t-2}{t+2} \cdot 1$
$= \dfrac{t-2}{t+2}$

47. $\dfrac{6-x}{x-6} = \dfrac{-(-6+x)}{x-6}$
$= \dfrac{-1(x-6)}{x-6}$
$= -1 \cdot \dfrac{x-6}{x-6}$
$= -1 \cdot 1$
$= -1$

49. $\dfrac{a-b}{b-a} = \dfrac{-(-a+b)}{b-a}$
$= \dfrac{-1(b-a)}{b-a}$
$= -1 \cdot \dfrac{b-a}{b-a}$
$= -1 \cdot 1$
$= -1$

51. $\dfrac{6t-12}{2-t} = \dfrac{-6(-t+2)}{2-t}$
$= \dfrac{-6(2-t)}{2-t}$
$= \dfrac{-6(2-t)}{2-t}$
$= -6$

53. $\dfrac{x^2-1}{1-x} = \dfrac{(x+1)(x-1)}{-1(-1+x)}$
$= \dfrac{(x+1)(x-1)}{-1(x-1)}$
$= \dfrac{(x+1)\cancel{(x-1)}}{-1\cancel{(x-1)}}$
$= -(x+1)$
$= -x-1$

55. $\dfrac{4x^3}{3x}\cdot\dfrac{14}{x}=\dfrac{4x^3\cdot 14}{3x\cdot x}$ Multiplying the numerators and the denominators

$=\dfrac{4\cdot x\cdot x\cdot x\cdot 14}{3\cdot x\cdot x}$ Factoring the numerator and the denominator

$=\dfrac{4\cdot \cancel{x}\cdot \cancel{x}\cdot x\cdot 14}{3\cdot \cancel{x}\cdot \cancel{x}}$ Removing a factor of 1

$=\dfrac{56x}{3}$ Simplifying

57. $\dfrac{3c}{d^2}\cdot\dfrac{4d}{6c^3}=\dfrac{3c\cdot 4d}{d^2\cdot 6c^3}$ Multiplying the numerators and the denominators

$=\dfrac{3\cdot c\cdot 2\cdot 2\cdot d}{d\cdot d\cdot 3\cdot 2\cdot c\cdot c\cdot c}$ Factoring the numerator and the denominator

$=\dfrac{\cancel{3}\cdot \cancel{c}\cdot \cancel{2}\cdot 2\cdot \cancel{d}}{\cancel{d}\cdot d\cdot \cancel{3}\cdot \cancel{2}\cdot \cancel{c}\cdot c\cdot c}$

$=\dfrac{2}{dc^2}$

59. $\dfrac{x+4}{x}\cdot\dfrac{x^2-3x}{x^2+x-12}=\dfrac{(x+4)(x^2-3x)}{x(x^2+x-12)}$

$=\dfrac{(x+4)(x)(x-3)}{x(x+4)(x-3)}$

$=1$

61. $\dfrac{a^2-9}{a^2}\cdot\dfrac{a^2-3a}{a^2+a-12}=\dfrac{(a-3)(a+3)(a)(a-3)}{a\cdot a(a+4)(a-3)}$

$=\dfrac{\cancel{(a-3)}(a+3)(\cancel{a})(a-3)}{\cancel{a}\cdot a(a+4)\cancel{(a-3)}}$

$=\dfrac{(a-3)(a+3)}{a(a+4)}$

63. $\dfrac{4a^2}{3a^2-12a+12}\cdot\dfrac{3a-6}{2a}=\dfrac{4a^2(3a-6)}{(3a^2-12a+12)2a}$

$=\dfrac{2\cdot 2\cdot a\cdot a\cdot 3\cdot (a-2)}{3\cdot (a-2)\cdot (a-2)\cdot 2\cdot a}$

$=\dfrac{\cancel{2}\cdot 2\cdot \cancel{a}\cdot a\cdot \cancel{3}\cdot \cancel{(a-2)}}{\cancel{3}\cdot \cancel{(a-2)}\cdot (a-2)\cdot \cancel{2}\cdot \cancel{a}}$

$=\dfrac{2a}{a-2}$

65. $\dfrac{t^4-16}{t^4-1}\cdot\dfrac{t^2+1}{t^2+4}$

$=\dfrac{(t^4-16)(t^2+1)}{(t^4-1)(t^2+4)}$

$=\dfrac{(t^2+4)(t+2)(t-2)(t^2+1)}{(t^2+1)(t+1)(t-1)(t^2+4)}$

$=\dfrac{\cancel{(t^2+4)}(t+2)(t-2)\cancel{(t^2+1)}}{\cancel{(t^2+1)}(t+1)(t-1)\cancel{(t^2+4)}}$

$=\dfrac{(t+2)(t-2)}{(t+1)(t-1)}$

67. $\dfrac{(x+4)^3}{(x+2)^3}\cdot\dfrac{x^2+4x+4}{x^2+8x+16}$

$=\dfrac{(x+4)^3(x^2+4x+4)}{(x+2)^3(x^2+8x+16)}$

$=\dfrac{(x+4)(x+4)(x+4)(x+2)(x+2)}{(x+2)(x+2)(x+2)(x+4)(x+4)}$

$=\dfrac{\cancel{(x+4)}\cancel{(x+4)}(x+4)\cancel{(x+2)}\cancel{(x+2)}}{\cancel{(x+2)}\cancel{(x+2)}(x+2)\cancel{(x+4)}\cancel{(x+4)}}$

$=\dfrac{x+4}{x+2}$

69. $\dfrac{5a^2-180}{10a^2-10}\cdot\dfrac{20a+20}{2a-12}=\dfrac{(5a^2-180)(20a+20)}{(10a^2-10)(2a-12)}$

$=\dfrac{5(a+6)(a-6)(2)(10)(a+1)}{10(a+1)(a-1)(2)(a-6)}$

$=\dfrac{5(a+6)\cancel{(a-6)}\cancel{(2)}\cancel{(10)}\cancel{(a+1)}}{\cancel{10}\cancel{(a+1)}(a-1)\cancel{(2)}\cancel{(a-6)}}$

$=\dfrac{5(a+6)}{a-1}$

71. Graph: $x+y=-1$

Find the x-intercept.

$x+0=-1$

$x=-1$

The x-intercept is $(-1,0)$.

Find the y-intercept.

$0+y=-1$

$y=-1$

The y-intercept is $(0,-1)$.

Find a third point as a check. Let $x=2$.

$2+y=-1$

$y=-3$

A third point is $(2,-3)$. Plot this point and the intercepts and draw the graph.

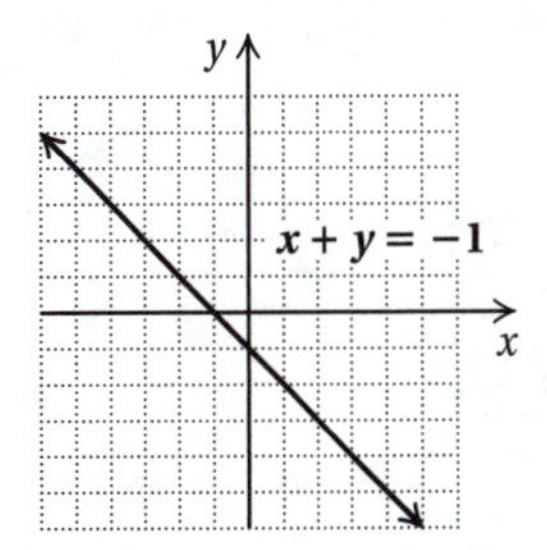

73. ***Familiarize***. Let x = the smaller even integer. Then $x+2$ = the larger even integer.

Translate. We reword the problem.

Smaller even integer times larger even integer is 360.

$x\cdot(x+2)=360$

Solve.

$$x(x+2)=360$$
$$x^2+2x=360$$
$$x^2+2x-360=0$$
$$(x+20)(x-18)=0$$
$$x+20=0 \quad \text{or} \quad x-18=0$$
$$x=-20 \quad \text{or} \quad x=18$$

Check. The solutions of the equation are -20 and 18. When $x=-20$, then $x+2=-18$ and $-20(-18)=360$. The numbers -20 and -18 are consecutive even integers which are solutions to the problem. When $x=18$, then $x+2=20$ and $18\cdot 20=360$. The numbers 18 and 20 are also consecutive even integers which are solutions to the problem.

State. We have two solutions, each of which consists of a pair of numbers: -20 and -18, and 18 and 20.

75. x^2-x-56

We look for a pair of numbers whose product is -56 and whose sum is -1. The numbers are -8 and 7.

$$x^2-x-56=(x-8)(x+7)$$

77. $x^5-2x^4-35x^3=x^3(x^2-2x-35)=x^3(x-7)(x+5)$

79.
$$\begin{aligned}16-t^4&=4^2-(t^2)^2 && \text{Difference of squares}\\ &=(4+t^2)(4-t^2)\\ &=(4+t^2)(2^2-t^2) && \text{Difference of squares}\\ &=(4+t^2)(2+t)(2-t)\end{aligned}$$

81. $x^2-9x+14$

We look for a pair of numbers whose product is 14 and whose sum is -9. The numbers are -2 and -7.

$$x^2-9x+14=(x-2)(x-7)$$

83.
$$\begin{aligned}&16x^2-40xy+25y^2\\ &=(4x)^2-2\cdot 4x\cdot 5y+(5y)^2 \quad \text{Trinomial square}\\ &=(4x-5y)^2\end{aligned}$$

85.
$$\begin{aligned}&\frac{x^4-16y^2}{(x^2+4y^2)(x-2y)}\\ &=\frac{(x^2+4y^2)(x+2y)(x-2y)}{(x^2+4y^2)(x-2y)}\\ &=\frac{\cancel{(x^2+4y^2)}(x+2y)\cancel{(x-2y)}}{\cancel{(x^2+4y^2)}\cancel{(x-2y)}(1)}\\ &=x+2y\end{aligned}$$

87.
$$\begin{aligned}&\frac{t^4-1}{t^4-81}\cdot\frac{t^2-9}{t^2+1}\cdot\frac{(t-9)^2}{(t+1)^2}\\ &=\frac{(t^2+1)(t+1)(t-1)(t+3)(t-3)(t-9)(t-9)}{(t^2+9)(t+3)(t-3)(t^2+1)(t+1)(t+1)}\\ &=\frac{\cancel{(t^2+1)}\cancel{(t+1)}(t-1)\cancel{(t+3)}\cancel{(t-3)}(t-9)(t-9)}{(t^2+9)\cancel{(t+3)}\cancel{(t-3)}\cancel{(t^2+1)}\cancel{(t+1)}(t+1)}\\ &=\frac{(t-1)(t-9)(t-9)}{(t^2+9)(t+1)},\ \text{or}\ \frac{(t-1)(t-9)^2}{(t^2+9)(t+1)}\end{aligned}$$

89.
$$\begin{aligned}&\frac{x^2-y^2}{(x-y)^2}\cdot\frac{x^2-2xy+y^2}{x^2-4xy-5y^2}\\ &=\frac{(x+y)(x-y)(x-y)(x-y)}{(x-y)(x-y)(x-5y)(x+y)}\\ &=\frac{\cancel{(x+y)}\cancel{(x-y)}\cancel{(x-y)}(x-y)}{\cancel{(x-y)}\cancel{(x-y)}(x-5y)\cancel{(x+y)}}\\ &=\frac{x-y}{x-5y}\end{aligned}$$

91.
$$\begin{aligned}\frac{5(2x+5)-25}{10}&=\frac{10x+25-25}{10}\\ &=\frac{10x}{10}\\ &=x\end{aligned}$$

You get the same number you selected.

To do a number trick, ask someone to select a number and then perform these operations. The person will probably be surprised that the result is the original number.

Exercise Set 14.2

1. The reciprocal of $\frac{4}{x}$ is $\frac{x}{4}$ because $\frac{4}{x}\cdot\frac{x}{4}=1$.

3. The reciprocal of x^2-y^2 is $\frac{1}{x^2-y^2}$ because $\frac{x^2-y^2}{1}\cdot\frac{1}{x^2-y^2}=1$.

5. The reciprocal of $\frac{1}{a+b}$ is $a+b$ because $\frac{1}{a+b}\cdot(a+b)=1$.

7. The reciprocal of $\frac{x^2+2x-5}{x^2-4x+7}$ is $\frac{x^2-4x+7}{x^2+2x-5}$ because $\frac{x^2+2x-5}{x^2-4x+7}\cdot\frac{x^2-4x+7}{x^2+2x-5}=1$.

9.
$$\begin{aligned}\frac{2}{5}\div\frac{4}{3}&=\frac{2}{5}\cdot\frac{3}{4} && \text{Multiplying by the reciprocal of the divisor}\\ &=\frac{2\cdot 3}{5\cdot 4}\\ &=\frac{2\cdot 3}{5\cdot 2\cdot 2} && \text{Factoring the denominator}\\ &=\frac{\cancel{2}\cdot 3}{5\cdot\cancel{2}\cdot 2} && \text{Removing a factor of 1}\\ &=\frac{3}{10} && \text{Simplifying}\end{aligned}$$

11.
$$\begin{aligned}\frac{2}{x}\div\frac{8}{x}&=\frac{2}{x}\cdot\frac{x}{8} && \text{Multiplying by the reciprocal of the divisor}\\ &=\frac{2\cdot x}{x\cdot 8}\\ &=\frac{2\cdot x\cdot 1}{x\cdot 2\cdot 4} && \text{Factoring the numerator and the denominator}\\ &=\frac{\cancel{2}\cdot\cancel{x}\cdot 1}{\cancel{x}\cdot\cancel{2}\cdot 4} && \text{Removing a factor of 1}\\ &=\frac{1}{4} && \text{Simplifying}\end{aligned}$$

13. $\frac{a}{b^2} \div \frac{a^2}{b^3} = \frac{a}{b^2} \cdot \frac{b^3}{a^2}$ Multiplying by the reciprocal of the divisor

$$= \frac{a \cdot b^3}{b^2 \cdot a^2}$$
$$= \frac{a \cdot b^2 \cdot b}{b^2 \cdot a \cdot a}$$
$$= \frac{\cancel{a} \cdot \cancel{b^2} \cdot b}{\cancel{b^2} \cdot \cancel{a} \cdot a}$$
$$= \frac{b}{a}$$

15. $\frac{a+2}{a-3} \div \frac{a-1}{a+3} = \frac{a+2}{a-3} \cdot \frac{a+3}{a-1}$

$$= \frac{(a+2)(a+3)}{(a-3)(a-1)}$$

17. $\frac{x^2-1}{x} \div \frac{x+1}{x-1} = \frac{x^2-1}{x} \cdot \frac{x-1}{x+1}$

$$= \frac{(x^2-1)(x-1)}{x(x+1)}$$
$$= \frac{(x-1)(x+1)(x-1)}{x(x+1)}$$
$$= \frac{(x-1)\cancel{(x+1)}(x-1)}{x\cancel{(x+1)}}$$
$$= \frac{(x-1)^2}{x}$$

19. $\frac{x+1}{6} \div \frac{x+1}{3} = \frac{x+1}{6} \cdot \frac{3}{x+1}$

$$= \frac{(x+1) \cdot 3}{6(x+1)}$$
$$= \frac{3(x+1)}{2 \cdot 3(x+1)}$$
$$= \frac{1 \cdot \cancel{3}\cancel{(x+1)}}{2 \cdot \cancel{3}\cancel{(x+1)}}$$
$$= \frac{1}{2}$$

21. $\frac{5x-5}{16} \div \frac{x-1}{6} = \frac{5x-5}{16} \cdot \frac{6}{x-1}$

$$= \frac{(5x-5) \cdot 6}{16(x-1)}$$
$$= \frac{5(x-1) \cdot 2 \cdot 3}{2 \cdot 8(x-1)}$$
$$= \frac{5\cancel{(x-1)} \cdot \cancel{2} \cdot 3}{\cancel{2} \cdot 8\cancel{(x-1)}}$$
$$= \frac{15}{8}$$

23. $\frac{-6+3x}{5} \div \frac{4x-8}{25} = \frac{-6+3x}{5} \cdot \frac{25}{4x-8}$

$$= \frac{(-6+3x) \cdot 25}{5(4x-8)}$$
$$= \frac{3(x-2) \cdot 5 \cdot 5}{5 \cdot 4(x-2)}$$
$$= \frac{3\cancel{(x-2)} \cdot \cancel{5} \cdot 5}{\cancel{5} \cdot 4\cancel{(x-2)}}$$
$$= \frac{15}{4}$$

25. $\frac{a+2}{a-1} \div \frac{3a+6}{a-5} = \frac{a+2}{a-1} \cdot \frac{a-5}{3a+6}$

$$= \frac{(a+2)(a-5)}{(a-1)(3a+6)}$$
$$= \frac{(a+2)(a-5)}{(a-1) \cdot 3 \cdot (a+2)}$$
$$= \frac{\cancel{(a+2)}(a-5)}{(a-1) \cdot 3 \cdot \cancel{(a+2)}}$$
$$= \frac{a-5}{3(a-1)}$$

27. $\frac{x^2-4}{x} \div \frac{x-2}{x+2} = \frac{x^2-4}{x} \cdot \frac{x+2}{x-2}$

$$= \frac{(x^2-4)(x+2)}{x(x-2)}$$
$$= \frac{(x-2)(x+2)(x+2)}{x(x-2)}$$
$$= \frac{\cancel{(x-2)}(x+2)(x+2)}{x\cancel{(x-2)}}$$
$$= \frac{(x+2)^2}{x}$$

29. $\frac{x^2-9}{4x+12} \div \frac{x-3}{6} = \frac{x^2-9}{4x+12} \cdot \frac{6}{x-3}$

$$= \frac{(x^2-9) \cdot 6}{(4x+12)(x-3)}$$
$$= \frac{(x-3)(x+3) \cdot 3 \cdot 2}{2 \cdot 2(x+3)(x-3)}$$
$$= \frac{\cancel{(x-3)}\cancel{(x+3)} \cdot 3 \cdot \cancel{2}}{\cancel{2} \cdot 2\cancel{(x+3)}\cancel{(x-3)}}$$
$$= \frac{3}{2}$$

31. $\frac{c^2+3c}{c^2+2c-3} \div \frac{c}{c+1} = \frac{c^2+3c}{c^2+2c-3} \cdot \frac{c+1}{c}$

$$= \frac{(c^2+3c)(c+1)}{(c^2+2c-3)c}$$
$$= \frac{c(c+3)(c+1)}{(c+3)(c-1)c}$$
$$= \frac{\cancel{c}\cancel{(c+3)}(c+1)}{\cancel{(c+3)}(c-1)\cancel{c}}$$
$$= \frac{c+1}{c-1}$$

33. $\dfrac{2y^2-7y+3}{2y^2+3y-2} \div \dfrac{6y^2-5y+1}{3y^2+5y-2}$

$= \dfrac{2y^2-7y+3}{2y^2+3y-2} \cdot \dfrac{3y^2+5y-2}{6y^2-5y+1}$

$= \dfrac{(2y^2-7y+3)(3y^2+5y-2)}{(2y^2+3y-2)(6y^2-5y+1)}$

$= \dfrac{(2y-1)(y-3)(3y-1)(y+2)}{(2y-1)(y+2)(3y-1)(2y-1)}$

$= \dfrac{\cancel{(2y-1)}(y-3)\cancel{(3y-1)}\cancel{(y+2)}}{\cancel{(2y-1)}\cancel{(y+2)}\cancel{(3y-1)}(2y-1)}$

$= \dfrac{y-3}{2y-1}$

35. $\dfrac{x^2-1}{4x+4} \div \dfrac{2x^2-4x+2}{8x+8} = \dfrac{x^2-1}{4x+4} \cdot \dfrac{8x+8}{2x^2-4x+2}$

$= \dfrac{(x^2-1)(8x+8)}{(4x+4)(2x^2-4x+2)}$

$= \dfrac{(x+1)(x-1)(2)(4)(x+1)}{4(x+1)(2)(x-1)(x-1)}$

$= \dfrac{\cancel{(x+1)}\cancel{(x-1)}\cancel{(2)}\cancel{(4)}(x+1)}{\cancel{4}\cancel{(x+1)}\cancel{(2)}(x-1)\cancel{(x-1)}}$

$= \dfrac{x+1}{x-1}$

37. ***Familiarize***. Let $s =$ Bonnie's score on the last test.

Translate. The average of the four scores must be at least 90. This means it must be greater than or equal to 90. We translate.

$$\frac{96+98+89+s}{4} \geq 90$$

Solve. We solve the inequality. First we multiply by 4 to clear the fraction.

$$4\left(\frac{96+98+89+s}{4}\right) \geq 4 \cdot 90$$

$$96+98+89+s \geq 360$$

$$283+s \geq 360$$

$$s \geq 77 \quad \text{Subtracting 283}$$

Check. We can do a partial check by substituting a value for s less than 77 and a value for s greater than 77.

For $s = 76$: $\dfrac{96+98+89+76}{4} = 89.75 < 90$

For $s = 78$: $\dfrac{96+98+89+78}{4} = 90.25 \leq 90$

Since the average is less than 90 for a value of s less than 77 and greater than or equal to 90 for a value greater than or equal to 77, the answer is probably correct.

State. The scores on the last test that will earn Bonnie an A are $\{s|s \geq 77\}$.

39. $(8x^3-3x^2+7)-(8x^2+3x-5) =$
$8x^3-3x^2+7-8x^2-3x+5 =$
$8x^3-11x^2-3x+12$

41. $(2x^{-3}y^4)^2 = 2^2(x^{-3})^2(y^4)^2$

$= 2^2x^{-6}y^8$ Multiplying exponents

$= 4x^{-6}y^8$ $(2^2 = 4)$

$= \dfrac{4y^8}{x^6}$ $\left(x^{-6} = \dfrac{1}{x^6}\right)$

43. $\left(\dfrac{2x^3}{y^5}\right)^2 = \dfrac{2^2(x^3)^2}{(y^5)^2}$

$= \dfrac{2^2x^6}{y^{10}}$ Multiplying exponents

$= \dfrac{4x^6}{y^{10}}$ $(2^2 = 4)$

45. $\dfrac{3a^2-5ab-12b^2}{3ab+4b^2} \div (3b^2-ab)$

$= \dfrac{3a^2-5ab-12b^2}{3ab+4b^2} \cdot \dfrac{1}{3b^2-ab}$

$= \dfrac{(3a+4b)(a-3b)}{b(3a+4b)\cdot b(3b-a)}$

$= \dfrac{(3a+4b)(-1)(3b-a)}{b(3a+4b)\cdot b(3b-a)}$

$= \dfrac{\cancel{(3a+4b)}(-1)\cancel{(3b-a)}}{b\cancel{(3a+4b)}\cdot b\cancel{(3b-a)}}$

$= -\dfrac{1}{b^2}$

47. $\dfrac{a^2b^2+3ab^2+2b^2}{a^2b^4+4b^4} \div (5a^2+10a)$

$= \dfrac{a^2b^2+3ab^2+2b^2}{a^2b^4+4b^4} \cdot \dfrac{1}{5a^2+10a}$

$= \dfrac{a^2b^2+3ab^2+2b^2}{(a^2b^4+4b^4)(5a^2+10a)}$

$= \dfrac{b^2(a^2+3a+2)}{b^4(a^2+4)(5a)(a+2)}$

$= \dfrac{b^2(a+1)(a+2)}{b^2\cdot b^2(a^2+4)(5a)(a+2)}$

$= \dfrac{\cancel{b^2}(a+1)\cancel{(a+2)}}{\cancel{b^2}\cdot b^2(a^2+4)(5a)\cancel{(a+2)}}$

$= \dfrac{a+1}{5ab^2(a^2+4)}$

Exercise Set 14.3

1. $12 = 2\cdot 2\cdot 3$

$27 = 3\cdot 3\cdot 3$

$\text{LCM} = 2\cdot 2\cdot 3\cdot 3\cdot 3$, or 108

3. $8 = 2\cdot 2\cdot 2$

$9 = 3\cdot 3$

$\text{LCM} = 2\cdot 2\cdot 2\cdot 3\cdot 3$, or 72

5. $6 = 2 \cdot 3$

$9 = 3 \cdot 3$

$21 = 3 \cdot 7$

$\text{LCM} = 2 \cdot 3 \cdot 3 \cdot 7$, or 126

7. $24 = 2 \cdot 2 \cdot 2 \cdot 3$

$36 = 2 \cdot 2 \cdot 3 \cdot 3$

$40 = 2 \cdot 2 \cdot 2 \cdot 5$

$\text{LCM} = 2 \cdot 2 \cdot 2 \cdot 3 \cdot 3 \cdot 5$, or 360

9. $10 = 2 \cdot 5$

$100 = 2 \cdot 2 \cdot 5 \cdot 5$

$500 = 2 \cdot 2 \cdot 5 \cdot 5 \cdot 5$

$\text{LCM} = 2 \cdot 2 \cdot 5 \cdot 5 \cdot 5$, or 500

(We might have observed at the outset that both 10 and 100 are factors of 500, so the LCM is 500.)

11. $24 = 2 \cdot 2 \cdot 2 \cdot 3$

$18 = 2 \cdot 3 \cdot 3$

$\text{LCD} = 2 \cdot 2 \cdot 2 \cdot 3 \cdot 3$, or 72

$$\frac{7}{24} + \frac{11}{18} = \frac{7}{2 \cdot 2 \cdot 2 \cdot 3} \cdot \frac{3}{3} + \frac{11}{2 \cdot 3 \cdot 3} \cdot \frac{2 \cdot 2}{2 \cdot 2}$$
$$= \frac{21}{2 \cdot 2 \cdot 2 \cdot 3 \cdot 3} + \frac{44}{2 \cdot 2 \cdot 2 \cdot 3 \cdot 3}$$
$$= \frac{65}{72}$$

13. $$\frac{1}{6} + \frac{3}{40}$$
$$= \frac{1}{2 \cdot 3} + \frac{3}{2 \cdot 2 \cdot 2 \cdot 5}$$

LCD is $2 \cdot 2 \cdot 2 \cdot 3 \cdot 5$, or 120

$$= \frac{1}{2 \cdot 3} \cdot \frac{2 \cdot 2 \cdot 5}{2 \cdot 2 \cdot 5} + \frac{3}{2 \cdot 2 \cdot 2 \cdot 5} \cdot \frac{3}{3}$$
$$= \frac{20 + 9}{2 \cdot 2 \cdot 2 \cdot 3 \cdot 5}$$
$$= \frac{29}{120}$$

15. $$\frac{1}{20} + \frac{1}{30} + \frac{2}{45}$$
$$= \frac{1}{2 \cdot 2 \cdot 5} + \frac{1}{2 \cdot 3 \cdot 5} + \frac{2}{3 \cdot 3 \cdot 5}$$

LCD is $2 \cdot 2 \cdot 3 \cdot 3 \cdot 5$, or 180

$$= \frac{1}{2 \cdot 2 \cdot 5} \cdot \frac{3 \cdot 3}{3 \cdot 3} + \frac{1}{2 \cdot 3 \cdot 5} \cdot \frac{2 \cdot 3}{2 \cdot 3} + \frac{2}{3 \cdot 3 \cdot 5} \cdot \frac{2 \cdot 2}{2 \cdot 2}$$
$$= \frac{9 + 6 + 8}{2 \cdot 2 \cdot 3 \cdot 3 \cdot 5}$$
$$= \frac{23}{180}$$

17. $6x^2 = 2 \cdot 3 \cdot x \cdot x$

$12x^3 = 2 \cdot 2 \cdot 3 \cdot x \cdot x \cdot x$

$\text{LCM} = 2 \cdot 2 \cdot 3 \cdot x \cdot x \cdot x$, or $12x^3$

19. $2x^2 = 2 \cdot x \cdot x$

$6xy = 2 \cdot 3 \cdot x \cdot y$

$18y^2 = 2 \cdot 3 \cdot 3 \cdot y \cdot y$

$\text{LCM} = 2 \cdot 3 \cdot 3 \cdot x \cdot x \cdot y \cdot y$, or $18x^2y^2$

21. $2(y-3) = 2 \cdot (y-3)$

$6(y-3) = 2 \cdot 3 \cdot (y-3)$

$\text{LCM} = 2 \cdot 3 \cdot (y-3)$, or $6(y-3)$

23. $t,\ t+2,\ t-2$

The expressions are not factorable, so the LCM is their product:

$\text{LCM} = t(t+2)(t-2)$

25. $x^2 - 4 = (x+2)(x-2)$

$x^2 + 5x + 6 = (x+3)(x+2)$

$\text{LCM} = (x+2)(x-2)(x+3)$

27. $t^3 + 4t^2 + 4t = t(t^2 + 4t + 4) = t(t+2)(t+2)$

$t^2 - 4t = t(t-4)$

$\text{LCM} = t(t+2)(t+2)(t-4) = t(t+2)^2(t-4)$

29. $a + 1 = a + 1$

$(a-1)^2 = (a-1)(a-1)$

$a^2 - 1 = (a+1)(a-1)$

$\text{LCM} = (a+1)(a-1)(a-1) = (a+1)(a-1)^2$

31. $m^2 - 5m + 6 = (m-3)(m-2)$

$m^2 - 4m + 4 = (m-2)(m-2)$

$\text{LCM} = (m-3)(m-2)(m-2) = (m-3)(m-2)^2$

33. $2 + 3x = 2 + 3x$

$4 - 9x^2 = (2+3x)(2-3x)$

$2 - 3x = 2 - 3x$

$\text{LCM} = (2+3x)(2-3x)$

35. $10v^2 + 30v = 10v(v+3) = 2 \cdot 5 \cdot v(v+3)$

$5v^2 + 35v + 60 = 5(v^2 + 7v + 12)$
$\qquad = 5(v+4)(v+3)$

$\text{LCM} = 2 \cdot 5 \cdot v(v+3)(v+4) = 10v(v+3)(v+4)$

37. $9x^3 - 9x^2 - 18x = 9x(x^2 - x - 2)$
$\qquad = 3 \cdot 3 \cdot x(x-2)(x+1)$

$6x^5 - 24x^4 + 24x^3 = 6x^3(x^2 - 4x + 4)$
$\qquad = 2 \cdot 3 \cdot x \cdot x \cdot x(x-2)(x-2)$

$\text{LCM} = 2 \cdot 3 \cdot 3 \cdot x \cdot x \cdot x(x-2)(x-2)(x+1) =$
$18x^3(x-2)^2(x+1)$

39. $x^5 + 4x^4 + 4x^3 = x^3(x^2 + 4x + 4)$
$\qquad = x \cdot x \cdot x(x+2)(x+2)$

$3x^2 - 12 = 3(x^2 - 4) = 3(x+2)(x-2)$

$2x + 4 = 2(x+2)$

$\text{LCM} = 2 \cdot 3 \cdot x \cdot x \cdot x(x+2)(x+2)(x-2)$
$\qquad = 6x^3(x+2)^2(x-2)$

41. $24w^4 = 2\cdot 2\cdot 2\cdot 3\cdot w\cdot w\cdot w\cdot w$

$w^2 = w\cdot w$

$10w^3 = 2\cdot 5\cdot w\cdot w\cdot w$

$w^6 = w\cdot w\cdot w\cdot w\cdot w\cdot w$

$\text{LCM} = 2\cdot 2\cdot 2\cdot 3\cdot 5\cdot w\cdot w\cdot w\cdot w\cdot w\cdot w = 120w^6$

43. $x^2 - 6x + 9 = x^2 - 2\cdot x\cdot 3 + 3^2$ Trinomial square

$= (x-3)^2$

45. $x^2 - 9 = x^2 - 3^2$ Difference of squares

$= (x+3)(x-3)$

47. $x^2 + 6x + 9 = x^2 + 2\cdot x\cdot 3 + 3^2$ Trinomial square

$= (x+3)^2$

49. $40x^3 = 2\cdot 2\cdot 2\cdot 5\cdot x\cdot x\cdot x$

$24x^4 = 2\cdot 2\cdot 2\cdot 3\cdot x\cdot x\cdot x\cdot x$

$\text{LCM} = 2\cdot 2\cdot 2\cdot 3\cdot 5\cdot x\cdot x\cdot x\cdot x = 120x^4$

$\text{GCF} = 2\cdot 2\cdot 2\cdot x\cdot x\cdot x = 8x^3$

$120x^4(8x^3) = 960x^7$

51. $20x^2 = 2\cdot 2\cdot 5\cdot x\cdot x$

$10x = 2\cdot 5\cdot x$

$\text{LCM} = 2\cdot 2\cdot 5\cdot x\cdot x = 20x^2$

$\text{GCF} = 2\cdot 5\cdot x = 10x$

$20x^2(10x) = 200x^3$

53. $10x^2 = 2\cdot 5\cdot x\cdot x$

$24x^3 = 2\cdot 2\cdot 2\cdot 3\cdot x\cdot x\cdot x$

$\text{LCM} = 2\cdot 2\cdot 2\cdot 3\cdot 5\cdot x\cdot x\cdot x = 120x^3$

$\text{GCF} = 2\cdot x\cdot x = 2x^2$

$120x^3(2x^2) = 240x^5$

55. The product of the LCM and the GCF is the product of the two expressions.

Exercise Set 14.4

1. $\frac{5}{8} + \frac{3}{8} = \frac{5+3}{8} = \frac{8}{8} = 1$

3. $\frac{1}{3+x} + \frac{5}{3+x} = \frac{1+5}{3+x} = \frac{6}{3+x}$

5. $\frac{4x+6}{2x-1} + \frac{5-8x}{-1+2x} = \frac{4x+6}{2x-1} + \frac{5-8x}{2x-1}$ $(-1+2x = 2x-1)$

$= \frac{4x+6+5-8x}{2x-1}$

$= \frac{-4x+11}{2x-1}$

7. $\frac{2}{x} + \frac{5}{x^2} = \frac{2}{x} + \frac{5}{x\cdot x}$ LCD $= x\cdot x$, or x^2

$= \frac{2}{x}\cdot\frac{x}{x} + \frac{5}{x\cdot x}$

$= \frac{2x+5}{x^2}$

9. $\left.\begin{array}{l}6r = 2\cdot 3\cdot r\\ 8r = 2\cdot 2\cdot 2\cdot r\end{array}\right\}\text{LCD} = 2\cdot 2\cdot 2\cdot 3\cdot r, \text{ or } 24r$

$\frac{5}{6r} + \frac{7}{8r} = \frac{5}{6r}\cdot\frac{4}{4} + \frac{7}{8r}\cdot\frac{3}{3}$

$= \frac{20+21}{24r}$

$= \frac{41}{24r}$

11. $\left.\begin{array}{l}xy^2 = x\cdot y\cdot y\\ x^2y = x\cdot x\cdot y\end{array}\right\}\text{LCD} = x\cdot x\cdot y\cdot y, \text{ or } x^2y^2$

$\frac{4}{xy^2} + \frac{6}{x^2y} = \frac{4}{xy^2}\cdot\frac{x}{x} + \frac{6}{x^2y}\cdot\frac{y}{y}$

$= \frac{4x+6y}{x^2y^2} = \frac{2(2x+3y)}{x^2y^2}$

13. $\left.\begin{array}{l}9t^3 = 3\cdot 3\cdot t\cdot t\cdot t\\ 6t^2 = 2\cdot 3\cdot t\cdot t\end{array}\right\}\text{LCD} = 2\cdot 3\cdot 3\cdot t\cdot t\cdot t, \text{ or } 18t^3$

$\frac{2}{9t^3} + \frac{1}{6t^2} = \frac{2}{9t^3}\cdot\frac{2}{2} + \frac{1}{6t^2}\cdot\frac{3t}{3t}$

$= \frac{4+3t}{18t^3}$

15. LCD $= x^2y^2$ (See Exercise 11.)

$\frac{x+y}{xy^2} + \frac{3x+y}{x^2y} = \frac{x+y}{xy^2}\cdot\frac{x}{x} + \frac{3x+y}{x^2y}\cdot\frac{y}{y}$

$= \frac{x(x+y)+y(3x+y)}{x^2y^2}$

$= \frac{x^2+xy+3xy+y^2}{x^2y^2}$

$= \frac{x^2+4xy+y^2}{x^2y^2}$

17. The denominators do not factor, so the LCD is their product, $(x-2)(x+2)$.

$\frac{3}{x-2} + \frac{3}{x+2} = \frac{3}{x-2}\cdot\frac{x+2}{x+2} + \frac{3}{x+2}\cdot\frac{x-2}{x-2}$

$= \frac{3(x+2)+3(x-2)}{(x-2)(x+2)}$

$= \frac{3x+6+3x-6}{(x-2)(x+2)}$

$= \frac{6x}{(x-2)(x+2)}$

19. $\left.\begin{array}{l}3x = 3\cdot x\\ x+1 = x+1\end{array}\right\}\text{LCD} = 3x(x+1)$

$\frac{3}{x+1} + \frac{2}{3x} = \frac{3}{x+1}\cdot\frac{3x}{3x} + \frac{2}{3x}\cdot\frac{x+1}{x+1}$

$= \frac{9x+2(x+1)}{3x(x+1)}$

$= \frac{9x+2x+2}{3x(x+1)}$

$= \frac{11x+2}{3x(x+1)}$

21. $\left.\begin{array}{l}x^2-16=(x+4)(x-4)\\ x-4=x-4\end{array}\right\}\text{LCD}=(x+4)(x-4)$

$$\begin{aligned}\frac{2x}{x^2-16}+\frac{x}{x-4}&=\frac{2x}{(x+4)(x-4)}+\frac{x}{x-4}\cdot\frac{x+4}{x+4}\\&=\frac{2x+x(x+4)}{(x+4)(x-4)}\\&=\frac{2x+x^2+4x}{(x+4)(x-4)}\\&=\frac{x^2+6x}{(x+4)(x-4)}\\&=\frac{x(x+6)}{(x+4)(x-4)}\end{aligned}$$

23. $\frac{5}{z+4}+\frac{3}{3z+12}=\frac{5}{z+4}+\frac{3}{3(z+4)}$ LCD $=3(z+4)$

$$\begin{aligned}&=\frac{5}{z+4}\cdot\frac{3}{3}+\frac{3}{3(z+4)}\\&=\frac{15+3}{3(z+4)}=\frac{18}{3(z+4)}\\&=\frac{3\cdot 6}{3(z+4)}=\frac{\cancel{3}\cdot 6}{\cancel{3}(z+4)}\\&=\frac{6}{z+4}\end{aligned}$$

25. $\frac{3}{x-1}+\frac{2}{(x-1)^2}$ LCD $=(x-1)^2$

$$\begin{aligned}&=\frac{3}{x-1}\cdot\frac{x-1}{x-1}+\frac{2}{(x-1)^2}\\&=\frac{3(x-1)+2}{(x-1)^2}\\&=\frac{3x-3+2}{(x-1)^2}\\&=\frac{3x-1}{(x-1)^2}\end{aligned}$$

27. $\frac{4a}{5a-10}+\frac{3a}{10a-20}=\frac{4a}{5(a-2)}+\frac{3a}{2\cdot 5(a-2)}$

LCD $=2\cdot 5(a-2)$

$$\begin{aligned}&=\frac{4a}{5(a-2)}\cdot\frac{2}{2}+\frac{3a}{2\cdot 5(a-2)}\\&=\frac{8a+3a}{10(a-2)}\\&=\frac{11a}{10(a-2)}\end{aligned}$$

29. $\frac{x+4}{x}+\frac{x}{x+4}$ LCD $=x(x+4)$

$$\begin{aligned}&=\frac{x+4}{x}\cdot\frac{x+4}{x+4}+\frac{x}{x+4}\cdot\frac{x}{x}\\&=\frac{(x+4)^2+x^2}{x(x+4)}\\&=\frac{x^2+8x+16+x^2}{x(x+4)}\\&=\frac{2x^2+8x+16}{x(x+4)}=\frac{2(x^2+4x+8)}{x(x+4)}\end{aligned}$$

31. $\frac{4}{a^2-a-2}+\frac{3}{a^2+4a+3}$

$$=\frac{4}{(a-2)(a+1)}+\frac{3}{(a+3)(a+1)}$$

LCD $=(a-2)(a+1)(a+3)$

$$\begin{aligned}&=\frac{4}{(a-2)(a+1)}\cdot\frac{a+3}{a+3}+\frac{3}{(a+3)(a+1)}\cdot\frac{a-2}{a-2}\\&=\frac{4(a+3)+3(a-2)}{(a-2)(a+1)(a+3)}\\&=\frac{4a+12+3a-6}{(a-2)(a+1)(a+3)}\\&=\frac{7a+6}{(a-2)(a+1)(a+3)}\end{aligned}$$

33. $\frac{x+3}{x-5}+\frac{x-5}{x+3}$ LCD $=(x-5)(x+3)$

$$\begin{aligned}&=\frac{x+3}{x-5}\cdot\frac{x+3}{x+3}+\frac{x-5}{x+3}\cdot\frac{x-5}{x-5}\\&=\frac{(x+3)^2+(x-5)^2}{(x-5)(x+3)}\\&=\frac{x^2+6x+9+x^2-10x+25}{(x-5)(x+3)}\\&=\frac{2x^2-4x+34}{(x-5)(x+3)}=\frac{2(x^2-2x+17)}{(x-5)(x+3)}\end{aligned}$$

35. $\frac{a}{a^2-1}+\frac{2a}{a^2-a}$

$$=\frac{a}{(a+1)(a-1)}+\frac{2a}{a(a-1)}$$

LCD $=a(a+1)(a-1)$

$$\begin{aligned}&=\frac{a}{(a+1)(a-1)}\cdot\frac{a}{a}+\frac{2a}{a(a-1)}\cdot\frac{a+1}{a+1}\\&=\frac{a^2+2a(a+1)}{a(a+1)(a-1)}=\frac{a^2+2a^2+2a}{a(a+1)(a-1)}\\&=\frac{3a^2+2a}{a(a+1)(a-1)}=\frac{a(3a+2)}{a(a+1)(a-1)}\\&=\frac{\cancel{a}(3a+2)}{\cancel{a}(a+1)(a-1)}=\frac{3a+2}{(a+1)(a-1)}\end{aligned}$$

37. $\frac{7}{8}+\frac{5}{-8}=\frac{7}{8}+\frac{5}{-8}\cdot\frac{-1}{-1}$

$$\begin{aligned}&=\frac{7}{8}+\frac{-5}{8}\\&=\frac{7+(-5)}{8}\\&=\frac{2}{8}=\frac{\cancel{2}\cdot 1}{4\cdot\cancel{2}}\\&=\frac{1}{4}\end{aligned}$$

39. $\frac{3}{t}+\frac{4}{-t}=\frac{3}{t}+\frac{4}{-t}\cdot\frac{-1}{-1}$
$=\frac{3}{t}+\frac{-4}{t}$
$=\frac{3+(-4)}{t}$
$=\frac{-1}{t}$
$=-\frac{1}{t}$

41. $\frac{2x+7}{x-6}+\frac{3x}{6-x}=\frac{2x+7}{x-6}+\frac{3x}{6-x}\cdot\frac{-1}{-1}$
$=\frac{2x+7}{x-6}+\frac{-3x}{x-6}$
$=\frac{(2x+7)+(-3x)}{x-6}$
$=\frac{-x+7}{x-6}$, or $\frac{7-x}{x-6}$, or $\frac{x-7}{6-x}$

43. $\frac{y^2}{y-3}+\frac{9}{3-y}=\frac{y^2}{y-3}+\frac{9}{3-y}\cdot\frac{-1}{-1}$
$=\frac{y^2}{y-3}+\frac{-9}{y-3}$
$=\frac{y^2+(-9)}{y-3}$
$=\frac{y^2-9}{y-3}$
$=\frac{(y+3)(y-3)}{y-3}$
$=\frac{(y+3)\cancel{(y-3)}}{1\cancel{(y-3)}}$
$=y+3$

45. $\frac{b-7}{b^2-16}+\frac{7-b}{16-b^2}=\frac{b-7}{b^2-16}+\frac{7-b}{16-b^2}\cdot\frac{-1}{-1}$
$=\frac{b-7}{b^2-16}+\frac{b-7}{b^2-16}$
$=\frac{(b-7)+(b-7)}{b^2-16}$
$=\frac{2b-14}{b^2-16}=\frac{2(b-7)}{(b+4)(b-4)}$

47. $\frac{a^2}{a-b}+\frac{b^2}{b-a}=\frac{a^2}{a-b}+\frac{b^2}{b-a}\cdot\frac{-1}{-1}$
$=\frac{a^2}{a-b}+\frac{-b^2}{a-b}$
$=\frac{a^2+(-b^2)}{a-b}$
$=\frac{a^2-b^2}{a-b}$
$=\frac{(a+b)(a-b)}{a-b}$
$=\frac{(a+b)\cancel{(a-b)}}{1\cancel{(a-b)}}$
$=a+b$

49. $\frac{x+3}{x-5}+\frac{2x-1}{5-x}+\frac{2(3x-1)}{x-5}$
$=\frac{x+3}{x-5}+\frac{2x-1}{5-x}\cdot\frac{-1}{-1}+\frac{2(3x-1)}{x-5}$
$=\frac{x+3}{x-5}+\frac{1-2x}{x-5}+\frac{2(3x-1)}{x-5}$
$=\frac{(x+3)+(1-2x)+(6x-2)}{x-5}$
$=\frac{5x+2}{x-5}$

51. $\frac{2(4x+1)}{5x-7}+\frac{3(x-2)}{7-5x}+\frac{-10x-1}{5x-7}$
$=\frac{2(4x+1)}{5x-7}+\frac{3(x-2)}{7-5x}\cdot\frac{-1}{-1}+\frac{-10x-1}{5x-7}$
$=\frac{2(4x+1)}{5x-7}+\frac{-3(x-2)}{5x-7}+\frac{-10x-1}{5x-7}$
$=\frac{(8x+2)+(-3x+6)+(-10x-1)}{5x-7}$
$=\frac{-5x+7}{5x-7}$
$=\frac{-1(5x-7)}{5x-7}$
$=\frac{-1\cancel{(5x-7)}}{\cancel{5x-7}}$
$=-1$

53. $\frac{x+1}{(x+3)(x-3)}+\frac{4(x-3)}{(x-3)(x+3)}+\frac{(x-1)(x-3)}{(3-x)(x+3)}$
$=\frac{x+1}{(x+3)(x-3)}+\frac{4(x-3)}{(x-3)(x+3)}+\frac{(x-1)(x-3)}{(3-x)(x+3)}\cdot\frac{-1}{-1}$
$=\frac{x+1}{(x+3)(x-3)}+\frac{4(x-3)}{(x-3)(x+3)}+\frac{-1(x^2-4x+3)}{(x-3)(x+3)}$
$=\frac{(x+1)+(4x-12)+(-x^2+4x-3)}{(x+3)(x-3)}$
$=\frac{-x^2+9x-14}{(x+3)(x-3)}$

55. $\dfrac{6}{x-y}+\dfrac{4x}{y^2-x^2}$

$=\dfrac{6}{x-y}+\dfrac{4x}{(y-x)(y+x)}$

$=\dfrac{6}{x-y}+\dfrac{4x}{(y-x)(y+x)}\cdot\dfrac{-1}{-1}$

$=\dfrac{6}{x-y}+\dfrac{-4x}{(x-y)(x+y)}$

$[-1(y-x)=x-y; y+x=x+y]$

LCD $=(x-y)(x+y)$

$=\dfrac{6}{x-y}\cdot\dfrac{x+y}{x+y}+\dfrac{-4x}{(x-y)(x+y)}$

$=\dfrac{6(x+y)-4x}{(x-y)(x+y)}$

$=\dfrac{6x+6y-4x}{(x-y)(x+y)}$

$=\dfrac{2x+6y}{(x-y)(x+y)}$

$=\dfrac{2(x+3y)}{(x-y)(x+y)}$

57. $\dfrac{4-a}{25-a^2}+\dfrac{a+1}{a-5}$

$=\dfrac{4-a}{25-a^2}\cdot\dfrac{-1}{-1}+\dfrac{a+1}{a-5}$

$=\dfrac{a-4}{a^2-25}+\dfrac{a+1}{a-5}$

$=\dfrac{a-4}{(a+5)(a-5)}+\dfrac{a+1}{a-5}$

LCD $=(a+5)(a-5)$

$=\dfrac{a-4}{(a+5)(a-5)}+\dfrac{a+1}{a-5}\cdot\dfrac{a+5}{a+5}$

$=\dfrac{a-4}{(a+5)(a-5)}+\dfrac{(a+1)(a+5)}{(a+5)(a-5)}$

$=\dfrac{(a-4)+(a+1)(a+5)}{(a+5)(a-5)}$

$=\dfrac{a-4+a^2+6a+5}{(a+5)(a-5)}$

$=\dfrac{a^2+7a+1}{(a+5)(a-5)}$

59. $\dfrac{2}{t^2+t-6}+\dfrac{3}{t^2-9}$

$=\dfrac{2}{(t+3)(t-2)}+\dfrac{3}{(t+3)(t-3)}$

LCD $=(t+3)(t-2)(t-3)$

$=\dfrac{2}{(t+3)(t-2)}\cdot\dfrac{t-3}{t-3}+\dfrac{3}{(t+3)(t-3)}\cdot\dfrac{t-2}{t-2}$

$=\dfrac{2(t-3)+3(t-2)}{(t+3)(t-2)(t-3)}$

$=\dfrac{2t-6+3t-6}{(t+3)(t-2)(t-3)}$

$=\dfrac{5t-12}{(t+3)(t-2)(t-3)}$

61. $(x^2+x)-(x+1)=x^2+x-x-1=x^2-1$

63. $(2x^4y^3)^{-3}=\dfrac{1}{(2x^4y^3)^3}=\dfrac{1}{2^3(x^4)^3(y^3)^3}=\dfrac{1}{8x^{12}y^9}$

65. $\left(\dfrac{x^{-4}}{y^7}\right)^3=\dfrac{(x^{-4})^3}{(y^7)^3}=\dfrac{x^{-12}}{y^{21}}=\dfrac{1}{x^{12}y^{21}}$

67. $y=\dfrac{1}{2}x-5=\dfrac{1}{2}x+(-5)$

The y-intercept is $(0,-5)$. We find two other pairs.

When $x=2$, $y=\dfrac{1}{2}\cdot2-5=1-5=-4$.

When $x=4$, $y=\dfrac{1}{2}\cdot4-5=2-5=-3$.

x	y
0	−5
2	−4
4	−3

Plot these points, draw the line they determine, and label the graph $y=\dfrac{1}{2}x-5$.

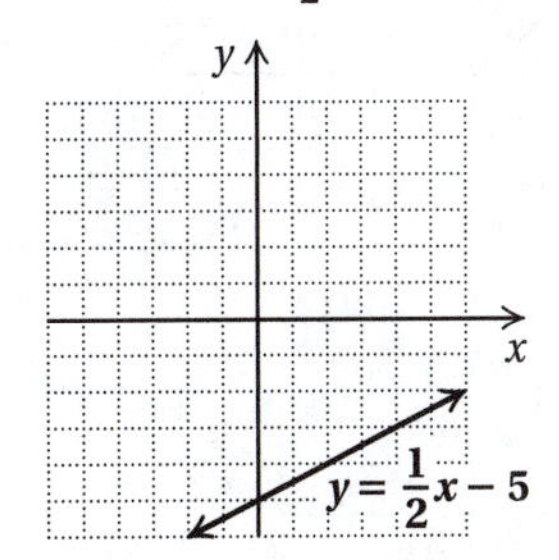

69. $y=3$

Any ordered pair $(x,3)$ is a solution. The variable y must be 3, but x can be any number we choose. A few solutions are listed below. Plot these points and draw the line.

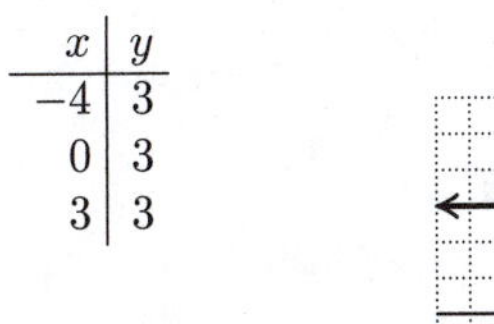

x	y
−4	3
0	3
3	3

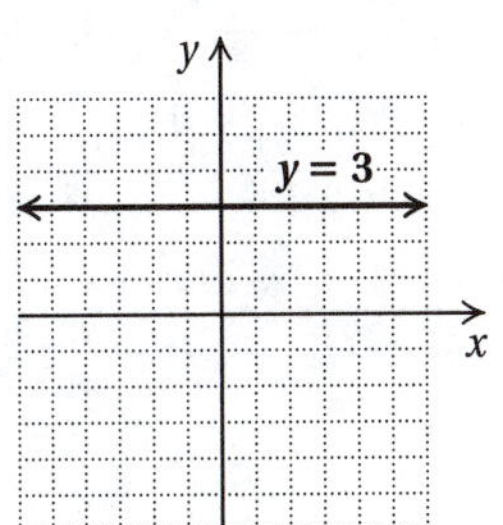

71.
$$\begin{aligned} 3x-7&=5x+9 \\ -2x-7&=9 \quad \text{Subtracting } 5x \\ -2x&=16 \quad \text{Adding } 7 \\ x&=-8 \quad \text{Dividing by } -2 \end{aligned}$$

The solution is -8.

73.
$$x^2-8x+15=0$$
$$(x-3)(x-5)=0$$
$$x-3=0 \text{ or } x-5=0 \quad \text{Principle of zero products}$$
$$x=3 \text{ or } x=5$$

The solutions are 3 and 5.

75. To find the perimeter we add the lengths of the sides:

$$\frac{y+4}{3}+\frac{y+4}{3}+\frac{y-2}{5}+\frac{y-2}{5} \quad \text{LCD} = 3\cdot 5$$
$$=\frac{y+4}{3}\cdot\frac{5}{5}+\frac{y+4}{3}\cdot\frac{5}{5}+\frac{y-2}{5}\cdot\frac{3}{3}+\frac{y-2}{5}\cdot\frac{3}{3}$$
$$=\frac{5y+20+5y+20+3y-6+3y-6}{3\cdot 5}$$
$$=\frac{16y+28}{15}$$

To find the area we multiply the length and the width:

$$\left(\frac{y+4}{3}\right)\left(\frac{y-2}{5}\right)=\frac{(y+4)(y-2)}{3\cdot 5}=\frac{y^2+2y-8}{15}$$

77.
$$\frac{5}{z+2}+\frac{4z}{z^2-4}+2$$
$$=\frac{5}{z+2}+\frac{4z}{(z+2)(z-2)}+\frac{2}{1} \quad \text{LCD} = (z+2)(z-2)$$
$$=\frac{5}{z+2}\cdot\frac{z-2}{z-2}+\frac{4z}{(z+2)(z-2)}+\frac{2}{1}\cdot\frac{(z+2)(z-2)}{(z+2)(z-2)}$$
$$=\frac{5z-10+4z+2(z^2-4)}{(z+2)(z-2)}$$
$$=\frac{5z-10+4z+2z^2-8}{(z+2)(z-2)}=\frac{2z^2+9z-18}{(z+2)(z-2)}$$
$$=\frac{(2z-3)(z+6)}{(z+2)(z-2)}$$

79.
$$\frac{3z^2}{z^4-4}+\frac{5z^2-3}{2z^4+z^2-6}$$
$$=\frac{3z^2}{(z^2+2)(z^2-2)}+\frac{5z^2-3}{(2z^2-3)(z^2+2)}$$
$$\text{LCD} = (z^2+2)(z^2-2)(2z^2-3)$$
$$=\frac{3z^2}{(z^2+2)(z^2-2)}\cdot\frac{2z^2-3}{2z^2-3}+\frac{5z^2-3}{(2z^2-3)(z^2+2)}\cdot\frac{z^2-2}{z^2-2}$$
$$=\frac{6z^4-9z^2+5z^4-13z^2+6}{(z^2+2)(z^2-2)(2z^2-3)}$$
$$=\frac{11z^4-22z^2+6}{(z^2+2)(z^2-2)(2z^2-3)}$$

Exercise Set 14.5

1. $\frac{7}{x}-\frac{3}{x}=\frac{7-3}{x}=\frac{4}{x}$

3. $\frac{y}{y-4}-\frac{4}{y-4}=\frac{y-4}{y-4}=1$

5.
$$\frac{2x-3}{x^2+3x-4}-\frac{x-7}{x^2+3x-4}$$
$$=\frac{2x-3-(x-7)}{x^2+3x-4}$$
$$=\frac{2x-3-x+7}{x^2+3x-4}$$
$$=\frac{x+4}{x^2+3x-4}$$
$$=\frac{x+4}{(x+4)(x-1)}$$
$$=\frac{\cancel{(x+4)}\cdot 1}{\cancel{(x+4)}(x-1)}$$
$$=\frac{1}{x-1}$$

7.
$$\frac{a-2}{10}-\frac{a+1}{5}=\frac{a-2}{10}-\frac{a+1}{5}\cdot\frac{2}{2} \quad \text{LCD} = 10$$
$$=\frac{a-2}{10}-\frac{2(a+1)}{10}$$
$$=\frac{(a-2)-2(a+1)}{10}$$
$$=\frac{a-2-2a-2}{10}$$
$$=\frac{-a-4}{10}$$

9.
$$\frac{4z-9}{3z}-\frac{3z-8}{4z}=\frac{4z-9}{3z}\cdot\frac{4}{4}-\frac{3z-8}{4z}\cdot\frac{3}{3}$$
$$\text{LCD} = 3\cdot 4\cdot z, \text{ or } 12z$$
$$=\frac{16z-36}{12z}-\frac{9z-24}{12z}$$
$$=\frac{16z-36-(9z-24)}{12z}$$
$$=\frac{16z-36-9z+24}{12z}$$
$$=\frac{7z-12}{12z}$$

11. $\dfrac{4x+2t}{3xt^2}-\dfrac{5x-3t}{x^2t}$ LCD $=3x^2t^2$

$=\dfrac{4x+2t}{3xt^2}\cdot\dfrac{x}{x}-\dfrac{5x-3t}{x^2t}\cdot\dfrac{3t}{3t}$

$=\dfrac{4x^2+2tx}{3x^2t^2}-\dfrac{15xt-9t^2}{3x^2t^2}$

$=\dfrac{4x^2+2tx-(15xt-9t^2)}{3x^2t^2}$

$=\dfrac{4x^2+2tx-15xt+9t^2}{3x^2t^2}$

$=\dfrac{4x^2-13xt+9t^2}{3x^2t^2}$

13. $\dfrac{5}{x+5}-\dfrac{3}{x-5}$ LCD $=(x+5)(x-5)$

$=\dfrac{5}{x+5}\cdot\dfrac{x-5}{x-5}-\dfrac{3}{x-5}\cdot\dfrac{x+5}{x+5}$

$=\dfrac{5x-25}{(x+5)(x-5)}-\dfrac{3x+15}{(x+5)(x-5)}$

$=\dfrac{5x-25-(3x+15)}{(x+5)(x-5)}$

$=\dfrac{5x-25-3x-15}{(x+5)(x-5)}$

$=\dfrac{2x-40}{(x+5)(x-5)}$

$=\dfrac{2(x-20)}{(x+5)(x-5)}$

15. $\dfrac{3}{2t^2-2t}-\dfrac{5}{2t-2}$

$=\dfrac{3}{2t(t-1)}-\dfrac{5}{2(t-1)}$ LCD $=2t(t-1)$

$=\dfrac{3}{2t(t-1)}-\dfrac{5}{2(t-1)}\cdot\dfrac{t}{t}$

$=\dfrac{3}{2t(t-1)}-\dfrac{5t}{2t(t-1)}$

$=\dfrac{3-5t}{2t(t-1)}$

17. $\dfrac{2s}{t^2-s^2}-\dfrac{s}{t-s}$ LCD $=(t-s)(t+s)$

$=\dfrac{2s}{(t-s)(t+s)}-\dfrac{s}{t-s}\cdot\dfrac{t+s}{t+s}$

$=\dfrac{2s}{(t-s)(t+s)}-\dfrac{st+s^2}{(t-s)(t+s)}$

$=\dfrac{2s-(st+s^2)}{(t-s)(t+s)}$

$=\dfrac{2s-st-s^2}{(t-s)(t+s)}$

$=\dfrac{s(2-t-s)}{(t-s)(t+s)}$

19. $\dfrac{y-5}{y}-\dfrac{3y-1}{4y}=\dfrac{y-5}{y}\cdot\dfrac{4}{4}-\dfrac{3y-1}{4y}$ LCD $=4y$

$=\dfrac{4y-20}{4y}-\dfrac{3y-1}{4y}$

$=\dfrac{4y-20-(3y-1)}{4y}$

$=\dfrac{4y-20-3y+1}{4y}$

$=\dfrac{y-19}{4y}$

21. $\dfrac{a}{x+a}-\dfrac{a}{x-a}$ LCD $=(x+a)(x-a)$

$=\dfrac{a}{x+a}\cdot\dfrac{x-a}{x-a}-\dfrac{a}{x-a}\cdot\dfrac{x+a}{x+a}$

$=\dfrac{ax-a^2}{(x+a)(x-a)}-\dfrac{ax+a^2}{(x+a)(x-a)}$

$=\dfrac{ax-a^2-(ax+a^2)}{(x+a)(x-a)}$

$=\dfrac{ax-a^2-ax-a^2}{(x+a)(x-a)}$

$=\dfrac{-2a^2}{(x+a)(x-a)}$

23. $\dfrac{11}{6}-\dfrac{5}{-6}=\dfrac{11}{6}-\dfrac{5}{-6}\cdot\dfrac{-1}{-1}$

$=\dfrac{11}{6}-\dfrac{-5}{6}$

$=\dfrac{11-(-5)}{6}$

$=\dfrac{11+5}{6}$

$=\dfrac{16}{6}$

$=\dfrac{8}{3}$

25. $\dfrac{5}{a}-\dfrac{8}{-a}=\dfrac{5}{a}-\dfrac{8}{-a}\cdot\dfrac{-1}{-1}$

$=\dfrac{5}{a}-\dfrac{-8}{a}$

$=\dfrac{5-(-8)}{a}$

$=\dfrac{5+8}{a}$

$=\dfrac{13}{a}$

27. $\dfrac{4}{y-1}-\dfrac{4}{1-y}=\dfrac{4}{y-1}-\dfrac{4}{1-y}\cdot\dfrac{-1}{-1}$

$=\dfrac{4}{y-1}-\dfrac{4(-1)}{(1-y)(-1)}$

$=\dfrac{4}{y-1}-\dfrac{-4}{y-1}$

$=\dfrac{4-(-4)}{y-1}$

$=\dfrac{4+4}{y-1}$

$=\dfrac{8}{y-1}$

29. $\dfrac{3-x}{x-7}-\dfrac{2x-5}{7-x}=\dfrac{3-x}{x-7}-\dfrac{2x-5}{7-x}\cdot\dfrac{-1}{-1}$

$=\dfrac{3-x}{x-7}-\dfrac{(2x-5)(-1)}{(7-x)(-1)}$

$=\dfrac{3-x}{x-7}-\dfrac{5-2x}{x-7}$

$=\dfrac{(3-x)-(5-2x)}{x-7}$

$=\dfrac{3-x-5+2x}{x-7}$

$=\dfrac{x-2}{x-7}$

31. $\dfrac{a-2}{a^2-25}-\dfrac{6-a}{25-a^2}=\dfrac{a-2}{a^2-25}-\dfrac{6-a}{25-a^2}\cdot\dfrac{-1}{-1}$

$=\dfrac{a-2}{a^2-25}-\dfrac{(6-a)(-1)}{(25-a^2)(-1)}$

$=\dfrac{a-2}{a^2-25}-\dfrac{a-6}{a^2-25}$

$=\dfrac{(a-2)-(a-6)}{a^2-25}$

$=\dfrac{a-2-a+6}{a^2-25}$

$=\dfrac{4}{a^2-25}=\dfrac{4}{(a+5)(a-5)}$

33. $\dfrac{4-x}{x-9}-\dfrac{3x-8}{9-x}=\dfrac{4-x}{x-9}-\dfrac{3x-8}{9-x}\cdot\dfrac{-1}{-1}$

$=\dfrac{4-x}{x-9}-\dfrac{8-3x}{x-9}$

$=\dfrac{(4-x)-(8-3x)}{x-9}$

$=\dfrac{4-x-8+3x}{x-9}$

$=\dfrac{2x-4}{x-9}=\dfrac{2(x-2)}{x-9}$

35. $\dfrac{5x}{x^2-9}-\dfrac{4}{3-x}$

$=\dfrac{5x}{(x+3)(x-3)}-\dfrac{4}{3-x}$ $x-3$ and $3-x$ are opposites

$=\dfrac{5x}{(x+3)(x-3)}-\dfrac{4}{3-x}\cdot\dfrac{-1}{-1}$

$=\dfrac{5x}{(x+3)(x-3)}-\dfrac{-4}{x-3}$ LCD $=(x+3)(x-3)$

$=\dfrac{5x}{(x+3)(x-3)}-\dfrac{-4}{x-3}\cdot\dfrac{x+3}{x+3}$

$=\dfrac{5x}{(x+3)(x-3)}-\dfrac{-4x-12}{(x+3)(x-3)}$

$=\dfrac{5x-(-4x-12)}{(x+3)(x-3)}$

$=\dfrac{5x+4x+12}{(x+3)(x-3)}$

$=\dfrac{9x+12}{(x+3)(x-3)}$

$=\dfrac{3(3x+4)}{(x+3)(x-3)}$

37. $\dfrac{t^2}{2t^2-2t}-\dfrac{1}{2t-2}$

$=\dfrac{t^2}{2t(t-1)}-\dfrac{1}{2(t-1)}$ LCD $=2t(t-1)$

$=\dfrac{t^2}{2t(t-1)}-\dfrac{1}{2(t-1)}\cdot\dfrac{t}{t}$

$=\dfrac{t^2}{2t(t-1)}-\dfrac{t}{2t(t-1)}$

$=\dfrac{t^2-t}{2t(t-1)}$

$=\dfrac{t(t-1)}{2t(t-1)}$

$=\dfrac{\cancel{t}\cancel{(t-1)}(1)}{2\cancel{t}\cancel{(t-1)}}$

$=\dfrac{1}{2}$

39. $\dfrac{x}{x^2+5x+6}-\dfrac{2}{x^2+3x+2}$

$=\dfrac{x}{(x+3)(x+2)}-\dfrac{2}{(x+2)(x+1)}$

LCD $=(x+3)(x+2)(x+1)$

$=\dfrac{x}{(x+3)(x+2)}\cdot\dfrac{x+1}{x+1}-\dfrac{2}{(x+2)(x+1)}\cdot\dfrac{x+3}{x+3}$

$=\dfrac{x^2+x}{(x+3)(x+2)(x+1)}-\dfrac{2x+6}{(x+3)(x+2)(x+1)}$

$=\dfrac{x^2+x-(2x+6)}{(x+3)(x+2)(x+1)}$

$=\dfrac{x^2+x-2x-6}{(x+3)(x+2)(x+1)}$

$=\dfrac{x^2-x-6}{(x+3)(x+2)(x+1)}$

$=\dfrac{(x-3)(x+2)}{(x+3)(x+2)(x+1)}$

$=\dfrac{(x-3)\cancel{(x+2)}}{(x+3)\cancel{(x+2)}(x+1)}$

$=\dfrac{x-3}{(x+3)(x+1)}$

41. $\dfrac{3(2x+5)}{x-1}-\dfrac{3(2x-3)}{1-x}+\dfrac{6x+1}{x-1}$

$=\dfrac{3(2x+5)}{x-1}-\dfrac{3(2x-3)}{1-x}\cdot\dfrac{-1}{-1}+\dfrac{6x-1}{x-1}$

$=\dfrac{3(2x+5)}{x-1}-\dfrac{-3(2x-3)}{x-1}+\dfrac{6x-1}{x-1}$

$=\dfrac{(6x+15)-(-6x+9)+(6x-1)}{x-1}$

$=\dfrac{6x+15+6x-9+6x-1}{x-1}$

$=\dfrac{18x+5}{x-1}$

43. $\dfrac{x-y}{x^2-y^2}+\dfrac{x+y}{x^2-y^2}-\dfrac{2x}{x^2-y^2}$

$=\dfrac{x-y+x+y-2x}{x^2-y^2}$

$=\dfrac{0}{x^2-y^2}$

$=0$

45. $\dfrac{2(x-1)}{2x-3}-\dfrac{3(x+2)}{2x-3}-\dfrac{x-1}{3-2x}$

$=\dfrac{2(x-1)}{2x-3}-\dfrac{3(x+2)}{2x-3}-\dfrac{x-1}{3-2x}\cdot\dfrac{-1}{-1}$

$=\dfrac{2(x-1)}{2x-3}-\dfrac{3(x+2)}{2x-3}-\dfrac{1-x}{2x-3}$

$=\dfrac{(2x-2)-(3x+6)-(1-x)}{2x-3}$

$=\dfrac{2x-2-3x-6-1+x}{2x-3}$

$=\dfrac{-9}{2x-3}$

47. $\dfrac{10}{2y-1}-\dfrac{6}{1-2y}+\dfrac{y}{2y-1}+\dfrac{y-4}{1-2y}$

$=\dfrac{10}{2y-1}-\dfrac{6}{1-2y}\cdot\dfrac{-1}{-1}+\dfrac{y}{2y-1}+\dfrac{y-4}{1-2y}\cdot\dfrac{-1}{-1}$

$=\dfrac{10}{2y-1}-\dfrac{-6}{2y-1}+\dfrac{y}{2y-1}+\dfrac{4-y}{2y-1}$

$=\dfrac{10-(-6)+y+4-y}{2y-1}$

$=\dfrac{10+6+y+4-y}{2y-1}$

$=\dfrac{20}{2y-1}$

49. $\dfrac{a+6}{4-a^2}-\dfrac{a+3}{a+2}+\dfrac{a-3}{2-a}$

$=\dfrac{a+6}{(2+a)(2-a)}-\dfrac{a+3}{2+a}+\dfrac{a-3}{2-a}$

$a+2=2+a$; LCD $=(2+a)(2-a)$

$=\dfrac{a+6}{(2+a)(2-a)}-\dfrac{a+3}{2+a}\cdot\dfrac{2-a}{2-a}+\dfrac{a-3}{2-a}\cdot\dfrac{2+a}{2+a}$

$=\dfrac{(a+6)-(a+3)(2-a)+(a-3)(2+a)}{(2+a)(2-a)}$

$=\dfrac{a+6-(-a^2-a+6)+(a^2-a-6)}{(2+a)(2-a)}$

$=\dfrac{a+6+a^2+a-6+a^2-a-6}{(2+a)(2-a)}$

$=\dfrac{2a^2+a-6}{(2+a)(2-a)}$

$=\dfrac{(2a-3)(a+2)}{(2+a)(2-a)}$

$=\dfrac{(2a-3)\cancel{(2+a)}}{\cancel{(2+a)}(2-a)}$

$=\dfrac{2a-3}{2-a}$

51. $\frac{2z}{1-2z}+\frac{3z}{2z+1}-\frac{3}{4z^2-1}$

$=\frac{2z}{1-2z}\cdot\frac{-1}{-1}+\frac{3z}{2z+1}-\frac{3}{4z^2-1}$

$=\frac{-2z}{2z-1}+\frac{3z}{2z+1}-\frac{3}{(2z-1)(2z+1)}$

LCD $=(2z-1)(2z+1)$

$=\frac{-2z}{2z-1}\cdot\frac{2z+1}{2z+1}+\frac{3z}{2z+1}\cdot\frac{2z-1}{2z-1}-\frac{3}{(2z-1)(2z+1)}$

$=\frac{(-4z^2-2z)+(6z^2-3z)-3}{(2z-1)(2z+1)}$

$=\frac{2z^2-5z-3}{(2z-1)(2z+1)}$

$=\frac{(z-3)(2z+1)}{(2z-1)(2z+1)}$

$=\frac{(z-3)\cancel{(2z+1)}}{(2z-1)\cancel{(2z+1)}}$

$=\frac{z-3}{2z-1}$

53. $\frac{1}{x+y}-\frac{1}{x-y}+\frac{2x}{x^2-y^2}$

$=\frac{1}{x+y}-\frac{1}{x-y}+\frac{2x}{(x+y)(x-y)}$

LCD $=(x+y)(x-y)$

$=\frac{1}{x+y}\cdot\frac{x-y}{x-y}-\frac{1}{x-y}\cdot\frac{x+y}{x+y}+\frac{2x}{(x+y)(x-y)}$

$=\frac{x-y-(x+y)+2x}{(x+y)(x-y)}$

$=\frac{x-y-x-y+2x}{(x+y)(x-y)}$

$=\frac{2x-2y}{(x+y)(x-y)}$

$=\frac{2(x-y)}{(x+y)(x-y)}$

$=\frac{2\cancel{(x-y)}}{(x+y)\cancel{(x-y)}}$

$=\frac{2}{x+y}$

55. $\frac{x^8}{x^3}=x^{8-3}=x^5$

57. $(a^2b^{-5})^{-4}=a^{2(-4)}b^{-5(-4)}=a^{-8}b^{20}=\frac{b^{20}}{a^8}$

59. $\frac{66x^2}{11x^5}=\frac{6\cdot\cancel{11}\cdot\cancel{x^2}}{\cancel{11}\cdot\cancel{x^2}\cdot x^3}=\frac{6}{x^3}$

61. $\frac{4}{7}+3x=\frac{1}{2}x-\frac{3}{14}$, LCD is 14

$14\left(\frac{4}{7}+3x\right)=14\left(\frac{1}{2}x-\frac{3}{14}\right)$

$14\cdot\frac{4}{7}+14\cdot 3x=14\cdot\frac{1}{2}x-14\cdot\frac{3}{14}$

$8+42x=7x-3$

$8+35x=-3$

$35x=-11$

$x=-\frac{11}{35}$

The solution is $-\frac{11}{35}$.

63. The shaded area has dimensions $x-6$ by $x-3$. Then the area is $(x-6)(x-3)$, or $x^2-9x+18$.

65. $\frac{2x+11}{x-3}\cdot\frac{3}{x+4}+\frac{2x+1}{4+x}\cdot\frac{3}{3-x}$

$=\frac{6x+33}{(x-3)(x+4)}+\frac{6x+3}{(4+x)(3-x)}$

$=\frac{6x+33}{(x-3)(x+4)}+\frac{6x+3}{(4+x)(3-x)}\cdot\frac{-1}{-1}$

$=\frac{6x+33}{(x-3)(x+4)}+\frac{-6x-3}{(x+4)(x-3)}$

$=\frac{6x+33-6x-3}{(x-3)(x+4)}$

$=\frac{30}{(x-3)(x+4)}$

67. $\frac{x}{x^4-y^4}-\left(\frac{1}{x+y}\right)^2$

$=\frac{x}{(x^2+y^2)(x+y)(x-y)}-\frac{1}{(x+y)^2}$

LCD $=(x^2+y^2)(x+y)^2(x-y)$

$=\frac{x}{(x^2+y^2)(x+y)(x-y)}\cdot\frac{x+y}{x+y}-\frac{1}{(x+y)^2}\cdot\frac{(x^2+y^2)(x-y)}{(x^2+y^2)(x-y)}$

$=\frac{x(x+y)-(x^2+y^2)(x-y)}{(x^2+y^2)(x+y)^2(x-y)}$

$=\frac{x^2+xy-(x^3-x^2y+xy^2-y^3)}{(x^2+y^2)(x+y)^2(x-y)}$

$=\frac{x^2+xy-x^3+x^2y-xy^2+y^3}{(x^2+y^2)(x+y)^2(x-y)}$

69. Let $l =$ the length of the missing side.

$$\frac{a^2-5a-9}{a-6}+\frac{a^2-6}{a-6}+l=2a+5$$
$$\frac{2a^2-5a-15}{a-6}+l=2a+5$$
$$l=2a+5-\frac{2a^2-5a-15}{a-6}$$
$$l=\left(2a+5\right)\cdot\frac{a-6}{a-6}-\frac{2a^2-5a-15}{a-6}$$
$$l=\frac{2a^2-7a-30}{a-6}-\frac{2a^2-5a-15}{a-6}$$
$$l=\frac{2a^2-7a-30-(2a^2-5a-15)}{a-6}$$
$$l=\frac{2a^2-7a-30-2a^2+5a+15}{a-6}$$
$$l=\frac{-2a-15}{a-6}$$

The length of the missing side is $\frac{-2a-15}{a-6}$.

Now find the area.

$$A=\frac{1}{2}\cdot b\cdot h$$
$$A=\frac{1}{2}\left(\frac{-2a-15}{a-6}\right)\left(\frac{a^2-6}{a-6}\right)$$
$$A=\frac{(-2a-15)(a^2-6)}{2(a-6)^2}\text{, or}$$
$$A=\frac{-2a^3-15a^2+12a+90}{2a^2-24a+72}$$

Chapter 14 Mid-Chapter Review

1. False; the reciprocal of $\frac{3-w}{w+2}$ is $\frac{w+2}{3-w}$, not $\frac{w-3}{w+2}$.

2. True; we must avoid *denominators* of 0. The numerator has no bearing on whether a rational expression is defined.

3. True; see pages 1053 and 1061 in the text.

4. False; x is a factor of the numerator, but it is a *term* of the denominator.

5. True; see page 1049 in the text.

6.
$$\frac{x-1}{x-2}-\frac{x+1}{x+2}-\frac{x-6}{4-x^2}$$
$$=\frac{x-1}{x-2}-\frac{x+1}{x+2}-\frac{x-6}{4-x^2}\cdot\frac{-1}{-1}$$
$$=\frac{x-1}{x-2}-\frac{x+1}{x+2}-\frac{6-x}{x^2-4}$$
$$=\frac{x-1}{x-2}-\frac{x+1}{x+2}-\frac{6-x}{(x-2)(x+2)}$$
$$=\frac{x-1}{x-2}\cdot\frac{x+2}{x+2}-\frac{x+1}{x+2}\cdot\frac{x-2}{x-2}-\frac{6-x}{(x-2)(x+2)}$$
$$=\frac{x^2+x-2}{(x-2)(x+2)}-\frac{x^2-x-2}{(x-2)(x+2)}-\frac{6-x}{(x-2)(x+2)}$$
$$=\frac{x^2+x-2-(x^2-x-2)-(6-x)}{(x-2)(x+2)}$$
$$=\frac{x^2+x-2-x^2+x+2-6+x}{(x-2)(x+2)}$$
$$=\frac{3x-6}{(x-2)(x+2)}$$
$$=\frac{3(x-2)}{(x-2)(x+2)}=\frac{x-2}{x-2}\cdot\frac{3}{x+2}=\frac{3}{x+2}$$

7. $\frac{t^2-16}{3}$

Since the denominator is the constant 3, there are no numbers for which the expression is not defined.

8. $\frac{x-8}{x^2-11x+24}$

Set the denominator equal to 0 and solve for x.

$$x^2-11x+24=0$$
$$(x-3)(x-8)=0$$
$$x-3=0 \quad or \quad x-8=0$$
$$x=3 \quad or \quad x=8$$

The expression is not defined for the numbers 3 and 8.

9. $\frac{7}{2w-7}$

Set the denominator equal to 0 and solve for x.

$$2w-7=0$$
$$2w=7$$
$$w=\frac{7}{2}$$

The expression is not defined for the number $\frac{7}{2}$.

10.
$$\frac{x^2+2x+3}{x^2-9}=\frac{(x+1)(x+3)}{(x+3)(x-3)}$$
$$=\frac{x+3}{x+3}\cdot\frac{x+1}{x-3}$$
$$=1\cdot\frac{x+1}{x-3}$$
$$=\frac{x+1}{x-3}$$

11. $\dfrac{6y^2+12y-48}{3y^2-9y+6}=\dfrac{6(y^2+2y-8)}{3(y^2-3y+2)}$

$=\dfrac{2\cdot 3(y+4)(y-2)}{3(y-2)(y-1)}$

$=\dfrac{3(y-2)}{3(y-2)}\cdot\dfrac{2(y+4)}{y-1}$

$=1\cdot\dfrac{2(y+4)}{y-1}$

$=\dfrac{2(y+4)}{y-1}$

12. $\dfrac{r-s}{s-r}=\dfrac{-1(-r+s)}{s-r}=\dfrac{-1(s-r)}{s-r}$

$=-1\cdot\dfrac{s-r}{s-r}$

$=-1\cdot 1=-1$

13. The reciprocal of $-x+3$ is $\dfrac{1}{-x+3}$, because $-x+3\cdot\dfrac{1}{-x+3}=1$. Since $-x+3$ can be written equivalently as $3-x$, we can also write the reciprocal as $\dfrac{1}{3-x}$.

14. $x^2-100=(x+10)(x-10)$

$10x^3=2\cdot 5\cdot x\cdot x\cdot x$

$x^2-20x+100=(x-10)(x-10)$

$\text{LCM}=2\cdot 5\cdot x\cdot x\cdot x(x+10)(x-10)(x-10)=$

$10x^3(x+10)(x-10)^2$

15. $\dfrac{a^2-a-2}{a^2-a-6}\div\dfrac{a^2-2a}{2a+a^2}=\dfrac{a^2-a-2}{a^2-a-6}\cdot\dfrac{2a+a^2}{a^2-2a}$

$=\dfrac{(a-2)(a+1)(a)(2+a)}{(a-3)(a+2)(a)(a-2)}$

$=\dfrac{\cancel{(a-2)}(a+1)\cancel{(a)}\cancel{(2+a)}}{(a-3)\cancel{(a+2)}\cancel{(a)}\cancel{(a-2)}}$

$=\dfrac{a+1}{a-3}$

16. $\dfrac{3y}{y^2-7y+10}-\dfrac{2y}{y^2-8y+15}$

$=\dfrac{3y}{(y-2)(y-5)}-\dfrac{2y}{(y-3)(y-5)}$,

LCD is $(y-2)(y-5)(y-3)$

$=\dfrac{3y}{(y-2)(y-5)}\cdot\dfrac{y-3}{y-3}-\dfrac{2y}{(y-3)(y-5)}\cdot\dfrac{y-2}{y-2}$

$=\dfrac{3y(y-3)-2y(y-2)}{(y-2)(y-5)(y-3)}$

$=\dfrac{3y^2-9y-2y^2+4y}{(y-2)(y-5)(y-3)}$

$=\dfrac{y^2-5y}{(y-2)(y-5)(y-3)}$

$=\dfrac{y(y-5)}{(y-2)(y-5)(y-3)}=\dfrac{y-5}{y-5}\cdot\dfrac{y}{(y-2)(y-3)}$

$=\dfrac{y}{(y-2)(y-3)}$

17. $\dfrac{x^2}{x-11}+\dfrac{121}{11-x}=\dfrac{x^2}{x-11}+\dfrac{121}{11-x}\cdot\dfrac{-1}{-1}$

$=\dfrac{x^2}{x-11}+\dfrac{-121}{x-11}$

$=\dfrac{x^2-121}{x-11}$

$=\dfrac{(x+11)(x-11)}{x-11}=\dfrac{x+11}{1}\cdot\dfrac{x-11}{x-11}$

$=x+11$

18. $\dfrac{x^2-y^2}{(x-y)^2}\cdot\dfrac{1}{x+y}=\dfrac{(x+y)(x-y)\cdot 1}{(x-y)^2(x+y)}$

$=\dfrac{\cancel{(x+y)}\cancel{(x-y)}\cdot 1}{\cancel{(x-y)}(x-y)\cancel{(x+y)}}$

$=\dfrac{1}{x-y}$

19. $\dfrac{3a-b}{a^2b}+\dfrac{a+2b}{ab^2}$, LCD is a^2b^2

$=\dfrac{3a-b}{a^2b}\cdot\dfrac{b}{b}+\dfrac{a+2b}{ab^2}\cdot\dfrac{a}{a}$

$=\dfrac{(3a-b)b+(a+2b)a}{a^2b^2}$

$=\dfrac{3ab-b^2+a^2+2ab}{a^2b^2}$

$=\dfrac{a^2+5ab-b^2}{a^2b^2}$

20. $\dfrac{5x}{x^2-4}-\dfrac{3}{x}+\dfrac{4}{x+2}$

$=\dfrac{5x}{(x+2)(x-2)}-\dfrac{3}{x}+\dfrac{4}{x+2}$, LCD is $x(x+2)(x-2)$

$=\dfrac{5x}{(x+2)(x-2)}\cdot\dfrac{x}{x}-\dfrac{3}{x}\cdot\dfrac{(x+2)(x-2)}{(x+2)(x-2)}+\dfrac{4}{x+2}\cdot\dfrac{x(x-2)}{x(x-2)}$

$=\dfrac{5x^2-3(x^2-4)+4x(x-2)}{x(x+2)(x-2)}$

$=\dfrac{5x^2-3x^2+12+4x^2-8x}{x(x+2)(x-2)}$

$=\dfrac{6x^2-8x+12}{x(x+2)(x-2)}$

$=\dfrac{2(3x^2-4x+6)}{x(x+2)(x-2)}$

21. $\dfrac{2}{x-2}\div\dfrac{1}{x+3}=\dfrac{2}{x-2}\cdot\dfrac{x+3}{1}=\dfrac{2(x+3)}{x-2}$

Answer E is correct.

22. $\dfrac{1}{x+3}-\dfrac{2}{x-2}=\dfrac{1}{x+3}\cdot\dfrac{x-2}{x-2}-\dfrac{2}{x-2}\cdot\dfrac{x+3}{x+3}$

$=\dfrac{x-2-2(x+3)}{(x+3)(x-2)}$

$=\dfrac{x-2-2x-6}{(x+3)(x-2)}$

$=\dfrac{-x-8}{(x+3)(x-2)}$

Answer A is correct.

23. $\dfrac{2}{x-2}-\dfrac{1}{x+3}=\dfrac{2}{x-2}\cdot\dfrac{x+3}{x+3}-\dfrac{1}{x+3}\cdot\dfrac{x-2}{x-2}$
$=\dfrac{2(x+3)-(x-2)}{(x-2)(x+3)}$
$=\dfrac{2x+6-x+2}{(x-2)(x+3)}$
$=\dfrac{x+8}{(x-2)(x+3)}$

Answer D is correct.

24. $\dfrac{1}{x+3}\div\dfrac{2}{x-2}=\dfrac{1}{x+3}\cdot\dfrac{x-2}{2}=\dfrac{x-2}{2(x+3)}$

Answer B is correct.

25. $\dfrac{2}{x-2}+\dfrac{1}{x+3}=\dfrac{2}{x-2}\cdot\dfrac{x+3}{x+3}+\dfrac{1}{x+3}\cdot\dfrac{x-2}{x-2}$
$=\dfrac{2(x+3)+x-2}{(x-2)(x+3)}$
$=\dfrac{2x+6+x-2}{(x-2)(x+3)}$
$=\dfrac{3x+4}{(x-2)(x+3)}$

Answer F is correct.

26. $\dfrac{2}{x-2}\cdot\dfrac{1}{x+3}=\dfrac{2}{(x-2)(x+3)}$

Answer C is correct.

27. If the numbers have a common factor, their product contains that factor more than the greatest number of times it occurs in any one factorization. In this case, their product is not their least common multiple.

28. Yes; consider the product $\dfrac{a}{b}\cdot\dfrac{c}{d}=\dfrac{ac}{bd}$. The reciprocal of the product is $\dfrac{bd}{ac}$. This is equal to the product of the reciprocals of the two original factors: $\dfrac{b}{a}\cdot\dfrac{d}{c}=\dfrac{bd}{ac}$.

29. Although multiplying the denominators of the expressions being added results in a common denominator, it is often not the *least* common denominator. Using a common denominator other than the LCD makes the expressions more complicated, requires additional simplification after the addition has been performed, and leaves more room for error.

30. Their sum is zero. Another explanation is that $-\left(\dfrac{1}{3-x}\right)=\dfrac{1}{-(3-x)}=\dfrac{1}{x-3}$.

31. $\dfrac{x+3}{x-5}$ is undefined for $x=5$, $\dfrac{x-7}{x+1}$ is undefined for $x=-1$, and $\dfrac{x+1}{x-7}$ $\left(\text{the reciprocal of } \dfrac{x-7}{x+1}\right)$ is undefined for $x=7$.

32. The binomial is a factor of the trinomial.

Exercise Set 14.6

1. $\dfrac{4}{5}-\dfrac{2}{3}=\dfrac{x}{9}$, LCM $=45$

$$45\left(\frac{4}{5}-\frac{2}{3}\right)=45\cdot\frac{x}{9}$$
$$45\cdot\frac{4}{5}-45\cdot\frac{2}{3}=45\cdot\frac{x}{9}$$
$$36-30=5x$$
$$6=5x$$
$$\frac{6}{5}=x$$

Check:

$$\begin{array}{c|c} \frac{4}{5}-\frac{2}{3} & \frac{x}{9} \\ \hline \frac{4}{5}-\frac{2}{3} \ ? & \frac{6/5}{9} \\ \frac{12}{15}-\frac{10}{15} & \frac{6}{5}\cdot\frac{1}{9} \\ \frac{2}{15} & \frac{2}{15} \end{array} \quad \text{TRUE}$$

This checks, so the solution is $\dfrac{6}{5}$.

3. $\dfrac{3}{5}+\dfrac{1}{8}=\dfrac{1}{x}$, LCM $=40x$

$$40x\left(\frac{3}{5}+\frac{1}{8}\right)=40x\cdot\frac{1}{x}$$
$$40x\cdot\frac{3}{5}+40x\cdot\frac{1}{8}=40x\cdot\frac{1}{x}$$
$$24x+5x=40$$
$$29x=40$$
$$x=\frac{40}{29}$$

Check:

$$\begin{array}{c|c} \frac{3}{5}+\frac{1}{8} & \frac{1}{x} \\ \hline \frac{3}{5}+\frac{1}{8} \ ? & \frac{1}{40/29} \\ \frac{24}{40}+\frac{5}{40} & 1\cdot\frac{29}{40} \\ \frac{29}{40} & \frac{29}{40} \end{array} \quad \text{TRUE}$$

This checks, so the solution is $\dfrac{40}{29}$.

5.
$$\frac{3}{8}+\frac{4}{5}=\frac{x}{20},\ \text{LCM}=40$$
$$40\left(\frac{3}{8}+\frac{4}{5}\right)=40\cdot\frac{x}{20}$$
$$40\cdot\frac{3}{8}+40\cdot\frac{4}{5}=40\cdot\frac{x}{20}$$
$$15+32=2x$$
$$47=2x$$
$$\frac{47}{2}=x$$

Check:

$\frac{3}{8}+\frac{4}{5}=\frac{x}{20}$	
$\frac{3}{8}+\frac{4}{5}$?	$\frac{\frac{47}{2}}{20}$
$\frac{15}{40}+\frac{32}{40}$	$\frac{47}{2}\cdot\frac{1}{20}$
$\frac{47}{40}$	$\frac{47}{40}$ TRUE

This checks, so the solution is $\frac{47}{2}$.

7.
$$\frac{1}{x}=\frac{2}{3}-\frac{5}{6},\ \text{LCM}=6x$$
$$6x\cdot\frac{1}{x}=6x\left(\frac{2}{3}-\frac{5}{6}\right)$$
$$6x\cdot\frac{1}{x}=6x\cdot\frac{2}{3}-6x\cdot\frac{5}{6}$$
$$6=4x-5x$$
$$6=-x$$
$$-6=x$$

Check:

$\frac{1}{x}=\frac{2}{3}-\frac{5}{6}$	
$\frac{1}{-6}$?	$\frac{2}{3}-\frac{5}{6}$
$-\frac{1}{6}$	$\frac{4}{6}-\frac{5}{6}$
	$-\frac{1}{6}$ TRUE

This checks, so the solution is -6.

9.
$$\frac{1}{6}+\frac{1}{8}=\frac{1}{t},\ \text{LCM}=24t$$
$$24t\left(\frac{1}{6}+\frac{1}{8}\right)=24t\cdot\frac{1}{t}$$
$$24t\cdot\frac{1}{6}+24t\cdot\frac{1}{8}=24t\cdot\frac{1}{t}$$
$$4t+3t=24$$
$$7t=24$$
$$t=\frac{24}{7}$$

Check:

$\frac{1}{6}+\frac{1}{8}=\frac{1}{t}$	
$\frac{1}{6}+\frac{1}{8}$?	$\frac{1}{24/7}$
$\frac{4}{24}+\frac{3}{24}$	$1\cdot\frac{7}{24}$
$\frac{7}{24}$	$\frac{7}{24}$ TRUE

This checks, so the solution is $\frac{24}{7}$.

11.
$$x+\frac{4}{x}=-5,\ \text{LCM}=x$$
$$x\left(x+\frac{4}{x}\right)=x(-5)$$
$$x\cdot x+x\cdot\frac{4}{x}=x(-5)$$
$$x^2+4=-5x$$
$$x^2+5x+4=0$$
$$(x+4)(x+1)=0$$
$$x+4=0\quad\text{or}\quad x+1=0$$
$$x=-4\quad\text{or}\quad x=-1$$

Check:

$x+\frac{4}{x}=-5$	
$-4+\frac{4}{-4}$?	-5
$-4-1$	
-5	TRUE

$x+\frac{4}{x}=-5$	
$-1+\frac{4}{-1}$?	-5
$-1-4$	
-5	TRUE

Both of these check, so the two solutions are -4 and -1.

13.
$$\frac{x}{4}-\frac{4}{x}=0,\ \text{LCM}=4x$$
$$4x\left(\frac{x}{4}-\frac{4}{x}\right)=4x\cdot 0$$
$$4x\cdot\frac{x}{4}-4x\cdot\frac{4}{x}=4x\cdot 0$$
$$x^2-16=0$$
$$(x+4)(x-4)=0$$
$$x+4=0\quad\text{or}\quad x-4=0$$
$$x=-4\quad\text{or}\quad x=4$$

Check:

$\frac{x}{4}-\frac{4}{x}=0$	
$\frac{-4}{4}-\frac{4}{-4}$?	0
$-1-(-1)$	
$-1+1$	
0	TRUE

$\frac{x}{4}-\frac{4}{x}=0$	
$\frac{4}{4}-\frac{4}{4}$?	0
$1-1$	
0	TRUE

Both of these check, so the two solutions are -4 and 4.

15. $\frac{5}{x} = \frac{6}{x} - \frac{1}{3}$, LCM $= 3x$

$$3x \cdot \frac{5}{x} = 3x\left(\frac{6}{x} - \frac{1}{3}\right)$$
$$3x \cdot \frac{5}{x} = 3x \cdot \frac{6}{x} - 3x \cdot \frac{1}{3}$$
$$15 = 18 - x$$
$$-3 = -x$$
$$3 = x$$

Check:

$$\frac{5}{x} = \frac{6}{x} - \frac{1}{3}$$
$$\frac{5}{3} \ ? \ \frac{6}{3} - \frac{1}{3}$$
$$\frac{5}{3} \quad \text{TRUE}$$

This checks, so the solution is 3.

17. $\frac{5}{3x} + \frac{3}{x} = 1$, LCM $= 3x$

$$3x\left(\frac{5}{3x} + \frac{3}{x}\right) = 3x \cdot 1$$
$$3x \cdot \frac{5}{3x} + 3x \cdot \frac{3}{x} = 3x \cdot 1$$
$$5 + 9 = 3x$$
$$14 = 3x$$
$$\frac{14}{3} = x$$

Check:

$$\frac{5}{3x} + \frac{3}{x} = 1$$
$$\frac{5}{3 \cdot (14/3)} + \frac{3}{(14/3)} \ ? \ 1$$
$$\frac{5}{14} + \frac{9}{14}$$
$$\frac{14}{14}$$
$$1 \quad \text{TRUE}$$

This checks, so the solution is $\frac{14}{3}$.

19. $\frac{t-2}{t+3} = \frac{3}{8}$, LCM $= 8(t+3)$

$$8(t+3)\left(\frac{t-2}{t+3}\right) = 8(t+3)\left(\frac{3}{8}\right)$$
$$8(t-2) = 3(t+3)$$
$$8t - 16 = 3t + 9$$
$$5t = 25$$
$$t = 5$$

Check:

$$\frac{t-2}{t+3} = \frac{3}{8}$$
$$\frac{5-2}{5+3} \ ? \ \frac{3}{8}$$
$$\frac{3}{8} \quad \text{TRUE}$$

This checks, so the solution is 5.

21. $\frac{2}{x+1} = \frac{1}{x-2}$, LCM $= (x+1)(x-2)$

$$(x+1)(x-2) \cdot \frac{2}{x+1} = (x+1)(x-2) \cdot \frac{1}{x-2}$$
$$2(x-2) = x+1$$
$$2x - 4 = x + 1$$
$$x = 5$$

This checks, so the solution is 5.

23. $\frac{x}{6} - \frac{x}{10} = \frac{1}{6}$, LCM $= 30$

$$30\left(\frac{x}{6} - \frac{x}{10}\right) = 30 \cdot \frac{1}{6}$$
$$30 \cdot \frac{x}{6} - 30 \cdot \frac{x}{10} = 30 \cdot \frac{1}{6}$$
$$5x - 3x = 5$$
$$2x = 5$$
$$x = \frac{5}{2}$$

This checks, so the solution is $\frac{5}{2}$.

25. $\frac{t+2}{5} - \frac{t-2}{4} = 1$, LCM $= 20$

$$20\left(\frac{t+2}{5} - \frac{t-2}{4}\right) = 20 \cdot 1$$
$$20\left(\frac{t+2}{5}\right) - 20\left(\frac{t-2}{4}\right) = 20 \cdot 1$$
$$4(t+2) - 5(t-2) = 20$$
$$4t + 8 - 5t + 10 = 20$$
$$-t + 18 = 20$$
$$-t = 2$$
$$t = -2$$

This checks, so the solution is -2.

27.

$$\frac{5}{x-1} = \frac{3}{x+2},$$
$$\text{LCD} = (x-1)(x+2)$$
$$(x-1)(x+2)\cdot\frac{5}{x-1} = (x-1)(x+2)\cdot\frac{3}{x+2}$$
$$5(x+2) = 3(x-1)$$
$$5x+10 = 3x-3$$
$$2x = -13$$
$$x = -\frac{13}{2}$$

This checks, so the solution is $-\frac{13}{2}$.

29.

$$\frac{a-3}{3a+2} = \frac{1}{5}, \text{ LCM} = 5(3a+2)$$
$$5(3a+2)\cdot\frac{a-3}{3a+2} = 5(3a+2)\cdot\frac{1}{5}$$
$$5(a-3) = 3a+2$$
$$5a-15 = 3a+2$$
$$2a = 17$$
$$a = \frac{17}{2}$$

This checks, so the solution is $\frac{17}{2}$.

31.

$$\frac{x-1}{x-5} = \frac{4}{x-5}, \text{ LCM} = x-5$$
$$(x-5)\cdot\frac{x-1}{x-5} = (x-5)\cdot\frac{4}{x-5}$$
$$x-1 = 4$$
$$x = 5$$

The number 5 is not a solution because it makes a denominator zero. Thus, there is no solution.

33.

$$\frac{2}{x+3} = \frac{5}{x}, \text{ LCM} = x(x+3)$$
$$x(x+3)\cdot\frac{2}{x+3} = x(x+3)\cdot\frac{5}{x}$$
$$2x = 5(x+3)$$
$$2x = 5x+15$$
$$-15 = 3x$$
$$-5 = x$$

This checks, so the solution is -5.

35.

$$\frac{x-2}{x-3} = \frac{x-1}{x+1}, \text{ LCM} = (x-3)(x+1)$$
$$(x-3)(x+1)\cdot\frac{x-2}{x-3} = (x-3)(x+1)\cdot\frac{x-1}{x+1}$$
$$(x+1)(x-2) = (x-3)(x-1)$$
$$x^2-x-2 = x^2-4x+3$$
$$-x-2 = -4x+3$$
$$3x = 5$$
$$x = \frac{5}{3}$$

This checks, so the solution is $\frac{5}{3}$.

37.

$$\frac{1}{x+3} + \frac{1}{x-3} = \frac{1}{x^2-9},$$
$$\text{LCM} = (x+3)(x-3)$$
$$(x+3)(x-3)\left(\frac{1}{x+3}+\frac{1}{x-3}\right) = (x+3)(x-3)\cdot\frac{1}{(x+3)(x-3)}$$
$$(x-3)+(x+3) = 1$$
$$2x = 1$$
$$x = \frac{1}{2}$$

This checks, so the solution is $\frac{1}{2}$.

39.

$$\frac{x}{x+4} - \frac{4}{x-4} = \frac{x^2+16}{x^2-16},$$
$$\text{LCM} = (x+4)(x-4)$$
$$(x+4)(x-4)\left(\frac{x}{x+4}-\frac{4}{x-4}\right) = (x+4)(x-4)\cdot\frac{x^2+16}{(x+4)(x-4)}$$
$$x(x-4)-4(x+4) = x^2+16$$
$$x^2-4x-4x-16 = x^2+16$$
$$x^2-8x-16 = x^2+16$$
$$-8x-16 = 16$$
$$-8x = 32$$
$$x = -4$$

The number -4 is not a solution because it makes a denominator zero. Thus, there is no solution.

41.

$$\frac{4-a}{8-a} = \frac{4}{a-8} \qquad 8-a \text{ and } a-8 \text{ are opposites}$$
$$\frac{4-a}{8-a}\cdot\frac{-1}{-1} = \frac{4}{a-8}$$
$$\frac{a-4}{a-8} = \frac{4}{a-8}, \text{ LCM} = a-8$$
$$(a-8)\left(\frac{a-4}{a-8}\right) = (a-8)\left(\frac{4}{a-8}\right)$$
$$a-4 = 4$$
$$a = 8$$

The number 8 is not a solution because it makes a denominator zero. Thus, there is no solution.

43. $$2 - \frac{a-2}{a+3} = \frac{a^2-4}{a+3}, \text{ LCM} = a+3$$

$$(a+3)\left(2 - \frac{a-2}{a+3}\right) = (a+3) \cdot \frac{a^2-4}{a+3}$$
$$2(a+3) - (a-2) = a^2 - 4$$
$$2a + 6 - a + 2 = a^2 - 4$$
$$0 = a^2 - a - 12$$
$$0 = (a-4)(a+3)$$
$$a - 4 = 0 \text{ or } a + 3 = 0$$
$$a = 4 \text{ or } \quad a = -3$$

Only 4 checks, so the solution is 4.

45. $$\frac{x+1}{x+2} = \frac{x+3}{x+4},$$
$$\text{LCM} = (x+2)(x+4)$$

$$(x+2)(x+4)\left(\frac{x+1}{x+2}\right) = (x+2)(x+4)\left(\frac{x+3}{x+4}\right)$$
$$(x+4)(x+1) = (x+2)(x+3)$$
$$x^2 + 5x + 4 = x^2 + 5x + 6$$
$$4 = 6 \quad \text{Subtracting } x^2 \text{ and } 5x$$

We get a false equation, so the original equation has no solution.

47. $$4a - 3 = \frac{a+13}{a+1}, \text{ LCM} = a+1$$

$$(a+1)(4a-3) = (a+1) \cdot \frac{a+13}{a+1}$$
$$4a^2 + a - 3 = a + 13$$
$$4a^2 - 16 = 0$$
$$4(a+2)(a-2) = 0$$
$$a + 2 = 0 \quad \text{or} \quad a - 2 = 0$$
$$a = -2 \quad \text{or} \quad a = 2$$

Both of these check, so the two solutions are -2 and 2.

49. $$\frac{4}{y-2} - \frac{2y-3}{y^2-4} = \frac{5}{y+2},$$
$$\text{LCM} = (y+2)(y-2)$$

$$(y+2)(y-2)\left(\frac{4}{y-2} - \frac{2y-3}{(y+2)(y-2)}\right) = (y+2)(y-2) \cdot \frac{5}{y+2}$$
$$4(y+2) - (2y-3) = 5(y-2)$$
$$4y + 8 - 2y + 3 = 5y - 10$$
$$2y + 11 = 5y - 10$$
$$21 = 3y$$
$$7 = y$$

This checks, so the solution is 7.

51. A rational expression is a quotient of two polynomials.

53. Two expressions are reciprocals of each other if their product is 1.

55. To find the LCM, use each factor the greatest number of times that it appears in any one factorization.

57. The quotient rule asserts that when dividing with exponential notation, if the bases are the same, keep the base and subtract the exponent of the denominator from the exponent of the numerator.

59. $$\frac{x}{x^2+3x-4} + \frac{x+1}{x^2+6x+8} = \frac{2x}{x^2+x-2}$$
$$\frac{x}{(x+4)(x-1)} + \frac{x+1}{(x+4)(x+2)} = \frac{2x}{(x+2)(x-1)}$$
$$x(x+2) + (x+1)(x-1) = 2x(x+4)$$

Multiplying by the LCM, $(x+4)(x-1)(x+2)$

$$x^2 + 2x + x^2 - 1 = 2x^2 + 8x$$
$$2x^2 + 2x - 1 = 2x^2 + 8x$$
$$2x - 1 = 8x$$
$$-1 = 6x$$
$$-\frac{1}{6} = x$$

This checks, so the solution is $-\frac{1}{6}$.

61. Left to the student

Exercise Set 14.7

1. ***Familiarize***. The job takes Mandy 4 hours working alone and Omar 5 hours working alone. Then in 1 hour Mandy does $\frac{1}{4}$ of the job and Omar does $\frac{1}{5}$ of the job. Working together, they can do $\frac{1}{4} + \frac{1}{5}$, or $\frac{9}{20}$ of the job in 1 hour. In two hours, Mandy does $2\left(\frac{1}{4}\right)$ of the job and Omar does $2\left(\frac{1}{5}\right)$ of the job. Working together they can do $2\left(\frac{1}{4}\right) + 2\left(\frac{1}{5}\right)$, or $\frac{9}{10}$ of the job in 2 hours. In 3 hours they can do $3\left(\frac{1}{4}\right) + 3\left(\frac{1}{5}\right)$, or $1\frac{7}{20}$ of the job which is more of the job then needs to be done. The answer is somewhere between 2 hr and 3 hr.

Translate. If they work together t hours, then Mandy does $t\left(\frac{1}{4}\right)$ of the job and Omar does $t\left(\frac{1}{5}\right)$ of the job. We want some number t such that

$$t\left(\frac{1}{4}\right) + t\left(\frac{1}{5}\right) = 1, \text{ or } \frac{t}{4} + \frac{t}{5} = 1.$$

Solve. We solve the equation.

$$\frac{t}{4} + \frac{t}{5} = 1, \text{ LCM} = 20$$
$$20\left(\frac{t}{4} + \frac{t}{5}\right) = 20 \cdot 1$$
$$20 \cdot \frac{t}{4} + 20 \cdot \frac{t}{5} = 20$$
$$5t + 4t = 20$$
$$9t = 20$$
$$t = \frac{20}{9}, \text{ or } 2\frac{2}{9}$$

Check. The check can be done by repeating the computations. We also have a partial check in that we expected from our familiarization step that the answer would be between 2 hr and 3 hr.

State. Working together, it takes them $2\frac{2}{9}$ hr to complete the job.

3. ***Familiarize***. The job takes Vern 45 min working alone and Nina 60 min working alone. Then in 1 minute Vern does $\frac{1}{45}$ of the job and Nina does $\frac{1}{60}$ of the job. Working together, they can do $\frac{1}{45}+\frac{1}{60}$, or $\frac{7}{180}$ of the job in 1 minute. In 20 minutes, Vern does $\frac{20}{45}$ of the job and Nina does $\frac{20}{60}$ of the job. Working together, they can do $\frac{20}{45}+\frac{20}{60}$, or $\frac{7}{9}$ of the job. In 30 minutes, they can do $\frac{30}{45}+\frac{30}{60}$, or $\frac{7}{6}$ of the job which is more of the job than needs to be done. The answer is somewhere between 20 minutes and 30 minutes.

Translate. If they work together t minutes, then Vern does $t\left(\frac{1}{45}\right)$ of the job and Nina does $t\left(\frac{1}{60}\right)$ of the job. We want some number t such that

$$t\left(\frac{1}{45}\right)+t\left(\frac{1}{60}\right)=1, \text{ or } \frac{t}{45}+\frac{t}{60}=1.$$

Solve. We solve the equation.

$$\begin{aligned}\frac{t}{45}+\frac{t}{60}&=1, \text{ LCM}=180\\ 180\left(\frac{t}{45}+\frac{t}{60}\right)&=180\cdot 1\\ 180\cdot\frac{t}{45}+180\cdot\frac{t}{60}&=180\\ 4t+3t&=180\\ 7t&=180\\ t&=\frac{180}{7}, \text{ or } 25\frac{5}{7}\end{aligned}$$

Check. The check can be done by repeating the computations. We also have a partial check in that we expected from our familiarization step that the answer would be between 20 minutes and 30 minutes.

State. It would take them $25\frac{5}{7}$ minutes to complete the job working together.

5. ***Familiarize***. The job takes Peggyann 9 hours working alone and Matthew 7 hours working alone. Then in 1 hour Peggyann does $\frac{1}{9}$ of the job and Matthew does $\frac{1}{7}$ of the job. Working together they can do $\frac{1}{9}+\frac{1}{7}$, or $\frac{16}{63}$ of the job in 1 hour. In two hours, Peggyann does $2\left(\frac{1}{9}\right)$ of the job and Matthew does $2\left(\frac{1}{7}\right)$ of the job. Working together they can do $2\left(\frac{1}{9}\right)+2\left(\frac{1}{7}\right)$, or $\frac{32}{63}$ of the job in two hours. In five hours they can do $5\left(\frac{1}{9}\right)+5\left(\frac{1}{7}\right)$, or $\frac{80}{63}$, or $1\frac{17}{63}$ of the job which is more of the job than needs to be done. The answer is somewhere between 2 hr and 5 hr.

Translate. If they work together t hours, Peggyann does $t\left(\frac{1}{9}\right)$ of the job and Matthew does $t\left(\frac{1}{7}\right)$ of the job. We want some number t such that

$$t\left(\frac{1}{9}\right)+t\left(\frac{1}{7}\right)=1, \text{ or } \frac{t}{9}+\frac{t}{7}=1.$$

Solve. We solve the equation.

$$\begin{aligned}\frac{t}{9}+\frac{t}{7}&=1, \text{ LCM}=63\\ 63\left(\frac{t}{9}+\frac{t}{7}\right)&=63\cdot 1\\ 63\cdot\frac{t}{9}+63\cdot\frac{t}{7}&=63\\ 7t+9t&=63\\ 16t&=63\\ t&=\frac{63}{16}, \text{ or } 3\frac{15}{16}\end{aligned}$$

Check. The check can be done by repeating the computations. We also have a partial check in that we expected from our familiarization step that the answer would be between 2 hr and 5 hr.

State. Working together, it takes them $3\frac{15}{16}$ hr to complete the job.

7. ***Familiarize***. Let t = the number of minutes it takes Nicole and Glen to weed the garden, working together.

Translate. We use the work principle.

$$t\left(\frac{1}{50}\right)+t\left(\frac{1}{40}\right)=1, \text{ or } \frac{t}{50}+\frac{t}{40}=1$$

Solve. We solve the equation.

$$\begin{aligned}\frac{t}{50}+\frac{t}{40}&=1, \text{ LCM}=200\\ 200\left(\frac{t}{50}+\frac{t}{40}\right)&=200\cdot 1\\ 200\cdot\frac{t}{50}+200\cdot\frac{t}{40}&=200\\ 4t+5t&=200\\ 9t&=200\\ t&=\frac{200}{9}, \text{ or } 22\frac{2}{9}\end{aligned}$$

Check. In $\frac{200}{9}$ min, the portion of the job done is $\frac{1}{50}\cdot\frac{200}{9}+\frac{1}{40}\cdot\frac{200}{9}=\frac{4}{9}+\frac{5}{9}=1$. The answer checks.

State. It would take $22\frac{2}{9}$ min to weed the garden if Nicole and Glen worked together.

9. ***Familiarize***. Let t = the number of minutes it would take the two machines to make one copy of the report, working together.

Translate. We use the work principle.

$$t\left(\frac{1}{10}\right)+t\left(\frac{1}{6}\right)=1, \text{ or } \frac{t}{10}+\frac{t}{6}=1$$

Solve. We solve the equation.

$$\frac{t}{10}+\frac{t}{6}=1, \text{ LCM} = 30$$

$$30\left(\frac{t}{10}+\frac{t}{6}\right)=30\cdot 1$$

$$30\cdot\frac{t}{10}+30\cdot\frac{t}{6}=30$$

$$3t+5t=30$$

$$8t=30$$

$$t=\frac{15}{4}, \text{ or } 3\frac{3}{4}$$

Check. In $\frac{15}{4}$ min, the portion of the job done is

$\frac{1}{10}\cdot\frac{15}{4}+\frac{1}{6}\cdot\frac{15}{4}=\frac{3}{8}+\frac{5}{8}=1$. The answer checks.

State. It would take the two machines $3\frac{3}{4}$ min to make one copy of the report, working together.

11. ***Familiarize***. We complete the table shown in the text.

$d = r \cdot t$

	Distance	Speed	Time	
Car	150	r	t	$\rightarrow 150 = r(t)$
Truck	350	$r+40$	t	$\rightarrow 350 = (r+40)t$

Translate. We apply the formula $d = rt$ along the rows of the table to obtain two equations:

$150 = rt$,

$350 = (r+40)t$

Then we solve each equation for t and set the results equal:

Solving $150 = rt$ for t: $t = \frac{150}{r}$

Solving $350 = (r+40)t$ for t: $t = \frac{350}{r+40}$

Thus, we have

$$\frac{150}{r}=\frac{350}{r+40}.$$

Solve. We multiply by the LCM, $r(r+40)$.

$$r(r+40)\cdot\frac{150}{r}=r(r+40)\cdot\frac{350}{r+40}$$

$$150(r+40)=350r$$

$$150r+6000=350r$$

$$6000=200r$$

$$30=r$$

Check. If r is 30 km/h, then $r+40$ is 70 km/h. The time for the car is 150/30, or 5 hr. The time for the truck is 350/70, or 5 hr. The times are the same. The values check.

State. The speed of Sarah's car is 30 km/h, and the speed of Rick's truck is 70 km/h.

13. ***Familiarize***. We complete the table shown in the text.

$d = r \cdot t$

	Distance	Speed	Time
Freight	330	$r-14$	t
Passenger	400	r	t

Translate. From the rows of the table we have two equations:

$330 = (r-14)t$,

$400 = rt$

We solve each equation for t and set the results equal:

Solving $330 = (r-14)t$ for t: $t = \frac{330}{r-14}$

Solving $400 = rt$ for t: $t = \frac{400}{r}$

Thus, we have

$$\frac{330}{r-14}=\frac{400}{r}.$$

Solve. We multiply by the LCM, $r(r-14)$.

$$r(r-14)\cdot\frac{330}{r-14}=r(r-14)\cdot\frac{400}{r}$$

$$330r=400(r-14)$$

$$330r=400r-5600$$

$$-70r=-5600$$

$$r=80$$

Then substitute 80 for r in either equation to find t:

$$t=\frac{400}{r}$$

$$t=\frac{400}{80} \quad \text{Substituting 80 for } r$$

$$t=5$$

Check. If $r = 80$, then $r-14 = 66$. In 5 hr the freight train travels $66\cdot 5$, or 330 mi, and the passenger train travels $80\cdot 5$, or 400 mi. The values check.

State. The speed of the passenger train is 80 mph. The speed of the freight train is 66 mph.

15. ***Familiarize***. We let r represent the speed going. Then $2r$ is the speed returning. We let t represent the time going. Then $t-3$ represents the time returning. We organize the information in a table.

$d = r \cdot t$

	Distance	Speed	Time
Going	120	r	t
Returning	120	$2r$	$t-3$

Translate. The rows of the table give us two equations:

$120 = rt$,

$120 = 2r(t-3)$

We can solve each equation for r and set the results equal:

Solving $120 = rt$ for r: $r = \dfrac{120}{t}$

Solving $120 = 2r(t-3)$ for r: $r = \dfrac{120}{2(t-3)}$, or

$$r = \frac{60}{t-3}$$

Then $\dfrac{120}{t} = \dfrac{60}{t-3}$.

Solve. We multiply on both sides by the LCM, $t(t-3)$.

$$t(t-3)\cdot\frac{120}{t} = t(t-3)\cdot\frac{60}{t-3}$$
$$120(t-3) = 60t$$
$$120t - 360 = 60t$$
$$-360 = -60t$$
$$6 = t$$

Then substitute 6 for t in either equation to find r, the speed going:

$$r = \frac{120}{t}$$
$$r = \frac{120}{6} \quad \text{Substituting 6 for } t$$
$$r = 20$$

Check. If $r = 20$ and $t = 6$, then $2r = 2\cdot 20$, or 40 mph and $t - 3 = 6 - 3$, or 3 hr. The distance going is $6\cdot 20$, or 120 mi. The distance returning is $40\cdot 3$, or 120 mi. The numbers check.

State. The speed going is 20 mph.

17. ***Familiarize***. Let r = Kelly's speed, in km/h, and t = the time the bicyclists travel, in hours. Organize the information in a table.

	Distance	Speed	Time
Hank	42	$r-5$	t
Kelly	57	r	t

Translate. We can replace the t's in the table above using the formula $t = d/r$.

	Distance	Speed	Time
Hank	42	$r-5$	$\frac{42}{r-5}$
Kelly	57	r	$\frac{57}{r}$

Since the times are the same for both bicyclists, we have the equation

$$\frac{42}{r-5} = \frac{57}{r}.$$

Solve. We first multiply by the LCD, $r(r-5)$.

$$r(r-5)\cdot\frac{42}{r-5} = r(r-5)\cdot\frac{57}{r}$$
$$42r = 57(r-5)$$
$$42r = 57r - 285$$
$$-15r = -285$$
$$r = 19$$

If $r = 19$, then $r - 5 = 14$.

Check. If Hank's speed is 14 km/h and Kelly's speed is 19 km/h, then Hank bicycles 5 km/h slower than Kelly. Hank's time is 42/14, or 3 hr. Kelly's time is 57/19, or 3 hr. Since the times are the same, the answer checks.

State. Hank travels at 14 km/h, and Kelly travels at 19 km/h.

19. ***Familiarize***. Let r = Ralph's speed, in km/h. Then Bonnie's speed is $r + 3$. Also set t = the time, in hours, that Ralph and Bonnie walk. We organize the information in a table.

	Distance	Speed	Time
Ralph	7.5	r	t
Bonnie	12	$r+3$	t

Translate. We can replace the t's in the table shown above using the formula $t = d/r$.

	Distance	Speed	Time
Ralph	7.5	r	$\frac{7.5}{r}$
Bonnie	12	$r+3$	$\frac{12}{r+3}$

Since the times are the same for both walkers, we have the equation

$$\frac{7.5}{r} = \frac{12}{r+3}.$$

Solve. We first multiply by the LCD, $r(r+3)$.

$$r(r+3)\cdot\frac{7.5}{r} = r(r+3)\cdot\frac{12}{r+3}$$
$$7.5(r+3) = 12r$$
$$7.5r + 22.5 = 12r$$
$$22.5 = 4.5r$$
$$5 = r$$

If $r = 5$, then $r + 3 = 8$.

Check. If Ralph's speed is 5 km/h and Bonnie's speed is 8 km/h, then Bonnie walks 3 km/h faster than Ralph. Ralph's time is 7.5/5, or 1.5 hr. Bonnie's time is 12/8, or 1.5 hr. Since the times are the same, the answer checks.

State. Ralph's speed is 5 km/h, and Bonnie's speed is 8 km/h.

21. ***Familiarize***. Let t = the time it takes Evan to drive to town and organize the given information in a table.

	Distance	Speed	Time
Evan	15	r	t
Hobart	20	r	$t+1$

Translate. We can replace the r's in the table above using the formula $r = d/t$.

	Distance	Speed	Time
Evan	15	$\frac{15}{t}$	t
Hobart	20	$\frac{20}{t+1}$	$t+1$

Since the speeds are the same for both riders, we have the equation

$$\frac{15}{t} = \frac{20}{t+1}.$$

Solve. We multiply by the LCD, $t(t+1)$.

$$t(t+1)\cdot\frac{15}{t} = t(t+1)\cdot\frac{20}{t+1}$$
$$15(t+1) = 20t$$
$$15t+15 = 20t$$
$$15 = 5t$$
$$3 = t$$

If $t = 3$, then $t+1 = 3+1$, or 4.

Check. If Evan's time is 3 hr and Hobart's time is 4 hr, then Hobart's time is 1 hr more than Evan's. Evan's speed is 15/3, or 5 mph. Hobart's speed is 20/4, or 5 mph. Since the speeds are the same, the answer checks.

State. It takes Evan 3 hr to drive to town.

23. $\frac{60 \text{ students}}{18 \text{ teachers}} = \frac{60}{18}$ students/teacher $= \frac{10}{3}$ students/teacher

25. $\frac{4.6 \text{ km}}{2 \text{ hr}} = 2.3$ km/h

27. ***Familiarize.*** A 120-lb person should eat at least 44 g of protein each day, and we wish to find the minimum protein required for a 180-lb person. We can set up ratios. We let p = the minimum number of grams of protein a 180-lb person should eat each day.

Translate. If we assume the rates of protein intake are the same, the ratios are the same and we have an equation.

$$\begin{array}{l}\text{Protein} \rightarrow \\ \text{Weight} \rightarrow\end{array} \frac{44}{120} = \frac{p}{180} \begin{array}{l}\leftarrow \text{Protein} \\ \leftarrow \text{Weight}\end{array}$$

Solve. We solve the proportion.

$$360\cdot\frac{44}{120} = 360\cdot\frac{p}{180} \quad \text{Multiplying by the LCM, 360}$$
$$3\cdot 44 = 2\cdot p$$
$$132 = 2p$$
$$66 = p$$

Check. $\frac{44}{120} = \frac{4\cdot 11}{4\cdot 30} = \frac{\cancel{4}\cdot 11}{\cancel{4}\cdot 30} = \frac{11}{30}$ and

$\frac{66}{180} = \frac{6\cdot 11}{6\cdot 30} = \frac{\cancel{6}\cdot 11}{\cancel{6}\cdot 30} = \frac{11}{30}$. The ratios are the same.

State. A 180-lb person should eat a minimum of 66 g of protein each day.

29. ***Familiarize.*** 10 cc of human blood contains 1.2 grams of hemoglobin, and we wish to find how many grams of hemoglobin are contained in 16 cc of the same blood. We can set up ratios. Let H = the amount of hemoglobin in 16 cc of the same blood.

Translate. Assuming the two ratios are the same, we can translate to a proportion.

$$\begin{array}{l}\text{Grams} \rightarrow \\ \text{cm}^3 \rightarrow\end{array} \frac{H}{16} = \frac{1.2}{10} \begin{array}{l}\leftarrow \text{Grams} \\ \leftarrow \text{cm}^3\end{array}$$

Solve. We solve the proportion.

We multiply by 16 to get H alone.

$$16\cdot\frac{H}{16} = 16\cdot\frac{1.2}{10}$$
$$H = \frac{19.2}{10}$$
$$H = 1.92$$

Check.

$\frac{1.92}{16} = 0.12 \qquad \frac{1.2}{10} = 0.12$

The ratios are the same.

State. 16 cc of the same blood would contain 1.92 grams of hemoglobin.

31. ***Familiarize.*** Let h = the amount of honey, in pounds, that 35,000 trips to flowers would produce.

Translate. We translate to a proportion.

$$\begin{array}{l}\text{Honey} \rightarrow \\ \text{Trips} \rightarrow\end{array} \frac{1}{20,000} = \frac{h}{35,000} \begin{array}{l}\leftarrow \text{Honey} \\ \leftarrow \text{Trips}\end{array}$$

Solve. We solve the proportion.

$$35,000\cdot\frac{1}{20,000} = 35,000\cdot\frac{h}{35,000}$$
$$1.75 = h$$

Check. $\frac{1}{20,000} = 0.00005$ and $\frac{1.75}{35,000} = 0.00005$.

The ratios are the same.

State. 35,000 trips to gather nectar will produce 1.75 lb of honey.

33. ***Familiarize.*** The ratio of the weight of copper to the weight of zinc in a U.S. penny is $\frac{1}{39}$, and we wish to find how much copper is needed if 50 kg of zinc is being turned into pennies. We can set up a second ratio to go with the one we already have. Let C = the amount of copper needed, in kg, if 50 kg of zinc is being turned into pennies.

Translate. We translate to a proportion.

$$\frac{1}{39} = \frac{C}{50}$$

Solve. We solve the proportion.

$$50\cdot\frac{1}{39} = 50\cdot\frac{C}{50}$$
$$\frac{50}{39} = C, \text{ or}$$
$$1\frac{11}{39} = C$$

Check. $\dfrac{50/39}{50} = \dfrac{1}{39}$, so the ratios are the same.

State. $1\dfrac{11}{39}$ kg of copper is needed if 50 kg of zinc is turned into pennies.

35. (a) $\dfrac{118}{439} \approx 0.269$

Howard's batting average was 0.269.

(b) Let $h =$ the number of hits Howard would get in the 162-game season. We translate to a proportion and solve it.

$$\frac{118}{114} = \frac{h}{162}$$
$$162 \cdot \frac{118}{114} = 162 \cdot \frac{h}{162}$$
$$168 \approx h$$

Howard would get 168 hits in the 162-game season.

(c) Let $h =$ the number of hits Howard would get if he batted 700 times. We translate to a proportion and solve it.

$$\frac{118}{439} = \frac{h}{700}$$
$$700 \cdot \frac{118}{439} = 700 \cdot \frac{h}{700}$$
$$188 \approx h$$

Howard would get 188 hits if he batted 700 times.

37. Let $h =$ the head circumference, in inches. We translate to a proportion and solve it.

$$\frac{6\frac{3}{4}}{21\frac{1}{5}} = \frac{7}{h}$$
$$6\frac{3}{4} \cdot h = 21\frac{1}{5} \cdot 7$$
$$\frac{27}{4} \cdot h = \frac{106}{5} \cdot 7$$
$$h = \frac{4}{27} \cdot \frac{106}{5} \cdot 7$$
$$h \approx 22$$

The head circumference is 22 in.

Now let $c =$ the head circumference, in centimeters. We translate to a proportion and solve it.

$$\frac{6\frac{3}{4}}{53.8} = \frac{7}{c}$$
$$\frac{6.75}{53.8} = \frac{7}{c} \quad \left(6\frac{3}{4} = 6.75\right)$$
$$6.75 \cdot c = 53.8 \cdot 7$$
$$c = \frac{53.8 \cdot 7}{6.75}$$
$$c \approx 55.8$$

The head circumference is 55.8 cm.

39. Let $h =$ the hat size. We translate to a proportion and solve it.

$$\frac{6\frac{3}{4}}{21\frac{1}{5}} = \frac{h}{22\frac{4}{5}}$$
$$6\frac{3}{4} \cdot 22\frac{4}{5} = 21\frac{1}{5} \cdot h$$
$$\frac{27}{4} \cdot \frac{114}{5} = \frac{106}{5} \cdot h$$
$$\frac{5}{106} \cdot \frac{27}{4} \cdot \frac{114}{5} = h$$
$$7.26 \approx h$$
$$7\frac{1}{4} \approx h$$

The hat size is $7\frac{1}{4}$.

Now let $c =$ the head circumference, in centimeters. We translate to a proportion and solve it. We use the hat size found above in the translation.

$$\frac{6\frac{3}{4}}{53.8} = \frac{7\frac{1}{4}}{c}$$
$$\frac{6.75}{53.8} = \frac{7.25}{c}$$
$$6.75 \cdot c = 53.8 \cdot 7.25$$
$$c = \frac{53.8 \cdot 7.25}{6.75}$$
$$c \approx 57.8$$

The head circumference is 57.8 cm. (Answers may vary slightly depending on when rounding occurs.)

41. Let $h =$ the hat size. We translate to a proportion and solve it.

$$\frac{6\frac{3}{4}}{53.8} = \frac{h}{59.8}$$
$$\frac{6.75}{53.8} = \frac{h}{59.8}$$
$$59.8 \cdot \frac{6.75}{53.8} = h$$
$$7.5 \approx h, \text{ or}$$
$$7\frac{1}{2} \approx h$$

The hat size is $7\frac{1}{2}$.

Now let $c =$ the head circumference, in inches. We translate to a proportion and solve it. We use the hat size found above in the translation.

$$\frac{6\frac{3}{4}}{21\frac{1}{5}} = \frac{7\frac{1}{2}}{c}$$

$$6\frac{3}{4} \cdot c = 21\frac{1}{5} \cdot 7\frac{1}{2}$$

$$\frac{27}{4} \cdot c = \frac{106}{5} \cdot \frac{15}{2}$$

$$c = \frac{4}{27} \cdot \frac{106}{5} \cdot \frac{15}{2}$$

$$c \approx 23.6, \text{ or}$$

$$c \approx 23\frac{3}{5}$$

The head circumference is $23\frac{3}{5}$ in.

43. ***Familiarize***. The ratio of trout tagged to the total trout population, P, is $\frac{112}{P}$. Of the 82 trout checked later, 32 were tagged. The ratio of trout tagged to trout checked is $\frac{32}{82}$.

Translate. Assuming the two ratios are the same, we can translate to a proportion.

Trout tagged originally $\longrightarrow$ / Trout population $\longrightarrow$ $\dfrac{112}{P} = \dfrac{32}{82}$ $\longleftarrow$ Tagged trout caught later / $\longleftarrow$ Trout caught later

Solve. We solve the equation.

$$82P \cdot \frac{112}{P} = 82P \cdot \frac{32}{82} \quad \text{Multiplying by the LCM, } 82P$$

$$82 \cdot 112 = P \cdot 32$$

$$9184 = 32P$$

$$287 = P$$

Check.

$$\frac{112}{287} \approx 0.390 \text{ and } \frac{32}{82} \approx 0.390.$$

The ratios are the same.

State. The trout population is 287.

45. ***Familiarize***. A sample of 144 firecrackers contained 9 duds, and we wish to find how many duds could be expected in a sample of 3200 firecrackers. We can set up ratios, letting $d =$ the number of duds expected in a sample of 3200 firecrackers.

Translate. Assuming the rates of occurrence of duds are the same, we can translate to a proportion.

Duds $\rightarrow$ / Sample size $\rightarrow$ $\dfrac{9}{144} = \dfrac{d}{3200}$ $\leftarrow$ Duds / $\leftarrow$ Sample size

Solve. We solve the equation. We multiply by 3200 to get d alone.

$$3200 \cdot \frac{9}{144} = 3200 \cdot \frac{d}{3200}$$

$$\frac{28,800}{144} = d$$

$$200 = d$$

Check.

$$\frac{9}{144} = 0.0625 \text{ and } \frac{200}{3200} = 0.0625$$

The ratios are the same.

State. You would expect 200 duds in a sample of 3200 firecrackers.

47. ***Familiarize***. The ratio of the weight of an object on Mars to the weight of an object on earth is 0.4 to 1.

a) We wish to find how much a 12-ton rocket would weigh on Mars.

b) We wish to find how much a 120-lb astronaut would weigh on Mars.

We can set up ratios. We let $r =$ the weight of a 12-ton rocket and $a =$ the weight of a 120-lb astronaut on Mars.

Translate. Assuming the ratios are the same, we can translate to proportions.

a) Weight on Mars $\rightarrow$ / Weight on earth $\rightarrow$ $\dfrac{0.4}{1} = \dfrac{r}{12}$ $\leftarrow$ Weight on Mars / $\leftarrow$ Weight on earth

b) Weight on Mars $\rightarrow$ / Weight on earth $\rightarrow$ $\dfrac{0.4}{1} = \dfrac{a}{120}$ $\leftarrow$ Weight on Mars / $\leftarrow$ Weight on earth

Solve. We solve each proportion.

a) $\dfrac{0.4}{1} = \dfrac{r}{12}$ b) $\dfrac{0.4}{1} = \dfrac{a}{120}$

$12(0.4) = r$ $120(0.4) = a$

$4.8 = r$ $48 = a$

Check. $\dfrac{0.4}{1} = 0.4$, $\dfrac{4.8}{12} = 0.4$, and $\dfrac{48}{120} = 0.4$.

The ratios are the same.

State. a) A 12-ton rocket would weigh 4.8 tons on Mars.

b) A 120-lb astronaut would weigh 48 lb on Mars.

49. We write a proportion and then solve it.

$$\frac{b}{6} = \frac{7}{4}$$

$$b = \frac{7}{4} \cdot 6 \quad \text{Multiplying by 6}$$

$$b = \frac{42}{4}$$

$$b = \frac{21}{2}, \text{ or } 10.5$$

(Note that the proportions $\frac{6}{b} = \frac{4}{7}$, $\frac{b}{7} = \frac{6}{4}$, or $\frac{7}{b} = \frac{4}{6}$ could also be used.)

51. We write a proportion and then solve it.

$$\frac{4}{f} = \frac{6}{4}$$
$$4f \cdot \frac{4}{f} = 4f \cdot \frac{6}{4}$$
$$16 = 6f$$
$$\frac{8}{3} = f \quad \text{Simplifying}$$

(One of the following proportions could also be used: $\frac{f}{4} = \frac{4}{6}, \frac{4}{f} = \frac{9}{6}, \frac{f}{4} = \frac{6}{9}, \frac{4}{9} = \frac{f}{6}, \frac{9}{4} = \frac{6}{f}$)

53. We write a proportion and then solve it.

$$\frac{h}{7} = \frac{10}{6}$$
$$h = \frac{10}{6} \cdot 7 \quad \text{Multiplying by 7}$$
$$h = \frac{70}{6}$$
$$h = \frac{35}{3} \quad \text{Simplifying}$$

(Note that the proportions $\frac{7}{h} = \frac{6}{10}, \frac{h}{10} = \frac{7}{6}$, or $\frac{10}{h} = \frac{6}{7}$ could also be used.)

55. We write a proportion and then solve it.

$$\frac{4}{10} = \frac{6}{l}$$
$$10l \cdot \frac{4}{10} = 10l \cdot \frac{6}{l}$$
$$4l = 60$$
$$l = 15 \text{ ft}$$

(One of the following proportions could also be used: $\frac{4}{6} = \frac{10}{l}, \frac{10}{4} = \frac{l}{6}$, or $\frac{6}{4} = \frac{l}{10}$)

57. The equation is $y = -2x + b$. We use the point $\left(\frac{1}{2}, \frac{3}{4}\right)$ to find b.

$$\frac{3}{4} = -2 \cdot \frac{1}{2} + b$$
$$\frac{3}{4} = -1 + b$$
$$\frac{7}{4} = b$$

The equation is $y = -2x + \frac{7}{4}$.

59. $x^5 \cdot x^6 = x^{5+6} = x^{11}$

61. $x^{-5} \cdot x^{-6} = x^{-5+(-6)} = x^{-11} = \frac{1}{x^{11}}$

63. Graph: $y = 2x - 6$.

We select some x-values and compute y-values.

If $x = 1$, then $y = 2 \cdot 1 - 6 = -4$.

If $x = 3$, then $y = 2 \cdot 3 - 6 = 0$.

If $x = 5$, then $y = 2 \cdot 5 - 6 = 4$.

x	y	(x, y)
1	-4	$(1, -4)$
3	0	$(3, 0)$
5	4	$(5, 4)$

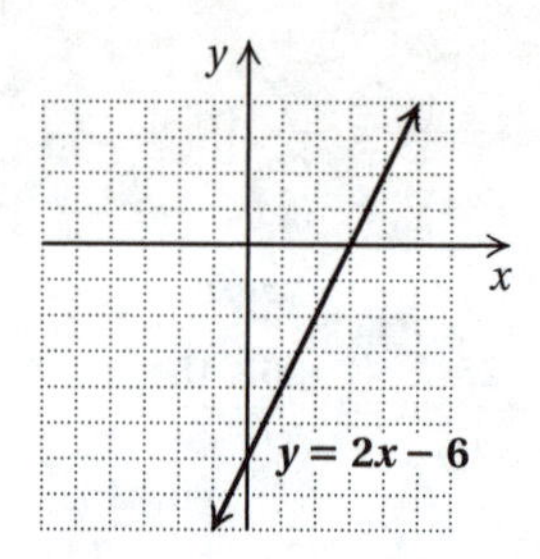

65. Graph: $3x + 2y = 12$.

We can replace either variable with a number and then calculate the other coordinate. We will find the intercepts and one other point.

If $y = 0$, we have:

$$3x + 2 \cdot 0 = 12$$
$$3x = 12$$
$$x = 4$$

The x-intercept is $(4, 0)$.

If $x = 0$, we have:

$$3 \cdot 0 + 2y = 12$$
$$2y = 12$$
$$y = 6$$

The y-intercept is $(0, 6)$.

If $y = -3$, we have:

$$3x + 2(-3) = 12$$
$$3x - 6 = 12$$
$$3x = 18$$
$$x = 6$$

The point $(6, -3)$ is on the graph.

We plot these points and draw a line through them.

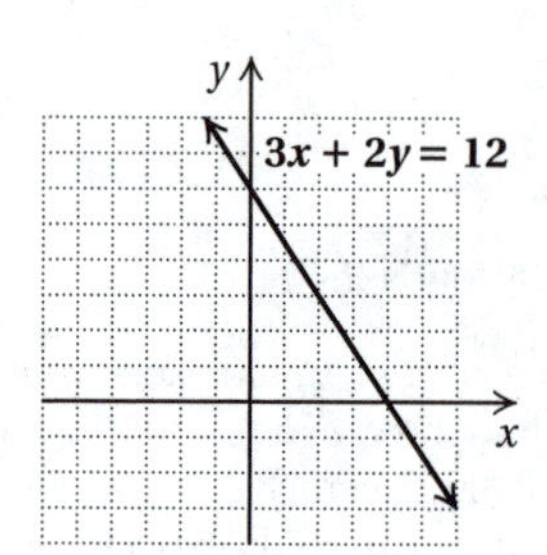

67. Graph: $y = -\frac{3}{4}x + 2$

We select some x-values and compute y-values. We use multiples of 4 to avoid fractions.

If $x = -4$, then $y = -\frac{3}{4}(-4) + 2 = 5$.

If $x = 0$, then $y = -\frac{3}{4} \cdot 0 + 2 = 2$.

If $x = 4$, then $y = -\frac{3}{4} \cdot 4 + 2 = -1$.

x	y	(x,y)
-4	5	$(-4,5)$
0	2	$(0,2)$
4	-1	$(4,-1)$

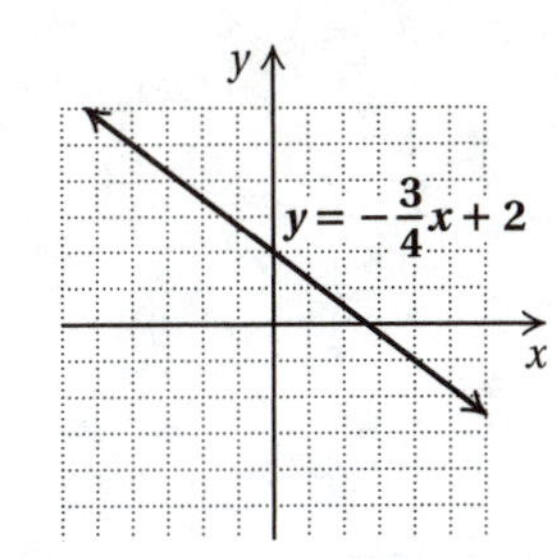

69. ***Familiarize***. Let $t =$ the time it would take for Ann to complete the report working alone. Then $t+6 =$ the time it would take Betty to complete the report working alone. In 1 hour they would complete $\frac{1}{t}+\frac{1}{t+6}$ of the report and in 4 hours they would complete $4\left(\frac{1}{t}+\frac{1}{t+6}\right)$, or $\frac{4}{t}+\frac{4}{t+6}$ of the report.

Translate. In 4 hours one entire job is done, so we have

$$\frac{4}{t}+\frac{4}{t+6}=1.$$

Solve. We solve the equation.

$$\frac{4}{t}+\frac{4}{t+6}=1,\ \text{LCM}=t(t+6)$$
$$t(t+6)\left(\frac{4}{t}+\frac{4}{t+6}\right)=t(t+6)\cdot 1$$
$$t(t+6)\cdot\frac{4}{t}+t(t+6)\cdot\frac{4}{t+6}=t^2+6t$$
$$4(t+6)+4t=t^2+6t$$
$$4t+24+4t=t^2+6t$$
$$0=t^2-2t-24$$
$$0=(t-6)(t+4)$$

$t-6=0$ *or* $t+4=0$

$t=6$ *or* $t=-4$

Check. The time cannot be negative, so we check only 6. If it takes Ann 6 hr to complete the report, then it would take Betty $6+6$, or 12 hr, to complete the report. In 4 hr Ann does $4\cdot\frac{1}{6}$, or $\frac{2}{3}$, of the report, Betty does $4\cdot\frac{1}{12}$, or $\frac{1}{3}$, of the report, and together they do $\frac{2}{3}+\frac{1}{3}$, or 1 entire job. The answer checks.

State. It would take Ann 6 hr and Betty 12 hr to complete the report working alone.

71. ***Familiarize***. Let $t =$ the number of minutes after 5:00 at which the hands of the clock will first be together. While the minute hand moves through t minutes, the hour hand moves through $t/12$ minutes. At 5:00 the hour hand is on the 25-minute mark. We wish to find when a move of the minute hand through t minutes is equal to $25+t/12$ minutes.

Translate. We use the last sentence of the familiarization step to write an equation.

$$t=25+\frac{t}{12}$$

Solve. We solve the equation.

$$t=25+\frac{t}{12}$$
$$12\cdot t=12\left(25+\frac{t}{12}\right)$$
$$12t=300+t \qquad \text{Multiplying by 12}$$
$$11t=300$$
$$t=\frac{300}{11}\ \text{or}\ 27\frac{3}{11}$$

Check. At $27\frac{3}{11}$ minutes after 5:00, the minute hand is at the $27\frac{3}{11}$-minutes mark and the hour hand is at the $25+\frac{27\frac{3}{11}}{12}$-minute mark. Simplifying $25+\frac{27\frac{3}{11}}{12}$, we get $25+\frac{\frac{300}{11}}{12}=25+\frac{300}{11}\cdot\frac{1}{12}=25+\frac{25}{11}=25+2\frac{3}{11}=27\frac{3}{11}$. Thus, the hands are together.

State. The hands are first together $27\frac{3}{11}$ minutes after 5:00.

73.

$$\frac{t}{a}+\frac{t}{b}=1,\ \text{LCM}=ab$$
$$ab\left(\frac{t}{a}+\frac{t}{b}\right)=ab\cdot 1$$
$$ab\cdot\frac{t}{a}+ab\cdot\frac{t}{b}=ab$$
$$bt+at=ab$$
$$t(b+a)=ab$$
$$t=\frac{ab}{b+a}$$

Exercise Set 14.8

1.

$$\frac{1+\frac{9}{16}}{1-\frac{3}{4}} \qquad \text{LCM of the denominators is 16.}$$
$$=\frac{1+\frac{9}{16}}{1-\frac{3}{4}}\cdot\frac{16}{16} \qquad \text{Multiplying by 1 using } \frac{16}{16}$$
$$=\frac{\left(1+\frac{9}{16}\right)16}{\left(1-\frac{3}{4}\right)16} \qquad \text{Multiplying numerator and denominator by 16}$$
$$=\frac{1(16)+\frac{9}{16}(16)}{1(16)-\frac{3}{4}(16)}$$
$$=\frac{16+9}{16-12}$$
$$=\frac{25}{4}$$

3. $\dfrac{1-\frac{3}{5}}{1+\frac{1}{5}}$

$= \dfrac{1\cdot\frac{5}{5}-\frac{3}{5}}{1\cdot\frac{5}{5}+\frac{1}{5}}$ Getting a common denominator in numerator and in denominator

$= \dfrac{\frac{5}{5}-\frac{3}{5}}{\frac{5}{5}+\frac{1}{5}}$

$= \dfrac{\frac{2}{5}}{\frac{6}{5}}$ Subtracting in numerator; adding in denominator

$= \dfrac{2}{5}\cdot\dfrac{5}{6}$ Multiplying by the reciprocal of the divisor

$= \dfrac{2\cdot 5}{5\cdot 2\cdot 3}$

$= \dfrac{\not{2}\cdot\not{5}\cdot 1}{\not{5}\cdot\not{2}\cdot 3}$

$= \dfrac{1}{3}$

5. $\dfrac{\frac{1}{2}+\frac{3}{4}}{\frac{5}{8}-\frac{5}{6}} = \dfrac{\frac{1}{2}\cdot\frac{2}{2}+\frac{3}{4}}{\frac{5}{8}\cdot\frac{3}{3}-\frac{5}{6}\cdot\frac{4}{4}}$ Getting a common denominator in numerator and in denominator

$= \dfrac{\frac{2}{4}+\frac{3}{4}}{\frac{15}{24}-\frac{20}{24}}$

$= \dfrac{\frac{5}{4}}{\frac{-5}{24}}$ Adding in numerator; subtracting in denominator

$= \dfrac{5}{4}\cdot\dfrac{24}{-5}$ Multiplying by the reciprocal of the divisor

$= \dfrac{5\cdot 4\cdot 6}{4\cdot(-1)\cdot 5}$

$= \dfrac{\not{5}\cdot\not{4}\cdot 6}{\not{4}\cdot(-1)\cdot\not{5}}$

$= -6$

7. $\dfrac{\frac{1}{x}+3}{\frac{1}{x}-5}$ LCM of the denominators is x.

$= \dfrac{\frac{1}{x}+3}{\frac{1}{x}-5}\cdot\dfrac{x}{x}$ Multiplying by 1 using $\dfrac{x}{x}$

$= \dfrac{\left(\frac{1}{x}+3\right)x}{\left(\frac{1}{x}-5\right)x}$

$= \dfrac{\frac{1}{x}\cdot x+3\cdot x}{\frac{1}{x}\cdot x-5\cdot x}$

$= \dfrac{1+3x}{1-5x}$

9. $\dfrac{4-\frac{1}{x^2}}{2-\frac{1}{x}}$ LCM of the denominators is x^2.

$= \dfrac{4-\frac{1}{x^2}}{2-\frac{1}{x}}\cdot\dfrac{x^2}{x^2}$

$= \dfrac{\left(4-\frac{1}{x^2}\right)x^2}{\left(2-\frac{1}{x}\right)x^2}$

$= \dfrac{4\cdot x^2-\frac{1}{x^2}\cdot x^2}{2\cdot x^2-\frac{1}{x}\cdot x^2}$

$= \dfrac{4x^2-1}{2x^2-x}$

$= \dfrac{(2x+1)(2x-1)}{x(2x-1)}$ Factoring numerator and denominator

$= \dfrac{(2x+1)\cancel{(2x-1)}}{x\cancel{(2x-1)}}$

$= \dfrac{2x+1}{x}$

11. $$\frac{8+\frac{8}{d}}{1+\frac{1}{d}} = \frac{8\cdot\frac{d}{d}+\frac{8}{d}}{1\cdot\frac{d}{d}+\frac{1}{d}}$$

$$= \frac{\frac{8d+8}{d}}{\frac{d+1}{d}}$$

$$= \frac{8d+8}{d}\cdot\frac{d}{d+1}$$

$$= \frac{8(d+1)(d)}{d(d+1)}$$

$$= \frac{8\cancel{(d+1)}\cancel{(d)}}{\cancel{d}\cancel{(d+1)}(1)}$$

$$= 8$$

13. $$\frac{\frac{x}{8}-\frac{8}{x}}{\frac{1}{8}+\frac{1}{x}}$$ LCM of the denominators is $8x$.

$$= \frac{\frac{x}{8}-\frac{8}{x}}{\frac{1}{8}+\frac{1}{x}}\cdot\frac{8x}{8x}$$

$$= \frac{\left(\frac{x}{8}-\frac{8}{x}\right)8x}{\left(\frac{1}{8}+\frac{1}{x}\right)8x}$$

$$= \frac{\frac{x}{8}(8x)-\frac{8}{x}(8x)}{\frac{1}{8}(8x)+\frac{1}{x}(8x)}$$

$$= \frac{x^2-64}{x+8}$$

$$= \frac{(x+8)(x-8)}{x+8}$$

$$= \frac{\cancel{(x+8)}(x-8)}{1\cancel{(x+8)}}$$

$$= x-8$$

15. $$\frac{1+\frac{1}{y}}{1-\frac{1}{y^2}} = \frac{1\cdot\frac{y}{y}+\frac{1}{y}}{1\cdot\frac{y^2}{y^2}-\frac{1}{y^2}}$$

$$= \frac{\frac{y+1}{y}}{\frac{y^2-1}{y^2}}$$

$$= \frac{y+1}{y}\cdot\frac{y^2}{y^2-1}$$

$$= \frac{(y+1)y\cdot y}{y(y+1)(y-1)}$$

$$= \frac{\cancel{(y+1)}\cancel{y}\cdot y}{\cancel{y}\cancel{(y+1)}(y-1)}$$

$$= \frac{y}{y-1}$$

17. $$\frac{\frac{1}{5}-\frac{1}{a}}{\frac{5-a}{5}}$$ LCM of the denominators is $5a$.

$$= \frac{\frac{1}{5}-\frac{1}{a}}{\frac{5-a}{5}}\cdot\frac{5a}{5a}$$

$$= \frac{\left(\frac{1}{5}-\frac{1}{a}\right)5a}{\left(\frac{5-a}{5}\right)5a}$$

$$= \frac{\frac{1}{5}(5a)-\frac{1}{a}(5a)}{a(5-a)}$$

$$= \frac{a-5}{5a-a^2}$$

$$= \frac{a-5}{-a(-5+a)}$$

$$= \frac{1\cancel{(a-5)}}{-a\cancel{(a-5)}}$$

$$= -\frac{1}{a}$$

19. $$\frac{\frac{1}{a}+\frac{1}{b}}{\frac{1}{a^2}-\frac{1}{b^2}}$$ LCM of the denominators is a^2b^2.

$$= \frac{\frac{1}{a}+\frac{1}{b}}{\frac{1}{a^2}-\frac{1}{b^2}}\cdot\frac{a^2b^2}{a^2b^2}$$

$$= \frac{\left(\frac{1}{a}+\frac{1}{b}\right)\cdot a^2b^2}{\left(\frac{1}{a^2}-\frac{1}{b^2}\right)\cdot a^2b^2}$$

$$= \frac{\frac{1}{a}\cdot a^2b^2+\frac{1}{b}\cdot a^2b^2}{\frac{1}{a^2}\cdot a^2b^2-\frac{1}{b^2}\cdot a^2b^2}$$

$$= \frac{ab^2+a^2b}{b^2-a^2}$$

$$= \frac{ab(b+a)}{(b+a)(b-a)}$$

$$= \frac{ab\cancel{(b+a)}}{\cancel{(b+a)}(b-a)}$$

$$= \frac{ab}{b-a}$$

21. $\dfrac{\dfrac{p}{q}+\dfrac{q}{p}}{\dfrac{1}{p}+\dfrac{1}{q}}$ LCM of the denominators is pq.

$$= \frac{\left(\frac{p}{q}+\frac{q}{p}\right)\cdot pq}{\left(\frac{1}{p}+\frac{1}{q}\right)\cdot pq}$$

$$= \frac{\frac{p}{q}\cdot pq+\frac{q}{p}\cdot pq}{\frac{1}{p}\cdot pq+\frac{1}{q}\cdot pq}$$

$$= \frac{p^2+q^2}{q+p}$$

23. $\dfrac{\dfrac{2}{a}+\dfrac{4}{a^2}}{\dfrac{5}{a^3}-\dfrac{3}{a}}$ LCD is a^3

$$= \frac{\frac{2}{a}+\frac{4}{a^2}}{\frac{5}{a^3}-\frac{3}{a}}\cdot\frac{a^3}{a^3}$$

$$= \frac{\frac{2}{a}\cdot a^3+\frac{4}{a^2}\cdot a^3}{\frac{5}{a^3}\cdot a^3-\frac{3}{a}\cdot a^3}$$

$$= \frac{2a^2+4a}{5-3a^2}$$

$$= \frac{2a(a+2)}{5-3a^2}$$

25. $$\frac{\frac{2}{7a^4}-\frac{1}{14a}}{\frac{3}{5a^2}+\frac{2}{15a}} = \frac{\frac{2}{7a^4}\cdot\frac{2}{2}-\frac{1}{14a}\cdot\frac{a^3}{a^3}}{\frac{3}{5a^2}\cdot\frac{3}{3}+\frac{2}{15a}\cdot\frac{a}{a}}$$

$$= \frac{\frac{4-a^3}{14a^4}}{\frac{9+2a}{15a^2}}$$

$$= \frac{4-a^3}{14a^4}\cdot\frac{15a^2}{9+2a}$$

$$= \frac{15\cdot \cancel{a^2}(4-a^3)}{14a^2\cdot \cancel{a^2}(9+2a)}$$

$$= \frac{15(4-a^3)}{14a^2(9+2a)}$$

27. $$\frac{\frac{a}{b}+\frac{c}{d}}{\frac{b}{a}+\frac{d}{c}} = \frac{\frac{a}{b}\cdot\frac{d}{d}+\frac{c}{d}\cdot\frac{b}{b}}{\frac{b}{a}\cdot\frac{c}{c}+\frac{d}{c}\cdot\frac{a}{a}}$$

$$= \frac{\frac{ad+bc}{bd}}{\frac{bc+ad}{ac}}$$

$$= \frac{ad+bc}{bd}\cdot\frac{ac}{bc+ad}$$

$$= \frac{ac(ad+bc)}{bd(bc+ad)}$$

$$= \frac{ac}{bd}\cdot\frac{ad+bc}{bc+ad}$$

$$= \frac{ac}{bd}\cdot 1$$

$$= \frac{ac}{bd}$$

29. $\dfrac{\dfrac{x}{5y^3}+\dfrac{3}{10y}}{\dfrac{3}{10y}+\dfrac{x}{5y^3}}$

Observe that, by the commutative law of addition, the numerator and denominator are equivalent, so the result is 1. We could also simplify this expression as follows:

$$\frac{\frac{x}{5y^3}+\frac{3}{10y}}{\frac{3}{10y}+\frac{x}{5y^3}} = \frac{\frac{x}{5y^3}+\frac{3}{10y}}{\frac{3}{10y}+\frac{x}{5y^3}}\cdot\frac{10y^3}{10y^3}$$

$$= \frac{\frac{x}{5y^3}\cdot 10y^3+\frac{3}{10y}\cdot 10y^3}{\frac{3}{10y}\cdot 10y^3+\frac{x}{5y^3}\cdot 10y^3}$$

$$= \frac{2x+3y^2}{3y^2+2x}$$

$$= 1$$

31. $$\frac{\frac{3}{x+1}+\frac{1}{x}}{\frac{2}{x+1}+\frac{3}{x}} = \frac{\frac{3}{x+1}+\frac{1}{x}}{\frac{2}{x+1}+\frac{3}{x}}\cdot\frac{x(x+1)}{x(x+1)}$$

$$= \frac{\frac{3}{x+1}\cdot x(x+1)+\frac{1}{x}\cdot x(x+1)}{\frac{2}{x+1}\cdot x(x+1)+\frac{3}{x}\cdot x(x+1)}$$

$$= \frac{3x+x+1}{2x+3(x+1)}$$

$$= \frac{4x+1}{2x+3x+3}$$

$$= \frac{4x+1}{5x+3}$$

33.

$$4-\frac{1}{6}x\ge -12$$
$$-\frac{1}{6}x\ge -16$$
$$-6\left(-\frac{1}{6}x\right)\le -6(-16)$$
$$x\le 96$$

The solution set is $\{x|x\le 96\}$.

35.

$$1.5x+19.2<4.2-3.5x$$
$$5x+19.2<4.2$$
$$5x<-15$$
$$x<-3$$

The solution set is $\{x|x<-3\}$.

37.

$$(2x^3-4x^2+x-7)+(4x^4+x^3+4x^2+x)$$
$$=4x^4+3x^3+2x-7$$

39. $p^2-10p+25=p^2-2\cdot p\cdot 5+5^2$ Trinomial square

$$=(p-5)^2$$

41. $50p^2-100=50(p^2-2)$ Factoring out the common factor

Since p^2-2 cannot be factored, we have factored completely.

43. ***Familiarize***. Let w = the width of the rectangle. Then $w+3$ = the length. Recall that the formula for the area of a rectangle is $A=lw$ and the formula for the perimeter of a rectangle is $P=2l+2w$.

Translate. We substitute in the formula for area.

$$10=lw$$
$$10=(w+3)w$$

Solve.

$$10=(w+3)w$$
$$10=w^2+3w$$
$$0=w^2+3w-10$$
$$0=(w+5)(w-2)$$
$$w+5=0 \quad\text{or}\quad w-2=0$$
$$w=-5 \quad\text{or}\quad w=2$$

Check. Since the width cannot be negative, we only check 2. If $w=2$, then $w+3=2+3$, or 5. Since $2\cdot 5=10$, the given area, the answer checks. Now we find the perimeter:

$$P=2l+2w$$
$$P=2\cdot 5+2\cdot 2$$
$$P=10+4$$
$$P=14$$

We can check this by repeating the calculation.

State. The perimeter is 14 yd.

45.

$$\frac{1}{\frac{2}{x-1}-\frac{1}{3x-2}}$$
$$=\frac{1}{\frac{2}{x-1}-\frac{1}{3x-2}}\cdot\frac{(x-1)(3x-2)}{(x-1)(3x-2)}$$
$$=\frac{(x-1)(3x-2)}{\left(\frac{2}{x-1}-\frac{1}{3x-2}\right)(x-1)(3x-2)}$$
$$=\frac{(x-1)(3x-2)}{\frac{2}{x-1}(x-1)(3x-2)-\frac{1}{3x-2}(x-1)(3x-2)}$$
$$=\frac{(x-1)(3x-2)}{2(3x-2)-(x-1)}$$
$$=\frac{(x-1)(3x-2)}{6x-4-x+1}$$
$$=\frac{(x-1)(3x-2)}{5x-3}$$

47.

$$1+\frac{1}{1+\frac{1}{1+\frac{1}{1+\frac{1}{x}}}}=1+\frac{1}{1+\frac{1}{1+\frac{1}{\frac{x+1}{x}}}}$$
$$=1+\frac{1}{1+\frac{1}{1+\frac{x}{x+1}}}$$
$$=1+\frac{1}{1+\frac{1}{\frac{x+1+x}{x+1}}}$$
$$=1+\frac{1}{1+\frac{1}{\frac{2x+1}{x+1}}}$$
$$=1+\frac{1}{1+\frac{x+1}{2x+1}}$$
$$=1+\frac{1}{\frac{2x+1+x+1}{2x+1}}$$
$$=1+\frac{1}{\frac{3x+2}{2x+1}}$$
$$=1+\frac{2x+1}{3x+2}$$
$$=\frac{3x+2+2x+1}{3x+2}$$
$$=\frac{5x+3}{3x+2}$$

Exercise Set 14.9

1. We substitute to find k.

$y = kx$

$36 = k \cdot 9$ Substituting 36 for y and 9 for x

$\frac{36}{9} = k$

$4 = k$ k is the variation constant.

The equation of the variation is $y = 4x$.

To find the value of y when $x = 20$ we substitute 20 for x in the equation of variation.

$y = 4x$

$y = 4 \cdot 20$

$y = 80$

The value of y is 80 when $x = 20$.

3. We substitute to find k.

$y = kx$

$0.8 = k \cdot 0.5$ Substituting 0.8 for y and 0.5 for x

$\frac{0.8}{0.5} = k$

$1.6 = k$ k is the variation constant.

The equation of the variation is $y = 1.6x$.

To find the value of y when $x = 20$ we substitute 20 for x in the equation of variation.

$y = 1.6x$

$y = 1.6(20)$

$y = 32$

The value of y is 32 when $x = 20$.

5. We substitute to find k.

$y = kx$

$630 = k \cdot 175$ Substituting 630 for y and 175 for x

$\frac{630}{175} = k$

$3.6 = k$ k is the variation constant.

The equation of the variation is $y = 3.6x$.

To find the value of y when $x = 20$ we substitute 20 for x in the equation of variation.

$y = 3.6x$

$y = 3.6(20)$

$y = 72$

The value of y is 72 when $x = 20$.

7. We substitute to find k.

$y = kx$

$500 = k \cdot 60$ Substituting 500 for y and 60 for x

$\frac{500}{60} = k$

$\frac{25}{3} = k$ k is the variation constant.

The equation of the variation is $y = \frac{25}{3}x$.

To find the value of y when $x = 20$ we substitute 20 for x in the equation of variation.

$y = \frac{25}{3}x$

$y = \frac{25}{3} \cdot 20$

$y = \frac{500}{3}$

The value of y is $\frac{500}{3}$ when $x = 20$.

9. ***Familiarize and Translate***. The problem states that we have direct variation between the variables P and H. Thus, an equation $P = kH$, $k > 0$, applies. As the number of hours increases, the paycheck increases.

Solve.

a) First find an equation of variation.

$P = kH$

$180 = k \cdot 15$ Substituting 180 for P and 15 for H

$\frac{180}{15} = k$

$12 = k$

The equation of variation is $P = 12H$.

b) Use the equation to find the pay for 35 hours work.

$P = 12H$

$P = 12(35)$ Substituting 35 for H

$P = 420$

Check. This check might be done by repeating the computations. We might also do some reasoning about the answer. The paycheck increased from \$180 to \$420. Similarly, the hours increased from 15 to 35.

State. a) The equation of variation is $P = 12H$.

b) For 35 hours work, the paycheck is \$420.

11. ***Familiarize and Translate***. The problem states that we have direct variation between the variables C and S. Thus, an equation $C = kS$, $k > 0$, applies. As the depth increases, the cost increases.

Solve.

a) First find an equation of variation.

$C = kS$

$67.5 = k \cdot 6$ Substituting 75 for C and 6 for S

$\frac{67.5}{6} = k$

$11.25 = k$

The equation of variation is $C = 11.25S$.

b) Use the equation to find the cost of filling the sandbox to a depth of 9 inches.

$C = 11.25S$

$C = 11.25(9)$ Substituting 9 for S

$C = 101.25$

Check. In addition to repeating the computations, we can also do some reasoning. The depth increased from 6 inches to 9 inches. Similarly, the cost increased from \$67.50 to \$101.25.

State. a) The equation of variation is $C = 11.25S$.

b) The sand will cost \$101.25.

13. ***Familiarize and Translate.*** The problem states that we have direct variation between the variables M and E. Thus, an equation $M = kE$, $k > 0$, applies. As the weight on earth increases, the weight on the moon increases.

Solve.

a) First find an equation of variation.

$$M = kE$$

$$32 = k \cdot 192 \quad \text{Substituting 32 for } M \text{ and 192 for } E$$

$$\frac{32}{192} = k$$

$$\frac{1}{6} = k$$

The equation of variation is $M = \frac{1}{6}E$.

b) Use the equation to find how much a 110-lb person would weigh on the moon.

$$M = \frac{1}{6}E$$

$$M = \frac{1}{6} \cdot 110 \qquad \text{Substituting 110 for } E$$

$$M = \frac{110}{6}, \text{ or } 18.\overline{3}$$

c) Use the equation to find how much a person who weighs 5 lb on the moon would weigh on Earth.

$$M = \frac{1}{6}E$$

$$5 = \frac{1}{6}E$$

$$30 = E \quad \text{Multiplying by 6}$$

Check. In addition to repeating the computations we can do some reasoning. When the weight on Earth decreased from 192 lb to 110 lb, the weight on the moon decreased from 32 lb to $18.\overline{3}$ lb. Similarly, when the weight on the moon decreased from 32 lb to 5 lb, the weight on Earth decreased from 192 lb to 30 lb.

State. a) The equation of variation is $M = \frac{1}{6}E$.

b) Elizabeth, who weighs 110 lb on Earth, would weigh $18.\overline{3}$ lb on the moon.

c) Jasmine, who weighs 5 lb on the moon, would weigh 30 lb on Earth.

15. ***Familiarize and Translate.*** The problem states that we have direct variation between the variables N and S. Thus, an equation $N = kS$, $k > 0$, applies. As the speed of the internal processor increases, the number of instructions increases.

Solve.

a) First find an equation of variation.

$$N = kS$$

$$2{,}000{,}000 = k \cdot 25 \quad \text{Substituting 2,000,000 for } N \text{ and 25 for } S$$

$$\frac{2{,}000{,}000}{25} = k$$

$$80{,}000 = k$$

The equation of variation is $N = 80{,}000S$.

b) Use the equation to find how many instructions the processor will perform at a speed of 200 megahertz.

$$N = 80{,}000S$$

$$N = 80{,}000 \cdot 200 \quad \text{Substituting 200 for } S$$

$$N = 16{,}000{,}000$$

Check. In addition to repeating the computations we can do some reasoning. The speed of the processor increased from 25 to 200 megahertz. Similarly, the number of instructions performed per second increased from 2,000,000 to 16,000,000.

State. a) The equation of variation is $N = 80{,}000S$.

b) The processor will perform 16,000,000 instructions per second running at a speed of 200 megahertz.

17. ***Familiarize and Translate.*** This problem states that we have direct variation between the variables S and W. Thus, an equation $S = kW$, $k > 0$, applies. As the weight increases, the number of servings increases.

Solve.

a) First find an equation of variation.

$$S = kW$$

$$70 = k \cdot 9 \quad \text{Substituting 70 for } S \text{ and 9 for } W$$

$$\frac{70}{9} = k$$

The equation of variation is $S = \frac{70}{9}W$.

b) Use the equation to find the number of servings from 12 kg of round steak.

$$S = \frac{70}{9}W$$

$$S = \frac{70}{9} \cdot 12 \qquad \text{Substituting 12 for } W$$

$$S = \frac{840}{9}$$

$$S = \frac{280}{3}, \text{ or } 93\frac{1}{3}$$

Check. A check can always be done by repeating the computations. We can also do some reasoning about the answer. When the weight increased from 9 kg to 12 kg, the number of servings increased from 70 to $93\frac{1}{3}$.

State. $93\frac{1}{3}$ servings can be obtained from 12 kg of round steak.

19. We substitute to find k.

$$y = \frac{k}{x}$$

$$3 = \frac{k}{25} \quad \text{Substituting 3 for } y \text{ and 25 for } x$$

$$25 \cdot 3 = k$$

$$75 = k$$

The equation of variation is $y = \frac{75}{x}$.

To find the value of y when $x = 10$ we substitute 10 for x in the equation of variation.

$$y = \frac{75}{x}$$

$$y = \frac{75}{10}$$

$$y = \frac{15}{2}, \text{ or } 7.5$$

The value of y is $\frac{15}{2}$, or 7.5, when $x = 10$.

21. We substitute to find k.

$$y = \frac{k}{x}$$

$$10 = \frac{k}{8} \quad \text{Substituting 10 for } y \text{ and 8 for } x$$

$$8 \cdot 10 = k$$

$$80 = k$$

The equation of variation is $y = \frac{80}{x}$.

To find the value of y when $x = 10$ we substitute 10 for x in the equation of variation.

$$y = \frac{80}{x}$$

$$y = \frac{80}{10}$$

$$y = 8$$

The value of y is 8 when $x = 10$.

23. We substitute to find k.

$$y = \frac{k}{x}$$

$$6.25 = \frac{k}{0.16} \quad \text{Substituting 6.25 for } y \text{ and 0.16 for } x$$

$$0.16(6.25) = k$$

$$1 = k$$

The equation of variation is $y = \frac{1}{x}$.

To find the value of y when $x = 10$ we substitute 10 for x in the equation of variation.

$$y = \frac{1}{x}$$

$$y = \frac{1}{10}$$

The value of y is $\frac{1}{10}$ when $x = 10$.

25. We substitute to find k.

$$y = \frac{k}{x}$$

$$50 = \frac{k}{42} \quad \text{Substituting 50 for } y \text{ and 42 for } x$$

$$42 \cdot 50 = k$$

$$2100 = k$$

The equation of variation is $y = \frac{2100}{x}$.

To find the value of y when $x = 10$ we substitute 10 for x in the equation of variation.

$$y = \frac{2100}{x}$$

$$y = \frac{2100}{10}$$

$$y = 210$$

The value of y is 210 when $x = 10$.

27. We substitute to find k.

$$y = \frac{k}{x}$$

$$0.2 = \frac{k}{0.3} \quad \text{Substituting 0.2 for } y \text{ and 0.3 for } x$$

$$0.06 = k$$

The equation of variation is $y = \frac{0.06}{x}$.

To find the value of y when $x = 10$ we substitute 10 for x in the equation of variation.

$$y = \frac{0.06}{x}$$

$$y = \frac{0.06}{10}$$

$$y = \frac{6}{1000} \quad \text{Multiplying } \frac{0.06}{10} \text{ by } \frac{100}{100}$$

$$y = \frac{3}{500}, \text{ or } 0.006$$

The value of y is $\frac{3}{500}$, or 0.006, when $x = 10$.

29. a) It seems reasonable that, as the number of hours of production increases, the number of compact-disc players produced will increase, so direct variation might apply.

b) ***Familiarize.*** Let H = the number of hours the production line is working, and let P = the number of compact-disc players produced. An equation $P = kH$, $k > 0$, applies. (See part (a)).

Translate. We write an equation of variation.

Number of players produced varies directly as hours of production. This translates to $P = kH$.

Solve.

a) First we find an equation of variation.

$P = kH$

$15 = k \cdot 8$ Substituting 8 for H and 15 for P

$\frac{15}{8} = k$

The equation of variation is $P = \frac{15}{8}H$.

b) Use the equation to find the number of players produced in 37 hr.

$P = \frac{15}{8}H$

$P = \frac{15}{8} \cdot 37$ Substituting 37 for H

$P = \frac{555}{8} = 69\frac{3}{8}$

Check. In addition to repeating the computations, we can do some reasoning. The number of hours increased from 8 to 37. Similarly, the number of compact disc players produced increased from 15 to $69\frac{3}{8}$.

State. About $69\frac{3}{8}$ compact-disc players can be produced in 37 hr.

31. a) It seems reasonable that, as the number of workers increases, the number of hours required to do the job decreases, so inverse variation might apply.

b) ***Familiarize.*** Let $T =$ the time required to cook the meal and $N =$ the number of cooks. An equation $T = k/N$, $k > 0$, applies. (See part (a)).

Translate. We write an equation of variation. Time varies inversely as the number of cooks. This translates to $T = \frac{k}{N}$.

Solve.

a) First find the equation of variation.

$T = \frac{k}{N}$

$4 = \frac{k}{9}$ Substituting 4 for T and 9 for N

$36 = k$

The equation of variation is $T = \frac{36}{N}$.

b) Use the equation to find the amount of time it takes 8 cooks to prepare the dinner.

$T = \frac{36}{N}$

$T = \frac{36}{8}$ Substituting 8 for N

$T = 4.5$

Check. The check might be done by repeating the computation. We might also analyze the results. The number of cooks decreased from 9 to 8, and the time increased from 4 hr to 4.5 hr. This is what we would expect with inverse variation.

State. It will take 8 cooks 4.5 hr to prepare the dinner.

33. ***Familiarize.*** The problem states that we have inverse variation between the variables N and P. Thus, an equation $N = k/P$, $k > 0$, applies. As the miles per gallon rating increases, the number of gallons required to travel the fixed distance decreases.

Translate. We write an equation of variation. Number of gallons varies inversely as miles per gallon rating. This translates to $N = \frac{k}{P}$.

Solve.

a) First find an equation of variation.

$N = \frac{k}{P}$

$20 = \frac{k}{14}$ Substituting 20 for N and 14 for P

$280 = k$

The equation is $N = \frac{280}{P}$.

b) Use the equation to find the number of gallons of gasoline needed for a car that gets 28 mpg.

$N = \frac{k}{P}$

$N = \frac{280}{28}$ Substituting 28 for P

$N = 10$

Check. In addition to repeating the computations, we can analyze the results. The number of miles per gallon increased from 14 to 28, and the number of gallons required decreased from 20 to 10. This is what we would expect with inverse variation.

State. a) The equation of variation is $N = \frac{280}{P}$.

b) A car that gets 28 mpg will need 10 gallons of gasoline to travel the fixed distance.

35. ***Familiarize.*** The problem states that we have inverse variation between the variables I and R. Thus, an equation $I = k/R$, $k > 0$, applies. As the resistance increases, the current decreases.

Translate. We write an equation of variation. Current varies inversely as resistance. This translates to $I = \frac{k}{R}$.

Solve.

a) First find an equation of variation.

$I = \frac{k}{R}$

$96 = \frac{k}{20}$ Substituting 96 for I and 20 for R

$1920 = k$

The equation of variation is $I = \frac{1920}{R}$.

b) Use the equation to find the current when the resistance is 60 ohms.

$I = \frac{1920}{R}$

$I = \frac{1920}{60}$ Substituting 60 for R

$I = 32$

Check. The check might be done by repeating the computations. We might also analyze the results. The resistance increased from 20 ohms to 60 ohms, and the current decreased from 96 amperes to 32 amperes. This is what we would expect with inverse variation.

State. a) The equation of variation is $I = \frac{1920}{R}$.

b) The current is 32 amperes when the resistance is 60 ohms.

37. ***Familiarize.*** The problem states that we have inverse variation between the variables m and n. Thus, an equation $m = k/n$, $k > 0$, applies. As the number of questions increases, the number of minutes allowed for each question decreases.

Translate. We write an equation of variation. Time allowed per question varies inversely as the number of questions.

Solve.

a) First find an equation of variation.

$m = \frac{k}{n}$

$2.5 = \frac{k}{16}$ Substituting 2.5 for m and 16 for n

$40 = k$

The equation of variation is $m = \frac{40}{n}$.

b) Use the equation to find the number of questions on a quiz when students have 4 min per question.

$m = \frac{40}{n}$

$4 = \frac{40}{n}$ Substituting 4 for m

$4n = 40$ Multiplying by n

$n = 10$ Dividing by 4

Check. The check might be done by repeating the computations. We might also analyze the results. The time allowed for each question increased from 2.5 min to 4 min, and the number of questions decreased from 16 to 10. This is what we would expect with inverse variation.

State. a) The equation of variation is $m = \frac{40}{n}$.

b) There would be 10 questions on a quiz for which students have 4 min per question.

39. ***Familiarize.*** The problem states that we have inverse variation between the variables A and d. Thus, an equation $A = k/d$, $k > 0$, applies. As the distance increases, the apparent size decreases.

Translate. We write an equation of variation. Apparent size varies inversely as the distance. This translates to $A = \frac{k}{d}$.

Solve.

a) First find an equation of variation.

$A = \frac{k}{d}$

$27.5 = \frac{k}{30}$ Substituting 27.5 for A and 30 for d

$825 = k$

The equation of variation is $A = \frac{825}{d}$.

b) Use the equation to find the apparent size when the distance is 100 ft.

$A = \frac{825}{d}$

$A = \frac{825}{100}$ Substituting 100 for d

$A = 8.25$

Check. The check might be done by repeating the computations. We might also analyze the results. The distance increased from 30 ft to 100 ft, and the apparent size decreased from 27.5 ft to 8.25 ft. This is what we would expect with inverse variation.

State. The flagpole will appear to be 8.25 ft tall when it is 100 ft from the observer.

41. $\left(4x-\frac{1}{4}\right)^2 = (4x)^2 - 2\cdot 4x\cdot \frac{1}{4} + \left(\frac{1}{4}\right)^2 = 16x^2 - 2x + \frac{1}{16}$

43.

$$\begin{array}{r} x^2 - 2x + 4 \\ x + 1 \\ \hline x^2 - 2x + 4 \\ x^3 - 2x^2 + 4x \\ \hline x^3 - x^2 + 2x + 4 \end{array}$$

45. $49x^2 - \frac{1}{16} = (7x)^2 - \left(\frac{1}{4}\right)^2 = \left(7x + \frac{1}{4}\right)\left(7x - \frac{1}{4}\right)$

47. $5x^2 + 8x - 21$

We will use the ac-method.

(1) There are no common factors (other than 1 or -1).

(2) Multiply 5 and -21: $5(-21) = -105$.

(3) Find a pair of factors of -105 whose sum is 8. The numbers we want are 15 and -7.

(4) Split the middle term: $8x = 15x - 7x$.

(5) Factor by grouping.

$$\begin{aligned} 5x^2 + 8x - 21 &= 5x^2 + 15x - 7x - 21 \\ &= 5x(x+3) - 7(x+3) \\ &= (x+3)(5x-7) \end{aligned}$$

49. $$\frac{x+2}{x+5} = \frac{x-4}{x-6}, \text{ LCM is } (x+5)(x-6)$$

$$(x+5)(x-6)\cdot\frac{x+2}{x+5} = (x+5)(x-6)\cdot\frac{x-4}{x-6}$$
$$(x-6)(x+2) = (x+5)(x-4)$$
$$x^2-4x-12 = x^2+x-20$$
$$-4x-12 = x-20 \quad \text{Subtracting } x^2$$
$$-5x = -8 \quad \text{Subtracting } x \text{ and adding 12}$$
$$x = \frac{8}{5}$$

The number $\frac{8}{5}$ checks and is the solution.

51. $x^2-25x+144=0$

$(x-9)(x-16)=0$

$x-9=0$ *or* $x-16=0$

$x=9$ *or* $x=16$

The solutions are 9 and 16.

53. $35x^2+8=34x$

$35x^2-34x+8=0$

$(7x-4)(5x-2)=0$

$7x-4=0$ *or* $5x-2=0$

$7x=4$ *or* $5x=2$

$x=\frac{4}{7}$ *or* $x=\frac{2}{5}$

The solutions are $\frac{4}{7}$ and $\frac{2}{5}$.

55. We do the divisions in order from left to right.

$$3^7 \div 3^4 \div 3^3 \div 3 = 3^3 \div 3^3 \div 3$$
$$= 1 \div 3$$
$$= \frac{1}{3}$$

57. $-5^2+4\cdot 6 = -25+4\cdot 6$

$= -25+24$

$= -1$

59.

X	Y1
1	.16667
2	.33333
3	.5
4	.66667
5	.83333
6	1
7	1.1667

X=1

The y-values become larger.

61. $P^2 = kt$

63. $P = kV^3$

Chapter 14 Concept Reinforcement

1. True; see page 1034 in the text.
2. False; this statement describes *inverse* variation.
3. True; $-(2-x) = -2+x = x-2$.
4. True; see page 1071 in the text.
5. False; $-(y+5) = -y-5 \neq y-5$.

Chapter 14 Important Concepts

1.
$$\frac{2x^2-2}{4x^2+24x+20} = \frac{2(x^2-1)}{4(x^2+6x+5)}$$
$$= \frac{2(x+1)(x-1)}{2\cdot 2(x+1)(x+5)}$$
$$= \frac{2(x+1)}{2(x+1)}\cdot\frac{x-1}{2(x+5)}$$
$$= \frac{x-1}{2(x+5)}$$

2.
$$\frac{2y^2+7y-15}{5y^2-45}\cdot\frac{y-3}{2y-3} = \frac{2y^2+7y-15)(y-3)}{(5y^2-45)(2y-3)}$$
$$= \frac{(2y-3)(y+5)(y-3)}{5(y^2-9)(2y-3)}$$
$$= \frac{\cancel{(2y-3)}(y+5)\cancel{(y-3)}}{5(y+3)\cancel{(y-3)}\cancel{(2y-3)}}$$
$$= \frac{y+5}{5(y+3)}$$

3.
$$\frac{b^2+3b-28}{b^2+5b-24} \div \frac{b-4}{b-3} = \frac{b^2+3b-28}{b^2+5b-24}\cdot\frac{b-3}{b-4}$$
$$= \frac{(b^2+3b-28)(b-3)}{(b^2+5b-24)(b-4)}$$
$$= \frac{(b+7)\cancel{(b-4)}\cancel{(b-3)}}{(b+8)\cancel{(b-3)}\cancel{(b-4)}}$$
$$= \frac{b+7}{b+8}$$

4.
$$\frac{x}{x-4}+\frac{2x-4}{4-x} = \frac{x}{x-4}+\frac{2x-4}{4-x}\cdot\frac{-1}{-1}$$
$$= \frac{x}{x-4}+\frac{-2x+4}{x-4}$$
$$= \frac{x-2x+4}{x-4}$$
$$= \frac{-x+4}{x-4}$$
$$= \frac{-1(x-4)}{x-4} = -1\cdot\frac{x-4}{x-4}$$
$$= -1$$

5. $$\frac{x}{x^2+x-2}-\frac{5}{x^2-1}$$
$$=\frac{x}{(x+2)(x-1)}-\frac{5}{(x+1)(x-1)},$$
LCM is $(x+2)(x-1)(x+1)$
$$=\frac{x}{(x+2)(x-1)}\cdot\frac{x+1}{x+1}-\frac{5}{(x+1)(x-1)}\cdot\frac{x+2}{x+2}$$
$$=\frac{x(x+1)-5(x+2)}{(x+2)(x-1)(x+1)}$$
$$=\frac{x^2+x-5x-10}{(x+2)(x-1)(x+1)}$$
$$=\frac{x^2-4x-10}{(x+2)(x-1)(x+1)}$$

6. $$\frac{1}{x}=\frac{2}{3-x},\ \text{LCM is } x(3-x)$$
$$x(3-x)\cdot\frac{1}{x}=x(3-x)\cdot\frac{2}{3-x}$$
$$3-x=2x$$
$$3=3x$$
$$1=x$$
The number 1 checks. It is the solution.

7. $$\frac{\frac{2}{5}-\frac{1}{y}}{\frac{3}{y}-\frac{1}{3}}\quad \text{(The LCM of 5, } y\text{, and 3 is } 15y\text{).}$$
$$=\frac{\frac{2}{5}-\frac{1}{y}}{\frac{3}{y}-\frac{1}{3}}\cdot\frac{15y}{15y}$$
$$=\frac{\frac{2}{5}\cdot 15y-\frac{1}{y}\cdot 15y}{\frac{3}{y}\cdot 15y-\frac{1}{3}\cdot 15y}$$
$$=\frac{6y-15}{45-5y}=\frac{3(2y-5)}{5(9-y)}$$

8. $y=kx$

$60=k(0.4)$ Substituting

$150=k$ Variation constant

The equation of variation is $y=150x$.

Substitute 2 for x and find y.

$$y=150x=150\cdot 2=300$$

The value of y is 300 when $x=2$.

9. $$y=\frac{k}{x}$$

$150=\frac{k}{1.5}$ Substituting

$225=k$ Variation constant

The equation of variation is $y=\frac{225}{x}$.

Substitute 10 for x and find y.

$$y=\frac{225}{x}=\frac{225}{10}=22.5$$

The value of y is 22.5 when $x=10$.

Chapter 14 Review Exercises

1. $\frac{3}{x}$

The denominator is 0 when $x=0$, so the expression is not defined for the replacement number 0.

2. $\frac{4}{x-6}$

To determine the numbers for which the rational expression is not defined, we set the denominator equal to 0 and solve:

$$x-6=0$$
$$x=6$$

The expression is not defined for the replacement number 6.

3. $\frac{x+5}{x^2-36}$

To determine the numbers for which the rational expression is not defined, we set the denominator equal to 0 and solve:

$$x^2-36=0$$
$$(x+6)(x-6)=0$$
$$x+6=0 \quad \textit{or} \quad x-6=0$$
$$x=-6 \quad \textit{or} \quad x=6$$

The expression is not defined for the replacement numbers −6 and 6.

4. $\frac{x^2-3x+2}{x^2+x-30}$

To determine the numbers for which the rational expression is not defined, we set the denominator equal to 0 and solve:

$$x^2+x-30=0$$
$$(x+6)(x-5)=0$$
$$x+6=0 \quad \textit{or} \quad x-5=0$$
$$x=-6 \quad \textit{or} \quad x=5$$

The expression is not defined for the replacement numbers −6 and 5.

5. $\frac{-4}{(x+2)^2}$

To determine the numbers for which the rational expression is not defined, we set the denominator equal to 0 and solve:

$$(x+2)^2=0$$
$$(x+2)(x+2)=0$$
$$x+2=0 \quad \textit{or} \quad x+2=0$$
$$x=-2 \quad \textit{or} \quad x=-2$$

The expression is not defined for the replacement number −2.

6. $\dfrac{x-5}{5}$

Since the denominator is the constant 5, there are no replacement numbers for which the expression is not defined.

7. $\dfrac{4x^2-8x}{4x^2+4x} = \dfrac{4x(x-2)}{4x(x+1)}$
$= \dfrac{4x}{4x} \cdot \dfrac{x-2}{x+1}$
$= 1 \cdot \dfrac{x-2}{x+1}$
$= \dfrac{x-2}{x+1}$

8. $\dfrac{14x^2-x-3}{2x^2-7x+3} = \dfrac{(2x-1)(7x+3)}{(2x-1)(x-3)}$
$= \dfrac{2x-1}{2x-1} \cdot \dfrac{7x+3}{x-3}$
$= 1 \cdot \dfrac{7x+3}{x-3}$
$= \dfrac{7x+3}{x-3}$

9. $\dfrac{(y-5)^2}{y^2-25} = \dfrac{(y-5)(y-5)}{(y+5)(y-5)}$
$= \dfrac{y-5}{y+5} \cdot \dfrac{y-5}{y-5}$
$= \dfrac{y-5}{y+5} \cdot 1$
$= \dfrac{y-5}{y+5}$

10. $\dfrac{a^2-36}{10a} \cdot \dfrac{2a}{a+6} = \dfrac{(a^2-36)(2a)}{10a(a+6)}$
$= \dfrac{(a+6)(a-6) \cdot 2 \cdot a}{2 \cdot 5 \cdot a \cdot (a+6)}$
$= \dfrac{\cancel{(a+6)}(a-6) \cdot \cancel{2} \cdot \cancel{a}}{\cancel{2} \cdot 5 \cdot \cancel{a} \cdot \cancel{(a+6)}}$
$= \dfrac{a-6}{5}$

11. $\dfrac{6t-6}{2t^2+t-1} \cdot \dfrac{t^2-1}{t^2-2t+1}$
$= \dfrac{(6t-6)(t^2-1)}{(2t^2+t-1)(t^2-2t+1)}$
$= \dfrac{6(t-1)(t+1)(t-1)}{(2t-1)(t+1)(t-1)(t-1)}$
$= \dfrac{6\cancel{(t-1)}\cancel{(t+1)}\cancel{(t-1)}}{(2t-1)\cancel{(t+1)}\cancel{(t-1)}\cancel{(t-1)}}$
$= \dfrac{6}{2t-1}$

12. $\dfrac{10-5t}{3} \div \dfrac{t-2}{12t} = \dfrac{10-5t}{3} \cdot \dfrac{12t}{t-2}$
$= \dfrac{(10-5t)(12t)}{3(t-2)}$
$= \dfrac{5(2-t) \cdot 3 \cdot 4t}{3(t-2)}$
$= \dfrac{5(-1)(t-2) \cdot 3 \cdot 4t}{3(t-2)}$ $\quad 2-t=-1(t-2)$
$= \dfrac{5(-1)\cancel{(t-2)} \cdot \cancel{3} \cdot 4t}{\cancel{3}\cancel{(t-2)} \cdot 1}$
$= -20t$

13. $\dfrac{4x^4}{x^2-1} \div \dfrac{2x^3}{x^2-2x+1} = \dfrac{4x^4}{x^2-1} \cdot \dfrac{x^2-2x+1}{2x^3}$
$= \dfrac{4x^4(x^2-2x+1)}{(x^2-1)(2x^3)}$
$= \dfrac{2 \cdot 2 \cdot x \cdot x^3(x-1)(x-1)}{(x+1)(x-1) \cdot 2 \cdot x^3}$
$= \dfrac{\cancel{2} \cdot 2 \cdot x \cdot \cancel{x^3}\cancel{(x-1)}(x-1)}{(x+1)\cancel{(x-1)} \cdot \cancel{2} \cdot \cancel{x^3}}$
$= \dfrac{2x(x-1)}{x+1}$

14. $3x^2 = 3 \cdot x \cdot x$
$10xy = 2 \cdot 5 \cdot x \cdot y$
$15y^2 = 3 \cdot 5 \cdot y \cdot y$
$\text{LCM} = 2 \cdot 3 \cdot 5 \cdot x \cdot x \cdot y \cdot y$, or $30x^2y^2$

15. $a-2 = a-2$
$4a-8 = 4(a-2)$
$\text{LCM} = 4(a-2)$

16. $y^2-y-2 = (y-2)(y+1)$
$y^2-4 = (y+2)(y-2)$
$\text{LCM} = (y-2)(y+1)(y+2)$

17. $\dfrac{x+8}{x+7} + \dfrac{10-4x}{x+7} = \dfrac{x+8+10-4x}{x+7} = \dfrac{-3x+18}{x+7} =$
$\dfrac{-3(x-6)}{x+7}$

18. $$\frac{3}{3x-9}+\frac{x-2}{3-x}=\frac{3}{3(x-3)}+\frac{x-2}{3-x}$$
$$=\frac{3}{3(x-3)}+\frac{x-2}{3-x}\cdot\frac{-1}{-1}$$
$$=\frac{3}{3(x-3)}+\frac{-1(x-2)}{-1(3-x)}$$
$$=\frac{3}{3(x-3)}+\frac{-x+2}{x-3}$$
$$=\frac{3}{3(x-3)}+\frac{-x+2}{x-3}\cdot\frac{3}{3}$$
$$=\frac{3}{3(x-3)}+\frac{-3x+6}{3(x-3)}$$
$$=\frac{3-3x+6}{3(x-3)}$$
$$=\frac{-3x+9}{3(x-3)}$$
$$=\frac{-3(x-3)}{3(x-3)}$$
$$=\frac{-1\cdot\cancel{3}(\cancel{x-3})}{1\cdot\cancel{3}(\cancel{x-3})}$$
$$=-1$$

19. $$\frac{2a}{a+1}+\frac{4a}{a^2-1}$$
$$=\frac{2a}{a+1}+\frac{4a}{(a+1)(a-1)},\text{ LCM is }(a+1)(a-1)$$
$$=\frac{2a}{a+1}\cdot\frac{a-1}{a-1}+\frac{4a}{(a+1)(a-1)}$$
$$=\frac{2a(a-1)+4a}{(a+1)(a-1)}$$
$$=\frac{2a^2-2a+4a}{(a+1)(a-1)}$$
$$=\frac{2a^2+2a}{(a+1)(a-1)}$$
$$=\frac{2a(a+1)}{(a+1)(a-1)}$$
$$=\frac{2a(\cancel{a+1})}{(\cancel{a+1})(a-1)}$$
$$=\frac{2a}{a-1}$$

20. $$\frac{d^2}{d-c}+\frac{c^2}{c-d}=\frac{d^2}{d-c}+\frac{c^2}{c-d}\cdot\frac{-1}{-1}$$
$$=\frac{d^2}{d-c}+\frac{-c^2}{d-c}$$
$$=\frac{d^2-c^2}{d-c}$$
$$=\frac{(d+c)(d-c)}{d-c}$$
$$=\frac{(d+c)(\cancel{d-c})}{(\cancel{d-c})\cdot 1}$$
$$=d+c$$

21. $$\frac{6x-3}{x^2-x-12}-\frac{2x-15}{x^2-x-12}=\frac{6x-3-(2x-15)}{x^2-x-12}$$
$$=\frac{6x-3-2x+15}{x^2-x-12}$$
$$=\frac{4x+12}{x^2-x-12}$$
$$=\frac{4(x+3)}{(x-4)(x+3)}$$
$$=\frac{4(\cancel{x+3})}{(x-4)(\cancel{x+3})}$$
$$=\frac{4}{x-4}$$

22. $$\frac{3x-1}{2x}-\frac{x-3}{x},\text{ LCM is }2x$$
$$=\frac{3x-1}{2x}-\frac{x-3}{x}\cdot\frac{2}{2}$$
$$=\frac{3x-1}{2x}-\frac{2(x-3)}{2x}$$
$$=\frac{3x-1-2(x-3)}{2x}$$
$$=\frac{3x-1-2x+6}{2x}$$
$$=\frac{x+5}{2x}$$

23. $$\frac{x+3}{x-2}-\frac{x}{2-x}=\frac{x+3}{x-2}-\frac{x}{2-x}\cdot\frac{-1}{-1}$$
$$=\frac{x+3}{x-2}-\frac{-x}{x-2}$$
$$=\frac{x+3-(-x)}{x-2}$$
$$=\frac{x+3+x}{x-2}$$
$$=\frac{2x+3}{x-2}$$

24. $$\frac{1}{x^2-25}-\frac{x-5}{x^2-4x-5}$$
$$=\frac{1}{(x+5)(x-5)}-\frac{x-5}{(x-5)(x+1)},$$
LCM is $(x+5)(x-5)(x+1)$
$$=\frac{1}{(x+5)(x-5)}\cdot\frac{x+1}{x+1}-\frac{x-5}{(x-5)(x+1)}\cdot\frac{x+5}{x+5}$$
$$=\frac{x+1}{(x+5)(x-5)(x+1)}-\frac{(x-5)(x+5)}{(x+5)(x-5)(x+1)}$$
$$=\frac{x+1-(x^2-25)}{(x+5)(x-5)(x+1)}$$
$$=\frac{x+1-x^2+25}{(x+5)(x-5)(x+1)}$$
$$=\frac{-x^2+x+26}{(x+5)(x-5)(x+1)}$$

25. $\frac{3x}{x+2} - \frac{x}{x-2} + \frac{8}{x^2-4}$

$= \frac{3x}{x+2} - \frac{x}{x-2} + \frac{8}{(x+2)(x-2)}$, LCM is $(x+2)(x-2)$

$= \frac{3x}{x+2}\cdot\frac{x-2}{x-2} - \frac{x}{x-2}\cdot\frac{x+2}{x+2} + \frac{8}{(x+2)(x-2)}$

$= \frac{3x(x-2)}{(x+2)(x-2)} - \frac{x(x+2)}{(x+2)(x-2)} + \frac{8}{(x+2)(x-2)}$

$= \frac{3x(x-2) - x(x+2) + 8}{(x+2)(x-2)}$

$= \frac{3x^2 - 6x - x^2 - 2x + 8}{(x+2)(x-2)}$

$= \frac{2x^2 - 8x + 8}{(x+2)(x-2)}$

$= \frac{2(x^2 - 4x + 4)}{(x+2)(x-2)}$

$= \frac{2(x-2)(x-2)}{(x+2)(x-2)}$

$= \frac{2\cancel{(x-2)}(x-2)}{(x+2)\cancel{(x-2)}}$

$= \frac{2(x-2)}{x+2}$

26. $\frac{\frac{1}{z}+1}{\frac{1}{z^2}-1}$ LCM of the denominators is z^2.

$= \frac{\frac{1}{z}+1}{\frac{1}{z^2}-1}\cdot\frac{z^2}{z^2}$

$= \frac{\left(\frac{1}{z}+1\right)z^2}{\left(\frac{1}{z^2}-1\right)z^2}$

$= \frac{\frac{1}{z}\cdot z^2 + 1\cdot z^2}{\frac{1}{z^2}\cdot z^2 - 1\cdot z^2}$

$= \frac{z+z^2}{1-z^2}$

$= \frac{z(1+z)}{(1+z)(1-z)}$

$= \frac{z\cancel{(1+z)}}{\cancel{(1+z)}(1-z)}$

$= \frac{z}{1-z}$

27. $\frac{\frac{c}{d}-\frac{d}{c}}{\frac{1}{c}+\frac{1}{d}}$

$= \frac{\frac{c}{d}\cdot\frac{c}{c}-\frac{d}{c}\cdot\frac{d}{d}}{\frac{1}{c}\cdot\frac{d}{d}+\frac{1}{d}\cdot\frac{c}{c}}$ Getting a common denominator in numerator and in denominator

$= \frac{\frac{c^2}{cd}-\frac{d^2}{cd}}{\frac{d}{cd}+\frac{c}{cd}}$

$= \frac{\frac{c^2-d^2}{cd}}{\frac{d+c}{cd}}$

$= \frac{c^2-d^2}{cd}\cdot\frac{cd}{d+c}$

$= \frac{(c^2-d^2)cd}{cd(d+c)}$

$= \frac{(c+d)(c-d)cd}{cd(d+c)}$

$= \frac{\cancel{(c+d)}(c-d)\cancel{cd}}{\cancel{cd}\,\cancel{(d+c)}\cdot 1}$

$= c-d$

28. $\frac{3}{y} - \frac{1}{4} = \frac{1}{y}$, LCM $= 4y$

$4y\left(\frac{3}{y}-\frac{1}{4}\right) = 4y\cdot\frac{1}{y}$

$4y\cdot\frac{3}{y} - 4y\cdot\frac{1}{4} = 4y\cdot\frac{1}{y}$

$12 - y = 4$

$-y = -8$

$y = 8$

This checks, so the solution is 8.

29. $\frac{15}{x} - \frac{15}{x+2} = 2$, LCM $= x(x+2)$

$x(x+2)\left(\frac{15}{x}-\frac{15}{x+2}\right) = x(x+2)\cdot 2$

$x(x+2)\cdot\frac{15}{x} - x(x+2)\cdot\frac{15}{x+2} = 2x(x+2)$

$15(x+2) - 15x = 2x(x+2)$

$15x + 30 - 15x = 2x^2 + 4x$

$30 = 2x^2 + 4x$

$0 = 2x^2 + 4x - 30$

$0 = 2(x^2 + 2x - 15)$

$0 = x^2 + 2x - 15$ Dividing by 2

$0 = (x+5)(x-3)$

$x + 5 = 0$ *or* $x - 3 = 0$

$x = -5$ *or* $x = 3$

Both numbers check. The solutions are -5 and 3.

30. ***Familiarize.*** Let t = the time the job would take if the crews worked together.

Translate. We use the work principle, substituting 9 for a and 12 for b.

$$\frac{t}{a}+\frac{t}{b}=1$$

$$\frac{t}{9}+\frac{t}{12}=1 \quad \text{Substituting}$$

Solve.

$$\frac{t}{9}+\frac{t}{12}=1, \text{ LCM} = 36$$

$$36\left(\frac{t}{9}+\frac{t}{12}\right)=36\cdot 1$$

$$36\cdot\frac{t}{9}+36\cdot\frac{t}{12}=36$$

$$4t+3t=36$$

$$7t=36$$

$$t=\frac{36}{7}, \text{ or } 5\frac{1}{7}$$

Check. In $\frac{36}{7}$ hr, the portion of the job done is $\frac{36}{7}\cdot\frac{1}{9}+\frac{36}{7}\cdot\frac{1}{12}=\frac{4}{7}+\frac{3}{7}=1.$ The answer checks.

State. The job would take $5\frac{1}{7}$ hr if the crews worked together.

31. ***Familiarize.*** Let r = the speed of the slower plane, in mph. Then $r+80$ = the speed of the faster plane. We organize the information in a table.

	Distance	Speed	Time
Slower plane	950	r	t
Faster plane	1750	$r+80$	t

Translate. We use the formula $d=rt$ in each row of the table to obtain two equations.

$$950=rt,$$

$$1750=(r+80)t$$

Since the times are the same, we solve each equation for t and set the results equal to each other.

$$950=rt, \text{ so } t=\frac{950}{r}.$$

$$1750=(r+80)t, \text{ so } t=\frac{1750}{r+80}.$$

Then we have

$$\frac{950}{r}=\frac{1750}{r+80}.$$

Solve.

$$\frac{950}{r}=\frac{1750}{r+80}, \text{ LCM} = r(r+80)$$

$$r(r+80)\cdot\frac{950}{r}=r(r+80)\cdot\frac{1750}{r+80}$$

$$950(r+80)=1750r$$

$$950r+76,000=1750r$$

$$76,000=800r$$

$$95=r$$

If $95=r$, then $r+80=95+80=175$.

Check. If the speeds are 95 mph and 175 mph, then the speed of the faster plane is 80 mph faster than the speed of the slower plane. At 95 mph, the slower plane travels 950 mi in 950/95, or 10 hr. At 175 mph, the faster plane travels 1750 mi in 1750/175, or 10 hr. The times are the same, so the answer checks.

State. The speed of the slower plane is 95 mph; the speed of the faster plane is 175 mph.

32. ***Familiarize.*** Let r = the speed of the slower train, in km/h. Then $r+40$ = the speed of the faster train. We organize the information in a table.

	Distance	Speed	Time
Slower train	60	r	t
Faster train	70	$r+40$	t

Translate. We use the formula $d=rt$ in each row of the table to obtain two equations.

$$60=rt,$$

$$70=(r+40)t$$

Since the times are the same, we solve each equation for t and set the results equal to each other.

$$60=rt, \text{ so } t=\frac{60}{r}.$$

$$70=(r+40)t, \text{ so } t=\frac{70}{r+40}.$$

Then we have

$$\frac{60}{r}=\frac{70}{r+40}.$$

Solve.

$$\frac{60}{r}=\frac{70}{r+40}, \text{ LCM} = r(r+40)$$

$$r(r+40)\cdot\frac{60}{r}=r(r+40)\cdot\frac{70}{r+40}$$

$$60(r+40)=70r$$

$$60r+2400=70r$$

$$2400=10r$$

$$240=r$$

If $r=240$, then $r+40=240+40=280$.

Check. If the speeds are 240 km/h and 280 km/h, then the speed of the faster train is 40 km/h faster than the speed of the slower train. At 240 km/h, the slower train travels 60 km in 60/240, or 1/4 hr. At 280 km/h, the faster train travels 70 km in 70/280, or 1/4 hr. Since the times are the same, the answer checks.

State. The speed of the slower train is 240 km/h; the speed of the faster train is 280 km/h.

33. ***Familiarize***. We can translate to a proportion, letting $d =$ the number of defective calculators that can be expected in a sample of 5000.

Translate.

$$\text{Number defective} \rightarrow \frac{8}{250} = \frac{d}{5000} \leftarrow \text{Number defective}$$
$$\text{Sample size} \rightarrow 250 \quad 5000 \leftarrow \text{Sample size}$$

Solve. We solve the proportion.

$$\frac{8}{250} = \frac{d}{5000}$$
$$5000 \cdot \frac{8}{250} = 5000 \cdot \frac{d}{5000}$$
$$160 = d$$

Check.

$$\frac{8}{250} = 0.032 \text{ and } \frac{160}{5000} = 0.032.$$

The ratios are the same, so the answer checks.

State. You would expect to find 160 defective calculators in a sample of 5000.

34. a) Let $x =$ the number of cups of onion that would be used. Then we can write and solve a proportion.

$$\frac{6}{13} = \frac{x}{2}$$
$$2 \cdot \frac{6}{13} = 2 \cdot \frac{x}{2}$$
$$\frac{12}{13} = x$$

Thus, $\frac{12}{13}$ cup of onion would be used.

b) Let $c =$ the number of cups of cheese that would be used. We can write and solve a proportion.

$$\frac{5}{7} = \frac{3}{c}$$
$$7c \cdot \frac{5}{7} = 7c \cdot \frac{3}{c}$$
$$5c = 21$$
$$c = \frac{21}{5}, \text{ or } 4\frac{1}{5}$$

Thus, $4\frac{1}{5}$ cups of cheese would be used.

c) Let $c =$ the number of cups of cheese that would be used. We can write and solve a proportion.

$$\frac{9}{14} = \frac{6}{c}$$
$$14c \cdot \frac{9}{14} = 14c \cdot \frac{6}{c}$$
$$9c = 84$$
$$c = \frac{84}{9} = \frac{28}{3}, \text{ or } 9\frac{1}{3}$$

Thus, $9\frac{1}{3}$ cups of cheese would be used.

35. ***Familiarize***. The ratio of blue whales tagged to the total blue whale population, P, is $\frac{500}{P}$. Of the 400 blue whales checked later, 20 were tagged. The ratio of blue whales tagged to blue whales checked is $\frac{20}{400}$.

Translate. Assuming the two ratios are the same, we can translate to a proportion.

$$\begin{array}{ccc} \text{Whales tagged originally} \longrightarrow & 500 & 20 \longleftarrow \text{Tagged whales caught later} \\ & \overline{} = \overline{} & \\ \text{Whale population} \longrightarrow & P & 400 \longleftarrow \text{Whales caught later} \end{array}$$

Solve. We solve the proportion.

$$400P \cdot \frac{500}{P} = 400P \cdot \frac{20}{400} \quad \text{Multiplying by the LCM, } 400P$$
$$400 \cdot 500 = P \cdot 20$$
$$200{,}000 = 20P$$
$$10{,}000 = P$$

Check.

$$\frac{500}{10{,}000} = \frac{1}{20} \text{ and } \frac{20}{400} = \frac{1}{20}.$$

The ratios are the same.

State. The blue whale population is about 10,000.

36. We write a proportion and solve it.

$$\frac{3.4}{8.5} = \frac{2.4}{x}$$
$$8.5x \cdot \frac{3.4}{8.5} = 8.5x \cdot \frac{2.4}{x}$$
$$3.4x = 20.4$$
$$x = 6$$

(Note that the proportions $\frac{8.5}{3.4} = \frac{x}{2.4}$, $\frac{8.5}{x} = \frac{3.4}{2.4}$, and $\frac{x}{8.5} = \frac{2.4}{3.4}$ could also be used.)

37. We substitute to find k.

$$y = kx$$
$$12 = k \cdot 4$$
$$3 = k$$

The equation of variation is $y = 3x$.

To find the value of y when $x = 20$ we substitute 20 for x in the equation of variation.

$$y = 3x$$
$$y = 3 \cdot 20$$
$$y = 60$$

The value of y is 60 when $x = 20$.

38. We substitute to find k.

$$y = kx$$
$$0.4 = k \cdot 0.5$$
$$\frac{0.4}{0.5} = k$$
$$\frac{4}{5} = k \quad \left(\frac{0.4}{0.5} = \frac{0.4}{0.5} \cdot \frac{10}{10} = \frac{4}{5}\right)$$

The equation of variation is $y = \frac{4}{5}x$.

To find the value of y when $x = 20$ we substitute 20 for x in the equation of variation.

$$y = \frac{4}{5}x$$
$$y = \frac{4}{5} \cdot 20$$
$$y = 16$$

The value of y is 16 when $x = 20$.

39. We substitute to find k.

$$y = \frac{k}{x}$$
$$5 = \frac{k}{6}$$
$$30 = k$$

The equation of variation is $y = \frac{30}{x}$.

To find the value of y when $x = 5$ we substitute 5 for x in the equation of variation.

$$y = \frac{30}{x}$$
$$y = \frac{30}{5}$$
$$y = 6$$

The value of y is 6 when $x = 5$.

40. We substitute to find k.

$$y = \frac{k}{x}$$
$$0.5 = \frac{k}{2}$$
$$1 = k$$

The equation of variation is $y = \frac{1}{x}$.

To find the value of y when $x = 5$ we substitute 5 for x in the equation of variation.

$$y = \frac{1}{x}$$
$$y = \frac{1}{5}$$

The value of y is $\frac{1}{5}$ when $x = 5$.

41. We substitute to find k.

$$y = \frac{k}{x}$$
$$1.3 = \frac{k}{0.5}$$
$$0.65 = k$$

The equation of variation is $y = \frac{0.65}{x}$.

To find the value of y when $x = 5$ we substitute 5 for x in the equation of variation.

$$y = \frac{0.65}{x}$$
$$y = \frac{0.65}{5}$$
$$y = 0.13$$

The value of y is 0.13 when $x = 5$.

42. ***Familiarize and Translate***. The problem states that we have direct variation between the variables P and H. Thus, an equation $P = kH$, $k > 0$, applies.

Solve.

a) First we find an equation of variation.

$$P = kH$$
$$165 = k \cdot 20$$
$$8.25 = k$$

The equation of variation is $P = 8.25H$.

b) Use the equation to find the pay for 35 hr of work.

$$P = 8.25H$$
$$P = 8.25(35)$$
$$P = 288.75$$

Check. We can repeat the computations. Also note that when the number of hours worked increased from 20 to 35, the pay increased from \$165.00 to \$288.75 so the answer seems reasonable.

State. The pay for 35 hr of work is \$288.75.

43. ***Familiarize and Translate***. Let M = the number of washing machines used and T = the time required to do the laundry. The problem states that we have inverse variation between T and M, so an equation $T = k/M$, $k > 0$, applies.

Solve.

a) First we find an equation of variation.

$$T = \frac{k}{M}$$
$$5 = \frac{k}{2}$$
$$10 = k$$

The equation of variation is $T = \frac{10}{M}$.

b) Use the equation to find the time required to do the laundry if 10 washing machines are used.

$$T = \frac{10}{M}$$
$$T = \frac{10}{10}$$
$$T = 1$$

Check. We can repeat the computations. Also note that as the number of washing machines increased from 5 to 10, the time required to do the laundry decreased from 2 hr to 1 hr, so the answer seems reasonable.

State. It would take 1 hr for 10 washing machines to do the laundry.

44. $\dfrac{3x^2 - 2x - 1}{3x^2 + x}$

Set the denominator equal to 0 and solve for x.

$$3x^2 + x = 0$$
$$x(3x + 1) = 0$$
$$x = 0 \quad or \quad 3x + 1 = 0$$
$$x = 0 \quad or \quad 3x = -1$$
$$x = 0 \quad or \quad x = -\frac{1}{3}$$

The expression is not defined for 0 and $-\dfrac{1}{3}$. Answer C is correct.

45. $\dfrac{1}{x-5} - \dfrac{1}{x+5}$, LCM is $(x-5)(x+5)$

$$= \frac{1}{x-5} \cdot \frac{x+5}{x+5} - \frac{1}{x+5} \cdot \frac{x-5}{x-5}$$
$$= \frac{x+5-(x-5)}{(x-5)(x+5)}$$
$$= \frac{x+5-x+5}{(x-5)(x+5)}$$
$$= \frac{10}{(x-5)(x+5)}$$

Answer A is correct.

46. $\dfrac{2a^2+5a-3}{a^2} \cdot \dfrac{5a^3+30a^2}{2a^2+7a-4} \div \dfrac{a^2+6a}{a^2+7a+12}$

$$= \frac{2a^2+5a-3}{a^2} \cdot \frac{5a^3+30a^2}{2a^2+7a-4} \cdot \frac{a^2+7a+12}{a^2+6a}$$
$$= \frac{(2a-1)(a+3)}{a^2} \cdot \frac{5a^2(a+6)}{(2a-1)(a+4)} \cdot \frac{(a+3)(a+4)}{a(a+6)}$$
$$= \frac{(2a-1)(a+3)(5a^2)(a+6)(a+3)(a+4)}{a^2(2a-1)(a+4)(a)(a+6)}$$
$$= \frac{\cancel{(2a-1)}(a+3) \cdot 5 \cdot \cancel{a^2}\,\cancel{(a+6)}(a+3)\cancel{(a+4)}}{\cancel{a^2}\,\cancel{(2a-1)}\cancel{(a+4)}(a)\cancel{(a+6)}}$$
$$= \frac{5(a+3)^2}{a}$$

47. $\dfrac{A+B}{B} = \dfrac{C+D}{D}$

$$\frac{A}{B} + \frac{B}{B} = \frac{C}{D} + \frac{D}{D}$$
$$\frac{A}{B} + 1 = \frac{C}{D} + 1$$
$$\frac{A}{B} = \frac{C}{D}$$

The two given proportions are equivalent.

Chapter 14 Discussion and Writing Exercises

1. No; when adding, no sign changes are required so the result is the same regardless of parentheses. When subtracting, however, the sign of each term of the expression being subtracted must be changed and parentheses are needed to make sure this is done.

2. Graph each side of the equation and determine the number of points of intersection of the graphs.

3. Canceling removes a factor of 1, allowing us to rewrite $a \cdot 1$ as a.

4. The larger the average gain per play, the smaller the number of plays required to go 80 yd. Thus, inverse variation applies.

5. Form a rational expression that has factors of $x+3$ and $x-4$ in the denominator.

6. If we multiply both sides of a rational equation by a variable expression in order to clear fractions, it is possible that the variable expression is equal to 0. Thus, an equivalent equation might not be produced.

Chapter 14 Test

1. $\dfrac{8}{2x}$

To determine the numbers for which the rational expression is not defined, we set the denominator equal to 0 and solve:

$$2x = 0$$
$$x = 0$$

The expression is not defined for the replacement number 0.

2. $\dfrac{5}{x+8}$

To determine the numbers for which the rational expression is not defined, we set the denominator equal to 0 and solve:

$$x + 8 = 0$$
$$x = -8$$

The expression is not defined for the replacement number -8.

3. $\dfrac{x-7}{x^2-49}$

To determine the numbers for which the rational expression is not defined, we set the denominator equal to 0 and solve:

$$x^2 - 49 = 0$$
$$(x+7)(x-7) = 0$$
$$x + 7 = 0 \quad or \quad x - 7 = 0$$
$$x = -7 \quad or \quad x = 7$$

The expression is not defined for the replacement numbers -7 and 7.

4. $\dfrac{x^2+x-30}{x^2-3x+2}$

To determine the numbers for which the rational expression is not defined, we set the denominator equal to 0 and solve:

$$x^2 - 3x + 2 = 0$$
$$(x-1)(x-2) = 0$$

$x-1=0$ *or* $x-2=0$

$x=1$ *or* $x=2$

The expression is not defined for the replacement numbers 1 and 2.

5. $\dfrac{11}{(x-1)^2}$

To determine the numbers for which the rational expression is not defined, we set the denominator equal to 0 and solve:

$$(x-1)^2=0$$
$$(x-1)(x-1)=0$$
$$x-1=0 \text{ or } x-1=0$$
$$x=1 \text{ or } x=1$$

The expression is not defined for the replacement number 1.

6. $\dfrac{x+2}{2}$

Since the denominator is the constant 2, there are no replacement numbers for which the expression is not defined.

7. $\dfrac{6x^2+17x+7}{2x^2+7x+3}=\dfrac{(2x+1)(3x+7)}{(2x+1)(x+3)}$

$=\dfrac{2x+1}{2x+1}\cdot\dfrac{3x+7}{x+3}$

$=1\cdot\dfrac{3x+7}{x+3}$

$=\dfrac{3x+7}{x+3}$

8. $\dfrac{a^2-25}{6a}\cdot\dfrac{3a}{a-5}=\dfrac{(a^2-25)(3a)}{6a(a-5)}$

$=\dfrac{(a+5)(a-5)\cdot 3\cdot a}{2\cdot 3\cdot a\cdot(a-5)}$

$=\dfrac{(a+5)\cancel{(a-5)}\cdot\cancel{3}\cdot\cancel{a}}{2\cdot\cancel{3}\cdot\cancel{a}\cdot\cancel{(a-5)}}$

$=\dfrac{a+5}{2}$

9. $\dfrac{25x^2-1}{9x^2-6x}\div\dfrac{5x^2+9x-2}{3x^2+x-2}$

$=\dfrac{25x^2-1}{9x^2-6x}\cdot\dfrac{3x^2+x-2}{5x^2+9x-2}$

$=\dfrac{(25x^2-1)(3x^2+x-2)}{(9x^2-6x)(5x^2+9x-2)}$

$=\dfrac{(5x+1)(5x-1)(3x-2)(x+1)}{3x(3x-2)(5x-1)(x+2)}$

$=\dfrac{(5x+1)\cancel{(5x-1)}\cancel{(3x-2)}(x+1)}{3x\cancel{(3x-2)}\cancel{(5x-1)}(x+2)}$

$=\dfrac{(5x+1)(x+1)}{3x(x+2)}$

10. $y^2-9=(y+3)(y-3)$

$y^2+10y+21=(y+3)(y+7)$

$y^2+4y-21=(y-3)(y+7)$

$\text{LCM}=(y+3)(y-3)(y+7)$

11. $\dfrac{16+x}{x^3}+\dfrac{7-4x}{x^3}=\dfrac{16+x+7-4x}{x^3}=\dfrac{23-3x}{x^3}$

12. $\dfrac{5-t}{t^2+1}-\dfrac{t-3}{t^2+1}=\dfrac{5-t-(t-3)}{t^2+1}$

$=\dfrac{5-t-t+3}{t^2+1}$

$=\dfrac{8-2t}{t^2+1}$

13. $\dfrac{x-4}{x-3}+\dfrac{x-1}{3-x}=\dfrac{x-4}{x-3}+\dfrac{x-1}{3-x}\cdot\dfrac{-1}{-1}$

$=\dfrac{x-4}{x-3}+\dfrac{-x+1}{x-3}$

$=\dfrac{x-4-x+1}{x-3}$

$=\dfrac{-3}{x-3}$

14. $\dfrac{x-4}{x-3}-\dfrac{x-1}{3-x}=\dfrac{x-4}{x-3}-\dfrac{x-1}{3-x}\cdot\dfrac{-1}{-1}$

$=\dfrac{x-4}{x-3}-\dfrac{-x+1}{x-3}$

$=\dfrac{x-4-(-x+1)}{x-3}$

$=\dfrac{x-4+x-1}{x-3}$

$=\dfrac{2x-5}{x-3}$

15. $\dfrac{5}{t-1}+\dfrac{3}{t}$, LCD is $t(t-1)$.

$=\dfrac{5}{t-1}\cdot\dfrac{t}{t}+\dfrac{3}{t}\cdot\dfrac{t-1}{t-1}$

$=\dfrac{5t}{t(t-1)}+\dfrac{3(t-1)}{t(t-1)}$

$=\dfrac{5t}{t(t-1)}+\dfrac{3t-3}{t(t-1)}$

$=\dfrac{5t+3t-3}{t(t-1)}$

$=\dfrac{8t-3}{t(t-1)}$

16. $\dfrac{1}{x^2-16}-\dfrac{x+4}{x^2-3x-4}$

$=\dfrac{1}{(x+4)(x-4)}-\dfrac{x+4}{(x-4)(x+1)}$, LCD is $(x+4)(x-4)(x+1)$.

$=\dfrac{1}{(x+4)(x-4)}\cdot\dfrac{x+1}{x+1}-\dfrac{x+4}{(x-4)(x+1)}\cdot\dfrac{x+4}{x+4}$

$=\dfrac{x+1-(x+4)(x+4)}{(x+4)(x-4)(x+1)}$

$=\dfrac{x+1-(x^2+8x+16)}{(x+4)(x-4)(x+1)}$

$=\dfrac{x+1-x^2-8x-16}{(x+4)(x-4)(x+1)}$

$=\dfrac{-x^2-7x-15}{(x+4)(x-4)(x+1)}$

17.
$$\frac{1}{x-1}+\frac{4}{x^2-1}-\frac{2}{x^2-2x+1}$$
$$=\frac{1}{x-1}+\frac{4}{(x+1)(x-1)}-\frac{2}{(x-1)(x-1)},$$
LCD is $(x+1)(x-1)(x-1)$
$$=\frac{1}{x-1}\cdot\frac{(x+1)(x-1)}{(x+1)(x-1)}+\frac{4}{(x+1)(x-1)}\cdot\frac{x-1}{x-1}-\frac{2}{(x-1)(x-1)}\cdot\frac{x+1}{x+1}$$
$$=\frac{(x+1)(x-1)+4(x-1)-2(x+1)}{(x+1)(x-1)(x-1)}$$
$$=\frac{x^2-1+4x-4-2x-2}{(x+1)(x-1)(x-1)}$$
$$=\frac{x^2+2x-7}{(x+1)(x-1)^2}$$

18. We multiply the numerator and the denominator by the LCM of the denominators, y^2.
$$\frac{9-\frac{1}{y^2}}{3-\frac{1}{y}}=\frac{9-\frac{1}{y^2}}{3-\frac{1}{y}}\cdot\frac{y^2}{y^2}$$
$$=\frac{\left(9-\frac{1}{y^2}\right)\cdot y^2}{\left(3-\frac{1}{y}\right)\cdot y^2}$$
$$=\frac{9\cdot y^2-\frac{1}{y^2}\cdot y^2}{3\cdot y^2-\frac{1}{y}\cdot y^2}$$
$$=\frac{9y^2-1}{3y^2-y}$$
$$=\frac{(3y+1)(3y-1)}{y(3y-1)}$$
$$=\frac{(3y+1)\cancel{(3y-1)}}{y\cancel{(3y-1)}}$$
$$=\frac{3y+1}{y}$$

19.
$$\frac{7}{y}-\frac{1}{3}=\frac{1}{4},\ \text{LCM is } 12y$$
$$12y\left(\frac{7}{y}-\frac{1}{3}\right)=12y\cdot\frac{1}{4}$$
$$12y\cdot\frac{7}{y}-12y\cdot\frac{1}{3}=12y\cdot\frac{1}{4}$$
$$84-4y=3y$$
$$84=7y$$
$$12=y$$

The number 12 checks, so it is the solution.

20.
$$\frac{15}{x}-\frac{15}{x-2}=-2,\ \text{LCM is } x(x-2)$$
$$x(x-2)\left(\frac{15}{x}-\frac{15}{x-2}\right)=x(x-2)(-2)$$
$$x(x-2)\cdot\frac{15}{x}-x(x-2)\cdot\frac{15}{x-2}=-2x(x-2)$$
$$15(x-2)-15x=-2x(x-2)$$
$$15x-30-15x=-2x^2+4x$$
$$-30=-2x^2+4x$$
$$2x^2-4x-30=0$$
$$2(x^2-2x-15)=0$$
$$x^2-2x-15=0\quad \text{Dividing by 2}$$
$$(x-5)(x+3)=0$$
$$x-5=0 \quad or \quad x+3=0$$
$$x=5 \quad or \quad x=-3$$

Both numbers check. The solutions are 5 and -3.

21. We substitute to find k.
$$y=kx$$
$$6=k\cdot 3$$
$$2=k$$

The equation of variation is $y=2x$.

To find the value of y when $x=25$ we substitute 25 for x in the equation of variation.
$$y=2x$$
$$y=2\cdot 25$$
$$y=50$$

The value of y is 50 when $x=25$.

22. We substitute to find k.
$$y=kx$$
$$1.5=k\cdot 3$$
$$0.5=k$$

The equation of variation is $y=0.5x$.

To find the value of y when $x=25$ we substitute 25 for x in the equation of variation.
$$y=0.5x$$
$$y=0.5(25)$$
$$y=12.5$$

The value of y is 12.5 when $x=25$.

23. We substitute to find k.
$$y=\frac{k}{x}$$
$$6=\frac{k}{3}$$
$$18=k$$

The equation of variation is $y=\frac{18}{x}$.

To find the value of y when $x = 100$ we substitute 100 for x in the equation of variation.

$$y = \frac{18}{x}$$
$$y = \frac{18}{100}$$
$$y = \frac{9}{50}$$

The value of y is $\frac{9}{50}$ when $x = 100$.

24. We substitute to find k.

$$y = \frac{k}{x}$$
$$11 = \frac{k}{2}$$
$$22 = k$$

The equation of variation is $y = \frac{22}{x}$.

To find the value of y when $x = 100$ we substitute 100 for x in the equation of variation.

$$y = \frac{22}{x}$$
$$y = \frac{22}{100}$$
$$y = \frac{11}{50}$$

The value of y is $\frac{11}{50}$ when $x = 100$.

25. ***Familiarize and Translate.*** The problem states that we have direct variation between the variables d and t. Thus, an equation $d = kt$, $k > 0$, applies.

Solve.

a) First find an equation of variation.

$$d = kt$$
$$60 = k \cdot \frac{1}{2}$$
$$120 = k$$

The equation of variation is $d = 120t$.

b) Use the equation to find the distance the train will travel in 2 hr.

$$d = 120t$$
$$d = 120 \cdot 2$$
$$d = 240$$

Check. We can repeat the computations. Also note that when the time increases from $\frac{1}{2}$ hr to 2 hr, the distance traveled increases from 60 km to 240 km so the answer seems reasonable.

State. The train will travel 240 km in 2 hr.

26. ***Familiarize and Translate.*** Let T = the time required to do the job and let M = the number of concrete mixers used. We have inverse variation between T and M so an equation $T = \frac{k}{M}$, $k > 0$, applies.

Solve.

a) First we find an equation of variation.

$$T = \frac{k}{M}$$
$$3 = \frac{k}{2}$$
$$6 = k$$

The equation of variation is $T = \frac{6}{M}$.

b) Use the equation to find the time required to do the job when 5 mixers are used.

$$T = \frac{6}{M}$$
$$T = \frac{6}{5}$$
$$T = 1\frac{1}{5}$$

Check. We can repeat the computations. Also note that as the number of mixers increased from 2 to 5, the time required to mix the concrete decreased from 3 hr to $1\frac{1}{5}$ hr, so the answer seems reasonable.

State. It would take $1\frac{1}{5}$ hr to do the job if 5 concrete mixers are used.

27. ***Familiarize.*** We can translate to a proportion, letting d = the number of defective spark plugs that would be expected in a sample of 500.

Translate.

$$\begin{array}{r} \text{Number defective} \rightarrow \\ \text{Sample size} \rightarrow \end{array} \frac{4}{125} = \frac{d}{500} \begin{array}{l} \leftarrow \text{Number defective} \\ \leftarrow \text{Sample size} \end{array}$$

Solve. We solve the proportion.

$$\frac{4}{125} = \frac{d}{500}$$
$$500 \cdot \frac{4}{125} = 500 \cdot \frac{d}{500}$$
$$16 = d$$

Check.

$$\frac{4}{125} = 0.032 \text{ and } \frac{16}{500} = 0.032.$$

The ratios are the same, so the answer checks.

State. You would expect to find 16 defective spark plugs in a sample of 500.

28. ***Familiarize.*** The ratio of zebras tagged to the total zebra population, P, is $\frac{15}{P}$. Of the 20 zebras checked later, 6 were tagged. The ratio of zebras tagged to zebras checked is $\frac{6}{20}$.

Translate. Assuming the two ratios are the same, we can translate to a proportion.

$$\begin{array}{c} \text{Zebras tagged} \\ \text{originally} \\ \text{Zebra} \\ \text{population} \end{array} \begin{array}{c} \longrightarrow \\ \\ \longrightarrow \end{array} \frac{15}{P} = \frac{6}{20} \begin{array}{c} \longleftarrow \\ \\ \longleftarrow \end{array} \begin{array}{c} \text{Tagged zebras} \\ \text{caught later} \\ \text{Zebras caught} \\ \text{later} \end{array}$$

Solve. We solve the proportion.

$$\frac{15}{P} = \frac{6}{20}$$
$$20P \cdot \frac{15}{P} = 20P \cdot \frac{6}{20}$$
$$300 = 6P$$
$$50 = P$$

Check.

$$\frac{15}{50} = 0.3 \text{ and } \frac{6}{20} = 0.3.$$

The ratios are the same, so the answer checks.

State. The zebra population is 50.

29. ***Familiarize***. Let t = the time, in minutes, required to copy the report using both copy machines working together.

Translate. We use the work principle, substituting 20 for a and 30 for b.

$$\frac{t}{a} + \frac{t}{b} = 1$$
$$\frac{t}{20} + \frac{t}{30} = 1$$

Solve.

$$\frac{t}{20} + \frac{t}{30} = 1, \text{ LCM is } 60$$
$$60\left(\frac{t}{20} + \frac{t}{30}\right) = 60 \cdot 1$$
$$60 \cdot \frac{t}{20} + 60 \cdot \frac{t}{30} = 60$$
$$3t + 2t = 60$$
$$5t = 60$$
$$t = 12$$

Check. In 12 min, the portion of the job done is

$$12 \cdot \frac{1}{20} + 12 \cdot \frac{1}{30} = \frac{3}{5} + \frac{2}{5} = 1.$$

The answer checks.

State. It would take 12 min to copy the report using both machines working together.

30. ***Familiarize***. Let r = Marilyn's speed, in km/h. Then $r + 20$ = Craig's speed. We organize the information in a table.

	Distance	Speed	Time
Marilyn	225	r	t
Craig	325	$r + 20$	t

Translate. We use the formula $d = rt$ in each row of the table to obtain two equations.

$$225 = rt,$$
$$325 = (r + 20)t$$

Since the times are the same, we solve each equation for t and set the results equal to each other.

$$225 = rt, \text{ so } t = \frac{225}{r}.$$
$$325 = (r + 20)t, \text{ so } t = \frac{325}{r + 20}.$$

Then we have

$$\frac{225}{r} = \frac{325}{r + 20}.$$

Solve.

$$\frac{225}{r} = \frac{325}{r + 20}, \text{ LCM is } r(r + 20)$$
$$r(r + 20) \cdot \frac{225}{r} = r(r + 20) \cdot \frac{325}{r + 20}$$
$$225(r + 20) = 325r$$
$$225r + 4500 = 325r$$
$$4500 = 100r$$
$$45 = r$$

If $r = 45$, then $r + 20 = 45 + 20 = 65$.

Check. If Marilyn's speed is 45 km/h and Craig's speed is 65 km/h, then Craig's speed is 20 km/h faster than Marilyn's. At 45 km/h, Marilyn travels 225 km in 225/45, or 5 hr. At 65 km/h, Craig travels 325 km in 325/65, or 5 hr. Since the times are the same, the answer checks.

State. The speed of Marilyn's car is 45 km/h and the speed of Craig's car is 65 km/h.

31. We write a proportion and solve it.

$$\frac{12}{9} = \frac{20}{x}$$
$$9x \cdot \frac{12}{9} = 9x \cdot \frac{20}{x}$$
$$12x = 180$$
$$x = 15$$

(Note that the proportions $\frac{9}{12} = \frac{x}{20}$, $\frac{12}{20} = \frac{9}{x}$, and $\frac{20}{12} = \frac{x}{9}$ could also be used.)

32.

$$\frac{2}{x-4} + \frac{2x}{x^2-16} = \frac{1}{x+4},$$
$$\text{LCM} = (x-4)(x+4)$$
$$(x-4)(x+4)\left(\frac{2}{x-4} + \frac{2x}{(x+4)(x-4)}\right) = (x-4)(x+4) \cdot \frac{1}{x+4}$$
$$2(x+4) + 2x = x - 4$$
$$2x + 8 + 2x = x - 4$$
$$4x + 8 = x - 4$$
$$3x = -12$$
$$x = -4$$

The number -4 is not a solution because it makes a denominator zero. Thus, there is no solution. The correct answer is D.

33. ***Familiarize***. Let r = the number of hours it would take Rema to do the job working alone. Then $r+6$ = the number of hours it would take Reggie to do the job working alone.

Translate. We use the work principle, substituting $2\frac{6}{7}$, or $\frac{20}{7}$, for t, r for a, and $r+6$ for b.

$$\frac{t}{a}+\frac{t}{b}=1$$

$$\frac{\frac{20}{7}}{r}+\frac{\frac{20}{7}}{r+6}=1$$

Solve.

$$\frac{\frac{20}{7}}{r}+\frac{\frac{20}{7}}{r+6}=1$$

$$\frac{20}{7}\cdot\frac{1}{r}+\frac{20}{7}\cdot\frac{1}{r+6}=1$$

$$\frac{20}{7r}+\frac{20}{7(r+6)}=1,\ \text{LCM is } 7r(r+6)$$

$$7r(r+6)\left(\frac{20}{7r}+\frac{20}{7(r+6)}\right)=7r(r+6)\cdot 1$$

$$7r(r+6)\cdot\frac{20}{7r}+7r(r+6)\cdot\frac{20}{7(r+6)}=7r(r+6)$$

$$20(r+6)+20r=7r^2+42r$$

$$20r+120+20r=7r^2+42r$$

$$40r+120=7r^2+42r$$

$$0=7r^2+2r-120$$

$$0=(7r+30)(r-4)$$

$7r+30=0$ *or* $r-4=0$

$7r=-30$ *or* $r=4$

$r=-\frac{30}{7}$ *or* $r=4$

Check. Since the time cannot be negative, $-\frac{30}{7}$ cannot be a solution. If $r=4$, then $r+6=4+6=10$. In $2\frac{6}{7}$ hr, or $\frac{20}{7}$ hr, then the portion of the job done is

$$\frac{\frac{20}{7}}{4}+\frac{\frac{20}{7}}{10}=\frac{20}{7}\cdot\frac{1}{4}+\frac{20}{7}\cdot\frac{1}{10}=\frac{5}{7}+\frac{2}{7}=1.$$

The answer checks.

State. Working alone it would take Rema 4 hr to do the job and it would take Reggie 10 hr.

34.

$$1+\cfrac{1}{1+\cfrac{1}{1+\cfrac{1}{a}}}=1+\cfrac{1}{1+\cfrac{1}{\frac{a}{a}+\frac{1}{a}}}$$

$$=1+\cfrac{1}{1+\cfrac{1}{\frac{a+1}{a}}}$$

$$=1+\cfrac{1}{1+1\cdot\frac{a}{a+1}}$$

$$=1+\cfrac{1}{1+\frac{a}{a+1}}$$

$$=1+\cfrac{1}{\frac{a+1}{a+1}+\frac{a}{a+1}}$$

$$=1+\cfrac{1}{\frac{a+1+a}{a+1}}$$

$$=1+\cfrac{1}{\frac{2a+1}{a+1}}$$

$$=1+1\cdot\frac{a+1}{2a+1}$$

$$=1+\frac{a+1}{2a+1}$$

$$=\frac{2a+1}{2a+1}+\frac{a+1}{2a+1}$$

$$=\frac{2a+1+a+1}{2a+1}$$

$$=\frac{3a+2}{2a+1}$$

Chapter 15

Systems of Equations

Exercise Set 15.1

1. We check by substituting alphabetically 1 for x and 5 for y.

$$\begin{array}{c|c} \multicolumn{2}{c}{5x - 2y = -5} \\ \hline 5 \cdot 1 - 2 \cdot 5 \ ? & -5 \\ 5 - 10 & \\ -5 & \text{TRUE} \end{array} \qquad \begin{array}{c|c} \multicolumn{2}{c}{3x - 7y = -32} \\ \hline 3 \cdot 1 - 7 \cdot 5 \ ? & -32 \\ 3 - 35 & \\ -32 & \text{TRUE} \end{array}$$

The ordered pair $(1, 5)$ is a solution of both equations, so it is a solution of the system of equations.

3. We check by substituting alphabetically 4 for a and 2 for b.

$$\begin{array}{c|c} \multicolumn{2}{c}{3b - 2a = -2} \\ \hline 3 \cdot 2 - 2 \cdot 4 \ ? & -2 \\ 6 - 8 & \\ -2 & \text{TRUE} \end{array} \qquad \begin{array}{c|c} \multicolumn{2}{c}{b + 2a = 8} \\ \hline 2 + 2 \cdot 4 \ ? & 8 \\ 2 + 8 & \\ 10 & \text{FALSE} \end{array}$$

The ordered pair $(4, 2)$ is not a solution of $b + 2a = 8$, so it is not a solution of the system of equations.

5. We check by substituting alphabetically 15 for x and 20 for y.

$$\begin{array}{c|c} \multicolumn{2}{c}{3x - 2y = 5} \\ \hline 3 \cdot 15 - 2 \cdot 20 \ ? & 5 \\ 45 - 40 & \\ 5 & \text{TRUE} \end{array} \qquad \begin{array}{c|c} \multicolumn{2}{c}{6x - 5y = -10} \\ \hline 6 \cdot 15 - 5 \cdot 20 \ ? & -10 \\ 90 - 100 & \\ -10 & \text{TRUE} \end{array}$$

The ordered pair $(15, 20)$ is a solution of both equations, so it is a solution of the system of equations.

7. We check by substituting alphabetically -1 for x and 1 for y.

$$\begin{array}{c|c} \multicolumn{2}{c}{x = -1} \\ \hline -1 \ ? & -1 \quad \text{TRUE} \end{array} \qquad \begin{array}{c|c} \multicolumn{2}{c}{x - y = -2} \\ \hline -1 - 1 \ ? & -2 \\ -2 & \text{TRUE} \end{array}$$

The ordered pair $(-1, 1)$ is a solution of both equations, so it is a solution of the system of equations.

9. We check by substituting alphabetically 18 for x and 3 for y.

$$\begin{array}{c|c} \multicolumn{2}{c}{y = \frac{1}{6}x} \\ \hline 3 \ ? & \frac{1}{6} \cdot 18 \\ & 3 \quad \text{TRUE} \end{array} \qquad \begin{array}{c|c} \multicolumn{2}{c}{2x - y = 33} \\ \hline 2 \cdot 18 - 3 \ ? & 33 \\ 36 - 3 & \\ 33 & \text{TRUE} \end{array}$$

The ordered pair $(18, 3)$ is a solution of both equations, so it is a solution of the system of equations.

11. We graph the equations.

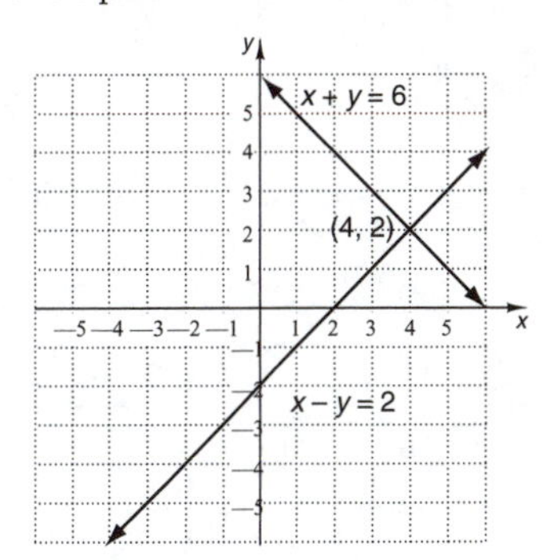

The point of intersection looks as if it has coordinates $(4, 2)$.

Check:

$$\begin{array}{c|c} \multicolumn{2}{c}{x - y = 2} \\ \hline 4 - 2 \ ? & 2 \\ 2 & \text{TRUE} \end{array} \qquad \begin{array}{c|c} \multicolumn{2}{c}{x + y = 6} \\ \hline 4 + 2 \ ? & 6 \\ 6 & \text{TRUE} \end{array}$$

The solution is $(4, 2)$.

13. We graph the equations.

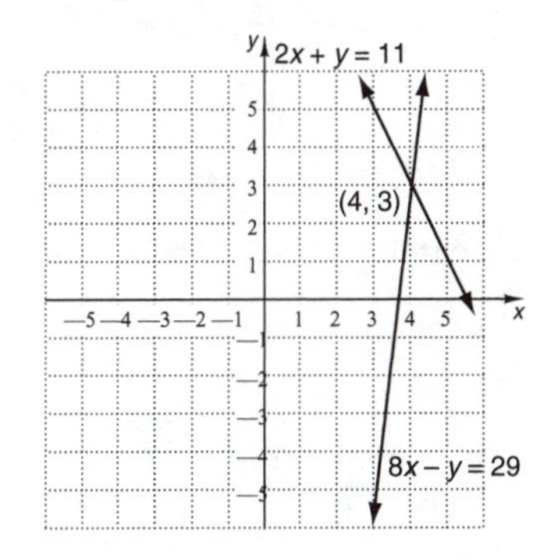

The point of intersection looks as if it has coordinates $(4, 3)$.

Check:

$$\begin{array}{c|c} \multicolumn{2}{c}{8x - y = 29} \\ \hline 8 \cdot 4 - 3 \ ? & 29 \\ 32 - 3 & \\ 29 & \text{TRUE} \end{array} \qquad \begin{array}{c|c} \multicolumn{2}{c}{2x + y = 11} \\ \hline 2 \cdot 4 + 3 \ ? & 11 \\ 8 + 3 & \\ 11 & \text{TRUE} \end{array}$$

The solution is $(4, 3)$.

15. We graph the equations.

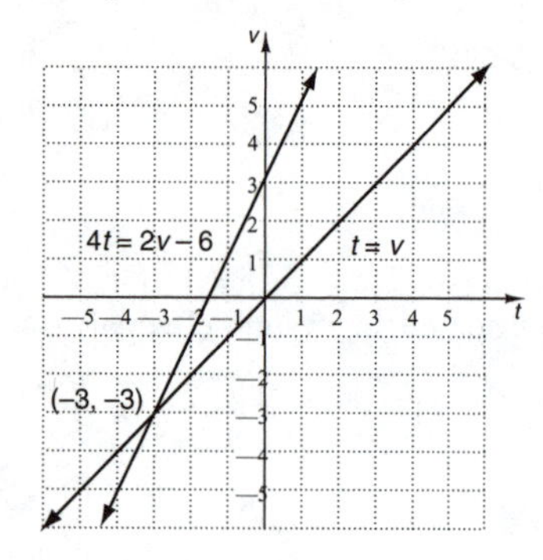

The point of intersection looks as if it has coordinates $(-3, -3)$.

Check:

$t = v$	
$-3 \;?\; -3$	TRUE

$4t = 2v - 6$		
$4(-3)$	$?\; 2(-3) - 6$	
-12	$-6 - 6$	
	-12	TRUE

The solution is $(-3, -3)$.

17. We graph the equations.

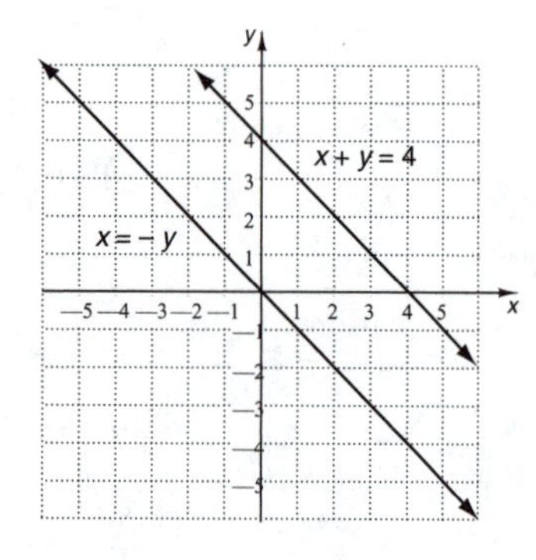

The lines are parallel. There is no solution.

19. We graph the equations.

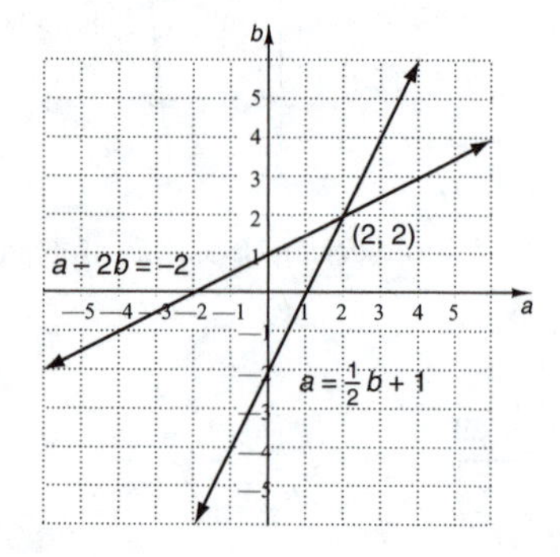

The point of intersection looks as if it has coordinates $(2, 2)$.

Check:

$a = \frac{1}{2}b + 1$		
2	$?\; \frac{1}{2} \cdot 2 + 1$	
	$1 + 1$	
	2	TRUE

$a - 2b = -2$		
$2 - 2 \cdot 2$	$?\; -2$	
$2 - 4$		
-2		TRUE

The solution is $(2, 2)$.

21. We graph the equations.

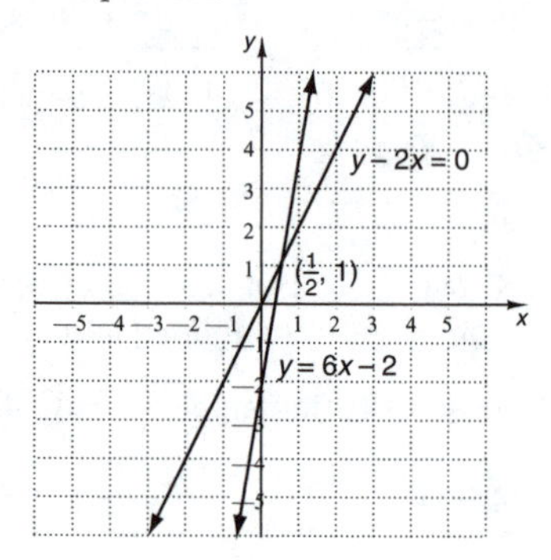

The point of intersection looks as if it has coordinates $\left(\frac{1}{2}, 1\right)$.

Check:

$y - 2x = 0$		
$1 - 2 \cdot \frac{1}{2}$	$?\; 0$	
$1 - 1$		
0		TRUE

$y = 6x - 2$		
1	$?\; 6 \cdot \frac{1}{2} - 2$	
	$3 - 2$	
	1	TRUE

The solution is $\left(\frac{1}{2}, 1\right)$.

23. We graph the equations.

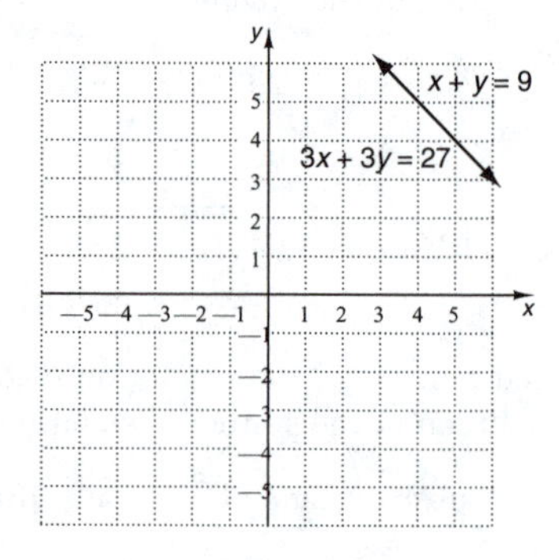

The lines coincide. The system has an infinite number of solutions.

25. We graph the equations.

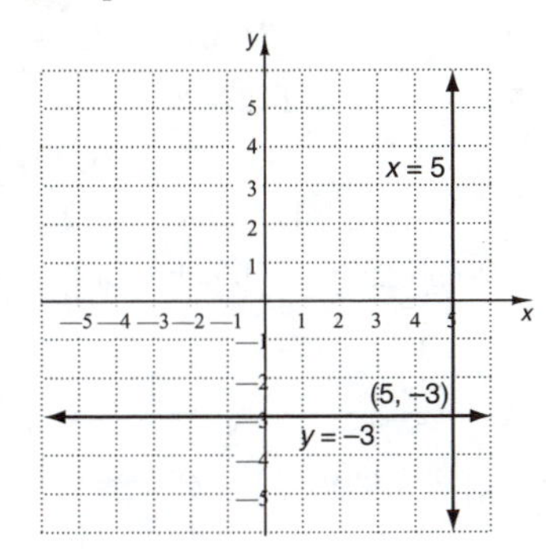

The point of intersection looks as if it has coordinates $(5, -3)$.

Check:

$x = 5$		$y = -3$	
$5 \ ? \ 5$	TRUE	$-3 \ ? \ -3$	TRUE

The solution is $(5, -3)$.

27. $\frac{1}{x} - \frac{1}{x^2} + \frac{1}{x+1}$, LCM is $x^2(x+1)$

$$= \frac{1}{x} \cdot \frac{x(x+1)}{x(x+1)} - \frac{1}{x^2} \cdot \frac{x+1}{x+1} + \frac{1}{x+1} \cdot \frac{x^2}{x^2}$$
$$= \frac{x(x+1) - (x+1) + x^2}{x^2(x+1)}$$
$$= \frac{x^2 + x - x - 1 + x^2}{x^2(x+1)}$$
$$= \frac{2x^2 - 1}{x^2(x+1)}$$

29. $\frac{x+2}{x-4} - \frac{x+1}{x+4}$, LCM is $(x-4)(x+4)$

$$= \frac{x+2}{x-4} \cdot \frac{x+4}{x+4} - \frac{x+1}{x+4} \cdot \frac{x-4}{x-4}$$
$$= \frac{(x+2)(x+4) - (x+1)(x-4)}{(x-4)(x+4)}$$
$$= \frac{x^2 + 6x + 8 - (x^2 - 3x - 4)}{(x-4)(x+4)}$$
$$= \frac{x^2 + 6x + 8 - x^2 + 3x + 4}{(x-4)(x+4)}$$
$$= \frac{9x + 12}{(x-4)(x+4)} = \frac{3(3x+4)}{(x-4)(x+4)}$$

31. The polynomial has exactly three terms, so it is a trinomial.

33. The polynomial has exactly one term, so it is a monomial.

35. $(2, -3)$ is a solution of $Ax - 3y = 13$. Substitute 2 for x and -3 for y and solve for A.

$$Ax - 3y = 13$$
$$A \cdot 2 - 3(-3) = 13$$
$$2A + 9 = 13$$
$$2A = 4$$
$$A = 2$$

$(2, -3)$ is a solution of $x - By = 8$. Substitute 2 for x and -3 for y and solve for B.

$$x - By = 8$$
$$2 - B(-3) = 8$$
$$2 + 3B = 8$$
$$3B = 6$$
$$B = 2$$

37. Answers may vary. Any two equations with a solution of $(6, -2)$ will do. One possibility is

$$x + y = 4,$$
$$x - y = 8.$$

39.-41. Left to the student

Exercise Set 15.2

1. $x = -2y$, (1)

$x + 4y = 2$ (2)

We substitute $-2y$ for x in Equation (2) and solve for y.

$x + 4y = 2$ Equation (2)

$-2y + 4y = 2$ Substituting

$2y = 2$ Collecting like terms

$y = 1$ Dividing by 2

Next we substitute 1 for y in either equation of the original system and solve for x.

$x = -2y$ Equation (1)

$x = -2 \cdot 1$

$x = -2$

We check the ordered pair $(-2, 1)$.

$x = -2y$		$x + 4y = 2$	
$-2 \ ? \ -2 \cdot 1$		$-2 + 4 \cdot 1 \ ? \ 2$	
	-2 TRUE	$-2 + 4$	
		2	TRUE

Since $(-2, 1)$ checks in both equations, it is the solution.

3. $y = x - 6$, (1)

$x + y = -2$ (2)

We substitute $x - 6$ for y in Equation (2) and solve for x.

$x + y = -2$ Equation (2)

$x + (x - 6) = -2$ Substituting

$2x - 6 = -2$ Collecting like terms

$2x = 4$ Adding 6

$x = 2$ Dividing by 2

Next we substitute 2 for x in either equation of the original system and solve for y. We choose Equation (1) since it has y alone on one side.

$$\begin{aligned} y &= x - 6 && \text{Equation (1)} \\ y &= 2 - 6 && \text{Substituting} \\ y &= -4 \end{aligned}$$

We check the ordered pair $(2, -4)$.

$$\begin{array}{c|c} \multicolumn{2}{c}{y = x - 6} \\ \hline -4 \ ? & 2 - 6 \\ & -4 \quad \text{TRUE} \end{array} \qquad \begin{array}{c|c} \multicolumn{2}{c}{x + y = -2} \\ \hline 2 + (-4) & ? \ -2 \\ -2 & \quad \text{TRUE} \end{array}$$

Since $(2, -4)$ checks in both equations, it is the solution.

5. $y = 2x - 5$, (1)
$3y - x = 5$ (2)

We substitute $2x - 5$ for y in Equation (2) and solve for x.

$$\begin{aligned} 3y - x &= 5 && \text{Equation (2)} \\ 3(2x - 5) - x &= 5 && \text{Substituting} \\ 6x - 15 - x &= 5 && \text{Removing parentheses} \\ 5x - 15 &= 5 && \text{Collecting like terms} \\ 5x &= 20 && \text{Adding 15} \\ x &= 4 && \text{Dividing by 5} \end{aligned}$$

Next we substitute 4 for x in either equation of the original system and solve for y.

$$\begin{aligned} y &= 2x - 5 && \text{Equation (1)} \\ y &= 2 \cdot 4 - 5 && \text{Substituting} \\ y &= 8 - 5 \\ y &= 3 \end{aligned}$$

We check the ordered pair $(4, 3)$.

$$\begin{array}{c|c} \multicolumn{2}{c}{y = 2x - 5} \\ \hline 3 \ ? & 2 \cdot 4 - 5 \\ & 8 - 5 \\ & 3 \quad \text{TRUE} \end{array} \qquad \begin{array}{c|c} \multicolumn{2}{c}{3y - x = 5} \\ \hline 3 \cdot 3 - 4 & ? \ 5 \\ 9 - 4 & \\ 5 & \quad \text{TRUE} \end{array}$$

Since $(4, 3)$ checks in both equations, it is the solution.

7. $x = y + 5$, (1)
$2x + y = 1$ (2)

We substitute $y + 5$ for x in Equation (2) and solve for y.

$$\begin{aligned} 2x + y &= 1 && \text{Equation (2)} \\ 2(y + 5) + y &= 1 && \text{Substituting} \\ 2y + 10 + y &= 1 && \text{Removing parentheses} \\ 3y + 10 &= 1 && \text{Collecting like terms} \\ 3y &= -9 && \text{Subtracting 10} \\ y &= -3 && \text{Dividing by 3} \end{aligned}$$

Next we substitute -3 for y in either equation of the original system and solve for x. We choose Equation (1) since it has x alone on one side.

$$\begin{aligned} x &= y + 5 && \text{Equation (1)} \\ x &= -3 + 5 && \text{Substituting} \\ x &= 2 \end{aligned}$$

We check the ordered pair $(2, -3)$.

$$\begin{array}{c|c} \multicolumn{2}{c}{x = y + 5} \\ \hline 2 \ ? & 3 + 5 \\ & 2 \quad \text{TRUE} \end{array} \qquad \begin{array}{c|c} \multicolumn{2}{c}{2x + y = 1} \\ \hline 2 \cdot 2 - 3 & ? \ 1 \\ 4 - 3 & \\ 1 & \quad \text{TRUE} \end{array}$$

Since $(2, -3)$ checks in both equations, it is the solution.

9. $x + y = 10$, (1)
$y = x + 8$ (2)

We substitute $x + 8$ for y in Equation (1) and solve for x.

$$\begin{aligned} x + y &= 10 && \text{Equation (1)} \\ x + (x + 8) &= 10 && \text{Substituting} \\ 2x + 8 &= 10 && \text{Collecting like terms} \\ 2x &= 2 && \text{Subtracting 8} \\ x &= 1 && \text{Dividing by 2} \end{aligned}$$

Next we substitute 1 for x in either equation of the original system and solve for y. We choose Equation (2) since it has y alone on one side.

$$\begin{aligned} y &= x + 8 && \text{Equation (2)} \\ y &= 1 + 8 && \text{Substituting} \\ y &= 9 \end{aligned}$$

We check the ordered pair $(1, 9)$.

$$\begin{array}{c|c} \multicolumn{2}{c}{x + y = 10} \\ \hline 1 + 9 & ? \ 10 \\ 10 & \quad \text{TRUE} \end{array} \qquad \begin{array}{c|c} \multicolumn{2}{c}{y = x + 8} \\ \hline 9 \ ? & 1 + 8 \\ & 9 \quad \text{TRUE} \end{array}$$

Since $(1, 9)$ checks in both equations, it is the solution.

11. $2x + y = 5$, (1)
$x = y + 7$ (2)

We substitute $y + 7$ for x in Equation (1) and solve for y.

$$\begin{aligned} 2x + y &= 5 && \text{Equation (1)} \\ 2(y + 7) + y &= 5 && \text{Substituting} \\ 2y + 14 + y &= 5 && \text{Removing parentheses} \\ 3y + 14 &= 5 && \text{Collecting like terms} \\ 3y &= -9 && \text{Subtracting 14} \\ y &= -3 && \text{Dividing by 3} \end{aligned}$$

Now we substitute -3 for y in Equation (2) and find x.

$$x = y + 7 = -3 + 7 = 4$$

The ordered pair $(4, -3)$ checks in both equations. It is the solution.

13. $x - y = 6$, (1)
$x + y = -2$ (2)

We solve Equation (1) for x.

$$\begin{aligned} x - y &= 6 && \text{Equation (1)} \\ x &= y + 6 && \text{Adding } y \qquad (3) \end{aligned}$$

We substitute $y+6$ for x in Equation (2) and solve for y.

$x+y=-2$ Equation (2)
$(y+6)+y=-2$ Substituting
$2y+6=-2$ Collecting like terms
$2y=-8$ Subtracting 6
$y=-4$ Dividing by 2

Now we substitute -4 for y in Equation (3) and compute x.

$x=y+6=-4+6=2$

The ordered pair $(2,-4)$ checks in both equations. It is the solution.

15. $y-2x=-6$, (1)
$2y-x=5$ (2)

We solve Equation (1) for y.

$y-2x=-6$ Equation (1)
$y=2x-6$ (3)

We substitute $2x-6$ for y in Equation (2) and solve for x.

$2y-x=5$ Equation (2)
$2(2x-6)-x=5$ Substituting
$4x-12-x=5$ Removing parentheses
$3x-12=5$ Collecting like terms
$3x=17$ Adding 12
$x=\frac{17}{3}$ Dividing by 3

We substitute $\frac{17}{3}$ for x in Equation (3) and compute y.

$y=2x-6=2\left(\frac{17}{3}\right)-6=\frac{34}{3}-\frac{18}{3}=\frac{16}{3}$

The ordered pair $\left(\frac{17}{3},\frac{16}{3}\right)$ checks in both equations. It is the solution.

17. $r-2s=0$, (1)
$4r-3s=15$ (2)

We solve Equation (1) for r.

$r-2s=0$ Equation (1)
$r=2s$ (3)

We substitute $2s$ for r in Equation (2) and solve for s.

$4r-3s=15$ Equation (2)
$4(2s)-3s=15$ Substituting
$8s-3s=15$ Removing parentheses
$5s=15$ Collecting like terms
$s=3$ Dividing by 5

Now we substitute 3 for s in Equation (3) and compute r.

$r=2s=2\cdot 3=6$

The ordered pair $(6,3)$ checks in both equations. It is the solution.

19. $2x+3y=-2$, (1)
$2x-y=9$ (2)

We solve Equation (2) for y.

$2x-y=9$ Equation (2)
$2x=9+y$ Adding y
$2x-9=y$ Subtracting 9 (3)

We substitute $2x-9$ for y in Equation (1) and solve for x.

$2x+3y=-2$ Equation (1)
$2x+3(2x-9)=-2$ Substituting
$2x+6x-27=-2$ Removing parentheses
$8x-27=-2$ Collecting like terms
$8x=25$ Adding 27
$x=\frac{25}{8}$ Dividing by 8

Now we substitute $\frac{25}{8}$ for x in Equation (3) and compute y.

$y=2x-9=2\left(\frac{25}{8}\right)-9=\frac{25}{4}-\frac{36}{4}=-\frac{11}{4}$

The ordered pair $\left(\frac{25}{8},-\frac{11}{4}\right)$ checks in both equations. It is the solution.

21. $x+3y=5$, (1)
$3x+5y=3$ (2)

We solve Equation (1) for x.

$x+3y=5$
$x=-3y+5$ Subtracting $3y$ (3)

We substitute $-3y+5$ for x in Equation (2) and solve for y.

$3x+5y=3$ Equation (2)
$3(-3y+5)+5y=3$ Substituting
$-9y+15+5y=3$ Removing parentheses
$-4y+15=3$ Collecting like terms
$-4y=-12$ Subtracting 15
$y=3$ Dividing by -4

Now we substitute 3 for y in Equation (3) and compute x.

$x=-3y+5=-3\cdot 3+5=-9+5=-4$

The ordered pair $(-4,3)$ checks in both equations. It is the solution.

23. $x-y=-3$, (1)
$2x+3y=-6$ (2)

We solve Equation (1) for x.

$x-y=-3$ Equation (1)
$x=y-3$ (3)

We substitute $y-3$ for x in Equation (2) and solve for y.

$2x+3y=-6$ Equation (2)

$2(y-3)+3y=-6$ Substituting

$2y-6+3y=-6$ Removing parentheses

$5y-6=-6$ Collecting like terms

$5y=0$ Adding 6

$y=0$ Dividing by 5

Now we substitute 0 for y in Equation (3) and compute x.

$x=y-3=0-3=-3$

The ordered pair $(-3,0)$ checks in both equations. It is the solution.

25. ***Familiarize***. We make a drawing. We let $l=$ the length and $w=$ the width, in inches.

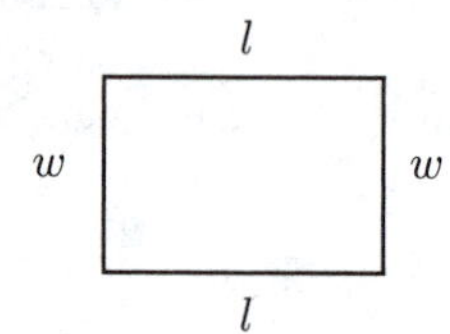

Translate. The perimeter is $2l+2w$. We translate the first statement.

The perimeter is 10 in.

$2l+2w = 10$

We translate the second statement.

The length is 2 in. more than the width.

$l = w+2$

The resulting system is

$2l+2w=10,$ (1)

$l=w+2.$ (2)

Solve. We solve the system. We substitute $w+2$ for l in Equation (1) and solve for w.

$2l+2w=10$ Equation (1)

$2(w+2)+2w=10$ Substituting

$2w+4+2w=10$ Removing parentheses

$4w+4=10$

$4w=6$ Collecting like terms

$w=\frac{3}{2}$, or $1\frac{1}{2}$

Now we substitute $\frac{3}{2}$ for w in Equation (2) and solve for l.

$l=w+2$ Equation (2)

$l=\frac{3}{2}+2$ Substituting

$l=\frac{3}{2}+\frac{4}{2}$

$l=\frac{7}{2}$, or $3\frac{1}{2}$

Check. A possible solution is a length of $\frac{7}{2}$, or $3\frac{1}{2}$ in. and a width of $\frac{3}{2}$, or $1\frac{1}{2}$ in. The perimeter would be $2\cdot\frac{7}{2}+2\cdot\frac{3}{2}$, or $7+3$, or 10 in. Also, the width plus 2 is $\frac{3}{2}+2$, or $\frac{7}{2}$, which is the length. These numbers check.

State. The length is $3\frac{1}{2}$ in., and the width is $1\frac{1}{2}$ in.

27. ***Familiarize***. Let $l=$ the length, in mi, and $w=$ the width, in mi. Recall that the perimeter of a rectangle with length l and width w is given by $2l+2w$.

Translate.

The perimeter is 1280 mi.

$2l+2w = 1280$

The width is the length less 90 mi.

$w = l - 90$

The resulting system is

$2l+2w=1280,$ (1)

$w=l-90.$ (2)

Solve. We solve the system. Substitute $l-90$ for w in the first equation and solve for l.

$2l+2(l-90)=1280$

$2l+2l-180=1280$

$4l-180=1280$

$4l=1460$

$l=365$

Now substitute 365 for l in Equation (2).

$w=l-90=365-90=275$

Check. If the length is 365 mi and the width is 275 mi, then the width is 90 mi less than the length and the perimeter is $2\cdot 365+2\cdot 275$, or $730+550$, or 1280. The answer checks.

State. The length is 365 mi, and the width is 275 mi.

29. ***Familiarize***. Let $l=$ the length in ft, and $w=$ the width, in ft. Recall that the perimeter of a rectangle with length l and width w is given by $2l+2w$.

Translate.

The perimeter is 120 ft.

$2l+2w = 120$

The length is twice the width.

$l = 2w$

The resulting system is

$2l+2w=120,$ (1)

$l=2w.$ (2)

Solve. We solve the system.

Substitute $2w$ for l in Equation (1) and solve for w.

$$2 \cdot 2w + 2w = 120 \quad (1)$$
$$4w + 2w = 120$$
$$6w = 120$$
$$w = 20$$

Now substitute 20 for w in Equation (2).

$$l = 2w \quad (2)$$
$$l = 2 \cdot 20 \quad \text{Substituting}$$
$$l = 40$$

Check. If the length is 40 ft and the width is 20 ft, the perimeter would be $2 \cdot 40 + 2 \cdot 20$, or $80 + 40$, or 120 ft. Also, the length is twice the width. These numbers check.

State. The length is 40 ft, and the width is 20 ft.

31. ***Familiarize***. Let l = the length and w = the width, in yards. The perimeter is $l + l + w + w$, or $2l + 2w$.

Translate.

The perimeter is 340 yd.

$$2l + 2w = 340$$

The length is 10 yd less than twice the width.

$$l = 2w - 10$$

The resulting system is

$$2l + 2w = 340, \quad (1)$$
$$l = 2w - 10. \quad (2)$$

Solve. We solve the system. We substitute $2w - 10$ for l in Equation (1) and solve for w.

$$2l + 2w = 340 \quad (1)$$
$$2(2w - 10) + 2w = 340$$
$$4w - 20 + 2w = 340$$
$$6w - 20 = 340$$
$$6w = 360$$
$$w = 60$$

Next we substitute 60 for w in Equation (2) and solve for l.

$$l = 2w - 10 = 2 \cdot 60 - 10 = 120 - 10 = 110$$

Check. The perimeter is $2 \cdot 110 + 2 \cdot 60$, or 340 yd. Also 10 yd less than twice the width is $2 \cdot 60 - 10 = 120 - 10 = 110$. The answer checks.

State. The length is 110 yd, and the width is 60 yd.

33. ***Familiarize***. We let x = the larger number and y = the smaller number.

Translate. We translate the first statement.

The sum of two numbers is 37.

$$x + y = 37$$

Now we translate the second statement.

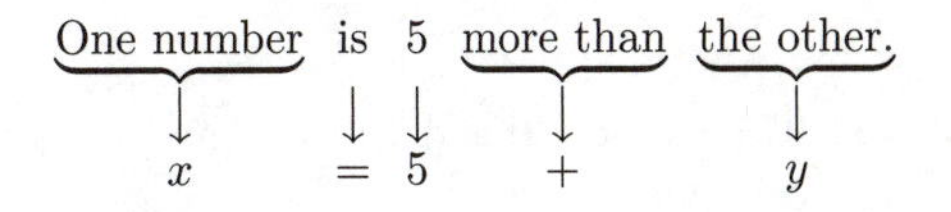

One number is 5 more than the other.

$$x = 5 + y$$

The resulting system is

$$x + y = 37, \quad (1)$$
$$x = 5 + y. \quad (2)$$

Solve. We solve the system of equations. We substitute $5 + y$ for x in Equation (1) and solve for y.

$$x + y = 37 \quad \text{Equation (1)}$$
$$(5 + y) + y = 37 \quad \text{Substituting}$$
$$5 + 2y = 37 \quad \text{Collecting like terms}$$
$$2y = 32 \quad \text{Subtracting 5}$$
$$y = 16 \quad \text{Dividing by 2}$$

We go back to the original equations and substitute 16 for y. We use Equation (2).

$$x = 5 + y \quad \text{Equation (2)}$$
$$x = 5 + 16 \quad \text{Substituting}$$
$$x = 21$$

Check. The sum of 21 and 16 is 37. The number 21 is 5 more than the number 16. These numbers check.

State. The numbers are 21 and 16.

35. ***Familiarize***. Let x = one number and y = the other.

Translate. We reword and translate.

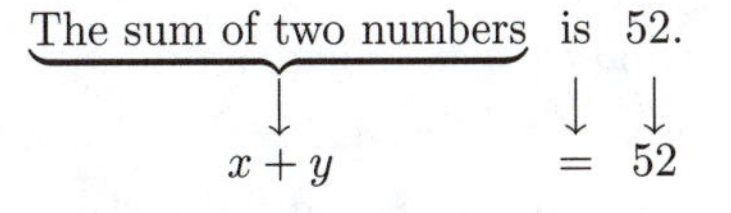

The sum of two numbers is 52.

$$x + y = 52$$

The difference of two numbers is 28.

$$x - y = 28$$

(The second statement could also be translated as $y - x = 28$.)

The resulting system is

$$x + y = 52, \quad (1)$$
$$x - y = 28. \quad (2)$$

Solve. We solve the system. First we solve Equation (2) for x.

$$x - y = 28 \quad \text{Equation (2)}$$
$$x = y + 28 \quad \text{Adding } y \quad (3)$$

We substitute $y + 28$ for x in Equation (1) and solve for y.

$$x + y = 52 \quad \text{Equation (1)}$$
$$(y + 28) + y = 52 \quad \text{Substituting}$$
$$2y + 28 = 52 \quad \text{Collecting like terms}$$
$$2y = 24 \quad \text{Subtracting 28}$$
$$y = 12 \quad \text{Dividing by 2}$$

Now we substitute 12 for y in Equation (3) and compute x.

$$x = y + 28 = 12 + 28 = 40$$

Check. The sum of 40 and 12 is 52, and their difference is 28. These numbers check.

State. The numbers are 40 and 12.

37. ***Familiarize.*** We let w = the width of the court, in ft, and l = the length, in ft. Recall that the perimeter of a rectangle with length l and width w is given by $2l + 2w$.

Translate.

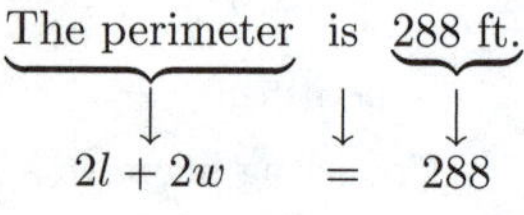

The length is 44 ft more than the width.

$$l = 44 + w$$

The resulting system is

$$2l + 2w = 288, \quad (1)$$
$$l = 44 + w. \quad (2)$$

Solve. We solve the system. Substitute $44 + w$ for l in the first equation and solve for w.

$$2(44 + w) + 2w = 288$$
$$88 + 2w + 2w = 288$$
$$88 + 4w = 288$$
$$4w = 200$$
$$w = 50$$

Now substitute 50 for w in Equation (2).

$$l = 44 + w = 44 + 50 = 94$$

Check. If the length is 94 ft and the width is 50 ft, then the length is 44 ft more than the width and the perimeter is $2 \cdot 94 + 2 \cdot 50$, or $188 + 100$, or 288 ft. The answer checks.

State. The length of the court is 94 ft, and the width is 50 ft.

39. ***Familiarize.*** We let x = the larger number and y = the smaller number.

Translate. We translate the first statement.

The difference between two numbers is 12.

$$x - y = 12$$

Now we translate the second statement.

Two times the larger number is five times the smaller.

$$2x = 5y$$

The resulting system is

$$x - y = 12, \quad (1)$$
$$2x = 5y. \quad (2)$$

Solve. We solve the system. First we solve Equation (1) for x.

$$x - y = 12 \quad \text{Equation (1)}$$
$$x = y + 12 \quad \text{Adding } y \quad (3)$$

We substitute $y + 12$ for x in Equation (2) and solve for y.

$$2x = 5y \quad \text{Equation (2)}$$
$$2(y + 12) = 5y \quad \text{Substituting}$$
$$2y + 24 = 5y \quad \text{Removing parentheses}$$
$$24 = 3y \quad \text{Subtracting } 2y$$
$$8 = y \quad \text{Dividing by 3}$$

Now we substitute 8 for y in Equation (3) and compute x.

$$x = y + 12 = 8 + 12 = 20$$

Check. The difference between 20 and 8 is 12. Two times 20, or 40, is five times 8. These numbers check.

State. The numbers are 20 and 8.

41. Graph: $2x - 3y = 6$

To find the x-intercept, let $y = 0$. Then solve for x.

$$2x - 3 \cdot 0 = 6$$
$$2x = 6$$
$$x = 3$$

The x-intercept is $(3, 0)$.

To find the y-intercept, let $x = 0$. Then solve for y.

$$2 \cdot 0 - 3y = 6$$
$$-3y = 6$$
$$y = -2$$

The y-intercept is $(0, -2)$.

We plot these points and draw the line.

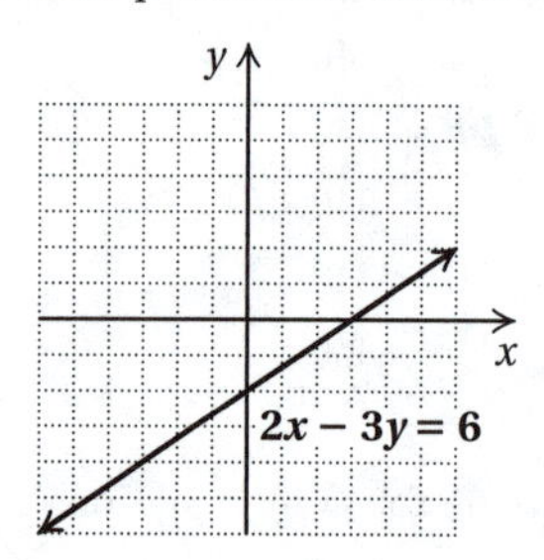

A third point should be used as a check. We let $x = -3$:

$$2(-3) - 3y = 6$$
$$-6 - 3y = 6$$
$$-3y = 12$$
$$y = -4$$

The point $(-3, -4)$ is on the graph, so our graph is probably correct.

43. Graph: $y = 2x - 5$

We select several values for x and compute the corresponding y-values.

When $x = 0$, $y = 2 \cdot 0 - 5 = 0 - 5 = -5$.

When $x = 2$, $y = 2 \cdot 2 - 5 = 4 - 5 = -1$.

When $x = 4$, $y = 2 \cdot 4 - 5 = 8 - 5 = 3$.

x	y	(x, y)
0	-5	$(0, -5)$
2	-1	$(2, -1)$
4	3	$(4, 3)$

We plot these points and draw the line connecting them.

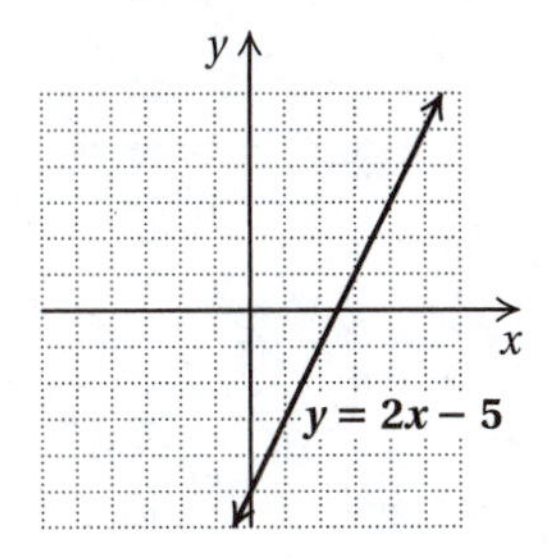

45. $6x^2 - 13x + 6$

The possibilities are $(x+\quad)(6x+\quad)$ and $(2x+\quad)(3x+\quad)$. We look for a pair of factors of the last term, 6, which produces the correct middle term. Since the last term is positive and the middle term is negative, we need only consider negative pairs. The factorization is $(2x - 3)(3x - 2)$.

47. $4x^2 + 3x + 2$

The possibilities are $(x+\quad)(4x+\quad)$ and $(2x+\quad)(2x+\quad)$. We look for a pair of factors of the last term, 2, which produce the correct middle term. Since the last term and the middle term are both positive, we need only consider positive pairs. We find that there is no possibility that works. The trinomial cannot be factored.

49. $\dfrac{x^{-2}}{x^{-5}} = x^{-2-(-5)} = x^3$

51. $x^{-2} \cdot x^{-5} = x^{-2+(-5)} = x^{-7} = \dfrac{1}{x^7}$

53. First put the equations in "$y =$" form by solving for y. We get

$$y_1 = x - 5,$$
$$y_2 = (-1/2)x + 7/2.$$

Then graph these equations in the standard window and use the INTERSECT feature from the CALC menu to find the coordinates of the point of intersection of the graphs. The solution is $(5.\overline{6}, 0.\overline{6})$.

55. First put the equations in "$y =$" form by solving for y. We get

$$y_1 = 2.35x - 5.97,$$
$$y_2 = (1/2.14)x + (4.88/2.14).$$

Then graph these equations in the standard window and use the INTERSECT feature from the CALC menu to find the coordinates of the point of intersection of the graphs. The solution is approximately $(4.38, 4.33)$.

57. ***Familiarize.*** Let $s =$ the perimeter of a softball diamond, in yards, and $b =$ the perimeter of a baseball diamond, in yards.

Translate.

$$\underbrace{\text{Perimeter of a softball diamond}}_{s} \ \underbrace{\text{is}}_{=} \ \underbrace{\tfrac{2}{3}}_{\frac{2}{3}} \ \underbrace{\text{of}}_{\cdot} \ \underbrace{\text{perimeter of a baseball diamond.}}_{b}$$

$$\underbrace{\text{The sum of the perimeters}}_{s+b} \ \underbrace{\text{is}}_{=} \ \underbrace{\text{200 yd.}}_{200}$$

The resulting system is

$$s = \frac{2}{3}b, \quad (1)$$
$$s + b = 200. \quad (2)$$

Solve. We solve the system of equations. We substitute $\frac{2}{3}b$ for s in Equation (2) and solve for b.

$$s + b = 200$$
$$\frac{2}{3}b + b = 200$$
$$\frac{5}{3}b = 200$$
$$\frac{3}{5} \cdot \frac{5}{3}b = \frac{3}{5} \cdot 200$$
$$b = 120$$

Next we substitute 120 for b in Equation (1) and solve for s.

$$s = \frac{2}{3}b = \frac{2}{3} \cdot 120 = 80$$

Each diamond has four sides of equal length, so we divide each perimeter by 4 to find the distance between bases in each sport. For the softball diamond the distance is $80/4$, or 20 yd. For the baseball diamond it is $120/4$, or 30 yd.

Check. The perimeter of the softball diamond, 80 yd, is $\frac{2}{3}$ of 120 yd, the perimeter of the baseball diamond. The sum of the perimeters is $80 + 120$, or 200 yd. We can also recheck the calculations of the distances between the bases. The answer checks.

State. The distance between bases on a softball diamond is 20 yd and the distance between bases on a baseball diamond is 30 yd.

Exercise Set 15.3

1.

$x - y = 7$	(1)
$x + y = 5$	(2)
$2x \quad = 12$	Adding
$x = 6$	Dividing by 2

Substitute 6 for x in either of the original equations and solve for y.

$$\begin{aligned} x+y&=5 \quad \text{Equation (2)}\\ 6+y&=5 \quad \text{Substituting}\\ y&=-1 \quad \text{Subtracting 6}\end{aligned}$$

Check:

$$\begin{array}{c|c} x-y=7 & \\ \hline 6-(-1) \; ? \; 7 & \\ 6+1 & \\ 7 & \text{TRUE} \end{array} \qquad \begin{array}{c|c} x+y=5 & \\ \hline 6+(-1) \; ? \; 5 & \\ 5 & \text{TRUE} \end{array}$$

Since $(6,-1)$ checks, it is the solution.

3. $$\begin{aligned} x+\ \ y&=8 \quad (1)\\ -x+2y&=7 \quad (2)\\ \hline 3y&=15 \quad \text{Adding}\\ y&=5 \quad \text{Dividing by 3}\end{aligned}$$

Substitute 5 for y in either of the original equations and solve for x.

$$\begin{aligned} x+y&=8 \quad \text{Equation (1)}\\ x+5&=8 \quad \text{Substituting}\\ x&=3\end{aligned}$$

Check:

$$\begin{array}{c|c} x+y=8 & \\ \hline 3+5 \; ? \; 8 & \\ 8 & \text{TRUE} \end{array} \qquad \begin{array}{c|c} -x+2y=7 & \\ \hline -3+2\cdot 5 \; ? \; 7 & \\ -3+10 & \\ 7 & \text{TRUE} \end{array}$$

Since $(3,5)$ checks, it is the solution.

5. $$\begin{aligned} 5x-y&=5 \quad (1)\\ 3x+y&=11 \quad (2)\\ \hline 8x\qquad&=16 \quad \text{Adding}\\ x&=2 \quad \text{Dividing by 8}\end{aligned}$$

Substitute 2 for x in either of the original equations and solve for y.

$$\begin{aligned} 3x+y&=11 \quad \text{Equation (2)}\\ 3\cdot 2+y&=11 \quad \text{Substituting}\\ 6+y&=11\\ y&=5\end{aligned}$$

Check:

$$\begin{array}{c|c} 5x-y=5 & \\ \hline 5\cdot 2-5 \; ? \; 5 & \\ 10-5 & \\ 5 & \text{TRUE} \end{array} \qquad \begin{array}{c|c} 3x+y=11 & \\ \hline 3\cdot 2+5 \; ? \; 11 & \\ 6+5 & \\ 11 & \text{TRUE} \end{array}$$

Since $(2,5)$ checks, it is the solution.

7. $$\begin{aligned} 4a+3b&=7 \quad (1)\\ -4a+\ \ b&=5 \quad (2)\\ \hline 4b&=12 \quad \text{Adding}\\ b&=3\end{aligned}$$

Substitute 3 for b in either of the original equations and solve for a.

$$\begin{aligned} 4a+3b&=7 \quad \text{Equation (1)}\\ 4a+3\cdot 3&=7 \quad \text{Substituting}\\ 4a+9&=7\\ 4a&=-2\\ a&=-\frac{1}{2}\end{aligned}$$

Check:

$$\begin{array}{c|c} 4a+3b=7 & \\ \hline 4\left(-\frac{1}{2}\right)+3\cdot 3 \; ? \; 7 & \\ -2+9 & \\ 7 & \text{TRUE} \end{array} \qquad \begin{array}{c|c} -4a+b=5 & \\ \hline -4\left(-\frac{1}{2}\right)+3 \; ? \; 5 & \\ 2+3 & \\ 5 & \text{TRUE} \end{array}$$

Since $\left(-\frac{1}{2},3\right)$ checks, it is the solution.

9. $$\begin{aligned} 8x-5y&=-9 \quad (1)\\ 3x+5y&=-2 \quad (2)\\ \hline 11x\qquad&=-11 \quad \text{Adding}\\ x&=-1\end{aligned}$$

Substitute -1 for x in either of the original equations and solve for y.

$$\begin{aligned} 3x+5y&=-2 \quad \text{Equation (2)}\\ 3(-1)+5y&=-2 \quad \text{Substituting}\\ -3+5y&=-2\\ 5y&=1\\ y&=\frac{1}{5}\end{aligned}$$

Check:

$$\begin{array}{c|c} 8x-5y=-9 & \\ \hline 8(-1)-5\left(\frac{1}{5}\right) \; ? \; -9 & \\ -8-1 & \\ -9 & \text{TRUE} \end{array} \qquad \begin{array}{c|c} 3x+5y=-2 & \\ \hline 3(-1)+5\left(\frac{1}{5}\right) \; ? \; -2 & \\ -3+1 & \\ -2 & \text{TRUE} \end{array}$$

Since $\left(-1,\frac{1}{5}\right)$ checks, it is the solution.

11. $$\begin{aligned} 4x-5y&=7\\ -4x+5y&=7\\ \hline 0&=14 \quad \text{Adding}\end{aligned}$$

We obtain a false equation, $0=14$, so there is no solution.

13. $$\begin{aligned} x+y&=-7, \quad (1)\\ 3x+y&=-9 \quad (2)\end{aligned}$$

We multiply on both sides of Equation (1) by -1 and then add.

$$\begin{aligned} -x-y&=7 \quad \text{Multiplying by } -1\\ 3x+y&=-9 \quad \text{Equation (2)}\\ \hline 2x\qquad&=-2 \quad \text{Adding}\\ x&=-1\end{aligned}$$

Substitute -1 for x in one of the original equations and solve for y.

$$x+y=-7 \quad \text{Equation (1)}$$
$$-1+y=-7 \quad \text{Substituting}$$
$$y=-6$$

Check:

$x+y=-7$	
$-1+(-6)\ ?\ -7$	
-7	TRUE

$3x+y=-9$	
$3(-1)+(-6)\ ?\ -9$	
$-3-6$	
-9	TRUE

Since $(-1,-6)$ checks, it is the solution.

15. $3x - y = 8$, (1)

$x + 2y = 5$ (2)

We multiply on both sides of Equation (1) by 2 and then add.

$$6x - 2y = 16 \quad \text{Multiplying by 2}$$
$$x + 2y = 5 \quad \text{Equation (2)}$$
$$7x = 21 \quad \text{Adding}$$
$$x = 3$$

Substitute 3 for x in one of the original equations and solve for y.

$$x+2y=5 \quad \text{Equation (2)}$$
$$3+2y=5 \quad \text{Substituting}$$
$$2y=2$$
$$y=1$$

Check:

$3x-y=8$	
$3\cdot 3-1\ ?\ 8$	
$9-1$	
8	TRUE

$x+2y=5$	
$3+2\cdot 1\ ?\ 5$	
$3+2$	
5	TRUE

Since $(3,1)$ checks, it is the solution.

17. $x - y = 5$, (1)

$4x - 5y = 17$ (2)

We multiply on both sides of Equation (1) by -4 and then add.

$$-4x + 4y = -20 \quad \text{Multiplying by } -4$$
$$4x - 5y = 17 \quad \text{Equation (2)}$$
$$-y = -3 \quad \text{Adding}$$
$$y = 3$$

Substitute 3 for y in one of the original equations and solve for x.

$$x-y=5 \quad \text{Equation (1)}$$
$$x-3=5 \quad \text{Substituting}$$
$$x=8$$

Check:

$x-y=5$	
$8-3\ ?\ 5$	
5	TRUE

$4x-5y=17$	
$4\cdot 8-5\cdot 3\ ?\ 17$	
$32-15$	
17	TRUE

Since $(8,3)$ checks, it is the solution.

19. $2w - 3z = -1$, (1)

$3w + 4z = 24$ (2)

We use the multiplication principle with both equations and then add.

$$8w - 12z = -4 \quad \text{Multiplying (1) by 4}$$
$$9w + 12z = 72 \quad \text{Multiplying (2) by 3}$$
$$17w = 68 \quad \text{Adding}$$
$$w = 4$$

Substitute 4 for w in one of the original equations and solve for z.

$$3w+4z=24 \quad \text{Equation (2)}$$
$$3\cdot 4+4z=24 \quad \text{Substituting}$$
$$12+4z=24$$
$$4z=12$$
$$z=3$$

Check:

$2w-3z=-1$	
$2\cdot 4-3\cdot 3\ ?\ -1$	
$8-9$	
-1	TRUE

$3w+4z=24$	
$3\cdot 4+4\cdot 3\ ?\ 24$	
$12+12$	
24	TRUE

Since $(4,3)$ checks, it is the solution.

21. $2a + 3b = -1$, (1)

$3a + 5b = -2$ (2)

We use the multiplication principle with both equations and then add.

$$-10a - 15b = 5 \quad \text{Multiplying (1) by } -5$$
$$9a + 15b = -6 \quad \text{Multiplying (2) by 3}$$
$$-a = -1 \quad \text{Adding}$$
$$a = 1$$

Substitute 1 for a in one of the original equations and solve for b.

$$2a+3b=-1 \quad \text{Equation (1)}$$
$$2\cdot 1+3b=-1 \quad \text{Substituting}$$
$$2+3b=-1$$
$$3b=-3$$
$$b=-1$$

Check:

$2a+3b=-1$	
$2\cdot 1+3(-1)\ ?\ -1$	
$2-3$	
-1	TRUE

$3a+5b=-2$	
$3\cdot 1+5(-1)\ ?\ -2$	
$3-5$	
-2	TRUE

Since $(1,-1)$ checks, it is the solution.

23. $x = 3y,\quad (1)$

$5x+14 = y \quad (2)$

We first get each equation in the form $Ax+By=C$.

$x-3y=0, \quad (1a) \quad$ Adding $-3y$

$5x-y=-14 \quad (2a) \quad$ Adding $-y-14$

We multiply by -5 on both sides of Equation (1a) and add.

$$\begin{array}{rl} -5x+15y=0 & \text{Multiplying by } -5 \\ 5x-\ \ y=-14 & \\ \hline 14y=-14 & \text{Adding} \\ y=-1 & \end{array}$$

Substitute -1 for y in Equation (1) and solve for x.

$$\begin{array}{rl} x-3y=0 & \\ x-3(-1)=0 & \text{Substituting} \\ x+3=0 & \\ x=-3 & \end{array}$$

Check:

$x-3y=0$	
$-3-3(-1)\ ?\ 0$	
$-3+3$	
0	TRUE

$5x-y=-14$	
$5(-3)-(-1)\ ?\ -14$	
$-15+1$	
-14	TRUE

Since $(-3,-1)$ checks, it is the solution.

25. $2x+5y=16,\quad (1)$

$3x-2y=5 \quad (2)$

We use the multiplication principle with both equations and then add.

$$\begin{array}{rl} 4x+10y=32 & \text{Multiplying (1) by 2} \\ 15x-10y=25 & \text{Multiplying (2) by 5} \\ \hline 19x\qquad\ =57 & \\ x=3 & \end{array}$$

Substitute 3 for x in one of the original equations and solve for y.

$$\begin{array}{rl} 2x+5y=16 & \text{Equation (1)} \\ 2\cdot 3+5y=16 & \text{Substituting} \\ 6+5y=16 & \\ 5y=10 & \\ y=2 & \end{array}$$

Check:

$2x+5y=16$	
$2\cdot 3+5\cdot 2\ ?\ 16$	
$6+10$	
16	TRUE

$3x-2y=5$	
$3\cdot 3-2\cdot 2\ ?\ 5$	
$9-4$	
5	TRUE

Since $(3,2)$ checks, it is the solution.

27. $p=32+q,\quad (1)$

$3p=8q+6 \quad (2)$

First we write each equation in the form $Ap+Bq=C$.

$p-q=32, \quad (1a) \quad$ Subtracting q

$3p-8q=6 \quad (2a) \quad$ Subtracting $8q$

Now we multiply both sides of Equation (1a) by -3 and then add.

$$\begin{array}{rl} -3p+\ 3q=-96 & \text{Multiplying by } -3 \\ 3p-\ 8q=6 & \text{Equation (2a)} \\ \hline -5q=-90 & \text{Adding} \\ q=18 & \end{array}$$

Substitute 18 for q in Equation (1) and solve for p.

$$\begin{array}{rl} p=32+q & \\ p=32+18 & \text{Substituting} \\ p=50 & \end{array}$$

Check:

$p-q=32$	
$50-18\ ?\ 32$	
32	TRUE

$3p-8q=6$	
$3\cdot 50-8\cdot 18\ ?\ 6$	
$150-144$	
6	TRUE

Since $(50,18)$ checks, it is the solution.

29. $3x-2y=10,\quad (1)$

$-6x+4y=-20 \quad (2)$

We multiply by 2 on both sides of Equation (1) and add.

$$\begin{array}{r} 6x-4y=20 \\ -6x+4y=-20 \\ \hline 0=0 \end{array}$$

We get an obviously true equation, so the system has an infinite number of solutions.

31. $0.06x+0.05y=0.07,$

$0.04x-0.03y=0.11$

We first multiply each equation by 100 to clear the decimals.

$6x+5y=7, \quad (1)$

$4x-3y=11 \quad (2)$

We use the multiplication principle with both equations of the resulting system.

$$\begin{array}{rl} 18x+15y=21 & \text{Multiplying (1) by 3} \\ 20x-15y=55 & \text{Multiplying (2) by 5} \\ \hline 38x\qquad\ =76 & \text{Adding} \\ x=2 & \end{array}$$

Substitute 2 for x in Equation (1) and solve for y.

$$6x + 5y = 7$$
$$6 \cdot 2 + 5y = 7$$
$$12 + 5y = 7$$
$$5y = -5$$
$$y = -1$$

Check:

$0.06x + 0.05y = 0.07$	
$0.06(2) + 0.05(-1)$? 0.07	
$0.12 - 0.05$	
0.07	TRUE

$0.04x - 0.03y = 0.11$	
$0.04(2) - 0.03(-1)$? 0.11	
$0.08 + 0.03$	
0.11	TRUE

Since $(2, -1)$ checks, it is the solution.

33. $\frac{1}{3}x + \frac{3}{2}y = \frac{5}{4}$,

$\frac{3}{4}x - \frac{5}{6}y = \frac{3}{8}$

First we clear the fractions. We multiply on both sides of the first equation by 12 and on both sides of the second equation by 24.

$$12\left(\frac{1}{3}x + \frac{3}{2}y\right) = 12 \cdot \frac{5}{4}$$
$$12 \cdot \frac{1}{3}x + 12 \cdot \frac{3}{2}y = 15$$
$$4x + 18y = 15$$

$$24\left(\frac{3}{4}x - \frac{5}{6}y\right) = 24 \cdot \frac{3}{8}$$
$$24 \cdot \frac{3}{4}x - 24 \cdot \frac{5}{6}y = 9$$
$$18x - 20y = 9$$

The resulting system is

$4x + 18y = 15$, (1)

$18x - 20y = 9$. (2)

We use the multiplication principle with both equations.

$72x + 324y = 270$ Multiplying (1) by 18

$-72x + 80y = -36$ Multiplying (2) by -4

$404y = 234$

$$y = \frac{234}{404}, \text{ or } \frac{117}{202}$$

Substitute $\frac{117}{202}$ for y in Equation (1) and solve for x.

$$4x + 18\left(\frac{117}{202}\right) = 15$$
$$4x + \frac{1053}{101} = 15$$
$$4x = \frac{462}{101}$$
$$x = \frac{1}{4} \cdot \frac{462}{101}$$
$$x = \frac{231}{202}$$

The ordered pair $\left(\frac{231}{202}, \frac{117}{202}\right)$ checks in both equations. It is the solution.

35. $-4.5x + 7.5y = 6$,

$-x + 1.5y = 5$

First we clear the decimals by multiplying by 10 on both sides of each equation.

$$10(-4.5x + 7.5y) = 10 \cdot 6$$
$$-45x + 75y = 60$$

$$10(-x + 1.5y) = 10 \cdot 5$$
$$-10x + 15y = 50$$

The resulting system is

$-45x + 75y = 60$, (1)

$-10x + 15y = 50$. (2)

We multiply both sides of Equation (2) by -5 and then add.

$-45x + 75y = 60$ Equation (1)

$50x - 75y = -250$ Multiplying by -5

$5x = -190$ Adding

$x = -38$

Substitute -38 for x in Equation (2) and solve for y.

$$-10x + 15y = 50$$
$$-10(-38) + 15y = 50$$
$$380 + 15y = 50$$
$$15y = -330$$
$$y = -22$$

The ordered pair $(-38, -22)$ checks in both equations. It is the solution.

37. Parallel lines have the same slope and different y-intercepts.

39. A solution of a system of two equations is an ordered pair that makes both equations true.

41. The graph of $y = b$ is a horizontal line.

43. The equation $y = mx+b$ is called the slope-intercept equation.

45.-53. Left to the student

55.-63. Left to the student

65. $3(x-y)=9,$

$x+y=7$

First we remove parentheses in the first equation.

$$3x - 3y = 9, \quad (1)$$
$$x + y = 7 \quad (2)$$

Then we multiply Equation (2) by 3 and add.

$$\begin{array}{rl} 3x - 3y &= 9 \\ 3x + 3y &= 21 \\ \hline 6x \quad &= 30 \\ x &= 5 \end{array}$$

Now we substitute 5 for x in Equation (2) and solve for y.

$$x+y=7$$
$$5+y=7$$
$$y=2$$

The ordered pair $(5,2)$ checks and is the solution.

67. $2(5a-5b)=10,$

$-5(6a+2b)=10$

First we remove parentheses.

$$10a - 10b = 10, \quad (1)$$
$$-30a - 10b = 10 \quad (2)$$

Then we multiply Equation (2) by -1 and add.

$$\begin{array}{rl} 10a - 10b &= 10 \\ 30a + 10b &= -10 \\ \hline 40a \quad &= 0 \\ a &= 0 \end{array}$$

Substitute 0 for a in Equation (1) and solve for b.

$$10\cdot 0 - 10b = 10$$
$$-10b = 10$$
$$b = -1$$

The ordered pair $(0,-1)$ checks and is the solution.

69. $y=-\frac{2}{7}x+3, \quad (1)$

$y=\frac{4}{5}x+3 \quad (2)$

Observe that these equations represent lines with different slopes and the same y-intercept. Thus, their point of intersection is the y-intercept, $(0,3)$ and this is the solution of the system of equations.

We could also solve this system of equations algebraically. First substitute $\frac{4}{5}x+3$ for y in Equation (1) and solve for x.

$$\frac{4}{5}x+3=-\frac{2}{7}x+3$$
$$35\left(\frac{4}{5}x+3\right)=35\left(-\frac{2}{7}x+3\right) \quad \text{Clearing fractions}$$
$$35\cdot\frac{4}{5}x+35\cdot 3=35\left(-\frac{2}{7}x\right)+35\cdot 3$$
$$28x+105=-10x+105$$
$$28x=-10x$$
$$38x=0$$
$$x=0$$

Now substitute 0 for x in one of the original equations and find y. We will use Equation (1).

$$y=-\frac{2}{7}x+3=-\frac{2}{7}\cdot 0+3=0+3=3$$

The ordered pair $(0,3)$ checks and is the solution.

71. $y=ax+b, \quad (1)$

$y=x+c \quad (2)$

Substitute $x+c$ for y in Equation (1) and solve for x.

$$y=ax+b$$
$$x+c=ax+b \quad \text{Substituting}$$
$$x-ax=b-c$$
$$(1-a)x=b-c$$
$$x=\frac{b-c}{1-a}$$

Substitute $\frac{b-c}{1-a}$ for x in Equation (2) and simplify to find y.

$$y=x+c$$
$$y=\frac{b-c}{1-a}+c$$
$$y=\frac{b-c}{1-a}+c\cdot\frac{1-a}{1-a}$$
$$y=\frac{b-c+c-ac}{1-a}$$
$$y=\frac{b-ac}{1-a}$$

The ordered pair $\left(\frac{b-c}{1-a},\frac{b-ac}{1-a}\right)$ checks and is the solution. This ordered pair could also be expressed as $\left(\frac{c-b}{a-1},\frac{ac-b}{a-1}\right)$.

Chapter 15 Mid-Chapter Review

1. False; a solution of a system of equations must make *both* equations true.

2. False; see pages 1124, 1125, and 1139 in the text.

3. True; the graphs of the equations are parallel lines, so there is no point of intersection.

4. True; the vertical line $x=a$ and the horizontal line $y=b$ intersect at the point (a,b).

5. $x + y = -1,\quad (1)$

$y = x - 3 \quad (2)$

$$x + (x - 3) = -1$$
$$2x - 3 = -1$$
$$2x = -1 + 3$$
$$2x = 2$$
$$x = 1$$

$$y = 1 - 3$$
$$y = -2$$

The solution is $(1, -2)$.

6.
$$2x - 3y = 7$$
$$x + 3y = -10$$
$$3x + 0y = -3$$
$$3x = -3$$
$$x = -1$$

$$-1 + 3y = -10$$
$$3y = -9$$
$$y = -3$$

The solution is $(-1, -3)$.

7.

$x + y = 1$	
$-4 + 5$? 1	
1	TRUE

$x = y - 9$	
-4 ? $5 - 9$	
	-4 TRUE

The ordered pair $(-4, 5)$ is a solution of both equations, so it is a solution of the system of equations.

8.

$x = y + 10$	
6 ? $-4 + 10$	
	6 TRUE

$x - y = 2$	
$6 - (-4)$? 2	
10	FALSE

The ordered pair $(6, -4)$ is not a solution of $x - y = 2$, so it is not a solution of the system of equations.

9.

$3x + 5y = 2$	
$3(-1) + 5 \cdot 1$? 2	
$-3 + 5$	
2	TRUE

$2x - y = -1$	
$2(-1) - 1$? -1	
$-2 - 1$	
-3	FALSE

The ordered pair $(-1, 1)$ is not a solution of $2x - y = -1$, so it is not a solution of the system of equations.

10.

$2x + y = 1$	
$2 \cdot 2 - 3$? 1	
$4 - 3$	
1	TRUE

$3x - 2y = 12$	
$3 \cdot 2 - 2(-3)$? 12	
$6 + 6$	
12	TRUE

The ordered pair $(2, -3)$ is a solution of both equations, so it is a solution of the system of equations.

11. We graph the equations.

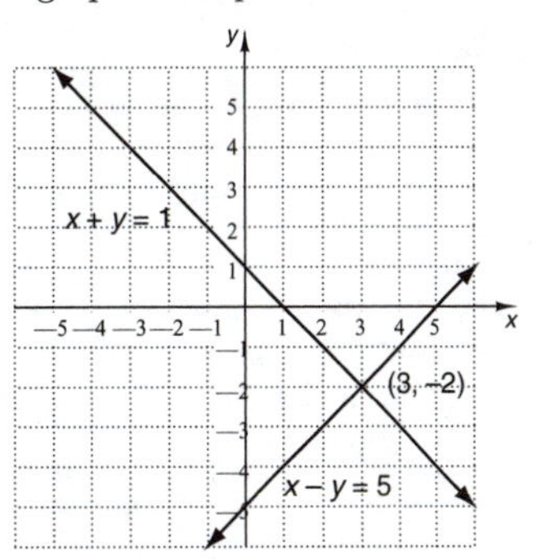

The point of intersection looks as if it has coordinates $(3, -2)$. The ordered pair $(3, -2)$ checks in both equations. It is the solution.

12. We graph the equations.

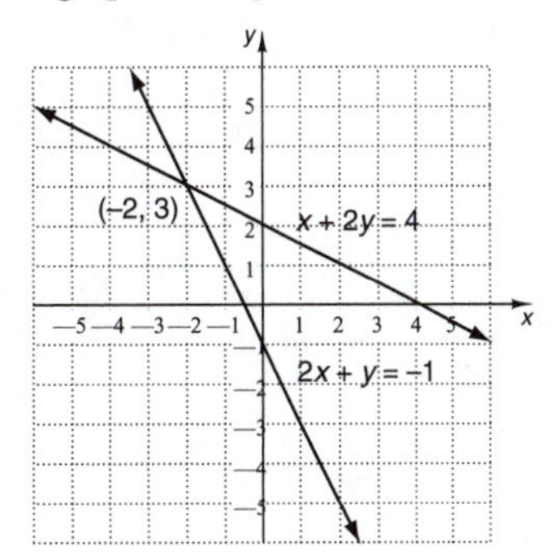

The point of intersection looks as if it has coordinates $(-2, 3)$. The ordered pair $(-2, 3)$ checks in both equations. It is the solution.

13. We graph the equations.

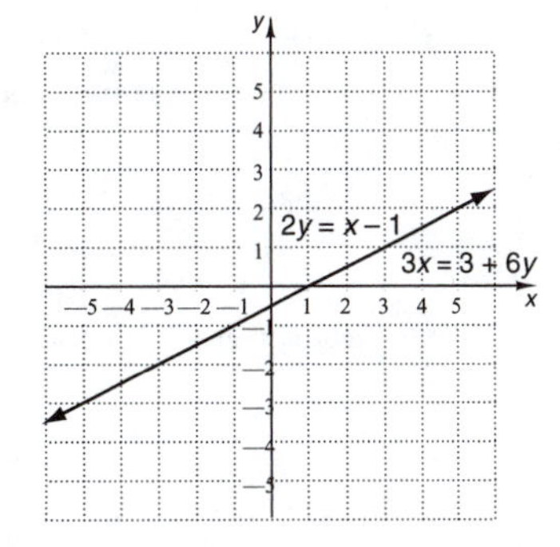

The lines coincide. The system has an infinite number of solutions.

14. We graph the equations.

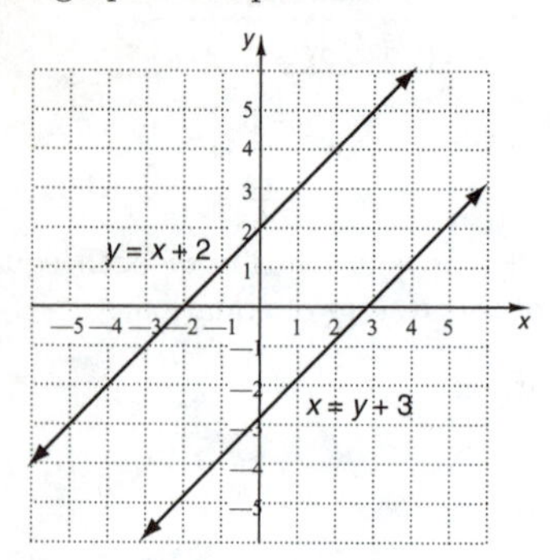

The lines are parallel. There is no solution.

15. $x + y = 2,\quad (1)$

$y = x - 8\quad (2)$

Substitute $x - 8$ for y in Equation (1) and solve for x.

$$\begin{aligned} x + y &= 2 \\ x + (x - 8) &= 2 \\ 2x - 8 &= 2 \\ 2x &= 10 \\ x &= 5 \end{aligned}$$

Now substitute 5 for x in Equation (2) and find y.

$$y = x - 8 = 5 - 8 = -3$$

The ordered pair $(5, -3)$ checks in both equations. It is the solution.

16. $x = y - 1,\quad (1)$

$2x - 5y = 1\quad (2)$

Substitute $y - 1$ for x in Equation (2) and solve for y.

$$\begin{aligned} 2x - 5y &= 1 \\ 2(y - 1) - 5y &= 1 \\ 2y - 2 - 5y &= 1 \\ -3y - 2 &= 1 \\ -3y &= 3 \\ y &= -1 \end{aligned}$$

Now substitute -1 for y in Equation (1) and find x.

$$x = y - 1 = -1 - 1 = -2$$

The ordered pair $(-2, -1)$ checks in both equations. It is the solution.

17. $x + y = 1,\quad (1)$

$3x + 6y = 1\quad (2)$

First solve Equation (1) for either x or y. We will choose to solve for y.

$$\begin{aligned} x + y &= 1 \\ y &= 1 - x\quad (3) \end{aligned}$$

Substitute $1 - x$ for y in Equation (2) and solve for x.

$$\begin{aligned} 3x + 6y &= 1 \\ 3x + 6(1 - x) &= 1 \\ 3x + 6 - 6x &= 1 \\ -3x + 6 &= 1 \\ -3x &= -5 \\ x &= \frac{5}{3} \end{aligned}$$

Substitute $\frac{5}{3}$ for x in Equation (3) and find y.

$$y = 1 - x = 1 - \frac{5}{3} = -\frac{2}{3}$$

The ordered pair $\left(\frac{5}{3}, -\frac{2}{3}\right)$ checks in both equations. It is the solution.

18. $2x + y = 2,\quad (1)$

$2x - y = -1\quad (2)$

First solve Equation (1) for y.

$$\begin{aligned} 2x + y &= 2 \\ y &= 2 - 2x\quad (3) \end{aligned}$$

Now substitute $2 - 2x$ for y in Equation (2) and solve for x.

$$\begin{aligned} 2x - y &= -1 \\ 2x - (2 - 2x) &= -1 \\ 2x - 2 + 2x &= -1 \\ 4x - 2 &= -1 \\ 4x &= 1 \\ x &= \frac{1}{4} \end{aligned}$$

Substitute $\frac{1}{4}$ for x in Equation (3) and find y.

$$y = 2 - 2x = 2 - 2 \cdot \frac{1}{4} = 2 - \frac{1}{2} = \frac{3}{2}$$

The ordered pair $\left(\frac{1}{4}, \frac{3}{2}\right)$ checks in both equations. It is the solution.

19.

$$\begin{array}{r} x + y = 3 \\ -x - y = 5 \\ \hline 0 = 8 \end{array}$$

We get a false equation. The system of equations has no solution.

20.

$$\begin{array}{rll} 3x - 2y &= 2, & (1) \\ 5x + 2y &= -2 & (2) \\ \hline 8x \quad &= 0 & \text{Adding} \\ x &= 0 & \end{array}$$

Substitute 0 for x in one of the original equations and solve for y.

$$\begin{aligned} 5x + 2y &= -2\quad (2) \\ 5 \cdot 0 + 2y &= -2 \\ 2y &= -2 \\ y &= -1 \end{aligned}$$

The ordered pair $(0,-1)$ checks in both equations. It is the solution.

21. $2x+3y=1,\quad (1)$

$3x+2y=-6\quad (2)$

Multiply the first equation by 2 and the second by -3 and then add.

$$\begin{array}{r} 4x+6y=2 \\ -9x-6y=18 \\ \hline -5x=20 \\ x=-4 \end{array}$$

Substitute -4 for x in one of the original equations and solve for y.

$$\begin{aligned} 2x+3y&=1 \quad (1) \\ 2(-4)+3y&=1 \\ -8+3y&=1 \\ 3y&=9 \\ y&=3 \end{aligned}$$

The ordered pair $(-4,3)$ checks in both equations. It is the solution.

22. $2x-3y=6\quad (1)$

$-4x+6y=-12\quad (2)$

Multiply Equation (1) by 2 and then add.

$$\begin{array}{r} 4x-6y=12 \\ -4x+6y=-12 \\ \hline 0=0 \end{array}$$

We get an equation that is true for all values of x and y. The system of equations has an infinite number of solutions.

23. ***Familiarize***. Let l = the length of the rug, in feet, and w = the width.

Translate. The formula for the perimeter of a rectangle gives us one equation.

$2l+2w=18$

The width is one foot shorter than the length, so we have a second equation.

$w=l-1$

The resulting system of equations is

$2l+2w=18,\quad (1)$

$w=l-1.\quad (2)$

Solve. First we substitute $l-1$ for w in Equation (1) and solve for l.

$$\begin{aligned} 2l+2w&=18 \\ 2l+2(l-1)&=18 \\ 2l+2l-2&=18 \\ 4l-2&=18 \\ 4l&=20 \\ l&=5 \end{aligned}$$

Now substitute 5 for l in Equation (2) and find w.

$w=l-1=5-1=4.$

Check. If the length is 5 ft and the width is 4 ft, the perimeter is $2\cdot 5+2\cdot 4=10+8=18$ ft. The width, 4 ft, is 1 ft shorter than the length, 5 ft. The answer checks.

State. The length is 5 ft, and the width is 4 ft.

24. ***Familiarize***. Let x and y represent the numbers.

Translate.

$$\underbrace{\text{The sum of the numbers}}_{x+y}\ \underset{=}{\text{is}}\ \underset{18}{18.}$$

$$\underbrace{\text{The difference of the numbers}}_{x-y}\ \underset{=}{\text{is}}\ \underset{86}{86.}$$

The resulting system of equations is

$x+y=18,\quad (1)$

$x-y=86.\quad (2)$

Solve. We add the equations.

$$\begin{array}{r} x+y=18 \\ x-y=86 \\ \hline 2x=104 \\ x=52 \end{array}$$

Now substitute 52 for x in one of the original equations and solve for y.

$$\begin{aligned} x+y&=18 \quad (1) \\ 52+y&=18 \\ y&=-34 \end{aligned}$$

Check. $52+(-34)=18$ and $52-(-34)=52+34=86$. The answer checks.

State. The numbers are 52 and -34.

25. ***Familiarize***. Let x = the larger number and y = the smaller number.

Translate.

$$\underbrace{\text{The difference of the numbers}}_{x-y}\ \underset{=}{\text{is}}\ \underset{4}{4.}$$

$$\underset{2}{\text{Two}}\ \underset{\cdot}{\text{times}}\ \underbrace{\text{the larger number}}_{x}\ \underset{=}{\text{is}}\ \underset{3}{\text{three}}\ \underset{\cdot}{\text{times}}\ \underbrace{\text{the smaller number.}}_{y}$$

The resulting system of equations is

$x-y=4,\quad (1)$

$2x=3y.\quad (2)$

Solve. First solve Equation (1) for x.

$$\begin{aligned} x-y&=4 \\ x&=y+4 \quad (3) \end{aligned}$$

Now substitute $y+4$ for x in Equation (2) and solve for y.

$$2x = 3y$$
$$2(y+4) = 3y$$
$$2y+8 = 3y$$
$$8 = y$$

Substitute 8 for y in Equation (3) and find x.

$$x = y+4 = 8+4 = 12$$

Check. $12-8 = 4$ and $2\cdot 12 = 24 = 3\cdot 8$. The answer checks.

State. The numbers are 12 and 8.

26. We know that the first coordinate of the point of intersection is 2. Substitute 2 for x in either $y = 3x-1$ or $y = 9-2x$ and find y, the second coordinate of the point of intersection.

27. The coordinates of the point of intersection of the graphs are not integers, so it is difficult to determine the solution from the graph.

28. The equations have the same coefficients of x and y but different constant terms. This means their graphs have the same slope but different y-intercepts. Thus, they have no points in common and the system of equations has no solution.

29. This is not the best approach, in general. If the first equation has x alone on one side, for instance, or if the second equation has a variable alone on one side, solving for y in the first equation is inefficient. This procedure could also introduce fractions in the computations unnecessarily.

Exercise Set 15.4

1. ***Familiarize***. Let a = the number of adults and c = the number of children. We organize the information in a table.

	Adults	Children	Total
Number of visitors	a	c	320
Price	\$3	\$2	
Money taken in	$3a$	$2c$	\$730

Translate. The first and last rows of the table give us two equations. The total number of visitors was 320.

$$a+c = 320$$

Total receipts were \$730.

$$3a+2c = 730$$

The resulting system of equations is

$$a + c = 320, \quad (1)$$
$$3a + 2c = 730. \quad (2)$$

Solve. Multiply Equation (1) by -2 and then add.

$$-2a - 2c = -640$$
$$3a + 2c = 730$$
$$\overline{a \quad = 90}$$

Now substitute 90 for a in one of the original equations and solve for c.

$$a+c = 320 \quad (1)$$
$$90+c = 320$$
$$c = 230$$

Check. The total number of visitors was $90+230$, or 320. The receipts from 90 adult tickets were $\$3\cdot 90$, or \$270, and the receipts from 230 children's tickets were $\$2\cdot 230$, or \$460. Then total receipts were \$270 + \$460, or \$730. The answer checks.

State. 90 adults and 230 children visited the exhibit.

3. ***Familiarize***. Let x = the number of 4×6 prints and y = the number of 5×7 prints. We organize the information in a table.

	4×6	5×7	Total
Number ordered	x	y	36
Price	\$0.10	\$0.60	
Amount paid	$0.1x$	$0.6y$	\$6.60

Translate. The first and last rows of the table give us two equations. Lucy ordered 36 prints.

$$x+y = 36$$

The total cost of the prints was \$6.60.

$$0.1x+0.6y = 6.60$$

We can multiply the second equation by 10 to clear the decimals. The resulting system of equations is

$$x + y = 36, \quad (1)$$
$$x + 6y = 66. \quad (2)$$

Solve. Multiply Equation (1) by -1 and then add.

$$-x - y = -36$$
$$x + 6y = 66$$
$$\overline{5y = 30}$$
$$y = 6$$

Substitute 6 for y in one of the original equations and solve for x.

$$x+y = 36 \quad (1)$$
$$x+6 = 36$$
$$x = 30$$

Check. A total of $30+6$, or 36, prints were ordered. The 30 4×6 prints cost $\$0.10\times 30$, or \$3, and the 6 5×7 prints cost $\$0.60\times 6$, or \$3.60. The total cost of the prints was \$3 + \$3.60, or \$6.60. The answer checks.

State. Lucy ordered 30 4×6 prints and 6 5×7 prints.

5. ***Familiarize***. Let x = the number of two-point baskets made and y = the number of three-point baskets made. Then $2x$ points were scored from two-point baskets and $3y$ points were scored from three-point baskets.

Translate. Since 40 baskets were made we have one equation: $x+y = 40$. Since 85 points were scored, we have a second equation: $2x+3y = 85$.

The resulting system is

$$x + y = 40, \quad (1)$$
$$2x + 3y = 85. \quad (2)$$

Solve. We use the elimination method. First we multiply both sides of Equation (1) by -2 and add.

$$-2x - 2y = -80$$
$$2x + 3y = 85$$
$$y = 5$$

Now we substitute 5 for y in Equation (1) and solve for x.

$$x + y = 40$$
$$x + 5 = 40$$
$$x = 35$$

Check. If 35 two-point baskets and 5 three-point baskets are made, then a total of $35 + 5$, or 40, baskets are made. The points scored are $2 \cdot 35 + 3 \cdot 5$, or $70 + 15$, or 85. The answer checks.

State. The Spurs made 35 two-point shots and 5 three-point shots.

7. ***Familiarize.*** Let x = the number of \$50 bonds and y = the number of \$100 bonds. Then the total value of the \$50 bonds is $50x$ and the total value of the \$100 bonds is $100y$.

Translate.

Total value of bonds is \$1250.

$$50x + 100y = 1250$$

Number of \$50 bonds is 7 more than number of \$100 bonds.

$$x = 7 + y$$

The resulting system is

$$50x + 100y = 1250, \quad (1)$$
$$x = 7 + y. \quad (2)$$

Solve. We use the substitution method, substituting $7+y$ for x in Equation (1).

$$50x + 100y = 1250 \quad (1)$$
$$50(7 + y) + 100y = 1250$$
$$350 + 50y + 100y = 1250$$
$$350 + 150y = 1250$$
$$150y = 900$$
$$y = 6$$

Now we substitute 6 for y in Equation (2) to find x.

$$x = 7 + y \quad (2)$$
$$x = 7 + 6 = 13$$

Check. If there are 13 \$50 bonds and 6 \$100 bonds, there are 7 more \$50 bonds than \$100 bonds. The total value of the bonds is $\$50 \cdot 13 + \$100 \cdot 6$, or $\$650 + \$600$, or \$1250. The answer checks.

State. Cassandra has 13 \$50 bonds and 6 \$100 bonds.

9. ***Familiarize.*** Let x = the number of cardholders tickets that were sold and y = the number of non-cardholders tickets. We arrange the information in a table.

	Card-holders	Non-card-holders	Total
Price	\$2.25	\$3	
Number sold	x	y	203
Money taken in	$2.25x$	$3y$	\$513

Translate. The last two rows of the table give us two equations. The total number of tickets sold was 203, so we have

$$x + y = 203.$$

The total amount of money collected was \$513, so we have

$$2.25x + 3y = 513.$$

We can multiply the second equation on both sides by 100 to clear decimals. The resulting system is

$$x + y = 203, \quad (1)$$
$$225x + 300y = 51,300. \quad (2)$$

Solve. We use the elimination method. We multiply on both sides of Equation (1) by -225 and then add.

$$-225x - 225y = -46,675 \quad \text{Multiplying by } -225$$
$$225x + 300y = 51,300$$
$$75y = 5625$$
$$y = 75$$

We go back to Equation (1) and substitute 75 for y.

$$x + y = 203$$
$$x + 75 = 203$$
$$x = 128$$

Check. The number of tickets sold was $128 + 75$, or 203. The money collected was $\$2.25(128) + \$3(75)$, or $\$288 + \$225$, or \$513. These numbers check.

State. 128 cardholders tickets and 75 non-cardholders tickets were sold.

11. ***Familiarize.*** We complete the table in the text. Note that x represents the number of liters of solution A to be used and y represents the number of liters of solution B.

Type of solution	A	B	Mixture
Amount of solution	x	y	100 L
Percent of acid	50%	80%	68%
Amount of acid in solution	$50\%x$	$80\%y$	$68\% \times 100$, or 68 L

Equation from first row: $x + y = 100$

Equation from third row: $50\%x + 80\%y = 68$

Translate. The first and third rows of the table give us two equations. Since the total amount of solution is 100 liters, we have

$x + y = 100.$

The amount of acid in the mixture is to be 68% of 100, or 68 liters. The amounts of acid from the two solutions are $50\%x$ and $80\%y$. Thus

$$50\%x + 80\%y = 68,$$
or $$0.5x + 0.8y = 68,$$
or $$5x + 8y = 680 \quad \text{Clearing decimals}$$

The resulting system is

$$x + y = 100, \quad (1)$$
$$5x + 8y = 680. \quad (2)$$

Solve. We use the elimination method. We multiply on both sides of Equation (1) by -5 and then add.

$$-5x - 5y = -500 \quad \text{Multiplying by } -5$$
$$5x + 8y = 680$$
$$3y = 180$$
$$y = 60$$

We go back to Equation (1) and substitute 60 for y.

$$x + y = 100$$
$$x + 60 = 100$$
$$x = 40$$

Check. We consider $x = 40$ and $y = 60$. The sum is 100. Now 50% of 40 is 20 and 80% of 60 is 48. These add up to 68. The numbers check.

State. 40 liters of solution A and 60 liters of solution B should be used.

13. ***Familiarize***. We let $x =$ the number of pounds of hay and $y =$ the number of pounds of grain that should be fed to the horse each day. We arrange the information in a table.

Type of feed	Hay	Grain	Mixture
Amount of feed	x	y	15
Percent of protein	6%	12%	8%
Amount of protein in mixture	$6\%x$	$12\%y$	$8\% \times 15$, or 1.2 lb

Translate. The first and last rows of the table give us two equations. The total amount of feed is 15 lb, so we have

$x + y = 15.$

The amount of protein in the mixture is to be 8% of 15 lb, or 1.2 lb. The amounts of protein from the two feeds are $6\%x$ and $12\%y$. Thus

$$6\%x + 12\%y = 1.2, \quad \text{or}$$
$$0.06x + 0.12y = 1.2, \quad \text{or}$$
$$6x + 12y = 120 \quad \text{Clearing decimals}$$

The resulting system is

$$x + y = 15, \quad (1)$$
$$6x + 12y = 120. \quad (2)$$

Solve. We use the elimination method. Multiply on both sides of Equation (1) by -6 and then add.

$$-6x - 6y = -90$$
$$6x + 12y = 120$$
$$6y = 30$$
$$y = 5$$

We go back to Equation (1) and substitute 5 for y.

$$x + y = 15$$
$$x + 5 = 15$$
$$x = 10$$

Check. The sum of 10 and 5 is 15. Also, 6% of 10 is 0.6 and 12% of 5 is 0.6, and $0.6 + 0.6 = 1.2$. These numbers check.

State. Brianna should feed her horse 10 lb of hay and 5 lb of grain each day.

15. ***Familiarize***. Let d represent the number of dimes and q the number of quarters. Then, $10d$ represents the value of the dimes in cents, and $25q$ represents the value of the quarters in cents. The total value is \$15.25, or 1525¢. The total number of coins is 103.

Translate.

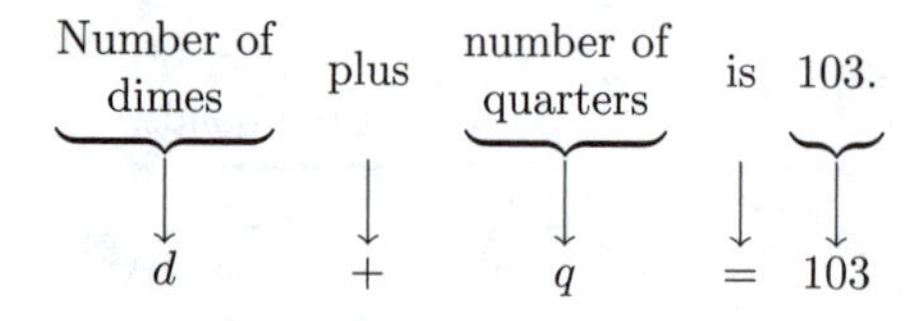

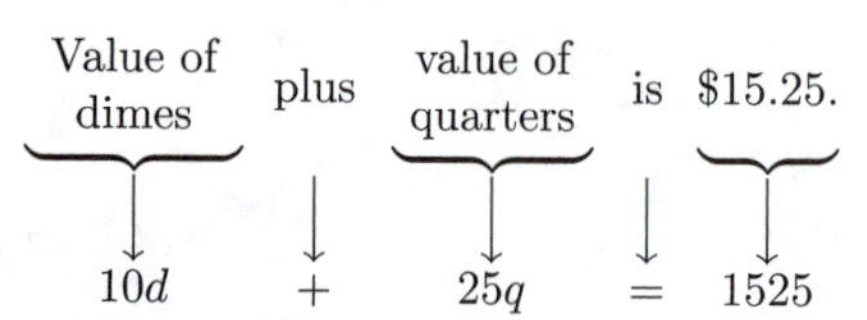

The resulting system is

$$d + q = 103, \quad (1)$$
$$10d + 25q = 1525. \quad (2)$$

Solve. We use the addition method. We multiply Equation (1) by -10 and then add.

$$-10d - 10q = -1030 \quad \text{Multiplying by } -10$$
$$10d + 25q = 1525$$
$$15q = 495 \quad \text{Adding}$$
$$q = 33$$

Now we substitute 33 for q in one of the original equations and solve for d.

$$d + q = 103 \quad (1)$$
$$d + 33 = 103 \quad \text{Substituting}$$
$$d = 70$$

Check. The number of dimes plus the number of quarters is $70+33$, or 103. The total value in cents is $10 \cdot 70+25 \cdot 33$, or $700+825$, or 1525. This is equal to \$15.25. This checks.

State. There are 70 dimes and 33 quarters.

17. Familiarize. We complete the table in the text. Note that x represents the number of pounds of Brazilian coffee to be used and y represents the number of pounds of Turkish coffee.

Type of coffee	Brazilian	Turkish	Mixture
Cost of coffee	\$19	\$22	\$20
Amount (in pounds)	x	y	300
Mixture	$19x$	$22y$	\$20(300), or \$6000

Equation from second row: $x+y=300$

Equation from third row: $19x+22y=6000$

Translate. The second and third rows of the table give us two equations. Since the total amount of the mixture is 300 lb, we have

$$x+y=300.$$

The value of the Brazilian coffee is $19x$ (x lb at \$19 per pound), the value of the Turkish coffee is $22y$ (y lb at \$22 per pound), and the value of the mixture is \$20(300) or \$6000. Thus we have

$$19x+22y=6000.$$

The resulting system is

$$x+y=300, \quad (1)$$
$$19x+22y=6000. \quad (2)$$

Solve. We use the elimination method. We multiply on both sides of Equation (1) by -19 and then add.

$$-19x-19y=-5700 \quad \text{Multiplying by } -19$$
$$\underline{19x+22y=6000}$$
$$3y=300$$
$$y=100$$

We go back to Equation (1) and substitute 100 for y.

$$x+y=300$$
$$x+100=300$$
$$x=200$$

Check. The sum of 200 and 100 is 300. The value of the mixture is $\$19(200)+\$22(100)$, or $\$3800+\$2200$, or \$6000. These values check.

State. 200 lb of Brazilian coffee and 100 lb of Turkish coffee should be used.

19. Familiarize. Let x and y represent the number of liters of 28%-fungicide solution and 40%-fungicide solution to be used in the mixture, respectively.

Translate. We organize the given information in a table.

Type of solution	28%	40%	36%
Amount of solution	x	y	300
Percent fungicide	28%	40%	36%
Amount of fungicide in solution	$0.28x$	$0.4y$	0.36(300), or 108

We get a system of equations from the first and third rows of the table.

$$x+\quad y=300,$$
$$0.28x+0.4y=108$$

Clearing decimals we have

$$x+\quad y=300, \quad (1)$$
$$28x+40y=10,800 \quad (2)$$

Solve. We use the elimination method. Multiply Equation (1) by -28 and add.

$$-28x-28y=-8400$$
$$\underline{28x+40y=10,800}$$
$$12y=2400$$
$$y=200$$

Now substitute 200 for y in Equation (1) and solve for x.

$$x+y=300$$
$$x+200=300$$
$$x=100$$

Check. The sum of 100 and 200 is 300. The amount of fungicide in the mixture is $0.28(100)+0.4(200)$, or $28+80$, or 108 L. These numbers check.

State. 100 L of the 28%-fungicide solution and 200 L of the 40%-fungicide solution should be used in the mixture.

21. Familiarize. We let $x=$ the number of pages in large type and $y=$ the number of pages in small type. We arrange the information in a table.

Size of type	Large	Small	Mixture (Book)
Words per page	830	1050	
Number of pages	x	y	12
Number of words	$830x$	$1050y$	11,720

Translate. The last two rows of the table give us two equations. The total number of pages in the document is 12, so we have

$$x+y=12.$$

The number of words on the pages with large type is $830x$ (x pages with 830 words per page), and the number of words on the pages with small type is $1050y$ (y pages with 1050 words per page). The total number of words is 11,720, so we have

$$830x+1050y=11,720.$$

The resulting system is

$$x+y=12, \quad (1)$$
$$830x+1050y=11,720. \quad (2)$$

Solve. We use the elimination method. We multiply on both sides of Equation (1) by -830 and then add.

$$\begin{aligned} -830x - 830y &= -9,960 \quad \text{Multiplying by } -830 \\ 830x + 1050y &= 11,720 \\ \hline 220y &= 1760 \\ y &= 8 \end{aligned}$$

We go back to Equation (1) and substitute 8 for y.

$$\begin{aligned} x+y &= 12 \\ x+8 &= 12 \\ x &= 4 \end{aligned}$$

Check. The sum of 4 and 8 is 12. The number of words in large type is $830 \cdot 4$, or 3320, and the number of words in small type is $1050 \cdot 8$, or 8400. Then the total number of words is $3320 + 8400$, or 11,720. These numbers check.

State. There were 4 pages in large type and 8 pages in small type.

23. ***Familiarize***. Let $x =$ the number of pounds of the 70% mixture and $y =$ the number of pounds of the 45% mixture to be used. We organize the information in the table.

Percent of cashews	70%	45%	
Amount	x	y	60
Mixture	$70\%x$, or $0.7x$	$45\%y$, or $0.45y$	60(60%) or 60(0.6) or 36

Translate. The last two rows of the table give us two equations. The total weight of the mixture is 60 lb, so we have

$$x+y=60.$$

The amount of cashews in the mixture is 36 lb, so we have

$$0.7x+0.45y=36, \text{ or}$$
$$70x+45y=3600 \quad \text{Clearing decimals}$$

The resulting system is

$$x+y=60, \quad (1)$$
$$70x+45y=3600. \quad (2)$$

Solve. We use the elimination method. We multiply Equation (1) by -45 and then add.

$$\begin{aligned} -45x - 45y &= -2700 \\ 70x + 45y &= 3600 \\ \hline 25x \qquad &= 900 \\ x &= 36 \end{aligned}$$

Next we substitute 36 for x in one of the original equations and solve for y.

$$\begin{aligned} x+y &= 60 \quad (1) \\ 36+y &= 60 \\ y &= 24 \end{aligned}$$

Check. The total weight of the mixture is 36 lb+24 lb, or 60 lb. The amount of cashews in the mixture is $0.7(36 \text{ lb}) + 0.45(24 \text{ lb})$, or 36 lb. Since 36 lb is 60% of 60 lb, the answer checks.

State. The new mixture should contain 36 lb of the 70% cashew mixture and 24 lb of the 45% cashew mixture.

25. ***Familiarize***. We arrange the information in a table. Let $a =$ the number of type A questions and $b =$ the number of type B questions.

Type of question	A	B	Mixture (Test)
Number	a	b	16
Time	3 min	6 min	
Value	10 points	15 points	
Mixture (Test)	$3a$ min, $10a$ points	$6b$ min, $15b$ points	60 min, 180 points

Translate. The table actually gives us three equations. Since the total number of questions is 16, we have

$$a+b=16.$$

The total time is 60 min, so we have

$$3a+6b=60.$$

The total number of points is 180, so we have

$$10a+15b=180.$$

The resulting system is

$$a+b=16, \quad (1)$$
$$3a+6b=60, \quad (2)$$
$$10a+15b=180. \quad (3)$$

Solve. We will solve the system composed of Equations (1) and (2) and then check to see that this solution also satisfies Equation (3). We multiply equation (1) by -3 and add.

$$\begin{aligned} -3a - 3b &= -48 \\ 3a + 6b &= 60 \\ \hline 3b &= 12 \\ b &= 4 \end{aligned}$$

Now we substitute 4 for b in Equation (1) and solve for a.

$$\begin{aligned} a+b &= 16 \\ a+4 &= 16 \\ a &= 12 \end{aligned}$$

Check. We consider $a = 12$ questions and $b = 4$ questions. The total number of questions is 16. The time required is $3 \cdot 12 + 6 \cdot 4$, or $36 + 24$, or 60 min. The total points are $10 \cdot 12 + 15 \cdot 4$, or $120 + 60$, or 180. These values check.

State. 12 questions of type A and 4 questions of type B were answered. Assuming all the answers were correct, the score was 180 points.

27. ***Familiarize***. Let $k =$ the age of the Kuyatt's house now and $m =$ the age of the Marconi's house now. Eight years ago the houses' ages were $k - 8$ and $m - 8$.

Translate. We reword and translate.

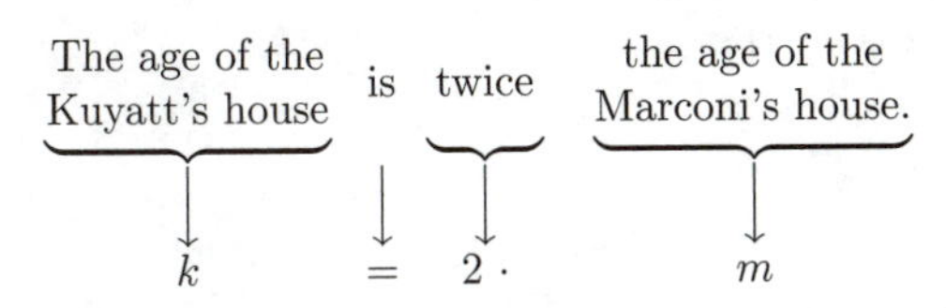

The age of the Kuyatt's house 8 years ago was three times the age of the Marconi's house 8 years ago.

$$k - 8 = 3 \cdot m - 8$$

The resulting system is

$$k = 2m, \qquad (1)$$

$$k - 8 = 3(m - 8). \qquad (2)$$

Solve. We use the substitution method. We substitute $2m$ for k in Equation (2) and solve for m.

$$k - 8 = 3(m - 8)$$

$$2m - 8 = 3(m - 8)$$

$$2m - 8 = 3m - 24$$

$$-8 = m - 24$$

$$16 = m$$

We find k by substituting 16 for m in Equation (1).

$$k = 2m$$

$$k = 2 \cdot 16$$

$$k = 32$$

Check. The age of the Kuyatt's house, 32 years, is twice the age of the Marconi's house, 16 years. Eight years ago, when the Kuyatt's house was 24 years old and the Marconi's house was 8 years old, the Kuyatt's house was three times as old as the Marconi's house. These numbers check.

State. The Kuyatt's house is 32 years old, and the Marconi's house is 16 years old.

29. ***Familiarize***. Let R = Randy's age now and M = Marie's age now. In twelve years their ages will be $R + 12$ and $M + 12$.

Translate. We reword and translate.

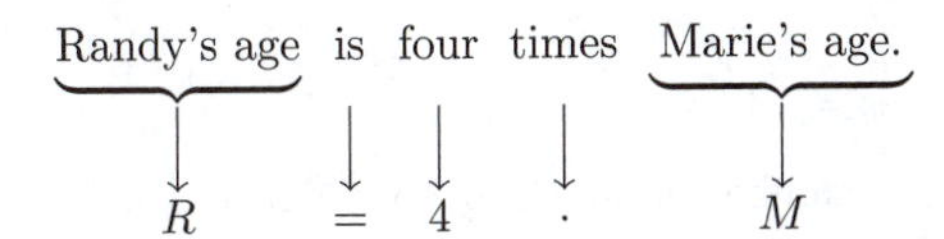

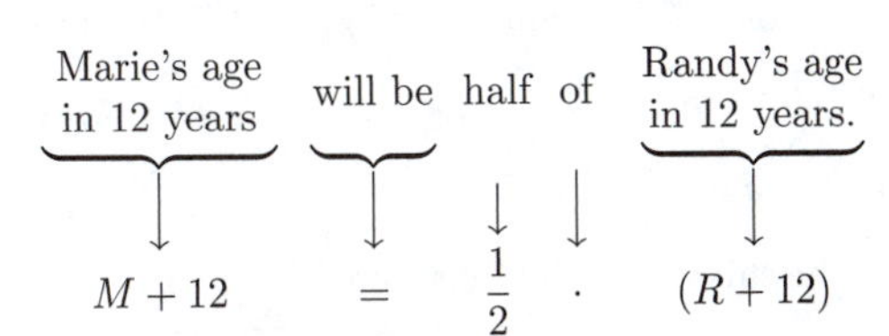

The resulting system is

$$R = 4M, \qquad (1)$$

$$M + 12 = \frac{1}{2}(R + 12). \qquad (2)$$

Solve. We use the substitution method. We substitute $4M$ for R in Equation (2) and solve for M.

$$M + 12 = \frac{1}{2}(R + 12)$$

$$M + 12 = \frac{1}{2}(4M + 12)$$

$$M + 12 = 2M + 6$$

$$12 = M + 6$$

$$6 = M$$

We find R by substituting 6 for M in Equation (1).

$$R = 4M$$

$$R = 4 \cdot 6$$

$$R = 24$$

Check. Randy's age now, 24, is 4 times 6, Marie's age. In 12 yr, when Randy will be 36 and Marie 18, Marie's age will be half of Randy's age. These numbers check.

State. Randy is 24 years old now, and Marie is 6.

31. ***Familiarize***. Let x = the smaller angle and y = the larger angle.

Translate. We reword the problem.

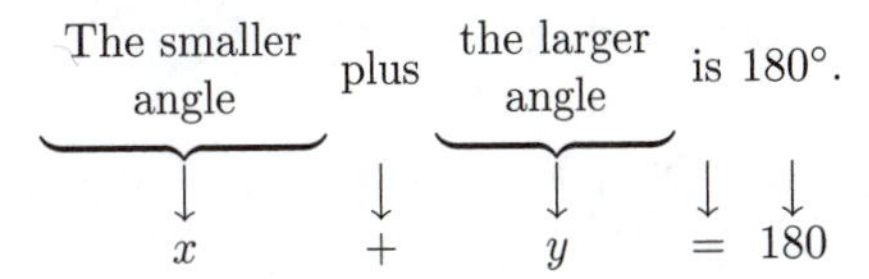

The larger angle is 30° more than 2 times the smaller angle.

$$y = 30 + 2 \cdot x$$

The resulting system is

$$x + y = 180,$$

$$y = 30 + 2x.$$

Solve. We solve the system. We will use the elimination method although we could also easily use the substitution method. First we get the second equation in the form $Ax + By = C$.

$$x + y = 180 \quad (1)$$

$$-2x + y = 30 \quad (2) \text{ Adding } -2x$$

Now we multiply Equation (2) by -1 and add.

$$\begin{aligned} x + y &= 180 \\ 2x - y &= -30 \\ \hline 3x \quad &= 150 \\ x &= 50 \end{aligned}$$

Then we substitute 50 for x in Equation (1) and solve for y.

$$x + y = 180 \quad \text{Equation (1)}$$

$$50 + y = 180 \quad \text{Substituting}$$

$$y = 130$$

Check. The sum of the angles is $50° + 130°$, or $180°$, so the angles are supplementary. Also, $30°$ more than two times the $50°$ angle is $30° + 2 \cdot 50°$, or $30° + 100°$, or $130°$, the other angle. These numbers check.

State. The angles are $50°$ and $130°$.

33. ***Familiarize***. We let $x =$ the larger angle and $y =$ the smaller angle.

Translate. We reword and translate the first statement.

$$\underbrace{\text{The sum of two angles}}_{x+y} \ \underset{=}{\text{is}} \ \underset{90}{90°}.$$

We reword and translate the second statement.

$$\underbrace{\text{The difference of two angles}}_{x-y} \ \underset{=}{\text{is}} \ \underset{34}{34°}.$$

The resulting system is

$$x + y = 90,$$
$$x - y = 34.$$

Solve. We solve the system.

$$\begin{aligned} x + y &= 90, \quad (1) \\ x - y &= 34 \quad (2) \\ \hline 2x &= 124 \quad \text{Adding} \\ x &= 62 \end{aligned}$$

Now we substitute 62 for x in Equation (1) and solve for y.

$$\begin{aligned} x + y &= 90 \quad \text{Equation (1)} \\ 62 + y &= 90 \quad \text{Substituting} \\ y &= 28 \end{aligned}$$

Check. The sum of the angles is $62° + 28°$, or $90°$, so the angles are complementary. The difference of the angles is $62° - 28°$, or $34°$. These numbers check.

State. The angles are $62°$ and $28°$.

35. ***Familiarize***. Let $x =$ the number of gallons of 87-octane gas and $y =$ the number of gallons of 93-octane gas that should be used. We arrange the information in a table.

Type of gas	87-octane	93-octane	Mixture
Amount of gas	x	y	18
Octane rating	87	93	89
Mixture	$87x$	$93y$	$18 \cdot 89$, or 1602

Translate. The first and last rows of the table give us a system of equations.

$$\begin{aligned} x + \quad y &= \ \ 18, \quad (1) \\ 87x + 93y &= 1602 \quad (2) \end{aligned}$$

Solve. We multiply Equation (1) by -87 and then add.

$$\begin{aligned} -87x - 87y &= -1566 \\ 87x + 93y &= 1602 \\ \hline 6y &= 36 \\ y &= 6 \end{aligned}$$

Then substitute 6 for y in Equation (1) and solve for x.

$$\begin{aligned} x + y &= 18 \\ x + 6 &= 18 \\ x &= 12 \end{aligned}$$

Check. The total amount of gas is 12 gal + 6 gal, or 18 gal. Also $87(12) + 93(6) = 1044 + 558 = 1602$. The answer checks.

State. 12 gal of 87-octane gas and 6 gal of 93-octane gas should be blended.

37. ***Familiarize***. Let $x =$ the number of ounces of Dr. Zeke's cough syrup and $y =$ the number of ounces of Vitabrite cough syrup that should be used. We organize the information in a table.

	Dr. Zeke's	Vitabrite	Mixture
Percent of alcohol	2%	5%	3%
Amount	x	y	80
Mixture	$2\%x$, or $0.02x$	$5\%x$, or $0.05y$	$3\% \cdot 80$, or $0.03 \cdot 80$, or 2.4

Translate. The last two rows of the table give us a system of equations.

$$x + y = 80,$$
$$0.02x + 0.05y = 2.4.$$

Clearing decimals, we have

$$\begin{aligned} x + y &= 80, \quad (1) \\ 2x + 5y &= 240. \quad (2) \end{aligned}$$

Solve. We use the elimination method. First we multiply Equation (1) by -2 and then add.

$$\begin{aligned} -2x - 2y &= -160 \\ 2x + 5y &= 240 \\ \hline 3y &= 80 \\ y &= 26\frac{2}{3} \end{aligned}$$

Substitute $26\frac{2}{3}$ for y in one of the original equations and solve for x.

$$\begin{aligned} x + y &= 80 \\ x + 26\frac{2}{3} &= 80 \\ x &= 53\frac{1}{3} \end{aligned}$$

Check. The number of ounces in the mixture is $53\frac{1}{3} + 26\frac{2}{3}$, or 80. The amount of alcohol in the mixture is $0.02\left(53\frac{1}{3}\right) + 0.05\left(26\frac{2}{3}\right)$, or 2.4 oz. Since 2.4 oz is 3% of 80 oz, the answer checks.

State. The mixture should contain $53\frac{1}{3}$ oz of Dr. Zeke's cough syrup and $26\frac{2}{3}$ oz of Vitabrite cough syrup.

39.
$$\begin{aligned} 25x^2 - 81 &= (5x)^2 - 9^2 \\ &= (5x + 9)(5x - 9) \end{aligned}$$

41. $4x^2 + 100 = 4(x^2 + 25)$

43. $y = -2x - 3$

The equation is in the form $y = mx + b$, so the y-intercept is $(0, -3)$.

To find the x-intercept, we let $y = 0$ and solve for x.

$$0 = -2x - 3$$
$$2x = -3$$
$$x = -\frac{3}{2}$$

The x-intercept is $\left(-\frac{3}{2}, 0\right)$.

We plot the intercepts and draw the line.

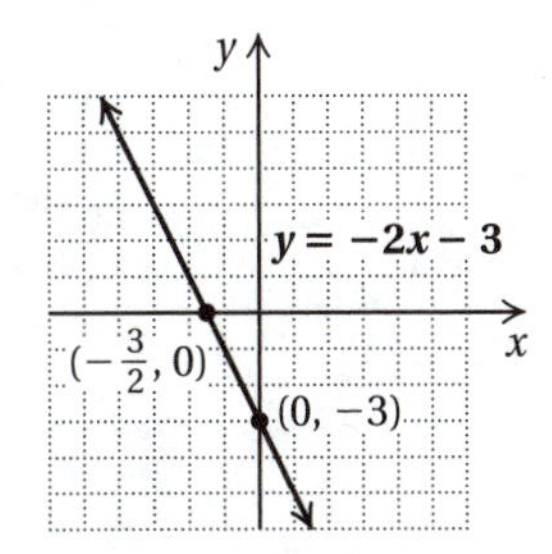

A third point should be used as a check.

For example, let $x = -3$. Then

$$y = -2(-3) - 3 = 6 - 3 = 3.$$

It appears that the point $(-3, 3)$ is on the graph, so the graph is probably correct.

45. $5x - 2y = -10$

To find the y-intercept, let $x = 0$ and solve for y.

$$5 \cdot 0 - 2y = -10$$
$$-2y = -10$$
$$y = 5$$

The y-intercept is $(0, 5)$.

To find the x-intercept, let $y = 0$ and solve for x.

$$5x - 2 \cdot 0 = -10$$
$$5x = -10$$
$$x = -2$$

The x-intercept is $(-2, 0)$.

We plot the intercepts and draw the line.

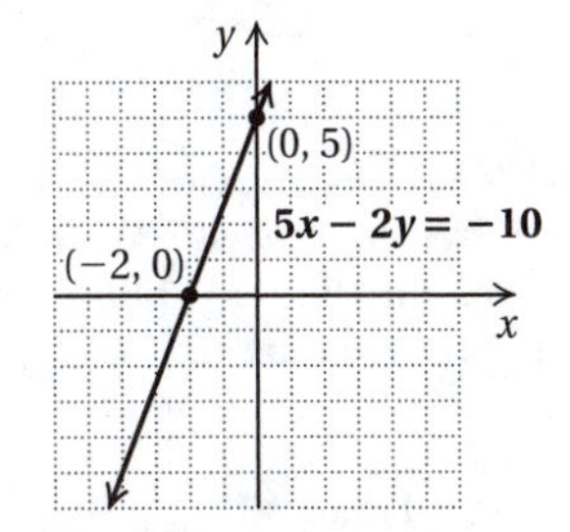

A third point should be used as a check. For example, let $x = -4$. then

$$5(-4) - 2y = -10$$
$$-20 - 2y = -10$$
$$-2y = 10$$
$$y = -5$$

It appears that the point $(-4, -5)$ is on the graph, so the graph is probably correct.

47. $$\frac{x^2 - 5x + 6}{x^2 - 4} = \frac{(x-3)(x-2)}{(x+2)(x-2)} = \frac{(x-3)\cancel{(x-2)}}{(x+2)\cancel{(x-2)}} = \frac{x-3}{x+2}$$

49. $$\frac{x-2}{x+3} - \frac{2x-5}{x-4} \quad \text{LCD is } (x+3)(x-4)$$

$$= \frac{x-2}{x+3} \cdot \frac{x-4}{x-4} - \frac{2x-5}{x-4} \cdot \frac{x+3}{x+3}$$

$$= \frac{(x-2)(x-4)}{(x+3)(x-4)} - \frac{(2x-5)(x+3)}{(x-4)(x+3)}$$

$$= \frac{x^2 - 6x + 8}{(x+3)(x-4)} - \frac{2x^2 + x - 15}{(x-4)(x+3)}$$

$$= \frac{x^2 - 6x + 8 - (2x^2 + x - 15)}{(x+3)(x-4)}$$

$$= \frac{x^2 - 6x + 8 - 2x^2 - x + 15}{(x+3)(x-4)}$$

$$= \frac{-x^2 - 7x + 23}{(x+3)(x-4)}$$

51. ***Familiarize***. We arrange the information in a table. Let x = the number of liters of skim milk and y = the number of liters of 3.2% milk.

Type of milk	4.6%	Skim	3.2% (Mixture)
Amount of milk	100 L	x	y
Percent of butterfat	4.6%	0%	3.2%
Amount of butterfat in milk	4.6% × 100, or 4.6 L	$0\% \cdot x$, or 0 L	$3.2\%y$

Translate. The first and third rows of the table give us two equations.

Amount of milk: $100 + x = y$

Amount of butterfat: $4.6 + 0 = 3.2\%y$, or $4.6 = 0.032y$.

The resulting system is

$$100 + x = y,$$
$$4.6 = 0.032y.$$

Solve. We solve the second equation for y.

$$4.6 = 0.032y$$
$$\frac{4.6}{0.032} = y$$
$$143.75 = y$$

We substitute 143.75 for y in the first equation and solve for x.

$$100 + x = y$$
$$100 + x = 143.75$$
$$x = 43.75$$

Check. We consider $x = 43.75$ L and $y = 143.75$ L. The difference between 143.75 L and 43.75 L is 100 L. There is no butterfat in the skim milk. There are 4.6 liters of butterfat in the 100 liters of the 4.6% milk. Thus there are 4.6 liters of butterfat in the mixture. This checks because 3.2% of 143.75 is 4.6.

State. 43.75 L of skim milk should be used.

53. ***Familiarize***. In a table we arrange the information regarding the solution after some of the 30% solution is drained and replaced with pure antifreeze. We let x represent the amount of the original (30%) solution remaining, and we let y represent the amount of the 30% mixture that is drained and replaced with pure antifreeze.

Type of solution	Original (30%)	Pure antifreeze	Mixture
Amount of solution	x	y	16
Percent of antifreeze	30%	100%	50%
Amount of antifreeze in solution	$0.3x$	$1 \cdot y$, or y	$0.5(16)$, or 8

Translate. The table gives us two equations.

Amount of solution: $x + y = 16$

Amount of antifreeze in solution: $0.3x + y = 8$, or $3x + 10y = 80$

The resulting system is

$$x + y = 16, \quad (1)$$
$$3x + 10y = 80. \quad (2)$$

Solve. We multiply Equation (1) by -3 and then add.

$$-3x - 3y = -48$$
$$3x + 10y = 80$$
$$7y = 32$$
$$y = \frac{32}{7}, \text{ or } 4\frac{4}{7}$$

Then we substitute $4\frac{4}{7}$ for y in Equation (1) and solve for x.

$$x + y = 16$$
$$x + 4\frac{4}{7} = 16$$
$$x = 11\frac{3}{7}$$

Check. When $x = 11\frac{3}{7}$ L and $y = 4\frac{4}{7}$ L, the total is 16 L. The amount of antifreeze in the mixture is $0.3\left(11\frac{3}{7}\right) + 4\frac{4}{7}$, or $\frac{3}{10} \cdot \frac{80}{7} + \frac{32}{7}$, or $\frac{24}{7} + \frac{32}{7} = \frac{56}{7}$, or 8 L. This is 50% of 16 L, so the numbers check.

State. $4\frac{4}{7}$ L of the original mixture should be drained and replaced with pure antifreeze.

55. ***Familiarize***. Let x = the tens digit and y = the ones digit. Then the number is $10x + y$.

Translate. The number is six times the sum of its digits, so we have

$$10x + y = 6(x + y)$$
$$10x + y = 6x + 6y$$
$$4x - 5y = 0.$$

The tens digit is 1 more than the ones digit so we have

$$x = y + 1.$$

The resulting system is

$$4x - 5y = 0, \quad (1)$$
$$x = y + 1. \quad (2)$$

Solve. First substitute $y + 1$ for x in Equation (1) and solve for y.

$$4x - 5y = 0 \quad (1)$$
$$4(y + 1) - 5y = 0$$
$$4y + 4 - 5y = 0$$
$$-y + 4 = 0$$
$$4 = y$$

Now substitute 4 for y in one of the original equations and solve for x.

$$x = y + 1 \quad (2)$$
$$x = 4 + 1$$
$$x = 5$$

Check. If the number is 54, then the sum of the digits is $5 + 4$, or 9, and $54 = 6 \cdot 9$. Also, the tens digit, 5, is one more than the ones digit, 4. The answer checks.

State. The number is 54.

Exercise Set 15.5

1. ***Familiarize***. We first make a drawing.

30 mph

Slower car t hours d miles

46 mph

Faster car t hours $d + 72$ miles

We let d = the distance the slower car travels. Then $d + 72$ = the distance the faster car travels. We call the time t. We complete the table in the text, filling in the distances as well as the other information.

$d = r \cdot t$

	Distance	Speed	Time
Slower car	d	30	t
Faster car	$d + 72$	46	t

Translate. We get an equation $d = rt$ from each row of the table. Thus we have

$$d = 30t, \quad (1)$$
$$d + 72 = 46t. \quad (2)$$

Solve. We use the substitution method. We substitute $30t$ for d in Equation (2).

$$d + 72 = 46t$$
$$30t + 72 = 46t \quad \text{Substituting}$$
$$72 = 16t \quad \text{Subtracting } 30t$$
$$4.5 = t \quad \text{Dividing by 16}$$

Check. In 4.5 hr the slower car travels 30(4.5), or 135 mi, and the faster car travels 46(4.5), or 207 mi. Since 207 is 72 more than 135, our result checks.

State. The cars will be 72 mi apart in 4.5 hr.

3. ***Familiarize.*** First make a drawing.

Station 72 mph
Slower train $t + 3$ hours d miles

Station 120 mph
Faster train t hours d miles

Trains meet here.

From the drawing we see that the distances are the same. Let's call the distance d. Let t represent the time for the faster train and $t+3$ represent the time for the slower train. We complete the table in the text.

	Distance (d)	Speed (r)	Time (t)
Slower train	d	72	$t + 3$
Faster train	d	120	t

Equation from first row: $d = 72(t + 3)$

Equation from second row: $d = 120t$

Translate. Using $d = rt$ in each row of the table, we get the following system of equations:

$$d = 72(t + 3), \quad (1)$$
$$d = 120t. \quad (2)$$

Solve. Substitute $120t$ for d in Equation (1) and solve for t.

$$d = 72(t + 3)$$
$$120t = 72(t + 3) \quad \text{Substituting}$$
$$120t = 72t + 216$$
$$48t = 216$$
$$t = \frac{216}{48}$$
$$t = 4.5$$

Check. When $t = 4.5$ hours, the faster train will travel 120(4.5), or 540 mi, and the slower train will travel 72(7.5), or 540 mi. In both cases we get the distance 540 mi.

State. In 4.5 hours after the second train leaves, the second train will overtake the first train. We can also state the answer as 7.5 hours after the first train leaves.

5. ***Familiarize.*** We first make a drawing.

With the current $r + 6$
4 hours d kilometers

Against the current $r - 6$
10 hours d kilometers

From the drawing we see that the distances are the same. Let d represent the distance. Let r represent the speed of the canoe in still water. Then, when the canoe is traveling with the current, its speed is $r + 6$. When it is traveling against the current, its speed is $r - 6$. We complete the table in the text.

	Distance (d)	Speed (r)	Time (t)
With current	d	$r + 6$	4
Against current	d	$r - 6$	10

Equation from first row: $d = (r + 6)4$

Equation from second row: $d = (r - 6)10$

Translate. Using $d = rt$ in each row of the table, we get the following system of equations:

$$d = (r + 6)4, \quad (1)$$
$$d = (r - 6)10 \quad (2)$$

Solve. Substitute $(r + 6)4$ for d in Equation (2) and solve for r.

$$d = (r - 6)10$$
$$(r + 6)4 = (r - 6)10 \quad \text{Substituting}$$
$$4r + 24 = 10r - 60$$
$$84 = 6r$$
$$14 = r$$

Check. When $r = 14$, $r + 6 = 20$ and $20 \cdot 4 = 80$, the distance. When $r = 14$, $r - 6 = 8$ and $8 \cdot 10 = 80$. In both cases, we get the same distance.

State. The speed of the canoe in still water is 14 km/h.

7. ***Familiarize.*** First make a drawing.

Passenger 96 km/h
$t - 2$ hours d kilometers

Freight 64 km/h
t hours d kilometers

Central City Clear Creek

From the drawing we see that the distances are the same. Let d represent the distance. Let t represent the time for the freight train. Then the time for the passenger train is $t-2$. We organize the information in a table.

$d \quad = \quad r \quad \cdot \quad t$

	Distance	Speed	Time
Passenger	d	96	$t-2$
Freight	d	64	t

Translate. From each row of the table we get an equation.

$d = 96(t-2), \quad (1)$

$d = 64t \quad (2)$

Solve. Substitute $64t$ for d in Equation (1) and solve for t.

$$d = 96(t-2)$$
$$64t = 96(t-2) \quad \text{Substituting}$$
$$64t = 96t - 192$$
$$192 = 32t$$
$$6 = t$$

Next we substitute 6 for t in one of the original equations and solve for d.

$$d = 64t \quad \text{Equation (2)}$$
$$d = 64 \cdot 6 \quad \text{Substituting}$$
$$d = 384$$

Check. If the time is 6 hr, then the distance the passenger train travels is $96(6-2)$, or 384 km. The freight train travels $64(6)$, or 384 km. The distances are the same.

State. It is 384 km from Central City to Clear Creek.

9. ***Familiarize***. We first make a drawing.

Downstream $r+6$

3 hours d miles

Upstream $r-6$

5 hours d miles

We let r represent the speed of the boat in still water and d represent the distance Antoine traveled downstream before he turned back. We organize the information in a table.

$d \quad = \quad r \quad \cdot \quad t$

	Distance	Speed	Time
Downstream	d	$r+6$	3
Upstream	d	$r-6$	5

Translate. Using $d = rt$ in each row of the table, we get the following system of equations:

$d = (r+6)3, \quad (1)$

$d = (r-6)5 \quad (2)$

Solve. Substitute $(r+6)3$ for d in Equation (2) and solve for r.

$$d = (r-6)5$$
$$(r+6)3 = (r-6)5 \quad \text{Substituting}$$
$$3r + 18 = 5r - 30$$
$$48 = 2r$$
$$24 = r$$

If $r = 24$, then $d = (r+6)3 = (24+6)3 = 30 \cdot 3 = 90$.

Check. If $r = 24$, then $r+6 = 24+6 = 30$ and $r-6 = 24-6 = 18$. If Antoine travels for 3 hr at 30 mph, then he travels $3 \cdot 30$, or 90 mi, downstream. If he travels for 5 hr at 18 mph, then he also travels $5 \cdot 18$, or 90 mi, upstream. Since the distances are the same, the answer checks.

State. (a) Antoine must travel at a speed of 24 mph.

(b) Antoine traveled 90 mi downstream before he turned back.

11. ***Familiarize***. We first make a drawing.

230 ft/min

Toddler $t+1$ min d ft

660 ft/min

Mother t min d ft

They meet here.

From the drawing we see that the distances are the same. Let's call the distance d. Let t = the time the mother runs. Then $t+1$ = the time the toddler runs. We arrange the information in a table.

$d \quad = \quad r \quad \cdot \quad t$

	Distance	Speed	Time
Toddler	d	230	$t+1$
Mother	d	660	t

Translate. Using $d = rt$ in each row of the table we get two equations.

$d = 230(t+1), \quad (1)$

$d = 660t \quad (2)$

Solve. Substitute $660t$ for d in Equation (1) and solve for t.

$$d = 230(t+1)$$
$$660t = 230(t+1) \quad \text{Substituting}$$
$$660t = 230t + 230$$
$$430t = 230$$
$$t = \frac{230}{430}, \text{ or } \frac{23}{43}$$

Check. When $t = \frac{23}{43}$ the toddler will travel $230\left(1\frac{23}{43}\right)$, or $230 \cdot \frac{66}{43}$, or $\frac{15,180}{43}$ ft and the mother will travel $660 \cdot \frac{23}{43}$, or $\frac{15,180}{43}$ ft. Since the distances are the same, our result checks.

State. The mother will overtake the toddler $\frac{23}{43}$ min after she starts running. We can also state the answer as $1\frac{23}{43}$ min after the toddler starts running.

13. ***Familiarize***. First make a drawing.

Home t hr 45 mph | $(2-t)$ hr 6 mph Work
Motorcycle distance | Walking distance

← 25 miles →

Let t represent the time the motorcycle was driven. Then $2-t$ represents the time the rider walked. We organize the information in a table.

$d = r \cdot t$

	Distance	Speed	Time
Motorcycling	Motorcycle distance	45	t
Walking	Walking distance	6	$2-t$
Total	25		

Translate. From the drawing we see that

Motorcycle distance + Walking distance = 25

Then using $d = rt$ in each row of the table we get

$$45t + 6(2-t) = 25$$

Solve. We solve this equation for t.

$$45t + 12 - 6t = 25$$
$$39t + 12 = 25$$
$$39t = 13$$
$$t = \frac{13}{39}$$
$$t = \frac{1}{3}$$

Check. The problem asks us to find how far the motorcycle went before it broke down. If $t = \frac{1}{3}$, then $45t$ (the distance the motorcycle traveled) $= 45 \cdot \frac{1}{3}$, or 15 and $6(2-t)$ (the distance walked) $= 6\left(2 - \frac{1}{3}\right) = 6 \cdot \frac{5}{3}$, or 10. The total of these distances is 25, so $\frac{1}{3}$ checks.

State. The motorcycle went 15 miles before it broke down.

15. $\frac{8x^2}{24x} = \frac{8}{24} \cdot \frac{x^2}{x} = \frac{1}{3} \cdot x^{2-1} = \frac{x}{3}$

17.
$$\frac{5a+15}{10} = \frac{5(a+3)}{5 \cdot 2} = \frac{\cancel{5}(a+3)}{\cancel{5} \cdot 2} = \frac{a+3}{2}$$

19. $\frac{2x^2-50}{x^2-25} = \frac{2(x^2-25)}{x^2-25} = \frac{2}{1} \cdot \frac{x^2-25}{x^2-25} = 2$

21.
$$\frac{x^2-3x-10}{x^2-2x-15} = \frac{(x-5)(x+2)}{(x-5)(x+3)} = \frac{\cancel{(x-5)}(x+2)}{\cancel{(x-5)}(x+3)} = \frac{x+2}{x+3}$$

23.
$$\frac{(x^2+6x+9)(x-2)}{(x^2-4)(x+3)} = \frac{(x+3)(x+3)(x-2)}{(x+2)(x-2)(x+3)} = \frac{\cancel{(x+3)}(x+3)\cancel{(x-2)}}{(x+2)\cancel{(x-2)}\cancel{(x+3)}} = \frac{x+3}{x+2}$$

25.
$$\frac{6x^2+18x+12}{6x^2-6} = \frac{6(x^2+3x+2)}{6(x^2-1)} = \frac{6(x+1)(x+2)}{6(x+1)(x-1)} = \frac{\cancel{6}\cancel{(x+1)}(x+2)}{\cancel{6}\cancel{(x+1)}(x-1)} = \frac{x+2}{x-1}$$

27. ***Familiarize***. We arrange the information in a table. Let d = the length of the route and t = Lindbergh's time. Note that 16 hr and 57 min $= 16\frac{57}{60}$ hr $= 16.95$ hr.

$d = r \cdot t$

	Distance	Speed	Time
Lindbergh	d	107.4	t
Hughes	d	217.1	$t - 16.95$

Translate. From the rows of the table we get two equations.

$$d = 107.4t, \quad (1)$$
$$d = 217.1(t - 16.95) \quad (2)$$

Solve. We substitute $107.4t$ for d in Equation (2) and solve for t.

$$d = 217.1(t - 16.95)$$
$$107.4t = 217.1(t - 16.95)$$
$$107.4t = 217.1t - 3679.845$$
$$-109.7t = -3679.845$$
$$t \approx 33.54$$

Now we go back to Equation (1) and substitute 33.54 for t.

$$d = 107.4t$$
$$d = 107.4(33.54)$$
$$d \approx 3602$$

Check. When $t \approx 33.54$, Lindbergh traveled $107.4(33.54) \approx 3602$ mi, and Hughes traveled $217.1(16.59) \approx 3602$ mi. Since the distances are the same, our result checks.

State. The route was 3602 mi long. (Answers may vary slightly due to rounding differences.)

29. ***Familiarize.*** We arrange the information in a table. Let's call the distance d. When the riverboat is traveling upstream its speed is $12 - 4$, or 8 mph. Its speed traveling downstream is $12 + 4$, or 16 mph.

$$d \quad = \quad r \quad \cdot \quad t$$

	Distance	Speed	Time
Upstream	d	8	Time upstream
Downstream	d	16	Time downstream
Total			1

Translate. From the table we see that (Time upstream) + (Time downstream) = 1. Then using $d = rt$, in the form $\frac{d}{r} = t$, in each row of the table we get

$$\frac{d}{8} + \frac{d}{16} = 1.$$

Solve. We solve the equation. The LCM is 16.

$$\frac{d}{8} + \frac{d}{16} = 1$$
$$16\left(\frac{d}{8} + \frac{d}{16}\right) = 16 \cdot 1$$
$$16 \cdot \frac{d}{8} + 16 \cdot \frac{d}{16} = 16$$
$$2d + d = 16$$
$$3d = 16$$
$$d = \frac{16}{3}, \text{ or } 5\frac{1}{3}$$

Check. When $d = \frac{16}{3}$,

(Time upstream) + (Time downstream)

$$= \frac{\frac{16}{3}}{8} + \frac{\frac{16}{3}}{16}$$
$$= \frac{16}{3} \cdot \frac{1}{8} + \frac{16}{3} \cdot \frac{1}{16}$$
$$= \frac{2}{3} + \frac{1}{3}$$
$$= 1 \text{ hr}$$

Thus the distance of $\frac{16}{3}$ mi, or $5\frac{1}{3}$ mi checks.

State. The pilot should travel $5\frac{1}{3}$ mi upstream before turning around.

Chapter 15 Concept Reinforcement

1. False; the ordered pairs must be solutions of both equations.

2. False; there are three possibilities: no solution, exactly one solution, or an infinite number of solutions. See page 1125 in the text.

3. The equations $y = ax + b$ and $y = ax - b$, $b \neq 0$, have the same slope (a) and different y-intercepts (b and $-b$). Thus, the lines are parallel and have no point of intersection. The given statement is true.

4. The system of equations consists of a horizontal line and a vertical line, so there is exactly one point of intersection. The ordered pair (a, b) makes both equations true, so it is the solution. We also know that the solution must lie on both lines so the x-coordinate must be a and the y-coordinate must be b. The given statement is true.

Chapter 15 Important Concepts

1.
$$\begin{array}{c|l} x + 3y = 1 & \\ \hline -2 + 3 \cdot 1 \ ? & 1 \\ -2 + 3 & \\ 1 & \text{TRUE} \end{array} \qquad \begin{array}{c|l} y = x + 3 & \\ \hline 1 \ ? & -2 + 3 \\ & 1 \quad \text{TRUE} \end{array}$$

The ordered pair $(-2, 1)$ is a solution of both equations, so it is a solution of the system of equations.

2. We graph the equations.

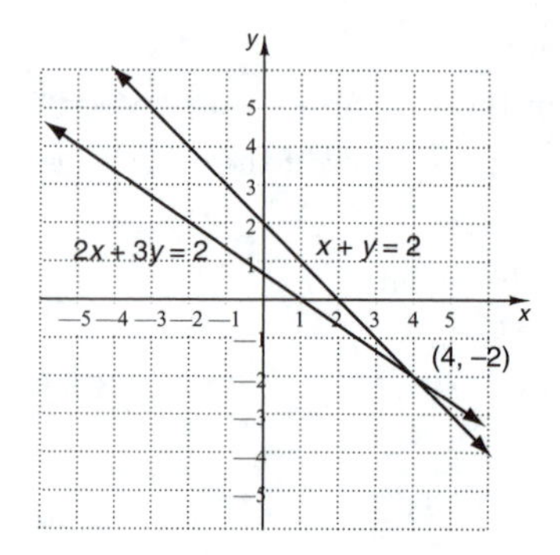

The point of intersection appears to be $(4, -2)$. This checks in both equations, so it is the solution.

3. $x + y = -1$, (1)
$2x + 5y = 1$ (2)

To solve this system of equations by the substitution method, we first solve one of the equations for one of the variables.

$$x + y = -1 \quad (1)$$
$$y = -x - 1 \quad (3)$$

Substitute $-x - 1$ for y in Equation (2) and solve for x.

$$2x + 5y = 1$$
$$2x + 5(-x - 1) = 1$$
$$2x - 5x - 5 = 1$$
$$-3x - 5 = 1$$
$$-3x = 6$$
$$x = -2$$

Now substitute -2 for x in Equation (3) and find y.

$$y = -x - 1 = -(-2) - 1 = 2 - 1 = 1$$

The ordered pair $(-2, 1)$ checks in both equations. It is the solution.

4. $3x + 2y = 6,\quad (1)$
$x - y = 7\quad (2)$

Multiply Equation (2) by 2 and then add.

$$\begin{aligned} 3x + 2y &= 6 \\ 2x - 2y &= 14 \\ \hline 5x \quad &= 20 \\ x &= 4 \end{aligned}$$

Now substitute 4 for x in one of the original equations and solve for y.

$$\begin{aligned} x - y &= 7 \quad (2) \\ 4 - y &= 7 \\ -y &= 3 \\ y &= -3 \end{aligned}$$

The ordered pair $(4, -3)$ checks in both equations. It is the solution.

Chapter 15 Review Exercises

1. We check by substituting alphabetically 6 for x and -1 for y.

$$\begin{array}{c|c} \multicolumn{2}{c}{x - y = 3} \\ \hline 6 - (-1) \ ? \ 3 & \\ 6 + 1 & \\ 7 & \text{FALSE} \end{array}$$

Since $(6, -1)$ is not a solution of the first equation, it is not a solution of the system of equations.

2. We check by substituting alphabetically 2 for x and -3 for y.

$$\begin{array}{c|c} \multicolumn{2}{c}{2x + y = 1} \\ \hline 2 \cdot 2 + (-3) \ ? \ 1 & \\ 4 - 3 & \\ 1 & \text{TRUE} \end{array} \qquad \begin{array}{c|c} \multicolumn{2}{c}{x - y = 5} \\ \hline 2 - (-3) \ ? \ 5 & \\ 2 + 3 & \\ 5 & \text{TRUE} \end{array}$$

The ordered pair $(2, -3)$ is a solution of both equations, so it is a solution of the system of equations.

3. We check by substituting alphabetically -2 for x and 1 for y.

$$\begin{array}{c|c} \multicolumn{2}{c}{x + 3y = 1} \\ \hline -2 + 3 \cdot 1 \ ? \ 1 & \\ -2 + 3 & \\ 1 & \text{TRUE} \end{array} \qquad \begin{array}{c|c} \multicolumn{2}{c}{2x - y = -5} \\ \hline 2(-2) - 1 \ ? \ -5 & \\ -4 - 1 & \\ -5 & \text{TRUE} \end{array}$$

The ordered pair $(-2, 1)$ is a solution of both equations, so it is a solution of the system of equations.

4. We check by substituting alphabetically -4 for x and -1 for y.

$$\begin{array}{c|c} \multicolumn{2}{c}{x - y = 3} \\ \hline -4 - (-1) \ ? \ 3 & \\ -4 + 1 & \\ -3 & \text{FALSE} \end{array}$$

Since $(-4, -1)$ is not a solution of the first equation, it is not a solution of the system of equations.

5. We graph the equations.

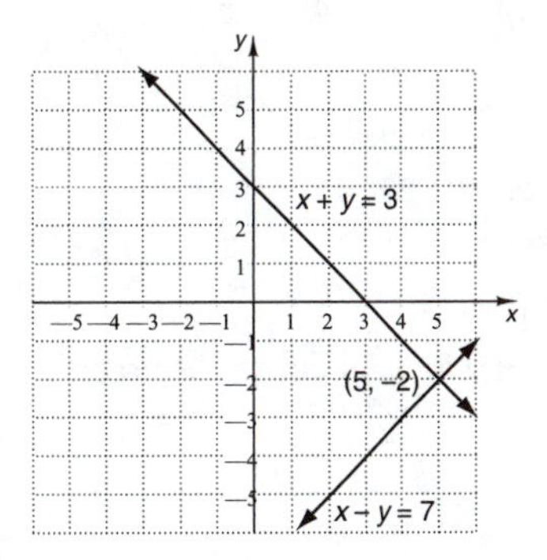

The point of intersection looks as if it has coordinates $(5, -2)$.

Check:

$$\begin{array}{c|c} \multicolumn{2}{c}{x + y = 3} \\ \hline 5 + (-2) \ ? \ 3 & \\ 3 & \text{TRUE} \end{array} \qquad \begin{array}{c|c} \multicolumn{2}{c}{x - y = 7} \\ \hline 5 - (-2) \ ? \ 7 & \\ 5 + 2 & \\ 7 & \text{TRUE} \end{array}$$

The solution is $(5, -2)$.

6. We graph the equations.

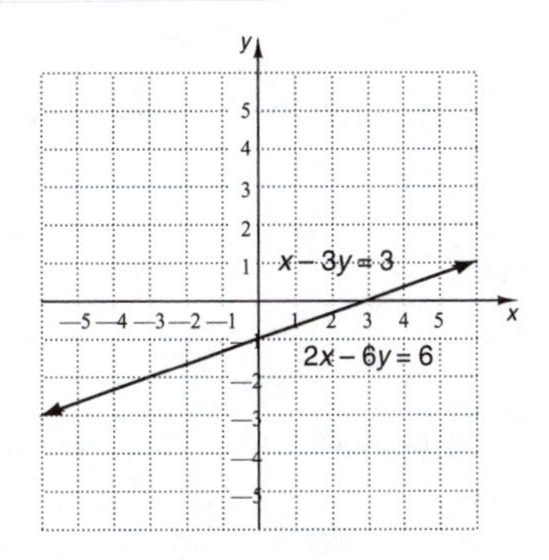

The graphs are the same. The system of equations has an infinite number of solutions.

7. We graph the equations.

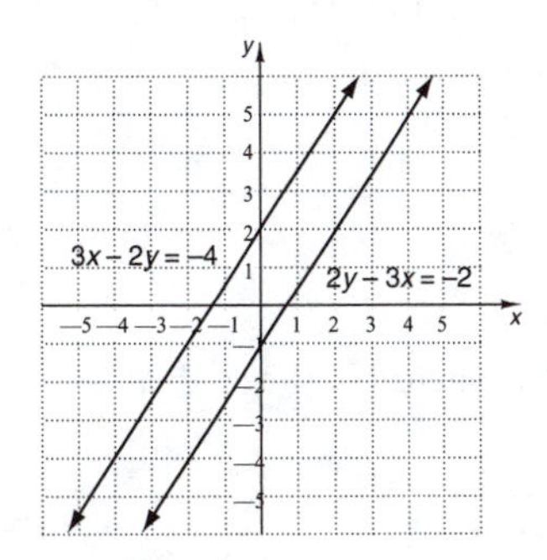

The lines are parallel. There is no solution.

8. $y = 5 - x,$ (1)

$3x - 4y = -20$ (2)

We substitute $5 - x$ for y in Equation (2) and solve for x.

$$3x - 4y = -20 \quad (2)$$
$$3x - 4(5 - x) = -20$$
$$3x - 20 + 4x = -20$$
$$7x - 20 = -20$$
$$7x = 0$$
$$x = 0$$

Next we substitute 0 for x in one of the original equations and solve for y.

$$y = 5 - x \quad (1)$$
$$y = 5 - 0$$
$$y = 5$$

The ordered pair $(0, 5)$ checks in both equations. It is the solution.

9. $x + y = 6,$ (1)

$y = 3 - 2x$ (2)

We substitute $3 - 2x$ for y in Equation (1) and solve for x.

$$x + y = 6 \quad (1)$$
$$x + (3 - 2x) = 6$$
$$-x + 3 = 6$$
$$-x = 3$$
$$x = -3$$

Now we substitute -3 for x in one of the original equations and solve for y.

$$y = 3 - 2x \quad (2)$$
$$y = 3 - 2(-3)$$
$$y = 3 + 6$$
$$y = 9$$

The ordered pair $(-3, 9)$ checks in both equations. It is the solution.

10. $x - y = 4,$ (1)

$y = 2 - x$ (2)

We substitute $2 - x$ for y in Equation (1) and solve for x.

$$x - y = 4 \quad (1)$$
$$x - (2 - x) = 4$$
$$x - 2 + x = 4$$
$$2x - 2 = 4$$
$$2x = 6$$
$$x = 3$$

Now substitute 3 for x in one of the original equations and solve for y.

$$y = 2 - x \quad (2)$$
$$y = 2 - 3$$
$$y = -1$$

The ordered pair $(3, -1)$ checks in both equations. It is the solution.

11. $s + t = 5,$ (1)

$s = 13 - 3t$ (2)

We substitute $13 - 3t$ for s in Equation (1) and solve for t.

$$s + t = 5 \quad (1)$$
$$(13 - 3t) + t = 5$$
$$13 - 2t = 5$$
$$-2t = -8$$
$$t = 4$$

Now substitute 4 for t in one of the original equations and solve for s.

$$s = 13 - 3t \quad (2)$$
$$s = 13 - 3 \cdot 4$$
$$s = 13 - 12$$
$$s = 1$$

The ordered pair $(1, 4)$ checks in both equations. It is the solution.

12. $x + 2y = 6,$ (1)

$2x + 3y = 8$ (2)

We solve Equation (1) for x.

$$x + 2y = 6 \quad (1)$$
$$x = -2y + 6 \quad (3)$$

We substitute $-2y + 6$ for x in Equation (2) and solve for y.

$$2x + 3y = 8 \quad (2)$$
$$2(-2y + 6) + 3y = 8$$
$$-4y + 12 + 3y = 8$$
$$-y + 12 = 8$$
$$-y = -4$$
$$y = 4$$

Now substitute 4 for y in Equation (3) and compute x.

$$x = -2y + 6 = -2 \cdot 4 + 6 = -8 + 6 = -2$$

The ordered pair $(-2, 4)$ checks in both equations. It is the solution.

13. $3x + y = 1,$ (1)

$x - 2y = 5$ (2)

We solve Equation (2) for x.

$$x - 2y = 5 \quad (1)$$
$$x = 2y + 5 \quad (3)$$

We substitute $2y + 5$ for x in Equation (1) and solve for y.

$$3x + y = 1 \quad (1)$$
$$3(2y + 5) + y = 1$$
$$6y + 15 + y = 1$$
$$7y + 15 = 1$$
$$7y = -14$$
$$y = -2$$

Now substitute -2 for y in Equation (3) and compute x.

$$x = 2y + 5 = 2(-2) + 5 = -4 + 5 = 1$$

The ordered pair $(1, -2)$ checks in both equations. It is the solution.

14. $x + y = 4,\quad (1)$

$2x - y = 5\quad (2)$

$3x \quad = 9$

$x = 3$

Substitute 3 for x in either of the original equations and solve for y.

$x + y = 4\quad (1)$

$3 + y = 4$

$y = 1$

The ordered pair $(3, 1)$ checks in both equations. It is the solution.

15. $x + 2y = 9,\quad (1)$

$3x - 2y = -5\quad (2)$

$4x \quad = 4$

$x = 1$

Substitute 1 for x in either of the original equations and solve for y.

$x + 2y = 9\quad (1)$

$1 + 2y = 9\quad (2)$

$2y = 8$

$y = 4$

The ordered pair $(1, 4)$ checks in both equations. It is the solution.

16. $x - y = 8,\quad (1)$

$2x - 2y = 7\quad (2)$

Multiply Equation (1) by -2 and then add.

$-2x + 2y = -16,\quad (1)$

$2x - 2y = 7\quad (2)$

$0 = -9$

We get a false equation. The system of equations has no solution.

17. $2x + 3y = 8,\quad (1)$

$5x + 2y = -2\quad (2)$

We use the multiplication principle with both equations and then add.

$4x + 6y = 16$ Multiplying (1) by 2

$-15x - 6y = 6$ Multiplying (2) by -3

$-11x \quad = 22$

$x = -2$

Substitute -2 for x in one of the original equations and solve for y.

$2x + 3y = 8\quad (1)$

$2(-2) + 3y = 8$

$-4 + 3y = 8$

$3y = 12$

$y = 4$

The ordered pair $(-2, 4)$ checks in both equations. It is the solution.

18. $5x - 2y = 2,\quad (1)$

$3x - 7y = 36\quad (2)$

We use the multiplication principle with both equations and then add.

$35x - 14y = 14$ Multiplying (1) by 7

$-6x + 14y = -72$ Multiplying (2) by -2

$29x \quad = -58$

$x = -2$

Substitute -2 for x in one of the original equations and solve for y.

$5x - 2y = 2\quad (1)$

$5(-2) - 2y = 2$

$-10 - 2y = 2$

$-2y = 12$

$y = -6$

The ordered pair $(-2, -6)$ checks in both equations. It is the solution.

19. $-x - y = -5,\quad (1)$

$2x - y = 4\quad (2)$

We multiply Equation (1) by -1 and then add.

$x + y = 5$

$2x - y = 4$

$3x \quad = 9$

$x = 3$

Substitute 3 for x in one of the original equations and solve for y.

$-x - y = -5\quad (1)$

$-3 - y = -5$

$-y = -2$

$y = 2$

The ordered pair $(3, 2)$ checks in both equations. It is the solution.

20. $6x + 2y = 4,\quad (1)$

$10x + 7y = -8\quad (2)$

We use the multiplication principle with both equations and then add.

$42x + 14y = 28$ Multiplying (1) by 7

$-20x - 14y = 16$ Multiplying (2) by -2

$22x \quad = 44$

$x = 2$

Substitute 2 for x in one of the original equations and solve for y.

$6x + 2y = 4\quad (1)$

$6 \cdot 2 + 2y = 4$

$12 + 2y = 4$

$2y = -8$

$y = -4$

The ordered pair $(2, -4)$ checks in both equations. It is the solution.

21. $-6x - 2y = 5,$ (1)

$12x + 4y = -10$ (2)

We multiply Equation (1) by 2 and then add.

$$\begin{array}{r} -12x - 4y = 10 \\ 12x + 4y = -10 \\ \hline 0 = 0 \end{array}$$

We get an obviously true equation, so the system has an infinite number of solutions.

22. $\frac{2}{3}x + y = -\frac{5}{3}$

$x - \frac{1}{3}y = -\frac{13}{3}$

First we multiply both sides of each equation to clear the fractions.

$$3\left(\frac{2}{3}x + y\right) = 3\left(-\frac{5}{3}\right)$$
$$3 \cdot \frac{2}{3}x + 3y = -5$$
$$2x + 3y = -5$$

$$3\left(x - \frac{1}{3}y\right) = 3\left(-\frac{13}{3}\right)$$
$$3x - 3 \cdot \frac{1}{3}y = -13$$
$$3x - y = -13$$

The resulting system is

$2x + 3y = -5,$ (1)

$3x - y = -13.$ (2)

Now we multiply Equation (2) by 3 and then add.

$$\begin{array}{r} 2x + 3y = -5 \\ 9x - 3y = -39 \\ \hline 11x \quad = -44 \end{array}$$
$$x = -4$$

Substitute -4 for x in Equation (1) and solve for y.

$$2x + 3y = -5$$
$$2(-4) + 3y = -5$$
$$-8 + 3y = -5$$
$$3y = 3$$
$$y = 1$$

The ordered pair $(-4, 1)$ checks in both equations. It is the solution.

23. ***Familiarize***. We make a drawing. We let $l =$ the length and $w =$ the width, in cm.

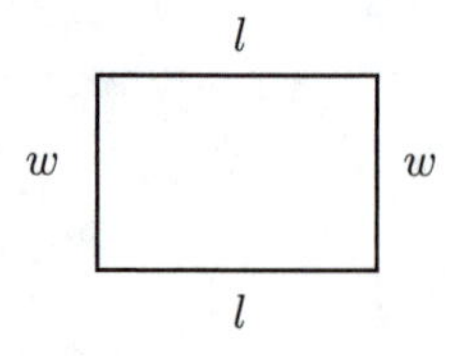

Translate. The perimeter is $2l + 2w$. We translate the first statement.

$$\underbrace{\text{The perimeter}}_{2l + 2w} \ \underbrace{\text{is}}_{=} \ \underbrace{96 \text{ cm.}}_{96}$$

We translate the second statement.

$$\underbrace{\text{The length}}_{l} \ \underbrace{\text{is}}_{=} \ \underbrace{27 \text{ cm}}_{27} \ \underbrace{\text{more than}}_{+} \ \underbrace{\text{the width.}}_{w}$$

The resulting system is

$2l + 2w = 96,$ (1)

$l = 27 + w.$ (2)

Solve. First we substitute $27 + w$ for l in Equation (1) and solve for w.

$$2l + 2w = 96 \quad (1)$$
$$2(27 + w) + 2w = 96$$
$$54 + 2w + 2w = 96$$
$$54 + 4w = 96$$
$$4w = 42$$
$$w = 10.5$$

Now we substitute 10.5 for w in Equation (2) and find l.

$$l = 27 + w = 27 + 10.5 = 37.5$$

Check. If the length is 37.5 cm and the width is 10.5 cm, then the perimeter is $2(37.5)+2(10.5)$, or $75+21$, or 96 cm. Also, the length is 27 cm more than the width. The answer checks.

State. The length of the rectangle is 37.5 cm, and the width is 10.5 cm.

24. ***Familiarize***. Let $x =$ the number of orchestra seats sold and $y =$ the number of balcony seats sold. We organize the information in a table.

	Orchestra	Balcony	Total
Price	\$25	\$18	
Number bought	x	y	508
Receipts	$25x$	$18y$	\$11,223

Translate. The last two rows of the table give us a system of equations.

$x + y = 508,$ (1)

$25x + 18y = 11,223.$ (2)

Solve. First we multiply Equation (1) by -18 and then add.

$$\begin{array}{r} -18x - 18y = -9144 \\ 25x + 18y = 11,223 \\ \hline 7x \quad = 2079 \end{array}$$
$$x = 297$$

Now we substitute 297 for x in Equation (1) and solve for y.

$$x + y = 508$$
$$297 + y = 508$$
$$y = 211$$

Check. The total number of tickets sold was $297 + 211$, or 508. The total receipts were $\$25 \cdot 297 + \$18 \cdot 211$, or $\$7425 + \$3798$, or \$11,223. The answer checks.

State. 297 orchestra seats and 211 balcony seats were sold.

25. ***Familiarize.*** Let $c =$ the number of liters of Clear Shine and $s =$ the number of liters of Sunstream window cleaner to be used in the mixture. We organize the information in a table.

Type of cleaner	Clear Shine	Sunstream	Mixture
Amount used	c	s	80
Percent of alcohol	30%	60%	45%
Amount of alcohol in solution	$0.3c$	$0.6s$	$45\% \times 80$, or 36 L

Translate. The first and third rows of the table give us a system of equations.

$$c + s = 80,$$
$$0.3c + 0.6s = 36$$

After we clear decimals we have

$$c + s = 80, \quad (1)$$
$$3c + 6s = 360. \quad (2)$$

Solve. First we multiply Equation (1) by -3 and then add.

$$-3c - 3s = -240$$
$$3c + 6s = 360$$
$$3s = 120$$
$$s = 40$$

Now we substitute 40 for s in Equation (1) and solve for c.

$$c + s = 80$$
$$c + 40 = 80$$
$$c = 40$$

Check. If 40 L of Clear Shine and 40 L of Sunstream are used, then there is 40 L + 40 L, or 80 L, of solution. The amount of alcohol in the solution is $0.3(40) + 0.6(40)$, or $12 + 24$, or 36 L. The answer checks.

State. 40 L of each window cleaner should be used.

26. ***Familiarize.*** Let $x =$ the weight of the Asian elephant and $y =$ the weight of the African elephant, in kg.

Translate.

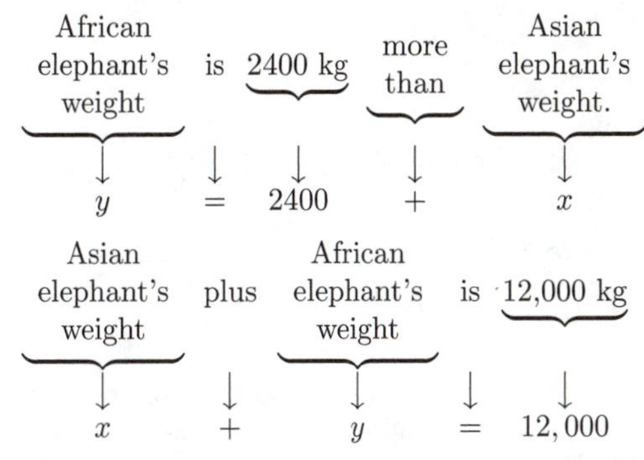

The resulting system is

$$y = 2400 + x, \quad (1)$$
$$x + y = 12,000. \quad (2)$$

Solve. First we substitute $2400 + x$ for y in Equation (2) and solve for x.

$$x + y = 12,000$$
$$x + (2400 + x) = 12,000$$
$$2x + 2400 = 12,000$$
$$2x = 9600$$
$$x = 4800$$

Now substitute 4800 for x in Equation (1) and find y.

$$y = 2400 + x = 2400 + 4800 = 7200$$

Check. If the Asian elephant weighs 7200 kg and the African elephant weighs 4800 kg, then the African elephant weighs 2400 kg more than the Asian elephant and their total weight is 4800 + 7200, or 12,000 kg. The answer checks.

State. The Asian elephant weighs 4800 kg, and the African elephant weighs 7200 kg.

27. ***Familiarize.*** Let $x =$ the number of pounds of peanuts and $y =$ the number of pounds of fancy nuts to be used. We organize the information in a table.

Type of nuts	Peanuts	Fancy	Mixture
Cost per pound	\$4.50	\$7.00	
Amount	x	y	13
Mixture	$4.5x$	$7y$	\$71

Translate. The last two rows of the table give us a system of equations.

$$x + y = 13,$$
$$4.5x + 7y = 71$$

After clearing decimals, we have

$$x + y = 13, \quad (1)$$
$$45x + 70y = 710. \quad (2)$$

Solve. First we multiply Equation (1) by -45 and then add.

$$\begin{aligned} -45x - 45y &= -585 \\ 45x + 70y &= 710 \\ \hline 25y &= 125 \\ y &= 5 \end{aligned}$$

Now substitute 5 for y in one of the original equations and solve for x.

$$\begin{aligned} x + y &= 13 \quad (1) \\ x + 5 &= 13 \\ x &= 8 \end{aligned}$$

Check. If 8 lb of peanuts and 5 lb of fancy nuts are used, the mixture weighs 13 lb. The value of the mixture is $\$4.50(8) + \$7.00(5) = \$36 + \$35 = \$71$. The answer checks.

State. 8 lb of peanuts and 5 lb of fancy nuts should be used.

28. ***Familiarize.*** Let $x =$ the number of gallons of 87-octane gas and $y =$ the number of gallons of 95-octane gas to be used. We arrange the information in a table.

Type of gas	87-octane	95-octane	Mixture
Amount	x	y	10
Octane rating	87	95	93
Mixture	$87x$	$95y$	$10 \cdot 93$, or 930

Translate. The first and last rows of the table give us a system of equations.

$$\begin{aligned} x + \quad y &= 10, \quad (1) \\ 87x + 95y &= 930 \quad (2) \end{aligned}$$

Solve. We multiply Equation (1) by -87 and then add.

$$\begin{aligned} -87x - 87y &= -870 \\ 87x + 95y &= 930 \\ \hline 8y &= 60 \\ y &= 7.5 \end{aligned}$$

Now substitute 7.5 for y in one of the original equations and solve for x.

$$\begin{aligned} x + y &= 10 \quad (1) \\ x + 7.5 &= 10 \\ x &= 2.5 \end{aligned}$$

Check. If 2.5 gal of 87-octane gas and 7.5 gal of 95-octane gas are used, then there are $2.5 + 7.5$, or 10 gal, of gas in the mixture. Also, $87(2.5) + 95(7.5) = 217.5 + 712.5 = 930$, so the answer checks.

State. 2.5 gal of 87-octane gas and 7.5 gal of 95-octane gas should be used.

29. ***Familiarize.*** Let $x =$ Jeff's age now and $y =$ his son's age now. In 13 yr their ages will be $x + 13$ and $y + 13$.

Translate.

Jeff's age is three times his son's age.

$$x = 3 \cdot y$$

In 13 yr,

Jeff's age will be two times his son's age.

$$x + 13 = 2 \cdot (y + 13)$$

The resulting system is

$$\begin{aligned} x &= 3y, \quad (1) \\ x + 13 &= 2(y + 13). \quad (2) \end{aligned}$$

Solve. First we substitute $3y$ for x in Equation (2) and solve for y.

$$\begin{aligned} x + 13 &= 2(y + 13) \\ 3y + 13 &= 2(y + 13) \\ 3y + 13 &= 2y + 26 \\ y + 13 &= 26 \\ y &= 13 \end{aligned}$$

Now substitute 13 for y in Equation (1) and find x.

$$x = 3y = 3 \cdot 13 = 39$$

Check. If Jeff is 39 years old and his son is 13 years old, then Jeff is three times as old as his son. In 13 yr Jeff's age will be $39 + 13$, or 52, his son's age will be $13 + 13$, or 26, and $52 = 2 \cdot 26$. The answer checks.

State. Jeff is 39 years old now, and his son is 13 years old.

30. ***Familiarize.*** Let $x =$ the measure of the larger angle and $y =$ the measure of the smaller angle. Recall that the sum of the measures of complementary angles is 90°.

Translate.

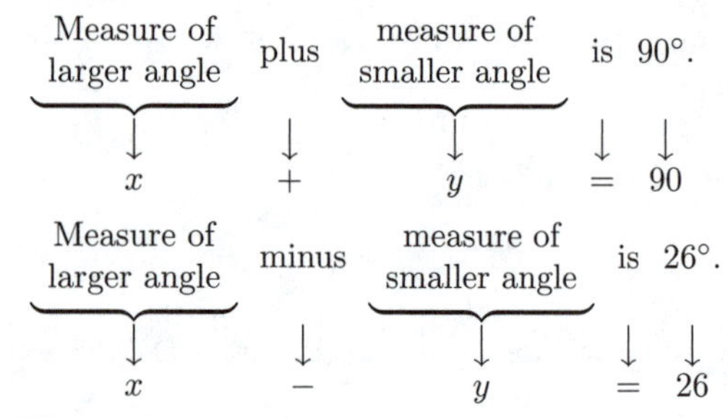

The resulting system is

$$\begin{aligned} x + y &= 90, \quad (1) \\ x - y &= 26. \quad (2) \end{aligned}$$

Solve. We add.

$$\begin{aligned} x + y &= 90 \\ x - y &= 26 \\ \hline 2x \quad &= 116 \\ x &= 58 \end{aligned}$$

Substitute 58 for x in one of the original equations and solve for y.

$$\begin{aligned} x + y &= 90 \quad (1) \\ 58 + y &= 90 \\ y &= 32 \end{aligned}$$

Check. $58° + 32° = 90°$ and $58° - 32° = 26°$, so the answer checks.

State. The measures of the angles are 58°and 32°.

31. ***Familiarize.*** Let $x =$ the measure of the larger angle and $y =$ the measure of the smaller angle. Recall that the sum of the measures of supplementary angles is 180°.

Translate.

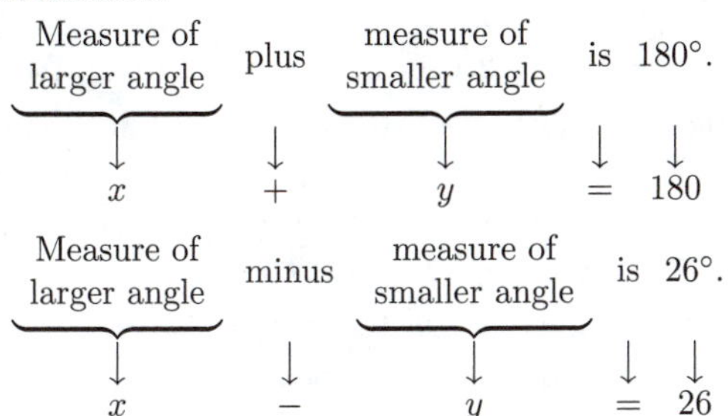

The resulting system is

$$x + y = 180, \quad (1)$$
$$x - y = 26. \quad (2)$$

Solve. We add.

$$x + y = 180$$
$$x - y = 26$$
$$2x \quad = 206$$
$$x = 103$$

Now substitute 103 for x in one of the original equations and solve for y.

$$x + y = 180 \quad (1)$$
$$103 + y = 180$$
$$y = 77$$

Check. $103° + 77° = 180°$ and $103° - 77° = 26°$; so the answer checks.

State. The measures of the angles are 103°and 77°.

32. ***Familiarize.*** Let $r =$ the speed of the airplane in still air, in km/h. Then $r + 15 =$ the speed with a 15 km/h tail wind and $r - 15 =$ the speed against a 15 km/h wind. We fill in the table in the text.

	Distance	Speed	Time
With wind	d	$r + 15$	4
Against wind	d	$r - 15$	5

Translate. We get an equation $d = rt$ from each row of the table.

$$d = (r + 15)4, \quad (1)$$
$$d = (r - 15)5 \quad (2)$$

Solve. We substitute $(r + 15)4$ for d in Equation (2).

$$d = (r - 15)5$$
$$(r + 15)4 = (r - 15)5$$
$$4r + 60 = 5r - 75$$
$$60 = r - 75$$
$$135 = r$$

Check. The plane's speed with the tail wind is $135 + 15$, or 150 km/h. At that speed, in 4 hr it will travel $150 \cdot 4$, or 600 km. The plane's speed against the wind is $135 - 15$, or 120 km/h. At that speed, in 5 hr it will travel $120 \cdot 5$, or 600 km. Since the distances are the same, the answer checks.

State. The speed of the airplane in still air is 135 km/h.

33. ***Familiarize.*** Let $t =$ the number of hours the slower car travels before the second car catches up to it. Then $t - 2 =$ the number of hours the faster car travels. We fill in the table in the text.

	Distance	Speed	Time
Slow car	d	55	t
Fast car	d	75	$t - 2$

Translate. We get an equation $d = rt$ from each row of the table.

$$d = 55t, \quad (1)$$
$$d = 75(t - 2) \quad (2)$$

Solve. We substitute $55t$ for d in Equation (2).

$$d = 75(t - 2)$$
$$55t = 75(t - 2)$$
$$55t = 75t - 150$$
$$-20t = -150$$
$$t = 7.5$$

Now substitute 7.5 for t in Equation (1) and find d.

$$d = 55t = 55(7.5) = 412.5$$

Check. From the calculation of d above, we see that the slow car travels 412.5 mi in 7.5 hr. The fast car travels $7.5 - 2$, or 5.5 hr. At a speed of 75 mph, it travels $75(5.5)$, or 412.5 mi. Since the distances are the same, the answer checks.

State. The second car catches up to the first car 412.5 mi from Phoenix.

34. $y = x - 2, \quad (1)$

$x - 2y = 6 \quad (2)$

Substitute $x - 2$ for y in Equation (2) and solve for x.

$$x - 2y = 6$$
$$x - 2(x - 2) = 6$$
$$x - 2x + 4 = 6$$
$$-x + 4 = 6$$
$$-x = 2$$
$$x = -2$$

Now substitute -2 for x in Equation (1) and find y.

$$y = x - 2 = -2 - 2 = -4$$

The ordered pair $(-2, -4)$ checks in both equations. It is the solution. The y-value of the solution is -4, so answer D is correct.

35. $3x + 2y = 5$, (1)
$x - y = 5$ (2)

Multiply Equation (2) by 2 and then add.

$$\begin{aligned} 3x + 2y &= 5 \\ 2x - 2y &= 10 \\ \hline 5x &= 15 \\ x &= 3 \end{aligned}$$

Now substitute 3 for x in one of the original equations and solve for y.

$$\begin{aligned} x - y &= 5 \quad (2) \\ 3 - y &= 5 \\ -y &= 2 \\ y &= -2 \end{aligned}$$

The ordered pair $(3, -2)$ checks in both equations. It is the solution. The x-value of the solution is 3, so answer A is correct.

36. We substitute 6 for x and 2 for y in each equation.

$$\begin{aligned} 2x - Dy &= 6 & Cx + 4y &= 14 \\ 2 \cdot 6 - D \cdot 2 &= 6 & C \cdot 6 + 4 \cdot 2 &= 14 \\ 12 - 2D &= 6 & 6C + 8 &= 14 \\ -2D &= -6 & 6C &= 6 \\ D &= 3 & C &= 1 \end{aligned}$$

37. $3(x - y) = 4 + x$, (1)
$x = 5y + 2$ (2)

Substitute $5y + 2$ for x in Equation (1) and solve for y.

$$\begin{aligned} 3(x - y) &= 4 + x \\ 3(5y + 2 - y) &= 4 + 5y + 2 \\ 3(4y + 2) &= 5y + 6 \\ 12y + 6 &= 5y + 6 \\ 7y + 6 &= 6 \\ 7y &= 0 \\ y &= 0 \end{aligned}$$

Now substitute 0 for y in Equation (2) and find x.

$$x = 5y + 2 = 5 \cdot 0 + 2 = 0 + 2 = 2$$

The solution is $(2, 0)$.

38. ***Familiarize***. Let $c =$ the compensation agreed upon for 12 months of work and let $h =$ the value of the horse. After 7 months, Stephanie would be owed $\frac{7}{12}$ of the compensation agreed upon for 12 months of work, or $\frac{7}{12}c$.

Translate.

Compensation for 12 months is a horse plus \$2400.

$$c = h + 2400$$

Compensation for 7 months is a horse plus \$1000.

$$\frac{7}{12}c = h + 1000$$

After clearing the fractions we have the following system of equations.

$$\begin{aligned} c &= h + 2400, \quad (1) \\ 7c &= 12h + 12{,}000 \quad (2) \end{aligned}$$

Solve. We substitute $h + 2400$ for c in Equation (2) and solve for h.

$$\begin{aligned} 7c &= 12h + 12{,}000 \\ 7(h + 2400) &= 12h + 12{,}000 \\ 7h + 16{,}800 &= 12h + 12{,}000 \\ 16{,}800 &= 5h + 12{,}000 \\ 4800 &= 5h \\ 960 &= h \end{aligned}$$

Check. If the value of the horse is \$960, then the compensation for 12 months of work is \$960 + \$2400, or \$3360; $\frac{7}{12}$ of \$3360 is \$1960. This is the value of the horse, \$960, plus \$1000, so the answer checks.

State. The value of the horse was \$960.

39. The line graphed in red contains the points $(0, 0)$ and $(3, 2)$. We find the slope:

$$m = \frac{2 - 0}{3 - 0} = \frac{2}{3}$$

The y-intercept is $(0, 0)$, so the equation of the line is $y = \frac{2}{3}x + 0$, or $y = \frac{2}{3}x$.

The line graphed in blue contains the points $(0, 5)$ and $(3, 2)$. We find the slope:

$$m = \frac{2 - 5}{3 - 0} = \frac{-3}{3} = -1$$

The y-intercept is $(0, 5)$, so the equation of the line is $y = -1 \cdot x + 5$, or $y = -x + 5$.

40. The line graphed in red contains the points $(-3, 0)$ and $(0, -3)$. We find the slope:

$$m = \frac{-3 - 0}{0 - (-3)} = \frac{-3}{3} = -1$$

The y-intercept is $(0, -3)$, so the equation of the line is $y = -1 \cdot x - 3$, or $y = -x - 3$, or $x + y = -3$.

The line graphed in blue contains the points $(0, 4)$ and $(4, 0)$. We find the slope:

$$m = \frac{0 - 4}{4 - 0} = \frac{-4}{4} = -1$$

The y-intercept is $(0, 4)$, so the equation of the line is $y = -1 \cdot x + 4$, or $y = -x + 4$, or $x + y = 4$.

41. ***Familiarize.*** Let x = the number of rabbits and y = the number of pheasants. Then the rabbits have a total of x heads and $4x$ feet; the pheasants have a total of y heads and $2y$ feet.

Translate.

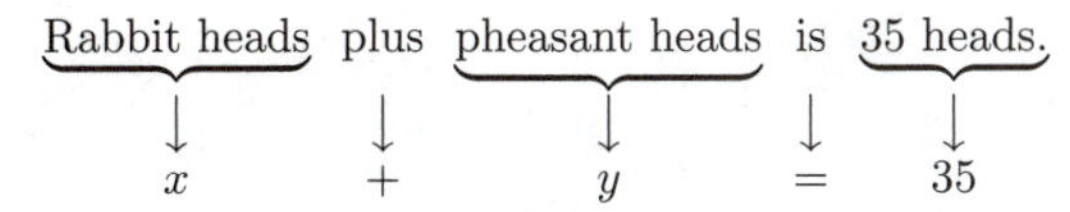

Rabbit feet plus pheasant feet is 94 feet.

$4x \quad + \quad 2y \quad = \quad 94$

The resulting system is

$$x + y = 35, \quad (1)$$
$$4x + 2y = 94. \quad (2)$$

Solve. First we multiply Equation (1) by -2 and then add.

$$\begin{aligned} -2x - 2y &= -70 \\ 4x + 2y &= 94 \\ \hline 2x \qquad &= 24 \\ x &= 12 \end{aligned}$$

Now substitute 12 for x in one of the original equations and solve for y.

$$x + y = 35 \quad (1)$$
$$12 + y = 35$$
$$y = 23$$

Check. If there are 12 rabbits and 23 pheasants, then there are $12+23$, or 35, heads and $4 \cdot 12 + 2 \cdot 23$, or $48+46$, or 94, feet. The answer checks.

State. There are 12 rabbits and 23 pheasants.

Chapter 15 Discussion and Writing Exercises

1. If we multiply the first equation by -2 we get the second equation. Thus the graphs of the equations will be the same, and the system of equations has an infinite number of solutions.

2. The multiplication principle might be used to obtain a pair of terms that are opposites. The addition principle is used to eliminate a variable. Once a variable has been eliminated, the multiplication and addition principles are also used to solve for the remaining variable and, after a substitution, are used again to find the variable that was eliminated.

3. Answers will vary.

4. A chart allows us to see the given information and missing information clearly and to see the relationships that yield equations.

Chapter 15 Test

1. We check by substituting alphabetically -2 for x and -1 for y.

$$\begin{array}{c|l} 2y - 3x & = 4 \\ \hline 2(-1) - 3(-2) & ?\ 4 \\ -2 + 6 & \\ 4 & \quad \text{TRUE} \end{array}$$

$$\begin{array}{c|l} x & = 4 + 2y \\ \hline -2 & ?\ 4 + 2(-1) \\ & 4 - 2 \\ & 2 \quad \text{FALSE} \end{array}$$

Although $(-2, -1)$ is a solution of the first equation, it is not a solution of the second equation and thus is not a solution of the system of equations.

2. We graph the equations.

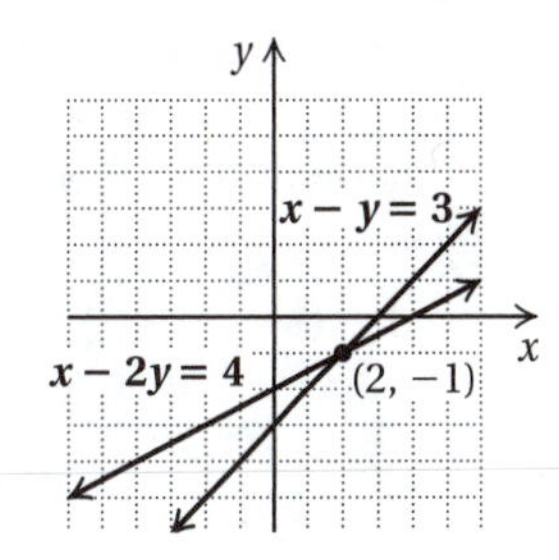

The point of intersection looks as if it has coordinates $(2, -1)$.

Check:

$$\begin{array}{c|l} x - y & = 3 \\ \hline 2 - (-1) & ?\ 3 \\ 2 + 1 & \\ 3 & \quad \text{TRUE} \end{array} \qquad \begin{array}{c|l} x - 2y & = 4 \\ \hline 2 - 2(-1) & ?\ 4 \\ 2 + 2 & \\ 4 & \quad \text{TRUE} \end{array}$$

The solution is $(2, -1)$.

3. $y = 6 - x, \quad (1)$
   $2x - 3y = 22 \quad (2)$

We substitute $6 - x$ for y in Equation (2) and solve for x.

$$\begin{aligned} 2x - 3y &= 22 \quad (2) \\ 2x - 3(6 - x) &= 22 \\ 2x - 18 + 3x &= 22 \\ 5x - 18 &= 22 \\ 5x &= 40 \\ x &= 8 \end{aligned}$$

Next we substitute 8 for x in one of the original equations and solve for y.

$$y = 6 - x \quad (1)$$
$$y = 6 - 8$$
$$y = -2$$

The ordered pair $(8, -2)$ checks in both equations. It is the solution.

4. $x + 2y = 5,\quad (1)$
$x + y = 2\quad (2)$

We solve Equation (1) for x.

$$x + 2y = 5 \quad (1)$$
$$x = -2y + 5 \quad (3)$$

We substitute $-2y + 5$ for x in Equation (2) and solve for y.

$$\begin{aligned} x + y &= 2 \quad (2)\\ -2y + 5 + y &= 2\\ -y + 5 &= 2\\ -y &= -3\\ y &= 3 \end{aligned}$$

Now substitute 3 for y in Equation (3) and compute x.

$$x = -2y + 5 = -2 \cdot 3 + 5 = -6 + 5 = -1.$$

The ordered pair $(-1, 3)$ checks in both equations. It is the solution.

5. $y = 5x - 2,\quad (1)$
$y - 2 = x\quad (2)$

Substitute $5x - 2$ for y in Equation (2) and solve for x.

$$\begin{aligned} y - 2 &= x \quad (2)\\ 5x - 2 - 2 &= x\\ 5x - 4 &= x\\ -4 &= -4x\\ 1 &= x \end{aligned}$$

Next we substitute 1 for x in Equation (1) and find y.

$$y = 5x - 2 = 5 \cdot 1 - 2 = 5 - 2 = 3.$$

The ordered pair $(1, 3)$ checks in both equations. It is the solution.

6.
$$\begin{aligned} x - y &= 6 \quad (1)\\ 3x + y &= -2 \quad (2)\\ \hline 4x &= 4\\ x &= 1 \end{aligned}$$

Substitute 1 for x in either of the original equations and solve for y.

$$\begin{aligned} 3x + y &= -2 \quad (2)\\ 3 \cdot 1 + y &= -2\\ 3 + y &= -2\\ y &= -5 \end{aligned}$$

The ordered pair $(1, -5)$ checks in both equations. It is the solution.

7. $\frac{1}{2}x - \frac{1}{3}y = 8,$
$\frac{1}{3}x - \frac{2}{9}y = 1$

First we multiply the first equation by 6 and the second equation by 9 to clear the fractions.

$$\begin{aligned} 6\left(\frac{1}{2}x - \frac{1}{3}y\right) &= 6 \cdot 8\\ 6 \cdot \frac{1}{2}x - 6 \cdot \frac{1}{3}y &= 48\\ 3x - 2y &= 48 \end{aligned}$$

$$\begin{aligned} 9\left(\frac{1}{3}x - \frac{2}{9}y\right) &= 9 \cdot 1\\ 9 \cdot \frac{1}{3}x - 9 \cdot \frac{2}{9}y &= 9\\ 3x - 2y &= 9 \end{aligned}$$

The resulting system is

$$3x - 2y = 48, \quad (1)$$
$$3x - 2y = 9. \quad (2)$$

We multiply Equation (1) by -1 and then add.

$$\begin{aligned} -3x + 2y &= -48\\ 3x - 2y &= 9\\ \hline 0 &= -39 \end{aligned}$$

We obtain a false equation. The system has no solution.

8. $-4x - 9y = 4,\quad (1)$
$6x + 3y = 1\quad (2)$

Multiply Equation (2) by 3 and then add.

$$\begin{aligned} -4x - 9y &= 4\\ 18x + 9y &= 3\\ \hline 14x &= 7\\ x &= \frac{1}{2} \end{aligned}$$

Now substitute $\frac{1}{2}$ for x in one of the original equations and solve for y.

$$\begin{aligned} 6x + 3y &= 1 \quad (2)\\ 6 \cdot \frac{1}{2} + 3y &= 1\\ 3 + 3y &= 1\\ 3y &= -2\\ y &= -\frac{2}{3} \end{aligned}$$

The ordered pair $\left(\frac{1}{2}, -\frac{2}{3}\right)$ checks in both equations. It is the solution.

9. $2x + 3y = 13,\quad (1)$
$3x - 5y = 10\quad (2)$

Multiply Equation (1) by 5 and Equation (2) by 3 and then add.

$$\begin{aligned} 10x + 15y &= 65\\ 9x - 15y &= 30\\ \hline 19x &= 95\\ x &= 5 \end{aligned}$$

Now substitute 5 for x in one of the original equations and solve for y.

$$2x + 3y = 13 \quad (1)$$
$$2 \cdot 5 + 3y = 13$$
$$10 + 3y = 13$$
$$3y = 3$$
$$y = 1$$

The ordered pair $(5, 1)$ checks in both equations. It is the solution.

10. ***Familiarize***. Let l = the length and w = the width, in yd. Recall that the perimeter of a rectangle is given by the formula $P = 2l + 2w$.

Translate.

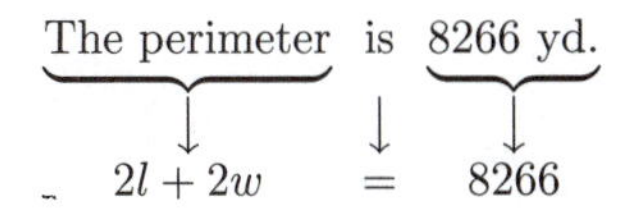

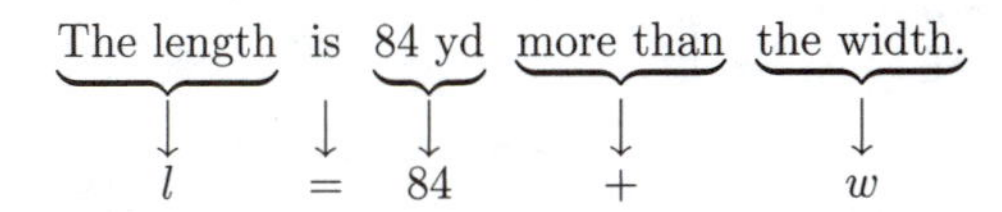

The resulting system is

$$2l + 2w = 8266, \quad (1)$$
$$l = 84 + w. \quad (2)$$

Solve. First we substitute $84 + w$ for l in Equation (1) and solve for w.

$$2l + 2w = 8266$$
$$2(84 + w) + 2w = 8266$$
$$168 + 2w + 2w = 8266$$
$$168 + 4w = 8266$$
$$4w = 8098$$
$$w = 2024.5$$

Now substitute 2024.5 for w in Equation (2) and find l.

$$l = 84 + w = 84 + 2024.5 = 2108.5$$

Check. If the length is 2108.5 yd and the width is 2024.5 yd, the length is 84 yd more than the width and the perimeter is $2(2108.5) + 2(2024.5)$, or $4217 + 4049$, or 8266 yd. The answer checks.

State. The length is 2108.5 yd and the width is 2024.5 yd.

11. ***Familiarize***. Let a = the amount of solution A and b = the amount of solution B in the mixture, in liters. We organize the information in a table.

Solution	A	B	Mixture
Amount	a	b	60 L
Percent of acid	25%	40%	30%
Amount of acid	$0.25a$	$0.4b$	$0.3(60)$ or 18 L

Translate. The first and third rows of the table give us a system of equations.

$$a + b = 60,$$
$$0.25a + 0.4b = 18$$

After clearing decimals, we have

$$a + b = 60, \quad (1)$$
$$25a + 40b = 1800. \quad (2)$$

Solve. First we multiply Equation (1) by -25 and then add.

$$-25a - 25b = -1500$$
$$25a + 40b = 1800$$
$$15b = 300$$
$$b = 20$$

Now we substitute 20 for b in one of the original equations and solve for a.

$$a + b = 60 \quad (1)$$
$$a + 20 = 60$$
$$a = 40$$

Check. If 40 L of solution A and 20 L of solution B are used, then there are $40 + 20$, or 60 L, in the mixture. The amount of acid in the mixture is $0.25(40) + 0.4(20)$, or $10 + 8$, or 18 L. The answer checks.

State. 40 L of solution A and 20 L of solution B should be used.

12. ***Familiarize***. Let r = the speed of the motorboat in still water, in km/h. Then the speed of the boat with the current is $r + 8$ and the speed against the current is $r - 8$. We organize the information in a table.

	Distance	Speed	Time
With current	d	$r+8$	2
Against current	d	$r-8$	3

Translate. We get an equation from each row of the table.

$$d = (r + 8)2 \quad (1)$$
$$d = (r - 8)3 \quad (2)$$

Solve. We substitute $(r + 8)2$ for d in Equation (2) and solve for r.

$$d = (r - 8)3 \quad (2)$$
$$(r + 8)2 = (r - 8)3$$
$$2r + 16 = 3r - 24$$
$$16 = r - 24$$
$$40 = r$$

Check. When $r = 40$, then $r + 8 = 40 + 8 = 48$ and in 2 hr at this speed the boat travels $48 \cdot 2$, or 96 km. Also, $r - 8 = 40 - 8 = 32$, and in 3 hr at this speed the boat travels $32 \cdot 3$, or 96 km. The distances are the same, so the answer checks.

State. The speed of the motorboat in still water is 40 km/h.

13. ***Familiarize***. Let c = the receipts from concessions and r = the receipts from the rides.

Translate.

Receipts from concessions plus receipts from rides is \$4275.

$$c + r = 4275$$

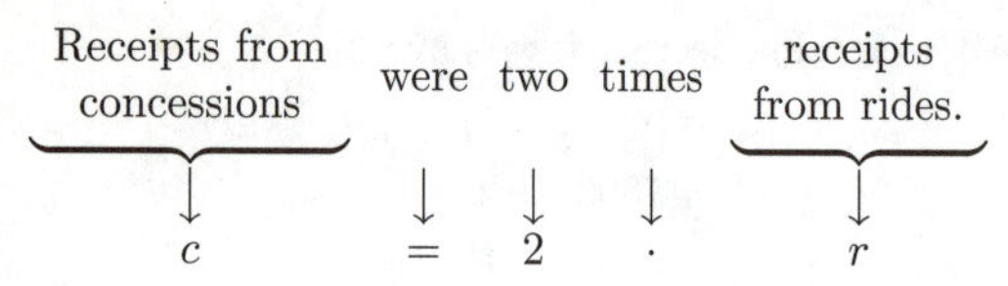

The resulting system is

$$c + r = 4275, \quad (1)$$
$$c = 2r. \quad (2)$$

Solve. First we substitute $2r$ for c in Equation (1) and solve for r.

$$c + r = 4275 \quad (1)$$
$$2r + r = 4275$$
$$3r = 4275$$
$$r = 1425$$

Now we substitute 1425 for r in one of the original equations and find c.

$$c = 2r = 2 \cdot 1425 = 2850$$

Check. If concessions brought in \$2850 and rides brought in \$1425, then the total receipts were \$2850 + \$1425, or \$4275. Also, $2 \cdot \$1425 = \$2850$, so concessions brought in twice as much as rides. The answer checks.

State. Concessions brought in \$2850; rides brought in \$1425.

14. ***Familiarize.*** Let $x =$ the number of acres of hay planted and $y =$ the number of acres of oats planted.

Translate.

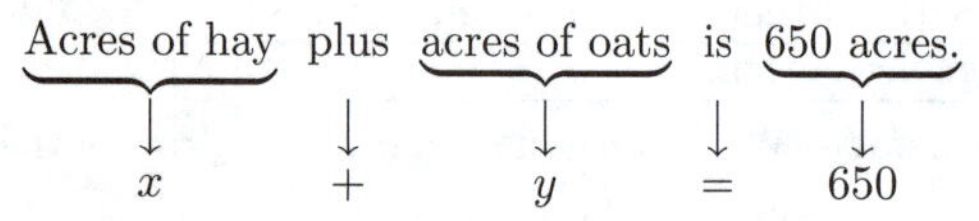

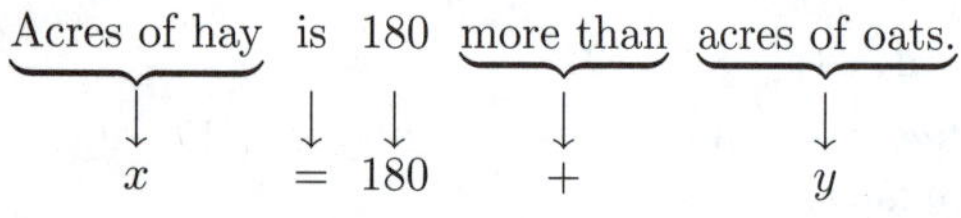

The resulting system is

$$x + y = 650, \quad (1)$$
$$x = 180 + y. \quad (2)$$

Solve. First we substitute $180 + y$ for x in Equation (1) and solve for y.

$$x + y = 650 \quad (1)$$
$$(180 + y) + y = 650$$
$$180 + 2y = 650$$
$$2y = 470$$
$$y = 235$$

Now substitute 235 for y in Equation (2) and find x.

$$x = 180 + y = 180 + 235 = 415$$

Check. If 415 acres of hay and 235 acres of oats are planted, a total of $415+235$, or 650 acres, is planted. Also, $235 + 180 = 415$, so 180 acres more of hay than of oats are planted. The answer checks.

State. 415 acres of hay and 235 acres of oats should be planted.

15. ***Familiarize.*** Let $x =$ the measure of the larger angle and $y =$ the measure of the smaller angle. Recall that the sum of the measures of supplementary angles is 180°.

Translate.

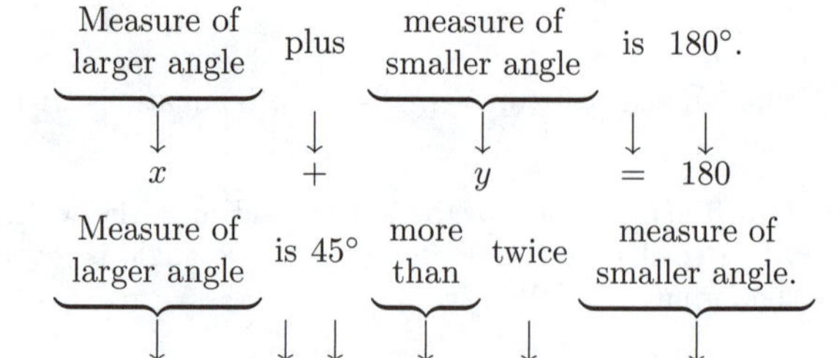

The resulting system is

$$x + y = 180, \quad (1)$$
$$x = 45 + 2y. \quad (2)$$

Solve. First we substitute $45 + 2y$ for x in Equation (1) and solve for y.

$$x + y = 180 \quad (1)$$
$$(45 + 2y) + y = 180$$
$$45 + 3y = 180$$
$$3y = 135$$
$$y = 45$$

Now substitute 45 for y in Equation (2) and find x.

$$x = 45 + 2y = 45 + 2 \cdot 45 = 45 + 90 = 135$$

Check. $135° + 45° = 180°$; and $135° = 45° + 2 \cdot 45°$, so the answer checks.

State. The measures of the angles are 135°and 45°.

16. ***Familiarize.*** Let $x =$ the number of gallons of 87-octane gas and $y =$ the number of gallons of 93-octane gas that should be used. We organize the information in a table.

Type of gas	87-octane	93-octane	Mixture
Amount	x	y	12
Octane rating	87	93	91
Mixture	$87x$	$93y$	$12 \cdot 91$, or 1092

Translate. The first and last rows of the table give us a system of equations.

$$x + y = 12, \quad (1)$$
$$87x + 93y = 1092 \quad (2)$$

Solve. We multiply Equation (1) by -87 and then add.

$$-87x - 87y = -1044$$
$$87x + 93y = 1092$$
$$\overline{\quad 6y = 48}$$
$$y = 8$$

Then substitute 8 for y in one of the original equations and solve for x.

$$x + y = 12$$
$$x + 8 = 12$$
$$x = 4$$

Check. If 4 gal of 87-octane gas and 8 gal of 93-octane gas are used, then there are $4+8$, or 12 gal, of gas in the mixture. Also $87 \cdot 4 + 93 \cdot 8 = 348 + 744 = 1092$, so the answer checks.

State. 4 gal of 87-octane gas and 8 gal of 93-octane gas should be used.

17. ***Familiarize***. Let $x =$ the number of minutes used in the \$2.95 plan and $y =$ the number of minutes used in the \$1.95 plan. Expressing 10¢ as \$0.10 and 15¢ as \$0.15, the \$2.95 plan costs $\$2.95 + \$0.10x$ per month and the \$1.95 plan costs $\$1.95 + \$0.15y$ per month.

Translate.

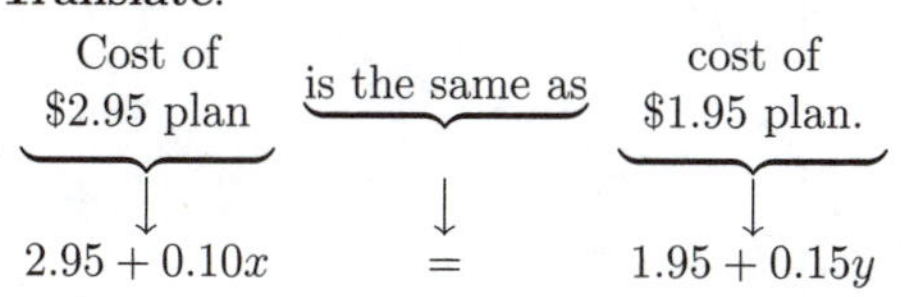

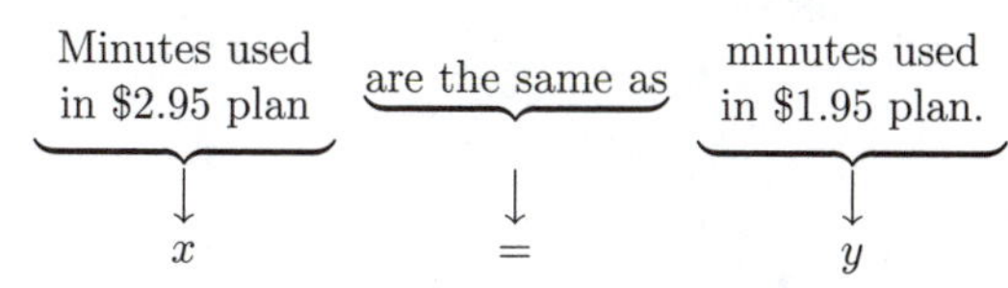

After clearing decimals we have the following system of equations.

$$295 + 10x = 195 + 15y, \quad (1)$$
$$x = y \quad (2)$$

Solve. We substitute x for y in Equation (1) and solve for x.

$$295 + 10x = 195 + 15y \quad (1)$$
$$295 + 10x = 195 + 15x$$
$$295 = 195 + 5x$$
$$100 = 5x$$
$$20 = x$$

Substitute 20 for x in Equation (2) to find y.

$$x = y$$
$$20 = y$$

Check. For 20 min the \$2.95 plan costs $\$2.95 + \$0.10(20)$, or \$4.95 per month and the \$1.95 plan costs $\$1.95 + \$0.15(20)$, or \$4.95 per month. The costs are the same, so the answer checks.

State. The two plans cost the same for 20 min.

18. ***Familiarize***. Let $t =$ the number of hours it will take the SUV to catch up with the car. Then $t + 2 =$ the number of hours the car travels. We organize the information in a table.

	Distance	Speed	Time
Car	d	55	$t+2$
SUV	d	65	t

Translate. We get an equation $d = rt$ from each row of the table.

$$d = 55(t+2), \quad (1)$$
$$d = 65t \quad (2)$$

Solve. Substitute $65t$ for d in Equation (1) and solve for t.

$$d = 55(t+2) \quad (1)$$
$$65t = 55(t+2)$$
$$65t = 55t + 110$$
$$10t = 110$$
$$t = 11$$

Check. At a speed of 55 mph for $11+2$, or 13 hr, the car travels $55 \cdot 13$, or 715 mi; at a speed of 65 mph for 11 hr, the SUV travels $65 \cdot 11$, or 715 mi. Since the distances are the same, the answer checks.

State. It will take the SUV 11 hr to catch up to the car.

19. $x - 2y = 4, \quad (1)$

$2x - 3y = 3 \quad (2)$

Multiply Equation (1) by -2 and then add.

$$\begin{aligned} -2x + 4y &= -8 \\ 2x - 3y &= 3 \\ \hline y &= -5 \end{aligned}$$

Now substitute -5 for y in one of the original equations and solve for x.

$$x - 2y = 4 \quad (1)$$
$$x - 2(-5) = 4$$
$$x + 10 = 4$$
$$x = -6$$

The ordered pair $(-6, -5)$ checks in both equations. It is the solution. Since both x and y are negative, answer D is correct.

20. Substitute -2 for x and 3 for y in each equation.

$$\begin{aligned} Cx - 4y &= 7 & 3x + Dy &= 8 \\ C(-2) - 4 \cdot 3 &= 7 & 3(-2) + D \cdot 3 &= 8 \\ -2C - 12 &= 7 & -6 + 3D &= 8 \\ -2C &= 19 & 3D &= 14 \\ C &= -\frac{19}{2} & D &= \frac{14}{3} \end{aligned}$$

21. ***Familiarize***. Let $a =$ the number of people ahead of Lily and $b =$ the number of people behind her. Then in the entire line there are the people ahead of her, the people behind her, and Lily herself, or $a + b + 1$.

Translate.

Number ahead of Lily | is | two | more than | number behind her.

$a \quad = \quad 2 \quad + \quad b$

Number in entire line | is | three | times | number behind her.

$a + b + 1 \quad = \quad 3 \quad \cdot \quad b$

The resulting system is

$$a = 2 + b, \quad (1)$$
$$a + b + 1 = 3b. \quad (2)$$

Solve. We substitute $2+b$ for a in Equation (2) and solve for b.

$$a+b+1=3b \quad (2)$$
$$(2+b)+b+1=3b$$
$$2b+3=3b$$
$$3=b$$

Now substitute 3 for b in Equation (1) and find a.

$$a=2+b=2+3=5$$

Check. If there are 5 people ahead of Lily and 3 people behind her, then the number of people ahead of her is two more than the number behind her. Also, in the entire line there are $5+1+3$, or 9 people, and $3 \cdot 3 = 9$. The answer checks.

State. There are 5 people ahead of Lily.

22. The line graphed in red contains the points $(-2,3)$ and $(3,0)$.

$$m = \frac{3-0}{-2-3} = \frac{3}{-5} = -\frac{3}{5}$$

Then the equation of the line is $y = -\frac{3}{5}x + b$. To find b we substitute the coordinates of either point for x and y in the equation. We use $(3,0)$.

$$y = -\frac{3}{5}x + b$$
$$0 = -\frac{3}{5} \cdot 3 + b$$
$$0 = -\frac{9}{5} + b$$
$$\frac{9}{5} = b$$

The equation of the line is $y = -\frac{3}{5}x + \frac{9}{5}$.

The line graphed in blue contains the points $(-2,3)$ and $(3,4)$.

$$m = \frac{3-4}{-2-3} = \frac{-1}{-5} = \frac{1}{5}$$

Then the equation is $y - \frac{1}{5}x + b$. To find b we substitute the coordinates of either point for x and y in the equation. We use $(3,4)$.

$$y = \frac{1}{5}x + b$$
$$4 = \frac{1}{5} \cdot 3 + b$$
$$4 = \frac{3}{5} + b$$
$$\frac{17}{5} = b$$

The equation of the line is $y = \frac{1}{5}x + \frac{17}{5}$.

23. The line graphed in red is a vertical line, so its equation is of the form $x = a$. The line contains the point $(3,-2)$, so the equation is $x = 3$.

The line graphed in blue is a horizontal line, so its equation is of the form $y = b$. The line contains the point $(3,-2)$, so the equation is $y = -2$.

Chapter 16

Radical Expressions and Equations

Exercise Set 16.1

1. The square roots of 4 are 2 and -2, because $2^2 = 4$ and $(-2)^2 = 4$.

3. The square roots of 9 are 3 and -3, because $3^2 = 9$ and $(-3)^2 = 9$.

5. The square roots of 100 are 10 and -10, because $10^2 = 100$ and $(-10)^2 = 100$.

7. The square roots of 169 are 13 and -13, because $13^2 = 169$ and $(-13)^2 = 169$.

9. The square roots of 256 are 16 and -16, because $16^2 = 256$ and $(-16)^2 = 256$.

11. $\sqrt{4} = 2$, taking the principal square root.

13. $\sqrt{9} = 3$, so $-\sqrt{9} = -3$.

15. $\sqrt{36} = 6$, so $-\sqrt{36} = -6$.

17. $\sqrt{225} = 15$, so $-\sqrt{225} = -15$.

19. $\sqrt{361} = 19$, taking the principal square root.

21. 2.236

23. 20.785

25. $\sqrt{347.7} \approx 18.647$, so $-\sqrt{347.7} \approx -18.647$.

27. 2.779

29. $\sqrt{8 \cdot 9 \cdot 200} = 120$, so $-\sqrt{8 \cdot 9 \cdot 200} = -120$.

31. a) We substitute 650 for P in the formula.

$W = 118.8\sqrt{650} \approx 3029$

The water flow is about 3029 GPM.

b) We substitute 1500 for P in the formula.

$W = 118.8\sqrt{1500} \approx 4601$

The water flow is about 4601 GPM.

33. Substitute 46 for V in the formula.

$T = 0.144\sqrt{46} \approx 0.977$ sec

35. Substitute 38 for V in the formula.

$T = 0.144\sqrt{38} \approx 0.888$ sec

37. The radicand is the expression under the radical, 200.

39. The radicand is the expression under the radical, $a - 4$.

41. The radicand is the expression under the radical, $t^2 + 1$.

43. The radicand is the expression under the radical, $\dfrac{3}{x+2}$.

45. No, because the radicand is negative

47. Yes, because the radicand is nonnegative

49. No, because the radicand is negative

51. $\sqrt{c^2} = c$ Since c is assumed to be nonnegative

53. $\sqrt{9x^2} = \sqrt{(3x)^2} = 3x$ Since $3x$ is assumed to be nonnegative

55. $\sqrt{(8p)^2} = 8p$ Since $8p$ is assumed to be nonnegative

57. $\sqrt{(ab)^2} = ab$

59. $\sqrt{(34d)^2} = 34d$

61. $\sqrt{(x+3)^2} = x + 3$

63. $\sqrt{a^2 - 10a + 25} = \sqrt{(a-5)^2} = a - 5$

65. $\sqrt{4a^2 - 20a + 25} = \sqrt{(2a-5)^2} = 2a - 5$

67. $\sqrt{121y^2 - 198y + 81} = \sqrt{(11y-9)^2} = 11y - 9$

69. ***Familiarize***. Let x and y represent the angles. Recall that supplementary angles are angles whose sum is 180°.

Translate. We reword the problem.

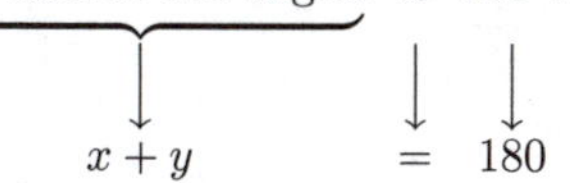

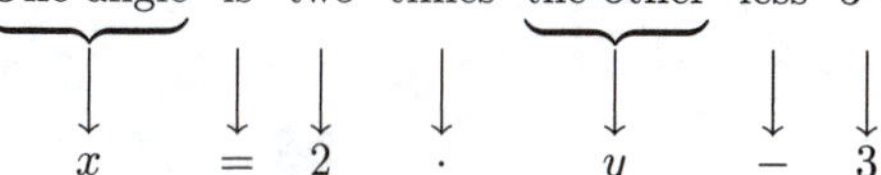

The resulting system is

$x + y = 180, \quad (1)$

$x = 2y - 3. \quad (2)$

Solve. We use substitution. We substitute $2y - 3$ for x in Equation (1) and solve for y.

$$(2y - 3) + y = 180$$
$$3y - 3 = 180$$
$$3y = 183$$
$$y = 61$$

We substitute 61 for y in Equation (2) to find x.

$$x = 2(61) - 3 = 122 - 3 = 119$$

Check. $61° + 119° = 180°$, so the angles are supplementary. Also, 3° less than twice 61° is $2 \cdot 61° - 3°$, or $122° - 3°$, or 119°. The numbers check.

State. The angles are 61° and 119°.

71. ***Familiarize***. This problem states that we have direct variation between F and I. Thus, an equation $F = kI$, $k > 0$, applies. As the income increases, the amount spent on food increases.

Translate. We write an equation of variation.

Amount spent on food varies directly as the income.

This translates to $F = kI$.

Solve.

a) First find an equation of variation.

$$F = kI$$

$$10,192 = k \cdot 39,200 \quad \text{Substituting 10,192 for } F \text{ and 39,200 for } I$$

$$\frac{10,192}{39,200} = k$$

$$0.26 = k$$

The equation of variation is $F = 0.26I$.

b) We use the equation to find how much a family spends on food when their income is \$41,000.

$$F = 0.26I$$

$$F = 0.26(\$41,000) \quad \text{Substituting \$41,000 for } I$$

$$F = \$10,660$$

Check. Let us do some reasoning about the answer. The income increased from \$39,200 to \$41,000. Similarly, the amount spend on food increased from \$10,192 to \$10,660. This is what we would expect with direct variation.

State. The amount spent on food is \$10,660.

73. $$\frac{x^2+10x-11}{x^2-1} \div \frac{x+11}{x+1}$$

$$= \frac{x^2+10x-11}{x^2-1} \cdot \frac{x+1}{x+11}$$

$$= \frac{(x^2+10x-11)(x+1)}{(x^2-1)(x+11)}$$

$$= \frac{(x+11)(x-1)(x+1)}{(x+1)(x-1)(x+11)}$$

$$= 1$$

75. To approximate $\sqrt{3}$, locate 3 on the x-axis, move up vertically to the graph, and then move left horizontally to the y-axis to read the approximation.

$\sqrt{3} \approx 1.7$ (Answers may vary.)

To approximate $\sqrt{5}$, locate 5 on the x-axis, move up vertically to the graph, and then move left horizontally to the y-axis to read the approximation.

$\sqrt{5} \approx 2.2$ (Answers may vary.)

To approximate $\sqrt{7}$, locate 7 on the x-axis, move up vertically to the graph, and then move left horizontally to the y-axis to read the approximation.

$\sqrt{7} \approx 2.6$ (Answers may vary.)

77. If $\sqrt{x^2} = 16$, then $x^2 = 256$ since $\sqrt{256} = 16$. Thus $x = 16$ or $x = -16$.

79. If $t^2 = 49$ then the values of t are the square roots of 49, 7 and -7.

Exercise Set 16.2

1. $\sqrt{12} = \sqrt{4 \cdot 3}$ 4 is a perfect square.

$= \sqrt{4}\,\sqrt{3}$ Factoring into a product of radicals

$= 2\sqrt{3}$ Taking the square root

3. $\sqrt{75} = \sqrt{25 \cdot 3}$ 25 is a perfect square.

$= \sqrt{25}\,\sqrt{3}$ Factoring into a product of radicals

$= 5\sqrt{3}$ Taking the square root

5. $\sqrt{20} = \sqrt{4 \cdot 5}$ 4 is a perfect square.

$= \sqrt{4}\,\sqrt{5}$ Factoring into a product of radicals

$= 2\sqrt{5}$ Taking the square root

7. $\sqrt{600} = \sqrt{100 \cdot 6}$ 100 is a perfect square.

$= \sqrt{100} \cdot \sqrt{6}$ Factoring into a product of radicals

$= 10\sqrt{6}$ Taking the square root

9. $\sqrt{486} = \sqrt{81 \cdot 6}$ 81 is a perfect square.

$= \sqrt{81} \cdot \sqrt{6}$ Factoring into a product of radicals

$= 9\sqrt{6}$ Taking the square root

11. $\sqrt{9x} = \sqrt{9 \cdot x} = \sqrt{9}\,\sqrt{x} = 3\sqrt{x}$

13. $\sqrt{48x} = \sqrt{16 \cdot 3 \cdot x} = \sqrt{16}\,\sqrt{3x} = 4\sqrt{3x}$

15. $\sqrt{16a} = \sqrt{16 \cdot a} = \sqrt{16}\,\sqrt{a} = 4\sqrt{a}$

17. $\sqrt{64y^2} = \sqrt{64}\,\sqrt{y^2} = 8y$, or

$\sqrt{64y^2} = \sqrt{(8y)^2} = 8y$

19. $\sqrt{13x^2} = \sqrt{13}\,\sqrt{x^2} = \sqrt{13} \cdot x$, or $x\sqrt{13}$

21. $\sqrt{8t^2} = \sqrt{2 \cdot 4 \cdot t^2} = \sqrt{4}\,\sqrt{t^2}\,\sqrt{2} = 2t\sqrt{2}$

23. $\sqrt{180} = \sqrt{36 \cdot 5} = 6\sqrt{5}$

25. $\sqrt{288y} = \sqrt{144 \cdot 2 \cdot y} = \sqrt{144}\,\sqrt{2y} = 12\sqrt{2y}$

27. $\sqrt{28x^2} = \sqrt{4 \cdot 7 \cdot x^2} = \sqrt{4}\,\sqrt{x^2}\,\sqrt{7} = 2x\sqrt{7}$

29. $\sqrt{x^2 - 6x + 9} = \sqrt{(x-3)^2} = x - 3$

31. $\sqrt{8x^2 + 8x + 2} = \sqrt{2(4x^2 + 4x + 1)} =$

$\sqrt{2(2x+1)^2} = \sqrt{2}\,\sqrt{(2x+1)^2} = \sqrt{2}\,(2x+1)$

33. $\sqrt{36y + 12y^2 + y^3} = \sqrt{y(36 + 12y + y^2)} =$

$\sqrt{y(6+y)^2} = \sqrt{y}\,\sqrt{(6+y)^2} = \sqrt{y}\,(6+y)$

35. $\sqrt{t^6} = \sqrt{(t^3)^2} = t^3$

37. $\sqrt{x^{12}} = \sqrt{(x^6)^2} = x^6$

39. $\sqrt{x^5} = \sqrt{x^4 \cdot x}$ One factor is a perfect square
$= \sqrt{x^4}\sqrt{x}$
$= \sqrt{(x^2)^2}\sqrt{x}$
$= x^2\sqrt{x}$

41. $\sqrt{t^{19}} = \sqrt{t^{18} \cdot t} = \sqrt{t^{18}}\sqrt{t} = \sqrt{(t^9)^2}\sqrt{t} = t^9\sqrt{t}$

43. $\sqrt{(y-2)^8} = \sqrt{[(y-2)^4]^2} = (y-2)^4$

45. $\sqrt{4(x+5)^{10}} = \sqrt{4[(x+5)^5]^2} = \sqrt{4}\sqrt{[(x+5)^5]^2} = 2(x+5)^5$

47. $\sqrt{36m^3} = \sqrt{36 \cdot m^2 \cdot m} = \sqrt{36}\sqrt{m^2}\sqrt{m} = 6m\sqrt{m}$

49. $\sqrt{8a^5} = \sqrt{2 \cdot 4 \cdot a^4 \cdot a} = \sqrt{2 \cdot 4 \cdot (a^2)^2 \cdot a} = \sqrt{4}\sqrt{(a^2)^2}\sqrt{2a} = 2a^2\sqrt{2a}$

51. $\sqrt{104p^{17}} = \sqrt{4 \cdot 26 \cdot p^{16} \cdot p} = \sqrt{4 \cdot 26 \cdot (p^8)^2 \cdot p} = \sqrt{4}\sqrt{(p^8)^2}\sqrt{26p} = 2p^8\sqrt{26p}$

53. $\sqrt{448x^6y^3} = \sqrt{64 \cdot 7 \cdot x^6 \cdot y^2 \cdot y} = \sqrt{64 \cdot 7 \cdot (x^3)^2 \cdot y^2 \cdot y} = \sqrt{64}\sqrt{(x^3)^2}\sqrt{y^2}\sqrt{7y} = 8x^3y\sqrt{7y}$

55. $\sqrt{3}\sqrt{18} = \sqrt{3 \cdot 18}$ Multiplying
$= \sqrt{3 \cdot 3 \cdot 6}$ Looking for perfect-square factors or pairs of factors
$= \sqrt{3 \cdot 3}\sqrt{6}$
$= 3\sqrt{6}$

57. $\sqrt{15}\sqrt{6} = \sqrt{15 \cdot 6}$ Multiplying
$= \sqrt{5 \cdot 3 \cdot 3 \cdot 2}$ Looking for perfect-square factors or pairs of factors
$= \sqrt{3 \cdot 3}\sqrt{5 \cdot 2}$
$= 3\sqrt{10}$

59. $\sqrt{18}\sqrt{14x} = \sqrt{18 \cdot 14x} = \sqrt{3 \cdot 3 \cdot 2 \cdot 2 \cdot 7 \cdot x} = \sqrt{3 \cdot 3}\sqrt{2 \cdot 2}\sqrt{7x} = 3 \cdot 2\sqrt{7x} = 6\sqrt{7x}$

61. $\sqrt{3x}\sqrt{12y} = \sqrt{3x \cdot 12y} = \sqrt{3 \cdot x \cdot 3 \cdot 4 \cdot y} = \sqrt{3 \cdot 3 \cdot 4 \cdot x \cdot y} = \sqrt{3 \cdot 3}\sqrt{4}\sqrt{x \cdot y} = 3 \cdot 2\sqrt{xy} = 6\sqrt{xy}$

63. $\sqrt{13}\sqrt{13} = \sqrt{13 \cdot 13} = 13$

65. $\sqrt{5b}\sqrt{15b} = \sqrt{5b \cdot 15b} = \sqrt{5 \cdot b \cdot 5 \cdot 3 \cdot b} = \sqrt{5 \cdot 5 \cdot b \cdot b \cdot 3} = \sqrt{5 \cdot 5}\sqrt{b \cdot b}\sqrt{3} = 5b\sqrt{3}$

67. $\sqrt{2t}\sqrt{2t} = \sqrt{2t \cdot 2t} = 2t$

69. $\sqrt{ab}\sqrt{ac} = \sqrt{ab \cdot ac} = \sqrt{a \cdot a \cdot b \cdot c} = \sqrt{a \cdot a}\sqrt{b \cdot c} = a\sqrt{bc}$

71. $\sqrt{2x^2y}\sqrt{4xy^2} = \sqrt{2x^2y \cdot 4xy^2} = \sqrt{2 \cdot x^2 \cdot y \cdot 4 \cdot x \cdot y^2} = \sqrt{4}\sqrt{x^2}\sqrt{y^2}\sqrt{2xy} = 2xy\sqrt{2xy}$

73. $\sqrt{18}\sqrt{18} = \sqrt{18 \cdot 18} = 18$

75. $\sqrt{5}\sqrt{2x-1} = \sqrt{5(2x-1)} = \sqrt{10x-5}$

77. $\sqrt{x+2}\sqrt{x+2} = \sqrt{(x+2)^2} = x+2$

79. $\sqrt{18x^2y^3}\sqrt{6xy^4} = \sqrt{18x^2y^3 \cdot 6xy^4} = \sqrt{3 \cdot 6 \cdot x^2 \cdot y^2 \cdot y \cdot 6 \cdot x \cdot y^4} = \sqrt{6 \cdot 6 \cdot x^2 \cdot y^6 \cdot 3 \cdot x \cdot y} = \sqrt{6 \cdot 6}\sqrt{x^2}\sqrt{y^6}\sqrt{3xy} = 6xy^3\sqrt{3xy}$

81. $\sqrt{50x^4y^6}\sqrt{10xy} = \sqrt{50x^4y^6 \cdot 10xy} = \sqrt{5 \cdot 10 \cdot x^4 \cdot y^6 \cdot 10 \cdot x \cdot y} = \sqrt{10 \cdot 10 \cdot x^4 \cdot y^6 \cdot 5 \cdot x \cdot y} = \sqrt{10 \cdot 10}\sqrt{x^4}\sqrt{y^6}\sqrt{5xy} = 10x^2y^3\sqrt{5xy}$

83. $\sqrt{99p^4q^3}\sqrt{22p^5q^2} = \sqrt{99p^4q^3 \cdot 22p^5q^2} = \sqrt{9 \cdot 11 \cdot p^4 \cdot q^2 \cdot q \cdot 2 \cdot 11 \cdot p^4 \cdot p \cdot q^2} = \sqrt{9 \cdot 11 \cdot 11 \cdot p^4 \cdot q^2 \cdot p^4 \cdot q^2 \cdot 2 \cdot q \cdot p} = 3 \cdot 11 \cdot p^2 \cdot q \cdot p^2 \cdot q\sqrt{2pq} = 33p^4q^2\sqrt{2pq}$

85. $\sqrt{24a^2b^3c^4}\sqrt{32a^5b^4c^7} = \sqrt{24a^2b^3c^4 \cdot 32a^5b^4c^7} = \sqrt{4 \cdot 2 \cdot 3 \cdot a^2 \cdot b^2 \cdot b \cdot c^4 \cdot 16 \cdot 2 \cdot a^4 \cdot a \cdot b^4 \cdot c^6 \cdot c} = \sqrt{4 \cdot 2 \cdot 2 \cdot 16 \cdot a^2 \cdot b^2 \cdot c^4 \cdot a^4 \cdot b^4 \cdot c^6 \cdot 3 \cdot b \cdot a \cdot c} = 2 \cdot 2 \cdot 4 \cdot a \cdot b \cdot c^2 \cdot a^2 \cdot b^2 \cdot c^3\sqrt{3abc} = 16a^3b^3c^5\sqrt{3abc}$

87.
$x - y = -6$ (1)
$x + y = 2$ (2)
$2x = -4$ Adding
$x = -2$

Now we substitute -2 for x in one of the original equations and solve for y.

$x + y = 2$ Equation (2)
$-2 + y = 2$ Substituting
$y = 4$

Since $(-2, 4)$ checks in both equations, it is the solution.

89.
$3x - 2y = 4,$ (1)
$2x + 5y = 9$ (2)

We will us the elimination method. We multiply on both sides of Equation (1) by 5 and on both sides of Equation (2) by 2. Then we add

$15x - 10y = 20$
$4x + 10y = 18$
$19x = 38$
$x = 2$

Now we substitute 2 for x in one of the original equations and solve for y.

$2x + 5y = 9$ Equation (2)
$2 \cdot 2 + 5y = 9$
$4 + 5y = 9$
$5y = 5$
$y = 1$

Since $(2, 1)$ checks, it is the solution.

91. ***Familiarize***. We organize the information in a table. The distance traveled downstream is the same as the distance traveled upstream. Let d = this distance, in miles. Let r = the speed at which Greg and Beth paddled. Then their speed downstream was $r+2$, and their speed upstream was $r-2$.

	Distance	Speed	Time
Downstream	d	$r+2$	2
Upstream	d	$r-2$	3

Translate. From each row of the table we get an equation $d = rt$.

$d = (r+2)2,$ (1)

$d = (r-2)3$ (2)

Solve. We solve using substitution. We will substitute $(r+2)2$ for d in equation (2).

$(r+2)2 = (r-2)3$ Substituting

$2r+4 = 3r-6$

$4 = r-6$ Subtracting $2r$

$10 = r$ Adding 6

Check. When $r = 10$, then $r+2 = 10+2$, or 12. The distance traveled at 12 mph for 2 hr is $12 \cdot 2$, or 24 mi. When $r = 10$, $r-2 = 8$. The distance traveled at 8 mph for 3 hr is $8 \cdot 3$, or 24 mi. The distances are the same so the answer checks.

State. Greg and Beth were paddling at a speed of 10 mph.

93. ***Familiarize***. Let a = the number of adults and c = the number of children who attended the fund-raiser. We organize the information in a table.

	Adults	Children	Total
Price	\$24	\$9	
Number	a	c	382
Amount taken in	$24a$	$9c$	\$6603

Translate. We get two equations from the last two rows of the table.

$a+c = 382,$ (1)

$24a+9c = 6603$ (2)

Solve. We will use the elimination method. First we multiply Equation (1) by -9 and then add.

$$\begin{aligned} -9a - 9c &= -3438 \\ 24a + 9c &= 6603 \\ \hline 15a \quad &= 3165 \\ a &= 211 \end{aligned}$$

Now substitute 211 for a in equation (1) and solve for c.

$a+c = 382$

$211+c = 382$

$c = 171$

Check. The total number in attendance was $211+171$, or 382. The amount taken in from 211 adults' tickets was $\$24 \cdot 211$, or \$5064. The amount taken in from 171 children's tickets was $\$9 \cdot 171$, or \$1539. The total amount taken in was $\$5064+\$1539$, or \$6603. The answer checks.

State. 211 adults and 171 children attended the fund-raiser.

95. $\sqrt{5x-5} = \sqrt{5(x-1)} = \sqrt{5}\sqrt{x-1}$

97. $\sqrt{x^2-36} = \sqrt{(x+6)(x-6)} = \sqrt{x+6}\sqrt{x-6}$

99. $\sqrt{x^3-2x^2} = \sqrt{x^2(x-2)} = \sqrt{x^2}\sqrt{x-2} = x\sqrt{x-2}$

101. $\sqrt{0.25} = \sqrt{(0.5)^2} = 0.5$

103. $\sqrt{\sqrt{\sqrt{256}}} = \sqrt{\sqrt{16}}$ $(\sqrt{256} = 16)$

$= \sqrt{4}$ $(\sqrt{16} = 4)$

$= 2$

105. $\sqrt{18(x-2)}\sqrt{20(x-2)^3} = \sqrt{18(x-2)\cdot 20(x-2)^3} = \sqrt{9\cdot 2\cdot 4\cdot 5(x-2)^4} = 3\cdot 2(x-2)^2\sqrt{2\cdot 5} = 6(x-2)^2\sqrt{10}$

107. $\sqrt{2^{109}}\sqrt{x^{306}}\sqrt{x^{11}} = \sqrt{2^{109}\cdot x^{306}\cdot x^{11}} = \sqrt{2^{109}x^{317}} = \sqrt{2^{108}\cdot 2\cdot x^{316}\cdot x} = 2^{54}x^{158}\sqrt{2x}$

109. $\sqrt{a}(\sqrt{a^3}-5) = \sqrt{a^4} - 5\sqrt{a} = a^2 - 5\sqrt{a}$

Exercise Set 16.3

1. $\dfrac{\sqrt{18}}{\sqrt{2}} = \sqrt{\dfrac{18}{2}} = \sqrt{9} = 3$

3. $\dfrac{\sqrt{108}}{\sqrt{3}} = \sqrt{\dfrac{108}{3}} = \sqrt{36} = 6$

5. $\dfrac{\sqrt{65}}{\sqrt{13}} = \sqrt{\dfrac{65}{13}} = \sqrt{5}$

7. $\dfrac{\sqrt{3}}{\sqrt{75}} = \sqrt{\dfrac{3}{75}} = \sqrt{\dfrac{1}{25}} = \dfrac{1}{5}$

9. $\dfrac{\sqrt{12}}{\sqrt{75}} = \sqrt{\dfrac{12}{75}} = \sqrt{\dfrac{4}{25}} = \dfrac{2}{5}$

11. $\dfrac{\sqrt{8x}}{\sqrt{2x}} = \sqrt{\dfrac{8x}{2x}} = \sqrt{4} = 2$

13. $\dfrac{\sqrt{63y^3}}{\sqrt{7y}} = \sqrt{\dfrac{63y^3}{7y}} = \sqrt{9y^2} = 3y$

15. $\sqrt{\dfrac{16}{49}} = \dfrac{\sqrt{16}}{\sqrt{49}} = \dfrac{4}{7}$

17. $\sqrt{\dfrac{1}{36}} = \dfrac{\sqrt{1}}{\sqrt{36}} = \dfrac{1}{6}$

19. $-\sqrt{\dfrac{16}{81}} = -\dfrac{\sqrt{16}}{\sqrt{81}} = -\dfrac{4}{9}$

21. $\sqrt{\frac{64}{289}} = \frac{\sqrt{64}}{\sqrt{289}} = \frac{8}{17}$

23. $\sqrt{\frac{1690}{1960}} = \sqrt{\frac{169 \cdot 10}{196 \cdot 10}} = \sqrt{\frac{169}{196} \cdot \frac{10}{10}} = \sqrt{\frac{169}{196} \cdot 1} =$

$\sqrt{\frac{169}{196}} = \frac{\sqrt{169}}{\sqrt{196}} = \frac{13}{14}$

25. $\sqrt{\frac{25}{x^2}} = \frac{\sqrt{25}}{\sqrt{x^2}} = \frac{5}{x}$

27. $\sqrt{\frac{9a^2}{625}} = \frac{\sqrt{9a^2}}{\sqrt{625}} = \frac{3a}{25}$

29. $\frac{\sqrt{50y^{15}}}{\sqrt{2y^{25}}} = \sqrt{\frac{50y^{15}}{2y^{25}}} = \sqrt{\frac{25}{y^{10}}} = \frac{\sqrt{25}}{\sqrt{y^{10}}} = \frac{5}{y^5}$

31. $\frac{\sqrt{7x^{23}}}{\sqrt{343x^5}} = \sqrt{\frac{7x^{23}}{343x^5}} = \sqrt{\frac{x^{18}}{49}} = \frac{\sqrt{x^{18}}}{\sqrt{49}} = \frac{x^9}{7}$

33. $\sqrt{\frac{2}{5}} = \sqrt{\frac{2}{5} \cdot \frac{5}{5}} = \sqrt{\frac{10}{25}} = \frac{\sqrt{10}}{\sqrt{25}} = \frac{\sqrt{10}}{5}$

35. $\sqrt{\frac{7}{8}} = \sqrt{\frac{7}{8} \cdot \frac{2}{2}} = \sqrt{\frac{14}{16}} = \frac{\sqrt{14}}{\sqrt{16}} = \frac{\sqrt{14}}{4}$

37. $\sqrt{\frac{1}{12}} = \sqrt{\frac{1}{12} \cdot \frac{3}{3}} = \sqrt{\frac{3}{36}} = \frac{\sqrt{3}}{\sqrt{36}} = \frac{\sqrt{3}}{6}$

39. $\sqrt{\frac{5}{18}} = \sqrt{\frac{5}{18} \cdot \frac{2}{2}} = \sqrt{\frac{10}{36}} = \frac{\sqrt{10}}{\sqrt{36}} = \frac{\sqrt{10}}{6}$

41. $\frac{3}{\sqrt{5}} = \frac{3}{\sqrt{5}} \cdot \frac{\sqrt{5}}{\sqrt{5}} = \frac{3\sqrt{5}}{5}$

43. $\sqrt{\frac{8}{3}} = \sqrt{\frac{8}{3} \cdot \frac{3}{3}} = \sqrt{\frac{24}{9}} = \frac{\sqrt{4 \cdot 6}}{\sqrt{9}} = \frac{\sqrt{4}\,\sqrt{6}}{\sqrt{9}} = \frac{2\sqrt{6}}{3}$

45. $\sqrt{\frac{3}{x}} = \sqrt{\frac{3}{x} \cdot \frac{x}{x}} = \sqrt{\frac{3x}{x^2}} = \frac{\sqrt{3x}}{\sqrt{x^2}} = \frac{\sqrt{3x}}{x}$

47. $\sqrt{\frac{x}{y}} = \sqrt{\frac{x}{y} \cdot \frac{y}{y}} = \sqrt{\frac{xy}{y^2}} = \frac{\sqrt{xy}}{\sqrt{y^2}} = \frac{\sqrt{xy}}{y}$

49. $\sqrt{\frac{x^2}{20}} = \sqrt{\frac{x^2}{20} \cdot \frac{5}{5}} = \sqrt{\frac{5x^2}{100}} = \frac{\sqrt{x^2 \cdot 5}}{\sqrt{100}} = \frac{\sqrt{x^2}\,\sqrt{5}}{\sqrt{100}} = \frac{x\sqrt{5}}{10}$

51. $\frac{\sqrt{7}}{\sqrt{2}} = \frac{\sqrt{7}}{\sqrt{2}} \cdot \frac{\sqrt{2}}{\sqrt{2}} = \frac{\sqrt{14}}{2}$

53. $\frac{\sqrt{9}}{\sqrt{8}} = \frac{\sqrt{9}}{\sqrt{8}} \cdot \frac{\sqrt{2}}{\sqrt{2}} = \frac{\sqrt{9 \cdot 2}}{\sqrt{16}} = \frac{3\sqrt{2}}{4}$

55. $\frac{\sqrt{3}}{\sqrt{2}} = \frac{\sqrt{3}}{\sqrt{2}} \cdot \frac{\sqrt{2}}{\sqrt{2}} = \frac{\sqrt{6}}{2}$

57. $\frac{2}{\sqrt{2}} = \frac{2}{\sqrt{2}} \cdot \frac{\sqrt{2}}{\sqrt{2}} = \frac{2\sqrt{2}}{2} = \sqrt{2}$

59. $\frac{\sqrt{5}}{\sqrt{11}} = \frac{\sqrt{5}}{\sqrt{11}} \cdot \frac{\sqrt{11}}{\sqrt{11}} = \frac{\sqrt{55}}{11}$

61. $\frac{\sqrt{7}}{\sqrt{12}} = \frac{\sqrt{7}}{\sqrt{12}} \cdot \frac{\sqrt{3}}{\sqrt{3}} = \frac{\sqrt{21}}{\sqrt{36}} = \frac{\sqrt{21}}{6}$

63. $\frac{\sqrt{48}}{\sqrt{32}} = \sqrt{\frac{48}{32}} = \sqrt{\frac{3}{2}} = \sqrt{\frac{3}{2} \cdot \frac{2}{2}} = \sqrt{\frac{6}{4}} = \frac{\sqrt{6}}{\sqrt{4}} = \frac{\sqrt{6}}{2}$

65. $\frac{\sqrt{450}}{\sqrt{18}} = \sqrt{\frac{450}{18}} = \sqrt{25} = 5$

67. $\frac{\sqrt{3}}{\sqrt{x}} = \frac{\sqrt{3}}{\sqrt{x}} \cdot \frac{\sqrt{x}}{\sqrt{x}} = \frac{\sqrt{3x}}{x}$

69. $\frac{4y}{\sqrt{5}} = \frac{4y}{\sqrt{5}} \cdot \frac{\sqrt{5}}{\sqrt{5}} = \frac{4y\sqrt{5}}{5}$

71. $\frac{\sqrt{a^3}}{\sqrt{8}} = \frac{\sqrt{a^3}}{\sqrt{8}} \cdot \frac{\sqrt{2}}{\sqrt{2}} = \frac{\sqrt{2a^3}}{\sqrt{16}} = \frac{\sqrt{a^2 \cdot 2a}}{\sqrt{16}} = \frac{a\sqrt{2a}}{4}$

73. $\frac{\sqrt{56}}{\sqrt{12x}} = \sqrt{\frac{56}{12x}} = \sqrt{\frac{14}{3x}} = \sqrt{\frac{14}{3x} \cdot \frac{3x}{3x}} = \sqrt{\frac{42x}{3x \cdot 3x}} =$

$\frac{\sqrt{42x}}{3x}$

75. $\frac{\sqrt{27c}}{\sqrt{32c^3}} = \sqrt{\frac{27c}{32c^3}} = \sqrt{\frac{27}{32c^2}} = \sqrt{\frac{27}{32c^2} \cdot \frac{2}{2}} = \sqrt{\frac{54}{64c^2}} =$

$\sqrt{\frac{9 \cdot 6}{64c^2}} = \frac{3\sqrt{6}}{8c}$

77. $\frac{\sqrt{y^5}}{\sqrt{xy^2}} = \sqrt{\frac{y^5}{xy^2}} = \sqrt{\frac{y^3}{x}} = \sqrt{\frac{y^3}{x} \cdot \frac{x}{x}} = \sqrt{\frac{xy^3}{x^2}} =$

$\sqrt{\frac{y^2 \cdot xy}{x^2}} = \frac{y\sqrt{xy}}{x}$

79. $\frac{\sqrt{45mn^2}}{\sqrt{32m}} = \sqrt{\frac{45mn^2}{32m}} = \sqrt{\frac{45n^2}{32}} = \sqrt{\frac{45n^2}{32} \cdot \frac{2}{2}} =$

$\sqrt{\frac{90n^2}{64}} = \frac{\sqrt{90n^2}}{\sqrt{64}} = \frac{\sqrt{9 \cdot n^2 \cdot 10}}{8} = \frac{3n\sqrt{10}}{8}$

81. $x = y + 2,$ (1)

$x + y = 6$ (2)

We substitute $y + 2$ for x in Equation (2) and solve for y.

$$(y + 2) + y = 6$$
$$2y + 2 = 6$$
$$2y = 4$$
$$y = 2$$

Substitute 2 for y in Equation (1) to find x.

$$x = 2 + 2 = 4$$

The ordered pair $(4, 2)$ checks in both equations. It is the solution.

83. $2x - 3y = 7$ (1)

$2x - 3y = 9$ (2)

We multiply Equation (2) by -1 and add.

$$\begin{array}{r} 2x - 3y = 7 \\ -2x + 3y = -9 \\ \hline 0 = -2 \end{array}$$

We get a false equation. The system of equations has no solution.

85.
$$\begin{aligned} x + y &= -7 \quad (1) \\ x - y &= 2 \quad (2) \\ \hline 2x &= -5 \quad \text{Adding} \\ x &= -\frac{5}{2} \end{aligned}$$

Substitute $-\frac{5}{2}$ for x in Equation (1) to find y.

$$\begin{aligned} x + y &= -7 \quad \text{Equation (1)} \\ -\frac{5}{2} + y &= -7 \quad \text{Substituting} \\ y &= -\frac{9}{2} \end{aligned}$$

The ordered pair $\left(-\frac{5}{2}, -\frac{9}{2}\right)$ checks in both equations. It is the solution.

87.
$$\begin{aligned} &\frac{x^2-49}{x+8} \div \frac{x^2-14x+49}{x^2+15x+56} \\ &= \frac{x^2-49}{x+8} \cdot \frac{x^2+15x+56}{x^2-14x+49} \\ &= \frac{(x^2-49)(x^2+15x+56)}{(x+8)(x^2-14x+49)} \\ &= \frac{(x+7)(x-7)(x+7)(x+8)}{(x+8)(x-7)(x-7)} \\ &= \frac{(x+7)\cancel{(x-7)}(x+7)\cancel{(x+8)}}{\cancel{(x+8)}\cancel{(x-7)}(x-7)} \\ &= \frac{(x+7)(x+7)}{x-7}, \text{ or } \frac{(x+7)^2}{x-7} \end{aligned}$$

89.
$$\begin{aligned} \frac{a^2-25}{6} \div \frac{a+5}{3} &= \frac{a^2-25}{6} \cdot \frac{3}{a+5} \\ &= \frac{(a^2-25) \cdot 3}{6(a+5)} \\ &= \frac{(a+5)(a-5) \cdot 3}{2 \cdot 3 \cdot (a+5)} \\ &= \frac{\cancel{(a+5)}(a-5) \cdot \cancel{3}}{2 \cdot \cancel{3} \cdot \cancel{(a+5)}} \\ &= \frac{a-5}{2} \end{aligned}$$

91. $(3x-7)(3x+7) = (3x)^2 - 7^2 = 9x^2 - 49$

93. 2 ft: $T \approx 2(3.14)\sqrt{\frac{2}{32}} = 6.28\sqrt{\frac{1}{16}} = 6.28\left(\frac{1}{4}\right) =$ 1.57 sec

8 ft: $T \approx 2(3.14)\sqrt{\frac{8}{32}} = 6.28\sqrt{\frac{1}{4}} = 6.28\left(\frac{1}{2}\right) =$ 3.14 sec

To find the period for a 10-in. pendulum, we first convert 10 in. to feet.

$$10 \text{ in.} = 10 \text{ in.} \times \frac{1 \text{ ft}}{12 \text{ in.}} = \frac{10}{12} \times \frac{\text{in.}}{\text{in.}} \times \text{ ft} = \frac{5}{6} \text{ ft}$$

Now we use the formula.

$$T \approx 2(3.14)\sqrt{\frac{5/6}{32}} = 6.28\sqrt{\frac{5}{6} \cdot \frac{1}{32}} = 6.28\sqrt{\frac{5}{192}} \approx 1.01 \text{ sec}$$

95. $\sqrt{\frac{5}{1600}} = \frac{\sqrt{5}}{\sqrt{1600}} = \frac{\sqrt{5}}{40}$

97. $\sqrt{\frac{3x^2y}{a^2x^5}} = \sqrt{\frac{3y}{a^2x^3}} = \sqrt{\frac{3y}{a^2x^3} \cdot \frac{x}{x}} = \sqrt{\frac{3xy}{a^2x^4}} =$
$\frac{\sqrt{3xy}}{\sqrt{a^2x^4}} = \frac{\sqrt{3xy}}{ax^2}$

99.
$$\begin{aligned} &\sqrt{\frac{1}{x^2} - \frac{2}{xy} + \frac{1}{y^2}}, \text{ LCD is } x^2y^2 \\ &= \sqrt{\frac{1}{x^2} \cdot \frac{y^2}{y^2} - \frac{2}{xy} \cdot \frac{xy}{xy} + \frac{1}{y^2} \cdot \frac{x^2}{x^2}} \\ &= \sqrt{\frac{y^2-2xy+x^2}{x^2y^2}} \\ &= \sqrt{\frac{(y-x)^2}{x^2y^2}} \\ &= \frac{\sqrt{(y-x)^2}}{\sqrt{x^2y^2}} \\ &= \frac{y-x}{xy} \end{aligned}$$

Chapter 16 Mid-Chapter Review

1. True; see page 1174 in the text.

2. For any nonnegative real number A, the principal square root of A^2 is A. The given statement is false.

3. The number 0 has only one square root 0. The given statement is false.

4. True; see page 1176 in the text.

5.
$$\begin{aligned} &\sqrt{3x^2-48x+192} \\ &= \sqrt{3(x^2-16x+64)} \\ &= \sqrt{3(x-8)^2} \\ &= \sqrt{3}\sqrt{(x-8)^2} \\ &= \sqrt{3}(x-8) \end{aligned}$$

6.
$$\begin{aligned} &\sqrt{30}\sqrt{40y} \\ &= \sqrt{30 \cdot 40y} \\ &= \sqrt{1200y} \\ &= \sqrt{100 \cdot 12 \cdot y} \\ &= \sqrt{100 \cdot 4 \cdot 3 \cdot y} \\ &= \sqrt{100}\sqrt{4}\sqrt{3y} \\ &= 10 \cdot 2\sqrt{3y} \\ &= 20\sqrt{3y} \end{aligned}$$

7. $\sqrt{18ab^2}\sqrt{14a^2b^4}$
$= \sqrt{18ab^2 \cdot 14a^2b^4}$
$= \sqrt{2 \cdot 3 \cdot 3 \cdot 2 \cdot 7 \cdot a^3 \cdot b^6}$
$= \sqrt{2^2 \cdot 3^2 \cdot 7 \cdot a^2 \cdot a \cdot b^6}$
$= \sqrt{2^2}\sqrt{3^2}\sqrt{a^2}\sqrt{b^6}\sqrt{7a}$
$= 2 \cdot 3 \cdot a \cdot b^3\sqrt{7a}$
$= 6ab^3\sqrt{7a}$

8. $\sqrt{\frac{3y^2}{44}} = \sqrt{\frac{3y^2}{2 \cdot 2 \cdot 11}} = \sqrt{\frac{3y^2}{2 \cdot 2 \cdot 11} \cdot \frac{11}{11}}$
$= \sqrt{\frac{33y^2}{2^2 \cdot 11^2}} = \frac{y\sqrt{33}}{2 \cdot 11} = \frac{y\sqrt{33}}{22}$

9. The square roots of 121 are 11 and -11, because $11^2 = 121$ and $(-11)^2 = 121$.

10. The radicand is the expression under the radical, $\frac{x-3}{7}$.

11. a) No, because the radicand is negative.
b) Yes, because the radicand is nonnegative.

12. $\sqrt{128r^7s^6} = \sqrt{64 \cdot 2 \cdot r^6 \cdot r \cdot s^6} = \sqrt{64r^6s^6}\sqrt{2r} = 8r^3s^3\sqrt{2r}$

13. $\sqrt{25(x-3)^2} = \sqrt{25}\sqrt{(x-3)^2} = 5(x-3)$

14. $\sqrt{\frac{1}{100}} = \frac{\sqrt{1}}{\sqrt{100}} = \frac{1}{10}$

15. $\sqrt{36} = 6$, so $-\sqrt{36} = -6$.

16. $-\sqrt{\frac{6250}{490}} = -\sqrt{\frac{625 \cdot 10}{49 \cdot 10}} = -\sqrt{\frac{625}{49}} = -\frac{\sqrt{625}}{\sqrt{49}} = -\frac{25}{7}$

17. $\sqrt{225} = 15$, taking the principal square root.

18. $\sqrt{(10y)^2} = 10y$

19. $\sqrt{4x^2 - 4x + 1} = \sqrt{(2x-1)^2} = 2x - 1$

20. $\sqrt{800x} = \sqrt{400 \cdot 2 \cdot x} = \sqrt{400}\sqrt{2x} = 20\sqrt{2x}$

21. $\frac{\sqrt{6}}{\sqrt{96}} = \sqrt{\frac{6}{96}} = \sqrt{\frac{6 \cdot 1}{6 \cdot 16}} = \sqrt{\frac{1}{16}} = \frac{\sqrt{1}}{\sqrt{16}} = \frac{1}{4}$

22. $\sqrt{32q^{11}} = \sqrt{16 \cdot 2 \cdot q^{10} \cdot q} = \sqrt{16q^{10}}\sqrt{2q} = 4q^5\sqrt{2q}$

23. $\sqrt{\frac{81}{x^2}} = \frac{\sqrt{81}}{\sqrt{x^2}} = \frac{9}{x}$

24. $\sqrt{25}\sqrt{25} = \sqrt{25 \cdot 25} = 25$

25. $\frac{\sqrt{18}}{\sqrt{98}} = \sqrt{\frac{18}{98}} = \sqrt{\frac{9 \cdot 2}{49 \cdot 2}} = \sqrt{\frac{9}{49}} = \frac{\sqrt{9}}{\sqrt{49}} = \frac{3}{7}$

26. $\frac{\sqrt{192x}}{\sqrt{3x}} = \sqrt{\frac{192x}{3x}} = \sqrt{\frac{3x \cdot 64}{3x \cdot 1}} = \sqrt{64} = 8$

27. $\sqrt{40c^2d^7}\sqrt{15c^3d^3} = \sqrt{40c^2d^7 \cdot 15c^3d^3} =$
$\sqrt{2 \cdot 2 \cdot 2 \cdot 5 \cdot c^2 \cdot d^6 \cdot d \cdot 3 \cdot 5 \cdot c^2 \cdot c \cdot d^2 \cdot d} =$
$\sqrt{2 \cdot 2 \cdot 5 \cdot 5 \cdot c^2 \cdot d^6 \cdot c^2 \cdot d^2 \cdot d \cdot d \cdot 2 \cdot 3 \cdot c} =$
$\sqrt{2 \cdot 2 \cdot 5 \cdot 5 \cdot c^2 \cdot d^6 \cdot c^2 \cdot d^2 \cdot d \cdot d}\sqrt{6c} =$
$2 \cdot 5 \cdot c \cdot d^3 \cdot c \cdot d \cdot d\sqrt{6c} = 10c^2d^5\sqrt{6c}$

28. $\sqrt{24x^5y^8z^2}\sqrt{60xy^3z} = \sqrt{24x^5y^8z^2 \cdot 60xy^3z} =$
$\sqrt{4 \cdot 2 \cdot 3 \cdot x^4 \cdot x \cdot y^8 \cdot z^2 \cdot 4 \cdot 3 \cdot 5 \cdot x \cdot y^2 \cdot y \cdot z} =$
$\sqrt{4 \cdot 3 \cdot 3 \cdot 4 \cdot x^4 \cdot x \cdot x \cdot y^8 \cdot y^2 \cdot z^2}\sqrt{2 \cdot 5 \cdot y \cdot z} =$
$2 \cdot 3 \cdot 2 \cdot x^2 \cdot x \cdot y^4 \cdot y \cdot z\sqrt{10yz} = 12x^3y^5z\sqrt{10yz}$

29. $\sqrt{2x}\sqrt{30y} = \sqrt{2x \cdot 30y} = \sqrt{2 \cdot x \cdot 2 \cdot 15 \cdot y} =$
$\sqrt{2 \cdot 2 \cdot x \cdot 15 \cdot y} = \sqrt{2 \cdot 2}\sqrt{x \cdot 15 \cdot y} = 2\sqrt{15xy}$

30. $\sqrt{21a}\sqrt{35a} = \sqrt{21a \cdot 35a} = \sqrt{3 \cdot 7 \cdot a \cdot 5 \cdot 7 \cdot a} =$
$\sqrt{7 \cdot 7 \cdot a \cdot a \cdot 3 \cdot 5} = \sqrt{7 \cdot 7 \cdot a \cdot a}\sqrt{3 \cdot 5} =$
$7a\sqrt{15}$

31. $\frac{\sqrt{3y^{29}}}{\sqrt{75y^5}} = \sqrt{\frac{3y^{29}}{75y^5}} = \sqrt{\frac{y^{24}}{25}} = \frac{\sqrt{y^{24}}}{\sqrt{25}} = \frac{y^{12}}{5}$

32. $\frac{x}{\sqrt{3}} = \frac{x}{\sqrt{3}} \cdot \frac{\sqrt{3}}{\sqrt{3}} = \frac{x\sqrt{3}}{3}$
$\sqrt{\frac{3}{x}} = \sqrt{\frac{3}{x} \cdot \frac{x}{x}} = \sqrt{\frac{3x}{x^2}} = \frac{\sqrt{3x}}{x}$
$\frac{3}{\sqrt{x}} = \frac{3}{\sqrt{x}} \cdot \frac{\sqrt{x}}{\sqrt{x}} = \frac{3\sqrt{x}}{x}$
$\frac{3x}{\sqrt{3}} = \frac{3x}{\sqrt{3}} \cdot \frac{\sqrt{3}}{\sqrt{3}} = \frac{3x\sqrt{3}}{3} = x\sqrt{3}$
$\frac{3}{\sqrt{3}} = \frac{3}{\sqrt{3}} \cdot \frac{\sqrt{3}}{\sqrt{3}} = \frac{3\sqrt{3}}{3} = \sqrt{3}$
$\sqrt{\frac{x}{3}} = \sqrt{\frac{x}{3} \cdot \frac{3}{3}} = \sqrt{\frac{3x}{9}} = \frac{\sqrt{3x}}{3}$

33. **The** square root of 10 is the principal, or positive, square root. **A** square root of 10 could refer to either the positive or the negative square root.

34. It is incorrect to take the square roots of the terms in the numerator individually; that is, $\sqrt{a+b}$ and $\sqrt{a} + \sqrt{b}$ are not equivalent. The following is correct:
$$\sqrt{\frac{9+100}{25}} = \frac{\sqrt{9+100}}{\sqrt{25}} = \frac{\sqrt{109}}{5}.$$

35. In general, $\sqrt{a^2 - b^2} \neq \sqrt{a^2} - \sqrt{b^2}$. In this case, let $x = 13$. Then
$\sqrt{x^2 - 25} = \sqrt{13^2 - 25} = \sqrt{169 - 25} = \sqrt{144} = 12$,
but $\sqrt{x^2} - \sqrt{25} = \sqrt{13^2} - \sqrt{25} = 13 - 5 = 8$.

36. 1. If necessary, rewrite the expression as $\sqrt{a}/\sqrt{b}$.
2. Simplify the numerator and denominator, if possible, by taking the square roots of perfect square factors.
3. Multiply by a form of 1 that produces an expression without a radical in the numerator.

Exercise Set 16.4

1. $7\sqrt{3}+9\sqrt{3}=(7+9)\sqrt{3}$
$=16\sqrt{3}$

3. $7\sqrt{5}-3\sqrt{5}=(7-3)\sqrt{5}$
$=4\sqrt{5}$

5. $6\sqrt{x}+7\sqrt{x}=(6+7)\sqrt{x}$
$=13\sqrt{x}$

7. $4\sqrt{d}-13\sqrt{d}=(4-13)\sqrt{d}$
$=-9\sqrt{d}$

9. $5\sqrt{8}+15\sqrt{2}=5\sqrt{4\cdot 2}+15\sqrt{2}$
$=5\cdot 2\sqrt{2}+15\sqrt{2}$
$=10\sqrt{2}+15\sqrt{2}$
$=25\sqrt{2}$

11. $\sqrt{27}-2\sqrt{3}=\sqrt{9\cdot 3}-2\sqrt{3}$
$=3\sqrt{3}-2\sqrt{3}$
$=(3-2)\sqrt{3}$
$=1\sqrt{3}$
$=\sqrt{3}$

13. $\sqrt{45}-\sqrt{20}=\sqrt{9\cdot 5}-\sqrt{4\cdot 5}$
$=3\sqrt{5}-2\sqrt{5}$
$=(3-2)\sqrt{5}$
$=1\sqrt{5}$
$=\sqrt{5}$

15. $\sqrt{72}+\sqrt{98}=\sqrt{36\cdot 2}+\sqrt{49\cdot 2}$
$=6\sqrt{2}+7\sqrt{2}$
$=(6+7)\sqrt{2}$
$=13\sqrt{2}$

17. $2\sqrt{12}+\sqrt{27}-\sqrt{48}=2\sqrt{4\cdot 3}+\sqrt{9\cdot 3}-\sqrt{16\cdot 3}$
$=2\cdot 2\sqrt{3}+3\sqrt{3}-4\sqrt{3}$
$=4\sqrt{3}+3\sqrt{3}-4\sqrt{3}$
$=(4+3-4)\sqrt{3}$
$=3\sqrt{3}$

19. $\sqrt{18}-3\sqrt{8}+\sqrt{50}=\sqrt{9\cdot 2}-3\sqrt{4\cdot 2}+\sqrt{25\cdot 2}$
$=3\sqrt{2}-3\cdot 2\sqrt{2}+5\sqrt{2}$
$=3\sqrt{2}-6\sqrt{2}+5\sqrt{2}$
$=(3-6+5)\sqrt{2}$
$=2\sqrt{2}$

21. $2\sqrt{27}-3\sqrt{48}+3\sqrt{12}=2\sqrt{9\cdot 3}-3\sqrt{16\cdot 3}+3\sqrt{4\cdot 3}$
$=2\cdot 3\sqrt{3}-3\cdot 4\sqrt{3}+3\cdot 2\sqrt{3}$
$=6\sqrt{3}-12\sqrt{3}+6\sqrt{3}$
$=(6-12+6)\sqrt{3}$
$=0\sqrt{3}$
$=0$

23. $\sqrt{4x}+\sqrt{81x^3}=\sqrt{4\cdot x}+\sqrt{81\cdot x^2\cdot x}$
$=2\sqrt{x}+9x\sqrt{x}$
$=(2+9x)\sqrt{x}$

25. $\sqrt{27}-\sqrt{12x^2}=\sqrt{9\cdot 3}-\sqrt{4\cdot 3\cdot x^2}$
$=3\sqrt{3}-2x\sqrt{3}$
$=(3-2x)\sqrt{3}$

27. $\sqrt{8x+8}+\sqrt{2x+2}=\sqrt{4(2x+2)}+\sqrt{2x+2}$
$=2\sqrt{2x+2}+1\sqrt{2x+2}$
$=(2+1)\sqrt{2x+2}$
$=3\sqrt{2x+2}$

29. $\sqrt{x^5-x^2}+\sqrt{9x^3-9}=\sqrt{x^2(x^3-1)}+\sqrt{9(x^3-1)}$
$=x\sqrt{x^3-1}+3\sqrt{x^3-1}$
$=(x+3)\sqrt{x^3-1}$

31. $4a\sqrt{a^2b}+a\sqrt{a^2b^3}-5\sqrt{b^3}$
$=4a\sqrt{a^2\cdot b}+a\sqrt{a^2\cdot b^2\cdot b}-5\sqrt{b^2\cdot b}$
$=4a\cdot a\sqrt{b}+a\cdot a\cdot b\sqrt{b}-5\cdot b\sqrt{b}$
$=4a^2\sqrt{b}+a^2b\sqrt{b}-5b\sqrt{b}$
$=(4a^2+a^2b-5b)\sqrt{b}$

33. $\sqrt{3}-\sqrt{\frac{1}{3}}=\sqrt{3}-\sqrt{\frac{1}{3}\cdot\frac{3}{3}}$
$=\sqrt{3}-\frac{\sqrt{3}}{3}$
$=\left(1-\frac{1}{3}\right)\sqrt{3}$
$=\frac{2}{3}\sqrt{3}$, or $\frac{2\sqrt{3}}{3}$

35. $5\sqrt{2}+3\sqrt{\frac{1}{2}}=5\sqrt{2}+3\sqrt{\frac{1}{2}\cdot\frac{2}{2}}$
$=5\sqrt{2}+\frac{3}{2}\sqrt{2}$
$=\left(5+\frac{3}{2}\right)\sqrt{2}$
$=\frac{13}{2}\sqrt{2}$, or $\frac{13\sqrt{2}}{2}$

37. $\sqrt{\frac{2}{3}}-\sqrt{\frac{1}{6}}=\sqrt{\frac{2}{3}\cdot\frac{3}{3}}-\sqrt{\frac{1}{6}\cdot\frac{6}{6}}$
$=\frac{\sqrt{6}}{3}-\frac{\sqrt{6}}{6}$
$=\left(\frac{1}{3}-\frac{1}{6}\right)\sqrt{6}$
$=\frac{1}{6}\sqrt{6}$, or $\frac{\sqrt{6}}{6}$

39. $\sqrt{3}(\sqrt{5}-1) = \sqrt{3}\,\sqrt{5} - \sqrt{3}\cdot 1$
$= \sqrt{15} - \sqrt{3}$

41. $(2+\sqrt{3})(5-\sqrt{7})$
$= 2\cdot 5 - 2\sqrt{7} + \sqrt{3}\cdot 5 - \sqrt{3}\sqrt{7}$ Using FOIL
$= 10 - 2\sqrt{7} + 5\sqrt{3} - \sqrt{21}$

43. $(2-\sqrt{5})^2$
$= 2^2 - 2\cdot 2\cdot\sqrt{5} + (\sqrt{5})^2$
Using $(A-B)^2 = A^2 - 2AB + B^2$
$= 4 - 4\sqrt{5} + 5$
$= 9 - 4\sqrt{5}$

45. $(\sqrt{2}+8)(\sqrt{2}-8)$
$= (\sqrt{2})^2 - 8^2$ Using $(A+B)(A-B) = A^2 - B^2$
$= 2 - 64$
$= -62$

47. $(\sqrt{6}-\sqrt{5})(\sqrt{6}+\sqrt{5})$
$= (\sqrt{6})^2 - (\sqrt{5})^2$ Using $(A+B)(A-B) = A^2 - B^2$
$= 6 - 5$
$= 1$

49. $(3\sqrt{5}-2)(\sqrt{5}+1)$
$= 3\sqrt{5}\,\sqrt{5} + 3\sqrt{5} - 2\sqrt{5} - 2$ Using FOIL
$= 3\cdot 5 + 3\sqrt{5} - 2\sqrt{5} - 2$
$= 15 + \sqrt{5} - 2$
$= 13 + \sqrt{5}$

51. $(\sqrt{x}-\sqrt{y})^2 = (\sqrt{x})^2 - 2\sqrt{x}\,\sqrt{y} + (\sqrt{y})^2$
Using $(A-B)^2 = A^2 - 2AB + B^2$
$= x - 2\sqrt{xy} + y$

53. We multiply by 1 using the conjugate of $\sqrt{3}-\sqrt{5}$, which is $\sqrt{3}+\sqrt{5}$, as the numerator and denominator.

$$\frac{2}{\sqrt{3}-\sqrt{5}} = \frac{2}{\sqrt{3}-\sqrt{5}}\cdot\frac{\sqrt{3}+\sqrt{5}}{\sqrt{3}+\sqrt{5}} \quad \text{Multiplying by 1}$$
$$= \frac{2(\sqrt{3}+\sqrt{5})}{(\sqrt{3}-\sqrt{5})(\sqrt{3}+\sqrt{5})} \quad \text{Multiplying}$$
$$= \frac{2\sqrt{3}+2\sqrt{5}}{(\sqrt{3})^2-(\sqrt{5})^2} = \frac{2\sqrt{3}+2\sqrt{5}}{3-5}$$
$$= \frac{2\sqrt{3}+2\sqrt{5}}{-2} = \frac{2(\sqrt{3}+\sqrt{5})}{-2}$$
$$= -(\sqrt{3}+\sqrt{5}) = -\sqrt{3}-\sqrt{5}$$

55. We multiply by 1 using the conjugate of $\sqrt{3}+\sqrt{2}$, which is $\sqrt{3}-\sqrt{2}$, as the numerator and denominator.

$$\frac{\sqrt{3}-\sqrt{2}}{\sqrt{3}+\sqrt{2}} = \frac{\sqrt{3}-\sqrt{2}}{\sqrt{3}+\sqrt{2}}\cdot\frac{\sqrt{3}-\sqrt{2}}{\sqrt{3}-\sqrt{2}} \quad \text{Multiplying by 1}$$
$$= \frac{(\sqrt{3}-\sqrt{2})^2}{(\sqrt{3}+\sqrt{2})(\sqrt{3}-\sqrt{2})}$$
$$= \frac{(\sqrt{3})^2 - 2\sqrt{3}\,\sqrt{2} + (\sqrt{2})^2}{(\sqrt{3})^2-(\sqrt{2})^2}$$
$$= \frac{3-2\sqrt{6}+2}{3-2} = \frac{5-2\sqrt{6}}{1}$$
$$= 5-2\sqrt{6}$$

57. We multiply by 1 using the conjugate of $\sqrt{10}+1$, which is $\sqrt{10}-1$, as the numerator and denominator.

$$\frac{4}{\sqrt{10}+1} = \frac{4}{\sqrt{10}+1}\cdot\frac{\sqrt{10}-1}{\sqrt{10}-1}$$
$$= \frac{4(\sqrt{10}-1)}{(\sqrt{10}+1)(\sqrt{10}-1)}$$
$$= \frac{4\sqrt{10}-4}{(\sqrt{10})^2-1^2} = \frac{4\sqrt{10}-4}{10-1}$$
$$= \frac{4\sqrt{10}-4}{9}$$

59. We multiply by 1 using the conjugate of $3+\sqrt{7}$, which is $3-\sqrt{7}$, as the numerator and denominator.

$$\frac{1-\sqrt{7}}{3+\sqrt{7}} = \frac{1-\sqrt{7}}{3+\sqrt{7}}\cdot\frac{3-\sqrt{7}}{3-\sqrt{7}}$$
$$= \frac{(1-\sqrt{7})(3-\sqrt{7})}{(3+\sqrt{7})(3-\sqrt{7})}$$
$$= \frac{3-\sqrt{7}-3\sqrt{7}+\sqrt{7}\,\sqrt{7}}{3^2-(\sqrt{7})^2}$$
$$= \frac{3-\sqrt{7}-3\sqrt{7}+7}{9-7} = \frac{10-4\sqrt{7}}{2}$$
$$= \frac{2(5-2\sqrt{7})}{2} = 5-2\sqrt{7}$$

61. We multiply by 1 using the conjugate of $4+\sqrt{x}$, which is $4-\sqrt{x}$, as the numerator and denominator.

$$\frac{3}{4+\sqrt{x}} = \frac{3}{4+\sqrt{x}}\cdot\frac{4-\sqrt{x}}{4-\sqrt{x}}$$
$$= \frac{3(4-\sqrt{x})}{(4+\sqrt{x})(4-\sqrt{x})}$$
$$= \frac{12-3\sqrt{x}}{4^2-(\sqrt{x})^2}$$
$$= \frac{12-3\sqrt{x}}{16-x}$$

63. We multiply by 1 using the conjugate of $8-\sqrt{x}$, which is $8+\sqrt{x}$, as the numerator and denominator.

$$\frac{3+\sqrt{2}}{8-\sqrt{x}} = \frac{3+\sqrt{2}}{8-\sqrt{x}}\cdot\frac{8+\sqrt{x}}{8+\sqrt{x}}$$
$$= \frac{(3+\sqrt{2})(8+\sqrt{x})}{(8-\sqrt{x})(8+\sqrt{x})}$$
$$= \frac{3\cdot 8+3\cdot\sqrt{x}+\sqrt{2}\cdot 8+\sqrt{2}\cdot\sqrt{x}}{8^2-(\sqrt{x})^2}$$
$$= \frac{24+3\sqrt{x}+8\sqrt{2}+\sqrt{2x}}{64-x}$$

65. We multiply by 1 using the conjugate of $1+\sqrt{a}$, which is $1-\sqrt{a}$, as the numerator and denominator.

$$\frac{\sqrt{a}-1}{1+\sqrt{a}} = \frac{\sqrt{a}-1}{1+\sqrt{a}}\cdot\frac{1-\sqrt{a}}{1-\sqrt{a}}$$
$$= \frac{(\sqrt{a}-1)(1-\sqrt{a})}{(1+\sqrt{a})(1-\sqrt{a})}$$
$$= \frac{\sqrt{a}\cdot 1-\sqrt{a}\cdot\sqrt{a}-1\cdot 1+1\cdot\sqrt{a}}{1^2-(\sqrt{a})^2}$$
$$= \frac{\sqrt{a}-a-1+\sqrt{a}}{1-a}$$
$$= \frac{2\sqrt{a}-a-1}{1-a}$$

67. We multiply by 1 using the conjugate of $\sqrt{a}-\sqrt{t}$, which is $\sqrt{a}+\sqrt{t}$, as the numerator and denominator.

$$\frac{4+\sqrt{3}}{\sqrt{a}-\sqrt{t}} = \frac{4+\sqrt{3}}{\sqrt{a}-\sqrt{t}}\cdot\frac{\sqrt{a}+\sqrt{t}}{\sqrt{a}+\sqrt{t}}$$
$$= \frac{(4+\sqrt{3})(\sqrt{a}+\sqrt{t})}{(\sqrt{a}-\sqrt{t})(\sqrt{a}+\sqrt{t})}$$
$$= \frac{4\cdot\sqrt{a}+4\cdot\sqrt{t}+\sqrt{3}\cdot\sqrt{a}+\sqrt{3}\cdot\sqrt{t}}{(\sqrt{a})^2-(\sqrt{t})^2}$$
$$= \frac{4\sqrt{a}+4\sqrt{t}+\sqrt{3a}+\sqrt{3t}}{a-t}$$

69.
$$3x+5+2(x-3) = 4-6x$$
$$3x+5+2x-6 = 4-6x$$
$$5x-1 = 4-6x$$
$$11x-1 = 4$$
$$11x = 5$$
$$x = \frac{5}{11}$$

The solution is $\frac{5}{11}$.

71.
$$x^2-5x = 6$$
$$x^2-5x-6 = 0$$
$$(x+1)(x-6) = 0$$
$$x+1=0 \quad \text{or} \quad x-6=0$$
$$x=-1 \quad \text{or} \quad x=6$$

The solutions are -1 and 6.

73. $\frac{7x^9}{27}\cdot\frac{9}{7x^3} = \frac{63x^9}{189x^3} = \frac{63}{189}x^{9-3} = \frac{1}{3}x^6$, or $\frac{x^6}{3}$

75. ***Familiarize***. Let a = the highest altitude of the Trail, in feet.

Translate.

Lowest altitude	plus	9990 ft	is	highest altitude
↓	↓	↓	↓	↓
4280	+	9990	=	a

Solve. We add on the left side.

$$4280+9990 = a$$
$$14,270 = a$$

Check. We can repeat the calculation. We can also estimate.

$$4280+9990 \approx 4300+10,000 = 14,300 \approx 14,270$$

The answer seems reasonable.

State. The highest altitude of the Trail is 14,270 ft.

77. $\sqrt{2^2+3^2} = \sqrt{4+9} = \sqrt{13}$

$\sqrt{2^2}+\sqrt{3^2} = 2+3 = 5$

79. Enter $y_1 = \sqrt{x}+\sqrt{3}$ and $y_2 = \sqrt{x+3}$ and look at a table of values. The values of y_1 and y_2 are different, so the given statement is not correct.

81.
$$\frac{3}{5}\sqrt{24}+\frac{2}{5}\sqrt{150}-\sqrt{96}$$
$$= \frac{3}{5}\sqrt{4\cdot 6}+\frac{2}{5}\sqrt{25\cdot 6}-\sqrt{16\cdot 6}$$
$$= \frac{3}{5}\cdot 2\sqrt{6}+\frac{2}{5}\cdot 5\sqrt{6}-4\sqrt{6}$$
$$= \frac{6}{5}\sqrt{6}+2\sqrt{6}-4\sqrt{6}$$
$$= \left(\frac{6}{5}+2-4\right)\sqrt{6}$$
$$= -\frac{4}{5}\sqrt{6}, \text{ or } -\frac{4\sqrt{6}}{5}$$

83. $\sqrt{10}+\sqrt{50} = \sqrt{10}+\sqrt{10}\,\sqrt{5} = \sqrt{10}(1+\sqrt{5})$

$\sqrt{10}+\sqrt{50} = \sqrt{10}+\sqrt{25\cdot 2} = \sqrt{10}+5\sqrt{2}$

$\sqrt{10}+\sqrt{50} = \sqrt{2}\,\sqrt{5}+\sqrt{2}\,\sqrt{25} =$

$\sqrt{2}(\sqrt{5}+\sqrt{25}) = \sqrt{2}(\sqrt{5}+5)$, or $\sqrt{2}(5+\sqrt{5})$

All three are correct.

Exercise Set 16.5

1.
$$\sqrt{x} = 6$$
$$(\sqrt{x})^2 = 6^2 \quad \text{Squaring both sides}$$
$$x = 36 \quad \text{Simplifying}$$

Check:

$\sqrt{x} = 6$	
$\sqrt{36}$? 6	
6	TRUE

The solution is 36.

3. $\sqrt{x} = 4.3$

$(\sqrt{x})^2 = (4.3)^2$ Squaring both sides

$x = 18.49$ Simplifying

Check: $\sqrt{x} = 4.3$

$\sqrt{18.49} \; ? \; 4.3$

$4.3 \;|\;$ TRUE

The solution is 18.49.

5. $\sqrt{y+4} = 13$

$(\sqrt{y+4})^2 = 13^2$ Squaring both sides

$y + 4 = 169$ Simplifying

$y = 165$ Subtracting 4

Check: $\sqrt{y+4} = 13$

$\sqrt{165+4} \; ? \; 13$

$\sqrt{169}$

$13 \;|\;$ TRUE

The solution is 165.

7. $\sqrt{2x+4} = 25$

$(\sqrt{2x+4})^2 = 25^2$ Squaring both sides

$2x + 4 = 625$ Simplifying

$2x = 621$ Subtracting 4

$x = \frac{621}{2}$ Dividing by 2

Check: $\sqrt{2x+4} = 25$

$\sqrt{2 \cdot \frac{621}{2} + 4} \; ? \; 25$

$\sqrt{621+4}$

$\sqrt{625}$

$25 \;|\;$ TRUE

The solution is $\frac{621}{2}$.

9. $3 + \sqrt{x-1} = 5$

$\sqrt{x-1} = 2$ Subtracting 3

$(\sqrt{x-1})^2 = 2^2$ Squaring both sides

$x - 1 = 4$

$x = 5$

Check: $3 + \sqrt{x-1} = 5$

$3 + \sqrt{5-1} \; ? \; 5$

$3 + \sqrt{4}$

$3 + 2$

$5 \;|\;$ TRUE

The solution is 5.

11. $6 - 2\sqrt{3n} = 0$

$6 = 2\sqrt{3n}$ Adding $2\sqrt{3n}$

$6^2 = (2\sqrt{3n})^2$ Squaring both sides

$36 = 4 \cdot 3n$

$36 = 12n$

$3 = n$

Check: $6 - 2\sqrt{3n} = 0$

$6 - 2\sqrt{3 \cdot 3} \; ? \; 0$

$6 - 2 \cdot 3$

$6 - 6$

$0 \;|\;$ TRUE

The solution is 3.

13. $\sqrt{5x-7} = \sqrt{x+10}$

$(\sqrt{5x-7})^2 = (\sqrt{x+10})^2$ Squaring both sides

$5x - 7 = x + 10$

$4x = 17$

$x = \frac{17}{4}$

Check: $\sqrt{5x-7} = \sqrt{x+10}$

$\sqrt{5 \cdot \frac{17}{4} - 7} \; ? \; \sqrt{\frac{17}{4} + 10}$

$\sqrt{\frac{85}{4} - \frac{28}{4}} \;|\; \sqrt{\frac{57}{4}}$

$\sqrt{\frac{57}{4}} \;|\;$ TRUE

The solution is $\frac{17}{4}$.

15. $\sqrt{x} = -7$

There is no solution. The principal square root of x cannot be negative.

17. $\sqrt{2y+6} = \sqrt{2y-5}$

$(\sqrt{2y+6})^2 = (\sqrt{2y-5})^2$

$2y + 6 = 2y - 5$

$6 = -5$

The equation $6 = -5$ is false; there is no solution.

19. $x - 7 = \sqrt{x-5}$

$(x-7)^2 = (\sqrt{x-5})^2$

$x^2 - 14x + 49 = x - 5$

$x^2 - 15x + 54 = 0$

$(x-9)(x-6) = 0$

$x - 9 = 0 \quad \text{or} \quad x - 6 = 0$

$x = 9 \quad \text{or} \quad x = 6$

Check: $x - 7 = \sqrt{x-5}$

$$\begin{array}{c|c} 9-7 \; ? & \sqrt{9-5} \\ 2 & \sqrt{4} \\ & 2 \quad \text{TRUE} \end{array}$$

$x - 7 = \sqrt{x-5}$

$$\begin{array}{c|c} 6-7 & \sqrt{6-5} \\ -1 & \sqrt{1} \\ & 1 \quad \text{FALSE} \end{array}$$

The number 9 checks, but 6 does not. The solution is 9.

21.

$$\begin{aligned} x - 9 &= \sqrt{x-3} \\ (x-9)^2 &= (\sqrt{x-3})^2 \\ x^2 - 18x + 81 &= x - 3 \\ x^2 - 19x + 84 &= 0 \\ (x-12)(x-7) &= 0 \end{aligned}$$

$x - 12 = 0$ or $x - 7 = 0$

$x = 12$ or $x = 7$

Check: $x - 9 = \sqrt{x-3}$

$$\begin{array}{c|c} 12-9 \; ? & \sqrt{12-3} \\ 3 & \sqrt{9} \\ & 3 \quad \text{TRUE} \end{array}$$

$x - 9 = \sqrt{x-3}$

$$\begin{array}{c|c} 7-9 \; ? & \sqrt{7-3} \\ -2 & \sqrt{4} \\ & 2 \quad \text{FALSE} \end{array}$$

The number 12 checks, but 7 does not. The solution is 12.

23.

$$\begin{aligned} 2\sqrt{x-1} &= x - 1 \\ (2\sqrt{x-1})^2 &= (x-1)^2 \\ 4(x-1) &= x^2 - 2x + 1 \\ 4x - 4 &= x^2 - 2x + 1 \\ 0 &= x^2 - 6x + 5 \\ 0 &= (x-5)(x-1) \end{aligned}$$

$x - 5 = 0$ or $x - 1 = 0$

$x = 5$ or $x = 1$

Both numbers check. The solutions are 5 and 1.

25.

$$\begin{aligned} \sqrt{5x+21} &= x + 3 \\ (\sqrt{5x+21})^2 &= (x+3)^2 \\ 5x + 21 &= x^2 + 6x + 9 \\ 0 &= x^2 + x - 12 \\ 0 &= (x+4)(x-3) \end{aligned}$$

$x + 4 = 0$ or $x - 3 = 0$

$x = -4$ or $x = 3$

Check: $\sqrt{5x+21} = x + 3$

$$\begin{array}{c|c} \sqrt{5(-4)+21} \; ? & -4+3 \\ \sqrt{1} & -1 \\ 1 & \quad \text{FALSE} \end{array}$$

$\sqrt{5x+21} = x + 3$

$$\begin{array}{c|c} \sqrt{5 \cdot 3+21} \; ? & 3+3 \\ \sqrt{36} & 6 \\ 6 & \quad \text{TRUE} \end{array}$$

The number 3 checks, but -4 does not. The solution is 3.

27.

$$\begin{aligned} \sqrt{2x-1} + 2 &= x \\ \sqrt{2x-1} &= x - 2 \qquad \text{Isolating the radical} \\ (\sqrt{2x-1})^2 &= (x-2)^2 \\ 2x - 1 &= x^2 - 4x + 4 \\ 0 &= x^2 - 6x + 5 \\ 0 &= (x-5)(x-1) \end{aligned}$$

$x - 5 = 0$ or $x - 1 = 0$

$x = 5$ or $x = 1$

Check: $\sqrt{2x-1} + 2 = x$

$$\begin{array}{c|c} \sqrt{2 \cdot 5 - 1} + 2 \; ? & 5 \\ \sqrt{10-1} + 2 & \\ \sqrt{9} + 2 & \\ 3 + 2 & \\ 5 & \quad \text{TRUE} \end{array}$$

$\sqrt{2x-1} + 2 = x$

$$\begin{array}{c|c} \sqrt{2 \cdot 1 - 1} + 2 \; ? & 1 \\ \sqrt{2-1} + 2 & \\ \sqrt{1} + 2 & \\ 1 + 2 & \\ 3 & \quad \text{FALSE} \end{array}$$

The number 5 checks, but 1 does not. The solution is 5.

29.

$$\begin{aligned} \sqrt{x^2+6} - x + 3 &= 0 \\ \sqrt{x^2+6} &= x - 3 \qquad \text{Isolating the radical} \\ (\sqrt{x^2+6})^2 &= (x-3)^2 \\ x^2 + 6 &= x^2 - 6x + 9 \\ -3 &= -6x \qquad \text{Adding } -x^2 \text{ and } -9 \\ \frac{1}{2} &= x \end{aligned}$$

Check: $\sqrt{x^2+6}-x+3=0$

$$\sqrt{\left(\frac{1}{2}\right)^2+6}-\frac{1}{2}+3 \ ? \ 0$$
$$\sqrt{\frac{25}{4}}-\frac{1}{2}+3$$
$$\frac{5}{2}-\frac{1}{2}+3$$
$$5 \quad \text{FALSE}$$

The number $\frac{1}{2}$ does not check. There is no solution.

31. $\sqrt{x^2-4}-x=6$

$\sqrt{x^2-4}=x+6$ Isolating the radical

$(\sqrt{x^2-4})^2=(x+6)^2$

$x^2-4=x^2+12x+36$

$-40=12x$ Adding $-x^2$ and -36

$-\frac{40}{12}=x$

$-\frac{10}{3}=x$

The number $-\frac{10}{3}$ checks. It is the solution.

33. $\sqrt{(p+6)(p+1)}-2=p+1$

$\sqrt{(p+6)(p+1)}=p+3$ Isolating the radical

$\left(\sqrt{(p+6)(p+1)}\right)^2=(p+3)^2$

$(p+6)(p+1)=p^2+6p+9$

$p^2+7p+6=p^2+6p+9$

$p=3$

The number 3 checks. It is the solution.

35. $\sqrt{4x-10}=\sqrt{2-x}$

$(\sqrt{4x-10})^2=(\sqrt{2-x})^2$

$4x-10=2-x$

$5x=12$ Adding 10 and x

$x=\frac{12}{5}$

Check: $\sqrt{4x-10}=\sqrt{2-x}$

$$\sqrt{4\cdot\frac{12}{5}-10} \ ? \ \sqrt{2-\frac{12}{5}}$$
$$\sqrt{\frac{48}{5}-10} \quad \Big| \quad \sqrt{-\frac{2}{5}}$$

Since $\sqrt{-\frac{2}{5}}$ does not represent a real number, there is no solution that is a real number.

37. $\sqrt{x-5}=5-\sqrt{x}$

$(\sqrt{x-5})^2=(5-\sqrt{x})^2$ Squaring both sides

$x-5=25-10\sqrt{x}+x$

$-30=-10\sqrt{x}$ Isolating the radical

$3=\sqrt{x}$ Dividing by -10

$3^2=(\sqrt{x})^2$ Squaring both sides

$9=x$

The number 9 checks. It is the solution.

39. $\sqrt{y+8}-\sqrt{y}=2$

$\sqrt{y+8}=\sqrt{y}+2$ Isolating one radical

$(\sqrt{y+8})^2=(\sqrt{y}+2)^2$ Squaring both sides

$y+8=y+4\sqrt{y}+4$

$4=4\sqrt{y}$ Isolating the radical

$1=\sqrt{y}$ Dividing by 4

$1^2=(\sqrt{y})^2$

$1=y$

The number 1 checks. It is the solution.

41. $\sqrt{x-4}+\sqrt{x+1}=5$

$\sqrt{x-4}=5-\sqrt{x+1}$ Isolating one radical

$(\sqrt{x-4})^2=(5-\sqrt{x+1})^2$

$x-4=25-10\sqrt{x+1}+x+1$

$-30=-10\sqrt{x+1}$ Isolating the radical

$3=\sqrt{x+1}$ Dividing by -10

$3^2=(\sqrt{x+1})^2$

$9=x+1$

$8=x$

The number 8 checks. It is the solution.

43. $\sqrt{x}-1=\sqrt{x-31}$

$(\sqrt{x}-1)^2=(\sqrt{x-31})^2$

$x-2\sqrt{x}+1=x-31$

$-2\sqrt{x}=-32$

$\sqrt{x}=16$

$(\sqrt{x})^2=16^2$

$x=256$

The number 256 checks. It is the solution.

45. Substitute 27,000 for h in the equation.

$D=\sqrt{2h}$

$D=\sqrt{2\cdot 27,000}$

$D=\sqrt{54,000}$

$D\approx 232$

You can see about 232 mi to the horizon.

47. Substitute 180 for D in the equation and solve for h.

$$D = \sqrt{2h}$$
$$180 = \sqrt{2h}$$
$$180^2 = (\sqrt{2h})^2$$
$$32,400 = 2h$$
$$16,200 = h$$

A pilot must fly 16,200 ft above sea level in order to see a horizon that is 180 mi away.

49. For 65 mph, substitute 64 for S in the equation and solve for x.

$$S = 2\sqrt{5x}$$
$$65 = 2\sqrt{5x}$$
$$65^2 = (2\sqrt{5x})^2$$
$$4225 = 4 \cdot 5x$$
$$4225 = 20x$$
$$211.25 = x$$

At 65 mph, a car will skid 211.25 ft.

For 75 mph, substitute 75 for S in the equation and solve for x.

$$S = 2\sqrt{5x}$$
$$75 = 2\sqrt{5x}$$
$$75^2 = (2\sqrt{5x})^2$$
$$5625 = 4 \cdot 5x$$
$$5625 = 20x$$
$$\frac{5625}{20} = x$$
$$281.25 = x$$

At 75 mph, a car will skid 281.25 ft.

51. Parallel lines have the same slope and different y-intercepts.

53. The number c is a principal square root of a if $c^2 = a$ and c is either zero or positive.

55. The quotient rule asserts that when dividing with exponential notation, if the bases are the same, keep the base and subtract the exponent of the denominator from the exponent of the numerator.

57. The quotient rule for radicals asserts that for any nonnegative number A and any positive number B, $\dfrac{\sqrt{A}}{\sqrt{B}} = \sqrt{\dfrac{A}{B}}$.

59.

$$\sqrt{5x^2+5} = 5$$
$$(\sqrt{5x^2+5})^2 = 5^2$$
$$5x^2+5 = 25$$
$$5x^2-20 = 0$$
$$5(x^2-4) = 0$$
$$5(x+2)(x-2) = 0$$

$$x+2 = 0 \quad \textit{or} \quad x-2 = 0$$
$$x = -2 \quad \textit{or} \quad x = 2$$

Both numbers check, so the solutions are 2 and -2.

61.

$$4+\sqrt{19-x} = 6+\sqrt{4-x}$$
$$\sqrt{19-x} = 2+\sqrt{4-x} \quad \text{Isolating one radical}$$
$$(\sqrt{19-x})^2 = (2+\sqrt{4-x})^2$$
$$19-x = 4+4\sqrt{4-x}+(4-x)$$
$$19-x = 4\sqrt{4-x}+8-x$$
$$11 = 4\sqrt{4-x}$$
$$11^2 = (4\sqrt{4-x})^2$$
$$121 = 16(4-x)$$
$$121 = 64-16x$$
$$57 = -16x$$
$$-\frac{57}{16} = x$$

$-\dfrac{57}{16}$ checks, so it is the solution.

63.

$$\sqrt{x+3} = \frac{8}{\sqrt{x-9}}$$
$$(\sqrt{x+3})^2 = \left(\frac{8}{\sqrt{x-9}}\right)^2$$
$$x+3 = \frac{64}{x-9}$$
$$(x-9)(x+3) = 64 \quad \text{Multiplying by } x-9$$
$$x^2-6x-27 = 64$$
$$x^2-6x-91 = 0$$
$$(x-13)(x+7) = 0$$
$$x-13 = 0 \quad \textit{or} \quad x+7 = 0$$
$$x = 13 \quad \textit{or} \quad x = -7$$

The number 13 checks, but -7 does not. The solution is 13.

65.-67. Left to the student

Exercise Set 16.6

1.

$$a^2+b^2 = c^2$$
$$8^2+15^2 = c^2 \quad \text{Substituting}$$
$$64+225 = c^2$$
$$289 = c^2$$
$$\sqrt{289} = c$$
$$17 = c$$

3.

$$a^2+b^2 = c^2$$
$$4^2+4^2 = c^2 \quad \text{Substituting}$$
$$16+16 = c^2$$
$$32 = c^2$$
$$\sqrt{32} = c \quad \text{Exact answer}$$
$$5.657 \approx c \quad \text{Approximation}$$

5.
$$\begin{aligned} a^2+b^2 &= c^2 \\ 5^2+b^2 &= 13^2 \\ 25+b^2 &= 169 \\ b^2 &= 144 \\ b &= 12 \end{aligned}$$

7.
$$\begin{aligned} a^2+b^2 &= c^2 \\ (4\sqrt{3})^2+b^2 &= 8^2 \\ 16\cdot 3+b^2 &= 64 \\ 48+b^2 &= 64 \\ b^2 &= 16 \\ b &= 4 \end{aligned}$$

9.
$$\begin{aligned} a^2+b^2 &= c^2 \\ 10^2+24^2 &= c^2 \\ 100+576 &= c^2 \\ 676 &= c^2 \\ 26 &= c \end{aligned}$$

11.
$$\begin{aligned} a^2+b^2 &= c^2 \\ 9^2+b^2 &= 15^2 \\ 81+b^2 &= 225 \\ b^2 &= 144 \\ b &= 12 \end{aligned}$$

13.
$$\begin{aligned} a^2+b^2 &= c^2 \\ a^2+1^2 &= (\sqrt{5})^2 \\ a^2+1 &= 5 \\ a^2 &= 4 \\ a &= 2 \end{aligned}$$

15.
$$\begin{aligned} a^2+b^2 &= c^2 \\ 1^2+b^2 &= (\sqrt{3})^2 \\ 1+b^2 &= 3 \\ b^2 &= 2 \\ b &= \sqrt{2} \quad \text{Exact answer} \\ b &\approx 1.414 \quad \text{Approximation} \end{aligned}$$

17.
$$\begin{aligned} a^2+b^2 &= c^2 \\ a^2+(5\sqrt{3})^2 &= 10^2 \\ a^2+25\cdot 3 &= 100 \\ a^2+75 &= 100 \\ a^2 &= 25 \\ a &= 5 \end{aligned}$$

19.
$$\begin{aligned} a^2+b^2 &= c^2 \\ (\sqrt{2})^2+(\sqrt{7})^2 &= c^2 \\ 2+7 &= c^2 \\ 9 &= c^2 \\ 3 &= c \end{aligned}$$

21. We use the drawing in the text, letting d represent the distance from P to R.

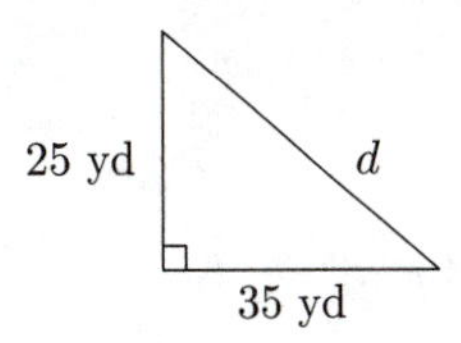

We know that $25^2+35^2=d^2$. We solve this equation.
$$\begin{aligned} 25^2+35^2 &= d^2 \\ 625+1225 &= d^2 \\ 1850 &= d^2 \\ \sqrt{1850} &= d \\ 43.012 &\approx d \end{aligned}$$

The distance from P to R is $\sqrt{1850}$ yd, or approximately 43.012 yd.

23. We use the drawing in the text, letting d represent the distance from b to c.

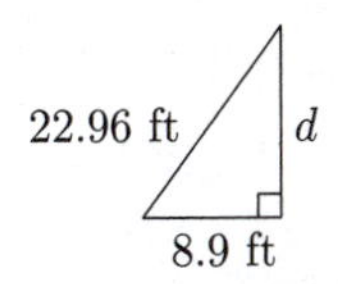

We know that $8.9^2+d^2=22.96^2$. We solve this equation.
$$\begin{aligned} 8.9^2+d^2 &= 22.96^2 \\ 79.21+d^2 &= 527.1616 \\ d^2 &= 447.9516 \\ d &= \sqrt{447.9516} \approx 21.2 \end{aligned}$$

The distance from b to c is approximately 21.2 ft.

25. We make a drawing, letting h represent the height of the top of the ladder.

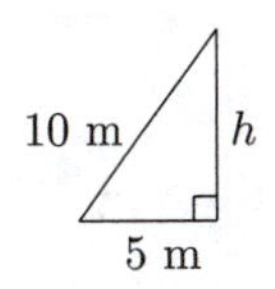

We know that $5^2+h^2=10^2$. We solve this equation.
$$\begin{aligned} 5^2+h^2 &= 10^2 \\ 25+h^2 &= 100 \\ h^2 &= 75 \\ h &= \sqrt{75} \approx 8.660 \end{aligned}$$

The top of the ladder is $\sqrt{75}$ m high, or approximately 8.660 m high.

27. We make a drawing. Let d represent the length of a diagonal of the field.

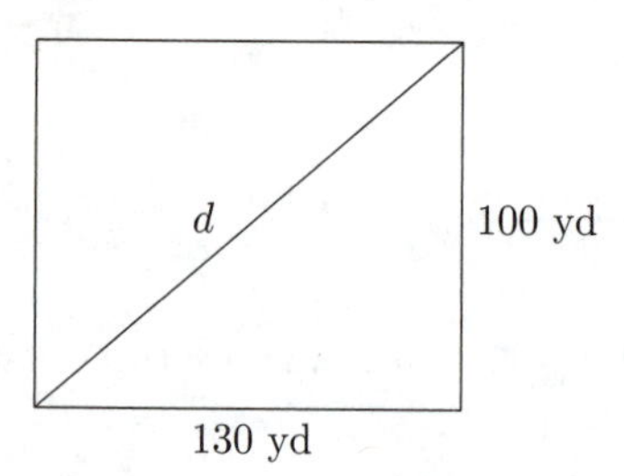

We know that $130^2 + 100^2 = d^2$. We solve this equation.

$$\begin{aligned} 130^2 + 100^2 &= d^2 \\ 16,900 + 10,000 &= d^2 \\ 26,900 &= d^2 \\ \sqrt{26,900} &= d \\ 164.012 &\approx d \end{aligned}$$

The length of a diagonal of the field is $\sqrt{26,900}$ yd, or approximately 164.012 yd.

29. $5x + 7 = 8y,$

$3x = 8y - 4$

$5x - 8y = -7 \quad (1)$ Rewriting

$3x - 8y = -4 \quad (2)$ the equations

We multiply Equation (2) by -1 and add.

$$\begin{aligned} 5x - 8y &= -7 \\ -3x + 8y &= 4 \\ \hline 2x \qquad &= -3 \\ x &= -\frac{3}{2} \end{aligned}$$

Substitute $-\frac{3}{2}$ for x in Equation (1) and solve for y.

$$\begin{aligned} 5x - 8y &= -7 \\ 5\left(-\frac{3}{2}\right) - 8y &= -7 \\ -\frac{15}{2} - 8y &= -7 \\ -8y &= \frac{1}{2} \\ y &= -\frac{1}{16} \end{aligned}$$

The ordered pair $\left(-\frac{3}{2}, -\frac{1}{16}\right)$ checks. It is the solution.

31. $3x - 4y = -11 \quad (1)$

$5x + 6y = 12 \quad (2)$

We multiply Equation (1) by 3 and Equation (2) by 2, and then we add.

$$\begin{aligned} 9x - 12y &= -33 \\ 10x + 12y &= 24 \\ \hline 19x \qquad &= -9 \\ x &= -\frac{9}{19} \end{aligned}$$

Substitute $-\frac{9}{19}$ for x in Equation (2) and solve for y.

$$\begin{aligned} 5x + 6y &= 12 \\ 5\left(-\frac{9}{19}\right) + 6y &= 12 \\ -\frac{45}{19} + 6y &= 12 \\ 6y &= \frac{273}{19} \quad \text{Adding } \frac{45}{19} \\ y &= \frac{273}{6 \cdot 19} \quad \text{Dividing by 6} \\ y &= \frac{91}{38} \quad \text{Simplifying} \end{aligned}$$

The ordered pair $\left(-\frac{9}{19}, \frac{91}{38}\right)$ checks. It is the solution.

33. Write the equation in the slope-intercept form.

$$\begin{aligned} 4 - x &= 3y \\ \frac{1}{3}(4 - x) &= y \\ \frac{4}{3} - \frac{1}{3}x &= y, \text{ or} \\ y &= -\frac{1}{3}x + \frac{4}{3} \end{aligned}$$

The slope is $-\frac{1}{3}$.

35.

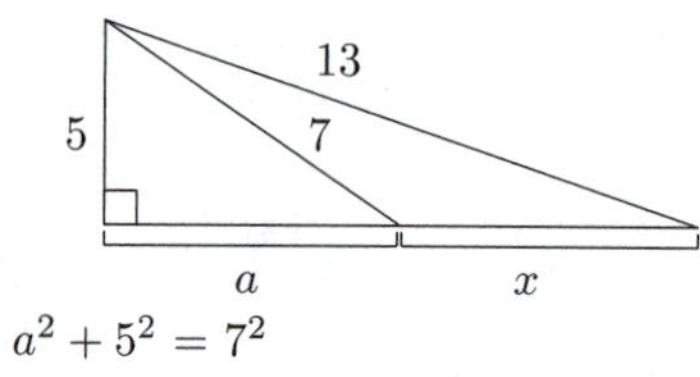

$$\begin{aligned} a^2 + 5^2 &= 7^2 \\ a^2 + 25 &= 49 \\ a^2 &= 24 \\ a &= \sqrt{24}, \text{ or } 2\sqrt{6} \end{aligned}$$

$$\begin{aligned} (a + x)^2 + 5^2 &= 13^2 \\ (2\sqrt{6} + x)^2 + 5^2 &= 13^2 \quad \text{Substituting } 2\sqrt{6} \text{ for } a \\ (2\sqrt{6} + x)^2 + 25 &= 169 \\ (2\sqrt{6} + x)^2 &= 144 \\ 2\sqrt{6} + x &= 12 \quad \text{Taking the principal square root} \\ x &= 12 - 2\sqrt{6} \\ x &\approx 7.101 \end{aligned}$$

Chapter 16 Concept Reinforcement

1. True; see page 1208 in the text
2. True; see page 1185 in the text.
3. True; see page 1207 in the text.
4. The statement is false. For example, the equation $2^2 = (-2)^2$ is true, but $2 = -2$ is not true.

Chapter 16 Important Concepts

1. The radicand is the expression under the radical, $y^2 - 3$.

2. a) $-\sqrt{-(-3)} = -\sqrt{3}$

   Yes; this expression represents a real number, because the radicand is positive.

   b) $\sqrt{-200}$

   No; the radicand is negative.

3. $\sqrt{1200y^2} = \sqrt{400 \cdot 3 \cdot y^2} = \sqrt{400}\sqrt{y^2}\sqrt{3} = 20y\sqrt{3}$

4. $\sqrt{175a^{12}b^9} = \sqrt{25 \cdot 7 \cdot a^{12} \cdot b^8 \cdot b} =$
   $\sqrt{25}\sqrt{a^{12}}\sqrt{b^8}\sqrt{7b} = 5a^6b^4\sqrt{7b}$

5. $\sqrt{8x^3y}\sqrt{12x^4y^3} = \sqrt{8x^3y \cdot 12x^4y^3} =$
   $\sqrt{4 \cdot 2 \cdot 4 \cdot 3 \cdot x^2 \cdot x \cdot x^4 \cdot y \cdot y^2 \cdot y} =$
   $\sqrt{4}\sqrt{4}\sqrt{x^2}\sqrt{x^4}\sqrt{y \cdot y}\sqrt{y^2}\sqrt{2 \cdot 3 \cdot x} =$
   $2 \cdot 2 \cdot x \cdot x^2 \cdot y \cdot y\sqrt{6x} = 4x^3y^2\sqrt{6x}$

6. $\dfrac{\sqrt{15b^7}}{\sqrt{5b^4}} = \sqrt{\dfrac{15b^7}{5b^4}} = \sqrt{3b^3} = \sqrt{3 \cdot b^2 \cdot b} =$
   $\sqrt{b^2}\sqrt{3 \cdot b} = b\sqrt{3b}$

7. $\sqrt{\dfrac{50}{162}} = \sqrt{\dfrac{2 \cdot 25}{2 \cdot 81}} = \sqrt{\dfrac{25}{81}} = \dfrac{\sqrt{25}}{\sqrt{81}} = \dfrac{5}{9}$

8. $\dfrac{2a}{\sqrt{50}} = \dfrac{2a}{\sqrt{50}} \cdot \dfrac{\sqrt{2}}{\sqrt{2}} = \dfrac{2a\sqrt{2}}{\sqrt{100}} = \dfrac{2a\sqrt{2}}{10} =$
   $\dfrac{2a\sqrt{2}}{2 \cdot 5} = \dfrac{2}{2} \cdot \dfrac{a\sqrt{2}}{5} = \dfrac{a\sqrt{2}}{5}$

9. $\sqrt{x^3 - x^2} + \sqrt{36x - 36} = \sqrt{x^2(x-1)} + \sqrt{36(x-1)} =$
   $\sqrt{x^2}\sqrt{x-1} + \sqrt{36}\sqrt{x-1} = x\sqrt{x-1} + 6\sqrt{x-1} =$
   $(x+6)\sqrt{x-1}$

10.
$$\begin{aligned}
&(\sqrt{13} - \sqrt{2})(\sqrt{13} + 2\sqrt{2})\\
&= \sqrt{13} \cdot \sqrt{13} + \sqrt{13} \cdot 2\sqrt{2} - \sqrt{2} \cdot \sqrt{13} - \sqrt{2} \cdot 2\sqrt{2}\\
&= 13 + 2\sqrt{26} - \sqrt{26} - 4\\
&= 9 + \sqrt{26}
\end{aligned}$$

11.
$$\begin{aligned}
\frac{5-\sqrt{2}}{9+\sqrt{2}} &= \frac{5-\sqrt{2}}{9+\sqrt{2}} \cdot \frac{9-\sqrt{2}}{9-\sqrt{2}}\\
&= \frac{5 \cdot 9 - 5 \cdot \sqrt{2} - \sqrt{2} \cdot 9 + \sqrt{2} \cdot \sqrt{2}}{9^2 - (\sqrt{2})^2}\\
&= \frac{45 - 5\sqrt{2} - 9\sqrt{2} + 2}{81 - 2}\\
&= \frac{47 - 14\sqrt{2}}{79}
\end{aligned}$$

12.
$$\begin{aligned}
x - 4 &= \sqrt{x-2}\\
(x-4)^2 &= (\sqrt{x-2})^2\\
x^2 - 8x + 16 &= x - 2\\
x^2 - 9x + 18 &= 0\\
(x-3)(x-6) &= 0\\
x - 3 = 0 \quad &or \quad x - 6 = 0\\
x = 3 \quad &or \quad x = 6
\end{aligned}$$

The number 3 does not check, but 6 does. The solution is 6.

13.
$$\begin{aligned}
12 - \sqrt{x} &= \sqrt{90 - x}\\
(12 - \sqrt{x})^2 &= (\sqrt{90-x})^2\\
144 - 24\sqrt{x} + x &= 90 - x\\
-24\sqrt{x} &= -54 - 2x\\
12\sqrt{x} &= 27 + x \qquad \text{Dividing by } -2\\
(12\sqrt{x})^2 &= (27+x)^2\\
144x &= 729 + 54x + x^2\\
0 &= x^2 - 90x + 729\\
0 &= (x-81)(x-9)\\
x - 81 = 0 \quad &or \quad x - 9 = 0\\
x = 81 \quad &or \quad x = 9
\end{aligned}$$

Both numbers check. The solutions are 81 and 9.

14.
$$\begin{aligned}
a^2 + b^2 &= c^2\\
a^2 + 32^2 &= 41^2\\
a^2 + 1024 &= 1681\\
a^2 &= 657\\
a &= \sqrt{657} \approx 25.632
\end{aligned}$$

Chapter 16 Review Exercises

1. The square roots of 64 are 8 and -8, because $8^2 = 64$ and $(-8)^2 = 64$.

2. The square roots of 400 are 20 and -20, because $20^2 = 400$ and $(-20)^2 = 400$.

3. $\sqrt{36} = 6$, taking the principal square root

4. $\sqrt{169} = 13$, so $-\sqrt{169} = -13$.

5. 1.732

6. 9.950

7. $\sqrt{320.12} \approx 17.892$, so $-\sqrt{320.12} \approx -17.892$.

8. 0.742

9. $\sqrt{\dfrac{47.3}{11.2}} \approx 2.055$, so $-\sqrt{\dfrac{47.3}{11.2}} \approx -2.055$.

10. 394.648

11. The radicand is the expression under the radical, $x^2 + 4$.

12. The radicand is the expression under the radical, $5ab^3$.

13. The radicand is the expression under the radical, $4-x$.

14. The radicand is the expression under the radical, $\dfrac{2}{y-7}$.

15. Yes, because the radicand is nonnegative

16. No, because the radicand is negative

17. No, because the radicand is negative

18. $\sqrt{(-3)(-27)} = \sqrt{81}$

Yes, because the radicand is positive.

19. $\sqrt{m^2} = m$

20. $\sqrt{(x-4)^2} = x-4$

21. $\sqrt{(16x)^2} = 16x$

22. $\sqrt{4p^2-12p+9} = \sqrt{(2p-3)^2} = 2p-3$

23. $\sqrt{48} = \sqrt{16\cdot 3} = \sqrt{16}\sqrt{3} = 4\sqrt{3}$

24. $\sqrt{32t^2} = \sqrt{16\cdot 2\cdot t^2} = \sqrt{16}\sqrt{t^2}\sqrt{2} = 4t\sqrt{2}$

25. $\sqrt{t^2-49} = \sqrt{(t+7)(t-7)} = \sqrt{t+7}\sqrt{t-7}$

26. $\sqrt{x^2+16x+64} = \sqrt{(x+8)^2} = x+8$

27. $\sqrt{x^8} = \sqrt{(x^4)^2} = x^4$

28. $\sqrt{75a^7} = \sqrt{25\cdot 3\cdot a^6\cdot a} = \sqrt{25}\sqrt{a^6}\sqrt{3a} = 5a^3\sqrt{3a}$

29. $\sqrt{3}\sqrt{7} = \sqrt{3\cdot 7} = \sqrt{21}$

30. $\sqrt{x-3}\sqrt{x+3} = \sqrt{(x-3)(x+3)} = \sqrt{x^2-9}$

31. $\sqrt{6}\sqrt{10} = \sqrt{6\cdot 10} = \sqrt{2\cdot 3\cdot 2\cdot 5} = \sqrt{2\cdot 2}\sqrt{3\cdot 5} = 2\sqrt{15}$

32. $\sqrt{5x}\sqrt{8x} = \sqrt{5x\cdot 8x} = \sqrt{5\cdot x\cdot 4\cdot 2\cdot x} =$
$\sqrt{4}\sqrt{x\cdot x}\sqrt{5\cdot 2} = 2x\sqrt{10}$

33. $\sqrt{5x}\sqrt{10xy^2} = \sqrt{5x\cdot 10xy^2} = \sqrt{5\cdot x\cdot 2\cdot 5\cdot x\cdot y^2} =$
$\sqrt{5\cdot 5}\sqrt{x\cdot x}\sqrt{y^2}\sqrt{2} = 5xy\sqrt{2}$

34. $\sqrt{20a^3b}\sqrt{5a^2b^2} = \sqrt{20a^3b\cdot 5a^2b^2} =$
$\sqrt{4\cdot 5\cdot a^2\cdot a\cdot b\cdot 5\cdot a^2\cdot b^2} =$
$\sqrt{4}\sqrt{5\cdot 5}\sqrt{a^2}\sqrt{a^2}\sqrt{b^2}\sqrt{ab} = 2\cdot 5\cdot a\cdot a\cdot b\sqrt{a\cdot b} =$
$10a^2b\sqrt{ab}$

35. $\sqrt{\dfrac{25}{64}} = \dfrac{\sqrt{25}}{\sqrt{64}} = \dfrac{5}{8}$

36. $\sqrt{\dfrac{49}{t^2}} = \dfrac{\sqrt{49}}{\sqrt{t^2}} = \dfrac{7}{t}$

37. $\dfrac{\sqrt{2c^9}}{\sqrt{32c}} = \sqrt{\dfrac{2c^9}{32c}} = \sqrt{\dfrac{c^8}{16}} = \dfrac{\sqrt{c^8}}{\sqrt{16}} = \dfrac{c^4}{4}$

38. $\sqrt{\dfrac{1}{2}} = \sqrt{\dfrac{1}{2}\cdot\dfrac{2}{2}} = \sqrt{\dfrac{2}{4}} = \dfrac{\sqrt{2}}{\sqrt{4}} = \dfrac{\sqrt{2}}{2}$

39. $\dfrac{\sqrt{x^3}}{\sqrt{15}} = \dfrac{\sqrt{x^3}}{\sqrt{15}}\cdot\dfrac{\sqrt{15}}{\sqrt{15}} = \dfrac{\sqrt{15\cdot x^2\cdot x}}{15} = \dfrac{x\sqrt{15x}}{15}$

40. $\sqrt{\dfrac{5}{y}} = \sqrt{\dfrac{5}{y}\cdot\dfrac{y}{y}} = \sqrt{\dfrac{5y}{y^2}} = \dfrac{\sqrt{5y}}{\sqrt{y^2}} = \dfrac{\sqrt{5y}}{y}$

41. $\dfrac{\sqrt{b^9}}{\sqrt{ab^2}} = \sqrt{\dfrac{b^9}{ab^2}} = \sqrt{\dfrac{b^7}{a}} = \sqrt{\dfrac{b^7}{a}\cdot\dfrac{a}{a}} = \sqrt{\dfrac{a\cdot b^6\cdot b}{a^2}} =$
$\dfrac{\sqrt{a\cdot b^6\cdot b}}{\sqrt{a^2}} = \dfrac{b^3\sqrt{ab}}{a}$

42. $\dfrac{\sqrt{27}}{\sqrt{45}} = \sqrt{\dfrac{27}{45}} = \sqrt{\dfrac{3}{5}} = \sqrt{\dfrac{3}{5}\cdot\dfrac{5}{5}} = \sqrt{\dfrac{15}{25}} = \dfrac{\sqrt{15}}{\sqrt{25}} = \dfrac{\sqrt{15}}{5}$

43. $\dfrac{\sqrt{45x^2y}}{\sqrt{54y}} = \sqrt{\dfrac{45x^2y}{54y}} = \sqrt{\dfrac{5x^2}{6}} = \sqrt{\dfrac{5x^2}{6}\cdot\dfrac{6}{6}} = \sqrt{\dfrac{30x^2}{36}} =$
$\dfrac{\sqrt{30x^2}}{\sqrt{36}} = \dfrac{x\sqrt{30}}{6}$

44. $10\sqrt{5}+3\sqrt{5} = (10+3)\sqrt{5} = 13\sqrt{5}$

45.
$$\begin{aligned}\sqrt{80}-\sqrt{45} &= \sqrt{16\cdot 5}-\sqrt{9\cdot 5}\\ &= 4\sqrt{5}-3\sqrt{5}\\ &= (4-3)\sqrt{5}\\ &= \sqrt{5}\end{aligned}$$

46.
$$\begin{aligned}3\sqrt{2}-5\sqrt{\frac{1}{2}} &= 3\sqrt{2}-5\sqrt{\frac{1}{2}\cdot\frac{2}{2}}\\ &= 3\sqrt{2}-5\cdot\frac{\sqrt{2}}{2}\\ &= 3\sqrt{2}-\frac{5}{2}\sqrt{2}\\ &= \left(3-\frac{5}{2}\right)\sqrt{2}\\ &= \frac{1}{2}\sqrt{2},\text{ or }\frac{\sqrt{2}}{2}\end{aligned}$$

47. $(2+\sqrt{3})^2 = 2^2+2\cdot 2\cdot\sqrt{3}+(\sqrt{3})^2 = 4+4\sqrt{3}+3 =$
$7+4\sqrt{3}$

48. $(2+\sqrt{3})(2-\sqrt{3}) = 2^2-(\sqrt{3})^2 = 4-3 = 1$

49.
$$\begin{aligned}\frac{4}{2+\sqrt{3}} &= \frac{4}{2+\sqrt{3}}\cdot\frac{2-\sqrt{3}}{2-\sqrt{3}}\\ &= \frac{4(2-\sqrt{3})}{(2+\sqrt{3})(2-\sqrt{3})}\\ &= \frac{8-4\sqrt{3}}{2^2-(\sqrt{3})^2} = \frac{8-4\sqrt{3}}{4-3}\\ &= \frac{8-4\sqrt{3}}{1} = 8-4\sqrt{3}\end{aligned}$$

50.
$$\begin{aligned}\sqrt{x-3} &= 7\\ (\sqrt{x-3})^2 &= 7^2\\ x-3 &= 49\\ x &= 52\end{aligned}$$

The number 52 checks. It is the solution.

51.
$$\begin{aligned}\sqrt{5x+3} &= \sqrt{2x-1}\\(\sqrt{5x+3})^2 &= (\sqrt{2x-1})^2\\5x+3 &= 2x-1\\3x+3 &= -1\\3x &= -4\\x &= -\frac{4}{3}\end{aligned}$$

The number $-\frac{4}{3}$ does not check. There is no solution.

52.
$$\begin{aligned}1+x &= \sqrt{1+5x}\\(1+x)^2 &= (\sqrt{1+5x})^2\\1+2x+x^2 &= 1+5x\\x^2-3x &= 0\\x(x-3) &= 0\end{aligned}$$

$x=0$ *or* $x-3=0$

$x=0$ *or* $x=3$

Both numbers check. The solutions are 0 and 3.

53.
$$\begin{aligned}\sqrt{x} &= \sqrt{x-5}+1\\(\sqrt{x})^2 &= (\sqrt{x-5}+1)^2\\x &= x-5+2\sqrt{x-5}+1\\x &= x-4+2\sqrt{x-5}\\4 &= 2\sqrt{x-5}\\2 &= \sqrt{x-5}\\2^2 &= (\sqrt{x-5})^2\\4 &= x-5\\9 &= x\end{aligned}$$

The number 9 checks. It is the solution.

54. a) We substitute 200 for L in the formula.

$$r = 2\sqrt{5\cdot 200} = 2\sqrt{1000} \approx 63 \text{ mph}$$

b) We substitute 90 for r and solve for L.

$$\begin{aligned}90 &= 2\sqrt{5L}\\90^2 &= (2\sqrt{5L})^2\\8100 &= 4\cdot 5L\\8100 &= 20L\\405 \text{ ft} &= L\end{aligned}$$

55.
$$\begin{aligned}a^2+b^2 &= c^2\\15^2+b^2 &= 25^2\\225+b^2 &= 625\\b^2 &= 400\\b &= 20\end{aligned}$$

56.
$$\begin{aligned}a^2+b^2 &= c^2\\1^2+(\sqrt{2})^2 &= c^2\\1+2 &= c^2\\3 &= c^2\\\sqrt{3} &= c \quad \text{Exact answer}\\1.732 &\approx c \quad \text{Approximation}\end{aligned}$$

57. First we subtract to find the vertical distance of the descent:

$$30,000 \text{ ft} - 20,000 \text{ ft} = 10,000 \text{ ft}$$

Let $c =$ the distance the plane travels during the descent. This is labeled "?" in the drawing in the text. We know that $10,000^2 + 50,000^2 = c^2$. We solve this equation.

$$\begin{aligned}10,000^2+50,000^2 &= c^2\\100,000,000+2,500,000,000 &= c^2\\2,600,000,000 &= c^2\\\sqrt{2,600,00,000} \text{ ft} &= c\\50,990 \text{ ft} &\approx c\end{aligned}$$

58. Let $d =$ the distance each brace reaches vertically. From the drawing in the text we see that $d^2+12^2 = 15^2$. We solve this equation.

$$\begin{aligned}d^2+12^2 &= 15^2\\d^2+144 &= 225\\d^2 &= 81\\d &= 9 \text{ ft}\end{aligned}$$

59.
$$\begin{aligned}x-2 &= \sqrt{4-9x}\\(x-2)^2 &= (\sqrt{4-9x})^2\\x^2-4x+4 &= 4-9x\\x^2+5x &= 0\\x(x+5) &= 0\end{aligned}$$

$x=0$ *or* $x+5=0$

$x=0$ *or* $x=-5$

Neither number checks. The equation has no solution. The correct choice is answer B.

60.
$$\begin{aligned}&(2\sqrt{7}+\sqrt{2})(\sqrt{7}-\sqrt{2})\\&= 2\sqrt{7}\cdot\sqrt{7}-2\sqrt{7}\cdot\sqrt{2}+\sqrt{2}\cdot\sqrt{7}-\sqrt{2}\sqrt{2}\\&= 2\cdot 7-2\sqrt{14}+\sqrt{14}-2\\&= 14-\sqrt{14}-2\\&= 12-\sqrt{14}\end{aligned}$$

The correct choice is answer C.

61. After $\frac{1}{2}$ hr, the car traveling east at 50 mph has traveled $\frac{1}{2}\cdot 50$, or 25 mi and the car traveling south at 60 mph has traveled $\frac{1}{2}\cdot 60$, or 30 mi. These distances are the legs of a right triangle. The length of the hypotenuse of the triangle is the distance that separates the cars. Let $d =$ this distance. We know that $25^2+30^2 = d^2$. We solve this equation.

$$\begin{aligned}25^2+30^2 &= d^2\\625+900 &= d^2\\1525 &= d^2\\\sqrt{1525} \text{ mi} &= d\\39.051 \text{ mi} &\approx d\end{aligned}$$

62.
$$A = \sqrt{a^2 + b^2}$$
$$A^2 = (\sqrt{a^2+b^2})^2$$
$$A^2 = a^2 + b^2$$
$$A^2 - a^2 = b^2$$

If $b^2 = A^2 - a^2$, then $b = \sqrt{A^2 - a^2}$ or $b = -\sqrt{A^2 - a^2}$, so we have $b = \pm\sqrt{A^2 - a^2}$.

63. Using the drawing in the text, let a = the hypotenuse of the triangle with legs 4 and x and let b = the hypotenuse of the triangle with legs 9 and x. Then we have

$$4^2 + x^2 = a^2, \text{ or } 16 + x^2 = a^2$$

and

$$9^2 + x^2 = b^2, \text{ or } 81 + x^2 = b^2.$$

Note that a and b are also legs of the large triangle with hypotenuse $4+9$, or 13. Then we have

$$a^2 + b^2 = 13^2, \text{ or } a^2 + b^2 = 169.$$

Adding the two equations containing x, we have

$$16 + x^2 = a^2$$
$$81 + x^2 = b^2$$
$$\overline{97 + 2x^2 = a^2 + b^2.}$$

We substitute $97+2x^2$ for a^2+b^2 in the equation pertaining to the large triangle.

$$a^2 + b^2 = 169$$
$$97 + 2x^2 = 169$$
$$2x^2 = 72$$
$$x^2 = 36$$
$$x = 6$$

Chapter 16 Discussion and Writing Exercises

1. It is important for the signs to differ to ensure that the product of the conjugates will be free of radicals.

2. Since $\sqrt{11-2x}$ cannot be negative, the statement $\sqrt{11-2x} = -3$ cannot be true for any value of x, including -1.

3. We often use the rules for manipulating exponents "in reverse" when simplifying radical expressions. For example, we might write x^5 as $x^4 \cdot x$ or y^6 and $(y^3)^2$.

4. No; consider the clapboard's height above ground level to be one leg of a right triangle. Then the length of the ladder is the hypotenuse of that triangle. Since the length of the hypotenuse must be greater than the length of a leg, a 28-ft ladder cannot be used to repair a clapboard that is 28 ft above ground level.

5. The square of a number is equal to the square of its opposite. Thus, while squaring both sides of a radical equation allows us to find the solutions of the original equation, this procedure can also introduce numbers that are not solutions of the original equation.

6. a) $\sqrt{5x^2} = \sqrt{5}\sqrt{x^2} = \sqrt{5}\cdot|x| = |x|\sqrt{5}$. The given statement is correct.

b) Let $b = 3$. Then $\sqrt{b^2-4} = \sqrt{3^2-4} = \sqrt{9-4} = \sqrt{5}$, but $b - 2 = 3 - 2 = 1$. The given statement is false.

c) Let $x = 3$. Then $\sqrt{x^2+16} = \sqrt{3^2+16} = \sqrt{9+16} = \sqrt{25} = 5$, but $x + 4 = 3 + 4 = 7$. The given statement is false.

Chapter 16 Test

1. The square roots of 81 are 9 and -9, because $9^2 = 81$ and $(-9)^2 = 81$.

2. $\sqrt{64} = 8$, taking the principal square root

3. $\sqrt{25} = 5$, so $-\sqrt{25} = -5$.

4. $\sqrt{116} \approx 10.770$

5. $\sqrt{87.4} \approx 9.349$, so $-\sqrt{87.4} \approx -9.349$.

6. $4\sqrt{5\cdot 6} \approx 21.909$

7. The radicand is the expression under the radical, $4 - y^3$.

8. Yes, because the radicand is nonnegative

9. No, because the radicand is negative

10. $\sqrt{a^2} = a$

11. $\sqrt{36y^2} = \sqrt{(6y)^2} = 6y$

12. $\sqrt{5}\sqrt{6} = \sqrt{5\cdot 6} = \sqrt{30}$

13. $\sqrt{x-8}\sqrt{x+8} = \sqrt{(x-8)(x+8)} = \sqrt{x^2-64}$

14. $\sqrt{27} = \sqrt{9\cdot 3} = \sqrt{9}\sqrt{3} = 3\sqrt{3}$

15. $\sqrt{25x-25} = \sqrt{25(x-1)} = \sqrt{25}\sqrt{x-1} = 5\sqrt{x-1}$

16. $\sqrt{t^5} = \sqrt{t^4\cdot t} = \sqrt{t^4}\sqrt{t} = \sqrt{(t^2)^2}\sqrt{t} = t^2\sqrt{t}$

17. $\sqrt{5}\sqrt{10} = \sqrt{5\cdot 10} = \sqrt{5\cdot 2\cdot 5} = \sqrt{5\cdot 5}\sqrt{2} = 5\sqrt{2}$

18. $\sqrt{3ab}\sqrt{6ab^3} = \sqrt{3ab\cdot 6ab^3} = \sqrt{3\cdot a\cdot b\cdot 2\cdot 3\cdot a\cdot b^2\cdot b} = \sqrt{3\cdot 3}\sqrt{a\cdot a}\sqrt{b\cdot b}\sqrt{b^2}\sqrt{2} = 3\cdot a\cdot b\cdot b\sqrt{2} = 3ab^2\sqrt{2}$

19. $\sqrt{\frac{27}{12}} = \sqrt{\frac{9}{4}} = \frac{\sqrt{9}}{\sqrt{4}} = \frac{3}{2}$

20. $\sqrt{\frac{144}{a^2}} = \frac{\sqrt{144}}{\sqrt{a^2}} = \frac{12}{a}$

21. $\sqrt{\frac{2}{5}} = \sqrt{\frac{2}{5}\cdot\frac{5}{5}} = \sqrt{\frac{10}{25}} = \frac{\sqrt{10}}{\sqrt{25}} = \frac{\sqrt{10}}{5}$

22. $\sqrt{\frac{2x}{y}} = \sqrt{\frac{2x}{y}\cdot\frac{y}{y}} = \sqrt{\frac{2xy}{y^2}} = \frac{\sqrt{2xy}}{y}$

23. $\frac{\sqrt{27}}{\sqrt{32}} = \sqrt{\frac{27}{32}} = \sqrt{\frac{27}{32}\cdot\frac{2}{2}} = \sqrt{\frac{54}{64}} = \frac{\sqrt{54}}{\sqrt{64}} = \frac{\sqrt{9\cdot 6}}{8} = \frac{3\sqrt{6}}{8}$

24. $\dfrac{\sqrt{35x}}{\sqrt{80xy^2}} = \sqrt{\dfrac{35x}{80xy^2}} = \sqrt{\dfrac{7}{16y^2}} = \dfrac{\sqrt{7}}{\sqrt{16y^2}} = \dfrac{\sqrt{7}}{4y}$

25. $3\sqrt{18} - 5\sqrt{18} = (3-5)\sqrt{18} = -2\sqrt{18} = -2\sqrt{9 \cdot 2} =$
$-2 \cdot 3\sqrt{2} = -6\sqrt{2}$

26. $\sqrt{5} + \sqrt{\dfrac{1}{5}} = \sqrt{5} + \sqrt{\dfrac{1}{5} \cdot \dfrac{5}{5}} = \sqrt{5} + \sqrt{\dfrac{5}{25}} =$
$\sqrt{5} + \dfrac{\sqrt{5}}{5} = \left(1 + \dfrac{1}{5}\right)\sqrt{5} = \dfrac{6}{5}\sqrt{5}$, or $\dfrac{6\sqrt{5}}{5}$

27. $(4 - \sqrt{5})^2 = 4^2 - 2 \cdot 4 \cdot \sqrt{5} + (\sqrt{5})^2 = 16 - 8\sqrt{5} + 5 =$
$21 - 8\sqrt{5}$

28. $(4 - \sqrt{5})(4 + \sqrt{5}) = 4^2 - (\sqrt{5})^2 = 16 - 5 = 11$

29. $\dfrac{10}{4-\sqrt{5}} = \dfrac{10}{4-\sqrt{5}} \cdot \dfrac{4+\sqrt{5}}{4+\sqrt{5}} = \dfrac{10(4+\sqrt{5})}{4^2-(\sqrt{5})^2} = \dfrac{40+10\sqrt{5}}{16-5} =$
$\dfrac{40+10\sqrt{5}}{11}$

30.
$$
\begin{aligned}
a^2 + b^2 &= c^2 \\
8^2 + 4^2 &= c^2 \\
64 + 16 &= c^2 \\
80 &= c^2 \\
\sqrt{80} &= c \quad \text{Exact answer} \\
8.944 &\approx c \quad \text{Approximation}
\end{aligned}
$$

31.
$$
\begin{aligned}
\sqrt{3x} + 2 &= 14 \\
\sqrt{3x} &= 12 \\
(\sqrt{3x})^2 &= 12^2 \\
3x &= 144 \\
x &= 48
\end{aligned}
$$
The number 48 checks. It is the solution.

32.
$$
\begin{aligned}
\sqrt{6x+13} &= x + 3 \\
(\sqrt{6x+13})^2 &= (x+3)^2 \\
6x + 13 &= x^2 + 6x + 9 \\
0 &= x^2 - 4 \\
0 &= (x+2)(x-2)
\end{aligned}
$$
$x + 2 = 0$ *or* $x - 2 = 0$
$x = -2$ *or* $x = 2$
Both numbers check. The solutions are -2 and 2.

33.
$$
\begin{aligned}
\sqrt{1-x} + 1 &= \sqrt{6-x} \\
(\sqrt{1-x} + 1)^2 &= (\sqrt{6-x})^2 \\
1 - x + 2\sqrt{1-x} + 1 &= 6 - x \\
2 - x + 2\sqrt{1-x} &= 6 - x \\
2\sqrt{1-x} &= 4 \\
\sqrt{1-x} &= 2 \\
(\sqrt{1-x})^2 &= 2^2 \\
1 - x &= 4 \\
-x &= 3 \\
x &= -3
\end{aligned}
$$
The number -3 checks. It is the solution.

34. a) Substitute 28,000 for h in the formula.
$D = \sqrt{2 \cdot 28,000} = \sqrt{56,000} \approx 237$ mi

b) Substitute 261 for D and solve for h.
$$
\begin{aligned}
261 &= \sqrt{2h} \\
261^2 &= (\sqrt{2h})^2 \\
68,121 &= 2h \\
34,060.5 &= h
\end{aligned}
$$
The airplane is 34,060.5 ft high.

35. Let $d =$ the length of a diagonal, in yd.
$$
\begin{aligned}
a^2 + b^2 &= c^2 \\
60^2 + 110^2 &= d^2 \\
3600 + 12,100 &= d^2 \\
15,700 &= d^2 \\
\sqrt{15,700} \text{ yd} &= d \\
125.300 \text{ yd} &\approx d
\end{aligned}
$$

36. $\sqrt{\dfrac{2a}{5b}} = \sqrt{\dfrac{2a}{5b} \cdot \dfrac{5b}{5b}} = \sqrt{\dfrac{10ab}{25b^2}} = \dfrac{\sqrt{10ab}}{\sqrt{25b^2}} = \dfrac{\sqrt{10ab}}{5b}$
Answer A is correct.

37. $\sqrt{\sqrt{\sqrt{625}}} = \sqrt{\sqrt{25}} = \sqrt{5}$

38. $\sqrt{y^{16n}} = \sqrt{(y^{8n})^2} = y^{8n}$

Chapter 17

Quadratic Equations

Exercise Set 17.1

1. $x^2 - 3x + 2 = 0$

This equation is already in standard form.

$a = 1,\ b = -3,\ c = 2$

3. $7x^2 = 4x - 3$

$7x^2 - 4x + 3 = 0$ Standard form

$a = 7,\ b = -4,\ c = 3$

5. $5 = -2x^2 + 3x$

$2x^2 - 3x + 5 = 0$ Standard form

$a = 2,\ b = -3,\ c = 5$

7. $x^2 + 5x = 0$

$x(x + 5) = 0$

$x = 0$ *or* $x + 5 = 0$

$x = 0$ *or* $x = -5$

The solutions are 0 and -5.

9. $3x^2 + 6x = 0$

$3x(x + 2) = 0$

$3x = 0$ *or* $x + 2 = 0$

$x = 0$ *or* $x = -2$

The solutions are 0 and -2.

11. $5x^2 = 2x$

$5x^2 - 2x = 0$

$x(5x - 2) = 0$

$x = 0$ *or* $5x - 2 = 0$

$x = 0$ *or* $5x = 2$

$x = 0$ *or* $x = \frac{2}{5}$

The solutions are 0 and $\frac{2}{5}$.

13. $4x^2 + 4x = 0$

$4x(x + 1) = 0$

$4x = 0$ *or* $x + 1 = 0$

$x = 0$ *or* $x = -1$

The solutions are 0 and -1.

15. $0 = 10x^2 - 30x$

$0 = 10x(x - 3)$

$10x = 0$ *or* $x - 3 = 0$

$x = 0$ *or* $x = 3$

The solutions are 0 and 3.

17. $11x = 55x^2$

$0 = 55x^2 - 11x$

$0 = 11x(5x - 1)$

$11x = 0$ *or* $5x - 1 = 0$

$x = 0$ *or* $5x = 1$

$x = 0$ *or* $x = \frac{1}{5}$

The solutions are 0 and $\frac{1}{5}$.

19. $14t^2 = 3t$

$14t^2 - 3t = 0$

$t(14t - 3) = 0$

$t = 0$ *or* $14t - 3 = 0$

$t = 0$ *or* $14t = 3$

$t = 0$ *or* $t = \frac{3}{14}$

The solutions are 0 and $\frac{3}{14}$.

21. $5y^2 - 3y^2 = 72y + 9y$

$2y^2 = 81y$

$2y^2 - 81y = 0$

$y(2y - 81) = 0$

$y = 0$ *or* $2y - 81 = 0$

$y = 0$ *or* $2y = 81$

$y = 0$ *or* $y = \frac{81}{2}$

The solutions are 0 and $\frac{81}{2}$.

23. $x^2 + 8x - 48 = 0$

$(x + 12)(x - 4) = 0$

$x + 12 = 0$ *or* $x - 4 = 0$

$x = -12$ *or* $x = 4$

The solutions are -12 and 4.

25. $5 + 6x + x^2 = 0$

$(5 + x)(1 + x) = 0$

$5 + x = 0$ *or* $1 + x = 0$

$x = -5$ *or* $x = -1$

The solutions are -5 and -1.

27. $18 = 7p + p^2$

$0 = p^2 + 7p - 18$

$0 = (p + 9)(p - 2)$

$p + 9 = 0$ *or* $p - 2 = 0$

$p = -9$ *or* $p = 2$

The solutions are -9 and 2.

29. $-15 = -8y + y^2$

$0 = y^2 - 8y + 15$

$0 = (y-5)(y-3)$

$y - 5 = 0$ *or* $y - 3 = 0$

$y = 5$ *or* $y = 3$

The solutions are 5 and 3.

31. $x^2 + 10x + 25 = 0$

$(x+5)(x+5) = 0$

$x + 5 = 0$ *or* $x + 5 = 0$

$x = -5$ *or* $x = -5$

The solution is -5.

33. $r^2 = 8r - 16$

$r^2 - 8r + 16 = 0$

$(r-4)(r-4) = 0$

$r - 4 = 0$ *or* $r - 4 = 0$

$r = 4$ *or* $r = 4$

The solution is 4.

35. $6x^2 + x - 2 = 0$

$(3x+2)(2x-1) = 0$

$3x + 2 = 0$ *or* $2x - 1 = 0$

$3x = -2$ *or* $2x = 1$

$x = -\frac{2}{3}$ *or* $x = \frac{1}{2}$

The solutions are $-\frac{2}{3}$ and $\frac{1}{2}$.

37. $3a^2 = 10a + 8$

$3a^2 - 10a - 8 = 0$

$(3a+2)(a-4) = 0$

$3a + 2 = 0$ *or* $a - 4 = 0$

$3a = -2$ *or* $a = 4$

$a = -\frac{2}{3}$ *or* $a = 4$

The solutions are $-\frac{2}{3}$ and 4.

39. $6x^2 - 4x = 10$

$6x^2 - 4x - 10 = 0$

$2(3x^2 - 2x - 5) = 0$

$2(3x-5)(x+1) = 0$

$3x - 5 = 0$ *or* $x + 1 = 0$

$3x = 5$ *or* $x = -1$

$x = \frac{5}{3}$ *or* $x = -1$

The solutions are $\frac{5}{3}$ and -1.

41. $2t^2 + 12t = -10$

$2t^2 + 12t + 10 = 0$

$2(t^2 + 6t + 5) = 0$

$2(t+5)(t+1) = 0$

$t + 5 = 0$ *or* $t + 1 = 0$

$t = -5$ *or* $t = -1$

The solutions are -5 and -1.

43. $t(t-5) = 14$

$t^2 - 5t = 14$

$t^2 - 5t - 14 = 0$

$(t+2)(t-7) = 0$

$t + 2 = 0$ *or* $t - 7 = 0$

$t = -2$ *or* $t = 7$

The solutions are -2 and 7.

45. $t(9+t) = 4(2t+5)$

$9t + t^2 = 8t + 20$

$t^2 + t - 20 = 0$

$(t+5)(t-4) = 0$

$t + 5 = 0$ *or* $t - 4 = 0$

$t = -5$ *or* $t = 4$

The solutions are -5 and 4.

47. $16(p-1) = p(p+8)$

$16p - 16 = p^2 + 8p$

$0 = p^2 - 8p + 16$

$0 = (p-4)(p-4)$

$p - 4 = 0$ *or* $p - 4 = 0$

$p = 4$ *or* $p = 4$

The solution is 4.

49. $(t-1)(t+3) = t - 1$

$t^2 + 2t - 3 = t - 1$

$t^2 + t - 2 = 0$

$(t+2)(t-1) = 0$

$t + 2 = 0$ *or* $t - 1 = 0$

$t = -2$ *or* $t = 1$

The solutions are -2 and 1.

51. $\frac{24}{x-2} + \frac{24}{x+2} = 5$

The LCM is $(x-2)(x+2)$.

$(x-2)(x+2)\left(\frac{24}{x-2} + \frac{24}{x+2}\right) =$

$(x-2)(x+2) \cdot 5$

$(x-2)(x+2) \cdot \frac{24}{x-2} + (x-2)(x+2) \cdot \frac{24}{x+2} =$

$5(x-2)(x+2)$

$24(x+2) + 24(x-2) =$

$5(x^2 - 4)$

$24x + 48 + 24x - 48 = 5x^2 - 20$

$48x = 5x^2 - 20$

$0 = 5x^2 - 48x - 20$

$0 = (5x+2)(x-10)$

$$5x+2=0 \quad or \quad x-10=0$$
$$5x=-2 \quad or \quad x=10$$
$$x=-\frac{2}{5} \quad or \quad x=10$$

Both numbers check. The solutions are $-\frac{2}{5}$ and 10.

53. $$\frac{1}{x}+\frac{1}{x+6}=\frac{1}{4}$$

The LCM is $4x(x+6)$.

$$4x(x+6)\left(\frac{1}{x}+\frac{1}{x+6}\right)=4x(x+6)\cdot\frac{1}{4}$$
$$4x(x+6)\cdot\frac{1}{x}+4x(x+6)\cdot\frac{1}{x+6}=x(x+6)$$
$$4(x+6)+4x=x(x+6)$$
$$4x+24+4x=x^2+6x$$
$$8x+24=x^2+6x$$
$$0=x^2-2x-24$$
$$0=(x-6)(x+4)$$

$$x-6=0 \quad or \quad x+4=0$$
$$x=6 \quad or \quad x=-4$$

Both numbers check. The solutions are 6 and -4.

55. $$1+\frac{12}{x^2-4}=\frac{3}{x-2}$$

The LCM is $(x+2)(x-2)$.

$$(x+2)(x-2)\left(1+\frac{12}{(x+2)(x-2)}\right)=(x+2)(x-2)\cdot\frac{3}{x-2}$$
$$(x+2)(x-2)\cdot 1+(x+2)(x-2)\cdot\frac{12}{(x+2)(x-2)}=3(x+2)$$
$$x^2-4+12=3x+6$$
$$x^2+8=3x+6$$
$$x^2-3x+2=0$$
$$(x-2)(x-1)=0$$

$$x-2=0 \quad or \quad x-1=0$$
$$x=2 \quad or \quad x=1$$

The number 1 checks, but 2 does not. (It makes the denominators x^2-4 and $x-2$ zero.) The solution is 1.

57. $$\frac{r}{r-1}+\frac{2}{r^2-1}=\frac{8}{r+1}$$

The LCM is $(r-1)(r+1)$.

$$(r-1)(r+1)\left(\frac{r}{r-1}+\frac{2}{(r-1)(r+1)}\right)=(r-1)(r+1)\cdot\frac{8}{r+1}$$
$$(r-1)(r+1)\cdot\frac{r}{r-1}+(r-1)(r+1)\cdot\frac{2}{(r-1)(r+1)}=8(r-1)$$
$$r(r+1)+2=8(r-1)$$
$$r^2+r+2=8r-8$$
$$r^2-7r+10=0$$
$$(r-5)(r-2)=0$$

$$r-5=0 \quad or \quad r-2=0$$
$$r=5 \quad or \quad r=2$$

Both numbers check. The solutions are 5 and 2.

59. $$\frac{x-1}{1-x}=-\frac{x+8}{x-8}$$

The LCM is $(1-x)(x-8)$.

$$(1-x)(x-8)\cdot\frac{x-1}{1-x}=(1-x)(x-8)\left(-\frac{x+8}{x-8}\right)$$
$$(x-8)(x-1)=-(1-x)(x+8)$$
$$x^2-9x+8=-(x+8-x^2-8x)$$
$$x^2-9x+8=-(-x^2-7x+8)$$
$$x^2-9x+8=x^2+7x-8$$
$$16=16x$$
$$1=x$$

The number 1 does not check. (It makes the denominator $1-x$ zero.) There is no solution.

61. $$\frac{5}{y+4}-\frac{3}{y-2}=4$$

The LCM is $(y+4)(y-2)$.

$$(y+4)(y-2)\left(\frac{5}{y+4}-\frac{3}{y-2}\right)=(y+4)(y-2)\cdot 4$$
$$5(y-2)-3(y+4)=4(y^2+2y-8)$$
$$5y-10-3y-12=4y^2+8y-32$$
$$2y-22=4y^2+8y-32$$
$$0=4y^2+6y-10$$
$$0=2(2y^2+3y-5)$$
$$0=2(2y+5)(y-1)$$

$$2y+5=0 \quad or \quad y-1=0$$
$$2y=-5 \quad or \quad y=1$$
$$y=-\frac{5}{2} \quad or \quad y=1$$

The solutions are $-\frac{5}{2}$ and 1.

63. ***Familiarize.*** We will use the formula

$$d=\frac{n^2-3n}{2},$$

where d is the number of diagonals and n is the number of sides.

Translate. We substitute 10 for n.

$$d=\frac{10^2-3\cdot 10}{2}$$

Solve. We do the computation.

$$d=\frac{10^2-3\cdot 10}{2}=\frac{100-30}{2}=\frac{70}{2}=35$$

Check. We can recheck our computation. We can also substitute 35 for d in the original formula and determine whether this yields $n=10$. Our result checks.

State. A decagon has 35 diagonals.

65. ***Familiarize.*** We will use the formula

$$d=\frac{n^2-3n}{2},$$

where d is the number of diagonals and n is the number of sides.

Translate. We substitute 14 for d.

$14 = \frac{n^2 - 3n}{2}$

Solve. We solve the equation.

$$\frac{n^2 - 3n}{2} = 14$$
$$n^2 - 3n = 28 \quad \text{Multiplying by 2}$$
$$n^2 - 3n - 28 = 0$$
$$(n-7)(n+4) = 0$$
$$n - 7 = 0 \quad or \quad n + 4 = 0$$
$$n = 7 \quad or \quad n = -4$$

Check. Since the number of sides cannot be negative, -4 cannot be a solution. To check 7, we substitute 7 for n in the original formula and determine if this yields $d = 14$. Our result checks.

State. The polygon has 7 sides.

67. $\sqrt{64} = 8$, taking the principal square root

69. $\sqrt{8} = \sqrt{4 \cdot 2} = \sqrt{4}\sqrt{2} = 2\sqrt{2}$

71. $\sqrt{20} = \sqrt{4 \cdot 5} = \sqrt{4}\sqrt{5} = 2\sqrt{5}$

73. $\sqrt{405} = \sqrt{81 \cdot 5} = \sqrt{81}\sqrt{5} = 9\sqrt{5}$

75. 2.646

77. 1.528

79.
$$4m^2 - (m+1)^2 = 0$$
$$4m^2 - (m^2 + 2m + 1) = 0$$
$$4m^2 - m^2 - 2m - 1 = 0$$
$$3m^2 - 2m - 1 = 0$$
$$(3m+1)(m-1) = 0$$
$$3m + 1 = 0 \quad or \quad m - 1 = 0$$
$$3m = -1 \quad or \quad m = 1$$
$$m = -\frac{1}{3} \quad or \quad m = 1$$

The solutions are $-\frac{1}{3}$ and 1.

81.
$$\sqrt{5}x^2 - x = 0$$
$$x(\sqrt{5}x - 1) = 0$$
$$x = 0 \quad or \quad \sqrt{5}x - 1 = 0$$
$$x = 0 \quad or \quad \sqrt{5}x = 1$$
$$x = 0 \quad or \quad x = \frac{1}{\sqrt{5}}, \text{ or } \frac{\sqrt{5}}{5}$$

The solutions are 0 and $\frac{\sqrt{5}}{5}$.

83. Graph $y_1 = 3x^2 - 7x$ and $y_2 = 20$. Then use the INTERSECT feature to find the first coordinate(s) of the point(s) of intersection. The solutions are 4 and approximately -1.7.

85. Graph $y_1 = 3x^2 + 8x$ and $y_2 = 12x + 15$. Then use the INTERSECT feature to find the first coordinate(s) of the point(s) of intersection. The solutions are 3 and approximately -1.7.

87. Graph $y_1 = (x-2)^2 + 3(x-2)$ and $y_2 = 4$. Then use the INTERSECT feature to find the first coordinate(s) of the point(s) of intersection. The solutions are -2 and 3.

89. Graph $y_1 = 16(x-1)$ and $y_2 = x(x+8)$. Then use the INTERSECT feature to find the first coordinate(s) of the point(s) of intersection. The solution is 4.

Exercise Set 17.2

1. $x^2 = 121$

$x = 11 \ or \ x = -11 \quad$ Principle of square roots

The solutions are 11 and -11.

3.
$$5x^2 = 35$$
$$x^2 = 7 \quad \text{Dividing by 5}$$
$$x = \sqrt{7} \ or \ x = -\sqrt{7} \quad \text{Principle of square roots}$$

The solutions are $\sqrt{7}$ and $-\sqrt{7}$.

5.
$$5x^2 = 3$$
$$x^2 = \frac{3}{5}$$
$$x = \sqrt{\frac{3}{5}} \quad or \quad x = -\sqrt{\frac{3}{5}} \quad \text{Principle of square roots}$$
$$x = \sqrt{\frac{3}{5} \cdot \frac{5}{5}} \quad or \quad x = -\sqrt{\frac{3}{5} \cdot \frac{5}{5}} \quad \text{Rationalizing denominators}$$
$$x = \frac{\sqrt{15}}{5} \quad or \quad x = -\frac{\sqrt{15}}{5}$$

The solutions are $\frac{\sqrt{15}}{5}$ and $-\frac{\sqrt{15}}{5}$.

7.
$$4x^2 - 25 = 0$$
$$4x^2 = 25$$
$$x^2 = \frac{25}{4}$$
$$x = \frac{5}{2} \ or \ x = -\frac{5}{2}$$

The solutions are $\frac{5}{2}$ and $-\frac{5}{2}$.

9.
$$3x^2 - 49 = 0$$
$$3x^2 = 49$$
$$x^2 = \frac{49}{3}$$
$$x = \frac{7}{\sqrt{3}} \quad or \quad x = -\frac{7}{\sqrt{3}}$$
$$x = \frac{7}{\sqrt{3}} \cdot \frac{\sqrt{3}}{\sqrt{3}} \quad or \quad x = -\frac{7}{\sqrt{3}} \cdot \frac{\sqrt{3}}{\sqrt{3}}$$
$$x = \frac{7\sqrt{3}}{3} \quad or \quad x = -\frac{7\sqrt{3}}{3}$$

The solutions are $\frac{7\sqrt{3}}{3}$ and $-\frac{7\sqrt{3}}{3}$.

11.
$$4y^2 - 3 = 9$$
$$4y^2 = 12$$
$$y^2 = 3$$
$$y = \sqrt{3} \quad or \quad y = -\sqrt{3}$$

The solutions are $\sqrt{3}$ and $-\sqrt{3}$.

13. $49y^2 - 64 = 0$
$$49y^2 = 64$$
$$y^2 = \frac{64}{49}$$
$$y = \frac{8}{7} \quad or \quad y = -\frac{8}{7}$$
The solutions are $\frac{8}{7}$ and $-\frac{8}{7}$.

15. $(x+3)^2 = 16$
$$x+3 = 4 \quad or \quad x+3 = -4 \quad \text{Principle of square roots}$$
$$x = 1 \quad or \quad x = -7$$
The solutions are 1 and -7.

17. $(x+3)^2 = 21$
$$x+3 = \sqrt{21} \quad or \quad x+3 = -\sqrt{21} \quad \text{Principle of square roots}$$
$$x = -3+\sqrt{21} \quad or \quad x = -3-\sqrt{21}$$
The solutions are $-3+\sqrt{21}$ and $-3-\sqrt{21}$, or $-3\pm\sqrt{21}$.

19. $(x+13)^2 = 8$
$$x+13 = \sqrt{8} \quad or \quad x+13 = -\sqrt{8}$$
$$x+13 = 2\sqrt{2} \quad or \quad x+13 = -2\sqrt{2}$$
$$x = -13+2\sqrt{2} \quad or \quad x = -13-2\sqrt{2}$$
The solutions are $-13+2\sqrt{2}$ and $-13-2\sqrt{2}$, or $-13\pm 2\sqrt{2}$.

21. $(x-7)^2 = 12$
$$x-7 = \sqrt{12} \quad or \quad x-7 = -\sqrt{12}$$
$$x-7 = 2\sqrt{3} \quad or \quad x-7 = -2\sqrt{3}$$
$$x = 7+2\sqrt{3} \quad or \quad x = 7-2\sqrt{3}$$
The solutions are $7+2\sqrt{3}$ and $7-2\sqrt{3}$, or $7\pm 2\sqrt{3}$.

23. $(x+9)^2 = 34$
$$x+9 = \sqrt{34} \quad or \quad x+9 = -\sqrt{34}$$
$$x = -9+\sqrt{34} \quad or \quad x = -9-\sqrt{34}$$
The solutions are $-9+\sqrt{34}$ and $-9-\sqrt{34}$, or $-9\pm\sqrt{34}$.

25. $\left(x+\frac{3}{2}\right)^2 = \frac{7}{2}$
$$x+\frac{3}{2} = \sqrt{\frac{7}{2}} \quad or \quad x+\frac{3}{2} = -\sqrt{\frac{7}{2}}$$
$$x = -\frac{3}{2}+\sqrt{\frac{7}{2}} \quad or \quad x = -\frac{3}{2}-\sqrt{\frac{7}{2}}$$
$$x = -\frac{3}{2}+\sqrt{\frac{7}{2}\cdot\frac{2}{2}} \quad or \quad x = -\frac{3}{2}-\sqrt{\frac{7}{2}\cdot\frac{2}{2}}$$
$$x = -\frac{3}{2}+\frac{\sqrt{14}}{2} \quad or \quad x = -\frac{3}{2}-\frac{\sqrt{14}}{2}$$
$$x = \frac{-3+\sqrt{14}}{2} \quad or \quad x = \frac{-3-\sqrt{14}}{2}$$
The solutions are $\frac{-3\pm\sqrt{14}}{2}$.

27. $x^2 - 6x + 9 = 64$
$$(x-3)^2 = 64 \quad \text{Factoring the left side}$$
$$x-3 = 8 \quad or \quad x-3 = -8 \quad \text{Principle of square roots}$$
$$x = 11 \quad or \quad x = -5$$
The solutions are 11 and -5.

29. $x^2 + 14x + 49 = 64$
$$(x+7)^2 = 64 \quad \text{Factoring the left side}$$
$$x+7 = 8 \quad or \quad x+7 = -8 \quad \text{Principle of square roots}$$
$$x = 1 \quad or \quad x = -15$$
The solutions are 1 and -15.

31. $x^2 - 6x - 16 = 0$
$$x^2 - 6x = 16 \quad \text{Adding 16}$$
$$x^2 - 6x + 9 = 16 + 9 \quad \text{Adding 9: } \left(\frac{-6}{2}\right)^2 = (-3)^2 = 9$$
$$(x-3)^2 = 25$$
$$x-3 = 5 \quad or \quad x-3 = -5 \quad \text{Principle of square roots}$$
$$x = 8 \quad or \quad x = -2$$
The solutions are 8 and -2.

33. $x^2 + 22x + 21 = 0$
$$x^2 + 22x = -21 \quad \text{Subtracting 21}$$
$$x^2 + 22x + 121 = -21 + 121 \quad \text{Adding 121: } \left(\frac{22}{2}\right)^2 = 11^2 = 121$$
$$(x+11)^2 = 100$$
$$x+11 = 10 \quad or \quad x+11 = -10 \quad \text{Principle of square roots}$$
$$x = -1 \quad or \quad x = -21$$
The solutions are -1 and -21.

35. $x^2 - 2x - 5 = 0$
$$x^2 - 2x = 5$$
$$x^2 - 2x + 1 = 5 + 1 \quad \text{Adding 1: } \left(\frac{-2}{2}\right)^2 = (-1)^2 = 1$$
$$(x-1)^2 = 6$$
$$x-1 = \sqrt{6} \quad or \quad x-1 = -\sqrt{6}$$
$$x = 1+\sqrt{6} \quad or \quad x = 1-\sqrt{6}$$
The solutions are $1\pm\sqrt{6}$.

37. $x^2 - 22x + 102 = 0$
$$x^2 - 22x = -102$$
$$x^2 - 22x + 121 = -102 + 121 \quad \text{Adding 121: } \left(\frac{-22}{2}\right)^2 = (-11)^2 = 121$$
$$(x-11)^2 = 19$$
$$x-11 = \sqrt{19} \quad or \quad x-11 = -\sqrt{19}$$
$$x = 11+\sqrt{19} \quad or \quad x = 11-\sqrt{19}$$
The solutions are $11\pm\sqrt{19}$.

39.
$$x^2 + 10x - 4 = 0$$
$$x^2 + 10x = 4$$
$$x^2 + 10x + 25 = 4 + 25 \quad \text{Adding 25: } \left(\frac{10}{2}\right)^2 = 5^2 = 25$$
$$(x+5)^2 = 29$$
$$x + 5 = \sqrt{29} \quad or \quad x + 5 = -\sqrt{29}$$
$$x = -5 + \sqrt{29} \quad or \quad x = -5 - \sqrt{29}$$
The solutions are $-5 \pm \sqrt{29}$.

41.
$$x^2 - 7x - 2 = 0$$
$$x^2 - 7x = 2$$
$$x^2 - 7x + \frac{49}{4} = 2 + \frac{49}{4} \quad \text{Adding } \frac{49}{4}: \left(\frac{-7}{2}\right)^2 = \frac{49}{4}$$
$$\left(x - \frac{7}{2}\right)^2 = \frac{8}{4} + \frac{49}{4} = \frac{57}{4}$$
$$x - \frac{7}{2} = \frac{\sqrt{57}}{2} \quad or \quad x - \frac{7}{2} = -\frac{\sqrt{57}}{2}$$
$$x = \frac{7}{2} + \frac{\sqrt{57}}{2} \quad or \quad x = \frac{7}{2} - \frac{\sqrt{57}}{2}$$
$$x = \frac{7+\sqrt{57}}{2} \quad or \quad x = \frac{7-\sqrt{57}}{2}$$
The solutions are $\frac{7 \pm \sqrt{57}}{2}$.

43.
$$x^2 + 3x - 28 = 0$$
$$x^2 + 3x = 28$$
$$x^2 + 3x + \frac{9}{4} = 28 + \frac{9}{4} \quad \text{Adding } \frac{9}{4}: \left(\frac{3}{2}\right)^2 = \frac{9}{4}$$
$$\left(x + \frac{3}{2}\right)^2 = \frac{121}{4}$$
$$x + \frac{3}{2} = \frac{11}{2} \quad or \quad x + \frac{3}{2} = -\frac{11}{2}$$
$$x = \frac{8}{2} \quad or \quad x = -\frac{14}{2}$$
$$x = 4 \quad or \quad x = -7$$
The solutions are 4 and -7.

45.
$$x^2 + \frac{3}{2}x - \frac{1}{2} = 0$$
$$x^2 + \frac{3}{2}x = \frac{1}{2}$$
$$x^2 + \frac{3}{2}x + \frac{9}{16} = \frac{1}{2} + \frac{9}{16} \quad \text{Adding } \frac{9}{16}: \left(\frac{3/2}{2}\right)^2 = \left(\frac{3}{4}\right)^2 = \frac{9}{16}$$
$$\left(x + \frac{3}{4}\right)^2 = \frac{17}{16}$$
$$x + \frac{3}{4} = \frac{\sqrt{17}}{4} \quad or \quad x + \frac{3}{4} = -\frac{\sqrt{17}}{4}$$
$$x = -\frac{3}{4} + \frac{\sqrt{17}}{4} \quad or \quad x = -\frac{3}{4} - \frac{\sqrt{17}}{4}$$
$$x = \frac{-3+\sqrt{17}}{4} \quad or \quad x = \frac{-3-\sqrt{17}}{4}$$
The solutions are $\frac{-3 \pm \sqrt{17}}{4}$.

47.
$$2x^2 + 3x - 17 = 0$$
$$\frac{1}{2}(2x^2 + 3x - 17) = \frac{1}{2} \cdot 0 \quad \text{Multiplying by } \frac{1}{2} \text{ to make the } x^2\text{-coefficient 1}$$
$$x^2 + \frac{3}{2}x - \frac{17}{2} = 0$$
$$x^2 + \frac{3}{2}x = \frac{17}{2}$$
$$x^2 + \frac{3}{2}x + \frac{9}{16} = \frac{17}{2} + \frac{9}{16} \quad \text{Adding } \frac{9}{16}: \left(\frac{3/2}{2}\right)^2 = \left(\frac{3}{4}\right)^2 = \frac{9}{16}$$
$$\left(x + \frac{3}{4}\right)^2 = \frac{145}{16}$$
$$x + \frac{3}{4} = \frac{\sqrt{145}}{4} \quad or \quad x + \frac{3}{4} = -\frac{\sqrt{145}}{4}$$
$$x = \frac{-3+\sqrt{145}}{4} \quad or \quad x = \frac{-3-\sqrt{145}}{4}$$
The solutions are $\frac{-3 \pm \sqrt{145}}{4}$.

49.
$$3x^2 + 4x - 1 = 0$$
$$\frac{1}{3}(3x^2 + 4x - 1) = \frac{1}{3} \cdot 0$$
$$x^2 + \frac{4}{3}x - \frac{1}{3} = 0$$
$$x^2 + \frac{4}{3}x = \frac{1}{3}$$
$$x^2 + \frac{4}{3}x + \frac{4}{9} = \frac{1}{3} + \frac{4}{9}$$
$$\left(x + \frac{2}{3}\right)^2 = \frac{7}{9}$$
$$x + \frac{2}{3} = \frac{\sqrt{7}}{3} \quad or \quad x + \frac{2}{3} = -\frac{\sqrt{7}}{3}$$
$$x = \frac{-2+\sqrt{7}}{3} \quad or \quad x = -\frac{-2-\sqrt{7}}{3}$$
The solutions are $\frac{-2 \pm \sqrt{7}}{3}$.

51.
$$2x^2 = 9x + 5$$
$$2x^2 - 9x - 5 = 0 \quad \text{Standard form}$$
$$\frac{1}{2}(2x^2 - 9x - 5) = \frac{1}{2} \cdot 0$$
$$x^2 - \frac{9}{2}x - \frac{5}{2} = 0$$
$$x^2 - \frac{9}{2}x = \frac{5}{2}$$
$$x^2 - \frac{9}{2}x + \frac{81}{16} = \frac{5}{2} + \frac{81}{16}$$
$$\left(x - \frac{9}{4}\right)^2 = \frac{121}{16}$$

$$x - \frac{9}{4} = \frac{11}{4} \quad or \quad x - \frac{9}{4} = -\frac{11}{4}$$
$$x = \frac{20}{4} \quad or \quad x = -\frac{2}{4}$$
$$x = 5 \quad or \quad x = -\frac{1}{2}$$

The solutions are 5 and $-\frac{1}{2}$.

53.
$$6x^2 + 11x = 10$$
$$6x^2 + 11x - 10 = 0 \qquad \text{Standard form}$$
$$\frac{1}{6}(6x^2 + 11x - 10) = \frac{1}{6} \cdot 0$$
$$x^2 + \frac{11}{6}x - \frac{5}{3} = 0$$
$$x^2 + \frac{11}{6}x = \frac{5}{3}$$
$$x^2 + \frac{11}{6}x + \frac{121}{144} = \frac{5}{3} + \frac{121}{144}$$
$$\left(x + \frac{11}{12}\right)^2 = \frac{361}{144}$$

$$x + \frac{11}{12} = \frac{19}{12} \quad or \quad x + \frac{11}{12} = -\frac{19}{12}$$
$$x = \frac{8}{12} \quad or \quad x = -\frac{30}{12}$$
$$x = \frac{2}{3} \quad or \quad x = -\frac{5}{2}$$

The solutions are $\frac{2}{3}$ and $-\frac{5}{2}$.

55. ***Familiarize.*** We will use the formula $s = 16t^2$.

Translate. We substitute 2684 for s.

$$2684 = 16t^2$$

Solve. We solve the equation.

$$2684 = 16t^2$$
$$\frac{2684}{16} = t^2 \qquad \text{Solving for } t^2$$
$$167.75 = t^2 \qquad \text{Dividing}$$
$$\sqrt{167.75} = t \quad or \quad -\sqrt{167.75} = t \qquad \text{Principle of square roots}$$
$$13.0 \approx t \quad or \quad -13.0 \approx t \qquad \text{Using a calculator and rounding to the nearest tenth}$$

Check. The number -13.0 cannot be a solution, because time cannot be negative in this situation. We substitute 13.0 in the original equation.

$$s = 16(13.0)^2 = 16(169) = 2704$$

This is close. Remember that we approximated a solution. Thus we have a check.

State. It takes about 13.0 sec for an object to fall to the ground from the top of the Burj Dubai.

57. ***Familiarize.*** We will use the formula $s = 16t^2$.

Translate. We substitute 1353 for s.

$$1353 = 16t^2$$

Solve. We solve the equation.

$$1353 = 16t^2$$
$$\frac{1353}{16} = t^2 \qquad \text{Solving for } t^2$$
$$84.5625 = t^2 \qquad \text{Dividing}$$
$$\sqrt{84.5625} = t \quad or \quad -\sqrt{84.5625} = t \qquad \text{Principle of square roots}$$
$$9.2 \approx t \quad or \quad -9.2 \approx t \qquad \text{Using a calculator and rounding to the nearest tenth}$$

Check. The number -9.2 cannot be a solution, because time cannot be negative in this situation. We substitute 9.2 in the original equation.

$$s = 16(9.2)^2 = 16(84.64) = 1354.24$$

This is close. Remember that we approximated a solution. Thus we have a check.

State. The fall would take approximately 9.2 sec.

59. The product rule asserts when multiplying with exponential notation, if the bases are the same, we keep the base and add the exponents.

61. The number -5 is not the principal square root of 25.

63. The quotient rule asserts that when dividing with exponential notation, if the bases are the same, we keep the base and subtract the exponent of the denominator from the exponent of the numerator.

65. The quotient rule for radicals asserts that for any nonnegative radicand A and positive number B, $\frac{\sqrt{A}}{\sqrt{B}} = \sqrt{\frac{A}{B}}$.

67. $x^2 + bx + 36$

The trinomial is a square if the square of one-half the x-coefficient is equal to 36. Thus we have:

$$\left(\frac{b}{2}\right)^2 = 36$$
$$\frac{b^2}{4} = 36$$
$$b^2 = 144$$
$$b = 12 \quad or \quad b = -12 \qquad \text{Principle of square roots}$$

69. $x^2 + bx + 128$

The trinomial is a square if the square of one-half the x-coefficient is equal to 128. Thus we have:

$$\left(\frac{b}{2}\right)^2 = 128$$
$$\frac{b^2}{4} = 128$$
$$b^2 = 512$$

$$b = \sqrt{512} \quad or \quad b = -\sqrt{512}$$
$$b = 16\sqrt{2} \quad or \quad b = -16\sqrt{2}$$

71. $x^2 + bx + c$

The trinomial is a square if the square of one-half the x-coefficient is equal to c. Thus we have:

$$\left(\frac{b}{2}\right)^2 = c$$
$$\frac{b^2}{4} = c$$
$$b^2 = 4c$$

$b = \sqrt{4c}$ *or* $b = -\sqrt{4c}$
$b = 2\sqrt{c}$ *or* $b = -2\sqrt{c}$

73. $4.82x^2 = 12,000$

$$x^2 = \frac{12,000}{4.82}$$

$x = \sqrt{\frac{12,000}{4.82}}$ *or* $x = -\sqrt{\frac{12,000}{4.82}}$ Principle of square roots

$x \approx 49.896$ *or* $x \approx -49.896$ Using a calculator and rounding

The solutions are approximately 49.896 and -49.896.

75. $\frac{x}{9} = \frac{36}{4x}$, LCM is $36x$

$36x \cdot \frac{x}{9} = 36x \cdot \frac{36}{4x}$ Multiplying by $36x$

$$4x^2 = 324$$
$$x^2 = 81$$

$x = 9$ *or* $x = -9$

Both numbers check. The solutions are 9 and -9.

Exercise Set 17.3

1. $x^2 - 4x = 21$

$x^2 - 4x - 21 = 0$ Standard form

We can factor.

$$x^2 - 4x - 21 = 0$$
$$(x-7)(x+3) = 0$$

$x - 7 = 0$ *or* $x + 3 = 0$
$x = 7$ *or* $x = -3$

The solutions are 7 and -3.

3. $x^2 = 6x - 9$

$x^2 - 6x + 9 = 0$ Standard form

We can factor.

$$x^2 - 6x + 9 = 0$$
$$(x-3)(x-3) = 0$$

$x - 3 = 0$ *or* $x - 3 = 0$
$x = 3$ *or* $x = 3$

The solution is 3.

5. $3y^2 - 2y - 8 = 0$

We can factor.

$$3y^2 - 2y - 8 = 0$$
$$(3y+4)(y-2) = 0$$

$3y + 4 = 0$ *or* $y - 2 = 0$
$3y = -4$ *or* $y = 2$
$y = -\frac{4}{3}$ *or* $y = 2$

The solutions are $-\frac{4}{3}$ and 2.

7. $4x^2 + 4x = 15$

$4x^2 + 4x - 15 = 0$ Standard form

We can factor.

$$4x^2 + 4x - 15 = 0$$
$$(2x-3)(2x+5) = 0$$

$2x - 3 = 0$ *or* $2x + 5 = 0$
$2x = 3$ *or* $2x = -5$
$x = \frac{3}{2}$ *or* $x = -\frac{5}{2}$

The solutions are $\frac{3}{2}$ and $-\frac{5}{2}$.

9. $x^2 - 9 = 0$ Difference of squares

$$(x+3)(x-3) = 0$$

$x + 3 = 0$ *or* $x - 3 = 0$
$x = -3$ *or* $x = 3$

The solutions are -3 and 3.

11. $x^2 - 2x - 2 = 0$

$a = 1,\ b = -2,\ c = -2$

We use the quadratic formula.

$$x = \frac{-(-2) \pm \sqrt{(-2)^2 - 4 \cdot 1 \cdot (-2)}}{2 \cdot 1}$$
$$x = \frac{2 \pm \sqrt{4+8}}{2}$$
$$x = \frac{2 \pm \sqrt{12}}{2} = \frac{2 \pm \sqrt{4 \cdot 3}}{2}$$
$$x = \frac{2 \pm 2\sqrt{3}}{2} = \frac{2(1 \pm \sqrt{3})}{2}$$
$$x = 1 \pm \sqrt{3}$$

The solutions are $1 + \sqrt{3}$ and $1 - \sqrt{3}$, or $1 \pm \sqrt{3}$.

13. $y^2 - 10y + 22 = 0$

$a = 1,\ b = -10,\ c = 22$

We use the quadratic formula.

$$y = \frac{-(-10) \pm \sqrt{(-10)^2 - 4 \cdot 1 \cdot 22}}{2 \cdot 1}$$
$$y = \frac{10 \pm \sqrt{100 - 88}}{2}$$
$$y = \frac{10 \pm \sqrt{12}}{2} = \frac{10 \pm \sqrt{4 \cdot 3}}{2}$$
$$y = \frac{10 \pm 2\sqrt{3}}{2} = \frac{2(5 \pm \sqrt{3})}{2}$$
$$y = 5 \pm \sqrt{3}$$

The solutions are $5 + \sqrt{3}$ and $5 - \sqrt{3}$, or $5 \pm \sqrt{3}$.

15. $x^2+4x+4=7$
$x^2+4x-3=0$ Adding -7 to get standard form

$a=1,\ b=4,\ c=-3$

We use the quadratic formula.

$$x=\frac{-4\pm\sqrt{4^2-4\cdot1\cdot(-3)}}{2\cdot1}=\frac{-4\pm\sqrt{16+12}}{2}$$

$$x=\frac{-4\pm\sqrt{28}}{2}=\frac{-4\pm\sqrt{4\cdot7}}{2}$$

$$x=\frac{-4\pm2\sqrt{7}}{2}=\frac{2(-2\pm\sqrt{7})}{2}$$

$$x=-2\pm\sqrt{7}$$

The solutions are $-2+\sqrt{7}$ and $-2-\sqrt{7}$, or $-2\pm\sqrt{7}$.

17. $3x^2+8x+2=0$

$a=3,\ b=8,\ c=2$

We use the quadratic formula.

$$x=\frac{-8\pm\sqrt{8^2-4\cdot3\cdot2}}{2\cdot3}=\frac{-8\pm\sqrt{64-24}}{6}$$

$$x=\frac{-8\pm\sqrt{40}}{6}=\frac{-8\pm\sqrt{4\cdot10}}{6}$$

$$x=\frac{-8\pm2\sqrt{10}}{6}=\frac{2(-4\pm\sqrt{10})}{2\cdot3}$$

$$x=\frac{-4\pm\sqrt{10}}{3}$$

The solutions are $\frac{-4+\sqrt{10}}{3}$ and $\frac{-4-\sqrt{10}}{3}$, or $\frac{-4\pm\sqrt{10}}{3}$.

19. $2x^2-5x=1$
$2x^2-5x-1=0$ Adding -1 to get standard form

$a=2,\ b=-5,\ c=-1$

We use the quadratic formula.

$$x=\frac{-(-5)\pm\sqrt{(-5)^2-4\cdot2\cdot(-1)}}{2\cdot2}=\frac{5\pm\sqrt{25+8}}{4}$$

$$x=\frac{5\pm\sqrt{33}}{4}$$

The solutions are $\frac{5+\sqrt{33}}{4}$ and $\frac{5-\sqrt{33}}{4}$, or $\frac{5\pm\sqrt{33}}{4}$.

21. $2y^2-2y-1=0$

$a=2,\ b=-2,\ c=-1$

We use the quadratic formula.

$$y=\frac{-(-2)\pm\sqrt{(-2)^2-4\cdot2\cdot(-1)}}{2\cdot2}=\frac{2\pm\sqrt{4+8}}{4}$$

$$y=\frac{2\pm\sqrt{12}}{4}=\frac{2\pm\sqrt{4\cdot3}}{4}$$

$$y=\frac{2\pm2\sqrt{3}}{4}=\frac{2(1\pm\sqrt{3})}{2\cdot2}$$

$$y=\frac{1\pm\sqrt{3}}{2}$$

The solutions are $\frac{1+\sqrt{3}}{2}$ and $\frac{1-\sqrt{3}}{2}$, or $\frac{1\pm\sqrt{3}}{2}$.

23. $2t^2+6t+5=0$

$a=2,\ b=6,\ c=5$

We use the quadratic formula.

$$t=\frac{-6\pm\sqrt{6^2-4\cdot2\cdot5}}{2\cdot2}=\frac{-6\pm\sqrt{36-40}}{4}$$

$$t=\frac{-6\pm\sqrt{-4}}{4}$$

Since square roots of negative numbers do not exist as real numbers, there are no real-number solutions.

25. $3x^2=5x+4$
$3x^2-5x-4=0$

$a=3,\ b=-5,\ c=-4$

We use the quadratic formula.

$$x=\frac{-(-5)\pm\sqrt{(-5)^2-4\cdot3\cdot(-4)}}{2\cdot3}=\frac{5\pm\sqrt{25+48}}{6}$$

$$x=\frac{5\pm\sqrt{73}}{6}$$

The solutions are $\frac{5+\sqrt{73}}{6}$ and $\frac{5-\sqrt{73}}{6}$, or $\frac{5\pm\sqrt{73}}{6}$.

27. $2y^2-6y=10$
$2y^2-6y-10=0$
$y^2-3y-5=0$ Multiplying by $\frac{1}{2}$ to simplify

$a=1,\ b=-3,\ c=-5$

We use the quadratic formula.

$$y=\frac{-(-3)\pm\sqrt{(-3)^2-4\cdot1\cdot(-5)}}{2\cdot1}=\frac{3\pm\sqrt{9+20}}{2}$$

$$y=\frac{3\pm\sqrt{29}}{2}$$

The solutions are $\frac{3+\sqrt{29}}{2}$ and $\frac{3-\sqrt{29}}{2}$, or $\frac{3\pm\sqrt{29}}{2}$.

29. $\frac{x^2}{x+3}-\frac{5}{x+3}=0$, LCM is $x+3$

$$(x+3)\left(\frac{x^2}{x+3}-\frac{5}{x+3}\right)=(x+3)\cdot0$$

$$x^2-5=0$$

$$x^2=5$$

$x=\sqrt{5}$ *or* $x=-\sqrt{5}$ Principle of square roots

Both numbers check. The solutions are $\sqrt{5}$ and $-\sqrt{5}$, or $\pm\sqrt{5}$.

31. $x+2=\frac{3}{x+2}$

$(x+2)(x+2)=(x+2)\cdot\frac{3}{x+2}$ Clearing the fraction

$$x^2+4x+4=3$$

$$x^2+4x+1=0$$

$a=1,\ b=4,\ c=1$

We use the quadratic formula.

$$x = \frac{-4 \pm \sqrt{4^2 - 4 \cdot 1 \cdot 1}}{2 \cdot 1} = \frac{-4 \pm \sqrt{16-4}}{2}$$
$$x = \frac{-4 \pm \sqrt{12}}{2} = \frac{-4 \pm \sqrt{4 \cdot 3}}{2}$$
$$x = \frac{-4 \pm 2\sqrt{3}}{2} = \frac{2(-2 \pm \sqrt{3})}{2}$$
$$x = -2 \pm \sqrt{3}$$

Both numbers check. The solutions are $-2 + \sqrt{3}$ and $-2 - \sqrt{3}$, or $-2 \pm \sqrt{3}$.

33. $\frac{1}{x} + \frac{1}{x+1} = \frac{1}{3}$, LCM is $3x(x+1)$

$$3x(x+1)\left(\frac{1}{x} + \frac{1}{x+1}\right) = 3x(x+1) \cdot \frac{1}{3}$$
$$3(x+1) + 3x = x(x+1)$$
$$3x + 3 + 3x = x^2 + x$$
$$6x + 3 = x^2 + x$$
$$0 = x^2 - 5x - 3$$

$a = 1,\ b = -5,\ c = -3$

We use the quadratic formula.

$$x = \frac{-(-5) \pm \sqrt{(-5)^2 - 4 \cdot 1 \cdot (-3)}}{2 \cdot 1} = \frac{5 \pm \sqrt{25+12}}{2}$$
$$x = \frac{5 \pm \sqrt{37}}{2}$$

The solutions are $\frac{5 + \sqrt{37}}{2}$ and $\frac{5 - \sqrt{37}}{2}$, or $\frac{5 \pm \sqrt{37}}{2}$.

35. $x^2 - 4x - 7 = 0$

$a = 1,\ b = -4,\ c = -7$

$$x = \frac{-(-4) \pm \sqrt{(-4)^2 - 4 \cdot 1 \cdot (-7)}}{2 \cdot 1}$$
$$x = \frac{4 \pm \sqrt{16+28}}{2} = \frac{4 \pm \sqrt{44}}{2}$$
$$x = \frac{4 \pm \sqrt{4 \cdot 11}}{2} = \frac{4 \pm 2\sqrt{11}}{2}$$
$$x = \frac{2(2 \pm \sqrt{11})}{2} = 2 \pm \sqrt{11}$$

Using a calculator, we have:

$2 + \sqrt{11} \approx 5.31662479 \approx 5.3$, and

$2 - \sqrt{11} \approx -1.31662479 \approx -1.3$.

The approximate solutions, to the nearest tenth, are 5.3 and -1.3.

37. $y^2 - 6y - 1 = 0$

$a = 1,\ b = -6,\ c = -1$

$$y = \frac{-(-6) \pm \sqrt{(-6)^2 - 4 \cdot 1 \cdot (-1)}}{2 \cdot 1}$$
$$y = \frac{6 \pm \sqrt{36+4}}{2} = \frac{6 \pm \sqrt{40}}{2}$$
$$y = \frac{6 \pm \sqrt{4 \cdot 10}}{2} = \frac{6 \pm 2\sqrt{10}}{2}$$
$$y = \frac{2(3 \pm \sqrt{10})}{2} = 3 \pm \sqrt{10}$$

Using a calculator, we have:

$3 + \sqrt{10} \approx 6.16227766 \approx 6.2$ and

$3 - \sqrt{10} \approx -0.1622776602 \approx -0.2$.

The approximate solutions, to the nearest tenth, are 6.2 and -0.2.

39. $4x^2 + 4x = 1$

$4x^2 + 4x - 1 = 0$ Standard form

$a = 4,\ b = 4,\ c = -1$

$$x = \frac{-4 \pm \sqrt{4^2 - 4 \cdot 4 \cdot (-1)}}{2 \cdot 4}$$
$$x = \frac{-4 \pm \sqrt{16+16}}{8} = \frac{-4 \pm \sqrt{32}}{8}$$
$$x = \frac{-4 \pm \sqrt{16 \cdot 2}}{8} = \frac{-4 \pm 4\sqrt{2}}{8}$$
$$x = \frac{4(-1 \pm \sqrt{2})}{4 \cdot 2} = \frac{-1 \pm \sqrt{2}}{2}$$

Using a calculator, we have:

$\frac{-1 + \sqrt{2}}{2} \approx 0.2071067812 \approx 0.2$ and

$\frac{-1 - \sqrt{2}}{2} \approx -1.207106781 \approx -1.2$.

The approximate solutions, to the nearest tenth, are 0.2 and -1.2.

41. $3x^2 - 8x + 2 = 0$

$a = 3,\ b = -8,\ c = 2$

$$x = \frac{-(-8) \pm \sqrt{(-8)^2 - 4 \cdot 3 \cdot 2}}{2 \cdot 3}$$
$$x = \frac{8 \pm \sqrt{64-24}}{6} = \frac{8 \pm \sqrt{40}}{6}$$
$$x = \frac{8 \pm \sqrt{4 \cdot 10}}{6} = \frac{8 \pm 2\sqrt{10}}{6}$$
$$x = \frac{2(4 \pm \sqrt{10})}{2 \cdot 3} = \frac{4 \pm \sqrt{10}}{3}$$

Using a calculator, we have:

$\frac{4 + \sqrt{10}}{3} \approx 2.387425887 \approx 2.4$ and

$\frac{4 - \sqrt{10}}{3} \approx 0.2792407799 \approx 0.3$.

The approximate solutions, to the nearest tenth, are 2.4 and 0.3.

43.
$$\begin{aligned}\sqrt{40} - 2\sqrt{10} + \sqrt{90} &= \sqrt{4 \cdot 10} - 2\sqrt{10} + \sqrt{9 \cdot 10}\\ &= \sqrt{4}\sqrt{10} - 2\sqrt{10} + \sqrt{9}\sqrt{10}\\ &= 2\sqrt{10} - 2\sqrt{10} + 3\sqrt{10}\\ &= (2 - 2 + 3)\sqrt{10}\\ &= 3\sqrt{10}\end{aligned}$$

45. $\sqrt{18}+\sqrt{50}-3\sqrt{8} = \sqrt{9\cdot 2}+\sqrt{25\cdot 2}-3\sqrt{4\cdot 2}$
$= \sqrt{9}\sqrt{2}+\sqrt{25}\sqrt{2}-3\sqrt{4}\sqrt{2}$
$= 3\sqrt{2}+5\sqrt{2}-3\cdot 2\sqrt{2}$
$= 3\sqrt{2}+5\sqrt{2}-6\sqrt{2}$
$= (3+5-6)\sqrt{2}$
$= 2\sqrt{2}$

47. $\sqrt{80}=\sqrt{16\cdot 5}=\sqrt{16}\sqrt{5}=4\sqrt{5}$

49. $\sqrt{9000x^{10}}=\sqrt{900\cdot 10\cdot x^{10}}=\sqrt{900}\sqrt{x^{10}}\sqrt{10}=30x^5\sqrt{10}$

51. $y=\frac{k}{x}$ Inverse variation

$235=\frac{k}{0.6}$ Substituting 0.6 for x and 235 for y

$141=k$ Constant of variation

$y=\frac{141}{x}$ Equation of variation

53. $5x+x(x-7)=0$
$5x+x^2-7x=0$
$x^2-2x=0$ We can factor.
$x(x-2)=0$

$x=0$ *or* $x-2=0$
$x=0$ *or* $x=2$

The solutions are 0 and 2.

55. $3-x(x-3)=4$
$3-x^2+3x=4$
$0=x^2-3x+1$ Standard form

$a=1,\ b=-3,\ c=1$

We use the quadratic formula.

$$x=\frac{-(-3)\pm\sqrt{(-3)^2-4\cdot 1\cdot 1}}{2\cdot 1}=\frac{3\pm\sqrt{9-4}}{2}$$

$$x=\frac{3\pm\sqrt{5}}{2}$$

The solutions are $\frac{3+\sqrt{5}}{2}$ and $\frac{3-\sqrt{5}}{2}$, or $\frac{3\pm\sqrt{5}}{2}$.

57. $(y+4)(y+3)=15$
$y^2+7y+12=15$
$y^2+7y-3=0$ Standard form

$a=1,\ b=7,\ c=-3$

We use the quadratic formula.

$$y=\frac{-7\pm\sqrt{7^2-4\cdot 1\cdot(-3)}}{2\cdot 1}=\frac{-7\pm\sqrt{49+12}}{2}$$

$$y=\frac{-7\pm\sqrt{61}}{2}$$

The solutions are $\frac{-7+\sqrt{61}}{2}$ and $\frac{-7-\sqrt{61}}{2}$, or $\frac{-7\pm\sqrt{61}}{2}$.

59. $x^2+(x+2)^2=7$
$x^2+x^2+4x+4=7$
$2x^2+4x+4=7$
$2x^2+4x-3=0$ Standard form

$a=2,\ b=4,\ c=-3$

We use the quadratic formula.

$$x=\frac{-4\pm\sqrt{4^2-4\cdot 2\cdot(-3)}}{2\cdot 2}=\frac{-4\pm\sqrt{16+24}}{4}$$

$$x=\frac{-4\pm\sqrt{40}}{4}=\frac{-4\pm\sqrt{4\cdot 10}}{4}$$

$$x=\frac{-4\pm 2\sqrt{10}}{4}=\frac{2(-2\pm\sqrt{10})}{2\cdot 2}$$

$$x=\frac{-2\pm\sqrt{10}}{2}$$

The solutions are $\frac{-2+\sqrt{10}}{2}$ and $\frac{-2-\sqrt{10}}{2}$, or $\frac{-2\pm\sqrt{10}}{2}$.

61. $x^2+x=1$

$x^2+x-1=0$

Graph $y=x^2+x-1$ and observe that the graph intersects the x-axis at two points. This indicates that the equation has two real-number solutions.

63.-69. Left to the student

Chapter 17 Mid-Chapter Review

1. The statement is true. See page 1238 in the text.

2. The solutions of $ax^2+bx+c=0$ are the first coordinates of the x-intercepts of the graph of $y=ax^2+bx+c$. The given statement is false.

3. $ax^2+bx=0$

$x(ax+b)=0$

$x=0$ *or* $ax+b=0$

$x=0$ *or* $ax=-b$

$x=0$ *or* $x=-\frac{b}{a}$

One solution of the equation is 0 and, because $b\neq 0$, then $-\frac{b}{a}\neq 0$, so the other solution is a non-zero number. The given statement is true.

4. $x^2-6x-2=0$

$x^2-6x=2$

$x^2-6x+9=2+9$

$(x-3)^2=11$

$x-3=\pm\sqrt{11}$

$x=3\pm\sqrt{11}$

5.
$$3x^2 = 8x - 2$$
$$3x^2 - 8x + 2 = 0$$
$$a = 3,\ b = -8,\ c = 2$$
$$x = \frac{-b \pm \sqrt{b^2 - 4ac}}{2a}$$
$$x = \frac{-(-8) \pm \sqrt{(-8)^2 - 4 \cdot 3 \cdot 2}}{2 \cdot 3}$$
$$x = \frac{8 \pm \sqrt{64 - 24}}{6} = \frac{8 \pm \sqrt{40}}{6} = \frac{8 \pm \sqrt{4 \cdot 10}}{6}$$
$$x = \frac{8 \pm 2\sqrt{10}}{6} = \frac{2(4 \pm \sqrt{10})}{2 \cdot 3} = \frac{4 \pm \sqrt{10}}{3}$$

6. $q^2 - 5q + 10 = 0$ Standard form

$a = 1,\ b = -5,\ c = 10$

7. $6 - x^2 = 14x + 2$

$0 = x^2 + 14x - 4$ Standard form

$a = 1,\ b = 14,\ c = -4$

8. $17z = 3z^2$

$0 = 3z^2 - 17x$ Standard form

$a = 3,\ b = -17,\ c = 0$

9. $16x = 48x^2$
$$0 = 48x^2 - 16x$$
$$0 = 16x(3x - 1)$$
$$16x = 0 \quad or \quad 3x - 1 = 0$$
$$x = 0 \quad or \quad 3x = 1$$
$$x = 0 \quad or \quad x = \frac{1}{3}$$
The solutions are 0 and $\frac{1}{3}$.

10.
$$x(x - 3) = 10$$
$$x^2 - 3x = 10$$
$$x^2 - 3x - 10 = 0$$
$$(x - 5)(x + 2) = 0$$
$$x - 5 = 0 \quad or \quad x + 2 = 0$$
$$x = 5 \quad or \quad x = -2$$
The solutions are 5 and -2.

11. $20x^2 - 20x = 0$
$$20x(x - 1) = 0$$
$$20x = 0 \quad or \quad x - 1 = 0$$
$$x = 0 \quad or \quad x = 1$$
The solutions are 0 and 1.

12.
$$x^2 = 14x - 49$$
$$x^2 - 14x + 49 = 0$$
$$(x - 7)(x - 7) = 0$$
$$x - 7 = 0 \quad or \quad x - 7 = 0$$
$$x = 7 \quad or \quad x = 7$$
The solution is 7.

13. $t^2 + 2t = 0$
$$t(t + 2) = 0$$
$$t = 0 \quad or \quad t + 2 = 0$$
$$t = 0 \quad or \quad t = -2$$
The solutions are 0 and -2.

14.
$$18w^2 + 21w = 4$$
$$18w^2 + 21w - 4 = 0$$
$$(3w + 4)(6w - 1) = 0$$
$$3w + 4 = 0 \quad or \quad 6w - 1 = 0$$
$$3w = -4 \quad or \quad 6w = 1$$
$$w = -\frac{4}{3} \quad or \quad w = \frac{1}{6}$$
The solutions are $-\frac{4}{3}$ and $\frac{1}{6}$.

15. $9y^2 - 5y^2 = 82y + 6y$
$$4y^2 = 88y$$
$$4y^2 - 88y = 0$$
$$4y(y - 22) = 0$$
$$4y = 0 \quad or \quad y - 22 = 0$$
$$y = 0 \quad or \quad y = 22$$
The solutions are 0 and 22.

16. $2(s - 3) = s(s - 3)$
$$2s - 6 = s^2 - 3s$$
$$0 = s^2 - 5s + 6$$
$$0 = (s - 3)(s - 2)$$
$$s - 3 = 0 \quad or \quad s - 2 = 0$$
$$s = 3 \quad or \quad s = 2$$
The solutions are 3 and 2.

17.
$$8y^2 - 40y = -7y + 35$$
$$8y^2 - 33y - 35 = 0$$
$$(8y + 7)(y - 5) = 0$$
$$8y + 7 = 0 \quad or \quad y - 5 = 0$$
$$8y = -7 \quad or \quad y = 5$$
$$y = -\frac{7}{8} \quad or \quad y = 5$$
The solutions are $-\frac{7}{8}$ and 5.

18. $x^2 + 2x - 3 = 0$
$$x^2 + 2x = 3$$
$$x^2 + 2x + 1 = 3 + 1$$
$$(x + 1)^2 = 4$$
$$x + 1 = 2 \quad or \quad x + 1 = -2$$
$$x = 1 \quad or \quad x = -3$$
The solutions are 1 and -3.

19. $x^2 - 9x + 6 = 0$

$$x^2 - 9x = -6$$
$$x^2 - 9x + \frac{81}{4} = -6 + \frac{81}{4}$$
$$\left(x - \frac{9}{2}\right)^2 = \frac{57}{4}$$
$$x - \frac{9}{2} = \frac{\sqrt{57}}{2} \quad or \quad x - \frac{9}{2} = -\frac{\sqrt{57}}{2}$$
$$x = \frac{9}{2} + \frac{\sqrt{57}}{2} \quad or \quad x = \frac{9}{2} - \frac{\sqrt{57}}{2}$$
$$x = \frac{9 + \sqrt{57}}{2} \quad or \quad x = \frac{9 - \sqrt{57}}{2}$$

The solutions are $\frac{9 + \sqrt{57}}{2}$ and $\frac{9 - \sqrt{57}}{2}$, or $\frac{9 \pm \sqrt{57}}{2}$.

20. $2x^2 = 7x + 8$

$$2x^2 - 7x = 8$$
$$\frac{1}{2}(2x^2 - 7x) = \frac{1}{2} \cdot 8$$
$$x^2 - \frac{7}{2}x = 4$$
$$x^2 - \frac{7}{2}x + \frac{49}{16} = 4 + \frac{49}{16}$$
$$\left(x - \frac{7}{4}\right)^2 = \frac{113}{16}$$
$$x - \frac{7}{4} = \frac{\sqrt{113}}{4} \quad or \quad x - \frac{7}{4} = -\frac{\sqrt{113}}{4}$$
$$x = \frac{7}{4} + \frac{\sqrt{113}}{4} \quad or \quad x = \frac{7}{4} - \frac{\sqrt{113}}{4}$$
$$x = \frac{7 + \sqrt{113}}{4} \quad or \quad x = \frac{7 - \sqrt{113}}{4}$$

The solutions are $\frac{7 + \sqrt{113}}{4}$ and $\frac{7 - \sqrt{113}}{4}$, or $\frac{7 \pm \sqrt{113}}{4}$.

21. $y^2 + 80 = 18y$

$$y^2 - 18y = -80$$
$$y^2 - 18y + 81 = -80 + 81$$
$$(y - 9)^2 = 1$$
$$y - 9 = 1 \quad or \quad y - 9 = -1$$
$$y = 10 \quad or \quad y = 8$$

The solutions are 10 and 8.

22. $t^2 + \frac{3}{2}t - \frac{3}{2} = 0$

$$t^2 + \frac{3}{2}t = \frac{3}{2}$$
$$t^2 + \frac{3}{2}t + \frac{9}{16} = \frac{3}{2} + \frac{9}{16}$$
$$\left(t + \frac{3}{4}\right)^2 = \frac{33}{16}$$
$$t + \frac{3}{4} = \frac{\sqrt{33}}{4} \quad or \quad t + \frac{3}{4} = -\frac{\sqrt{33}}{4}$$
$$t = -\frac{3}{4} + \frac{\sqrt{33}}{4} \quad or \quad t = -\frac{3}{4} - \frac{\sqrt{33}}{4}$$
$$t = \frac{-3 + \sqrt{33}}{4} \quad or \quad t = \frac{-3 - \sqrt{33}}{4}$$

The solutions are $\frac{-3 + \sqrt{33}}{4}$ and $\frac{-3 - \sqrt{33}}{4}$, or $\frac{-3 \pm \sqrt{33}}{4}$.

23. $x + 7 = -3x^2$

$$3x^2 + x = -7$$
$$\frac{1}{3}(3x^2 + x) = \frac{1}{3}(-7)$$
$$x^2 + \frac{1}{3}x = -\frac{7}{3}$$
$$x^2 + \frac{1}{3}x + \frac{1}{36} = -\frac{7}{3} + \frac{1}{36}$$
$$\left(x + \frac{1}{6}\right)^2 = -\frac{83}{36}$$

Because $-\frac{83}{36}$ has no real-number square root, the equation has no real-number solution.

24. $6x^2 = 384$

$$x^2 = 64$$
$$x = 8 \quad or \quad x = -8$$

The solutions are 8 and -8.

25. $5y^2 + 2y + 3 = 0$

$$a = 5, \ b = 2, \ y = 3$$
$$y = \frac{-b \pm \sqrt{b^2 - 4ac}}{2a}$$
$$y = \frac{-2 \pm \sqrt{2^2 - 4 \cdot 5 \cdot 3}}{2 \cdot 5} = \frac{-2 \pm \sqrt{4 - 60}}{10}$$
$$y = \frac{-2 \pm \sqrt{-56}}{10}$$

The radicand is negative, so there are no real-number solutions.

26. $6(x - 3)^2 = 12$

$$(x - 3)^2 = 2 \quad \text{Dividing by 6}$$
$$x - 3 = \sqrt{2} \quad or \quad x - 3 = -\sqrt{2}$$
$$x = 3 + \sqrt{2} \quad or \quad x = 3 - \sqrt{2}$$

The solutions are $3 + \sqrt{2}$ and $3 - \sqrt{2}$, or $3 \pm \sqrt{2}$.

27. $4x^2 + 4x = 3$

$$4x^2 + 4x - 3 = 0$$
$$(2x - 1)(2x + 3) = 0$$
$$2x - 1 = 0 \quad or \quad 2x + 3 = 0$$
$$2x = 1 \quad or \quad 2x = -3$$
$$x = \frac{1}{2} \quad or \quad x = -\frac{3}{2}$$

The solutions are $\frac{1}{2}$ and $-\frac{3}{2}$.

28. $8y^2 - 5 = 19$

$8y^2 = 24$

$y^2 = 3$ Dividing by 8

$y = \sqrt{3}$ *or* $y = -\sqrt{3}$

The solutions are $\sqrt{3}$ and $-\sqrt{3}$, or $\pm\sqrt{3}$.

29. $a^2 = a + 1$

$a^2 - a - 1 = 0$

$a = 1,\ b = -1,\ c = -1$

(Note that a is used in two different ways in this exercise.)

$$a = \frac{-b \pm \sqrt{b^2 - 4ac}}{2a}$$

$$a = \frac{-(-1) \pm \sqrt{(-1)^2 - 4 \cdot 1 \cdot (-1)}}{2 \cdot 1} = \frac{1 \pm \sqrt{1+4}}{2}$$

$$a = \frac{1 \pm \sqrt{5}}{2}$$

The solutions are $\frac{1+\sqrt{5}}{2}$ and $\frac{1-\sqrt{5}}{2}$, or $\frac{1 \pm \sqrt{5}}{2}$.

30. $(w-2)^2 = 100$

$w - 2 = 10$ *or* $w - 2 = -10$

$w = 12$ *or* $w = -8$

The solutions are 12 and -8.

31. $5m^2 + 2m = -3$

$5m^2 + 2m + 3 = 0$

$a = 5, b = 2, c = 3$

$$m = \frac{-b \pm \sqrt{b^2 - 4ac}}{2a}$$

$$m = \frac{-2 \pm \sqrt{2^2 - 4 \cdot 5 \cdot 3}}{2 \cdot 5} = \frac{-2 \pm \sqrt{4 - 60}}{10}$$

$$m = \frac{-2 \pm \sqrt{-56}}{10}$$

The radicand is negative, so there are no real-number solutions.

32. $\left(y - \frac{1}{2}\right)^2 = \frac{5}{4}$

$y - \frac{1}{2} = \frac{\sqrt{5}}{2}$ *or* $y - \frac{1}{2} = -\frac{\sqrt{5}}{2}$

$y = \frac{1}{2} + \frac{\sqrt{5}}{2}$ *or* $y = \frac{1}{2} - \frac{\sqrt{5}}{2}$

$y = \frac{1+\sqrt{5}}{2}$ *or* $y = \frac{1-\sqrt{5}}{2}$

The solutions are $\frac{1+\sqrt{5}}{2}$ and $\frac{1-\sqrt{5}}{2}$, or $\frac{1 \pm \sqrt{5}}{2}$.

33. $3x^2 - 75 = 0$

$3x^2 = 75$

$x^2 = 25$

$x = 5$ *or* $x = -5$

The solutions are 5 and -5.

34. $2x^2 - 2x - 5 = 0$

$a = 2,\ b = -2,\ c = -5$

$$x = \frac{-b \pm \sqrt{b^2 - 4ac}}{2a}$$

$$x = \frac{-(-2) \pm \sqrt{(-2)^2 - 4 \cdot 2 \cdot (-5)}}{2 \cdot 2} = \frac{2 \pm \sqrt{4+40}}{4}$$

$$x = \frac{2 \pm \sqrt{44}}{4} = \frac{2 \pm \sqrt{4 \cdot 11}}{4} = \frac{2 \pm 2\sqrt{11}}{4}$$

$$x = \frac{2(1 \pm \sqrt{11})}{2 \cdot 2} = \frac{1 \pm \sqrt{11}}{2}$$

The solutions are $\frac{1+\sqrt{11}}{2}$ and $\frac{1-\sqrt{11}}{2}$, or $\frac{1 \pm \sqrt{11}}{2}$.

35. $(x+2)^2 = -5$

Because -5 has no real-number square root, the equation has no real-number solution.

36. $y^2 - y - 8 = 0$

$a = 1,\ b = -1,\ c = -8$

$$y = \frac{-b \pm \sqrt{b^2 - 4ac}}{2a}$$

$$y = \frac{-(-1) \pm \sqrt{(-1)^2 - 4 \cdot 1 \cdot (-8)}}{2 \cdot 1} = \frac{1 \pm \sqrt{1+32}}{2}$$

$$y = \frac{1 \pm \sqrt{33}}{2}$$

Using a calculator, we have

$\frac{1+\sqrt{33}}{2} \approx 3.4$ and

$\frac{1-\sqrt{33}}{2} \approx -2.4$.

The solutions are approximately 3.4 and -2.4.

37. $2x^2 + 7x + 1 = 0$

$a = 2, b = 7, c = 1$

$$x = \frac{-b \pm \sqrt{b^2 - 4ac}}{2a}$$

$$x = \frac{-7 \pm \sqrt{7^2 - 4 \cdot 2 \cdot 1}}{2 \cdot 2} = \frac{-7 \pm \sqrt{49-8}}{4}$$

$$x = \frac{-7 \pm \sqrt{41}}{4}$$

Using a calculator, we have

$\frac{-7+\sqrt{41}}{4} \approx -0.1$ and

$\frac{-7-\sqrt{41}}{4} \approx -3.4$.

The solutions are approximately -0.1 and -3.4.

38. $x^2 - x - 6 = 0$

$(x-3)(x+2) = 0$

$x - 3 = 0$ *or* $x + 2 = 0$

$x = 3$ *or* $x = -2$

The equation has two real-number solutions. Answer A is correct.

39. $x^2 = -9$

Because -9 has no real-number square root, the equation has no real-number solutions. Answer B is correct.

40. $x^2 = 31$

$x = \sqrt{31}$ *or* $x = -\sqrt{31}$

The equation has two real-number solutions. Answer A is correct.

41. $x^2 = 0$

$x = 0$

The only solution is 0, so answer C is correct.

42. $x^2 - x + 6 = 0$

$a = 1, b = -1, c = 6$

$$x = \frac{-b \pm \sqrt{b^2 - 4ac}}{2a}$$

$$x = \frac{-(-1) \pm \sqrt{(-1)^2 - 4 \cdot 1 \cdot 6}}{2 \cdot 1} = \frac{1 \pm \sqrt{1 - 24}}{2}$$

$$x = \frac{1 \pm \sqrt{-23}}{2}$$

Because the radicand is negative, there are no real-number solutions. Answer B is correct.

43. $(x-3)^2 = 36$

$x - 3 = 6$ *or* $x - 3 = -6$

$x = 9$ *or* $x = -3$

The solutions are 9 and -3. Answer C is correct.

44. $$\frac{-24 \pm \sqrt{720}}{18} = \frac{-24 \pm \sqrt{144 \cdot 5}}{18} = \frac{-24 \pm 12\sqrt{5}}{18} =$$

$$\frac{6(-4 \pm 2\sqrt{5})}{6 \cdot 3} = \frac{-4 \pm 2\sqrt{5}}{3}$$

Answer B is correct.

45. Mark does not recognize that the $\pm$ sign yields two solutions, one in which the radical is added to 3 and the other in which the radical is subtracted from 3.

46. The addition principle should be used at the outset to get 0 on one side of the equation. Since this was not done, the principle of zero products was not used correctly.

47. The first coordinates of the x-intercepts of the graph of $y = (x-2)(x+3)$ are the solutions of the equation $(x-2)(x+3) = 0$.

48. The quadratic formula would not be the easiest way to solve a quadratic equation when the equation can be solved by factoring or by using the principle of square roots.

49. Answers will vary. Any equation of the form $ax^2+bx+c = 0$, where $b^2 - 4ac < 0$, will do. Then the graph of the equation $y = ax^2 + bx + c$ will not cross the x-axis.

50. If $x = -5$ or $x = 7$, then $x + 5 = 0$ or $x - 7 = 0$. Thus, the equation $(x+5)(x-7) = 0$, or $x^2 - 2x - 35 = 0$, has solutions -5 and 7.

Exercise Set 17.4

1.
$q = \frac{VQ}{I}$

$I \cdot q = I \cdot \frac{VQ}{I}$ Multiplying by I

$Iq = VQ$ Simplifying

$I = \frac{VQ}{q}$ Dividing by q

3.
$S = \frac{kmM}{d^2}$

$d^2 \cdot S = d^2 \cdot \frac{kmM}{d^2}$ Multiplying by d^2

$d^2S = kmM$ Simplifying

$\frac{d^2S}{kM} = m$ Dividing by kM

5.
$S = \frac{kmM}{d^2}$

$d^2 \cdot S = d^2 \cdot \frac{kmM}{d^2}$ Multiplying by d^2

$d^2S = kmM$ Simplifying

$d^2 = \frac{kmM}{S}$ Dividing by S

7.
$T = \frac{10t}{W^2}$

$W^2 \cdot T = W^2 \cdot \frac{10t}{W^2}$ Multiplying by W^2

$W^2T = 10t$

$W^2 = \frac{10t}{T}$ Dividing by T

$W = \sqrt{\frac{10t}{T}}$ Principle of square roots. Assume W is nonnegative.

9.
$A = at + bt$

$A = t(a+b)$ Factoring

$\frac{A}{a+b} = t$ Dividing by $a+b$

11.
$y = ax + bx + c$

$y - c = ax + bx$ Subtracting c

$y - c = x(a+b)$ Factoring

$\frac{y-c}{a+b} = x$ Dividing by $a+b$

13.
$\frac{t}{a} + \frac{t}{b} = 1$

$ab\left(\frac{t}{a} + \frac{t}{b}\right) = ab \cdot 1$ Multiplying by ab

$ab \cdot \frac{t}{a} + ab \cdot \frac{t}{b} = ab$

$bt + at = ab$

$bt = ab - at$ Subtracting at

$bt = a(b-t)$ Factoring

$\frac{bt}{b-t} = a$ Dividing by $b-t$

15.
$$\frac{1}{p}+\frac{1}{q}=\frac{1}{f}$$
$$pqf\left(\frac{1}{p}+\frac{1}{q}\right)=pqf\cdot\frac{1}{f}\quad\text{Multiplying by } pqf$$
$$pqf\cdot\frac{1}{p}+pqf\cdot\frac{1}{q}=pq$$
$$qf+pf=pq$$
$$qf=pq-pf\quad\text{Subtracting } pf$$
$$qf=p(q-f)\quad\text{Factoring}$$
$$\frac{qf}{q-f}=p\quad\text{Dividing by } q-f$$

17.
$$A=\frac{1}{2}bh$$
$$2\cdot A=2\cdot\frac{1}{2}bh\quad\text{Multiplying by 2}$$
$$2A=bh$$
$$\frac{2A}{h}=b\quad\text{Dividing by } h$$

19.
$$S=2\pi r(r+h)$$
$$S=2\pi r^2+2\pi rh\quad\text{Removing parentheses}$$
$$S-2\pi r^2=2\pi rh\quad\text{Subtracting } 2\pi r^2$$
$$\frac{S-2\pi r^2}{2\pi r}=h,\text{ or}\quad\text{Dividing by } 2\pi r$$
$$\frac{S}{2\pi r}-r=h$$

21.
$$\frac{1}{R}=\frac{1}{r_1}+\frac{1}{r_2}$$
$$Rr_1r_2\cdot\frac{1}{R}=Rr_1r_2\left(\frac{1}{r_1}+\frac{1}{r_2}\right)\quad\text{Multiplying by } Rr_1r_2$$
$$r_1r_2=Rr_1r_2\cdot\frac{1}{r_1}+Rr_1r_2\cdot\frac{1}{r_2}$$
$$r_1r_2=Rr_2+Rr_1$$
$$r_1r_2=R(r_2+r_1)\quad\text{Factoring}$$
$$\frac{r_1r_2}{r_2+r_1}=R\quad\text{Dividing by } r_2+r_1$$

23.
$$P=17\sqrt{Q}$$
$$\frac{P}{17}=\sqrt{Q}\quad\text{Isolating the radical}$$
$$\left(\frac{P}{17}\right)^2=(\sqrt{Q})^2\quad\text{Principle of squaring}$$
$$\frac{P^2}{289}=Q\quad\text{Simplifying}$$

25.
$$v=\sqrt{\frac{2gE}{m}}$$
$$v^2=\left(\sqrt{\frac{2gE}{m}}\right)^2\quad\text{Principle of squaring}$$
$$v^2=\frac{2gE}{m}$$
$$mv^2=2gE\quad\text{Multipying by } m$$
$$\frac{mv^2}{2g}=E\quad\text{Dividing by } 2g$$

27.
$$S=4\pi r^2$$
$$\frac{S}{4\pi}=r^2\quad\text{Dividing by } 4\pi$$
$$\sqrt{\frac{S}{4\pi}}=r\quad\text{Principle of square roots. Assume } r \text{ is nonnegative.}$$
$$\sqrt{\frac{1}{4}\cdot\frac{S}{\pi}}=r$$
$$\frac{1}{2}\sqrt{\frac{S}{\pi}}=r$$

29.
$$P=kA^2+mA$$
$$0=kA^2+mA-P\quad\text{Standard form}$$
$$a=k,\ b=m,\ c=-P$$
$$A=\frac{-b\pm\sqrt{b^2-4ac}}{2a}\quad\text{Quadratic formula}$$
$$A=\frac{-m\pm\sqrt{m^2-4\cdot k\cdot(-P)}}{2\cdot k}\quad\text{Substituting}$$
$$A=\frac{-m+\sqrt{m^2+4kP}}{2k}\quad\text{Using the positive root}$$

31.
$$c^2=a^2+b^2$$
$$c^2-b^2=a^2$$
$$\sqrt{c^2-b^2}=a\quad\text{Principle of square roots. Assume } a \text{ is nonnegative.}$$

33.
$$s=16t^2$$
$$\frac{s}{16}=t^2$$
$$\sqrt{\frac{s}{16}}=t\quad\text{Principle of square roots. Assume } t \text{ is nonnegative.}$$
$$\frac{\sqrt{s}}{4}=t$$

35.
$$A=\pi r^2+2\pi rh$$
$$0=\pi r^2+2\pi hr-A$$
$$a=\pi,\ b=2\pi h,\ c=-A$$
$$r=\frac{-b\pm\sqrt{b^2-4ac}}{2a}$$
$$r=\frac{-2\pi h\pm\sqrt{(2\pi h)^2-4\cdot\pi\cdot(-A)}}{2\cdot\pi}$$
$$r=\frac{-2\pi h+\sqrt{4\pi^2h^2+4\pi A}}{2\pi}\quad\text{Using the positive root}$$
$$r=\frac{-2\pi h+\sqrt{4(\pi^2h^2+\pi A)}}{2\pi}$$
$$r=\frac{-2\pi h+2\sqrt{\pi^2h^2+\pi A}}{2\pi}$$
$$r=\frac{2\left(-\pi h+\sqrt{\pi^2h^2+\pi A}\right)}{2\pi}$$
$$r=\frac{-\pi h+\sqrt{\pi^2h^2+\pi A}}{\pi}$$

37.
$$F = \frac{Av^2}{400}$$
$$400F = Av^2 \quad \text{Multiplying by 400}$$
$$\frac{400F}{A} = v^2 \quad \text{Dividing by } A$$
$$\sqrt{\frac{400F}{A}} = v \quad \text{Principle of square roots. Assume } v \text{ is nonnegative.}$$
$$\sqrt{400 \cdot \frac{F}{A}} = v$$
$$20\sqrt{\frac{F}{a}} = v$$

39.
$$c = \sqrt{a^2 + b^2}$$
$$c^2 = (\sqrt{a^2 + b^2})^2 \quad \text{Principle of squaring}$$
$$c^2 = a^2 + b^2$$
$$c^2 - b^2 = a^2$$
$$\sqrt{c^2 - b^2} = a \quad \text{Principle of square roots. Assume } a \text{ is nonnegative.}$$

41.
$$h = \frac{a}{2}\sqrt{3}$$
$$2h = a\sqrt{3}$$
$$\frac{2h}{\sqrt{3}} = a$$
$$\frac{2h\sqrt{3}}{3} = a \quad \text{Rationalizing the denominator}$$

43. $n = aT^2 - 4T + m$
$$0 = aT^2 - 4T + m - n$$
$$a = a,\ b = -4,\ c = m - n$$
$$T = \frac{-b \pm \sqrt{b^2 - 4ac}}{2a}$$
$$T = \frac{-(-4) \pm \sqrt{(-4)^2 - 4 \cdot a \cdot (m - n)}}{2 \cdot a}$$
$$T = \frac{4 + \sqrt{16 - 4a(m - n)}}{2a} \quad \text{Using the positive root}$$
$$T = \frac{4 + \sqrt{4[4 - a(m - n)]}}{2a}$$
$$T = \frac{4 + 2\sqrt{4 - a(m - n)}}{2a}$$
$$T = \frac{2\left(2 + \sqrt{4 - a(m - n)}\right)}{2 \cdot a}$$
$$T = \frac{2 + \sqrt{4 - a(m - n)}}{a}$$

45.
$$v = 2\sqrt{\frac{2kT}{\pi m}}$$
$$\frac{v}{2} = \sqrt{\frac{2kT}{\pi m}} \quad \text{Isolating the radical}$$
$$\left(\frac{v}{2}\right)^2 = \left(\sqrt{\frac{2kT}{\pi m}}\right)^2 \quad \text{Principle of squaring}$$
$$\frac{v^2}{4} = \frac{2kT}{\pi m}$$
$$\frac{v^2}{4} \cdot \frac{\pi m}{2k} = \frac{2kT}{\pi m} \cdot \frac{\pi m}{2k} \quad \text{Multiplying by } \frac{\pi m}{2k}$$
$$\frac{v^2 \pi m}{8k} = T$$

47. $3x^2 = d^2$
$$x^2 = \frac{d^2}{3} \quad \text{Dividing by 3}$$
$$x = \frac{d}{\sqrt{3}} \quad \text{Principle of square roots. Assume } x \text{ is nonnegative.}$$
$$x = \frac{d}{\sqrt{3}} \cdot \frac{\sqrt{3}}{\sqrt{3}} \quad \text{Rationalizing the denominator}$$
$$x = \frac{d\sqrt{3}}{3}$$

49.
$$N = \frac{n^2 - n}{2}$$
$$2N = n^2 - n \quad \text{Multiplying by 2}$$
$$0 = n^2 - n - 2N \quad \text{Finding standard form}$$
$$a = 1,\ b = -1,\ c = -2N$$
$$n = \frac{-b \pm \sqrt{b^2 - 4ac}}{2a}$$
$$n = \frac{-(-1) \pm \sqrt{(-1)^2 - 4 \cdot 1 \cdot (-2N)}}{2 \cdot 1} \quad \text{Substituting}$$
$$n = \frac{1 + \sqrt{1 + 8N}}{2} \quad \text{Using the positive root}$$

51.
$$S = \frac{a + b}{3b}$$
$$3b \cdot S = 3b \cdot \frac{a + b}{3b}$$
$$3bS = a + b$$
$$3bS - b = a$$
$$b(3S - 1) = a$$
$$b = \frac{a}{3S - 1}$$

53.
$$\frac{A - B}{AB} = Q$$
$$AB \cdot \frac{A - B}{AB} = AB \cdot Q$$
$$A - B = ABQ$$
$$A = ABQ + B$$
$$A = B(AQ + 1)$$
$$\frac{A}{AQ + 1} = B$$

55.
$$S = 180(n-2)$$
$$S = 180n - 360$$
$$S + 360 = 180n$$
$$\frac{S+360}{180} = n, \text{ or}$$
$$\frac{S}{180} + 2 = n$$

57.
$$A = P(1+rt)$$
$$A = P + Prt$$
$$A - P = Prt$$
$$\frac{A-P}{Pr} = t$$

59.
$$\frac{A}{B} = \frac{C}{D}$$
$$BD \cdot \frac{A}{B} = BD \cdot \frac{C}{D}$$
$$AD = BC$$
$$D = \frac{BC}{A}$$

61.
$$C = \frac{Ka-b}{a}$$
$$a \cdot C = a \cdot \frac{Ka-b}{a}$$
$$aC = Ka - b$$
$$aC - Ka = -b$$
$$a(C-K) = -b$$
$$a = \frac{-b}{C-K}, \text{ or}$$
$$a = \frac{b}{K-C}$$

63. $a^2 + b^2 = c^2$ Pythagorean equation
$4^2 + 7^2 = c^2$ Substituting
$16 + 49 = c^2$
$65 = c^2$
$\sqrt{65} = c$ Exact answer
$8.062 \approx c$ Approximate answer

65. $a^2 + b^2 = c^2$ Pythagorean equation
$4^2 + 5^2 = c^2$ Substituting
$16 + 25 = c^2$
$41 = c^2$
$\sqrt{41} = c$ Exact answer
$6.403 \approx c$ Approximate answer

67. $a^2 + b^2 = c^2$ Pythagorean equation
$2^2 + b^2 = (8\sqrt{17})^2$ Substituting
$4 + b^2 = 64 \cdot 17$
$4 + b^2 = 1088$
$b^2 = 1084$
$b = \sqrt{1084}$ Exact answer
$b \approx 32.924$ Approximate answer

69. We make a drawing. Let l = the length of the guy wire.

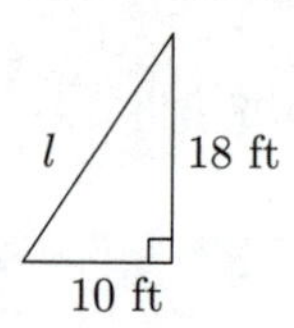

Then we use the Pythagorean equation.
$$10^2 + 18^2 = l^2$$
$$100 + 324 = l^2$$
$$424 = l^2$$
$\sqrt{424} = l$ Exact answer
$20.591 \approx l$ Approximation

The length of the guy wire is $\sqrt{424}$ ft ≈ 20.591 ft.

71. $\sqrt{3x} \cdot \sqrt{6x} = \sqrt{18x^2} = \sqrt{9 \cdot x^2 \cdot 2} = \sqrt{9}\sqrt{x^2}\sqrt{2} = 3x\sqrt{2}$

73. $3\sqrt{t} \cdot \sqrt{t} = 3\sqrt{t^2} = 3t$

75. a) $C = 2\pi r$
$$\frac{C}{2\pi} = r$$

b) $A = \pi r^2$
$A = \pi \cdot \left(\frac{C}{2\pi}\right)^2$ Substituting $\frac{C}{2\pi}$ for r
$$A = \pi \cdot \frac{C^2}{4\pi^2}$$
$$A = \frac{C^2}{4\pi}$$

Exercise Set 17.5

1. ***Familiarize.*** Using the drawing in the text, we have w = the width of the raspberry patch, in feet, and $3w + 7$ = the length.

Translate. We use the formula for the area of a rectangle.
$$w(3w+7) = 76.$$

Solve. We solve the equation.
$$w(3w+7) = 76$$
$$3w^2 + 7w = 76$$
$$3w^2 + 7w - 76 = 0$$
$$(3w+19)(w-4) = 0$$
$$3w + 19 = 0 \quad or \quad w - 4 = 0$$
$$3w = -19 \quad or \quad w = 4$$
$$w = -\frac{19}{3} \quad or \quad w = 4$$

Since the width cannot be negative, we consider only 4 as a possible solution.

If $w = 4$, then $3w + 7 = 3 \cdot 4 + 7 = 12 + 7 = 19$.

Check. 19 ft is 7 more than 3 times 4 ft. The area is $19 \cdot 4$, or 76 ft^2. The answer checks.

State. The length of the raspberry patch is 19 ft, and the width is 4 ft.

3. ***Familiarize.*** Using the drawing in the text, we let s and $s+8$ represent the lengths of the legs, in inches.

Translate. We use the Pythagorean equation.

$$s^2 + (s+8)^2 = (8\sqrt{13})^2$$

Solve. We solve the equation.

$$s^2 + (s+8)^2 = (8\sqrt{13})^2$$
$$s^2 + s^2 + 16s + 64 = 64 \cdot 13$$
$$2s^2 + 16s + 64 = 832$$
$$2s^2 + 16s - 768 = 0$$
$$s^2 + 8s - 384 = 0 \quad \text{Dividing by 2}$$
$$(s+24)(s-16) = 0$$
$$s+24 = 0 \quad or \quad s-16 = 0$$
$$s = -24 \quad or \quad s = 16$$

Since the length of a leg cannot be negative, we consider only 16 as a possible solution.

If $s = 16$, then $s + 8 = 16 + 8 = 24$.

Check. The 24-in. side is 8 in. longer than the 16-in. side. Also, $16^2 + 24^2 = 256 + 576 = 832 = (8\sqrt{13})^2$. The answer checks.

State. The lengths of the legs of the square are 16 in. and 24 in.

5. ***Familiarize.*** Using the drawing in the text, we have $w =$ the width, in yards, and $3w + 1 =$ the length.

Translate. We use the formula for the area of a rectangle.

$$w(3w+1) = 68$$

Solve. We solve the equation.

$$w(3w+1) = 68$$
$$3w^2 + w = 68$$
$$3w^2 + w - 68 = 0$$

We will use the quadratic equation with $a = 3$, $b = 1$, and $c = -68$.

$$w = \frac{-b \pm \sqrt{b^2 - 4ac}}{2a}$$
$$w = \frac{-1 \pm \sqrt{1^2 - 4 \cdot 3 \cdot (-68)}}{2 \cdot 3} = \frac{-1 \pm \sqrt{1+816}}{6}$$
$$w = \frac{-1 \pm \sqrt{817}}{6}$$

Using a calculator, we have

$$\frac{-1+\sqrt{817}}{6} \approx 4.6 \text{ and}$$
$$\frac{-1-\sqrt{817}}{6} \approx -4.9.$$

Since the width cannot be negative, we consider only 4.6 as a possible solution.

If $w \approx 4.6$, then $3w + 1 \approx 3(4.6) + 1 = 13.8 + 1 = 14.8$.

Check. A length of 14.8 yd is 1 yd longer than three times a width of 4.6 yd. Also, $14.8(4.6) = 68.08 \approx 68$, so the answer checks.

State. The length of the pool is about 14.8 yd, and the width is about 4.6 yd.

7. ***Familiarize.*** We first make a drawing. We let x represent the length. Then $x - 4$ represents the width.

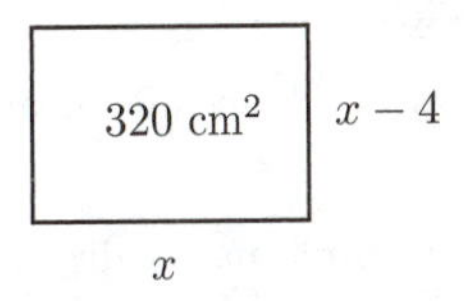

Translate. The area is length × width. Thus, we have two expressions for the area of the rectangle: $x(x-4)$ and 320. This gives us a translation.

$$x(x-4) = 320.$$

Solve. We solve the equation.

$$x^2 - 4x = 320$$
$$x^2 - 4x - 320 = 0$$
$$(x-20)(x+16) = 0$$
$$x - 20 = 0 \quad or \quad x + 16 = 0$$
$$x = 20 \quad or \quad x = -16$$

Check. Since the length of a side cannot be negative, -16 does not check. But 20 does check. If the length is 20, then the width is $20 - 4$, or 16. The area is 20×16, or 320. This checks.

State. The length is 20 cm, and the width is 16 cm.

9. ***Familiarize.*** We first make a drawing. We let x represent the length of one leg. Then $x + 2$ represents the length of the other leg.

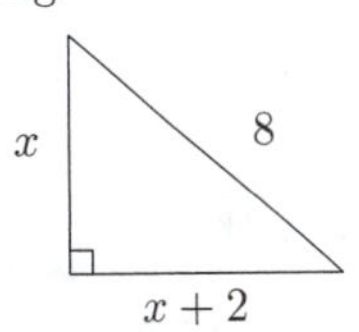

Translate. We use the Pythagorean equation.

$$x^2 + (x+2)^2 = 8^2.$$

Solve. We solve the equation.

$$x^2 + x^2 + 4x + 4 = 64$$
$$2x^2 + 4x + 4 = 64$$
$$2x^2 + 4x - 60 = 0$$
$$x^2 + 2x - 30 = 0 \quad \text{Dividing by 2}$$

$a = 1$, $b = 2$, $c = -30$

$$x = \frac{-2 \pm \sqrt{2^2 - 4 \cdot 1 \cdot (-30)}}{2 \cdot 1}$$
$$= \frac{-2 \pm \sqrt{4+120}}{2} = \frac{-2 \pm \sqrt{124}}{2}$$
$$= \frac{-2 \pm \sqrt{4 \cdot 31}}{2} = \frac{-2 \pm 2\sqrt{31}}{2}$$
$$= \frac{2(-1 \pm \sqrt{31})}{2} = -1 \pm \sqrt{31}$$

Using a calculator or Table 2 we find that $\sqrt{31} \approx 5.568$:

$$-1 + \sqrt{31} \approx -1 + 5.568 \quad or \quad -1 - \sqrt{31} \approx -1 - 5.568$$
$$\approx 4.6 \quad or \quad \approx -6.6$$

Check. Since the length of a leg cannot be negative, -6.6 does not check. But 4.6 does check. If the shorter leg is 4.6, then the other leg is 4.6 + 2, or 6.6. Then $4.6^2 + 6.6^2 = 21.16 + 43.56 = 64.72$ and using a calculator, $\sqrt{64.72} \approx 8.04 \approx 8$. Note that our check is not exact since we are using an approximation.

State. One leg is about 4.6 m, and the other is about 6.6 m long.

11. ***Familiarize.*** We first make a drawing. We let x represent the width and $x + 2$ the length.

20 in^2 x

$x + 2$

Translate. The area is length × width. We have two expressions for the area of the rectangle: $(x + 2)x$ and 20. This gives us a translation.

$(x + 2)x = 20.$

Solve. We solve the equation.

$$x^2 + 2x = 20$$
$$x^2 + 2x - 20 = 0$$

$a = 1,\ b = 2,\ c = -20$

$$x = \frac{-2 \pm \sqrt{2^2 - 4 \cdot 1 \cdot (-20)}}{2 \cdot 1}$$
$$= \frac{-2 \pm \sqrt{4 + 80}}{2} = \frac{-2 \pm \sqrt{84}}{2}$$
$$= \frac{-2 \pm \sqrt{4 \cdot 21}}{2} = \frac{-2 \pm 2\sqrt{21}}{2}$$
$$= \frac{2(-1 \pm \sqrt{21})}{2} = -1 \pm \sqrt{21}$$

Using a calculator or Table 2 we find that $\sqrt{21} \approx 4.583$:

$-1 + \sqrt{21} \approx -1 + 4.583$ *or* $-1 - \sqrt{21} \approx -1 - 4.583$
≈ 3.6 *or* ≈ -5.6

Check. Since the length of a side cannot be negative, -5.6 does not check. But 3.6 does check. If the width is 3.6, then the length is 3.6 + 2, or 5.6. The area is 5.6(3.6), or $20.16 \approx 20$. This checks.

State. The length is about 5.6 in., and the width is about 3.6 in.

13. ***Familiarize.*** We make a drawing and label it. We let w = the width of the rectangle and $2w$ = the length.

20 cm^2 w

$2w$

Translate. Recall that area = length × width. Then we have

$2w \cdot w = 20.$

Solve. We solve the equation.

$2w^2 = 20$
$w^2 = 10$ Dividing by 2
$w = \sqrt{10}$ *or* $w = -\sqrt{10}$ Principle of square roots
$w \approx 3.2$ *or* $w \approx -3.2$

Check. We know that -3.2 is not a solution of the original problem, because width cannot be negative. When $w \approx 3.2$, then $2w \approx 6.4$ and the area is about (6.4)(3.2), or 20.48. This checks, although the check is not exact since we used an approximation for $\sqrt{10}$.

State. The length is about 6.4 cm, and the width is about 3.2 cm.

15. ***Familiarize.*** Using the drawing in the text, we have x = the width of the frame, $20 - 2x$ = the width of the picture showing, and $25 - 2x$ = the length of the picture showing.

Translate. Recall that area = length × width. Then we have

$(25 - 2x)(20 - 2x) = 266.$

Solve. We solve the equation.

$500 - 90x + 4x^2 = 266$
$4x^2 - 90x + 234 = 0$
$2x^2 - 45x + 117 = 0$ Dividing by 2
$(2x - 39)(x - 3) = 0$

$2x - 39 = 0$ *or* $x - 3 = 0$
$2x = 39$ *or* $x = 3$
$x = 19.5$ *or* $x = 3$

Check. The number 19.5 cannot be a solution, because when $x = 19.5$ then $20 - 2x = -19$, and the width cannot be negative. When $x = 3$, then $20 - 2x = 20 - 2 \cdot 3$, or 14 and $25 - 2x = 25 - 2 \cdot 3$, or 19 and $19 \cdot 14 = 266$. This checks.

State. The width of the frame is 3 cm.

17. ***Familiarize.*** Referring to the drawing in the text, we complete the table.

	d	r	t
Upstream	40	$r - 3$	t_1
Downstream	40	$r + 3$	t_2
Total Time			14

Translate. Using $t = d/r$ and the rows of the table, we have

$$t_1 = \frac{40}{r - 3} \text{ and } t_2 = \frac{40}{r + 3}.$$

Since the total time is 14 hr, $t_1 + t_2 = 14$, and we have

$$\frac{40}{r - 3} + \frac{40}{r + 3} = 14.$$

Solve. We solve the equation. We multiply by $(r - 3)(r + 3)$, the LCM of the denominators.

$$(r - 3)(r + 3)\left(\frac{40}{r - 3} + \frac{40}{r + 3}\right) = (r - 3)(r + 3) \cdot 14$$
$$40(r + 3) + 40(r - 3) = 14(r^2 - 9)$$
$$40r + 120 + 40r - 120 = 14r^2 - 126$$
$$80r = 14r^2 - 126$$
$$0 = 14r^2 - 80r - 126$$
$$0 = 7r^2 - 40r - 63$$
$$0 = (7r + 9)(r - 7)$$

$$
\begin{aligned}
7r+9=0 \quad &or \quad r-7=0\\
7r=-9 \quad &or \quad r=7\\
r=-\frac{9}{7} \quad &or \quad r=7
\end{aligned}
$$

Check. Since speed cannot be negative, $-\frac{9}{7}$ cannot be a solution. If the speed of the boat is 7 km/h, the speed upstream is 7 − 3, or 4 km/h, and the speed downstream is 7 + 3, or 10 km/h. The time upstream is $\frac{40}{4}$, or 10 hr. The time downstream is $\frac{40}{10}$, or 4 hr. The total time is 14 hr. This checks.

State. The speed of the boat in still water is 7 km/h.

19. ***Familiarize.*** Let r = the speed of the stream. Then the boat's speed upstream is $8-r$ and the speed downstream is $8+r$. We fill in the table in the text.

	d	r	t
Upstream	60	$8-r$	t_1
Downstream	60	$8+r$	t_2
Total Time			16

Translate. Using $t=d/r$ and the rows of the table, we have

$$t_1=\frac{60}{8-r} \text{ and } t_2=\frac{60}{8+r}.$$

Since the total time is 16 hr, $t_1+t_2=16$, and we have

$$\frac{60}{8-r}+\frac{60}{8+r}=16.$$

Solve. We first multiply by $(8-r)(8+r)$, the LCM of the denominators.

$$
\begin{aligned}
(8-r)(8+r)\left(\frac{60}{8-r}+\frac{60}{8+r}\right)&=(8-r)(8+r)\cdot 16\\
60(8+r)+60(8-r)&=(64-r^2)16\\
480+60r+480-60r&=1024-16r^2\\
960&=1024-16r^2\\
16r^2-64&=0\\
16(r^2-4)&=0\\
16(r+2)(r-2)&=0
\end{aligned}
$$

$$
\begin{aligned}
r+2=0 \quad &or \quad r-2=0\\
r=-2 \quad &or \quad r=2
\end{aligned}
$$

Check. Since the speed of the stream cannot be negative, −2 cannot be a solution. If the speed of the stream is 2 km/h, the boat's speed upstream is 8 − 2 or 6 km/h, and the speed downstream is 8 + 2, or 10 km/h. The time upstream is $\frac{60}{6}$, or 10 hr. The time downstream is $\frac{60}{10}$, or 6 hr. The total time is 10+6, or 16 hr. The answer checks.

State. The speed of the stream is 2 km/h.

21. ***Familiarize.*** Let r represent the speed of the wind. Then the speed of the plane flying with the wind is $300+r$ and the speed against the wind is $300-r$. We fill in the table in the text.

	d	r	t
With wind	680	$300+r$	t_1
Against wind	520	$300-r$	t_2

Translate. Using $t=d/r$ and the rows of the table, we have

$$t_1=\frac{680}{300+r} \text{ and } t_2=\frac{520}{300-r}.$$

Since the total time is 4 hr, $t_1+t_2=4$, and we have

$$\frac{680}{300+r}+\frac{520}{300-r}=4.$$

Solve. We solve the equation. We multiply by $(300+r)(300-r)$, the LCM of the denominators.

$$
\begin{aligned}
(300+r)(300-r)\left(\frac{680}{300+r}+\frac{520}{300-r}\right)&=(300+r)(300-r)\cdot 4\\
680(300-r)+520(300+r)&=4(90,000-r^2)\\
204,000-680r+156,000+520r&=360,000-4r^2\\
360,000-160r&=360,000-4r^2\\
4r^2-160r&=0\\
4r(r-40)&=0
\end{aligned}
$$

$$
\begin{aligned}
4r=0 \quad &or \quad r-40=0\\
r=0 \quad &or \quad r=40
\end{aligned}
$$

Check. If $r=0$, then the speed of the wind is 0 km/h. That is, there is no wind. In this case the plane travels 680 km in 680/300, or $2\frac{4}{15}$ hr, and it travels 520 km in 520/300, or $1\frac{11}{15}$ hr. The total time is $2\frac{4}{15}+1\frac{11}{15}$, or 4 hr, so we have one solution. If the speed of the wind is 40 km/h, then the speed of the airplane with the wind is 300 + 40, or 340 km/h, and the speed against the wind is 300−40, or 260 km/h. The time with the wind is 680/340, or 2 hr, and the time against the wind is 520/260, or 2 hr. The total time is 2 hr + 2 hr, or 4 hr, so we have a second solution.

State. The speed of the wind is 0 km/h (There is no wind.) or 40 km/h.

23. ***Familiarize.*** We first make a drawing. We let r represent the speed of the boat in still water. Then $r-4$ is the speed of the boat traveling upstream and $r+4$ is the speed of the boat traveling downstream.

Upstream
$r-4$ mph
4 mi

Downstream
$r+4$ mph
12 mi

We summarize the information in a table.

	d	r	t
Upstream	4	$r-4$	t_1
Downstream	12	$r+4$	t_2
Total Time			2

Translate. Using $t = d/r$ and the rows of the table, we have

$$t_1 = \frac{4}{r-4} \text{ and } t_2 = \frac{12}{r+4}.$$

Since the total time is 2 hr, $t_1 + t_2 = 2$, and we have

$$\frac{4}{r-4} + \frac{12}{r+4} = 2.$$

Solve. We solve the equation. We multiply by $(r-4)(r+4)$, the LCM of the denominators.

$$\begin{aligned}(r-4)(r+4)\left(\frac{4}{r-4} + \frac{12}{r+4}\right) &= (r-4)(r+4)\cdot 2\\ 4(r+4) + 12(r-4) &= 2(r^2-16)\\ 4r + 16 + 12r - 48 &= 2r^2 - 32\\ 16r - 32 &= 2r^2 - 32\\ 0 &= 2r^2 - 16r\\ 0 &= 2r(r-8)\end{aligned}$$

$$\begin{aligned}2r = 0 \quad &or \quad r - 8 = 0\\ r = 0 \quad &or \quad\quad r = 8\end{aligned}$$

Check. If $r = 0$, then the speed upstream, $0-4$, would be negative. Since speed cannot be negative, 0 cannot be a solution. If the speed of the boat is 8 mph, the speed upstream is $8-4$, or 4 mph, and the speed downstream is $8+4$, or 12 mph. The time upstream is $\frac{4}{4}$, or 1 hr. The time downstream is $\frac{12}{12}$, or 1 hr. The total time is 2 hr. This checks.

State. The speed of the boat in still water is 8 mph.

25. ***Familiarize.*** We first make a drawing. We let r represent the speed of the stream. Then $9-r$ represents the speed of the boat traveling upstream and $9+r$ represents the speed of the boat traveling downstream.

Upstream
$9-r$ km/h
80 km

Downstream
$9+r$ km/h
80 km

We summarize the information in a table.

	d	r	t
Upstream	80	$9-r$	t_1
Downstream	80	$9+r$	t_2

Translate. Using $t = d/r$ and the rows of the table, we have

$$t_1 = \frac{80}{9-r} \text{ and } t_2 = \frac{80}{9+r}.$$

Since the total time is 18 hr, $t_1 + t_2 = 18$, and we have

$$\frac{80}{9-r} + \frac{80}{9+r} = 18.$$

Solve. We solve the equation. We multiply by $(9-r)(9+r)$, the LCM of the denominators.

$$\begin{aligned}(9-r)(9+r)\left(\frac{80}{9-r} + \frac{80}{9+r}\right) &= (9-r)(9+r)\cdot 18\\ 80(9+r) + 80(9-r) &= 18(81-r^2)\\ 720 + 80r + 720 - 80r &= 1458 - 18r^2\\ 1440 &= 1458 - 18r^2\\ 18r^2 &= 18\\ r^2 &= 1\end{aligned}$$

$r = 1$ *or* $r = -1$ Principle of square roots

Check. Since speed cannot be negative, -1 cannot be a solution. If the speed of the stream is 1 km/h, the speed upstream is $9-1$, or 8 km/h, and the speed downstream is $9+1$, or 10 km/h. The time upstream is $\frac{80}{8}$, or 10 hr. The time downstream is $\frac{80}{10}$, or 8 hr. The total time is 18 hr. This checks.

State. The speed of the stream is 1 km/h.

27. $$\begin{aligned}5\sqrt{2} + \sqrt{18} &= 5\sqrt{2} + \sqrt{9\cdot 2}\\ &= 5\sqrt{2} + \sqrt{9}\sqrt{2}\\ &= 5\sqrt{2} + 3\sqrt{2}\\ &= (5+3)\sqrt{2}\\ &= 8\sqrt{2}\end{aligned}$$

29. $$\begin{aligned}\sqrt{4x^3} - 7\sqrt{x} &= \sqrt{4\cdot x^2\cdot x} - 7\sqrt{x}\\ &= \sqrt{4}\sqrt{x^2}\sqrt{x} - 7\sqrt{x}\\ &= 2x\sqrt{x} - 7\sqrt{x}\\ &= (2x-7)\sqrt{x}\end{aligned}$$

31. $$\begin{aligned}\sqrt{2} + \sqrt{\frac{1}{2}} &= \sqrt{2} + \sqrt{\frac{1}{2}\cdot\frac{2}{2}}\\ &= \sqrt{2} + \sqrt{\frac{2}{4}}\\ &= \sqrt{2} + \frac{\sqrt{2}}{\sqrt{4}}\\ &= \sqrt{2} + \frac{\sqrt{2}}{2}\\ &= (1 + \frac{1}{2})\sqrt{2}\\ &= \frac{3}{2}\sqrt{2}, \text{or } \frac{3\sqrt{2}}{2}\end{aligned}$$

33. $$\begin{aligned}&\sqrt{24} + \sqrt{54} - \sqrt{48}\\ &= \sqrt{4\cdot 6} + \sqrt{9\cdot 6} - \sqrt{16\cdot 3}\\ &= \sqrt{4}\cdot\sqrt{6} + \sqrt{9}\cdot\sqrt{6} - \sqrt{16}\cdot\sqrt{3}\\ &= 2\sqrt{6} + 3\sqrt{6} - 4\sqrt{3}\\ &= 5\sqrt{6} - 4\sqrt{3}\end{aligned}$$

35. $8x = 4 - y$

To find the y-intercept, we let $x = 0$ and solve for y.

$$8 \cdot 0 = 4 - y$$
$$0 = 4 - y$$
$$y = 4$$

The y-intercept is $(0, 4)$.

To find the x-intercept, we let $y = 0$ and solve for x.

$$8x = 4 - 0$$
$$8x = 4$$
$$x = \frac{1}{2}$$

The x-intercept is $\left(\frac{1}{2}, 0\right)$.

37. ***Familiarize.*** The radius of a 12-in. pizza is $\frac{12}{2}$, or 6 in. The radius of a d-in. pizza is $\frac{d}{2}$ in. The area of a circle is πr^2.

Translate.

Area of d-in. pizza is Area of 12-in. pizza plus Area of 12-in. pizza

$$\pi\left(\frac{d}{2}\right)^2 = \pi \cdot 6^2 + \pi \cdot 6^2$$

Solve. We solve the equation.

$$\frac{d^2}{4}\pi = 36\pi + 36\pi$$
$$\frac{d^2}{4}\pi = 72\pi$$
$$\frac{d^2}{4} = 72 \quad \text{Dividing by } \pi$$
$$d^2 = 288$$

$$d = \sqrt{288} \ \ or \ \ d = -\sqrt{288}$$
$$d = 12\sqrt{2} \ \ or \ \ d = -12\sqrt{2}$$
$$d \approx 16.97 \ \ or \ \ d \approx -16.97 \quad \text{Using a calculator}$$

Check. Since the diameter cannot be negative, -16.97 is not a solution. If $d = 12\sqrt{2}$, or 16.97, then $r = 6\sqrt{2}$ and the area is $\pi(6\sqrt{2})^2$, or 72π. The area of the two 12-in. pizzas is $2 \cdot \pi \cdot 6^2$, or 72π. The value checks.

State. The diameter of the pizza should be $12\sqrt{2}$ in. $\approx$ 16.97 in.

The radius of a 16-in. pizza is $\frac{16}{2}$, or 8 in., so the area is $\pi(8)^2$, or 64π. We found that the area of two 12-in. pizzas is 72π and $72\pi > 64\pi$, so you get more to eat with two 12-in. pizzas than with a 16-in. pizza.

Exercise Set 17.6

1. $y = x^2 + 1$

We first find the vertex. The x-coordinate is

$$-\frac{b}{2a} = -\frac{0}{2 \cdot 1} = 0.$$

We substitute into the equation to find the second coordinate of the vertex.

$$y = x^2 + 1 = 0^2 + 1 = 1$$

The vertex is (0,1). This is also the y-intercept. The line of symmetry is $x = 0$, the y-axis.

We choose some x-values on both sides of the vertex and graph the parabola.

When $x = 1$, $y = 1^2 + 1 = 1 + 1 = 2$.

When $x = -1$, $y = (-1)^2 + 1 = 1 + 1 = 2$.

When $x = 2$, $y = 2^2 + 1 = 4 + 1 = 5$.

When $x = -2$, $y = (-2)^2 + 1 = 4 + 1 = 5$.

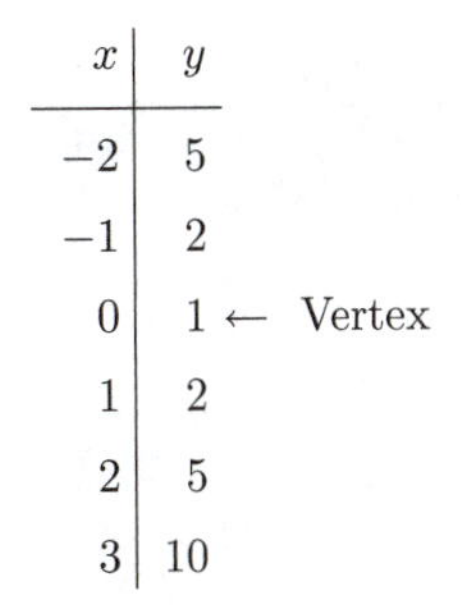

x	y	
-2	5	
-1	2	
0	1	← Vertex
1	2	
2	5	
3	10	

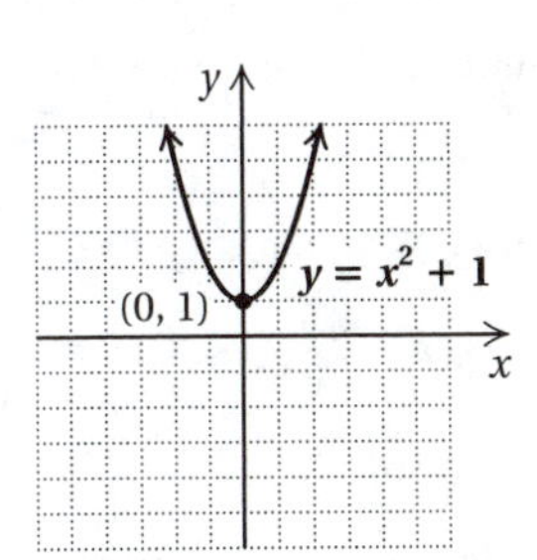

3. $y = -1 \cdot x^2$

Find the vertex. The x-coordinate is

$$-\frac{b}{2a} = -\frac{0}{2(-1)} = 0.$$

The y-coordinate is

$$y = -1 \cdot x^2 = -1 \cdot 0^2 = 0.$$

The vertex is (0,0). This is also the y-intercept. The line of symmetry is $x = 0$, the y-axis.

Choose some x-values on both sides of the vertex and graph the parabola.

When $x = -2$, $y = -1 \cdot (-2)^2 = -1 \cdot 4 = -4$.

When $x = -1$, $y = -1 \cdot (-1)^2 = -1 \cdot 1 = -1$.

When $x = 1$, $y = -1 \cdot 1^2 = -1 \cdot 1 = -1$.

When $x = 2$, $y = -1 \cdot 2^2 = -1 \cdot 4 = -4$.

x	y	
0	0	←Vertex
-2	-4	
-1	-1	
1	-1	
2	-4	

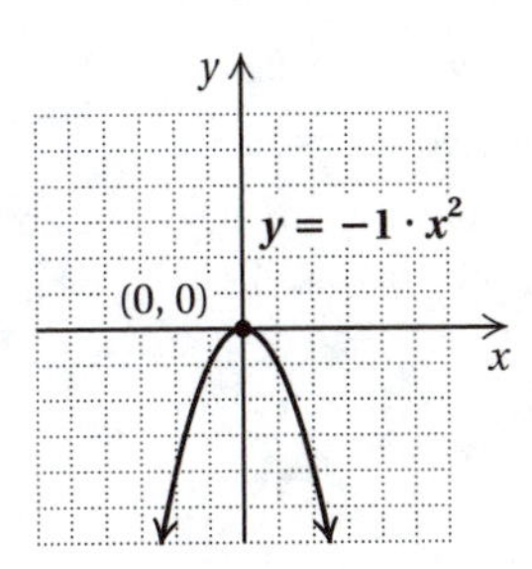

5. $y = -x^2 + 2x$

Find the vertex. The x-coordinate is

$$-\frac{b}{2a} = -\frac{2}{2(-1)} = -(-1) = 1.$$

The y-coordinate is

$$y = -x^2 + 2x = -(1)^2 + 2 \cdot 1 = -1 + 2 = 1.$$

The vertex is (1,1).

We choose some x-values on both sides of the vertex and graph the parabola. We make sure we find y when $x = 0$. This gives us the y-intercept.

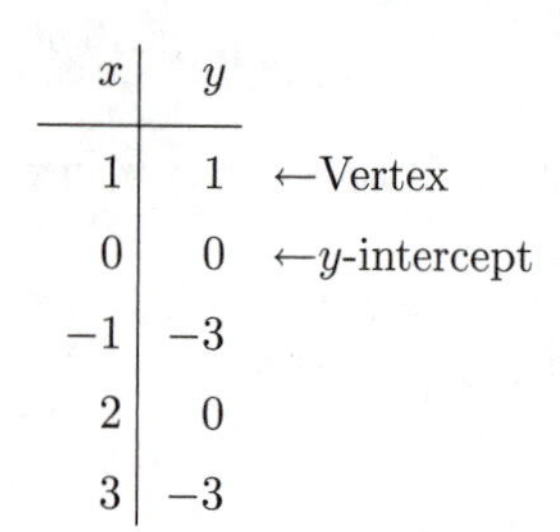

x	y	
1	1	←Vertex
0	0	←y-intercept
−1	−3	
2	0	
3	−3	

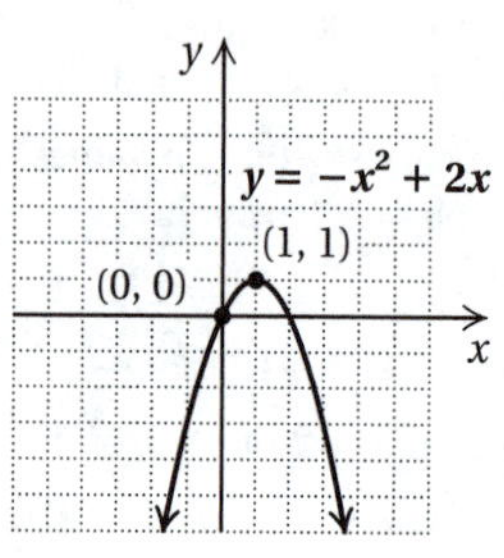

7. $y = 5 - x - x^2$, or $y = -x^2 - x + 5$

Find the vertex. The x-coordinate is

$$-\frac{b}{2a} = -\frac{-1}{2(-1)} = -\frac{1}{2}.$$

The y-coordinate is

$$y = 5 - x - x^2 = 5 - \left(-\frac{1}{2}\right) - \left(-\frac{1}{2}\right)^2 = 5 + \frac{1}{2} - \frac{1}{4} = \frac{21}{4}.$$

The vertex is $\left(-\frac{1}{2}, \frac{21}{4}\right)$.

We choose some x-values on both sides of the vertex and graph the parabola.

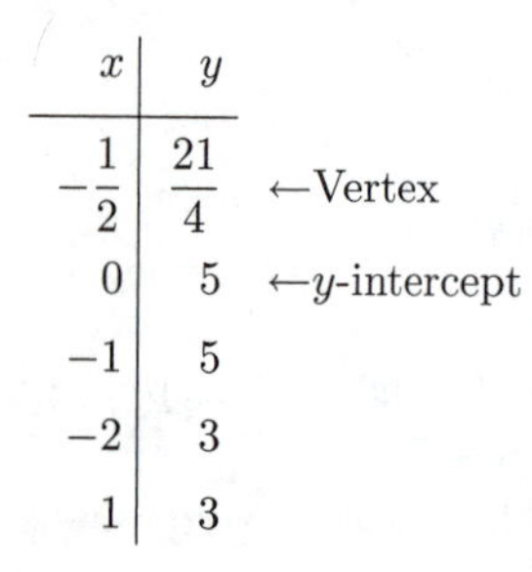

x	y	
$-\frac{1}{2}$	$\frac{21}{4}$	←Vertex
0	5	←y-intercept
−1	5	
−2	3	
1	3	

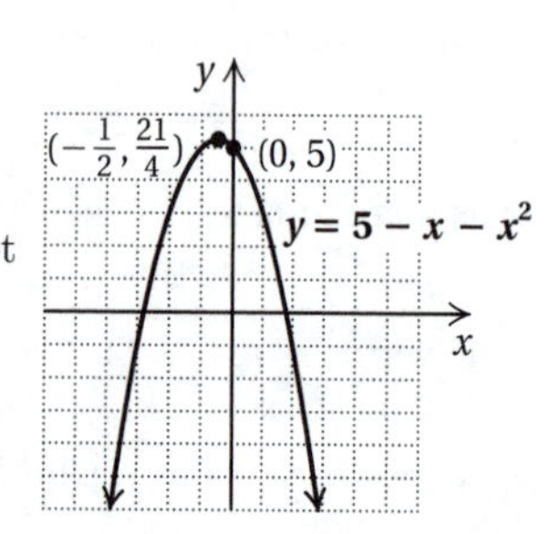

9. $y = x^2 - 2x + 1$

Find the vertex. The x-coordinate is

$$-\frac{b}{2a} = -\frac{-2}{2 \cdot 1} = -(-1) = 1.$$

The y-coordinate is

$$y = x^2 - 2x + 1 = 1^2 - 2 \cdot 1 + 1 = 1 - 2 + 1 = 0.$$

The vertex is $(1, 0)$.

We choose some x-values on both sides of the vertex and graph the parabola.

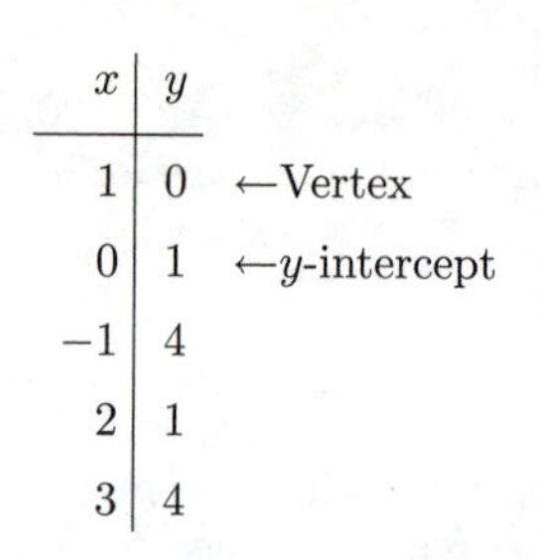

x	y	
1	0	←Vertex
0	1	←y-intercept
−1	4	
2	1	
3	4	

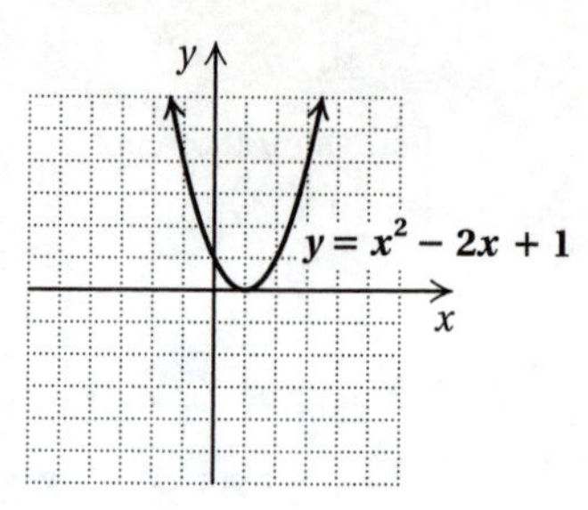

11. $y = -x^2 + 2x + 3$

Find the vertex. The x-coordinate is

$$-\frac{b}{2a} = -\frac{2}{2(-1)} = -(-1) = 1.$$

The y-coordinate is

$$y = -x^2 + 2x + 3 = -(1)^2 + 2 \cdot 1 + 3 = -1 + 2 + 3 = 4.$$

The vertex is $(1, 4)$.

We choose some x-values on both sides of the vertex and graph the parabola.

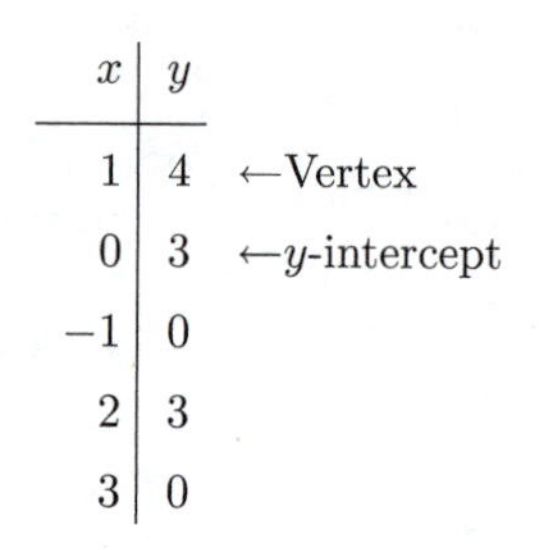

x	y	
1	4	←Vertex
0	3	←y-intercept
−1	0	
2	3	
3	0	

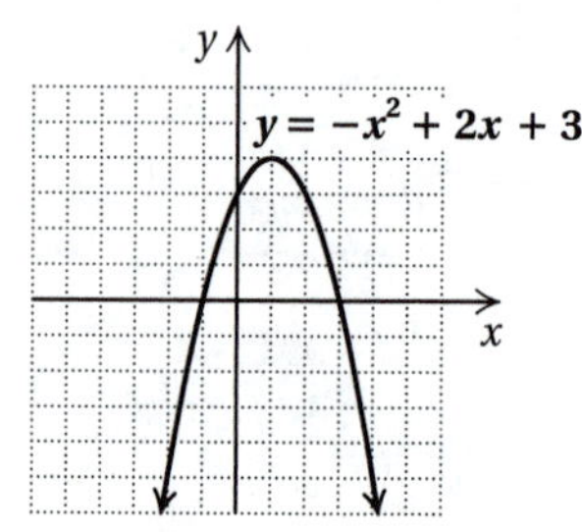

13. $y = -2x^2 - 4x + 1$

Find the vertex. The x-coordinate is

$$-\frac{b}{2a} = -\frac{-4}{2(-2)} = -1.$$

The y-coordinate is

$$y = -2x^2 - 4x + 1 = -2(-1)^2 - 4(-1) + 1 = -2 + 4 + 1 = 3.$$

The vertex is $(-1, 3)$.

We choose some x-values on both sides of the vertex and graph the parabola.

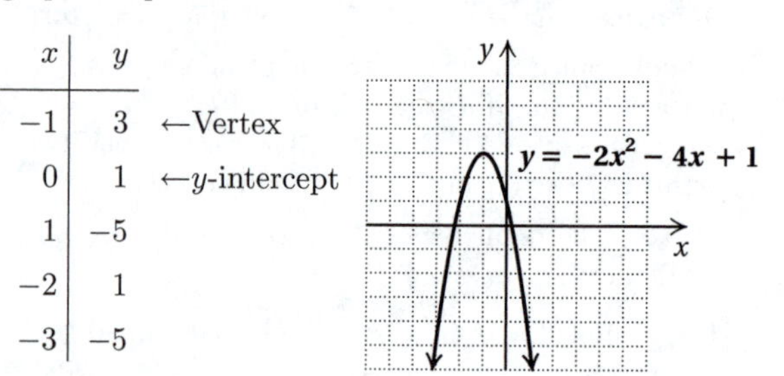

x	y	
−1	3	←Vertex
0	1	←y-intercept
1	−5	
−2	1	
−3	−5	

15. $y = 5 - x^2$, or $y = -x^2 + 5$

Find the vertex. The x-coordinate is

$$-\frac{b}{2a} = -\frac{0}{2(-1)} = 0.$$

The y-coordinate is

$$y = 5 - x^2 = 5 - 0^2 = 5.$$

The vertex is $(0, 5)$. This is also the y-intercept.

We choose some x-values on both sides of the vertex and graph the parabola.

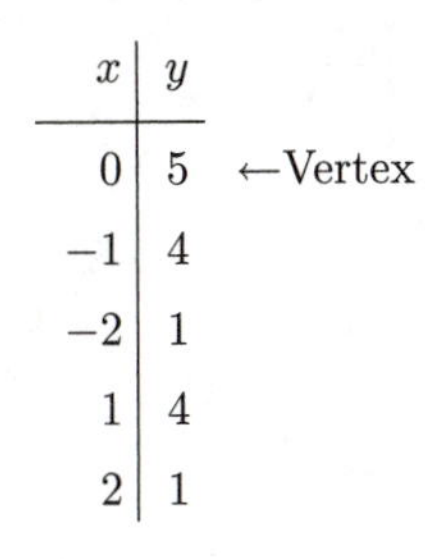

x	y	
0	5	←Vertex
−1	4	
−2	1	
1	4	
2	1	

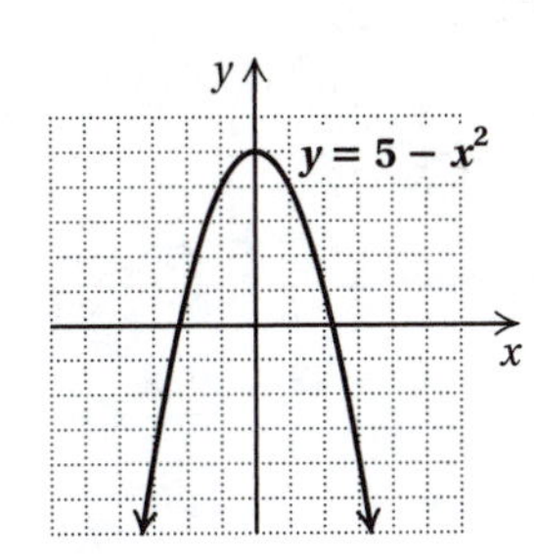

17. $y = \frac{1}{4}x^2$

Find the vertex. The x-coordinate is

$$-\frac{b}{2a} = -\frac{0}{2\left(\frac{1}{4}\right)} = 0.$$

The y-coordinate is

$$y = \frac{1}{4}x^2 = \frac{1}{4} \cdot 0^2 = 0.$$

The vertex is $(0, 0)$. This is also the y-intercept.

We choose some x-values on both sides of the vertex and graph the parabola.

x	y	
0	0	←Vertex
−2	1	
−4	4	
2	1	
4	4	

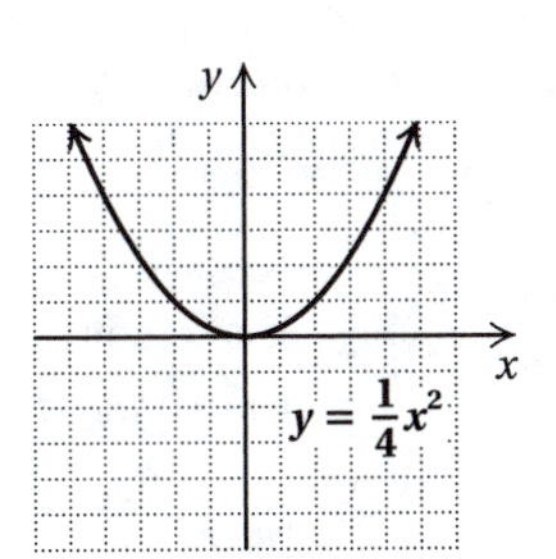

19. $y = -x^2 + x - 1$

Find the vertex. The x-coordinate is

$$-\frac{b}{2a} = -\frac{1}{2(-1)} = -\left(-\frac{1}{2}\right) = \frac{1}{2}.$$

The y-coordinate is

$$y = -x^2 + x - 1 = -\left(\frac{1}{2}\right)^2 + \frac{1}{2} - 1 = -\frac{1}{4} + \frac{1}{2} - 1 = -\frac{3}{4}.$$

The vertex is $\left(\frac{1}{2}, -\frac{3}{4}\right)$.

We choose some x-values on both sides of the vertex and graph the parabola.

x	y	
$\frac{1}{2}$	$-\frac{3}{4}$	←Vertex
0	−1	←y-intercept
−1	−3	
1	−1	
2	−3	

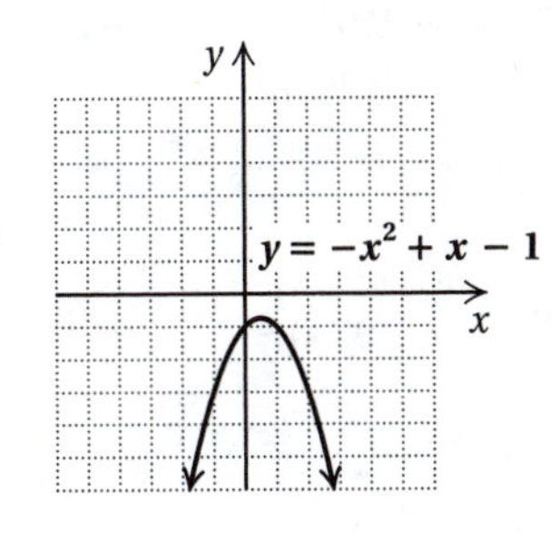

21. $y = -2x^2$

Find the vertex. The x-coordinate is

$$-\frac{b}{2a} = -\frac{0}{2(-2)} = 0.$$

The y-coordinate is

$$y = -2x^2 = -2 \cdot 0^2 = 0.$$

The vertex is $(0, 0)$. This is also the y-intercept.

We choose some x-values on both sides of the vertex and graph the parabola.

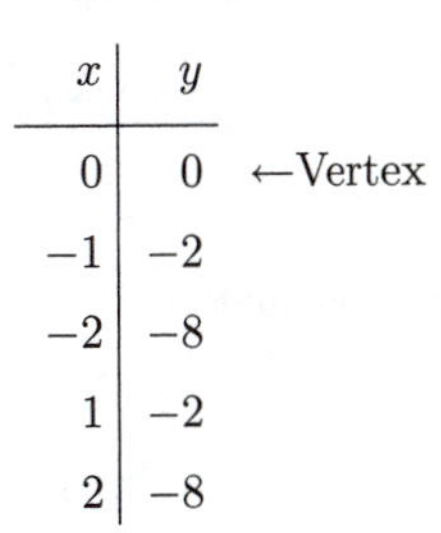

x	y	
0	0	←Vertex
−1	−2	
−2	−8	
1	−2	
2	−8	

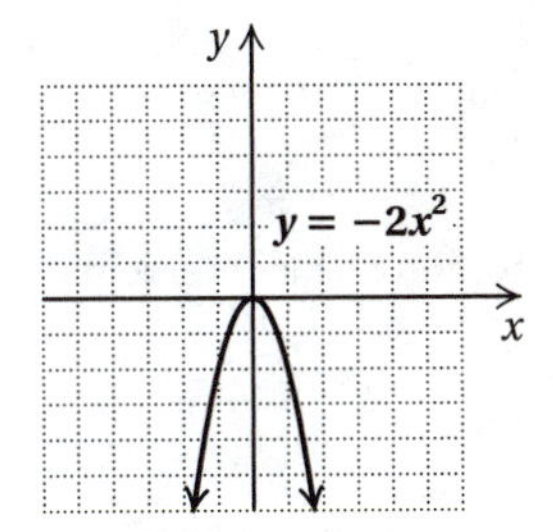

23. $y = x^2 - x - 6$

Find the vertex. The x-coordinate is

$$-\frac{b}{2a} = -\frac{-1}{2 \cdot 1} = -\left(-\frac{1}{2}\right) = \frac{1}{2}.$$

The y-coordinate is

$$y = x^2 - x - 6 = \left(\frac{1}{2}\right)^2 - \frac{1}{2} - 6 = \frac{1}{4} - \frac{1}{2} - 6 = -\frac{25}{4}.$$

The vertex is $\left(\frac{1}{2}, -\frac{25}{4}\right)$.

We choose some x-values on both sides of the vertex and graph the parabola.

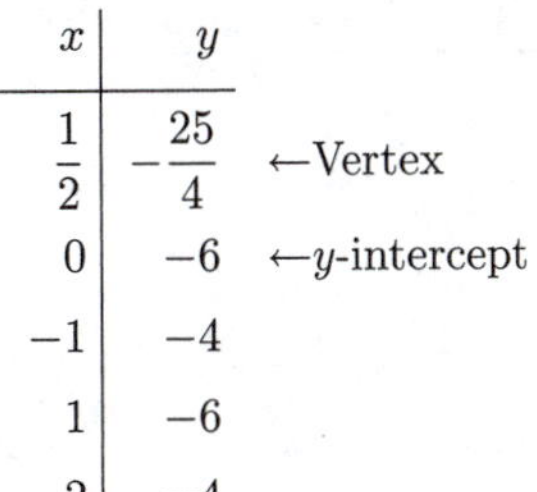

x	y	
$\frac{1}{2}$	$-\frac{25}{4}$	←Vertex
0	−6	←y-intercept
−1	−4	
1	−6	
2	−4	

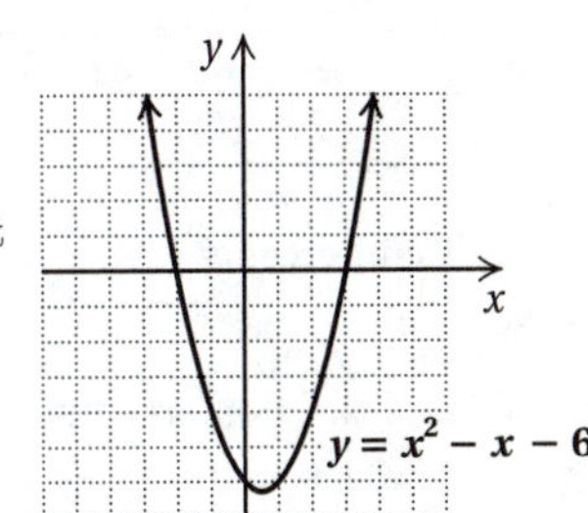

25. $y = x^2 - 2$

To find the x-intercepts we solve the equation $x^2 - 2 = 0$.

$$x^2 - 2 = 0$$
$$x^2 = 2$$

$x = \sqrt{2}$ *or* $x = -\sqrt{2}$ Principle of square roots

The x-intercepts are $(\sqrt{2}, 0)$ and $(-\sqrt{2}, 0)$.

27. $y = x^2 + 5x$

To find the x-intercepts we solve the equation $x^2 + 5x = 0$.

$$x^2 + 5x = 0$$
$$x(x + 5) = 0$$

$x = 0$ *or* $x + 5 = 0$

$x = 0$ *or* $x = -5$

The x-intercepts are $(0, 0)$ and $(-5, 0)$.

29. $y = 8 - x - x^2$

To find the x-intercepts we solve the equation $8 - x - x^2 = 0$.

$8 - x - x^2 = 0$

$x^2 + x - 8 = 0$ Standard form

$a = 1,\ b = 1,\ c = -8$

$$x = \frac{-1 \pm \sqrt{1^2 - 4 \cdot 1 \cdot (-8)}}{2 \cdot 1}$$

$$x = \frac{-1 \pm \sqrt{33}}{2}$$

The x-intercepts are $\left(\frac{-1+\sqrt{33}}{2}, 0\right)$ and $\left(\frac{-1-\sqrt{33}}{2}, 0\right)$.

31. $y = x^2 - 6x + 9$

To find the x-intercepts we solve the equation $x^2 - 6x + 9 = 0$.

$x^2 - 6x + 9 = 0$

$(x-3)(x-3) = 0$

$x - 3 = 0$ *or* $x - 3 = 0$

$x = 3$ *or* $x = 3$

The x-intercept is $(3, 0)$.

33. $y = -x^2 - 4x + 1$

To find the x-intercepts we solve the equation $-x^2 - 4x + 1 = 0$.

$-x^2 - 4x + 1 = 0$

$x^2 + 4x - 1 = 0$ Standard form

$a = 1,\ b = 4,\ c = -1$

$$x = \frac{-4 \pm \sqrt{4^2 - 4 \cdot 1 \cdot (-1)}}{2 \cdot 1}$$

$$x = \frac{-4 \pm \sqrt{20}}{2} = \frac{-4 \pm \sqrt{4 \cdot 5}}{2} = \frac{-4 \pm 2\sqrt{5}}{2}$$

$$x = \frac{2(-2 \pm \sqrt{5})}{2} = -2 \pm \sqrt{5}$$

The x-intercepts are $(-2 + \sqrt{5}, 0)$ and $(-2 - \sqrt{5}, 0)$.

35. $y = x^2 + 9$

To find the x-intercepts we solve the equation $x^2 + 9 = 0$.

$x^2 + 9 = 0$

$x^2 = -9$

The negative number -9 has no real-number square roots. Thus there are no x-intercepts.

37. $\sqrt{8} + \sqrt{50} + \sqrt{98} + \sqrt{128}$

$= \sqrt{4 \cdot 2} + \sqrt{25 \cdot 2} + \sqrt{49 \cdot 2} + \sqrt{64 \cdot 2}$

$= 2\sqrt{2} + 5\sqrt{2} + 7\sqrt{2} + 8\sqrt{2}$

$= 22\sqrt{2}$

39. $y = \dfrac{k}{x}$

$12.4 = \dfrac{k}{2.4}$ Substituting

$29.76 = k$ Variation constant

$y = \dfrac{29.76}{x}$ Equation of variation

41. $-\dfrac{1}{5} + \dfrac{7}{10} - \left(-\dfrac{4}{15}\right) + \dfrac{1}{60}$

$= -\dfrac{1}{5} + \dfrac{7}{10} + \dfrac{4}{15} + \dfrac{1}{60}$

$= -\dfrac{1}{5} \cdot \dfrac{12}{12} + \dfrac{7}{10} \cdot \dfrac{6}{6} + \dfrac{4}{15} \cdot \dfrac{4}{4} + \dfrac{1}{60}$

$= -\dfrac{12}{60} + \dfrac{42}{60} + \dfrac{16}{60} + \dfrac{1}{60}$

$= \dfrac{-12 + 42 + 16 + 1}{60}$

$= \dfrac{47}{60}$

43. a) We substitute 128 for H and solve for t:

$128 = -16t^2 + 96t$

$16t^2 - 96t + 128 = 0$

$16(t^2 - 6t + 8) = 0$

$16(t-2)(t-4) = 0$

$t - 2 = 0$ *or* $t - 4 = 0$

$t = 2$ *or* $t = 4$

The projectile is 128 ft from the ground 2 sec after launch and again 4 sec after launch. The graph confirms this.

b) We find the first coordinate of the vertex of the function $H = -16t^2 + 96t$:

$$-\frac{b}{2a} = -\frac{96}{2(-16)} = -\frac{96}{-32} = -(-3) = 3$$

The projectile reaches its maximum height 3 sec after launch. The graph confirms this.

c) We substitute 0 for H and solve for t:

$0 = -16t^2 + 96t$

$0 = -16t(t - 6)$

$-16t = 0$ *or* $t - 6 = 0$

$t = 0$ *or* $t = 6$

At $t = 0$ sec the projectile has not yet been launched. Thus, we use $t = 6$. The projectile returns to the ground 6 sec after launch. The graph confirms this.

45. $y = x^2 + 2x - 3$

$a = 1,\ b = 2,\ c = -3$

$b^2 - 4ac = 2^2 - 4 \cdot 1 \cdot (-3) = 4 + 12 = 16$

Since $b^2 - 4ac = 16 > 0$, the equation $x^2 + 2x - 3 = 0$ has two real-number solutions.

47. $y = -0.02x^2 + 4.7x - 2300$

$a = -0.02,\ b = 4.7,\ c = -2300$

$b^2 - 4ac = (4.7)^2 - 4(-0.02)(-2300) = 22.09 - 184 = -161.91$

Since $b^2 - 4ac = -161.91 < 0$, the equation $-0.02x^2 + 4.7x - 2300 = 0$ has no real solutions.

Exercise Set 17.7

1. Yes; each member of the domain is matched to only one member of the range.

3. Yes; each member of the domain is matched to only one member of the range.

5. No; a member of the domain is matched to more than one member of the range. In fact, each member of the domain is matched to 3 members of the range.

7. Yes; each member of the domain is matched to only one member of the range.

9. This correspondence is a function, because each class member has only one seat number.

11. This correspondence is a function, because each shape has only one number for its area.

13. This correspondence is not a function, because it is reasonable to assume that at least one person has more than one aunt.

The correspondence is a relation, because it is reasonable to assume that each person has at least one aunt.

15. $f(x) = x + 5$

a) $f(4) = 4 + 5 = 9$

b) $f(7) = 7 + 5 = 12$

c) $f(-3) = -3 + 5 = 2$

d) $f(0) = 0 + 5 = 5$

e) $f(2.4) = 2.4 + 5 = 7.4$

f) $f\left(\frac{2}{3}\right) = \frac{2}{3} + 5 = 5\frac{2}{3}$

17. $h(p) = 3p$

a) $h(-7) = 3(-7) = -21$

b) $h(5) = 3 \cdot 5 = 15$

c) $h(14) = 3 \cdot 14 = 42$

d) $h(0) = 3 \cdot 0 = 0$

e) $h\left(\frac{2}{3}\right) = 3 \cdot \frac{2}{3} = \frac{6}{3} = 2$

f) $h(-54.2) = 3(-54.2) = -162.6$

19. $g(s) = 3s + 4$

a) $g(1) = 3 \cdot 1 + 4 = 3 + 4 = 7$

b) $g(-7) = 3(-7) + 4 = -21 + 4 = -17$

c) $g(6.7) = 3(6.7) + 4 = 20.1 + 4 = 24.1$

d) $g(0) = 3 \cdot 0 + 4 = 0 + 4 = 4$

e) $g(-10) = 3(-10) + 4 = -30 + 4 = -26$

f) $g\left(\frac{2}{3}\right) = 3 \cdot \frac{2}{3} + 4 = 2 + 4 = 6$

21. $f(x) = 2x^2 - 3x$

a) $f(0) = 2 \cdot 0^2 - 3 \cdot 0 = 0 - 0 = 0$

b) $f(-1) = 2(-1)^2 - 3(-1) = 2 + 3 = 5$

c) $f(2) = 2 \cdot 2^2 - 3 \cdot 2 = 8 - 6 = 2$

d) $f(10) = 2 \cdot 10^2 - 3 \cdot 10 = 200 - 30 = 170$

e) $f(-5) = 2(-5)^2 - 3(-5) = 50 + 15 = 65$

f) $f(-10) = 2(-10)^2 - 3(-10) = 200 + 30 = 230$

23. $f(x) = |x| + 1$

a) $f(0) = |0| + 1 = 0 + 1 = 1$

b) $f(-2) = |-2| + 1 = 2 + 1 = 3$

c) $f(2) = |2| + 1 = 2 + 1 = 3$

d) $f(-3) = |-3| + 1 = 3 + 1 = 4$

e) $f(-10) = |-10| + 1 = 10 + 1 = 11$

f) $f(22) = |22| + 1 = 22 + 1 = 23$

25. $f(x) = x^3$

a) $f(0) = 0^3 = 0$

b) $f(-1) = (-1)^3 = -1$

c) $f(2) = 2^3 = 8$

d) $f(10) = 10^3 = 1000$

e) $f(-5) = (-5)^3 = -125$

f) $f(-10) = (-10)^3 = -1000$

27. $F(x) = 2.75x + 71.48$

a) $$\begin{aligned} F(32) &= 2.75(32) + 71.48 \\ &= 88 + 71.48 \\ &= 159.48 \text{ cm} \end{aligned}$$

b) $$\begin{aligned} F(30) &= 2.75(30) + 71.48 \\ &= 82.5 + 71.48 \\ &= 153.98 \text{ cm} \end{aligned}$$

29. $$\begin{aligned} P(d) &= 1 + \frac{d}{33} \\ P(20) &= 1 + \frac{20}{33} = 1\frac{20}{33} \text{ atm} \\ P(30) &= 1 + \frac{30}{33} = 1\frac{10}{11} \text{ atm} \\ P(100) &= 1 + \frac{100}{33} = 1 + 3\frac{1}{33} = 4\frac{1}{33} \text{ atm} \end{aligned}$$

31. $$\begin{aligned} W(d) &= 0.112d \\ W(16) &= 0.112(16) = 1.792 \text{ cm} \\ W(25) &= 0.112(25) = 2.8 \text{ cm} \\ W(100) &= 0.112(100) = 11.2 \text{ cm} \end{aligned}$$

33. Graph $f(x) = 3x - 1$

Make a list of function values in a table.

When $x = -1$, $f(-1) = 3(-1) - 1 = -3 - 1 = -4$.

When $x = 0$, $f(0) = 3 \cdot 0 - 1 = 0 - 1 = -1$.

When $x = 2$, $f(2) = 3 \cdot 2 - 1 = 6 - 1 = 5$.

x	$f(x)$
-1	-4
0	-1
2	5

Plot these points and connect them.

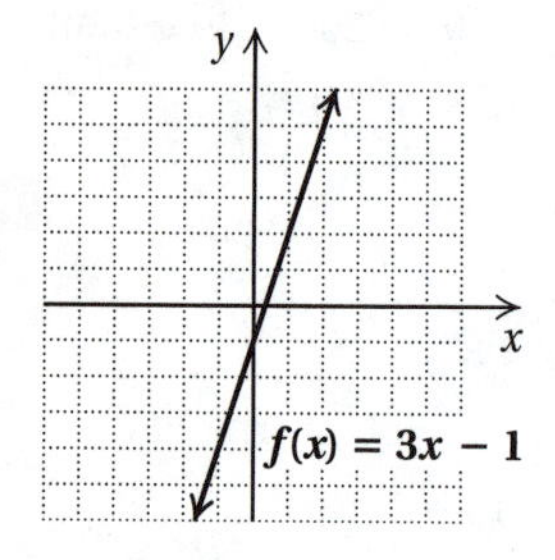

35. Graph $g(x) = -2x + 3$

Make a list of function values in a table.

When $x = -1$, $g(-1) = -2(-1) + 3 = 2 + 3 = 5$.

When $x = 0$, $g(0) = -2 \cdot 0 + 3 = 0 + 3 = 3$.

When $x = 3$, $g(3) = -2 \cdot 3 + 3 = -6 + 3 = -3$.

x	$g(x)$
-1	5
0	3
3	-3

Plot these points and connect them.

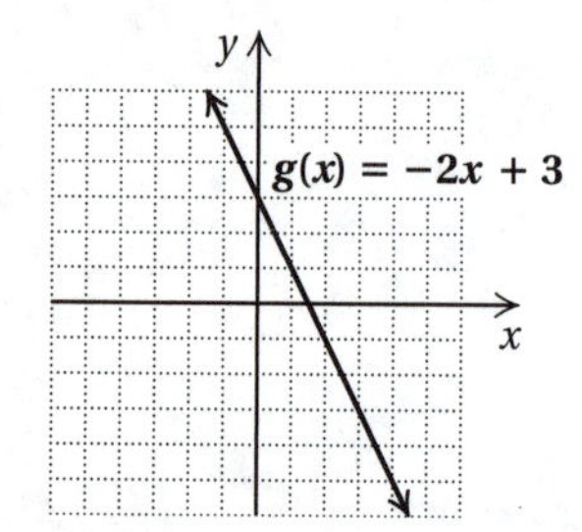

37. Graph $f(x) = \frac{1}{2}x + 1$.

Make a list of function values in a table.

When $x = -2$, $f(-2) = \frac{1}{2}(-2) + 1 = -1 + 1 = 0$.

When $x = 0$, $f(0) = \frac{1}{2} \cdot 0 + 1 = 0 + 1 = 1$.

When $x = 4$, $f(4) = \frac{1}{2} \cdot 4 + 1 = 2 + 1 = 3$.

x	$f(x)$
-2	0
0	1
4	3

Plot these points and connect them.

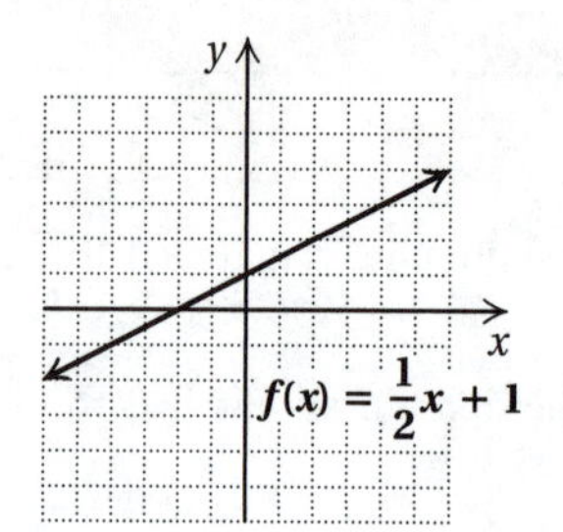

39. Graph $f(x) = 2 - |x|$.

Make a list of function values in a table.

When $x = -4$, $f(-4) = 2 - |-4| = 2 - 4 = -2$.

When $x = 0$, $f(0) = 2 - |0| = 2 - 0 = 2$.

When $x = 3$, $f(3) = 2 - |3| = 2 - 3 = -1$.

x	$f(x)$
-4	-2
0	2
3	-1

Plot these points and connect them.

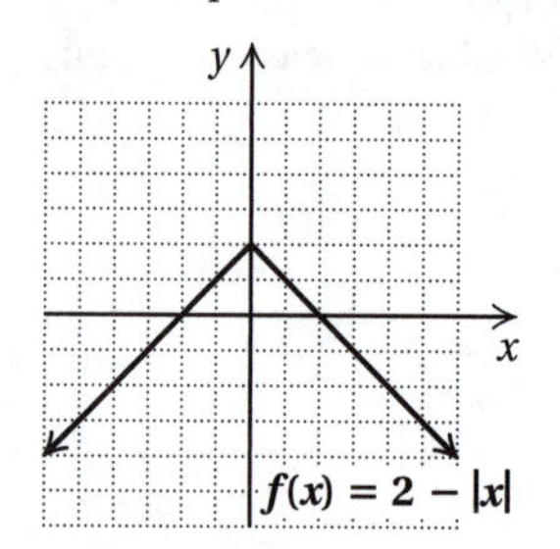

41. Graph $f(x) = x^2$.

Recall from Section 17.6 that the graph is a parabola. Make a list of function values in a table.

When $x = -2$, $f(-2) = (-2)^2 = 4$.

When $x = -1$, $f(-1) = (-1)^2 = 1$.

When $x = 0$, $f(0) = 0^2 = 0$.

When $x = 1$, $f(1) = 1^2 = 1$.

When $x = 2$, $f(2) = 2^2 = 4$.

x	$f(x)$
-2	4
-1	1
0	0
1	1
2	4

Plot these points and connect them.

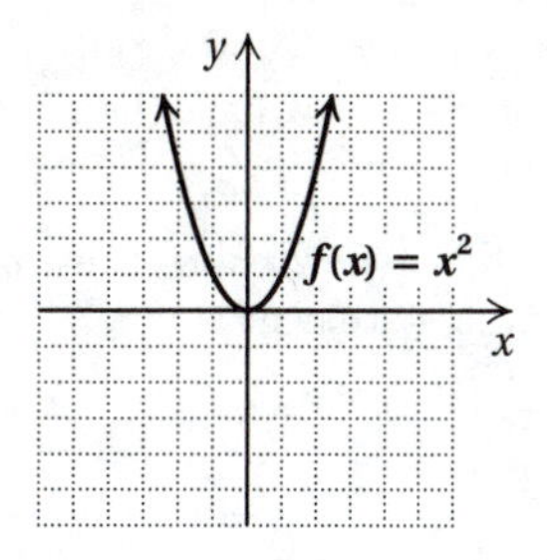

43. Graph $f(x) = x^2 - x - 2$.

Recall from Section 17.6 that the graph is a parabola. Make a list of function values in a table.

When $x = -1$, $f(-1) = (-1)^2 - (-1) - 2 = 1 + 1 - 2 = 0$.

When $x = 0$, $f(0) = 0^2 - 0 - 2 = -2$.

When $x = 1$, $f(1) = 1^2 - 1 - 2 = 1 - 1 - 2 = -2$.

When $x = 2$, $f(2) = 2^2 - 2 - 2 = 4 - 2 - 2 = 0$.

x	$f(x)$
-1	0
0	-2
1	-2
2	0

Plot these points and connect them.

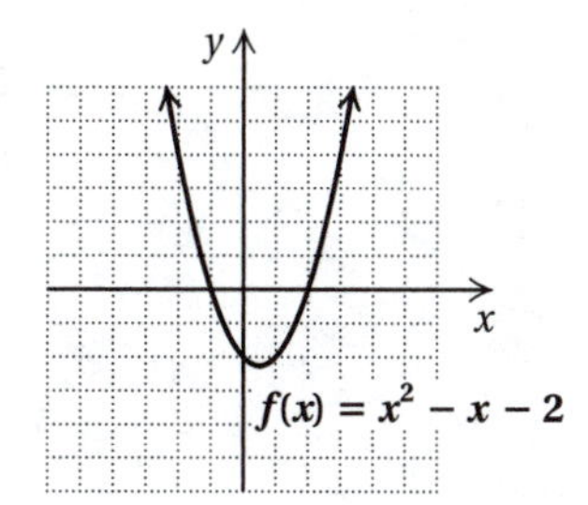

45. We can use the vertical line test:

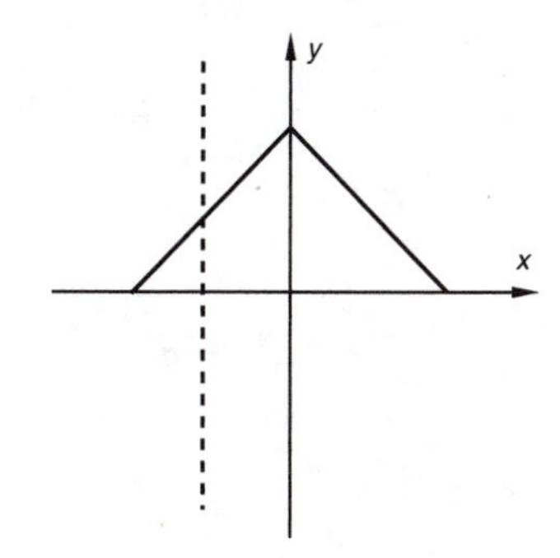

Visualize moving this vertical line across the graph. No vertical line will intersect the graph more than once. Thus, the graph is a graph of a function.

47. We can use the vertical line test:

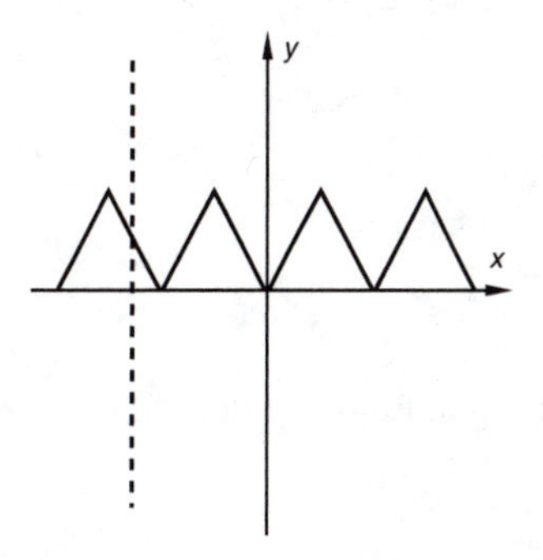

Visualize moving this vertical line across the graph. No vertical line will intersect the graph more than once. Thus, the graph is a graph of a function.

49. We can use the vertical line test.

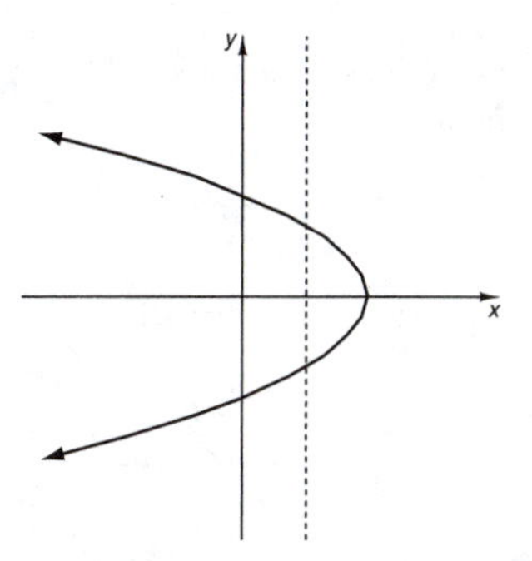

It is possible for a vertical line to intersect the graph more than once. Thus this is not the graph of a function.

51. We can use the vertical line test.

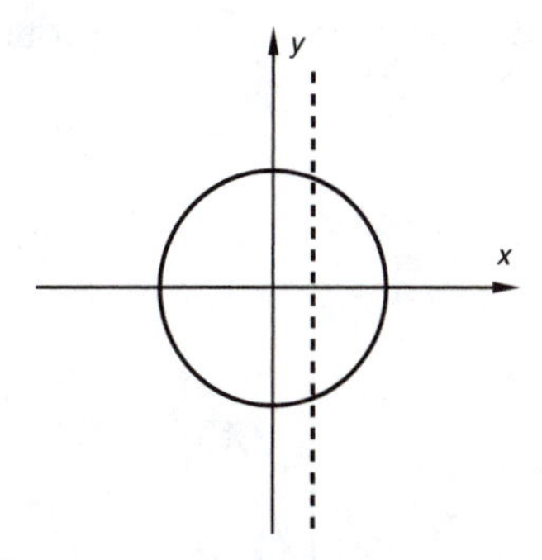

It is possible for a vertical line to intersect the graph more than once. Thus this is not a graph of a function.

53. Locate the point that is directly above 225. Then estimate its second coordinate by moving horizontally from the point to the vertical axis. The rate is about 75 per 10,000 men.

55. The first equation is in slope-intercept form:

$$y = \frac{3}{4}x - 7,\ m = \frac{3}{4}$$

We write the second equation in slope-intercept form.

$$3x + 4y = 7$$
$$4y = -3x + 7$$
$$y = -\frac{3}{4}x + \frac{7}{4},\ m = -\frac{3}{4}$$

Since the slopes are different, the equations do not represent parallel lines.

57. $2x - y = 6$, (1)

$4x - 2y = 5$ (2)

We solve Equation (1) for y.

$2x - y = 6$ (1)

$2x - 6 = y$ Adding y and -6

Substitute $2x - 6$ for y in Equation (2) and solve for x.

$$4x - 2y = 5 \quad (2)$$
$$4x - 2(2x - 6) = 5$$
$$4x - 4x + 12 = 5$$
$$12 = 5$$

We get a false equation, so the system has no solution.

59. Graph $g(x) = x^3$.

Make a list of function values in a table. Then plot the points and connect them.

x	$g(x)$
-2	-8
-1	-1
0	0
1	1
2	8

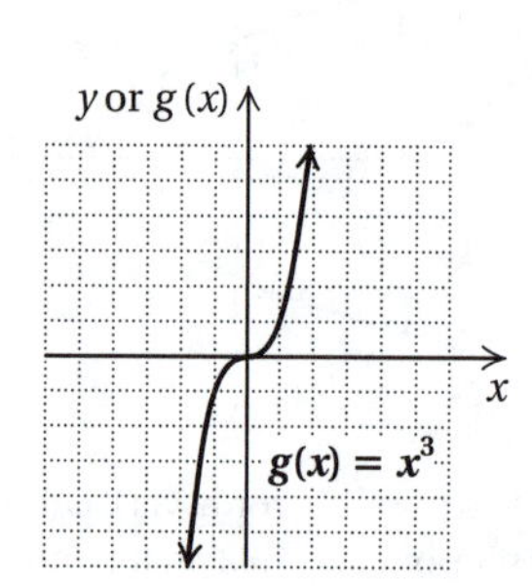

61. Graph $g(x) = |x| + x$.

Make a list of function values in a table. Then plot the points and connect them.

x	$f(x)$
-3	0
-2	0
-1	0
0	0
1	2
2	4
3	6

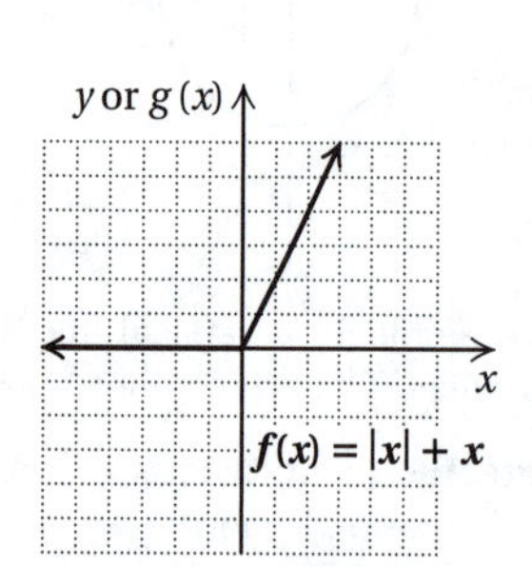

Chapter 17 Concept Reinforcement

1. A graph represents a function if it is not possible to draw a vertical line that intersects the graph more than once. The given statement is false.

2. When $x = 0$, $y = a \cdot 0^2 + b \cdot 0 + c = c$, so all graphs of quadratic equations, $y = ax^2 + bx + c$, have the y-intercept $(0, c)$. The given statement is true.

3. If $a < 0$, then the graph opens down and, thus, q is the largest value of y. The given statement is true.

4. True; see page 1272 in the text.

Chapter 17 Important Concepts

1. $\dfrac{3}{x+2} + \dfrac{1}{x} = \dfrac{5}{4}$

We multiply by the LCM, $4x(x+2)$.

$$4x(x+2)\left(\frac{3}{x+2} + \frac{1}{x}\right) = 4x(x+2) \cdot \frac{5}{4}$$
$$4x(x+2) \cdot \frac{3}{x+2} + 4x(x+2) \cdot \frac{1}{x} = x(x+2) \cdot 5$$
$$4x \cdot 3 + 4(x+2) = 5x^2 + 10x$$
$$12x + 4x + 8 = 5x^2 + 10x$$
$$16x + 8 = 5x^2 + 10x$$
$$0 = 5x^2 - 6x - 8$$
$$0 = (5x+4)(x-2)$$

$5x + 4 = 0$ *or* $x - 2 = 0$

$5x = -4$ *or* $x = 2$

$x = -\dfrac{4}{5}$ *or* $x = 2$

The solutions are $-\dfrac{4}{5}$ and 2.

2. $7x^2 - 3 = 8$

$7x^2 = 11$

$x^2 = \dfrac{11}{7}$

$x = \sqrt{\dfrac{11}{7}}$ *or* $x = -\sqrt{\dfrac{11}{7}}$

$x = \sqrt{\dfrac{11}{7} \cdot \dfrac{7}{7}}$ *or* $x = -\sqrt{\dfrac{11}{7} \cdot \dfrac{7}{7}}$

$x = \dfrac{\sqrt{77}}{7}$ *or* $x = -\dfrac{\sqrt{77}}{7}$

The solutions are $\dfrac{\sqrt{77}}{7}$ and $-\dfrac{\sqrt{77}}{7}$, or $\pm\dfrac{\sqrt{77}}{7}$.

3. $x^2 - 4x + 1 = 0$

$x^2 - 4x = -1$

$x^2 - 4x + 4 = -1 + 4$

$(x-2)^2 = 3$

$x - 2 = \sqrt{3}$ *or* $x - 2 = -\sqrt{3}$

$x = 2 + \sqrt{3}$ *or* $x = 2 - \sqrt{3}$

The solutions are $2 + \sqrt{3}$ and $2 - \sqrt{3}$, or $2 \pm \sqrt{3}$.

4. $4y^2 = 6y + 3$

$4y^2 - 6y - 3 = 0$

$a = 4,\ b = -6,\ c = -3$

$$y = \frac{-b \pm \sqrt{b^2 - 4ac}}{2a}$$

$$y = \frac{-(-6) \pm \sqrt{(-6)^2 - 4 \cdot 4 \cdot (-3)}}{2 \cdot 4} = \frac{6 \pm \sqrt{36 + 48}}{8}$$

$$y = \frac{6 \pm \sqrt{84}}{8} = \frac{6 \pm \sqrt{4 \cdot 21}}{8} = \frac{6 \pm 2\sqrt{21}}{8}$$

$$y = \frac{2(3 \pm \sqrt{21})}{2 \cdot 4} = \frac{3 \pm \sqrt{21}}{4}$$

The solutions are $\dfrac{3 \pm \sqrt{21}}{4}$.

5. $y = x^2 - 4x + 2$.

We first find the vertex. The x-coordinate is

$$-\frac{b}{2a} = -\frac{-4}{2 \cdot 1} = 2.$$

We substitute into the equation to find the second coordinate of the vertex.

$$y = 2^2 - 4 \cdot 2 + 2 = 4 - 8 + 2 = -2$$

The vertex is $(2, -2)$.

We choose some x-values on both sides of the vertex and graph the parabola.

x	y	
-2	14	
-1	7	
0	2	⟵ x-intercept
1	-1	
2	-2	⟵ vertex
3	-1	
4	2	

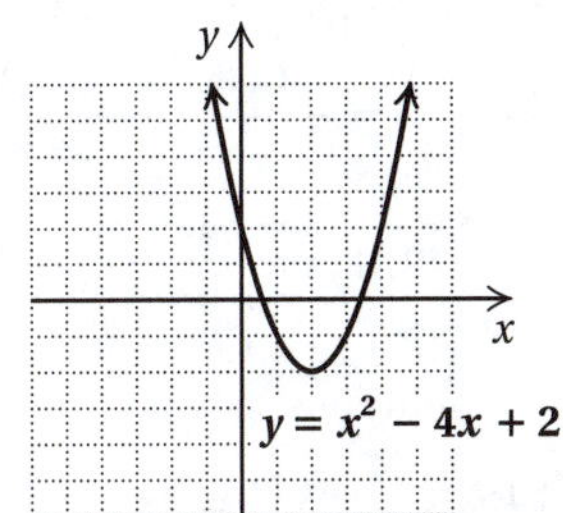

6. The correspondence is a function because each member of the domain is matched to only one member of the range.

7. a) $h(5) = \frac{1}{5} \cdot 5^2 + 5 - 1 = \frac{1}{5} \cdot 25 + 5 - 1 = 5 + 5 - 1 = 9$

b) $f(0) = -3 \cdot 0 - 4 = 0 - 4 = -4$

Chapter 17 Review Exercises

1. $8x^2 = 24$

$x^2 = 3$

$x = \sqrt{3}$ *or* $x = -\sqrt{3}$

The solutions are $\sqrt{3}$ and $-\sqrt{3}$.

2. $40 = 5y^2$

$8 = y^2$

$y = \sqrt{8}$ *or* $y = -\sqrt{8}$

$y = 2\sqrt{2}$ *or* $y = -2\sqrt{2}$

The solutions are $2\sqrt{2}$ and $-2\sqrt{2}$.

3. $5x^2 - 8x + 3 = 0$

$(5x - 3)(x - 1) = 0$

$5x - 3 = 0$ *or* $x - 1 = 0$

$5x = 3$ *or* $x = 1$

$x = \frac{3}{5}$ *or* $x = 1$

The solutions are $\frac{3}{5}$ and 1.

4. $3y^2 + 5y = 2$

$3y^2 + 5y - 2 = 0$

$(3y - 1)(y + 2) = 0$

$3y - 1 = 0$ *or* $y + 2 = 0$

$3y = 1$ *or* $y = -2$

$y = \frac{1}{3}$ *or* $y = -2$

The solutions are $\frac{1}{3}$ and -2.

5. $(x + 8)^2 = 13$

$x + 8 = \sqrt{13}$ *or* $x + 8 = -\sqrt{13}$

$x = -8 + \sqrt{13}$ *or* $x = -8 - \sqrt{13}$

The solutions are $-8 \pm \sqrt{13}$.

6. $9x^2 = 0$

$x^2 = 0$

$x = 0$

The solution is 0.

7. $5t^2 - 7t = 0$

$t(5t - 7) = 0$

$t = 0$ *or* $5t - 7 = 0$

$t = 0$ *or* $5t = 7$

$t = 0$ *or* $t = \frac{7}{5}$

The solutions are 0 and $\frac{7}{5}$.

8. $x^2 - 2x - 10 = 0$

$a = 1,\ b = -2,\ c = -10$

$$x = \frac{-(-2) \pm \sqrt{(-2)^2 - 4 \cdot 1 \cdot (-10)}}{2 \cdot 1}$$

$$x = \frac{2 \pm \sqrt{4 + 40}}{2}$$

$$x = \frac{2 \pm \sqrt{44}}{2} = \frac{2 \pm \sqrt{4 \cdot 11}}{2}$$

$$x = \frac{2 \pm 2\sqrt{11}}{2} = \frac{2(1 \pm \sqrt{11})}{2}$$

$$x = 1 \pm \sqrt{11}$$

The solutions are $1 \pm \sqrt{11}$.

9. $9x^2 - 6x - 9 = 0$

$a = 9,\ b = -6,\ c = -9$

$$x = \frac{-(-6) \pm \sqrt{(-6)^2 - 4 \cdot 9 \cdot (-9)}}{2 \cdot 9}$$

$$x = \frac{6 \pm \sqrt{36 + 324}}{18}$$

$$x = \frac{6 \pm \sqrt{360}}{18} = \frac{6 \pm \sqrt{36 \cdot 10}}{18}$$

$$x = \frac{6 \pm 6\sqrt{10}}{18} = \frac{6(1 \pm \sqrt{10})}{3 \cdot 6}$$

$$x = \frac{1 \pm \sqrt{10}}{3}$$

The solutions are $\frac{1 \pm \sqrt{10}}{3}$.

10. $x^2 + 6x = 9$

$x^2 + 6x - 9 = 0$

$a = 1,\ b = 6,\ c = -9$

$$x = \frac{-6 \pm \sqrt{6^2 - 4 \cdot 1 \cdot (-9)}}{2 \cdot 1}$$

$$x = \frac{-6 \pm \sqrt{36 + 36}}{2}$$

$$x = \frac{-6 \pm \sqrt{72}}{2} = \frac{-6 \pm \sqrt{36 \cdot 2}}{2}$$

$$x = \frac{-6 \pm 6\sqrt{2}}{2} = \frac{2(-3 \pm 3\sqrt{2})}{2}$$

$$x = -3 \pm 3\sqrt{2}$$

The solutions are $-3 \pm 3\sqrt{2}$.

11. $1 + 4x^2 = 8x$

$4x^2 - 8x + 1 = 0$

$a = 4,\ b = -8,\ c = 1$

$$x = \frac{-(-8) \pm \sqrt{(-8)^2 - 4 \cdot 4 \cdot 1}}{2 \cdot 4}$$

$$x = \frac{8 \pm \sqrt{64 - 16}}{8}$$

$$x = \frac{8 \pm \sqrt{48}}{8} = \frac{8 \pm \sqrt{16 \cdot 3}}{8}$$

$$x = \frac{8 \pm 4\sqrt{3}}{8} = \frac{4(2 \pm \sqrt{3})}{2 \cdot 4}$$

$$x = \frac{2 \pm \sqrt{3}}{2}$$

The solutions are $\frac{2 \pm \sqrt{3}}{2}$.

12. $6 + 3y = y^2$

$0 = y^2 - 3y - 6$

$a = 1,\ b = -3,\ c = -6$

$$y = \frac{-(-3) \pm \sqrt{(-3)^2 - 4 \cdot 1 \cdot (-6)}}{2 \cdot 1}$$

$$y = \frac{3 \pm \sqrt{9 + 24}}{2} = \frac{3 \pm \sqrt{33}}{2}$$

The solutions are $\frac{3 \pm \sqrt{33}}{2}$.

13. $3m = 4 + 5m^2$

$0 = 5m^2 - 3m + 4$

$a = 5,\ b = -3,\ c = 4$

$$m = \frac{-(-3) \pm \sqrt{(-3)^2 - 4 \cdot 5 \cdot 4}}{2 \cdot 5}$$

$$m = \frac{3 \pm \sqrt{9 - 80}}{10} = \frac{3 \pm \sqrt{-71}}{10}$$

Since the radicand is negative, there are no real-number solutions.

14. $3x^2 = 4x$

$3x^2 - 4x = 0$

$x(3x - 4) = 0$

$x = 0$ *or* $3x - 4 = 0$

$x = 0$ *or* $3x = 4$

$x = 0$ *or* $x = \frac{4}{3}$

The solutions are 0 and $\frac{4}{3}$.

15.
$$\frac{15}{x} - \frac{15}{x+2} = 2, \text{ LCM is } x(x+2)$$
$$x(x+2)\left(\frac{15}{x} - \frac{15}{x+2}\right) = x(x+2)(2)$$
$$x(x+2)\cdot\frac{15}{x} - x(x+2)\cdot\frac{15}{x+2} = 2x(x+2)$$
$$15(x+2) - 15x = 2x^2 + 4x$$
$$15x + 30 - 15x = 2x^2 + 4x$$
$$30 = 2x^2 + 4x$$
$$0 = 2x^2 + 4x - 30$$
$$0 = x^2 + 2x - 15 \quad \text{Dividing by 2}$$
$$0 = (x+5)(x-3)$$

$x + 5 = 0$ *or* $x - 3 = 0$

$x = -5$ *or* $x = 3$

Both numbers check. The solutions are -5 and 3.

16.
$$x + \frac{1}{x} = 2, \text{ LCM is } x$$
$$x\left(x + \frac{1}{x}\right) = x \cdot 2$$
$$x \cdot x + x \cdot \frac{1}{x} = 2x$$
$$x^2 + 1 = 2x$$
$$x^2 - 2x + 1 = 0$$
$$(x-1)^2 = 0$$
$$x - 1 = 0$$
$$x = 1$$

The number 1 checks. It is the solution.

17. $x^2 - 4x + 2 = 0$

$x^2 - 4x = -2$

$x^2 - 4x + 4 = -2 + 4$ Adding 4: $\left(\frac{-4}{2}\right)^2 = 4$

$(x-2)^2 = 2$

$x - 2 = \sqrt{2}$ *or* $x - 2 = -\sqrt{2}$

$x = 2 + \sqrt{2}$ *or* $x = 2 - \sqrt{2}$

The solutions are $2 \pm \sqrt{2}$.

18. $3x^2 - 2x - 5 = 0$

$\frac{1}{3}(3x^2 - 2x - 5) = \frac{1}{3} \cdot 0$

$x^2 - \frac{2}{3}x - \frac{5}{3} = 0$

$x^2 - \frac{2}{3}x = \frac{5}{3}$

$x^2 - \frac{2}{3}x + \frac{1}{9} = \frac{5}{3} + \frac{1}{9}$ Adding $\frac{1}{9}$:

$$\left[\frac{1}{2}\left(-\frac{2}{3}\right)\right]^2 = \left(-\frac{1}{3}\right)^2 = \frac{1}{9}$$

$\left(x - \frac{1}{3}\right)^2 = \frac{16}{9}$

$x - \frac{1}{3} = \frac{4}{3}$ *or* $x - \frac{1}{3} = -\frac{4}{3}$

$x = \frac{5}{3}$ *or* $x = -1$

The solutions are $\frac{5}{3}$ and -1.

19. From Exercise 17, we know the solutions are $\frac{5 \pm \sqrt{17}}{2}$.

Using a calculator, we have

$\frac{5 + \sqrt{17}}{2} \approx 4.6$ and $\frac{5 - \sqrt{17}}{2} \approx 0.4$.

20. $4y^2 + 8y + 1 = 0$

$a = 4$, $b = 8$, $c = 1$

$$y = \frac{-8 \pm \sqrt{8^2 - 4 \cdot 4 \cdot 1}}{2 \cdot 4}$$
$$y = \frac{-8 \pm \sqrt{64 - 16}}{8} = \frac{-8 \pm \sqrt{48}}{8}$$

Using a calculator, we have $\frac{-8 + \sqrt{48}}{8} \approx -0.1$ and $\frac{-8 - \sqrt{48}}{8} \approx -1.9$.

21.
$$V = \frac{1}{2}\sqrt{1 + \frac{T}{L}}$$
$$V^2 = \left(\frac{1}{2}\sqrt{1 + \frac{T}{L}}\right)^2$$
$$V^2 = \frac{1}{4}\left(1 + \frac{T}{L}\right)$$
$$4 \cdot V^2 = 4 \cdot \frac{1}{4}\left(1 + \frac{T}{L}\right)$$
$$4V^2 = 1 + \frac{T}{L}$$
$$4V^2 - 1 = \frac{T}{L}$$
$$L(4V^2 - 1) = L \cdot \frac{T}{L}$$
$$L(4V^2 - 1) = T$$

22. $y = 2 - x^2$, or $y = -x^2 + 2$

Find the vertex. The x-coordinate is

$$-\frac{b}{2a} = -\frac{0}{2(-1)} = 0.$$

The y-coordinate is

$$y = 2 - x^2 = 2 - 0^2 = 2.$$

The vertex is $(0, 2)$. This is also the y-intercept.

We choose some x-values on both sides of the vertex and graph the parabola.

x	y	
0	2	←Vertex
−1	1	
−2	−2	
1	1	
2	−2	

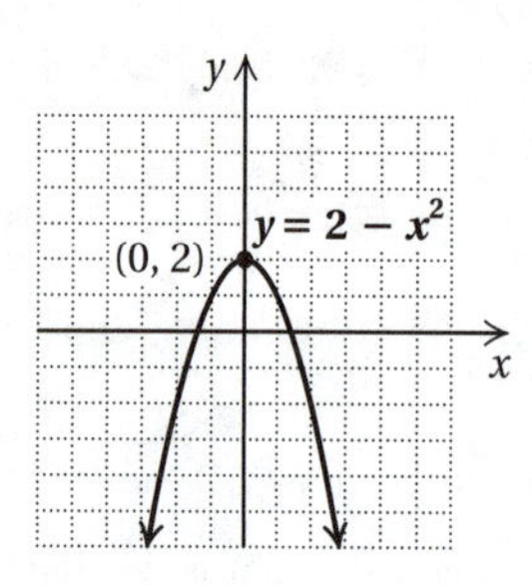

23. $y = x^2 - 4x - 2$

Find the vertex. The x-coordinate is

$$-\frac{b}{2a} = -\frac{-4}{2\cdot 1} = -(-2) = 2.$$

The y-coordinate is

$$y = 2^2 - 4\cdot 2 - 2 = 4 - 8 - 2 = -6.$$

The vertex is $(2, -6)$.

We choose some x-values on both sides of the vertex and graph the parabola.

x	y	
2	−6	← Vertex
0	−2	← y-intercept
−1	3	
3	5	
5	3	

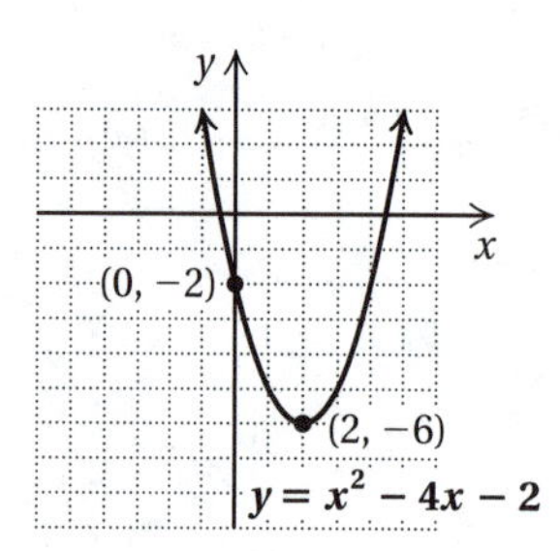

24. $y = 2 - x^2$

To find the x-intercepts we solve the equation $2 - x^2 = 0$.

$$2 - x^2 = 0$$
$$2 = x^2$$
$$x = \sqrt{2} \quad or \quad x = -\sqrt{2}$$

The x-intercepts are $(\sqrt{2}, 0)$ and $(-\sqrt{2}, 0)$.

25. $y = x^2 - 4x - 2$

To find the x-intercepts we solve the equation $x^2 - 4x - 2 = 0$.

$$x = \frac{-(-4) \pm \sqrt{(-4)^2 - 4\cdot 1(-2)}}{2\cdot 1}$$

$$x = \frac{4 \pm \sqrt{16+8}}{2} = \frac{4+\sqrt{24}}{2}$$

$$x = \frac{4 \pm \sqrt{4\cdot 6}}{2} = \frac{4 \pm 2\sqrt{6}}{2}$$

$$x = \frac{2(2\pm\sqrt{6})}{2} = 2 \pm \sqrt{6}$$

The x-intercepts are $(2-\sqrt{6}, 0)$ and $(2+\sqrt{6}, 0)$.

26. ***Familiarize***. Using the labels on the drawing in the text, we let a and $a+3$ represent the lengths of the legs, in cm.

Translate. We use the Pythagorean equation.

$$a^2 + (a+3)^2 = 5^2$$

Solve.

$$a^2 + (a+3)^2 = 5^2$$
$$a^2 + a^2 + 6a + 9 = 25$$
$$2a^2 + 6a + 9 = 25$$
$$2a^2 + 6a - 16 = 0$$
$$a^2 + 3a - 8 = 0 \quad \text{Dividing by 2}$$

We use the quadratic formula.

$$a = \frac{-3 \pm \sqrt{3^2 - 4\cdot 1\cdot(-8)}}{2\cdot 1}$$

$$a = \frac{-3 \pm \sqrt{9+32}}{2} = \frac{-3\pm\sqrt{41}}{2}$$

Using a calculator, we have

$$a = \frac{-3-\sqrt{41}}{2} \approx -4.7 \text{ and } a = \frac{-3+\sqrt{41}}{2} \approx 1.7.$$

Check. Since the length of a leg cannot be negative, -4.7 cannot be a solution. If $a \approx 1.7$, then $a + 3 \approx 4.7$ and $(1.7)^2 + (4.7)^4 = 24.98 \approx 25 = 5^2$. The answer checks.

State. The lengths of the legs are about 1.7 cm and 4.7 cm.

27. ***Familiarize***. Using the labels on the drawing in the text, we let s and $s - 5$ represent the lengths of the legs, in ft.

Translate. We use the Pythagorean equation.

$$s^2 + (s-5)^2 = 25^2$$

Solve.

$$s^2 + (s-5)^2 = 25^2$$
$$s^2 + s^2 - 10s + 25 = 625$$
$$2s^2 - 10s + 25 = 625$$
$$2s^2 - 10s - 600 = 0$$
$$s^2 - 5s - 300 = 0 \quad \text{Dividing by 2}$$
$$(s-20)(s+15) = 0$$
$$s - 20 = 0 \quad or \quad s + 15 = 0$$
$$s = 20 \quad or \quad s = -15$$

Check. Since the length of a leg of the triangle cannot be negative, -15 cannot be a solution. If $s = 20$, then $s - 5 = 20 - 5 = 15$ and $20^2 + 15^2 = 625 = 25^2$. The answer checks.

State. The height of the ramp is 15 ft.

28. ***Familiarize***. We will use the formula $s = 16t^2$.

Translate. We substitute 1053 for s.

$$1053 = 16t^2$$

Solve.

$$1053 = 16t^2$$
$$\frac{1053}{16} = t^2$$
$$65.8125 = t^2$$
$$t = \sqrt{65.8125} \quad or \quad t = -\sqrt{65.8125}$$
$$t \approx 8.1 \quad or \quad t \approx -8.1$$

Check. Time cannot be negative in this application, so -8.1 cannot be a solution. We check 8.1.

$$16(8.1)^2 = 1049.76 \approx 1053$$

The answer is close, so we have a check. (Remember that we rounded the value of t.)

State. It would take about 8.1 sec for an object to fall to the ground from the top of the Royal Gorge Bridge.

29. $f(x) = 2x - 5$

$f(2) = 2 \cdot 2 - 5 = 4 - 5 = -1$

$f(-1) = 2(-1) - 5 = -2 - 5 = -7$

$f(3.5) = 2(3.5) - 5 = 7 - 5 = 2$

30. $g(x) = |x| - 1$

$g(1) = |1| - 1 = 1 - 1 = 0$

$g(-1) = |-1| - 1 = 1 - 1 = 0$

$g(-20) = |-20| - 1 = 20 - 1 = 19$

31. $C(p) = 15p$

$C(180) = 15 \cdot 180 = 2700$ calories

32. $g(x) = 4 - x$

We find some function values.

When $x = -1$, $g(-1) = 4 - (-1) = 4 + 1 = 5$

When $x = 0$, $g(0) = 4 - 0 = 4$.

When $x = 3$, $g(3) = 4 - 3 = 1$.

x	$g(x)$
-1	5
0	4
3	1

Plot these points and connect them.

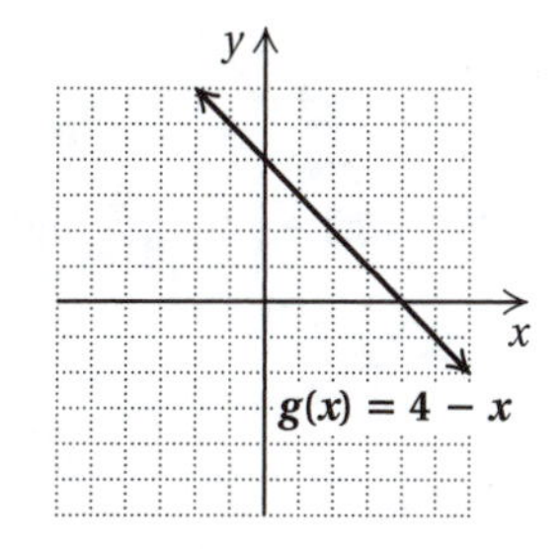

33. $f(x) = x^2 - 3$

Recall from Section 17.6 that the graph is a parabola. We find some function values.

When $x = -3$, $f(-3) = (-3)^2 - 3 = 9 - 3 = 6$.

When $x = -1$, $f(-1) = (-1)^2 - 3 = 1 - 3 = -2$.

When $x = 0$, $f(0) = 0^2 - 3 = 0 - 3 = -3$.

When $x = 1$, $f(1) = 1^2 - 3 = 1 - 3 = -2$.

When $x = 2$, $f(2) = 2^2 - 3 = 4 - 3 = 1$.

x	$f(x)$
-3	6
-1	-2
0	-3
1	-2
2	1

Plot these points and connect them.

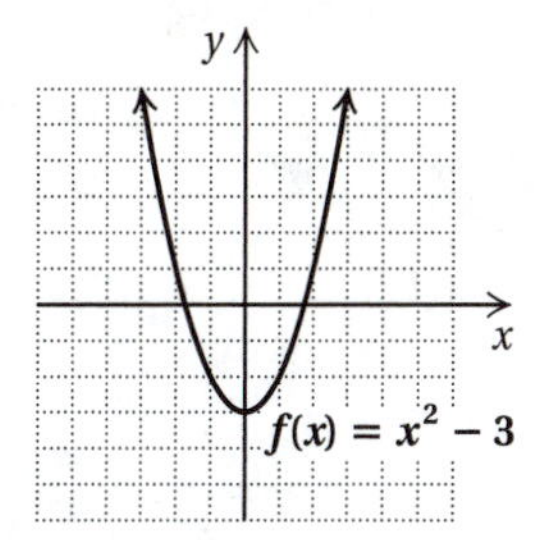

34. $h(x) = |x| - 5$

We find some function values.

When $x = -4$, $h(-4) = |-4| - 5 = 4 - 5 = -1$.

When $x = -2$, $h(-2) = |-2| - 5 = 2 - 5 = -3$.

When $x = 0$, $h(0) = |0| - 5 = 0 - 5 = -5$.

When $x = 1$, $h(1) = |1| - 5 = 1 - 5 = -4$.

When $x = 3$, $h(3) = |3| - 5 = 3 - 5 = -2$.

x	$h(x)$
-4	-1
-2	-3
0	-5
1	-4
3	-2

Plot these points and connect them.

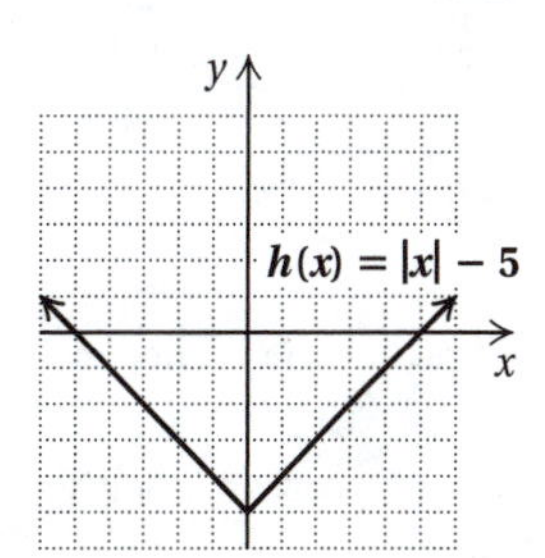

35. $f(x) = x^2 - 2x + 1$

Recall from Section 17.6 that the graph is a parabola.

When $x = -1$, $f(-1) = (-1)^2 - 2(-1) + 1 = 1 + 2 + 1 = 4$.

When $x = 0$, $f(0) = 0^2 - 2 \cdot 0 + 1 = 0 - 0 + 1 = 1$.

When $x = 1$, $f(1) = 1^2 - 2 \cdot 1 + 1 = 1 - 2 + 1 = 0$.

When $x = 2$, $f(2) = 2^2 - 2 \cdot 2 + 1 = 4 - 4 + 1 = 1$.

When $x = 3$, $f(3) = 3^2 - 2 \cdot 3 + 1 = 9 - 6 + 1 = 4$.

x	$f(x)$
-1	4
0	1
1	0
2	1
3	4

Plot these points and connect them.

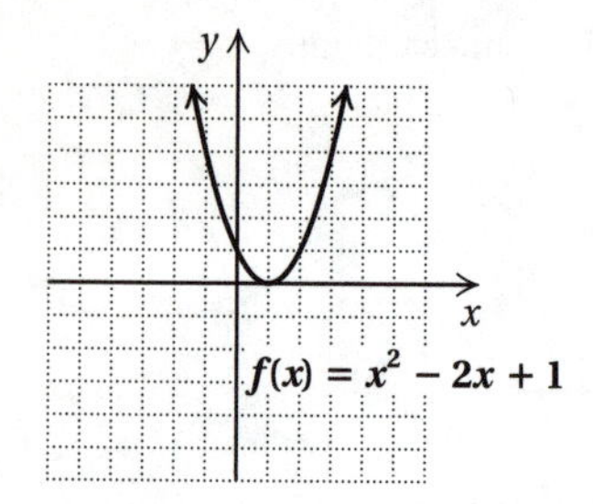

36. It is possible for a vertical line to intersect the graph more than once, so this is not the graph of a function.

37. No vertical line will intersect the graph more than once, so this is the graph of a function.

38. $40x - x^2 = 0$

$x(40 - x) = 0$

$x = 0$ *or* $40 - x = 0$

$x = 0$ *or* $40 = x$

Answer D is correct.

39. $\frac{1}{2}c^2 + c - \frac{1}{2} = 0$

$a = \frac{1}{2}$, $b = 1$, $c = -\frac{1}{2}$

(Note that we are using c in two different ways in this exercise.)

$$c = \frac{-b \pm \sqrt{b^2 - 4ac}}{2a}$$

$$c = \frac{-1 \pm \sqrt{1^2 - 4 \cdot \frac{1}{2} \cdot (-\frac{1}{2})}}{2 \cdot \frac{1}{2}} = \frac{-1 \pm \sqrt{1+1}}{1}$$

$c = -1 \pm \sqrt{2}$

Answer A is correct.

40. ***Familiarize.*** Let x = the first integer. Then $x + 1$ = the second integer.

Translate. If the numbers are positive, then $(x+1)^2$ is larger than x^2 and we have:

$$\underbrace{\text{Square of larger number}}_{(x+1)^2} \quad \underbrace{\text{minus}}_{-} \quad \underbrace{\text{square of smaller number}}_{x^2} \quad \underbrace{\text{is}}_{=} \quad \underbrace{63.}_{63}$$

If the numbers are negative, then x^2 is larger than $(x+1)^2$ and we have:

$$\underbrace{\text{Square of smaller number}}_{x^2} \quad \underbrace{\text{minus}}_{-} \quad \underbrace{\text{square of larger number}}_{(x+1)^2} \quad \underbrace{\text{is}}_{=} \quad \underbrace{63.}_{63}$$

Solve. We solve each equation.

$$(x+1)^2 - x^2 = 63$$
$$x^2 + 2x + 1 - x^2 = 63$$
$$2x + 1 = 63$$
$$2x = 62$$
$$x = 31$$

If $x = 31$, then $x + 1 = 31 + 1 = 32$.

$$x^2 - (x+1)^2 = 63$$
$$x^2 - (x^2 + 2x + 1) = 63$$
$$x^2 - x^2 - 2x - 1 = 63$$
$$-2x - 1 = 63$$
$$-2x = 64$$
$$x = -32$$

If $x = -32$, then $x + 1 = -32 + 1 = -31$.

Check. 31 and 32 are consecutive integers and $32^2 - 31^2 = 1024 - 961 = 63$. Also, -32 and -31 are consecutive integers and $(-32)^2 - (-31)^2 = 1024 - 961 = 63$. Both pairs of numbers check.

State. The integers are 31 and 32 or -32 and -31.

41. ***Familiarize.*** The area of a square with side s is s^2; the area of a circle with radius 5 in. is $\pi \cdot 5^2$, or 25π in^2.

Translate.

$$\underbrace{\text{Area of square}}_{s^2} \quad \underbrace{\text{equals}}_{=} \quad \underbrace{\text{area of circle}}_{25\pi}.$$

Solve.

$s^2 = 25\pi$

$s = 5\sqrt{\pi}$ *or* $s = -5\sqrt{\pi}$

Check. Since the length of a side of the square cannot be negative, $-5\sqrt{\pi}$ cannot be a solution. If the length of a side of the square is $5\sqrt{\pi}$, then the area of the square is $(5\sqrt{\pi})^2 = 25\pi$. Since this is also the area of the circle, $5\sqrt{\pi}$ checks.

State. $s = 5\sqrt{\pi}$ in. ≈ 8.9 in.

42. $x - 4\sqrt{x} - 5 = 0$

Let $u = \sqrt{x}$. Then $u^2 = x$. Substitute u for $\sqrt{x}$ and u^2 for x and solve for u.

$u^2 - 4u - 5 = 0$

$(u+1)(u-5) = 0$

$u + 1 = 0$ *or* $u - 5 = 0$

$u = -1$ *or* $u = 5$

Now we substitute $\sqrt{x}$ for u and solve for x.

$\sqrt{x} = -1$ has no real-number solutions.

If $u = 5$, then we have:

$\sqrt{x} = 5$

$(\sqrt{x})^2 = 5^2$

$x = 25$

The number 25 checks. It is the solution.

43. The graph of $y = (x+3)^2$ contains the points $(-4, 1)$ and $(-2, 1)$, so the solutions of $(x+3)^2 = 1$ are -4 and -2.

44. The graph of $y = (x+3)^2$ contains the points $(-5, 4)$ and $(-1, 4)$, so the solutions of $(x+3)^2 = 4$ are -5 and -1.

45. The graph of $y = (x+3)^2$ contains the points $(-6, 9)$ and $(0, 9)$, so the solutions of $(x+3)^2 = 9$ are -6 and 0.

46. The graph of $y = (x+3)^2$ contains the point $(-3, 0)$, so the solution of $(x+3)^2 = 0$ is -3.

Chapter 17 Discussion and Writing Exercises

1. The second line should be $x+6 = \sqrt{16}$ *or* $x+6 = -\sqrt{16}$. Then we would have

$$x+6 = 4 \quad or \quad x+6 = -4$$
$$x = -2 \quad or \quad x = -10$$

Both numbers check, so the solutions are -2 and -10.

2. No; since each input has exactly one output, the number of outputs cannot exceed the number of inputs.

3. Find the average, v, of the x-coordinates of the x-intercepts. Then the equation of the line of symmetry is $x = v$. The number v is also the first coordinate of the vertex. Substitute this value for x in the equation of the parabola to find the y-coordinate of the vertex.

4. If $a > 0$ the graph opens up. If $a < 0$, the graph opens down.

5. The solutions will be rational numbers because each is the solution of a linear equation of the form $bx + a = 0$.

Chapter 17 Test

1.
$$7x^2 = 35$$
$$x^2 = 5$$
$$x = \sqrt{5} \quad or \quad x = -\sqrt{5}$$

The solutions are $\sqrt{5}$ and $-\sqrt{5}$.

2.
$$7x^2 + 8x = 0$$
$$x(7x+8) = 0$$
$$x = 0 \quad or \quad 7x + 8 = 0$$
$$x = 0 \quad or \quad 7x = -8$$
$$x = 0 \quad or \quad x = -\frac{8}{7}$$

The solutions are 0 and $-\frac{8}{7}$.

3.
$$48 = t^2 + 2t$$
$$0 = t^2 + 2t - 48$$
$$0 = (t+8)(t-6)$$
$$t + 8 = 0 \quad or \quad t - 6 = 0$$
$$t = -8 \quad or \quad t = 6$$

The solutions are -8 and 6.

4.
$$3y^2 - 5y = 2$$
$$3y^2 - 5y - 2 = 0$$
$$(3y+1)(y-2) = 0$$
$$3y + 1 = 0 \quad or \quad y - 2 = 0$$
$$3y = -1 \quad or \quad y = 2$$
$$y = -\frac{1}{3} \quad or \quad y = 2$$

The solutions are $-\frac{1}{3}$ and 2.

5.
$$(x-8)^2 = 13$$
$$x - 8 = \sqrt{13} \quad or \quad x - 8 = -\sqrt{13}$$
$$x = 8 + \sqrt{13} \quad or \quad x = 8 - \sqrt{13}$$

The solutions are $8 \pm \sqrt{13}$.

6.
$$x^2 = x + 3$$
$$x^2 - x - 3 = 0$$
$$a = 1,\ b = -1,\ c = -3$$
$$x = \frac{-(-1) \pm \sqrt{(-1)^2 - 4 \cdot 1 \cdot (-3)}}{2 \cdot 1}$$
$$x = \frac{1 \pm \sqrt{1+12}}{2}$$
$$x = \frac{1 \pm \sqrt{13}}{2}$$

The solutions are $\frac{1 \pm \sqrt{13}}{2}$.

7.
$$m^2 - 3m = 7$$
$$m^2 - 3m - 7 = 0$$
$$a = 1,\ b = -3,\ c = -7$$
$$m = \frac{-(-3) \pm \sqrt{(-3)^2 - 4 \cdot 1 \cdot (-7)}}{2 \cdot 1}$$
$$m = \frac{3 \pm \sqrt{9+28}}{2}$$
$$m = \frac{3 \pm \sqrt{37}}{2}$$

The solutions are $\frac{3 \pm \sqrt{37}}{2}$.

8.
$$10 = 4x + x^2$$
$$0 = x^2 + 4x - 10$$
$$a = 1,\ b = 4,\ c = -10$$
$$x = \frac{-4 \pm \sqrt{4^2 - 4 \cdot 1 \cdot (-10)}}{2 \cdot 1}$$
$$x = \frac{-4 \pm \sqrt{16+40}}{2} = \frac{-4 \pm \sqrt{56}}{2}$$
$$x = \frac{-4 \pm \sqrt{4 \cdot 14}}{2} = \frac{-4 \pm 2\sqrt{14}}{2}$$
$$x = \frac{2(-2 \pm \sqrt{14})}{2} = -2 \pm \sqrt{14}$$

The solutions are $-2 \pm \sqrt{14}$.

9.
$$3x^2 - 7x + 1 = 0$$
$$a = 3,\ b = -7,\ c = 1$$
$$x = \frac{-(-7) \pm \sqrt{(-7)^2 - 4 \cdot 3 \cdot 1}}{2 \cdot 3}$$
$$x = \frac{7 \pm \sqrt{49-12}}{6} = \frac{7 \pm \sqrt{37}}{6}$$

The solutions are $\frac{7 \pm \sqrt{37}}{6}$.

10.
$$x - \frac{2}{x} = 1, \text{ LCM is } x$$
$$x\left(x - \frac{2}{x}\right) = x \cdot 1$$
$$x \cdot x - x \cdot \frac{2}{x} = x$$
$$x^2 - 2 = x$$
$$x^2 - x - 2 = 0$$
$$(x-2)(x+1) = 0$$
$$x - 2 = 0 \quad or \quad x + 1 = 0$$
$$x = 2 \quad or \quad x = -1$$

Both numbers check. The solutions are 2 and -1.

11.
$$\frac{4}{x} - \frac{4}{x+2} = 1, \text{ LCM is } x(x+2)$$
$$x(x+2)\left(\frac{4}{x} - \frac{4}{x+2}\right) = x(x+2) \cdot 1$$
$$x(x+2) \cdot \frac{4}{x} - x(x+2) \cdot \frac{4}{x+2} = x(x+2)$$
$$4(x+2) - 4x = x^2 + 2x$$
$$4x + 8 - 4x = x^2 + 2x$$
$$8 = x^2 + 2x$$
$$0 = x^2 + 2x - 8$$
$$0 = (x+4)(x-2)$$
$$x + 4 = 0 \quad or \quad x - 2 = 0$$
$$x = -4 \quad or \quad x = 2$$

Both numbers check. The solutions are -4 and 2.

12. $x^2 - 4x - 10 = 0$
$$x^2 - 4x = 10$$
$$x^2 - 4x + 4 = 10 + 4 \quad \text{Adding 4: } \left(\frac{-4}{2}\right)^2 = (-2)^2 = 4$$
$$(x-2)^2 = 14$$
$$x - 2 = \sqrt{14} \quad or \quad x - 2 = -\sqrt{14}$$
$$x = 2 + \sqrt{14} \quad or \quad x = 2 - \sqrt{14}$$

The solutions are $2 \pm \sqrt{14}$.

13. From Exercise 12 we know that the solutions of the equation are $2 \pm \sqrt{14}$.

Using a calculator, we have

$2 - \sqrt{14} \approx -1.7$ and $2 + \sqrt{14} \approx 5.7$.

14. $d = an^2 + bn$
$$0 = an^2 + bn - d$$

We will use the quadratic formula with $a = a$, $b = b$, and $c = -d$.
$$n = \frac{-b \pm \sqrt{b^2 - 4 \cdot a \cdot (-d)}}{2 \cdot a}$$
$$n = \frac{-b + \sqrt{b^2 + 4ad}}{2a} \quad \text{Using the positive square root}$$

15. To find the x-intercepts we solve the following equation.
$$-x^2 + x + 5 = 0$$
$$x^2 - x - 5 = 0 \quad \text{Standard form}$$
$$a = 1,\ b = -1,\ c = -5$$
$$x = \frac{-(-1) \pm \sqrt{(-1)^2 - 4 \cdot 1 \cdot (-5)}}{2 \cdot 1}$$
$$x = \frac{1 \pm \sqrt{1 + 20}}{2} = \frac{1 \pm \sqrt{21}}{2}$$

The x-intercepts are $\left(\frac{1 - \sqrt{21}}{2}, 0\right)$ and $\left(\frac{1 + \sqrt{21}}{2}, 0\right)$.

16. $y = 4 - x^2$, or $y = -x^2 + 4$

Find the vertex. The x-coordinate is
$$-\frac{b}{2a} = -\frac{0}{2(-1)} = 0.$$

The y-coordinate is
$$y = 4 - x^2 = 4 - 0^2 = 4.$$

The vertex is $(0, 4)$. This is also the y-intercept.

We choose some x-values on both sides of the vertex and graph the parabola.

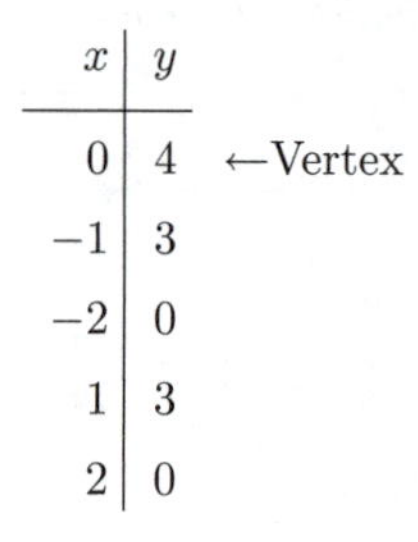

x	y	
0	4	←Vertex
−1	3	
−2	0	
1	3	
2	0	

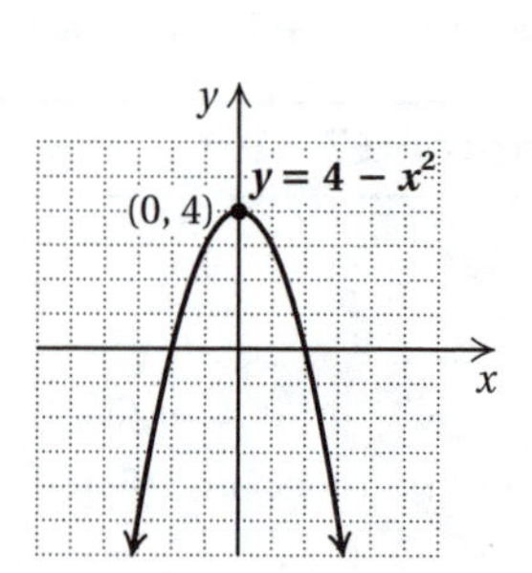

17. $y = -x^2 + x + 5$

Find the vertex. The x-coordinate is
$$-\frac{b}{2a} = -\frac{1}{2(-1)} = -\left(-\frac{1}{2}\right) = \frac{1}{2}.$$

The y-coordinate is
$$y = -\left(\frac{1}{2}\right)^2 + \frac{1}{2} + 5 = -\frac{1}{4} + \frac{1}{2} + 5 = \frac{21}{4}.$$

The vertex is $\left(\frac{1}{2}, \frac{21}{4}\right)$.

We choose some x-values on both sides of the vertex and graph the parabola.

x	y	
$\frac{1}{2}$	$\frac{21}{4}$	← Vertex
0	5	← y-intercept
−2	−1	
−1	3	
2	3	
3	−1	

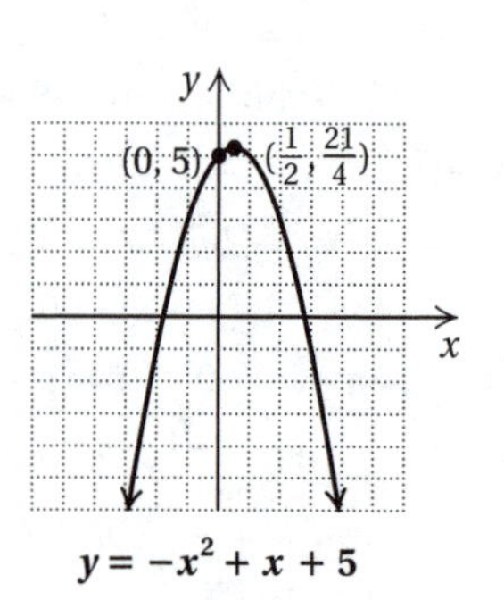

18. $f(x) = \frac{1}{2}x + 1$

$f(0) = \frac{1}{2} \cdot 0 + 1 = 0 + 1 = 1$

$f(1) = \frac{1}{2} \cdot 1 + 1 = \frac{1}{2} + 1 = 1\frac{1}{2}$

$f(2) = \frac{1}{2} \cdot 2 + 1 = 1 + 1 = 2$

19. $g(t) = -2|t| + 3$

$g(-1) = -2|-1| + 3 = -2 \cdot 1 + 3 = -2 + 3 = 1$

$g(0) = -2|0| + 3 = -2 \cdot 0 + 3 = 0 + 3 = 3$

$g(3) = -2|3| + 3 = -2 \cdot 3 + 3 = -6 + 3 = -3$

20. ***Familiarize.*** Using the labels on the drawing in the text, we let l and $l-4$ represent the length and width of the rug, respectively, in meters. Recall that the area of a rectangle is (length) × (width).

Translate.

$$\underbrace{\text{The area}}_{l(l-4)} \quad \underbrace{\text{is}}_{=} \quad \underbrace{16.25 \text{ m}^2}_{16.25}.$$

Solve.

$$l(l-4) = 16.25$$
$$l^2 - 4l = 16.25$$
$$l^2 - 4l - 16.25 = 0$$

The factorization of $l^2 - 4l - 16.25$ is not readily apparent so we will use the quadratic formula with $a = 1$, $b = -4$, and $c = -16.25$.

$$l = \frac{-(-4) \pm \sqrt{(-4)^2 - 4 \cdot 1 \cdot (-16.25)}}{2 \cdot 1}$$

$$l = \frac{4 \pm \sqrt{16 + 65}}{2} = \frac{4 \pm \sqrt{81}}{2}$$

$$l = \frac{4 \pm 9}{2}$$

$$l = \frac{4 - 9}{2} = \frac{-5}{2} = -\frac{5}{2}, \text{ or } -2.5$$

or

$$l = \frac{4 + 9}{2} = \frac{13}{2} = 6.5$$

Check. Since the length cannot be negative, -2.5 cannot be a solution. If $l = 6.5$, then $l - 4 = 6.5 - 4 = 2.5$ and $6.5(2.5) = 16.25$. Thus, the width is 4 m less than the length and the area is 16.25 m^2. The answer checks.

State. The length of the rug is 6.5 m, and the width is 2.5 m.

21. ***Familiarize.*** Let r = the speed of the boat in still water. Then $r - 2$ = the speed upstream and $r + 2$ = the speed downstream. We organize the information in a table.

	d	r	t
Upstream	44	$r-2$	t_1
Downstream	52	$r+2$	t_2

Translate. Using $t = d/r$ and the rows of the table, we have

$$t_1 = \frac{44}{r-2} \text{ and } t_2 = \frac{52}{r+2}.$$

Since the total time is 4 hr, $t_1 + t_2 = 4$, and we have

$$\frac{44}{r-2} + \frac{52}{r+2} = 4.$$

Solve. We solve the equation. We multiply by $(r-2)(r+2)$, the LCM of the denominators.

$$(r-2)(r+2)\left(\frac{44}{r-2} + \frac{52}{r+2}\right) = (r-2)(r+2) \cdot 4$$
$$44(r+2) + 52(r-2) = 4(r^2 - 4)$$
$$44r + 88 + 52r - 104 = 4r^2 - 16$$
$$96r - 16 = 4r^2 - 16$$
$$0 = 4r^2 - 96r$$
$$0 = 4r(r - 24)$$

$4r = 0$ *or* $r - 24 = 0$

$r = 0$ *or* $r = 24$

Check. The boat cannot travel upstream if its speed in still water is 0 km/h. If the speed of the boat in still water is 24 km/h, then it travels at a speed of $24 - 2$, or 22 km/h, upstream and $24 + 2$, or 26 km/h, downstream. At 22 km/h the boat travels 44 km in 44/22, or 2 hr. At 26 km/h it travels 52 km in 52/26, or 2 hr. The total time is $2 + 2$, or 4 hr, so the answer checks.

State. The speed of the boat in still water is 24 km/h.

22. In 2012, $t = 2012 - 1940 = 72$.

$R(72) = 30.18 - 0.06(72) = 30.18 - 4.32 = 25.86$

We predict that the record will be 25.86 min in 2012.

23. $h(x) = x - 4$

We find some function values.

When $x = -1$, $h(-1) = -1 - 4 = -5$.

When $x = 2$, $h(2) = 2 - 4 = -2$.

When $x = 5$, $h(5) = 5 - 4 = 1$.

x	$h(x)$
-1	-5
2	-2
5	1

Plot these points and connect them.

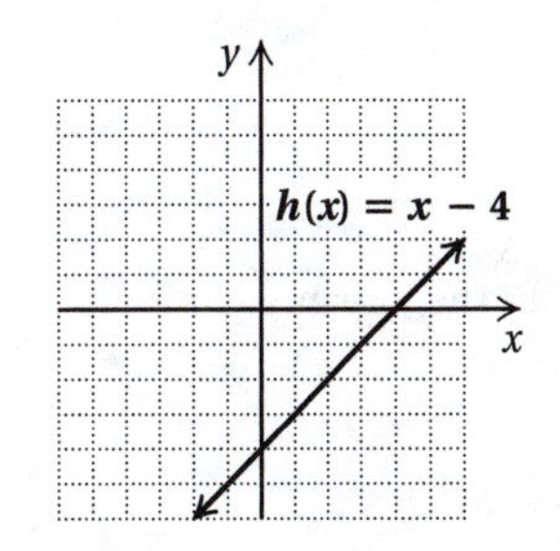

24. $g(x) = x^2 - 4$

Recall from Section 17.6 that the graph is a parabola. We find some function values.

When $x = -3$, $g(-3) = (-3)^2 - 4 = 9 - 4 = 5$.

When $x = -1$, $g(-1) = (-1)^2 - 4 = 1 - 4 = -3$.

When $x = 0$, $g(0) = 0^2 - 4 = 0 - 4 = -4$.

When $x = 2$, $g(2) = 2^2 - 4 = 4 - 4 = 0$.

When $x = 3$, $g(3) = 3^2 - 4 = 9 - 4 = 5$.

x	$g(x)$
-3	5
-1	-3
0	-4
2	0
3	5

Plot these points and connect them.

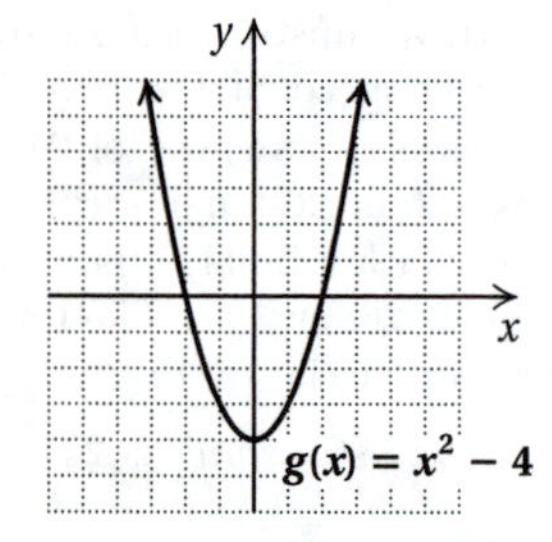

25. No vertical line will intersect the graph more than once, so this is the graph of a function.

26. It is possible for a vertical line to intersect the graph more than once, so this is not the graph of a function.

27. $g(x) = -2x - x^2$

$g(-8) = -2(-8) - (-8)^2 = 16 - 64 = -48$

Answer D is correct.

28. ***Familiarize***. We make a drawing. Let s = the length of a side of the square, in feet. Then $s + 5$ = the length of a diagonal.

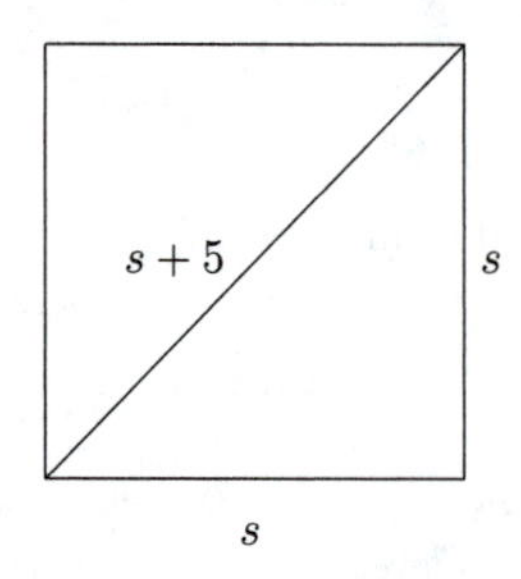

Translate. We use the Pythagorean equation.

$s^2 + s^2 = (s + 5)^2$

Solve.

$$s^2 + s^2 = (s + 5)^2$$
$$2s^2 = s^2 + 10s + 25$$
$$s^2 - 10s - 25 = 0$$

We use the quadratic formula with $a = 1$, $b = -10$, and $c = -25$.

$$s = \frac{-(-10) \pm \sqrt{(-10)^2 - 4 \cdot 1 \cdot (-25)}}{2 \cdot 1}$$
$$s = \frac{10 \pm \sqrt{100 + 100}}{2} = \frac{10 \pm \sqrt{200}}{2}$$
$$s = \frac{10 \pm \sqrt{100 \cdot 2}}{2} = \frac{10 \pm 10\sqrt{2}}{2}$$
$$s = \frac{2(5 \pm 5\sqrt{2})}{2} = 5 \pm 5\sqrt{2}$$

Check. Since $5 - 5\sqrt{2}$ is negative, it cannot be the length of a side of the square. If the length of a side is $5+5\sqrt{2}$ ft, then $(5+5\sqrt{2})^2+(5+5\sqrt{2})^2 = 25+50\sqrt{2}+50+25+50\sqrt{2}+50 = 150+100\sqrt{2}$. The length of a diagonal is $5+5\sqrt{2}+5$ ft, or $10 + 5\sqrt{2}$, and $(10 + 5\sqrt{2})^2 = 100 + 100\sqrt{2} + 50 = 150 + 100\sqrt{2}$. Since these lengths satisfy the Pythagorean equation, the answer checks.

State. The length of a side of the square is $5 + 5\sqrt{2}$ ft.

29. $x - y = 2$, (1)

$xy = 4$ (2)

Solve Equation (1) for y.

$$x - y = 2$$
$$-y = -x + 2$$
$$y = x - 2 \quad (3)$$

Substitute $x - 2$ for y in Equation (2) and solve for x.

$$xy = 4$$
$$x(x - 2) = 4$$
$$x^2 - 2x = 4$$
$$x^2 - 2x - 4 = 0$$

We use the quadratic formula with $a = 1$, $b = -2$, and $c = -4$.

$$x = \frac{-(-2) \pm \sqrt{(-2)^2 - 4 \cdot 1 \cdot (-4)}}{2 \cdot 1}$$
$$x = \frac{2 \pm \sqrt{4 + 16}}{2} = \frac{2 \pm \sqrt{20}}{2}$$
$$x = \frac{2 \pm \sqrt{4 \cdot 5}}{2} = \frac{2 \pm 2\sqrt{5}}{2}$$
$$x = \frac{2(1 \pm \sqrt{5})}{2} = 1 \pm \sqrt{5}$$

Now substitute these values for x in Equation (3) and solve for the corresponding y-values.

For $x = 1 + \sqrt{5}$: $y = x - 2 = 1 + \sqrt{5} - 2 = -1 + \sqrt{5}$.

For $x = 1 - \sqrt{5}$: $y = x - 2 = 1 - \sqrt{5} - 2 = -1 - \sqrt{5}$.

The solutions are $(1+\sqrt{5}, -1+\sqrt{5})$ and $(1-\sqrt{5}, -1-\sqrt{5})$.